W9-BIO-184

BIOLOGY OF THE Invertebrates

sixth edition

JAN A. PECHENIK
Tufts University

Boston Burr Ridge, IL Dubuque, IA New York San Francisco St. Louis
Bangkok Bogotá Caracas Kuala Lumpur Lisbon London Madrid Mexico City
Milan Montreal New Delhi Santiago Seoul Singapore Sydney Taipei Toronto

The McGraw-Hill Companies

BIOLOGY OF THE INVERTEBRATES, SIXTH EDITION

Published by McGraw-Hill, a business unit of The McGraw-Hill Companies, Inc., 1221 Avenue of the Americas, New York, NY 10020.

This book is printed on acid-free paper.

2 3 4 5 6 7 8 9 0 QPD/QPD 0 9

ISBN 978-0-07-302826-2
MHID 0-07-302826-6

Publisher: *Janice Roerig-Blong*
Executive Editor: *Patrick E. Reidy*
Director of Development: *Kristine Tibbetts*
Senior Managing Editor: *Faye M. Schilling*
Lead Production Supervisor: *Sandy Ludovissy*
Associate Design Coordinator: *Brenda A. Rolwes*
Cover Designer: *Studio Montage, St. Louis, Missouri*
(USE) Cover Image: © *James Gritz/Getty Images*
Senior Photo Research Coordinator: *Lori Hancock*
Compositor: *S4Carlisle Publishing Services*
Typeface: *10.5/12 Minion*
Printer: *Quebecor World Dubuque, IA*

Library of Congress Cataloging-in-Publication Data

Pechenik, Jan A.
Biology of the invertebrates / Jan A. Pechenik. – 6th ed.
p. cm.
Includes index.
ISBN 978-0-07-302826-2 — ISBN 0-07-302826-6 (hard copy : alk. paper)
1. Invertebrates. I. Title.

QL362.P43 2010
592—dc22

2008048546

www.mhhe.com

To Oliver

My favorite son,
now and always.

And
to all who love
invertebrates
and find them
worthy of study.

List of Research Focus Boxes

3.1 *Phenotypic Plasticity and Behavioral Complexity among Ciliates* **54**

4.1 *Histoincompatibility in Sponges* **80**

6.1 *Control of Nutrient Transfer in Symbiotic Associations* **112**

6.2 *Chemical Defenses on Coral Reefs* **126**

7.1 *Ctenophore Feeding Biology* **142**

8.1 *Control of Schistosomiasis* **165**

10.1 *Adaptive Value of Rotifer Morphology* **192**

11.1 *Assessing Phylogenetic Relationships* **206**

12.1 *Gastropod Torsion* **230**

12.2 *Bacterial Symbiosis* **252**

12.3 *Cephalopod Behavior* **264**

13.1 *Living with Sulfides* **310**

13.2 *Unexpected Responses to Pollution* **320**

14.1 *Halting the Spread of Malaria* **362**

14.2 *Honeybee Navigation* **368**

14.3 *Copepod Feeding* **386**

15.1 *The Energetics of Onychophoran Feeding* **426**

16.1 *Nematode Development* **440**

18.1 *Chaetognath Prey Capture* **464**

19.1 *Colonial Metabolism* **484**

20.1 *Influence of Pollutants* **517**

24.1 *Costs of Delaying Metamorphosis in the Field* **578**

Brief Contents

1 Introduction and Environmental Considerations **1**

2 Invertebrate Classification and Relationships **7**

3 The Protists **37**

4 The Poriferans and Placozoans **79**

5 Introduction to the Hydrostatic Skeleton **97**

6 The Cnidarians **101**

7 The Ctenophores **137**

8 The Platyhelminths **149**

9 The Mesozoans: Possible Flatworm Relatives **179**

10 The Gnathifera: Rotifers, Acanthocephalans, and Two Smaller Groups **183**

11 The Nemertines **203**

12 The Molluscs **215**

13 The Annelids **295**

14 The Arthropods **341**

15 Two Phyla of Likely Arthropod Relatives: Tardigrades and Onychophorans **421**

16 The Nematodes **431**

17 Four Phyla of Likely Nematode Relatives: Nematomorpha, Priapulida, Kinorhyncha, and Loricifera **451**

18 Three Phyla of Uncertain Affiliation: Gastrotricha, Chaetognatha, and Cycliophora **459**

19 The "Lophophorates" (Phoronids, Brachiopods, Bryozoans) and Entoprocts **473**

20 The Echinoderms **497**

21 The Hemichordates **529**

22 The Xenoturbellids: Deuterostomes at Last? **537**

23 The Nonvertebrate Chordates **539**

24 Invertebrate Reproduction and Development—An Overview **555**

Expanded Contents

List of Research Focus Boxes ***iv***
Preface ***xi***
Guide to the Major Animal Groups ***xvi***
The Geologic Time Scale ***xvii***

1

Introduction and Environmental Considerations 1

Introduction: The Importance of Research on Invertebrates **1**
Environmental Considerations **2**
Air Is Dry, Water Is Wet **2**
Water Is the "Universal Solvent" **3**
Water Is Denser than Air **4**
Water Has Thermal Stability **4**
Problems with the Aquatic Life **4**
Origins and Diversity of Life **6**

2

Invertebrate Classification and Relationships 7

Introduction **7**
Classification by Cell Number, Embryology, and Body Symmetry **9**
Classification by Developmental Pattern **9**
Classification by Evolutionary Relationship **16**
Inferring Evolutionary Relationships **19**
Why Determine Evolutionary Trees? **22**
How Evolutionary Relationships Are Determined **25**
Uncertainty about Evolutionary Relationships **30**
Time Machines: Help from the Fossil Record **31**
Classification by Habitat and Lifestyle **31**

3

The Protists 37

Introduction **37**
The Protozoa **38**
General Characteristics **39**
Protozoan Locomotory Systems **42**
Structure and Function of Cilia **42**
Structure and Function of Flagella **43**
Structure and Function of Pseudopodia **43**
Protozoan Reproduction **44**
Protozoan Feeding **46**
General Summary **46**
The Alveolates **46**
Phylum Ciliophora **46**
Patterns of Ciliation **47**
Other Morphological Features **47**
Reproductive Characteristics **49**
Ciliate Lifestyles **52**
Phylum Dinozoa (= Dinoflagellata) **54**
Phylum Apicomplexa **58**
A Brief Aside on Microsporideans and Myxozoans **61**
Amoeboid Protozoans **62**
The Amoebozoans **62**
The Arcellinids and Related Test-Bearing Amoebae **63**
The Rhizaria **65**
Flagellated Protozoans **68**
The Phytoflagellated Protozoans **68**
The Zooflagellated Protozoans—Free-Living Forms **68**
The Zooflagellated Protozoans—Parasitic Forms **68**
Two Interesting Transitional Forms **72**

4

The Poriferans and Placozoans 79

Phylum Porifera **79**
- Introduction **79**
- General Characteristics **81**
- Poriferan Diversity **86**
- Other Features of Poriferan Biology: Reproduction and Development **89**

Phylum Placozoa **91**
- Introduction **91**
- General Characteristics **91**

5

Introduction to the Hydrostatic Skeleton 97

6

The Cnidarians 101

Introduction and General Characteristics **101**

Subphylum Medusozoa **104**

Class Scyphozoa **104**

Class Cubozoa **108**

Class Hydrozoa **110**
- Order Hydroida **110**
- Order Siphonophora **113**
- The Hydrocorals **114**

One Group of Non-Medusozoans **115**
- Class Myxozoa (= Myxospora) **115**

Class Anthozoa **118**
- Subclass Hexacorallia (= Zoantharia) **120**
- Subclass Octocorallia (= Alcyonaria) **124**

7

The Ctenophores 137

Introduction and General Characteristics **137**

Ctenophore Diversity **143**
- Class Tentaculata **144**
- Class Nuda **146**

8

The Platyhelminths 149

Introduction and General Characteristics **149**

Class Turbellaria **151**

Class Cestoda **157**

Class Monogenea **160**

Class Trematoda **161**
- The Digeneans **161**
- The Aspidogastreans (= Aspidobothreans) **170**

9

The Mesozoans: Possible Flatworm Relatives 179

The Mesozoans **179**
- Class Orthonectida **180**
- Class Rhombozoa **181**

10

The Gnathifera: Rotifers, Acanthocephalans, and Two Smaller Groups 183

Introduction **183**

Phylum Rotifera **184**

Introduction and General Characteristics **184**
- Class Seisonidea **188**
- Class Bdelloidea **188**
- Class Monogononta **190**

Other Features of Rotifer Biology **195**
- Digestive System **195**
- Nervous and Sensory Systems **195**
- Excretion and Water Balance **196**

Phylum Acanthocephala **196**

Phylum Gnathostomulida **198**

Phylum Micrognathozoa **199**

11

The Nemertines 203

Introduction and General Characteristics **203**

Other Features of Nemertine Biology **210**
- Classification **210**

Protection from Predators 210
Reproduction and Development 210

12 The Molluscs 215

Introduction and General Characteristics 215
Class Polyplacophora 218
Class Aplacophora 222
Class Monoplacophora 222
Class Gastropoda 224
The Prosobranchs 229
The Opisthobranchia 235
The Pulmonata 237
Class Bivalvia (= Pelecypoda) 238
Subclass Protobranchia 239
The Lamellibranchs 241
Subclass Anomalodesmata 254
Class Scaphopoda 255
Class Cephalopoda 256
Other Features of Molluscan Biology 265
Reproduction and Development 265
Circulation, Blood Pigments, and Gas Exchange 266
Nervous System 266
Digestive System 269
Excretory System 271

13 The Annelids 295

Introduction 295
Phylum Annelida 295
General Annelid Characteristics 295
Class Polychaeta 297
Polychaete Reproduction 300
Family Siboglinidae (formerly the Pogonophora) 304
General Characteristics 304
Siboglinid Reproduction and Development 311
The Echiurans 312
The Sipunculans 314
Class Clitellata 318
Subclass Oligochaeta 318
Oligochaete Reproduction 322
Subclass Hirudinea 322
Leech Reproduction 325
Other Features of Annelid Biology 325
Digestive System 325
Nervous System and Sense Organs 327
Circulatory System 328

14 The Arthropods 341

Introduction and General Characteristics 341
The Exoskeleton 342
The Hemocoel 342
Molting 343
Nerves and Muscles 344
The Circulatory System 344
Arthropod Visual Systems 345
Arthropod Reproduction 349
Classification and Evolutionary Relationships 349
Subphylum Trilobitomorpha 350
Class Trilobita 351
Subphylum Chelicerata 352
Class Merostomata 352
Class Arachnida 352
Class Pycnogonida (= Pantopoda) 357
Subphylum Mandibulata 358
Class Myriapoda 358
Superclass Hexapoda 360
Class Insecta 360
Class Crustacea 373
Subclass Malacostraca 374
Subclass Branchiopoda 379
Subclass Ostracoda 381
Subclass Copepoda 381
Subclass Pentastomida 382
Subclass Cirripedia 389
Crustacean Development 389
Other Features of Arthropod Biology 392
Digestion 392
Excretion 395
Blood Pigments 396

15

Two Phyla of Likely Arthropod Relatives: Tardigrades and Onychophorans 421

Introduction and General Characteristics **421**
Phylum Tardigrada **422**
Phylum Onychophora **424**

16

The Nematodes 431

Introduction and General Characteristics **431**
Body Coverings and Body Cavities **432**
Musculature, Internal Pressure, and Locomotion **434**
Organ Systems and Behavior **436**
Reproduction and Development **438**
Parasitic Nematodes **438**
Beneficial Nematodes **444**

17

Four Phyla of Likely Nematode Relatives: Nematomorpha, Priapulida, Kinorhyncha, and Loricifera 451

Introduction **451**
Phylum Nematomorpha **452**
Phylum Priapulida **454**
Phylum Kinorhyncha (= Echinoderida) **454**
Phylum Loricifera **456**

18

Three Phyla of Uncertain Affiliation: Gastrotricha, Chaetognatha, and Cycliophora 459

Introduction **459**
Phylum Gastrotricha **459**
Phylum Chaetognatha **461**
General Characteristics and Feeding **461**
Chaetognath Reproduction **463**
Chaetognath Lifestyles and Behavior **463**
Chaetognath Relationships **465**
Phylum Cycliophora **467**

19

The "Lophophorates" (Phoronids, Brachiopods, Bryozoans) and Entoprocts 473

Introduction and General Characteristics **473**
Phylum Phoronida **475**
Phylum Brachiopoda **476**
Phylum Bryozoa (= Ectoprocta; = Polyzoa) **480**
Class Phylactolaemata **482**
Class Gymnolaemata **484**
Class Stenolaemata **487**
Other Features of "Lophophorate" Biology **488**
Reproduction **488**
Digestion **490**
Nervous System **490**
Phylum Entoprocta (= Kamptozoa) **490**

20

The Echinoderms 497

Introduction and General Characteristics **497**
Class Crinoidea **500**
Class Stelleroidea **503**
Subclass Ophiuroidea **503**
Subclass Asteroidea **505**
Class Echinoidea **509**
Class Holothuroidea **513**
Other Features of Echinoderm Biology **518**
Reproduction and Development **518**
Nervous System **520**

21

The Hemichordates 529

Introduction and General Characteristics **529**

Class Enteropneusta **530**

Class Pterobranchia **534**

22

The Xenoturbellids: Deuterostomes at Last? 537

Phylum Xenoturbellida **537**

23

The Nonvertebrate Chordates 539

Introduction and General Characteristics **539**

Subphylum Tunicata (= Urochordata) **540**

- Class Ascidiacea **540**
- Class Larvacea (= Appendicularia) **543**
- Class Thaliacea **544**

Other Features of Urochordate Biology **547**

- Reproduction **547**
- Excretory and Nervous Systems **547**

Subphylum Cephalochordata (= Acrania) **548**

24

Invertebrate Reproduction and Development— An Overview 555

Introduction **555**

Asexual Reproduction **556**

Sexual Reproduction **558**

- Patterns of Sexuality **558**
- Gamete Diversity **561**
- Getting the Gametes Together **561**
- Larval Forms **567**

Dispersal as a Component of the Life-History Pattern **577**

Glossary of Frequently Used Terms **587**

Index **590**

Preface

About This Book

Invertebrate zoology is a fascinating but enormous field. More than 98% of all known animal species are invertebrates, and that proportion is increasing with time as more species are described. Invertebrates are distributed among at least 30 phyla and a mind-boggling number of classes, subclasses, orders, and families. The degree of morphological and functional diversity found within some groups, even within single orders, can overwhelm the beginning student. The enormity of the field and the great range of potential approaches to the subject make invertebrate biology challenging both to teach and to learn. In preparing the sixth edition of this book, I have endeavored to make the tasks of both teaching and learning easier, and even enjoyable, while nevertheless presenting the latest thinking in the field.

Too many people think of invertebrate zoology as an exercise in memorizing terms and, perhaps, interesting but trivial stories about interesting but irrelevant animals. Too many people think of invertebrate zoology as an outdated field in which everything is already known. I hope that this book alters those perceptions, by presenting invertebrate zoology as a lively and modern area of ongoing and worthwhile biological inquiry.

Like previous editions, the sixth is designed as a nonintimidating, readable introduction to the biology of each group, emphasizing those characteristics that set each group apart from all others. The book is intended to serve as the foundation for further learning—in lecture, laboratory, field, and library—a foundation that is largely manageable by students. Instructors are then free to embellish and expand on that foundation to suit any desired focus: taxonomy and phylogeny, behavior, conservation, diversity of form and function, physiology, ecology, or current research in any of those areas. The book whets the student's appetite and provides the required background, buying instructors the time to discuss more fully whatever they feel are the most interesting and important aspects of the field. This is the guiding principle in my decisions about the level of detail to provide: I generalize wherever possible to build a firm foundation without intimidation or confusion. Given a chance, the animals themselves soon win most students over.

The most difficult part of writing this book has been deciding what to leave out. My decisions have been reached largely by reading many dozens of research articles on the biology of each group and determining the specific terminology and level of background information that students will need to read those papers. Although all phyla are covered, I have aimed for conciseness, not exhaustiveness, and have emphasized unifying principles rather than the diversity found within each group. Students are best prepared to encounter the diversity of form and function in lecture, laboratory, and field, once they have mastered basic concepts and terminology. I provide a sense of the ecological diversity encountered within each group in a Taxonomic Detail section at the end of most chapters, a section that also adds to the value of the book for reference.

The text remains somewhat biased toward functional morphology, bringing animals to life for students and preparing them to make careful observations of living animals in the laboratory and in the field. Most chapters contain a section entitled *Topics for Further Discussion and Investigation,* highlighting many of the major research questions that have been and are being addressed for the animals covered in each chapter. For each topic, I have selected references from the primary literature that should be intellectually accessible to any interested, beginning student once he or she has read the relevant textbook chapter. I have had to exclude many excellent papers because they were too advanced, were published in less widely distributed journals, or were review papers rather than primary journal articles. The topics I have chosen, along with the accompanying references, could be used as a basis for lectures, class discussion, term papers, research proposals, or other writing assignments, or simply as a convenient way of easing students into the original literature by having them investigate topics that excite their curiosity. Students gain little by reading and memorizing predigested summaries of the primary literature; they gain much by reading and discussing that literature. Similarly, I have again decided against adding "end of chapter summaries," arguing that students will learn far more of lasting value by writing and discussing their own summaries than by memorizing mine.

The excitement of invertebrate biology is found in the primary research literature, and a major goal of this book is to motivate and prepare students to read that literature—both the recent literature and that of past decades, before "synapomorphy" and "lophotrochozoa" were common in the literature, and when most workers still thought that Echiura was a phylum and that myxozoans were members of the Protozoa. The *Research Focus Boxes* scattered throughout the book are based on individual papers drawn from the primary literature to illustrate the range of questions that biologists have been asking about invertebrates and the variety of approaches that have been used to address those questions. My goal here is to prepare students to read the primary literature by focusing on how questions are formulated, how data are collected and interpreted, and how each study typically leads to further questions. Students interested in the topic of a particular Research Focus Box might wish to read the original paper on which that Focus Box was based and then use that Focus Box as a model for summarizing other papers on related topics. For Focus Boxes based on older research articles, interested students may wish to follow the topic forward in time, using an indexing service such as the Web of Science. Many instructors now ask their students to write Research Focus Boxes modeled on the ones in this book; such assignments can teach students much of lasting value, particularly if they include guided revision—not an onerous chore for either student or instructor since each Focus Box is only a few pages long. This assignment allows students to put their growing vocabulary to immediate use, and it doesn't seem to hurt, either, that students feel they're learning useful skills in the process.

As with previous editions, chapters are self-contained and can be assigned in whatever order best suits the organization of any particular course, once the introductory chapters have been covered. In my own course I cover some introductory material and then begin with annelids and other protostomes. Within each chapter, the material has been arranged in manageable, readable units for the convenience of both student and instructor. For example, a section entitled *Introduction and General Characteristics* might be assigned prior to a lecture on a particular group of organisms, while a section called *Feeding and Digestion* might best be assigned prior to the accompanying laboratory session; it is always easier to assign additional sections of a text than to tell students what *not* to read within a larger section. I find that periodic scheduled quizzes, designed to reward students for doing the assigned reading, provide excellent motivation. The final chapter (Chapter 24) brings together all of the major phyla by considering general principles of invertebrate reproduction and development, providing students an opportunity to reminisce about all the animals they have encountered during the term and to begin moving beyond phylum-by-phylum compartmentalization toward synthesis.

This edition has a somewhat greater phylogenetic orientation than its predecessors, but it still avoids prolonged phylogenetic discussion. There is still no consensus concerning many invertebrate interrelationships: Developmental studies, molecular techniques, and cladistic analyses continue to revolutionize our thinking about evolutionary relationships, or at least to challenge many treasured assumptions. Annelids and arthropods may or may not be closely related, segmentation may be a derived rather than ancestral character in molluscs, the ancestral mollusc may have more closely resembled a bivalve than a gastropod, nematodes may be more closely related to arthropods than to rotifers, insects may have evolved from crustacean ancestors, and phoronids may be modified brachiopods. Similarly, opisthobranch molluscs and polychaete annelids may not be valid monophyletic groups, and nemertine worms, long considered acoelomates, may actually be unusual coelomates with no direct flatworm affinities. Acoels may be primitive but may not belong in the phylum Platyhelminthes. For some workers, even the definition of what it means to be a protostome has shifted substantially over the past few years, and now relies less on anatomical and developmental criteria. Indeed, the usefulness of morphology and ultrastructure in inferring phylogenetic relationships has been seriously challenged in recent years. Lophophorates, for example, appear to be deuterostomes based on morphological and developmental criteria, but group unambiguously with protostomes in most molecular analyses. Similarly, molecular data now align chaetognaths with protostomes, despite their many deuterostome developmental characteristics. The relationships between cephalochordates, chordates, and echinoderms are also uncertain: Some recent molecular data suggest that cephalochordates are more closely related to echinoderms than to other chordates, implying that some key chordate features were present in the ancestral deuterostome and later lost in the evolution of echinoderms and hemichordates. And molecular analyses, often in concert with careful ultrastructural studies, have mostly destroyed the idea that pseudopodia and flagella inform us about protozoan relationships.

Clearly, invertebrate systematics is a work in progress. Although the phylogenetic atmosphere is charged with excitement, beginning students typically view textbook discussion of such controversies as simply another set of facts to be memorized. For this reason, such issues are best treated in lecture, where they can be used to animate class discussion. Indeed, the controversies surrounding phylogenetic speculation are what make phylogeny interesting, and can be used to make the animals themselves interesting. Chapter 2 introduces students to the range of approaches used in reconstructing phylogenies, and includes substantial discussion of cladistic analysis and the promise and potential pitfalls associated with the incorporation of molecular data. The book provides the foundation upon which instructors and students can build.

With increasing acceptance of the Phylocode, taxonomic rankings (e.g., class, subclass, order, family) are likely to be abandoned in the near future; nevertheless, I retain them here to make the hierarchical arrangements clearer to students, and easier to discuss. Whenever possible, I provide "Defining Characteristics" as each new animal group is introduced, to help students keep track of features separating each group of animals from other groups at the same taxonomic level. In essence, these Defining Characteristics are synapomorphies. For some groups there are no clear Defining Characteristics, or characteristics that have been proposed are too controversial to be included at the present time.

Most chapters conclude with a section entitled "Search the Web," guiding students to particularly good websites associated with the group under discussion. I have listed only those sites that speak with verifiable authority and that are likely to be around and updated for a number of years.

Changes for the Sixth Edition

When I finished the second edition of this book in 1984, I thought that it might need to be revised again in another 10–12 years. A series of careful developmental studies in concert with the increasing acceptance and use of cladistic methodology and molecular data in phylogenetic analyses have made the past 25 years far more exciting than I had imagined they would be, and the changes to this edition are substantial, despite the passage of only 4 years since I completed the previous edition.

Most chapters have been revised to reflect new discoveries and expanding research areas, including research of commercial importance and environmental relevance, and I now note the incipient emergence of The Phylocode, which, if widely accepted, will revolutionize systematics.

Remarkably, all of the rather dramatic organizational changes that I made for the fourth and fifth editions have held up, and indeed have generally been bolstered by additional evidence. The pogonophorans, for example, for many years treated as a separate phylum, are now widely accepted as modified annelids (members of the polychaete family Siboglinidae) based upon morphological, developmental, and molecular data; additional data also now support the inclusion of both sipunculans and echiurans among annelids. And support has generally increased for separating protostomes into at least 2 great groups—the Ecdysozoa and the Lophotrochozoa—although there is not yet complete agreement about exactly which animals each group contains.

Nevertheless, some impressive changes are new to this edition. A new order of insects has been added (the Mantophasmatodea), for example, the first new order of living insects to be established in nearly 100 years, and the medusa-lacking Staurozoa are now a separate class of cnidarians. The phyla Echiura and Sipuncula have been provisionally abandoned for this edition, since these animals—along with pogonophorans—now seem to be *bona fide* annelids. I have also updated phylogenetic relationships for some other animals, including the cladocerans and other branchiopods. Molecular data confirm that the collembolids, proturans, and diplurans are hexapods, but that they are not insects. The former grouping, Apterygota (=Thysanura), which contained all primitively wingless insects, has therefore been abandoned. The gnathostomulids have come to settle comfortably as close relatives of rotifers and acanthocephalans within the new clade Gnathifera, and a new deuterostome phylum, the Xenoturbellida, has been proposed for the ever-enigmatic worm *Xenoturbella bocki*, bringing the total number of extant deuterostome phyla to 4; as pointed out in another recent paper, however (Wallberg et al., 2007. *Zool. Scripta* 36: 509–23), the earlier (2006!) work did not include gene sequences from acoel or nemertodermatid flatworms, leaving the door open to potential changes in affiliation. Also new to this edition, molecular data now align chaetognaths with protostomes, despite their many deuterostome developmental characteristics, and the centipedes and millipedes (myriapods) have been unexpectedly linked with chelicerates, although through molecular rather than morphological analyses. And the myxozoans, long a most perplexing group of simple parasitic worms, have seemingly come to rest among the cnidarians, substantially increasing the diversity of cnidarian body plans.

Nematode systematics has undergone a thorough revolution in the past decade. For example, the old nematode orders Secernentea and Adenophorea are abandoned, as members of the Secernentea appear to have evolved from within the Adenophorea. Similarly, members of the former nematode order Strongylida (the strongylids) clearly evolved from rhabditid ancestors and so have been folded into the order Rhabditida.

I have updated the terminology used in discussing the siboglinids (formerly pogonophorans): Perviates are now called frenulates, and the Obturata is now the Vestimentifera.[1]

Chapter 3 has been seriously revised to incorporate the startling advances and uncertainties in our understanding of protist relationships, largely caused by recently published molecular data. The revision reflects the tentative consensus reached recently by the International Society of Protistologists (formerly the Society of Prozoologists). Among many other changes, the foraminiferans and radiolarians are now united in the Rhizaria, and the commercially important parasite *Perkinsus* is now classified with the Dinozoa rather than with the Apicomplexa.

1. Halanych, K. M. 2004. Molecular phylogeny of siboglinid annelids (a.k.a. pogonophorans): a review. *Hydriobiol.* 535–36 xvi (b): 297–307.

By popular request, I have moved the material on reproduction and development from the end of Chapter 14 (arthropods) into appropriate sections of the main presentation. I have also added the enigmatic subclass Facetotecta to the *Taxonomic Detail* at the end of that chapter. I also have a little more to say about insect flight, and about structure and function of compound eyes.

I have made such major changes only where they are supported by substantial argument and data, usually from at least several leading workers publishing in major journals, and have been careful in each chapter to indicate my sources.

Many new drawings and photographs have been added for this edition, largely taken from the recent research literature, and many older figures have been redrawn or otherwise improved. The current edition deals somewhat more conspicuously with biomedical relevance, biological invasions, habitat degradation, and other contemporary issues. I have written one new *Research Focus Box,* added new *Topics for Further Discussion and Investigation* to some chapters, and included many new references to the recent primary literature, including many published as recently as 2007 and 2008. I have also added some wonderful new websites to the Search the Web sections that close each chapter, and have written one new invertebrate riddle; readers are invited to submit their answers (and their own riddles) to me directly.

The paradigm shifts that have occurred over the past 20 years have been truly remarkable, and there is no end in sight. If insects really evolved from crustacean ancestors, for example, as recent molecular and morphological studies indicate, that will require a substantial redefinition of what it means to be a crustacean. Similarly, there is now substantial molecular evidence that oligochaetes and leeches evolved from polychaete ancestors, which would essentially make the Polychaeta the equivalent of the Annelida, and a recent study (2007) using expressed sequence tags has supported the inclusion of Ectoprocta and Entoprocta within a single phylum, the Bryozoa, bringing us back to a classification established over 100 years ago. Finally, although recent molecular studies using individual molecules continue to add support for the Ecdysozoa-Lophotrochozoa dichotomy, as noted earlier, a recently reported whole-genome study of 9 eukaryotic species does not support that arrangement, but favors instead the older Articulata hypothesis in which arthropods are more closely related to annelids than to nematodes.

As more molecular data from more species and more genes from each species are collected, we should see increasing stability in the accepted arrangements. There is some hope: Chapter 18, formerly Five Phyla of Uncertain Affiliation, is now Three Phyla of Uncertain Affiliation.

As always, I welcome constructive criticism from all readers, both instructors and students.

Jan.Pechenik@tufts.edu

Acknowledgments

It is a great pleasure to thank the reviewers who have helped to make this a better text.

Bill Biggers
Wilkes University

Ronald Dimock
Wake Forest University

William Kirby-Smith
Duke University

Robert Knowlton
George Washington University

Kenneth McCravy
Western Illinois University

Michele Nishiguchi
New Mexico University–Las Cruces

Owen Sholes
Assumption College

William Woods
Tufts University

I would also like to thank the many people who contributed their expertise to previous editons of the text:

Frank E. Anderson, *Southern Illinois University;* **Christopher J. Bayne,** *Oregon State University;* **John F. Belshe,** *Central Missouri State University;* **Yehuda Benayahu,** *Tel Aviv University;* **Robert Bieri,** *Antioch University;* **Chip Biernbaum,** *College of Charleston;* **Brian L. Bingham,** *Western Washington University;* **Susan Bornstein-Forst,** *Marian College;* **Kenneth J. Boss,** *Harvard University;* **Barbara C. Boyer,** *Union College;* **Robert H. Brewer,** *Trinity College;* **Maria Byrne,** *University of Sydney*

C. Bradford Calloway, *Harvard University;* **Ron Campbell,** *University of Massachusetts at Dartmouth;* **John C. Clamp,** *North Carolina Central University;* **Clayton B. Cook,** *Bermuda Biological Station for Research;* **John O. Corliss,** *University of Maryland;* **Bruce Coull,** *University of South Carolina*

Ferenc A. deSzalay, *Kent State University;* **Don Diebei,** *Memorial University of Newfoundland;* **Ronald V. Dimock, Jr.,** *Wake Forest University;* **William G. Dyer,** *Southern Illinois University, Carbondale*

David A. Evans, *Kalamazoo College*

Daphne G. Fautin, *University of Kansas;* **Paul Fell,** *Connecticut College;* **Ben Foote,** *Kent State University;* **Bernard Fried,** *Lafayette College;* **Peter Funch,** *University of Århus*

Audrey Gabel, *Black Hills State University;* **James R. Garey,** *University of South Florida;* **Ann Grens,** *Indiana University, South Bend*

Kenneth M. Halanych, *Woods Hole Oceanographic Institution;* **G. Richard Harbison,** *Woods Hole Oceanographic Institution;* **Norman Hecht,** *Tufts University;* **Gordon Hendler,** *Natural History Museum of Los Angeles County;* **Anson H. Hines,** *Smithsonian Environmental Research Center;* **Edward S. Hodgson,** *Tufts University;* **Alan R. Holyoak,** *Manchester College;* **Duane Hope,** *Smithsonian Institution;* **William D. Hummon,** *Ohio University, Athens*

William W. Kirby-Smith, *Duke University;* **Robert E. Knowlton,** *George Washington University*

John Lawrence, *University of South Florida;* **William Layton,** *Dartmouth Medical School;* **Herbert W. Levi,** *Harvard University;* **Gail M. Lima,** *Illinois Wesleyan University;* **B. Staffan Lindgren,** *University of Northern British Columbia*

James McClintock, *University of Alabama at Birmingham;* **Rachel Ann Merz,** *Swarthmore College;* **Nancy Milburn,** *Tufts University*

Diane R. Nelson, *East Tennessee State University;* **Claus Nielsen,** *University of Copenhagen;* **Jon Norenburg,** *Smithsonian Institution*

Steve Palumbi, *University of Hawaii;* **Lloyd Peck,** *British Antarctic Survey;* **John F. Pilger,** *Agnes State College;* **William J. Pohley,** *Franklin College of Indiana;* **Gary Polis,** *Vanderbilt University;* **Rudolph Prins,** *Western Kentucky University*

Mary Rice, *Smithsonian Institute;* **Robert Robertson,** *The Academy of Natural Sciences;* **Pamela Roe,** *California State University, Stanislaus;* **Frank Romano,** *Jacksonville State University*

Amelia H. Scheltema, *Woods Hole Oceanographic Institution;* **J. Malcolm Shick,** *University of Maine;* **Owen D. V. Sholes,** *Assumption College;* **Terry Snell,** *University of Tampa;* **Eve C. Southward,** *Plymouth Marine Laboratory (U.K.)*

John Tibbs, *University of Montana, Missoula*

Steve Vogel, *Duke University*

J. Evan Ward, *Salisbury State University;* **Daniel Wickham,** *University of California–Bodega Marine Laboratory;* **Edward O. Wilson,** *Harvard University;* **Jon D. Witman,** *Marine Science Institute–Northeastern University;* **Robert Woollacott,** *Harvard University*

Russel L. Zimmer, *University of Southern California*

I am also indebted to many other colleagues around the world who cheerfully provided their stunning photographs for this and previous editions, answered specific queries in person or by telephone and e-mail, or commented on drafts of new Research Focus Boxes that I based on their original research papers. Thanks also to Jim Carlton for drawing my attention to the correct placement of air ferns (they are hydrozoans, not bryozoans).

I am grateful to all of these people: their devotion to, enthusiasm for, and knowledge about invertebrate biology is both inspiring and humbling.

I am happy to thank my editors Patrick Reidy and Faye Schilling at McGraw-Hill for all their help, and Wendy Langerud and Roxanne Klaas at S4Carlisle Publishing Services who saw the manuscript so competently through production.

Finally, I am happy to thank Steve Wainwright for getting me off to a good start. Chapters 1 and 24 of this book are based on papers I wrote as an undergraduate for his Animal Diversity course at Duke University.

2008 Guide to the Major Animal Groups
(aschelminth groups are shaded blue)
(P = Platyzoa)

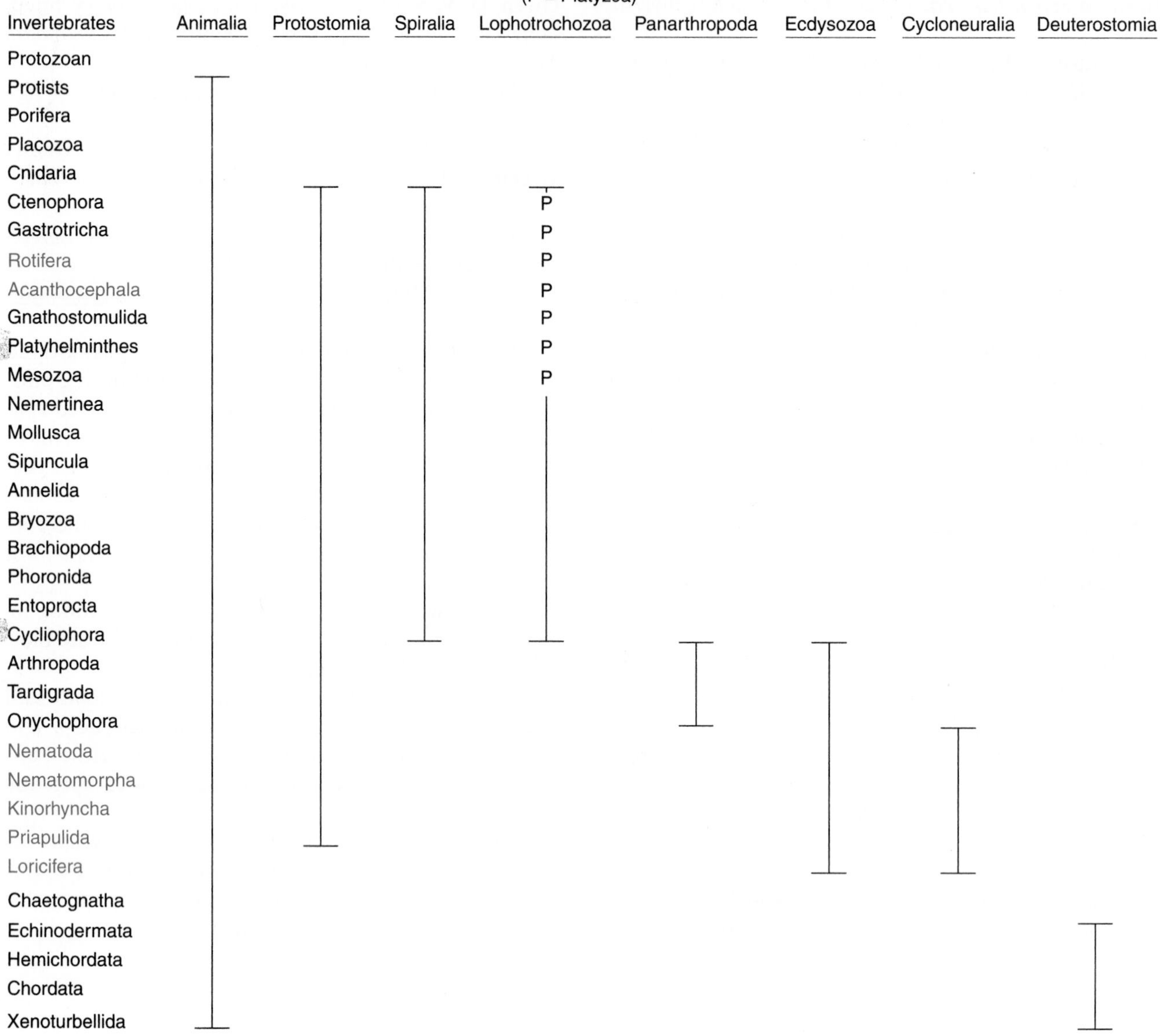

The Geologic Timescale (MYA)
(major global extinctions are shaded blue)

Eons	Eras	Periods	Epochs	MYA	Estimated % of genera going extinct*
		Quaternary	Holocene Pleistocene		
				1.6	
Phanerozoic	Cenozoic	Tertiary	Pliocene		
				5	
			Miocene		
				24	
			Oligocene		
				37	
			Eocene		
				58	
			Paleocene		
				66	47% ± 4%
	Mesozoic	Cretaceous			
				190	
		Jurassic			
				205	53% ± 4%
		Triassic			
				250	82% ± 4%
	Paleozoic	Permian			
				290	
		Pennsylvanian (Late Carboniferous)			
				325	
		Mississippian (Early Carboniferous)			
				355	57% ± 3%
		Devonian			
				410	
		Silurian			
				438	60% ± 4%
		Ordovician			
				510	
		Cambrian			
				544	
		Precambrian			

*Based on Sepkoski, J. J. Jr. 1996. In: O. H. Walliser, ed. *Global Events and Event Stratigraphy.* Springer: Berlin, pp. 35–51.

MYA = Million of years ago

I

Introduction and Environmental Considerations

Introduction: The Importance of Research on Invertebrates

> "We need invertebrates, but they don't need us."
> E. O. Wilson. 1987. The little things that run the world (the importance and conservation of invertebrates).
> *Conservation Biology* 1:344–346

It surprises me that everyone doesn't want to learn more about invertebrates. Much fascinating and important research has been conducted and continues to be conducted using invertebrates. Many diseases of humans and of the animals and plants upon which we depend are caused by invertebrates, either directly or indirectly, and invertebrates play critical roles in most food webs in all habitats. Studies on various invertebrate species have taught us much of what we presently know about: the control of gene expression, mitosis, meiosis, and regeneration; the design of gene regulatory networks in embryonic development; aging, programmed cell death, wound repair, and regeneration; the mechanisms of pattern formation during embryonic development; the control and consequences of phenotypic plasticity, in which a single genotype can produce different phenotypes under different environmental conditions; the evolutionary history of hemoglobin and ecdysteroid function; fertilization and chemoreception; the transmission of nerve impulses; the biochemical basis of learning and memory; the biology of vision; and the biochemical and genetic basis for predisposition to some major diseases (e.g., type II diabetes). Much of what we know about the mechanisms by which genetic diversity originates is maintained and transmitted to succeeding generations through the study of invertebrates, as are many basic principles of animal behavior, development, physiology, ecology, and evolution. Similarly, molecular studies on various invertebrate species are rapidly increasing our understanding of the genetic basis for evolutionary shifts in morphology and life history, including the possible role of horizontal gene transfer, in which sets of genes may be transferred intact from one species to another, and the role that such transfer may play in evolution.

In addition, modern research on invertebrates is helping to unravel the story of how immune recognition systems evolved and how they work. Interest in certain invertebrates as biological agents for controlling various agricultural pests and as sources of unique chemicals of potential biomedical and commercial importance also is increasing. Some of the substances isolated from marine sponges, for example, promise to be potent antitumor agents, and others isolated from certain spiders and venomous snails are providing neurobiologists with highly specific chemical probes for studying key aspects of nerve and muscle function, such as how ion channels are opened and closed. Still other substances derived from invertebrates show considerable promise as instant adhesives (glues produced by onychophorans and barnacles and by some spider and bivalve species, for example) and anticorrosion agents (e.g., barnacle cements); thousands of novel compounds have already been isolated from marine organisms. Detailed studies of crustacean and insect navigation and locomotion, and how that locomotion is controlled and coordinated, may lead to the design of new robots, both flying and crawling, macro and micro; studies on the optical properties of certain sponge fibers may lead to the manufacture of more effective fiber optic cables; and detailed studies of how echinoderms form their remarkable calcite crystals may have similarly sophisticated engineering applications.

Invertebrates also have become widely used to evaluate and monitor pollutant stress in aquatic environments, and the rapid loss of invertebrate species from both terrestrial and aquatic habitats is gaining increasing attention in biodiversity studies.

The recently documented phenomenon Colony Collapse Disorder, in which hundreds of thousands of honeybees simply abandon their hives and disappear, is worrisome: apples, almonds and approximately 90 other crops in the United States depend on honeybees for pollination, accounting for some $14 billion in annual sales. Bumblebees, which pollinate some 15% of commercial crops in the U.S., are also in serious decline. Similarly ominous is the recently documented increase in the acidity of seawater in the world's oceans, as discussed in the next section.

Finally, there is a growing concern about the increased spread of various invertebrate species into nonnative habitats, and increasing attention is being paid to the mechanisms of transport and to the ecological impact of such biological invasions. Of course, nobody yet knows the consequences of continued pollution, biological invasions, and global climate change on food web function, in either aquatic or terrestrial food-webs; probably there is only one way to do the experiment, and we're all participating.

This book, together with lectures and laboratory sessions, opens the door to the great and growing literature of invertebrate zoology.

Environmental Considerations

The organisms considered in this book are grouped into more than 30 phyla. The members of almost half of these phyla are entirely marine, and the members of the remaining phyla are found primarily in marine and, to a lesser extent, freshwater habitats. Excluding the arthropods, invertebrates have generally been far less successful in invading terrestrial environments. Even those invertebrate species that are terrestrial as adults often have aquatic developmental stages. It is worthwhile, therefore, to consider some of the physical properties of water—both fresh and salt—to discover why so many species are aquatic for all or part of their lives. The physical properties of salt water, freshwater, and air play major roles in determining the structural, physiological, and behavioral characteristics displayed by animals living in various habitats.

Air Is Dry, Water Is Wet

Air is dry, whereas water is wet. As trivial as this statement may seem, the repercussions with respect to morphology, respiratory physiology, nitrogen metabolism, and reproductive biology are tremendous, as seen in Table 1.1.

Since aquatic organisms are in no danger of drying out, gas exchange can be accomplished across the general body surface. Thus, the body walls of aquatic invertebrates are generally thin and water permeable, and any specialized respiratory structures that exist may be external and in direct contact with the surrounding medium. Gills, which can be structurally quite complex, are simply vascularized extensions of the outer body wall. These extensions increase the surface area available for gas exchange and, if they are especially thin walled, may also increase the efficiency of respiration (measured as the volume of gas exchanged per unit time per unit area).

In contrast to the minimal complexity required for aquatic respiratory systems, terrestrial organisms must cope with potential desiccation (dehydration). Terrestrial species relying on simple diffusion of gases through unspecialized body surfaces must have some means of maintaining a moist outer body surface, as by the secretion of mucus in earthworms. Truly terrestrial invertebrates generally have a water-impermeable outer body covering that prevents rapid dehydration. Gas exchange in such species must be accomplished through specialized, internal respiratory structures.

The union of sperm and egg, and the subsequent development of a zygote, can be achieved far more simply by aquatic invertebrates than by terrestrial species. Marine organisms, in particular, may shed sperm and eggs freely into the environment. Because the gametes and embryos of marine species are not subject to dehydration or to osmotic stress, fertilization and development can be completed entirely in the water. Fertilization in the terrestrial environment, on the other hand, must be internal to avoid dehydration of gametes; terrestrial

Table 1.1 Summary of the Different Lifestyles Possible in the Two Major Environments, Aquatic and Terrestrial, as They Reflect Differences in the Physical Properties of Water and Air

Property	Water	Air
Humidity	High: Exposed respiratory surfaces; external fertilization*; external development; excretion of ammonia	Low: Internalized respiratory surfaces; internal fertilization; protected development; excretion of urea and uric acid
Density	High: Rigid skeletal supports unnecessary; filter-feeding lifestyle possible; external fertilization; dispersing developmental stages*	Low: Rigid skeletal supports necessary; must move to find food; internal fertilization; sedentary developmental stages
Compressibility	Low: Transmits pressure changes uniformly and effectively	High: Less effective at transmitting pressure changes
Specific heat	High: Temperature stability	Low: Wide fluctuations in ambient temperature
Oxygen solubility	Low: 5–6 ml O_2/liter of water	High: 210 ml O_2/liter of air
Viscosity	High: Organisms sink slowly; greater frictional resistance to movement	Low: Faster rates of falling; less frictional resistance to movement
Rate of oxygen diffusion	Low: Animal must move (or must move water) for gas exchange	High: (about 10,000 times higher than in water)
Nutrient content	High: Salts and nutrients available through absorption directly from water for all life stages*; adults may make minimal nutrient investment per egg*	Low: No nutrients available via direct absorption from air; adults supply eggs with all nutrients and salts needed for development
Light-extinction coefficient	High: Animals may be far removed from sites of surface-water primary production	Low: Animals never far from sites of primary production

*Signifies features that are especially characteristic of marine invertebrates and uncommon among freshwater invertebrates.

invertebrates therefore require more complex reproductive systems than do their marine counterparts. Successful fertilization of terrestrial eggs often involves complex reproductive behaviors as well.

Ammonia is the basic end product of amino acid metabolism in all organisms, regardless of habitat. Ammonia is usually very toxic, largely through its effects on cellular respiration. Even a small accumulation of ammonia in the tissues and blood is detrimental to individuals of most species. However, few terrestrial organisms can afford the luxury of constantly eliminating ammonia as it is produced, since the water required to flush out the ammonia is in short supply. As an adaptation to life on land, terrestrial organisms usually incorporate ammonia into less toxic compounds (urea and uric acid), which can then be excreted in a smaller amount of water. This detoxification of ammonia requires additional biochemical pathways and an increased expenditure of energy. Aquatic invertebrates, on the other hand, can simply use the surrounding water to dilute away metabolic ammonia as it is produced. Moreover, because water is wet, ammonia may be excreted by simple diffusion across the general body surface of many aquatic invertebrates. In contrast, all terrestrial animals require complex excretory systems.

Water Is the "Universal Solvent"

Water is a remarkably versatile solvent. The benefits to aquatic invertebrates are both direct and indirect. First of all, aquatic animals can potentially take up dissolved nutrients (including amino acids, carbohydrates, and salts) directly from the surrounding water by diffusion or by active uptake. In particular, dissolved salts and organic water-soluble nutrients may be taken up directly from water by developing embryos and larvae. Embryos of terrestrial organisms must be supplied (by their parents) with all food and salts needed for development and must be protected from desiccation as well. Second, as an indirect benefit to aquatic invertebrates, suspension in a nutrient-containing, wet medium permits primary producers to take the form of small (typically less than 25 μm [micrometers]), suspended, single-celled organisms (**phytoplankton**); roots are not mandatory. Phytoplankton cells can attain high concentrations in water, and can easily be harvested and ingested by many suspension-feeding aquatic herbivores, including the developmental stages of many invertebrate species.

Water Is Denser than Air

Water is far denser than air, a fact that has profound consequences for invertebrates. For example, a rigid skeletal support system is not required in water, since the medium itself is supportive; water also supports delicate anatomical structures, such as gill filaments, that would collapse and cease to function properly in air. For the same reason, animals can often move with greater efficiency in water than in air, expending less energy to progress a given distance. Indeed, many aquatic invertebrate species expend virtually no energy at all for movement—they simply don't move. How do such animals feed without the ability to move? Because water is wet and dense, microscopic free-floating "plants" and animals (phytoplankton and **zooplankton,** respectively) live in suspension; this enables many other aquatic animals to make their livings "sitting down," capturing food particles directly from the medium as it flows past the stationary animal. Often, some energy must be expended to move water past the animal's feeding structures, but the animal need not use energy in a search for food. Such a **suspension-feeding** existence, quite commonly encountered in aquatic environments, seems to have been exploited only by web-building spiders in the terrestrial habitat. Potential food particles simply do not occur in high concentrations in the dry, unsupportive air.

External fertilization and external development of embryos, so commonly encountered among marine invertebrates, are made possible as much by water's high density as by its wetness; the water supports both sperm and eggs and the embryo itself as it develops. In many groups of marine invertebrates, external fertilization and/or external larval development is the rule rather than the exception. Because little energy may be required to remain afloat in the aquatic medium, developmental stages (e.g., embryos, larvae) of aquatic invertebrates often serve as the dispersal stages for sedentary adults—exactly the opposite of the situation encountered among most terrestrial animals.

Water Has Thermal Stability

One additional advantage of water as a biological environment is its relatively high temperature stability with respect to air. Water has a high specific heat; that is, the number of calories required to heat 1 g (gram) of water 1°C is considerably greater than that required to raise the temperature of 1 g of most other substances by the same 1°C. Because of its high specific heat, water is slow to cool and slow to heat up; water temperature is relatively insensitive to short-term fluctuations in air temperature. Over a 24-hour period, air temperatures at midlatitudes may vary by 20°C or more. For reasonably large volumes of water, local surface temperatures will probably not vary by more than 1°C to 2°C over the same time interval.

Differences in seasonal temperature fluctuations are even more striking. Near Cape Cod, Massachusetts, for example, local seawater temperature may vary between approximately 5°C during the winter and 20°C during the summer: a seasonal range of about 15°C. Air temperatures, on the other hand, fluctuate between approximately –25°C and 40°C: a seasonal range of 65°C. Even in small lakes and ponds, the annual range of water temperatures is much smaller than that of air temperatures in the same geographic area. Because the rates of all chemical reactions, including those associated with organismal metabolism, are altered by temperature, wide fluctuations in temperature (especially those occurring over short time intervals) are highly stressful to most invertebrates. Invertebrates living in thermally variable environments require biochemical, physiological, and/or behavioral adaptations not required by organisms living in more stable, aquatic habitats.

Problems with the Aquatic Life

Life in water does pose some problems. Light is extinguished over a much shorter distance in water than in air, so most aquatic **primary production** (fixation of carbon from carbon dioxide into carbohydrates, generally by photosynthesizing plants, algae, and phytoplankton) is limited to the upper 20 m to 50 m (meters) or so. Moreover, water's oxygen-carrying capacity, volume for volume, is only about 2.5% that of air. An additional problem for aquatic organisms is that the time required for a given molecule to diffuse across a given distance in water is much, much greater than the time required for the same molecule to diffuse across the same distance in air: Indeed, oxygen moves more than 300,000 times faster in air than in water! An organism sitting completely still in motionless water would have a severe gas-exchange problem once the fluid immediately in contact with the respiratory surface had given up all available oxygen (and/or had become saturated with carbon dioxide). On the other hand, even the slightest movement of the water surrounding an animal's respiratory surface enhances gas exchange significantly. **Sessile** (nonmotile) organisms living in areas of significant water-current velocity thus benefit in terms of gas exchange as well as nutrient replenishment. Sessile animals living in still water invariably have some means of creating water flow over their respiratory surfaces.

Potential difficulties are also created by the greater density and viscosity of water. Water is about 800 times denser and about 50 times more viscous than air. **Viscosity** essentially measures the extent to which the molecules of a fluid stick to each other. In contrast, density, which was referred to earlier in discussing the benefits of life in water, is a measure of mass per unit volume. Because of water's greater density and viscosity, animals swimming through it or facing a current experience far more frictional resistance (called **drag**) than they would experience in air. For large animals moving quickly (or facing a fast-moving current), the greater drag is due primarily to the greater density of water, whereas small animals moving slowly (or facing a slow-moving current) are affected mainly by water's greater viscosity. Because viscosity increases much more dramatically in water than in air for any given decline in temperature, small, slow-moving aquatic organisms experience noticeably greater frictional resistance in swimming as temperature falls.

Indeed, small organisms—which really must swim slowly because of their small size—live in a world dominated by viscous forces, a world in which **Reynolds numbers** (essentially a ratio of inertial to viscous forces) are very low. It is difficult for us to imagine what life is like in such a world, a world in which there is no such thing as "gliding" to a stop; instead, as soon as propulsion stops, the animal stops. Moreover, in a world of low Reynolds numbers, water tends to move primarily around rather than through bristly appendages; in such a world, rake-shaped objects behave much like solid paddles, so that they cannot readily filter food particles from the water. Clearly, animals operating at low Reynolds numbers are subjected to some physical selective pressures quite unlike those acting on larger, faster-moving organisms; even such basic biological functions as locomotion and suspension-feeding may require specialized physiological and behavioral adaptations, adaptations that often seem somewhat peculiar and counterintuitive.[1]

The fact that water is the so-called universal solvent creates another problem that should be particularly acute for aquatic invertebrates. Many of our industrial and agricultural waste products are water soluble, and we insert fantastic amounts of such pollutants into aquatic ecosystems each year. Aquatic animals must live in particularly intimate contact with these pollutants. Consider, for example, that the gas exchange surfaces of aquatic invertebrates are always in direct contact with the surrounding fluid. Consider also that many aquatic invertebrates are small, so that the surface area across which pollutants can diffuse is high relative to the animal's body volume. Free-living embryonic and larval stages, so common in aquatic invertebrate life cycles, would seem especially vulnerable to pollutant insult, partly because of their high-surface-area:volume ratios and partly because they are undergoing such complex and critical developmental processes. Indeed, for any given toxicant, developmental stages typically suffer adverse effects at only one-tenth to one-hundredth the concentration required to affect adults of the same species to the same degree.

Carbon dioxide, like other gases, is also water soluble. Scientists estimate that the oceans have absorbed about a third of our excess CO_2 emissions over the past 50 years or so. Incredibly, this uptake of CO_2 has overwhelmed the bicarbonate buffering system of seawater—one of the most remarkable things to have occurred in my lifetime—and lowered ocean pH by about 0.1 pH unit. The pH of seawater is expected to continue declining (by between 0.14 and 0.35 pH units) for the rest of the century. Rising acidity should eventually interfere with the ability of many marine organisms to calcify. The consequences of continued acidification will be interesting to document: organisms like foraminiferans, corals, sea urchins, snails, clams, and the developmental stages of such creatures as sea urchins, snails, and clams—all of which secrete calcium carbonate supporting or protective structures—should be especially vulnerable.

Organisms living in freshwater face several difficulties unique to the freshwater environment. For one thing, most bodies of freshwater are ultimately ephemeral, with smaller ponds and lakes being subject to drying up at yearly or even more frequent intervals. Most marine invertebrates are not faced with such a high degree of habitat unreliability. Second, the internal body fluids of freshwater organisms are always higher in osmotic concentration than is the surrounding medium; that is, freshwater organisms are **hyperosmotic** to their surroundings, and water tends to diffuse inward along the osmotic concentration gradient. Some freshwater animals have reduced surface permeability to water, reducing the magnitude of this inflow. Complete impermeability to water is not possible, however, because respiratory surfaces must remain permeable for gas exchange to occur. Thus, all freshwater animals must be capable of constantly expelling large volumes of incoming freshwater. In contrast, marine invertebrates are approximately in **osmotic equilibrium** with the medium in which they live; that is, the concentration of solutes in their body fluids matches that of the surrounding seawater.

Also, because salts are relatively rare in the freshwater medium (by definition of freshwater), most of the salts necessary for embryonic development must be supplied to the egg by the mother. By contrast, all salts required for the differentiation and growth of marine embryos are readily available in the surrounding medium.

The relative paucity of salts in freshwater has additional ramifications for animals living in it. Freshwater organisms, which must constantly expel incoming water, often possess sophisticated physiological mechanisms for reclaiming precious salts from the urine before the urine leaves the body; they also must possess mechanisms for replacing any salt loss that does occur.

Finally, the pH of salt water is far less variable, both spatially and temporally, than is the pH of freshwater. The pH of salt water is maintained near 8.1 by the tremendous buffering capacity of the bicarbonate ion. Most freshwater environments lack such a high buffering capacity, and the pH of the water is therefore far more sensitive to local fluctuation of acid and base content.

---

1. See *Topic for Further Discussion and Investigation* at the end of the chapter.

Origins and Diversity of Life

From such considerations of the properties of air, salt water, and freshwater, it is easy to understand why life must have originated in the ocean. The specialized physiological and/or morphological adaptations essential for existence on land or in freshwater are not required for the relatively simple and generally less stressful existence possible in the marine environment. Once life arose, various **preadaptations** eventually evolved that made a transition from saltwater environments to other habitats possible. Such preadaptations for terrestrial and freshwater life apparently arose rarely in many groups of animals and not at all in others. Not surprisingly, most phyla are still best represented in the ocean, both in terms of species numbers and in terms of the diversity of body plans and lifestyles.

Topic for Further Discussion and Investigation

In what ways are small invertebrates adapted to life at low Reynolds numbers?

Vogel, S. 1994. *Life in Moving Fluids,* 2d ed. Princeton, N.J.: Princeton University Press.

Vogel, S. 2003. *Comparative Biomechanics: Life's Physical World.* Princeton, N.J.: Princeton University Press.

2

Invertebrate Classification and Relationships

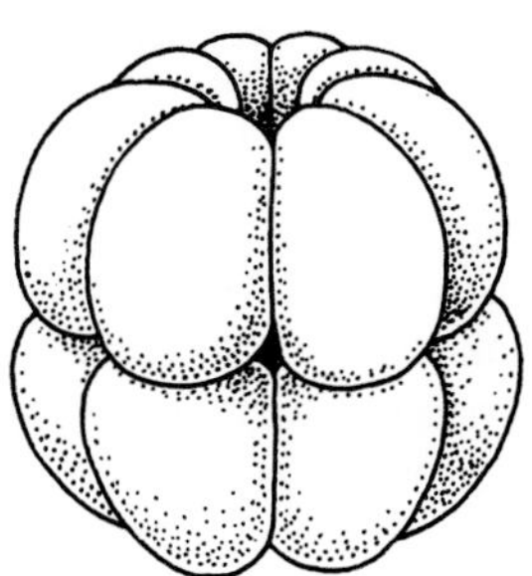

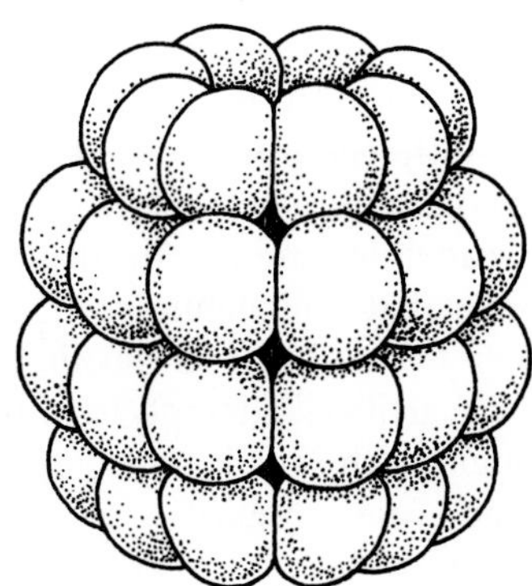

"According to my opinion, (which I give every one leave to hoot at . . .) classification consists in grouping beings according to their actual *relationship*, i.e., their consanguinity, or descent from common stocks."
Charles Darwin (1843)

Introduction

At one time, presumably, there were no animals. The marvelous variety of animal life-forms seen today and in the fossil record must have evolved gradually, beginning over 3 billion years ago; the Earth itself is over 4.5 billion years old. More than one million animal species have now been described and named, but at least another 4 to 10 million species probably await discovery and description; many of these will undoubtedly become extinct without being discovered. Probably several hundred million other species were here previously but are now extinct.

Multicellular life seems to have taken quite a long time to evolve from single-celled ancestral forms: Fossils of the earliest known unicellular eukaryotes (see Chapter 3) are about 1.8 to 1.9 billion years old, but the oldest known fossils of multicellular animals (called **metazoans**) or their burrows are no more than 543 to 635 million years old, members of the so-called Ediacaran fauna first discovered in South Australia. Moreover, none of those Ediacaran animals had shells, bones, or other hard parts, and their relationship to modern animals, if any, is unclear.[1] The first sizable metazoans that are clearly related to modern animals appear abruptly in the Cambrian period about 543 mya (million years ago). The best-studied invertebrate fossils are from the Burgess Shale of British Columbia, first discovered only in 1909 but formed some 525 mya in the Cambrian. Many of these animals were soft-bodied and others had hard parts, but their most conspicuous feature

1. See *Topics for Further Discussion and Investigation*, no. 3, at the end of the chapter.

is their substantial diversity. A similar fauna was discovered more recently in China, from older sedimentary rocks formed in the early Cambrian, about 540 mya. This amazingly sudden appearance and apparently rapid diversification of complex animals over several millions of years has been called the **Cambrian explosion.**

There is now some evidence that the Cambrian explosion reflects an incomplete fossil record.[2] For example, what may be cnidarian-like, echinoderm-like, and arthropod-like metazoan embryos have recently been described from southern China in rocks formed about 580 mya (Fig. 2.1), suggesting that forms related to modern animals existed at least 40 million years before the recorded Cambrian explosion.[3] More dramatically, some recent molecular studies suggest that most basic animal body plans existed at least 100 million years before any were preserved as fossils. This suggestion is based on differences in the amino acid sequences of particular proteins or differences in the nucleotide sequences of particular genes (e.g., cytochrome *c*) that are widespread among various animal groups, coupled with estimates of how long it should have taken for the proteins or underlying gene sequences to have diverged that far from each other. If the interpretations of these data are correct, the basic animal groups may have begun diverging as long ago as 1 billion years, but without leaving any historical record for the first 400 to 500 million years of their evolution. Possibly these early animals were simply too small and lacking in hard parts to be fossilized. Perhaps it was the gradual increase in atmospheric oxygen above some critical concentration, due to increased photosynthetic activity, that permitted larger body sizes and hard, impermeable outer body coverings to evolve, creating novel opportunities for fossilization. Or, perhaps the particular environmental conditions needed for fossil formation simply did not exist before about 600 mya. If the molecular data are correct, the explosion of animal body plans recorded in the Cambrian period reflects an increase in the numbers and kinds of fossilizable animals, not the sudden invention of new animal designs. Or perhaps the molecular analyses are misleading and there really was an explosion of animal body plans somewhere around 540 mya, attributable perhaps to dramatically increased pressures of predation and competition.

In any event, nearly all of today's major animal phyla are represented among the Cambrian fossils formed some 525–540 mya; without ancestral stages and stages that are intermediate between the various animal groups, the fossil record provides no clues about how these phyla are related to each other. Studying the extensive Cambrian and post-Cambrian fossil record can tell us something about evolution *since* the Cambrian explosion, but nothing about the ancestors from which these fossilized animals evolved. However, if we make the very reasonable assumption that all animals have ancestral forms in common, and that as animals evolved from those common ancestors they became less and less alike, we can infer evolutionary relationships, with varying degrees of certainty. Such inferences are based on morphological, developmental, physiological, biochemical, and genetic similarities and differences among animal groups. In the next few sections, we'll look at some of those key traits.

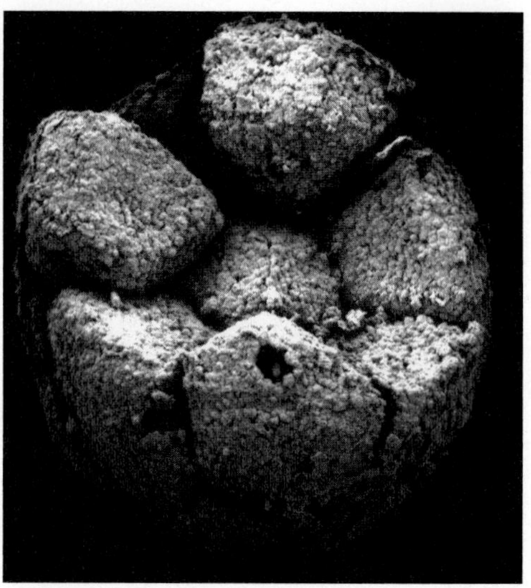

Figure 2.1
What appears to be a multicellular fossilized embryo (~ 500 μm diameter) from deposits in southern China formed about 580 million years ago. If this is truly the embryo of a bilaterally symmetrical, multicellular animal then a diversity of multicellular animal life undoubtedly existed long before the Burgess Shale record of the Cambrian explosion. But recent evidence suggests that the "embryos" could instead be clusters of dividing bacteria.* Stay tuned.
Courtesy Shuhai Xiao and Ed Seling. From Xiao et al. 1998. *Nature* 391:553–58. Reprinted with permission from *Nature*, © Macmillan Magazines Ltd. Harvard University, © President and Fellows of Harvard College. * Bailey, J.V. et al. 2007. *Nature* 445: 198–201.

Before we can consider the evolutionary interrelationships among different groups of organisms, we must sort the millions of animal species into categories, which can be done only after determining the degrees of similarity and difference that will define each category. It is important to keep in mind that all classification schemes are, at least in part, artificial attempts to impose order. As we will see throughout this book, many organisms do not fit cleanly into any one group; it is relatively simple to decide upon the categories to be used but often far more difficult to determine the category to which a given organism belongs. Once the organisms are assigned to taxonomic categories, it becomes possible to consider the evolutionary relationships among and within those categories. In this chapter, we will consider some of the schemes that have been developed to sort animals into groups, and to then deduce the evolutionary relationships among and within those groups.

2. See *Topics for Further Discussion and Investigation,* no. 6, at the end of this chapter.

3. See *Topics for Further Discussion and Investigation,* no. 8, at the end of this chapter.

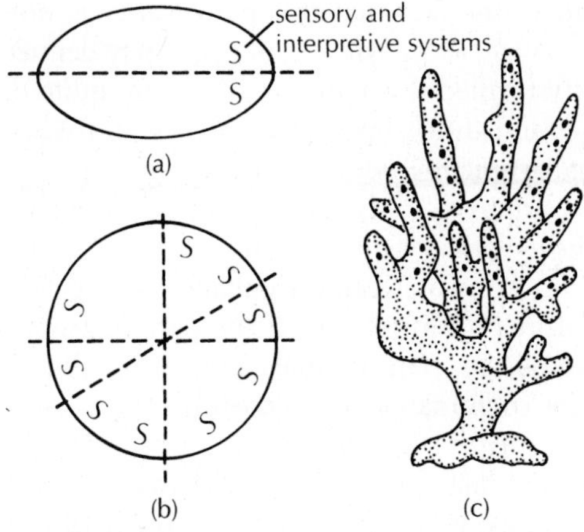

Figure 2.2
Various types of body symmetry. (a) Bilateral symmetry. (b) Radial symmetry. (c) Asymmetrical body plan of a marine sponge.

Classification by Cell Number, Embryology, and Body Symmetry

Invertebrates have been categorized in many ways. One of the most basic divisions is based upon whether individuals are single celled or composed of many cells. True animals are multicellular, generally diploid organisms that each develop from a blastula; these organisms are referred to collectively as the Metazoa, or as **metazoans.** Other invertebrates are considered either **unicellular** (single celled) or **acellular** (without cells)—a distinction discussed further in Chapter 3—and do not develop from anything resembling a metazoan embryo. As we will see in the next several chapters, the point at which an association of cells can be viewed as composing a multicellular organism is not always clear-cut. It is widely agreed that multicellular life evolved from some unicellular organism. Thus, there has been considerable interest in trying to determine how many times multicellularity arose, and from which unicellular ancestors it arose.

Animals may also be classified according to their general body form. Most metazoans show one of two types of body symmetry (Fig. 2.2a,b), at least superficially. Animals like ourselves are **bilaterally symmetrical,** possessing right and left sides that are approximate mirror images of each other. Bilateral symmetry is highly correlated with **cephalization,** which is the concentration of nervous and sensory tissues and organs at one end of an animal, resulting in distinct anterior and posterior ends. For an animal that shows cephalization, two mirror images can be produced only when a slice is made parallel to the animal's long (anterior-posterior) axis, with the cut passing down the midline. Any cut perpendicular to this midline, even when passing through the animal's center, creates two dissimilar pieces. This is not so for a **radially symmetrical** organism. Such an animal can be divided into two approximately equal halves by any cut that passes through its center. Thus, most animals belong to either the Radiata or the Bilateria. Asymmetrical invertebrates—those having no ordered pattern to their gross morphology—are uncommon (Fig. 2.2c).

Once again, what seems to be straightforward on the surface is never quite so simple when dealing with actual animals. Many species whose external appearances are the epitome of uncontroversial radial symmetry have asymmetrical internal anatomies. Some sea anemones, for example, are internally bilaterally symmetrical, and even show patterns of gene expression during development that resemble those of other bilateral animals.[4] Perhaps it would have been better to group animals based on degree of cephalization rather than on the basis of body symmetry. I bow, however, to historical precedent.

Classification by Developmental Pattern

Developmental pattern has long played a pivotal role in classification schemes, and in deducing evolutionary relationships, as discussed in the next several sections. Multicellular invertebrates have for many years been divided into two groups based upon the number of distinguishable germ layers formed during embryogenesis. **Germ layers** are groups of cells that behave as a unit during the early stages of embryonic development and give rise to distinctly different tissue and/or organ systems in the adult. In **diploblastic** animals (*diplo* = Greek: double), only 2 distinct germ layers form during or following the movement of cells into the embryo's interior. The outermost layer of cells is called the **ectoderm** (*ecto* = G: outer; *derm* = G: skin) and the innermost layer of cells is called the **endoderm** (*endo* = G: inner). Members of only a few phyla (notably the Cnidaria) are generally considered to be diploblastic (Fig. 2.10). Most metazoans are instead **triploblastic** (*triplo* = G: triple). During the ontogeny of triploblastic animals, cells of either the ectoderm or, more usually, the endoderm give rise to a third germ layer, the **mesoderm** (*meso* = G: middle). This mesodermal layer of tissue always lies between the outer ectodermal tissue and the inner endodermal tissue.

The absence of a distinct, embryonic, third tissue layer does not mean that the adult of a diploblastic species will lack the tissues that are derived from this layer in adults of a triploblastic species. Muscular elements, for example, derive from the mesodermal layer in triploblastic animals, but diploblastic adults also have musculature, despite the absence of a morphologically or behaviorally distinct group of cells that can be termed mesoderm in the early embryo.

Triploblastic animals have been further classified into 3 basic plans of body construction, based on whether they have an internal body cavity independent of the digestive

4. Finnerty *et al.*, 2004. *Science* 304:1335–37; Matus, D. Q. *et al.*, 2006. *Proc. Natl. Acad. Sci.* 103:11195–200.

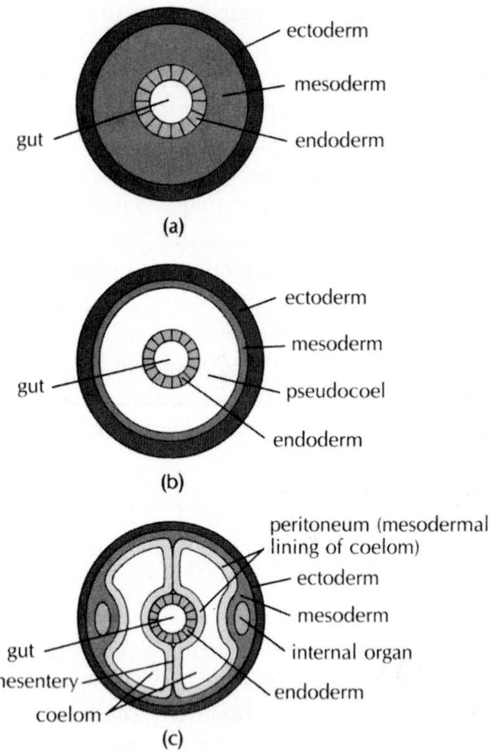

Figure 2.3
(a) Diagrammatic cross section through the body of an acoelomate. The space between the gut and the outer body wall musculature is completely filled with tissue derived from embryonic mesoderm. (b) Cross section through the body of a pseudocoelomate. The gut derives entirely from endoderm and is therefore not lined with mesoderm. (c) Cross section through the body of a coelomate. The entire coelomic space is bordered by tissue derived from embryonic mesoderm.

tract (gut) and on how this cavity forms during embryogenesis. Although their importance as phylogenetic clues has diminished considerably in recent years, or at least has become more controversial, these developmental characteristics have long played central roles in arguments about how triploblastic metazoans are related to each other. Triploblastic animals lacking an internal body cavity are said to be **acoelomate** (*a* = G: without; *coelom* = G: a hollow space). Characteristically, the region lying between the outer body wall and the gut of acoelomates is solid, being occupied by mesoderm (Fig. 2.3a), and there is no trace of an internal body cavity during embryological development.

In a second group of animals, the region between the outer body wall musculature and the endoderm of the gut is a fluid-filled cavity (Figs. 2.3b, 2.4a); in some species, this cavity is derived from the **blastocoel,** an internal space that develops in the embryo prior to gastrulation [(Figs. 2.4b(1), c(2), and 2.5a)]. This type of body cavity is termed a **pseudocoel,** and the organism housing it is said to be a **pseudocoelomate.** The name is a bit misleading. The *pseudo* is not intended to disparage the *coel;* the body cavity is genuine. The *pseudo* prefix merely draws attention to the fact that this body cavity is not a true coelom, which, as we will see, is a precisely defined internal cavity formed through one of several quite different processes and always lined completely with tissue derived from embryonic mesoderm.

This brings us to the third group of triploblastic animals, those with a true **coelom:** an internal, fluid-filled body cavity lying between the gut and the outer body wall musculature and lined with tissue derived from embryonic mesoderm. The animals possessing such a body cavity are **coelomates** (or **eucoelomates;** *eu* = G: true, proper). Coelom formation may occur by either of 2 quite dissimilar mechanisms a characteristic that has long been used to assign coelomates to one of two major subgroups: protostomes or deuterostomes. Among **protostomes,** coelom formation occurs by gradual enlargement of a split in the mesoderm (Fig. 2.4b). This process is termed **schizocoely** (*schizo* = G: split). Among **deuterostomes,** on the other hand, the coelom typically forms through evagination of the archenteron into the embryonic blastocoel (Figs. 2.4c and 2.5). Because the coelom of deuterostomes forms from a part of what eventually becomes the gut, coelom formation in this group of animals is termed **enterocoely** (*entero* = G: gut).

Whether the coelom forms by schizocoely or enterocoely, the end result is similar. The organism is left with a fluid-filled internal body cavity lying between the gut and the outer body wall musculature, and unlike the cavity of pseudocoelomates, this cavity is lined by a mesodermally derived epithelium. The fact that internal cavities develop by any of 3 distinctly different mechanisms (enterocoely, schizocoely, or persistence of embryonic blastocoel) suggests that such cavities have been independently evolved at least 3 times. If so, the selective pressures favoring the evolution of internal body cavities must have been substantial.

Indeed, selective advantages are easy to imagine. For example, with an internal body cavity the gut is somewhat independent from muscular, locomotory activities of the body wall. Also, the animal gains internal space into which can bulge digestive organs, gonads, and developing embryos, and an internal fluid that can serve to distribute oxygen, nutrients, and hormones or neurosecretory substances throughout the body, facilitating the evolution of larger body sizes. Perhaps most significantly, fluid-filled body cavities can lead to more effective locomotory systems, as discussed in Chapter 5.

To summarize, triploblastic animals can be acoelomate, pseudocoelomate, or coelomate, depending on whether they possess an internal, fluid-filled body cavity and on whether this body cavity is lined by mesodermally derived tissue. These are 3 distinctive types of organization. However, what they tell us about the evolutionary relationships among animals in the different categories, or even among animals within each of the categories is uncertain; in particular, it has become increasingly obvious that coelomic cavities can be lost as well as gained,

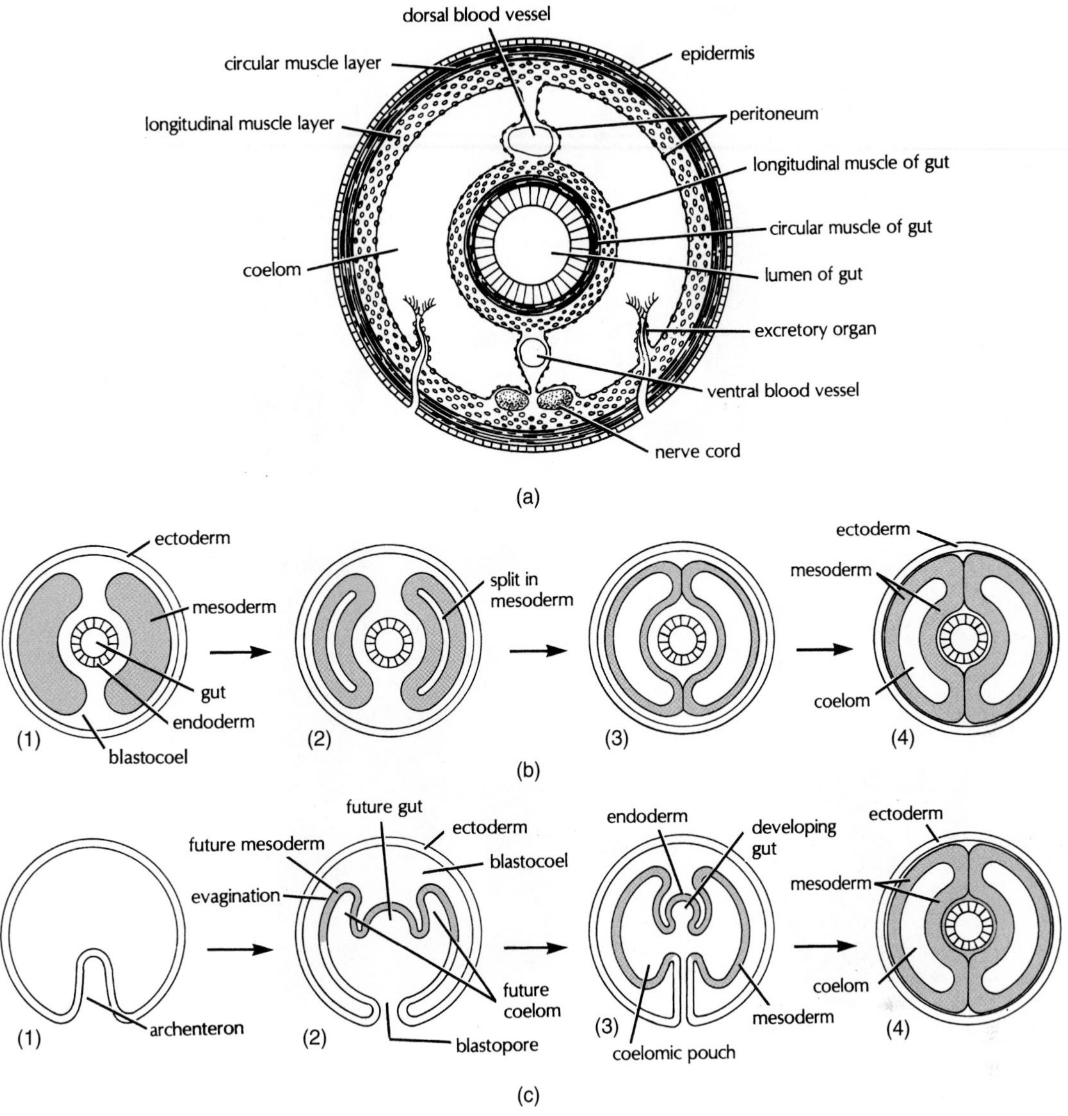

Figure 2.4

(a) Detailed cross section through the body of a coelomate. The tissues bordering the coelomic space include the musculature of the gut; the mesenteries, which suspend the various organs in the coelom; and the peritoneum, which lines the coelomic cavity. (b) Coelom formation by schizocoely—that is, by an actual split, or schism, in the mesodermal tissue. (c) Coelom formation by enterocoely, in which the archenteron evaginates into the embryonic blastocoel.

that acoelomates are unlikely to have evolved from a single common ancestor, and that the acoelomate body plan may not be *primitive* (i.e., closest to the ancestral triploblastic condition). Some workers have suggested that the earliest triploblastic animals were coelomate, and that the acoelomate condition may thus represent a number of independent losses of the body cavity.

In contrast, the distinction between protostomes and deuterostomes seems secure; the validity of these 2 groups has so far been largely upheld by molecular data, suggesting that protostome species are indeed more closely related to each other than to any deuterostome species.

But mode of coelom formation is only one of several characteristics distinguishing protostomes from deuterostomes. In fact, the terms *protostome* and *deuterostome* were actually coined to reflect differences in the embryonic origin of the mouth (*stoma* = G: mouth). Among the protostomes, the mouth (and sometimes the anus) forms from the blastopore (the opening from the outside into the archenteron); hence, the term *protostome,* meaning "first mouth,"—since the mouth forms from the first opening that appears during embryonic development. Among the deuterostomes, the mouth never develops from the blastopore: Although the blastopore may give rise to the anus, as in some protostomes, the deuterostome mouth always forms as a second, novel opening elsewhere on the embryo—hence, the term *deuterostome,* meaning "second mouth."

What other characteristics distinguish protostomes from deuterostomes? In addition to differing in the mode of coelom formation and in the embryological origin of the mouth, they also typically differ in the number of

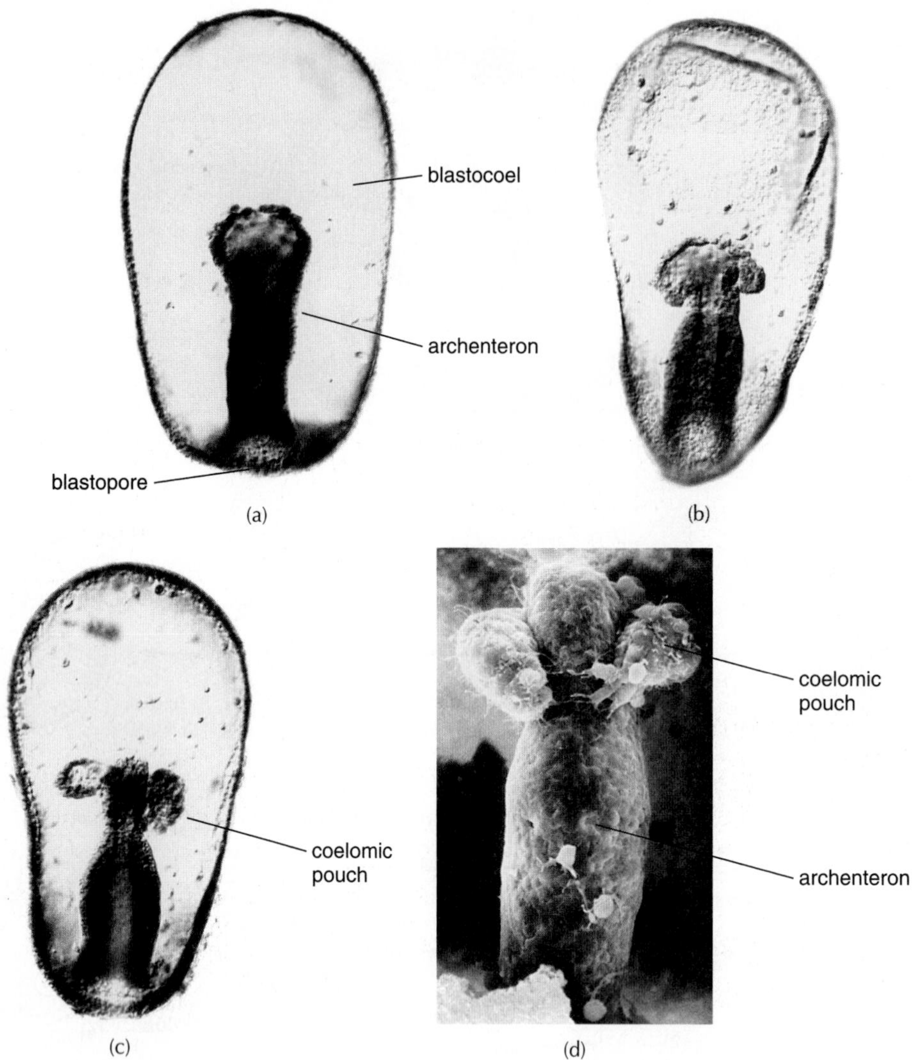

Figure 2.5
(a–c) Enterocoely in a sea star, showing the coelomic pouches forming from the sides of the archenteron and splitting off. (d) The coelomic pouches are clearly shown in this scanning electron micrograph. The pouches gradually enlarge to form the coelom.
(a–d) Courtesy of B. J. Crawford, from Crawford and Chia, 1978 *Journal of Morphology* 157:99. Reprinted by permission of Wiley-Liss, Inc., a subsidiary of John Wiley & Sons, Inc.

coelomic cavities formed as they develop. Among protostomes, the number of coelomic cavities is highly variable: For example, an annelid worm can have as many coelomic cavities as it has segments—hundreds in some species. Among deuterostomes, however, the original coelomic cavity generally subdivides to form 3 pairs of coelomic pouches (i.e., the deuterostome coelom is commonly *tripartite*). Protostomes and deuterostomes may also differ with respect to the orientation of the spindle axes of the cells during cleavage, the point in development at which cell fates become irrevocably fixed, and how the mesoderm originates.

Cleavage is often referred to as being either radial or spiral, depending on the orientation of the mitotic spindles relative to the egg axis. Generally, yolk is asymmetrically distributed within eggs, and the nucleus occurs in, or moves to, the region of lower yolk density. This is the **animal pole,** and it is here that the polar bodies are given off during meiosis. The opposite end of the egg is termed the **vegetal** (not vegetable!) **pole.**

In **radial cleavage** (deuterostomes), the spindles of a given cell, and thus the cleavage planes, are oriented either parallel or perpendicular to the animal-vegetal axis. Thus, daughter cells derived from a division in which the cleavage plane is parallel to the animal-vegetal axis end up lying in the same plane as the original mother cell (Fig. 2.6a,b). The two daughter cells resulting from a division perpendicular to the animal-vegetal axis come to lie directly one atop the other, with the center of the upper cell lying directly over the center of the underlying cell (Figs. 2.6c–f and 2.7).

In contrast, the spindle axes of cells undergoing **spiral cleavage** are oriented (after the first two cleavages) at 45° angles to the animal-vegetal axis (Fig. 2.6j–k). Moreover, the division line does not necessarily pass through the center of the dividing cell. As a result, by the eight-cell stage we often see a group of smaller cells (**micromeres**)

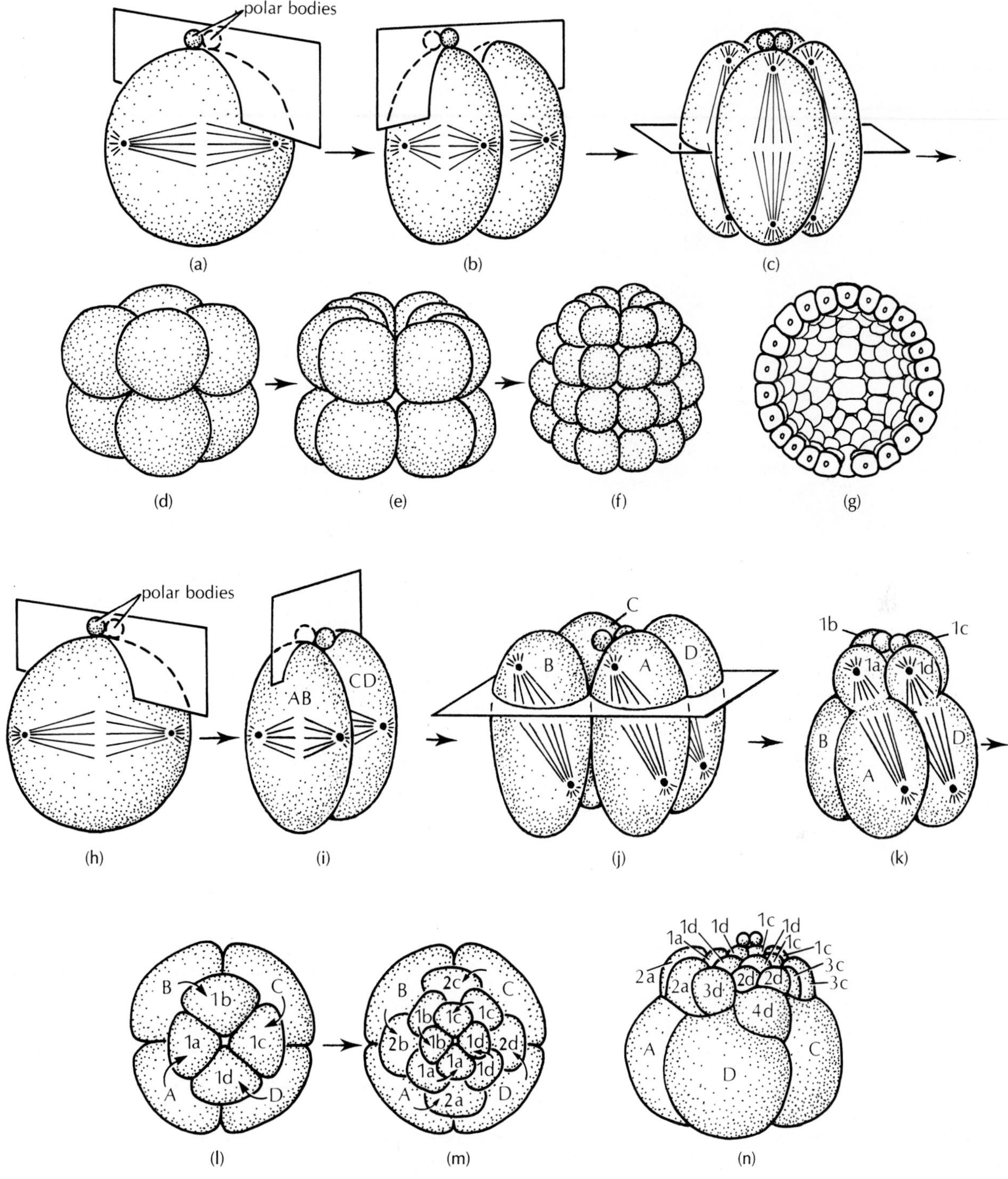

Figure 2.6

(a–g) Radial cleavage, as seen in the sea cucumber *Synapta digitata.* In (g), part of the embryo has been removed to reveal the blastocoel. (h–n) Spiral cleavage. The first two cleavages (h,i) are identical with those seen in radially cleaving embryos, forming 4 large blastomeres (j). The cleavage plane during the next cleavage, however, is oblique to the animal-vegetal axis of the embryo and does not pass through the center of a given cell (k). This produces a ring of smaller cells (micromeres) lying between the underlying larger cells (macromeres), as shown in (l). The lettering system illustrated was devised by embryologist E. B. Wilson in the late 1800s to make possible a discussion of particular cell origins and fates. The number preceding a letter indicates the cleavage in which a particular micromere was formed. Capital letters refer to macromeres, while lowercase letters refer to micromeres. With each subsequent cleavage, the macromeres divide to form one daughter macromere and one daughter micromere, while the micromeres divide to form 2 daughter micromeres. The 32-cell embryo of the marine snail *Crepidula fornicata* is shown in (n). Note the 4d cell, from which most of the mesodermal tissue of protostomes will ultimately derive.

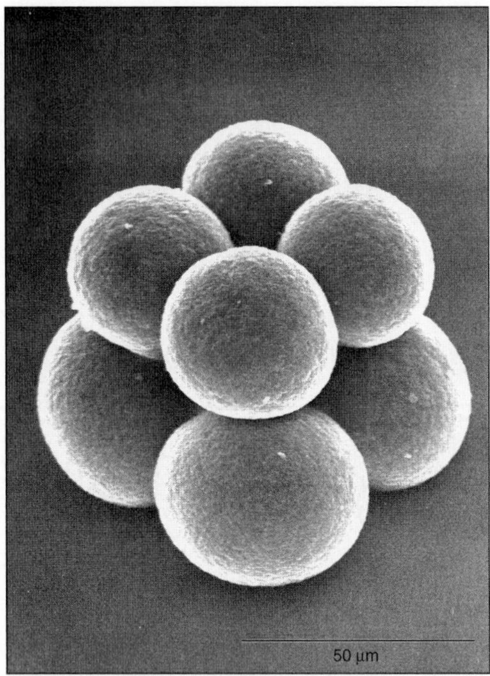

Figure 2.7

Scanning electron micrograph of radial cleavage (8-cell stage) in the cephalochordate *Branchiostoma belcheri* (Chapter 23).
Courtesy of N. Kajita. From Hirakow, R., and N. Kajita. 1990. *J. Morphol.* 203:331–44. Reprinted by permission of Wiley-Liss, Inc., a subsidiary of John Wiley & Sons, Inc.

lying in the spaces between the underlying larger cells (**macromeres**) (Fig. 2.6k–m). Cell division continues in this fashion, with the cleavage planes always oblique to the polar axis of the embryo.

Cleaving embryos of protostomes and deuterostomes also typically differ with respect to when their cells become fully committed to a particular fate. Among deuterostomes, one can separate the cells of a two-cell or four-cell embryo, and each cell will typically develop into a small but complete and fully functional animal. Thus, deuterostomes are said to show **indeterminate** (or **regulative**) cleavage; each cell retains—sometimes as late as the eight-cell stage—the capacity to differentiate the entire organism if that cell loses contact with its associates. Among most protostomes, in contrast, the developmental potential of each cell is irrevocably determined at the first cleavage; separate the blastomeres of a two-celled protostome embryo and each cell will, in most species, give rise only to a short-lived, malformed monster. Protostome cleavage is therefore said to be **determinate** or **mosaic.** Species with determinate development can never produce identical twins, which in deuterostomes arise from the natural separation of blastomeres during early cleavage. Interestingly, both protostomes (e.g., many annelids) and deuterostomes (e.g., many echinoderms) can regenerate body parts as adults, regardless of whether their development is determinate or indeterminate.

A further difference between the two groups of coelomates concerns the source of mesoderm. Among protostomes, much of the mesodermal tissue derives from a single cell of the 64-cell embryo, located at the edge of the blastopore. This is not true of deuterostomes, which produce mesoderm from the walls of the archenteron.

During their first 1 or 2 cleavages, the embryos of some protostomes form **polar lobes** (not to be confused with polar bodies, which arise during meiosis). A polar lobe is a conspicuous bulge of cytoplasm that forms prior to cell division and that contains no nuclear material. After cell division is complete, the bulge is resorbed into the single daughter cell to which it is still attached (Fig. 2.8). Although the functional significance of this phenomenon for the embryo is still not fully understood,[5] polar-lobe formation has provided developmental biologists with an intriguing system through which to study the role of cytoplasmic factors in determining cell fate. In the basic experiment, the fully formed polar lobe is detached from an embryo, and the development of the lobeless embryo is subsequently monitored. Polar-lobe formation is characteristic of only some protostome species (some annelids and some molluscs), but it is never encountered among deuterostomes.

Finally, the ciliary bands involved in feeding and locomotion among deuterostome larvae (and adults) are typically monociliated (Fig. 2.9a), while those found among larval protostomes are typically composed of multiciliated cells (Fig. 2.9b). Details of particle capture also differ among protostome and deuterostome larvae (Fig. 2.9c).

The developmental features distinguishing ideal protostomes from ideal deuterostomes are summarized in Table 2.1. Figure 2.10 shows where each animal phylum fits into the framework discussed so far in this section. The number following each listing gives the page on which the group is first discussed.

Unfortunately, biologists often find it far simpler to construct logical classification systems than to neatly distribute animals within them. As faith in the phylogenetic significance of body cavities has diminished, the definition of "protostome" has been broadened to include both acoelomate and pseudocoelomate animals (e.g., flatworms and nematodes, respectively). Few protostome species exhibit all of the other listed protostome characteristics. Not all protostomes exhibit spiral, determinate cleavage, for example.[6] Similarly, although the blastopore generally becomes the mouth during protostome development, in some protostome species it becomes instead the anus, as in deuterostomes. And some deuterostome species (e.g., the sea squirts) show the fully determinate cleavage pattern typically associated with protostomes, while at least one protostome species (a tardigrade)[7] shows the indeterminate cleavage pattern typically associated with deuterostomes.

5. Henry, J. Q., K. J. Perry, and M. Q. Martindale. 2006. *Devel. Biol.* 297:295–307.

6. Matus, D. Q., *et al.*, 2006. *Current Biol.* 16:575–76; Hausdorf, B. *et al.*, 2007. *Molec. Biol. Evol.* 24:2723–29.

7. Hejnol, A., and R. Schnabel. 2005. *Development* 132:1349–61.

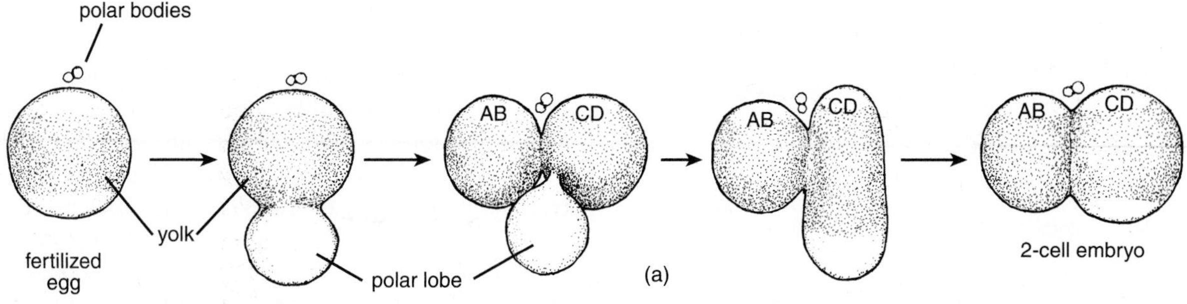

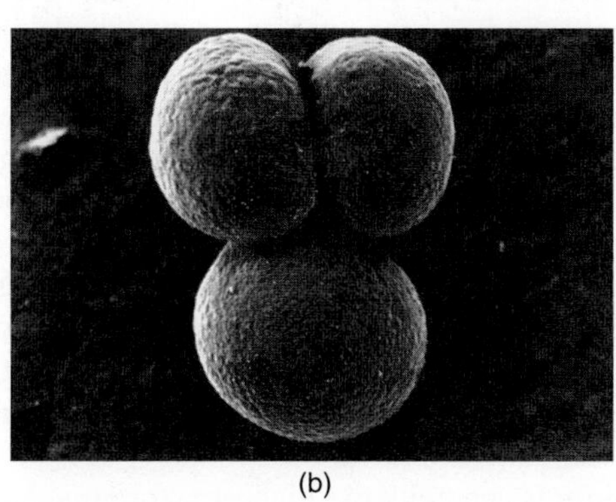
(b)

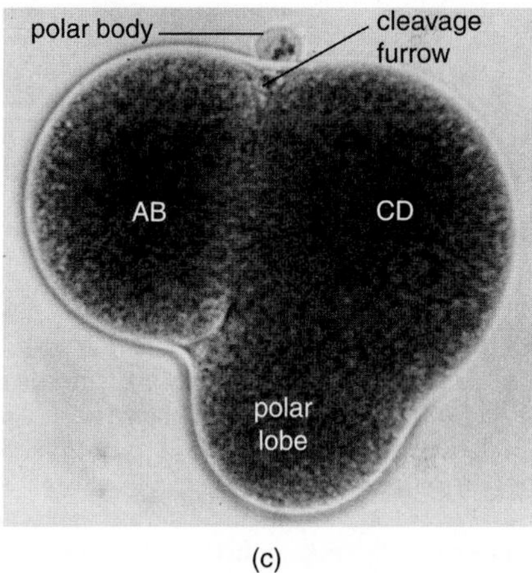

(c)

Figure 2.8

(a) Polar lobe formation during the development of a protostome. Following resorption of the polar lobe, the two blastomeres are clearly unequal in size because the cytoplasm held within the polar lobe does not participate directly in the process of cleavage. (b) Scanning electron micrograph of the two-celled embryo of the marine snail *Nassarius reticulatus*. The polar lobe (at bottom) is nearly equal in size to the blastomere into which it will be resorbed. (c) Polar-lobe formation during first cleavage in the blue mussel, *Mytilus edulis*, as seen with light microscopy. The newly formed cleavage furrow is visible between the AB and CD blastomeres. The polar lobe is clearly affiliated with only one of the daughter cells (the CD blastomere). A polar body (a product of meiotic division prior to cleavage) can be seen at the animal pole of the embryo.

(b) Courtesy of M. R. Dohmen. (c) © Carolina Biological Supply Company/PhotoTake.

Table 2.1 Summary of the Developmental Characteristics of Idealized Protostomous and Deuterostomous Coelomates

Developmental Characteristic	Protostomes	Deuterostomes
Mouth origin	From blastopore	Never from blastopore
Coelom formation	Schizocoely	Enterocoely
Arrangement of coelomic cavities	Variable in number	Generally in 3 pairs
Mesoderm origin	4d cell (Fig. 2.6n)	Other
Cleavage pattern	Spiral, determinate	Radial, indeterminate
Polar-lobe formation	Present in some species	Not present in any species
Larval ciliary bands	Compound cilia from multiciliated cells	Simple cilia, one cilium per cell
	Downstream particle capture	Upstream particle capture

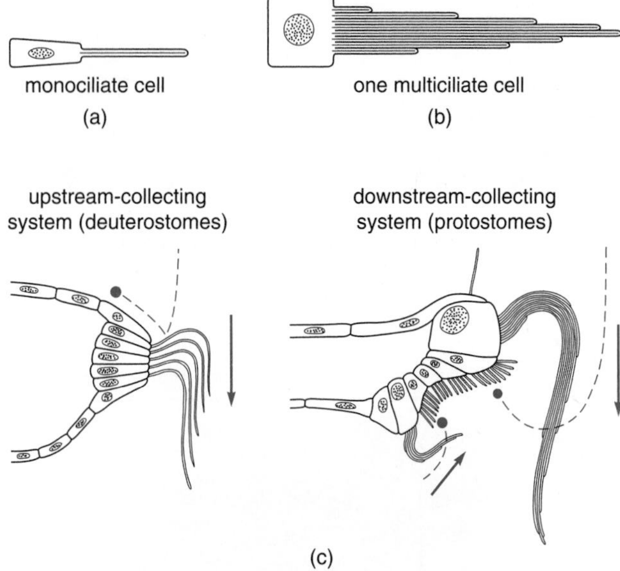

Figure 2.9

(a) A monociliated cell, characteristic of ciliated deuterostome larvae. (b) A multiciliated cell with a compound cilium, typical of protostome larvae. (c) The upstream and downstream particle-collection systems characterizing the ciliated larvae of all deuterostomes and at least most protostomes, respectively. In the upstream-collection system, individual cilia temporarily reverse their direction of beat—probably in response to mechanical contact with food particles—so that particles are directed to the upstream surface of the ciliary band.

From C. Nielsen, 1987. *Acta Zoologica* (Stockholm), 68:205–62. Reprinted by permission.

Finally, some species exhibit a combination of protostome and deuterostome characteristics. Because all animal groups have had ancestral forms in common at some time during their evolution, because evolution is an ongoing process, and because embryos as well as adults are subject to the modifying forces of natural selection, some species are likely to have developmental characteristics that fall outside the mainstream. In any event, those species exhibiting entirely or primarily protostome characteristics are most likely to be more closely related to each other than to those species exhibiting purely deuterostome characteristics.

Classification by Evolutionary Relationship

Probably the most familiar classification scheme is the taxonomic framework established about 250 years ago (1758) by Carolus Linnaeus. The system is hierarchical; that is, one category contains less inclusive groups, which in turn contain still less inclusive groups, and so on:

Kingdom
 Phylum
  Class
   Order
    Family
     Genus
      species

Many subgroups are superimposed upon this basic framework. One encounters among arthropods, for example, subclasses within classes, suborders within orders, infraorders within suborders, and even sections within infraorders, and families are grouped together within superfamilies. Any named group of organisms (e.g., sea urchins, banana slugs) that is sufficiently distinct to be assigned to such a category is called a **taxon.**

The members of any given taxon show a high degree of similarity—morphological, developmental, biochemical, genetic, and sometimes behavioral—and are presumed to be more closely related to each other than to the members of any other taxon at the same taxonomic level. The members of a particular order of snails, for example, are all presumed to have evolved from a single ancestor that is not an ancestor of snails in other orders. Similarly, all the members of any particular phylum are presumed to have evolved from a single ancestral form. Such groups, at every taxonomic level, are said to be **monophyletic** (G: single-tribed). Most modern workers now agree that all monophyletic groups must also include all descendants of the originating ancestor. A group that does not do so is said to be **paraphyletic.** By this definition, the invertebrates form a paraphyletic group, since their vertebrate descendants are excluded.

Phylum is generally the highest taxonomic level that will concern us in this text. Invertebrate animals are presently distributed among at least 23 phyla (32 phyla in this textbook), each representing a unique body plan, and unicellular invertebrates (protists) are distributed among still more phyla. The distribution of described species among the various animal phyla is summarized in Figure 2.11. Note that the percentage of species contained within our own phylum—the phylum Chordata—is quite small (no more than about 5% of all described species) and that this phylum contains both invertebrates and vertebrates.

Remarkably, based on existing fossil evidence, no new phylum-level body plans have arisen in the past 600 million years, despite substantial radiation following each of the 5 major and about 10 smaller extinctions that took place during that time. In the most devastating extinction event to date, 251 million years ago at the Permian-Triassic boundary, nearly 95% of existing species-level animal diversity was lost. In the subsequent 250 million years many new species evolved, often representing new orders and classes, but no new phylum-level body plans seem to have appeared. It is possible, of course, that some groups with no fossil record are of more recent origin.

The category of **species** has particular biological significance, although a single, precise, functional definition has not been found. Theoretically, the members of one species are reproductively isolated from members of all other species. The species, therefore, forms a pool of genetic material that only members of that species have access to and that is isolated from the gene pool of all other species.

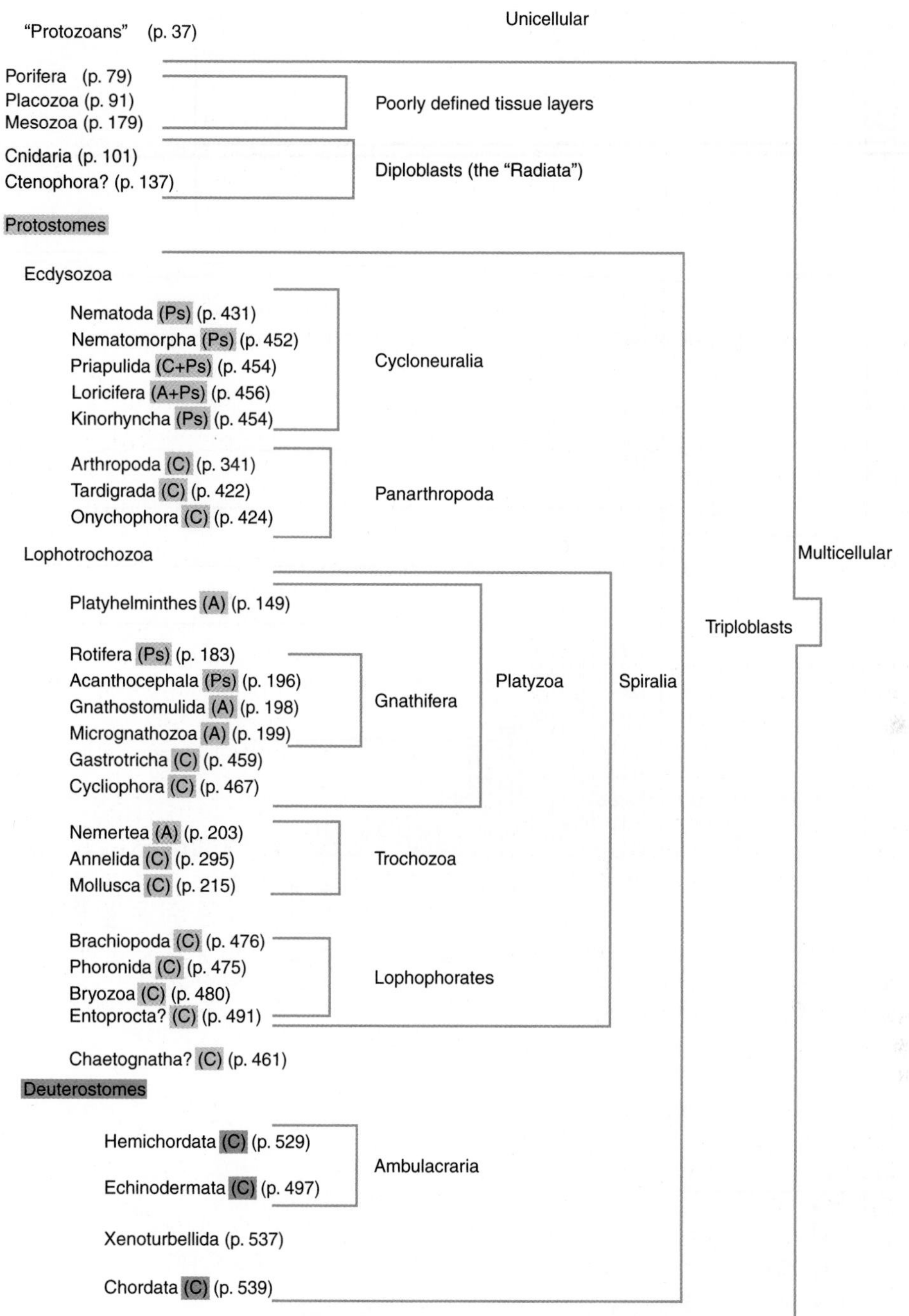

Figure 2.10

A modern arrangement of triploblastic animal groups according to the factors discussed in this chapter. The two major animal groupings, protostome and deuterostome are shaded blue, as are the indications of body cavity types: A = acoelomate, Ps = pseudocoelomate, C = coelomate. A "+" indicates that different members of a group belong to different categories (e.g., "C+Ps" means that the group includes both coelomates and pseudocoelomates). No body cavity type designated indicates uncertainty for that group. Some groupings (e.g., Ecdysozoa) are still controversial, and there is not yet complete agreement about membership in the major protostome groups Cycloneuralia and Lophotrochozoa. In some schemes, the lophophorate phyla are included within the Trochozoa. Based on several recent sources.

The scientific name of a species is binomial (has two parts): the **generic name** and the **specific name.** The generic and specific names (i.e., the **species name**) are usually italicized in print and underlined in writing. The generic name begins with a capital letter, but the specific name does not. For example, the proper scientific name for one of the common shallow-water marine snails found off Cape Cod, Massachusetts, is *Crepidula fornicata.* Related species are *Crepidula plana* and *Crepidula convexa.* Once the generic name is spelled out, it may be abbreviated when used subsequently, as long as no confusion results (if an author is referring to the 2 genera

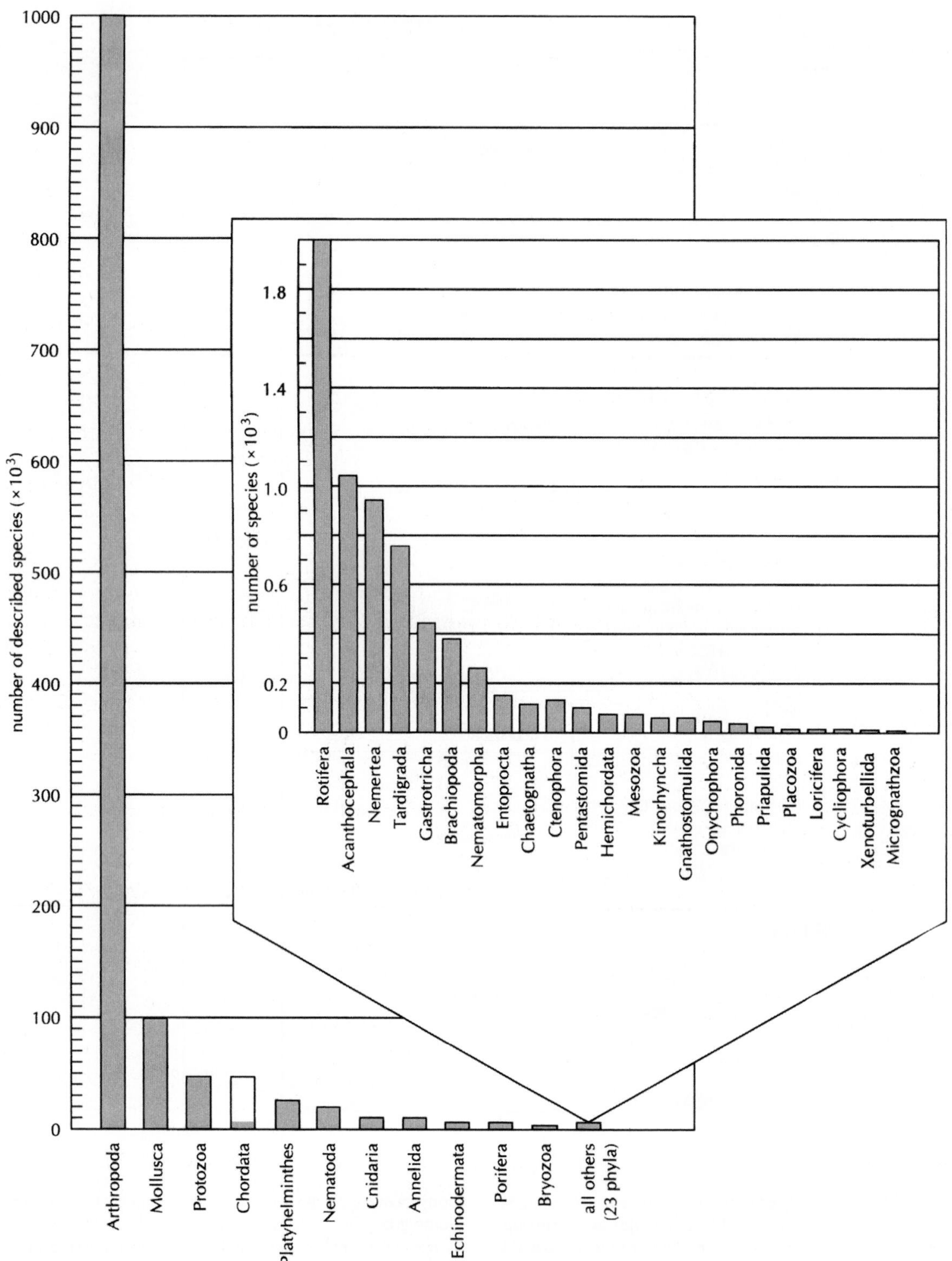

Figure 2.11

Graphic representation of the distribution of described species among the 33 major groups of invertebrates. In this text, protozoans are divided among more than one dozen phyla. Metazoan phyla containing fewer than 2000 described species are presented in the inset. Note the different scale on the Y-axis of the inset. The open (unshaded) area of the bar labeled "Chordata" represents vertebrate species. All other species in all other phyla are invertebrates.

Crepidula and *Conus*, for example, neither genus name can be abbreviated as "*C.*"). Thus, *Crepidula fornicata, C. plana,* and *C. convexa* are common shallow-water marine gastropods found near Woods Hole, Massachusetts. They all belong to the phylum Mollusca and are contained within the class Gastropoda, family Calyptraeidae. The family Calyptraeidae contains other genera besides *Crepidula;* the class Gastropoda contains other families besides the Calyptraeidae; and the phylum Mollusca contains other classes besides the Gastropoda. The taxonomic classification system is indeed hierarchical.

The name of the person who first described the organism often follows the species name. It is capitalized, but not italicized. A barnacle common along the coast of the southeastern United States, for example, is *Balanus amphitrite* Darwin, first described by Charles Darwin. Linnaeus's name is often abbreviated as L., since he is associated with the descriptions of so many species. If the organism was originally described as being in a different genus than the one in which it is currently placed, the describer's name is enclosed within parentheses. Thus, the snail *Ilyanassa obsoleta* (Say) was described by a man named Say, who originally assigned the species to another genus (the genus *Nassa*). This snail was later determined to be sufficiently dissimilar from other members of the genus *Nassa* to warrant its assignment to a different genus. Occasionally, a person's name is followed by a date, identifying the year in which the species was first described. For example, the shrimplike animal known as "*Euphausia superba,* Dana 1858" was first described by Dana in 1858, and it has remained in the genus *Euphausia* since it was originally named.

The system just described has been with us for so long that it seems obvious . . . and permanent. In 1998, however, a group of influential biologists met with the intention of replacing it. The proposed replacement is called **The PhyloCode.** Unlike the Linnaean system, The PhyloCode—although it remains hierarchical—is rankless: There will be no classes, orders, or families. It promises to revolutionize systematics.

The following two quotes give a hint of what may lie ahead:

> "We argue that taxon names under the Linnaean system are unclear in meaning and provide unstable group-name associations . . . . Furthermore, the Linnaean rank assignments lack justification and invite unwarranted comparisons across taxa." F. Pleijel and G. W. Rouse. 2003. *J. Zool. Syst. Evol. Res.* 41:162.

> "The semantic changes advocated by the PhyloCode, which uses a conceptually flawed topology of evolutionary trees, will hopelessly augment confusion and disdain for taxonomy." L. Margulis. 2005. *Amer. Scient.* 93:290.

Stay tuned (see *Search the Web* at the end of this chapter).

Inferring Evolutionary Relationships

> "The whole field of metazoan phylogenetics is 'up for grabs' with conflicting data and interpretations appearing in the literature constantly." F. R. Schram. 2003. *Proc. 18th Int. Congr. Zool.:* 359–68.

An ideal taxonomic classification scheme reflects degrees of phylogenetic relatedness; that is, all members of a given taxonomic group should have descended from a single ancestral species and thus be more closely related to each other than to the members of any other group.

Biologists have long made logical, reasoned guesses about the origins of various animal groups, based upon detailed studies of developmental patterns, studies of morphological and biochemical characteristics, and careful examination of animals preserved in the fossil record. Comparative molecular analyses of protein structure and of DNA and ribosomal RNA (rRNA)sequences among species have altered some of these views substantially.[8] Ferreting out probable relationships is no easy task.[9] In part, the difficulty concerns the relative importance of phenotypic similarities among taxa, phenotypic differences among taxa, and the degree to which one is willing to admit (and deal with the fact) that phenotype may be a very misleading indicator of underlying genetic similarities and differences. Through the process of **convergence,** distantly related animals may come to resemble each other rather closely. Features that resemble each other through convergence are referred to as **analogous,** as opposed to homologous. For example, the eye of an octopus (a cephalopod mollusc) is remarkably like that of a human, but these visual organs are believed to be analogues, not homologues, and do not indicate any close evolutionary relationship between vertebrates and molluscs. Which features indicate evolutionary closeness and which do not? Should we try to make this distinction? How can we know if we've decided correctly?

Moreover, in the evolutionary process, structures sometimes become less complex rather than more complex. Suppose, for example, you discover a new species of wingless insect. How can you tell whether this species evolved before insect wings evolved or whether it instead descended from a winged ancestor and lost the wings over time? It is often very difficult to determine which of 2 character states is the original (*primitive,* or *plesiomorphic*) condition and which is the advanced (*derived,* or *apomorphic*) condition.

Until very recently, evolutionary relationships have been deduced entirely through anatomical and ultrastructural studies, with phenotypes serving as reflections of the underlying genotypes. During the past 20 years or so, however, biochemical and molecular studies have allowed us to examine genotypic diversity directly. Particularly remarkable are recent interspecific comparisons of nucleotide sequences of genes coding for ribosomal RNA

8. See *Topics for Further Discussion and Investigation,* no. 5.

9. See *Topics for Further Discussion and Investigation,* nos. 1 and 2.

(rRNA), comparisons made feasible through development of the polymerase chain reaction (PCR) in the mid-1980s. The PCR permits biologists to very quickly and inexpensively generate many copies of specific DNA sequences; a billion copies of a single DNA molecule can be obtained in a few hours, producing sufficient material for analysis.

Molecular studies often produce some remarkable and surprising results, results that differ considerably from those of earlier, organismal studies. These results are frequently controversial; in some cases, there is considerable disagreement among workers about the procedures used to prepare and analyze the data, and about how the results of molecular studies should be interpreted, as discussed later in this chapter. But even before molecular biologists joined the fray, proposed phylogenetic relationships were controversial. A variety of phylogenetic trees have been proposed over the years. Six of these "dendrograms" are illustrated in Figure 2.12. None of the proposed schemes represents idle speculation; all reflect hard work and detailed and careful reasoning. The oldest scheme (Fig. 2.12a) assumes that all multicellular animals descended from some form of single-celled protist, most likely a colonial flagellate, and presents sponges (phylum Porifera) as the earliest experiments in multicellularity with no close relationship to any other existing phyla.

The hypothesized relationships among annelids, arthropods, and molluscs differ considerably among the different viewpoints; compare, for example, Figure 2.12a, d, and f. The more closely you look at the different schemes, the more fascinating the comparisons become; it is well worth returning to Figure 2.12 at intervals as you read the rest of this book. Fig. 2.12c represents one particularly widespread current view, with all acoelomate and pseudocoelomate animals folded into the Protostomia, and the protostomes divided into two major clades: the Ecdysozoa (molting animals) and the Lophotrochozoa (see also Fig. 2.10). Studies of highly conserved *Hox* gene insertions and deletions (so-called "signature sequences") have added strong support for these ecdysozoan and lophotrochozoan clades. Just as I was completing the manuscript for this revision, however, a colleague showed me a recent paper entitled, "Ecdysozoan clade rejected by genome-wide analysis of rare amino acid replacements."[10] Everything, it seems, is still up for grabs.

At least some of the differences among the various schemes may be attributed to insufficient data. As additional information about the various groups is gradually obtained, the evidence in favor of one scheme over some others may become more compelling, or additional modifications may be proposed.

The assignment of a given animal or group of animals to a particular position within the taxonomic hierarchy is not an irrevocable event. Studies of an animal's early development, for example, can reveal new information about the nature of the organism's internal body cavity, information that may affiliate that organism with an entirely different group of animals from those with which it was previously grouped. Controversies can also diminish—or increase—when data from the fossil record are added to data from extant species. Or a detailed study might call the usefulness of particular characters into question. If, for example, a certain embryonic cleavage pattern arose only once in evolution, then those animals that develop in this particular way must be closely related. But if evidence is found that this particular pattern evolved independently in several animal groups, then that trait conveys little, if any, phylogenetic information.

Classifications also change when biologists discover organisms having characteristics not shared with any existing groups. For example, 2 arthropod classes (the Remipedia and Tantulocarida, p. 417) and 3 small but remarkably distinct phyla of recently discovered marine animals called loriciferans (p. 456), micrognathozoans, and cycliophorans (p. 467) have been established in the last 25 years or so. Cycliophorans were first described in 1995, and micrognathozoans in 2000 (p. 467).

Sometimes classifications change when biologists reexamine previously studied material, or acquire new material. A small but fascinating group of gutless worms, for example, the pogonophorans (p. 304) were originally characterized as unquestionable deuterostomes, based on adult morphology. Years later, specimens with a small additional body part were obtained—the posterior part of the animal had detached unnoticed from previous specimens—and the animals were quickly reclassified as a phylum of protostomes. Indeed, largely on the basis of features of that small terminal portion, pogonophorans have recently been incorporated into the phylum Annelida, a group that contains earthworms and leeches. Such placement has now been supported by molecular data.

Finally, molecular studies comparing selected gene sequences—and, more recently, by analysis of entire genomes—among representatives of different groups are quickly altering our understanding of many invertebrate relationships. While molecular data often support previous conclusions based on morphology and developmental pattern, such as the monophyly of living animals and the distinction between protostomes and deuterostomes, they frequently suggest relationships quite different from those based on other criteria. Where molecular data produce phylogenies very different from those based on morphology, decisions will have to be made about which evidence is more likely to be correct. And it is worth noting that molecular studies, as powerful as they are, will never resolve all phylogenetic issues, no matter how sophisticated these studies become. For one thing, when species diversified too rapidly, molecular studies are unable to resolve the order of divergence. Moreover, molecular studies will never be able to tell us the precise sequence of steps that took place as one form gave rise to another or what selective pressures brought about these morphological changes. And molecular studies can never tell us what ancestral, unfossilized animals looked like. Perhaps molecular, paleontological,

10. Rogozin, I. B., Y. I. Wolf, L. Carmel, and E. V. Koonin. 2007. *Mol. Biol. Evol.* 24:1080–90.

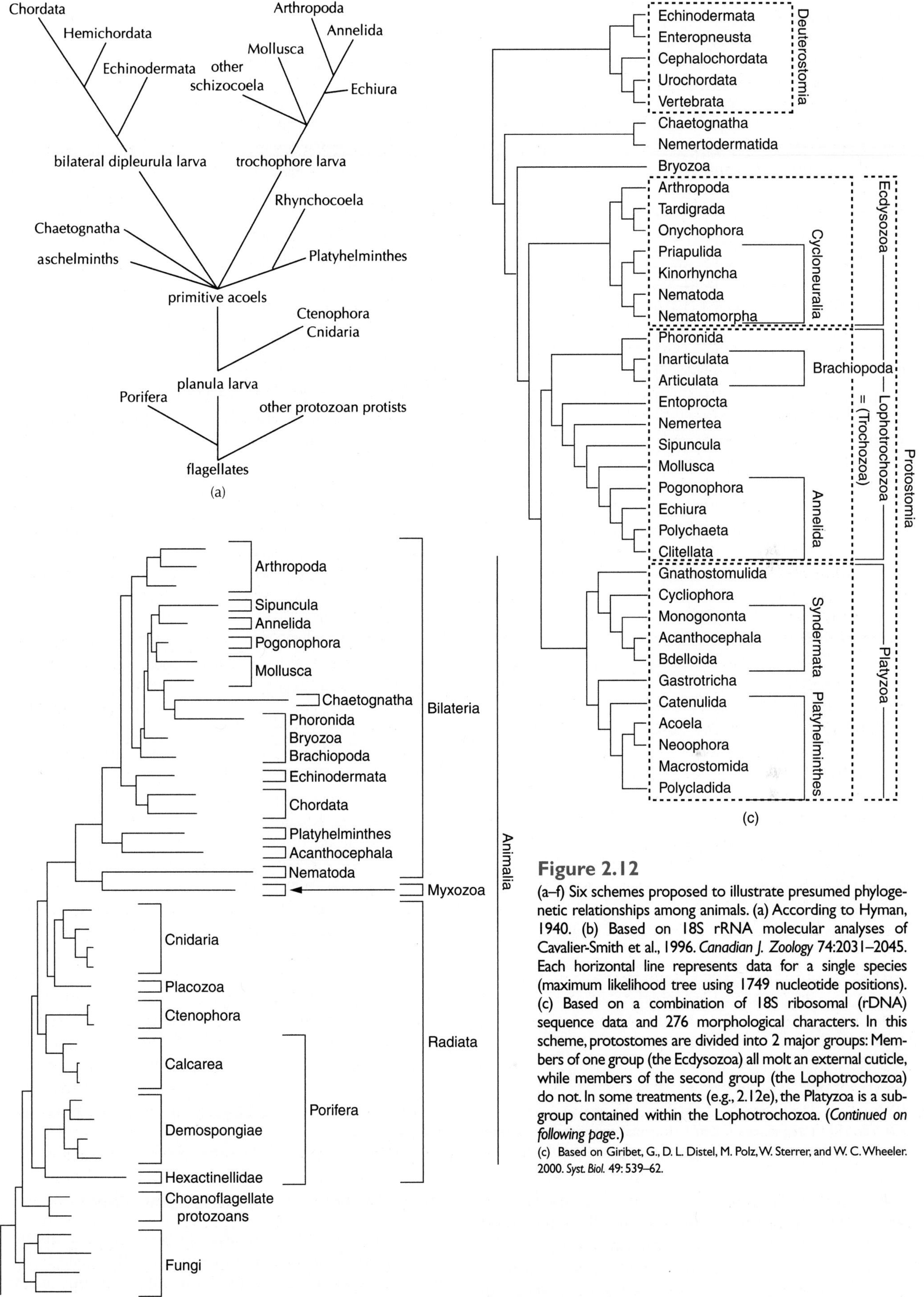

Figure 2.12

(a–f) Six schemes proposed to illustrate presumed phylogenetic relationships among animals. (a) According to Hyman, 1940. (b) Based on 18S rRNA molecular analyses of Cavalier-Smith et al., 1996. *Canadian J. Zoology* 74:2031–2045. Each horizontal line represents data for a single species (maximum likelihood tree using 1749 nucleotide positions). (c) Based on a combination of 18S ribosomal (rDNA) sequence data and 276 morphological characters. In this scheme, protostomes are divided into 2 major groups: Members of one group (the Ecdysozoa) all molt an external cuticle, while members of the second group (the Lophotrochozoa) do not. In some treatments (e.g., 2.12e), the Platyzoa is a subgroup contained within the Lophotrochozoa. (*Continued on following page*.)

(c) Based on Giribet, G., D. L. Distel, M. Polz, W. Sterrer, and W. C. Wheeler. 2000. *Syst. Biol.* 49: 539–62.

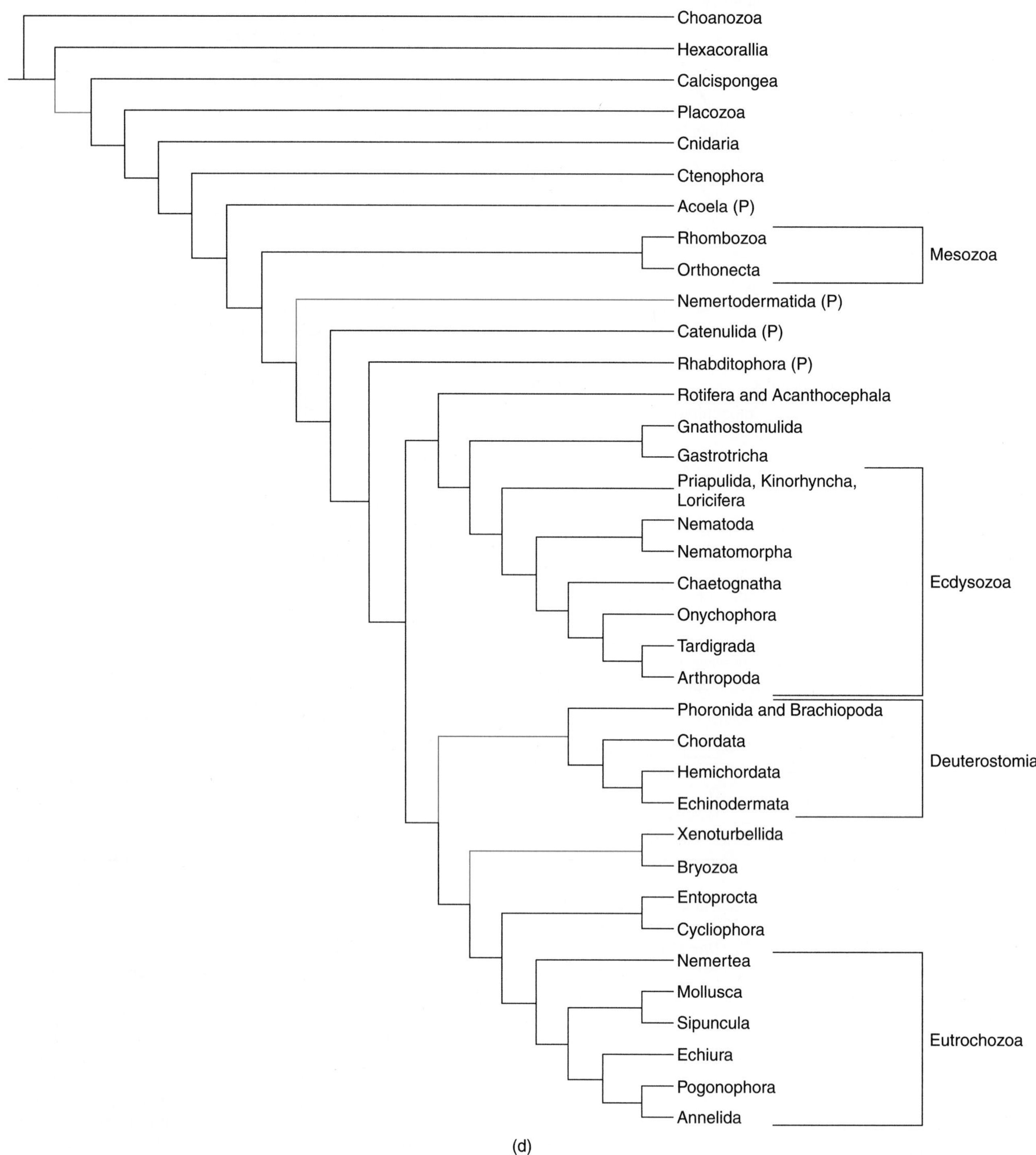

Figure 2.12 *Continued*
(d) Based on a combination of 276 morphological characters and 151 18S rRNA sequences (from 151 animal species). (P) indicates the 4 platyhelminth flatworm groups. Blue shading indicates components in which morphological and molecular evidence are especially in disagreement.
(d) Modified from J. Zrzavý, S. Mihulka, P. K. Kepka, A. Bezděk, and D. Tietz, 1998. *Cladistics* 14:249–85.

ecological, and morphological evidence can be used in concert to deduce relationships, but we will still need to decide how much weight to give each line of evidence when the different approaches imply different evolutionary scenarios.

Phylogenetic relationships have been argued about for over 150 years. Such arguments will likely continue long into the future.

Why Determine Evolutionary Trees?

One goal of classification schemes is simply to facilitate discussions about different groups of animals, and ideally to arrange those groups in the correct evolutionary context. But knowing with certainty the precise pattern of evolutionary change that gave rise to the present

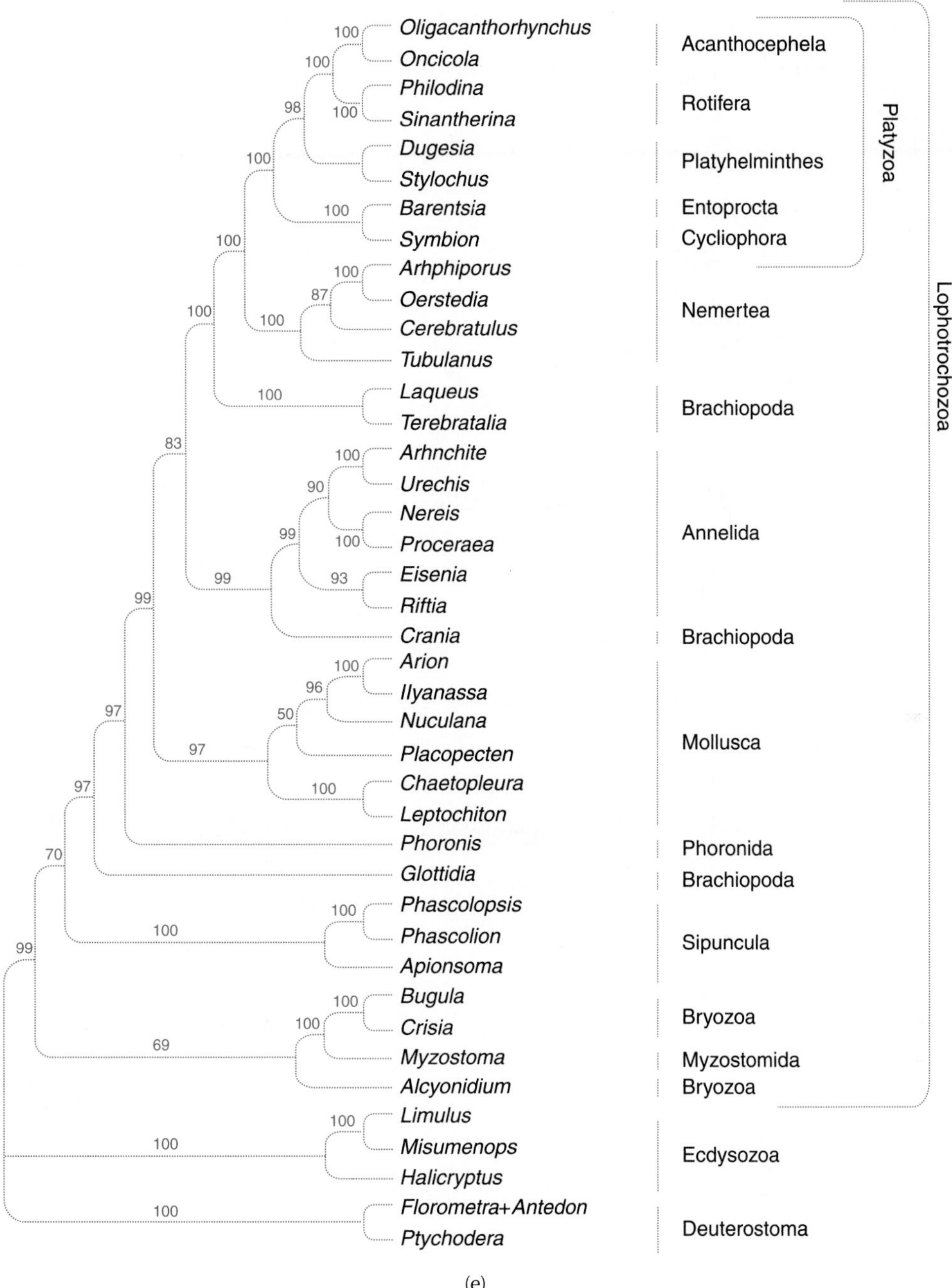

Figure 2.12 *Continued*
(e) Based on 18S and 28S ribosomal gene sequence data for 36 lophotrochozoan taxa. Numbers suggest the degree of confidence in particular branching points. (*Continued on following page.*)
(e) Based on Passamaneck, Y., and K. M. Halanych. 2006. *Molec. Phylog. Evol.* 40: 20–28.

diversity of animal form would give us far more than a convenient and stable classification system. Finding one species of coral, for example, that produces a particular defensive compound of great biomedical potential, we might know which other species were most likely to synthesize related compounds. We would also be better able to understand the sequence of genetic changes involved in body plan evolution, and would be able to tell with certainty how many times certain traits had evolved independently within any particular group of animals.

For example, Figure 2.13 shows one recent hypothesis regarding the evolutionary relationships among 37 species of stick insect (order Phasmida), a group in which individuals mimic—both morphologically and behaviorally—a variety of sticks and leaves. About 40% of all known stick-insect

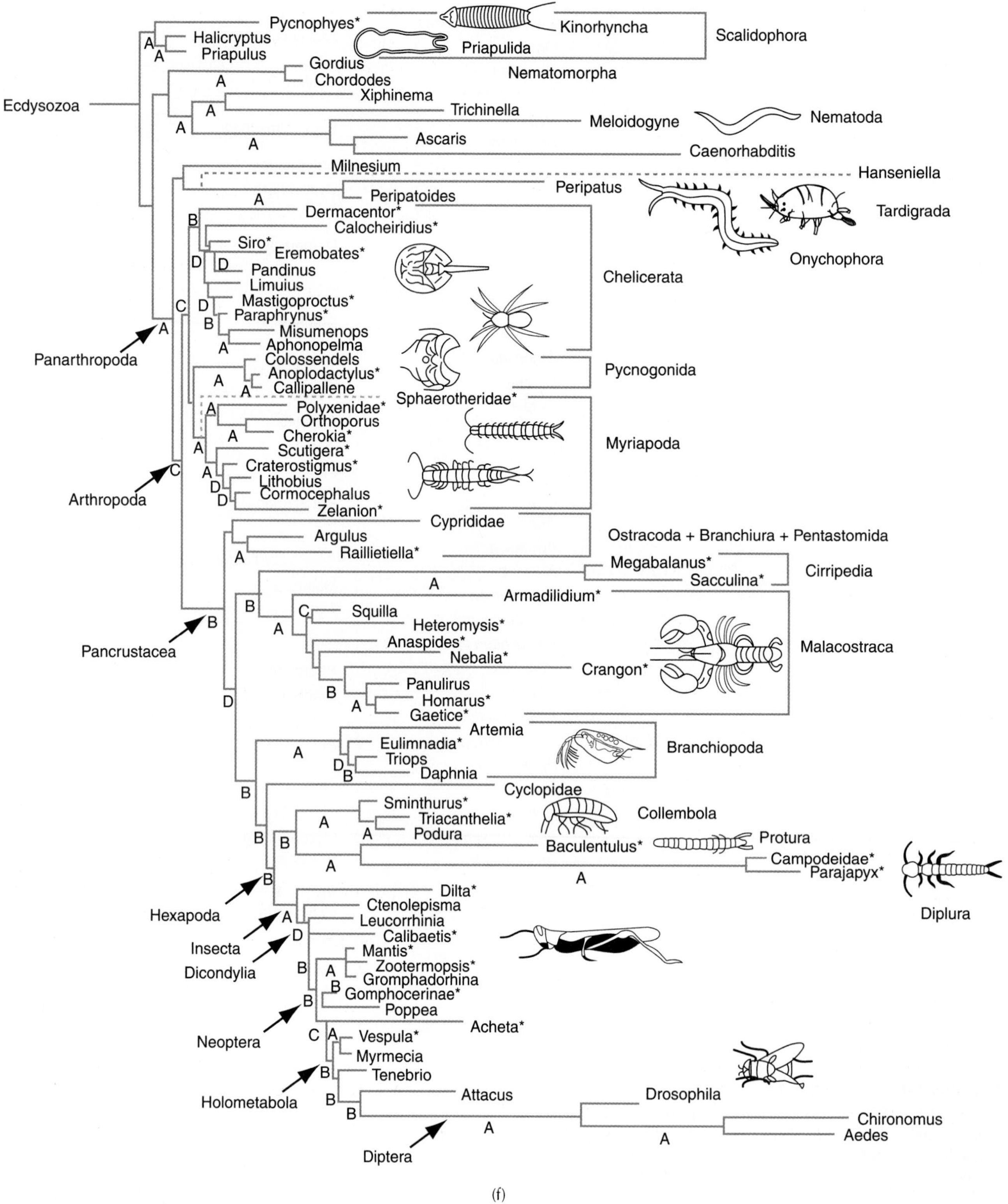

Figure 2.12 *Continued*

(f) Detail of ecdysozoan interrelationships, based on nearly complete 28S and 18S rRNA genes.

(f) Based on Mallat, J., and G. Giribet. 2006. *Molec. Phylog. Evol.* 40: 772–94.

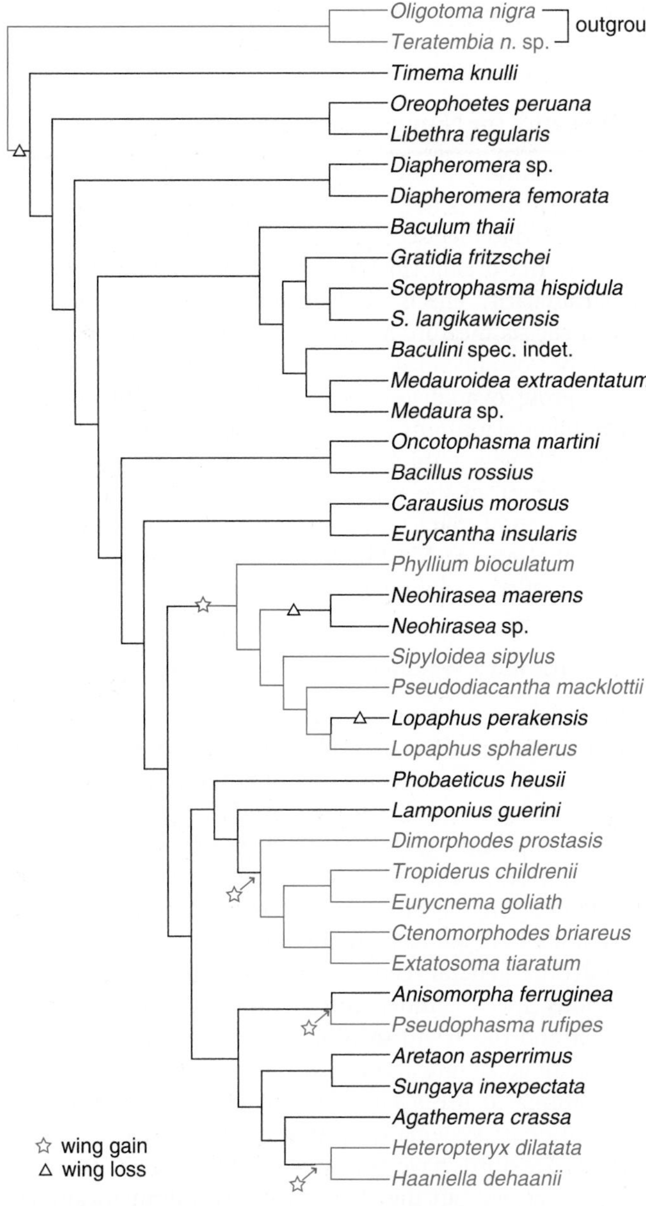

Figure 2.13

A phylogenetic hypothesis for the evolutionary relationship among 23 species of stick insects, suggesting 4 independent instances of wing gain and 3 instances of wing loss. The 2 winged species at the top of the figure belong to different insect orders, and serve here as the "outgroup" (see "How Evolutionary Relationships Are Determined," right column).

From Whiting, M. F., S. Bradler, and T. Maxwell. 2003. Loss and recovery of wings in stick insects, *Nature* 421:264–67.

species (about 1200 species) are fully winged, but the other 60% (about 1800 species) have either reduced hind-wings or no wings at all. In the figure, species with full or reduced wings are indicated with blue shading.

A convincing phylogeny for these animals can tell us much about the evolution of wings within the group. If the scenario shown in Figure 2.13 is correct, then the earliest stick insects lacked wings, and they diversified into many different species in that wingless condition. We also see that wings must have later appeared at least 4 times independently (indicated by the 4 stars), from at least 4 different wingless ancestors. Wings were then lost at least 2 times in subsequent evolution (indicated by the 2 triangles near the center of the figure). The most remarkable aspect of the scenario presented is that the wings of stick insects in the different groups seem fully homologous with each other and with the wings of other insect species, implying that the genetic instructions for wing development were maintained unexpressed but unaltered for many thousands of generations in wingless stick insects before being reactivated. This is the first published support for the idea that wings can be re-evolved in insect lineages that have lost them.

Similar arguments are being made for the evolution of life histories, behaviors, parasitic associations, morphological features, and biochemical or physiological attributes in a wide range of other animal groups. Thus, there is a lot riding on our ability to convincingly ascertain exactly how animals are related to each other.

How Evolutionary Relationships Are Determined

> "Even with a consistent method, the best tree need not be the correct tree." R. Raff et al. 1997. *Ann. Rev. Ecol. Syst.* 25:351–75.

> ". . . it is very difficult to distinguish true progress in our understanding of metazoan macroevolution from mere change of opinions with the passage of time." R. Jenner, 2003. Unleashing the force of cladistics? Metazoan phylogenetics and hypothesis testing. *Integr. Comp. Biol.* 43:207–18.

If all living metazoans evolved from a single ancestral form many millions of years ago, then all animals are related to each other: No matter how distant, there must be some genealogical connection between flatworms, snails, squid, annelid worms, insects, lobsters, sea urchins, and baleen whales. Trace your own ancestry far enough back, and you must find an invertebrate in your family tree.

Trying to unravel the evolutionary connections among the major animal groups is one of life's greatest puzzles and presents a great intellectual challenge. In particular, there are many difficulties in deciding how best to go about arranging and sorting the puzzle pieces, and in judging the accuracy of the picture that emerges when the sorting is done. These and related difficulties are discussed in the next few pages. Far more detailed discussions are found in the references listed at the end of this chapter.

Charles Darwin originally referred to what we now call evolution as "descent with modification." The members of any species tend to resemble each other from one generation to the next, as long as there is random mating within the gene pool. But if some individuals become reproductively isolated from other members of the species, they can evolve in quite different directions, particularly if they face different selective pressures—different temperature or salinity regimes, for example, or different sorts of predators or food resources. If species have gradually acquired differences—physical, physiological, biochemical, behavioral, genetic—at constant rates, and if they have continually evolved to resemble their ancestors less and less over time, then it should be easy to deduce evolutionary relationships. But animals do not evolve in so straightforward a way, which opens the door for both sophisticated creativity and controversy.

The centerpiece of any phylogenetic detective work is **homology.** Morphological features that share a common evolutionary origin are said to be **homologous;** our cranium, for example, is homologous with that of cats, dogs, frogs, and whales—the cranium in such animals has a single, common evolutionary origin. Any differences in homologous features among different animal groups reflect descent from ancestors with modification. In many cases, homologous features develop through similar pathways controlled by the same genetic instructions. If you can recognize homology when you see it, evolutionary puzzles should be easily solved. If you can safely assume, for example, that spiral cleavage evolved only once, then spiral cleavage is a homologous trait in all groups that exhibit it: All spirally cleaving animals have descended from a common ancestor and must be more closely related to each other than to animals that show any other cleavage pattern. But what if spiral cleavage is *not* homologous in all groups? Suppose that when eggs cleave there are only a few ways for the daughter cells to sit in stable relation to each other, and that the spiral pattern formed by adjacent cells simply represents one particularly stable geometric arrangement. In that case, different animal groups are likely to have independently converged upon spiral cleavage as an especially successful way to initiate development: Cleavage pattern then misleads us in our thinking about evolutionary relationships, and the molluscs, annelids, flatworms, and other spirally cleaving animals need not be closely related. Similarly, if coelomic cavities evolved only once, then coelomates form a monophyletic group and we then face the issue of determining whether protostomes evolved from deuterostomes or vice versa. But if coelomic cavities originated independently 2 or more times in different ancestral species, then the coelomate condition conveys only a very garbled phylogenetic message at best.

Even very complex morphologies can independently evolve from very different ancestors to give a close resemblance by *convergence,* as discussed earlier. It is often difficult to decide whether features that look similar in different animal groups are homologous or not.

The second particularly thorny issue concerns the direction, or **polarity,** of evolutionary change. Even if two characters are considered to be homologous, there is the question of which one represents the original, or *ancestral,* state and which represents the more advanced, or *derived* state. Issues of homology and polarity are at the source of most current debates among systematists. There are three basic approaches to deducing evolutionary relatedness, as described in the following pages.

Phenetics (or Numerical Taxonomy)

One solution to the homology/polarity dilemma is to assume that it is not possible to ascertain either with certainty, and to then set about establishing taxonomic groups that reflect overall similarity alone, regardless of whether that similarity reflects common ancestry or not. In practice, pheneticists measure as many characters as possible—the number of appendages on the head, for example—from each group of animals under study, and then apply complex computer algorithms to determine which groups are most alike and which are most different from each other. The main appeal of the approach is in not having to grapple with issues of character homology and polarity, but it now has little support. Most biologists *want* a classification system to reflect evolutionary relationships.

Evolutionary Systematics (Classical Taxonomy)

Evolutionary systematics has been practiced for over 100 years. In contrast to pheneticists, the evolutionary systematist wrestles with issues of homology at the outset of an analysis, and also decides which characters are most likely to hold the greatest amount of phylogenetic information; other characters are given less weight (underweighted) in the analysis or ignored altogether. Once what are believed to be homologous characters are used to deduce general relationships, the extent to which the various species under consideration differ from each other and the extent to which they resemble each other are both taken into account in constructing the final classification. To use a familiar non-invertebrate example, evolutionary systematists put birds in a separate class, the class Aves. Birds have clearly evolved from ancient reptilian ancestors, but they have evolved so dramatically far from those ancestors they deserve status as a separate class. The other, more reptile-like descendants of that same ancestor are grouped in a separate class, the Reptilia. It makes intuitive sense to form groups of similar-looking species and to exclude species that look very different, but as you will see below, all systematists do not share this feeling, in large part because the classical approach often leads to the formation of paraphyletic groups. The Reptilia, for example, is paraphyletic because it excludes some descendants of the original reptilian ancestor: the birds, and the mammals. The evolutionary systematist is not troubled by paraphyletic groupings.

Constructing classifications and evolutionary trees by this method is painstakingly slow, and requires decades of experience working with the animals being categorized. Intuition and logic play important roles in all decisions made. Major objections to this process are that it lacks objectivity and a rigorously standardized methodology, and that outsiders have difficulty arguing with the results.

Table 2.2 Cladistic Analysis: Some Common Terms Defined

clade: a group of organisms that includes the most recent common ancestor of all its members and all descendants of that ancestor; every valid clade forms a "monophyletic" group (see monophyletic taxon).

cladogenesis: the splitting of a single lineage into 2 or more distinct lineages (*klados* = G: twig, or branch).

anagenesis: change occurring within a lineage.

cladogram: the pictorial representation of branching sequences that are characterized by particular changes in key morphological or molecular characteristics (character states).

homologous characters, homology: characters that have the same evolutionary origin from a common ancestor, often coded for by the same genes. Homology is the basis for all decisions about evolutionary relationships among species.

taxon: any named group of organisms, such as jellyfish or sea urchins or slippershell snails (*Crepidula fornicata*); plural = **taxa.**

monophyletic taxon: a group of species that evolved from a single ancestor and includes all descendants of that ancestor. By definition, every valid clade must form a monophyletic taxon.

parsimony: a principle stating that, in the absence of other evidence, one should always accept the least complex scenario.

polarity: the direction of evolutionary change.

ancestral (primitive) state: the character state exhibited by the ancestor from which current members of a clade have evolved. Also called the **"plesiomorphic"** state.

derived state: an altered state, modified from the original, or ancestral condition. Also called the **"apomorphic"** state.

apomorphy: any derived or specialized character.

pleisiomorphy: any ancestral or primitive character.

synapomorphy: a derived character that is shared by the most recent common ancestor and by 2 or more descendants of that ancestor. In cladistic methodology, synapomorphies define clades; that is, they determine which species (or other groups) are most closely related to each other. Essentially, synapomorphies are homologous characters that define clades.

autapomorphy: a derived character possessed by only one descendant of an ancestor, and thus of no use in discerning relationships among other descendants.

homoplasy: the independent acquisition of similar characteristics (character states) from different ancestors through convergence or parallelism. Such homoplastic events create the illusion of homology.

paraphyletic grouping: a group of species sharing an immediate ancestor but not including all descendants of that ancestor.

polyphyletic grouping: an incorrect grouping containing species that descended from 2 or more different ancestors. Members of polyphyletic groups do not all share the same immediate ancestor. Members of polyphyletic groups may resemble each other because of the independent evolution of similar traits by different ancestors.

bootstrapping: a technique for evaluating the reliability of a branch of a phylogenetic tree by resampling some number of characters from the original data set (with replacement) at random. Each resampling thus creates a new data set, with some values duplicated and others omitted. The bootstrap value given for each branch shows the percentage of resamplings (typically 500–1000) that recover that branch. For example, in Figure 2.13, bootstrap values were at least 95% at most branch points, meaning that those branching patterns were recovered in a least 95 of every 100 simulations.

Bayesian inference: a statistical technique used to infer the probability that a particular phylogenetic hypothesis is correct.

jackknifing: a technique for evaluating the reliability of a branch of a phylogenetic tree by deleting some percentage of information (e.g., base-pair position information) at random and then rerunning the analysis. The jackknife value given for each branch shows the percentage of such resamplings (typically 500–1000) that recover that branch.

node: a branching point on a cladogram

Cladistics (Phylogenetic Systematics)

This approach to deducing evolutionary relationships has gained a large and enthusiastic following over the past 30 years or so. Although not universally accepted, cladistic procedures have become so widely used and so widely discussed (and argued about) that I will go into the philosophical and procedural underpinnings in more detail than I have for the other two approaches, to facilitate class discussions and individual forays into the burgeoning primary literature. Unfortunately, the field of cladistics is terminologically well endowed. In this treatment, I use as few of those terms as possible; for those wishing to read and discuss the relevant literature, I define the most important terms in Table 2.2.

Among cladists, phylogenies are constructed and assessed using one of several well-defined procedures and a number of widely available, highly sophisticated computer

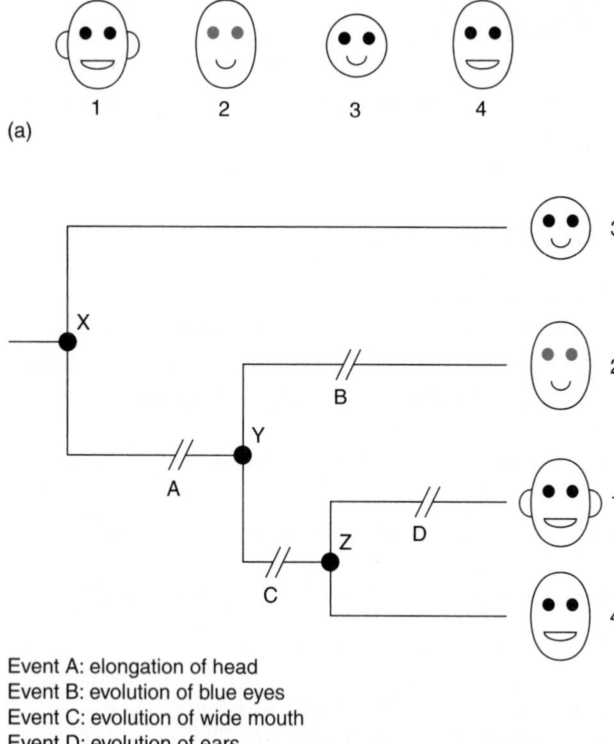

Figure 2.14

Cladistics in action. (a) Imaginary animals are placed into 4 groups according to differences in head characteristics as shown. Individuals differ in head shape, eye color, mouth shape, and the presence or absence of ears. (b) Evolutionary relationships among the 4 groups as determined by cladistic principles. Letters represent the evolution of uniquely defining characteristics (synapomorphies) shared by all descendants of the individual in which the characters first evolved. Each filled circle represents an ancestral species that gave rise to the groups stemming from that ancestor.

programs. The only characters of importance in establishing evolutionary relationships are so-called **synapomorphies,** shared characters derived from a common ancestor in which the characters originated. The astute reader will notice that this resembles the definition of homologues as well. However, the cladist is interested only in homologous characters that are *not* present in any earlier ancestors; only evolutionary novelties are used to construct cladistic classifications and to infer evolutionary relationships. In some cases a single morphological characteristic can suffice to define a major evolutionary event. The presence of a water vascular system derived from the central coelomic chamber (mesocoel) during embryogenesis, for example, uniquely defines the echinoderms. A particularly dramatic departure from the classical systematist's approach is the cladist's insistence that valid taxa include all descendants of an ancestor. As mentioned earlier, for example, cladists cannot recognize the Reptilia as a valid taxonomic category because it does not include birds and mammals, which have evolved from reptilian ancestors.

Let us consider a simple example of cladistics in action, with 4 groups of imaginary animals. Distinguishing characteristics are shown in Figure 2.14. We will assume here that these characteristics are genetically determined. The direction of evolutionary change (**polarity**) is first determined by comparison with a closely related taxon (the **outgroup**) that lies outside the taxa being studied; this outgroup's characteristics are assumed to represent the ancestral condition. (In Figure 2.13, the 2 species at the top of the figure served as the outgroup for that analysis: They belong to a different insect order, the Embioptera.) Suppose that through comparison with an imaginary outgroup, we assume that the ancestor to all 4 of the groups shown in Figure 2.14 had a round head, black eyes, a thin mouth, and no ears. Thus, we begin by assuming that blue eyes evolved from black-eyed precursors in the immediate ancestor to animals in Group 2. If this assumption is wrong, then our conclusions will be wrong, too.

Figure 2.14b presents a dendrogram (now called a **cladogram;** *clados* = G: a branch) showing the least complex, most **parsimonious** way of explaining the evolutionary history of these groups. Groups 1 and 4 are said to be **sister groups,** both derived from the same ancestor (as represented by the node labeled Z), an animal that was not ancestral to members of the other 2 groups. By the same reasoning, Group 2 is the sister group to the combination of Groups 1 and 4: Members of both groups are descended from a common ancestor represented by node Y. Try drawing alternative evolutionary scenarios. You will find that of all possible alternatives, the one illustrated does indeed require the fewest evolutionary changes; to derive Group 4 from Group 2, for example, would require a reversion of eye color back to the ancestral black-eyed condition and the independent evolution of a wide mouth. The scenario shown in Figure 2.13 was also established on this principle of maximum parsimony, although it is supported by other techniques as well.

Published analyses consider dozens of different characters. Many of those characters give conflicting signals, some leading in one direction with others pointing in other directions, because not all similarities between different animal groups are caused by homology. Some similarities arise through convergence or parallelism, but all we see are the similarities, not how they arose.

Consider the example in Figure 2.15. Two new characters have been added: Members of 2 groups now have noses and members of 2 groups have hair. Assuming that the characteristics of Group 1 represent the primitive, or ancestral condition, 3 very different but perfectly reasonable cladograms can be constructed. Two of these are presented in Figure 2.15b and c. In cladogram b, animals 3 and 4 form sister groups, whereas in cladogram c, animals 2 and 4 form sister groups. In the third scenario, which you can work out for yourself, animals 2 and 3 are sister groups. In other words, we can't really tell how these different groups of animals are related. By the principle of parsimony, we favor the first scenario (Fig. 2.15b) because

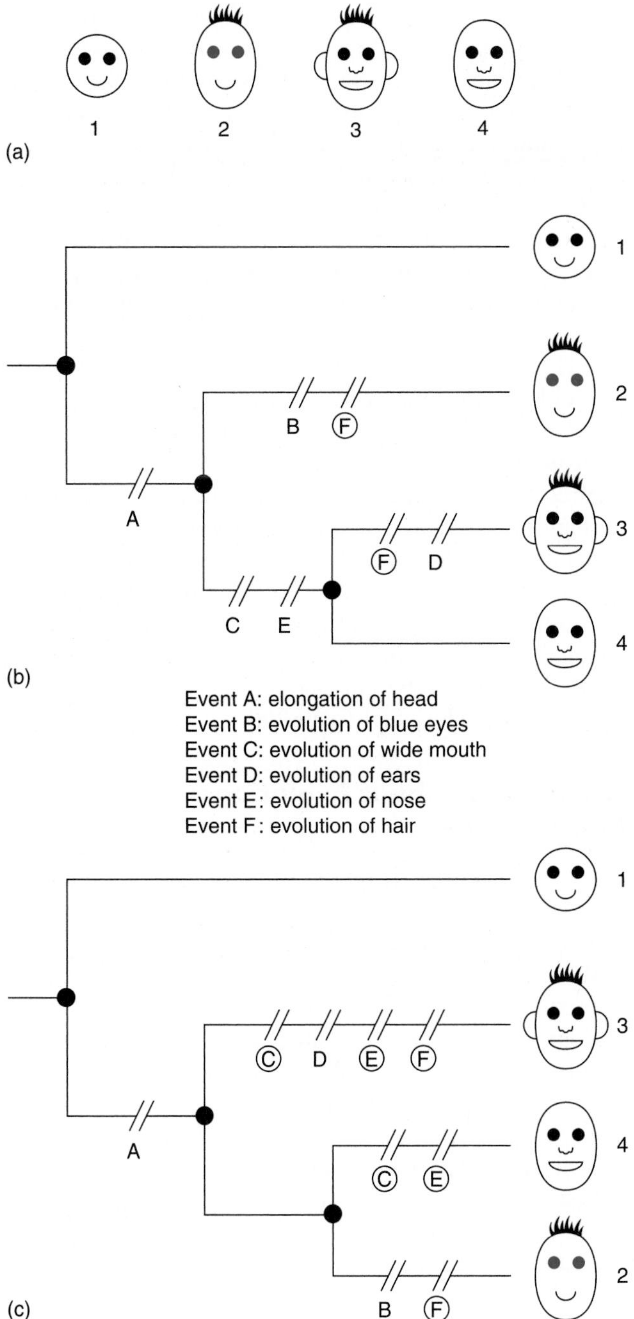

Figure 2.15

The difficulties created by undetected convergent events. (a) As in Figure 2.14, 4 groups of animals differ in head characteristics as shown. Now, however, different sets of characters suggest different evolutionary relationships among the animal groups. Two of these relationships are depicted in (b) and (c). Capital letters indicate key evolutionary events. Circled letters indicate events that must have occurred independently in at least two different ancestors.

it requires only 7 evolutionary steps (count them); the second scenario requires 9 steps. How many steps are required for the third scenario?

Note that the cladograms in Figure 2.15b and c both postulate convergent evolutionary events (indicated by circled letters). In cladogram b, hair evolved once in Group 2 and a second time, independently, in Group 3. Convergence plays an even larger role in cladogram c; noses, wide mouths, and hair each evolved twice (separately in Groups 3 and 4 for noses and mouths, and separately in Groups 2 and 3 for hair). Other scenarios are possible, too. In Figure 2.15c, we could imagine hair evolving once, along with elongation of the head, and then being lost in the ancestor giving rise to the animals in Group 4. This could in fact be the actual sequence of evolutionary events, of course, but we eliminate it as being a relatively unparsimonious explanation. When different characters give conflicting signals like this, the scenarios supported by a majority of characters and involving the fewest number of evolutionary transformations are selected for further consideration.

One can test the robustness of particular trees in a variety of ways—using different computer algorithms, for example, and seeing how much the resulting trees resemble each other, or by repeatedly subsampling different elements of the data set and seeing whether the trees change a great deal or only a little as different aspects of the data are randomly included or excluded from the data set (see "bootstrapping" and "jackknifing" in Table 2.2). If branching patterns remain stable when different components of the data set are randomly added or removed, that branching pattern is obviously more convincing than if the patterns change with every subsampling of the data. In this sense, the phylogenetic hypotheses that each tree represents are testable. But even with additional data, there is really no way to determine which of several competing trees is the correct one. If convergence is relatively uncommon, then cladistic analyses of morphological characters may produce an accurate tree. But many believe that convergent evolution has been very common, which seems especially likely as many different body plan features are apparently controlled by similar sets of genes in widely different animal groups. If convergence is indeed rampant, then the best trees may well be incorrect trees, because branching patterns will be based on false homologies.

The data contained in DNA molecules may present our best hope of eventually resolving such dilemmas, as discussed in the following text.

Cladistic Treatment of Molecular Data

One great appeal of cladistic methodology is that it accommodates molecular data. Assuming that the gene sequences for particular proteins in different organisms are homologous, each base in the sequence conveys potential information about evolutionary relationships. Consider the short stretch of DNA shown for 4 groups of animals in Figure 2.16. In actual practice, biologists work with sequences that are hundreds or tens of thousands of base pairs long. Figure 2.13, for example, is based mostly upon data from 18S (about 1800 base pairs) and 28S (about 3500–4000 base pairs) ribosomal RNA. If we are convinced that we are starting at exactly the same point in the same gene sequence for each group, then we can read from left to right

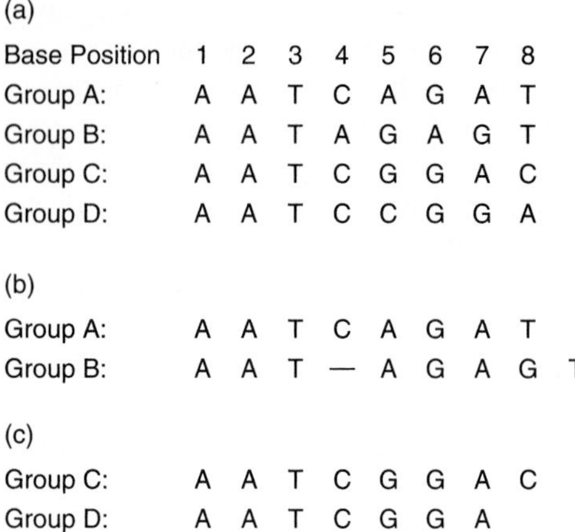

(a)

Base Position	1	2	3	4	5	6	7	8
Group A:	A	A	T	C	A	G	A	T
Group B:	A	A	T	A	G	A	G	T
Group C:	A	A	T	C	G	G	A	C
Group D:	A	A	T	C	C	G	G	A

(b)

Group A:	A	A	T	C	A	G	A	T	
Group B:	A	A	T	—	A	G	A	G	T

(c)

Group C:	A	A	T	C	G	G	A	C
Group D:	A	A	T	C	G	G	A	

Figure 2.16

Four hypothetical short stretches of a particular gene sequence (e.g., 18S rRNA) taken from 4 different species. Each letter represents a different nucleotide base (A = adenine; G = guanine; C = cytosine; T = thymine). Sequences from different species must be correctly aligned before they can be compared.

in our simplified example, checking to see if the bases are the same or different at each position in the sequence. Each difference represents a separate evolutionary event. In Figure 2.16a, for example, all 4 groups have identical bases at positions 1, 2, and 3; these first 3 positions offer no phylogenetically useful information. But animals in Group B have a different base at position 4, an adenine instead of a cytosine. This represents an evolutionary event—a mutation in the gene sequence that occurred only in the immediate ancestor to the animals in Group B if the animals in Group B represent the derived condition, or a mutation that occurred in the immediate ancestor of the other 3 groups. Some of these mutations will alter animal form or function, creating the materials through which natural selection works. Other changes will have no phenotypic effects because the genetic code is degenerate (several triplets code for the same amino acid) or because the mutation occurs in a noncoding region of the DNA or is compensated for (in diploid organisms) by lack of mutation in the homologous chromosome. In our example, position 4 appears to be phylogenetically informative.

But wait: How do we know that the sequences for the 4 groups are correctly aligned? Looking more carefully, we see that after position 3, only one of the remaining bases from Group B animals agrees with those in Group A. Perhaps the original base in position 4 was deleted in the ancestor to Group B; deletions are *bona fide* mutational events. So let us redraw the sequence data assuming a deletion event at position 4 (Fig. 2.16b). Now we get much better agreement between base pairs in sequences A and B. Similarly, if we assume for Group D (Fig. 2.16a) that the cytosine at position 5 was not present in the ancestor, but was in fact added through mutation (an insertion), we would realign the sequences for Group D as shown in Figure 2.16c. Now there are no mismatches between the bases in this short DNA sequence.

All sequences must be correctly aligned before the bases can be compared. Otherwise you won't be comparing homologous bases; if there is an insertion or deletion that has not been properly adjusted for, you will be comparing apples and oranges (or apples and iPods). Once sequences are aligned to the satisfaction of the biologist doing the study, ancestral states are determined through comparison with an outgroup, and the rest of the analysis proceeds much as for analyses using morphological criteria, except that now we are working with hundreds or thousands of characters for each animal group: Each base is a distinct character.

These simple examples give you an idea of the sorts of problems one must consider when conducting cladistic analyses. Some of the startling findings based upon cladistic analyses of molecular data are summarized in Figures 2.12b–f. Compare, for example, the relationships in Figures 2.12a and c for arthropods, molluscs, and annelids. Over the past 10–15 years, molecular data have caused biologists to question many long-standing assumptions about invertebrate evolution. A more detailed example of the use of molecular data in evaluating relationships between major animal groups is given in Research Focus Box 11.1 (page 206).

Uncertainty about Evolutionary Relationships

There is always the temptation to embrace new technology and complex procedures as giving definitive answers. Indeed, the approach outlined here using cladistic analysis and molecular data is so seductive that one is immediately tempted to draw firm conclusions from the cladograms that are generated; the procedures are so sophisticated and so logical that it is easy to forget that the products are working hypotheses. In fact, many biologists have serious reservations about the cladistic approach, and there are many additional controversies within the cladistic community. Even with morphological data, the relationships deduced by computer vary depending on the outgroup used, the characters included in or omitted from the analysis, the computer algorithm used to examine the data, and sometimes even the order in which data are entered into the calculations. The central assumption of parsimony itself periodically falls under attack: Why assume that animals evolve from one form to another in the fewest number of steps? In fact, what we know about the evolutionary process suggests this must often not be the case. There are other ways (Distance, Maximum Likelihood, and Bayesian methods) to construct phylogenetic trees that do not maximize parsimony, but these methods are subject to different criticisms. And is it valid to assume that the characteristics of the outgroup used for comparison always represent the ancestral character states? Even the simplest of animals, and their genomes, have been evolving for hundreds of millions of years.

Molecular data carry their own set of additional complications. Some parts of a given molecule are more likely to change than other parts of the same molecule, and to change at widely different rates within the same animal group, some types of mutations occur with greater frequency than others, different molecules (mitochondrial DNA and 18S rRNA, for example) evolve at different rates within the same animal groups, and homologous molecules evolve at different rates in different animal groups. Of particular interest, gene sequences that evolve unusually rapidly produce longer branches that tend to group closely together—not because the associated animals are close relatives, but only because of the rapid evolution of the sequences in question: "long branches attract." Distinguishing between evolutionary affinity and such "long branch attraction" problems has not been easy. I already mentioned the additional problems associated with sequence alignment. Also, each position in a molecule can change more than once: Guanine can be replaced by adenine, only to be later replaced by guanine, erasing any sign of the molecular evolution that occurred. And, as with morphological features, molecules can also come to resemble each other through convergence. Indeed, with only 4 possible bases for each position in a DNA molecule, some similarity through convergence is quite likely. Researchers have developed very sophisticated ways of addressing all of these problems, but each manipulation introduces more uncertainty into the analyses.

A major appeal of cladistics and its associated numerical methods is that the data and the methodology are completely up-front and open to examination and criticism. It is also very democratic, not requiring years of experience working with the animals under consideration. Whether this is a step forward or a step backward is an issue for class discussion. But acceptance of the cladistic approach is certain to keep growing. Indeed, cladistics may offer the best hope for forging an eventual consensus about the evolutionary relationships among phyla, particularly as more molecular data from more species are incorporated, and as our understanding about the interpretation of those data improves. But whether cladistics eventually reveals the TRUTH about evolutionary relationships may be impossible to ever evaluate. Even if all biologists should eventually agree on any particular set of relationships, there is really no way to test the accuracy of those conclusions, barring the invention of time travel.

Time Machines: Help from the Fossil Record

Probably the closest we will ever get to time travel is the fossil record. For many years there was great controversy about the evolutionary relationship between insects and crustaceans (crabs, lobsters, shrimp, barnacles, etc.). Crustacean appendages are typically biramous (two-branched), while those of insects are exclusively uniramous (single-branched). Strong arguments were made that the crustacean ancestor bore only biramous appendages, and that one branch from each appendage was lost in the evolution of insects. Other biologists advanced equally strong arguments supporting the independent evolution of crustaceans and insects from separate ancestors, with the insect ancestor never having had biramous appendages. A report of fossilized insects bearing biramous appendages in 1992 laid this long-standing controversy to rest. Subsequently, we have learned that the branching of appendages as they develop is under a very simple genetic control mechanism; in hindsight it seems obvious that the uniramous insects could easily have evolved from ancestors with biramous appendages.

Molecular data cannot tell us what any hypothesized ancestor looked like; the controversy about the relationships between insects and crustaceans was resolved by physical evidence from the fossil record, not by sophisticated technology or complex computer algorithms. Similarly, when one finds a 570-million-year-old metazoan embryo, we can know with certainty that complex multicellular animals of some kind existed at that time. Molecular data are leading us to question many evolutionary scenarios and many of the assumptions on which those scenarios are based, and will undoubtedly tell us very exciting things about the genetic changes underlying evolutionary shifts in form and function. Additional fossil material will probably resolve at least some key phylogenetic issues, but issues resolvable by new fossil discoveries are mostly limited to those within phyla, not among phyla: Informative fossils from before the Cambrian explosion will likely remain very limited. The changes that have occurred in DNA sequences as animals have evolved probably offer our best chance of ever fully resolving issues about evolutionary interrelationships among the various animal phyla.

Classification by Habitat and Lifestyle

Animals also may be categorized to reflect degrees of ecological similarity rather than the closeness of evolutionary relationship. For example, one group of animals may be **terrestrial,** living on land, while another is **marine,** living in the ocean. Marine animals, in turn, may be **intertidal** (living between the physical limits of high and low tides and thus exposed to air periodically); **subtidal** (living below the low-tide line and thus exposed to air only under extreme conditions, if ever); or **open ocean** creatures. In addition, animals may be **mobile** (capable of locomotion), **sessile** (immobile), or, perhaps, **sedentary** (exhibiting only limited locomotory capabilities). Some aquatic organisms may be able to move but have negligible locomotory powers with respect to the movement of the medium in which they live; such individuals are said to be **planktonic** (G: forced to drift or wander).

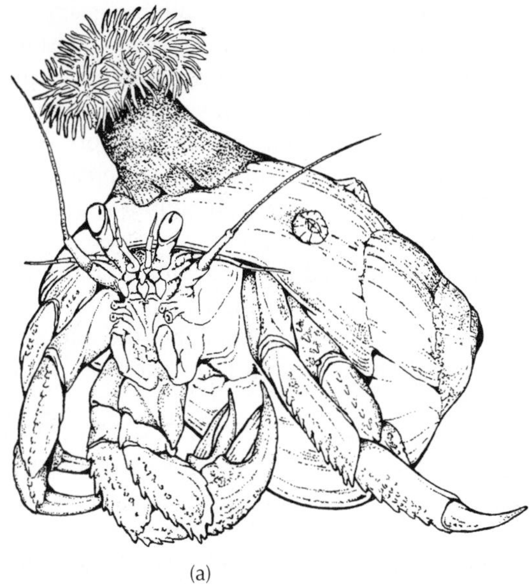

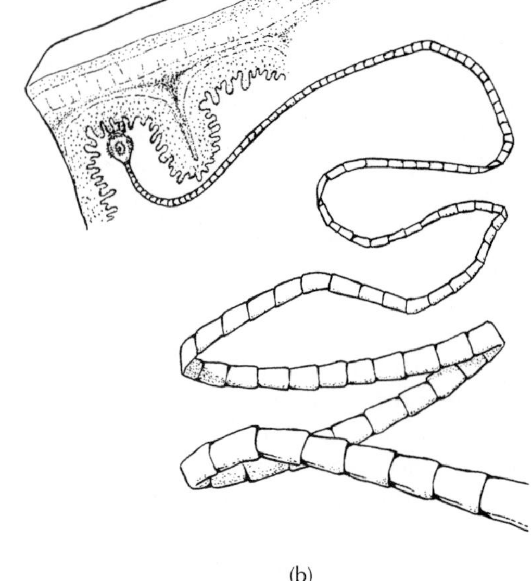

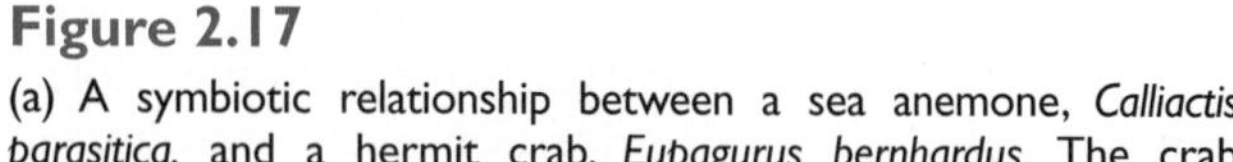

Figure 2.17

(a) A symbiotic relationship between a sea anemone, *Calliactis parasitica,* and a hermit crab, *Eupagurus bernhardus.* The crab deliberately places the anemones on its shell. (b) A tapeworm, *Taenia solium,* shown attached to the intestinal wall of its vertebrate host.
(a) After Hardy. (b) After Villee.

Animals are often categorized as to how they feed or what they eat. For example, some species are **herbivores** (plant eaters), while others are **carnivores** (flesh eaters). Some species remove small food particles from the surrounding medium **(suspension feeders),** while others ingest sediment, digesting the organic component as the sediment moves through the digestive tract **(deposit feeders).**

Members of one species frequently live in intimate association with those of another species. These **symbiotic associations,** or **symbioses,** frequently relate to the feeding biology of one or both of the participants (**symbionts**) in the association (Fig. 2.17). **Ectosymbionts** live near or on the body of the other participant, while **endosymbionts** live within the body of the other participant. When both symbionts benefit, the relationship is said to be **mutualistic,** or an example of **mutualism.** When the benefit accrues to only one of the symbionts and the other is neither benefited nor harmed, the relationship is one of **commensalism,** and the benefiting member is the **commensal.** Last, some animals are **parasites;** that is, they depend upon their **host** for continuation of the species, generally subsisting on either the blood or the tissues of the host. A parasite may or may not substantially impair the host's activities. The essence of parasitism is that the parasite is metabolically dependent upon the host and that the association is obligate for the parasite.

The boundaries between parasitism, mutualism, commensalism, and predation are not always distinct. For example, a parasite that eventually kills its host essentially becomes a predator. A parasite that produces a metabolic end product from which the host benefits borders on being mutualistic. Indeed, transitional forms in the process of evolving from one type of relationship to another are not uncommon. Such transitional forms make tidy categorization of animals into human-made schemes difficult; definitions of some categories have been modified by various workers in an attempt to improve the fit, but every rule seems to have an exception.

Topics for Further Discussion and Investigation

1. Historically, there have been 3 major approaches to animal classification: **phenetics** (based entirely on degree of overall anatomical and biochemical similarity, without regard to whether the similarities reflect homology or convergence), **cladistics** or **phylogenetic classification** (based entirely upon recency of common descent inferred from the mutual possession of particular specialized, derived morphological traits called **synapomorphies**), and **evolutionary classification** (which attempts to consider both ancestry and the degree to which organisms have subsequently diverged from the ancestral form). Discuss the advantages and disadvantages inherent in any 2 of these 3 approaches to the inferring of phylogenetic relationships.

Ax, P. 1987. *The Phylogenetic System: The Systematization of Organisms on the Basis of Their Phylogenesis.* New York: John Wiley & Sons.

Bock, W. 1965. Review of Hennig, *Phylogenetic Systematics. Evolution* 22:646.

Bock, W. J. 1982. Biological classification. In *Synopsis and Classification of Living Organisms,* vol. 2, edited by S. P. Parker. New York: McGraw-Hill, 1068–71.

Cavalier-Smith, T. 1998. A revised six-kingdom system of life. *Biol. Rev.* 73:203.

Cunningham, C. W., K. E. Omland, and T. H. Oakley. 1998. Reconstructing ancestral character states: A critical reappraisal. *Trends Ecol. Evol.* (TREE) 13:361.

Estabrook, G. F. 1986. Evolutionary classification using convex phenetics. *Syst. Zool.* 35:560.

Mayr, E. 1965. Numerical phenetics and taxonomic theory. *Syst. Zool.* 14:73.

Mayr, E., and P. D. Ashlock. 1991. *Principles of Systematic Zoology,* 2d ed. New York: McGraw-Hill.

Moore, J., and P. Willmer. 1997. Convergent evolution in invertebrates. *Biol. Rev.* 72:1–60.

Panchen, A. L. 1992. *Classification, Evolution, and the Nature of Biology.* New York: Cambridge University Press.

Smith, A. B. 1993. *Systematics and the Fossil Record: Documenting Evolutionary Patterns.* Cambridge, Mass.: Blackwell Scientific Publications.

2. What are the characteristics of the "ideal" classification system, and why is this ideal so difficult to attain?

Benton, M. J. 2000. Stems, nodes, crown clades, and rank-free lists: Is Linnaeus dead? *Biol. Rev.* 75:633–48.

Erwin, D. H. 1991. Metazoan phylogeny and the Cambrian radiation. *Trends Ecol. Evol.* 6:131.

Farris, J. 1982. Simplicity and informativeness in systematics and phylogeny. *Syst. Zool.* 31:413.

Hall, B. K. 1994. *Homology: The Hierarchical Basis of Comparative Biology.* New York: Academic Press.

Jenner, R. A. 2004. When molecules and morphology clash: Reconciling conflicting phylogenies of the Metazoa by considering secondary character loss. *Evol. Dev.* 6:372–78.

Mayr, E. 1974. Cladistic analysis or cladistic classification? *Zool. Syst. Evol.-forsch.* 12:94. (Reprinted in E. Mayr. 1976. *Evolution and the Diversity of Life—Selected Essays.* Cambridge, Mass.: Harvard University Press, 433–76.)

Page, M. D., and P. H. Harvey. 1988. Recent developments in the analysis of comparative data. *Q. Rev. Biol.* 63:413.

3. The fossil record shows a remarkable radiation of animal body plans beginning about 543 mya in a geologic time period termed the Cambrian. The earliest known metazoans—from the so-called Ediacaran fauna collected originally in the Ediacara hills of South Australia in the 1940s and more recently from Newfoundland—are somewhat older (about 550–575 mya). To what extent do the Ediacaran and Cambrian faunas differ, and what might account for those differences?

Dzik, J. 2003. Anatomical information content in the Ediacaran fossils and their possible zoological affinities. *Integr. Comp. Biol.* 43:114.

McMenamin, M. A. S. 1987. The emergence of animals. *Sci. Amer.* 256(4):94.

Moore, J. A. 1990. A conceptual framework for Biology, Part II. *Amer. Zool.* 30:752.

Morris, S. C. 1989. Burgess shale faunas and the Cambrian explosion. *Science* 246:339.

Narbonne, G. M. 2004. Modular construction of early Ediacaran complex life forms. *Science* 305:1141.

4. Molecular analyses allow us a direct look at differences in the genetic structure among species, potentially overcoming some of the problems associated with morphological studies. Moreover, if the chosen molecules are not subject to selection and if mutation rates are constant, degrees of difference between the molecules of 2 different species should indicate the amount of time elapsed since the species diverged from their common ancestor. Why haven't molecular data been able to resolve all previous phylogenetic controversies?

Archibald, J. K., M. E. Mort, and D. J. Crawford. 2003. Bayesian inference of phylogeny: A non-technical primer. *Taxon.* 52:187–91.

Bromham, L. 2003. What can DNA tell us about the Cambrian explosion? *Integr. Comp. Biol.* 43:148.

Ciccarelli, F. D., T. Doerks, C. von Mering, C. J. Creevey, B. Snel, and P. Bork. 2006. Toward automatic reconstruction of a highly resolved tree of life. *Science* 311:1283–87.

Cunningham, C. W., K. E. Omland, and T. H. Oakley. 1998. Reconstructing ancestral character states: A critical reappraisal. *Trends Ecol. Evol.* 13:361.

Dopazo, H., and J. Dopazo. 2005. Genome-scale evidence of the nematode-arthropod clade. *Genome Biol.* 6:R41.

Moore, J., and P. Willmer. 1997. Convergent evolution in invertebrates. *Biol. Rev.* 72:1–60.

Raff, R. 1996. The shape of life: Genes, development, and the evolution of animal form. Chapter 4: *Molecular Phylogeny: Dissecting the Metazoan Radiation.* Univ. Chicago Press.

Raff, R. A., C. R. Marshall, and J. M. Turbeville. 1994. Using DNA sequences to unravel the Cambrian radiation of the animal phyla. *Ann. Rev. Ecol. Syst.* 25:351–75.

Rand, D. M. 1994. Thermal habit, metabolic rate and the evolution of mitochondrial DNA. *Trends Ecol. Evol.* 9:125.

Rogozin, I. B., Y. I. Wolf, L. Carmel, and E. V. Koonin. 2007. Ecdysozoan clade rejected by genome-wide analysis of rare amino acid replacements. *Mol. Biol. Evol.* 24:1080–90.

Rokas, A., and P. W. H. Holland. 2000. Rare genomic changes as a tool for phylogenetics. *Trends Ecol. Evol.* 15:454.

Smythe, A. B., M. J. Sanderson, and S. A. Nadler. 2006. Nematode small subunit phylogeny correlates with alignment parameters. *Syst. Biol.* 55:972–92.

Valentine, J. W., D. H. Erwin, and D. Jablonski. 1996. Developmental evolution of metazoan bodyplans: The fossil evidence. *Devel. Biol.* 173:373–81.

Whitfield, J. 2007. Linnaeus at 300: We are family. *Nature* 446: 247–49.

5. Compare and contrast the relationships among phyla depicted in any three parts of Figure 2.12. For example, note that molluscs (such as snails, clams, and squid) are shown as having an immediate ancestor in common with annelids (such as earthworms and leeches) in (b) but as evolving independently of annelids in (a).

6. Was there a Cambrian explosion?

Ayala, F. J., A. Rzhetsky, and F. J. Ayala. 1998. Origin of the metazoan phyla: Molecular clocks confirm paleontological estimates. *Proc. Natl. Acad. Sci.* 95:606–11.

Brasier, M. 1998. From deep time to late arrivals. *Nature* 395:547–48.

Bromham, L. D., and M. D. Hendy. 2000. Can fast early rates reconcile molecular dates with the Cambridge explosion? *Proc. Royal Soc. London B* 267:1041.

Budd, G. E. 2003. The Cambrian fossil record and the origin of the phyla. *Integr. Comp. Biol.* 43:157.

Conway Morris, S. 2000. The Cambrian "explosion": Slow-fuse or megatonnage? *Proc. Nat. Acad. Sci.* 97:4426.

Cooper, A., and R. Fortey. 1998. Evolutionary explosions and the phylogenetic fuse. *Trends Ecol. Evol.* (TREE) 13:151–56.

Davidson, E. H., K. J. Peterson, and R. A. Cameron. 1995. Origin of bilaterian body plans: Evolution of developmental regulatory mechanisms. *Science* 270:1319.

Grotzinger, J. P., S. A. Bowring, B. Z. Saylor, and A. J. Kaufman. 1995. Biostratigraphic and geochronologic constraints on early animal evolution. *Science* 270:598.

Jensen, S., J. G. Gehling, and M. L. Droser. 1998. Ediacara-type fossils in Cambrian sediments. *Nature* 393:567–69.

Morris, S. C. 1993. The fossil record and the early evolution of the Metazoa. *Nature* 361:219–25.

Narbonne, G. M. 2004. Modular construction of early Ediacaran complex life forms. *Science* 305:1141–44.

Regier, J. C., J. W. Shultz, and R. E. Kambic. 2005. Pancrustacean phylogeny: Hexapods are terrestrial crustaceans and maxillopods are not monophyletic. *Proc. R. Soc. B.* 272:395–401.

7. For many decades, meticulous embryological studies have been undertaken in an attempt to unravel evolutionary connections among animal groups. Why is it so difficult to interpret such data?

Guralnick, R. P., and D. R. Lindberg. 2002. Cell lineage data and Spiralian evolution: A reply to Nielsen and Meier. *Evolution* 56:2558–60.

Nielsen, C., and R. Meier. 2002. What cell lineages tell us about the evolution of Spiralia remains to be seen. *Evolution* 56:2554–57.

8. Fossilized embryos are rare in the fossil record. What seem to be fossilized animal embryos were recently discovered in rocks that predate the Cambrian explosion by about 40 million years. What makes us think that these specimens really are fossilized animal embryos? What arguments suggest that they are not?

Bailey, J. V., S. B. Joye, K. M. Kalanetra, B. E. Flood, and F. A. Corsetti. 2007. Evidence of giant sulphur bacteria in Neoproterozoic phosphorites. *Nature* 445:198–201.

Bengtson, S., and G. Budd. 2004. Comment on "Small bilaterian fossils from 40–55 million years before the Cambrian." *Science* 306:1291a.

Chen, J.-Y., P. Oliveri, E. Davidson, and D. J. Bottjer. 2004. Response to comment on "Small bilaterian fossils from 40–55 million years before the Cambrian." *Science* 306:1291b.

Donoghue, P. C. J. 2007. Embryonic identity crisis. *Nature* 445: 155–56.

Donoghue, P. C. J., S. Bengtson, X.-P. Dong, M. J. Gostling, T. Huldtgren, J. A. Cunningham, C. Yin, Z. Yue, F. Peng, and M. Stamanoni. 2006. Synchrotron X-ray tomographic microscopy of fossil embryos. *Nature* 442:680–83.

Raff. E. C., J. T. Villinski, F. R. Turner, P. C. J. Donoghue, and R. A. Raff. 2006. Experimental taphonomy shows the feasibility of fossil embryos. *Proc. Natl. Acad. Sci.* 103:5846–51.

General References about Metazoan Origins and Evolution

Adoutte, A., G. Balavoine, N. Lartillot, O. Lespinet, B. Prud'homme, and R. de Rosa. 2000. The new animal phylogeny: Reliability and implications. *Proc. Nat. Acad. Sci.* 97:4453–56.

Aguinaldo, A. M. A., J. M. Turbeville, L. S. Linford, M. C. Rivera, J. R. Garey, R. A. Raff, and J. A. Lake. 1997. Evidence for a clade of nematodes, arthropods and other moulting animals. *Nature* 387:489.

Benton, M. J. 2000. Stems, nodes, crown clades, and rank-free lists: Is Linnaeus dead? *Biol. Rev.* 75:633–48.

Budd, G. E., and S. Jensen. 2000. A critical reappraisal of the fossil record of the bilaterian phyla. *Biol. Rev.* 75:253–95.

Cameron, C. B., J. R. Garey, and B. J. Swalla. 2000. Evolution of the chordate body plan: New insights from phylogenetic analyses of deuterostome phyla. *Proc. Natl. Acad. Sci.* 97:4469–74.

Cavalier-Smith, T. 1998. A revised six-kingdom system of life. *Biol. Rev.* 73:203–66.

Donoghue, P. C. J., and M. J. Benton. 2007. Rocks and clocks: Calibrating the Tree of Life using fossils and molecules. *Trends Ecol. Evol.* 22:424–31.

Garey, J. R. 2002. The lesser-known protostome taxa: An introduction and a tribute to Robert P. Higgins. *Integr. Comp. Biol.* 42:611–18.

Gilbert, S. F., J. M. Opitz, and R. A. Raff. 1996. Resynthesizing evolutionary and developmental biology. *Devel. Biol.* 173:357–72.

Halanych, K., J. D. Bacheller, A. M. Aguinaldo, S. M. Liva, D. M. Hillis, and J. A. Lake. 1995. Evidence from 18S ribosomal DNA that the lophophorates are protostome animals. *Science* 267: 1641–43.

Halanych, K. 2004. The new view of animal phylogeny. *Ann. Rev. Ecol. Syst.* 35:229–56.

Jenner, R. A. 2006. Challenging perceived wisdoms: Some contributions of the new microscopy to the new animal phylogeny. *Integr. Comp. Biol.* 46:93–103.

Lake, J. A. 1988. Tracing origins with molecular sequences: Metazoan and eukaryotic beginnings. *TIBS* (Trends in Biochemical Sciences) 16:46–50.

Moore, P. 1990. *Invertebrate Relationships: Patterns in Animal Evolution.* Cambridge Univ. Press, 400 pp.

Morris, S. C. 1998. *The Crucible of Creation: The Burgess Shale and the Rise of Animals.* Oxford Univ. Press, 242 pp.

Nielsen, C. 2001. *Animal Evolution: Interrelationships of the Living Phyla.*, 2nd ed. Oxford Univ. Press, 576 pp.

Pennisi, E., and W. Roush. 1997. Developing a new view of evolution. *Science* 277:34–37.

Peterson, K. J., and D. J. Eernisse. 2001. Animal phylogeny and the ancestry of bilaterians: Inferences from morphology and 18S rDNA gene sequences. *Evolution and Development* 3:170–205.

Raff, R. A. 1996. *The Shape of Life: Genes, Development, and the Evolution of Animal Form.* Univ. Chicago Press, 520 pp.

Schmidt-Rhaesa, A., U. Ehlers, T. Bartolomaeus, C. Lemburg, and J. R. Garey. 1998. The phylogenetic position of the Arthropoda. *J. Morphol.* 238:263–85.

Smith, A. B., and K. J. Peterson. 2002. Dating the time of origin of major clades: Molecular clocks and the fossil record. *Ann. Rev. Earth and Planet. Sci.* 30:65–88.

Valentine, J. W. 2001. Defining phyla: Evolutionary pathways to metazoan body plans. *Evol. Devel.* 3:432–42.

Valentine, J. W. 2004. *On the origin of phyla.* Univ. Chicago Press, 639 pp.

General References about Phylogenetic Analysis

Baldauf, S. L. 2003. Phylogeny for the faint of heart: A tutorial. *Trends in Genetics* 29:345–51.

Benton, M. J. 2000. Stems, nodes, crown clades, and rank-free lists: Is Linnaeus dead? *Biol. Rev.* 75:633–48.

Bryant, H. N., and P. D. Cantino. 2002. A review of criticisms of phylogenetic nomenclature: Is taxonomic freedom the fundamental issue? *Biol. Rev.* 77:39–55. [a spirited defense of the proposed PhyloCode system]

Cunningham, C. W., K. E. Omland, and T. H. Oakley. 1998. Reconstructing ancestral character states: A critical reappraisal. *TREE* 13:361–66.

Freeman, S., and J. C. Herron. 1998. *Evolutionary Analysis.* Prentice Hall, Upper Saddle River, New Jersey.

Gaffney, E. S., L. Dingus, and M. K. Smith. 1995. Why cladistics? *Natural History.* June:33–35.

Huelsenbeck, J. P., F. Ronquist, R. Nielsen, and J. P. Bollback. 2001. Bayesian inference of phylogeny and its impact on evolutionary biology. *Science* 294:2310–14.

Jenner, R. A., and F. R. Schram. 1999. The grand game of metazoan phylogeny: Rules and strategies. *Biol. Rev.* 4:121–42.

Kitching, I. J., P. L. Forey, C. J. Humphries, and D. M. Williams. 1998. *Cladistics: The Theory and Practice of Parsimony Analysis,* 2nd ed. Oxford Univ. Press, New York, 228 pp.

Kolaczkowski, B., and J. W. Thornton. 2004. Performance of maximum parsimony and likelihood phylogenetics when evolution is heterogeneous. *Nature* 431:980–84.

Minelli, A. 1993. *Biological Systematics: The State of the Art.* Chapman & Hall, New York, 387 pp.

Page, R. D. M., and E. C. Holmes. 1998. *Molecular evolution: A phylogenetic approach.* (See especially Chapter 2, "Trees.") Blackwell Science, Inc., Maiden, Mass. 352 pp.

Sanderson, M. J., and H. B. Shaffer. 2002. Troubleshooting molecular phylogenetic analyses. *Ann. Rev. Ecol. Syst.* 33:49–72.

Swofford, D. L., G. J. Olsen, P. J. Waddell, and D. M. Hillis. 1996. Phylogenetic inference. In D. M. Hillis, C. Moritz, and B. K. Mable (eds.), *Molecular Systematics,* 2nd ed. Sunderland, Mass.: Sinauer Associates, pp. 407–514.

Search the Web

1. http://evolution.berkeley.edu/evolibrary/article/phylogenetics_01

   Click on "Journey into Phylogenetic Systematics" for an introduction to the principles and implications of cladistic analysis, produced at the University of California at Berkeley.

2. http://tolweb.org/tree/phylogeny.html

   The Tree of Life, containing information about phylogeny and biodiversity, and orchestrated by D. R. Maddison at the University of Arizona.

3. http://www.ucmp.berkeley.edu/cambrian/camb.html

   This site, produced by the University of California Museum of Paleontology, concerns the Cambrian explosion. It includes excellent photographs of Burgess Shale fossils.

4. http://www.ohiou.edu/phylocode

   This is the PhyloCode website, giving a detailed description of the new system's goals and the rules by which it will operate. For contrasting views on this enterprise, do a Google search on the term "phylocode debate."

5. http://www.mhhe.com/biosci/pae/zoology/animalphylogenetics/

   This site offers an excellent tutorial on molecular phylogenetics, written by C. Leon Harris and offered through McGraw-Hill Publishers.

6. http://www.ucmp.berkeley.edu/IB181/VPL/Phylo/Phylo2.html

   A brief but informative introduction to cladistics, written by N. C. Arens, C. Strömberg, and A. Thompson.

7. http://stri.discoverlife.org

   Click on "Tree of Life" at the left side of the opening page to see a detailed overview of relationships among all life-forms, and then on the group of interest. The site is sponsored by the Polistes Foundation and operated by the Smithsonian Tropical Research Institute, University of Georgia, and the Missouri Botanical Garden. It also includes detailed identification guides to species.

8. http://bioinf.ncl.ac.uk/molsys/lectures.html

   This site presents an excellent tutorial (including Powerpoint and PDF files) on the methods of molecular systematics. Based on a course organized by Professor T. Martin Embleyis.

9. http://animaldiversity.ummz.umich.edu/site/index.html

   This is the Animal Diversity Website, from the University of Michigan Museum of Zoology, giving a largely up-to-date classification, with illustrations, of many animal groups.

10. www.tolweb.org/Eukaryotes/3

    This is the Tree of Life site, providing classifications for all organisms, including protozoans, with accompanying photographs and drawings.

11. http://evolution.berkeley.edu/evolibrary/article/phylogenetics_01

    This site, from the University of California at Berkeley, provides a good introduction to constructing and using phylogenetic trees.

3

The Protists

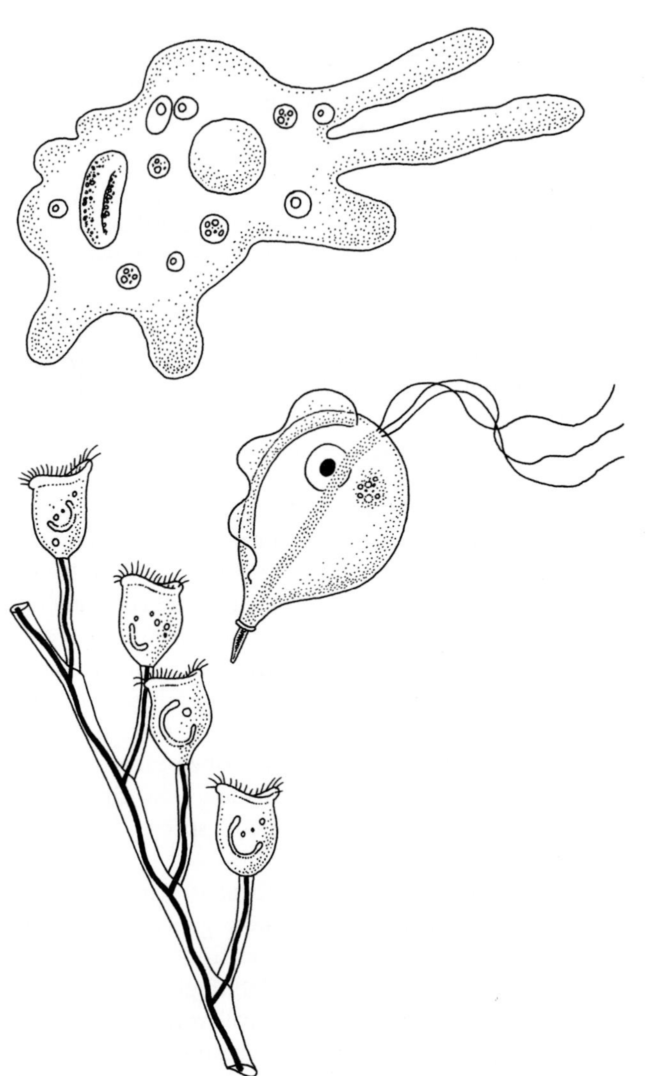

Introduction

Protists (*proto* = Greek: first) are unicellular eukaryotes (*eu* = G: true; *karyo* = G: nuclei) that blur the distinction between animals and plants. Indeed, all algal, plant, and animal life—that is, all multicellular life—must have evolved from protist ancestors. As with mitochondria and the other membrane-bound organelles that characterize eukaryotes, chloroplasts have been acquired by many unicellular organisms through ancient symbiotic relationships. Thus, while some protists ingest solid food particles, others photosynthesize. Still others live amidst decaying plant and animal matter—or live as parasites—and feed by taking up dissolved organic material and other nutrients across their surfaces. Some species are capable of 2 or even all 3 nutritional modes, either simultaneously or at different times. For this reason, about 140 years ago the great German scientist Ernst Haeckel suggested placing all of the troublesome organisms that weren't clearly plants or animals within a new kingdom, the Protista (*protist* = G: the very first). Eventually, the Protista came to include everything except plants, animals, fungi, and bacteria. But the diversity of protist ultrastructure, life cycles, lifestyles, and evolutionary trajectories proved too extreme for a single kingdom to support. Indeed, protists are now distributed among all kingdoms. In consequence, in the classification scheme adopted for this edition[1] there is no longer a formal taxonomic category called the Protista, although "protist" still serves as a useful general term for referring to this remarkable collection of unicellular eukaryotes. This chapter is concerned primarily with the animal-like (heterotrophic) protists, commonly called "protozoans."

1. Based largely on Adl, S. M., A. G. B. Simpson, M. A. Farmer, *et al.* 2005. *J. Eukaryot. Microbiol.* 52: 399–451; Keeling, P. J., G. Buerger, D. G. Durnford, B. F. Lang, *et al.* 2005. *Trends Ecol. Evol.* 20: 670–76.

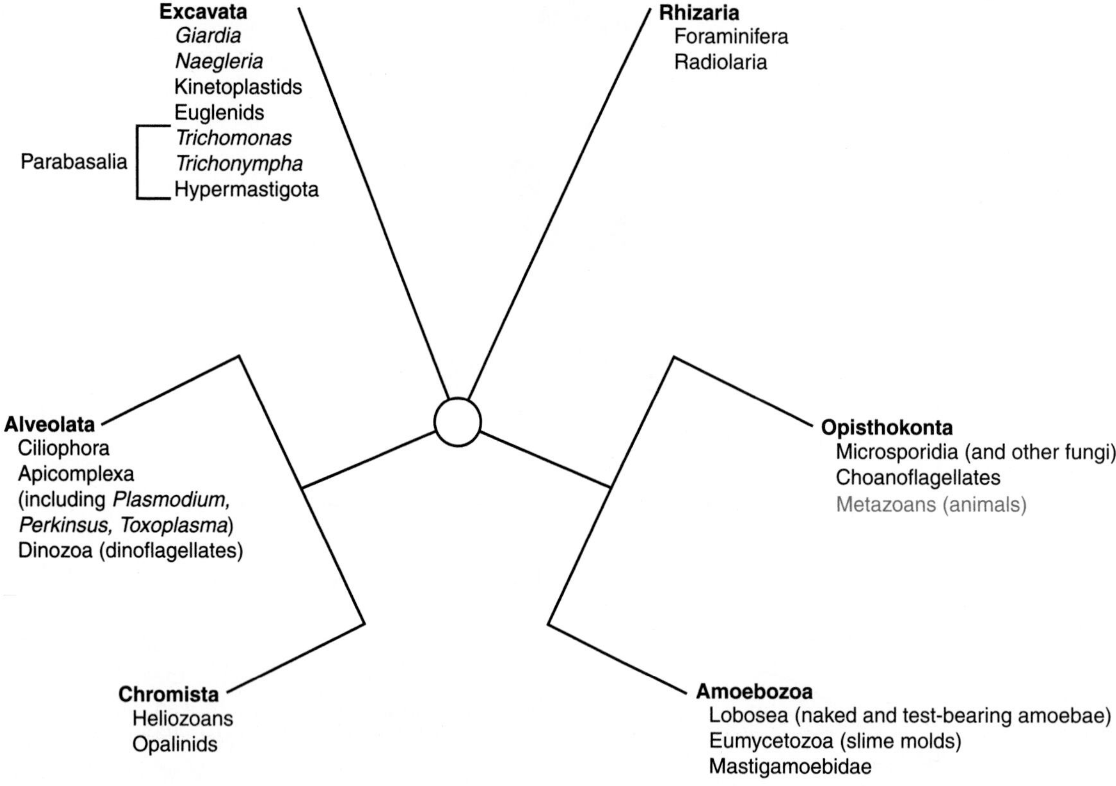

Figure 3.1
Current taxonomic placement of the organisms discussed in this chapter, based largely on molecular data and ultrastructural evidence. The rest of this book concerns animals ("metazoans," shaded blue), just one of many groups of opisthokonts (only 3 groups are shown here). Compiled from several sources.

The Protozoa

Proto • zoa
(G: the very first • animals)
prō-tō-zō´-ah

Defining Characteristics:[2] 1) All are unicellular eukaryotes lacking collagen and chitinous cell walls; 2) all are nonphotosynthetic in the primitive condition

---

> "We are frustratingly trapped between existing classifications of protists that are recognized to be faulty and some future scheme(s) not yet available."
> John D. Corliss

> "The traditional taxonomy of protozoa, based on the categories amoebae, flagellates, ciliates, and sporozoa, has almost completely broken down, and a new phylogenetic taxonomy is struggling to be born."
> James D. Berger

> "The names of many protist groups and the genera they include have been changed so many times that the classification scheme is unclear, and it is difficult to determine which names apply." Sina M. Adl

The Protozoa has generally been abandoned as a formal taxonomic category. Indeed, the former Society of Protozoologists is now the International Society for Protistologists! The organisms described in this chapter are currently distributed among a variety of groups (Fig. 3.1), whose interrelationships are still unclear. Many dozens of protozoan species have not yet been placed into any of these categories. At least 30,000 extant species have been described, with hundreds of new species being described each year. Tens or hundreds of thousands of new species descriptions are anticipated. At least another 44,000 species have been described from fossilized remains.

Like other eukaryotes, protozoans harbor distinct nuclei and other membrane-bound organelles, but unlike animals, protozoans—like other protists—never develop from a blastula embryonic stage. Although protozoans are not animals, all animals seem to have evolved from flagellated protozoan ancestors. This suggestion, first proposed in part because most animals produce flagellated sperm, is well supported by current molecular data. Protozoans, then, bridge the gap not only between animals and plants, but also between unicellular and multicellular organisms.

2. Characteristics distinguishing the members of this kingdom from members of other kingdoms.

Protozoan classification has long been uncertain and controversial, and nucleotide sequence and other molecular data are changing the scenarios of protozoan interrelationships at a rapid rate. Precise evolutionary relationships are difficult to determine for most groups of organisms, but the problem is especially great for protozoans. The fossil record for protozoans is mostly limited to a few groups, and what does exist is not particularly helpful in deducing relationships; members of some rather sophisticated protozoan groups possessing hard parts are encountered as 600-million-year-old fossils and have forms that are virtually identical to those found in some genera today. The small size of most individuals makes structural studies difficult, although they are slowly proceeding (and at times altering substantially our understanding of evolutionary relationships). As with all organisms, however, structural similarities need not imply close evolutionary relationships; similar structures often arise independently in different, unrelated groups of organisms in response to similar selective pressures, a phenomenon known as **convergent evolution** (see Chapter 2).

Controversy abounds at the most fundamental levels of protozoan organization. For many years, protozoans were placed into one of 4 general categories: amoeboid forms, flagellated forms, ciliated forms, and spore-forming parasitic forms. Both molecular and ultrastructural data now indicate that those categories are artificial, not accurately reflecting evolutionary relationships. It seems, for example, that amoeboid forms have evolved many times, mostly from a variety of flagellated ancestors. Although it seems clear that the old system must be abandoned, there is only now the beginning of a consensus about what should take its place. Of particular interest, hundreds of unicellular species lack mitochondria. Whether these species represent a primitive condition before the advent of mitochondria, or whether they evolved from protozoan ancestors that had mitochondria is not completely clear. The information presented here should be read with the understanding that there is as yet no agreement about evolutionary relationships among and within the various protozoan groups. Clearly, this is an exciting time to be a protistologist.

General Characteristics

Protozoans absolutely defy tidy categorization, demonstrating a tremendous range of sizes, morphologies, ultrastructural characteristics, nutritional modes, behavioral and physiological diversity, and genetic diversity. Indeed, in many respects, the diversity of form and function encountered among protozoans rivals that encountered among all invertebrate animals combined. Discussing all of the protozoans within one chapter is analogous to attempting a single coherent discussion of annelids, molluscs, arthropods, echinoderms, flatworms, and gastrotrichs!

Although most people never see protozoans, few other groups of organisms rival them in economic and scientific importance. Protozoans play major roles both in primary production and in decomposition, and may serve as a major food source for many invertebrates and, indirectly, for many vertebrates as well.[3] In addition, protozoans cause a number of human ailments, including malaria, African sleeping sickness, and dysentery, and a variety of devastating diseases of poultry, sheep, cattle, cabbage, and other human food. Protozoans also have provided biologists with outstanding material for genetic, physiological, developmental, and ecological studies. Remarkably, these ubiquitous and important organisms have been known for only about 330 years, their discovery awaiting the invention of the microscope and the patience of the Dutch draper and amateur scientist Antony van Leeuwenhoek, who discovered protozoans in 1674.

Protozoans occur wherever there is moisture. Free-living species are found in both marine and freshwater habitats, and in moist soil. Many protozoans live in close association with other protozoans, with animals, or with plants, either as commensals or as parasites. Indeed, every major group of protozoans contains at least some parasitic species, and the members of several major groups are exclusively parasitic. Most known protozoans are microscopic, usually 5–250 μm (micrometers) long; researchers have recently begun discovering many additional species in the size range of bacteria (about 0.5–2 μm). The largest protozoans rarely exceed 6–7 mm (millimeters). About 82,000 protozoan species have been described so far, more than half of which are known only as fossils. Probably, most extant protozoan species have yet to be described.

Protozoans are clearly among the simplest of life-forms. Nevertheless, in many respects protozoans are as complex as any multicellular animal **(metazoan),** and a great deal of protozoan biology remains poorly understood despite much sophisticated research. Protozoans are not composed of individual cells; in essence, each protozoan *is* an individual cell. This is a major characteristic that sets protozoans apart from animals, and is one of the few characteristics applicable to nearly all protozoans. Remarkably, this single cell functions as a complete organism, doing all of the basic things—feeding, digesting, locomoting, behaving, and reproducing—that multicellular animals do. Complexity arises from the specialization of organelles rather than from the specialization of individual cells and tissues. In addition, most protozoans lack specialized circulatory, respiratory, and excretory (waste-removal) structures. The surface area of their bodies is high relative to body volume, so that gases can be exchanged and soluble wastes removed by simple diffusion across the entire exposed body surface. The cytoplasm of a few species contains hemoglobin, although its role in gas exchange has not yet been demonstrated. The relatively high surface area of protozoans may also facilitate active uptake of dissolved nutrients from the surrounding fluids.

The entire protozoan body is bounded by a **plasmalemma** (cell membrane) that is structurally and chemically identical to that of multicellular organisms.

3. See *Topics for Further Discussion and Investigation,* no. 7, at the end of the chapter.

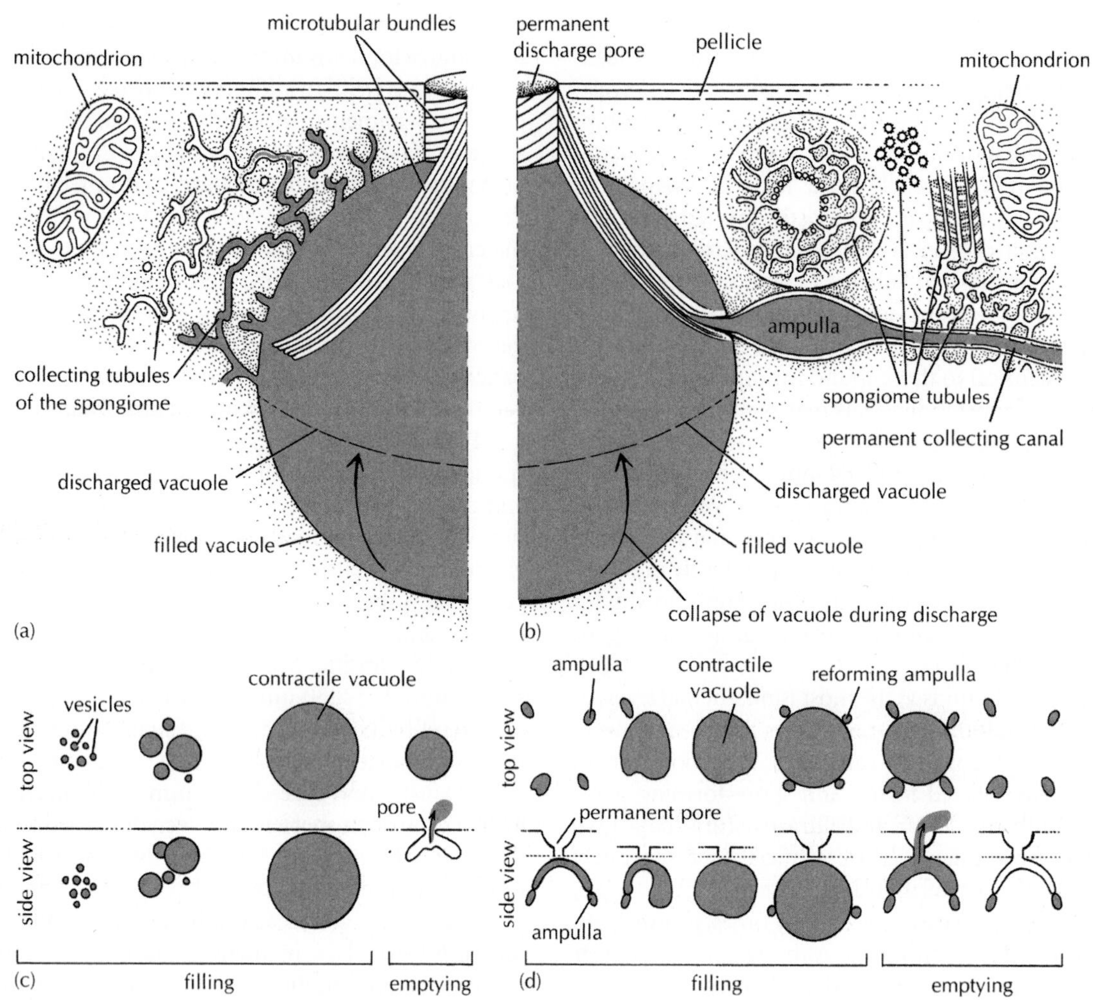

Figure 3.2

(a,b) Diagrammatic illustration of two types of contractile vacuole system encountered among protozoans. The two differ largely in the complexity of the spongiome and in other aspects of the fluid collection system. Other contractile vacuole complexes (not illustrated) may lack ampullae, permanent pores, and the associated bundles of microtubules. (c,d) The behavior of two types of contractile vacuole complexes during filling and emptying; arrows represent the flow of liquid out of the emptying vacuole. The vacuole is viewed from above in the upper series (discharge emerges perpendicular to the plane of the page) and from the side in the lower series. In (c), a permanent pore is lacking, and the vacuole forms through the fusion of many small, fluid-filled vesicles; the vacuole disappears completely following discharge. In (d), the contractile vacuole is filled by conspicuous ampullae and discharges through a permanent pore. The ampullae may begin to refill before the vacuole discharges its fluid to the outside.

From Patterson, in *Biological Review of the Philosophical Society,* 55:1, 1980. Copyright © 1980 Cambridge University Press, New York. Reprinted by permission.

The cytoplasm bounded by the plasmalemma resembles that of animal cells, except that it is often differentiated into a clear, gelatinous outer region, the **ectoplasm,** and an inner, more fluid region, the **endoplasm.** Within the cytoplasm are organelles and other components typical of metazoan cells, including nuclei, nucleoli, chromosomes, Golgi bodies, endoplasmic reticula (with and without ribosomes), lysosomes, centrioles, mitochondria, and in some individuals, chloroplasts. In a sense, then, the typical protozoan is a single-celled organism, although this cell is functionally more versatile than any single cell found within any multicellular animal.

In addition to the organelles encountered typically in multicellular animals, many protozoans contain organelles not generally found among metazoans. These organelles are, most notably, contractile vacuoles, trichocysts, and toxicysts. Many protozoans are also characterized by highly complex arrays of microtubules and microfilaments. Still other highly specialized organelles are restricted to various small groups of protozoans.

Contractile vacuoles are organelles that expel excess water from the cytoplasm. Apparently, fluid is collected from the cytoplasm by a system of membranous vesicles and tubules called the **spongiome.**[4] The collected fluid is transferred to a contractile vacuole and subsequently discharged to the outside through a pore in the plasmalemma (Fig. 3.2). Contractile vacuoles are most commonly seen among freshwater protozoan

4. See *Topics for Further Discussion and Investigation*, no. 3.

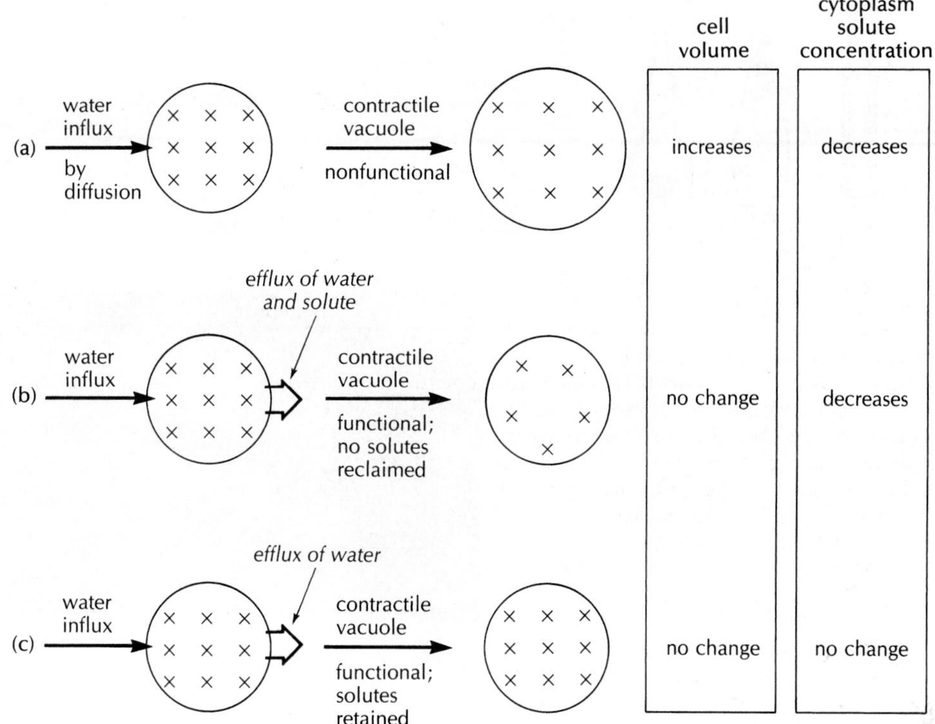

Figure 3.3

Diagrammatic illustration of the functional significance of contractile vacuoles in freshwater protozoans. The vacuole system functions both to maintain body volume and to maintain proper solute concentrations within the cell (X represents solute). (a) No functioning contractile vacuole; water diffuses inward, swelling the cell and diluting internal solute. (b) The contractile vacuole pumps out cytoplasmic fluid, including dissolved solute; cell volume is maintained, but solute is continuously lost. (c) The contractile vacuole pumps out fluid containing little solute, maintaining both cell volume and osmotic concentration.

species because concentrations of dissolved solutes in the cytoplasm of freshwater protozoa are much higher than solute concentrations in the surrounding medium. Thus, water continuously diffuses into the cytoplasm across the plasmalemma, in proportion to the magnitude of the solute concentration gradient (i.e., the **osmotic gradient**). Without some form of compensating mechanism, water would keep diffusing into the protozoan until either the osmotic gradient across the plasmalemma was reduced to zero or the protozoan burst. Moreover, cells can function over only a narrow range of internal solute concentrations, and protozoans are no exception; even a small dilution of the cytoplasmic solute concentration could be disabling.

Clearly, the contractile vacuole must function not only to prevent swelling (by ridding the body of fluid as fast as water crosses the plasmalemma), but also to maintain a physiologically acceptable solute concentration within the cell (Fig. 3.3). In other words, the contractile vacuole must function both in **volume regulation** (maintaining a constant body volume) and in **osmotic regulation** (maintaining a constant intracellular solute concentration).[5] The contractile vacuole achieves the latter result by pumping out of the cell a fluid that is dilute relative to the surrounding cytoplasm in the cell; that is, the solute is essentially separated from the water and retained within the cell before the remaining fluid is expelled. The mechanism by which water is separated from the cytoplasm is not known, nor is the mechanism by which the vacuole fluid is discharged to the outside, although researchers are making some progress in these studies. Contractile vacuoles typically fill and discharge several to many times per minute, and a single individual may possess several contractile vacuoles.

Other interesting organelles called **trichocysts** develop within membrane-bound vesicles in the cytoplasm and eventually come to lie along the periphery of the protozoan. The trichocysts themselves are elongated capsules that can be triggered by a variety of mechanical and/or chemical stimuli to discharge a long, thin filament (Fig. 3.4). This discharge occurs within several thousandths of a second and is thought to be initiated through an osmotic mechanism involving a rapid influx of water. The adaptive significance of trichocysts is unknown; likely possibilities are that they protect against predation or anchor the animal during feeding. Related structures called **toxicysts** are clearly involved in predation; filaments discharged from toxicysts paralyze prey

5. See *Topics for Further Discussion and Investigation,* no. 4.

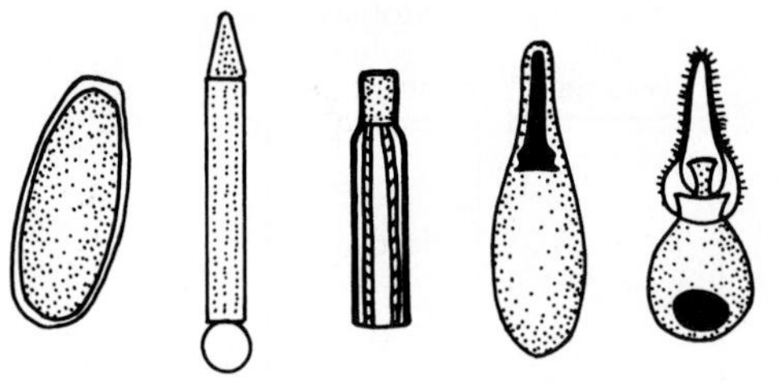

(a)

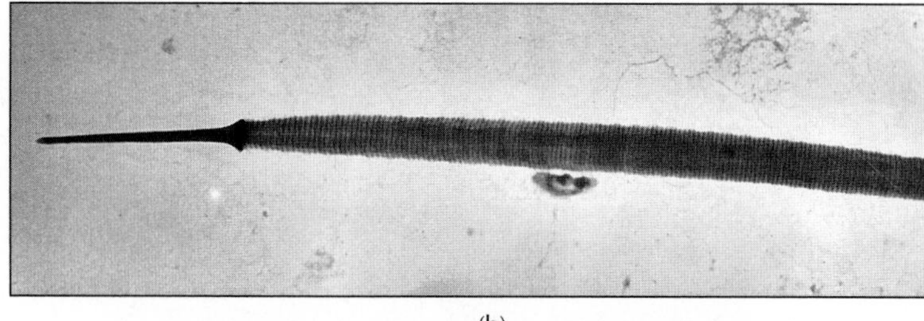

(b)

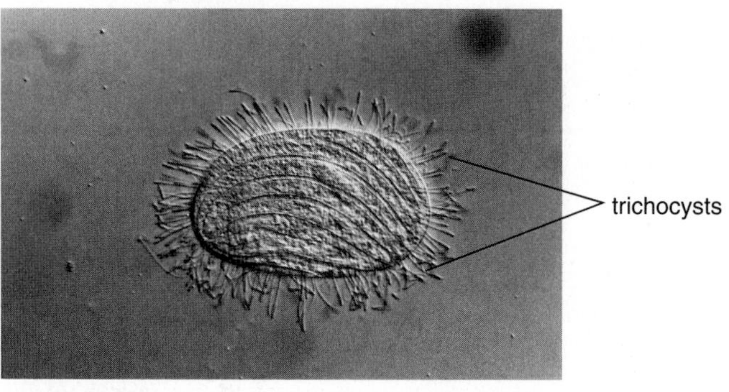

(c)

Figure 3.4

(a) Undischarged extrusomes of different types. At least 12 morphologically distinct types of extrusome have been described. Some expel mucus (left); others eject filaments of various lengths. Paralytic toxins are injected by one class of extrusome (toxicysts, not shown). Others (far right) are used primarily for adhering to prey during food capture. (b) Transmission electron micrograph of a trichocyst that has been discharged from *Paramecium* sp. (c) A ciliate, *Pseudomicrothorax dubis,* with its trichocysts extended.

(a) From Corliss, in *American Zoologist* 19:573, 1979. Copyright © 1979 American Society of Zoologists, Thousand Oaks, California. (b) Courtesy of M. A. Jakus, National Institutes of Health. (c) From S. Eperon and R. K. Peck. *The Journal of Protozoology* 35:280–86, fig. 1, p. 282, Allen Press, Inc. © 1988 Dr. Edna Kaneshiro, editor.

and initiate digestion. At least 10 other types of **extrusomes** (organelles capable of ejectability) have so far been described from protozoans.

Protozoan Locomotory Systems

Most free-living protozoans move, using either cilia, flagella, or flowing cytoplasmic fingers called **pseudopodia** (*pseudo* = G: false; *podium* = G: foot). The members of some species possess different locomotory systems at different stages of the life cycle or under different environmental conditions. As mentioned earlier, it has become all too clear that similar locomotory systems among different protozoan groups do not always indicate close evolutionary connections.

Structure and Function of Cilia

The ultrastructure of cilia is remarkably uniform among all organisms in which cilia occur, from protozoans to vertebrates, probably reflecting evolution by common descent. Although modifications to the basic plan are found sporadically, cilia are always cylindrical, with each cilium arising from a **basal body (kinetosome).** Within the cilium are a number of long rods called **microtubules,** composed of a protein known as **tubulin.** Tubulin is extremely similar to the actin of metazoan striated muscle. A cross section of a cilium's kinetosome in the region beneath the outer body surface shows a ring of 9 groups of microtubules, with 3 microtubules to a group (Fig. 3.5, lower right). The A, or innermost microtubule of each group, is physically connected to the C, or outermost microtubule of an adjacent group, via a thin filament. Additional filaments connect the A microtubule to a central tubule, like the spokes of a wheel. This configuration of microtubules changes somewhat near the distal end of the cilium.

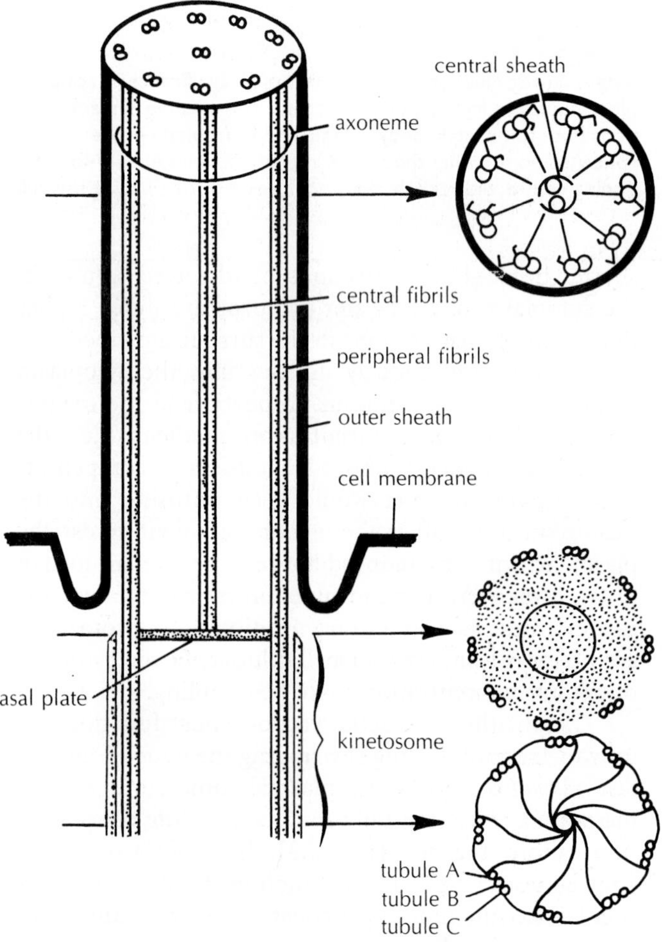

Figure 3.5

Ciliary ultrastructure. The appearance of the cilium in cross section changes along its length as indicated.

After Sherman and Sherman; after Wells.

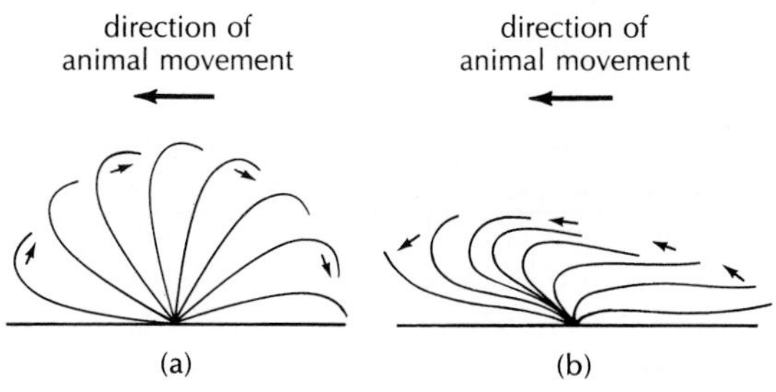

Figure 3.6

(a) Power and (b) recovery strokes in isolated cilia. Arrows indicate the direction of motion of the cilium. Note that more of the ciliary surface area is involved in pushing against the water in (a) than in (b). Thus, the recovery stroke does less work and does not undo all the work of the power stroke.

A cross section through a cilium external to the body surface shows a ring of 9 groups of microtubules, with only two microtubules per group; the C microtubule is not found (Fig. 3.5, upper right). Instead, a pair of **dynein arms** projects outward from each A microtubule toward the B tubule of the neighboring pair of microtubules. The primary protein component of these arms (**dynein**) is similar in some respects to muscle myosin. Like myosin, dynein possesses the ability to cleave ATP, releasing chemical energy. One pair of microtubules is located centrally within the cilium. These two microtubules form the central shaft of the cilium, and they are often surrounded by a membrane constituting a **central sheath.** Distinct filaments extend from the A microtubule of each outer doublet in toward this central sheath. The entire microtubular complex, consisting of the nine doublet microtubules and the inner pair of single microtubules, is termed the **axoneme.**

Although many details of ciliary operation remain to be elucidated, present evidence indicates that ciliary bending is achieved through the differential sliding of some groups of adjacent microtubules relative to others within the same cilium. The energy for this sliding appears to derive from the ATPase activity of the dynein arms, and the sliding itself appears to involve an interaction between tubulin and dynein, in a manner highly reminiscent of the interaction between the actin and myosin filaments of metazoan striated muscle. As we will see later in this chapter, microtubules play a variety of roles, both direct and indirect, in the locomotion of most cells—even in those lacking cilia.

The rhythmic and coordinated bending and recovery of cilia enable both locomotion and food collection. Ciliates are the fastest moving of all protozoans, achieving speeds of up to 2 mm per second. In addition, ciliary activity is probably important in ridding the surroundings of wastes and in continually bringing oxygenated water into contact with the general body surface.

To be effective—that is, to create unidirectional movement of an organism in water or unidirectional flow of water past an organism—the **power stroke** and the **recovery stroke** of a cilium must differ in form. By analogy, if you are swimming the breaststroke and move your arm forward to the front of your head by the same path and with the same force that you used to bring the arm back to your side, you will move backward by about the same distance that you moved forward during your power stroke. The cilium faces a similar dilemma. During ciliary locomotion, the power stroke, as the name implies, does work against the environment. The cilium is outstretched for maximum resistance as it bends downward toward the body (Fig. 3.6a). During the recovery stroke, the cilium bends in such a manner as to considerably reduce resistance, doing less work against the environment (Fig. 3.6b). Thus, the recovery stroke does not undo the work of the power stroke. In addition, the cilium moves fastest in the power stroke, increasing the force of the power stroke relative to that of the recovery stroke. By analogy, more force is required to move your arm through the water quickly than slowly.

Structure and Function of Flagella

Flagella are structurally much like cilia. In cross section, flagella exhibit the same characteristic arrangement of 9 pairs of microtubules ringing a pair of central microtubules over most of their length (see Fig. 3.5). Like cilia, flagella are produced from basal bodies, and the movement of flagella is believed to involve the sliding of microtubules in relation to each other. The cross-sectional appearance of the basal body of a flagellum resembles that previously described for cilia. Indeed, the structural and functional similarities between cilia and flagella are so striking—and differ so much from those of bacterial flagella—that eukaryote cilia and flagella are sometimes grouped together under the descriptive heading **undulipodia.**

Unlike cilia, however, flagella often bear numerous, external hairlike projections (**mastigonemes**) along their length. Presumably, these mastigonemes increase the effective surface area of the flagellum, thus increasing the power that it can generate when it moves through the water. Flagella are usually longer than cilia, and the typical flagellate bears many fewer flagella than the typical ciliate has cilia. Unlike with cilia, several waves of movement may be in progress simultaneously along a single flagellum, and the wave may be initiated at the tip of the flagellum rather than at the base, pulling the organism forward. Flagellate locomotion can be fairly rapid, up to about 200 μm per second. This is only about one-tenth the speed attained by many ciliates but is about 40 times that reached by the fastest amoebae, which typically move using pseudopodia, as described next.

Structure and Function of Pseudopodia

Pseudopodia are characteristic of amoebae. They are used both in feeding and locomotion[6] and come in several forms, as illustrated in Figures 3.7 and 3.9. **Lobopodia**

6. See *Topics for Further Discussion and Investigation,* no. 1.

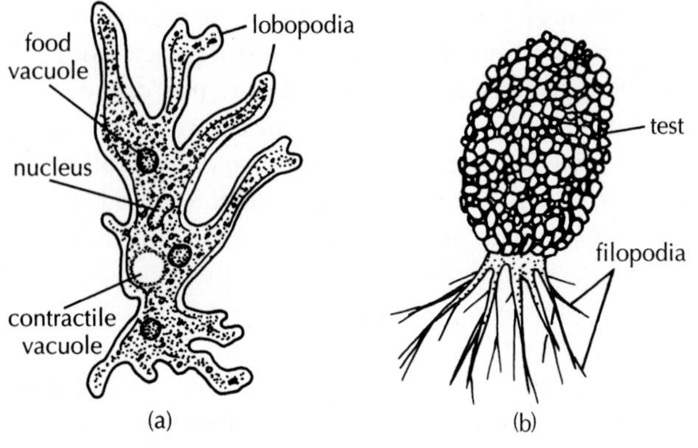

Figure 3.7

(a) A naked sarcodinid, *Amoeba proteus,* with lobopodia. (b) *Pseudodifflugia* sp., with filopodia.

(a) From Hyman, *The Invertebrates,* Vol. III. Copyright © 1951 McGraw-Hill Book Company, New York. Reprinted by permission. (b) After Pennak.

are typical of the familiar amoeba (Fig. 3.7a); they are broad with rounded tips, like fingers (*lobo* = G: a lobe), and bear a distinctly clear ectoplasmic area, called the **hyaline cap,** near each tip. In contrast, **filopodia** (*filo* = L: a thread) lack a hyaline cap, are very slender, and often branch (Fig. 3.7b). With both types, the organism flows into the advancing pseudopodium, a process called **cytoplasmic streaming.**

The amoeboid body is thus formless, lacking permanent anterior, posterior, or lateral surfaces (Figs. 3.7a, 3.26, p. 63); pseudopodia can generally form at practically any point on the body surface. This is also true of parasitic amoeboid species, including the ubiquitous agent of amoebic dysentery, *Entamoeba histolytica.* The mechanism by which pseudopodia form and change shape is not certain, although it seems clear that movement involves a controlled transition of cytoplasm between the gelatinous, ectoplasmic form **(gel)** and the more fluid, endoplasmic form **(sol).** The factors coordinating this transformation in different parts of the cell are not fully understood, although the hypotheses have become wonderfully complex during the past decade. The transformation may involve the interaction of actin and myosin molecules, both abundant in amoeba cytoplasm. A model of sliding actin filaments has been proposed, in which actin and myosin interact in a way that resembles their interaction in the muscle tissue of multicellular animals. Other hypotheses minimize the role of myosin in the gel-sol-gel transitions and instead emphasize the potential role of selective actin polymerization and depolymerization. In one model (Fig. 3.8), localized actin disassembly creates an area of increased osmotic pressure, which draws water from the more central region of the body to the periphery, forming a pseudopod. The actin is then repolymerized to form a bracing network, fixing the pseudopod's shape until the next bout of depolymerization. Whatever the mechanism, pseudopodial locomotion is extremely slow, usually less than 300 μm per minute (1.8 cm per h).

Pseudopodia can also come in very elaborate, anastomosing forms called **reticulopodia** (*reticularia* = L: a network), in which extremely thin filaments branch and coalesce repeatedly in highly complex patterns and exhibit a characteristic and striking bidirectional streaming; reticulopodia are especially characteristic of foraminiferans (Fig. 3.9), a group discussed later in the chapter (p. 65). Finally, some pseudopodia radiate outwards along thin spines, or along equally thin rods composed of microtubules (Fig. 3.31, p. 67). These **axopodia** (*axo* = G: an axle; *podia* = G: foot) are used more for food collection than for locomotion.

Protozoan Reproduction

As single-celled organisms, protozoans necessarily lack gonads. Although sexual reproduction occurs in most groups, asexual (*a* = G: without) reproduction is also common among all groups of protozoans and is the only form of reproduction reported for many species. By definition, asexual reproduction does not generate new genotypes. Protozoans reproduce asexually through **fission,** a controlled mitotic replication of chromosomes and splitting of the parent into 2 or more parts. Indeed, asexual reproduction among protozoans has long been exploited by biologists as a general model for studies of mitosis. **Binary fission** occurs when the protozoan splits into 2 individuals (Fig. 3.10). In **multiple fission,** many nuclear divisions precede the rapid differentiation of the cytoplasm into many distinct individuals. In **budding,** a portion of the parent breaks off and differentiates to form a new, complete individual. In some multinucleate species, the parent simply divides in two in the absence of any mitotic division, the original nuclei being distributed between the 2 daughter cells. This process is termed **plasmotomy** (*tomy* = G: to cut).

In keeping with the widespread occurrence of asexual reproductive capabilities, many protozoans possess a great capacity for regeneration. One example of this capacity is the phenomenon of encystment and excystment exhibited by many freshwater and parasitic species. During **encystment,** the organism dedifferentiates substantially. The individual loses its distinctive surface features, including cilia and flagella, and becomes rounded. The contractile vacuole(s) then pumps out all excess water from the cytoplasm, and a covering (often gelatinous) is secreted. This covering soon hardens to form a protective **cyst.** In this encysted form, the quiescent individual can withstand long periods (several to many years in some species) of exposure to what would otherwise be intolerable environmental conditions of acidity, dryness, thermal stress, and food or oxygen deprivation. Winds, animals, and other vectors may disperse such encysted individuals widely. Once conditions improve, **excystment** quickly ensues and all former internal and external structures are regenerated.

(a) (b)

(c) (d)

Figure 3.8

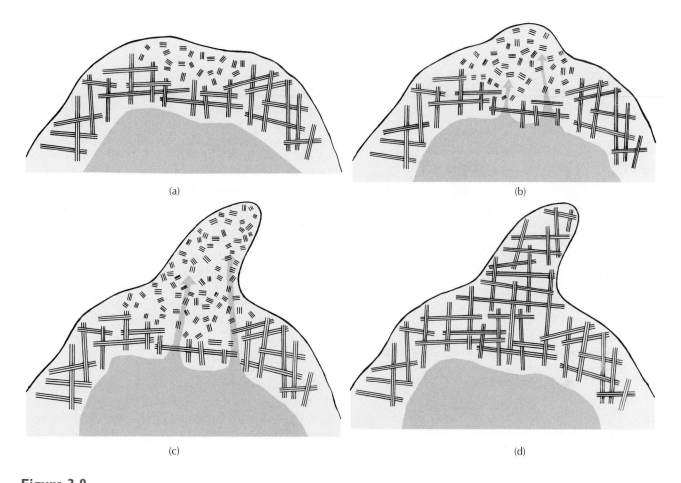

Hypothesized mechanism of pseudopod formation during amoeboid locomotion. (a) Localized breakdown of the actin network increases osmotic concentration in that part of the cytoplasm. (b) Fluid from the interior of the cell moves toward the periphery along the osmotic gradient, forming a pseudopod (c). (d) Actin is repolymerized, re-forming a stabilizing network of filaments.

From T. P. Stossel, "How Cells Crawl" in *American Scientist* 78:408–23, 1990. © American Scientist. Reprinted by permission.

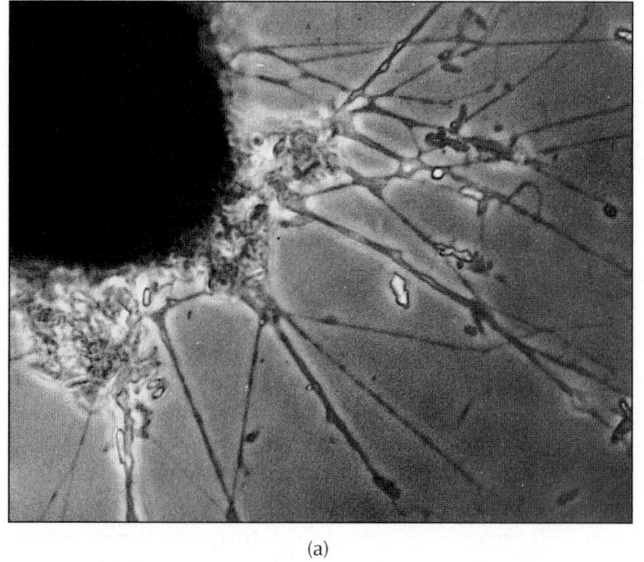

(a) (b)

Figure 3.9

(a) Reticulopodia projecting from a foraminiferan (*Allogromia laticollaris*) with a very translucent, non-calcareous test. (b) Detail of reticulopodia from *Astrammina rara* showing complex patterns of branching and coalescence.

(a) Courtesy of John J. Lee. (b) Courtesy of Sam Bowser, from Lee, J. J., G. F. Leedale, and P. Bradbury. 2000. *An Illustrated Guide to the Protozoa,* 2nd ed., Society of Protozoologists.

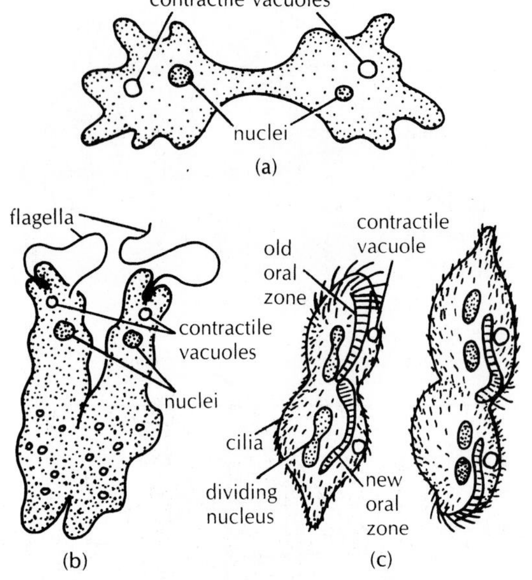

Figure 3.10

Binary fission among protozoans. (a) In an amoeba. (b) In a flagellate. (c) In a ciliate.

(a) After Pennak. (b,c) After Hyman; after Gregory.

Patterns of sexual reproduction will be discussed as appropriate on a group-by-group basis, as few generalizations are possible.

Protozoan Feeding

Food particles are digested internally among protozoans[7]: Since they are single-celled organisms, digestion is, of necessity, entirely intracellular. Ingested food particles generally become surrounded by membrane (similar to the plasmalemma), forming a distinct **food vacuole** or **phagosome** (*phago* = G: eat; *soma* = G: body). These vacuoles move about in the fluid cytoplasm of the cell as the vacuolar contents are digested enzymatically. By feeding protozoans food stained with pH-sensitive chemicals, researchers have determined that the contents of the vacuoles first become quite acidic and later become strongly basic. As in other organisms, including humans, digestion requires exposing the food to a series of enzymes, each with a specific role to play and each with a narrowly defined pH optimum. The controlled changes of pH that occur within protozoan food vacuoles allow for the sequential disassembly of foods by a series of different enzymes, despite the absence of a digestive tract *per se*. Once solubilized, nutrients move across the vacuole wall and into the endoplasm of the cell. Indigestible solid wastes are commonly discharged to the outside through an opening in the plasma membrane.

7. See *Topics for Further Discussion and Investigation*, no. 6.

General Summary

Clearly, protozoans are unlike animals in many features of their biology, but impressively complex nevertheless. Protozoans are also intriguing in another important respect: Even the distinction between unicellular and multicellular is not always easily made. Although most protozoans occur as single, one-celled individuals, many species are **colonial;** that is, a single individual divides asexually to form a colony of attached, genetically identical individuals. Usually, the individuals of a protozoan colony are morphologically and functionally identical. However, the individuals comprising the colonies of several species show a degree of structural and functional differentiation that is very reminiscent of that encountered within some groups of metazoans. Protozoans thus bridge the gap not only between plants and animals but also between unicellular and multicellular life-forms. A discussion of major protozoan groups follows.

The Alveolates

This is a large and diverse group of organisms (comprising the Alveolata) possessing very distinctive membrane-bound sacs, called **alveoli** (*alveolus* = Latin: a cavity), just below the outer cell membrane (see Fig. 3.16, p. 50). Molecular studies support this grouping. The 3 phyla included are the Ciliophora (ciliates), the Dinozoa (dinoflagellates), and the Apicomplexa, a remarkable group of parasitic species.

Phylum Ciliophora

Phylum Cilio • phora
(G: cilia bearing)
sil´-ē-ō-for´-ah

Defining Characteristics: 1) Body externally ciliated in at least some life cycle stages; 2) individual cilia are connected below the body surface through a complex cord of fibers (the infraciliature); 3) all individuals possess at least one micronucleus and at least one macronucleus

Ciliates show the highest degree of subcellular specialization encountered among protozoans and thus are considerably advanced from the primitive protozoan condition. Nevertheless, we will begin with the ciliates because they are predominately free-living and, compared with other protozoan groups, are relatively uniform in basic body plan. Moreover, ciliates clearly form a monophyletic group, with all members derived from a common ancestor. The phylum contains about 3500 described species. Only a few species are known as fossils.

First among the unique features of this phylum is the presence of external ciliation in at least some stage of the life cycle of nearly all species.

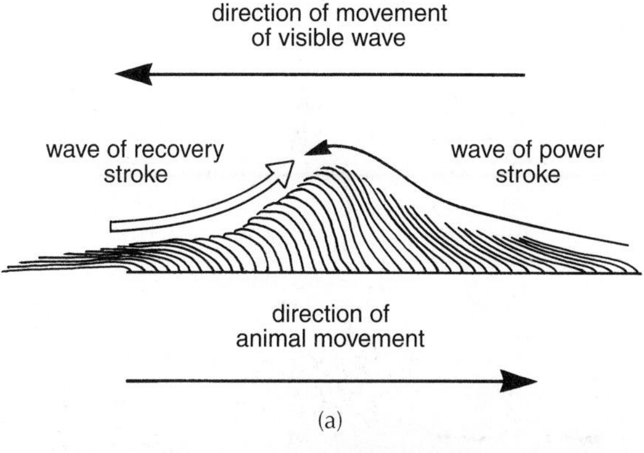

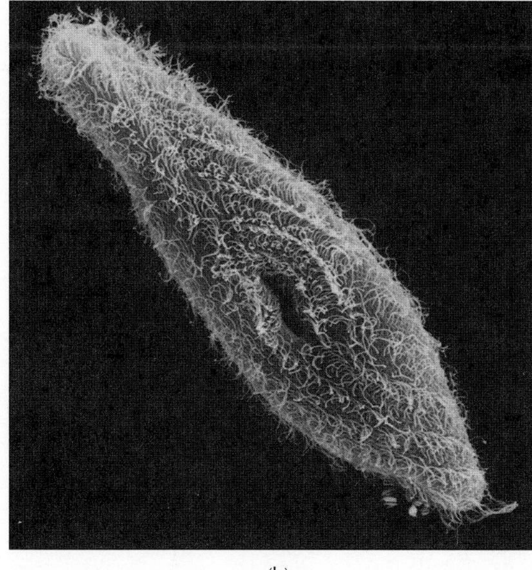

Figure 3.11

(a) A metachronal wave passing along a row of cilia. (b) A ciliate (*Paramecium sonneborni*), showing metachronal waves of ciliary activity. Note the oral opening near the middle of the body. The paramecium is 40 μm in length.

(a) After Wells; after Sleigh. (b) From K. J. Aufderheide et al. 1983. *The Journal of Protozoology* 30:128.

Generally, a large part of the body surface is covered by distinct rows of cilia, with movement achieved by the coordinated, **metachronal** beating of the cilia in each row (Fig. 3.11). In metachronal beating, the power and recovery strokes of a cilium are begun immediately following the initiation of the comparable strokes of an adjacent cilium. The direction of metachronal beating may be quickly reversed in response to chemical, mechanical, or other stimuli, permitting mobile organisms to quickly reverse direction and perhaps escape undesirable situations. As mentioned earlier, ciliates are the fastest moving of all protozoans.

Patterns of Ciliation

Basic ciliary structure and function has already been described (pp. 42–43). Among ciliates, individual cilia, seen external to the cell body, are associated with each other through a complex **infraciliature** below the body surface (Figs. 3.12 and 3.13). A striated fibril, called a **kinetodesmos,** extends from each kinetosome (**basal body**) in the direction of an adjacent cilium of the same row. Thus, running along the right side of each row of basal bodies is a cord of fibers, termed the **kinetodesmata.** Direct microtubular connections between adjacent kinetosomes have rarely been demonstrated. Each of the kinetosomes has its own array of microtubular, microfibrillar, and other organelles, giving ciliate bodies a more complex cytoarchitecture than is found in many other protozoans and in any metazoan.

This infraciliature is found only within the Ciliophora and is encountered in the adults of all ciliate species, even when those adults lack external ciliation. The structure of this infraciliature is one of the primary tools used to distinguish the different ciliate species and to assess the degree to which different species are related. The functional significance of the complex structure of the infraciliature has not yet been conclusively demonstrated.[8]

Cilia cover virtually the entire body of some species, but they are reduced or otherwise modified to various degrees in others. In some species, groups of cilia are functionally associated in such a way as to form discrete organelles. One such organelle is the so-called **undulating membrane,** a flattened sheet of cilia that moves as a single unit (Fig. 3.14a). A second commonly encountered ciliary organelle is termed a **membranelle;** here, a smaller number of cilia in several adjacent rows appear to lean toward each other, forming, in effect, a 2-dimensional triangular tooth (Fig. 3.14a, b). In addition, cilia may form a discrete bundle (**cirrus),** which tapers to a point toward the tip (Figs. 3.14a, c and 3.15). The cilia comprising such organelles are structurally identical to those that function as individuals, and no permanent physical attachments between the cilia comprising undulatory membranes, membranelles, or cirri have been observed. The mechanism by which their activities are so closely coordinated remains uncertain.

Besides the infraciliature and general placement and pattern of body ciliation, other characteristics of major taxonomic significance are the position, ultrastructure, and pattern of ciliation of the oral region. The mouth opening, called the **cytostome** (*cyto* = G: cell; *stoma* = G: mouth), may be located anteriorly, laterally, or ventrally on the body. Often, the cytostome is preceded by one or more preoral chambers, whose complexity varies greatly among major groups of ciliate species.

Other Morphological Features

One other morphological feature particularly characteristic of ciliates is the covering of the body by an often complex series of membranes, forming a **pellicle** (Fig. 3.16). The pellicle may be rigid or highly flexible, depending upon how the membranes are organized, and may serve a

8. See *Topics for Further Discussion and Investigation,* no. 2.

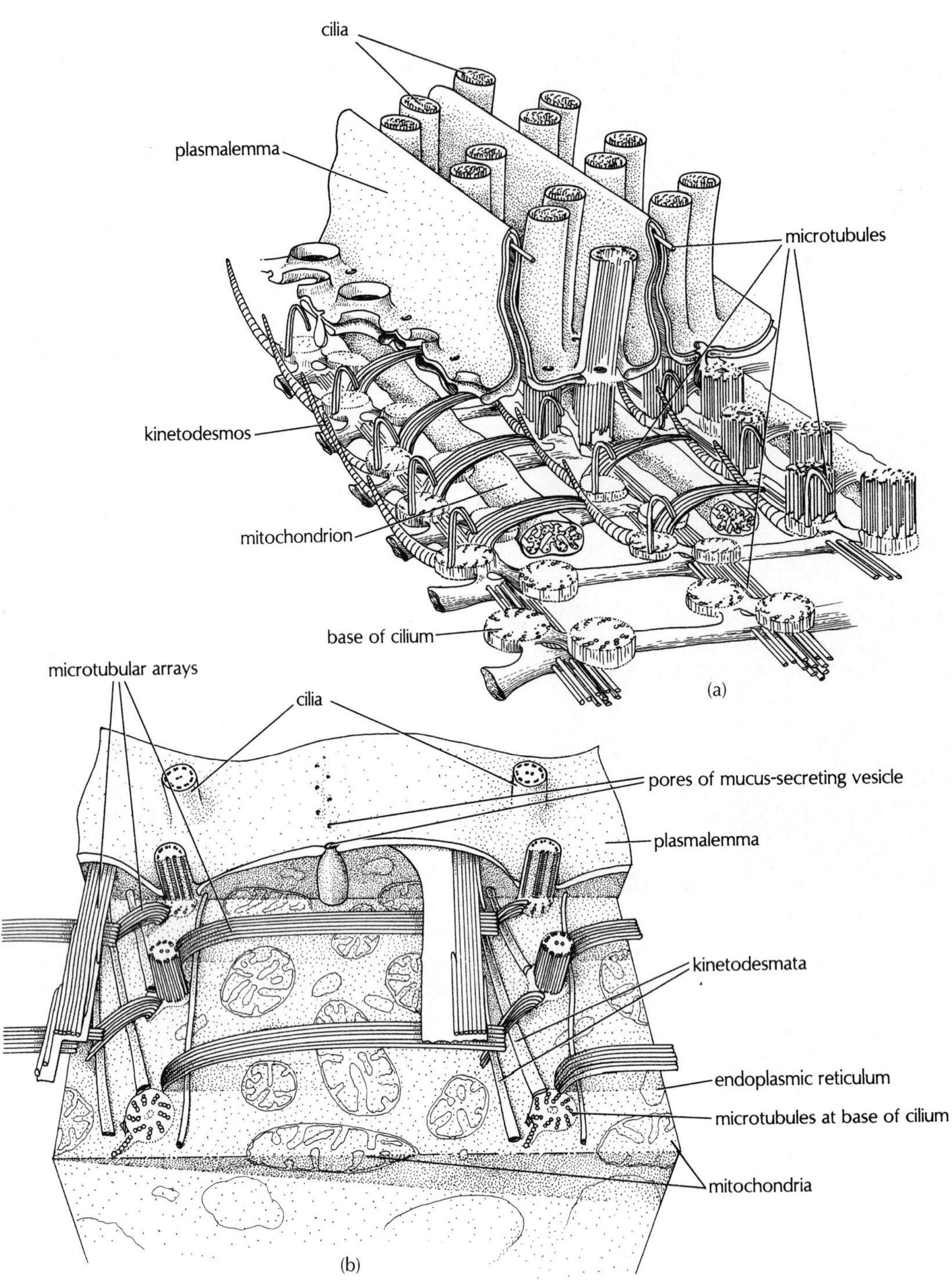

Figure 3.12
The complex infraciliature of ciliates. (a) *Conchophthirus* sp. (b) *Tetrahymena pyriformis*. These ultrastructural details were unknown before the advent of the electron microscope.

(a) From Corliss, in *American Zoologist* 19:573, 1979. Copyright © 1979 American Society of Zoologists. Reprinted by permission. (b) From R. D. Allen, "Fine Structure, Reconstruction and Possible Functions of Components of the Cortex of *Tetrahymena Pyriformis*" in *Journal of Protozoology* 14:553, 1967. Copyright © 1967 Society of Protozoologists, Lawrence, Kansas. Reprinted by permission.

supportive function in some species, helping to maintain shape. Trichocysts are characteristically associated with the ciliate pellicle. The inner membranes, lying beneath the single plasmalemma enveloping the body, form a series of elongated, flattened vesicles called **alveoli** (Fig. 3.16). Cilia project to the outside between adjacent alveoli.

In keeping with their unusually high level of structural complexity, only ciliates have permanent excretory pores associated with their contractile vacuoles. These excretory openings are maintained by arrays of microtubules. The system for collecting cytoplasmic fluid and transferring it to the contractile vacuole is also particularly

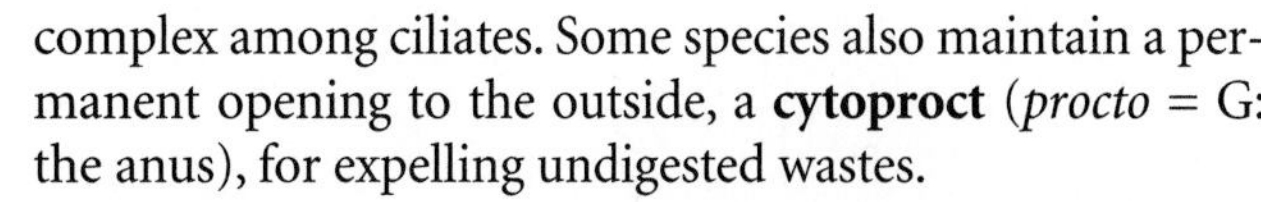

(a) (b)

Figure 3.13

(a) *Climacostomum* sp., stained to reveal the infraciliature. (b) Detail of 3 rows (note arrows) of kinetosomes (basal bodies). A row of kinetosomes, with their associated cilia and kinetodesmata, is termed a kinety.

(a,b) From C. F. Dubochet et al. 1979. *The Journal of Protozoology* 26:218.

complex among ciliates. Some species also maintain a permanent opening to the outside, a **cytoproct** (*procto* = G: the anus), for expelling undigested wastes.

Reproductive Characteristics

All ciliates possess 2 types of nuclei; that is, the nuclei of every ciliate are **dimorphic** (of 2 kinds). The ciliates are thus **heterokaryotic** (*hetero* = G: different; *karyo* = G: nucleus), while all other cells, including other protozoans, are **monomorphic** (one kind) or **homokaryotic** (*homo* = G: alike). Every ciliate contains one or more large nuclei, termed **macronuclei,** and one or more smaller nuclei, termed **micronuclei.** Micronuclei are often more abundant than macronuclei within a given individual; individuals in some species possess more than 80 micronuclei.

The macronucleus is polyploid (contains both DNA and RNA) and is involved both in the day-to-day operations of the protozoan and in differentiation and regeneration. Macronuclear shape varies markedly among

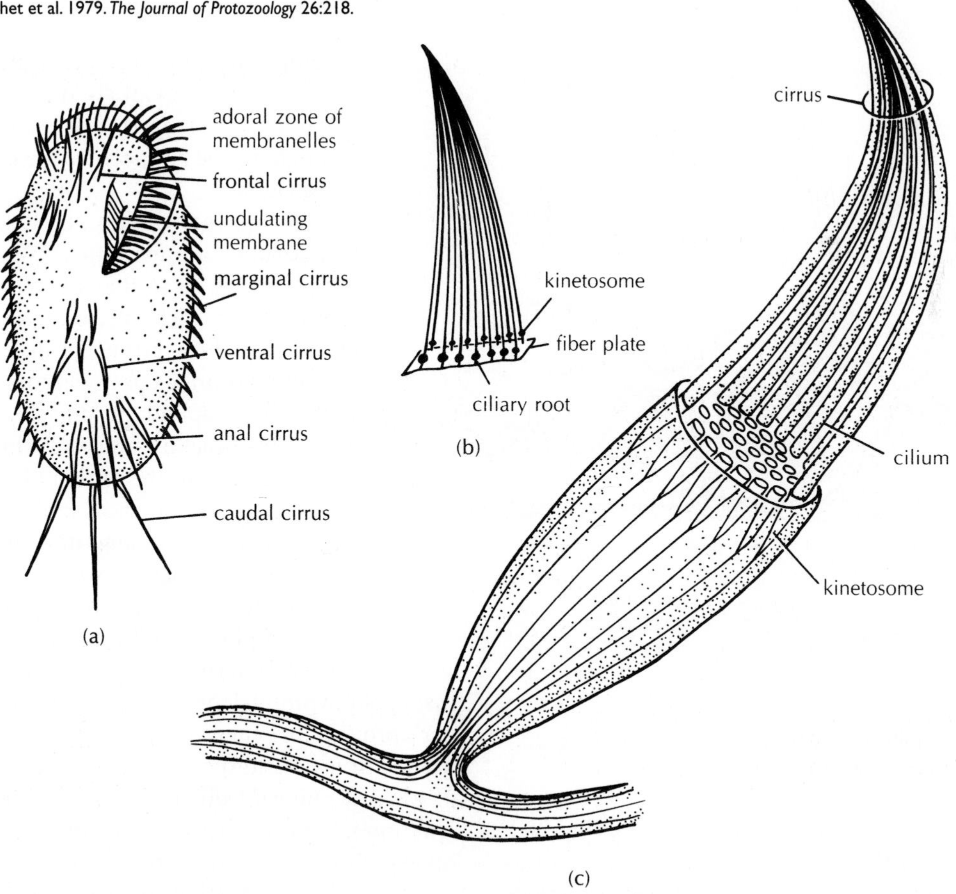

Figure 3.14

(a) *Stylonychia* sp., showing several ciliary organelles: undulating membranes, membranelles, and cirri. (b) Diagrammatic illustration of membranelle structure. (c) Diagrammatic representation of the structure of a single cirrus. Cirri typically contain between 24 and 36 individual cilia.

(a,b) After Kudo. (c) Sherman/Sherman, *The Invertebrates: Function and Form,* 2/e, © 1976, p. 15. Reprinted by permission of Prentice Hall, Upper Saddle River, New Jersey.

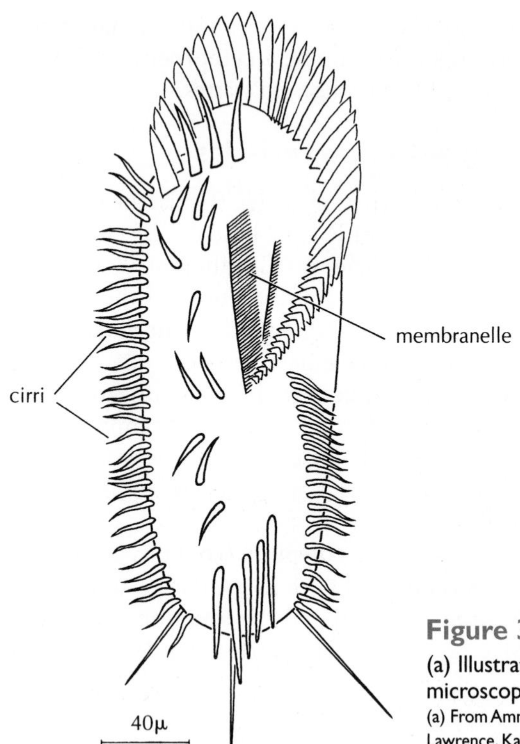

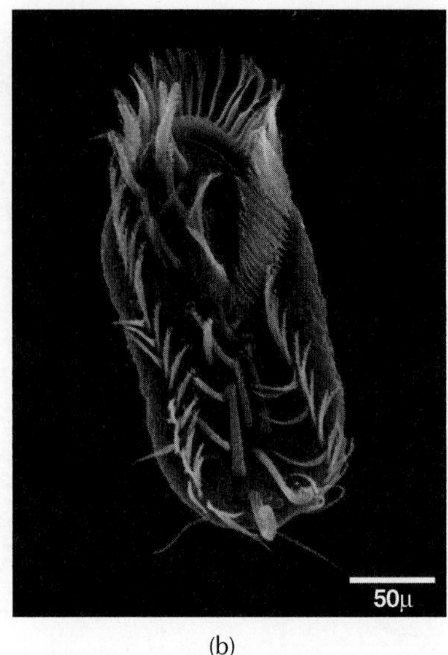

Figure 3.15

(a) Illustration of *Stylonychia lemnae*. (b) The same organism, seen with the scanning electron microscope.

(a) From Ammermann and Schlegel, in *Journal of Protozoology* 30:290, 1983. Copyright © 1983 Society of Protozoologists, Lawrence, Kansas. Reprinted by permission. (b) From Ammermann and Schlegel, 1983. *The Journal of Protozoology* 30:290.

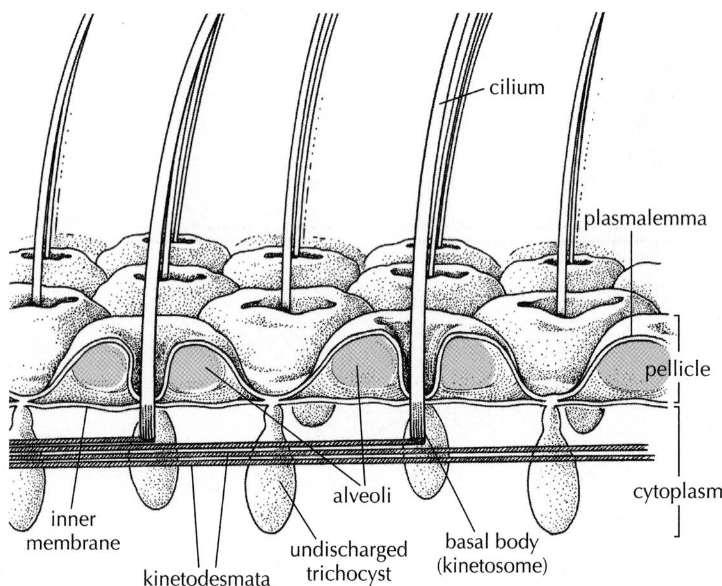

Figure 3.16

Illustration of pellicle, showing alveoli.

After Corliss; after Sherman and Sherman.

species. Ciliates cannot live without their macronuclei, but they can live without their micronuclei. Micronuclei, however, are essential for sexual reproduction, while the macronucleus plays no direct role in ciliate sexual activities.

Sexuality among ciliates never involves gamete formation. Rather, their primary sexual activity invariably involves a process called **conjugation** (Fig. 3.17), typically a temporary physical association between 2 "consenting" individuals during which genetic material is exchanged. The exchange takes place through a tube connecting the cytoplasm of the 2 individuals.

During conjugation, the macronuclei disintegrate and the micronuclei, which are diploid, divide by meiosis so that 4 haploid **pronuclei** are formed from each micronucleus. Typically, all but one of these pronuclei degenerate, and the remaining pronucleus undergoes mitosis to form 2 identical haploid pronuclei. One of these 2 pronuclei somehow migrates through the cytoplasmic tube into the other individual. Exchange of pronuclei is reciprocal so that each individual winds up with one migratory pronucleus. Each migratory pronucleus subsequently fuses with the stay-at-home, stationary pronucleus, restoring the diploid condition through the formation of a **synkaryon** (i.e., a nucleus formed by the fusion of the pronucleus from one individual with its partner's pronucleus). Once the 2 conjugants separate, following the transfer and fusion of micronuclei, the synkaryon of each exconjugant divides mitotically from one to several times. Some of the products form micronuclei, while others give rise to macronuclei. Cytoplasmic divisions (i.e., actual reproduction of individuals) may follow, resulting in several individual offspring that are genetically dissimilar to the parental conjugants. All but one of the micronuclei disintegrate preceding these "distributive" divisions. The remaining micronucleus then divides mitotically, so that one is received by each of the offspring.

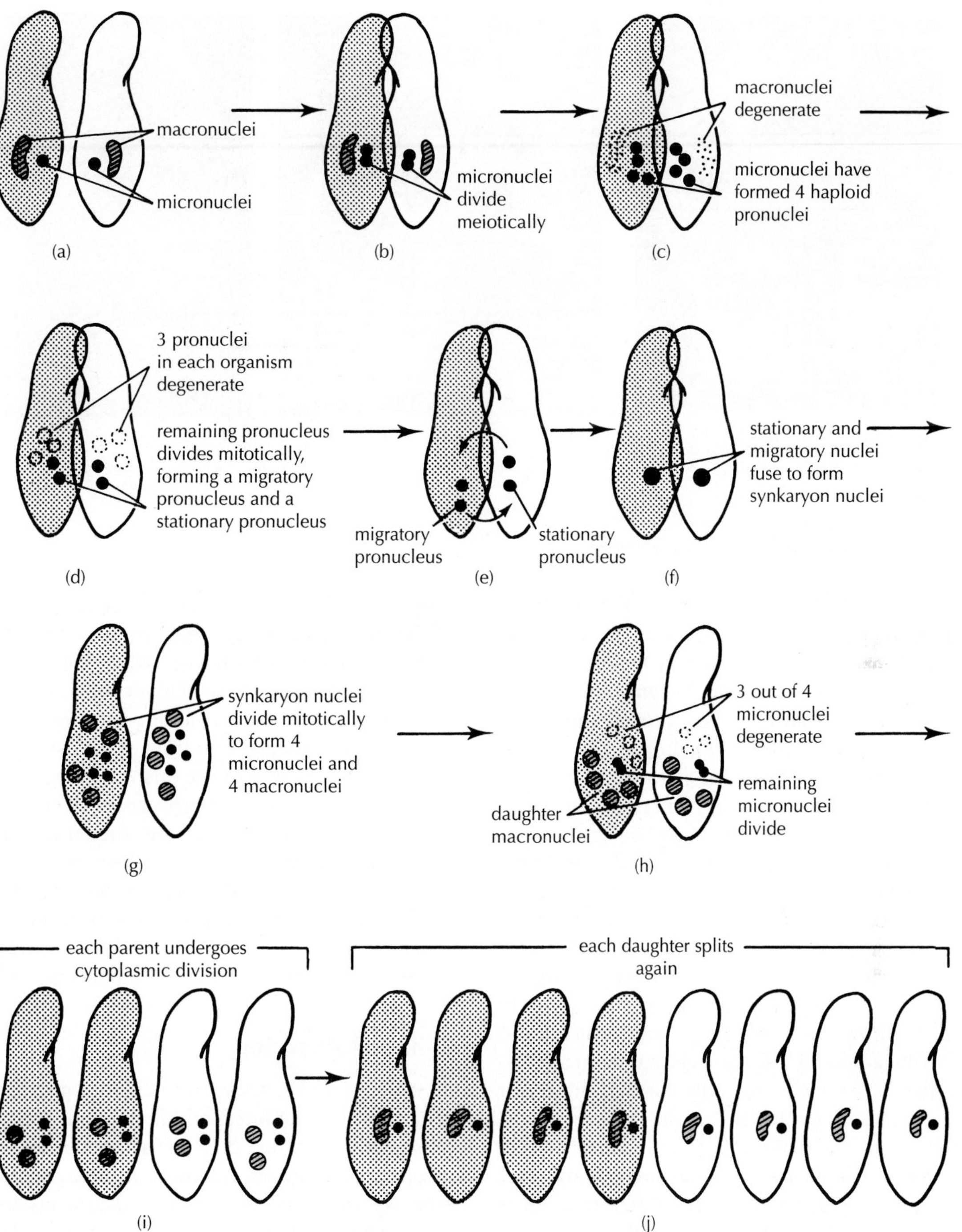

Figure 3.17

Conjugation in *Paramecium caudatum.* (a) Two individuals of different mating types come together. (b) A tubular cytoplasmic bridge forms between the 2 conjugants, and the micronuclei divide meiotically. (c) The macronuclei degenerate and the micronuclei divide mitotically, forming 4 pronuclei in each individual. (d) In each organism, 3 pronuclei degenerate and the remaining pronucleus undergoes another mitotic division, forming one migratory pronucleus and one stationary pronucleus. (e) The migratory pronuclei are exchanged through a cytoplasmic bridge and fuse with the stationary pronuclei (f). (g,h) The exconjugants separate, while a series of nuclear divisions and degenerations produce 4 macronuclei and a pair of micronuclei. (i,j) Cytoplasmic divisions produce 4 daughter individuals from each parent.

Although sexual dimorphism *per se* appears to be lacking in ciliates, the species that have been well studied do exhibit different **mating types.** These mating types, in turn, belong to separate varieties called **syngens.** For example, *Paramecium aurelia* has 16 to 18 known syngens, each containing several different mating types. Conjugation occurs only between individuals of different mating types and only between individuals within one syngen. In the case of *P. aurelia,* the various syngens have now been formally recognized as separate taxonomic species, each with its own set of mating types.

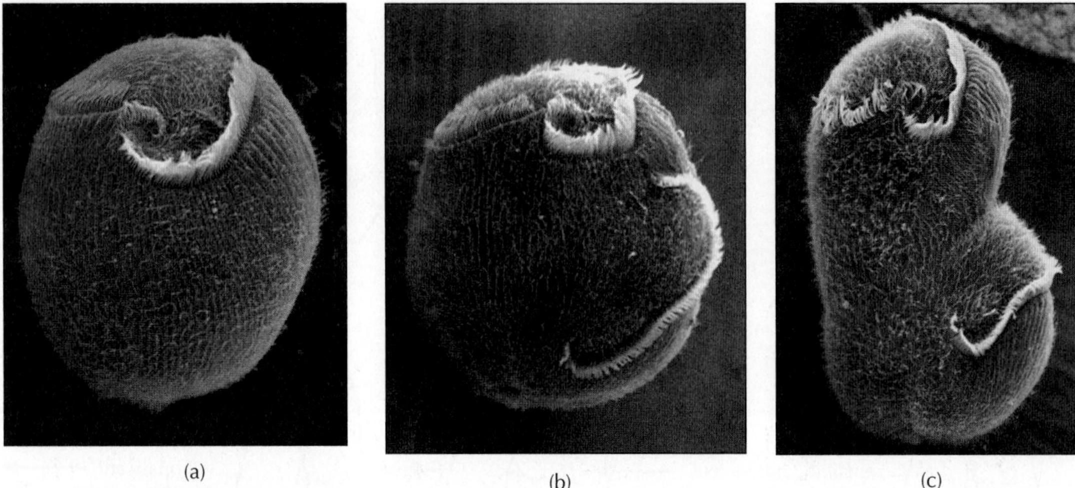

Figure 3.18
Binary fission in *Stentor coeruleus*.
(a–c) From D. R. Diener et al., 1979. *The Journal of Protozoology* 30:84, fig. 2a–d.

All ciliates are capable of asexual reproduction, as are all other protozoans. Among ciliates, asexual reproduction takes the form of transverse binary fission. The cell becomes bisected perpendicular to its long axis, so that complete individuals result from the anterior and posterior halves of the parent (Fig. 3.18). In contrast, binary fission in other groups of protozoa is longitudinal (i.e., parallel to the long axis), producing unicellular offspring that are mirror images of each other. By convention, offspring produced by binary fission are called **daughters;** this should not be taken as a reference to sexuality.

During binary fission in ciliates, the micronuclei divide mitotically and redistribute throughout the cytoplasm. The macronuclei elongate but do not undergo mitosis. In some cases, all macronuclei fuse prior to elongation, forming one very large macronucleus. A cleavage furrow gradually forms, dividing the macronucleus and the body itself into anterior and posterior halves. The detached, posterior half of the body must then regenerate all external and internal structures that have been appropriated by the anterior half and vice versa. In many species, much of this differentiation is completed prior to cell division. Commonly, all parental organelles are eventually replaced in both daughters. Such differentiation is, as always, under the control of the macronucleus. Current studies of pattern formation during ciliate reproduction may increase our understanding of pattern formation mechanisms in metazoan development.

One other form of nuclear reorganization is commonly encountered among ciliates. In this process, called **autogamy,** a form of sexual activity (in which new genotypes may be formed) takes place with the involvement of only a single individual! In some respects, events are reminiscent of the preliminaries to conjugation. The macronucleus (or macronuclei) degenerates while the micronucleus (or micronuclei) undergoes meiosis to form pronuclei (Fig. 3.19). Meiosis is followed by several mitotic divisions of the pronuclei. Two of these pronuclei fuse, forming a zygote nucleus (a **synkaryon**), while the remaining division products disintegrate. Subsequent events, including the formation of a new macronucleus, resemble those of conjugation following separation of the conjugants.

Periodic renewal of the macronucleus appears to be essential for a number of protozoan species. Laboratory cultures will eventually die out after many generations of asexual reproduction if some form of nuclear reorganization is prevented from occurring periodically. Macronuclear regeneration thus appears to have a rejuvenating effect on the animals, which would otherwise apparently exhibit senescence and, ultimately, death.

Ciliate Lifestyles

A variety of lifestyles and surprisingly sophisticated behaviors are exhibited within the phylum Ciliophora (Research Focus Box 3.1, p. 54). About 65% of ciliate species are free-living, and most of these are motile. Some other species form temporary attachments to living or nonliving substrates for feeding purposes, while others are permanently attached (i.e., **sessile**) and may form colonies. Although all ciliates have a distinct pellicle, some sessile species also produce a rigid, protective encasement termed a **test** or **lorica** (*lorica* = L: armor). Probably the best-known examples of such test-forming ciliate species are encountered among tintinnids, folliculinids (Fig. 3.20), and peritrichs (not illustrated). Folliculinids are quite remarkable organisms in that, although they are permanently attached to solid substrate as adults, they can dedifferentiate to a "larval" free-living form, abandon their encasement, relocate at a distance from the original site, secrete a new case, and resume the adult lifestyle. In fact, some species seem to dedifferentiate and then undergo fission, with the posterior half of the organism remaining

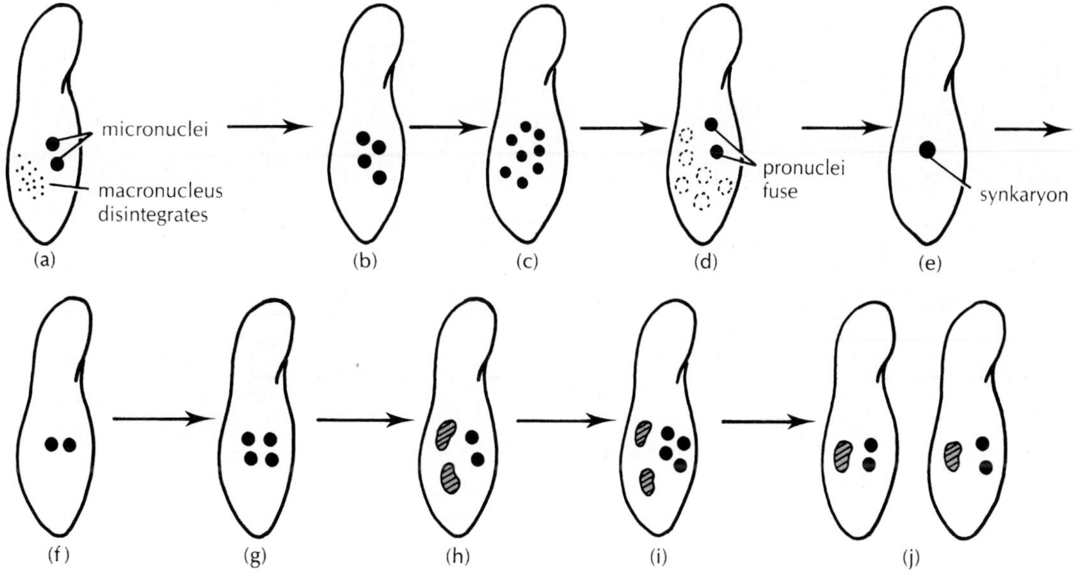

Figure 3.19

Diagrammatic illustration of autogamy. (a) Degeneration of macronucleus. (b–d) Micronuclei replicate meiotically and then mitotically; selective degeneration of haploid pronuclei follows. (e) Two pronuclei fuse to form a synkaryon. (f–i) A series of mitotic divisions produces macronuclei and micronuclei. (j) Cytoplasmic division produces 2 individuals, each with the proper number of micronuclei and macronuclei.

behind to redifferentiate at the original location while the anterior half of the organism leaves to establish itself elsewhere!

About one-third of all ciliate species are symbiotic in or on a variety of other invertebrates (including crustaceans, molluscs, bryozoans, and annelids) and vertebrates (including the digestive tracts of humans, producing an ulcerative condition of the intestine, and the skin of fishes, causing the disease known to freshwater fish aficionados as "ick"). Some species are commensals, attaching to the outer surface of organisms such as crabs, or within the body of such diverse hosts as cattle, horses, sheep, frogs, and cockroaches, and some species are parasitic. Many species, for example, parasitize fish, and several species exclusively parasitize the gonads of sea stars (phylum Echinodermata), particularly the males.

Most free-living ciliates are **holozoic;** that is, they ingest particulate foods. Some species may be **raptorial;** that is, they hunt and ingest living prey. Still others may be passive suspension feeders, primarily ingesting clumps of bacteria, unicellular algae, other protozoans, or small metazoans.

The mouths of some raptorial species can expand to body width or larger, permitting the ingestion of prey that are large with respect to the ingestor. Members of one species, *Didinium nasutum,* about 125 µm long, can ingest prey (exclusively members of the ciliate genus *Paramecium*) 2 to 3 times larger than themselves (Fig. 3.21).

Sedentary suspension feeders, in contrast, are relatively passive. Food particles are carried to the mouth by water currents generated by ciliary activity in the mouth region. Commonly, the suspension-feeding species have stalks, as in members of the genus *Vorticella.* The stalks of such species may contain a **spasmoneme**—a helical, membrane-bound bundle of contractile fibers (Fig. 3.22f, g). Contraction of the spasmoneme fibers results in a very rapid shortening (and coiling) of the stalk and constitutes the organism's only escape response. Other sessile species, such as *Stentor* spp. (Fig. 3.22a), lack stalks, but contractile fibers in the body proper often permit extensive shape changes to occur.

One group of usually sessile ciliates, the suctorians, merit special attention in that they, of all free-living ciliates, seem to have diverged furthest from the basic body plan. The body may attach to a substrate directly or by means of a long stalk (Fig. 3.22h). This stalk is never contractile. Adult suctorians lack cilia, although the immature stages are ciliated and free-swimming. The systems of kinetodesmata of the immature dispersive stage are retained even after external ciliation is lost during maturation. No cytostome is ever found, even in immature stages. Besides the infraciliature, the only clear indication that adult suctorians are indeed ciliates is the presence of both macronuclei and micronuclei. In contrast to what we find among other ciliates, suctorians reproduce asexually not by binary fission, but by simply budding off small parts of the adult. These buds differentiate into ciliated swimming forms that disperse, attach elsewhere, and differentiate to adulthood. Also, sexual reproduction involves the total fusion of the 2 conjugants rather than the temporary union described previously. One individual typically differentiates into a **microconjugant,** which then swims to a partner. Conjugants also fuse in other sessile ciliates.

Since adult suctorians lack both a conventional cytostome and cilia, food collection and ingestion clearly must take place by methods atypical of ciliates in

RESEARCH FOCUS BOX 3.1

Phenotypic Plasticity and Behavioral Complexity among Ciliates

Grønlien, H. K., T. Berg, and A. M. Lovlie. 2002. In the polymorphic ciliate *Tetrahymena vorax,* the non-selective phagocytosis seen in microstomes changes to a highly selective process in macrostomes. *J. Exp. Biol.* 205:2089–97.

In the presence of bacteria and other similarly small food particles, the ciliate *Tetrahymena vorax* has a normal, small oral apparatus anteriorly for capturing particulate food non-selectively. However, when the related ciliate *Tetrahymena thermophila* is present, the normal oral apparatus of *T. vorax* is quickly resorbed and replaced within a few hours by a substantially larger food-collecting pouch; the body also becomes generally larger (Focus Table 3.1). That is, the body shape is phenotypically plastic, with the drastic remodeling being triggered by chemicals released unwittingly by the future prey. These "remodeled" individuals are called macrostomes (G: large mouths), while the original forms are called microstomes (G: small mouths) (Focus Figure 3.1). The transformed individuals of *T. vorax* then become carnivores on individuals of *T. thermophila.* In this paper, Grønlien et al. (2002) ask whether macrostomes of *T. vorax* capture only cells of *T. thermophila.* That is, is the cell remodeling of *T. vorax* a specific adaptation for predation upon *T. thermophila*?

To address this question, the researchers compared the rates of feeding by macrostomes of *T. vorax* on individuals of *T. thermophila* and on inert latex beads of similar size. To make the comparison fair, the researchers first artificially removed the cilia from the protozoan prey (i.e., they "deciliated" them). The beads and immobilized protozoan prey were then offered to about 500 macrostomes per container alone or in combination at an optimal concentration of 10^6 items per ml. After 30 minutes the experiment was ended by adding the preservative formaldehyde; the number of predators that had captured cells and the number that had captured beads were then estimated. In 3 of these experiments,

Focus Table 3.1 Characteristics of *Tetrahymena vorax* and Its Protozoan Prey, *Tetrahymena thermophila.*

Organism	Dimensions (μm)
T. vorax normal phase (microstome)	60 × 20
T. vorax remodeled phase (macrostome)	120 × 80
T. thermophila	35 × 25

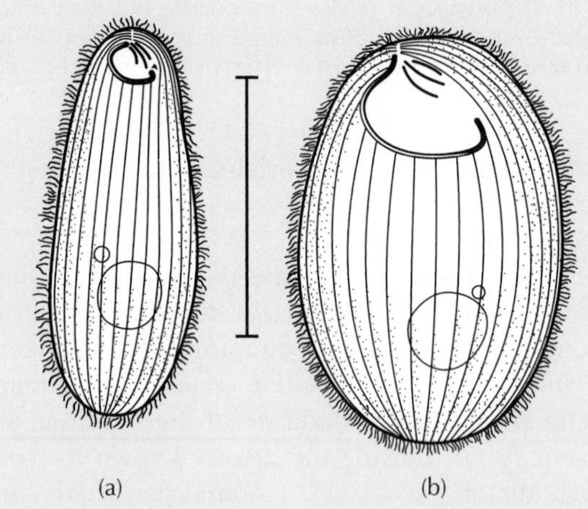

Focus Figure 3.1
Cells of the ciliated protozoan *Tetrahymena paravorax.* (a) Microstome form. (b) Macrostome form.
From Lee, J. J., et al. 2000. *An Illustrated Guide to the Protozoa,* 2nd ed. Lawrence, Kansas: Society of Protozoologists.

general. Small prey are captured by tentacles, which stud the general body surface (Fig. 3.22h). Some of these tentacles are hollow and bear knobs with individual "mouth" openings at their ends. Numerous membrane-bound organelles called **haptocysts** stud these knobs. When an appropriate prey organism contacts the tentacles, it triggers discharge of the haptocysts. The haptocysts then penetrate the prey's body surface, presumably by enzymatic secretion, and help maintain contact between prey and suctorian tentacle. Prey cytoplasm is then drawn through the lumen of the hollow tentacles into the suctorian body (Fig. 3.22i). These are ingenious little suckers indeed.

Phylum Dinozoa (= Dinoflagellata)

(*dino* = G: whirling; *zoa* = G: animal)

Dinoflagellates, or dinozoans, occur in both freshwater and marine habitats. Each individual bears 2 structurally distinctive flagella, one located within a longitudinal groove (termed the **sulcus**) and the other located in a transverse groove (termed the **girdle**) encircling the body (Fig. 3.23a, b). The position and orientation of the 2 grooves are major tools in distinguishing among dinoflagellate species. The orientation and beating patterns of the 2 flagella cause dinoflagellate cells to whirl the number of beads or protozoans in food vacuoles or oral pouches was determined for 100 predators. In a separate set of studies, one macrostome was placed in a drop of water with either 5 microstomes (6 replicates) or a mixture of 25 microstomes and 5 individuals of *T. thermophila* (another 6 replicates).

The results were striking (Focus Figure 3.2). Remodeled predators (macrostomes) consumed both deciliated individuals of *T. thermophila* and latex beads if offered only one or the other. But when both potential prey were offered together (Focus Figure 3.2), the predators preferentially ate individuals of *T. thermophila*. Similarly, when given no choice of prey, the macrostomes of *T. vorax* consumed nontransformed members (microstomes) of its own species. However, when macrostomes were offered a choice of microstomes or *T. thermophila*, they ate exclusively the individuals of *T. thermophila*, even when microstomes dominated by a ratio of 5:1.

The data imply that the remodeling caused specifically by the presence of *T. thermophila* is not just a morphological transformation but a behavioral transformation as well, with the modified ciliate switching from nonselective feeding to highly selective feeding on the species that induced the transformation. How do you think the predator might be recognizing the prey? How would you test your hypotheses?

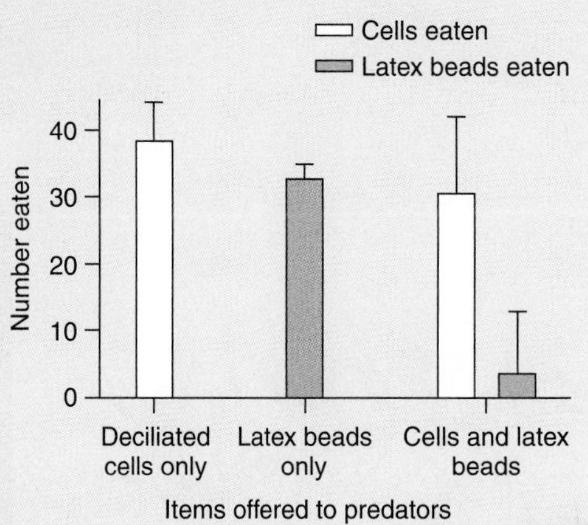

Focus Figure 3.2

The percentage of remodeled *Tetrahymena vorax* cells that captured either deciliated *T. thermophila* (open bars) or latex beads of similar size (blue bars). Each bar represents the mean of 3 experiments, with 100 *T. vorax* cells counted in each experiment.

in very distinctive patterns when they swim, a trait reflected in the phylum name. The entire individual is covered by an intricate array of cellulose plates that are secreted within **alveolar sacs** that lie just below the plasma membrane. As indicated earlier, it is these alveolar sacs that unite dinoflagellates with ciliates and with the next group to be discussed, the Apicomplexa. About 2000 dinoflagellate species exist today.

Dinoflagellates are best known for commonly exhibiting **bioluminescence** (biochemical production of light) and for occasionally producing highly toxic "red tides." These are dense aggregations of certain dinoflagellate species whose neurotoxins (saxitoxin and related compounds) kill fish and crustaceans and accumulate in the tissues of otherwise edible clams, mussels, and oysters, causing diarrhetic shellfish poisoning. Various benthic dinoflagellates produce a similarly potent neurotoxin that accumulates in certain warm-water fish. This "ciguatera poisoning" can kill people that eat the fish, although the fish themselves are somehow unaffected.

But perhaps the most infamous of dinoflagellates currently is *Pfiesteria piscicida*, a species with a fabulously complex life cycle and whose toxic secretions have apparently been responsible for many massive fish kills in recent years in estuarine waters. The life cycle includes flagellated, amoeboid, and encysted stages.

Figure 3.20

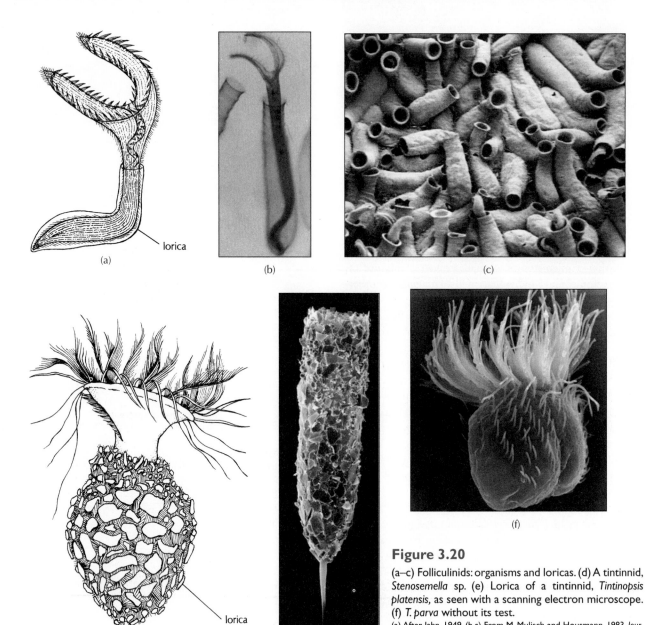

(a–c) Folliculinids: organisms and loricas. (d) A tintinnid, *Stenosemella* sp. (e) Lorica of a tintinnid, *Tintinopsis platensis,* as seen with a scanning electron microscope. (f) *T. parva* without its test.

(a) After Jahn, 1949. (b,c) From M. Mulisch and Housmann, 1983. *Journal of Protozoology* 30:97–104. (d) From Corliss, in *American Zoologist* 19:573, 1979. Copyright © 1979 American Society of Zoologists. Reprinted by permission. (e,f) From K. Gold, 1979. *Journal of Protozoology* 28:415.

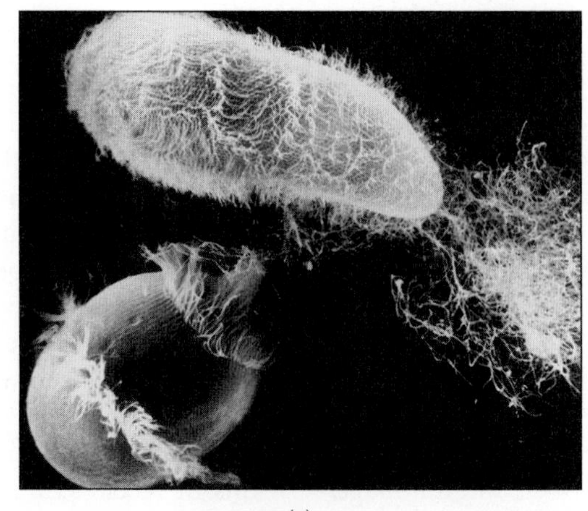

(a)

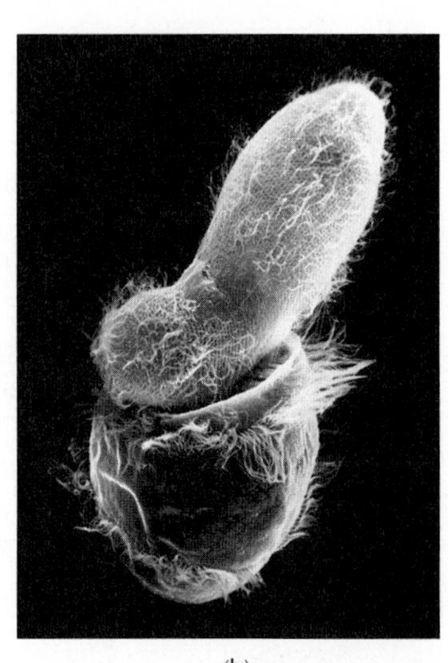

(b)

Figure 3.21

(a) This *Paramecium multimicronucleatum* has discharged numerous trichocysts in response to attack by another ciliate, *Didinium nasutum.* (b) The persistent *D. nasutum* begins to ingest its prey.

(a,b) Courtesy of Gregory Antipa, San Francisco State University.

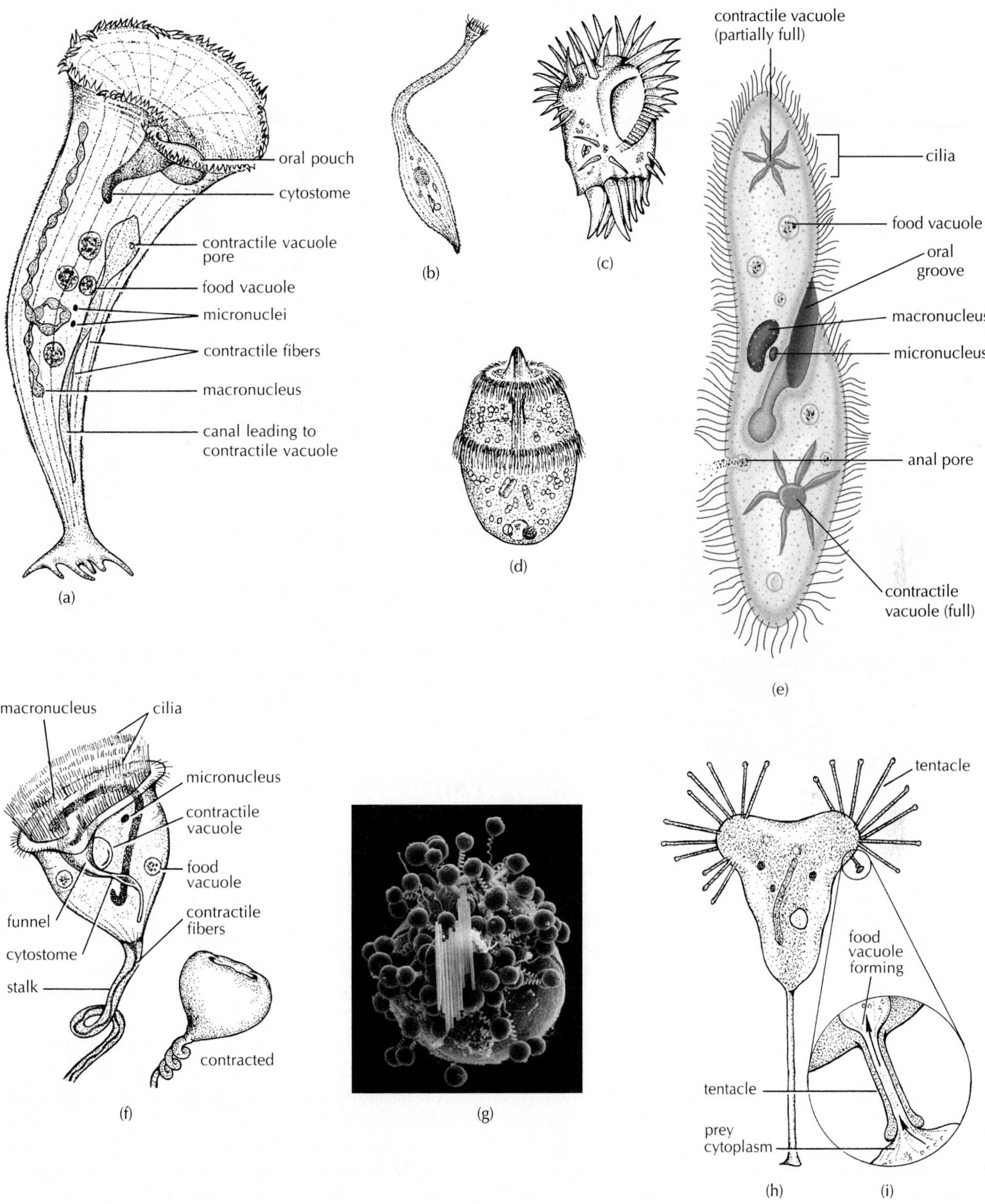

Figure 3.22

Ciliate diversity. (a) *Stentor.* (b) *Tracheloraphis kahli.* (c) *Diophyrs scutum.* (d) *Didinium nasutum.* (e) *Paramecium caudatum.* (f) *Vorticella.* (g) Vorticellids covering the shell of a larval marine bivalve. Note the coiled, springlike stalks. The larval shell is approximately 200 µm in length. (h) A suctorian, *Tokophyra quadripartita.* The body is approximately 100–175 µm long. (i) Prey cytoplasm being moved through a hollow tentacle of the predaceous suctorian. Haptocysts aid in maintaining contact between prey and predator.

(a) After Sherman and Sherman. (b,c) Based on Bayard H. McConnaughey and Robert Zottoli, *Introduction to Marine Biology,* 4th ed., redrawn from T. Fenchel, 1969, "Ecology of Marine Microbenthos IV" *Ophelia* 6:1–182, July. (d) From Hyman after Blochmann. (e) From R. L. Wallace, W. K. Taylor, and J. R. Litton, 1989. *Invertebrate Zoology,* 4th ed., Macmillan Publ. (f) After Sherman and Sherman. (g) Courtesy of C. Bradford Calloway and R. D. Turner. (h) From Hyman; after Kent.

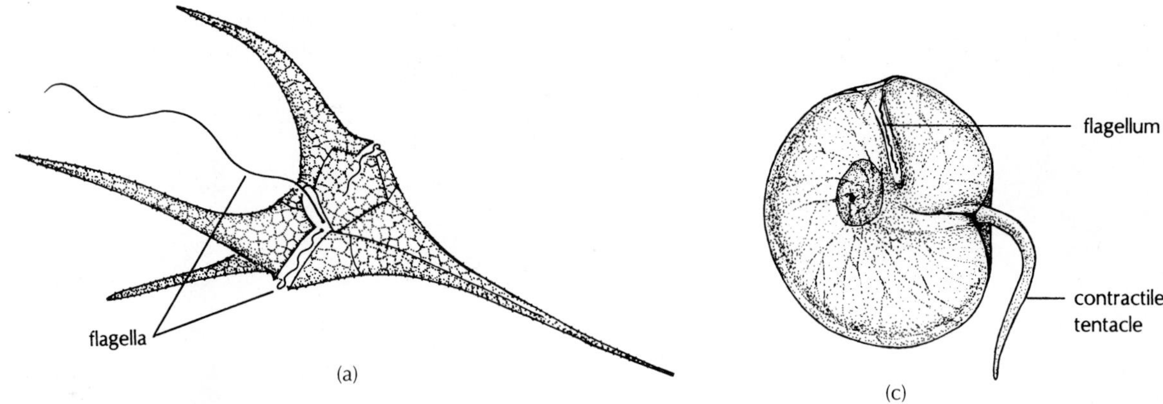

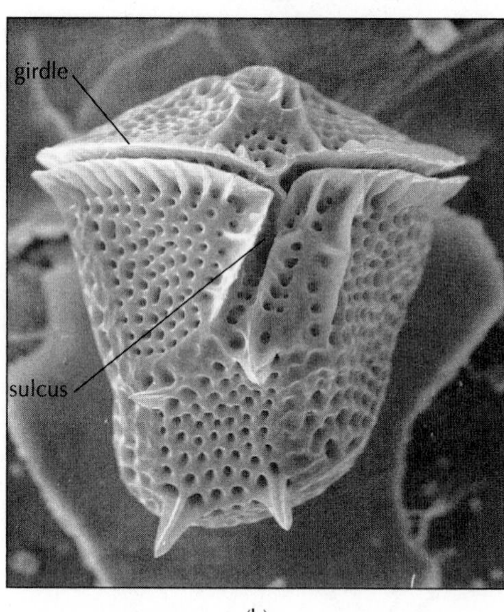

(b)

Figure 3.23

Dinoflagellates. (a) *Ceratium hirundinella,* an armored dinoflagellate. Like many other dinoflagellates, the members of this genus are encased in sturdy cellulose plates. (b) *Triadinium* sp. an armored dinoflagellate. Note the 2 conspicuous flagella grooves, one longitudinal and the other transverse. (c) *Noctiluca scintillans,* an unarmored dinoflagellate. Unlike other dinoflagellates, these possess only a single, very short flagellum and a long, sticky, contractile tentacle that catches particulate food. Members of this genus are strikingly bioluminescent.

(a,c) After Pennak. (b) Courtesy of J. J. Lee. Copyright © 1983 Sinauer Associates, Sunderland, MA. Reprinted by permission.

About half of all dinoflagellate species contain chlorophyll and photosynthesize. Individuals of some of these autotrophic species live intracellularly as important symbionts within some foraminiferans and in the tissues of various multicellular marine invertebrates, including reef-forming corals. In such cases, the photosynthesizing activities of the dinoflagellates (now referred to as **zooxanthellae,** mostly members of the genus *Symnodinium*) contribute significantly to the nutritional needs of their host. Other species, such as the bioluminescent *Noctiluca* (Fig. 3.23c), lack chlorophyll and feed only as heterotrophs, ingesting particulate foods through phagocytosis. At least some species in this genus may be important predators of larval bivalves and other zooplankton. Finally, members of some dinoflagellate species parasitize invertebrates, vertebrates, or other protozoans. Indeed, one such dinoflagellate (*Hematodinium perezi*) may be responsible for the huge recent decline in blue crab populations along the U.S. East Coast. Apparently, the parasite ingests the blue crab's oxygen-binding protein, hemacyanin, so that heavily infested crabs eventually die of asphyxiation. Similarly, members of the genus *Perkinsus* (previously thought to be an apicomplexan—see next section) commonly parasitize oysters and other molluscs, and are widely associated with molluscan mortality in both field and aquaculture situations.

Phylum Apicomplexa

Defining Characteristic: Infective stages possess a distinctive cluster of microtubules and organelles (the "apical complex") at one end of the cell in certain stages of the life cycle

The over 6000 described members of phylum Apicomplexa are all endoparasites in animals, and as such must deal with a number of very interesting problems faced by all parasites in all phyla. One is the difficulty of perpetuating the species from host to host. Typically, parasites can reach adulthood in only one type of host, called the **definitive host,** but need assistance in getting their offspring from one definitive host to another. Thus, one or more **intermediate hosts** commonly serve as **vectors**—agents of transfer between definitive hosts (Fig. 3.24). Not surprisingly, many parasites exhibit remarkable adaptations that increase their ability to locate or attract proper intermediate and definitive hosts. Parasites must also be able to adapt to the different physiological requirements they encounter in different hosts, and be able to evade host immune responses. The apicomplexans present a fine opportunity to consider many of these issues, which are discussed further later in this chapter (p. 70) and again in Chapter 8.

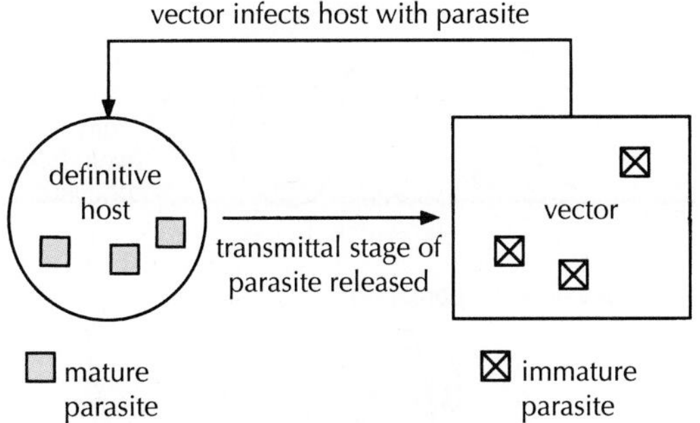

Figure 3.24
Diagram of a parasitic life cycle in which an invertebrate vector transfers the parasite from one definitive host to another. The parasite cannot attain sexual maturity within the body of the vector.

The previous name for the phylum, Sporozoa, has fallen out of favor, in part because many of the species included do not produce "spores," and in part because the Sporozoa included some groups (such as the myxozoans and microsporidians, p. 61) that are now known to be unrelated.

At 2 or 3 stages in the apicomplexan life cycle most individuals possess a structurally complicated grouping of microtubules and organelles at one end of the cell; the phylum name, Apicomplexa, formally acknowledges this "apical complex." The phylum includes 2 particularly important groups of protozoans, the gregarines and the coccidians. In both groups, adults completely lack locomotory organelles, although flagellated gametes are produced in many apicomplexan species. The gregarines parasitize insects and other invertebrates, while the coccidians parasitize both vertebrates and invertebrates, with the invertebrate serving as an intermediate host in the life cycle. A number of coccidian species are blood-cell parasites of humans. Most conspicuous among these are members of the genus *Plasmodium,* the agent of malaria. Four *Plasmodium* species transmit human malaria, with *P. falciparum* being the most deadly of the four.

Malaria has been called the most important disease in the world today: Probably more than 2 billion people living in (and visiting) tropical and semitropical areas are at risk, and at least 500 million people become seriously ill with malaria each year. Between 2 and 3 million people—mostly young children—die of malaria each year; indeed, one person dies of malaria every 10–12 seconds. (For comparison, cancer affects only about 10 million people worldwide.) Adults generally survive infection but are severely disabled.

In typical parasite fashion, apicomplexan life cycles are complicated affairs, including sexual reproduction by means of gamete fusion, asexual reproduction through fission, and the production of resistant or infective spores. The life cycle of the malarial parasite is diagrammed in Figure 3.25. Female mosquitoes (3 species in the genus *Anopheles*) are excellent vectors of malaria, easily transmitting it from one person to another, because they must continually ingest blood from vertebrate hosts to nourish their developing embryos.

To begin the cycle, haploid **gametocytes,** quiescent within the red blood cells of their human host, are removed from the human bloodstream through the bite of a female mosquito. Within the mosquito, gametocytes quickly emerge from the red blood cells and mature into either male or female gametes. The female gametes are fertilized within about one-half hour of the mosquito's blood meal. The motile, diploid zygotes (called **ookinetes**) move to the mosquito's stomach, encyst in the stomach wall, and grow. Each encysted ookinete **(oocyst)** now undergoes meiosis and a large number of fission events, the products of which mature into infective haploid **sporozoites.** As many as 10,000 sporozoites are produced asexually from an individual zygote.

Mature sporozoites migrate to the insect's salivary glands, to be injected into a human host at the next blood meal. A single mosquito bite injects 10–20 sporozoites, which begin infecting human liver cells (hepatocytes); becoming intracellular removes them from direct attack by the host immune system. The sporozoite forms a resistant cyst, which is followed by a remarkable number of asexual divisions; a single fissioning sporozoite (now called a **schizont**) can give rise, within about 1 week, to nearly 40,000 genetically identical offspring called **merozoites.** The merozoites rupture the liver cell housing them, enter the bloodstream, and immediately invade the host's red blood cells. Soon after entering a blood cell, the parasite somehow alters the cell's surface chemistry so that the cell sticks to the tissue lining the blood vessel. This removes the red blood cell (and its parasite) from general circulation, preventing destruction of that blood cell by the immune surveillance system within the host's spleen: The cells simply do not circulate through the spleen. Accumulations of infected cells in the blood vessels of the host's brain can lead to severe neurological impairment or death.

Within the red blood cell, merozoite development follows 1 of 2 pathways: Either the merozoite multiplies asexually to form 10–20 new merozoites over the next 48 hours, or it differentiates into a gametocyte. The new merozoites lyse the red blood cell and quickly invade new red blood cells for another round of asexual replication and red blood cell lysis (leading to debilitating anemia in the host) or for differentiation into a gametocyte. The fever and chills characterizing malarial infection result from toxins released during the synchronous rupture of infected red blood cells, once every 48 hours for most human malarial species. Gametocytes can continue the life cycle only if the blood cells they reside in are ingested by a mosquito; if not ingested, they perish within several weeks.

Let's summarize some key elements of this life cycle. Only the haploid gametocytes, living within human red blood cells, infect the mosquito vector. Only the haploid sporozoites, which accumulate in certain of the mosquito's salivary gland cells, can infect humans.

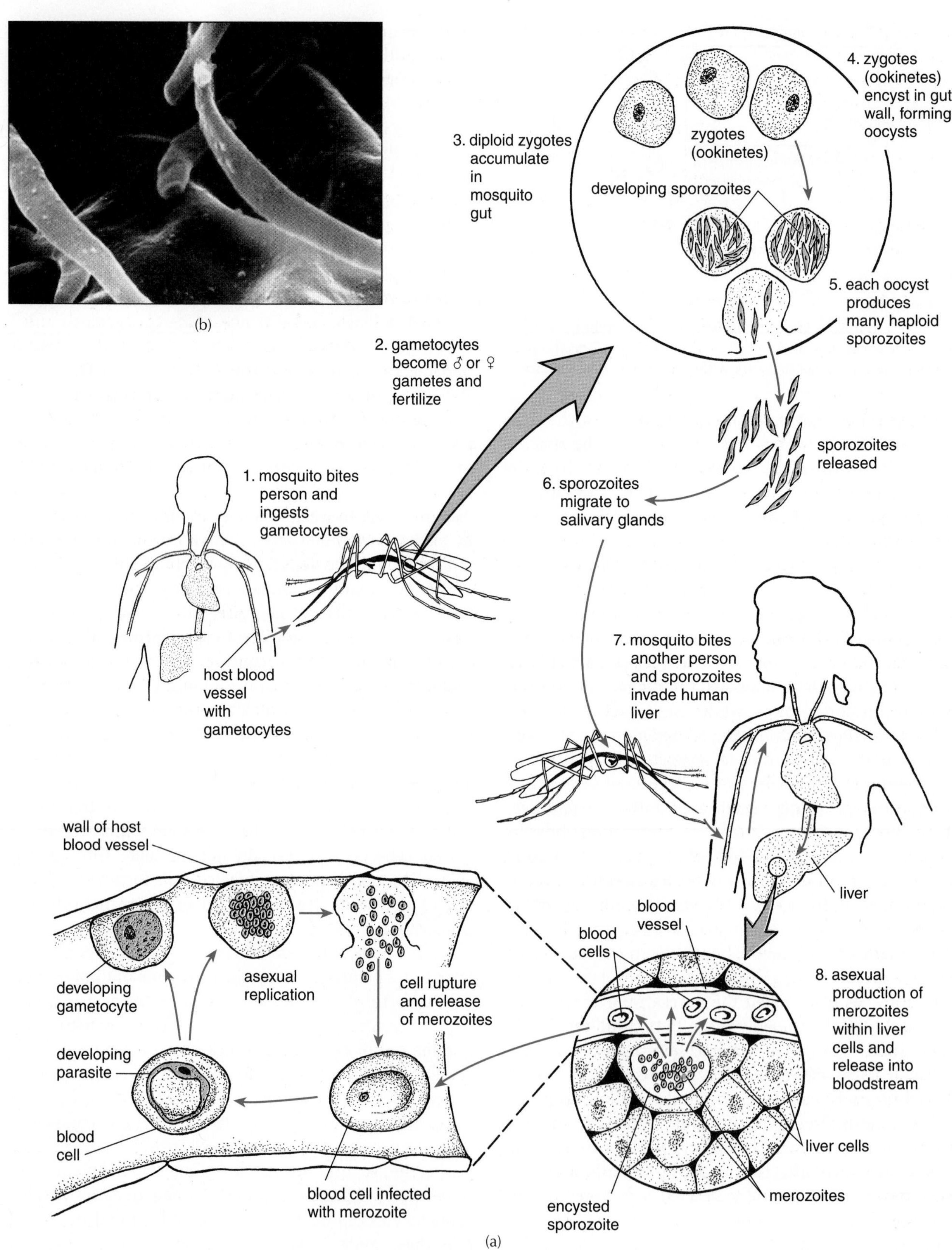
(b)
1. mosquito bites person and ingests gametocytes
host blood vessel with gametocytes
2. gametocytes become ♂ or ♀ gametes and fertilize
3. diploid zygotes accumulate in mosquito gut
zygotes (ookinetes)
4. zygotes (ookinetes) encyst in gut wall, forming oocysts
developing sporozoites
5. each oocyst produces many haploid sporozoites
sporozoites released
6. sporozoites migrate to salivary glands
7. mosquito bites another person and sporozoites invade human liver
liver
blood vessel
blood cells
8. asexual production of merozoites within liver cells and release into bloodstream
liver cells
merozoites
encysted sporozoite
wall of host blood vessel
developing gametocyte
asexual replication
cell rupture and release of merozoites
developing parasite
blood cell
blood cell infected with merozoite
(a)

Figure 3.25

(a) The life cycle of the malarial parasite, *Plasmodium* spp. (phylum Apicomplexa). (b) Sporozoites escaping from an oocyst on the gut wall of the mosquito vector *Anopheles gambiae*.

(b) Courtesy of R. E. Sinden. From Cox, F. E. G. 1992. *Nature* 359:361–62. © Macmillan Magazines Ltd.

Pathological effects in humans are caused only by the red blood cell (merozoite) stage of the life cycle. Note that the parasite is haploid for most of its life; a diploid stage occurs only in the mosquito vector.

Past efforts to control malaria have focused on eliminating mosquito populations with DDT and other pesticides and on controlling human infection with the drug chloroquine. Over the past 30 years or so, the treated mosquito populations have become resistant to the pesticides, and the parasites have become resistant to the chloroquine; in consequence, the incidence of malarial infection is increasing rapidly.

Here are some of the other potential control strategies that are currently being investigated: (1) Produce antibodies that will successfully attack the complex antigenic surfaces of the merozoite stage parasites during their very limited period of exposure in the bloodstream or that will attack the modified antigenic surfaces of infected red blood cells. This approach will be exceedingly difficult, as the merozoites vary the protein components of their surface coats with great frequency, as do the trypanosomes discussed earlier in this chapter, and vary the proteins expressed on the surfaces of the red blood cells they infect. (2) Produce a vaccine from sporozoites extracted from irradiated mosquitoes. (3) Prevent the parasite from invading host liver cells (hepatocytes), by somehow disarming the specific sporozoite surface proteins that recognize and bind to receptors on the liver cells. (4) Prevent the diploid zygotes (ookinetes) from producing the chitinase that enables them to penetrate the chitinous membrane lining the mosquito gut and invade the stomach wall. (5) Genetically engineer mosquitoes that will prevent gametocytes from maturing in the mosquito vector or that will produce antibodies against the surface proteins expressed on zygotes and oocysts, and then develop the means of spreading those genes throughout all populations of all relevant mosquito species. (6) Immunize patients with sporozoites modified so that they are unable to express genes essential for maturing in the host liver or invading human red blood cells.[9] (7) Genetically alter the bacteria or protozoans living symbiotically with the mosquito vector, so that the symbiont will kill or disable the parasite. (8) Prevent sporozoites from invading the mosquito's salivary gland cells, by disrupting the binding sites on the parasite that interact with mosquito salivary gland receptors. (9) Disable the vasodilators and anticlotting and antiplatelet compounds found in mosquito saliva, which prevent normal vertebrate defenses against blood loss. This could be done by engineering mosquitos that can't secrete saliva and then developing ways to spread those genes through mosquito populations.

One other potential approach takes advantage of the requirement of *Plasmodium* for large amounts of glucose. After invading a red blood cell, the parasite somehow inserts or activates supplemental transport proteins in the cell's plasma membrane, increasing the flux of glucose into the cell by about 40 times. Perhaps researchers can develop drugs that will deactivate that pathway or prevent it from becoming activated, possibly starving the parasite to death within the blood cell.

Additional efforts to control the disease may focus on reversing resistance of the parasites to chloroquine and developing new antimalarials. A promising substitute compound (artemisinin, the active ingredient in the Chinese herb qinghoa), produced naturally by sweet wormwood plants, is now being synthesized successfully in the laboratory; once in the mosquito vector, the chemical breaks down to form free radicals that seem to inhibit the action of one or more of the parasite's key metabolic proteins.

The phylum Apicomplexa contains many other commercially and medically noteworthy species. For example, *Toxoplasma gondii* is a particularly widespread and common parasite of warm-blooded mammals. Humans typically become contaminated from cats, which release tremendous numbers of encysted eggs in their feces. As the parasite can be deadly to developing fetuses, pregnant women should avoid contact with cats and litter boxes. The parasite *Pneumocystis carinii* may also be a member of this group, although many now believe it is more likely a fungus. It causes a fatal form of pneumonia in people with suppressed immune systems and is especially common among patients with AIDS (acquired immune deficiency syndrome).

A Brief Aside on Microsporideans and Myxozoans

Microsporideans superficially resemble the apicomplexans in several respects, and in fact were until recently placed with apicomplexans in a single phylum, the Sporozoa. As with apicomplexans, all microsporan species are intracellular parasites, and the life cycle includes production of dispersive or resistant (chitinous) spores. The life cycle includes an amoeboid stage that invades the host's tissues. Some species infect the eggs of their host, thereby infecting the host's offspring. They are unable, however, to infect sperm, because sperm contain so little cytoplasm; if they end up in a male host, they somehow cause the male to develop into a female . . . and then infect "her" eggs! Recent molecular analyses indicate that most microsporans are actually degenerate fungi, not protozoans.

Myxozoans are some 1300 species of extracellular, spore-forming parasites that were also formerly grouped with the apicomplexans and microsporans in the phylum Sporozoa. Recent molecular data, however, indicate that myxozoans are not protozoans at all, but rather highly modified cnidarians. Thus, they are now discussed in Chapter 6. I include this aside primarily to give you a sense of the revolution in "protozoan" taxonomy that has taken place over the past 10 years or so.

9. Mueller, A.-K., M. Labaied, S. H. I. Kappe, and K. Matuschewski. 2005. *Nature* 433: 164–67.

Amoeboid Protozoans

The organisms to be discussed in this section were formerly grouped together as the Sarcodina, which were themselves formerly grouped with the flagellates and some others (the opalinids, p. 72) in the phylum Sarcomastigophora (G: bearers of flesh or whips). Neither grouping, it turns out, is monophyletic.

Some 56,000 amoeboid species have been described to date, with about 44,000 of those species known only as fossils. None of the species have cilia or infraciliature, none have a pellicle, and none conjugate; where sexual reproduction is known to occur, gamete formation is usually involved. Some species produce flagellated gametes, which has intriguing phylogenetic implications, although no adult stages possess flagella. In addition, amoeboid protozoans generally contain a single type of nucleus. A few species are multinucleate, but the nuclei are always monomorphic, in contrast to the distinctly dimorphic macronuclei and micronuclei found among ciliates.

Contractile vacuoles are present, particularly among freshwater species, although they are not fixed in position as they are in ciliates but move freely within the cytoplasm. As with ciliates, amoebae lack specialized sensory organelles. The typical amoeba is characterized above all by cytoplasmic extensions of the body called **pseudopodia,** or "pseudopods" (*pseudo* = G: false; *pod* = G: a foot), used for locomotion, food collection, or both, as discussed previously (pp. 43–44). Body shapes range from completely amorphous and ever changing (*amoeba* = G: change) to highly structured, with elaborate, rigid skeletal supports. Members of most species feed on small particulate material exclusively, although some sarcodinids house photosynthesizing endosymbiotic algae.

Most amoeboid protozoans are free-living, although about 2% of known species parasitize vertebrate and invertebrate hosts, including other protozoans. Amoebic dysentery (caused by *Entamoeba histolytica,* as mentioned earlier, p. 44) is probably the best-known human ailment caused by a member of this group; probably more than 500,000 people are infected worldwide at any given time, and up to 100,000 people die from the infection yearly. Some amoeboid species are parasites within the bodies of other parasites; that is, they are **hyperparasites.**

All amoeboid protozoan species reproduce asexually, chiefly by binary fission or multiple fission. In addition, sexual reproduction has been described for many species. In some cases, a single individual essentially becomes a gamete, which then fuses with another individual. In other cases, flagellated gametes are formed, pairs of which fuse to form zygotes. Encystment is common among amoeboid protozoans, especially in freshwater and parasitic species, providing a means of withstanding unfavorable environmental circumstances.

Relationships among amoeboid protozoans are still being actively investigated and much debated. Both molecular and ultrastructural evidence now indicate that the amoeboid form evolved many times from different flagellated ancestors, so that the presence of pseudopodia—regardless of form—is now considered to be phylogenetically uninformative. Certainly the amoeboid creatures do not form a monophyletic group. This creates problems for both students and textbook writers: Many of these organisms clearly group together, but it is often unclear what larger taxon the groups belong to, and how the different groups are related to each other. Dozens of amoebozoan genera cannot yet be placed with confidence into any major taxonomic grouping.

The amoeboid groups discussed in the next several pages fall into 2 major categories. The first 2 groups (the naked and test-bearing amoebae) belong to the phylum Lobosea, in the larger grouping Amoebozoa. The other amoeboid groups discussed here, including the foraminiferans and radiolarians, are now tentatively placed within the Rhizaria.

The Amoebozoans

This major grouping contains thousands of described species exhibiting amoeboid movement. Most members of the Amoebozoa are free-living, although the group does include some parasitic species (such as the infamous *Entamoeba histolytica* just mentioned. Nearly all possess branching tubular mitochondrial cristae, so that they are sometimes referred to as the "naked ramicristate amoebae" (*ramus* = L: branching).

The Naked Amoebae (Gymnamoebae)

(*gymno* = G: naked, bare)
jim´-nah-mē´-bē

Defining Characteristic: Individuals possess shapeless ("amoeboid") bodies, with wide, blunt (lobose) or thin, pointy pseudopodia

These members of the Amoebozoa are surrounded only by their own cell membranes—that is, they are naked—and use pseudopodia of various shapes to move and/or feed (Fig. 3.26) (see pp. 43–44). The body shape changes continuously as the organisms move. Most naked amoebae are free-living, and very common in moist soil, freshwater, estuaries, and the sea, at concentrations of thousands of individuals per cm^3. About 200 species have been described.

All free-living naked amoebae are heterotrophic, capturing food with their pseudopodia.[10] Commonly, ingestion involves **phagocytosis** (*phago* = G: to eat). In this process, lobopodia or filopodia advance on both sides of, and often on top of, the intended prey, forming a **food cup.** The tips of the pseudopodia soon come together and fuse, thereby internalizing the prey. Amoebozoans typically feed on bacteria, other protozoans, and unicellular algae (e.g., diatoms). In **pinocytosis** (*pino* = G: to drink),

10. See *Topics for Further Discussion and Investigation,* no. 6.

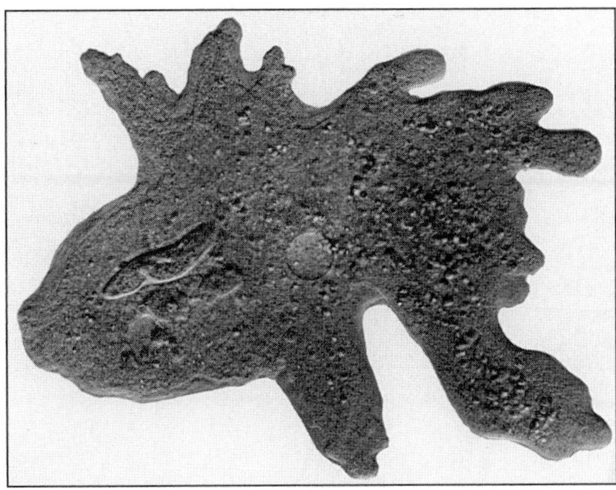

Figure 3.26
Chaos sp., a naked sarcodinid that can attain lengths of up to 5 mm. Note the numerous lobopodia extending in various directions about the periphery of the organism. Also note the ingested *Paramecium* to the left of the (central) contractile vacuole.

much smaller pseudopods are formed, capturing extremely small particulates or fluids rich in dissolved organic matter. Although none of these organisms are parasitic, many species harm vertebrates. Some species of *Acanthoamoebae,* for example, which abound in soil and freshwater, cause eye infections in humans.

The Arcellinids and Related Test-Bearing Amoebae

Many amoebae either secrete a single-chambered, proteinaceous or siliceous covering about themselves, called a **test** (*test* = L: a tile, shell), or cement fine sand particles or detritus to themselves to form such coverings (Fig. 3.27). The body is thus given form, with pseudopodia projecting through a single opening in the test. Testate amoebae are common in soil, lakes, and rivers, and as associates of sphagnum mosses. Modern soils can contain as many as several hundred million individuals per m^2. Fossilized tests of shelled amoebae have been found in freshwater sediments deposited some 500 million years ago.

Most test-bearing amoebae are in the Amoebozoa, and at least 75% of these testate species are placed in the group Arcellinida. All have lobopodia (p. 43); species are distinguished primarily on the basis of their test characteristics (Fig. 3.27).

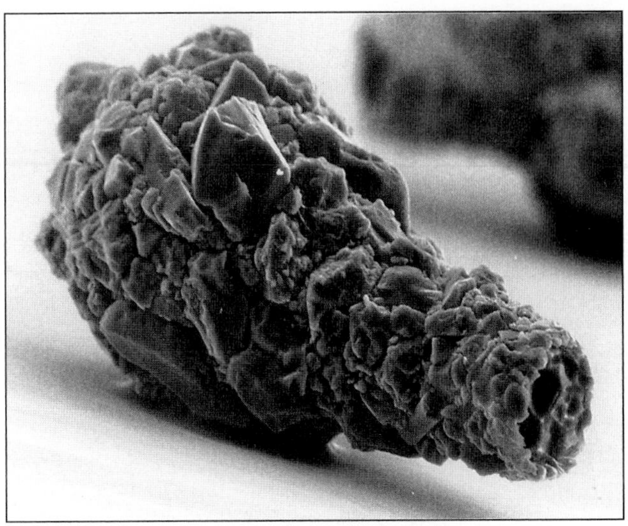

Figure 3.27
Test of the freshwater amoeba *Difflugia gassowskii,* about 125 μm long. This specimen was collected from moist floodplain sediments in western Germany.
Courtesy of Ralf Meisterfeld.

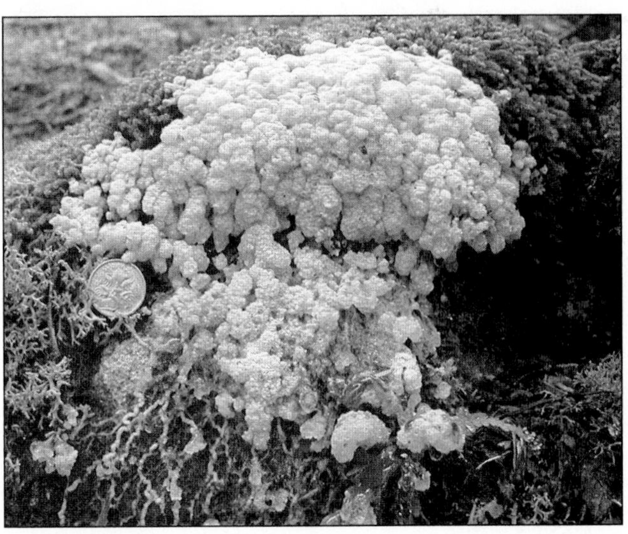

Figure 3.28
Slime mold, South Island of New Zealand. The coin at left is about 2 cm across.
Photo by J. Pechenik.

The Slime Molds

The cellular and acellular slime molds, members of the Eumycetozoa (*myceto* = G: a fungus; i.e., the true fungus animals) are also members of the Amoebozoa (see Fig. 3.1). For reasons that will soon become apparent, the slime molds have often been classified as plants or as fungi. Nevertheless, for much of their lives, the slime molds exist as individual amoeboid cells, moving and feeding on particulate matter by means of lobopodia. Most species are harmless and are commonly found among decaying vegetation in terrestrial habitats (Fig 3.28), although some species are plant pathogens of considerable commercial importance. In all cases, the amoeboid individuals eventually aggregate (Fig. 3.29a). For the cellular slime molds, commonly called "the social amoebae," aggregation clearly results from chemical communication between cells. Probably in response to dwindling food supplies, some of the amoebae secrete cAMP, which attracts other amoebae to form a large multinuclear mass—essentially a giant, multinucleate amoeba, or "slug." Whereas the acellular slime molds form a true **plasmodium,** in which cell membranes are lost to form a large, immobile syncytial mass, the cellular slime molds form what

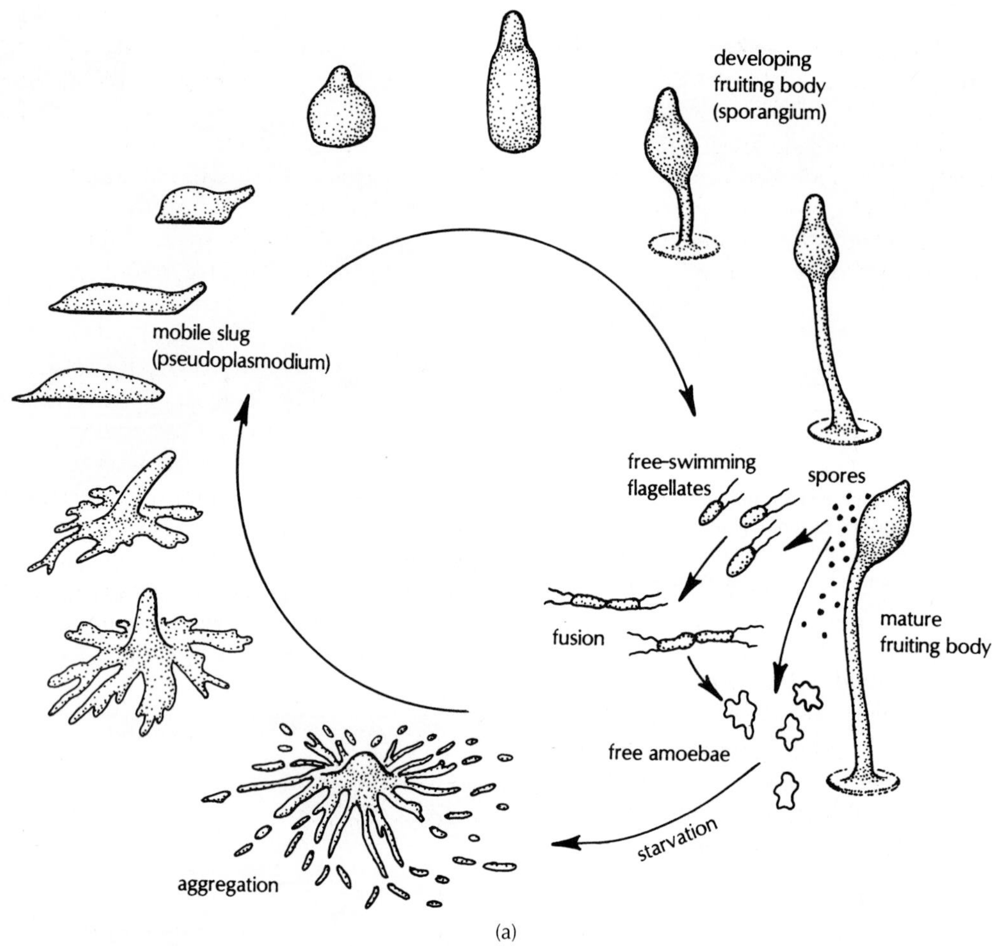

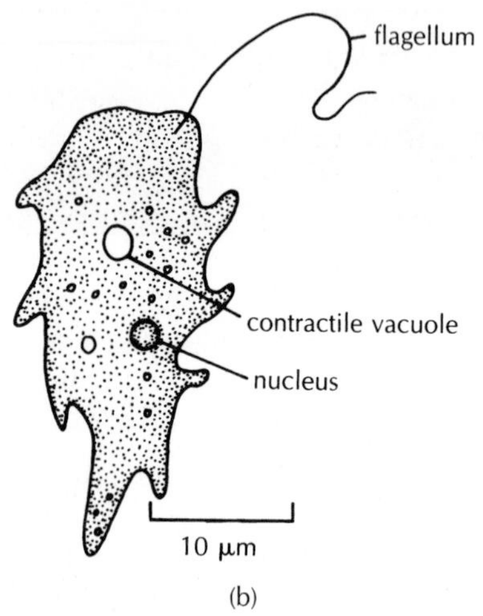

Figure 3.29

(a) The life cycle of a cellular slime mold. *Dictyostelium* is one of the best-studied slime molds. Note that the spores may give rise to swimming flagellates, which eventually fuse and form amoebae, or may germinate amoebae directly, depending upon the species. The slug stage is typically formed from about 100,000 individual amoebae. (b) *Mastigamoeba longifilum*.

Adapted from *Biological Science*, Third Edition, by William T. Keeton, illustrated by Paula DiSanto Bensadoun, by permission of W. W. Norton & Company, Inc. Copyright © 1980, 1979, 1978, 1972, 1967 by W.W. Norton & Company, Inc.

is termed a **pseudoplasmodium,** in which individual cell membranes persist. The pseudoplasmodium is motile, and it continues to move and feed in standard amoeboid fashion for some time (Fig. 3.29a). The slug of one species, *Dictyostelium discoideum,* consists of as many as 100,000 distinct cells.

If two pseudoplasmodia or plasmodia of the same species come into contact with each other, they often fuse, creating an even larger "individual." A decrease in nutrient availability causes the pseudoplasmodium or plasmodium to develop funguslike **sporangia,** or fruiting bodies. Highly resistant spores are released from these sporangia, and only appropriate environmental conditions will bring about germination of these spores. An amoeboid form or a flagellated form may emerge from a spore, depending upon the species. In some cases, the amoeboid individual subsequently transforms into a flagellated individual, which proceeds to eat and undergo repeated binary fission. Eventually, the flagellates give rise to amoebae, which feed and ultimately aggregate to form a multinucleate plasmodium or pseudoplasmodium that then differentiates to produce another round of resistant spores.

Mastigamoebae

One of the intriguing characteristics of the acellular slime molds just discussed is that the life history of some species includes both amoeboid and flagellated stages. A similarly intriguing group are the mastigamoebids (Fig. 3.29b). These organisms possess both locomotory flagella and pseudopodia, either simultaneously or in succession. These species are clearly at the crossroads of flagellate-amoeba organization and provide further evidence of a close evolutionary relationship between the 2 forms. Mastigamoebae are most commonly encountered in stagnant water in both marine and freshwater environments. Recent molecular data, based on study of small subunit rDNA, indicate that these organisms are close relatives of the Eumycetozoa just discussed.

The Rhizaria

(*rhiza* = G: root)
Rye-zahr´-ee-uh

Rhizarians possess splender pseudopodia (called filopodia), which are in some species supported by microtubules. The pseudopodia may be simple or branching, sometimes forming highly complex networks (**reticulopodia:** Figs. 3.30g, 3.9). This large group, the Rhizaria, was first proposed in 2002, based exclusively on molecular data—its members are so far not united by any morphological or biochemical features. The two major rhizarian groups to be discussed here are the foraminiferans and the radiolarians.

Phylum Foraminifera

Phylum Foramini • fera
(L: bearing openings)
for´-am-ih-nif´-fer-ah

Defining Characteristics: 1) Individuals secrete multichambered tests, generally of calcium carbonate; 2) pseudopodia (reticulopodia) emerge through pores in the test and branch extensively to form dense networks

Foraminiferans (often referred to unofficially as "forams"), are primarily marine and live on bottom sediments as benthic organisms; a small number of species are planktonic. Forams are among the most abundant protozoans found in marine and brackish waters. There are no parasitic species, although some are ectocommensal. Foraminiferans secrete multichambered tests, typically of calcium carbonate (Fig. 3.30). The individual chambers of these tests are often demarcated from each other by perforated septa. Thin, extensively branching pseudopodia called **reticulopodia,** project from minute openings in the test (Figs. 3.9, 3.30g), forming dense pseudopodial networks (*reticul* = L: a network) used primarily for food capture. Repeated extension and shortening of the pseudopodia also permit slow crawling over the ocean bottom. Foraminiferans feed on a remarkable variety of food, including other protists, small metazoans, fungi, bacteria, and organic detritus. In addition, some species house photosynthesizing symbionts, including dinoflagellates, and others can take up dissolved organic material from seawater. As the world's oceans continue to acidify, due to continued absorption of the huge quantities of CO_2 released into the air by humans, forams may be under increasing threat, as the low acidity may prevent them from fully calcifying their tiny shells.

Foraminiferan tests (representing more than 30,000 species!) abound in the fossil record. The oldest specimens date from the Early Cambrian, about 543 million years ago, although recent molecular clock calculations suggest that foraminiferans were diversifying for 150–600 million years before this. Although tests of present-day species rarely exceed 9 mm in diameter, with those of most species being less than 1 mm, fossilized tests of up to 15 cm in diameter have been found—quite a remarkable size for a protozoan. The extensive fossilized remains of foraminiferans have taken on considerable economic importance; for example, cement and blackboard chalk are foraminiferan-containing products. In addition, certain foraminiferan fossils are used as fairly reliable indicators of likely places to drill for oil.

Phylum Radiozoa (= Radiolaria)

Radio • zoa
(L: ray animals)
rā´-dē-ō-zō´-ah

Defining Characteristic: Body is divided into distinct intracapsular and extracapsular zones separated by a perforated membrane or capsule

The members of this phylum, called radiolarians and acantharians, support their pseudopodia with thin, radiating microtubules that give a spiny, rayed appearance to many species (Fig. 3.31a, b). The complexity and symmetry of their rigid infrastructure make many radiolarians exceedingly beautiful. The pseudopodia and their microtubular supports are called **axopodia** (*axo* = G: an axle). All radiozoans also possess a rigid endoskeleton composed either of silica (in radiolarians) or strontium sulfate (in acantharians). Because of their siliceous (silica-containing) skeletons, radiolarians, like the foraminiferans and shelled amoebae, are prominent in the fossil record; of the approximately 8700 described radiolarian species, about 100 are extant and 7700 are known only as fossils. In contrast, acantharians have left no fossil record, as the strontium sulfate supports degrade soon after the organism dies. Only about 160 acantharian species have been described.

In contrast to the mostly bottom-dwelling Foraminifera, all radiolarians and most acantharians are planktonic organisms, passively carried about by ocean currents. Many species possess symbiotic algae and so meet at least some of their nutritional requirements through photosynthesis. They also feed as carnivores, capturing microscopic prey with the cytoplasm that flows along their axopods.

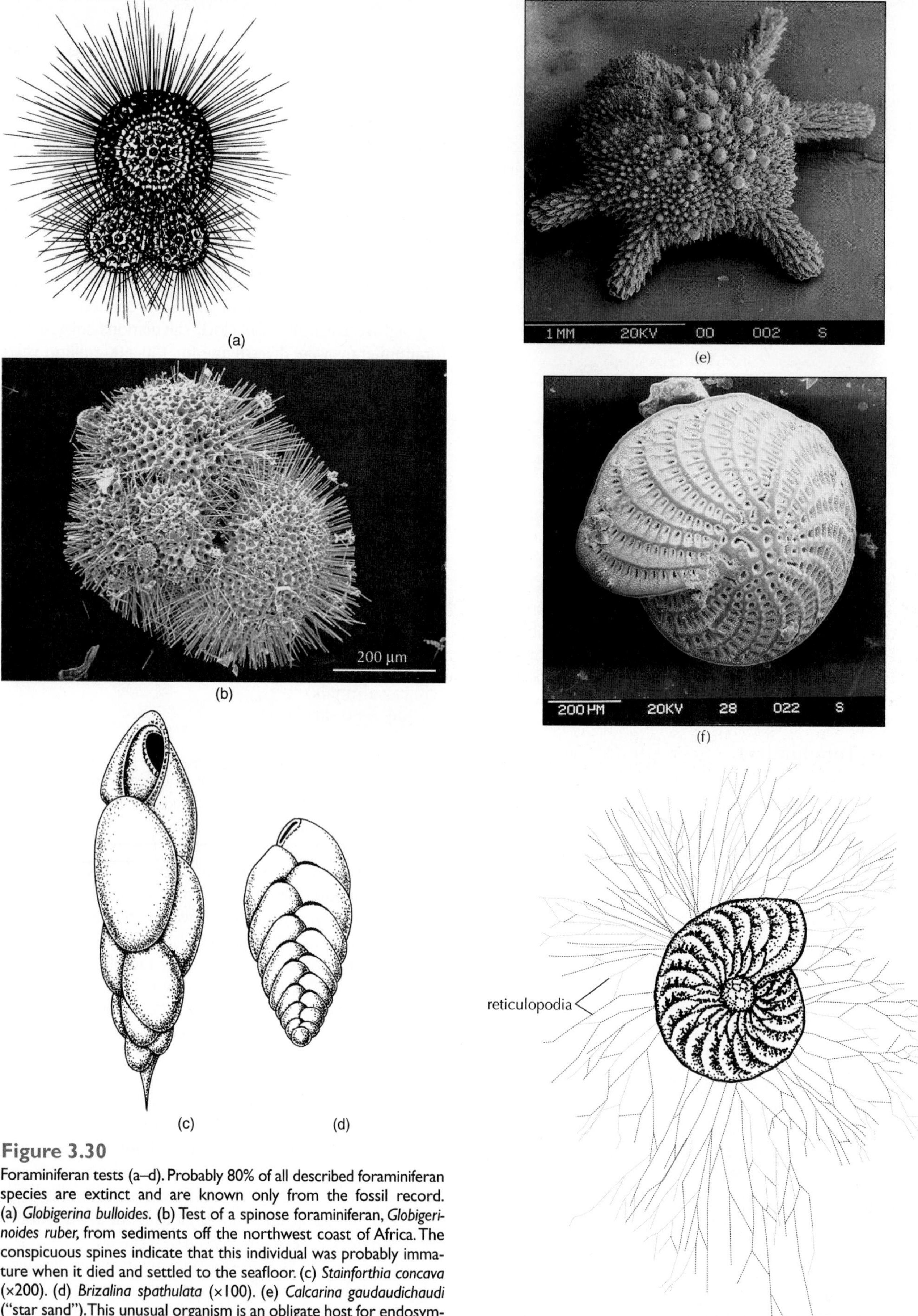

Figure 3.30

Foraminiferan tests (a–d). Probably 80% of all described foraminiferan species are extinct and are known only from the fossil record. (a) *Globigerina bulloides.* (b) Test of a spinose foraminiferan, *Globigerinoides ruber,* from sediments off the northwest coast of Africa. The conspicuous spines indicate that this individual was probably immature when it died and settled to the seafloor. (c) *Stainforthia concava* (×200). (d) *Brizalina spathulata* (×100). (e) *Calcarina gaudaudichaudi* ("star sand"). This unusual organism is an obligate host for endosymbiotic diatoms; it has a very complex internal canal system that bathes the symbiotic algae. (f) *Elphidium crispum.* This species incorporates chloroplasts from captured diatoms. (g) The foraminiferan *Polystomella strigillata,* with reticulopodia.

(a) After Kingsley. (b) Courtesy of Leslie Sautter. (c, d) From Haynes, Foraminifera. Copyright © 1981 Macmillan Accounts and Administration, Ltd., Hampshire, England. (e, f) Courtesy of John J. Lee. (g) After Kingsley; after Whiteley.

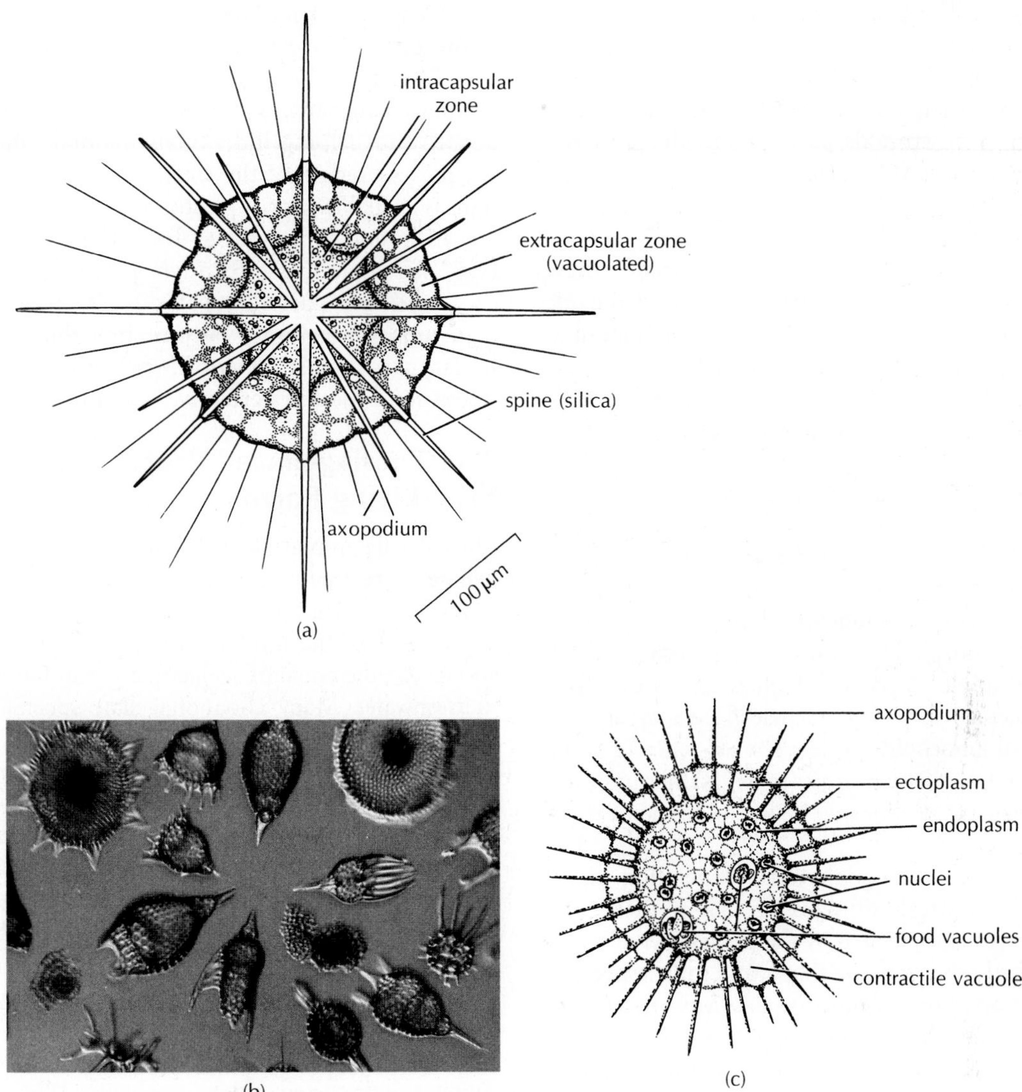

Figure 3.31
(a) A radiolarian, *Acanthometra elasticum.* (b) Radiolarian skeletons, composed of silica. (c) A heliozoan, *Actinosphaerium* sp.
(a) After Farmer. (b) © Eric Gravé/Photo Researchers. (c) After Beck and Braithwaite.

The radiolarian body is generally spherical and is divided into an **intracapsular zone** and an **extracapsular zone** by a perforated, spherical membrane or capsule. Food vacuole formation and digestion occur in the extracapsular region. The nucleus is contained in the intracapsular zone. The acantharian body is also composed of distinct layers, with the innermost layer containing a number of small nuclei.

An Amoeboid Misfit: The Heliozoans

Phylum Helio • zoa
(G: sun animals)
hēl´-ē-ō-zō´-ah

Defining Characteristic: Body is divided into distinct inner and outer regions, but the regions are not separated by any physical boundary; axonemes have numerous microtubules arranged as hexagons or triangles

Like the radiolarians and acantharians, heliozoans (also called "centrohelids") are primarily floating organisms with axopodia (Fig. 3.31c), but they are largely restricted to freshwater habitats. As with the radiozoans, heliozoan bodies are demarcated into (1) a frothy outer region of ectoplasm in which digestion occurs and (2) a less highly vacuolated inner region of endoplasm containing the nucleus. However, heliozoans do not show a distinct physical boundary (i.e., a capsular membrane) between the two regions. The axopodia, which usually project through a thick layer of silica scales, serve primarily to capture food items; in some species, they also function in locomotion. In such cases, a coordinated retraction and reextension of axopods permit individuals to roll slowly over a surface. Microtubules located in the cores of the axopods are apparently involved in mediating the changes in axopod length.

Long thought to be related to radiolarians, these organisms have now been provisionally placed outside the Rhizaria in the Chromisata, a group that is thought to be closely related to the alveolates (ciliates, apicomplexans, and dinoflagellates) discussed earlier (see Fig. 3.1). Fewer than 100 heliozoan species have been described.

Flagellated Protozoans

Unlike amoebae, flagellated protozoans possess one to many flagella (see p. 43), so that the 2 groups differ dramatically in both structure and function. Remarkably, however, these dramatic morphological differences between adult flagellates and adult amoebae are apparently misleading: Current molecular data suggest that flagellates are the ancestors of all amoeboid species. The existence of some very intriguing transitional forms (p. 72) further indicates a close evolutionary relationship between the members of these 2 groups.

Flagellates (or mastigophorans, depending upon whether one's classical affinities are Latin or Greek, respectively) are characterized by the possession of a pellicle, giving the body a definite shape, and most especially by the possession of one or more flagella. Most species are free-living and motile. A cytostome is present in some species, but its morphology is never as complex as that encountered within the Ciliophora. Contractile vacuoles may be present, particularly in freshwater species; if present, their position is fixed within the cell cytoplasm, in contrast to the contractile vacuoles of sarcodinid species.

Flagellates include photosynthesizing, particle-feeding, and parasitic species. To simplify discussion, I have grouped flagellates according to lifestyle. One group—the dinoflagellates—has already been discussed (pp. 54–58), as its members are now believed to be more closely related to the ciliates and apicomplexans than to other flagellated organisms (see Fig. 3.1). Many other flagellated species are not presented here; *Volvox* and its relatives, for example, are now usually placed in the kingdom Plantae.

The Phytoflagellated Protozoans

Some flagellated protozoans such as *Euglena* (Fig. 3.32b) contain chlorophyll, obtain their energy directly from sunlight, and rely exclusively on carbon dioxide as a carbon source. Other photosynthesizing species use light as an energy source but require various dissolved organic compounds as well. Neither type of organism has a mouth or forms food vacuoles. Together, these two groups comprise the flagellated plantlike protozoans.

Many phytoflagellate species bear a red, cup-shaped, photosensitive organelle called a **stigma** (Fig. 3.32b). This is one of the few true sensory organelles known among protozoans. Both flagella and stigma are clearly adaptive in helping individuals maintain themselves in the narrow region of the water column where sufficient light is available for net photosynthesis to occur.

Although some species are completely **autotrophic** (i.e., self-nourishing through photosynthesis), many can feed on particulate foods if necessary. Some euglenids, for example, become holozoic if maintained in darkness for a sufficient time. These individuals then ingest solid food in perfectly good animal-like fashion.

The Zooflagellated Protozoans—Free-Living Forms

Most strictly animal-like flagellate species (i.e., the **zooflagellates**) are free-living in freshwater, saltwater, or soil. Many zooflagellate species have been assigned to the Cercozoa. One of the most interesting groups of free-living species are the choanoflagellates, a group found primarily in freshwater. Many choanoflagellate species are sessile, being permanently attached to a substrate (Fig. 3.33). Each individual bears a single flagellum, which extends for part of its length through a cylindrical network (collar) of closely spaced protoplasmic strands called **microvilli.** Flagellar movements create feeding currents; small food particles stick to the microvillar collar and are ingested. Commonly, individuals are stalked and/or embedded in a gelatinous secretion. Most species are colonial and immobile. Members of the genus *Proterospongia* form (planktonic) colonies of up to several hundred cells; these colonies may bear a striking resemblance to primitive sponges, as discussed in the next chapter. Whether this similarity reflects a true phylogenetic relationship between the unicellular flagellates and the multicellular sponges or whether the similarities are a product of independent, convergent evolution is not certain, but it has been suggested that choanoflagellates are either the ancestors of animals or the sister group to animals. They have, in fact, been tentatively grouped with fungi and animals as "opisthokonts" (see *Taxonomic Detail,* p. 77). Another possibility, of course, is that choanoflagellates evolved *from* sponges.[11]

The Zooflagellated Protozoans—Parasitic Forms

About 25% of zooflagellate species are parasitic or commensal with plants, invertebrates, and vertebrates, including humans. Species in these parasitic groups often exhibit levels of structural and functional complexity not observed in other flagellates. Life cycles are especially complex, as they include adaptations for perpetuating the species from host to host as discussed earlier (pp. 58–59).

11. Maldonado, M. 2004. *Invert. Biol.* 123: 1–22.

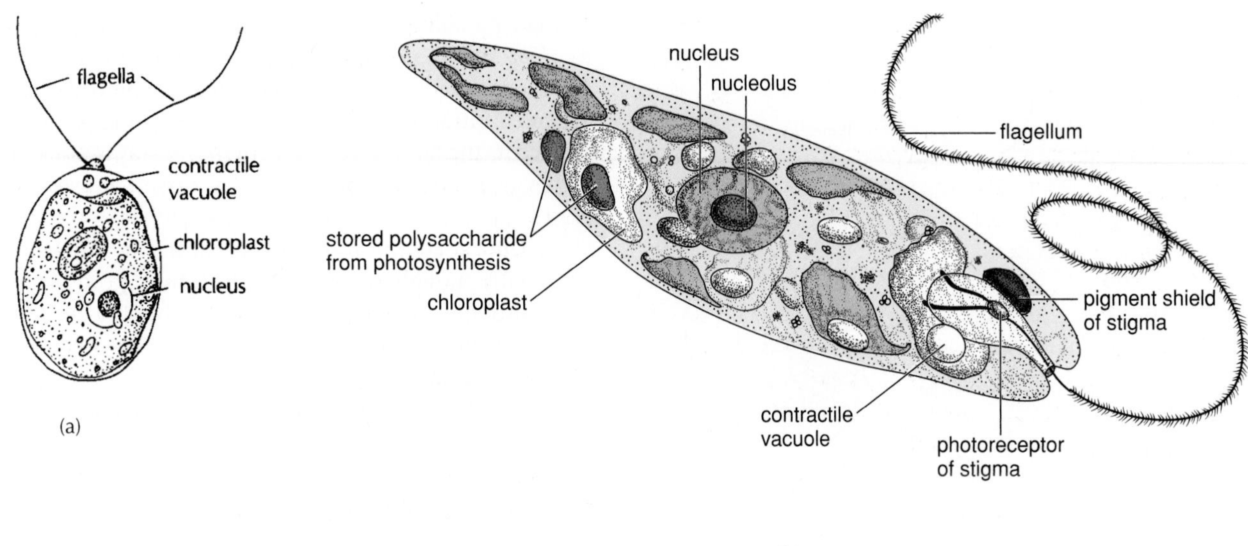

Figure 3.32

Flagellate diversity: phytoflagellates. (a) *Chlamydomonas.* (b) *Euglena.*
(a) After Pennak. (b) Redrawn from Purves and Orians, *Life: The Science of Biology,* 2d ed. 1983 Sinauer Associates, Sunderland, MA.

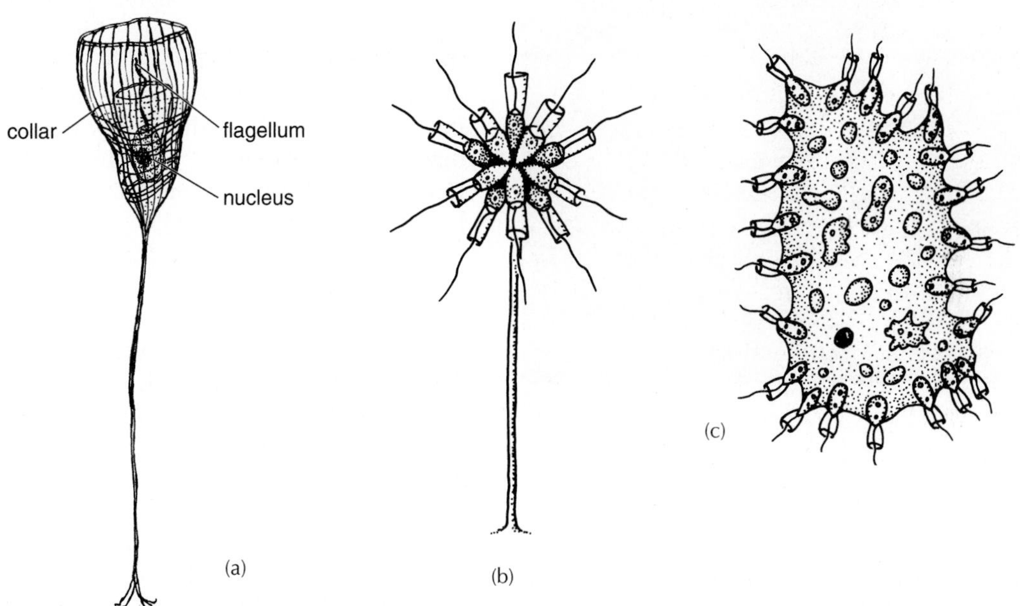

Figure 3.33

Flagellate diversity: zooflagellates. (a–c) Choanoflagellates: (a) *Stephanoeca campanula* (b) *Codosiga botrytis,* a colonial species. (c) *Proterospongia,* another colonial species, with the individuals embedded in a thick, gelatinous matrix.

(a) From H. A. Thomsen, K. R. Buck, and F. P. Chavez, in *Ophelia* 33:131–64, 1991. Copyright © 1991 Ophelia Publications, Helsingør, Denmark. Reprinted by permission. (b) After Kingsley. (c) From Hyman; after Kent.

Trypanosomes (phylum Euglenozoa, members of the Excavata—see Fig. 3.1) are flagellated parasites of special interest to humans (Fig. 3.34). Various species are pathogens in flowering plants, in people, and in cattle, sheep, goats, horses, and other domesticated animals. A key morphological characteristic is the possession of a dark-staining disc of DNA, called a **kinetoplast** (Fig. 3.34a), located within a single large mitochondrion. Trypanosomes, with the tsetse fly as vector, are responsible for the well-known African sleeping sickness and for many other human ailments. *Leishmania donovani,* for example, causes extreme disfigurement and death (with mortality

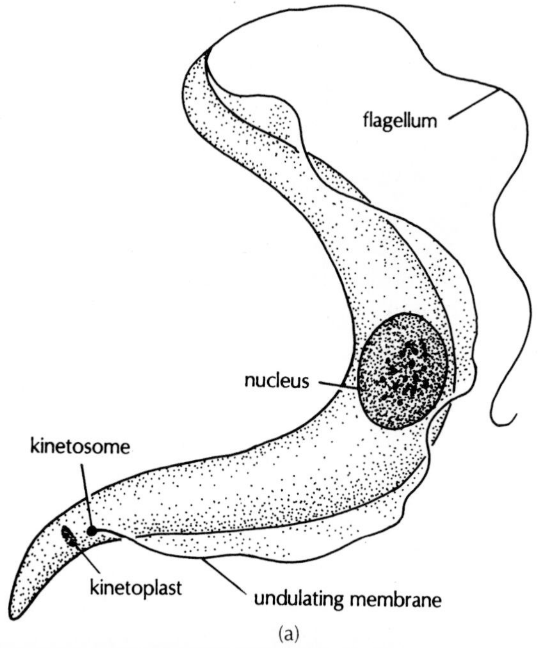

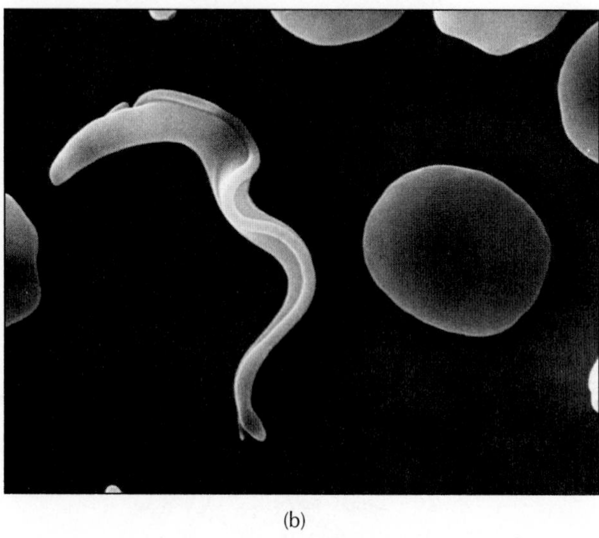
(b)

Figure 3.34

Flagellate diversity: trypanosomes. (a) *Trypanosoma lewisi.* Note the undulating membrane (not to be confused with the ciliate structure of the same name). (b) Scanning electron micrograph (5500 × magnification) of an African trypanosome (*Trypanosoma brucei brucei*) among host red blood cells. This trypanosome is a close relative of the subspecies causing African sleeping sickness.

(a) Sherman/Sherman, *The Invertebrates: Function and Form,* 2/e, © 1976, p. 9. Reprinted by permission of Prentice Hall, Upper Saddle River, New Jersey. (b) From Donelson, J. E. 1988. *The Biology of Parasitism.* pp. 371–400, fig. 1, 372. Reprinted by permission of John Wiley & Sons, Inc.

approaching 95%) in many areas of the world. About 12 million people are affected. Sand flies serve as vectors for the transmission of leishmanial parasites from human to human. Dogs also serve as suitable definitive hosts, functioning, in a sense, as reservoirs for later human infestation. The human defense system recognizes these trypanosomes as foreign beings, and the trypanosomes are quickly phagocytosed by appropriate cells (macrophages) of the human reticuloendothelial system. Remarkably, however, *L. donovani* is not then digested within these cells but instead succeeds in greatly increasing in numbers through repeated intracellular binary fission. How the parasite circumvents the host defense system in this way is not known. Other trypanosomes are equally successful at evading the host's humoral immune response, either by changing the chemical composition of their antigenic surface coats frequently enough to prevent the host from producing specific antibody in effective concentrations or by quickly invading specific tissues that are reasonably well isolated from circulating antibody.

The trick of changing surface antigens has been especially well studied in African trypanosomes, *Trypanosoma* spp. Each individual is covered with a single protein. One would think that the human immune system could easily deal with such a target, and that it should be a simple matter for scientists to produce antibodies against that protein or to help the host produce such antibodies. Unfortunately for us, however, the biochemical nature of the surface antigen changes as the parasite population multiplies within the host. Initially, all the individuals in an infective population have the same external protein composition. The host immune system turns on and begins producing antibodies against that surface protein; within a few days, most of the invaders are, in fact, destroyed. But for a few individuals in the population, a new gene is activated during a division cycle so that a new population begins to develop, one composed of individuals that present a new antigenic surface—one not targeted by the circulating antibody (Fig. 3.35). By the time the host immune system has produced sufficient antibodies to handle the second parasite population, a new variant with a third type of surface coating has appeared, and yet another population of trypanosomes quickly grows (Fig. 3.35). In this way, the trypanosomes stay one step ahead of the humoral immune system. One trypanosome has been estimated to hold over 1000 different genes and pseudogenes coding for surface proteins. Since each gene can be activated more than once, the parasite has more than enough gene combinations to maintain an infection for the life of the host. The molecular mechanisms responsible for turning the various genes on and off are currently under intensive study; preventing the parasite from changing its coat would seem an excellent way of combating the disease. At least 15 million people—and innumerable livestock—become infected by various trypanosomes yearly.

Assuming that natural selection ultimately promotes associations that do not kill the host (an obvious detriment to continued existence of the parasite population), the most highly evolved associations between parasitic zooflagellates and their hosts must be nonpathogenic in nature. The parabasalids (phylum Parabasala) represent one such group of primarily commensal flagellates, containing about 450 species. Nonpathogenic species generally attract less research attention (and funding) than do pathogenic species, but some species wreak enough havoc upon humans to have merited careful study. *Trichomonas vaginalis,* for example, is a typically small protozoan parasitizing the human vagina, prostate, and urethra. *T. vaginalis* generally causes little or no discomfort in men, but it often produces considerable inflammation and irritation in women. The disease is readily transmittable, sexually and otherwise (through contact with toilet seats and towels, for example).

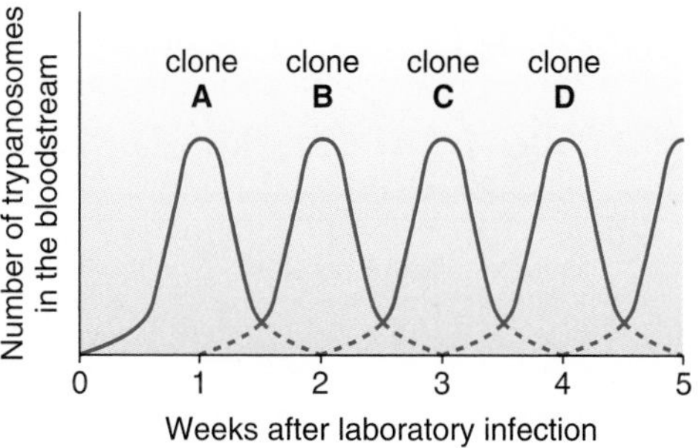

Figure 3.35

The cyclic appearance of trypanosomes in the host bloodstream. The test animal was initially injected with a single trypanosome, which gave rise to other individuals (clone *a*) by asexual replication; all of these individuals express a single surface antigen. Most of these individuals are destroyed by the host immune system within a few days, by which time a few trypanosomes have begun producing a new surface antigen, one not recognized by the circulating antibody produced in response to clone *a.* Clones *b, c,* and *d* represent trypanosomes expressing 3 distinctly different surface antigens. With the periodic expression of new surface antigens, the host can never eliminate the infection.

Based on Donelson, J. E., in *The Biology of Parasitism.*

Probably the most morphologically complex flagellate species are the hypermastigotes (phylum Parabasala, class Hypermastigota) living commensally in the guts of termites, cockroaches, and wood roaches (Fig. 3.36). A great many individuals can easily be obtained for study by squeezing out some of the fluid from the guts of termites. One species alone (*Trichonympha campanula*) may account for up to one-third of the biomass of an individual termite! The species in this group are generally large (for flagellates), typically attaining body lengths of several hundred micrometers. The bodies are commonly divisible into several morphologically and functionally distinct regions and typically bear numerous (up to several thousand!) flagella.

Perhaps the most intriguing feature of the biology of these hypermastigid protozoans is the mutual dependence between host and flagellate. Several species inhabiting termite guts are capable of digesting cellulose, something that most animals, including termites, cannot do. If the termite is deprived of its flagellates, it soon dies, even though it never stops ingesting wood. Similarly, wood-eating roaches soon die if deprived of their flagellate gut fauna. The flagellates, on the other hand, gain from their

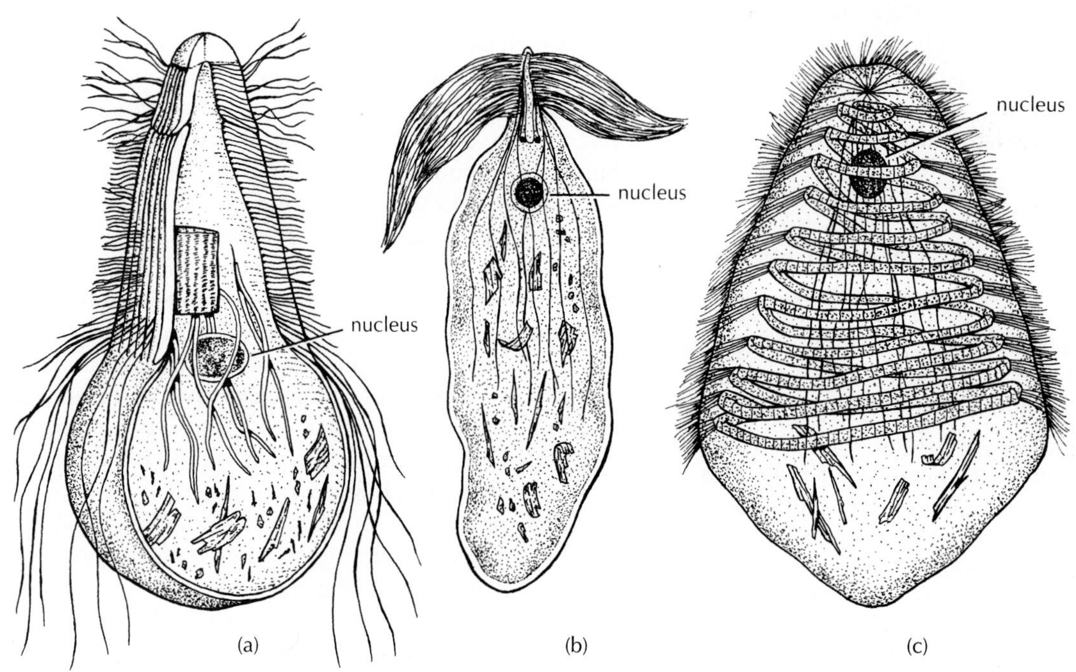

Figure 3.36

Symbiotic flagellates, the most morphologically complex of all flagellate species. (a) *Trichonympha collaris* (about 150 μm), taken from the intestine of a termite. (b) *Rhynchonympha tarda,* from the gut of a wood roach. (c) *Macrospironympha xylopletha,* from the gut of a wood roach.

(a) From Hyman; after Kirby. (b,c) From Hyman; after Cleveland.

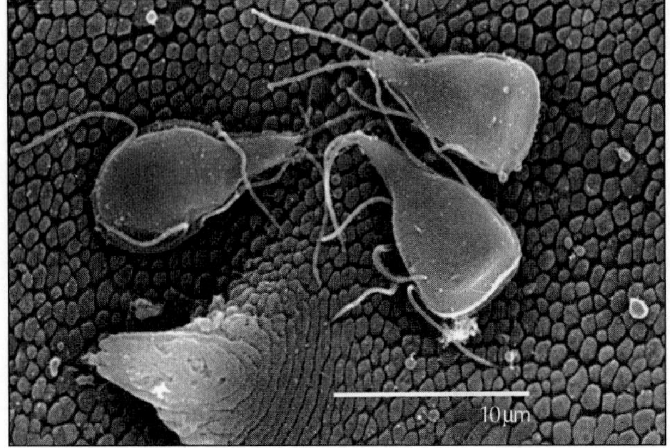

Figure 3.37

Scanning electron micrograph of *Giardia* sp. (3 individuals) attached to the outer body wall (tegument) of the parasitic flatworm *Echinostoma caproni.* Both parasites were taken from the small intestine of an experimentally infected golden hamster.

Courtesy of S.W. B. Irwin, University of Ulster at Jordanstown, Northern Ireland.

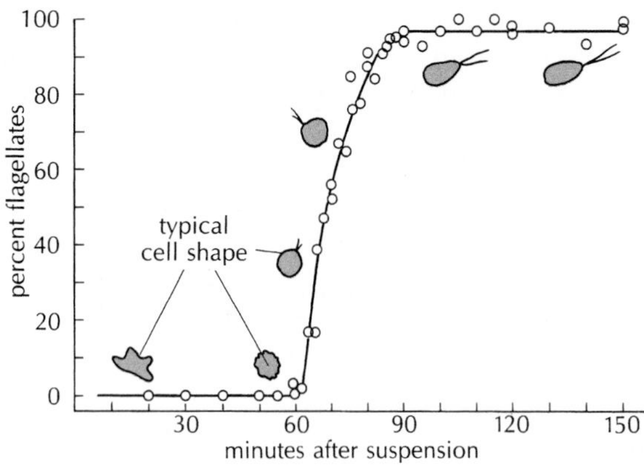

Figure 3.38

The transformation of *Naegleria gruberi* from an amoeboid form to a flagellated form. The transformation was initiated at 25°C by moving the amoebae from the surfaces on which they had been living, into suspension in a test tube. The total number of cells in the culture did not change appreciably during the experiment.

From Fulton and Dingle, in *Developmental Biology* 15:165, 1967. Copyright © 1967 Academic Press, Inc. Reprinted by permission.

host's protection from desiccation and predation, and rely upon the host for a source of cellulose to digest; cellulose is the primary carbon source for these protozoans.

Some particularly unpleasant flagellates are contained in the genus *Giardia* (Fig. 3.37). One species, *G. lamblia,* causes intense nausea, cramps, and diarrhea in its human hosts, a condition that may take several months to dissipate. The infection is readily transmitted through drinking water contaminated with infected feces. Many sadder-but-wiser hikers have discovered that even water in pristine wilderness areas may be so contaminated. Remarkably, *Giardia* and its close relatives lack mitochondria; whether this is a primitive condition or a derived condition related to its lifestyle remains to be seen.

Two Interesting Transitional Forms

Many protozoans offer fascinating links between amoeboid and flagellated forms. Some amoeboid species, for example, release flagellated gametes in the course of sexual reproduction (p. 62), some dinoflagellate species (p. 54) have amoeboid stages in the life cycle, and some slime molds have flagellated stages in the life cycle (pp. 63–64). Moreover, individuals of some other species are "amoebo-flagellates"; that is, they are basically flagellated amoebae.

One interesting transitional species is *Naegleria gruberi,* an apparently primitive protist that lacks mitochondria and typical Golgi bodies and whose ribosomes exhibit a number of prokaryotic features. *N. gruberi* and its relatives are commonly referred to as *amoebomastigotes* or *amoeboflagellates* because they are clearly neither full-fledged amoebae nor full-fledged flagellates. Under normal laboratory conditions, *N. gruberi* is a completely convincing, amorphous, crawling amoeba. Lobopodia form anywhere on the cell surface for locomotion and phagocytosis, and the single contractile vacuole has no fixed position within the body. However, a variety of stimuli, including salinity changes or simply maintaining the animals in suspension, quickly induce a complete transformation of the amoebae into equally convincing flagellates.[12] A pellicle forms, the contractile vacuole becomes fixed in one position, and functional flagella appear. The entire transformation, which takes place within about 1.5 hours, has been the subject of considerable study by developmental biologists (Fig. 3.38). At least one species of *Naegleria (N. fowleri)* recently has been shown to be a facultative parasite of humans. It probably enters through the nasal passages while the unlucky victim is swimming and then migrates to the brain, killing the host within a week. Fortunately, fewer than 100 cases have been reported in the United States since the disease was first recognized about 25 years ago.

Opalinids also bridge a gap, at least superficially, but this time between flagellates and ciliates. Opalinids are exclusively parasitic or commensal within the guts or recta of frogs, toads, and a few fishes. The bodies of opalinids bear numerous rows of cilia (with an associated infraciliature). However, opalinids lack a mouth and the dimorphic nuclei that characterize ciliates, and they exhibit sexual reproduction by gamete formation rather than by conjugation. They have recently been placed along with heliozoans and a variety of other interesting organisms within the Chromista (see Fig. 3.1).

12. See *Topics for Further Discussion and Investigation,* no. 5.

Taxonomic Summary

The Protozoan Protists

The Alveolate Protozoans (Alveolata)

Phylum Ciliophora
Phylum Dinozoa (= Dinoflagellata)—dinoflagellates
Phylum Apicomplexa (≈ Sporozoa)

The Amoebozoa

The Gymnamoebae—naked amoebae
The arcellanids and other test-bearing amoebae
The mastigamoebids
The eumycetozoans—the cellular and acellular slime molds

The Rhizaria

Phylum Foraminifera—the foraminiferans
Phylum Radiozoa—the radiolarians and acantharians
Phylum Heliozoa—the heliozoans
Many amoeboflagellate and zooflagellate species

The Excavata

Giardia
Naegleria
Phylum Parabasala
Class Trichomonada
Class Hypermastigia
Phylum Euglenida
Order Kinetoplastea—the kinetoplastids (e.g., *Bodo, Trypanosoma*)
The Chromista
Heliozoans
Opalinids
The Opisthokonta
Microsporidia
The choanoflagellates
(Metazoans)

Topics for Further Discussion and Investigation

1. The past several decades have seen considerable advances in our understanding of amoeboid locomotory mechanisms. What aspects of their locomotory biology remain puzzling?

Allen, R. D., D. Francis, and R. Zeh. 1971. Direct test of the positive pressure gradient theory of pseudopod extension and retraction in amoebae. *Science* 174:1237.

Cullen, K. J., and R. D. Allen. 1980. A laser microbeam study of amoeboid movement. *Exp. Cell Res.* 128:353.

Edds, K. T. 1975. Motility in *Echinosphaerium nucleofilum.* II. Cytoplasmic contractility and its molecular basis. *J. Cell Biol.* 66:156.

Fukui, Y., T. J. Lynch, H. Brzeska, and E. D. Korn. 1989. Myosin I is located at the leading edges of locomoting *Dictyostelium* amoebae. *Nature* 341:328.

Fukui, Y., T. Q. P. Uyeda, C. Kitayama and S. Inoue. 2000. How well can an amoeba climb? *Proc. Nat. Acad. Sci.* U. S. A. 97:10020.

Heath, J., and B. Holifield. 1991. Actin alone in lamellipodia. *Nature* 352:107.

Lazowski, K., and L. Kuznicki. 1991. Influence of light of different colors on motile behavior and cytoplasmic streaming in *Amoeba proteus. Acta Protozool.* 30:73.

Lombardi, M. L., D. A. Knecht, M. Dembo, and J. Lee. 2007. Traction force microscopy in *Dictyostelium* reveals distinct roles for myosin II motor and actin-crosslinking activity in polarized cell movement. *J. Cell Sci.* 120: 1624–34.

Stossel, T. P. 1989. From signal to pseudopod: How cells control cytoplasmic actin assembly. *J. Biol. Chem.* 264:18261.

Travis, J. L., J. F. X. Kenealy, and R. D. Allen. 1983. Studies on the motility of the Foraminifera. II. The dynamic microtubular cytoskeleton of the reticulopodial network of *Allogromia laticollaris. J. Cell Biol.* 97:1668.

2. Discuss the coordination of ciliary beating in free-living ciliates. What is the evidence that the infraciliature is or is not involved in this coordination?

Naitoh, Y., and R. Eckert. 1969. Ciliary orientation: Controlled by cell membrane or by intracellular fibrils? *Science* 166:1633.

Tamm, S. L. 1972. Ciliary motion in *Paramecium:* A scanning electron microscope study. *J. Cell Biol.* 55:250.

3. Discuss what we do and do not understand about how contractile vacuoles function.

Ahmad, M. 1979. The contractile vacuole of *Amoeba proteus.* III. Effects of inhibitors. *Canadian J. Zool.* 57:2083.

Organ, A. E., E. C. Bovee, and T. L. Jahn. 1972. The mechanisms of the water expulsion vesicle of the ciliate *Tetrahymena pyriformis. J. Cell Biol.* 55:644.

Patterson, D. J., and M. A. Sleigh. 1976. Behavior of the contractile vacuole of *Tetrahymena pyriformis* W: A redescription with comments on the terminology. *J. Protozool.* 23:410.

Riddick, D. H. 1968. Contractile vacuole in the amoeba, *Pelomyxa carolinensis. Amer. J. Physiol.* 215:736.

Stock, C., H. K. Grønlien, R. D. Allen, and Y. Naitoh. 2002. Osmoregulation in *Paramecium: In situ* ion gradients permit water to cascade through the cytosol to the contractile vacuole. *J. Cell Sci.* 115:2339.

Wigg, D., E. C. Bovee, and T. L. Jahn. 1967. The evacuation mechanism of the water expulsion vesicle ("contractile vacuole") of *Amoeba proteus. J. Protozool.* 14:104.

4. What is the evidence that contractile vacuoles play a role in osmotic regulation?

Cronkite, D. L., J. Neuman, D. Walker, and S. K. Pierce. 1991. The response of contractile and noncontractile vacuoles of *Paramecium calkinsi* to widely varying salinities. *J. Protozool.* 38:565.

Hampton, J. R., and J. R. L. Schwartz. 1976. Contractile vacuole function in *Pseudocohnilembus persalinus:* Responses to variation in ion and total solute concentration. *Comp. Biochem. Physiol.* 55A:1.

Schmidt-Nielson, B., and C. R. Schrauger. 1963. *Amoeba proteus:* Studying the contractile vacuole by micropuncture. *Science* 139:606.

Stock, C., R. D. Allen, and Y. Naitoh. 2001. How external osmolarity affects the activity of the contractile vacuole complex, the cytosolic osmolarity and the water permeability of the plasma membrane in *Paramecium multimicronucleatum. J. Exp. Biol.* 204: 291–304.

5. Several protozoans can quickly transform from one body form to another. Describe the morphological changes involved and discuss the environmental factors that initiate these dramatic transformations.

Fulton, C., and A. D. Dingle. 1967. Appearance of the flagellate phenotype in populations of *Naegleria. Devel. Biol.* 15:165.

Nelson, E. M. 1978. Transformation in *Tetrahymena thermophila.* Development of an inducible phenotype. *Devel. Biol.* 66:17.

Willmer, E. N. 1956. Factors which influence the acquisition of flagella by the amoeba, *Naegleria gruberi. J. Exp. Biol.* 33:583.

6. Investigate some of the morphological, behavioral, and physiological adaptations for nutrient acquisition among free-living protozoans.

Alexander, S. P., and T. E. DeLaca. 1987. Feeding adaptations of the foraminiferan *Cibicides refulgens* living epizoically and parasitically on the antarctic scallop *Adamussium colbecki. Biol. Bull.* 173:136.

Bowers, B., and T. E. Olszewski. 1983. *Acanthamoeba* discriminates internally between digestible and indigestible particles. *J. Cell Biol.* 97:317.

Bowser, S. S., T. E. DeLaca, and C. L. Rieder. 1986. Novel extracellular matrix and microtubule cables associated with pseudopodia of *Astrammina rara,* a carnivorous Antarctic foraminifer. *J. Ultra. Mol. Res.* 94:149.

Mast, S. O., and F. M. Root. 1916. Observations on amoeba feeding on rotifers, nematodes and ciliates, and their bearing on the surface tension theory. *J. Exp. Zool.* 21:33.

Salt, G. W. 1968. The feeding of *Amoeba proteus* on *Paramecium aurelia. J. Protozool.* 15:275.

Stoecker, D. K., A. E. Michaels, and L. H. Davis. 1987. Large proportion of marine planktonic ciliates found to contain functional chloroplasts. *Nature* 326:790.

Sundermann, C. A., J. J. Paulin, and H. W. Dickerson. 1986. Recognition of prey by suctoria: The role of cilia. *J. Protozool.* 33:473.

7. How does one document the relative importance of microscopic protozoans in aquatic food webs?

Banse, K. 1982. Cell volumes, maximal growth rates of unicellular algae and ciliates and the role of ciliates in the marine pelagial. *Limnol. Oceanogr.* 27:1059.

Carlough, L. A., and J. L. Meyer. 1989. Protozoans in 2 southeastern blackwater rivers and their importance to trophic transfer. *Limnol. Oceanogr.* 34:163.

Dupuy, C., M. Ryckaert, S. Le Gall, and H. J. Hartmann. 2007. Seasonal variations in planktonic community structure and production in an Atlantic coastal pond: The importance of nanoflagellates. *Microbiol. Ecol.* 53: 537–48.

Fukami, K., A Watanabe, S. Fujita, K. Yamaoka, and T. Nishijima. 1999. Predation on naked protozoan microzooplankton by fish larvae. *Marine Ecol. Progr. Ser.* 185:285.

Porter, K. G., E. B. Sherr, B. F. Sherr, M. Pace, and R. W. Sanders. 1985. Protozoa in planktonic food-webs. *J. Protozool.* 32:409.

8. This chapter presented several examples of ways that parasitic protozoans confound the sophisticated immune systems of their hosts. Investigate some of the ways that free-living ciliates outfox their would-be predators.

Fyda, J., G. Kennaway, K. Adamus, and A. Warren. 2006. Ultrastuctural events in the predator-induced defense response of *Colpidium kleini* (Ciliophora : Hymenostomatia). *Acta Protozool.* 45: 461–64.

Kuhlmann, H. W., and K. Heckmann. 1985. Interspecific morphogens regulating prey-predator relationships in protozoa. *Science* 227:1347.

Kusch, J. 1993. Predator-induced morphological changes in *Euplotes* (Ciliata): Isolation of the inducing substance released from *Stenostomum sphagnetorum* (Turbellaria). *J. Exp. Zool.* 265:613.

Washburn, J. O., M. E. Gross, D. R. Mercer, and J. R. Anderson. 1988. Predator-induced trophic shift of a free-living ciliate: Parasitism of mosquito larvae by their prey. *Science* 240:1193.

9. Despite their lack of organs and tissues, free-living protozoans often show surprisingly complex behaviors. Discuss the evidence that free-living amoebae can orient to chemical signals.

Bailey, G. B., G. J. Leitch, and D. B. Day. 1985. Chemotaxis by *Entamoeba histolytica. J. Protozool.* 32:341.

Biron, D., P. Libros, D. Sagi, D. Mirelman, and E. Moses. 2001. 'Midwives' assist dividing amoebae. *Nature* 410:430.

Fisher, P. R., R. Merkl, and G. Gerisch. 1989. Quantitative analysis of cell motility and chemotaxis in *Dictyostelium discoideum* by using an image processing system and a novel chemotaxis chamber providing stationary chemical gradients. *J. Cell Biol.* 108:973.

Kuhlmann, H.-W., C. Brünen-Nieweler, and K. Heckmann. 1997. Pheromones of the ciliate *Euplotes octocarinatus* not only induce conjugation but also function as chemoattractants. *J. Exp. Zool.* 277:38.

Tani, T., and Y. Naitoh. 1999. Chemotactic responses of *Dictyostelium discoideum* amoebae to a cyclic AMP concentration gradient: Evidence to support a spatial mechanism for sensing cyclic AMP. *J. Exp. Biol.* 202:1.

Zaki, M., N. Andrew, and R. H. Insall. 2006. *Entamoeba histolytica* cell movement: A central role for self-generated chemokines and chemorepellents. *Proc. Nat. Acad. Sci. U.S.A.* 103: 18751–56.

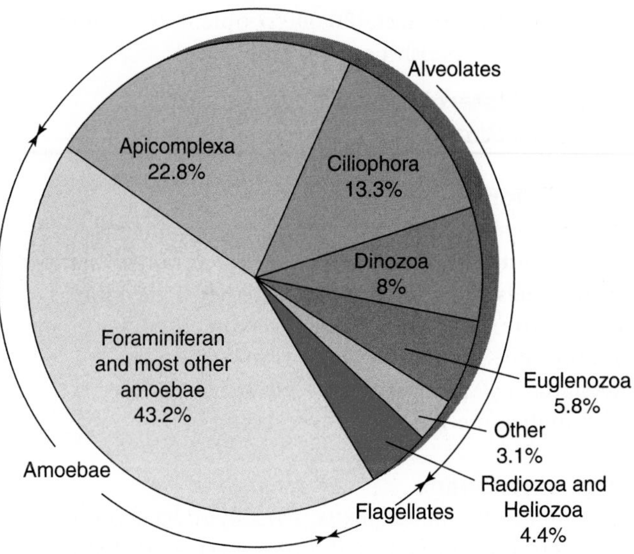

Taxonomic Detail*

Kingdom Protozoa

The Alveolata

All members possess flattened membranous sacs (alveoli) beneath the outer cell membrane. The mitochondrial cristae are tubular, as in most other eukaryotes.

Phylum Ciliophora
Blepharisma, Didinium, Euplotes, Folliculina, Paramecium, Stentor, Stylonychia, Tetrahymena, Tintinopsis, Tokophyra, Vorticella. The ciliates, about 3500 described species.

Phylum Dinozoa (= Dinoflagellata)
Alexandrium (= Protogonyaulax), Amphidinium, Ceratium, Dinophysis, Gonyaulax, Gymnodimium, Noctiluca, Perkinsus, Prorocentrum, Pyrocystis, Zooanthella. The dinoflagellates. Extant species, about 2000.

Phylum Apicomplexa
Babesia, Diplospora, Eimeria, Gigaductus, Gregarina, Haplosporidium, Monocystis, Paramyxa, Plasmodium, Sarcocystis, Toxoplasma. A highly diverse group of parasites, largely restricted to invertebrate hosts. About 6000 species.

The Excavata

These protozoans all possess disc-shaped mitochondrial cristae, and most members also have a deep, ventral feeding groove, after which the group is named (Excavata, as in "excavation").

Phylum Parabasalia
These flagellates are mostly symbiotic, living in hosts ranging from humans to termites and wood roaches. Instead of mitochondria, they possess hydrogenosomes, which may have been derived from mitochondria or may reflect an ancient, independent symbiotic incorporation.

Class Trichomonadida
Dientamoeba, Trichomonas

Class Hypermastigia
Holomastigotes, Lophomonas, Trichonympha, Spironympha. All are intestinal symbionts of termites, cockroaches, and woodroaches, with each individual bearing hundreds to thousands of flagella.

Phylum Euglenozoa

Class Euglenida

Order Euglenia—*Euglena, Peranema, Ploeotia.* Most of the approximately 1000 species are photosynthetic, but some are parasitic and some feed on particulate food. Euglinids are mostly found in freshwater habitats, although some species occupy marine or brackish waters.

Order Kinetoplastea—*Bodo, Leishmania, Leptomonas, Trypanosoma.* This group of flagellates (1 to 2 flagella per individual) includes free-living, symbiotic, and parasitic species. Parasitic species infect animals, flowering plants, and other protozoans. All members have a prominent body of massed DNA within the mitochondrion called a kinetoplast, and a unique microtubular cytoskeleton. A number of trypanosome species infect and debilitate, or kill, humans, livestock, and a variety of other mammals. *Leishmania* does similar harm by causing massive skin

---

* Based on Adl, S. M., A. G. B. Simpson, M. A. Farmer, *et al.* 2005. *J. Eukaryot. Microbiol.* 52: 399–451. Keeling, P. J., G. Buerger, D. G. Durnford, B. F. Lang, *et al.* 2005. *Trends Ecol. Evol.* 20: 670–76.

lesions: At least 400,000 people in 67 countries contract leishmaniasis each year.

Phylum Heterolobsea

Naegleria, Percolomonas, Vahlkampfia.

The diplomonads

Enteromonas, Giardia, Spironucleus, Trigomonas. A coherent group of flagellates of uncertain affiliation. Individuals lack mitochondria. Widely believed to derive from amitochondrial ancestors, these flagellates, it now appears, have in fact lost mitochondria secondarily. Most species are intestinal symbionts, although some species are free-living.

The oxymonads

Oxymonas, Polymastimastix, Pyrsonympha. Another coherent group of flagellates of uncertain affiliation. All species are intestinal symbionts of termites and wood-eating roaches, with one species (*Monocercomonoides*) also living in vertebrate guts.

The Amoebozoa

Most members have branching ("ramicristate"), tubular mitochondria. Includes the naked amoebae (gymnamoebae; e.g., *Acanthamoeba, Amoeba, Chaos, Entamoeba*), mostly free-living individuals with lobose pseudopodia, although some species are obligate pathogens of humans and other animals. Usually without flagellated stages in the life cycle. Concentrations of up to 4200 amoebae per cm^3 of sand have been reported from intertidal beaches.

The arcellanids

Arcella, Difflugia, Pentagonia. This group contains about 75% of all shelled amoebae, which protrude their pseudopodia through a single opening on the test.

The xenophyophoreans

Aschemonella, Homogammina, Psammetta. All are deep-water (500–8000+ meters depth) marine amoeboid protozoans that cement foreign particles into large tests, up to 25 cm long. They are among the largest known protists. For many years they were thought to be either foraminiferans or sponges. About 50 described species.

Phylum Mastigamoididae

Mastigamoeba, Mastigella. All individuals have amoeboid bodies, in some cases exhibiting functional pseudopodia, with the addition of one or more flagella. Marine and freshwater. None have mitochondria or Golgi bodies, but this seems to be a secondary loss rather than the primitive condition. Mastigamoebae are especially common in stagnant water rich in organic matter.

Phylum Eumycetozoa

Prostelium, Echinostelium, Dictyostelium. The cellular and acellular slime molds, commonly found among decaying vegetation.

The Rhizaria

This is a tremendously diverse group of protozoans defined exclusively by molecular characteristics. It includes nonphotosynthetic amoebae, amoeboflagellates, and a very large number of zooflagellate species of great abundance and probable ecological importance, particularly in soil and freshwater habitats. Most have tubular mitochondrial cristae (as do the alveolates and many other eukaryotes).

Phylum Foraminifera

Allogramia, Ammonia, Elphidium, Globigerina, Spirillina. The foraminiferans, most of which secrete calcium carbonate tests. Pseudopodia protrude through numerous pores. Most forams are microscopic, but individuals in some species can exceed 2 cm in diameter. In marine habitats shallower than about 2000 meters (at greater depths calcium carbonate solubilizes readily) they form the sediments known as *Globigerina* ooze. A small number of foraminiferan species are planktonic, occurring at concentrations of about 1 individual per liter of seawater. Some species form symbiotic relationships with algae (zooxanthellae), similar to those formed by reef-building corals and some other invertebrate metazoans. Forams have left an extensive fossil record; species composition can be used to date ancient sediments. In addition, the ratio of ^{18}O to ^{16}O in ancient shells of some species is used to determine past climate changes.

Phylum Radiolaria

All members possess radiating microtubular supports, called axopodia.

Class Polycystinea

Acanthodesmia, Heliodiscus, Hexacontium, Astrosphaera, Spumella, Staurolonche, Triplecta. Another group of small (about 30–250 μm) marine predators, subdivided into spherical forms (Spumellaria) and nonspherical forms (Nassellaria). Polycystines may be solitary or colonial. Unlike phaeodareans, many polycystine species harbor algal symbionts. The skeleton is composed of silica, providing an extensive fossil record. Formerly grouped with Phaeodarea as radiolarians.

Class Acantharia

Acanthocolla, Acanthometra, Acanthospira, Acanthostaurus, Heliolithium, Stauracon. All are marine and free-living micro-predators, with a skeleton of strontium sulfate. Like the phaeodarea and polycystines discussed earlier, actinarians exhibit axopodia. Formerly grouped with phaeodarea and polycystines, with which they share the presence of axopodia, it now appears that axopodia may have been independently acquired in the ancestors of these groups.

Phylum Cercozoa

Class Phaeodarea
Aulacantha, Aulosphaera, Castenella, Gymnosphaera, Phaeodina. These microscopic predators secrete ornate shells of biogenic opal, and range in size from about 50 to 250 μm. They are entirely marine, living to depths of at least 8000 m, and were previously placed in the Radiolaria. Little is yet known about their precise roles in marine food webs. None harbor algal symbionts.

The Chromista

Phylum Stramenopiles

The "heliozoans"
Actinosphaerium, Actinophyrs. There is a growing belief that heliozoans are polyphyletic and that Heliozoa should be abandoned as a formal category. Many "heliozoans" are now in the class Actinophyridae.

The opalinids (class Opalinata)
Opalina, Proteromonas. Flagellated protozoans known mostly from the intestines of reptiles, fishes, frogs, and other cold-blooded vertebrates.

The Opisthokonta

This recently established group contains the true fungi and their protozoan relatives, and the multicellular animals with their protist relatives. All opisthokonts have a posterior (*opistho* = G: behind) flagellum (*kont* = G: flagellum) in some part of the life cycle.

The choanoflagellates (phylum Choanomonada)
Acanthoeca, Codosiga, Diaphanoeca, Monosiga, Proterospongia. Free-living protozoans with a single flagellum surrounded by a basket-like collar composed of siliceous filaments. Many species form colonies, from which both metazoans and fungi may have evolved.

Phylum Fungi

Phylum Microsporidia
Buxtehudea, Loma, Metchnikovella, Microfilum, Nosema. Gene sequence data indicate that microsporidians are degenerate fungi. If so, their lack of mitochondria reflects a secondary loss from ancestors that had them. The members of one genus (*Metchnikovella*) all are intracellular parasites of gregarine protozoans (phylum Apicomplexa).

The Myxozoans
Chloromyxum, Myxidium. Molecular data now show that the members of this group, which are all parasitic, mostly in fishes, are highly degenerate metazoans, probably most closely related to jellyfish and other cnidarians.

General References about Protozoans

Baldauf, S. L. 2003.The deep roots of eukaryotes. *Science* 300:1703–6.

Boardman, R. S., A. H. Cheetham, and A. J. Rowell, eds. 1987. *Fossil Invertebrates.* Palo Alto, Calif.: Blackwell Scientific, 67–91.

Cavalier-Smith, T. 1998. A revised 6-kingdom system of life. *Biol. Rev.* 73:203–66.

Cheng, T. C. 1986. *General Parasitology,* 2d ed. New York: Academic Press.

Coombs, G. H., K. Vickerman, M. A. Sleigh, and A. Warren. 1998. *Evolutionary Relationships among Protozoa.* Chapman & Hall, London.

Fenchel, T. 1987. *Ecology of Protozoa: The Biology of Free-Living Phagotrophic Protists.* New York: Springer-Verlag.

Hyman, L. H. 1949. *The Invertebrates, Vol. 1. Protozoa through Ctenophora.* New York: McGraw-Hill.

Keeling, P. J., G. Burger, D. G. Durnford, B. F. Lang, R.W. Lee, R. E. Pearlman, A. J. Roger, and M. W. Gray. 2005. The tree of eukaryotes. *Trends Ecol. Evol.* 20: 670–76.

Lee, J. J., G. F. Leedale, and P. Bradbury (eds). 2000. *An Illustrated Guide to the Protozoa,* 2nd ed. Lawrence, Kansas: Society of Protozoologists.

Matthews, B. E. 1998. *An Introduction to Parasitology.* New York: Cambridge University Press.

Parker, S. P., ed. 1982. *Classification and Synopsis of Living Organisms,* vol. 1. New York: McGraw-Hill, 491–637.

Roberts, L. S., J. Janovy, and P. Schmidt. 2004. *Foundations of Parasitology,* 7th ed. New York: McGraw-Hill.

Sleigh, Michael A. 1989. *Protozoa and Other Protists.* London: Edward Arnold.

Stechmann, A., and T. Cavalier-Smith. 2002. Rooting the eukaryote tree by using a derived gene fusion. *Science* 297:89–91.

Zimmer, C. 2000. *Parasite Rex: Inside the Bizarre World of Nature's Most Dangerous Creatures.* New York: The Free Press.

Search the Web

1. www.ucmp.berkeley.edu/help/taxaform.html

   Click on "Protista" to see excellent photographs and drawings of a variety of protozoan species, with accompanying text. The site is offered by the University of California at Berkeley.

2. www.who.int

   The latest information about the spread and control of malaria and other parasitic diseases, provided by the World Health Organization. Once at the site, try searching under "malaria," "*Giardia,*" or "*Entamoeba.*"

3. www.malaria.org

   This is the home page for Malaria Foundation International, containing current information about the spread and control of malaria.

4. www.cfsan.fda.gov

   This site is maintained by the U.S. Food and Drug Administration; it includes current information on disease-causing protozoans. Under "Program Areas," select "Bad Bug Book" under the heading "Foodborne illness" and then click on entries for *Giardia lamblia, Entamoeba histolytica,* and *Acanthamoeba.*

5. www.pfiesteria.org/pfiesteria/

   This site, offered by the Center for Applied Aquatic Ecology at North Carolina State University, provides detailed information about the toxic dinoflagellate *Pfiesteria.*

6. www.ucmp.berkeley.edu/protista/slimemolds.html

   This site provides photographs and information about slime molds, along with links to related sites.

7. www.micrographia.com/index.htm

   This site provides exquisite photographs of protozoans and many other organisms along with detailed instructions for making good photomicrographs. Click on "Specimen Galleries" to access the list of organisms available. Provided by Micrographia.

8. http://www.tolweb.org/Eukaryotes

   This address on the Tree of Life site provides an excellent introduction to eukaryotes, including a very readable discussion of the relationships between metazoans and protists.

9. http://protist.i.hosei.ac.jp/Protist_menuE.html

   This site, provided by the Japanese Protist Information Server, provides thousands of photographs of various protists. Click on the group of interest under the heading "Digital Specimen Archives."

10. www.bio.uscd.edu.au/Protsvil/

    From the University of Sydney, Australia, this site includes videos of protozoans and a variety of outside links.

4

The Poriferans and Placozoans

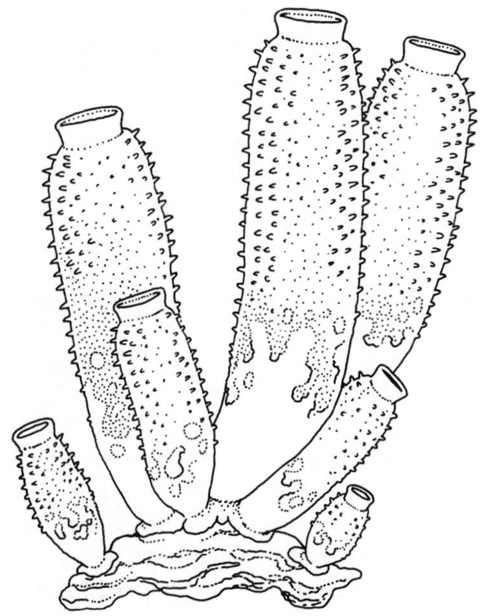

Sponges and placozoans are discussed together in this chapter primarily because they constitute the simplest of multicellular animals (metazoans). The 2 groups may not be closely related.

Phylum Porifera

Phylum Pori • fera
(Latin: pore bearing)
por-i´-fer-ah

Defining Characteristic:[1] Microvillar collars surround flagella, with units arising from either single cells or syncytia

Introduction

Less than 2% of all sponge species are found in freshwater, the remaining 98% of the species being marine. There are no terrestrial sponges. The approximately 7000 to 15,000 sponge species currently with us are perhaps more remarkable for the characteristics they lack than for those they possess. Like placozoans and unlike almost any other metazoan, sponges lack nerves and have no true musculature. All sponges feed on food particles suspended in the water. No specialized reproductive, digestive, respiratory, sensory, or excretory organs are found in this group, either; indeed, no organs are found at all. Often, sponges are amorphous, asymmetrical creatures, although there are some very beautiful exceptions to this generalization. No sponge has anything corresponding to anterior, posterior, or oral surfaces. Moreover, only a few different cell types are encountered within any given individual. These cells are functionally independent to

1. Characteristics distinguishing the members of this phylum from members of other phyla.

RESEARCH FOCUS BOX 4.1

Histoincompatibility in Sponges

Hildemann, W. H., I. S. Johnson, and P. L. Jokiel. 1979. Immunocompetence in the lowest metazoan phylum: Transplantation immunity in sponges. *Science* 204:420–22, 1979.

To qualify as having an immune system, (1) the organism (or the cells of that organism) must evidence some form of antagonism toward foreign substances, (2) the antagonism must be specific toward that substance, and (3) future responses should be altered by the first response—that is, the system must "remember" the prior encounter. Forty years ago, invertebrates were thought to lack immune systems. This is no longer the case: The ability to distinguish between self and nonself has been clearly demonstrated for many nonvertebrate animal groups, including sea urchins and sea stars, insects, crustaceans, annelid worms, clams and snails, sea anemones, and corals. Sponge cells of any given species have long been known to avoid fusing with cells of other sponge species, but can the cells of a sponge discriminate between sponge cells of another individual of the *same* species? Can those cells show a true immune response?

Hildemann et al. (1979) studied intraspecific (among individuals of one species) tissue incompatibility of the tropical sponge *Callyspongia diffusa*, a large, purple sponge that grows upward like long fingers arising from a common base on coral rock in Hawaii. They chose to study this species because, although they frequently observed tissue fusion between adjacent fingers or pieces of a single sponge colony in the field, they never observed such grafts between adjacent *C. diffusa* colonies, suggesting that the sponge tissues might be capable of intraspecific self-recognition.

To examine this possibility in the laboratory, they wired pieces of sponge (each approximately 2 cm by 8 cm) to heavy plastic plates, placing two pieces of sponge from one colony (isogeneic treatments) or pieces from 2 different colonies collected from different locations (allogeneic treatments) close together on each plate. The researchers then kept the sponges in the laboratory at about 27°C and monitored the fate of each sponge daily.

Sponge fragments in the isogeneic treatments (paired from the same colony) always fused together within a few days, showing complete tissue compatibility. In marked contrast, tissues in the allogeneic treatments never fused. In fact, the researchers observed pronounced toxic responses between the tissues of these incompatible sponges; within 2 weeks, tissues in contact with the neighboring sponge fragment lost their purple coloration and died (Focus Table 4.1). The sponges showed no such reaction to the plastic holders or to the wire they were tied down with; the response was clearly one of histoincompatibility: specific, intense tissue disharmony.

the extent that an entire sponge can be dissociated into its constituent cells in the laboratory, with no long-term impact. The cells dedifferentiate to an amoeboid form, reaggregate, and at least in some species, can redifferentiate to reform the sponge.[2] Just as remarkably, for such a rudimentary animal, one sponge can distinguish between its own cells and those of different individuals in the same species (Research Focus Box 4.1). Such non-self-recognition (i.e., alloincompatibility) has now been reported from at least 25 sponge species.[3]

Probably the most surprising aspect of sponge biology is that sponges work so well with so little. They are often major components of aquatic communities and compete impressively with other sedentary metazoans for food and, especially, space. They also provide homes for a great variety of animals from many other phyla, and for a variety of bacteria and cyanobacteria.

Considering the lack of organs and systems in sponges,[4] one can wonder whether sponges are truly multicellular animals or whether they represent a highly evolved form of colonial living by individual cells. In fact, some freshwater colonial protozoans (the choanoflagellates, p. 68) bear very definite morphological similarities to the simplest sponges, and as recently as 100 years ago it was suggested that sponges be classified as colonial protozoans. However, recent molecular data support a common evolutionary origin for sponges and more structurally complex animals: Sponges now seem comfortably seated in the animal kingdom, although their exact placement remains uncertain. Long considered an "evolutionary dead-end," it now seems from molecular analyses that while demosponges and glass sponges may well have given rise to no further lineages, the calcareous sponges may in fact be the ancestors of other multicellular animals, making the phylum Porifera paraphyletic (see definition

2. See *Topics for Further Discussion and Investigation,* no. 5, at the end of the chapter.

3. See *Topics for Further Discussion and Investigation,* no. 7.

4. See *Topics for Further Discussion and Investigation,* no. 4.

Focus Table 4.1 Reaction Times of *Callyspongia diffusa* Fragments to Each Other

1 Source of Individuals Tested	2 Days to React in First Test (median ± one standard deviation)	3 Number of Pairs Tested	4 Days to React in Second Test (median ± one standard deviation)	5 Number of Pairs Tested
A & B	9.0 ± 1.9	24	3.8 ± 0.9	10
A & C	8.9 ± 6.9	30	4.2 ± 1.3	13
B & C	7.2 ± 2.2	21	4.0 ± 1.2	11

Note: For each test, sponges collected from 3 different locations (A, B, or C) were grafted together (in pairs) in intimate soft tissue contact. Tissue death at points of contact was clearly visible in a little over 1 week, on average (column 2). When the pairs were regrafted after a 12-day separation, the tissue-toxic responses occurred even more rapidly (column 4). For example, the median response for individuals collected from locations A and B required about 9 days in the first encounter but less than 4 days in the second encounter. The faster second response clearly was a consequence of prior experience, indicating that these sponges have an immune memory system.

Even more remarkably, when the incompatible sponge fragments were permitted to "rest" apart for 12 days and were then reexposed to each other at new tissue locations, the toxic interactions recurred significantly more rapidly than before (Focus Table 4.1), suggesting an immune memory: The second response was clearly modified by the first.

All 200 intercolony challenges were marked by distinct tissue incompatibility responses in the first trial and accelerated responses in the second. The tissue of each sponge was clearly capable of recognizing the differences between many subtly different surface molecules characterizing individual sponges of the same species. The genetic loci coding for histocompatibility molecules (i.e., those responsible for making the self/nonself distinction) at the cell surface of sponges must be highly polymorphic, coded for by perhaps hundreds of different genes at particular loci. This ability to discriminate so finely between self and nonself most certainly plays a key role in helping an individual sponge genotype to compete successfully for space.

Clearly, even sponges have a spectacularly sensitive immunorecognition system capable of distinguishing even minor chemical differences among allogeneic surface markers. Further studies of this system in sponges may tell us something about the evolutionary origin of cell-mediated immunity in all animals and, perhaps, about the functioning of the major histocompatibility complex of the human immune system.

in Chapter 2, p. 27).[5] Their fossil record extends back about 580 million years.

General Characteristics

In its simplest form, a sponge is a fairly rigid, perforated bag, whose inner surface is lined with flagellated cells. The empty space of this bag is called the **spongocoel.** The flagellated cells lining the spongocoel are called **choanocytes** (literally, "funnel cells"), or **collar cells** in recognition of the cylindrical arrangement of the cytoplasmic extensions (collars) surrounding the proximal portion of each flagellum (Fig. 4.1). The collars of sponge choanocytes are almost certainly homologous with those of protozoan choanoflagellates. These collar cells perform the following functions:

1. They generate currents that help maintain circulation of seawater within and through the sponge.
2. They capture small food particles.
3. They capture incoming sperm for fertilization.

5. Nichols, S., and G. Wörheide. 2005. *Integr. Comp. Biol.* 45: 333–34.

Adjacent to the choanocyte layer is a gelatinous, nonliving layer of material called the **mesohyl** layer (*meso* = G: middle; *hyl* = G: stuff, matter). Although the mesohyl is acellular and nonliving, it contains live cells. Amorphous, amoeboid cells called **archaeocytes** wander throughout the mesohyl by typical cytoplasmic streaming, which involves the formation of pseudopodia as in amoeboid protozoans. Archaeocytes perform a number of essential functions within sponges and can develop into more specialized cell types when necessary. Archaeocytes are responsible for digesting food particles captured by the choanocytes, so that digestion is entirely intracellular. Some archaeocytes also store digested food material. In addition, archaeocytes may give rise to both sperm (which are flagellated) and eggs, although gametes may also arise through morphological modification of existing choanocytes. They also probably play an active role in non-self-recognition reactions in response to contact with other sponges. Finally, archaeocytes play a role in eliminating wastes and, in addition, can become specialized to secrete the supporting elements located in the

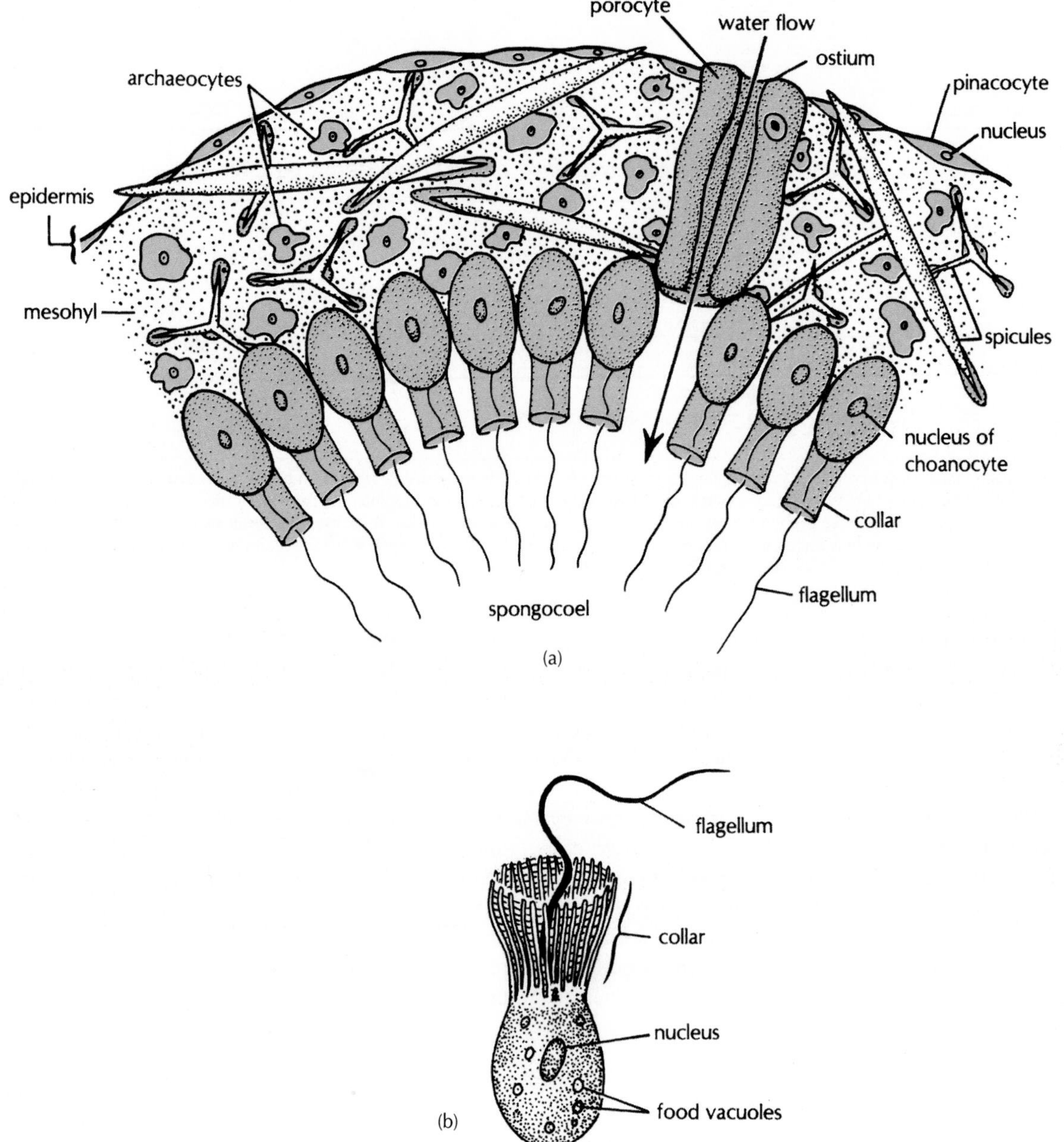

Figure 4.1
(a) Diagrammatic illustration of the body wall of a sponge. (b) Detail of a choanocyte.

(a) From Hyman, *The Invertebrates*, Vol. III. Copyright © 1951 McGraw-Hill Book Company, New York. Reprinted by permission. (b) After Rasmont.

mesohyl layer. These support elements may be calcareous or siliceous **spicules,** or they may be fibers composed of a collagenous protein called **spongin.** The cells secreting spicules are termed **sclerocytes,** and those producing spongin fibers are termed **spongocytes**[6] (Fig. 4.2). Both of these cell types are derived from archaeocytes. Clearly, archaeocyte cells are quite versatile. The spicules and fibers secreted by the sclerocytes and spongocytes are of great importance (1) to systematists, as an indispensable factor in species identification, and (2) to sponges, which generally depend upon the support elements for maintaining shape and, possibly, for discouraging predation.

At certain times of the year, many freshwater sponge species (and a few marine species) produce dormant structures called **gemmules.** To begin the process, archaeocytes accumulate nutrients by phagocytizing other cells and then cluster together within the sponge. Certain cells surrounding each cluster secrete a thick, protective covering; the gemmule consists of the cluster plus its surrounding capsule (Fig. 4.3). Gemmules are typically far more resistant to desiccation, freezing, and anoxia (lack of oxygen) than are the sponges that produce them (Fig. 4.4). In fact, the gemmules of many species must spend several months at low temperature before they become capable of hatching; that is, they require a period of **vernalization.** Under appropriate environmental conditions, the

6. See *Topics for Further Discussion and Investigation*, no. 1.

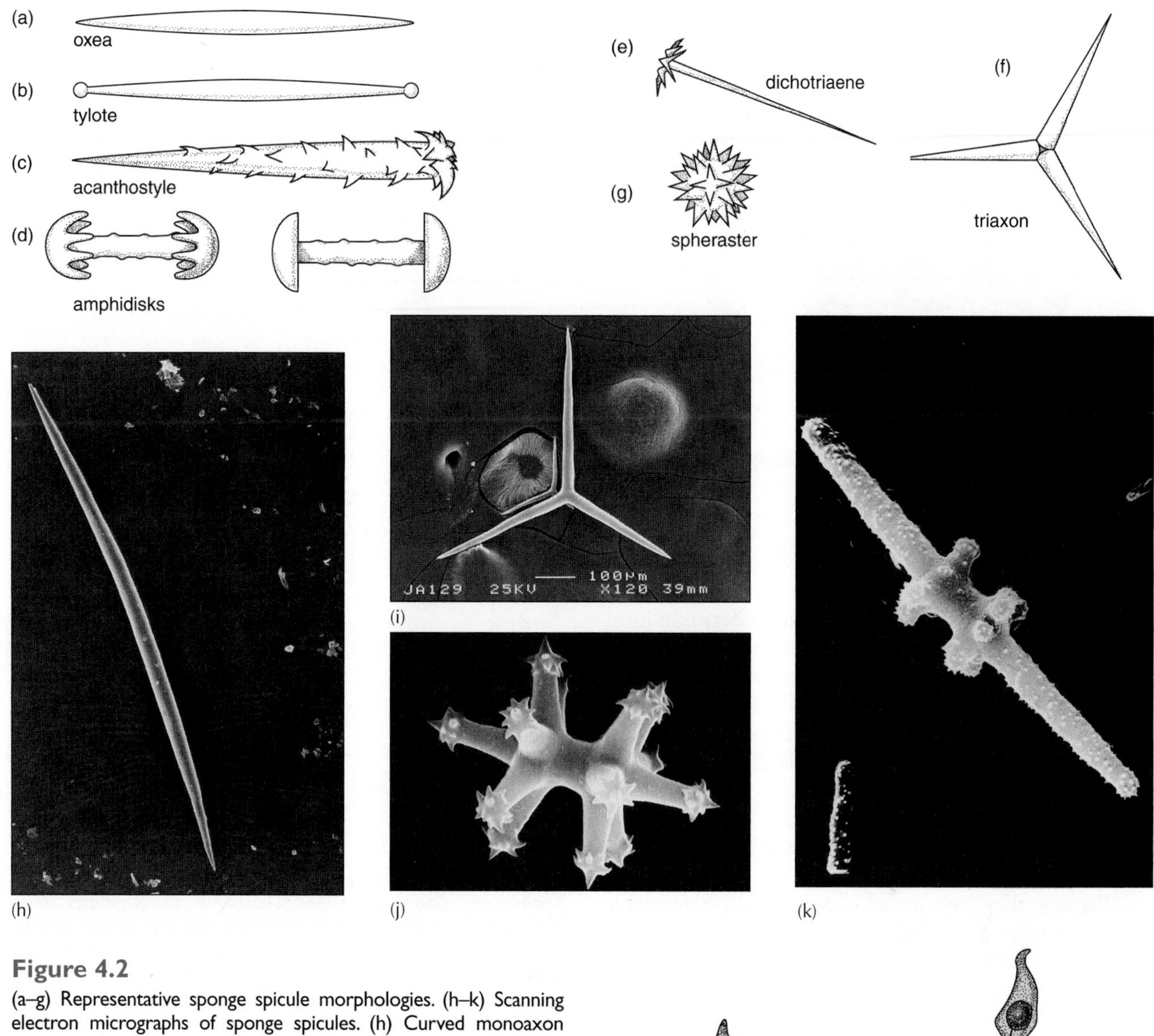

Figure 4.2

(a–g) Representative sponge spicule morphologies. (h–k) Scanning electron micrographs of sponge spicules. (h) Curved monoaxon spicule from *Sycon* sp. (i) Triradiate spicule from *Sycon* sp. (j) Small aphiaster from *Alectona wallichii.* (k) Fusiform amphiaster from *Alectona wallichii.* (l–n) The production of a sponge spicule by a binucleate sclerocyte. The spicule is initiated between the two nuclei. After the spicule is completed, the cells will wander off into the mesoglea.

(a–n) Based on Beck/Braithwaite, *Invertebrate Zoology, Laboratory Workbook,* 3/e, 1968. (h,i) Courtesy of Micha Ilan and J. Aizenberg. From Ilan, M., J. Aizenberg, and O. Gilor, 1996. *Proc. Roy. Soc.* London B 263:133–39, figs. 1c–d. (j,k) Courtesy of G. Bavestrello. From Bavestrello, G., B. Calcinai, C. Cerrano, and M. Sera, 1998. *J. Mar. Biol.* Ass. U.K. 78:59–73. (l, m, n,) From Hyman, after Woodman.

living cells leave the gemmule (hatch) through a narrow opening and differentiate to form a functional sponge.[7] Gemmule formation thus allows sponges to withstand unfavorable environmental conditions by entering a stage of developmental arrest, a period of dormancy. In light of some of the differences between freshwater and marine environments mentioned in Chapter 1, the adaptive significance of gemmule formation by freshwater sponges should be particularly apparent. Also, since each sponge produces many gemmules, gemmule formation can be an effective means of asexual reproduction, resulting in numerous genetically identical offspring. Gemmules of some sponge species have hatched successfully after 25 years of storage.

Instead of forming gemmules, some freshwater and marine species undergo pronounced tissue regression during unfavorable periods, with the sponge becoming reduced to little more than a compact cellular mass with an outer protective covering. The cells reactivate when environmental conditions improve, regenerating all of the structures present before the regression.

7. See *Topics for Further Discussion and Investigation,* no. 2.

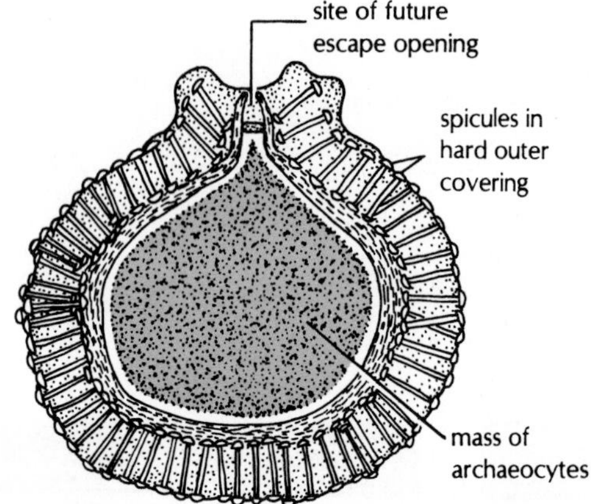

Figure 4.3

The gemmule of a freshwater sponge, *Ephydatia,* as seen in cross section.
From Hyman; after Evans.

The outer, epithelial-like layer of most sponges is composed of flattened contractile cells called **pinacocytes** (Fig. 4.5), forming a layer called the **pinacoderm.** In most animals, epithelial cells rest on a collagenous sheet of extracellular matrix that they secrete, called the **basal lamina** (or **basement membrane**); pinacocytes lack a basal lamina. The pinacocytes also line the incurrent canals and the spongocoel in places where choanocytes are lacking. Contraction of pinacocytes enables sponges to undergo minor to major shape changes[8] and also may play a role in regulating water flow through the sponge by varying the diameter of the incurrent openings.

Because they lack muscles, nerves, and deformable bodies, sponges are utterly dependent upon water flow for food, gas exchange, waste removal, and the dissemination and collection of sperm. Partly as a consequence of choanocyte activity and partly as a consequence of sponge architecture (Fig. 4.6),[9] water flows into the spongocoel through narrow openings **(ostia)** and exits the spongocoel through larger openings **(oscula).** Ostia are always numerous on the sponge's body, but there may be only one osculum present per individual. The name Porifera attests to the supreme importance of all these openings in sponge morphology and physiology.

The large size of many sponges—the members of some species exceed 1 m (meter) in height or diameter—attests to the large volumes of water filtered for food: probably dozens of liters per sponge per day.[10] However, rapid flow past a food-capturing surface is incompatible with efficient particle capture. Ideally, the water would slow down within the choanocyte chambers, allowing time for efficient particle removal, and then speed up on the way out of the sponge to dissipate wastes (and sperm) effectively. In fact, this is exactly what happens, not because of differences in flagellar activity in different parts of the sponge, but simply because of internal sponge architecture. Table 4.1 presents water flow data for a sponge that filters 0.18 cm^3 of water per second, which works out to a respectable 0.65 liters of water per hour. This sponge bears a large number of ostia (940,000!); because incoming water enters through all these available openings, water enters through each ostium at a leisurely 0.057 cm per second. However, the ostia lead into a mind-boggling number of flagellated chambers, nearly 29 million

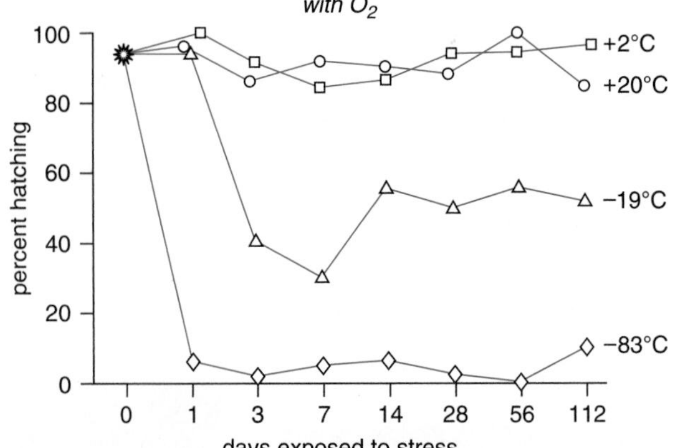

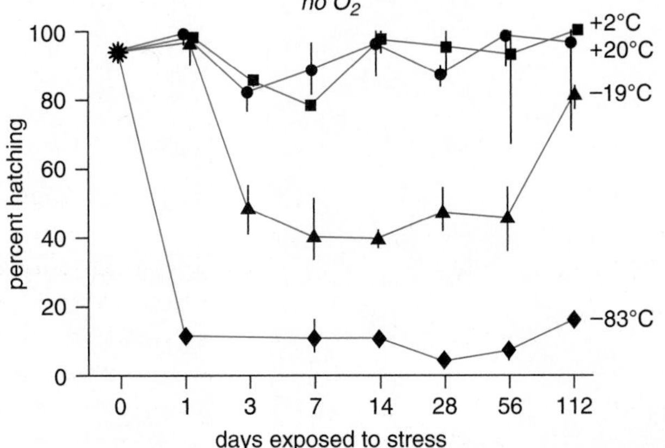

Figure 4.4

The effect of low temperature and oxygen deprivation (anoxia) on survival of sponge gemmules. Thousands of gemmules were obtained from a single specimen of the freshwater sponge *Ephydatia muelleri.* Some gemmules were sealed in glass ampules in well-oxygenated water, while others were sealed in glass ampules purged with nitrogen to displace all oxygen from the water. Each ampule contained about 30 gemmules; about 3,500 gemmules were used in the entire experiment. The ampules were then distributed among 4 temperature treatments. On 7 dates over the next 16 weeks, at least one ampule from each treatment was opened and the percent of gemmules hatching within 4 weeks at 20°C was determined. Data for gemmules maintained under anoxic conditions generally show the means and ranges for 3 replicates of 30 gemmules per sample. There was no replication for gemmules in the other treatment group. The gemmules were impressively tolerant of low temperature—in general, 5–18% hatched even after having spent 16 weeks at −83°C—and they survived equally well in the presence and absence of oxygen.
From H. M. Reiswig and T. L. Miller, 1998. *Invert. Biol.* 117:1–8.

8. See *Topics for Further Discussion and Investigation,* no. 4.
9. See *Topics for Further Discussion and Investigation,* nos. 3 and 4.
10. See *Topics for Further Discussion and Investigation,* no. 10.

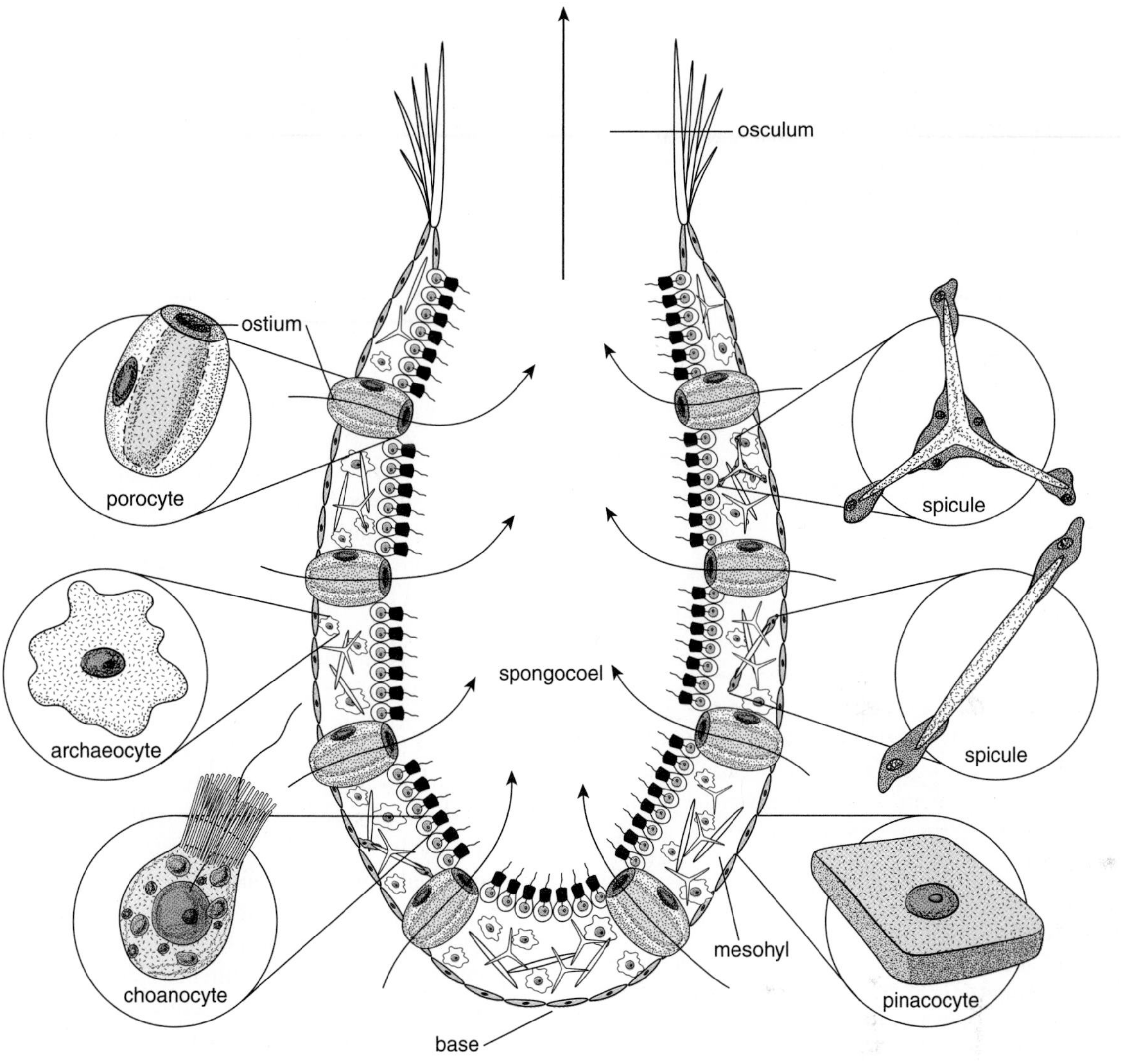

Figure 4.5
Diagrammatic representation of a simple (asconoid) sponge, illustrating its various cellular and structural components. In asconoids, the incurrent canal is simply a tube passing through a modified pinacocyte, called a *porocyte*. Note that 6 cells are involved in producing a triradiate spicule.
Based on Sherman/Sherman, *The Invertebrates: Function and Form,* 2/e, 1976, p. 45.

in this case. Although these chambers are each very small, multiply their individual cross-sectional area by 29 million and you get a surprisingly large total cross-sectional area, as indicated in the table. Since the total volume of water entering the sponge must also be the volume moving through and leaving the sponge in the same amount of time (note that for each row of the table, total area multiplied by water velocity always equals about 0.18 cm^3/sec), water velocity must decline dramatically inside the choanocyte chambers, as indicated, increasing the opportunity for food particle or gamete capture. Our own blood capillary system operates similarly in increasing the efficiency of gas exchange between blood and tissues, but sponges were clearly there first, a nice example of convergent morphological evolution in the face of similar selective pressures. Once particles (and oxygen) have been removed in the flagellated chambers, water velocity again increases, simply due to the decreased total cross-sectional area across which the water now flows; to move the same volume of water through a smaller cross-sectional area of the sponge per time, velocity must increase.

Natural products chemists have become extremely interested in sponges in recent years, as sources of novel chemical compounds produced uniquely by particular sponge species to protect against predation and to discourage the larvae of barnacles and other "fouling" organisms from attaching to their surfaces. Some of these chemicals are produced by the sponges themselves and some are produced by bacterial and cyanobacterial (blue-green algal) symbionts. Some of the chemicals released may not discourage fouling by the larvae of other animals directly, but rather by modifying the species' composition

Table 4.1 Water Transport Characteristics for a Marine Leuconoid Sponge

Anatomical Feature	Individual Cross-Sectional Area (cm^2)	Approximate No. Per Sponge	Total Cross-Sectional Area (cm^2)	Water Velocity (cm/sec)
Ostia	3.33×10^{-6}	940,000	3.14	0.057
Flagellated chambers	7.06×10^{-6}	2.88×10^{7}	203.0	8.69×10^{-4}
Osculum	0.034	1.0	0.034	5.1

The sponge on which the data are based had a total volume of 2.4 cm^3. *From LaBarbera, M., and S. Vogel. 1982.* Amer. Scient. *70:54–60.*

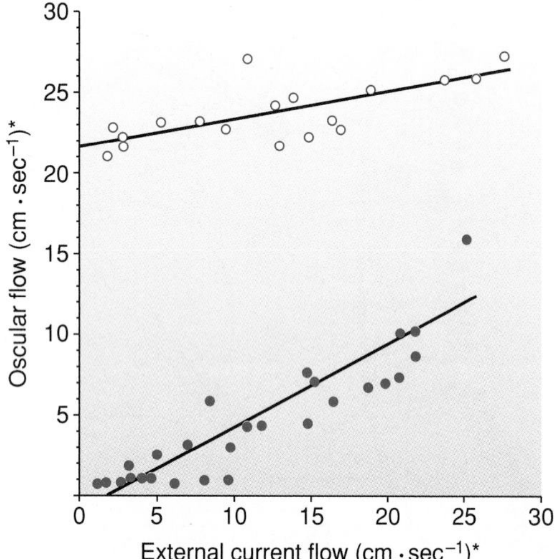

Figure 4.6
Influence of morphology on water flow through the marine sponge *Haliclona viridis.* (•) Velocity of water leaving sponge oscula for undisturbed sponges. (o) Data for sponges whose choanocytes were inactivated by immersing sponges in freshwater for several minutes. The data show that, in a current, water flows through the sponge with or without the help of choanocyte activity. In a current of 15 cm/sec, for example, water moves through an inactivated sponge at nearly 30% of the rate measured in a fully functional sponge, simply because of the way the sponge's architecture takes advantage of surrounding water flow. All flow measurements were made in the field, by positioning one tiny flow-sensitive probe next to a sponge and another in the center of its osculum.
Based on Steven Vogel, in *Proceedings of the National Academy of Science* 74:2069–71, 1977.
*centimeters per second

of adjacent bacterial communities.[11] Some of the chemicals have potential biomedical applications, and others may prove useful in discouraging larvae from attaching to ship hulls and other underwater surfaces.

Poriferan Diversity

Evolutionary advances in sponge morphology imply selection for maximizing current flow through the spongocoel and increasing the amount of surface area available for food collection. There are 3 basic levels of sponge construction: asconoid, syconoid, and leuconoid, in order of increasing complexity. Each form simply reflects an increased degree of evagination of the choanocyte layer away from the spongocoel, increasing the extent of flagellated surface area enclosed by the sponge (Fig. 4.7). Most sponge species are of leuconoid construction.

Representative sponges are illustrated in Figure 4.9. Sponges exist in a tremendous variety of colors and shapes, ranging from encrusting forms only a few millimeters wide to elaborate upright forms over 1 m high. Although most sponges are immobile, members of a few sponge species can move at speeds of several millimeters per hour, presumably by the coordinated cytoplasmic movements of individual, amoeboid cells.[12]

Sponges are distributed among the following 3 classes, based largely upon the chemical composition and morphology of the support elements: Calcarea, Demospongiae, and Hexactinellida. Until recently there was a fourth class of poriferan, the Sclerospongiae, containing 16 species, all of leuconoid construction and all restricted to obscure caves and crevices in coral reefs. Unlike that of most other sponges, the sclerosponges secrete a substantial supporting mass of calcium carbonate (Fig. 4.8) in addition to the more normal microscopic spicules made of calcium carbonate, silica, and spongin. About 10 years ago, similarly solid calcareous skeletons were described in at least one species within the Calcarea and at least one species within the Demospongiae, suggesting that this particular character does not define a unique class of sponges. Recent comparisons of 28S ribosomal RNA gene sequences support this suggestion: The sclerosponges included in the analysis clearly belonged in either the Calcarea or the Demospongiae.[13] The ability to secrete massive calcium carbonate supports has apparently evolved several times independently in different sponge groups. In consequence, the Sclerospongiae is abandoned as a formal taxonomic category, and we are back to 3 classes.

11. Lee, O. O., L. H. Yang, X. C. Li, J. R. Pawlik, and P. Y. Qian. 2007. *Mar. Ecol. Progr. Ser.* 339: 25–40.

12. See *Topics for Further Discussion and Investigation,* no. 4.

13. Chombard, C., N. Boury-Esnault, A. Tillier, and J. Vacelet. 1997. *Biol. Bull.* 193:359–67.

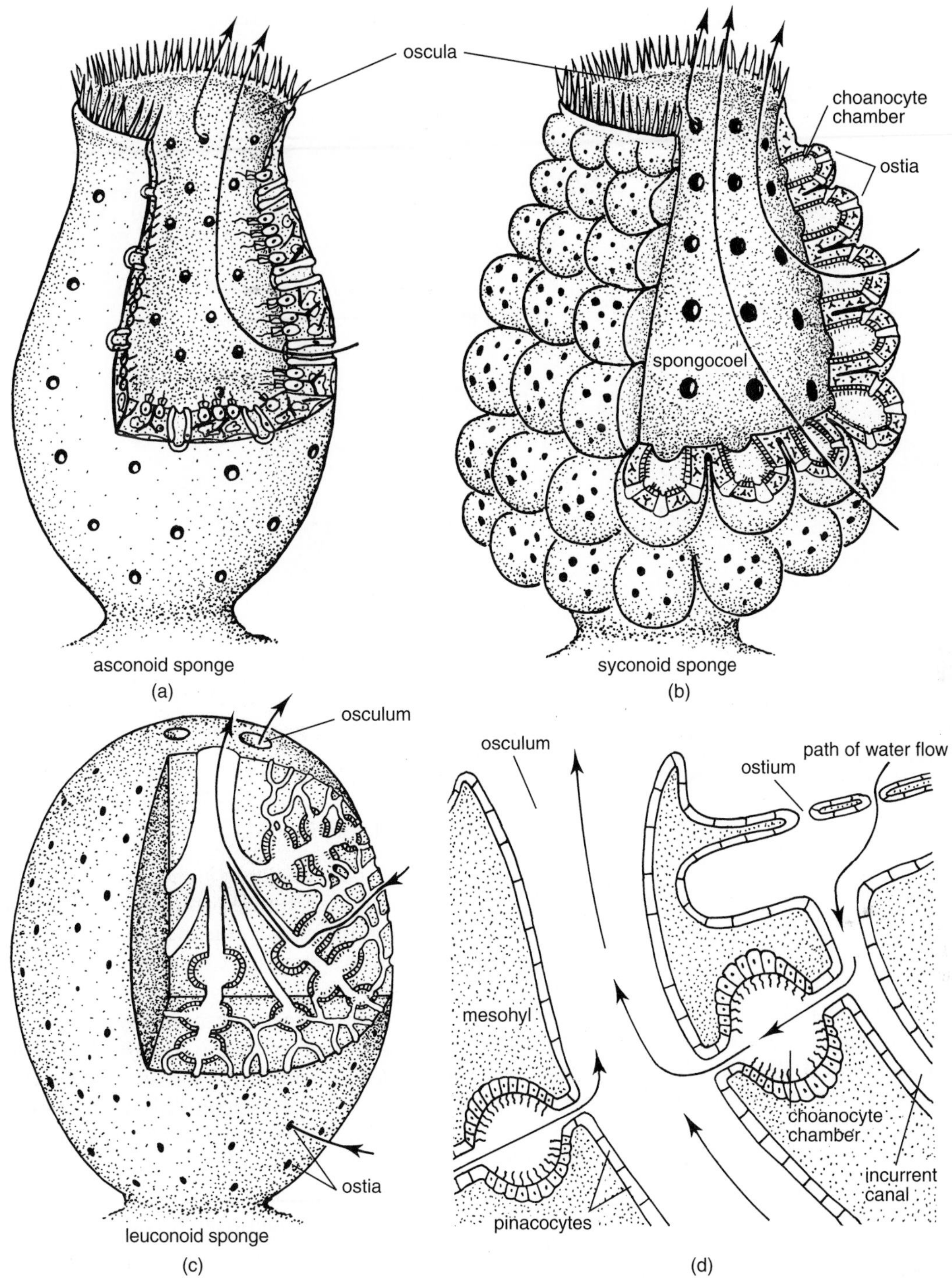

Figure 4.7

Diagrammatic illustrations of the different levels of complexity of sponge architecture. Arrows indicate the direction of water flow: in at the ostia and out at the oscula. (a) Asconoid sponge. (b) Syconoid sponge. (c) Leuconoid sponge. (d) Detail of water circulation in a leuconoid sponge.

(a–c) From Bayer and Owre, *The Free-Living Lower Invertebrates.* Reprinted by permission of the authors. From Johnston and Hildemann, in *The Reticuloendothelial System,* Cohen and Siegel, eds. Copyright © 1982 Plenum Publishing Corporation, New York. Reprinted by permission. (d) Redrawn from Johnston and Hildemann, in *The Reticuloendothelial System,* Cohen and Siegel, eds. Copyright © 1982 Plenum Publishing Corporation, New York. Reprinted by permission.

Class Calcarea

Members of the class Calcarea bear spicules composed only of calcium carbonate ($CaCO_3$). Representatives of all 3 types of construction occur in this class. Indeed, the only living asconoid forms are found among the Calcarea.

Class Demospongiae

Members of the largest class (containing at least 80% of all sponge species), the Demospongiae, are nearly all of leuconoid construction. The supporting spicules and fibers of the Demospongiae may be composed of spongin and/or

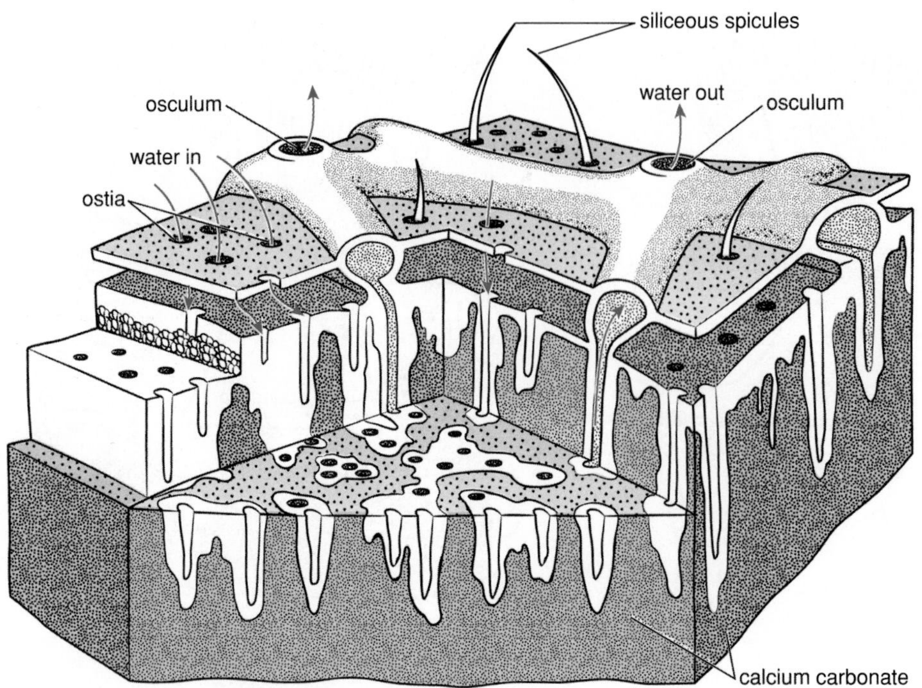

Figure 4.8

Schematic representation of a sclerosponge ("coralline sponge"), collected off the coast of Jamaica. Note the unusually narrow choanocyte chambers and the massive calcareous skeleton.
Based on Willenz and Hartman in *Marine Biology* 103:387–401, 1989.

silica but never of $CaCO_3$. Recent work[14] shows that the skeletal fibers of demosponges (and of the glass sponges to be discussed shortly) also contain chitin. A small number of species possess neither fibers nor spicules. All freshwater sponges (fewer than 300 species) are found in this class. Interestingly, these freshwater species possess contractile vacuoles, which are organelles specialized for eliminating water from cytoplasm; they are found elsewhere only among protozoans (see pp. 40–41).

A bizarre group of deep sea sponges was described in 1995; the spicules of these "cladorhizid" sponges place them convincingly within the Demospongiae, although their morphology and feeding biology are so aberrant, one wonders if they really are sponges! They possess no ostia and no oscula—no internal canal system at all, in fact—and no choanocytes. Remarkably, they feed as carnivores, entrapping small (mostly less than 1 mm) swimming crustaceans on numerous long, thin filaments that cover much of the body surface. The captured prey are gradually surrounded by adjacent epithelial cells and the growth of new filaments, and are digested externally within a few days. Additional nutrition may be provided by symbiotic bacteria, as described elsewhere in this book for bivalved molluscs (p. 238) and "pogonophoran" annelids (p. 304).[15]

14. Ehrlich, H., M. Maldonado, K. D. Spindler, C. Eckert, T. Hanke, R. Born, C. Goebel, P. Simon, S. Heinemann, and H. Worch. 2007. *J. Exp. Zool.* 308B: 347–56.

15. See *Topics for Further Discussion and Investigation,* no. 8.

Class Hexactinellida

Finally, sponges whose bodies are supported entirely by interconnected 6-rayed spicules of silica and chitin are placed in the class Hexactinellida. These sponges, known as the glass sponges, are marvels of structural complexity and symmetry (Fig. 4.9g). The members of some species live in soft sediment, anchored by tufts of spicules, while members of other species live attached to solid substratum. Hexactinellid canal systems may be either syconoid or leuconoid, but hexactinellids stand apart from all other sponges in that the outer layer is syncytial (having many nuclei contained within a single plasma membrane) rather than cellular and lacks contractile elements; thus, there is no pinacoderm layer. The inner, flagellated layer is also syncytial, again setting the glass sponges apart from all others. Some workers argue that the hexactinellids, therefore, belong in a separate phylum, but most are content to place them in a separate subphylum within the Porifera.

Scientists have recently found[16] that the long (5–15 cm), thin (about 50 μm in diameter) silica fibers secreted by these sponges at their base have light-guiding properties superior to those found in commercially produced fiber optic cables, and are less prone to fracture. Not surprisingly, there is growing commercial interest in determining how the sponges fabricate these spicules.

16. Sundar, V. C. *et al.* 2003. Fibre-optical features of a glass sponge. *Nature* 4424: 899–900.

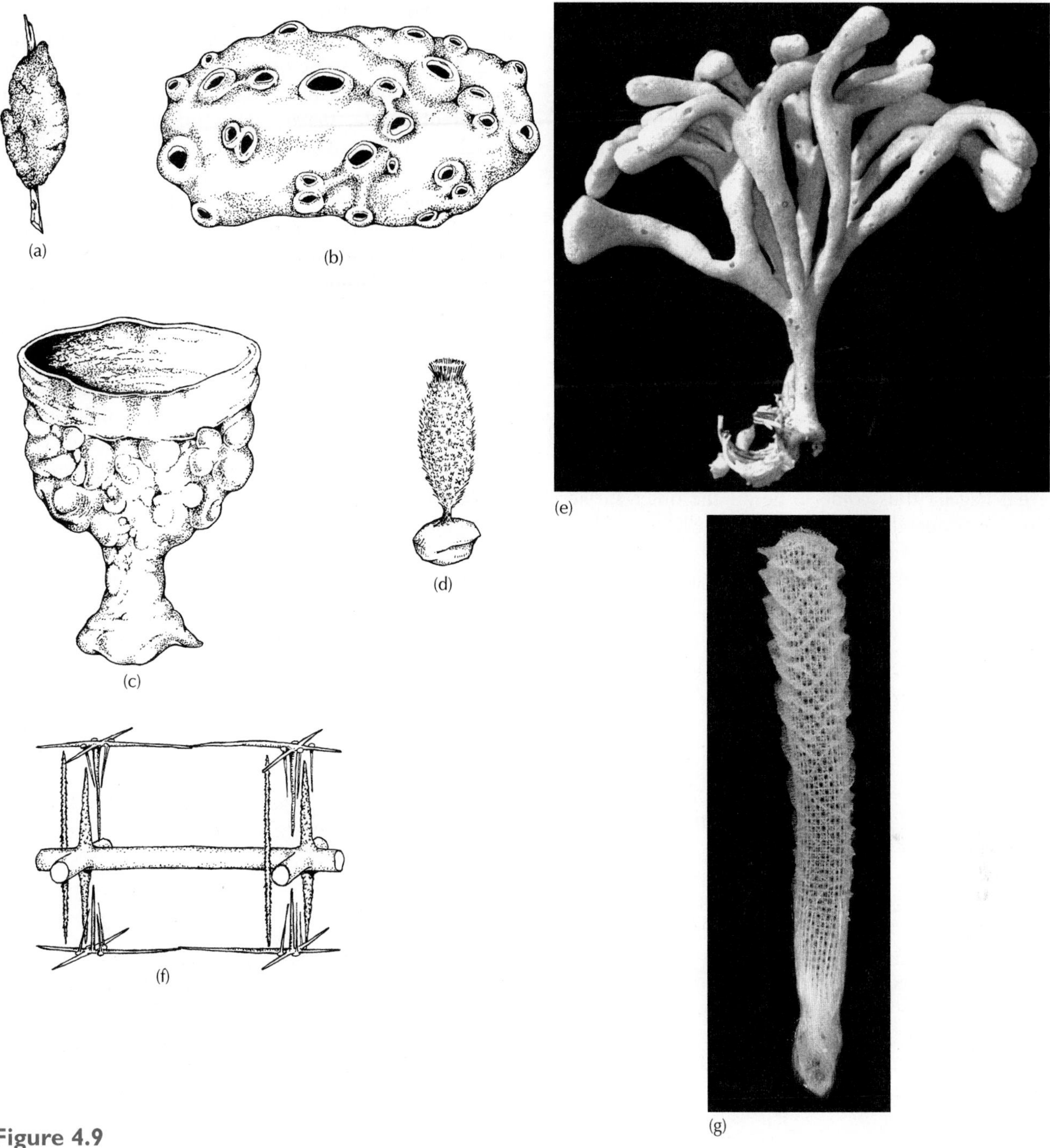

Figure 4.9

Sponge diversity. (a) A freshwater species encrusting a twig. (b) An encrusting marine sponge. (c) Neptune's goblet sponge, *Poterion neptuni.* (d) A syconoid sponge, *Sycon* sp. (e) A fingered sponge, *Haliclona oculata.* (g) *Euplectella* sp., Venus's flower basket. This is a member of the Hexactinellida. The spicules are 6-rayed and composed of silica, as shown in (f).

(a) After Hyman. (b,c) After Pimentel. (d) After MacGinitie and MacGinitie. (e) Courtesy of J. A. Kaandorp, from *Fractal Modelling: Growth and Form in Biology.* Springer-Verlag, Berlin, New York, 1994. (f) Barnes, *Invertebrate Zoology,* 4th ed. Orlando: W. B. Saunders Company, 1980. (g) Courtesy of James Sumich.

Other Features of Poriferan Biology: Reproduction and Development

Sponges reproduce asexually, either by fragmentation or through the production of gemmules or buds, and sexually, through the production of eggs and sperm. Many sponge species are **hermaphroditic,** with a single individual producing both types of gametes. Fertilization and early development are typically internal, a surprising elaboration for what is generally considered a rather unsophisticated animal. In at least some species, choanocytes capture the incoming sperm, dedifferentiate to amoeboid form, and then transport the sperm to the mesohyl, where the eggs are fertilized.

Most sponge species retain (i.e., "brood") the developing embryos for a time, releasing them through the oscula as swimming larvae. A much smaller number of species are oviparous; the newly fertilized eggs (or the gametes themselves) are shed into the seawater, so that embryonic development is completely external.

Development of sponge embryos differs from that of other metazoan embryos. In calcareous sponges and in some demosponges the embryo develops into a hollow blastula (a **coeloblastula;** *coelo* = G: hollow). In some calcareous sponges (including members of the genus *Grantia*), rapidly dividing cells at one end of the embryo soon become flagellated, with the flagella directed into the blastocoel rather than toward the outside of the embryo. The other cells of the embryo divide more slowly and remain unflagellated. The blastocoel opens to the outside in the middle of this group of relatively large, slowly dividing cells, and as development continues the embryo demonstrates a process called **inversion** and turns inside out through this opening (Fig. 4.10a). The internal flagella thus come to lie on the outer surface, where they propel the larva forward (flagellated end leading) once the individual is discharged through the excurrent canal system. Even after turning inside out, the embryo remains only one cell thick, so that inversion is not to be confused with gastrulation.

The hollow, swimming sponge larva just described, flagellated at one end only, is termed an **amphiblastula** (Fig. 4.10a,b). In some other calcareous sponge species, the initially hollow embryo (coeloblastula) becomes solid (a **stereoblastula;** *stereo* = G: solid) as numerous cells detach from the wall of the blastula and completely fill the blastocoel.

In contrast, the embryos of most demosponges develop directly into stereoblastulae, which then differentiate to form extensively flagellated **parenchymella** (also called **parenchymula** larvae) (Fig. 4.10c); each cell bears one flagellum, as in other sponge larvae already described. In some species, particularly those living in freshwater, these parenchymella larvae possess spicules and choanocytes, and even rudimentary canal systems. Hexactinellid larvae also become highly differentiated, possessing both siliceous spicules and choanocyte chambers; only cells in the larval midsection are flagellated, and each cell bears a number of flagella.

Sponge larvae are typically incapable of feeding, and swim for less than 24 hours before metamorphosing. Before losing the ability to swim, the larvae attach to a substrate. During the ensuing process of metamorphosis, cells from various parts of the embryo undergo extensive migrations

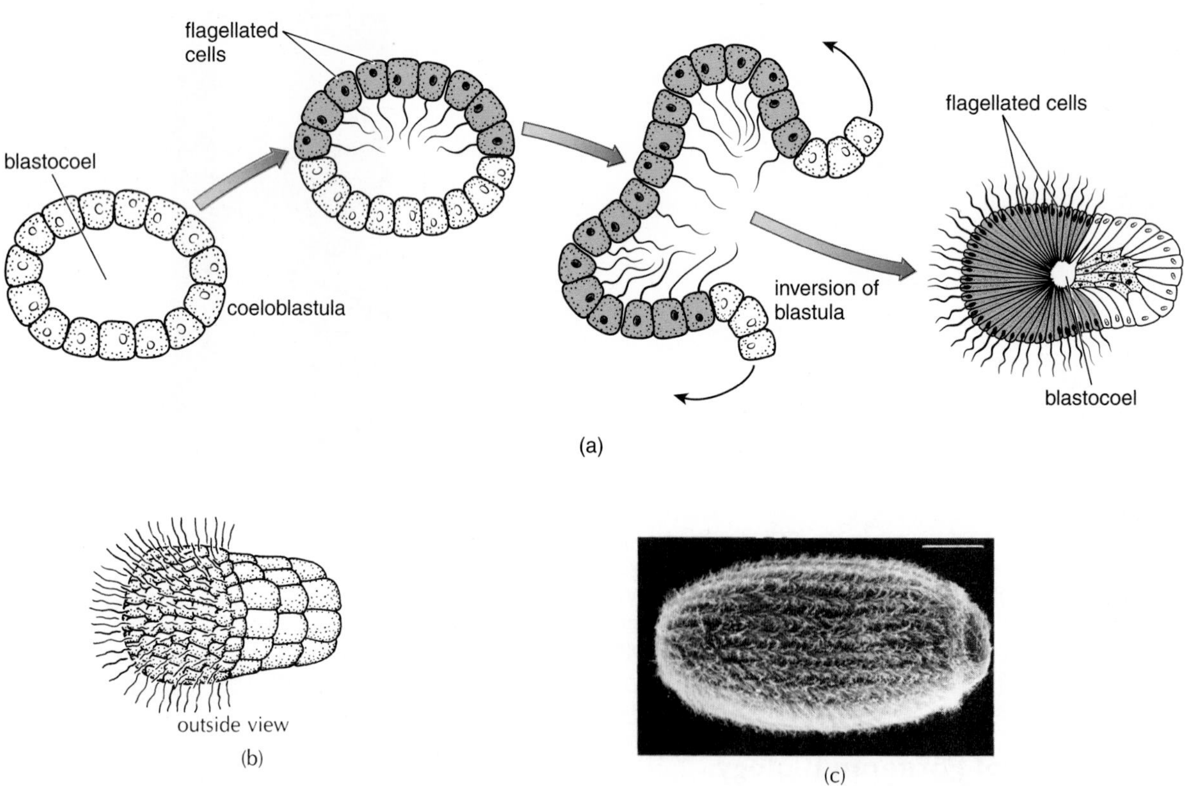

Figure 4.10
(a) Development of amphiblastula larva, seen in cross section. (b) Typical amphiblastula larva, external view. This free-swimming larva will attach to a substrate before undergoing further development. (c) Scanning electron micrograph of a parenchymella larva, *Halichondria* sp. The larva is about 300 μm long.

(a) Based on P. E. Fell 1997. In S. F. Gilbert and A. M. Raunio. *Embryology: Constructing the Organism,* p. 46, Inc. (b) After Hammer. (c) Courtesy of Claus Nielson, from C. Nielson 1987, *Acta Zoologica* 68:205–62, Permission of Royal Swedish Academy of Sciences.

and begin to, or continue to, differentiate to form the future adult sponge. In at least some calcareous and demosponge species, choanocytes develop directly from the flagellated cells of the larva. In some other sponge species, the flagella are degraded and choanocytes then differentiate *de novo* from archaeocytes. In at least one freshwater species, the flagellated cells first dedifferentiate into amoeboid cells, which later redifferentiate into choanocytes.

With respect to their mode of reproduction, sponges are unusual in that both marine and freshwater species have a free-swimming larval stage in their development. Such a free-swimming dispersal stage is often suppressed in the life histories of other freshwater invertebrates, as noted in Chapter 1. How freshwater parenchymella larvae withstand the osmotic stress associated with life in freshwater has not been determined.

Phylum Placozoa

Phylum Placo • zoa
(G: flat plate animal)
plack-ō-zō´-ah

Defining Characteristic: Multicellular, amorphous, mobile, flagellated animals lacking a body cavity, digestive system, and nervous system and composed of 2 layers of epithelial cells

Introduction

Although only a single placozoan species has been described, *Trichoplax adhaerens* (Fig. 4.11), recent findings of surprisingly high amounts of genetic diversity among individuals suggest that there may be more than one species after all. *Trichoplax* has been collected from marine aquariums and from shallow-water marine habitats around the world. Although placozoans have been known since 1883, their biology has been only poorly explored.

General Characteristics

Like sponges, placozoans have no front or back and no right or left, and they lack organs and tissues. There is no digestive system, no nervous system, and no true musculature. Placozoans, like sponges, additionally lack specialized sensory structures and, as with sponges, cells that are artificially dissociated will reaggregate to re-form a functioning animal. However, unlike sponges, placozoans are fully mobile. They are apparently planktonic for part of their lives but are most often seen gliding across hard substrates on thousands of motile flagella, changing shape markedly, amoeba-style, as they travel. The animals rarely get much larger than a few millimeters across in laboratory culture, and individuals collected from the field are even smaller: no more than about 1/10 that size.

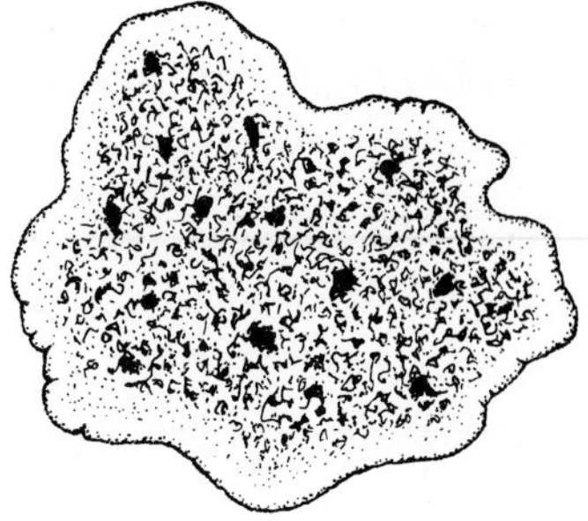

Figure 4.11
The placozoan *Trichoplax adhaerens*. This individual was approximately 0.5 mm in its longest dimension.

Placozoans are flat, with 2 distinct layers of epithelial cells, each layer containing perhaps a thousand or so cells. As with sponges, the epithelial cells lack a basal lamina. The ventral layer is composed of columnar cells, each bearing a single flagellum. Associated glandular cells apparently secrete digestive enzymes beneath the animal as it sits atop the algae and protozoans on which it apparently feeds; digestion seems to be entirely extracellular, as there is no mouth and no sign of phagocytosis. The much thinner, upper layer of the animal bears flagellated cells but no gland cells. In a sense, the upper layer is ectodermal, while the lower layer, because of its involvement in digesting food and absorbing nutrients, is endodermal. Between the upper and lower cell layers is a fluid-filled space containing a dense network of fibrous cells that may be contractile.

Asexual reproduction—by budding, fragmentation, or binary fission—occurs commonly in the laboratory. Also, individuals have no difficulty regenerating pieces that are cut off. Apparently, placozoans also reproduce sexually, although we don't yet know many details. Generally, each individual produces one oocyte. While fertilization has never been seen, some individual placozoans have produced embryos; unfortunately, the embryos have never developed beyond 64 cells in the laboratory.

The relationship between placozoans and other metazoans is obscure. Some suggest that placozoans are related in some way to sponges, while other studies have suggested a closer relationship to cnidarians, a group discussed in Chapter 6. Recent studies of mitochondrial genome sequences[17] place placozoans at the very root of the metazoan tree, that is, as the oldest (most basal) group of metazoans, whereas recent studies of large and small subunit ribosomal RNA gene sequences[18] do not. If placozoans are secondarily simplified from more complex animals, then placozoan evolution must reflect loss of the nervous system, and other degenerative events. Recent molecular data both support and do not support this scenario of placozoans as degenerate cnidarians.

17. Dellaporta, S. L., A. Xu, S. Sagasser, W. Jakob, M. A. Moreno, L. W. Buss, and B. Schierwater. 2006. *Proc. Natl. Acad. Sci. U.S.A.* 103: 8751–56.

18. da Silva, F. B., V. C. Muschner, and S. L. Bonatto. 2007. *Genet. Molec. Biol.* 30: 127–32.

Taxonomic Summary

Phylum Porifera—the sponges
  Class Calcarea
  Class Demospongiae
  Class Hexactinellida—the glass sponges
Phylum Placozoa

Topics for Further Discussion and Investigation

1. How are sponge spicules secreted by sclerocytes?

Dendy, A. 1926. Origin, growth, and arrangement of sponge spicules. *Q. J. Microsc. Sci.* 70:1.

Ilan, M., J. Aizenberg, and O. Gilor. 1996. Dynamics and growth patterns of calcareous sponge spicules. *Proc. Royal Soc. London B* 263:133.

Ledger, P. W., and W. C. Jones. 1977. Spicule formation in the calcareous sponge *Sycon ciliatum. Cell Tissue Res.* 181:553.

Uriz, M. J. 2006. Mineral skeletogenesis in sponges. *Can. J. Zool.* 84: 322–56.

2. Investigate the environmental control of hatching from sponge gemmules.

Benfey, T. J., and H. M. Reiswig. 1982. Temperature, pH, and photoperiod effects upon gemmule hatching in the freshwater sponge, *Ephydatia mülleri* (Porifera, Spongillidae). *J. Exp. Zool.* 221:13.

Fell, P. E. 1992. Salinity tolerance of the gemmules of *Eunapius fragilis* (Leidy) and the inhibition of germination by various salts. *Hydrobiol.* 242: 33–39.

3. How does sponge architecture contribute to increased flow of water through a sponge and to increased feeding efficiency?

LaBarbera, M., and S. Vogel. 1982. The design of fluid transport systems in organisms. *Amer. Scient.* 70:54.

Reiswig, H. M. 1975. The aquiferous systems of three marine Demospongiae. *J. Morphol.* 145:493.

Vogel, S. 1974. Current-induced flow through the sponge, *Halichondria. Biol. Bull.* 147:443.

Vogel, S. 1978. Organisms that capture currents. *Sci. Amer.* 234:108.

4. No one has ever demonstrated nerve tissue in any poriferan. Nevertheless, in some species there is evidence of cooperation among different areas of the sponge, resulting in locomotion and in the regulation of water flow through the animal. What is the evidence for such internal coordination, and through what mechanisms might it be accomplished in the absence of nerve cells?

Lawn, I. D., G. O. Mackie, and G. Silver. 1981. Conduction system in a sponge. *Science* 211:1169.

Leys, S. P., and R. W. Meech. 2006. Physiology of coordination in sponges. *Can. J. Zool.* 84: 288–306.

Leys, S. P., G. O. Mackie, and R. W. Meech. 1999. Impulse conduction in a sponge. *J. Exp. Biol.* 202:1139.

Mackie, G. O., I. D. Lawn, and M. Pavans de Deccaty. 1983. Studies on hexactinellid sponges. II. Excitability conduction and coordinated responses in *Rhabdocalyptus dawsoni* (Lambe, 1873). *Phil. Trans. Royal Soc. London* B 301:401.

Nickel, M. 2004. Kinetics and rhythm of body contractions in the sponge *Tethya wilhelma* (Porifera: Demospongiae). *J. Exp. Biol.* 207: 4515–24.

Nickel, M. 2006. Like a "rolling stone": Quantitative analysis of the body movement and skeletal dynamics of the sponge *Tethya wilhelma. J. Exp. Biol.* 209: 2839–46.

5. How do dissociated sponge cells recognize each other in the reaggregation process?

Galtsoff, P. S. 1925. Regeneration after dissociation (an experimental study on sponges). I. Behavior of dissociated cells of *Microciona prolifera* under normal and altered conditions. *J. Exp. Zool.* 42:183.

Humphreys, T. 1963. Chemical dissolution and in vitro reconstruction of sponge cell adhesions. I. Isolation of and functional demonstration of the components involved. *Devel. Biol.* 8:27.

McClay, D. R. 1974. Cell aggregation: Properties of cell surface factors from five species of sponge. *J. Exp. Zool.* 188:89.

Misevic G. N., Y. Guerardel, L. T. Sumanovski, M. C. Slomianny, M. Demarty, C. Ripoll, Y. Karamanos, E. Maes, O. Popescu, and G. Strecker. 2004. Molecular recognition between glyconectins as an adhesion self-assembly pathway to multicellularity. *J. Biol. Chem.* 279: 15579–90.

Spiegel, M. 1954. The role of specific surface antigens in cell adhesion. I. The reaggregation of sponge cells. *Biol. Bull.* 107:130.

Wilson, H. V., and J. T. Penney. 1930. The regeneration of sponges (*Microciona*) from dissociated cells. *J. Exp. Zool.* 56:73.

6. Sponges compete for space with a variety of organisms, including macroalgae, corals, bryozoans, and other sponges. In addition, sponges must be subject to considerable predation, especially by gastropods, crustaceans, and fishes. Finally, the great surface area of sponges would seem to make them ideal for the settlement and growth of other sedentary organisms. Lacking specialized organs and behaviors, sponges compete and protect themselves by chemical means. Discuss the evidence for chemical defense among sponges.

Becerro, M. A., V. J. Paul, and J. Starmer. 1998. Intracolonial variation in chemical defenses of the sponge *Cacospongia* sp. and its consequences on generalist fish predators and the specialist nudibranch predator *Glossodoris pallida. Marine Ecol. Progr. Ser.* 168:187.

Chanas, B., and J. R. Pawlik. 1996. Does the skeleton of a sponge provide a defense against predatory reef fish? *Oecologia* 107:225.

Furrow, F. B., C. D. Amsler, J. B. McClintock, and B. J. Baker. 2003. Surface sequestration of chemical feeding deterrents in the Antarctic sponge *Latrunculia apicalis* as an optimal defense against sea star spongivory. *Marine Biol.* 143:443.

Hill, M. S., N. A. Lopez, and K. A. Young. 2005. Anti-predator defenses in western North Atlantic sponges with evidence of enhanced defense through interactions between spicules and chemicals. *Mar. Ecol. Progr. Ser.* 291: 93–102.

Lee, O. O., L. H. Yang, X. C. Li, J. R. Pawlik, and P. Y. Qian. 2007. Surface bacterial community, fatty acid profile, and antifouling activity of two congeneric sponges from Hong Kong and the Bahamas. *Mar. Ecol. Progr. Ser.* 339: 25–40.

Marin, A., M. D. López, M A. Esteban, J. Meseguer, J. Muñoz, and A. Fontana. 1998. Anatomical and ultrastructural studies of chemical defense in the sponge *Dysidea fragilis. Marine Biol.* 131:639.

McClintock, J. B. 1987. Investigation of the relationship between invertebrate predation and biochemical composition, energy content, spicule armament and toxicity of benthic sponges at McMurdo Sound, Antarctica. *Marine Biol.* 94:479.

Porter, J. W., and N. M. Targett. 1988. Allelochemical interactions between sponges and corals. *Biol. Bull.* 175:230.

Swearingen, D. C. III, and J. R. Pawlik. 1998. Variability in the chemical defense of the sponge *Chondrilla nucula* against predatory reef fishes. *Marine Biol.* 131:619.

Uriz, M. J., X. Turon, M. A. Becerro, and J. Galera. 1996. Feeding deterrence in sponges. The role of toxicity, physical defenses, energetic contents, and life-history stage. *J. Exp. Marine Biol. Ecol.* 205:187.

7. Describe the immune recognition responses of sponges that come into contact with other sponges.

Amano, S. 1990. Self and non-self recognition in a calcareous sponge, *Leucandra abratsbo. Biol. Bull.* 179:272.

Ilan, M., and Y. Loya. 1990. Ontogenetic variation in sponge histocompatibility responses. *Biol. Bull.* 179:279.

Jokiel, P. L., and C. M. Bigger. 1994. Aspects of histocompatibility and regeneration in the solitary reef coral *Fungia scutaria. Biol. Bull.* 186:72.

McGhee, K. E. 2006. The importance of life-history stage and individual variation in the allorecognition system of a marine sponge. *J. Exp. Mar. Biol. Ecol.* 333:241–50.

Müller, W. E. G., and I. M. Müller. 2003. Origin of the metazoan immune system: Identification of the molecules and their functions in sponges. *Integr. Comp. Biol.* 43:281.

Van der Vyver, G., S. Holvoet, and P. DeWint. 1990. Variability of the immune response in freshwater sponges. *J. Exp. Zool.* 254:215.

8. How convincing is the evidence that cladorhizid sponges supplement their carnivorous habit with nutrients provided by methane-oxidizing bacterial symbionts?

Vacelet, J., A. Fiala-Médioni, C. R. Fisher, and N. Boury-Esnault. 1996. Symbiosis between methane-oxidizing bacteria and a deep-sea carnivorous cladorhizid sponge. *Marine Ecol. Progr. Ser.* 145:77.

9. Based upon your knowledge of sponge biology and the properties of air and water, why are there no terrestrial sponges?

10. Because of the large volumes of water passing through their bodies daily, sponges may be exposed to substantial amounts of dissolved toxic chemicals. Discuss the potential use of sponges as monitors of chemical pollutant stress in marine and freshwater ecosystems.

Batel, R., N. Bihari, B. Rinkevich, J. Dapper, H. Schäcke, H. C. Schröder, and W. E. G. Müller. 1993. Modulation of organotin-induced apoptosis by the water pollutant methyl mercury in a human lymphoblastoid tumor cell line and a marine sponge. *Marine Ecol. Progr. Ser.* 93:245.

Müller, W. E. G., C. Koziol, B. Kurelec, J. Dapper, R. Batel, and B. Rinkevich. 1995. Combinatory effects of temperature stress and nonionic organic pollutants on stress protein (hsp70) gene expression in the fresh water sponge *Ephydatia fluviatilis. Environ. Toxicol. Chem.* 14:1203.

Schröder, H. C., S. M. Efremova, B. A. Margulis, I. V. Guzhova, V. B. Itskovich, and W. E. G. Müller. 2006. Stress response in Baikalian sponges exposed to pollutants. *Hydrobiol.* 568 (Suppl. 1): 277–87.

Schröder, H. C., K. Shostak, V. Gamulin, M. Lacorn, A. Skorokhod, V. Kavsan, and W. E. G. Müller. 2000. Purification, cDNA cloning and expression of a cadmium-inducible cysteine-rich metallothionein-like protein from the marine sponge *Suberites domuncula. Marine Ecol. Progr. Ser.* 200:149.

Taxonomic Detail

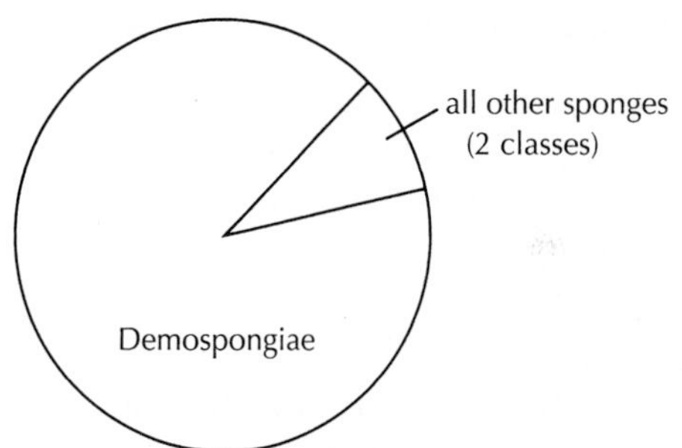

Subkingdom Parazoa

Phylum Porifera

Subphylum Cellularia (all members possess distinct choanocytes)

Class Demospongiae

This class contains more than 90% of all existing sponge species, including all freshwater species. Support elements are never calcareous; they are instead composed of silica, spongin, or both, apparently along with chitin. All members are of leuconoid construction. Species show a variety of forms—thin and encrusting, erect and branching, multilobed, spherical, tubular—and a single individual may exceed 2 m in diameter. Sixty-five families.

Family Clionidae

Cliona. Boring sponges: Marine sponges that excavate burrows in calcareous material, such as mollusc shells and coral. Etching chemicals are released at the tips of specialized surface cells; these secretions dissolve only the edges of calcareous chips, which then fall to the seafloor, contributing up to 40% of the reef sediment in some places. Species in this family are distributed from shallow water to depths exceeding 2100 m.

Family Spongiidae
Spongia, Hippospongia. All commercial sponges come from these two genera. All species in this family are marine. They live in all waters, from the tropics to the arctic and antarctic.

Family Haliclonidae
Haliclona. Members of this family are among the most commonly found and widely distributed of all shallow-water sponges, although some species reach depths of nearly 2500 m.

Family Halichondriidae
Halichondria. These encrusting, marine sponges are common in shallow water.

Family Clathriidae
Microciona. The first experiments on sponge regeneration were performed on *Microciona prolifera* during the early 1900s.

Family Callyspongiidae
Callyspongia. The species in this group are common in shallow tropical oceans. Species may be upright, encrusting, branching, tubular, or vase-shaped; some are massive.

Family Spongillidae
Spongilla, Eunapius, Ephydatia, Heteromeyenia. This widespread group contains most of the approximately 300 freshwater sponge species, which usually encrust solid surfaces in ponds, streams, rivers, and lakes. An individual sponge may exceed 1 m in diameter. A few species are found in brackish (i.e., slightly salty) waters. *Ephydatia* is the best-studied sponge genus in the world.

Family Lubomirskiidae
This group contains freshwater species restricted to Lake Baikal in Siberia. None of these species produce gemmules.

Family Mycalidae
Mycale

Family Plakinidae
Corticium, Plakina, Plakortis. These unusual sponges have no supporting skeleton, lacking both spicules and spongin fibers. Along with one other family (the Oscarellidae), they are included in a separate subclass of the Demospongiae, the Homoscleromorpha.

Family Cladorhizidae
Asbestopluma, Cladorhiza. These are mostly deep-sea sponges, living to depths as great as 8840 m, although one species has now been found in a shallow-water Mediterranean cave. Unlike other sponges, these animals lack choanocytes and internal water canal systems. They apparently feed as carnivores, passively entrapping small crustaceans that swim by, and may also gain nutrients from the activities of symbiotic bacteria.

Class Calcarea
All species are marine and have spicules exclusively composed of calcium carbonate. Species in this class may have asconoid, syconoid, or leuconoid grades of construction. Sixteen families.

Family Leucosoleniidae
Leucosolenia. All members of this family are marine, asconoid, and contained in a single genus. Species are found from the intertidal zone to depths exceeding 2400 m.

Family Grantiidae
Grantia (= *Scypha*). Species are all marine. Members of this family are distributed from the intertidal zone to depths of about 2200 m.

The Sclerosponges
Acanthochaetetes, Astrosclera, Stromatospongia. Most sclerosponges live in deep water on coral reefs or in caves, crevices, or tunnels within the reefs. All species are of leuconoid complexity. The body is supported by a thick layer of calcium carbonate in addition to spicules of silica and fibers of spongin, giving rise to the common name "coralline sponges." Recent morphological and molecular analyses suggest that sclerosponge species do not have a common, unique origin; members of the former class Sclerospongiae are currently assigned either to the Calcarea or to the Demospongiae.

Subphylum Symplasma (epithelial and "choanocyte" tissues are syncytial)

Class Hexactinellida—the glass sponges
Skeletal supports of all species are composed of 6-sided spicules of silica and chitin; many spicules are fused to form long, thin, glasslike strands. Sixteen families.

Family Euplectellidae
Euplectella—Venus's flower basket. Individual sponges often harbor a single pair of shrimp, one of each sex, that enter the sponge when small and reach reproductive adulthood imprisoned together within the spongocoel. The species in this family are found at depths ranging from 100 m to over 5200 m.

General References about the Sponges and Placozoans

Bergquist, P. R. 1978. *Sponges.* Berkeley, Calif.: Univ. Calif. Press.

Fell, P. E. 1997. Poriferans, the sponges. In *Embryology: Reconstructing the Organism,* edited by S. F. Gilbert and A. M. Raunio. Sunderland, MA: Sinauer Associates, Inc., pp. 39–54.

Harrison, F. W., and J. A. Westfall, eds. 1991. *Microscopic Anatomy of the Invertebrates, Vol. 2. Placozoa, Porifera, Cnidaria, and Ctenophora.* New York: Wiley-Liss, pp. 13–27 (placozoans), 29–89 (sponges).

Hyman, L. 1940. *The Invertebrates, Vol. 1. Protozoa through Ctenophora.* New York: McGraw-Hill.

Leys, S. P., G. O. Mackie, and H. M. Reiswig. 2007. The biology of glass. *Adv. Mar. Biol.* 52: 1–145.

Miller, D. J., and E. E. Ball. 2005. Animal evolution: the enigmatic phylum Placozoa revisited. *Current Biology* 15: R26–R28.

Nichols, S., and G. Wörheide. 2005. Sponges: New views of old animals. *Integr. Comp. Biol.* 45:333–34. (This is the lead paper for a series of sponge-related papers presented at a symposium.)

Simpson, T. L. 1984. *The Cell Biology of Sponges.* New York: Springer-Verlag.

Thorpe, J. H., and A. P. Covich. 2001. *Ecology and Classification of North American Freshwater Invertebrates,* 2nd ed. New York: Academic Press.

Wulff, J. L. 2006. Ecological interactions of marine sponges. *Can. J. Zool.* 84:146–66.

Search the Web

1. www.ucmp.berkeley.edu/porifera/porifera.html

   Information about the anatomy, ecology, life cycles, systematics, and fossil record of sponges; includes excellent color slides of many sponge species.

2. www.mbl.edu

   Choose "Biological Bulletin" under Quick Links, and then "Other Biological Bulletin Publications." Next select "Keys to Marine Invertebrates of the Woods Hole Region," and then choose "Porifera." This brings up a taxonomic key to the marine sponges around Woods Hole, Massachusetts; includes a glossary of sponge terminology and summarizes techniques for identifying sponges.

3. http://biodidac.bio.uottawa.ca

   Choose "Organismal Biology," "Animalia," and then "Porifera" or "Placozoa" for photographs and drawings, including sectional material.

4. http://www.ucmp.berkeley.edu/phyla/placozoa/placozoa.html

   This site provides information on placozoans, images, and links to other sites.

5. http://www.tolweb.org/tree/

   Search on the terms "Placozoa" and "Porifera" to find relevant information, images, and links to other sites.

5

Introduction to the Hydrostatic Skeleton

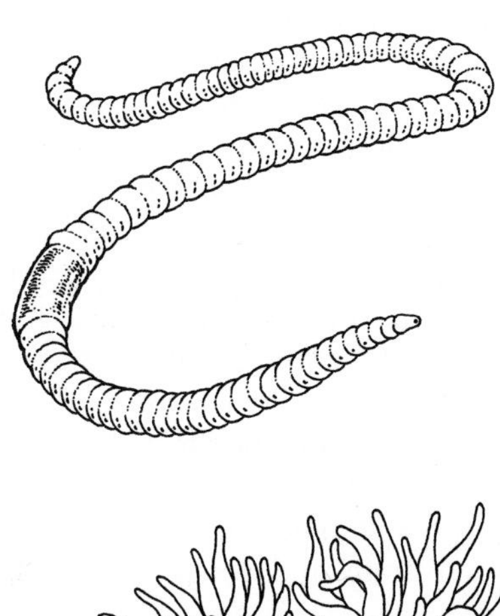

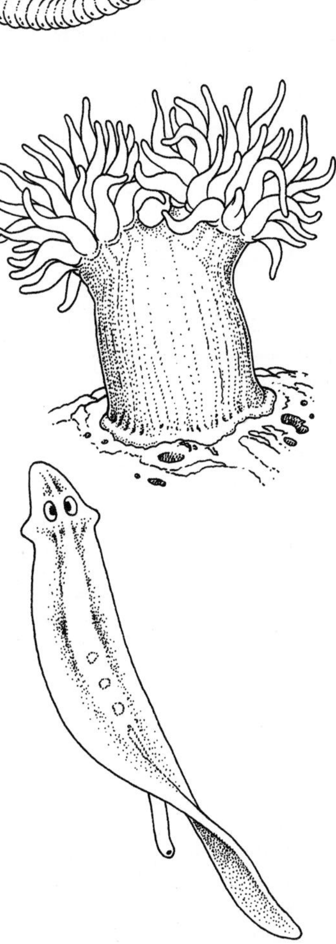

The word *skeleton* invariably conjures up an image of the articulated bones hanging in the corner of the high school biology classroom, or perhaps in the corner of the general practitioner's office. However, jointed bones are only one form of skeletal system. Nearly all multicellular animals, even invertebrates, require a skeleton for movement. The only exceptions to this rule are those small, aquatic metazoans that may move exclusively by cilia. A functional definition of the word *skeleton* is

> A solid or fluid system permitting muscles to be stretched back to their original length following a contraction. Such a system may or may not have protective and supportive functions as well.

A skeletal system is essential simply because muscles are capable of only 2 of the 3 activities required for repeated movements: Muscles can shorten or relax, but they cannot actively extend themselves. To bend your arm at the elbow, one set of muscles, the biceps, must contract. This contraction of the biceps not only causes your arm to bend at the elbow, but also serves to stretch another muscle in your arm, the triceps (Fig. 5.1). The triceps can now contract, making it possible for you to reextend your arm. Reextension of your arm, in turn, serves to stretch the biceps. The bones in your arm have functioned in these movements as the vehicle through which the triceps and biceps take turns stretching each other back to precontraction length; that is, the muscles **antagonize** (act against) each other, making controlled, repeatable movement possible. In vertebrates, the mutual antagonism of muscles is mediated through a solid skeleton. A rigid skeletal system is essential in a terrestrial environment, in part because the skeleton must also serve to support the body in a nonsupportive medium (see Chapter 1). Aquatic organisms are supported by the medium in which they live, so a rigid skeletal system is not required.

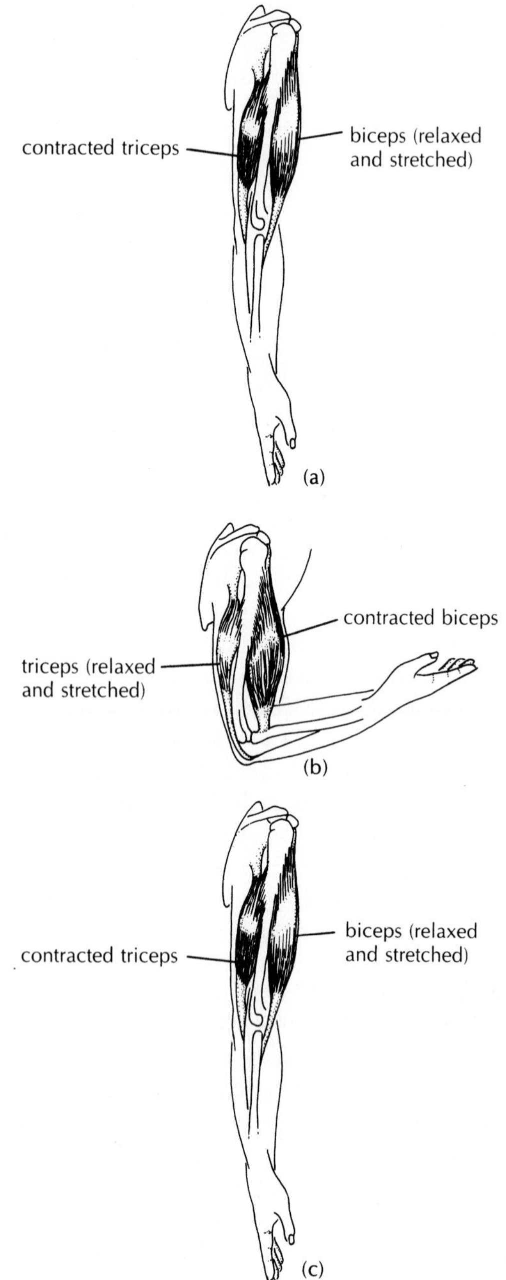

Figure 5.1
Antagonistic interaction between the biceps and triceps in the human arm. Contraction of the biceps (a) results not only in movement of the arm (b), but also in stretching of the opposing muscle, the triceps. Contraction of the triceps then returns the arm to its initial position (c) and stretches the biceps. In vertebrates, muscle pairs antagonize each other through a rigid skeleton, which is internal and jointed.

In fact, fluid can serve as the vehicle through which sets of muscles interact. Such **hydrostatic skeletons** are common among invertebrates. The basic hydrostatic skeleton requires:

1. the presence of a cavity housing an incompressible fluid that transmits pressure changes uniformly in all directions;
2. that this cavity be surrounded by a flexible outer body membrane, so that the outer body wall can be deformed;
3. that the volume of fluid in the cavity remain constant; and
4. that the animal be capable of forming temporary attachments to the substrate, if progressive locomotion is to occur on or within a substrate.

Let us assume that these 4 attributes are met in the hypothetical organism shown in Figure 5.2. This cylindrical being is equipped with **longitudinal muscles** only. If this animal attaches at point *X* (as shown in Fig. 5.2a) and then contracts its musculature, the increase in internal hydrostatic pressure will deform the outer body wall, resulting in a shorter, fatter animal (Fig. 5.2b). This animal can regain its initial shape only if it is surrounded by a stiff, elastic covering that will spring back to its original shape upon relaxation of the longitudinal musculature. Such a stiff covering could be difficult to deform in the first place and is not commonly encountered among unjointed invertebrates.

Instead, we add a second set of muscles (**circular muscles**) to our hypothetical animal. Forward locomotion then results from the series of contractions illustrated in Figure 5.3. In (a), the circular muscles are contracted, and the longitudinal muscles are stretched. The longitudinal muscles now contract while the circular muscles relax, producing the shorter, wider animal of Figure 5.3b (and generating powerful radial forces in the process). In (b), the animal releases its anterior attachment to the substrate and forms a new temporary

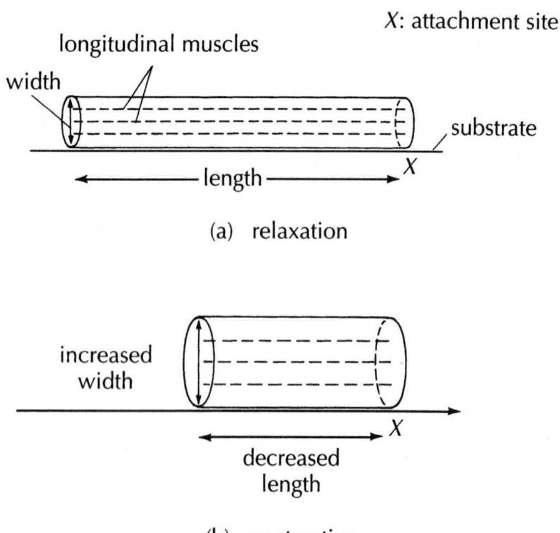

Figure 5.2
(a,b) Shape changes possible in a worm-like organism equipped with only longitudinal muscles. Because the fluid volume is constant, a change in the width of the hypothetical animal must be compensated for by a change in length, brought about by the increase in internal hydrostatic pressure during muscle contraction. Similarly, a shortening of the worm is accompanied by an increase in width, as seen in (b).

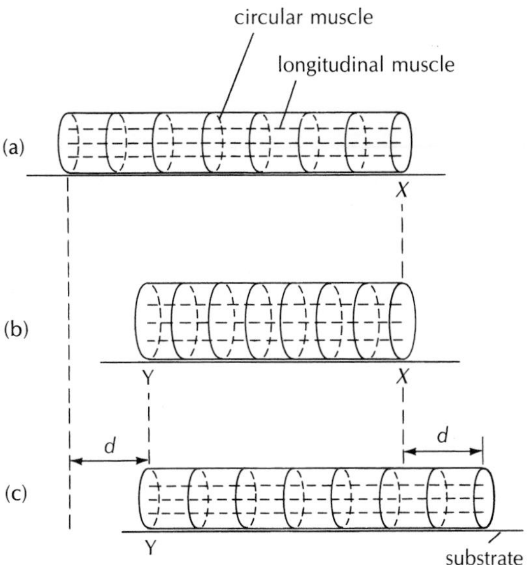

Figure 5.3

Locomotion in a hypothetical worm possessing both circular and longitudinal muscles and a continuous, fluid-filled internal body cavity of constant volume. (a) The animal attaches to the substrate at point *X*, relaxes its circular muscles, and contracts its longitudinal muscles. The contraction increases the hydrostatic pressure within the body cavity because the volume of the cavity cannot be decreased and the fluid within the cavity is incompressible. This pressure is relieved by permitting the circular muscles to stretch. (b) The animal releases its anterior attachment, forms an attachment at point *Y*, and relaxes its longitudinal muscles. (c) The circular muscles have contracted, causing another increase in pressure, which, in turn, is relieved as the longitudinal muscles are extended. The animal has now advanced by distance *d*, and regained its initial body shape.

attachment posteriorly. The circular muscles then contract while the longitudinal muscles relax, thrusting the animal's anterior end forward. In Figure 5.3c, we see that the animal has advanced by a distance *d* and has regained its initial shape, ready to repeat the cycle of muscular contractions.

From this discussion, it is obvious that one addition must be made to the previous list of requirements for a functional hydrostatic skeleton:

5. the presence of a deformable but elastic covering or the presence of at least 2 sets of muscles that can act against each other.

Clearly, the skeletal system in our hypothetical organism is a fluid. Temporary increases in internal pressure are caused by the contraction of one set of muscles, and this temporary pressure increase elongates another set of muscles. I emphasize that the internal pressure increase is temporary; elongation of the opposing set of muscles relieves the pressure. The essentially incompressible fluid thus makes possible the mutual antagonism of the 2 sets of muscles, resulting in repeatable locomotory movements. Hydrostatic skeletons play a role in the movements made by representatives of nearly every animal phylum. Researchers at a number of institutions are currently trying to develop flexible, shape-changing robots based on these principles.

Topics for Further Discussion and Investigation

1. Which features of a hydrostatic skeleton do sponges (Chapter 4) possess? Which features do sponges lack?

2. Fluid-rich deformable cells can act as hydrostatic skeletal systems. Examples are the deformable muscle cells of squid tentacles ("muscular hydrostats," Chapter 12) and the parenchyma of turbellarian flatworms (Chapter 8). What sequence of muscle contractions and relaxations is likely to be involved in (a) elongating a squid tentacle, (b) mediating flatworm locomotion via pedal waves, and (c) flatworm locomotion via "looping"?

6

The Cnidarians

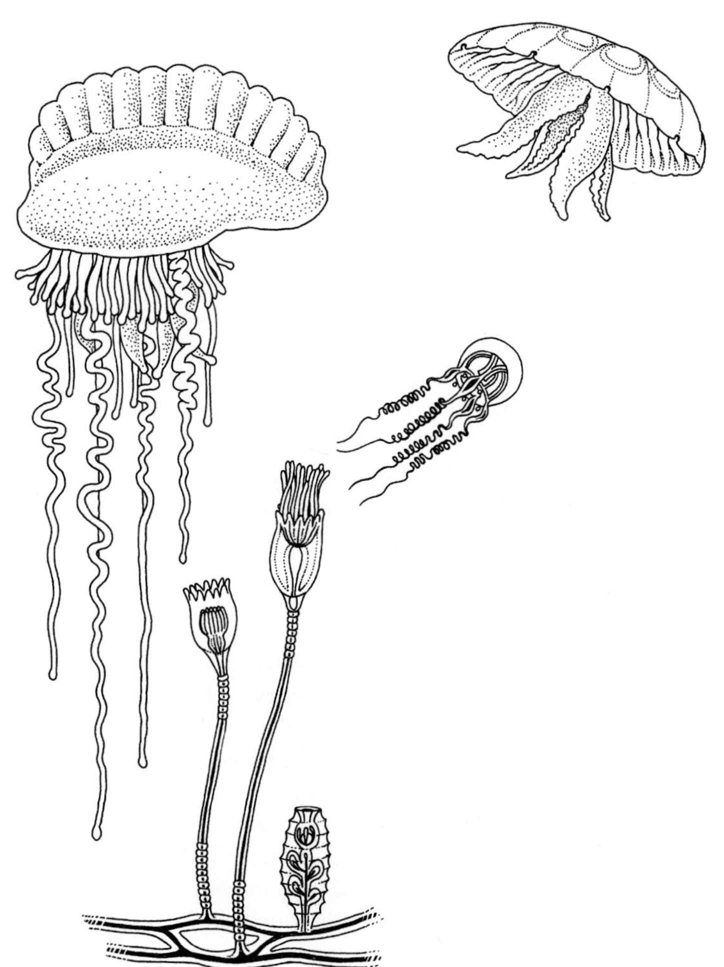

Introduction and General Characteristics

Phylum Cnidaria (= Coelenterata)
(Greek: a stinging thread [G: hollow gut])
nī-dare′-ē-ah (sih-len′-ter-ah′-tah)

Defining Characteristics:[1] 1) Secretion of complex intracellular organelles called cnidae (nematocysts); 2) planula larvae in the life cycle

The phylum Cnidaria contains over 11,000 species, including such animals as the sea anemones, corals, jellyfish, freshwater *Hydra,* and the Portuguese man-of-war. Only about 0.2% of the species live in freshwater; the rest are marine. Two strikingly different body plans are found among cnidarians: a **medusa** form, which resembles a gelatinous saucer or upside-down cup and generally swims (Fig. 6.5); and a **polyp** form, which has a tubular body and is generally stationary (Fig. 6.13). In many species, each body form is present in a different part of the life cycle, and in some species both are represented simultaneously in one individual. Many people have representatives of the polyp stage in their homes without knowing it: Colonies of these marine species (hydrozoans, see p. 110) are dried, dyed, and sold commercially as "air ferns." No wonder they never need to be watered!

Despite these organisms' structural and functional diversity, there is no doubt regarding their membership in a single phylum. Cnidarians have a basic radial symmetry and possess only 2 layers of living tissue (the epidermis and the gastrodermis). All cnidarians possess a gelatinous layer, the **mesoglea,** located between the epidermis and the gastrodermis. Although the mesoglea is itself nonliving, it may

1. Characteristics distinguishing the members of this phylum from members of other phyla.

contain living cells derived from embryonic ectoderm; amoebocytes in the mesoglea probably play roles in digestion, nutrient transport and storage, wound repair, and antibacterial defense. All cnidarians have tentacles surrounding the mouth and only a single opening to the digestive system. All cnidarians secrete cnidae.

Cnidae[2] (G: a nettle, a stinging thread), unique to the members of this phylum, are remarkable organelles secreted within cells called **cnidoblasts** (or **nematoblasts**) and discharged with explosive force for a variety of functions. Of the 3 major categories of cnidae, **nematocysts** (literally, "thread-bags") are the most widespread and the best studied. Over 30 types of nematocyst have been described; many different types often occur within a single individual. Cnidae are among the most complex intracellular secretion products known.

Each cnida consists of a rounded, proteinaceous capsule, with an opening at one end that is often occluded by a hinged operculum. Within the sac is a long, hollow, coiled tube. During discharge, the hollow tube shoots out explosively from the sac, turning inside out as it goes (Fig. 6.1). The entire process requires only about 3 ms (milliseconds). Discharge is triggered by a combination of chemical and tactile stimulation, generally perceived through a cluster of modified cilia (the **cnidocil**) that projects from the cnidoblast and by surface chemoreceptors on specialized nearby cells.[3] Each cnida can be discharged only once.

The primary force behind the actual expulsion of the cnidal filament is osmotic pressure, although the exact firing mechanism remains uncertain. One hypothesis is that discharge results from a sudden, dramatic increase in osmotic concentration within the capsular fluid. An alternative hypothesis is that the osmotic concentration is high at all times, the filament simply discharging when the capsule operculum is somehow opened. Perhaps different types of cnidae are operated by different mechanisms.

Within a given individual, cnidae may be specialized for wrapping around small objects, sticking to surfaces, penetrating surfaces, or secreting proteinaceous toxins, some of which are among the most deadly toxins known. Cnidae function in food collection, defense, and, to some extent, locomotion. They are especially abundant on the feeding tentacles of all species and within the digestive cavity of some species. The functional significance of cnidae lies in their great number per square millimeter of body surface, rather than in their individual size; cnidal capsules rarely exceed 50 μm (micrometers) in diameter and none exceed 100 μm. Cnidal morphology is often important in making species identifications.

2. In some treatments, "cnidae" and "nematocysts" are equivalent terms; it is becoming more common, however, to use the term "nematocyst" only in reference to the most widely distributed and best-studied of the 3 categories of cnidae.

3. See *Topics for Further Discussion and Investigation,* nos. 6–8, at the end of the chapter.

For those contemplating reincarnation, a major drawback to life as a cnidarian would seem to be the absence of an anus. All undigested food material passes out through the same opening through which the food enters: the mouth. This is not particularly appetizing, from the human point of view, but the shortcomings of life without an anus are not merely aesthetic. The sequential disassembly of particulate food material that occurs in an open-ended tubular gut is not possible in the cnidarian digestive system and, indeed, the animal must expel the undigested remains of one meal before it can ingest more food. Moreover, the animal's movements are generally accompanied by physical distortion of the digestive cavity, including partial or complete expulsion of the contained fluid. Extensive movement is therefore not conducive to leisurely, thorough digestion. Finally, gonadal development often takes place within the digestive cavity, and the gametes or embryos must be released into this cavity before being expelled to the exterior through the mouth.

Cnidarians are primarily carnivorous, although some soft-coral species will also eat phytoplankton. In many species, individuals obtain additional nutrients through the photosynthesizing activities of unicellular algae living symbiotically in their tissues. In particular, endosymbiotic algae characterize all reef-building (**hermatypic**) corals.[4]

In contrast to members of the Porifera and Placozoa (Chapter 4), cnidarians possess *bona fide* nerves and muscles. However, they have no central nervous system. Their nervous system consists instead of a network of nerve cells (neurons) and their processes (neurites), which generally synapse on one another repeatedly before terminating at a neuromuscular junction (Fig. 6.2). Although nerve impulses may cross certain synapses in one direction only, many synapses permit impulses to pass in both directions. Moreover, a given cell body may give rise to 2 or more neurites, radiating in different directions. Thus, a nerve impulse received by one neuron may proceed in several directions at once.

With such a nerve network, stimulation of a given sensory cell in the epithelium results in an outward spread of excitation over the animal's entire body (Fig. 6.2b). The amount of surface area of the cnidarian that is affected by stimulating a given nerve cell increases in proportion to the frequency of stimulation.

In addition to this slow-conducting nerve network, a second, fast-conducting nerve network generally underlies the epithelium. These cells are less branched (bipolar as opposed to multipolar) than are the cells in the slow-conducting network, so that signal transmission is more directed. Furthermore, neurites in the 2 nets differ in size: Nerves in the fast-conducting network are of greater diameter, allowing more rapid conduction of nerve impulses. Ultrastructural studies indicate that these "giant" fibers (perhaps 1 μm to 5 μm in diameter) can

4. See *Topics for Further Discussion and Investigation,* no. 3.

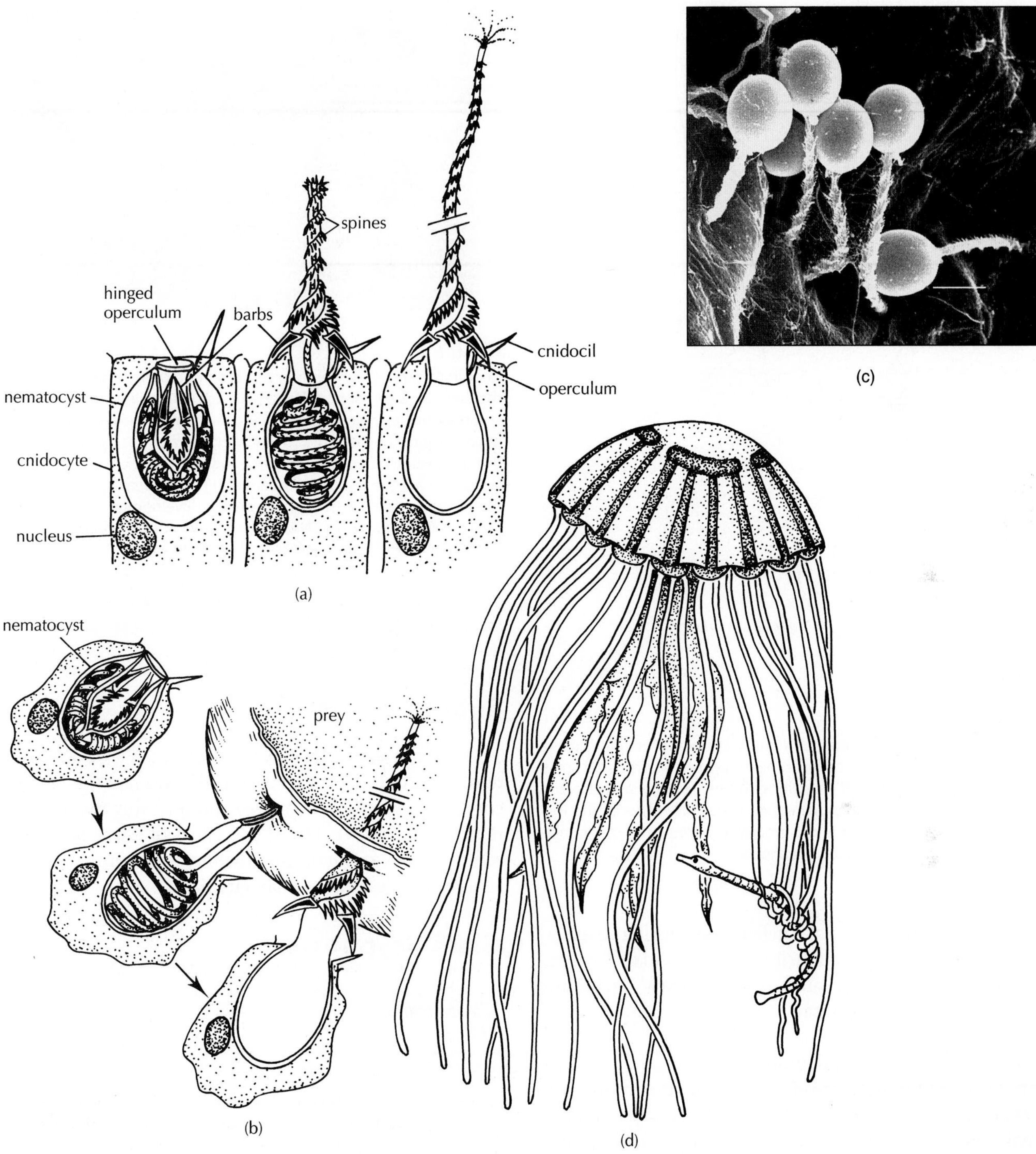

Figure 6.1

(a) Stages in the discharge of a nematocyst, stimulated by chemical and/or physical contact with the cnidocil. The mechanism of discharge remains unclear, although it is apparently mediated by an inflow of water along a concentration gradient. (b) Penetration of the nematocyst filament into another animal. Spines on the filament are exposed as the filament emerges (see part [a]), cutting through the prey tissues. (c) Six nematocysts fired by the scyphozoan jellyfish *Cyanea capillata* penetrating human skin. (Length of scale bar = 10 μm.) (d) Jellyfish (medusa) capturing prey. Nematocysts commonly inject toxins that paralyze the prey prior to ingestion.

(b) Hardy, *The Open Sea: Its Natural History.* Boston: Houghton Mifflin Company, 1965.
(c) Courtesy of Thomas Heeger. From T. Heeger *et al.*, 1992. *Marine Biology* 113:669–78. © Springer-Verlag.

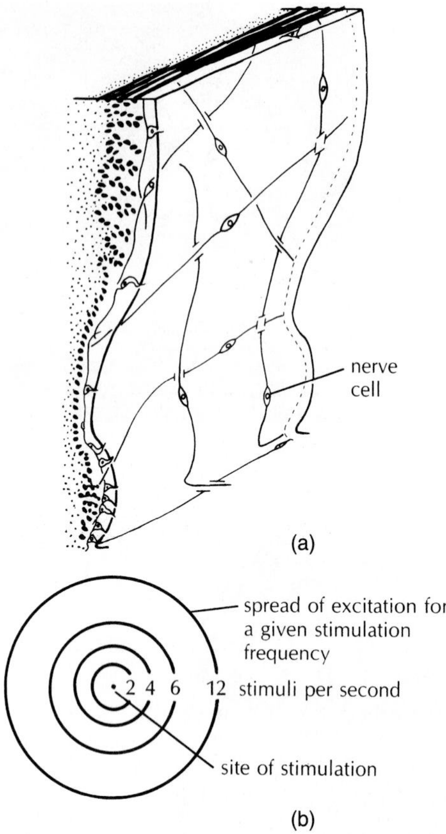

Figure 6.2
(a) Diagrammatic illustration of a cnidarian nerve net. The nerve cells synapse with each other repeatedly. Nerve impulses may cross the synapses in both directions. (b) The surface area affected by stimulation of a cnidarian nerve cell varies directly with the frequency of stimulation; that is, the greater the frequency of stimulation, the greater the surface area affected.
(a) *After Bullock and Horridge.*

arise through the fusion of smaller fibers during development (Fig. 6.3). Although their nervous sytem differs dramatically from that of most other animals, cnidarians often exhibit complex behaviors.

Because cnidarians are diploblastic animals, their musculature cannot (by definition of *diploblastic*) have its origins in embryonic mesoderm. Instead, the muscle layers are composed of numerous ectodermal and endodermal cells that possess elongated, contractile bases anchored in the mesoglea; these cells are termed *epitheliomuscular cells* or *nutritive-muscular cells,* respectively (Fig. 6.4). Depending upon the orientation of the contractile bases, the cells form layers of either longitudinal musculature (running from the base to the tip of a tentacle, for example) or circular musculature (running around the circumference of a sea anemone's body column, for example). In addition to their contractile function, nutritive-muscular cells also capture small food particles and then digest them intracellularly. Although epitheliomuscular cells characterize all cnidarians, they are not a uniquely cnidarian feature; they are found sporadically among members of several other phyla.

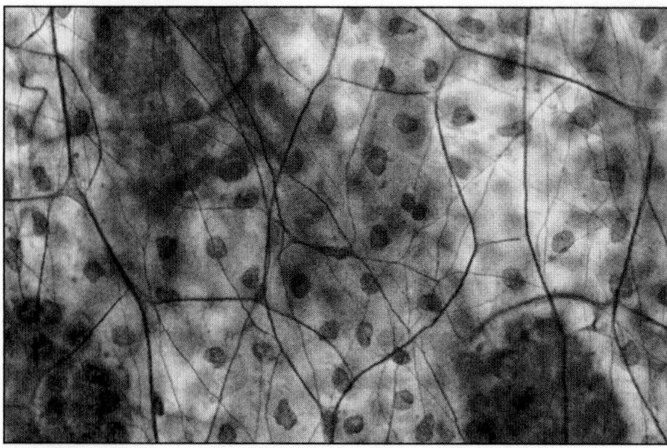

Figure 6.3
The cnidarian nerve net. Nerve cells in epithelial tissue of *Velella,* the by-the-wind sailor (class Hydrozoa), were impregnated with silver to make their outlines stand out when viewed with a light microscope. The network of smaller nerves (less than 1 μm in diameter) seems to be synaptic, and the additional network of larger nerves (up to 5 μm in diameter) may be syncytial, owing to the fusion of smaller nerve cells during development.
Courtesy of G. O. Mackie. From Mackie, Singla, and Arkett, 1988. *J. Morphol.* 198:15–23, Permission of Wiley-Liss, Inc., a subsidiary of John Wiley & Sons, Inc.

Cnidarians lack gills and other specialized respiratory structures; gases diffuse across all exposed epidermal and gastrodermal surfaces.

Subphylum Medusozoa[5]

This recently established but well-accepted grouping contains all cnidarians except for the sea anemones and corals (class Anthozoa, p. 118) and probably the parasitic myxozoans (pp. 61, 115). Unlike that of any other metazoan so far studied, the medusozoan mitochondrial genome is linear rather than circular.

Class Scyphozoa

Class Scypho • zoa
(G: cup animals)
sky´-fō-zo´-ah

Defining Characteristic:[6] Asexual replication by strobilation

The Scyphozoa contains only a few hundred species, all of which are marine and many of which are quite large (up to about 2 m across). Some former scyphozoans have recently been moved to a separate, new class (see *Taxonomic Detail,* p.131). The mesoglea layer of scyphozoans is thick and has the consistency of firm gelatin; thus, scyphozoans are known collectively as *jellyfish.*

5. Classification is based on Marques, A. C., and A. G. Collins. 2004. *Invert. Biol.* 123: 23–42; Collins, A. G., P. Cartwright, C. S. McFadden, and B. Schierwater. 2005. *Integr. Comp. Biol.* 45: 585–94.

6. Characteristics distinguishing the members of this class from those of other classes within the phylum.

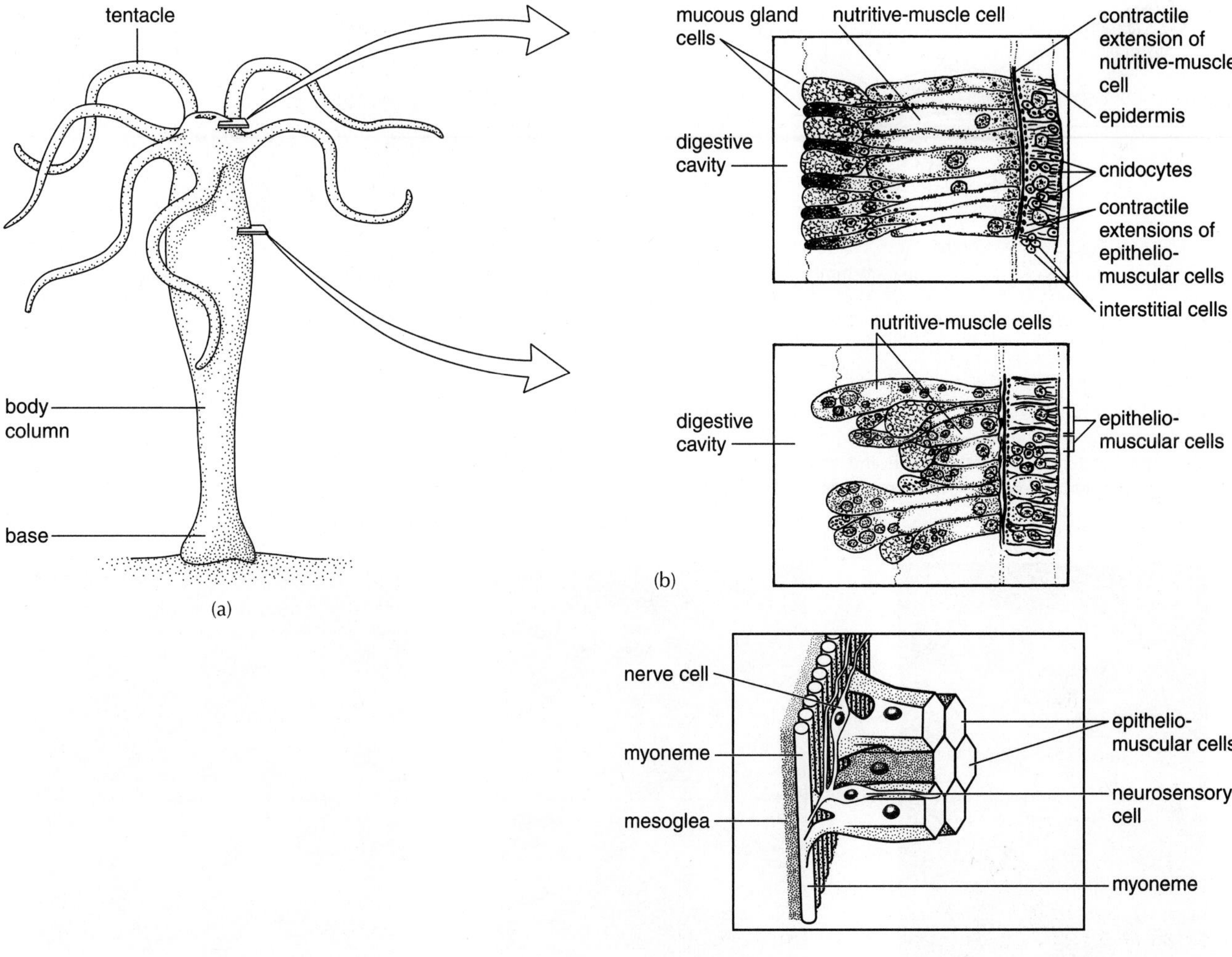

Figure 6.4

(a) Body wall of a freshwater *Hydra* seen in cross section from two regions of the body. Note that, as illustrated in (b), the contractile bases of epitheliomuscular cells form a layer of longitudinal musculature extending up and down the body column, whereas the contractile bases of the nutritive muscle cells form a layer of circular muscles extending around the body column. **Cnidocytes** are mature cnidoblasts, containing fully formed and functional cnidae. The **interstitial cells** include cnidoblasts and other, as yet, undifferentiated cells. (c) Epitheliomuscular cells of a cnidarian. The columnar (or sometimes cylindrical) upper portions of the cells form the outer epidermis, while the elongated contractile bases form the musculature. A portion of the nerve net is also shown.

(c) Based on G. O. Mackie and L. M. Pussano, in *Journal of General Physiology* 52:600–608, 1968.

Jellyfish morphology is described as **medusoid.** The body is in the form of an inverted cup, with nematocyst-studded tentacles extending downward from the cup, or **bell** (Fig. 6.5). The mouth is borne at the end of a muscular cylinder known as the **manubrium.**

Members of most scyphozoan species can swim actively, by contracting muscles and exploiting the mechanical properties of the mesoglea. When the muscle fibers of the swimming bell contract, the volume of fluid enclosed under the bell decreases. Water is forcefully expelled from under the bell as a consequence, and the animal is propelled in the opposite direction (Fig. 6.6). The muscular contraction deforms the elastic mesoglea so that when the musculature is relaxed, the mesoglea "pops" back to its normal shape. This, of course, pulls the jellyfish downward as the volume enclosed by the swimming bell increases. Net forward movement of the animal occurs primarily because the speed with which the bell contracts exceeds the speed with which the bell recoils to its resting state. Note that water is forced out from under the swimming bell, not through the manubrium. Also note that the mesoglea functions here as a skeletal system, stretching the contracted muscles when they relax. Recent research[7] shows that the mesoglea can also function as an oxygen store, to be exploited by the animal whenever surrounding oxygen concentrations are low.

Scyphozoans are characterized by a well-developed system of fluid-filled **gastrovascular canals,** ultimately connecting to the mouth through the manubrium

7. Thuesen, E.V., L. D. Rutherford, Jr., P. L. Brommer, K. Garrison, M.A. Gutowska, and T. Towanda. 2005. *J. Exp. Biol.* 208: 2475–82.

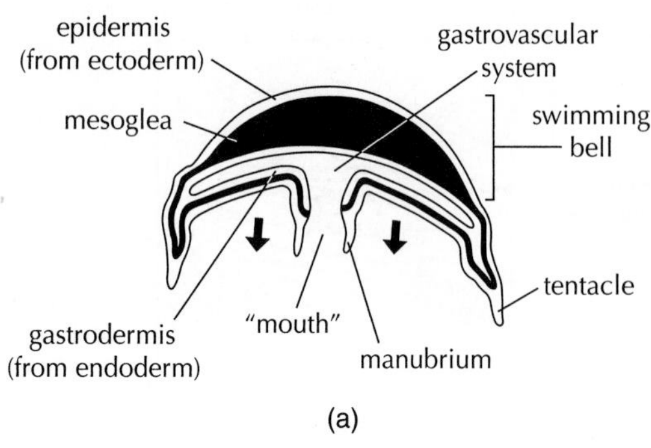

gastrovascular
canal
stomach
manubrium
circular
muscles
tentacle
oral arms
mouth opening
(b)

Figure 6.5

(a) A medusa, seen in longitudinal section. Note the very thick layer of mesoglea and the single opening to the gastrovascular cavity; this opening serves as both mouth and anus. Arrows indicate water leaving when bell musculature contracts. (b) Lateral view of a medusa, showing the gastrovascular canal system and the arrangement of tentacles, oral arms, and musculature of the swimming bell.

(a) After Russell-Hunter.

(a)

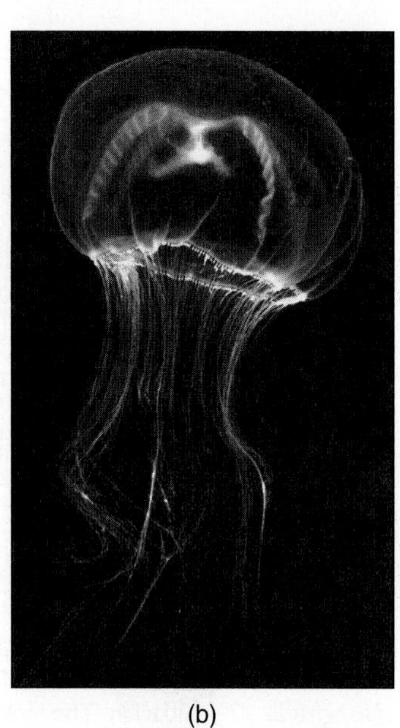

(b)

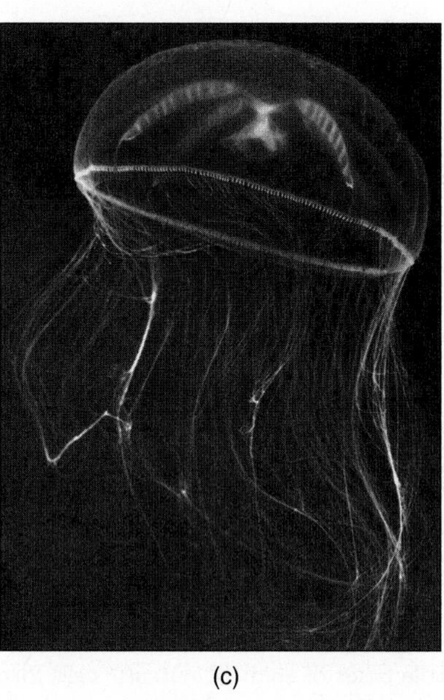

(c)

Figure 6.6

Locomotion of the medusa stage of *Mitrocoma cellularia.* The bell has a diameter of about 70 mm. As muscle contractions force water out from under the swimming bell, the animal is propelled in the opposite direction (upward, in photographs). Efficient jet propulsion relies on the incompressibility of water; fluid must leave the bell as it cannot "hide" under the bell by compression. The animal shown is actually a hydrozoan rather than a scyphozoan; the velum is seen clearly as a sheet protruding from the bell in (a). (See page 111 for a discussion of hydrozoan medusae.) (a) Power stroke nearly completed; note that a bolus of expelled water can be seen pushing the tentacles outward about midway along their length. (b) Beginning of recovery period; swimming bell is expanding; note that the bolus of ejected water has moved farther down the tentacles. (c) Bell fully relaxed and expanded, ready for the next contraction.

(a–c) Courtesy of Claudia E. Mills.

(Fig. 6.7). Food particles captured by nematocysts on the tentacles and/or oral arms are ingested at the mouth and conveyed to the stomach through the manubrium. Food is then distributed among 4 **gastric pouches,** which contain short, nematocyst-bearing tentacles (**gastric filaments**) that secrete an array of digestive enzymes. The partially digested food particles are then phagocytosed and digestion is completed intracellularly, a typical feature of cnidarian biology. Fluid within the gut is circulated by cilia lining the walls of the gastrovascular canals. The gastrovascular canals are believed to function in circulating oxygen and carbon dioxide (the "vascular" part of *gastrovascular*), as well as in distributing nutrients (the "gastro" part of the term).

Some scyphozoans also obtain nutrients from certain unicellular algae (**zooxanthellae**) that live symbiotically in the jellyfish tissues.

Although neurologically unsophisticated in comparison with most other metazoans, scyphozoans nevertheless show a variety of fairly sophisticated behaviors, including

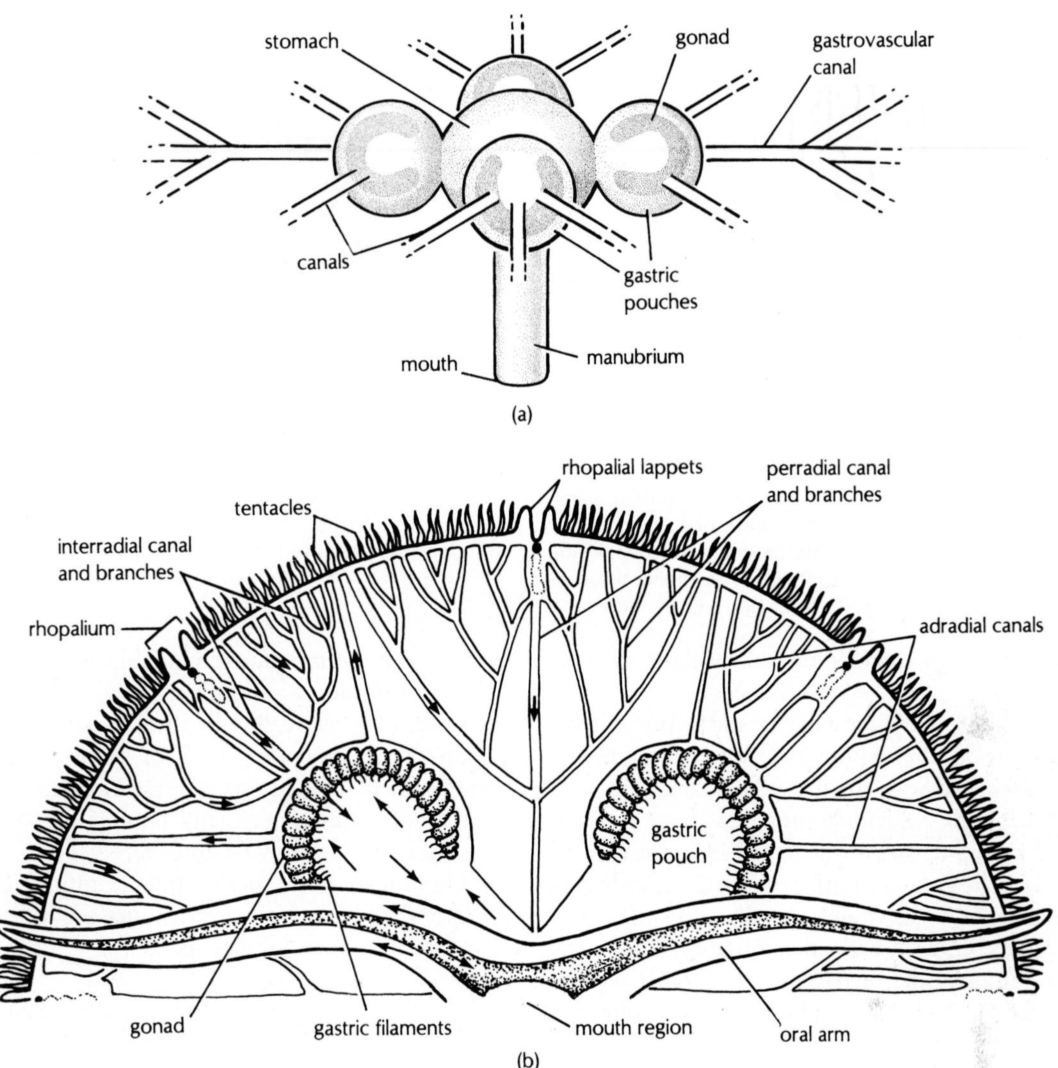

Figure 6.7

Detail of the scyphozoan gastrovascular canal system. (a) Lateral view of the moon jelly, *Aurelia* sp. (b) Oral view. Cilia lining the canal system draw water in through the mouth to the gastric pouches. From the pouches, water circulates to the periphery of the bell through a complex series of narrow canals, as shown by the arrows. Details of the rhopalia, which are sensory organs, are shown in Figure 6.8.
(b) After Hickman.

periodic vertical migrations from surface waters to deeper waters and back again, and the temporary formation of breeding aggregations. As befits any mobile organism, scyphozoan medusae are equipped with fairly sophisticated sensory receptors, implying that the nervous system is able to process and integrate a variety of sensory inputs. Sensory systems include balance organs **(statocysts),** simple light receptors **(ocelli),** and, in some species, touch receptors **(sensory lappets).** The statocysts and ocelli are contained within club-shaped structures called **rhopalia,** which are distributed along the margins of the swimming bell (Figs. 6.7b and 6.8). Dense aggregations of nerve tissue are found associated with the rhopalia. These ganglia act as pacemakers, triggering the rhythmic contraction of the swimming bell.[8]

8. See *Topics for Further Discussion and Investigation,* no. 11.

Statocysts operate on a beautifully simple principle. Tubular pieces of tissue (the **rhopalia**) hang freely at several locations around the margins of the swimming bell. Each of these rhopalia is adjacent to (but not in continuous contact with) sensory cilia. Also, each tube is weighted at the free end with a spherical calcareous mass (the **statolith**). If the animal tilts in a particular direction, those statocysts on the lowermost margin press against their respective cilia (Fig. 6.8b), causing the associated nerve cells to generate action potentials. The rhopalium/statocyst system thus provides a mechanism through which the animal can be informed of its physical orientation—that is, whether the body is horizontal or tilted—and the jellyfish can alter its posture accordingly, through stronger contractions of the musculature on one side of the bell.

The non-image-forming ocelli (light receptors) are also found along the bell margin. An **ocellus** is simply a

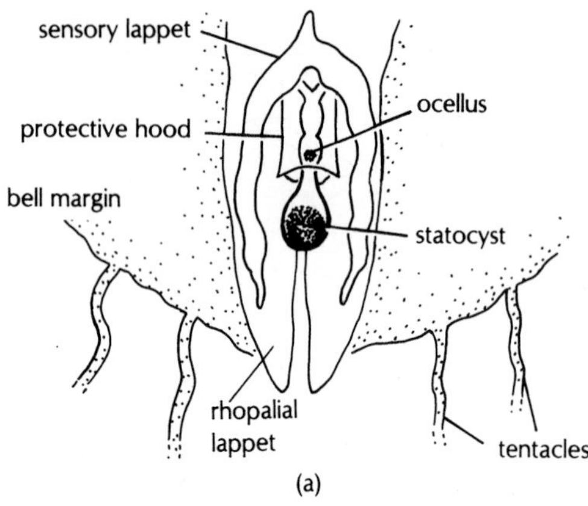

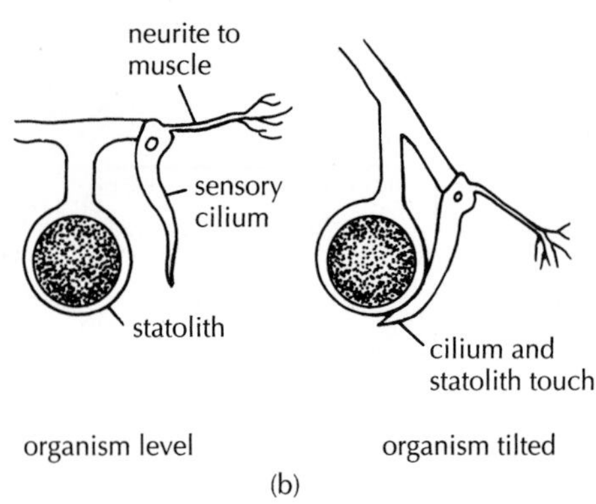

Figure 6.8

(a) Detail of one rhopalium from the scyphozoan *Aurelia aurita.* The species illustrated has a statocyst, a simple ocellus, and a pair of specialized sensory lappets associated with each rhopalium. The sensory lappets are believed to function as chemical receptors. (b) The principle of statocyst operation. When the animal tilts sideways, the statolith swings against a sensory cilium, initiating a nerve signal to the appropriate muscles; these muscle contractions then restore the animal to proper orientation.

(a) After Hyman. (b) After Wells.

small area, often cup shaped, backed by light-sensitive pigment.

Scyphozoan life cycles are a diagnostic feature of their biology. Gonads develop within gastrodermal tissue and are closely associated with the gastric pouches (Fig. 6.7). With few exceptions, individual medusae are either male or female; that is, the sexes are generally separate, and the species is said to be **gonochoristic** (*gono* = G: reproductive organs; *chorist* = G: separate), or **dioecious** (G: two houses). This contrasts with the situation frequently encountered among other invertebrates, in which a given individual may be both male and female, either simultaneously or in sequence. Such species are said to be **hermaphroditic** or **monoecious** (G: a single house).

A **planula** larva eventually results from the union between sperm and egg, as in other cnidarians. This larva typically has the form of a heavily ciliated, microscopic sausage. The nonfeeding planula larva soon attaches to a substrate and transforms into a small polypoid individual called a **scyphistoma** (Fig. 6.9). This **polyp** form has the same two-layered construction (plus mesoglea layer) as the medusa, but the mesoglea layer is substantially thinner in the polyp than in the medusa morph. The scyphistoma is sessile and lacks ocelli and statocysts. It is a feeding individual, with the mouth oriented away from the substrate.

As the scyphistoma grows, it may produce additional scyphistomae asexually by budding. Eventually, a process called **strobilation** takes place in most species. During strobilation, the body column of a scyphistoma subdivides transversely, forming numerous modules that are stacked on top of each other like hotel ashtrays (Fig. 6.9). Each module eventually breaks away from the stack as a swimming **ephyra.** As it swims, each ephyra gradually grows and changes in physical appearance, becoming an adult scyphozoan.

Reflect for a moment upon this life cycle. Scyphozoans use the relatively inconspicuous polyp morph to achieve something quite remarkable: From a single fertilized egg producing a unique genotype (or **genet**), a large number of genetically identical, sexually reproducing medusae are generated. Each of these independent but genetically identical units is termed a **ramet** (Fig. 6.9).[9] A similar phenomenon occurs among the parasitic flatworms, as discussed in Chapter 8.

In at least a few species, free-swimming ephyrae can differentiate back into scyphistoma polyps under unfavorable laboratory conditions.[10]

Class Cubozoa

> "The wife was screaming with pain, and walked with the assistance of her husband—who was also in pain—to the beach, where she became irrational and unable to stand. She then became 'quiet and pale'." S. K. Sutherland and J. Tibballs. 2001. *Australian Animal Toxins.* Oxford Univ. Press.

Class Cubo • zoa
(G: cube animals)
cube´-ō-zō´-ah

Defining Characteristics: 1) Medusa with boxlike body; 2) rhopalia bear complex, lensed eyes

The members of the small (about 20 species) but interesting class Cubozoa are called *cubomedusae.* Recent molecular

9. See *Topics for Further Discussion and Investigation,* no. 17.

10. Piraino, S., D. de Vito, J. Schmich, J. Bouillon, and F. Boero. 2004. *Can. J. Zool.* 82: 1748–54.

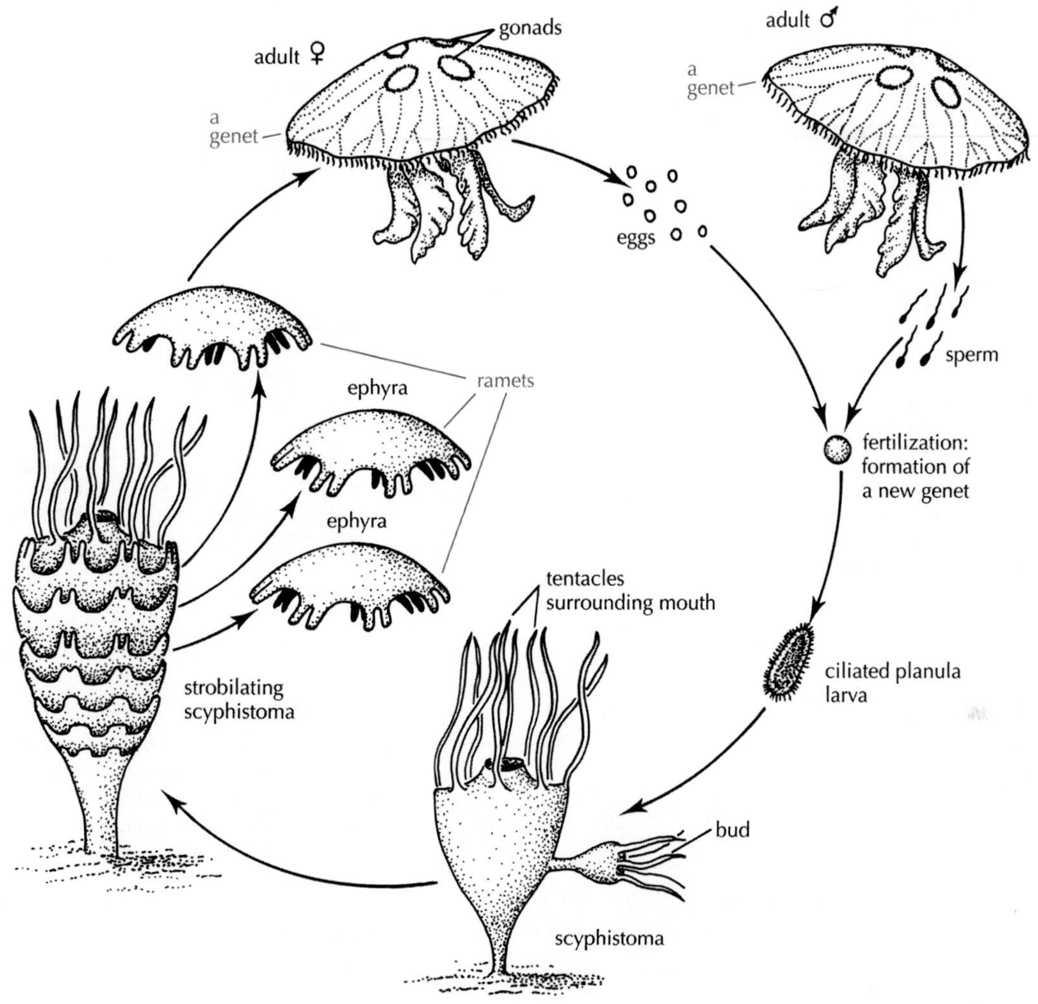

Figure 6.9

Life cycle of the moon jelly *Aurelia aurita*, a typical scyphozoan. The polyp stage (scyphistoma) is small—often only a few millimeters in length—and is often found hanging downward from the undersides of underwater rock ledges. The scyphistoma shows modular growth, both by budding new scyphistomae and by budding to produce ephyrae, which are ramets of the original genet.

data suggest that they are the most derived cnidarians. As in scyphozoans, the medusa (jellyfish) stage dominates the cubozoan life cycle. The cubomedusa is typically only a few centimeters in its largest dimension and extremely transparent. The "cubo" part of the name refers to the fact that all members of the class have a cuboidal swimming bell: The bell is actually square in transverse section. Each individual bears 4 tentacles, or 4 clusters of tentacles, emerging from the 4 corners of the bell, near the 4 rhopalia (Fig. 6.10a). An animal only 2–3 cm in diameter can have tentacles over 30 cm long. These tentacles boast highly virulent nematocysts that have earned cubomedusae the well-deserved epithet "sea wasp." Cubomedusae mostly eat small fish, often killing individuals much larger than themselves. However, some cubomedusan species can kill humans, inflicting considerable pain along the way.

Cubomedusae are unusually active and strong swimmers for jellyfish and possess an unusually well-developed nervous system and remarkably complex eyes (Fig. 6.10b,c) that can probably form images.[11] Unlike that of scyphozoans, the swimming bell of sea wasps curves inward at the lower edge, restricting the size of the opening through which water is expelled when the bell contracts. This increases the force with which the water exits the bell, and thus the amount of jet propulsion obtained. The effect is similar to that obtained by putting a nozzle on a garden hose: Greater propulsive force results from pushing the same volume of water out through a smaller opening in the same amount of time.

Cubozoans also differ from the true jellyfish (scyphozoans) in that the polyp stage does not strobilate. Rather, the polyp resulting from a single planula larva buds off more polyps, each of which develops into a single medusa. The asexual production of genetically identical

11. Nilsson, D.-E., L. Gislén, M. M. Coates, C. Skogh, and A. Garm. 2005. *Nature* 435: 201–4.

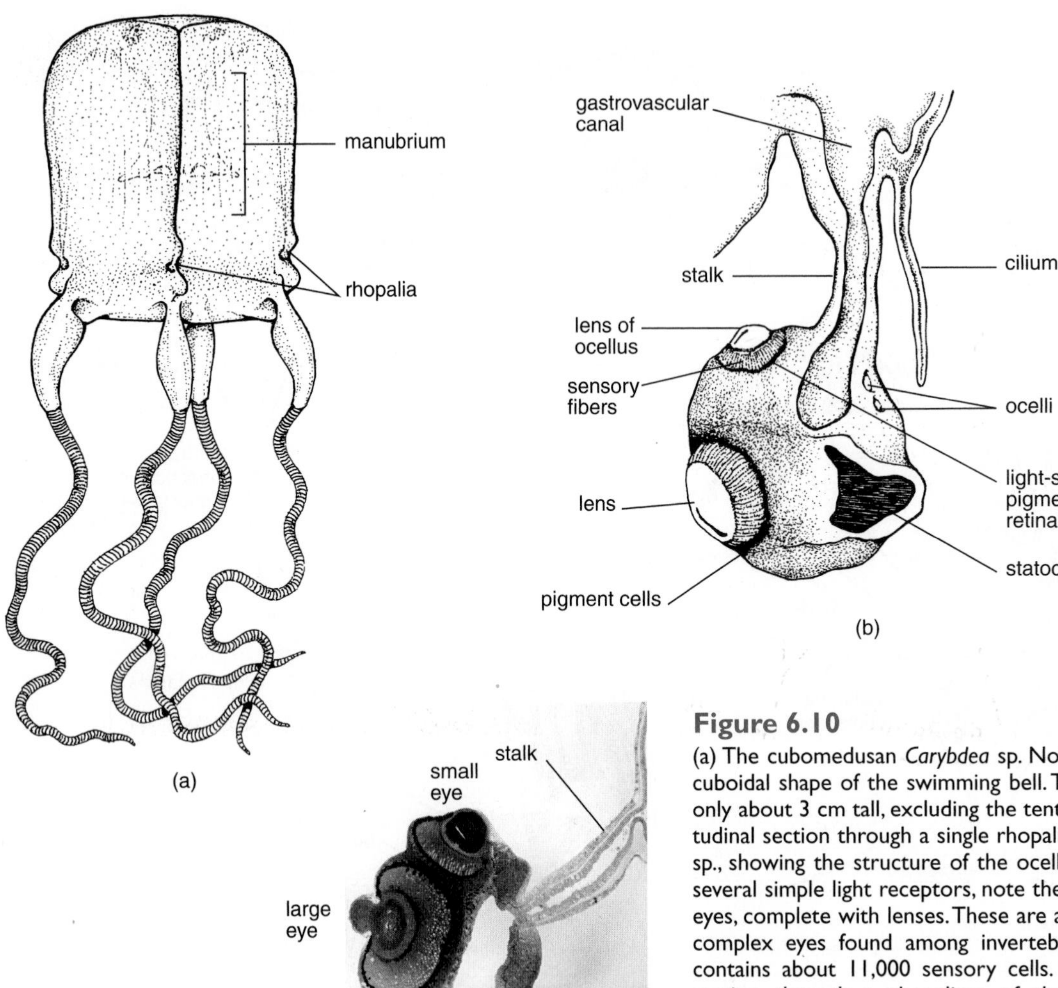

Figure 6.10

(a) The cubomedusan *Carybdea* sp. Note the distinctly cuboidal shape of the swimming bell. This individual is only about 3 cm tall, excluding the tentacles. (b) Longitudinal section through a single rhopalium of *Carybdea* sp., showing the structure of the ocelli. In addition to several simple light receptors, note the more complex eyes, complete with lenses. These are among the most complex eyes found among invertebrates. Each eye contains about 11,000 sensory cells. (c) Histological section through a rhopalium of the cubomedusan *Tripedalia cystophora.*

(b) After Bayre and Owre: after Mayer, and after Conant. (c) From J. Piatigorsky, et al., 1989. In *Journal of Comparative Physiology* (A) 164:577–87, 1989. Copyright © Springer-Verlag, New York. Reprinted by permission.

individuals (ramets) characterizing scyphozoans is again achieved in the cubozoan life cycle, but through a different vehicle.

Cubozoans are restricted to tropical and subtropical areas; they can be quite common in those waters at certain times of the year.

Class Hydrozoa

Class Hydro • zoa
(G: water animals)
hī´-drō-zō´-ah

Members of the Hydrozoa are characterized by generally greater representation of the polyp morph in the life cycle than is the case for scyphozoans, although the polyp and medusa morphs are about equally prominent in a number of hydrozoan species, and the medusa morph dominates in a few others. In contrast to other cnidarians, the gastrodermal tissue of hydrozoans lacks nematocysts, and no cells are found within the mesoglea; nematocysts are restricted to the epidermis. The class Hydrozoa comprises 3 major orders (and several smaller ones not discussed here) containing fewer than 3000 mostly marine species.

Order Hydroida

Although most members of the Hydroida (hī-droid´-ah) are marine, a number of freshwater species also exist—*Hydra,* for example (see Fig. 6.13a). Hydroids are generally medusoid as adults; that is, the sexual stage of the life cycle resembles that found among the Scyphozoa, and like scyphozoan medusae, those of hydrozoans are commonly called jellyfish. While the best-known species live in surface waters, recent studies

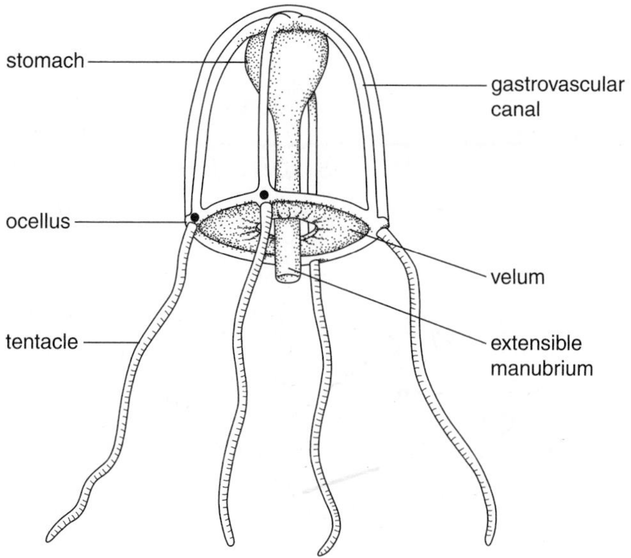

Figure 6.11
Diagrammatic illustration of a typical hydrozoan medusa. Note the conspicuous velum, through which water is forcefully expelled when the musculature of the swimming bell contracts. The water is expelled from under the swimming bell, as in scyphozoan medusae. The velum extends toward, but does not fuse with, the manubrium.

using submersibles and underwater cameras have revealed a number of jellyfish, both hydromedusae and scyphomedusae, living on or just above the ocean bottom several hundred to more than 1000 m below the water's surface.

As in scyphozoans, the mesoglea layer of the hydrozoan medusa is thick, the mouth is borne at the end of a manubrium, and ocelli and statocysts are present. The sense organs may be found at the base of the tentacles, as in scyphozoans, or between the tentacles. All medusae are gonochoristic, a given individual being either male or female but never both. However, hydrozoan medusae tend to be much smaller than scyphozoan medusae (generally only a few centimeters or less across), and they usually possess a shelf of tissue (the **velum**) extending inward from the edge of the swimming bell toward the manubrium (Fig. 6.11). The presence of the velum causes water to be ejected from under the swimming bell through a narrower opening, and thus with greater velocity, when the musculature contracts. The effect is virtually identical to that achieved by the inwardly directed edge of the swimming bell of cubomedusae (p. 105). Scyphozoan medusae lack a velum.

As in the Scyphozoa, the planula larvae of hydrozoans (Fig. 6.12) develop from fertilized eggs, and the planula typically metamorphoses into a sessile polypoid individual lacking both statocysts and ocelli. Hydrozoan polyps are structurally and functionally more complex than are the scyphistomae of the scyphozoan life cycle. The hydrozoan genus most familiar to readers is probably the freshwater *Hydra*, a rather atypical hydrozoan. In

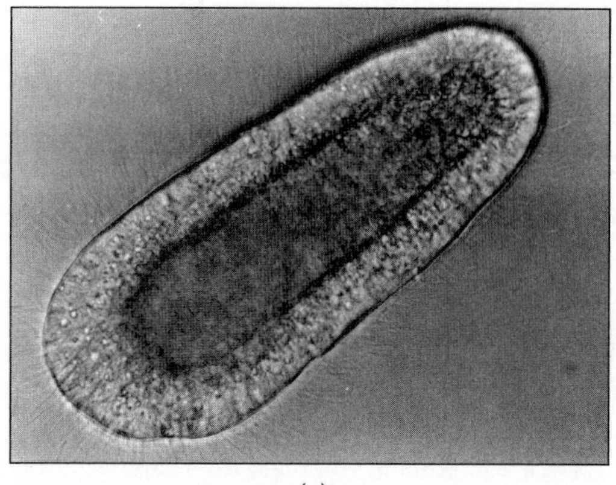

(a)

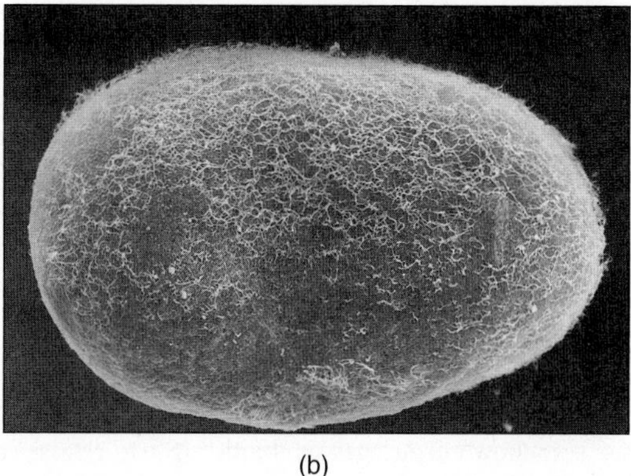

(b)

Figure 6.12
(a) The planula larva of a hydrozoan, *Mitrocomella polydiademata*. (b) Ciliated planula of the Red Sea soft coral *Dendronephthya hemprichi*.
(a) Courtesy of Vicki J. Martin. (b) Courtesy of Yehuda Benayahu, Dept. of Zoology, Tel Aviv University.

Hydra, each polyp is a separate, distinct being, completely responsible for its own welfare (Fig. 6.13a). Some species of *Hydra* (and some other hydrozoans) harbor unicellular green algae, called **zoochlorellae,** in their tissues. Both the host and the algae benefit from the relationship, but the details of the interaction are still being discovered (Research Focus Box 6.1).[12] Unlike that of most other hydrozoans, the *Hydra* life cycle lacks a medusa stage. Also, most other hydrozoans are colonial in the polyp stage of the life cycle; that is, a single planula usually gives rise to a large number of polyps, called **zooids** (or **modules** of the colony), all of which are interconnected and share a continuous gastrovascular cavity (Fig. 6.13b). The zooids are often connected to each other, or to a substrate, by means of a rootlike **stolon.** The oral end of a polyp (i.e., the end bearing the mouth and tentacles) is called the **hydranth.**

The stolon and stalks of the colony are commonly encased in a transparent protective tube, known as the **perisarc,** composed of polysaccharide, protein, and

12. See *Topics for Further Discussion and Investigation,* no. 5.

RESEARCH FOCUS BOX 6.1

Control of Nutrient Transfer in Symbiotic Associations

Douglas, A. E., 1987. "The influence of host contamination on maltose release by symbiotic *Chlorella*." *Limnology and Oceanography* 32:1363–65.

Symbiotic algae supplement the diets of their invertebrate hosts with substantial quantities of carbohydrates and other energy-rich compounds. Although the relationship between cnidarians and their symbiotic algae has been studied intently for more than 25 years, much is still not understood about the process of nutrient exchange between algae and host. One especially intriguing issue is whether the cnidarian host can regulate the release of nutrients from the algal cells. After all, free-living algae retain most of their photosynthetic end products for their own use in respiration, growth, and cell division. When algal cells (*Chlorella* sp.) are isolated from the tissues of freshwater hydrozoans (*Hydra* spp.) and cultured in the laboratory, the algal cells release little carbohydrate to the surrounding medium. Yet, when living symbiotically, the algae release to the host about 60% of the carbohydrate they manufacture, usually as maltose. These observations imply that host tissue somehow encourages symbiotic algae to release carbohydrate that they would normally retain for their own use.

To examine this tantalizing prospect, Angela Douglas (1987) concentrated algal cells from cnidarian tissue by grinding up (**homogenizing**) green hydra and centrifuging the suspension; the preparation at the bottom of the centrifuge tube contained algal cells and host tissue. Douglas hypothesized that, if host tissue stimulates algae to release the products of their photosynthetic activity, such release should occur in the centrifuge tube, and that removing host tissue from the preparation should decrease the rate at which photosynthates are released.

When the crude preparation (algae and host cells) was incubated in the presence of radioactive carbon dioxide ($^{14}CO_2$ provided as ^{14}C-sodium bicarbonate) and adequate lighting, algal cells incorporated the ^{14}C into carbohydrates (Focus Table 6.1, row 1, column 1), indicating that the cells were actively photosynthesizing. More importantly, the algae released maltose into the medium at high rates (Focus Table 6.1, row 1, column 2), as predicted. In contrast, when host tissue in the hydra homogenates was solubilized with a detergent (SDS: sodium dodecyl sulfate), the algae released little maltose, even though they continued to photosynthesize at high rates (Focus Table 6.1, row 2). This comparison strongly suggests that host cells regulate the release of carbohydrates from the algae.

As further support for the hypothesis of regulation by hosts, Douglas was able to decrease the rate of maltose release

chitin (Fig. 6.14a). The perisarc may or may not extend upward to encase the hydranth of a polyp, depending on the species. The perisarc surrounding the hydranth is known as a **hydrotheca** (Figs. 6.13c and 6.14a), and the hydroid is said to be **thecate** (as opposed to **athecate,** the Greek prefix *a* meaning "not," or "without"; Fig. 6.14b).

Several structurally and functionally distinct modules are often present in a single hydroid colony; such colonies are **dimorphic** (consisting of 2 types of modules) or **polymorphic** (consisting of more than 2 types of modules). Modules specialized for feeding are called **gastrozooids.** Gastrozooids collect small animals using the tentacles (which are densely clothed in nematocysts) and ingest the prey through the single opening into the gastrovascular cavity. Digestion is extracellular in the gastrovascular cavity, and then becomes intracellular as the partially digested food is distributed throughout the colony, largely through rhythmic contractions of the muscular polyps.

Medusoids are produced asexually by budding from various regions of the polyp colony. A single hydrozoan colony typically produces either male or female medusae, but in some species a colony can apparently produce medusae of both sexes. Often, the medusoids derive from a particular type of module called a **gonozooid** (Fig. 6.14a). Some gonozooids lack tentacles and so are incapable of feeding; they are specialized for producing medusoids and must depend upon other members of the colony for nutrition. In some hydrozoans, the medusae eventually break free of the polyp colony, swim off, and commingle gametes with other medusae as already discussed. More commonly, however, the gamete-producing medusoid morph remains attached to the hydrozoan colony. No free-swimming medusa stage exists in such species. In fact, the medusa morph may be little more than a mass of gonadal tissue. Even so, it is best to think of the medusa as the adult. The polyp stage, no matter how conspicuous, is then best thought of as a prepubescent juvenile that produces an independent adult stage asexually. In a few hydrozoan species, free-living medusae bud off more, genetically identical medusae asexually, either from the manubrium or from the tentacular bulbs, before getting on with the business at hand. And as with some scyphozoan medusae, the medusae of some hydrozoan species can "reverse" their

Focus Table 6.1 Effect of Several Treatments on Rates of Photosynthesis and Maltose Release by Symbiotic Zoochlorellae

Treatment	1 Rate of ^{14}C Incorporation (isotope counts per minute per 10,000 cells per hour)	2 Rate of Maltose Release (fmol per cell per hour)
Untreated	2250	1.30
Detergent	2150	Undetectable
Density gradation centrifugation	2380	Undetectable

Note: Zoochlorellae in this study were concentrated by homogenizing green hydra in a physiologically suitable solution, centrifuging the resulting suspension, and collecting the sedimented cells. (fmol = femtomole [10^{-18} moles]).

when she removed host tissue from her preparation by an entirely different method: centrifuging the homogenate in the presence of a density gradient, to cleanly separate components differing in density—in this case, separating the algal cells from the host tissue. Once again, maltose release by algae in the absence of host tissue was barely detectable, even though the algae were clearly photosynthesizing (Focus Table 6.1, row 3). In both treatments, purified algal cells incorporated ^{14}C into carbohydrates as fast as nonpurified preparations (Focus Table 6.1, column 1), indicating that the treatments did not harm the algae.

These data suggest that something in host tissue (a "host release factor") regulates the release of carbohydrates from the symbiotic algae in green hydra. Similar findings have been reported for marine cnidarians. A reasonable next goal is to understand how hosts exert these regulatory effects on their algal guests. How would you proceed?

Data from A. E. Douglas, "The influence of host contamination on maltose release by symbiotic *Chlorella.*" *Limnology and Oceanography* 32:1363–65, 1987. Copyright © 1987 American Society of Limnology and Oceanography, Inc., Seattle, Washington.

ontogeny under stressful conditions (e.g., when starving or experiencing substantial changes in salinity or temperature), developing back into polyps[13]; thus, the complete genetic program for forming the polyp stage is retained by the medusa and can be reactivated under certain circumstances.

Hydrozoan colonies commonly contain fingerlike modules specialized for defense (**dactylozooids;** *dactylus* = G: finger), as well as modules specialized for feeding and reproduction (Figs. 6.14b and 6.15). The dactylozooids are heavily studded with nematocysts. Dactylozooids never possess mouths and thus, like some of the highly specialized gonozooids, they depend upon the gastrozooids for food collection. All modules in a given colony, no matter how polymorphic the colony, are originally derived from a single planula larva, which serves as the founding member of a new genet.

Many hydrozoans form species-specific symbiotic associations with other animals, including fish, sea urchins, sea squirts (ascidians), polychaete annelids, gastropods, bivalves, crustaceans, bryozoans, sponges, sea anemones, and even other hydrozoans (Figs. 6.14b and 6.15a).

Order Siphonophora

The epitome of hydrozoan polymorphism is reached within the order Siphonophora (sī-fon´-ō-for´-ah). The siphonophores, which include the Portuguese man-of-war, are free-floating hydrozoan colonies in which medusoid and polypoid morphs are present simultaneously in a number of different incarnations. Modified medusae serve as modules modified to propel the colony through the water by jet propulsion **(nectophores),** or as leaflike defensive modules (**bracts,** or **phyllozooids;** *phyllo* = G: leaf). Some species exhibit gas-filled (carbon monoxide!) floats called **pneumatophores,** which may also derive from basic medusoid architecture. The mesoglea layer is much reduced or entirely absent in pneumatophores. Nectophores lack both mouth and tentacles. Bracts are well-endowed with nematocysts. The polyp morph is represented by gastrozooids, gonozooids, and dactylozooids (Fig. 6.16).

13. Piraino, S., D. de Vito, J. Schmich, J. Bouillon, and F. Boero. 2004. *Can. J. Zool.* 82: 1748–54.

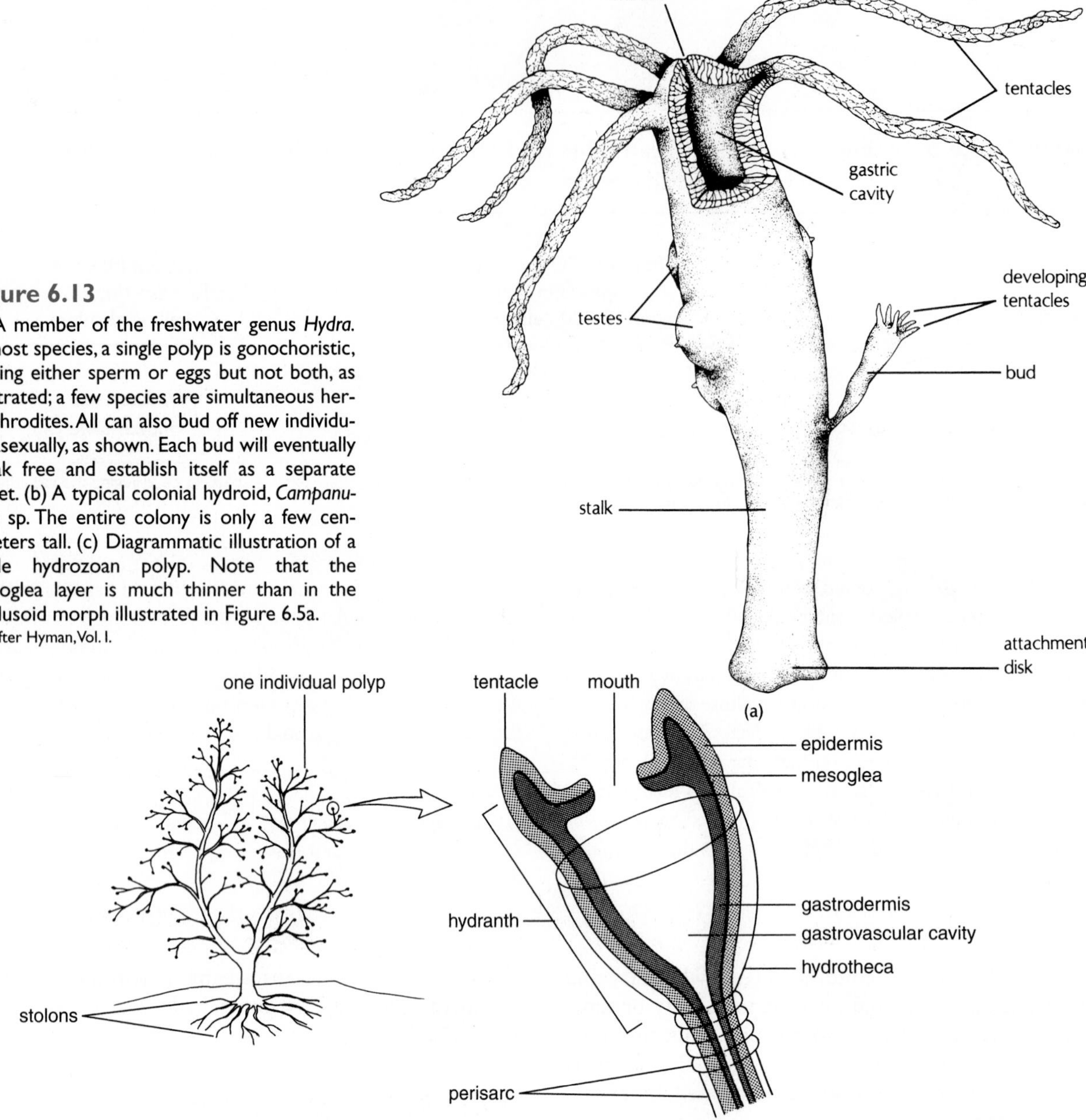

Figure 6.13
(a) A member of the freshwater genus *Hydra*. In most species, a single polyp is gonochoristic, making either sperm or eggs but not both, as illustrated; a few species are simultaneous hermaphrodites. All can also bud off new individuals asexually, as shown. Each bud will eventually break free and establish itself as a separate ramet. (b) A typical colonial hydroid, *Campanularia* sp. The entire colony is only a few centimeters tall. (c) Diagrammatic illustration of a single hydrozoan polyp. Note that the mesoglea layer is much thinner than in the medusoid morph illustrated in Figure 6.5a.
(b) After Hyman, Vol. I.

Each gastrozooid has a single tentacle associated with it; elongated, nematocyst-bearing structures **(tentillae)** may project from these tentacles. Dactylozooids may also have associated tentacles. All siphonophore tentacles are highly retractile, and the nematocysts are often very toxic, even to humans.

Modules within a colony often occur in clusters, called **cormidia,** arranged upon a long stem. Each cormidium typically contains gonozooids, dactylozooids, phyllozooids (bracts), and gastrozooids.

One or more morphs may be absent in some groups of siphonophores. For example, the Portuguese man-of-war lacks nectophores (Fig. 6.16c); the animals are moved by wind and water currents. Some other species (Fig. 6.16b) lack pneumatophores.

All siphonophores, like other hydrozoans, are voracious carnivores.

The Hydrocorals

The hydrocorals are sometimes placed in a single order, the Hydrocorallina (hī-drō-cor-ah-līn´-ah), and sometimes distributed among 2 separate orders, as in the *Taxonomic Detail* at the end of this chapter. Either way, we are talking about a small number of species, all of which are colonial and secrete a substantial calcareous skeleton. Hydrocorals are largely restricted to warm waters. The dactylozooids are especially abundant and potent in many species; the common name "fire coral" is well deserved. These animals are not true corals, however. The

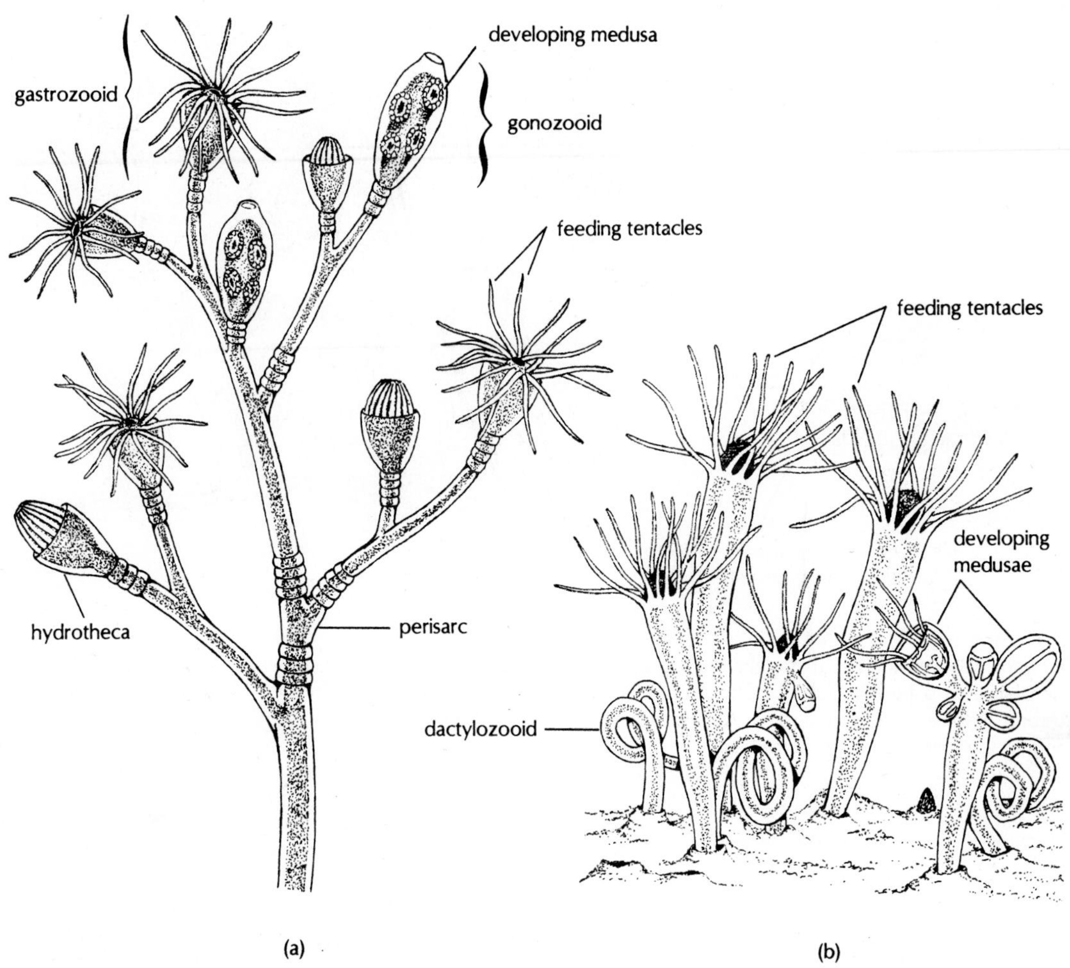

Figure 6.14
(a) A thecate, marine hydroid, *Obelia commissuralis,* showing specialized reproductive and feeding polyps (gonozooids and gastrozooids, respectively). This species commonly forms a pale, bushy covering on pilings and floats in protected harbors. (b) An athecate hydroid, *Podocoryne carnea.* This species possesses individuals specialized for protection (dactylozooids). *P. carnea* is commonly encountered inside the openings of snail shells occupied by marine hermit crabs.
(a) After McConnaughey and Zottoli; after Nutting. (b) After McConnaughey and Zottoli; after Fraser.

true corals are contained within a different class of cnidarians, the Anthozoa, as discussed shortly.

One Group of Non-Medusozoans

Class Myxozoa (= Myxospora)

mix´-ō-zō´-ah

Myxozoans are some 2000 species of extracellular, spore-forming parasites classified until recently as protozoans. The infective spore serves to disperse the parasite to a new host. Myxozoans mostly infect fish, with aquatic annelids apparently serving as intermediate hosts; the complete life cycle has not been worked out for any myxozoan species. As with another group of organisms until recently thought to be protozoans, the microsporidians (p. 61), myxozoans possess distinctive organelles called **polar filaments;** individual myxozoans usually bear several of these filaments, and each is coiled tightly within specialized **polar capsules;** these capsules may occupy much of the space within the spore (Fig. 6.17a,b). Moreover, each spore is often multicellular, containing 2 or more amoeboid sporoplasms. The sporoplasms are released by rupture of the spore, rather than by discharge through an everted polar filament.

So what are these parasites doing in this chapter? A number of biologists had argued for some time that myxozoans were really very degenerate multicellular animals, not protozoans at all. Molecular data collected over the past 10 years or so have supported that contention. One of the best-studied (although atypical) myxozoans, *Buddenbrockia plumatellae* (Fig. 6.17c), was thought for several years to be a peculiar, degenerate bryozoan, but it turns out that the DNA samples leading to that idea had been contaminated by tissue from their bryozoan host. With the contaminated DNA excluded, *Buddenbrockia* groups

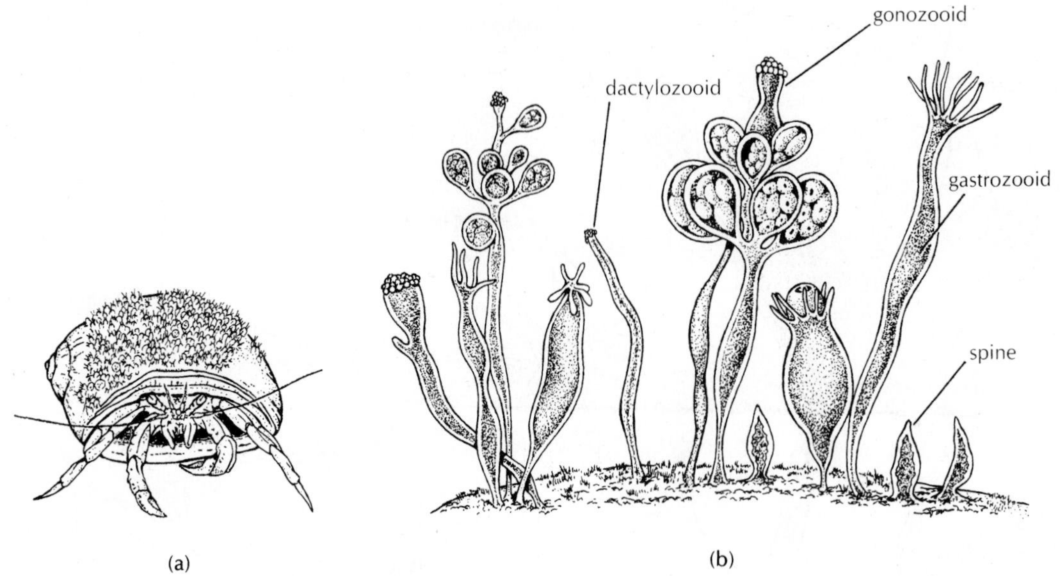

Figure 6.15

(a) Colony of *Hydractinia echinata* encrusting the outside surface of a snail shell inhabited by a hermit crab. (b) Detail of *H. echinata* colony, illustrating gastrozooids, gonozooids, dactylozooids, and individuals modified to form sharp protective spines.

(a) From R. I. Smith, *Key to Marine Invertebrates of the Woods Hole Region.* Copyright © 1964 Marine Biological Laboratory, Woods Hole, MA. Reprinted by permission. (b) After Bayer and Owre.

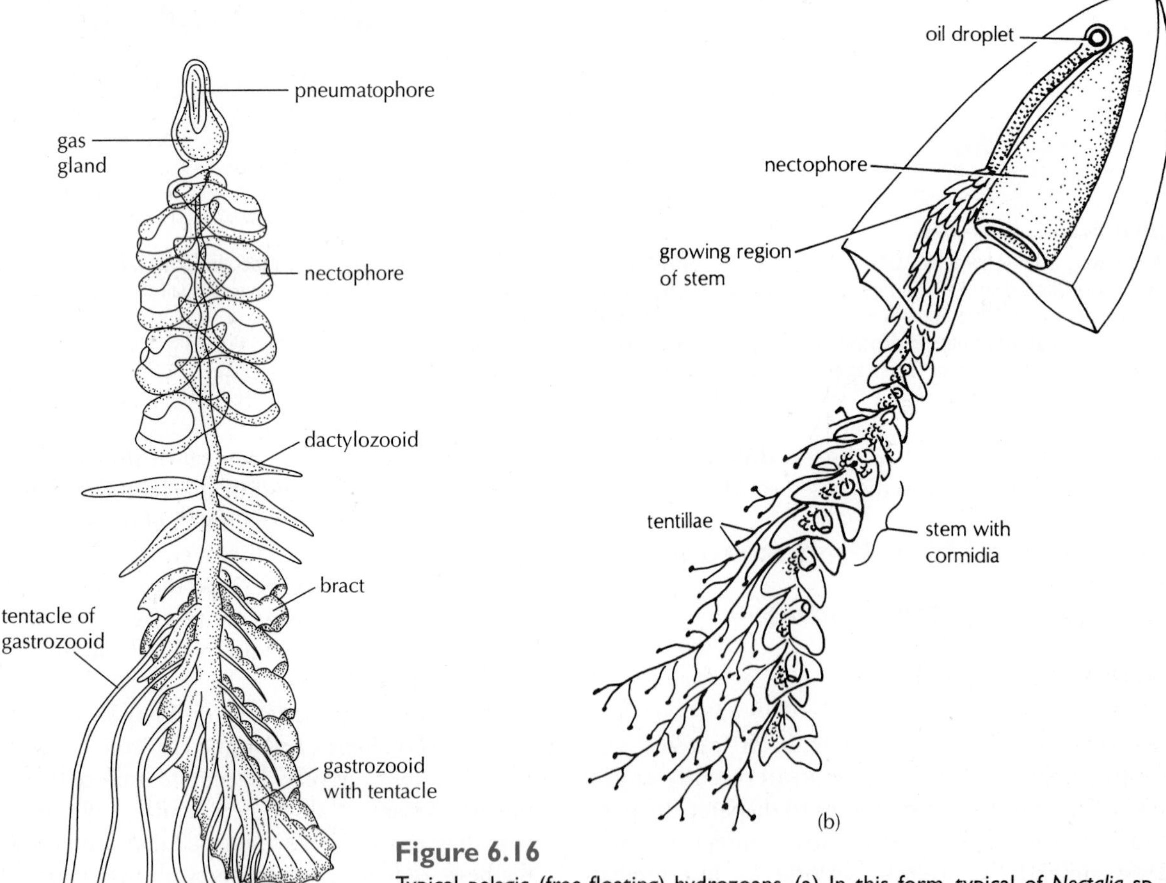

Figure 6.16

Typical pelagic (free-floating) hydrozoans. (a) In this form, typical of *Nectalia* sp., the pneumatophore is relatively small. Nectophore contraction provides jet propulsion. (b) *Muggiaea* sp., a siphonophore that lacks a pneumatophore. The largest member of the colony is a nectophore. Note the long stem with cormidia (clusters of gonozooids, bracts, gastrozooids, and dactylozooids).

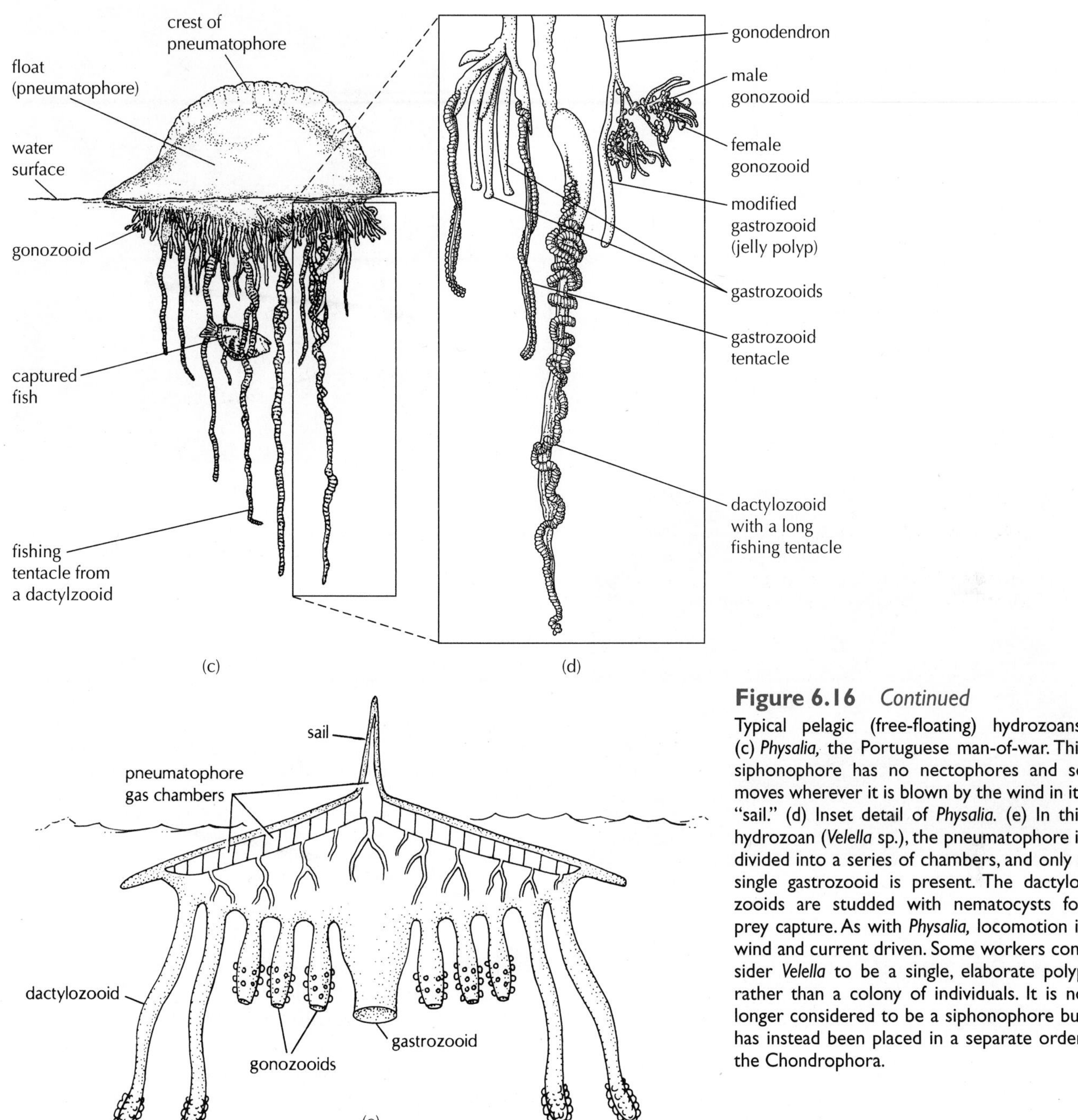

Figure 6.16 *Continued*
Typical pelagic (free-floating) hydrozoans. (c) *Physalia,* the Portuguese man-of-war. This siphonophore has no nectophores and so moves wherever it is blown by the wind in its "sail." (d) Inset detail of *Physalia.* (e) In this hydrozoan (*Velella* sp.), the pneumatophore is divided into a series of chambers, and only a single gastrozooid is present. The dactylozooids are studded with nematocysts for prey capture. As with *Physalia,* locomotion is wind and current driven. Some workers consider *Velella* to be a single, elaborate polyp rather than a colony of individuals. It is no longer considered to be a siphonophore but has instead been placed in a separate order, the Chondrophora.

definitively with the cnidarians, as close relatives (the sister group in fact) of medusazoans.[14] One interesting consequence is that the so-called "polar filaments" of myxozoans must actually be nematocysts, as a number of workers had suspected for a long time based on ultrastructural and developmental studies. It is not yet clear how myxozoans are related to other members of the phylum.

Unlike most other myxozoans, *Buddenbrockia plumatellae* is **vermiform** (worm-shaped) (Fig. 6.17c) and contains 4 blocks of radially arranged longitudinal muscles that it uses to generate sinusoidal waves for locomotion, as do nematodes. Even so, *Buddenbrockia* has no well-defined nervous system or sense organs, and lacks a gut. As suggested recently,[15] it seems reasonable to think that a saclike spore-producing form will be found in the life cycle eventually, something that has possibly already been described as a separate myxozoan species.

14. Jiménez-Guri, E., H. Philippe, B. Okamura, and P. W. H. Holland. 2007. *Science* 317: 116–18.

15. Monteiro, A. S., B. Okamura, and P. W. H. Holland. 2002. *Molec. Biol. Evol.* 19: 968–71.

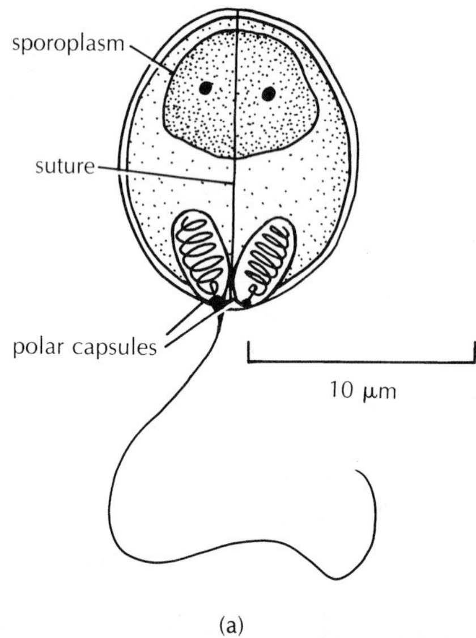

(a)

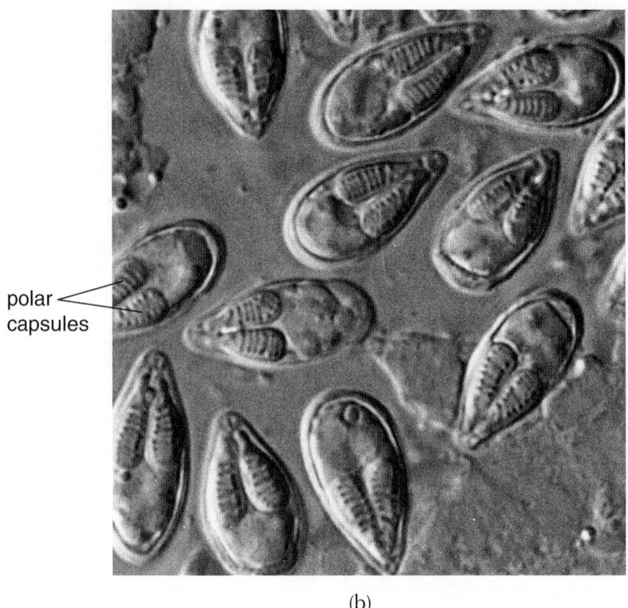

(b)

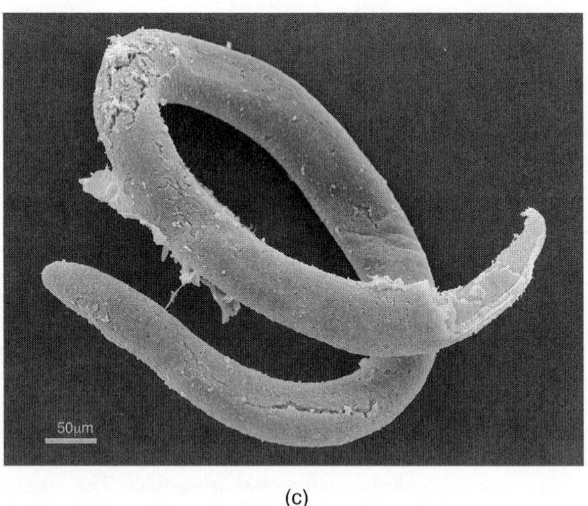

(c)

Figure 6.17

(a) Myxosporan spore with 2 polar capsules, one with an extruded filament. Infective sporozoites emerge from the spore following rupture along the suture. (b) Spores of the myxozoan *Myxobolus* sp. parasitizing a minnow. (c) The atypical, worm-shaped myxozoan *Buddenbrockia plumatellae*, the species most responsible for the current placement of myxozoans within the Cnidaria.

(a) From John N. Farmer, *The Protozoa*. Copyright © 1980 The C.V. Mosby Company (b) From Current et al., 1979. Myxosoma funduli Kudo (Myxosporida) in Fundulus kansae: ultrastructure of the plasmodium wall and of sporgensis. *Journal of Protozool.* 26: 574–83. (c) Courtesy of P. W. H. Holland, from Jiménez-Guri et al., 2007. *Science* 317: 116–18.

Class Anthozoa

Class Antho • zoa
(G: flower animals)
an-thō-zoō´-ah

Defining Characteristics: 1) Absence (loss?) of a medusa stage (or any trace of one); 2) absence (loss?) of operculum and cnidocil; 3) mitochondrial DNA is circular (as in most eukaryotes) rather than linear (as in other cnidarians); 4) presence of ciliated groove (siphonoglyph) in the pharyngeal wall leading from the mouth; 5) coelenteron partitioned by distinct sheets of tissue (mesenteries/septa)

Anthozoans (including the sea anemones and the corals) consist of about 6000 species, nearly 70% of all cnidarian species described. Anthozoans form the sister group to the Medusozoa. All anthozoans are marine and all exploit the polyp body form and lifestyle exclusively; no trace of the medusa morph appears in the life cycle, raising some evolutionary questions that have intrigued biologists for over 100 years: In particular, was the original cnidarian a polyp-shaped animal, with medusae evolving later; or was the ancestral cnidarian a medusa, with the polyp form evolving later and leading, eventually, to the elimination of medusae in all anthozoans and many hydrozoans? And were cnidarians originally benthic, or originally planktonic? These issues remain unresolved despite many decades of active inquiry and considerable debate.[16] The 2 most commonly discussed evolutionary scenarios are summarized in Figure 6.18.

Lacking a medusa, gametes are produced directly by the anthozoan polyp. A planula larva develops from the fertilized egg and metamorphoses to form another polyp. The planula larvae of some anthozoan species can feed on phytoplankton and other microscopic food particles; no feeding planulae have been found in other cnidarian groups.

16. See *Topics for Further Discussion and Investigation,* no. 1.

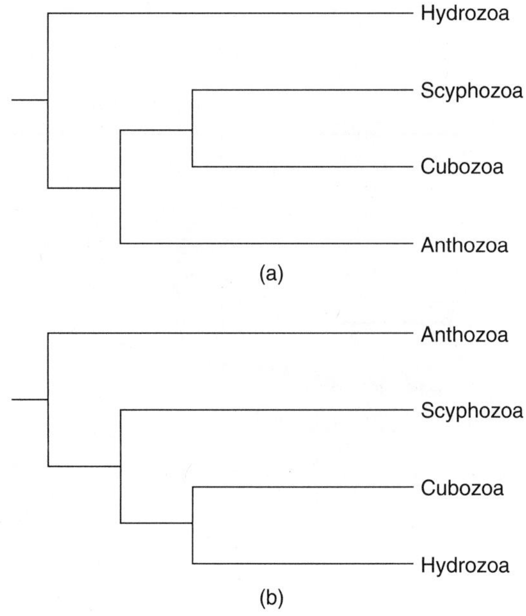

Figure 6.18
Two major hypotheses about cnidarian evolution. In (a) the original cnidarians would have had both polyps and medusae in the life cycle, while in (b) the polyp would have been the primary body plan, with the medusa form evolving later. Current evidence favors option (b).

Many anthozoan species also reproduce asexually, often through longitudinal or transverse **fission** or through a process of **pedal laceration,** in which parts of the pedal disc (foot) detach from the rest of the animal and gradually differentiate to form a new single-module ramet.[17] In coral reef species, fragmentation resulting from violent storms can also lead to substantial asexual replication.

As with most hydrozoans and scyphozoans, anthozoans are primarily carnivorous; they capture food using nematocyst-studded tentacles and transfer it to a central mouth opening. However, anthozoan polyps differ from those of hydrozoans in several respects. The anthozoan mouth opens into a tubular pharynx rather than directly into the gastrovascular cavity, and one or 2 discrete, ciliated grooves, called **siphonoglyphs,** typically extend down the pharynx from the mouth (Fig. 6.19). The anthozoan gastrovascular cavity is partitioned by numerous sheets of tissue called **mesenteries** or **septa,** whereas no mesenteries are found in the gastrovascular cavity of hydrozoan polyps. These infoldings of gastroderm and mesoglea greatly increase the surface area available for secreting digestive enzymes and absorbing nutrients. Mesenteries that extend far enough from the body wall into the gastrovascular cavity to actually attach to the pharynx are called **primary** or **complete mesenteries.** Those extending only partway into the gastrovascular cavity are termed **incomplete mesenteries.** The free edge of the incomplete mesenteries is trilobed, ciliated, and studded with nematocysts, cells that secrete digestive enzymes, and cells that phagocytose bacteria (Fig. 6.19d). Additionally, the mesenteries contain thick, longitudinal retractor muscles (Fig. 6.19b) and bear the gonads. Anthozoans are usually gonochoristic, but some species are sequential hermaphrodites.

Internally, near the base of an anthozoan, thin filaments called **acontia** extend in some species from the middle lobe of the mesenteries (Fig. 6.19a); they are loaded with nematocysts and secretory cells and can be extended outside the body through small pores in the body wall. Acontia are used both offensively and defensively, and they may also function in digestion.

A number of anemone species possess rings of small, spherical bulges extending around the circumference of the body column just below the tentacles. These hollow **acrorhagi** can be extended a substantial distance from the body column, possibly by forcing fluid into them from the gastrovascular cavity, with which they are continuous. Acrorhagi are covered with very potent nematocysts; they are used in defending a territory against invasion by other anemones, even different genets of the same species (Fig. 6.20). Some anemone species that lack acrorhagi have, instead, tentacles that are specialized for fighting. These **catch tentacles** are analogous to acrorhagi, functioning in aggressive encounters among individuals.[18]

Anthozoan tissues contain both circular and longitudinal muscle fibers (Fig. 6.21). Provided that the animal keeps its mouth closed by contracting appropriate sphincter muscles, the seawater in the gastrovascular cavity can serve as a hydrostatic skeleton (Chapter 5). For example, by closing the mouth, relaxing the longitudinal musculature, and contracting the circular muscles of the body wall, the animal becomes long and thin as the longitudinal muscles are stretched by the elevated pressure within the gastrovascular cavity (Fig. 6.22). Then, by contracting the longitudinal muscles on one side of the body and relaxing those on the other side, the animal can bend to one side, provided that the circular muscles are not permitted to stretch. In an emergency, all the longitudinal muscles can be contracted while the mouth is open, causing the animal to flatten considerably as the fluid of the gastrovascular cavity is expelled. The resulting shape has been referred to as the "bubble gum on a rock" disguise (Fig. 6.22c). Reinflation is rather slow, being dependent upon the activity of the cilia lining the siphonoglyphs; these cilia "drive" water back into the gastrovascular cavity.

Many anthozoan species can move from place to place under their own power, although usually very slowly. A fast-moving anemone might achieve a speed of several millimeters per minute. Some anemones are commonly found on the backs of more mobile invertebrates and are

17. See *Topics for Further Discussion and Investigation,* no. 10.

18. See *Topics for Further Discussion and Investigation,* no. 9.

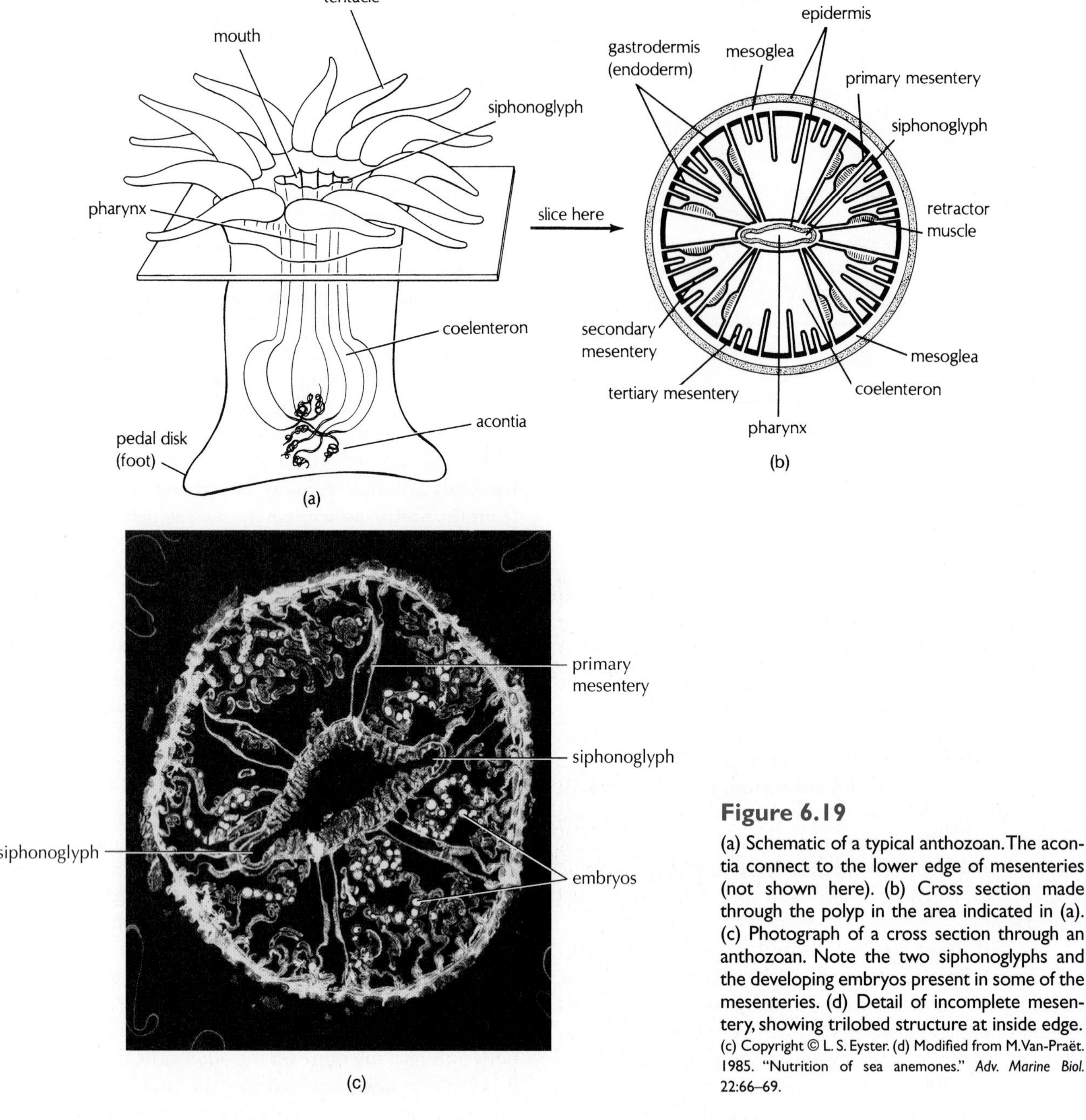

Figure 6.19

(a) Schematic of a typical anthozoan. The acontia connect to the lower edge of mesenteries (not shown here). (b) Cross section made through the polyp in the area indicated in (a). (c) Photograph of a cross section through an anthozoan. Note the two siphonoglyphs and the developing embryos present in some of the mesenteries. (d) Detail of incomplete mesentery, showing trilobed structure at inside edge.
(c) Copyright © L. S. Eyster. (d) Modified from M. Van-Praët. 1985. "Nutrition of sea anemones." *Adv. Marine Biol.* 22:66–69.

transported adventitiously. A few species can "swim" for short distances in response to predators,[19] but most anemones are basically stay-at-homes.

Some anthozoans show a decided bilateral symmetry, and that symmetry seems to be generated by the same molecular mechanisms (*Hox* gene expression) that generate bilateral symmetry in other animals.[20] This raises the tantalizing possibility that bilateral symmetry arose within the Cnidaria, before the split that gave rise to what are traditionally recognized as bilateral animals.

One of these bilaterally symmetrical anemones, *Nematostella vectensis,* is the first cnidarian to have its genome completely sequenced, the sequencing being completed in 2006.

Subclass Hexacorallia (= Zoantharia)

Subclass Hexa • corallia (= Zoantharia)
(G: six)
hex-ah-cor-ahl´-ē-ah (zō-an-thair´-ē-ah)

Hexacorallians possess numerous tentacles around the mouth opening (usually in some multiple of 6; *hexa* = G: 6) and 6 pairs of primary mesenteries (Fig. 6.19b). One pair of siphonoglyphs is associated with the pharynx. Many species in this subclass are **solitary** (i.e., they are independent ramets rather than colonies of connected modules)

19. See *Topics for Further Discussion and Investigation,* no. 2.

20. Finnerty, J. R., *et al.,* 2004. *Science* 304: 1335–37.

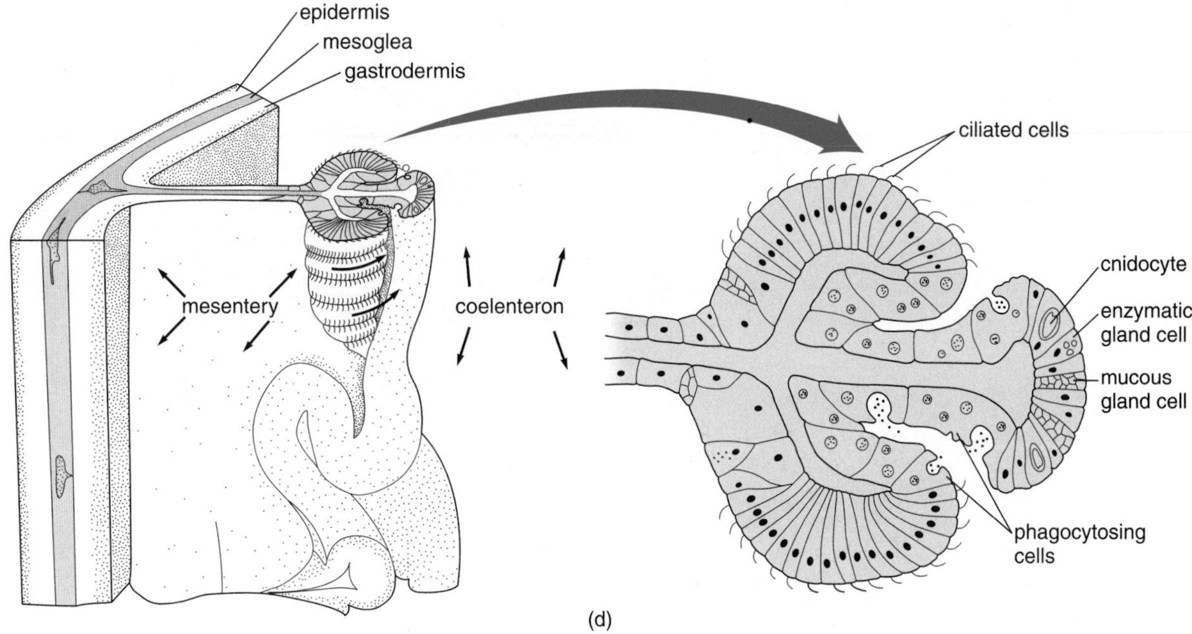

Figure 6.19 *Continued*

Figure 6.20
A sea anemone, *Anthopleura krebsi*. The individual on the left is displaying numerous stubby acrorhagi below its thinner feeding tentacles. The acrorhagi are used by the anemone to defend territory against invasion by neighboring anemones.

Courtesy of C. H. Bigger, from Bigger, 1980 *Biological Bulletin* 159:117 fig. 1, p. 120. Permission of Biological Bulletin.

and lack any specialized protective covering. These species are the sea anemones, as described in the class description. The other species in this subclass tend to be colonial; unlike the colonial hydrozoans, however, these anthozoans are never polymorphic. The best known of these colonial species are the true (or stony) corals, which secrete substantial external calcium carbonate skeletons. These species are also called scleractinian corals (*sclero* = G: hard), named

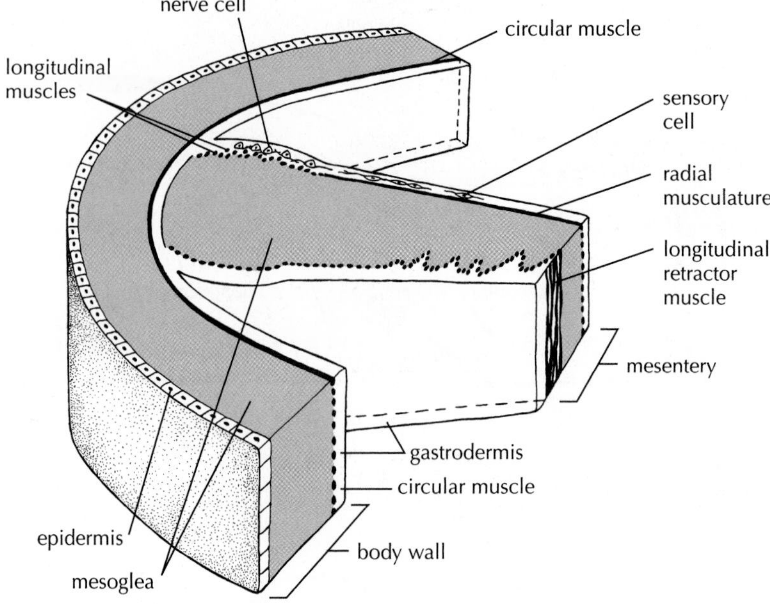

Figure 6.21
Diagrammatic illustration of the musculature and tissue layers in the body wall and on the mesentery (septum) of a typical anthozoan.
After Bullock and Horridge.

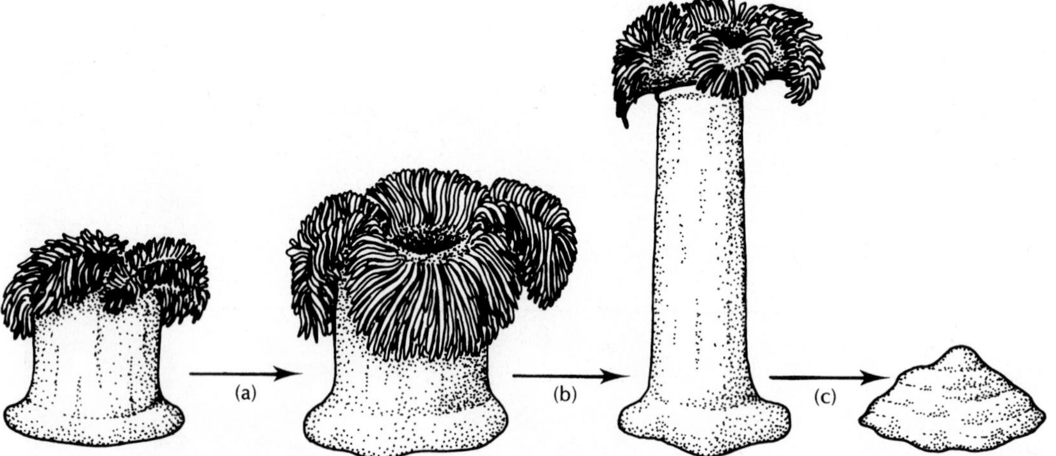

Figure 6.22
Shape changes in the sea anemone *Metridium senile*. (a) The animal is slowly inflating, using the cilia on the siphonoglyph to drive water into the gastrovascular cavity. (b) By closing the mouth, relaxing the longitudinal muscles, and contracting the circular muscles, the anemone increases in height but decreases in body width. (c) By opening the mouth and contracting the longitudinal muscles, most of the fluid in the gastrovascular cavity is rapidly expelled. To regain its original shape, the animal must pump water back into the coelenteron by ciliary action, a much slower process.
After Batham and Pantin.

for the order in which they are placed, the order Scleractinia (Figs. 6.23 and 6.25a,b). Scleractinian corals may be reef-building **(hermatypic)** or not **(ahermatypic)**.

Most hermatypic corals are restricted to clear, warm waters. Coral reefs are especially abundant in tropical areas of the Indo-Pacific, forming chains of islands and other structures of massive proportions. The Great Barrier Reef along Australia's northeast shoreline is more than 2,000 km (kilometers) long and 145 km wide. This reef is one of the world's most diverse, most complex ecosystems. Although other organisms (particularly calcareous red algae, foraminiferans, shelled molluscs, certain tube-dwelling polychaetes, and bryozoans) make significant contributions to reef structure and stability, anthozoans typically play the major role in constructing these spectacular habitats.[21] Coral calcification adds about 10 kg of additional $CaCO_3$ (calcium carbonate) per m^2 of reef yearly. Recent years have seen growing concern about the effects of disease, global warming, increasing seawater acidity, ozone depletion, increased commercial exploitation, and increased pollutant and sewage discharge on the health and stability of coral reef ecosystems.[22] In 1998, for

21. See *Topics for Further Discussion and Investigation,* no. 13.

22. See *Topics for Further Discussion and Investigation,* no. 15.

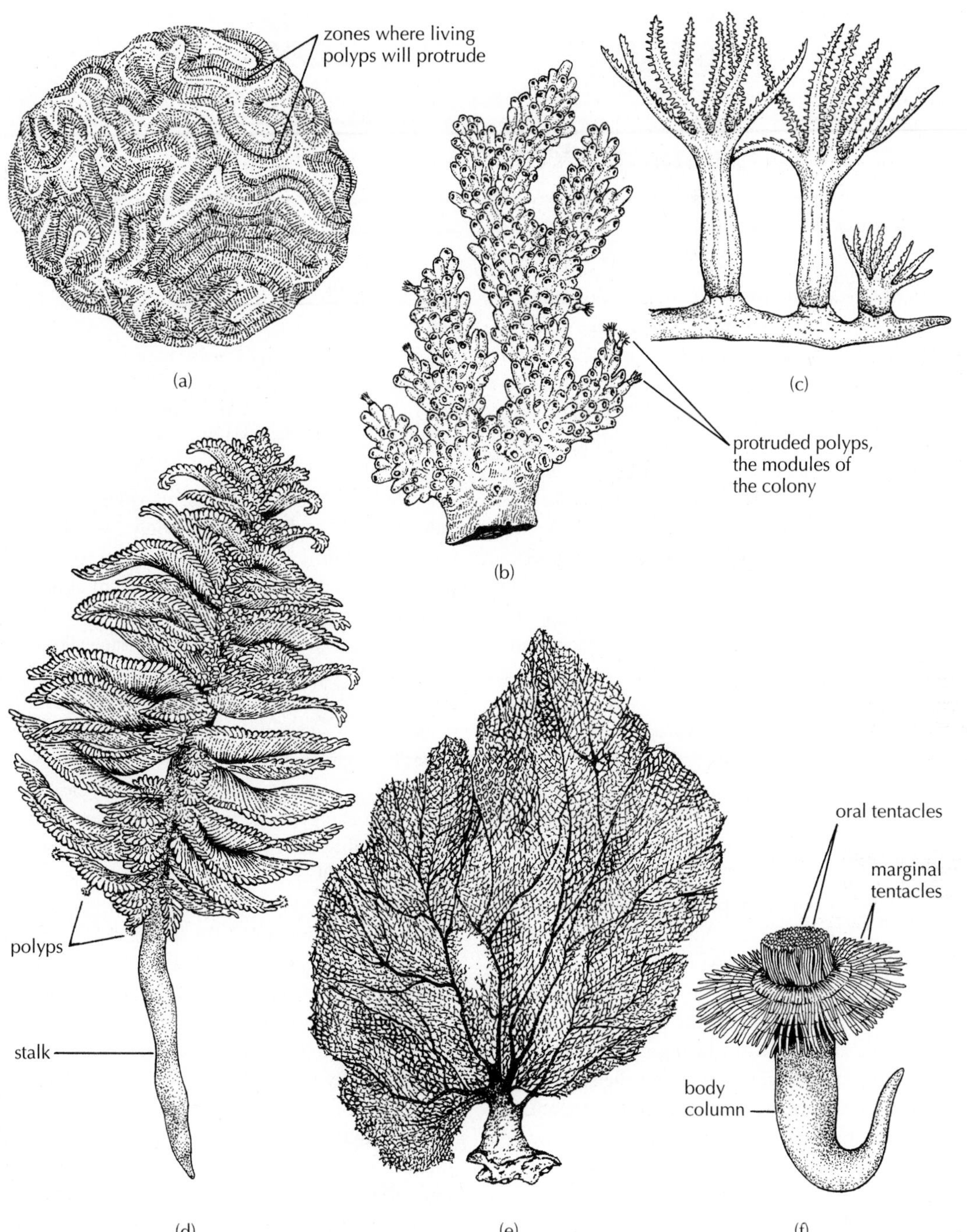

Figure 6.25

Anthozoan diversity. (a) The hexacoral, *Heliastra heliopora,* a brain coral. (b) Another hexacoral, *Astraea pallida,* with some of its polyps expanded and others retracted into the protective calcareous base of the colony. (c) An octocoral, *Clavularia,* showing the pinnate tentacles characteristic of the subclass Octocorallia. (d) *Pennatula* sp., the sea pen, another octocoral. (e) The sea fan *Gorgonia* sp., another octocoral. (f) A burrowing hexacoral anthozoan, *Cerianthus* sp., taken from the muddy substrate in which it lives. These animals (f) are structurally similar to sea anemones except that they lack a basal disc and have the tentacles arranged in two rings, as shown. Longitudinal muscles are especially well developed, permitting rapid withdrawal into mucus-lined burrows. Only the tentacles are generally visible at the surface of the sediment.
(a,b) After Kingsley. (c) After Gohar. (d) After Kolliker. (e) After Bayer and Owre.

The tissues of octocorallians have long been known to accumulate a variety of unusual biochemicals derived from fatty-acid metabolism.[24] These compounds are not directly involved in the metabolic processes of the corals, but seem instead to protect the coral from predation and overgrowth by other organisms (Research Focus Box 6.2). Recent research indicates that some of these chemicals kill certain mammalian tumors, increasing interest in cnidarians as potential sources of effective anticancer agents.

24. See *Topics for Further Discussion and Investigation,* no. 12.

RESEARCH FOCUS BOX 6.2

Chemical Defenses on Coral Reefs

O'Neal, W., and J. R. Pawlik. 2002. A reappraisal of the chemical and physical defenses of Carribbean gorgonian corals against predatory fishes. *Marine Ecol. Progr. Ser.* 240: 117–26.

Gorgonian corals are among the most conspicuous animals on Caribbean coral reefs, reaching densities as high as 25 colonies per m^2. Oddly enough for such conspicuous and soft-bodied organisms, few predators seem to eat them. This apparent invulnerability suggests that gorgonians are defended chemically or by their calcareous spicules, or both. Another possibility is that gorgonians are just too low in nuritional value to be of interest to most predators. In this paper, O'Neal and Pawlik considered these hypotheses for 32 gorgonian species living in the Bahamas and Key Largo.

The basic approach in this study was to offer tissue extracts or spicules from gorgonians to fish predators (the common wrasse, *Thalassoma bifasciatum*), and to then monitor how much the predators ate. To test for chemical or structural defenses, it's important to first remove size, shape, color, and behavior from the prey. This was done by making crude organic extracts of prey tissue; pieces were removed from the growing tips of gorgonian colonies (so that no colonies were killed during the study) and extracted in organic solvents. The resulting organic extracts were then incorporated into gelatinous pellets of uniform size and consistency, about 3–5 mm long, being careful to maintain the natural ratio of tissue concentration per unit volume. A homogenate of squid mantle tissue was included in each pellet, to mimic the nutritional value of gorgonian tissue and to make the pellets more appealing to the predatory wrasses. Similar pellets were also made in which gorgonian spicules rather than tissue extracts were added to pellets; chlorine bleach was used to remove gorgonian tissue from spicules, and thus to isolate the spicules. Again, the researchers made certain that spicule concentrations in the food pellets (per volume of pellet) were similar to those found in living gorgonians. Control pellets contained only squid mantle tissue without any gorgonian additives, and fish responses were included in the study only if the fish ate a control pellet before and after each assay.

Finally, the researchers also measured the per ml protein content, energy content, and inorganic ("ash") content of gorgonian samples, to see if any of these factors could explain differences in the palatability of the different gorgonian species tested.

Clearly, most of the gorgonians tested were chemically rather than physically defended—only 2 species (*B. asbestinum* [encrusting form] and *P. citrina*) were protected at all by their spicules when those spicules were present at natural concentrations (Focus Fig. 6.1, left side). In contrast, pellets with organic extracts of the gorgonian tissues alone were rarely eaten (Focus Fig. 6.1, right side). This dominance of chemical over physical defenses has also been recently demonstrated in 65 gorgonian species in the Indo Pacific, and 7 species in Guam.

Perhaps surprisingly, palatability to wrasses did not correlate with energy content of the prey (shown by the intensity of shading in Focus Fig. 6.1). For example, fish ate on average 3.5 pellets made from tissue extracts of *Iciligorgia schrammi* (Focus Fig. 6.1 right side), a gorgonian with low energy content, and about the same number of pellets made from *Muricea pendula,* a species with a much higher energy content. Similarly, palatability did not correlate with protein or ash content (data not shown here).

Note that in these studies the researchers tested gorgonian palatability to members of only one fish species; members of some other fish species might not be so readily deterred by the chemical defenses, or might be deterred as much or more by the spicules. How would you decide what other fish species, or what non-fish predators, to test? If spicules are not defensive, what do you suppose they do? How would you test your hypotheses? Finally, earlier studies

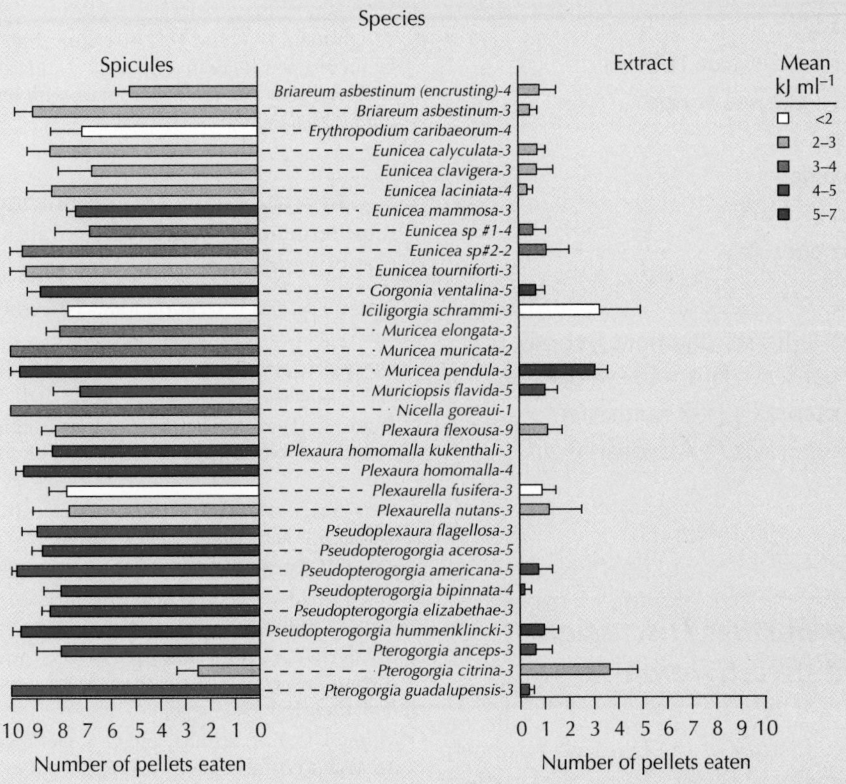

Focus Figure 6.1

The influence of tissue, spicules, and energy content on the palatability of 31 gorgonian species to blue wrasses *(Thalassoma bifasciatum)*. Each fish was offered 10 pellets in each assay; the number of assays conducted is indicated by the number adjacent to each bar. Each assay was conducted with material taken from a different gorgonian colony. Each bar represents the mean number of pellets consumed for a particular gorgonian species. Pellets that were not eaten were rejected by the fish; typically the pellets were tasted and then ignored, or taken into the mouth and then spat out. Darker coloration within bars represents higher energy content of gorgonian tissue (ranging from about 1.5–7 kJ ml^{-1})*.

*kilojoules per milliliter

have shown that spicules *do* protect against predators if they are present in pellets at higher than normal concentrations. If so, why do you think that spicules aren't found at those higher concentrations in gorgonian tissues? Can you think of any way to test your hypothesis?

Taxonomic Summary

Phylum Cnidaria (= Coelenterata)
Subphylum Medusozoa
- Class Scyphozoa—the true jellyfish
- Class Cubozoa—the sea wasps
- Class Hydrozoa
  - Order Hydroida
  - Order Siphonophora
  - Order Hydrocorallina
- Class Myxozoa
- Class Anthozoa—the sea anemones, corals, sea whips, sea pens, sea fans, and sea pansies
  - Subclass Hexacorallia (= Zoantharia)
  - Subclass Octocorallia (= Alcyonaria)

Topics for Further Discussion and Investigation

1. The evolutionary history of the Cnidaria has been argued about for many decades. Were the earliest cnidarians benthic or planktonic, were they polypoid or medusoid? Argue convincingly either that anthozoans are the most primitive, or the most advanced cnidarians. What does each scenario suggest about the evolution and loss of key characteristics within the phylum?

Bridge, D., C. W. Cunningham, R. DeSalle, and L. W. Buss. 1995. Class-level relationships in the phylum Cnidaria: Molecular and morphological evidence. *Molec. Biol. Evol.* 12:679.

Hyman, L. H. 1940. *The Invertebrates,* vol. I. New York: McGraw-Hill, 632–41.

Moore, J., and P. Willmer. 1997. Convergent evolution in invertebrates. *Biol. Rev.* 72:1.

Rees, W. J., ed. 1966. *The Cnidaria and Their Evolution.* New York: Academic Press.

2. Investigate the morphological and functional adaptations for swimming or burrowing encountered among some members of the Hexacorallia and Octocorallia, respectively.

Lawn, I. D., and D. M. Ross. 1982. The behavioural physiology of the swimming sea anemone *Boloceroides mcmurrichi. Proc. Royal Soc. London* 216B:315.

Mariscal, R. N., E. J. Conklin, and C. H. Bigger. 1977. The ptychocyst, a major new category of cnida used in tube construction by a cerianthid anemone. *Biol. Bull.* 152:392.

Robson, E. A. 1961. Some observations on the swimming behavior of the anemone *Stomphia coccinea. J. Exp. Biol.* 38:343.

Ross, D. M., and L. Sutton. 1967. Swimming sea anemones of Puget Sound: Swimming of *Actinostola* new species in response to *Stomphia coccinea. Science* 155:1419.

Sund, P. N. 1958. A study of the muscular anatomy and swimming behaviour of the sea anemone *Stomphia coccinea. Q. J. Microsc. Sci.* 99:401.

Weightman, J. O., and D. J. Arsenault. 2002. Predator classification by the sea pen *Ptilosarcus gurneyi* (Cnidaria): role of waterborne chemical cues and physical contact with predatory sea stars. *Canadian J. Zool.* 80:185.

3. To what extent are the nutritional and respiratory requirements of hermatypic corals and other anthozoans met by resident zooxanthellae?

Battey, J. F., and J. S. Patton. 1987. Glycerol translocation in *Condylactis gigantea. Marine Biol.* 95:37.

Davies, P. S. 1991. Effect of daylight variations on the energy budgets of shallow-water corals. *Marine Biol.* 108:137.

Dunn, K. W. 1988. The effect of host feeding on the contribution of endosymbiotic algae to the growth of green hydra. *Biol. Bull.* 175:193.

Fabricius, K. E., Y. Benayahu, and A. Genin. 1995. Herbivory in asymbiotic soft corals. *Science* 268:90.

Kevin, K. M., and R. C. L. Hudson. 1979. The role of zooxanthellae in the hermatypic coral *Plesiastrea urvellei* (Milne Edwards and Haime) from cold waters. *J. Exp. Marine Biol. Ecol.* 36:157.

Kinzie, R. A. III, and G. S. Chee. 1979. The effect of different zooxanthellae on the growth of experimentally reinfected hosts. *Biol. Bull.* 156:315.

Meyer, J. L., E. T. Schultz, and G. S. Helfman. 1983. Fish schools: An asset to corals. *Science* 220:1047.

Muscatine, L., and J. W. Porter. 1977. Reef corals: Mutualistic symbiosis adapted to nutrient-poor environments. *BioScience* 27:454.

Wethey, D. C., and J. W. Porter. 1976. Sun and shade differences in productivity of reef corals. *Nature (London)* 262:281.

4. Investigate the behavioral, morphological, or biochemical characteristics of cnidarians that increase the photosynthetic contributions of their symbiotic zooxanthellae.

Dykens, J. A., and J. M. Schick. 1984. Photobiology of the symbiotic sea anemone, *Anthopleura elegantissima:* Defenses against photodynamic effects, and seasonal photoacclimatization. *Biol. Bull.* 167:383.

Gates, R. D., O. Hoegh-Guldberg, M. J. McFall-Ngai, K. Y. Bil, and L. Muscatine. 1995. Free amino acids exhibit anthozoan "host factor" activity: They induce the release of photosynthate from symbiotic dinoflagellates *in vitro. Proc. Natl. Acad. Sci.* USA 92:7430.

Gates, R. D., K. Y. Bil, and L. Muscatine. 1999. The influence of an anthozoan "host factor" on the physiology of a symbiotic dinoflagellate. *J. Exp. Marine Biol. Ecol.* 232:241.

Grant, A. J., M. Rémond, and R. Hinde. 1998. Low molecular-weight factor from *Plesiastrea versipora* (Scleractinia) that modifies release and glycerol metabolism of isolated symbiotic algae. *Marine Biol.* 130:553.

Lasker, H. R. 1979. Light-dependent activity patterns among reef corals: *Montastrea cavernosa. Biol. Bull.* 156:196.

Schlicter, D., H. W. Fricke, and W. Weber. 1986. Light harvesting by wavelength transformation in a symbiotic coral of the Red Sea twilight zone. *Marine Biol.* 91:403.

Sebens, K. P., and K. DeRiemer. 1977. Diel cycles of expansion and contraction in coral reef anthozoans. *Marine Biol.* 43:247.

Sutton, D. C., and O. Hoegh-Guldberg. 1990. Host-zooxanthella interactions in four temperate marine invertebrate symbioses: Assessment of effect of host extracts on symbionts. *Biol. Bull.* 178:175.

5. Notwithstanding the obvious nutritional contributions of photosynthesizing symbionts to the cnidarian host, the algal cells may themselves exploit their situation within the host in obtaining nutrients for themselves. When the cnidarian captures and digests particulate food from the surrounding seawater, the resulting nutrients become potentially accessible to the intracellular algae. To what extent do zooxanthellae and zoochlorellae compete with their host for nutrients obtained by the host's feeding activities?

Blanquet, R. S., D. Emanuel, and T. A. Murphy. 1988. Suppression of exogenous alanine uptake in isolated zooxanthellae by cnidarian host homogenate fractions: Species and symbiosis specificity. *J. Exp. Marine Biol. Ecol.* 117:1.

Cook, C. B., C. F. D'Elia, and G. Muller-Parker. 1988. Host feeding and nutrient sufficiency for zooxanthellae in the sea anemone *Aiptasia pallida. Marine Biol.* 98:253.

McAuley, P. J. 1987. Quantitative estimation of movement of an amino acid from host to *Chlorella* symbionts in green hydra. *Biol. Bull.* 173:504.

McDermott, A. M., and R. S. Blanquet. 1991. Glucose and glycerol uptake by isolated zooxanthellae from *Cassiopea xamachana:* Transport mechanisms and regulation by host homogenate fractions. *Marine Biol.* 108:129.

Rees, T. A. V. 1991. Are symbiotic algae nutrient deficient? *Proc. Royal Soc. London B* 243:227.

Steen, R. G. 1986. Evidence for heterotrophy by zooxanthellae in symbiosis with *Aiptasia pulchella. Biol. Bull.* 170:267.

Thorington, G., and L. Margulis. 1981. *Hydra viridis:* Transfer of metabolites between *Hydra* and symbiotic algae. *Biol. Bull.* 160:175.

6. The nematocyst is a morphologically and functionally complex structure. Through what process are nematocysts formed by the cnidoblasts?

Skaer, R. J. 1973. The secretion and development of nematocysts in a siphonophore. *J. Cell Sci.* 13:371.

7. What is the relative importance of chemical versus physical stimuli in triggering nematocyst discharge?

Conklin, E. J., and R. N. Mariscal. 1976. Increase in nematocyst and spirocyst discharge in a sea anemone in response to mechanical stimulation. In *Coelenterate Ecology and Behavior,* edited by G. O. Mackie. New York: Plenum, 549–58.

Grosvenor, W., and G. Kass-Simon. 1987. Feeding behavior in *Hydra.* I. Effects of *Artemia* homogenate on nematocyst discharge. *Biol. Bull.* 173:527.

Kawaii, S., K. Yamashita, N. Nakai, and N. Fusetani. 1997. Intracellular calcium transients during nematocyst discharge in actinulae of the hydroid *Tubularia mesembryanthemum. J. Exp. Zool.* 278:299.

Thorington, G. U., and D. A. Hessinger. 1990. Control of cnida discharge. III. Spirocysts are regulated by three classes of chemoreceptors. *Biol. Bull.* 178:74.

Thorington, G. U., and D. A. Hessinger. 1998. Effect mechanisms of discharging cnidae: II. A nematocyst release response in the sea anemone tentacle. *Biol. Bull.* 195:145.

Watson, G. M., and D. A. Hessinger. 1994. Antagonistic frequency tuning of hair bundles by different chemoreceptors regulates nematocyst discharge. *J. Exp. Biol.* 187:57.

8. To what extent is nematocyst discharge under direct nervous control?

Aerne, B. L., R. P. Stidwill, and P. Tardent. 1991. Nematocyst discharge in *Hydra* does not require the presence of nerve cells. *J. Exp. Zool.* 258:137.

Lubbock, R. 1979. Chemical recognition and nematocyte excitation in a sea anemone. *J. Exp. Biol.* 83:283.

Mire-Thibodeaux, P., and G. M. Watson. 1993. Direct monitoring of intracellular calcium ions in sea anemone tentacles suggests regulation of nematocyst discharge by remote, rare epidermal cells. *Biol. Bull.* 185:335.

Pantin, C. F. A. 1942. The excitation of nematocysts. *J. Exp. Biol.* 19:294.

Ross, D. M., and L. Sutton. 1964. Inhibition of the swimming response by food and of nematocyst discharge during swimming in the sea anemone *Stomphia coccinea. J. Exp. Biol.* 41:751.

Ruch, R. J., and C. B. Cook. 1984. Nematocyst inactivation during feeding in *Hydra littoralis. J. Exp. Biol.* 111:31.

Sandberg, D. M., P. Kanciruk, and R. N. Mariscal. 1971. Inhibition of nematocyst discharge correlated with feeding in a sea anemone, *Calliactis tricolor* (Leseur). *Nature (London)* 232:263.

9. Sea anemones, corals, and hydrozoans must compete for space with each other, as well as with members of many other animal (and plant or algal) groups. What are the roles of acrorhagi, acontia, and catch tentacles in mediating the competition for space among sea anemones and corals? To what extent does the ability to distinguish self from nonself contribute to competitive ability? Through what mechanisms are such distinctions made?

Ayre, D. J. 1982. Inter-genotype aggression in the solitary sea anemone *Actinia tenebrosa. Marine Biol.* 68:199.

Bigger, C. H. 1980. Interspecific and intraspecific acrorhagial aggressive behavior among sea anemones: A recognition of self and non-self. *Biol. Bull.* 159:117.

Chadwick, N. E. 1987. Interspecific aggressive behavior of the corallimorpharian *Corynactis californica* (Cnidaria: Anthozoa): Effects on sympatric corals and sea anemones. *Biol. Bull.* 173:110.

Chornesky, E. A. 1983. Induced development of sweeper tentacles on the reef coral *Agaricia agaricites:* A response to direct competition. *Biol. Bull.* 165:569.

Francis, L. 1973. Intraspecific aggression and its effect on the distribution of *Anthopleura elegantissima* and some related anemones. *Biol. Bull.* 144:73.

Fukui, Y. 1986. Catch tentacles in the sea anemone *Haliplanella luciae.* Role as organs of social behavior. *Marine Biol.* 91:245.

Kramer, A., and L. Francis. 2004. Predation resistance and nematocyst scaling for *Metridium senile* and *M. farcimen. Biol. Bull.* 207:130–40.

Lange R. G., M. H. Dick, and W. A. Müller. 1992. Specificity and early ontogeny of historecognition in the hydroid *Hydractinia. J. Exp. Zool.* 262:307

Langmead, O., and N. E. Chadwich-Furhman. 1999. Marginal tentacles of the corallimorpharian *Rhodactis rhodostsoma*. 2. Induced development and long-term effects on coral competitors. *Marine Biol.* 134:491.

Miles, J. S. 1991. Inducible agonistic structures in the tropical corallimorpharian, *Discosoma sanctithomae. Biol. Bull.* 180:406.

Nozawa, Y., and Y. Loya. 2005. Genetic relationship and maturity state of the allorecognition system affect contact reactions in juvenile *Seriatopora* corals. *Mar. Ecol. Progr. Ser.* 286:115–23.

Salter-Cid, L., and C. H. Bigger. 1991. Alloimmunity in the gorgonian coral *Swiftia exserta. Biol. Bull.* 181:127.

Sauer, K. P., M. Muller, and M. Weber. 1986. Alloimmune memory for glycoprotein recognition molecules in sea anemones competing for space. *Marine Biol.* 92:73.

Sebens, K. P., and J. S. Miles. 1988. Sweeper tentacles in a gorgonian octocoral: Their function in competition for space. *Biol. Bull.* 175:378.

Turner, V. L. G., S. M. Lynch, L. Paterson, J. L. León-Cortés, and J. P. Thorpe. 2003. Aggession as a function of genetic relatedness in the sea anemone *Actinia equina* (Anthozoa: Actiniaria). *Marine Ecol. Progr. Ser.* 247:85.

10. Some form of asexual reproduction is encountered among all 3 classes of cnidarians. What are the adaptive benefits of asexual versus sexual reproduction?

Grassle, J. F., and J. M. Schick, eds. 1979. Ecology of asexual reproduction in animals. *Amer. Zool.* 19:667.

Kramarsky-Winter, E., M. Fine, and Y. Loya. 1997. Coral polyp expulsion. *Nature* 387:137.

Lirman, D. 2000. Fragmentation in the branching coral *Acropora palmata* (Lamarck): growth, survivorship, and reproduction of colonies and fragments. *J. Exp. Marine Biol. Ecol.* 251:41.

11. How are swimming and other behaviors mediated by the cnidarian nervous system?

Leonard, J. L. 1982. Transient rhythms in the swimming activity of *Sarsia tubulosa* (Hydrozoa). *J. Exp. Biol.* 96:181.

Lerner, J., S. A. Meleon, I. Waldron, and R. M. Factor. 1971. Neural redundancy and regularity of swimming beats in scyphozoan medusae. *J. Exp. Biol.* 55:177.

Mackie, G. O. 1990. Giant axons and control of jetting in the squid *Loligo* and the jellyfish *Aglantha. Canadian J. Zool.* 68:799.

Mackie, G. O., and R. W. Meech. 1995. Central circuitry in the jellyfish *Aglantha digitale.* I. The relay system. *J. Exp. Biol.* 198:2261.

Mire, P. 1998. Evidence for stretch-regulation of fission in a sea anemone. *J. Exp. Zool.* 282:344.

Sawyer, S. J., H. B. Dowse, and J. M. Shick. 1994. Neurophysiological correlates of the behavioral response to light in the sea anemone *Anthopleura elegantissima. Biol. Bull.* 186:195.

12. To what extent do cnidarians use chemical and structural defenses against predation?

Fenical, W., and J. R. Pawlik. 1991. Defensive properties of secondary metabolites from the Caribbean gorgonian coral *Erythropodium caribaeorum. Marine Ecol. Progr. Ser.* 75:1.

Harvell C. D., W. Fenical, and C. H. Greene. 1988. Chemical and structural defenses of Caribbean gorgonians (*Pseudopterogorgia* spp.) I. Development of an *in situ* feeding assay. *Marine Ecol. Progr. Ser.* 49:287.

La Barre, S. C., J. C. Coll, and P. W. Sammarco. 1986. Defensive strategies of soft corals (Coelenterata: Octocorallia) of the Great Barrier Reef. II. The relationship between toxicity and feeding deterrence. *Biol. Bull.* 171:565.

O'Neal, W., and J. R. Pawlik. 2002. A reappraisal of the chemical and physical defenses of Caribbean gorgonian corals against predatory fishes. *Marine Ecol. Progr. Ser.* 240:117.

Puglisi, M. P., V. J. Paul, J. Biggs, and M. Slattery. 2002. Co-occurrence of chemical and structural defenses in the gorgonian corals of Guam. *Marine Ecol. Progr. Ser.* 239:105.

Sammarco, P. W., S. La Barre, and J. C. Coll. 1987. Defensive strategies of soft corals (Coelenterata: Octocorallia) of the Great Barrier Reef. III. The relationship between ichthyotoxicity and morphology. *Oecologia* (Berlin) 74:93.

Stachowicz, J. J., and N. Lindquist. 2000. Hydroid defenses against predators: the importance of secondary metabolites versus nematocysts. *Oecologia* 124:280.

13. What factors influence growth rates and growth forms of hermatypic corals?

Al-Horani, F. A., S. M. Al-Moghrabi, and D. de Beer. 2003. The mechanisms of calcification and its relation to photosynthesis and respiration in the scleractinian coral *Galaxea fascicularis. Marine Biol.* 142:419.

Dunstan, P. 1975. Growth and form in the reef-building coral *Montastrea annularis. Marine Biol.* 33:101.

Goreau, T. F., N. I. Goreau, and T. J. Goreau. 1979. Corals and coral reefs. *Sci. Amer.* 241:124.

Tambutté, É., D. Allemand, E. Mueller, and J. Jaubert. 1996. A compartmental approach to the mechanism of calcification in hermatypic corals. *J. Exp. Biol.* 199:1029.

14. How does an economist resemble a sexually mature cnidarian?

15. To what extent is human activity influencing the incidence of coral bleaching and bacterial infection?

Ben-Haim, Y., and E. Rosenberg. 2002. A novel *Vibrio* sp. pathogen of the coral *Pocillopora damicornis. Marine Biol.* 141:47.

Bunkley-Williams, L., and E. H. Williams Jr. 1990. Global assault on coral reefs. *Nat. Hist.* (April):46.

Garrison, V. H., E. A. Shinn, W. T. Foreman, D. W. Griffin, C. W. Holmes, C. A. Kellogg, M. S. Majewski, L. L. Richardson, K. B. Ritchie, and G. W. Smith. 2003. African and Asian dust: From desert soils to coral reefs. *BioScience* 53:469.

Gleason, D. F., and G. M. Wellington. 1993. Ultraviolet radiation and coral bleaching. *Nature* 365:836.

Jompa, J., and L. J. McCook. 2002. The effects of nutrients and herbivory on competition between a hard coral (*Porites cylindrica*) and a brown alga (*Lobophora variegata*). *Limnol. Oceanogr.* 47:527–34.

Kushmaro, A., Y. Loya, M. Fine, and E. Rosenberg. 1996. Bacterial infection and coral bleaching. *Nature* 380:396.

Rowan, R., N. Knowlton, A. Baker, and J. Jara. 1997. Landscape ecology of algal symbionts creates variation in episodes of coral bleaching. *Nature* 388:265.

Saxby, T., W. C. Dennison, and O. Hoegh-Guldberg. 2003. Photosynthetic responses of the coral *Montipora digitata* to cold temperature stress. *Marine Ecol. Progr. Ser.* 248:85.

16. Coral bleaching, in which symbiotic dinoflagellates (zooxanthellae) are actively expelled from their coral hosts, is a common response to elevated temperature. What is the potential for the coral-zooxanthellae system to adapt or acclimate to increasing water temperatures over time?

Baker, A. C., C. J. Starger, T. R. McClanahan, and P. W. Glynn. 2004. Corals' adaptive response to climate change. *Nature* 430: 741.

Goulet, T. L. 2007. Most scleractinian corals and octocorals host a single symbiotic zooxanthella clade. *Mar. Ecol. Progr. Ser.* 335: 43–248.

Trapido-Rosenthal, H., S. Zielke, R. Owen, L. Buxton, B. Boeing, R. Bhagooli, and J. Archer. 2005. Increased zooxanthellae nitric oxide synthase activity is associated with coral bleaching. *Biol. Bull.* 208: 3–6.

Visram, S., and A. E. Douglas. 2007. Resilience and acclimation to bleaching stressors in the scleractinian coral *Porites cylindrica*. *J. Exp. Mar. Biol. Ecol.* 349: 35–44.

17. Why would it be ambiguous to speak of "individuals" for species that exhibit fission, budding, and other forms of asexual replication?

18. Why is the term "jellyfish" an inadequate descriptor for Scyphozoans?

Taxonomic Detail

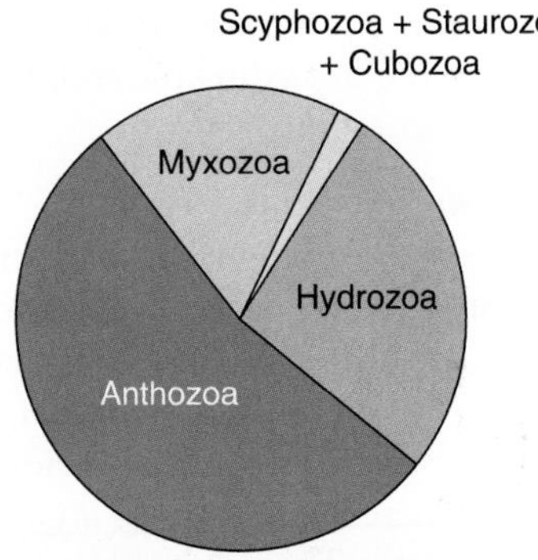

Phylum Cnidaria (= Coelenterata)

This phylum contains approximately 11,200 species, distributed among 6 classes. The phylogenetic relationships among some of the classes are still uncertain.

Subphylum Medusozoa

This group contains all cnidarians except for the anthozoans and probably the Myxozoa.

Class Staurozoa

Order Stauromedusae

Haliclystus. Unlike other scyphozoans, the medusae in this order are always sessile and develop directly from the scyphistoma stage without strobilation. Thus, only one juvenile arises from each scyphistoma; each planula larva can, however, bud off other, genetically identical planulae, so that numerous, genetically identical juveniles can still be produced from each fertilized egg. The planulae are nonciliated, creeping creatures, unlike those of other cnidarians. Adults attach to firm substrates, such as macroalgae and stones, using a central adhesive disc. Some species can move from place to place, but none can swim. Two families containing about 50 species.

Order Coronatae

Stephanoscyphus. This order contains over 24 mostly large, deep-water species. Members of the genus *Stephanoscyphus* are unique among scyphozoans in that the scyphistoma secretes a chitinous perisarc about itself. The medusae may reproduce sexually within the perisarc, never leaving the tube. The members of at least one species produce ephyrae that are said to develop directly into planulae, which then settle and attach to a substrate. Seven families.

Class Scyphozoa

The true jellyfish (about 125 species, all marine), in which medusae are generated from polyps, through strobilation. Four orders.

Order Semaeostomeae

Aurelia—the moon jelly. *Cyanea*—the lion's mane jellyfish, which produces the largest of all known medusae; individuals can reach 2 m in diameter and bear tentacles up to 30 m long. Contact with the oral arms or the tentacles extending from the circumference of the swimming bell brings a painful sting to unwary swimmers along the east coast of the United States. Certain species of fish and crustaceans often live in close association with these jellyfish, as commensals, external parasites, or predators. *Aurelia aurita* can reach concentrations of 300–600 medusae per m^3. *Chrysaora*—the sea nettle. Sea nettles are bothersome to summer swimmers along the Atlantic coast of North America. A huge, bright red jellyfish (currently named "Big Red!") was discovered in 2003, living at depths of about 600–1450 m in the Pacific Ocean. The swimming bell grows to at least 0.9 m in diameter, and instead of numerous thin tentacles the animal bears 4–7 thick arms. Three families.

Order Rhizostomeae

Cassiopea, Rhizostoma, Stomolophus. Instead of a central mouth, there are many small mouths opening into a complex canal system formed by the fusion of oral arms over the original mouth opening. Jellyfish in the genus *Cassiopea* grow to 30 cm or so in diameter, and all harbor symbiotic unicellular algae (zooxanthellae), which give the animals a distinctive greenish brown color. Often called the upside-down jellyfish, members of this genus typically lie upside down, pulsating on the substrate, presumably farming their zooxanthellae by maximizing their exposure to sunlight. Ten families.

Class Cubozoa

One order

Order Cubomedusae

Carybdea, Chironex, Tripedalia—the sea wasps. The jellyfish contained in this order are mostly small with a highly transparent swimming bell that is essentially square in transverse section. Each animal has 4 tentacles; a 2.5-cm individual can have tentacles up to 30 cm long. Sea wasps swim actively, up to 6 m per minute, and are common in all tropical and subtropical oceans. Their sting is exceedingly painful and can be fatal to humans. *Chironex fleckeri* has been called the most dangerous venomous animal in the world. Individuals range from thimble-sized to basketball sized, with tentacles up to 15 feet long; the species is found only in parts of the Pacific Ocean. Two families, 20 species.

Class Hydrozoa

The medusae often have unknown polyp stages and vice versa, so the taxonomy is likely to be reorganized considerably once more life cycles have been worked out. Seven orders, about 3000 species.

Order Hydroida

This group contains most of the species in the class Hydrozoa. Three suborders.

Suborder Anthomedusae

Hydranths lack a chitinous covering. Free-living medusae, if produced, are always bell shaped. *Bougainvillia, Sarsia, Eudendrium, Pennaria, Hydractinia, Podocoryne, Stylactis, Tubularia, Hydra* (freshwater species with no medusa stage in the life cycle), *Cordylophora* (a medusa-producing hydroid from rivers and brackish waters). This suborder contains most of the species in the class. Species in the genus *Tubularia* produce **actinula** larvae (see also order Trachylina below). *Hydractinia* forms pinkish, bushy, highly polymorphic colonies covering marine gastropod shells inhabited by hermit crabs. Thirty-two families.

Suborder Leptomedusae

Aequorea, Campanularia, Clytia, Eugymnanthea, Obelia, Sertularia. Hydranths possess a chitinous covering. Medusae, when present, are always flat and never bell shaped. All species are marine. The colonial hydroid *Clytia gracilis* is planktonic, sometimes occurring in dense aggregations. Members of the genus *Eugymnanthea* live exclusively in the mantle cavity of some oysters, mussels, and other bivalves.

Suborder Limnomedusae

Craspedacusta (the common and widespread freshwater jellyfish), *Gonionemus, Limnocodium, Pochella.* Most species occur only in freshwater, with some groups restricted to lakes in Africa and India. A few species in the group are marine, with highly restricted distributions. Species of *Pochella*, for example, are exclusively marine and live only on the rim of tubes built by certain sedentary polychaete annelids (sabellid polychaetes). The medusae produced by at least some species of *Gonionemus* have adhesive pads on the marginal tentacles; they use these pads to adhere to algae and sea grasses between bouts of swimming.

Order Milleporina

Millepora—fire coral. The few species in this order are all colonial and restricted to coral reefs, and all host symbiotic zooxanthellae. The nematocysts are highly irritating to humans. All milleporine species secrete substantial calcium carbonate skeletons, making significant contributions to reef mass. The members of this and the next order, the Stylasterina, are sometimes combined to form a single order, the Hydrocorallina.

Order Stylasterina

The hydrocorals. *Allopora Stylaster.* These widespread, colonial, marine hydrozoans secrete substantial calcium carbonate skeletons, although they, like the milleporines previously discussed, are not true corals. The species in this group are found in both warm and cold waters, and at all depths. The members of this order are sometimes combined with the members of the preceding order to form a single order, the Hydrocorallina.

Order Trachylina

Liriope. Most species are marine and restricted to warm open-ocean waters. Planulae derived from the sexual adventures of the medusa stage develop further into highly specialized, tentacled, swimming **actinula** larvae, which eventually develop directly into medusae. There is typically no attached polyp stage in the life cycle. Three suborders.

Suborder Narcomedusae

In this group, the actinulae bud off additional actinulae asexually before developing to the medusa stage. Although most species are marine, one species is an internal parasite of freshwater fish in the former Soviet Union, making it the only known parasitic adult cnidarian.

Order Siphonophora

Physalia—the Portuguese man-of-war. All species form marine, polymorphic colonies that spend their lives swimming or floating in the water. The nematocysts inject highly toxic, paralyzing secretions, making the species in this group dangerous to humans, especially small children.

Order Chondrophora

Porpita—the blue button; *Velella*—by-the-wind sailor. All species are marine, and individuals consist of what most workers believe is a single, large polyp with a gas-filled float that maintains the animal at the ocean surface. Some species also possess a "sail," allowing the animal to move by wind power. *Velella* is often found associated with the gastropod snail *Janthina janthina,* which eats the "sailor" as they travel together. In all members of the order, the floating polyps bud off small, free-swimming medusae, which represent the sexual stage of the life cycle.

Order Actinulida

Halammohydra. Individuals of all species in this group are marine, solitary (i.e., never forming colonies), and very small, typically less than 1.5 mm long. All members live interstitially, in the spaces between sand grains. There is no free-living medusa stage in the life cycle, and the polyp somewhat resembles the actinula stage of the trachylinids. The entire body is ciliated, permitting individuals to swim from place to place. Fewer than 10 species have been described.

Two Groups of Non-Medusozoans

Class Myxozoa

More than 2000 species have been described, distributed among about 20 families. All are internal parasites—primarily in fish, but to a much lesser extent in some bryozoans, amphibians, and reptiles as well. Members of most species are specific for particular hosts, and for particular tissues within those hosts. Myxozoans may eventually be incorporated into the Medusozoa, but their relationships to the members of that subphylum are still uncertain.

Subclass Myxosporea

This group includes all but 2 of the myxozoan species.

Family Myxosomatidae. *Myxobolus, Sphaerospora.* This is generally regarded as one of the most important of myxozoan families; all members parasitize fish.

Subclass Malacosporea (*Buddenbrockia, Tetracapsuloides*)

The group contains only 2 known species, both of which parasitize freshwater bryozoans. *Buddenbrockia plumatellae* is worm-shaped, with 4 blocks of radially arranged longitudinal muscles that it uses to generate sinusoidal waves for locomotion within its bryozoan host.

Class Anthozoa

All species are marine. About two-thirds of all described cnidarian species are contained in this class. Fifteen orders, 6000 species.

Subclass Alcyonaria (= Octocorallia)

Eight orders.

Order Stolonifera

Tubipora—organ-pipe coral. These are found only on coral reefs of the tropical Indo-Pacific.

Order Gorgonacea

Gorgonia—sea fans; *Plexaura, Plexaurella; Briareum*—deadman's fingers; *Leptogorgia*—sea whips; *Pseudopterogorgia*—sea feathers. Eighteen families. Most species are tropical, and all form colonies. Most colonies are supported by a firm, central, proteinaceous skeleton of gorgonin.

Order Alcyonacea

The soft corals. *Alcyonium, Sinularia.* These colonial anthozoans lack a rigid skeleton (calcareous or proteinaceous) but do have calcareous spicules (sclerites) embedded in the tissues. Species occur in all oceans, especially in the tropics.

Order Pennatulacea

Ptilosarcus—sea pens; *Renilla*—sea pansies. All species in this group are marine and adapted for life on soft substrates. A single, long axial polyp extends the length of the colony, with additional, much smaller polyps typically occurring on side branches (except in *Renilla*). The base of the axial polyp anchors the colony into the substrate. Sea pens stand erect. Sea pansies, which lie flat on the substrate, typically exhibit strong bioluminescence in the dark.

Subclass Zoantharia (= Hexacorallia)

Seven orders

Order Actiniaria

The sea anemones. *Bunodactis, Aiptasia, Anthopleura, Actinia, Diadumene, Calliactis, Bunodosoma, Edwardsia, Haliplanella, Metridium, Nematostella, Sagartia, Stomphia.* This large group contains about 800 species, distributed among 41 families. There are no colonial species in this order. Some members of one genus, *Stomphia,* can swim, although awkwardly, if attacked by predators. After detaching from the substrate, the anemone swims short distances by violently contracting the longitudinal body wall muscles on one side and then the other, bending the body from side to side. *Nematostella vectensis* (the starlet sea anemone), the first cnidarian to have its genome completely sequenced (completed in 2006), has become a leading model in studies of early metazoan evolution.

Order Corallimorpharia

Corynactis. Species in this order occur worldwide in shallow and deep water, in the tropics, in the temperate zone, and even at the poles. Members of all species are solitary, rather than colonial, and lack rigid skeletal elements.

Order Scleractinia

The true (stony, or hard) corals. The hermatypic (reef-building) species all form symbiotic relationships with unicellular algae (zooxanthellae or zoochlorellae). Many species commonly propagate asexually from branches that break off during storms. *Acropora*—staghorn and elkhorn corals, the most important reef-building species; *Astrangia; Agaricia; Fungia*—mushroom corals, forming the largest known solitary coral polyps, up to 1 m in diameter; *Porites*—another major reef builder; *Stylophora; Siderastraea; Pocillopora; Oculina; Montastraea; Diploria*—brain coral.

Order Zoanthinaria (= Zoanthidea)

Zoanthus. Polyps may be solitary or colonial, but they never secrete a solid skeleton. Three families.

Order Ceriantharia

Cerianthus. The species in this group are all solitary, living in tubes buried vertically in soft sediments; only the animal's oral disc is visible to the snorkeler. The body column musculature is largely ectodermal, allowing the animal to withdraw rapidly into its tube when attacked. The tubes are made by specialized nematocysts called **ptychocysts**, unique to members of this order. Unlike other anthozoans, the oral disc of cerianthids bears 2 distinct whorls of tentacles. Three families.

Order Ptychodactiaria

These anemonelike anthozoans are all restricted to cold waters in and near the Arctic and Antarctic oceans.

Order Antipatharia

Antipathes. The black or thorny corals. Most members of this order are deep-water tropical species. They are commonly exploited for jewelry making. The skeletal support system is composed of chitinous fibrils and protein that can be bent and shaped when heated. This axial skeleton was once thought to have medicinal properties (*anti* = Latin: against; *pathes* = L: disease).

General References about the Cnidarians

Arai, M. N. 1997. *A Functional Biology of Scyphozoa.* London: Chapman & Hall.

Blackstone, N. W., and R. E. Steele. 2005. Introduction to the symposium. *Integr. Comp. Biol.* 45:583–84. (This paper is followed by a series of papers from the 2004 symposium, "Model systems for the basal metazoans: cnidarians, ctenophores, and placozoans," 4 of which deal exclusively with cnidarian biology: fission, the genetics of self, non-self recognition, programmed cell death (apoptosis), and the symbiotic relationship between zooxanthellae and anthozoans).

Darling, J. A., A. R. Reitzel, P. M. Burton, M. E. Mazza, J. F. Ryan, J. C. Sullivan, and J. R. Finnerty. 2005. Rising starlet: The starlet sea anemone *Nematostella vectensis. BioEssays* 27: 11–21.

Halstead, B. W., P. S. Auerbach, and D. R. Campbell. 1990. *A Color Atlas of Dangerous Marine Animals.* Boca Raton, Florida: CRC Press, Inc.

Harrison, F. W., and J. A., eds. 1991. *Microscopic Anatomy of Invertebrates, Vol. 2. Placazoa, Porifera, Cnidaria, and Ctenophora.* New York: Wiley-Liss.

Hessinger, D. A., and H. M. Lenhoff, eds. 1988. *The Biology of Nematocysts.* San Diego, Calif.: Academic Press.

Jiménez-Guri, E. B. Okamura, and P. W. H. Holland. 2007. Origin and evolution of a myxozoan worm. *Integr. Comp. Biol.* 47:752–58.

Lesser, M. P. 2004. Experimental biology of coral reef ecosystems. *J. Exp. Mar. Biol. Ecol.* 300:217–52.

Mackie, G. O., ed. 1976. *Coelenterate Ecology and Behavior.* New York: Plenum Press.

Purcell, J. E., S. Uye, and W. T. Lo. 2007. Anthropogenic causes of jellyfish blooms and their direct consequences for humans: A review. *Mar. Ecol. Progr. Ser.* 350:153–74.

Roberts, L. S., and J. Janovy, Jr. 2004. *Foundations of Parasitology,* 7th ed. McGraw-Hill Publishers.

Shick, J. M. 1991. *A Functional Biology of Sea Anemones.* New York: Chapman & Hall.

Sutherland, S. K., and J. Tibballs. 2001. *Australian Animal Toxins.* New York: Oxford University Press.

Thorpe, J. H., and A. P. Covich. 2001. *Ecology and Classification of North American Freshwater Invertebrates,* 2nd ed. New York: Academic Press.

Tops, S., A. Curry, and B. Okamura. 2005. Diversity and systematics of the Malacosporea (Myxozoa). *Invert. Biol.* 124:285–95.

Search the Web

1. www.mbl.edu

   This site is maintained by the Marine Biological Laboratories in Woods Hole, MA. Under "Quick Links," choose "Biological Bulletin," and then "Other Biological Bulletin Publications." Next select "Keys to Marine Invertebrates of the Woods Hole Region," and then choose "Cnidaria." This brings up a taxonomic key, including a glossary of terms appropriate to the taxon. Once at the site, click on the cnidarian class of interest.

   Also, from the same home page, you can instead choose "Marine Organisms" under "Quick Links" and then select the "Flescher Slide Collection" from the Image Databases menu to see color photographs of various cnidarians common in the waters near Woods Hole, MA, along with explanatory captions.

2. http://biodidac.bio.uottawa.ca

   Choose "Organismal Biology," "Animalia," and then "Cnidaria" for photographs and drawings, including illustrations of sectioned material.

3. www.ucmp.berkeley.edu/cnidaria/cnidaria.html

   Information about all cnidarian groups, and color images of many representatives. The site is operated through the University of California at Berkeley.

4. http://tolweb.org/tree/phylogeny.html

   This brings you to the Tree of Life site, run by the University of Arizona. Use the search terms "Cnidaria" and "Myxozoa" to find information about cnidarian features and phylogenetic relationships, along with excellent images.

5. http://faculty.washington.edu/cemills/

   This site, written by Claudia Mills, provides information, along with arresting photographs, about a number of interesting cnidarians. It includes a detailed discussion about the bioluminescent jellyfish *Aequorea victorea* from Puget Sound, Washington. This jellyfish is the source of several markers that are now widely used in biomedical research.

6. http://animaldiversity.ummz.umich.edu/animalia.html

   Search under "Cnidaria" to find interesting information about selected species in each class. This site is provided by the Museum of Zoology at the University of Michigan.

7

The Ctenophores

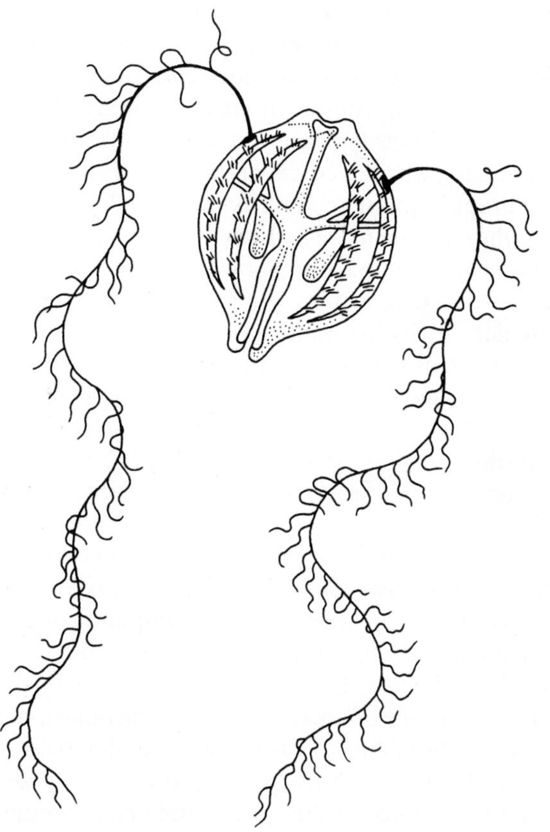

Introduction and General Characteristics

Phylum Cteno • phora
(Greek: comb bearing)
teen´-ō-for´-ah (or, teen-ahf´-or-ah)

Defining Characteristics:[1] 1) Plates of fused cilia arranged in rows; 2) adhesive prey-capturing cells (colloblasts)

Nearly all ctenophores are predators, and most are nearly transparent. Only one parasitic species is known. The 150 or so described species are exclusively marine, and most are **planktonic;** that is, members of most species are weak swimmers, carried about by ocean currents. Ctenophores can be major predators upon larval fish and other zooplankton, at times playing major roles in estuarine ecology and influencing the success of commercial fisheries.

As a particularly dramatic example of their impact, the invasive species *Mnemiopsis leidyi* has devastated the anchovy fishery in the Black Sea in the approximately 25 years since its accidental introduction, in part by consuming fish eggs and larvae and in part by competing with them for food, by consuming other zooplankton that the fish feed on. The Caspian Sea, where the same species was introduced about 8 years ago—probably through the discharge of ship ballast water—has already seen a dramatic decline in fish catches; predators on those fish, including the endemic Caspian seal, are now also endangered. *M. leidyi* has also now been reported from the Baltic Sea (2006), where it threatens in particular the cod-fish industry, and even more recently (2007) from the North Sea.

1. Characteristics distinguishing the members of this phylum from members of other phyla.

The body architecture of ctenophores somewhat resembles that of cnidarian medusae. Like that of a medusa, the ctenophore body consists of an outer epidermis, an inner gastrodermis, and a thick, gelatinous middle mesoglea layer. Both groups show a basic radial symmetry, with oral and aboral surfaces.

The digestive systems are similar in the two groups as well. The mouth leads into a pharynx (also called a **stomodeum**), which serves as a site of extracellular digestion, and thence through a stomach into a series of **gastrovascular canals,** where digestion is completed intracellularly. A functional excretory system has not been documented for either group, nor are any specialized respiratory organs found. The ctenophore nervous system takes the form of a subepidermal nerve network, as in many cnidarians. Moreover, at least one ctenophore species has a planula larval stage in its life history. In addition, ctenophore tentacles are solid, as in most cnidarians, rather than hollow. Lastly, ctenophores may be diploblastic, forming—as in cnidarians—no distinct mesodermal tissue layer during embryogenesis. There is disagreement on this point: Muscle-forming cells do segregate early in ctenophore development and some workers regard these as true mesodermal cells, although the source of these cells during embryological development requires further scrutiny; if they are truly mesodermal, ctenophores would differ from cnidarians in qualifying for triploblastic status. One ctenophore species is known to use nematocysts in prey capture, but these nematocysts are obtained through the ingestion of cnidarian medusae; the nematocysts are not manufactured by the ctenophore.

Ctenophores have left a poor fossil record, so that our best chance of unraveling their evolutionary origins probably lies in the molecular realm. Although there may be an evolutionary relationship between the Ctenophora and Cnidaria based upon the morphological similarities just discussed, ctenophores are clearly not cnidarians. For one thing, the symmetry of ctenophores is more accurately described as *biradial* rather than radial; there are only 2 ways to slice ctenophores and end up with equivalent halves. Radial symmetry in itself is probably not convincing evidence of relatedness anyway; the basic radial symmetry of ctenophores could well be a secondary adaptation to planktonic existence. I have already mentioned an important embryological difference in the formation of the musculature. In addition, polymorphism, an almost diagnostic characteristic of hydrozoans and scyphozoans, is never encountered among the Ctenophora, and no ctenophores are colonial. And whereas all cnidarians have monociliated cells (one cilium per cell), ctenophore cells are always multiciliated, each bearing two or more cilia. Moreover, the type of musculature differs among the members of the 2 groups, as do the swimming mechanism, the system of maintaining balance, the mechanism and mode of food capture, the means of eliminating solid wastes, the nature of sexuality, and several aspects of embryonic development. Each of these characteristics will be considered in sequence.

The muscles of ctenophores develop from amoeboid cells found within the mesoglea. Thus, the resulting muscle fibers actually reside in the mesoglea layers, supporting the view that ctenophores are triploblastic. In contrast, the musculature of cnidarians is found within the gastrodermis and, to a lesser extent, within the epidermis. Moreover, ctenophores have genuine smooth muscle tissue, and in fact they lack the myoepithelial cells that are so characteristic of cnidarian musculature. The first known giant (up to 6 cm long) smooth muscle fibers have recently been isolated from 2 ctenophore species, *Mnemiopsis leidyi* and *Beroë* sp.; muscle preparations from these species should provide excellent material for general studies of smooth muscle biology.[2]

In most coastal ctenophore species, the musculature plays little or no direct role in locomotion. Instead, swimming is accomplished by the activity of many bands of partially fused, remarkably long cilia. Each band is called a **ctene** because of its resemblance to a comb (*ctene* = G: a comb). This explains the name of the phylum ("comb bearer") and the common reference to its members as "comb jellies;" ctenophores are unique among animals in having their cilia fused into ctenes. The ctenes are typically organized into 8 distinct rows, which are equally spaced about the body. These **comb rows,** or **costae,** extend from the oral to the aboral surface of the animal (Fig. 7.1) and are strikingly iridescent. The power stroke of the cilia comprising each ctene is toward the aboral surface, so that the typical ctenophore swims mouth first. In summary, whereas cnidarian medusae swim by means of jet propulsion, ctenophore locomotion depends largely upon the coordinated activities of the partially fused cilia in the various comb rows.[3] Some open-ocean ctenophore species use ctene activity for feeding but use vigorous muscular activity in escaping from predators.[4]

The intensity of activity in the different comb rows is under the control of a single **apical sense organ,** located at the aboral end of the ctenophore (Figs. 7.1a and 7.2). The **statolith,** a single sphere of calcium carbonate ($CaCO_3$), sits atop 4 tufts of fused cilia called **balancers,** or **springs.** Each balancer may consist of several hundred cilia. A ciliated groove radiates from each balancer and bifurcates to service two adjacent comb rows. Experiments have shown these grooves to be agents of nerve impulse conduction from the apical sense organ to the ctenes of the comb rows. If the animal becomes tilted, the statolith presses against one of the balancers more than the others, causing the cilia of the comb rows associated with that balancer to increase their beat frequency until a satisfactory body orientation is restored. If the apical sense organ is surgically removed, the ctenophore continues to swim. However, the surgery obliterates any coordination of ciliary beating in

2. See *Topics for Further Discussion and Investigation,* no. 3, at the end of the chapter.

3. See *Topics for Further Discussion and Investigation,* no. 1.

4. See *Topics for Further Discussion and Investigation,* no. 4.

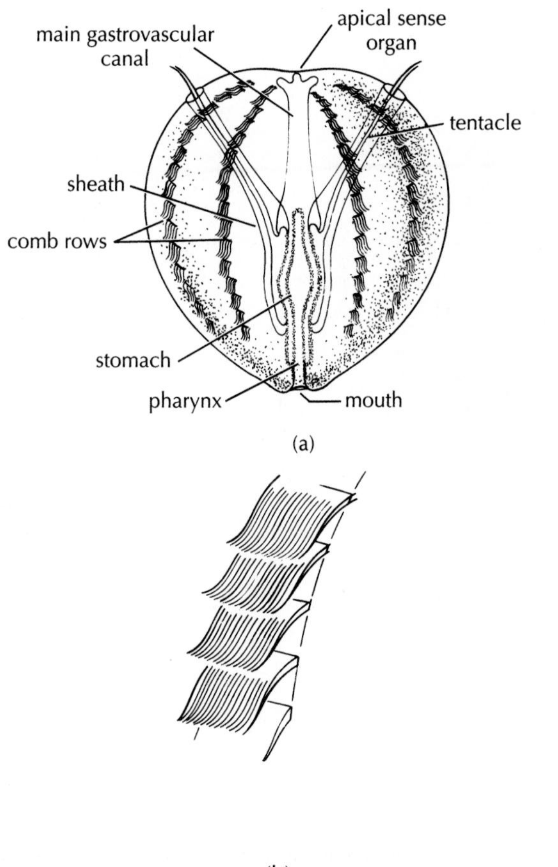

Figure 7.1

(a) External anatomy of a ctenophore, *Pleurobrachia* sp., showing the comb rows and several other anatomical features. The tentacles have been withdrawn into their sheaths. (b) Detail of a comb row, showing 4 ctenes. Each ctene is made up of thousands of long, compound cilia.
(a) After Hyman. (b) After Hardy.

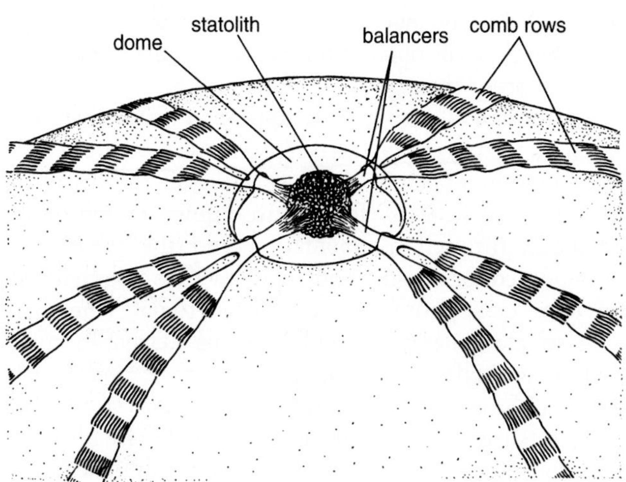

Figure 7.2

Detail of an apical (aboral) sense organ and its transparent covering. Tilting of the animal causes the statolith to press against particular balancers, stimulating activity in the associated comb rows and thus restoring the animal's orientation.

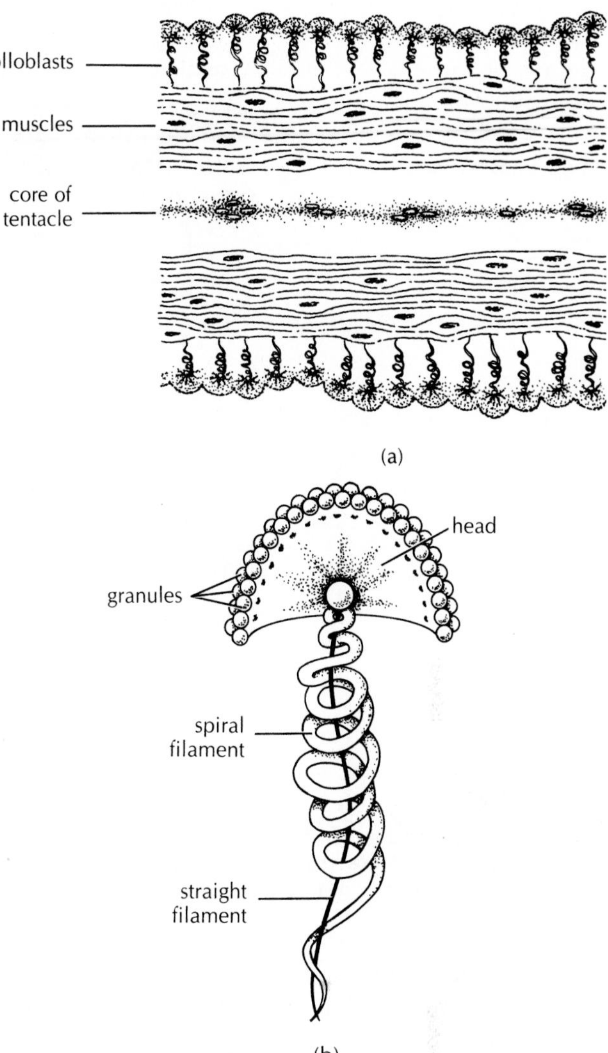

Figure 7.3

(a) Longitudinal section through a tentacle. (b) A single colloblast. The spiral filament is contractile. The head contains granules that, when discharged, produce a sticky secretion that traps prey.
(a) From Hyman; after Hertwig. (b) From Hyman; after Komai.

different comb rows, attesting to the synchronizing role of the apical sense organ.

The epidermal nerve net also seems to play a role in coordinating the activities of the ctenes. For example, mechanical stimulation of the oral end of the animal results in a sudden reversal of the ciliary beat in all comb rows. This response is observed even if the apical sense organ has been surgically removed.

As with medusae, many ctenophore species capture their prey using tentacles (see Research Focus Box 7.1, p. 142). Unlike the tentacles of cnidarians, however, ctenophore tentacles can be completely retracted into proximal pits or sheaths. Moreover, with one exception, the tentacle and/or the general epidermis of ctenophores is studded not with nematocysts but rather with quite different structures called **colloblasts** (Fig. 7.3). Each colloblast cell consists of a bulbous, sticky head connected to a long,

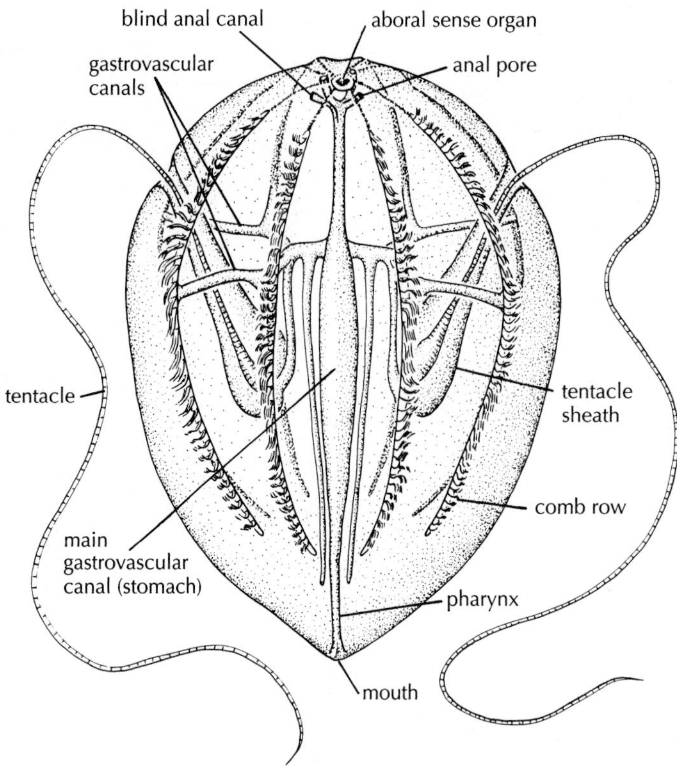

Figure 7.4

Diagrammatic illustration of the ctenophore *Pleurobrachia* sp., the sea gooseberry. Note that the main gastrovascular canal terminates as 4 branches near the aboral (apical) sense organ. Two of these branches end blindly, but the other two connect to the outside by means of small anal pores. All other gastrovascular canals end blindly.
After Bayer and Owre; after Hardy.

straight filament and a spiral, contractile filament. Prey organisms become stuck to the tentacles, which are then retracted. In many species, the body rotates to bring the mouth in contact with the tentacles once the food items are brought within range. The tentacles of some species may be extended more than 100 times the length of the body and retracted in seconds.

In other species, the food-catching function of the tentacles is much reduced. Instead, the body surface area is increased by lateral compression, and major areas of the body are coated with a sticky mucus and with colloblast cells. The body itself thus becomes the major organ of food collection, and the small tentacles found in some of these species merely aid in transporting food to the mouth. As mentioned earlier, most ctenophores eat small crustaceans (phylum Arthropoda) and the larval stages of various fish and shellfish species, including some of commercial importance, such as oysters. Some ctenophores are carnivorous on other ctenophores or on gelatinous animals in other groups (especially the Cnidaria and the Urochordata). In addition, some ctenophore species can also eat phytoplankton and ciliated protozoans,[5] so they have the potential to influence food web dynamics in a variety of ways, not just by consuming zooplankton.

5. Scolardi, K. M., K. L. Daly, E. A. Pakhomov, and J. J. Torres. 2006. *Mar. Ecol. Progr. Ser.* 317: 111–26.

The digestive systems of ctenophores and cnidarians differ in one interesting respect. As discussed earlier, the digestive tract of a medusa has but one opening, which serves as both mouth and anus. In ctenophores, on the other hand, 4 **digestive canals** lead from the roof of the stomach to the animal's aboral surface (Fig. 7.4). Although 2 of the digestive canals terminate as blind sacs, the other 2 canals open to the outside. Undigested wastes are discharged through these **anal pores.**

Although all ctenophores appear to have substantial regenerative powers, asexual reproduction is rare within the Ctenophora. Only a few species (all of which are members of the Platyctenida) are known to reproduce asexually, through fragmentation and the subsequent development of missing body parts by each fragment. All but a few ctenophore species described to date (members of 2 lobate genera) are simultaneous hermaphrodites; that is, a single individual has both male and female gonads. The gonads are located on the walls of some or all of the gastrovascular canals, so that gametes are liberated into the digestive tract and are commonly discharged through the mouth. Eggs are usually fertilized externally (i.e., in the surrounding seawater), except in some platyctene species that fertilize their eggs internally.

Ctenophore development differs in several respects from that of cnidarians. In particular, ctenophore cleavage is highly determinate; cell fates are fixed at the first cell

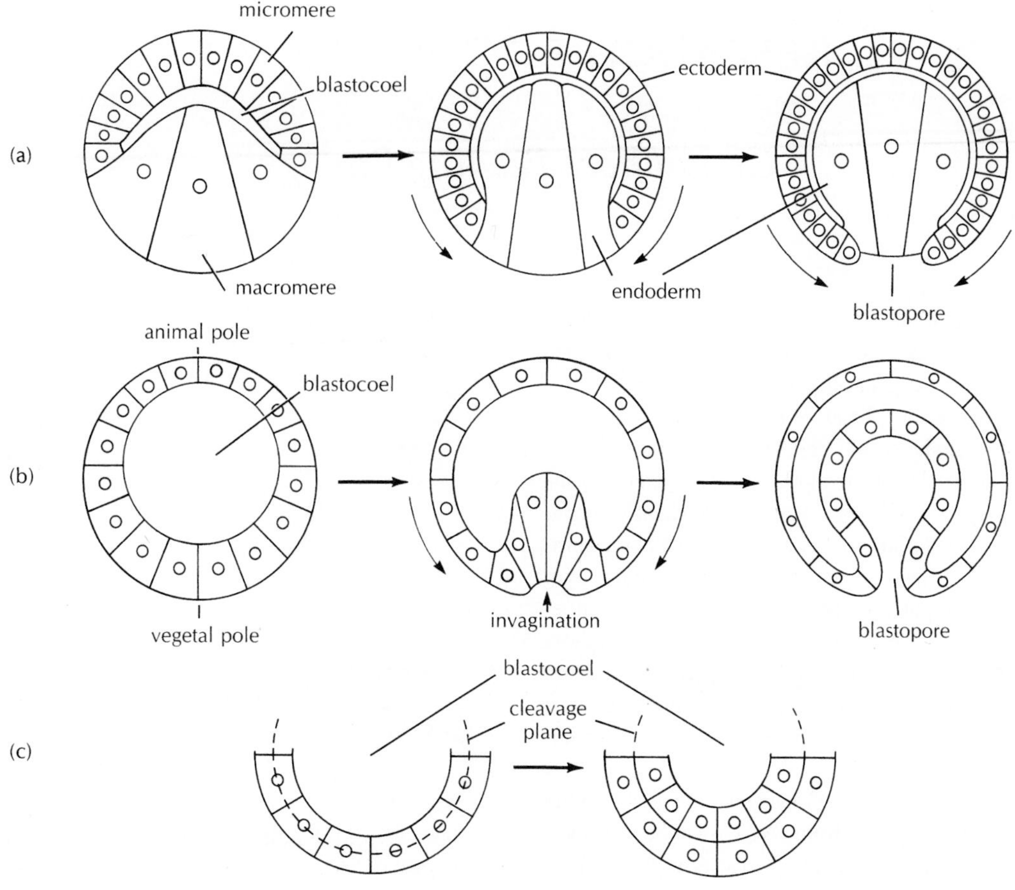

Figure 7.5

Patterns of gastrulation in ctenophores and cnidarians. (a) Gastrulation by epiboly, in which the smaller cells (micromeres) grow over larger cells (macromeres). (b) Gastrulation by invagination, in which a group of adjacent cells indents into the blastocoel. (c) Gastrulation by delamination, in which the second cell layer is formed by mitotic division. In ctenophores, gastrulation is by epiboly or by invagination. In marked contrast, cnidarians typically gastrulate by delamination or ingression (in which cells on the outside of the embryo migrate inward).

division. Cell fates of cnidarian embryos become fixed later in development. Moreover, the mechanism of **gastrulation** in ctenophores (i.e., the formation of distinct inner and outer germ layers) is quite unlike that found in most cnidarians. Among ctenophores, gastrulation is achieved either by **epiboly,** a process in which a sheet of micromeres spreads over what were the adjacent macromeres (Fig. 7.5a), or by **invagination,** in which groups of cells push into the blastocoelic space (Fig. 7.5b). Although gastrulation by invagination occurs among cnidarians, epiboly does not, and most cnidarians gastrulate by a process of **delamination** (Fig. 7.5c). In delamination, the cells of the blastula divide with the cleavage plane approximately parallel to the surface of the embryo. Thus, the cells essentially divide into the blastocoel, forming an inner and outer cell layer, between which the mesoglea is later secreted. Gastrulation in still other cnidarians is by **ingression,** in which certain cells become detached from their neighbors and simply move into the blastocoel, creating a second layer of cells.

Ctenophore embryos, in contrast to cnidarian embryos, rarely develop into ciliated planula larvae. Rather, the ctenophore embryo usually develops directly into a miniature ctenophore called a **cydippid.** The cydippid is approximately spherical in shape; is endowed with 8 comb rows, a fully formed apical sense organ, and a pharynx; and usually bears a pair of branched tentacles. In many species, the cydippid closely resembles the adult. In the substantially modified benthic ctenophore species (See Fig. 7.7d, p. 145), however, the cydippid undergoes a gradual, but considerable alteration in morphology to attain the adult form, and so can be considered a true larval stage that metamorphoses.

Certain ctenophore and cnidarian characteristics are compared in Table 7.1.

Nearly all ctenophores are **bioluminescent.**[6] Unlike iridescence, in which colors are generated by the diffraction of incident light, bioluminescence results from a chemical reaction in which much of the excess energy is given off as light rather than as heat. Although the particular form of the reaction taking place in ctenophores seems to be peculiar to the members of this phylum, the phenomenon of bioluminescence is not. Indeed, at least some species from most major animal phyla display some form

6. See *Topics for Further Discussion and Investigation,* no. 5.

RESEARCH FOCUS BOX 7.1

Ctenophore Feeding Biology

Greene, C. H., M. R. Landry, and B. C. Monger. 1986. "Foraging behavior and prey selection by the ambush entangling predator *Pleurobrachia bachei*." *Ecology*, 67:1493–1501.

Ctenophore population densities can become quite high, particularly in coastal waters. Individual species can reach concentrations of several hundred individuals per m^3. As carnivores, ctenophores are therefore likely to have substantial impact on the abundance and population dynamics of other zooplankton. In particular, ctenophores may severely deplete the numbers of copepods (Focus Fig. 7.1), which also serve as major food sources for the larvae and adults of many commercially important fish species. Quantifying the importance of ctenophores in regulating the sizes and dynamics of copepod populations requires detailed knowledge of ctenophore feeding biology. Do ctenophores preferentially select certain prey species or certain life-history stages of particular prey, or do they feed nonselectively on whatever they happen to catch? In addition, at what rates do they feed?

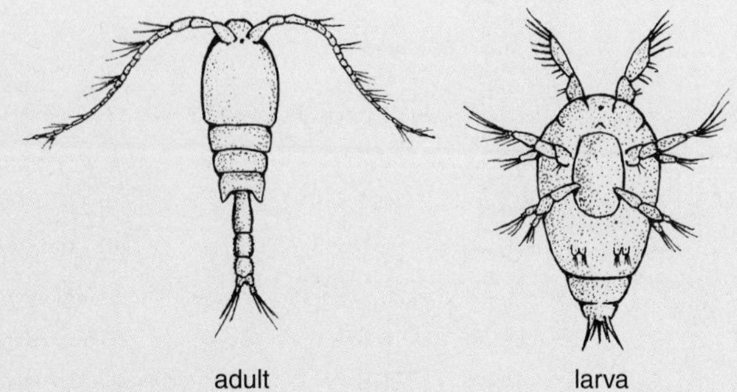

Focus Figure 7.1
Adult and larval stages of a typical planktonic copepod (not to scale). Copepods are crustaceans (phylum Arthropoda). Adults are typically less than one mm long, and larvae are much smaller.

The study by Greene et al. (1986) documents the ability of the small (at most a few centimeters in diameter), coastal ctenophore *Pleurobrachia bachei* to feed selectively on adults of the following copepod species: *Acartia clausi, Calanus pacificus,* and *Pseudocalanus* sp. The authors also examined feeding on different larval and juvenile stages of *C. pacificus.*

In the basic experiment, the researchers set up a number of 3.8 l (liter) jars of seawater and added one ctenophore and either 25 or 50 copepod prey to each. The ctenophores were allowed to feed in the dark for 12 hours. Like other tentaculate ctenophore species, *P. bachei* drifts passively in the water for long periods, trailing its long fishing tentacles behind. When prey are ensnared by colloblasts, the ctenophore draws the successful tentacle across the mouth and ingests the prey. At the end of 12 hours, the number of copepods remaining in each container was determined. There was no decline in copepod concentration in control jars (which contained copepods but no ctenophores), so that the disappearance of copepods in experimental treatments must have reflected predation by the ctenophores.

For comparative purposes, the data on rates of copepod disappearance were expressed in terms of the volume of water each ctenophore must have processed each hour to have captured and eaten the recorded number of copepods ingested. For example, if the jar initially contained 215 copepods in 3.8 l of seawater and the ctenophore ate 50 copepods over the next 12 hours, the average concentration

Table 7.1 Similarities and Differences between the 2 Phyla of Diploblastic Gelatinous Animals

Characteristic	Ctenophores	Cnidarians
Cleavage	Determinate	Indeterminate
Gastrulation	Epiboly or invagination	Delamination, ingression, or invagination
Common developmental stage	Cydippid	Planula
Digestive system	Gastrovascular canals	Gastrovascular canals
Nematocysts	None (unless "borrowed")	Present
Colloblasts	Present	None
Sexuality	Typically hermaphroditic	Typically gonochoristic
Musculature	Within mesoglea	Within gastrodermis
Ciliation	Multiciliated cells	Monociliated cells

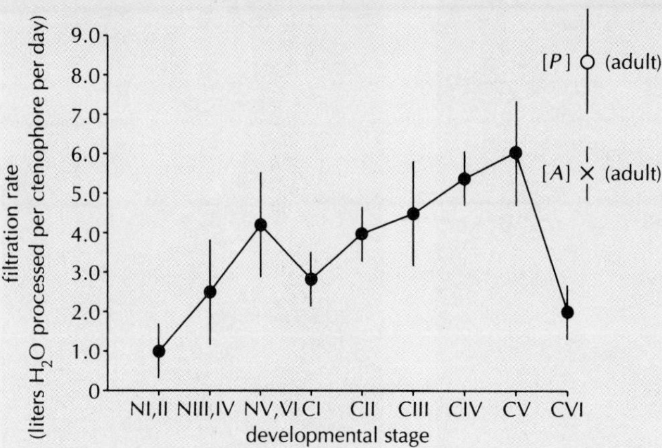

Focus Figure 7.2

Laboratory feeding rates of the ctenophore *Pleurobrachia bachei* on copepods of different developmental stages (*Calanus pacificus:* filled-in circles) or of different species (*Acartia clausi* adults: X; *Pseudocalanus* sp. adults: open circle). Developmental stages NI–NVI are larvae of *C. pacificus,* while stages CI–CVI are juveniles of *C. pacificus.*

of copepods in the jar during that time was 50 copepods per liter ([215 + 165]/2 copepods ÷ 3.8 l). To have eaten 50 copepods, the ctenophore must have filtered one liter of seawater in the 12-hour period.

The results of the feeding studies clearly demonstrate that, at least in the laboratory, this ctenophore species feeds selectively (Focus Fig. 7.2). Of particular interest, feeding rates on the various larval and juvenile stages of *C. pacificus* varied unpredictably from stage to stage. One might have expected feeding rates to be consistently higher for older—larger and faster-swimming—prey, but the data for *C. pacificus* do not show such a continuous increase in feeding rate (Focus Fig 7.2). Moreover, feeding rates on adults of different copepod species also differed markedly. Ctenophores fed at the highest rates on *Pseudocalanus* sp., processing seawater at rates up to 8.4 l daily (Focus Fig. 7.2).

Additional experiments demonstrated that the feeding-rate differences shown when ctenophores were feeding on only one prey species at a time were maintained when the ctenophores were fed mixtures of the prey species. This clearly demonstrates that the members of this ctenophore species can feed selectively in the laboratory, ingesting more of one prey type than another even when both prey types are present at equal concentrations.

Does this mean that ctenophores actively *select* certain prey over others? An alternative explanation is that some prey are simply less likely to be captured and ingested; that is, differential predation might reflect differences in prey behavior rather than deliberate choices made by the predator. In particular, differences in prey susceptibility could reflect (1) specific avoidance behaviors exhibited by the copepods; (2) differential ability to escape from the tentacles after capture; or (3) different swimming speeds for copepods of different species and different developmental stages (the faster one moves, the sooner one is likely to run into trouble). The authors went on to examine these possibilities. How do you suppose they designed this part of their study?

The laboratory experiments just described suggest that ctenophore predation does not have equal impact on all copepod species or on all life-history stages of any one copepod species. In the field, of course, ctenophores would be free to extend their tentacles to lengths not permitted in small glass jars, and copepods would have greater opportunities to avoid capture. Can you think of a way to test for differential predation by ctenophores under more natural field conditions?

of bioluminescence. The functional significance of bioluminescence is often unclear. Mate location and species recognition, the luring of prey, and the startling of would-be predators are possibilities that may apply to the bioluminescing members of some phyla. Some species may use bioluminescence to avoid detection by visual predators, producing light of ambient intensity. This would break up the silhouette of the animal when observed by potential predators from below, helping the lighted form blend into the surroundings. The adaptive significance of bioluminescence among ctenophores has not been examined, although the possibility of mate recognition can probably be ruled out by the lack of distinct photoreceptors and by the likely inability of the nervous system to process such complex information. Regardless of its significance, the bioluminescence, together with the iridescence of the comb rows, the delicacy of the form, and the grace of movement, make living ctenophores among the most glorious of animals to observe.

Ctenophore Diversity

Most ctenophore species live in the open sea far from coastlines, so their biology is largely unstudied. The recent switch from collecting samples with nets towed from fast-moving vessels to *in situ* observations and careful collection by divers and by submersibles has brought to light numerous species never before known, species much too fragile to have withstood the trauma of what were formerly considered standard collection techniques. As many as one dozen new species were captured a few years ago during a single 2-week expedition. At present, ctenophore species

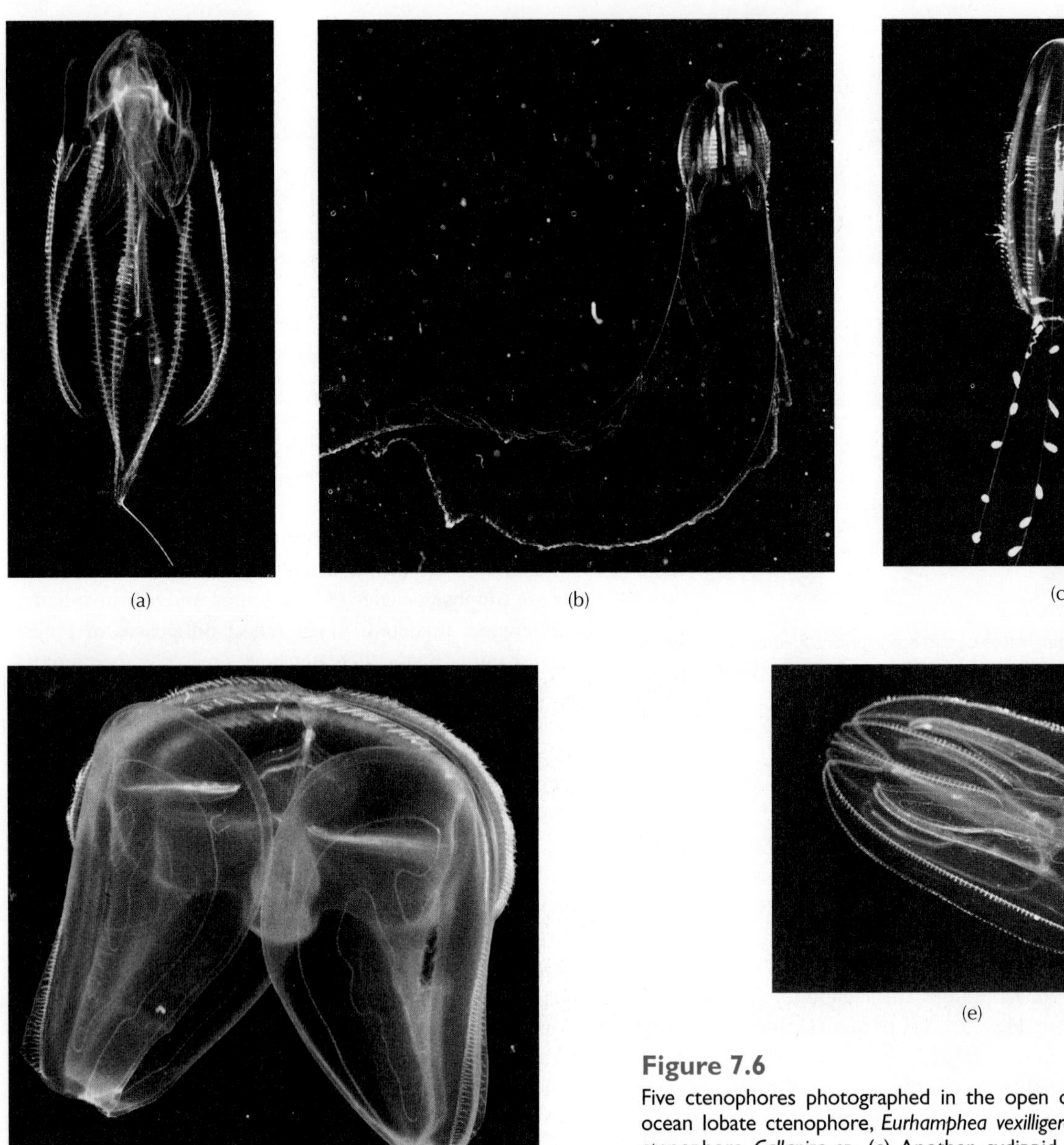

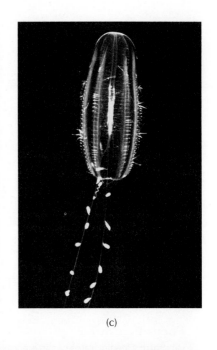

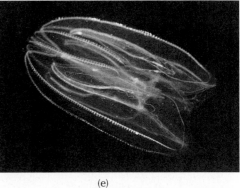

(a) (b) (c) (d) (e)

Figure 7.6

Five ctenophores photographed in the open ocean. (a) An open ocean lobate ctenophore, *Eurhamphea vexilligera.* (b) The cydippid ctenophore *Callanira* sp. (c) Another cydippid ctenophore, *Euplokamis dunlapae* (body length approximately 1.4 cm). (d) *Ocyropsis maculata,* a lobate ctenophore. (e) A coastal lobate ctenophore, *Mnemiopsis macrydi.*

(a,b) Courtesy of G. R. Harbison and M. Jones. (c) Courtesy of Claudia E. Mills. From Mackie, G. O., et al. 1992. *Biol. Bull.* 182:248–56. (d,e) By Anne Rudloe, Gulf Specimen Marine Laboratories, Inc.

are divided into 2 classes, primarily on the basis of whether they possess conspicuous tentacles as adults and/or as cydippids. Representative adults are shown in Figure 7.6.

Class Tentaculata

Much of tentaculate evolution seems to be a story of modifying the mechanism of prey capture.[7] Some tentaculate species closely resemble the cydippid larvae already described, except, of course, that functional gonads are present. Long, retractable tentacles are well developed throughout life, and food is captured exclusively by these few tentacles and their side branches. These ctenophores comprise the order Cydippida (Figs. 7.6b, c and 7.7a). In other species, the body is somewhat compressed laterally, only 4 of the comb rows are fully developed, and the tentacles are generally much reduced in length. Large **oral lobes,** covered with mucus and colloblasts, constitute the primary food collection surfaces. These are the lobate ctenophores (order Lobata, Figs. 7.6d, e and 7.7b, f). Muscular activity of the

7. See *Topics for Further Discussion and Investigation,* no. 2.

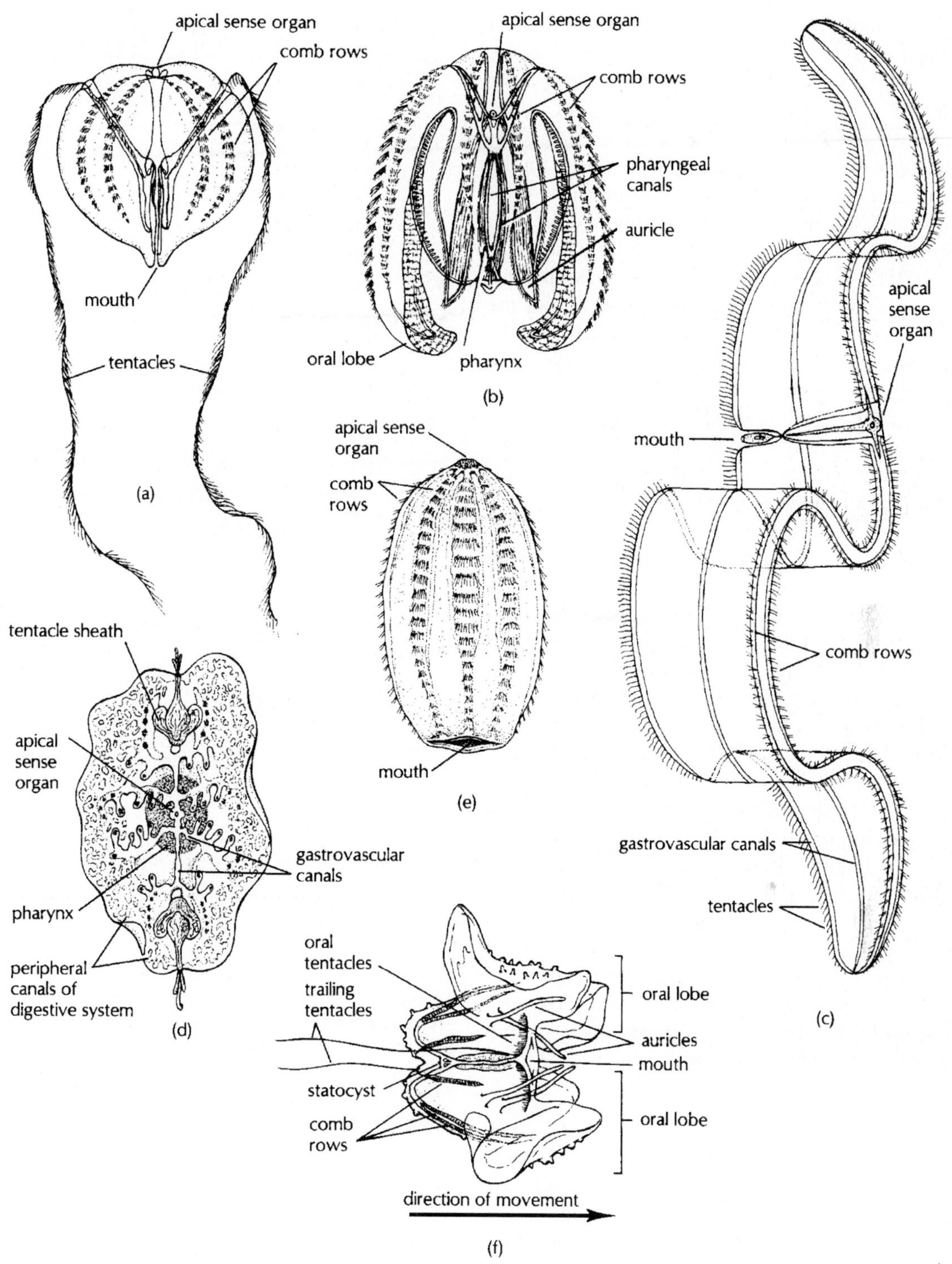

Figure 7.7

Ctenophore diversity. (a) The cydippid *Pleurobrachia* sp., approximately twice natural size. (b) A lobate ctenophore, *Mnemiopsis leidyi.* (c) A member of the order Cestida, *Cestum veneris,* commonly known as Venus's girdle. These animals attain lengths of about 1.5 m. (d) A member of the Platyctenida, *Coeloplana mesnili,* viewed from above. The animal is about 6 cm in length and spends its life in association with certain corals in the Indo-Pacific. (e) A member of the class Nuda, *Beroë* sp. The members of this class lack tentacles throughout life. (f) *Leucothea* sp., an open-ocean lobate ctenophore. The highly fragile lobes break into pieces at the slightest physical disturbance, so that these animals can be studied only by diving in their midst.

(a,b) After Hyman. (c) From Hyman; after Mayer. (d) From Hyman, after Dawydoff. (e) After Hardy. (f) From Matsumoto and Hamner, *Marine Biology* 97:551–58, 1988.

2 oral lobes aids locomotion in some species. Two pairs of ciliated, paddle- or tentacle-like structures called **auricles** (Fig. 7.7b), unique to lobate ctenophores, apparently assist in prey capture.

In another group, the cestids, the body is so compressed laterally that it now forms a long ribbon, with the mouth and apical sense organ on opposite sides at its midpoint (Fig. 7.7c). Swimming is accomplished through a combination of ctene activity and sinuous, muscular movements of the body; only 4 of the comb rows are well developed in adults. Despite the great surface area of the body, prey are captured not by the ribbon itself, but by the numerous short tentacles extending along the extensive oral edge of the ctenophore.

Thus, selection for increased food-collection capabilities seems to have favored a redistribution of colloblasts, together with an increased body surface area, in one group of tentaculate ctenophores—the Lobata—and an increased number of feeding tentacles in another group—the Cestida. Both adaptations have involved lateral compression of the body in comparison with the cydippid form, which is believed to represent the more primitive condition. In the fourth and final major order of tentaculate ctenophores, the Platyctenida, the body is compressed in a different plane (Fig. 7.7d). Specifically, the oral and aboral surfaces have moved toward each other, so the body forms a flattened plate (*platy* = G: flat). The bottom of the plate is formed largely by the pharynx, which is extensively and permanently everted. Some species simply float in the water. Others spend their time creeping slowly over solid substrates, apparently using both pharyngeal cilia and muscular contractions for locomotion. The only non-planktonic ctenophores are members of this order. Comb rows may be present, although these are reduced or absent in the adults of many species. Some locomotion may also be accomplished by muscular flapping of the lateral lobes of the body. The adults generally bear 2 long tentacles.

Class Nuda

The members of the relatively small class Nuda are contained in a single order, the Beroida. No individuals in this group have tentacles (even in the cydippid stage of development) or oral lobes. All 8 comb rows, however, are well developed (Fig. 7.7e). Prey, including other ctenophores, are captured and engulfed by muscular lips surrounding the mouth. The mouth can be widened to accommodate prey substantially larger than the predator. **Macrocilia,** consisting of thousands of 9 + 2 axonemes enclosed by a single membrane, are located just inside the mouth; these macrocilia are used as teeth, to chop especially large prey into bite-sized pieces.

Taxonomic Summary

Phylum Ctenophora
- Class Tentaculata
  - Order Cydippida
  - Order Lobata
  - Order Cestida
  - Order Platyctenida
- Class Nuda
  - Order Beroida

Topics for Further Discussion and Investigation

1. What role does mechanical interaction between adjacent ctenes play in coordinating the beating of cilia in a comb row?

Tamm, S. L. 1973. Mechanisms of ciliary coordination in ctenophores. *J. Exp. Biol.* 59:231.

Tamm, S. L. 1983. Motility and mechanosensitivity of macrocilia in the ctenophore *Beroë. Nature* 305:430.

2. Compare and contrast the feeding biology of lobate, cydippid, and cestid ctenophores.

Costello, J. H., and R. Coverdale. 1998. Planktonic feeding and evolutionary significance of the lobate body plan within the Ctenophora. *Biol. Bull.* 195:247.

Costello, J. H., R. Loftus, and R. Waggett. 1999. Influence of prey detection on capture success for the ctenophore *Mnemiopsis leidyi* feeding upon adult *Acartia tonsa* and *Oithona colcarva* copepods. *Marine Ecol. Progr. Ser.* 191:207.

Hamner, W. M., S. W. Strand, G. I. Matsumoto, and P. P. Hamner. 1987. Ethological observations on foraging behavior of the ctenophore *Leucothea* sp. in the open sea. *Limnol. Oceanogr.* 32:645.

Harbison, G. R., L. P. Madin, and N. R. Swanberg. 1978. On the natural history and distribution of oceanic ctenophores. *Deep-Sea Res.* 25:233.

Main, R. J. 1928. Observations of the feeding mechanism of the ctenophore, *Mnemiopsis leidyi. Biol. Bull.* 55:69.

Matsumoto, G. I., and G. R. Harbison. 1993. *In situ* observations of foraging, feeding, and escape behavior in 3 orders of oceanic ctenophores: Lobata, Cestida, and Beroida. *Marine Biol.* 117:279.

Reeve, M. R., M. A. Walter, and T. Ikeda. 1978. Laboratory studies of ingestion and food utilization in lobate and tentaculate ctenophores. *Limnol. Oceanogr.* 23:740.

Sullivan, L. J., and D. J. Gifford. 2004. Diet of the larval ctenophore *Mnemiopsis leidyi* A. Agassiz (Ctenophora, Lobata). *J. Plankton. Res.* 26: 417–31.

Swanberg, N. 1974. The feeding behavior of *Beroë ovata. Marine Biol.* 24:69.

3. The feeding behavior of ctenophores commonly involves a surprisingly complex series of rapid and apparently coordinated muscle contractions. How do ctenophores achieve such coordination without a central nervous system?

Bilbaut, A., M.-L. Hernandez-Nicaise, C. A. Leech, and R. W. Meech. 1988. Membrane currents that govern smooth muscle contraction in a ctenophore. *Nature* 331:533.

4. Discuss the roles of cilia and musculature in the escape responses exhibited by planktonic ctenophores.

Kreps, T. A., J. E. Purcell, and K. B. Heidelberg. 1997. Escape of the ctenophore *Mnemiopsis leidyi* from the scyphomedusa predator *Chrysaora quinquecirrha*. *Marine Biol.* 128:441.

Mackie, G. O., C. E. Mills, and C. L. Singla. 1992. Giant axons and escape swimming in *Euplokamis dunlapae* (Ctenophora: Cydippida). *Biol. Bull.* 182:248.

Matsumoto, G. I., and G. R. Harbison. 1993. *In situ* observations of foraging, feeding, and escape behavior in 3 orders of oceanic ctenophores: Lobata, Cestida, and Beroida. *Marine Biol.* 117:279.

5. How can one conclusively demonstrate that a particular ctenophore species is *not* capable of bioluminescing?

Haddock, S. H. D., and J. F. Case. 1995. Not all ctenophores are bioluminescent: *Pleurobrachia*. *Biol. Bull.* 189:356.

Taxonomic Detail

Phylum Ctenophora

The approximately 150 described species are distributed among 7 orders and 19 families.

Class Tentaculata

Four orders

Order Cydippida

Five families

Family Pleurobrachiidae. *Pleurobrachia.* This family is represented over a very wide geographic range, from polar to tropical waters and from coastal to open-ocean habitats.

Family Euplokidae. *Euplokamis*

Family Mertensidae. *Mertensia, Callianira*

Order Platyctenida

These peculiar, flattened ctenophores are restricted to shallow waters in the tropics and at the poles. Unlike other ctenophores, they reproduce asexually—by breaking off pieces of the body—as well as sexually. Fertilization is commonly internal, and the developing animal always passes through a perfectly normal cydippid stage. Four families.

Family Ctenoplanidae. *Ctenoplana.* Most individuals are smaller than about 2 cm, and all are found only in tropical waters. Although capable of swimming, they spend most of their time crawling over various benthic substrates.

Family Coeloplanidae. *Coeloplana, Vallicula.* These animals reach lengths of about 6 cm and occur in shallow water throughout the world. Some species are active swimmers, while others drift in the water or creep along the bottom. The comb rows are completely lost during development so that the animal looks superficially more like a flatworm than a ctenophore. Some species are commensal with various other benthic organisms.

Order Lobata

All members are planktonic and all have aural lobes and auricles. Six families.

Family Bolinopsidae. *Bolinopsis, Mnemiopsis.* All species live in coastal waters, with representatives in all oceans from the tropics to the poles. The tentacles are long when animals are in the cydippid stage but short and inconspicuous in adults. Individuals may reach about 15 cm in height. *Mnemiopsis leidyi* (Fig. 7.7b) was accidently introduced into the Black Sea in the early 1980s, probably transported in cargo ship ballast water from the coasts of North or South America, or from the Caribbean. It has subsequently caused the near collapse of the Black Sea anchovy fishery, by competing for food with the fish, eating their eggs and larvae, and clogging fishing nets. The species is currently spreading into the Mediterranean, Baltic, and North Seas.

Family Ocyropsidae. *Ocyropsis.* These common, tropical, open-ocean ctenophores swim using a combination of ctene activity and vigorous flapping of the oral lobes. Early juveniles have short tentacles, but these are reduced or completely lost by adulthood. All members of the genus are gonochoristic.

Family Leucotheidae. *Leucothea.* The oral lobes of these open-ocean ctenophores are especially large and fragile; even moderate water turbulence will cause the lobes to break apart. The auricles are extremely long and thin. Representatives are present in all oceans, from tropical to temperate regions.

Order Cestida

Cestum, Velamen. All individuals are pelagic. Although most are smaller than about 15 cm, members of the genus *Cestum* may grow to exceed 2 m. Development includes a normal, tentacled cydippid, which gradually elongates to adult form. Although the animals are propelled by the ctene rows when feeding, they can also wriggle like eels to escape from potential predators. The members of this order are found worldwide, with especially high concentrations in the tropics. One family (Cestidae).

Order Ganeshida

Ganesha. This order contains 2 species of pelagic ctenophores whose members possess branched tentacles but lack both auricles and oral lobes. Some workers think these ctenophores may be the developmental stages of lobate ctenophores.

Order Thalassocalycida

Thalassocalyce. This order contains a single species of pelagic, very fragile ctenophore whose members have oral lobes and tentacles but lack auricles and tentacular sheaths. The mouth and pharynx are borne on a central stalk. Individuals reach about 25 cm and have been collected from as deep as 1000 m.

Class Nuda

One order

Order Beroida

Beroë. These ctenophores are all pelagic and lack tentacles. They feed by engulfing their prey with a large, highly distensible mouth or by biting large prey into pieces with sharp-edged macrociliary bundles. These animals feed on other gelatinous animals, including salps (see Chapter 22) and other ctenophores. They occur in all the world's oceans, from the poles to the tropics, and reach lengths of about 30 cm. In 1997, they appeared in the Black Sea where they have been feeding voraciously on the previously introduced lobate ctenophore *Mnemiopsis leidyi.* One family (Beroidae).

General References about the Ctenophores

Haddock, S. H. D. 2007. Comparative feeding behavior of planktonic ctenophores. *Integr. Comp. Biol.* 47: 847–53.

Harbison, G. R., L. P. Madin, and N. R. Swanberg. 1978. On the natural history and distribution of oceanic ctenophores. *Deep-Sea Res.* 25:233–56.

Harrison, F. W., and J. A. Westfall, eds. 1991. *Microscopic Anatomy of Invertebrates,* vol. 2. New York: Wiley-Liss.

Hyman, L. 1940. *The Invertebrates, Vol. 1. Protozoa through Ctenophora.* New York: McGraw-Hill.

Morris, S. C., et al., eds. 1985. *The Origins and Relationships of Lower Invertebrates. Systematics Association, Special Vol. 28.* Oxford: Clarendon Press, 78–100.

Parker, S. P., ed. 1982. *Classification and Synopsis of Living Organisms,* vol. 1. New York: McGraw-Hill, 707–15.

Search the Web

1. www.mbl.edu/html/KEYS/INVERTS/5/start.html

   This site is run by the Marine Biological Laboratory, Woods Hole, MA. Under "Quick Links," choose "Biological Bulletin," and then "Other Biological Bulletin Publications." Next select "Keys to Marine Invertebrates of the Woods Hole Region," and then choose "Cnidaria." This includes a key to ctenophore species common in Woods Hole, along with references about the group.

2. www.imagequest3d.com/pages/general/news/blackseajellies/blackseajellies.htm

   This website describes the invasion of the Black Sea by ctenophores, with excellent photographs of the major players.

3. www.ucmp.berkeley.edu/cnidaria/ctenophora.html

4. http://tolweb.org/tree/phylogeny.html

   This brings you to the Tree of Life site, run by the University of Arizona. Use the search term "Ctenophora."

8

The Platyhelminths

Introduction and General Characteristics

Phylum Platy • helminthes
(Greek: flatworm)
plat´-ē-hel-min´-thēs

The platyhelminths are a group of some 34,000 described species, with—at present—no uniquely defining characters (synapomorphies). The group includes one class of mostly free-living individuals (the turbellarians) and 3 classes of exclusively parasitic individuals (the monogeneans, trematodes, and cestodes) that are believed to have evolved from free-living turbellarian ancestors; more than 80% of all described platyhelminth species are parasites. It has been estimated that another 36,500 species await description. All flatworms are acoelomate, triploblastic, and bilaterally symmetrical. Indeed, the free-living forms are usually considered the most primitive (i.e., **basal**) bilateral animals, and the first group to have evolved a true mesoderm; all coelomate animals may ultimately have evolved from flatworm-like ancestors.

On the other hand, there is also compelling evidence and logical argument suggesting that the flatworm acoelomate condition is a secondary phenomenon; that is, that the flatworms descended from coelomate ancestors. Possibly, the coelomic space of some flatworm ancestor filled in as an adaptive accompaniment to the evolution of reduced-body size, for example, or perhaps the acoelomates arose from a coelomate developmental stage when an embryo became sexually mature before developing a coelomic space. Both are reasonable possibilities. The fossil record is extremely limited for flatworms and thus of no help in resolving the issue. The oldest known flatworm fossil, approximately 40 million years old, was found just a few years ago, preserved in fossilized tree resin (amber).[1] In modern analyses, flatworms are sometimes grouped with

1. Poinar, G., Jr. 2003. *Invert. Biol.* 122: 308–12.

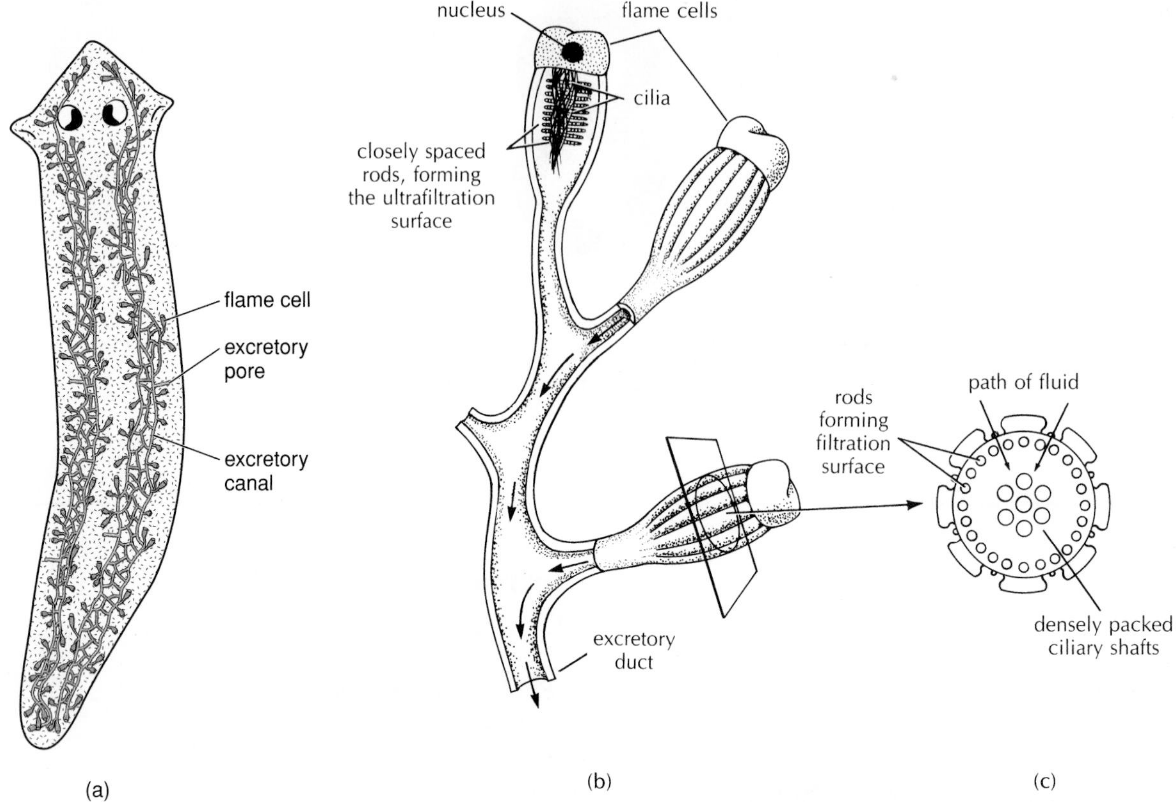

Figure 8.1

(a) The elaborately branching excretory system of a freshwater free-living flatworm. (b) Detail of several flame cells emptying into a common collecting duct. Arrows indicate direction of fluid flow. (c) Cross section of a flame cell at the level of the ciliary bundle.

(a) Based on Schmidt-Nielson, *Animal Physiology: Adaptation and Environment*, 3d ed.

molluscs, annelids, bryozoans, and members of a number of other phyla as **lophotrochozoans** (see Chapter 2, p. 17), but the controversy regarding the primitive or derived nature of the platyhelminth acoelomate condition remains unresolved. As discussed later (p. 154), some flatworms may be secondarily acoelomate while the acoelomate condition may be original in others; if so, the phylum Platyhelminthes contains at least 2 groups of animals without a common immediate ancestor, making it polyphyletic and invalid (see Chapter 2). The debate is not yet over.[2]

Although flatworms are acoelomate, their development is, in other respects, protostome-like: Cleavage is spiral and, at least in some species, determinate,[3] and the mouth forms before the anus, and forms from the blastopore. Most species have a conspicuous anterior brain, which is connected to at least one pair of longitudinal nerve cords. In the most advanced species, only a single pair of nerve cords is present, and these cords are always located ventrally. The mesodermal layer of the embryo develops into a loose collection of cells known as **parenchyma** tissue. This tissue occupies the entire space between the outer body wall and the endoderm of the gut.

2. See *Topics for Further Discussion and Investigation*, no. 9, at the end of this chapter.

3. See *Topics for Further Discussion and Investigation*, no. 6.

Like cnidarians, most platyhelminths have no anus; food enters, and unmetabolized wastes leave the digestive system, through a single opening.

Perhaps the most conspicuous feature of platyhelminths is that they are flat. They have no specialized respiratory organs and no specialized circulatory system, although a very few species possess hemoglobin. Gas exchange is accomplished by simple diffusion across the body surface. The rate at which such exchange can occur (milliliters of oxygen transported from the surrounding medium into the tissues per unit of time) depends upon several factors: the oxygen concentration gradient across the body wall, the body wall's permeability to gas, the thickness of the body wall, and the total exposed surface area across which diffusion can occur. By being flat, the flatworms achieve a high surface area relative to their enclosed volume, and a sufficient amount of gas exchange can occur to support an active lifestyle despite the lack of gills and an internal circulatory system.

Metabolic wastes probably move out of flatworms mostly by diffusing across the general body surface. Again, being flat is helpful in this regard. In addition, most platyhelminths contain a series of specialized organs called **protonephridia** (G: first kidney). The typical protonephridium consists of a group of cilia projecting into a fine-meshed cup (Fig. 8.1). The beating of the cilia within the cup has been likened to the flickering of a flame; thus, the common name for this type of cell is **flame cell.** Other protonephridia take the form of **solenocytes,** in which a single flagellum is found within the cup. In both cases, the mesh cup is attached to a long, convoluted tubule that connects to the outside of the animal through a single excretory pore. The inner workings of protonephridia are not fully understood. Apparently, the beating of the flagella or cilia creates a negative pressure, drawing fluid through the mesh cup and into the protonephridial tubule. The liquid entering the protonephridium is thus ultrafiltered: Large molecules (e.g., proteins) are excluded by their inability to pass through the openings in the cup wall. Measurements made on the liquid entering and leaving protonephridia indicate that the chemical composition of the fluid changes as it moves along the tubule to the excretory pore: Ions may be selectively absorbed or secreted, and the water content may be altered as well. Thus, flatworm protonephridia likely play an important role in regulating ionic and water balance (osmoregulation), in addition to their possible role in eliminating metabolic wastes, including ammonia, urea, and amino acids.

The vast majority of flatworm species, in all 4 classes, are **simultaneous hermaphrodites;** that is, each individual can, at any one time, function as both a female and a male. In consequence, sperm exchange and egg fertilization can occur when any individual encounters another individual of the same species. Individuals generally cannot fertilize themselves; some exceptions are discussed shortly.

The evolutionary interrelationships among the major flatworm groups are much debated, and platyhelminth systematics is decidedly unsettled in consequence. There is even some disagreement about whether platyhelminths form a monophyletic group. I have chosen to use the simplest possible system for this presentation, with all "turbellarian" flatworms placed within a single class, the Turbellaria. In all other flatworm groups (the exclusively parasitic Cestoda, Monogenea, Digenea, and Aspidogastrea), the larval epidermis is replaced at metamorphosis by a living, nonciliated, syncytial **tegument,** a possible homology suggesting that parasitism arose once in an ancestor to all of these classes and that they therefore be grouped within a single clade, the Neodermata (G: new skin). A leading view of evolutionary relationships among the major platyhelminth groups to be discussed in this chapter is shown in Figure 8.2.

Class Turbellaria

Only about 16% of all flatworm species are turbellarians (ter-bel-air´-ē-ans). The high surface-area-to-volume ratio of turbellarians makes them especially prone to dehydration in air, so most turbellarians live in aquatic environments. Most of the 3000 turbellarian species are free-living, but about 150 species are commensal or parasitic with other invertebrates. Parasitic turbellarians are

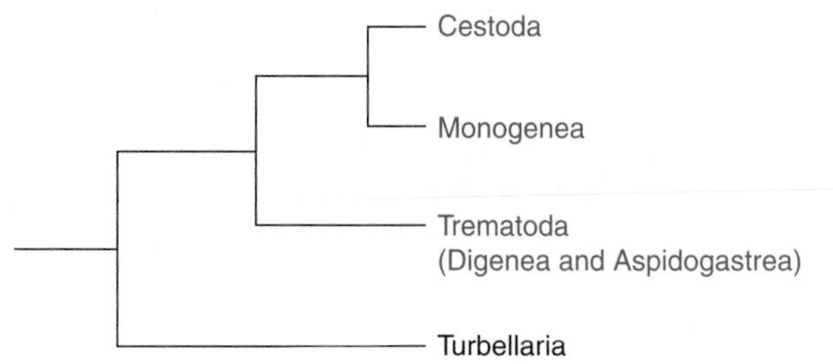

Figure 8.2
A leading view of the evolutionary relationships among the major platyhelminth groups. Summarized from several sources. Blue shading indicates groups contained within the Neodermata.

believed to have evolved independently of the Neodermata, from at least one different ancestor. Most turbellarian species are marine; a number of species are found in freshwater; and a few species are considered terrestrial, although these are restricted to very humid areas. Individuals are typically less than 1 cm long, regardless of habitat, although members of some terrestrial and marine species are considerably longer.

The nervous system consists of a coelenterate-style, diffuse nerve net in the most primitive turbellarian species. Increasing compactness of the system is associated with advancement in the class, culminating in the possession of a cerebral ganglion—a primitive but distinct brain—and from 1 to 3 (rarely 4) pairs of longitudinal nerve cords (Fig. 8.3). Such an advanced nervous system also characterizes the parasitic members of the phylum. Turbellarians typically bear one or more pairs of eyes anteriorly, along with a variety of cells that sense chemicals (such as potential foods), pressure changes (such as those produced by water currents), and mechanical stimuli. Individuals in about 10% of the species also have statocysts, which provide feedback regarding body orientation.

Most aquatic turbellarian species are **benthic;** that is, they live in or on the ocean, lake, pond, or river bottom. The body's outer surface is ciliated, often more so on the ventral surface than on the dorsal surface. Most species move at least partly by secreting mucus from the ventral surface and beating the ventral cilia within this viscous mucus. As a consequence of being flat, increased size is accompanied by a substantial increase in the amount of surface area in contact with the substrate over which the animal is moving. Thus, an increased number of cilia in contact with the substrate compensates for the increased weight of a larger animal, and the ability to move need not suffer as the animal grows. In contrast to the monociliated condition of cnidarians and sponges, each flatworm epidermal cell is multiciliated, bearing several to many cilia.

The locomotion of many individuals involves subtle waves of muscular contraction along the animal's ventral surface. These **pedal waves** are unidirectional, moving from the anterior of the worm posteriorly. As a wave of contraction moves down the length of the body, small portions of the ventral surface are pulled up and away

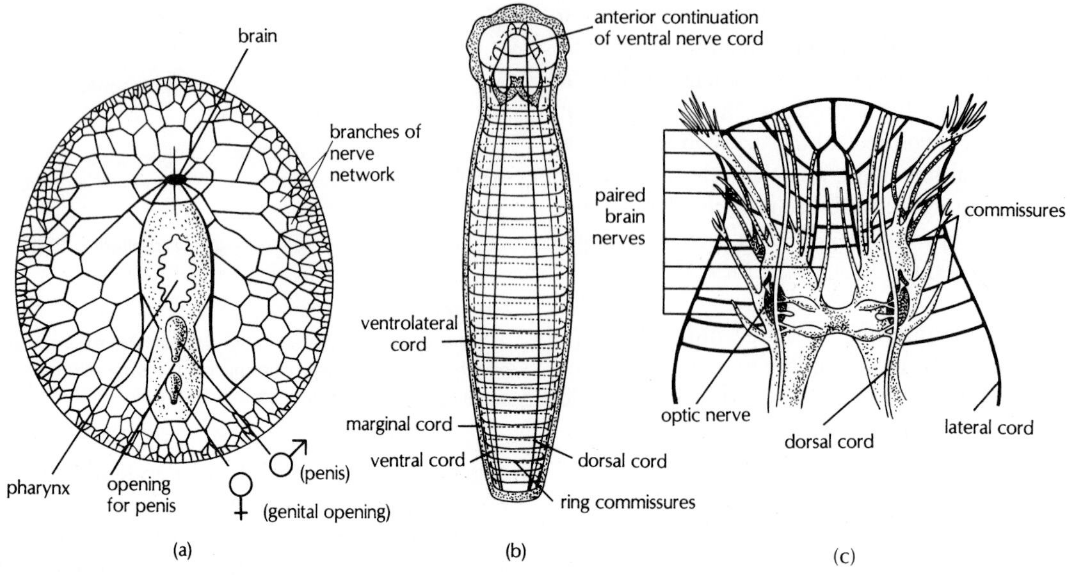

Figure 8.3

Turbellarian nervous system. (a) Nerve net system in *Planocera.* (b) Longitudinal nerve cord system in *Bothrioplana.* (c) Detail of the brain of *Crenobia.*

(a) After Lang. (b) After Reisinger; after Micoletzky. (c) From Bayer and Owre, *The Free-Living Lower Invertebrates.* Reprinted by permission of the authors.

from the substrate. Circular muscles contract just in advance of the wave, squeezing and thrusting the body forward, and longitudinal muscles contract just behind the wave, pulling the body in the direction of locomotion (Fig. 8.4b). The magnitude of the muscle contractions involved in generating pedal waves is quite small, and a number of waves are generally progressing down the body simultaneously. Thus, the progress of the animal along the substrate is very graceful and nearly indistinguishable from that powered entirely by ciliary action.

The musculature of the body wall includes fibers running longitudinally, circumferentially, dorsoventrally, and diagonally (Fig. 8.4a). All of this musculature is brought into play in the locomotory movements of some turbellarians. This movement, called **looping,** is quite pronounced. The individual attaches at the anterior end, pulls the posterior forward by contracting longitudinal muscles, attaches at the posterior end, releases the anterior end, and then thrusts the body forward by contracting circular muscles (Fig. 8.5).

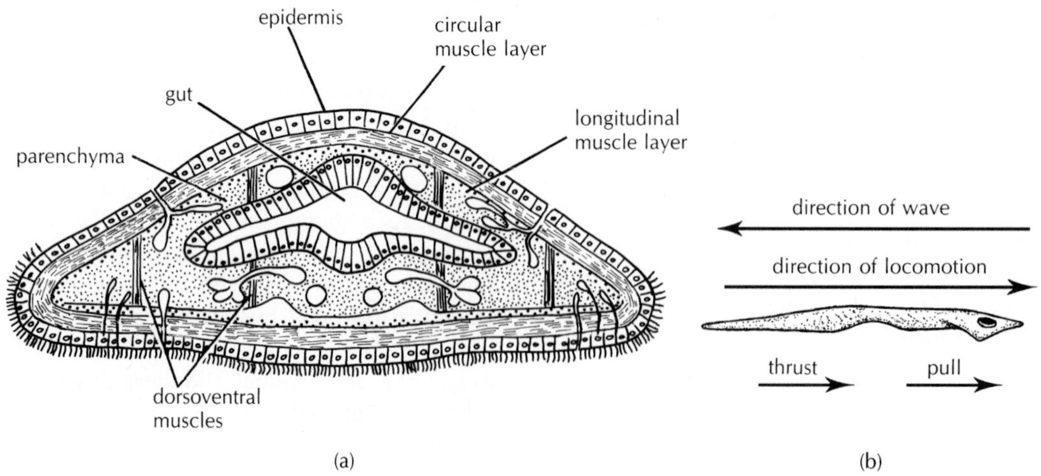

Figure 8.4

(a) Diagrammatic illustration of a turbellarian in cross section, showing the arrangement of muscle layers. (b) Locomotion by pedal waves in a turbellarian flatworm. A single wave is shown traversing the body's ventral surface. A small portion of the body is both thrust and pulled forward as the wave of dorsoventral contraction lifts that region of the body away from the substrate. The thrusting and pulling forces are generated by the regions of the body that are anchored against the substrate, on either side of the contraction. In most species, many waves of contraction travel down the body simultaneously.

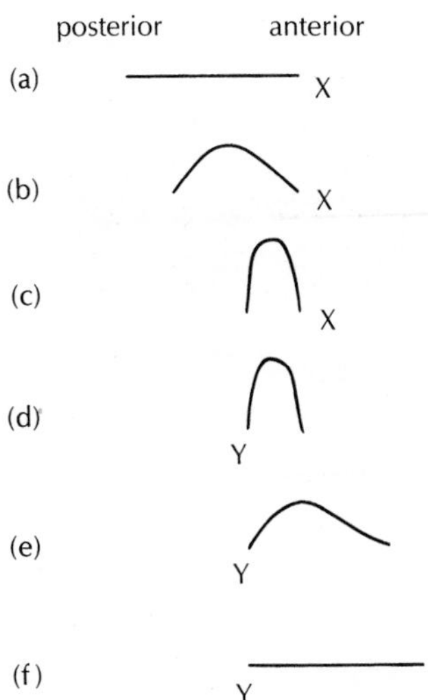

Figure 8.5

Diagrammatic illustration of looping by turbellarians. Once an attachment to the substrate is made at point X (a), contraction of the longitudinal musculature brings the rear of the animal anteriorly (b,c). The anterior attachment is released, and a new attachment is made at the posterior end (d) at point Y, allowing the head of the animal to be thrust forward (e,f).

For effective looping locomotion the flatworm must be able to adhere locally to the substrate, to prevent sliding backward while the pulling and pushing forces are being generated. On the other hand, attachments to the substrate must be only temporary if the animal is to progress forward. Flatworms typically possess a large number of paired secretory cells (**duo-glands**) located on the ventral surface and opening to the exterior. One cell of each pair seems to produce a viscous glue, while the other cell presumably secretes a chemical that breaks this attachment to the substrate (Fig. 8.6).

A number of turbellarian species can swim, either through ciliary activity or by vigorous, controlled waves of contraction of the body wall musculature. Some benthic species probably swim only when environmental conditions deteriorate, while a number of shallow-water species appear to swim routinely, remaining benthic only during low tide. Some small (< 1 mm) acoel flatworm species (see below) are routinely found in plankton samples taken in warm, oceanic surface waters, suggesting that these flatworms may be permanently planktonic.

The body surface of many species bears numerous aggregations of small, cylindrical **rhabdites** and related **rhabdoids** (Fig. 8.6). These structures are uniquely turbellarian, although their function is uncertain; they release a thick mucus that coats the animal's body, possibly in response to attempted predation or to desiccation.

The turbellarian digestive system is fairly simple, although the details vary considerably among species. Indeed, differences in the structure of the mouthparts and the gut, along with differences in reproductive morphology, are key elements in dividing the class into its 12 constituent orders.

Some species bear a simple mouth opening on the ventral surface and have no well-formed gut cavity; these are the acoel flatworms (order Acoela; *a* = G: without; *coel* = G: a cavity). In these flatworms, food is essentially thrust into a densely packed mass of specialized digestive cells (Fig. 8.7a). Acoel flatworms are exclusively marine and bear a superficial morphological resemblance to the planula larvae of

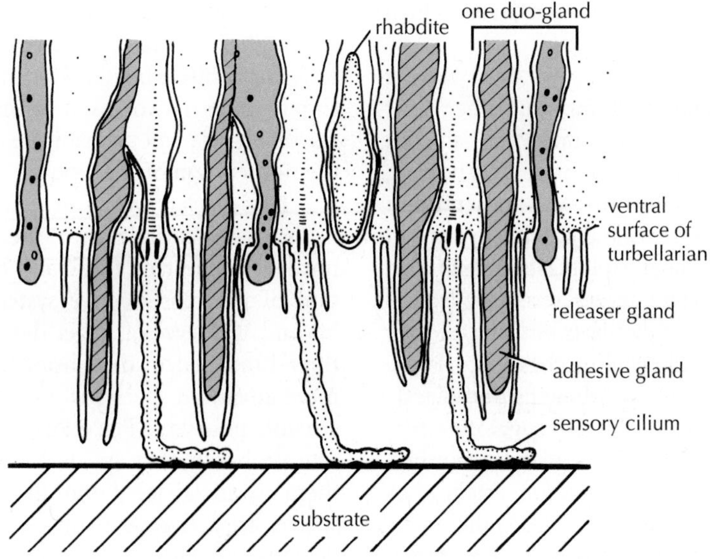

Figure 8.6

Diagrammatic illustration of the ventral surface of a free-living turbellarian flatworm, showing the arrangement of the duo-gland system. Adhesive glands produce a chemical that attaches part of the animal to a substrate; the releaser glands secrete a chemical that dissolves the attachment as appropriate. Rodlike rhabdites, as illustrated, are encountered in the epidermis of most turbellarians; a defensive function has been suggested.

Modified from Tyler. 1976. *Zoomorphology* 84:1.

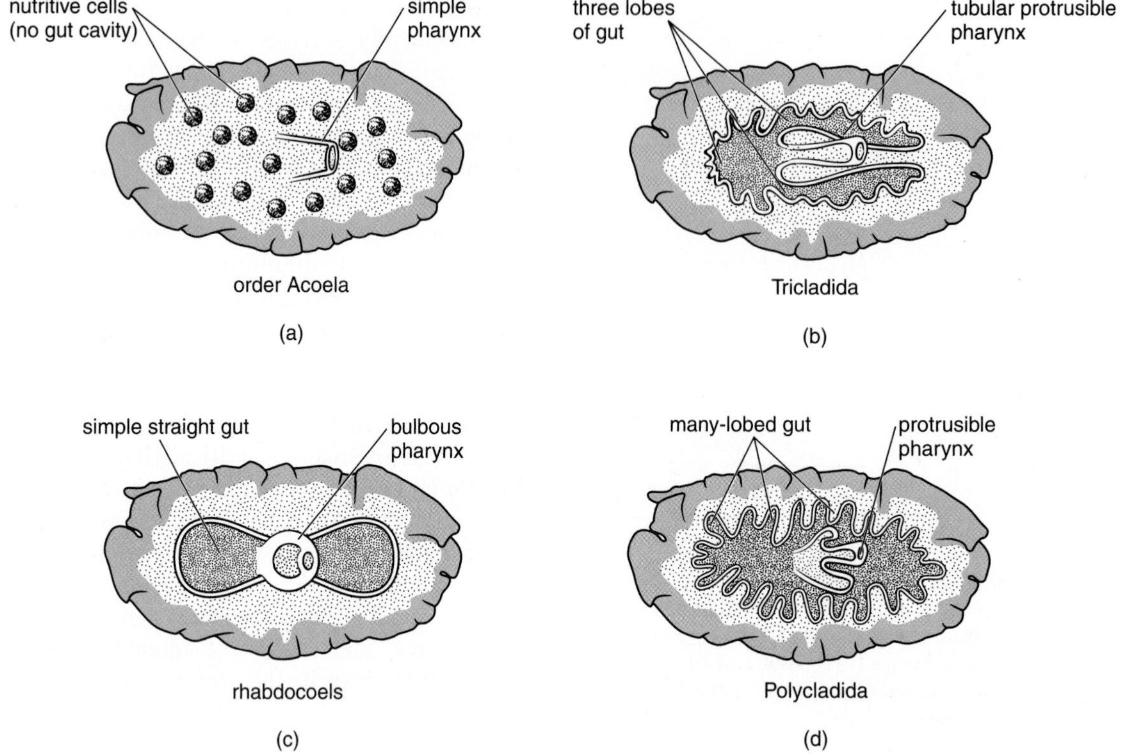

Figure 8.7
Morphological diversity of the turbellarian digestive system, ranging from (a) no well-defined gut; (b) a gut that is unbranched; (c) a three-branched gut; (d) a multibranched gut.
(a–d) Based on W. D. Russell-Hunter, *A Life of Invertebrates*. 1979.

cnidarians. In fact, some zoologists suspect that both cnidarians and flatworms evolved from a planula-like ancestor, in which case the acoel turbellarians would resemble the most primitive triploblastic metazoans, from which all other metazoans are presumably derived.

Support for this attractive scenario has been waxing and waning practically from month to month. Some recent studies suggest that acoels are not primitive at all, but instead have descended from more elaborate, gut-bearing (and possibly even coelomate) ancestors, while other recent studies restore flatworms in general, and acoel flatworms in particular, to central status as basal triploblasts. As hinted at in the introduction to this chapter, several recent molecular studies present acoels as basal triploblasts but suggest they don't belong in the phylum Platyhelminthes.[4] Acoels are also nearly unique in continually withdrawing and digesting damaged ciliated epidermal cells, again suggesting that they may not belong in the phylum Platyhelminthes. Indeed, recent work using a series of genes for certain RNA molecules (microRNAs, abbreviated "miRNAs") that regulate gene expression by disabling or destabilizing target mRNAs provides further support for separating the acoels from other flatworms. Triclad and polyclad flatworms (Fig. 8.8, top 2 lines) seem to express more of these genes than do any other animals, and in fact express at least 6 unique miRNAs (light blue shading). In contrast, only 5 of the 28 miRNAs considered were found among acoel flatworms (Fig. 8.8, second line from the bottom, dark blue shading). The data provide strong evidence that acoels are indeed primitive bilateral animals, with no close relationship to other flatworms. If confirmed with additional studies, the Acoela will probably be an independent phylum by the next edition of this book.

Turbellarian evolution seems to have been largely a tale of increasing complexity of the digestive system and of the means of acquiring food, and of an increasing specialization of the reproductive system. The guts of flatworms beyond the level of acoel development may be straight, three-branched, or multibranched (Fig. 8.7). The mouth of more advanced species is often borne at the end of a **protrusible pharynx** (Fig. 8.9b); other species possess a separate proboscis that spears food items and then transfers them to an adjacent mouth opening. Most turbellarian species are active carnivores, although some species ingest detritus and algae,[5] and a number harbor algal symbionts. Digestion of food is initially extracellular through secretion of enzymes. Particles are later phagocytosed, and digestion is completed intracellularly.

4. See General References on page 178.

5. See *Topics for Further Discussion and Investigation,* no. 1.

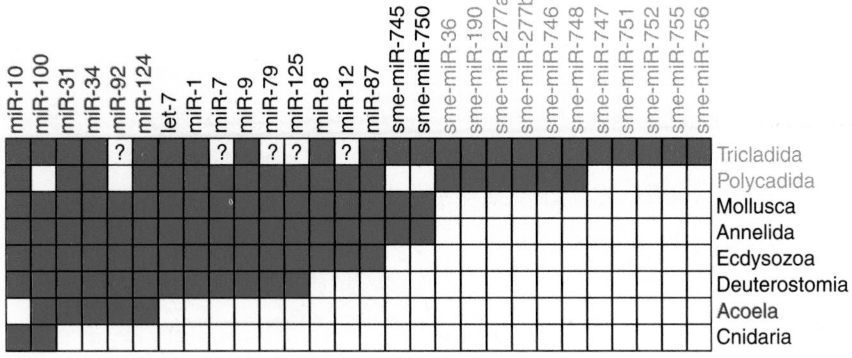

Figure 8.8

Expression of microRNAs (as detected by Nothern blots) for flatworms and a variety of other metaozoans. Light blue shading indicates expression unique to triclad and polyclad flatworms, while dark blue shading indicates genes expressed by acoel flatworms.

Based on Sempere, L. F., O. P. Martinez, C. Cole, J. Baguña, and K. J. Peterson. 2007. Phylogenetic distribution of microRNAs supports the basal position of acoel flatworms and the polyphyly of Platyhelminthes. *Evol. Devel.* 9: 409–415.

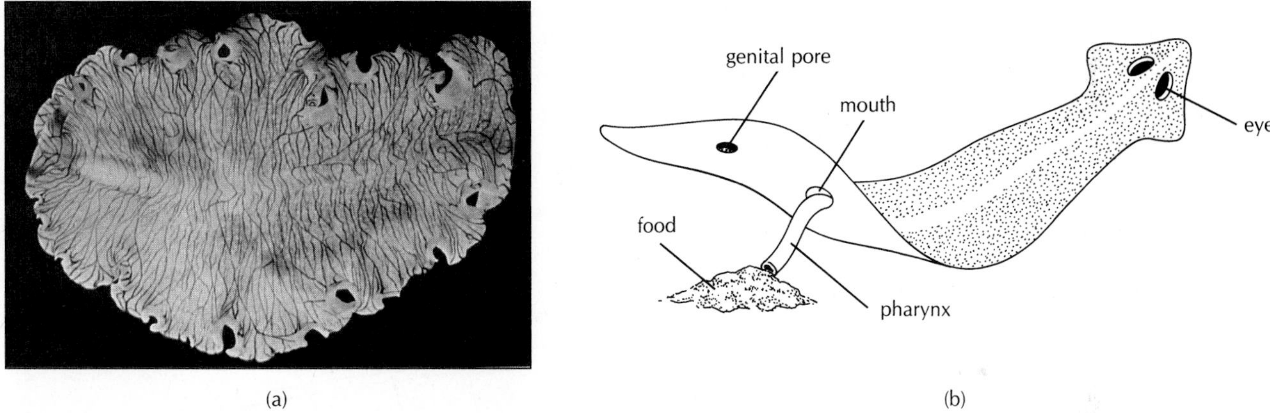

Figure 8.9

(a) A marine polyclad flatworm, *Pseudoceros crozieri*. This individual was nearly 4 cm long. (b) Typical flatworm (*Dugesia*) with pharynx extended for feeding.

(a) Courtesy of Anne Rudloe, Gulf Specimen Marine Laboratories, Inc.

The most structurally and physiologically sophisticated system found among turbellarians is the reproductive system. Both male and female reproductive organs are found within a single individual, and the male system is particularly complex (Fig. 8.10). When triclad flatworms mate, each individual typically inserts its penis into the female opening of the other member of the pair, so sperm transfer is reciprocal (Fig. 8.11a). For most other turbellarians, copulation may occur by hypodermic impregnation, in which the stylets of the penis pierce the body of the partner. The eggs of each animal are released after fertilization and generally develop directly into miniature flatworms within a protective capsule; in most turbellarian species, there is no free-living larval stage in the life cycle. In several marine species, however, the developing embryo gives rise to a short-lived, microscopic free-swimming larval stage, most commonly a **Müller's larva** (Fig. 8.11b).

Many turbellarian species possess remarkable regenerative powers that go far beyond the ability to repair wounds,[6] as illustrated in Figure 8.12. Regeneration is accomplished through the activities of **neoblasts,** undifferentiated cells—unique to turbellarians—with remarkably versatile developmental plasticity. In addition, many turbellarian species reproduce routinely by asexual fission, an ability closely linked to their great regenerative capabilities. Developmental biologists have studied these phenomena intently; certain freshwater triclads in particular have long served as models for studying the processes controlling cellular differentiation. It is not yet clear why the members of some turbellarian species can regenerate while others cannot. Note that having cell fates largely fixed at the first cleavage (a defining characteristic for protostomes) does not necessarily limit later regenerative capacity.

6. See *Topics for Further Discussion and Investigation,* no. 5.

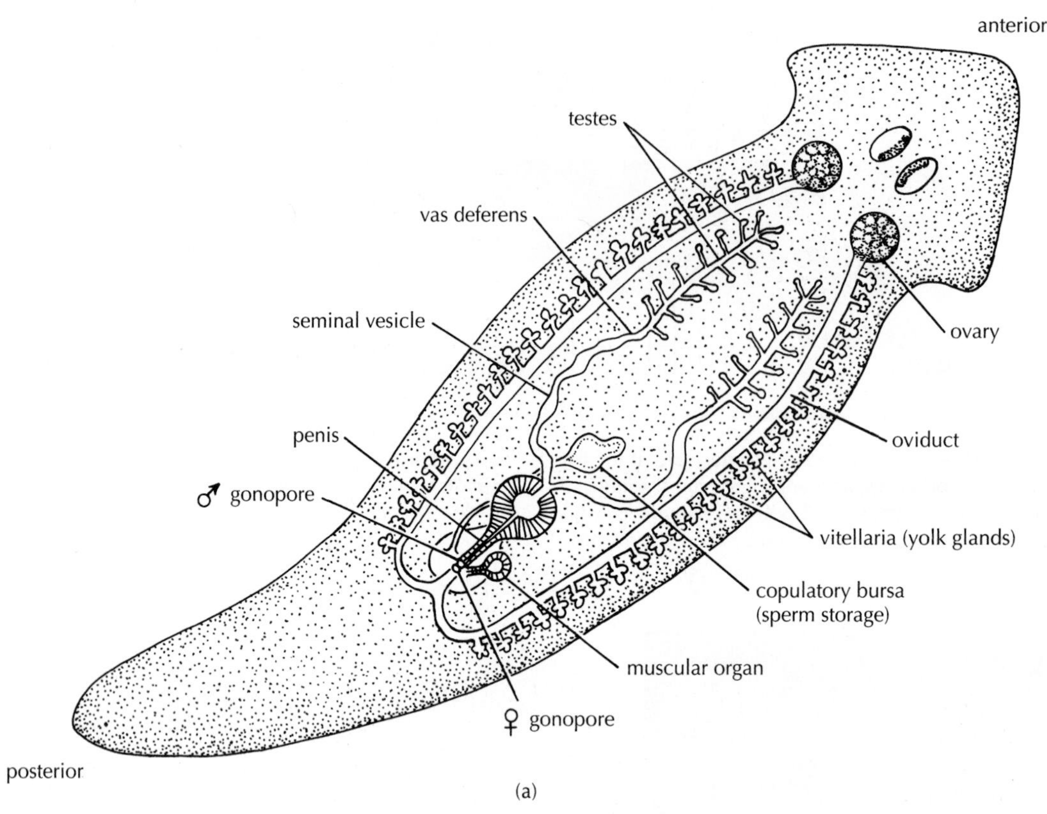

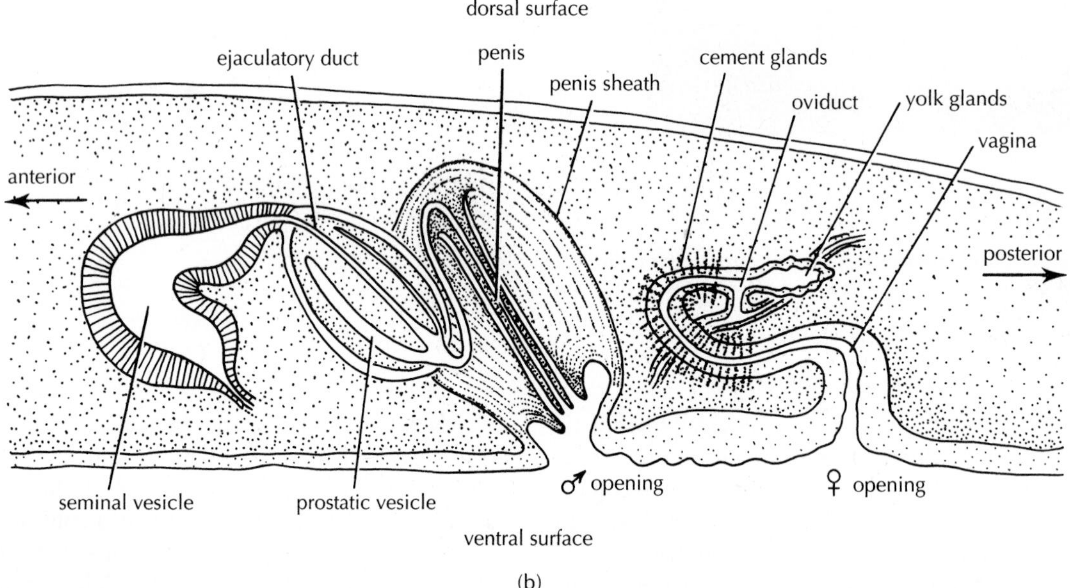

Figure 8.10

(a) Triclad turbellarian reproductive system. Note the presence of both male and female reproductive organs in a single individual. The male reproductive system consists of the testes, prostate gland, and the ducts (vas deferens and seminal vesicle) that conduct sperm from the testes to the ejaculatory duct of the muscular penis. Sperm accumulate in the seminal vesicle prior to copulation. The female reproductive tract includes the ovaries, the oviducts, and the gonopore, through which fertilized eggs are released. As eggs move down the oviduct toward the gonopore, they become surrounded by nutritive yolk cells manufactured by the vitellaria. The copulatory bursa serves to receive and store sperm contributed by the partner. Some species possess an additional sperm storage organ, the seminal receptacle. After copulation, cement glands coat the emerging ova with a sticky substance that glues them to a variety of solid substrates, including macroalgae and other vegetation. (b) Detail of copulatory apparatus and associated structures.

(a,b) After Steinmann.

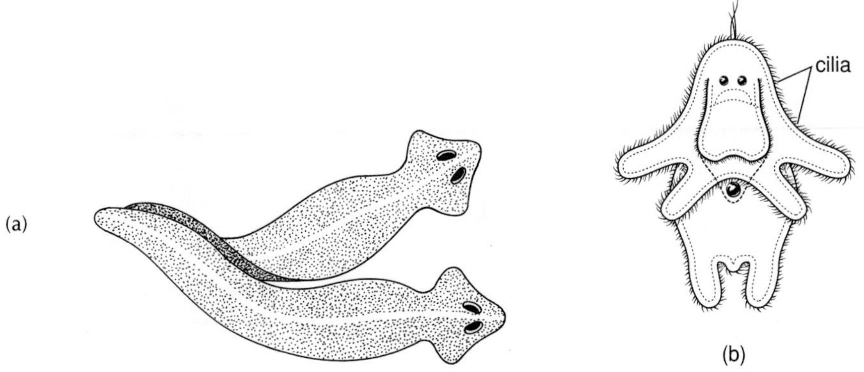

Figure 8.11

(a) Copulating flatworms. The penis of each individual is inserted into the female opening of the mate. (b) Müller's larva of a marine flatworm.

(b) Based on E. E. Ruppert, "A review of metamorphosis of turbellarian larvae" in *Settlement and Metamorphosis of Marine Invertebrate Larvae,* Chia and Rice, eds. 1978.

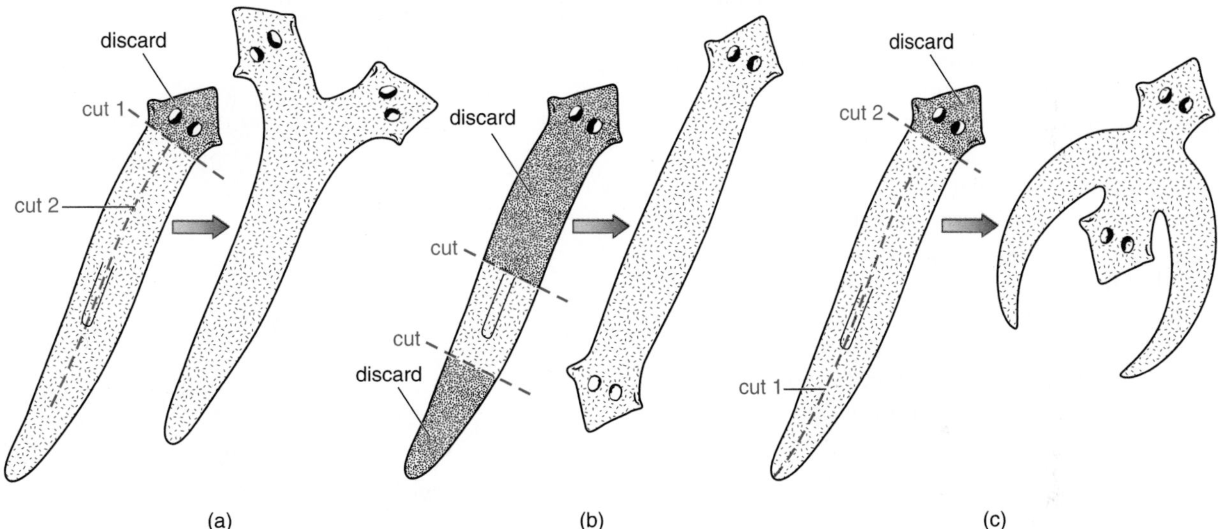

Figure 8.12

Three experiments illustrating the considerable powers of regeneration possessed by many free-living flatworms. Cuts are made using a clean razor blade. Blue shading indicates discarded body parts.
Based on Sherman/Sherman, *The Invertebrates: Function and Form,* 2e. 1976,

Class Cestoda

Defining Characteristics:[7] 1) Small anterior hooked attachment organ (scolex); 2) division of body into segments (proglottids) arising from anterior end, behind the scolex; 3) absence (loss) of digestive tract

---

Members of the class Cestoda (ses-to´-dah)—most of which are members of the subclass Eucestoda and commonly known as tapeworms—are all internal parasites; that is, they are **endoparasitic** (*endo* = G: within). They are primarily parasites of vertebrates, inhabiting various regions of the host digestive tract. As many as 135 million people worldwide are estimated to have tapeworm infections.

Cestodes are strikingly different from turbellarians, the differences reflecting an extremely high degree of cestode specialization for an endoparasitic existence. Instead of the ciliated epidermis characterizing turbellarians, a nonciliated **tegument** covers the cestodes. The tegument contains numerous nuclei, but these are not separated by cell membranes; that is, the tegument is **syncytial.** The outer surface is outfolded into numerous cytoplasmic projections, vastly increasing the amount of exposed surface area across which nutrients can be taken up from the host's gut. Indeed, the cestode must receive all of its nutrients in this manner, as it has no mouth or digestive tract of its own at any point in its life cycle.

7. *Characteristics distinguishing the members of this class from members of other classes within the phylum.*

Although lacking a mouth, cestodes do have an anterior end, which in most species takes the form of a **scolex.** The scolex is studded with hooks and/or suckers that are used to maintain position within the host's gut. The relatively few cestode species that lack a scolex are mostly placed within a second subclass, the Cestodaria (Fig. 8.13a).

The essence of tapeworm existence, however, really lies just posterior to the scolex, in a region known as the **neck.** A seemingly endless series of sections called **proglottids** bud from the neck of most cestodes at a rate of several per day (Fig. 8.13b). Only a small number of cestode species, mostly members of the subclass Cestodaria (Fig. 8.13a), do not produce proglottids.

Each proglottid is involved primarily with the process of sexual reproduction. Not only is each tapeworm a simultaneous hermaphrodite, but so is each proglottid; that is, each proglottid contains both male and female reproductive systems (Fig. 8.14). In some species, each proglottid may contain numerous ovaries and as many as 1000 distinct testes. The tapeworms might best be thought of as franchisers of eggs and sperm. And like the best franchising operations, tapeworms daily turn out a large amount of product, qualitatively similar from proglottid to proglottid.

Each proglottid contains perhaps 50,000 eggs. The eggs are generally fertilized by sperm from a neighboring cestode, but they can be fertilized by sperm from the same individual—or, in fact, from the same proglottid. Since proglottids are rarely more than 3–5 mm long and the total length of a cestode may exceed 10–12 m, perhaps 2000 to 4000 proglottids are produced per individual in many species, and an incredible number of fertilized eggs (tens of thousands, or even hundreds of thousands) are produced per day per cestode.

In some species, the posteriormost proglottids break off periodically. In other species, mature proglottids burst open, releasing the fertilized eggs into the host's gut. Either way, the eggs leave the host's body along with the feces. Generally, the fertilized eggs cannot take up residence in the **definitive** (final) **host** immediately; they must first enter an **intermediate host** or, in some species, a series of intermediate hosts. Different cestode species require different intermediate hosts, which include both vertebrates and invertebrates.

When a fertilized cestode egg is ingested by the appropriate intermediate host, an **oncosphere** larva commonly hatches out. Each oncosphere has muscles, flame cells, and most significantly, 3 pairs of hooks with which it attaches to the wall of the host's digestive tract (*oncus* = G: a barb or hook). The oncosphere then **lyses** (dissolves) its way through the intestinal wall, taking up residence as an encysted form in the coelomic space or in specific organs and tissues of the host. Among tapeworms in the genus *Taenia,* an oncosphere typically produces a single resting individual called a **cysticercus,** or **bladder worm.** In some species, however, the resting stage divides asexually many times within the intermediate host to produce a large, sometimes deadly, **hydatid cyst.** Such asexual replication is especially common among taeniids (family Taeniidae), but is rarely encountered among other cestode species.

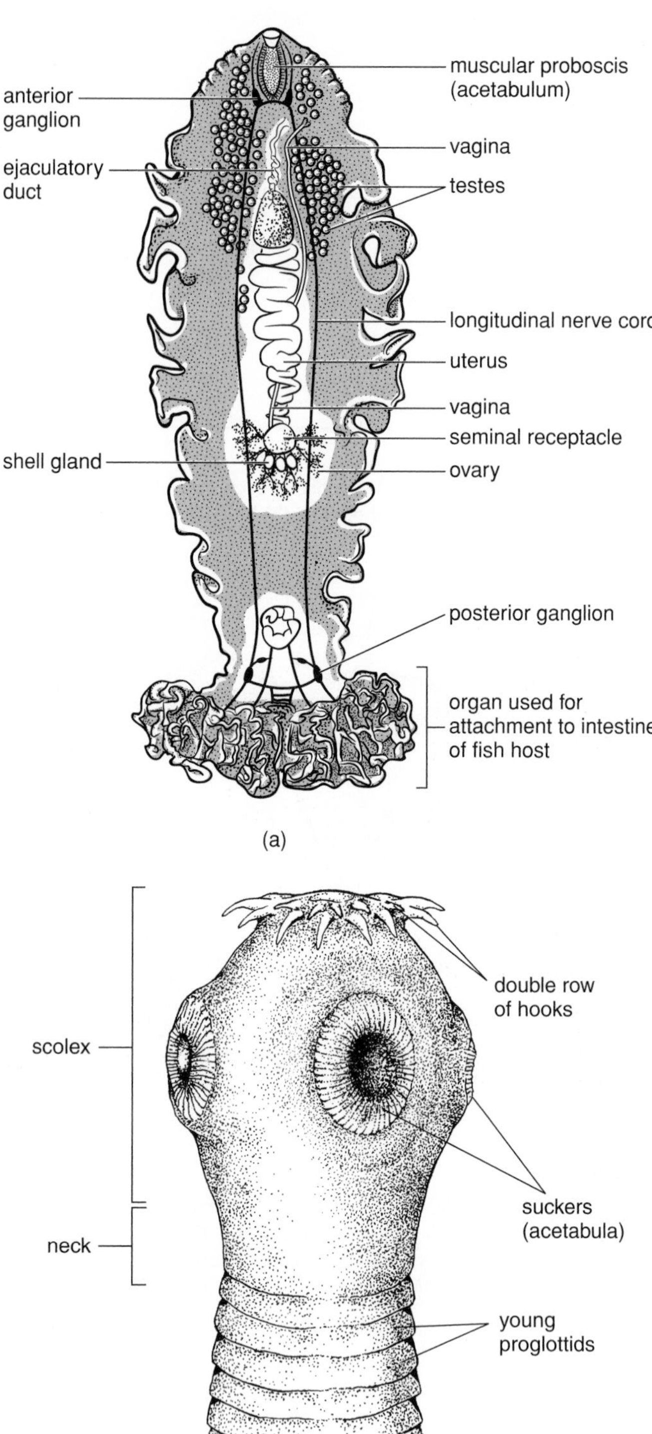

Figure 8.13

(a) *Gyrocotyle fimbriata,* a member of the small subclass Cestodaria. (b) Anterior end of the common tapeworm *Taenia* sp., magnified 12×.

(a) Based on T. Cheng (after Lynch, 1945), *General Parasitology,* 2d ed. 1986.

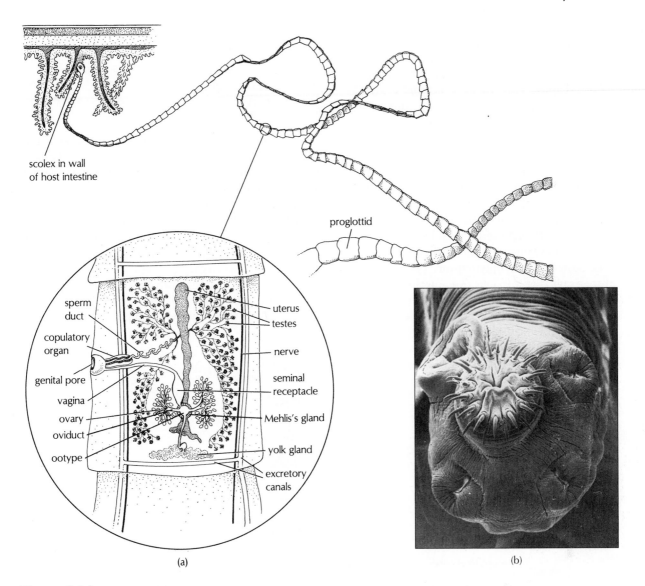

Figure 8.14

(a) *Taenia solium,* the pork tapeworm, attached to the intestinal wall of its host. Note that the entire proglottid is dedicated to the task of reproduction. Eggs leave the ovary and are fertilized either in transit to or within the ootype (the region where the oviducts join) by sperm feeding in from the seminal receptacle. Fertilized eggs are then enclosed individually in protective capsules, possibly made from secretions of a gland (Mehlis's gland) surrounding the ootype. (b) Scanning electron micrograph of the anterior end of *Taenia hydatigena,* magnified 170×.

(b) From D. W. Featherston, *International Journal for Parasitology* 5:615, fig. 1, © 1975 Pergamon Press. Reprinted with permission.

Further development is arrested until the intermediate host is eaten by a different host, which may be the final host or another intermediate host. Thus, the complete sequence from egg to adult is achieved only if the fertilized cestode eggs reach the appropriate intermediate hosts, which in turn must be ingested by the appropriate final, or definitive, host (Fig. 8.15). Among the vertebrates, fishes, cows, pigs, dogs, and sometimes birds may serve as intermediate hosts. Humans often serve as acceptable final and intermediate hosts. (Think twice before eating undercooked beef, pork, or fish and before letting a dog lick your face!) Most invertebrate intermediate hosts are arthropods.

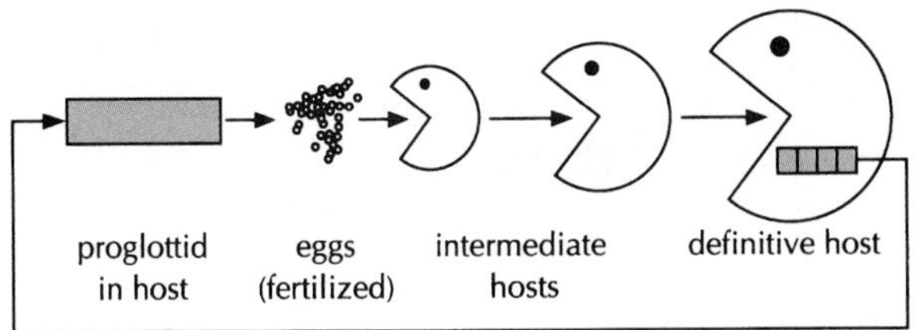

Figure 8.15

Diagrammatic illustration of the typical cestode life cycle. The parasite cannot attain sexual maturity until the definitive host is located.

Class Monogenea

Defining Characteristics: 1) Posterior attachment organ (haptor [= opisthaptor]) including sucker and complex attachment hooks and sclerites; 2) larva (oncomiracidium) bearing 3 bands of cilia and usually 1 or 2 pairs of eyes

Monogenetic flatworms are usually parasitic on the skin or gills of fishes; that is, most are **ectoparasites** (*ecto* = G: outside). This is the flatworm's sole host, to which it attaches primarily by means of suckers, hooks, and complex sclerites located at its posterior end. The highly specialized posterior attachment organ is called the **haptor** (*hapto* = G: fastened) (= **opisthaptor;** *opistho* = G: behind, at the rear) (Fig 8.16a, b). An anterior adhesive organ (often called the **prohaptor**), consisting of suckers and adhesive glands, aids attachment. There are no intermediate hosts, so the life cycle generally involves the following stages:

(1) Sexual maturity reached in or on fish → (2) Egg production →
(3) Larval stage **(oncomiracidium)** → (4) attachment to fish

Most monogenean species show a very high level of host specificity, and typically occupy highly specific sites within a host; members of one monogenean species live only at the base of a fish's gill filaments, for example, while members of another species are found only near the tips of the same filaments. Only about 8000 species have been described so far, but about 25,000 species are thought to exist, which would make this the most speciose group in the Platyhelminthes.

The taxonomic status of the monogeneans has been uncertain for some time. Long thought to be trematodes (flukes), which they closely resemble as adults, some

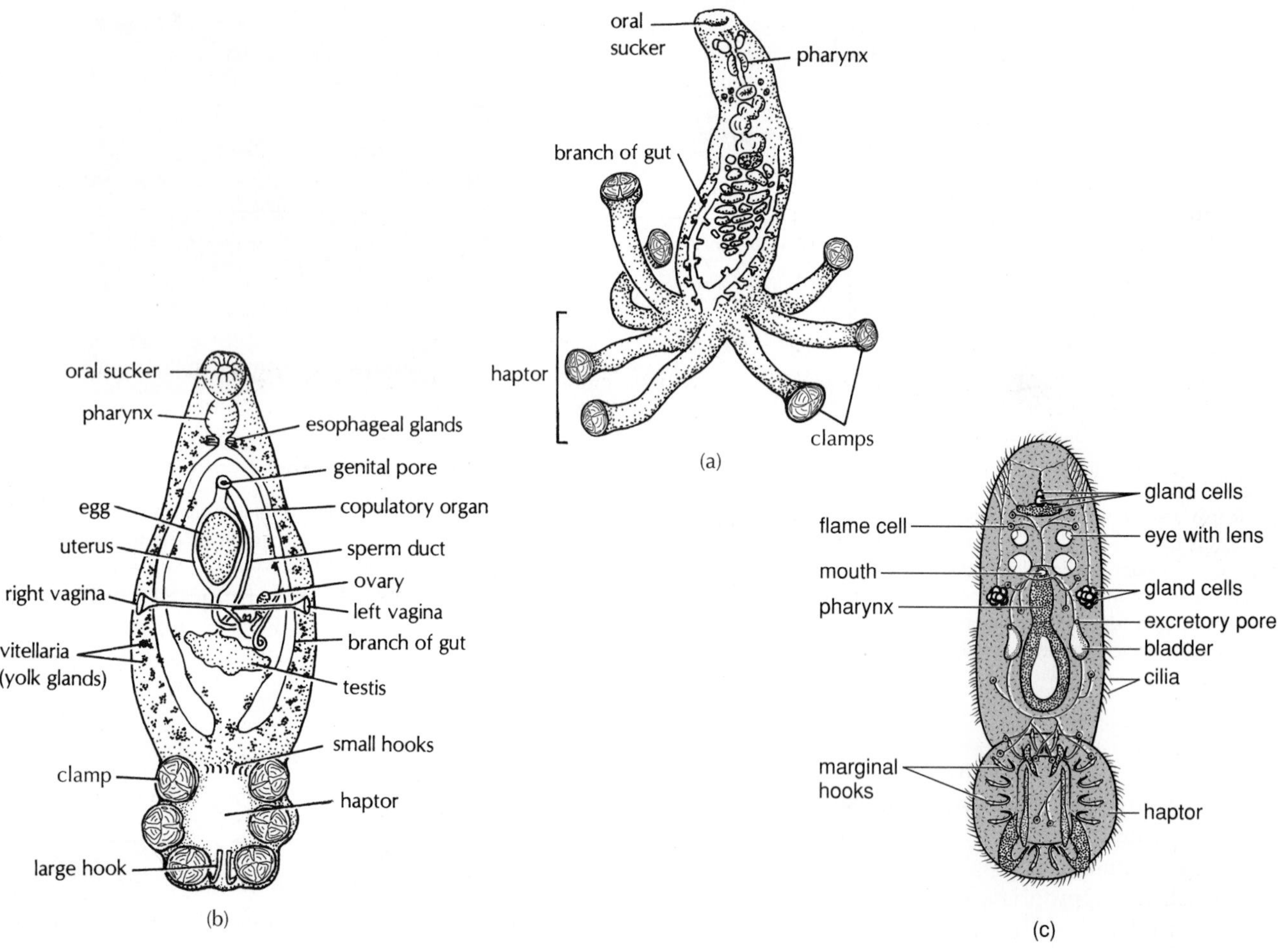

Figure 8.16

Monogenean diversity. (a) *Choricotyle louisianensis,* a monogenetic trematode, taken from the gills of a fish. Note the complex, multi-piece haptor. (b) *Polystomoidella oblongum,* another monogenetic trematode, taken from the urinary bladder of a turtle. Note the single-unit haptor of *P. oblongum,* studded with suckers for securing attachment to the host tissues. (c) Oncomiracidium larva of the monogenean *Entobdella soleae* seen in ventral view. (Larval length is about 240 μm.) Note the posterior haptor and 3 conspicuous bands of cilia.

(a) From Hyman, after Looss. (b) From Noble and Noble; after Cable. (c) Based on Frederick W. Harrison, John O. Corliss, and Jane A. Westfall, Eds., *Microscopic Anatomy of Invertebrates,* Vol. 3.

workers have argued that monogeneans are more closely related to the cestodes, largely on the basis of the similarity between the cestode oncosphere and the monogenean **oncomiracidium** larva (Fig. 8.16c). Others have argued that the monogeneans are sufficiently dissimilar from either cestodes or trematodes to merit categorization as a separate class, but some ultrastructural studies suggest that monogeneans do not form a monophyletic group. Molecular data presently link the monogeneans with the cestodes as a separate class of flatworms, the Monogenea (see Fig. 8.2). In that case, the morphological similarities between adult monogeneans and trematodes reflect strongly convergent evolution. Certainly the life cycles of the monogeneans have little in common with those of the digenetic trematodes, as discussed next.

Class Trematoda

All members of the Trematoda (trem-ah-tō´-dah) are parasitic, and most reach adulthood only as parasites in or on vertebrates. A parasitic existence poses a number of problems, none of them trivial (see also pp. 58–61). The successful parasite must

1. reproduce within the definitive host
2. get fertilized eggs or embryos out of the host
3. contact and recognize a new, appropriate host;[8]
4. obtain entrance into the host
5. locate the appropriate environment within the host
6. maintain position within the host
7. withstand what is often a rather anaerobic (oxygen-poor) environment
8. avoid digestion or attack by the host's immune system;[9] and
9. avoid killing the host, at least until reproduction has been achieved.

These problems are faced by all parasites, but perhaps the most remarkable adaptations to numbers 3 and 4 are encountered among the trematodes also called "flukes."

The outer body layer of adult trematodes, like that of the cestodes and monogeneans, is an unciliated syncytial tegument. In other respects, the trematode body more closely resembles that of a turbellarian. Trematodes have a mouth opening and a blind-ended digestive tract that is, with a few exceptions, bilobed. The body is never segmented (Fig. 8.17). The parasite ingests the host's tissues and blood through its mouth.

Schistosomiasis, an often deadly disease prominent in many regions of the world, results from an infection by trematodes known as "blood flukes" (Fig. 8.17d). Presently, more than 200 million persons in 77 countries are estimated to suffer from schistosomiasis, making it the second most prevalent disease in the world, next to malaria (which is caused by a protozoan—see Chapter 3). Schistosomiasis kills about 800,000 people yearly, in part by inducing cancer,[10] and infections in livestock cause hundreds of millions of dollars in economic damage.

The flukes are placed into one major group with over 4100 species, the **digenetic trematodes** (also called "digeneans"), and one much smaller group with fewer than 100 species, the aspidogastreans (also called aspidobothreans).

The Digeneans

As their name implies (*di* = G: two; *gena* = G: birth), digenetic flukes always require at least one intermediate host before reaching the final host. Unlike the passive process found among the cestodes, host location among digenetic trematodes is generally an active process, mediated by highly specialized, free-living larval stages.

Among the digenetic trematodes, each fertilized egg generally gives rise to a single, free-living, ciliated **miracidium** larva (Fig. 8.18). The gutless miracidium is then either eaten by, or locates and bores into, an intermediate host, which is almost always a mollusc and most commonly a snail. Miracidia produced by a given trematode species must generally enter a particular species of molluscan host to develop further. Indeed, molecular biologists may eventually eradicate schistosomiasis by inserting into the appropriate snail species genes that will prevent schistosome miracidia from penetrating or developing within those hosts.

Host penetration is accomplished by secretions from several glands located anteriorly in the miracidium (Fig. 8.18a). The miracidium then dedifferentiates into a **mother sporocyst** stage, in which most of the miracidial structures (including the external cilia) are lost, except for the protonephridia. The mother sporocyst lives in the molluscan blood circulatory system (the hemocoel). As no mouth is present at this stage of development, the mother sporocyst grows within the host by taking up dissolved nutrients from the surrounding body fluids. Within each sporocyst are numerous balls of cells. Each of these **germ balls** develops either into another sporocyst (the daughter sporocysts) or into another larval stage, the **redia.** The rediae or daughter sporocysts migrate to their host's digestive gland or gonad. The rediae are active feeders, possessing a mouth and a functional, blind-ended gut (Fig. 8.18a), whereas the daughter sporocysts do not feed (although they probably absorb soluble nutrients from their surroundings). Within each redia larva, or within each daughter sporocyst, are numerous germ balls, each of which develops into yet another anatomically distinct larval stage, the **cercaria.** The cercariae leave the "mother" by means of a birth canal if escaping from redia or by lysing through the body wall if escaping from sporocysts, and then lyse

8. See *Topics for Further Discussion and Investigation,* no. 2.

9. See *Topics for Further Discussion and Investigation,* no. 3.

10. Mayer, D.A., and B. Fried. 2007. *Adv. Parasitol.* 65: 239–96.

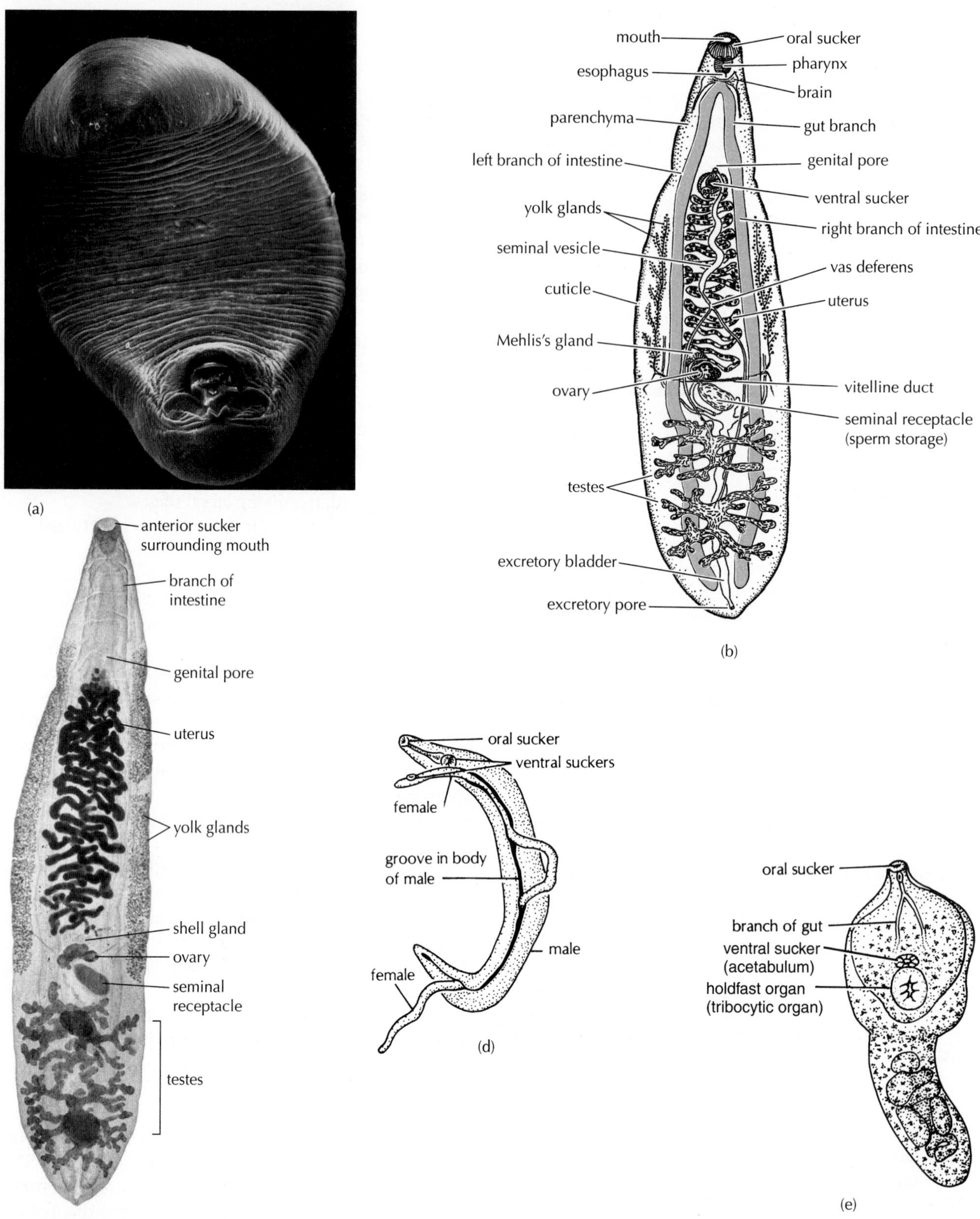

Figure 8.17

Digenetic trematode diversity. (a) Scanning electron micrograph showing external morphology of *Zygocotyle lunata,* taken from a mouse. The oral sucker is at the top of the photograph. (b) The Chinese liver fluke, *Opisthorchis sinensis,* seen in ventral view. Note the large percentage of the body devoted to reproduction. Mehlis's gland, also called the shell gland, is a conspicuous feature of the female reproductive tract; its function in trematodes is uncertain. (c) The Chinese liver fluke *Opisthorchis sinensis*. Note the two-branched gut and the proportion of the body devoted to reproduction. (d) The blood fluke *Schistosoma haematobium*, a gonochoristic digenetic species. During copulation, the female lies in a specialized groove in the body of the male. This and several other related species cause in humans the disease known as "schistosomiasis." "Swimmer's itch" is caused by members of the same family. (e) The digenetic trematode *Neodiplostomum paraspathula*, seen in ventral view. The tribocytic organ releases a variety of digestive enzymes that solubilize host tissue at the point of attachment.

(a) Courtesy of S.W. B. Irwin, University of Ulster, Northern Ireland. (b) From Brown, 1950. *Selected Invertebrate Types.* John Wiley & Sons. (c) Carolina Biological Supply Company/ PhotoTake; (d) From Hyman, after Looss. (e) From Noble and Noble; after Noble.

miracidium hatching from egg capsule

penetration gland

apical gland

miracidium

sporocyst

developing redia

germ balls

germ balls

cilia

(a)

redia

pharynx

intestinal caecum

developing cercariae

germ ball

anterior sucker

intestine

penetration glands

acetabulum

tail

cercariae

rediae

cercariae

cercariae

(b)

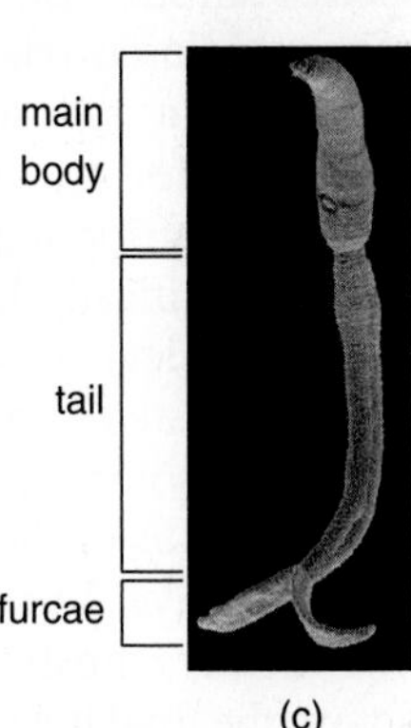

(c)

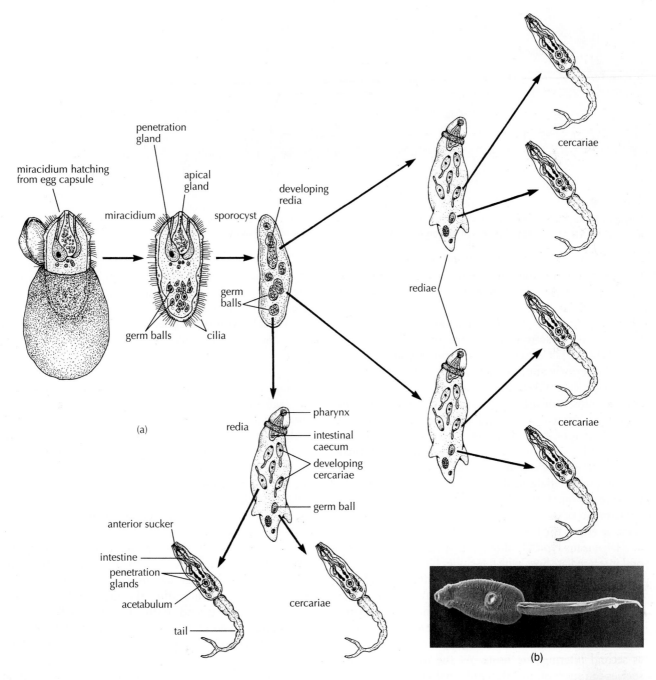

Figure 8.18

(a) Replication of larval stages in the trematode life cycle. Large numbers of cercariae are contained within each redia. Many rediae are, in turn, contained within each sporocyst. All of the larvae derived from a single fertilized egg are produced asexually and are therefore genetically identical to each other. The miracidium and cercaria stages are commonly free-swimming. The sporocyst and redia stages occur within an intermediate host, usually a snail. Note the germ balls in the miracidium and sporocyst stages; these will develop into redia larvae. Similarly, the germ balls within the rediae are future cercariae. (b) Scanning electron micrograph of the cercaria stage of *Echinostoma trivolvis*, a digenetic trematode, seen in ventral view. (Total length is about 0.5 mm.) The tail is unforked in this and many other species. Note the prominent oral and ventral suckers. (c) Scanning electron micrograph of a *Schistosoma mansoni* cercaria.

(a) Modified after Noble and Noble; after Cheng, T. C. 1986. *General Parasitology,* 2d ed. Academic Press. (b) Courtesy B. Fried and M. A. Haseeb. From Fried and Haseeb 1991. *Microscopic Anatomy of Invertebrates,* Vol. 3. pp. 141–209. Reprinted by permission of Wiley-Liss, Inc., a subsidiary of John Wiley & Sons, Inc. (c) Courtesy of D. W. Halton. From G. R., Mair, A. G., Maule, B., Fried, T. A., Day, and D. W. Halton. 2003. "Organization of the musculature of the schistosome cercariae," *J. Parasitol.* 89:623–625.

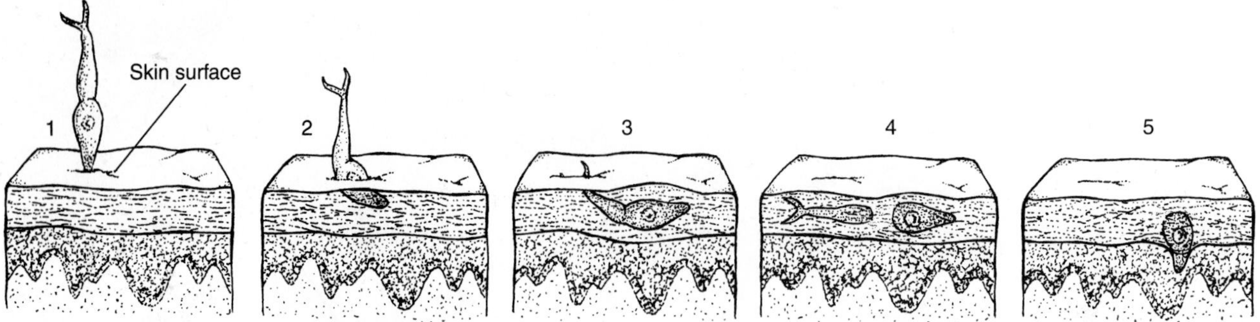

Figure 8.19

Sequential stages in penetration of a *Schistosoma mansoni* cercaria into a vertebrate host. Note loss of tail in step 4.

Modified from Ginetsinskaya, T. A. 1988. *Trematodes, Their Life Cycles, Biology and Evolution.* New Delhi: Amerind, Publ. Co., Pvt., Ltd., 59, Fig. 96.

their way out of the intermediate host or, in some species, wait within the intermediate host until that host is eaten by a predator.

Free-swimming cercaria larvae are nonciliated, but they can swim actively by means of a muscular tail (Fig. 8.18a, b). Cercariae usually possess at least one sucker anteriorly and, in many species, also have a ventral sucker (Fig. 8.18a, b). The cercaria larvae of some species transform into adults once they encounter the next host, attach to it, and enzymatically penetrate the host tissues. In other cases, the cercaria encysts on submerged vegetation likely to be eaten by the definitive host. In most species, however, the next stop is another animal intermediate host. In any event, penetration by the cercaria larva into the next host is accompanied by detachment of the cercarial tail, either before or after the cercaria enters the host's body (Fig. 8.19).

Digeneans typically show little specificity for the second intermediate host, which includes both vertebrate and invertebrate species representing most animal phyla. Even a single trematode species can often make use of a wide range of second intermediate hosts. One species in the genus *Echinostoma,* for example, can use either certain freshwater snails or the larval stage of certain amphibians as second intermediate hosts. Yet the miracidia of that species can successfully infect only snails of one species or of several closely related species.

Once in the second intermediate host, the cercaria transforms, in many trematode species, into an encysted waiting stage, the **metacercaria,** in which most of the specifically larval organs degenerate. The adult trematode develops only when the second intermediate host is ingested by an appropriate definitive host, which may take some time; metacercariae may survive for many months in the intermediate host while remaining fully infective. In some cases, one or more additional intermediate hosts may be involved in the life cycle before the final host is reached. Once eaten by the definitive host, the juvenile trematode migrates through the digestive tract to take up residence in the proper, species-specific location, where adulthood is reached. The sexually mature trematode produces eggs at a frightening rate; an individual adult schistosome can release more than 3000 fertilized eggs daily.

What I have just described is a generalized digenean life cycle; details vary considerably among species (Figs. 8.20 and 8.21).

Note that in contrast to the ectoparasitic lifestyle of monogeneans, digenetic trematodes, like the cestodes, are exclusively **endoparasitic,** living within host tissue (*endo* = G: within). The digenean adult has a mouth and a blind-ended gut, usually 2-branched, and often has both an anterior and a ventral sucker for maintaining position within the host (see Fig. 8.17). Most species are hermaphroditic, although some have separate, anatomically distinct sexes; these species are gonochoristic (Research Focus Box 8.1). Most of the body is taken up by the reproductive system.

The life cycle of digenetic trematodes is, to say the least, highly complex: Several hosts are required for completion of the life cycle; the hosts often occupy very different habitats, complicating the movement from one host to the next; and both the intermediate larval stages and the adult stage require specific host species (Fig. 8.20), although specificity for the first intermediate host is always far greater than that for either the second intermediate or definitive host. The free-living larval stages are nonfeeding and can generally remain alive and infective for only a few hours, during which time the next host animal must be located (Fig. 8.22). Clearly, the probability that any given fertilized egg will reach adulthood is very small. Therein lies the adaptive significance of producing tremendous numbers of dispersal stages. Cestodes achieve a high rate of offspring production mainly by franchising egg and sperm manufacture and by adding on new franchisees (i.e., proglottids) daily. Among the digenetic trematodes, a high rate of offspring production is accomplished through the geometric multiplication of larval stages.[11] Each female generally has only a single ovary, and each fertilized egg develops into

11. See *Topics for Further Discussion and Investigation,* no. 7.

RESEARCH FOCUS BOX 8.1

Control of Schistosomiasis

Eveland, L. K., and M. A. Haseeb. 1989. *Schistosoma mansoni:* Onset of chemoattraction in developing worms. *Experientia* 45:309–10.

Schistosomiasis is one of the major diseases in the world today, affecting an estimated 200 million people (Fig. 8.20c, d). Many biologists are therefore seeking ways to eliminate it or at least to control its spread. Since the pathological effects of schistosome infection in humans are caused by the eggs, the parasites should be eliminated before egg laying begins. One widely attempted approach is to develop vaccines that attack schistosome-specific surface proteins on juveniles. Another approach is to control populations of the intermediate host, a freshwater snail. A third approach is to interfere with the efficacy of the schistosome digestive system, preventing growth and sexual maturation by interfering with key digestive enzymes. A fourth approach, and the focus of the paper by Eveland and Haseeb (1989), is to take advantage of the gonochoristic sexuality of schistosomes—unique among trematodes—and somehow interfere with the essential process of mate location; eggs cannot be fertilized and released until a female worm comes to lie in the specialized groove of the male (Fig. 8.17d). Previous research has indicated that males and females locate each other by chemical means. Eveland and Haseeb's research defines the age at which worms first begin to attract each other.

Eveland and Haseeb conducted their study using a strain of *Schistosoma mansoni,* one of several species in the genus causing schistosomiasis. Cercaria larvae were obtained from the freshwater snail *Biomphalaria glabrata* (Gastropoda: Pulmonata) and applied to the shaved skin of several mice. From 20 to 28 days after the mice were infected, the researchers killed the mice and recovered the juvenile parasites to test for chemoattractive capacity.

For each bioassay, 1 male and 1 female schistosome were pipetted 6–15 mm apart into a glass container, and their subsequent movements toward or away from each other were recorded on videotape. The distance between the 2 worms was measured every 3 minutes for 30 minutes, and the percent change in amount of physical separation (the "migration index") was calculated for each time interval. At least 2 pairs, and as many as 4 pairs, of worms were tested in each of the age groups examined.

Clearly, worms became chemically attractive to each other sometime between 21 and 23 days after infecting the host (Focus Fig. 8.1); before that time, males and females actually seemed to repel each other. Identifying the factors responsible for the repulsion could be as useful as determining the factors responsible for the later attraction. In both cases, the data suggest the feasibility of interfering with schistosome reproductive success by preventing worms from locating mates. This would not only aid the host, by preventing egg laying and the associated inflammatory responses, but would also reduce the spread of the disease to other individuals, by reducing the output of infective eggs into the environment.

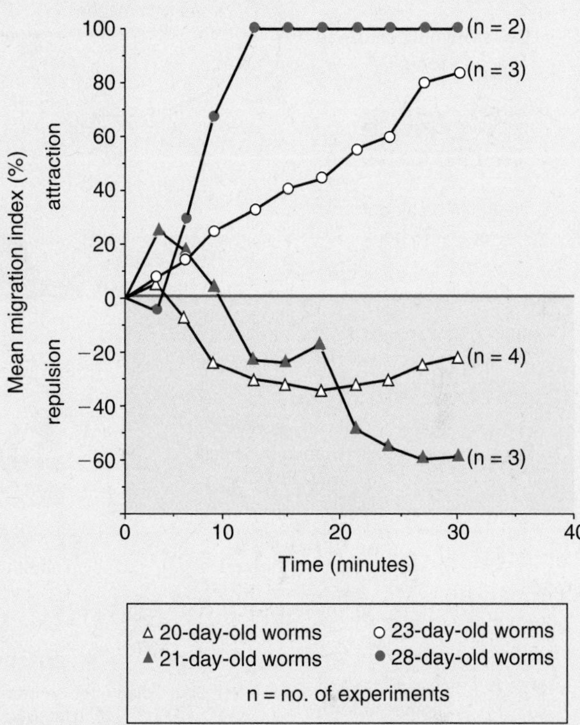

Focus Figure 8.1

Influence of schistosome age on chemoattractive capacity in 30-minute bioassays. Day 0 is the day on which cercaria larvae of *Schistosoma mansoni* were allowed to infect a mouse host. The data indicate mean percent changes in the distance between male and female worms; each experiment was repeated 2 to 4 times. A value of 0 indicates no net movement of worms toward or away from each other.

Based on L. K. Eveland and M. A. Haseeb, "Schistosoma mansoni: Onset of Chemoattraction in Developing Worms" in *Experientia* 45:309–10. 1989.

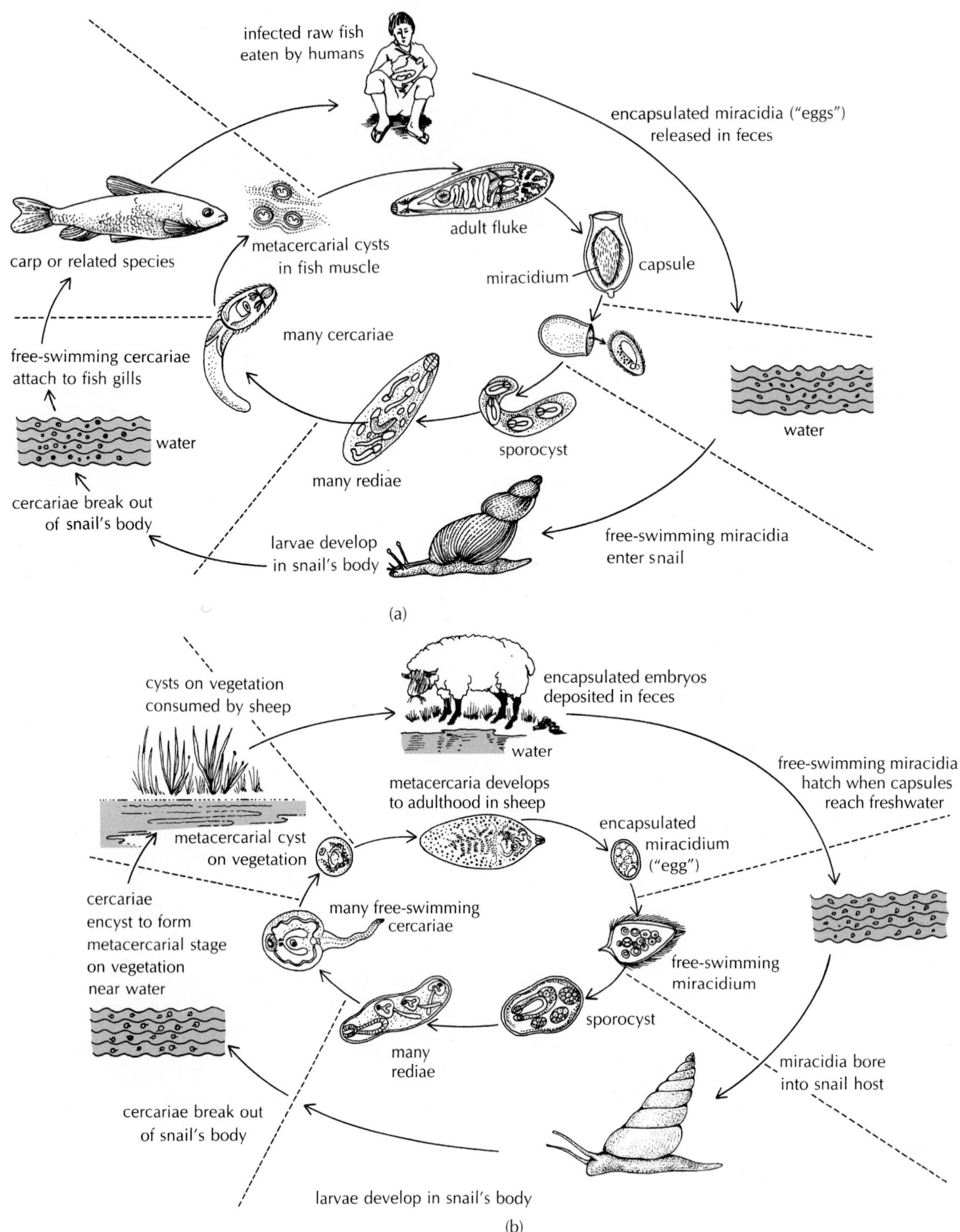

Figure 8.20

Fluke life cycles. The species are all digenetic trematodes. (a) Chinese human liver fluke *Opisthorchis sinensis*. There are 2 intermediate hosts (a snail, usually in the genus *Bithynia* and a fish, of the family Cyprinidae). This fluke can reach sexual maturity only in humans, who thus serve as the definitive hosts. Encapsulated miracidia pass out of the definitive host's body along with feces. The miracidium only hatches following ingestion by the snail. Each miracidium becomes a single sporocyst, which produces (asexually) many rediae, which, in turn, each produce (asexually) many cercariae. These leave the snail and swim freely in the surrounding water. Upon encountering an appropriate fish host, the cercariae bore in and encyst within the fish muscle. The life cycle reaches completion only when humans eat raw or undercooked fish. Adulthood is reached within the human bile duct. (b) The sheep liver fluke *Fasciola hepatica*. This parasite has a snail as the sole intermediate host. Infected sheep deposit fluke eggs along with feces. Free-swimming miracidia hatch from the eggs if the eggs reach freshwater. The miracidia bore into an appropriate snail host; sporocyst, redia, and cercaria stages follow within the snail. The cercariae leave the snail and encyst to form a metacercarial resting stage on emergent vegetation. Adulthood is reached if the encysted stage is ingested by grazing sheep. (c) *Schistosoma mansoni*, one of the species causing schistosomiasis in humans. Adults are only about one cm long. Members of this genus are atypical trematodes in that the adults are gonochoristic, with the adult female nestling within a specialized groove in the body of the larger male. Fertilized eggs (perhaps 300 per day per female) leave the human host with fecal material, and miracidia hatch if the feces contact freshwater. The miracidia must locate and bore into the tissues of the single intermediate host, a snail in the genus *Biomphalaria*. Each miracidium then develops into a single sporocyst,

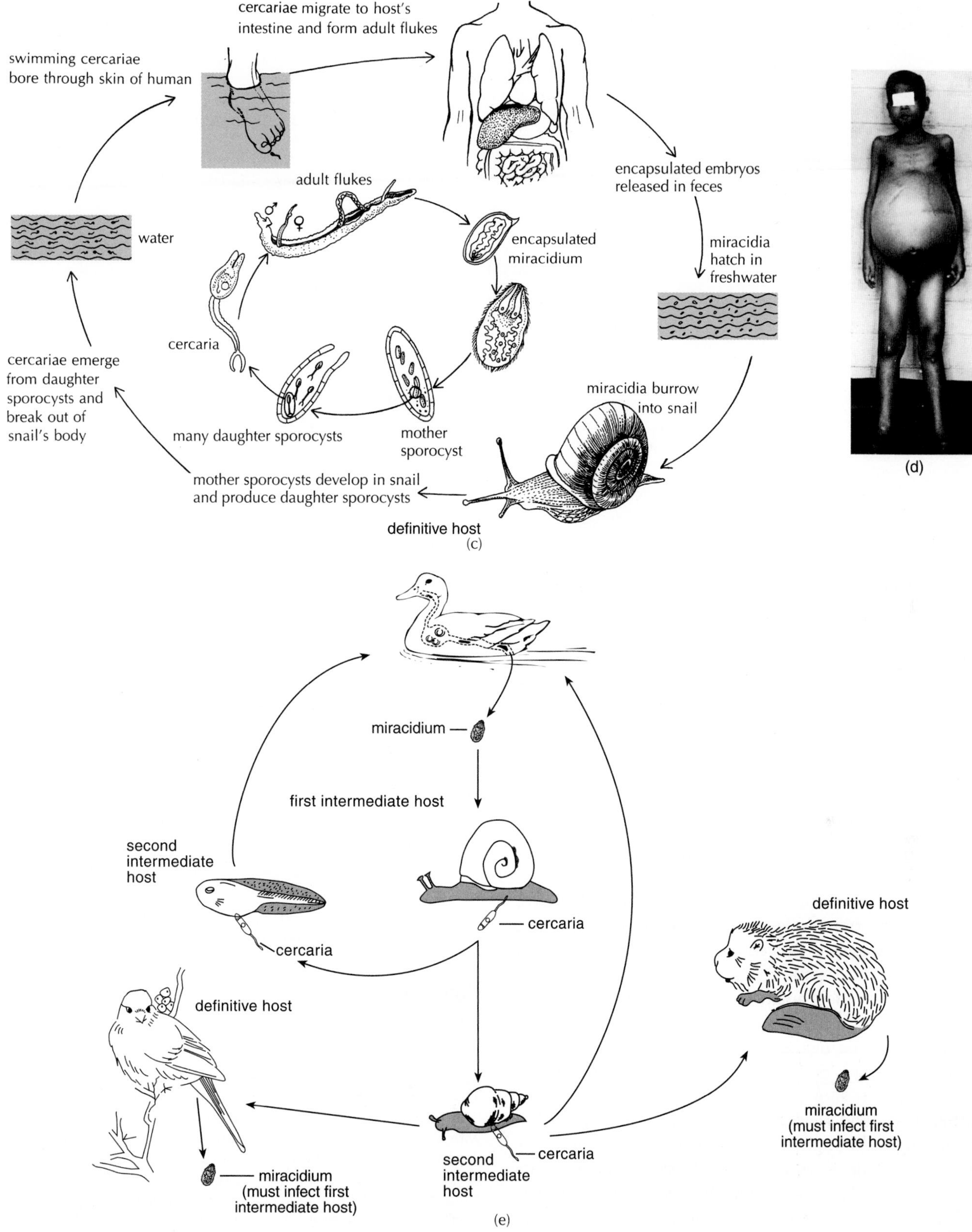

Figure 8.20 *Continued*

which then produces many daughter sporocysts. Many cercaria larvae emerge from each daughter sporocyst and exit the snail. If the free-swimming cercaria contacts a human, it bores through the skin and migrates into the circulatory system, within which it travels through the heart, to the lungs, and thence to the kidneys, where it feeds and grows. The parasite eventually reaches maturity within the blood vessels of the host intestine, where it may live for years. Pathology results mainly from inflammation caused by eggs becoming entrapped in host tissues. (d) This boy has the distended belly typical of infection by *Schistosoma japonicum.* (e) Generalized life cycle of a 37-collar-spined trematode *Echinostoma* spp. Adults live in intestines and bile ducts of many vertebrate hosts, especially aquatic birds and mammals. The adults bear 37 spines on a collar surrounding the oral sucker.

(d) Courtesy Centers for Disease Control, Department of Health and Human Services, Atlanta, GA. (e) From Huffman and Fried, "Echinostoma and Echinostomiasis," in *Advances in Parasitology* 29:224, 1990. Copyright © Academic Press Inc., Orlando, Florida. Reprinted by permission.

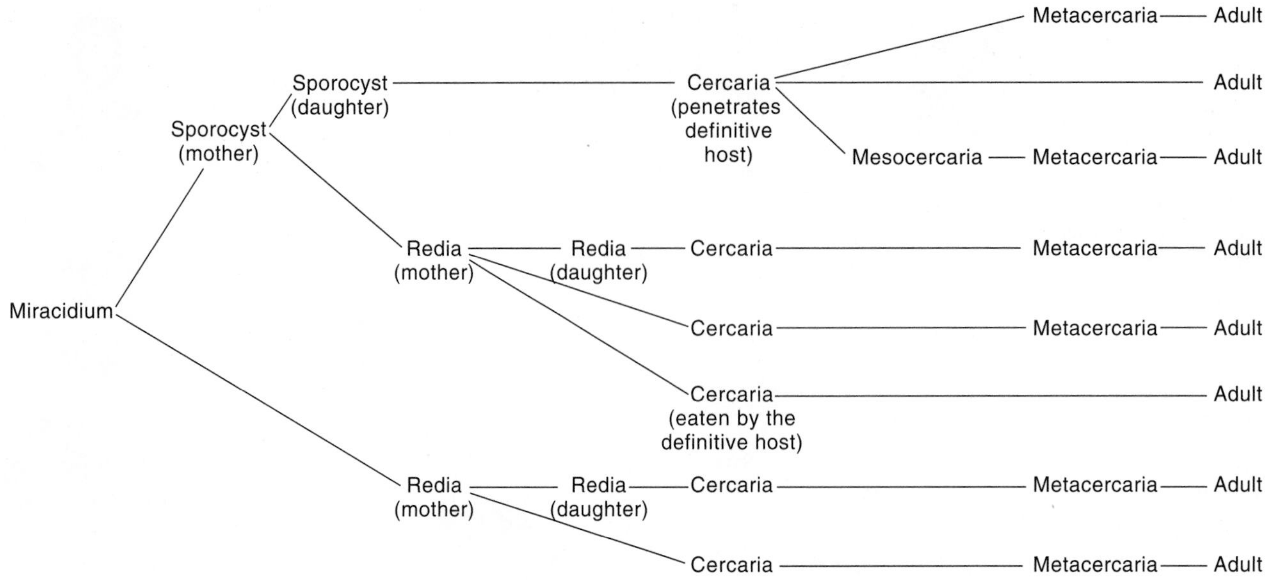

Figure 8.21
Variation in digenetic trematode life cycles. Note that mother sporocysts produce either daughter sporocysts or rediae but never both in the same species.

only a single miracidium larva. Yet, within each miracidium are the germ balls of the redia stage (or of an additional sporocyst stage), within which are the germ balls of the cercaria stage. A single miracidium generally gives rise to, on average, tens of thousands of cercariae—many hundreds of thousands for some schistosome species—all of which are genetically identical to each other and to the original fertilized egg. A moderately infected intermediate host may release thousands of cercaria larvae per day for many years.

Several remarkable behavioral adaptations that have arisen among trematodes increase the likelihood that a given larval stage will indeed make it to the next required host. Some schistosome species, for example, emerge as cercariae from their aquatic snail hosts at only certain times of the day, times when the target vertebrate host is most likely to be in the water. But perhaps one of the most wonderful adaptations for facilitating parasite transmission is shown by a liver fluke found in sheep, cattle, pigs, and several other terrestrial vertebrates (including—although rarely—humans). The species is *Dicrocoelium dendriticum,* and we will go through its story in some detail, to review the events of the trematode life cycle. The life cycle of *D. dendriticum* is remarkable in many respects, including the fact that it takes place entirely on land.

The adult liver flukes are hermaphroditic and deposit their eggs into the bile ducts of their host. The eggs, which are less than 50 μm (micrometers) in diameter, leave the host with the feces. The eggs will not hatch until they are eaten by one particular species, or possibly one of several species, of land snail. Miracidia emanating

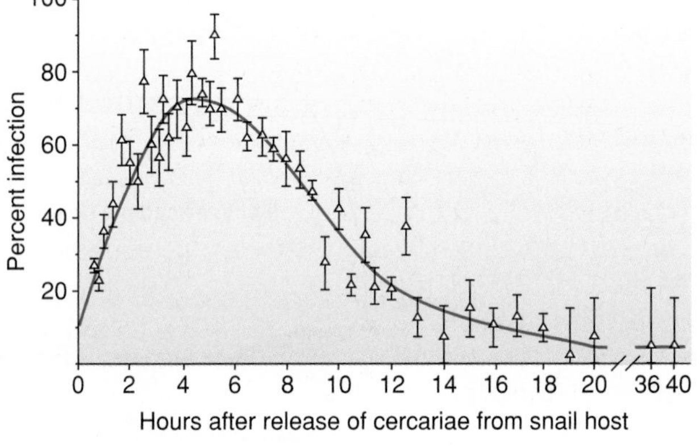

Figure 8.22
Functional life span for cercaria larvae of the digenetic trematode *Plagiorchis elegans.* To determine infective capability, cercariae of different ages were exposed to a suitable second intermediate host, larvae of the mosquito *Aedes aegypti,* for 15 minutes at 20°C. The number of metacercariae in each host was counted 24 hours later. Each point shows the mean infectivity to 10 mosquito larvae, with 5 cercariae tested per replicate. Error bars show one standard error about the mean. Ability of cercariae to infect the mosquito host declined abruptly about 6 hours after cercariae emerged from the first intermediate gastropod host.
Based on C.A. Lowenberger and M. E. Rau, in *Parasitology* 109. 1994.

from the fertilized eggs lyse their way through the gut wall of the snail host and migrate to a specific region of the snail's digestive system, where they gradually transform to sporocysts. Each mother sporocyst produces many daughter sporocysts, each of which, in turn, gives birth to large numbers of cercariae; this species has no redia stage. The cercariae—each about 600 μm long—migrate to the "lung" of the snail, where groups of them become engulfed in mucus. These mucous balls are expelled from the snail to the outside. The outer surface of each slime ball dries to form a water-resistant outer coat, while the cercariae persist in the watery environment within.

From here, the tale becomes even less believable. The slime balls are routinely collected and ingested by several species of ant. Within the ant, the cercariae encyst in various tissues, including the brain, to form the metacercarial stage, which somehow succeeds in altering the ant's behavior. Each evening, the ants are now obliged to crawl upward on a blade of grass and to bite down firmly upon its tip. They apparently remain in this position, unable to open their jaws, until the air temperature rises during the following day. Meanwhile, they are especially vulnerable to grazing by local herbivores, which tend to feed primarily during the evening and early morning. Only if the ants are ingested by an appropriate host can the life cycle be completed. Adult flukes of this species can reside in the bile ducts and gallbladders of many mammals, including horses, sheep, cattle, dogs, rabbits, pigs, and humans.

Exactly how the trematode metaceraria is able to influence ant behavior is not known, but it is easy to imagine how this ability would be selected for once it appeared. Those metacercariae possessing this capability would have a greater chance of infecting the definitive host and leaving offspring. Some proportion of those offspring would also possess the capacity to alter the behavior of the ant intermediate host, since the capability is genetically determined, and gradually the trait would become a species-specific characteristic.

The life cycle of this particular trematode is perhaps more remarkable than most. Equally remarkable is the patience and persistence of the people who have pieced this and other trematode life cycles together.[12]

The Evolution of Digenean Life Cycles

What could account for the evolution of the complex, seemingly roundabout digenean life cycle? Let's summarize the key features: (1) The life cycle typically includes 5 distinct and unique developmental stages (miracidium, sporocyst, redia, cercaria, and metacercaria) and typically 2 or more intermediate hosts. (2) The first intermediate host almost always must be a mollusc: Of the approximately 10,000 known trematode species, all but 2 require a gastropod or bivalved mollusc as the first intermediate host. (3) Species specificity for the first intermediate host is extremely high. (4) Specificity for subsequent intermediate hosts and for the definitive host is far less. (5) The digenean life cycle includes asexual replication of developmental stages, something not found in any turbellarian species, and that asexual replication occurs entirely within the molluscan intermediate host. (6) Finally, most species can reach adulthood only in a vertebrate host, usually after a period of encystment in the environment or in a previous host.

The evolutionary origin of such a life cycle presents a most intriguing puzzle. It is easy to imagine that the present parasitic relationship began as a benign, commensal one between a free-living flatworm (probably a rhabdocoel) and another animal. But with what other animal was the flatworm initially associated? Which was the original host: the mollusc or the vertebrate? If the relationship developed before the evolution of vertebrates (about 500 million years ago), the answer to that question is obvious; but how long ago *did* the first association develop? And was the original parasite the larval stage of the flatworm or the adult stage? In what order were the 3 standard hosts most likely acquired? Was asexual replication a part of the ancestral life cycle before the group became parasitic, or did it evolve later, as an adaptation to parasitic life? Would that the answers were as readily available as the questions.

Although there has been considerable speculation and much lively debate on these issues, no definitive answers are possible. But it is always a wonderful intellectual challenge to try to come up with hypotheses that others have difficulty refuting![13] Certainly the evolution of trematode life cycles poses such a challenge. One particularly attractive scenario[14] is that originally the flatworm larva parasitized a molluscan host. This would explain the almost complete present specificity for a molluscan first intermediate host and the high degree to which that host tolerates the parasite, assuming that, over evolutionary time, parasites and their hosts tend to become better adapted to each other. It is easy to imagine that the adult flatworm remained free-living for a time. Eventually, asexual replication within the molluscan host evolved, and the free-living cercaria developed the ability to encyst on vegetation, perhaps as an adaptation to environmental stresses; cercariae of some trematode species still encyst on vegetation rather than within a host. The ingestion of those free-living cysts by some vertebrate could have then led to the adult stage becoming parasitic in what is now the definitive vertebrate host; if so, the second intermediate host would have become incorporated into the life cycle later, probably as a vehicle for increasing throughput of the parasite to the vertebrate. This could account for the particularly wide range of acceptable second intermediate hosts even within individual trematode species: Essentially all animals

12. See *Topics for Further Discussion and Investigation,* no. 4.

13. See *Topics for Further Discussion and Investigation,* no. 10.

14. Based on Ginetsinskaya, T. A. 1988. *Trematodes, Their Life Cycles, Biology and Evolution.* New Delhi: Amerind Publishing Company.

routinely eaten by the final hosts can serve as acceptable intermediate hosts. Again, this is not the only possible pathway to the present trematode life cycle, but it seems a very reasonable one.

The Aspidogastreans (= Aspidobothreans)

Defining Characteristic: Large ventral sucker divided by septa, generally forming a row of suckers

As promised, we will now consider one last, small group of trematodes. The aspidogastrean (as-pid′-o-gas-tray′-un) trematodes, a mere 80 species, bear similarities to monogeneans and digeneans, but they fit neatly into neither group. Most species have a simple life cycle involving a single host, as in the Monogenea. That host is always a mollusc (freshwater mussels or gastropods) but otherwise, aspidogastreans generally show very little host specificity. However, aspidogastreans never have the posterior attachment organ (haptor) characterizing monogenean species. Instead, the entire ventral surface of the body is modified to form a powerful attachment sucker (Fig. 8.23). Also unlike monogeneans, some species require an intermediate host to complete the life cycle; these species reach adulthood in fishes or turtles, using the mollusc as an intermediate host. In this respect, they resemble the digenetic trematodes, but unlike the digenetic life cycle, the developmental stages never exhibit asexual replication within the intermediate molluscan host.

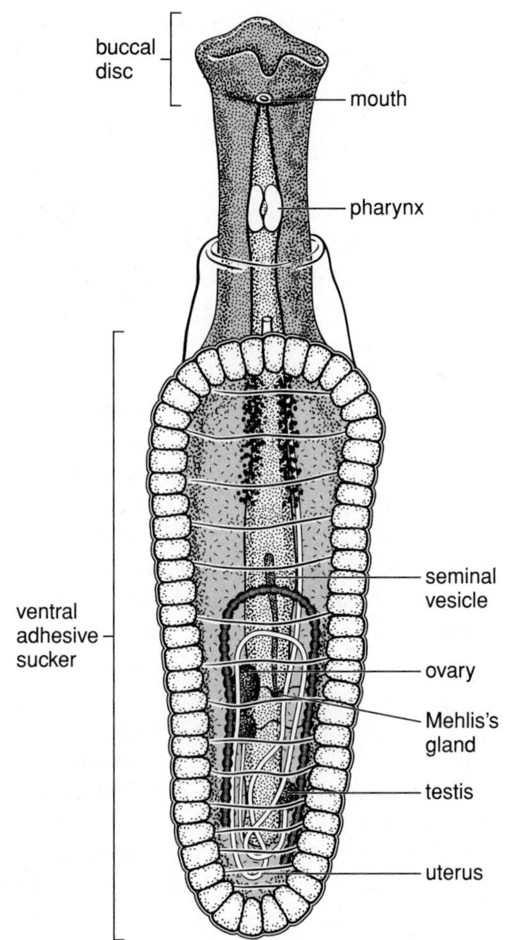

Figure 8.23

Young juvenile of the aspidogastrean *Cotylogaster occidentalis* taken from the gill of a freshwater snail or bivalve. Note the single large ventral sucker divided into a series of compartments.

Based on B. Fried and M.A. Haseeb. In Frederick W. Harrison, John O. Corliss, and Jane A. Westfall, Eds., *Microscopic Anatomy of Invertebrates,* Vol. 3. 1992.

Taxonomic Summary

Phylum Platyhelminthes—the flatworms
- Class Turbellaria—the free-living flatworms
- Class Cestoda
  - Subclass Cestodaria
  - Subclass Eucestoda—the tapeworms
- Class Monogenea
- Class Trematoda—the flukes
  - Subclass Digenea
  - Subclass Aspidogastrea

Topics for Further Discussion and Investigation

1. Discuss the feeding biology of turbellarian flatworms.

Calow, P., and D. A. Read. 1981. Transepidermal uptake of the amino acid leucine by freshwater triclads. *Comp. Biochem. Physiol.* 69A:443.

Dumont, H. J., and I. Carels. 1987. Flatworm predator (*Mesostoma* cf. *lingua*) releases a toxin to catch planktonic prey (*Daphnia magna*). *Limnol. Oceanogr.* 32:699.

Jennings, J. B. 1989. Epidermal uptake of nutrients in an unusual turbellarian parasitic in the starfish *Coscinasterias calamaria* in Tasmanian waters. *Biol. Bull.* 176:327.

Jennings, J. B., and J. I. Phillips. 1978. Feeding and digestion in three entosymbiotic graffillid rhabdocoels from bivalve and gastropod molluscs. *Biol. Bull.* 155:542.

2. How do trematode larvae locate and penetrate their future hosts, and then find the location appropriate for reaching adulthood within that host?

Bartoli, P., M. Bourgeay-Causse, and C. Combes. 1997. Parasite transmission via a vitamin supplement. *BioScience* 47:251.

Blankespoor, H. D., and H. van der Schalie. 1976. Attachment and penetration of miracidia observed by scanning electron microscopy. *Science* 191:291.

Curtis, L. A. 1993. Parasite transmission in the intertidal zone: Vertical migrations, infective stages, and snail trails. *J. Exp. Marine Biol. Ecol.* 173:197.

Feiler, W., and W. Haas. 1988. Host-finding in *Trichobilharzia ocellata* cercariae: Swimming and attachment to the host. *Parasitology* 96:493.

Fried, B., and B. W. King. 1989. Attraction of *Echinostoma revolutum* cercariae to *Biomphalaria glabrata* dialysate. *J. Parasitol.* 75:55.

Hass, W., M. Gui, B. Haberl, and M. Ströbel. 1991. Miracidia of *Schistosoma japonicum:* Approach and attachment to the snail host. *J. Parasitol.* 77:509.

Mason, P. R. 1977. Stimulation of the activity of *Schistosoma mansoni* miracidia by snail-conditioned water. *J. Parasitol.* 75:325.

Roberts, T. M., S. Ward, and E. Chernin. 1979. Behavioral responses of *Schistosoma mansoni* miracidia in concentration gradients of snail-conditioned water. *J. Parasitol.* 65:41.

Salafsky, B., Y-S. Wang, A. C. Fusco, and J. Antonacci. 1984. The role of essential fatty acids and prostaglandins in cercarial penetration (*Schistosoma mansoni*). *J. Parasitol.* 70:656.

Sukhdeo, M. V. K. 1997. Earth's third environment: The worm's eye view. *BioScience* 47:141.

Wilson, R. A. N., and J. Dennison. 1970. Short chain fatty acids as stimulants of turning activity by miracidia of *Fasciola hepatica. Comp. Biochem. Physiol.* 32:511.

3. How do flatworm parasites evade the host immune system?

Clegg, J. A., S. R. Smithers, and R. J. Terry. 1971. Acquisition of human antigens by *Schistosoma mansoni* during cultivation "in vitro." *Nature* 232:653.

Damian, R. T. 1967. Common antigens between adult *Schistosoma mansoni* and the laboratory mouse. *J. Parasitol.* 53:60.

Dineen, J. K. 1963. Immunological aspects of parasitism. *Nature* 197:268.

Hopkins, C. A., and H. E. Stallard. 1974. Immunity to intestinal tapeworms: The rejection of *Hymenolepis citelli* by mice. *Parasitology* 69:63.

Smithers, S. R., R. J. Terry, and D. J. Hockley. 1969. Host antigens in schistosomiasis. *Proc. Royal Soc. London* 171B:483.

Wakelin, D. 1997. Parasites and the immune system. *BioScience* 47:32.

4. How do researchers go about documenting the complex life histories of trematode flatworms?

Reversat, J., R. Leducq, R. Marin, and F. Renaud. 1991. A new methodology for studying parasite specificity and life cycles of trematodes. *Int. J. Parasitol.* 21:467.

Shinn, G. L. 1985. Infection of new hosts by *Anoplodium hymanae,* a turbellarian flatworm (Neorhabdocoela, Umagillidae) inhabiting the coelom of the sea cucumber *Stichopus californicus. Biol. Bull.* 169:199.

Stunkard, H. W. 1941. Specificity and host-relations in the trematode genus *Zoogonus. Biol. Bull.* 81:205.

Stunkard, H. W. 1964. Studies on the trematode genus *Renicola:* Observations on the life-history, specificity, and systematic position. *Biol. Bull.* 126:467.

Stunkard, H. W. 1980. The morphology, life-history, and taxonomic relations of *Lepocreadium aveolatum* (Linton, 1900) Stunkard, 1969 (Trematoda: Digenea). *Biol. Bull.* 158:154.

5. Flatworms are capable of considerable regeneration of lost body parts, including the head. Investigate regeneration in turbellarian flatworms.

Baguna, J., E. Salo, and C. Auladell. 1989. Regeneration and pattern formation in planarians. III. Evidence that neoblasts are totipotent stem cells and the source of blastema cells. *Development* 107:77.

Best, J. B., A. B. Goodman, and A. Pigon. 1969. Fissioning in planarians: Control by brain. *Science* 164:565.

Egger, B., P. Ladurner, K. Nimeth, R. Gschwentner, and R. Rieger. 2006. The regeneration capacity of the flatworm *Macrostomum lignano*—on repeated regeneration, rejuvenation, and the minimal size needed for regeneration. *Devel. Genes Evol.* 216:565–77.

Goldsmith, E. D. 1940. Regeneration and accessory growth in planarians. II. Initiation of the development of regenerative and accessory growths. *Physiol. Zool.* 13:43.

Mead, R. W. 1985. Proportioning and regeneration in fissioned and unfissioned individuals of the planarian *Dugesia tigrina. J. Exp. Zool.* 235:45.

Nentwig, M. R. 1978. Comparative morphological studies of head development after decapitation and after fission in the planarian *Dugesia dorotocephala. Trans. Amer. Microsc. Soc.* 97:297.

Nishimura K., Y. Kitamura, T. Inoue, Y. Umesono, S. Sano, K. Yoshimoto, M. Inden, K. Takata, T. Taniguchi, S. Shimohama, and K. Agata. 2007. Reconstruction of dopaminergic neural network and locomotion function in planarian regenerates. *Devel. Neurobiol.* 67:1059–78.

Saló, E., and J. Baguña. 2002. Regeneration in planarians and other worms: New findings, new tools, and new perspectives. *J. Exp. Zool.* 292:528.

6. How can one determine whether cleavage is determinate (= mosaic) or indeterminate (= regulative) in spirally cleaving embryos?

Boyer, B. C. 1990. The role of the first quartet micromeres in the development of the polyclad *Hoploplana inquilina. Biol. Bull.* 177:338.

7. Discuss the similarities and differences in the life cycles of digenean flatworms and members of the cnidarian class Scyphozoa. Or, compare the life cycle of digenetic trematodes with that of malarial parasites in the protozoan phylum Apicomplexa (pp. 58–61).

8. Like the flatworms, many members of the phylum Cnidaria lack specialized respiratory structures and blood circulatory systems. How are cnidarians able to meet their gas exchange requirements without being flat?

9. The acoelomate condition of flatworms is of considerable interest to those attempting to work out the phylogenetic interrelationships among the animal phyla. If we can safely assume that the flatworm body plan preceded that of the coelomates and pseudocoelomates, then it is very likely that arthropods, molluscs, annelids, and other major metazoan groups evolved from flatworm ancestors. If we cannot, the multicellular ancestors of coelomates must lie elsewhere. Debate the issue of whether the acoelomate condition is primitive or advanced.

Hyman, L. H. 1951. *The Invertebrates: Platyhelminthes and Rhynchocoela,* vol. 2. New York: McGraw-Hill, 52–219.

Rieger, R. M. 1985. The phylogenetic status of the acoelomate organization with the Bilateria: A histological perspective. S. C. Morris et al., eds. *The Origins and Relationships of Lower Invertebrates.* Oxford: Clarendon Press, 101–22.

Ruiz-Trillo, I., M. Riutort, D. T. J. Littlewood, E. A. Herniou, and V. J. Baguna. 1999. Acoel flatworms: Earliest extant metazoans, not members of the Platyhelminthes. *Science* 283:1919–23.

Sempere, L. F., P. Martinez, C. Cole, J. Baguña, and K. J. Peterson. 2007. Phylogenetic distribution of microRNAs supports the basal position of acoel flatworms and the polyphyly of Platyhelminthes. *Evol. Devel.* 9:409–15.

Smith, J. P. S. III, and S. Tyler. 1985. The acoel turbellarians: Kingpins of metazoan evolution or a specialized offshoot? S. C. Morris et al., eds. *The Origins and Relationships of Lower Invertebrates.* Oxford: Clarendon Press, 123–42.

Telford, M. J., A. E. Lockyer, C. Carwright-Finch, and D. T. J. Littlewood. 2003. Combined large and small subunit ribosomal RNA phylogenies support a basal position of the acoelomorph flatworms. *Proc. Royal Soc. London* B 270:1077–83.

10. The order in which the various hosts were acquired in the evolution of digenetic trematode life cycles is very controversial. Here are the 3 most plausible scenarios:

Scenario	Stage I	Stage II	Stage III
A	molluscan host	acquire 2d intermediate host	acquire vertebrate host
B	molluscan host	acquire vertebrate host	acquire 2d intermediate host
C	vertebrate host	acquire molluscan host	acquire 2d intermediate host

Suggest one sort of evidence that would support each scenario and one sort of evidence that would argue against each scenario.

11. What does a trematode cercaria have in common with a long-staying house guest?

Taxonomic Detail

Phylum Platyhelminthes

The approximately 34,000 species are distributed among 4 classes. Members of the Neodermata are shown in blue.

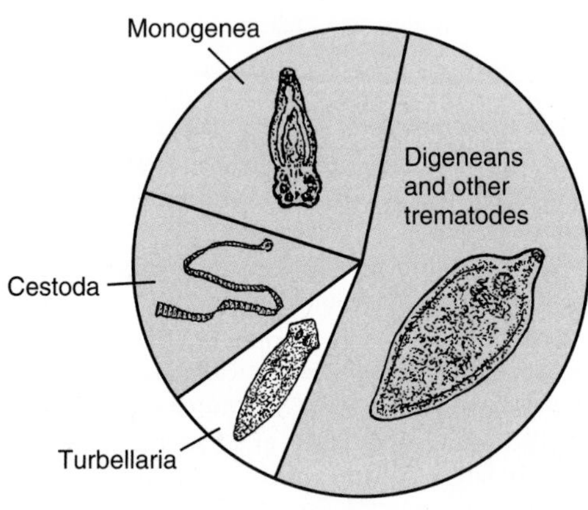

Class Turbellaria[15]

The approximately 3000 described species in this class are distributed among 12 orders and about one hundred twenty families. Probably another 10,000 or more species await description.

Order Acoela

This order contains about 200 species, all of which are marine. These worms have no permanent digestive cavity, and they have a statocyst housing a single statolith. There are no distinct gonads. Most individuals are smaller than about 1 mm, and none are larger than several millimeters. Most species occur in or on marine sediments, but a few species (e.g., *Ectocotyla paguris* and *Avagina* spp.) are commensal with other invertebrates, particularly echinoderms, and a few species are planktonic, spending their lives swimming in the water. Some benthic species within the genus *Convoluta* harbor algal symbionts within the parenchyma and display a rhythmic "sunning" behavior that promotes algal photosynthesis. At least a few other acoel species thrive in sediments low in oxygen, living on the energy gained from oxidizing hydrogen sulfide gas.

15. In many recent treatments, based on ultrastructural studies and cladistic analyses, the Turbellaria has been abandoned as a formal taxonomic category. There is still much uncertainty, however, about the evolutionary interrelationships among many platyhelminth groups. Until there is greater stability, I will retain here the older, simpler classification scheme.

Swimming acoels can be common components of warm, ocean surface plankton; these flatworms always house algal symbionts. At least one acoel species shows indeterminate cleavage rather than the determinate pattern characteristic of protostomes.

Order Nemertodermatida

Nemertoderma. These free-living flatworms resemble the acoels in many respects (including a 9 + 2 microtubular array in the sperm flagellum; most other flatworms show a 9 + 1 arrangement), but they possess a permanent gut cavity, a statocyst with 2 statoliths rather than one and have sperm with a single flagellum rather than 2. Recent molecular studies suggest that these worms are, like acoel flatworms, primitively acoelomate and representing the condition of the most primitive bilateral animals; however, there appears to be no close evolutionary relationship between the two groups. One family, containing only 10–11 species.

Order Rhabdocoela

This group originally contained all flatworms with linear guts, but its members are now distributed among the following 3 new orders: Catenulida, Macrostomida, and Neorhabdocoela. Catenulids and macrostomids are thought to be closely related to the acoel flatworms, while neorhabdocoels are believed to be considerably more advanced. All turbellarian species with straight guts are still referred to collectively as rhabdocoels, but the term no longer has formal taxonomic status.

Order Catenulida

Catenula, Stenostomum (freshwater flatworms), *Rectonectes* (marine, interstitial flatworms). Most of the 70 species occur only in freshwater lakes and ponds, although a few species are marine. Some species can use their cilia to swim short distances. Members of the freshwater species reproduce primarily by asexual means: Organ systems are replicated, and the worms then divide into 2 individuals. The worms in this order are extremely fragile.

Order Macrostomida

Microstomum, Macrostomum. Most species are interstitial, living between sand grains either in freshwater or marine sediments. Most species show numerous adhesive glands and rhabdites, but none have statocysts. Members of the genus *Microstomum* are somehow able to sequester the nematocysts (defensive structures) of the freshwater hydras on which they feed and use them for their own defense. Three families, containing nearly 200 species.

Order Lecithoepitheliata

Prorhynchus, Geocentrophora. The freshwater and terrestrial species contained in this small order are common worldwide. The order also includes a few deep-water, marine species. Two families.

Order Polycladida

This large group of flatworms contains no parasitic species, and all but one species (which occurs only in freshwater) are exclusively marine. Polyclads exhibit a gut that is highly branched (*poly* = G: many; *clad* = G: branch). Individuals are typically several centimeters long; most marine flatworms visible with the unaided eye belong to this order. Some can swim short distances by undulating the lateral margins of the body. Twenty-nine families.

Suborder Acotylea

Seventeen families, many with no more than 12 species.

Family Stylochidae. *Stylochus*—the oyster leech. A number of species prey on juvenile and adult oysters, making them commercially important shellfish pests. They burrow through the oyster's shell enzymatically and then destroy the adductor muscles holding the oyster's 2 shell valves closed. The oyster's tissues are then easy prey. Other species are commensal with other invertebrates, including hermit crabs and ascidians (sea squirts—phylum Chordata).

Family Discocelidae. *Adenoplana, Coronadena.* A widespread group in temperate and tropical marine environments.

Family Hoploplanidae. *Hoploplana.* At least one species is symbiotic in the mantle cavities of marine snails, although most species are free-living carnivores.

Family Planoceridae. A number of species in this family are transparent and pelagic throughout their lives.

Suborder Cotylea

Twelve families, many with fewer than 6 species. Some species are interstitial.

Family Pseudoceridae. *Pseudoceros.* This large family of more than 200 described species includes numerous, brightly colored, tropical flatworms. The margins of the body are always folded anteriorly to form a pair of pronounced tentacles (*pseudo* = G: false; *cera* = G: a horn). All species are marine.

Order Prolecithophora (= Holocoela)

Plagiostomum. This group contains both marine and freshwater species, most of which are free-living.

A few species are external parasites on certain crabs (decapod crustaceans). The sperm of all species are unusual in lacking flagella. Nine families.

Order Proseriata

Most species are marine; they are divided among eight families.

Family Monocelididae. *Monocelis.* This group contains about 60 species, several of which live in freshwater. Most are carnivorous, but some parasitize or live commensally with crabs.

Family Coelogynoporidae. This is a widely distributed group of marine interstitial flatworms.

Family Bothrioplanidae. This family contains only one species, *Bothrioplana semperi,* but it lives in restricted freshwater habitats (e.g., ephemeral ponds, groundwater, and drainage ditches) all over the world. It reproduces by parthenogenesis and has a generation time of only about 3 weeks.

Order Tricladida

All individuals exhibit a gut with 3 distinct branches (*tri* = G: three; *clad* = G: branch). The freshwater members of this order are known collectively as "planarians." Seven families.

Family Bdellouridae. *Bdelloura.* Most species are free-living, but the most famous members of this family are commensal on the book gills of horseshoe crabs; they attach to the host using sticky secretions and a specialized, posterior sucker. Females deposit their eggs on the host's gills.

Family Procerodidae. *Procerodes.* The members of this family live in a great variety of marine and freshwater habitats. Some species are free-living, while others are commensal with chitons, skates, or Japanese horseshoe crabs. Freshwater species occur in Tahiti and in Mexican caves.

Family Planariidae. *Planaria, Polycelis.* This familiar group of some 80 freshwater species occurs only in the Northern Hemisphere.

Family Dugesiidae. *Dugesia.* These flatworms are common in streams, lakes, and ponds and are frequently used in studies of regeneration. They differ from individuals in the family Planariidae, particularly in possessing eyes with multicellular retinas and pigment cups. Over 100 species have been described, many of which reproduce primarily (or exclusively) by asexual fission.

Family Bipaliidae. *Bipalium.* A sizable group (about 100 species) of terrestrial planarians. Individuals may attain lengths exceeding 0.5 m. *Bipalium adventitium,* a land planarian introduced accidentally to the United States from Southeast Asia, probably via imported horticultural plants, preys preferentially on earthworms, and can in fact drive local earthworm populations to extinction. Another species is common in greenhouses worldwide.

Family Geoplanidae. *Geoplana.* This is another large group of terrestrial triclads, including about 200 species.

Family Dendrocoelidae. This group includes both marine and freshwater species, including many blind cave dwellers. About 150 species are known, all from the Northern Hemisphere.

Order Neorhabdocoela

Four suborders, containing thirty-four families. All members bear straight guts and a single pair of ventral nerve cords. The members of the different suborders differ largely with respect to morphology of the male genitalia. Some species can use their external cilia to swim.

Suborder Dalyellioida

There are some indications that all of the parasitic groups (monogeneans, digeneans, aspidogastreans, and cestodes) may have originated among the members of this suborder. Eight families.

Family Dalyelliidae. *Castrella, Dalyellia, Microdalyellia.* The 145 species in this family are widely distributed geographically and are particularly abundant in freshwater lakes and ponds. The group also contains a few marine species and one species commensal with certain small crustaceans (amphipods). Individuals in several species harbor symbiotic zoochlorellae.

Family Provorticidae. *Provortex.* Most of the approximately 50 species are marine or freshwater, but one (*Archivortex*) is terrestrial, and another (*Oekiocolax*) parasitizes other marine turbellarians.

Family Graffillidae. *Graffilla, Paravortex.* All 7 species are marine and widespread. Several species are exclusively parasitic in bivalves or gastropods, while the others are free-living. *Paravortex* spp. are among the few hemoglobin-containing flatworms.

Family Umagillidae. *Anoplodium.* All approximately 35 species are parasitic, mostly in the guts or coelomic cavities of echinoderms.

Family Acholadidae. *Acholades.* The family contains a single species, parasitic in the tube feet of a particular sea-star species from

Tasmania. These turbellarians lack a mouth and digestive tract, apparently subsisting on dissolved organic nutrients obtained through the enzymatic digestion of adjacent sea-star tissue.

Family Hypoblepharinidae. *Hypoblepharina.* All 4 species lack eyes and seem to be commensal on amphipod crustaceans in the Antarctic.

Family Fecampiidae. *Fecampia, Kronborgia.* All 12 species are parasitic in the hemocoel of certain marine crustaceans. These flatworms lack mouths, digestive tracts, and eyes as adults, and sometimes in the larval stage as well. Adults leave the host to deposit eggs within a secreted cocoon (about 8 mm long in *Fecampia*). Members of the genus *Kronborgia,* reaching lengths of over 39 cm, castrate the host and then kill it upon exiting. Of small consolation to the host, the worm itself dies soon after egg laying. *Fecampia erythrocephala* also castrates at least some of its hosts (hermit crabs, caridean shrimp), and probably also kills at least some hosts (shore crabs) upon exiting.

Suborder Typhloplanoida

Promesostoma, Typhloplana. This group includes about 175 species, mostly free-living in marine, freshwater, or terrestrial habitats. One species is ectoparasitic, living only between the parapodia of one polychaete annelid species (*Nephtys scolopendroides*). Members of some marine species appear to be very active and powerful swimmers. Some species in the genus *Typhloplana* harbor symbiotic green algae (zoochlorellae). Nine families, most of which contain fewer than 12 species each.

Suborder Kalyptorhynchia

Most of the approximately 140 species live in marine sediments, although a few live in freshwater. Fifteen families, most of which contain fewer than 6 species each.

Suborder Temnocephalida

Temnocephala. All are freshwater commensals, living especially on the gills of decapod crustaceans. Members of a few species live with freshwater snails or turtles. Unlike that of most other turbellarians, the body is not externally ciliated. Most of the nearly 40 species occur only in South America, Australia, or New Zealand. These small (less than 2 mm long) flatworms typically crawl in leech-like fashion, using 2 or more anterior tentacles and a large, posterior, adhesive disc. They feed mostly on other smaller invertebrates rather than on host tissue. The anterior tentacles apparently function also in prey capture and defense. Two families.

Class Cestoda

The approximately 5000 species are divided among nearly sixty families. Two subclasses.

Subclass Cestodaria. *Amphilina, Gyrocotyle.* These gutless flatworms are endoparasites of fish and, to a lesser extent, turtles. The life cycles are poorly known, probably because none of the species has any economic or medical importance. Unlike most of the true tapeworms in the subclass Eucestoda, cestodarians lack a scolex and show no external segmentation, and the first larval stage has 10 hooks instead of 6. No individual houses more than a single set of male and female reproductive organs. Individuals of some species may exceed 35 cm in length. Three families.

Subclass Eucestoda. The true tapeworms are distributed among 12 orders, largely based on differences in scolex morphology. Fifty-six families.

Order Caryophyllidea

Archegetes, Caryophyllaeides. The species in this small group are unusual in that the body is unsegmented (no proglottids) and bears only a single set of male and female organs. Where known, the life cycles involve freshwater oligochaete worms as intermediate hosts. Adults parasitize freshwater fish and oligochaetes. One family.

Order Spathebothriidea

Diplocotyle, Spathebothrium. These cestodes are unusual in lacking a scolex or other attachment organs. Moreover, they show no external segmentation into proglottids, although the reproductive organs are arranged internally in a linear series. Adults parasitize marine teleosts. One family.

Order Trypanorhyncha

Grillotia, Lacistorhyncus. All species in this order parasitize the stomachs or spiral valves of sharks and other elasmobranchs. Sixteen families.

Order Pseudophyllidea

These widespread flatworms mostly parasitize marine and freshwater fishes, but some species infect toads, birds, amphibians, or humans, sometimes with severe consequences. Ten families.

Family Diphyllobothriidae. *Diphyllobothrium, Polygonoporus, Spirometra.* This family of some 60 species contains the largest of all tapeworms: *Polygonoporus giganticus,* a parasite of whales, reaches 30 m in length. *Diphyllobothrium latum* grows to 20 m in human hosts and contains many thousands of proglottids. Humans become infected by ingesting undercooked fish; abdominal pain and weight loss characterize the infection.

Order Tetraphyllidea

These tapeworms parasitize sharks and other elasmobranchs. Their life cycles are poorly explored. The order contains one of the few known gonochoristic tapeworms (genus *Dioecotaenia*). Four families.

Order Cyclophyllidea

This important and well-studied order includes most of the tapeworm species infecting people, pets, and livestock. All species have 4 suckers on the scolex. Fourteen families, containing several hundred species.

Family Triplotaeniidae. All species parasitize Australian marsupials exclusively.

Family Tetrabothriidae. The approximately 60 species in this family all parasitize marine birds and mammals, including whales and seals.

Family Dioecocestidae. This small group of bird parasites includes most of the few known gonochoristic tapeworm species. (See also order Tetraphyllidea.)

Family Taeniidae. *Taenia, Taeniarhynchus, Echinococcus.* This family of approximately 100 species is of great economic and medical importance. The group includes *Taenia solium,* which uses a pig as its intermediate host; humans eating infected, undercooked pork can develop severe neurological problems, including blindness and paralysis, which can lead to death. Adults of *T. solium* can reach 7 m in length. A more common human parasite is the beef tapeworm, *Taeniarhynchus* (= *Taenia*) *saginata,* currently infecting some 61 million to 77 million people worldwide. As the common name implies, infection is transmitted through consumption of infected, undercooked beef. Adults typically possess about 1000 proglottids, but they rarely exceed lengths of 60 cm (although individuals 3 m to 5 m long have been reported). The pathology associated with infection is far less severe than that associated with infection by the pork tapeworm, *T. solium.* Members of the related genus *Echinococcus* reach adulthood in dogs and are rarely larger than about 3 mm, possessing no more than 5 proglottids. However, in ruminants and humans, which serve as intermediate hosts, the larval stages multiply asexually to form fluid-filled, orange- or even grapefruit-sized **hydatid cysts,** each of which may house many thousands of scolices. The cysts must be removed surgically. Humans typically become infected from dogs, which often carry the eggs on their tongues.

Family Dilepididae. This is a major group of tapeworms, particularly common in birds and mammals, with arthropods as intermediate hosts. Members of *Dipylidium caninum* parasitize cats and dogs worldwide and infect humans (especially children) who ingest the intermediate hosts (fleas and lice). Infection causes no major medical difficulties.

Family Hymenolepididae. This group includes hundreds of species parasitizing birds and mammals, including humans. *Hymenolepis diminuta* is the major cestode model for medical studies. *Vampirolepis nana* infects humans but causes few medical problems; this species is unique among cestodes in needing no intermediate host.

Class Monogenea

This group has previously been considered a subclass within the class Trematoda. All individuals of all species are simultaneous hermaphrodites, and there are no intermediate hosts in the life cycle. Forty-four families, containing about 8000 named species with another 17,000 species awaiting description.

Suborder Monopisthocotylea

Gyrodactylus, Dactylogyrus, Protogyrodactylus, Trivitellina, Capsala. The species in this suborder bear a haptor (posterior attachment organ) with a single mass of hooks, suckers, or other attachment devices. Most parasitize marine or freshwater fishes (skin, gills, urinary tracts, respiratory tracts, or rectal cavities, in particular), but some species parasitize certain marine copepods (which are themselves parasites of fish), and others parasitize amphibians. Most species feed on host epidermal tissue and mucus. Fifteen families.

Suborder Polyopisthocotylea

Diplozoon, Discocotyle, Microcotyle, Polystoma. All members are characterized by having a haptor (posterior attachment organ) subdivided into at least 2 distinct parts. Most species parasitize the gills or skin of marine or freshwater fishes, but some parasitize the urinary tracts of amphibians or reptiles. Individuals in most species feed on blood. Twenty-nine families.

Class Trematoda

The approximately 8000 described species are distributed among more than 200 families. Probably another 5000 species await description. All members of this class are parasitic. Two subclasses.

Subclass Digenea. This subclass contains more than 99% of all trematode species. Five orders, one hundred sixty families.

Order Strigeidida

Strigea, Cotylurus, Alaria. The species in this group parasitize such diverse habitats as the digestive tracts of birds, crocodiles, turtles, snakes, and mammals; the mouth or esophagus of reptiles and birds; the scales of freshwater or estuarine fishes; the rectum of birds and mammals; and the circulatory systems of fishes, birds, and mammals. Some species require 4 successive hosts to complete the life cycle. Thirteen families, containing over 1350 species.

Family Schistosomatidae. *Austrobilharzia, Schistosoma, Trichobilharzia*—blood flukes. Adults are all parasitic in the blood vessels of mammals and birds. The life cycle requires only 2 hosts—a snail and a vertebrate—so there is no metacercaria stage. Neither is there any redia stage in the life history; cercaria larvae are produced directly from the sporocysts. Several species are responsible for the widespread and debilitating disease called schistosomiasis. Also, the cercaria larvae of some marine and some freshwater species penetrate the skin of humans despite their inability to survive within that host; this produces a short-lived (typically several weeks in duration) but frustratingly itchy dermatitis appropriately called "swimmer's itch." This family contains some of the only gonochoristic trematode species. Schistosomes are also unusual among flukes in living in blood vessels and lacking an encysted metacercarial stage.

Family Gymnophallidae. *Bartolius.* Gymnophallids use a variety of marine bivalve species as first and second intermediate hosts, parasitizing marine birds as adults.

Order Azygiida

Azygia, Hemiurus. Species (over 500) in this group commonly parasitize the coelomic cavity or various parts of the digestive tracts of marine and freshwater fishes. Some species in this order parasitize the respiratory tract or Eustachian tubes of amphibians. A variety of invertebrates and vertebrates, including barnacles, copepods, cnidarians, and fishes, serve as intermediate hosts. Twenty-six families, many of which contain only one or a few species. About 80% of all species are contained in just two families, the Hemiuridae and Didymozoidae.

Order Echinostomida

Aporchis, Echinostoma, Fasciola. Members of this large order of some 1360 species commonly parasitize the cloaca and intestines of birds and mammals (including dogs, cats, herbivores, and marine mammals); the respiratory systems of birds and other vertebrates; the intestines of fishes; and the lungs of turtles. People can become infected, but medical problems are minor. *Fasciola hepatica* releases about 25,000 fertilized eggs daily! Twenty-eight families.

Order Plagiorchiida

Plagiorchis, Paragonimus, Dicrocoelium, Nanophyetus. The life cycles often require 3 hosts. Adults parasitize a variety of hosts and habitats, including the gallbladder, kidney, liver, ovaries, rectum, and lungs of virtually all vertebrates (including fishes, snakes, amphibians, birds, and mammals); the intestines of marine and freshwater fishes; and the coelom and bladder of teleosts, elasmobranchs, amphibians, and turtles. Individuals of some species may reach 12 m in length. The cercariae of *Nanophyetus salmincola* encyst in salmonid and some other fishes, which serve as intermediate hosts. This fluke also carries the agent of rickettsial disease, which usually kills dogs that eat raw salmon or other infected fish. The lancet liver fluke, *Dicrocoelium dendriticum,* is unusual among trematodes in that its life cycle is entirely terrestrial; a land snail and an ant serve as intermediate hosts. Fifty-three families, containing many hundreds of species.

Order Opisthorchiida

This order contains about 700 species parasitizing a variety of organs in marine and freshwater fishes, reptiles, birds, and mammals. The life cycle always requires 3 hosts for completion. Seven families.

Family Opisthorchiidae. *Opisthorchis, Clonorchis sinensis*—the Chinese liver fluke. Adults live in the bile ducts of a human host, causing serious medical problems, including bile duct cancer. Twenty million people currently may be infected with species of *Opisthorchis* and *Clonorchis.* The family includes over 140 species.

Family Heterophyidae. *Cryptocotyle, Heterophyes, Haplorchis.* These worms are common intestinal parasites of birds and mammals, including dogs, cats, and humans. Infection in humans can cause intense intestinal pain; also, the eggs may collect in the heart, causing cardiac arrest.

Family Nasitrematidae. *Nasitrema.* The 10 species in this family all parasitize the nasal cavities of whales.

Subclass Aspidogastrea (= Aspidobothrea)

Aspidogaster, Stichocotyle. Most of the 80 species in this subclass are internal parasites of freshwater mussels, gastropods, fishes, or turtles. The entire ventral surface of the body is modified to form a powerful attachment sucker or series of suckers. Most species require an intermediate host to complete the life cycle; in this respect, they resemble the digenetic trematodes, but unlike the digenetic life cycle, the developmental stages never exhibit asexual replication within the intermediate host. Three families.

General References about the Platyhelminths and Their Evolutionary History

Bush, A. O., J. C. Fernández, G. W. Esch, and J. R. Seed. 2002. *Parasitism: The Diversity and Ecology of Animal Parasites.* New York: Cambridge University Press.

Cheng, T. C. 1967. Marine molluscs as hosts for symbioses, with a review of known parasites of commercially important species. *Adv. Marine Biol.* 5:1–424.

Ellis, C. H. Jr., and A. Fausto-Sterling. 1997. Platyhelminthes, the flatworms. In: Gilbert S. F., and A. M. Raunio, eds. *Embryology: Constructing the Organism.* Sunderland, MA: Sinauer Associates, Inc. Publishers.

Fried, B. and T. K. Graczyk, eds. 1997. *Advances in Trematode Biology.* New York: CRC Press.

Harrison, F. W., and B. J. Bogtish, eds. 1991. *Microscopic Anatomy of Invertebrates,* Vol. 3. New York: Wiley-Liss.

Higgins, R. P., and H. Thiel. 1988. *Introduction to the Study of Meiofauna.* Washington, D. C.: Smithsonian Institute.

Hyman, L. H. 1951. *The Invertebrates: Platyhelminthes and Rhynchocoela, the Acoelomate Bilateria.* New York: McGraw-Hill.

Kearn, G. C. 1998. *Parasitism and the Platyhelminths.* London: Chapman and Hall.

Moore, J. 2002. *Parasites and the Behaviour of Animals.* New York: Oxford University Press.

Roberts, L. S., J. Janovy, Jr., and P. Schmidt. 2004. *Foundations of Parasitology,* 7th edition. New York: McGraw-Hill.

Rohde, K. 1996. Robust phylogenies and adaptive radiations: A critical examination of methods used to identify key innovations. *Amer. Nat.* 148:481–500.

Ruiz-Trillo, I., M. Riutort, D. T. J. Littlewood, E. A. Herniou, and J. Baguña. 1999. Acoel flatworms: Earliest extant metazoans, not members of the Platyhelminthes. *Science* 283:1919–23.

Sempere, L. F., F. Lorenzo, P. Martinez, C. Cole, J. Baguña, and K. J. Peterson. 2007. Phylogenetic distribution of microRNA's supports the basal position of acoel flatworms and the polyphyly of Platyhelminthes. *Evol. Devel.* 9:409–15.

Telford, M. J., A. E. Lockyer, C. Carwright-Finch, and D. T. J. Littlewood. 2003. Combined large and small subunit ribosomal RNA phylogenies support a basal position of the acoelomorph flatworms. *Proc. Royal Soc.* London B 270:1077–83.

Toleo, R., J.-G. Esteban, and B. Fried. 2006. Immunology and pathology of intestinal trematodes in their definitive hosts. *Adv. Parasitol.* 63:285–365.

Search the Web

1. www.ucmp.berkeley.edu/platyhelminthes/platyhelminthes.html

   This site is maintained by the University of California at Berkeley. It includes connections to the web pages at other sites, including the University of Arizona Tree of Life and the World Health Organization.

2. http://tolweb.org/tree/phylogeny.html

   Searching under the term "Platyhelminthes" will bring you to the flatworm section of the Tree of Life site, maintained by the University of Arizona.

3. www.mbl.edu

   Choose "Biological Bulletin" under "Quick Links," and then "Other Biological Bulletin Publications." Next select "Keys to Marine Invertebrates of the Woods Hole Region," and then choose "Platyhelminthes." This brings up an identification key to the free-living turbellarian species found near Woods Hole. It also includes a section on how to collect and handle turbellarian flatworms.

4. vm.cfsan.fda.gov

   This site is maintained by the U. S. Food and Drug Administration. Under the heading "Program Areas," select "The Bad Bug Book" and click on entries for *Diphyllobothrium* spp. (tapeworms) and *Nanophyetus* spp. (flukes).

5. www.who.int/neglected_diseases/en/

   This site is maintained by the World Health Organization, and includes information about a variety of parasitic flatworms. Click on entries for *Dracunculiasis, Schistosomiasis, Lymphatic filariasis, Leishmaniasis,* and *Intestinal parasites* (includes the latest epidemiological data, infection trends, and maps showing geographical distribution of infection).

6. http://biodidac.bio.uottawa.ca

   Choose "Organismal Biology," "Animals," and then "Platyhelminthes" for photographs and drawings, including illustrations of sectioned material.

7. http://animaldiversity.ummz.umich.edu/animalia.html

   Under "Classification," select "Platyhelminthes" to find interesting information (and photos) about selected species in each class. This site is provided by the Museum of Zoology at the University of Michigan.

8. http://turbellaria.umaine.edu/

   This site gives an up-to-date listing of turbellarian taxonomy, courtesy of Seth Tyler, University of Maine.

9

The Mesozoans: Possible Flatworm Relatives

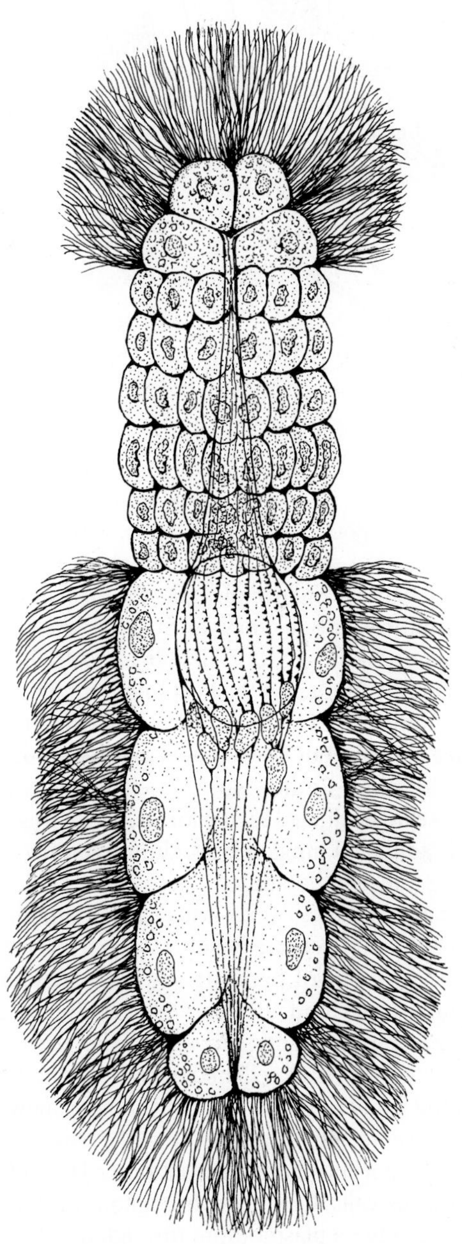

The Mesozoans

Phylum Meso • zoa
(Greek: middle animal)
mē´-sō-zō´-ah

The phylogenetic relationship between mesozoans and other multicellular animals is very uncertain. As the name "Mesozoa" implies, these animals are somewhere between the unicellular protists and the triploblastic flatworms in their level of organization. Mesozoans are definitely multicellular, typically with just 20–30 cells, but they are neither clearly diploblastic nor triploblastic animals: Their cells form at most 2 tissue layers, but these apparently correspond to ectoderm and mesoderm; there seems to be no endoderm, which constitutes the second tissue layer in diploblasts. Moreover, there is no collagenous connective tissue and the mesozoan epithelium lacks a basal lamina.

Like the cestodes and trematodes (phylum Platyhelminthes), all mesozoans parasitize other animals, and like the body of turbellarian flatworms, the outer body surface is ciliated in at least some parts of the life cycle (Figs. 9.1 and 9.2), with each cell bearing 2 or more cilia. Like the cestodes, mesozoans lack any trace of a mouth or digestive system and are without any specialized circulatory, respiratory, sensory, or nervous systems. In most other respects, however, their biology is both peculiar and unique. Probably the most remarkable aspect of their biology is that they develop intracellularly; that is, the reproductive cells and embryos develop *within* other cells. Such intracellular development is otherwise unknown among animals, and suggests that mesozoans represent an early branch of multicellular evolution separate from that leading to, or from, any other metazoans.

Alternatively, their simple morphology may reflect extreme degeneration from more complex ancestors as an adaptation to parasitism. Several recent analyses suggest that mesozoans are indeed related to platyhelminth flatworms (see, for example, Fig. 2.12d), although a few other

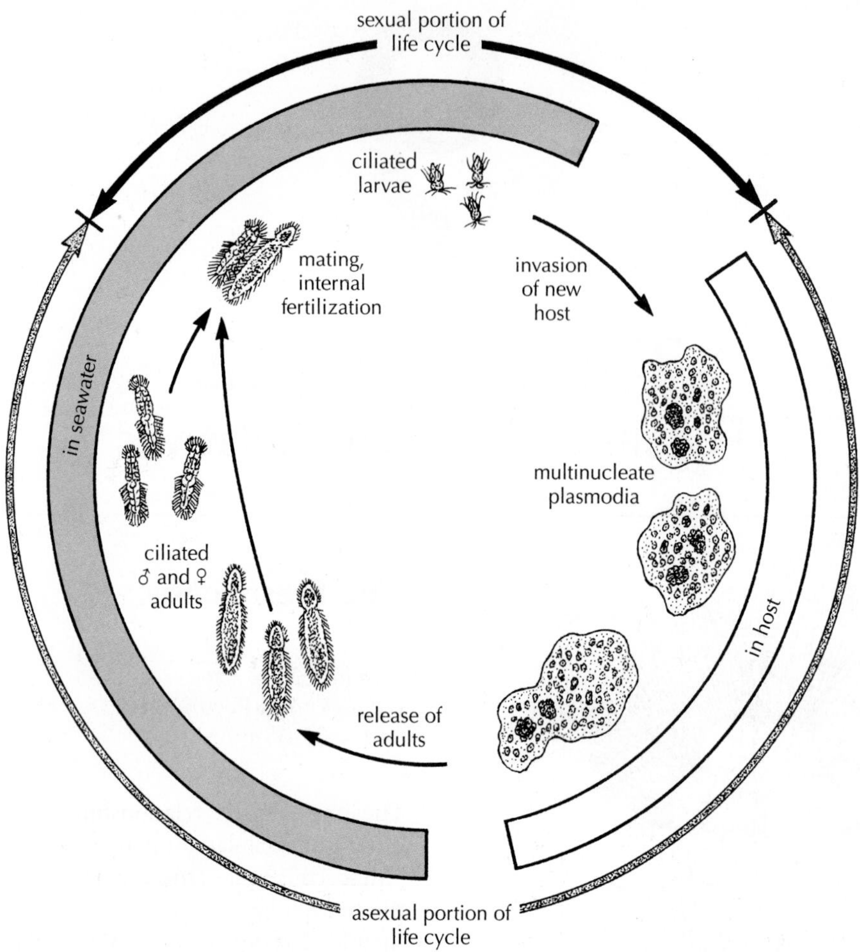

Figure 9.1
The orthonectid life cycle.

studies have suggested a closer relationship to nematode roundworms (Chapter 17). *Hox* gene sequence data suggest that at least some mesozoans (dicyemids, which parasitize cephalopod kidneys) group with molluscs, annelids, and a variety of other non-molting organisms as **lophotrochozoans** (see Fig. 2.12c). Mesozoan relationships to other phyla are just not clear. Neither is the relationship between the 2 mesozoan groups, the orthonectids and rhombozoans. The list of morphological and life history differences between orthonectids and rhombozoans exceeds the short list of similarities, suggesting that they have no close ancestor in common. Indeed, in a previous edition of this book they were each given phylum status. However, some recent analyses (e.g., Fig. 2.12d) indicate that they form a monophyletic group despite their differences, once again uniting them provisionally, as I have done here, within the single phylum Mesozoa.

Class Orthonectida

The orthonectids parasitize a variety of other marine invertebrates, including bivalved molluscs such as clams and oysters, polychaete annelids, echinoderms, and turbellarian flatworms. Individuals rarely exceed lengths of 300 μm, and most show gonochoristic sexuality, with a single individual possessing either male or female gonads but not both. Once in the host, juveniles take the form of asymmetrical, multinucleate but syncytial, amoeba-like individuals called **plasmodia** (Fig. 9.1). These live mostly in the host gonad, typically rendering the host impotent by castration. The plasmodia are juveniles in the sense that they cannot reproduce sexually, but they do not "mature" in the normal sense of the word. Rather, within itself, each plasmodium produces cellular, ciliated sexual forms, which might as well be called "adults." A single plasmodium typically produces both male and female adults. These small adults—rarely longer than about 150 μm—leave the plasmodium and then leave the invertebrate host, to swim freely in the sea. When 2 individuals of opposite sex meet, they juxtapose their genital pores and the male releases sperm into the female opening, fertilizing her eggs. The fertilized eggs develop into 2-layered, ciliated larval forms that leave through the female's genital opening and seek new hosts to infect. Upon entering the genital duct of a new invertebrate host, an orthonectid larva loses its outer layer of ciliated cells. The inner cells then move throughout the host gonad, and each becomes a plasmodium through mitotic division. Only about 20 orthonectid species have been described.

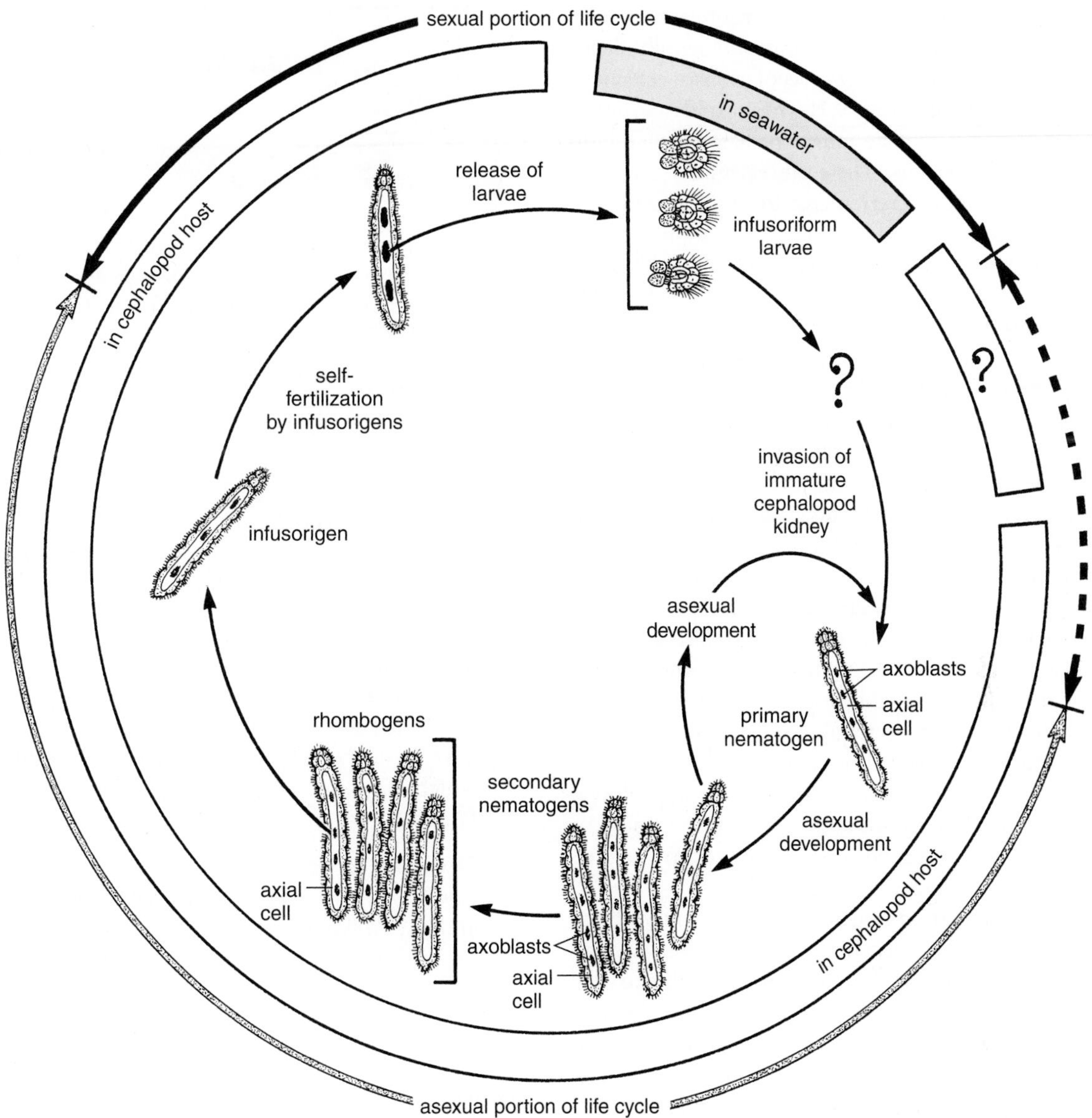

Figure 9.2
The rhombozoan life cycle.

Class Rhombozoa

The class Rhombozoa contains about 65 species. The rhombozoan life cycle little resembles that of orthonectids or other metazoans. The associated terminology is correspondingly cumbersome, but unavoidably so.

Rhombozoans all parasitize the nephridia (kidneys) and associated structures of cephalopod molluscs, particularly the cuttlefish and octopuses. The first stage found in the cephalopod host is called a **primary nematogen** (or **stem nematogen**). Like the bodies of adult orthonectids, rhombozoan nematogen bodies are vermiform (worm-shaped). Some individuals are 7.5 mm long, but even the largest animals have bodies composed of no more than about 30 cells. The 20 to 30 externally ciliated cells of the outer body layer surround one (or several) elongated **axial cell(s),** which forms a central shaft in the animal (Fig. 9.2). This central shaft is the animal's reproductive center, but it is not a gonad in the normal sense of the term; rather, reproduction is a function of other cells—up to about 100 in some species—that develop *within* this axial cell. Each intracellular **axoblast cell** (also called an **agamete**) develops into more asexually reproducing nematogens, which in turn develop more axoblasts that develop into more nematogens (Fig. 9.2). Once each nematogen attains a species-specific number of cells early in its development, all further growth results from increases in cell size rather than mitotic replication.

This asexual production of nematogens continues through several, or many, generations of nematogens until the kidney of the cephalopod host is filled; sexual maturation of the host may also play a role in ending nematogen production. At that point, either existing nematogens develop into a sexual stage, called the **rhombogen,** or the last generation of nematogens produces rhombogens from their axoblast cells. The rhombogen is morphologically identical to the nematogen and has its own set of axoblast cells internally, within its axial cell(s) (Fig. 9.2). These axoblasts, not the rhombogen itself, differentiate into nonciliated, sexual individuals called **infusorigens.** These are essentially the adult rhombozoans, but they never emerge from the rhombogen that produced them. To some extent, the situation is analogous to a human becoming sexually mature within the mother's uterus. To improve the analogy, the "child" would have to have arisen asexually, as would have the mother. Moreover, infusorigens are simultaneous hermaphrodites, producing both sperm and eggs by meiosis within a single individual. Surprisingly, there is no cross-fertilization; gametes produced by one infusorigen fertilize only other gametes produced by that same individual. (When it comes to rhombozoans, it does *not* "take 2 to tango.") Self-fertilization is also common among cestodes, as discussed in Chapter 8.

Following fertilization, the zygotes differentiate to form ciliated **infusoriform larvae** that escape from the rhombogen and then from the cephalopod host, emerging into the sea in host urine and then sinking to the bottom. The larvae do not seem capable of infecting a new cephalopod host directly and may require an intermediate host, although nothing is known about this part of the life cycle. The nephridia (kidneys) of another juvenile cephalopod are eventually invaded by something—we don't know by what or from where—that becomes a nematogen and goes on to produce more nematogens asexually.

Nematogen and rhombogen stages apparently obtain their nourishment exclusively from the host urine in which they live. The nutritive content may be high, but the oxygen concentration is extremely low, and the metabolism of nematogens and rhombogens is correspondingly anaerobic; in fact, these organisms apparently survive longer in the presence of cyanide than in the presence of oxygen.

Rhombozoans cause little detriment to their host, despite their extensive asexual proliferation in the host's kidney.

Taxonomic Summary

Phylum Mesozoa
- Class Orthonectida
- Class Rhombozoa—dicyemids and others

Topic for Further Discussion and Investigation

Compare and contrast the life cycle of a rhombozoan with that of a digenean flatworm (Chapter 8).

Taxonomic Detail

Phylum Mesozoa

The approximately 100 species in this phylum parasitize other marine invertebrates. Mesozoans are divided into 2 major groups, each of which may merit status as a separate phylum.

Class Orthonectida

Rhopalura, Stoecharthrum. All species in this small class (about 20 species) are endoparasitic in various invertebrates. Two families.

Class Rhombozoa

This group contains about 75 species, all of which parasitize the nephridial (kidney) system of cephalopods. Two orders.

Order Dicyemida

Dicyema, Pseudicyema. This order contains all but 2 of the known rhombozoan species. Most species have a poorly defined "head," termed a **calotte,** which may attach the parasite to host tissues. One family.

Order Heterocyemida

Conocyema, Microcyema. The nematogens lack cilia and have a syncytial outer cell layer. The life cycles for the 2 species in this order are incompletely described.

General References about the Mesozoans

Cheng, T. C. 1986. *Parasitology,* 2d ed. New York: Academic Press.

Furuya, H., F. G. Hochberg, and K. Tsuneki. 2003. Reproductive traits in dicyemids. *Marine Biol.* 142:693–706.

Garey, J. R., and A. Schmidt-Rhaesa. 1998. The essential role of "minor" phyla in molecular studies of animal evolution. *Amer. Zool.* 38:907–17.

Kobayashi, M., H. Furuya, and P. W. H. Holland. 1999. Dicyemids are higher animals. *Nature* 401:762.

Lapan, E. A., and H. Morowitz. 1972. The Mesozoa. *Sci. Amer.* 227:94–101.

Morris, S. C., et al., eds. 1985. *The Origins and Relationships of Lower Invertebrates. Systematics Association, Special Vol. 28.* Oxford: Clarendon Press.

Parker, S. P., ed. 1982. *Classification and Synopsis of Living Organisms,* vol. 1. New York: McGraw-Hill, 880–929.

Roberts, L. S., J. Janovy, Jr., and P. Schmidt. 2004. *Foundations of Parasitology,* 7th edition. McGraw-Hill.

Stunkard, H. W. 1954. The life history and systematic relations of the Mesozoa. *Q. Rev. Biol.* 29:230–44.

10

The Gnathifera: Rotifers, Acanthocephalans, and Two Smaller Groups

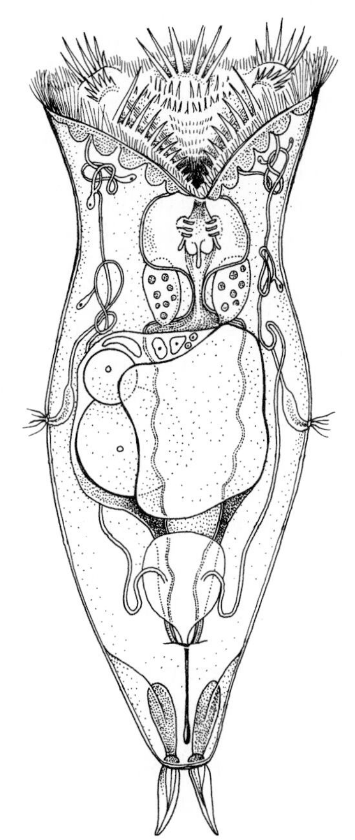

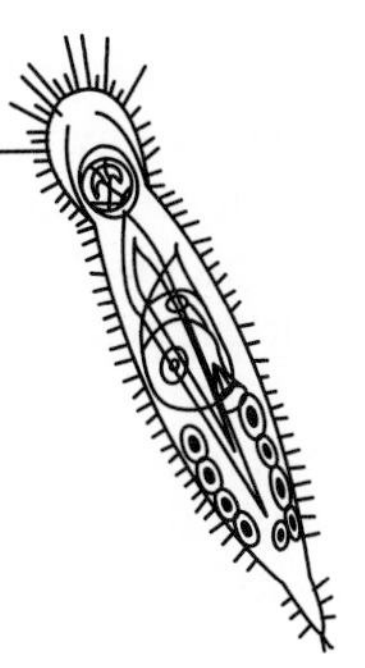

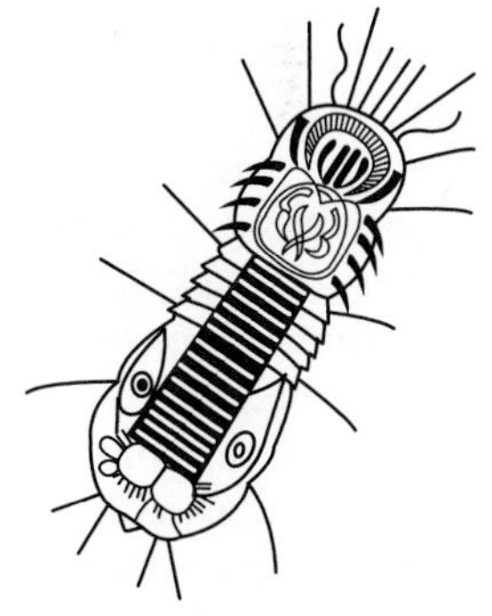

Gnatho = Greek: the jaw
nath-ih´-fur-ah

Defining Characteristic: All members possess pharyngeal jaws with a similar and complex ultrastructure

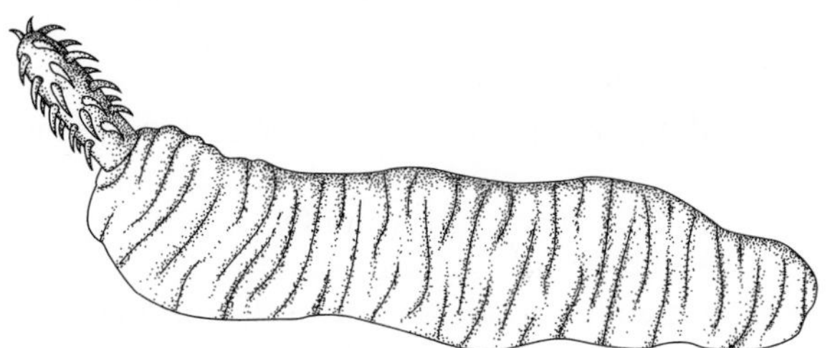

Introduction

The animals discussed in this chapter were formerly placed with nematodes and a number of other animal groups (gastrotrichs, kinorhynchs, and nematomorphs) in the phylum Aschelminthes. The similarities among at least some of these animals were increasingly understood to reflect convergence rather than common ancestry; thus, each has been awarded status as a separate phylum. The term "aschelminth" is still widely used in talking about these animals, however, reflecting a degree of organizational similarity. Although the precise relationships among aschelminths are still being worked out, they are presently split into 2 major groups—those animals that molt, and those that do not. The molting aschelminths are grouped together in the taxon Cycloneuralia (Fig. 2.10), while the non-molting aschelminths, discussed here, are placed in the taxon Gnathifera; a growing volume of evidence, both molecular and morphological, indicates that all 4 groups discussed in this chapter have evolved from a common ancestor.

As with other "aschelminths," the animals included in this chapter have a fluid-filled body cavity that forms neither by schizocoely nor enterocoely, and that is never lined by mesodermal epithelia; it is a **pseudocoel,** not a true coelom (Fig. 2.3). Traditionally the pseudocoel has been defined as a persistent blastocoel, but it apparently forms in other ways in many aschelminth species. In at least some groups, the pseudocoel may be a secondarily modified coelomic cavity.

Aschelminths also demonstrate a phenomenon called **eutely:** growth through increased cell size rather than increased cell number. Particular tissues and organs in juveniles and adults are composed of a species-specific, fixed number of cells. Probably because cell division ceases at the end of embryonic development, aschelminths are unable to regenerate lost body parts.

Most gnathiferan species are found within the first 2 groups discussed in this chapter, the largely free-living Rotifera and the completely parasitic Acanthocephala. Growing evidence from both morphological and molecular studies indicates that the 2 groups are closely related, but the exact relationship remains unclear. In particular, members of both groups have a syncytial epidermis with a unique type of non-molted skeletal structure in the cytoplasm. Some workers have started referring to a "phylum Syndermata" (*syn* = Greek: with; *dermata* = G: skin) that would encompass both groups, making rotifers and acanthocephalans classes within that phylum. Others[1] have suggested folding acanthocephalans within the Rotifera, and retaining Rotifera as the phylum name. In the absence of a consensus, I have retained the 2 groups as separate phyla for this edition.

Phylum Rotifera

Phylum Roti • fera
(Greek: wheel bearer)
rō-tih´-fer-ah

Defining Characteristics:[2] 1) Pharynx highly muscular and containing jaws (trophi) for grasping, crushing, or grinding prey, or attaching to a host; 2) toes with adhesive glands

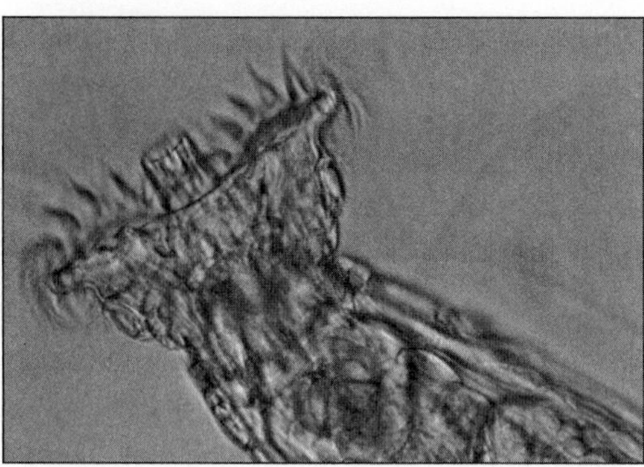

Figure 10.1
Light micrograph of rotifer corona, showing pattern of ciliation.
©Thomas E. Adams/Visuals Unlimited.

Introduction and General Characteristics

"I call it a Water Animal, because its Appearance as a living Creature is only in that Element. I give it also for Distinction Sake the Name of Wheeler, Wheel Insect, or Animal; from its being furnished with a Pair of Instruments, which in Figure and Motion appear much to resemble Wheels. It can, however, continue many Months out of Water, and dry as Dust; in which Condition its Shape is globular, its Bigness exceeds not a Grain of Sand, and no Signs of Life appear. Notwithstanding, being put into Water, in the Space of Half an Hour a languid Motion begins, the Globule turns itself about, lengthens by slow Degrees, becomes in the Form of a lively Maggot, and most commonly in a few Minutes afterwards puts out its Wheels, and swims vigorously through the Water in Search of Food."

So wrote a Mr. Baker, in a letter addressed to the President of the Royal Society in London, in 1744, on the subject of rotifers. Rotifers were so named because of the 2 ciliated anterior lobes present in many species. This ciliated surface is called the **corona** (Figs. 10.1 and 10.2). The coronal cilia do not beat in synchrony. Instead, each cilium is at a slightly earlier stage in the beat cycle than the preceding cilium in the sequence; that is, the cilia beat **metachronally.** A wave of ciliary beating therefore appears to pass around the periphery of the ciliated lobes, giving the impression of rotation. The degree and pattern of coronal ciliation vary considerably among species.

Approximately 1850 rotifer species have been described, mostly (about 95%) from freshwater environments, including lakes, ponds, and the moist surface films of mosses and other semiterrestrial vegetation. Some freshwater rotifers live **interstitially,** in the spaces between the sand grains of freshwater beaches. In fact, rotifers are generally considered one of the most characteristic groups of freshwater animals. Typically, 40 to 500

1. For example, Sørensen, M. V., and G. Giribet. 2006. *Molec. Phylog. Evol.* 40:61–72

2. Characteristics distinguishing the members of this phylum from members of other phyla.

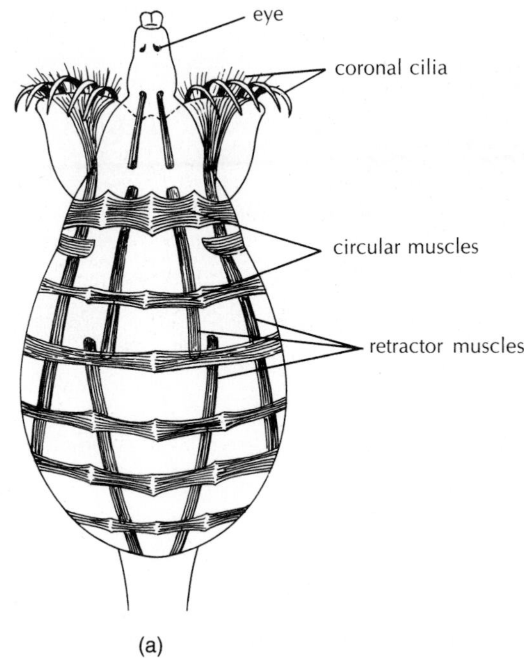

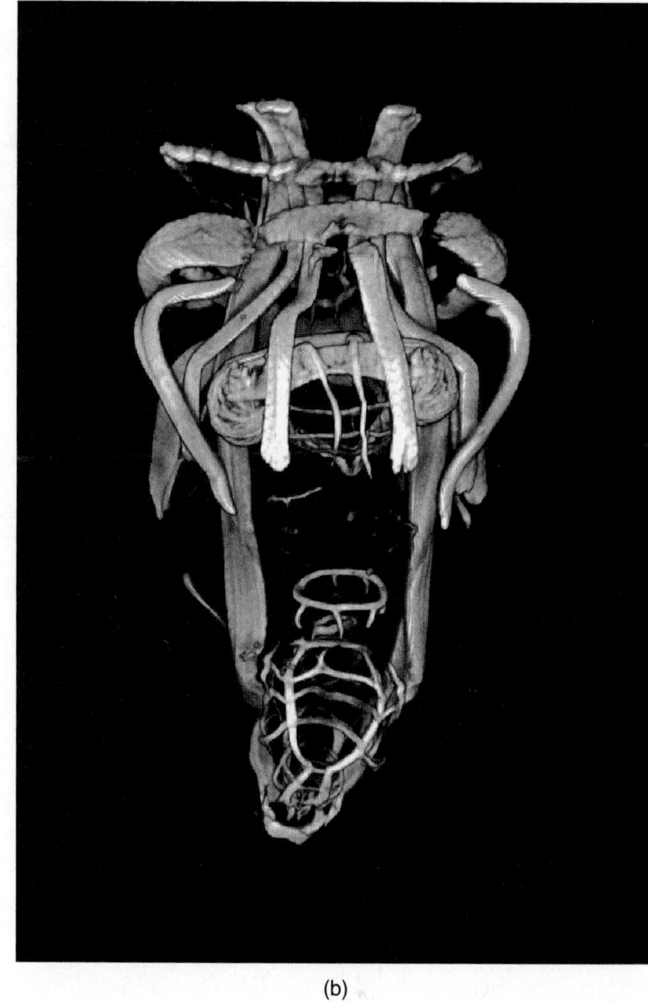

Figure 10.2

(a) Illustration of rotifer musculature (*Rotatoria* sp.). The musculature of a given individual rotifer may be smooth, striated, or a combination of both. (b) A computer-rendered dorsal view of the musculature of *Filinia novaezealandiae*. Muscles were stained with phalloidin (a fungus-derived peptide that binds to actin) and visualized with confocal laser scanning microscopy.

(a) From Hyman, *The Invertebrates,* Vol. III. Copyright © 1951 McGraw-Hill Book Company, New York. Reprinted by permission.
(b) Courtesy of Rick Hochberg, University of Massachusetts at Lowell. From Hochberg, R., and O.A. Gurbuz. 2007. Functional morphology of somatic muscles and anterolateral setae in *Filinia novaezealandiae* Shiel and Sanoamuang, 1993 (Rotifera). *Zoologischer Anzeiger* 246: 11–22.

individuals are found per liter of lake or pond water, with densities as high as 5000 individuals per liter having been recorded on several occasions. About 5% of rotifer species are found in shallow-water marine or estuarine environments; some of these marine species are interstitial.

Like most aschelminths, rotifers are pseudocoelomate and generally **eutelic;** mitotic divisions cease early in development, so that further increases in body size are due to an increase in cell size rather than in cell number, and the different organs and tissues are characterized by species-specific numbers of nuclei. The epidermis of rotifers is syncytial; that is, cell membranes between nuclei are incomplete. The syncytial epidermis produces a nonchitinous, intracellular "cuticle" that is never molted. This intracellular epidermal layer is supported by a complex web of protein filaments, something encountered elsewhere only among acanthocephalans. Rotifers possess both smooth and striated muscle fibers (Fig. 10.2), the latter being used for the rapid movement of spines and other appendages. Specialized respiratory and blood circulatory systems are lacking.

Most species are free-living and short-lived. Typically, rotifers live only 1–2 weeks, although individuals of a few species can survive for up to about 5 weeks. Some free-living species spend their adult lives permanently attached to a substrate, while others are capable of moving from place to place. Some rotifers are parasitic, but their hosts are always invertebrates, especially arthropods and annelids (Fig. 10.3). Because parasitic rotifers have minimal impact on humans, their life histories have never commanded the attention accorded those of most other parasitic animals. The free-living species have never been much in the public eye either. Most are only about 100–500 μm (micrometers) long, which is remarkably small for an adult metazoan. The largest individuals never exceed a length of 3 mm.

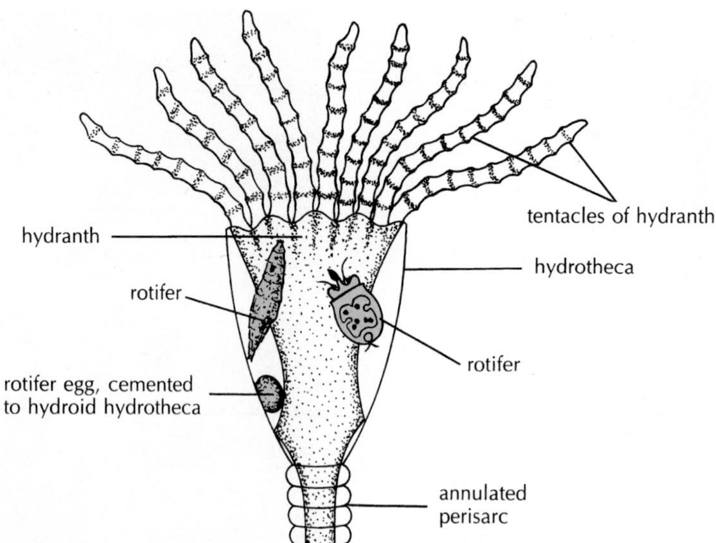

Figure 10.3

Cnidarian hydranth with rotifer parasites (*Proales gonothyraea*) in hydrotheca. The rotifer feeds on the cnidarian tissues and lays its eggs in the hydranth. Other species within this genus are parasites in or on protozoans, oligochaetes (Annelida), gastropod embryos (Mollusca), crustaceans (Arthropoda), and filamentous algae. Several other genera are common internal parasites of oligochaetes, leeches (Annelida), and slugs (Mollusca).
From Hyman; after Remane.

Interest in rotifer biology seems to be increasing somewhat, as several free-living rotifer species have been found to be good food sources for the rearing of some commercially important fish and crustaceans. Rotifers may also play important roles in determining aquatic community structure and in mediating the flow of energy through freshwater ecosystems. Quite recently, rotifers have been utilized as models for research on the process of **senescence** (aging) and as monitors of pollution.

Nonparasitic (free-living) rotifers feed on a variety of items. Most species are omnivores, selectively ingesting algae of appropriate size and chemical composition, small free-living animals **(zooplankton),** and detritus of appropriate size. Some other species feed on the intracellular juices of algae, and a number of species are carnivores, preying on a variety of animals smaller than themselves, including other rotifers.

Certain elements of the rotifer feeding and digestive system—notably the mastax and the trophi—are unique. The **mastax** is a prominent, muscular modification of the pharynx (Fig. 10.4). In some parasitic species, the mastax is modified for attachment to the host. Within the mastax of all species are found a number of rigid structures, the **trophi,** which are often used to grind food following ingestion (Fig. 10.5). In some species, the trophi are used to suck food in through the mouth; in other species, the trophi can be protruded from the mouth to grab and/or pierce prey. The structure and sculpturing of trophi vary considerably in different species, in accordance with how the trophi are used. They thus are important tools in species identification. Trophi ultrastructure is almost unique to rotifers; only the jaws of gnathostomulids (discussed later in this chapter) are structurally similar to those of rotifers, part of the evidence suggesting a close evolutionary relationship between the 2 groups.

Free-living rotifers swim by using the coronal cilia. In some species, particularly well-developed spines also contribute to a leaping sort of locomotion (Fig. 10.6). Some species are entirely planktonic; that is, individuals swim throughout their lives. Members of many other species stop swimming periodically and form temporary attachments to solid substrates by secreting a cementing substance from a pair of pedal glands. The pedal glands open to the outside through pores in the **toes** of the foot. Rotifer feet possess up to 4 toes, depending upon the species (Fig. 10.7). Once the rotifer attaches to the substrate, the coronal cilia generate water currents for respiration and food collection. In a number of planktonic and sedentary rotifers, the coronal cilia form 2 parallel bands, with a ciliated food groove lying between. The cilia of the 2 bands beat toward each other, sweeping particles into the food groove (Fig. 10.8). Cilia in the food groove then conduct these captured particles to the mouth for ingestion.

Many free-living species can move upon solid substrates between bouts of swimming. The foot and toes play a role in such locomotion by forming temporary attachments to the substrate. Once the attachment is made, the body can be elongated through the contraction of circular muscles, which are distributed about the body in discrete bands. Contracting the circular musculature elongates the body and stretches out the relaxed longitudinal musculature. The pseudocoel of these rotifers thus functions as a hydrostatic skeleton, permitting the mutual antagonism of the circular and longitudinal musculature through the generation of temporary increases in hydrostatic pressure.

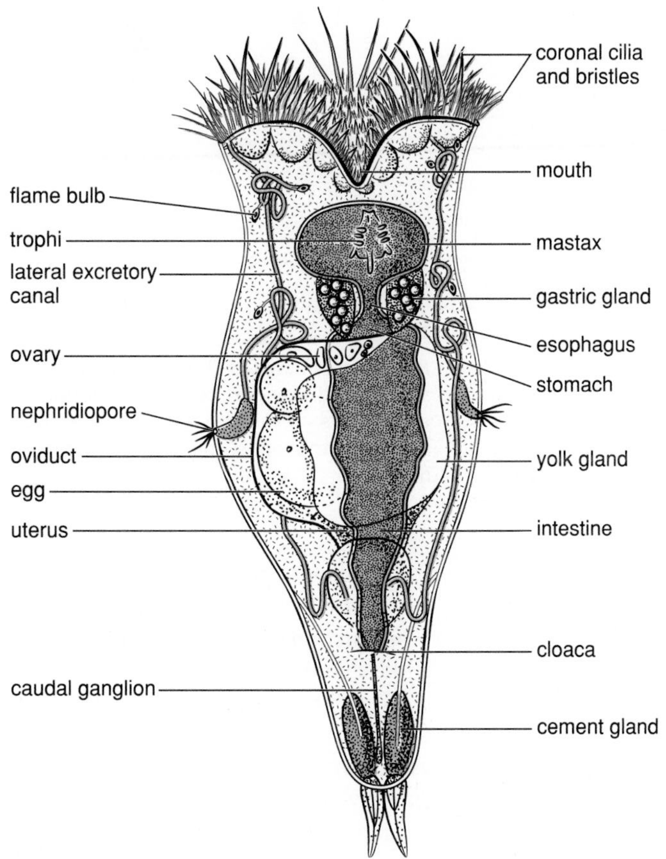

Figure 10.4

Epiphanes senta, a common free-living rotifer from freshwater. Note the placement of the mouth and feeding apparatus. The trophi/mastax complex is used to grind ingested food into smaller particles. In some species, the trophi can be protruded from the mouth for prey capture.

Based on Frank Brown, *Selected Invertebrate Types*.

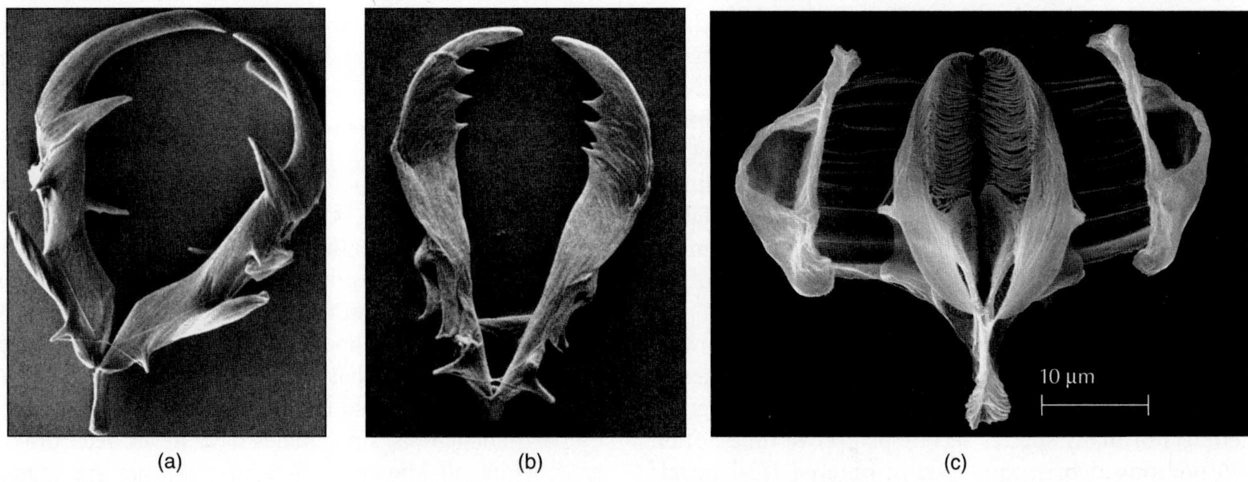

Figure 10.5

Scanning electron micrographs of rotifer trophi. (a) *Asplanchna sieboldi* (700×). (b) *Asplanchna priodonta* (1400×). (c) *Floscularia ringens.*

(a,b) Courtesy George Salt. From Salt et al, 1978. *Amer. Microsc. Soc. Trans.* 97:469. (c) Courtesy of Guilio Melone. From Fontaneto, D., Melone, G., and R. L. Wallace. 2003. Morphology of *Floscularia ringens* (Rotifera, Monogononta) from egg to adult. *Invert. Biol.* 122:231–40. With permission from the American Microscopical Society.

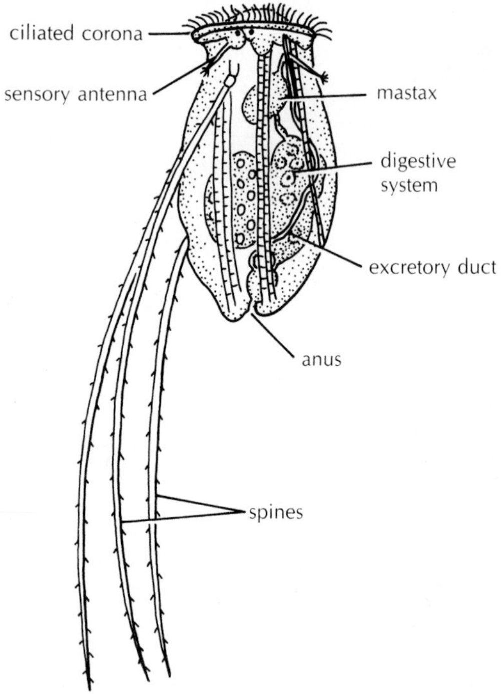

Figure 10.6

Filinia longiseta, a pelagic rotifer commonly found in lakes. Note the lack of a foot, the lack of toes, and the presence of conspicuous spines for swimming.
From Hyman; after Weber.

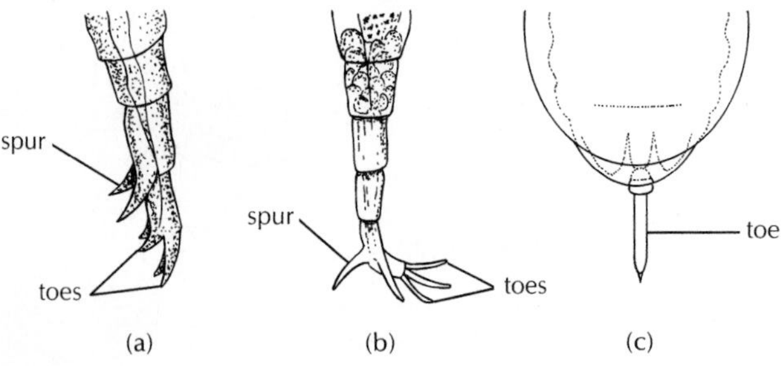

Figure 10.7

Rotifers may have 0, 1, 2, 3, or 4 toes on the foot, depending on the species, and may also bear a pair of nonsecretory protuberances called spurs. (a) *Philodina roseola,* with 4 toes and 2 spurs. (b) *Rotaria* sp., with 3 toes and 2 spurs. (c) *Monostyla* sp., with a single toe and no spurs.

As the rotifer's body elongates, the corona is withdrawn, turning the animal's anterior end into a suction-generating **proboscis** (Fig. 10.9). The proboscis can be applied to the substrate, forming a new attachment site. The posterior attachment is then broken, and the body becomes shorter and fatter through contraction of the longitudinal musculature and stretching of the relaxed circular musculature. Finally, the anterior attachment is released while the foot regains its grip on the substrate, and the body is again elongated. The animal can thus progress by **looping,** somewhat as described previously for free-living flatworms (phylum Platyhelminthes, class Turbellaria—Chapter 8) and as yet to be described for leeches (phylum Annelida, p. 321). Some rotifers exhibit extraordinary shape changes during locomotion, which are often facilitated by the foot and trunk of the body being divided into a number of sections that can be telescoped into each other, much like a collapsible drinking cup.

Other nonparasitic rotifers are **sessile** as adults; that is, they are incapable of locomotion. Pedal gland secretions attach these animals to a substrate permanently. Members of many species secrete protective tubes, often incorporating debris, sand grains, or even fecal pellets into their walls (Fig. 10.10). All species that are sessile as adults have free-swimming young resembling the young of planktonic species. Thus, even sedentary species are capable of locomotion—at least for a short time between adult generations.

Rotifer reproduction is unusual in a number of respects. As among nematodes and other aschelminths, reproduction by fission or fragmentation is unknown, again a likely consequence of eutely. Where asexual reproduction does occur among rotifers, it is by **parthenogenesis,** the development of unfertilized eggs. The details of sexual and asexual reproduction among rotifers are best discussed on a class-by-class basis. There are 3 classes in the phylum.

Class Seisonidea

The small class Seisonidea (sī-sun-id´-ē-ah) is composed exclusively of ectoparasites of marine crustaceans. As might be expected, the corona of these species is often greatly reduced in size (Fig. 10.11a). Reproduction seems to be exclusively sexual, and all individuals have separate sexes; that is, the members of this class are always **gonochoristic (dioecious).** Fertilization is internal, either through true copulation or by means of hypodermic impregnation of the female by the male. In the latter case, sperm are injected by the male into the pseudocoel of the female, from which they find their way into the ovary.

Class Bdelloidea

Members of the class Bdelloidea (del-oy´-dē-ah) are all free-living and mobile: there are no tube-dwelling, sessile species and no parasitic species. Most members are omnivorous suspension feeders, and the corona is correspondingly well developed and bilobed (Fig. 10.11b, c); one predatory species has recently been described. Reproduction among bdelloid rotifers appears to be exclusively by parthenogenesis, since males have never been discovered. Thus, all known rotifers in this class are female. Recent molecular evidence[3] suggests that bdelloid rotifers have persisted for millions of years without any exchange

3. See *Topics for Further Discussion and Investigation,* no. 6, at the end of the chapter.

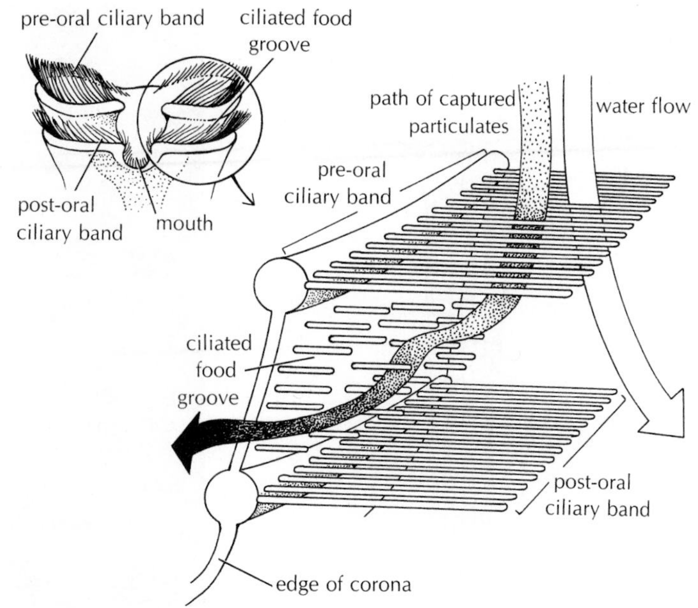

Figure 10.8

Double-banded ciliary system of suspension-feeding rotifers. As water is swept between the cilia of the pre-oral ciliary band, suspended food particles are captured and conducted to the mouth by the cilia of the food groove.

Modified from Strathmann et al., 1972. *Biological Bulletin* 142:505–19.

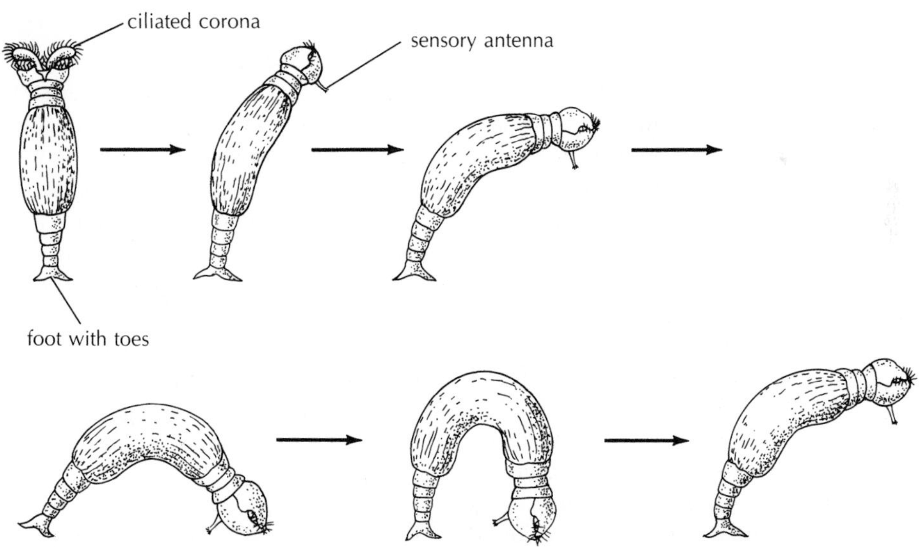

Figure 10.9

Looping locomotion of *Philodina roseola*. Contraction of circular muscles elongates the body, which is then bent toward the substrate by the differential contraction of longitudinal muscles. Once the anterior end of the animal grips the substrate, the foot releases its attachment and the rear of the body may be pulled forward. The body may then reattach posteriorly and extend forward again. Note that the corona is withdrawn during this sequence of movements.

After Harmer and Shipley.

of genetic material between individuals, a direct affront to the idea that the loss of sex leads inevitably to extinction. About 370 species have been described.

Bdelloid rotifers commonly inhabit environments that periodically expose the animals to physiologically stressful conditions, such as freezing, dehydration, or high temperatures. Species living in polar lakes and ponds, in temporary ponds, and among emergent mosses and lichens are typically capable of entering a state of extremely low metabolism, or **cryptobiosis.** Some species secrete a gelatinous covering during the early stages of drying. This covering then hardens to

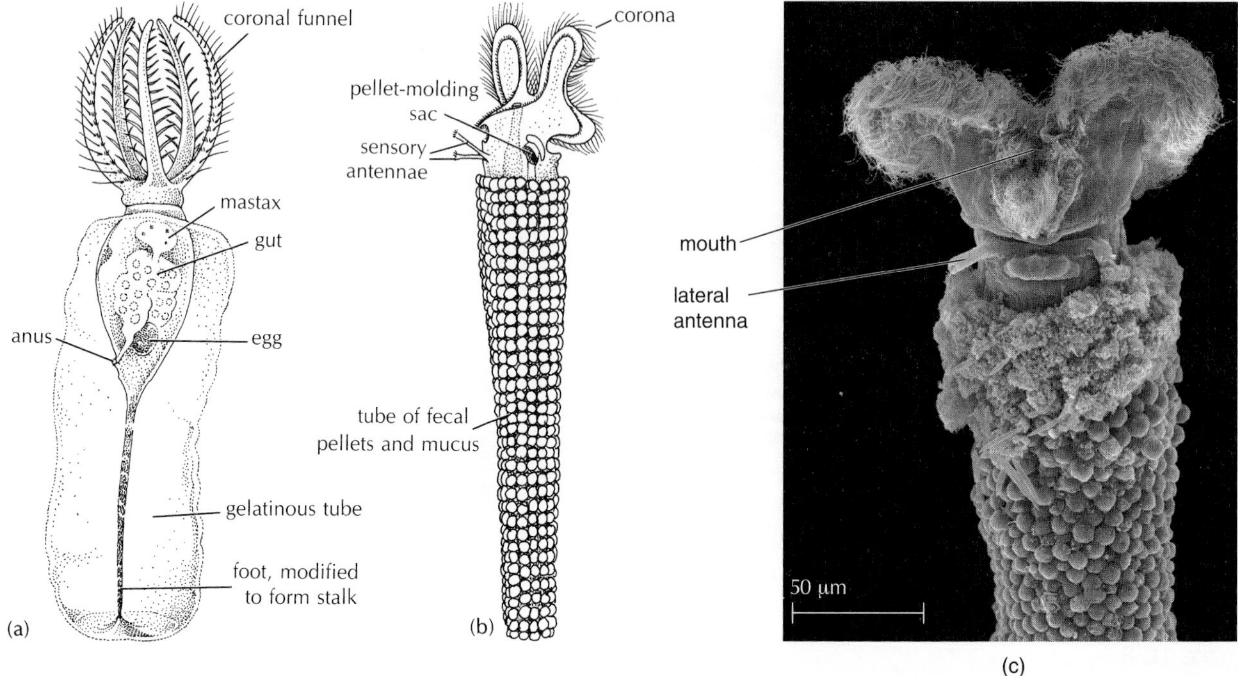

Figure 10.10

Two sedentary rotifer species: (a) *Stephanoceros fimbriatus.* (b) *Floscularia ringens,* showing region of pellet formation. (c) SEM of feeding adult of *F. ringens* in its tube.

(a) After Jurczyk; and after Jägersten. (b) After Pennak. (c) Courtesy of Guilio Melone. From Fontaneto, D., G. Melone, and R. L. Wallace. 2003. Morphology of *Floscularia ringens* (Rotifera, Monogononta) from egg to adult. *Invert. Biol.* 122:231–240. With permission from the American Microscopical Society.

form a **cyst** (Fig. 10.12). Like the tardigrades and nematodes (to be discussed in Chapters 15 and 16, respectively), rotifers in the cryptobiotic state can withstand environmental extremes, including extensive desiccation, for prolonged time periods. Indeed, some bdelloid rotifers have been successfully rehydrated following 20 years or more of desiccation! How cryptobiosis permits rotifers to withstand what would otherwise be lethal environmental conditions is not yet known.

Class Monogononta

Most rotifers belong to the class Monogononta (mah-nō-gon-on′-tah). Monogonont rotifers may be free-swimming or sessile. In fact, this class contains the only free-living sessile rotifers. The sessile rotifers are generally found attached to macroscopic plants, to filamentous algae, or to the tubes of other sessile rotifers of the same or of different species. Some sessile species use the corona to collect food particles as described previously, or with a modification of that theme. In other species, the corona is poorly ciliated or is nonciliated, but it may bear long spines surrounding a funnel-like anterior end. The spines can be moved to entrap small metazoans that come too close, and the captured prey are then forced into the mouth for ingestion. Thus, many members of this class are carnivores. Ciliation of the **buccal field** surrounding the mouth presumably aids ingestion.

Many sessile species live in protective tubes (Figs. 10.10 and 10.11f). In some instances, these are merely gelatinous secretions from specialized glands opening externally. In other species, particles are collected by the corona, coated with mucus, and cemented in place, thereby elongating the tube as the animal grows in size. In some species, fecal pellets are incorporated into the tube. In both free-swimming and sessile monogonont species, the cuticle is commonly thick and rigid, forming a protective **lorica.** Lorica production is encountered only among members of this class. Within a species, the shape of the lorica may be modified by environmental factors (Research Focus Box 10.1).

The reproductive pattern of monogonont rotifers is unique. Typically, monogononts reproduce by means of parthenogenesis. Females generally produce diploid eggs by mitosis, usually one at a time, and each egg develops into another diploid female—all in the absence of males. Such females are termed **amictic,** referring to the production of eggs without the "mixing in" of genes from any other individual. Young rotifers emerge from amictic eggs soon after the eggs are released; thus, amictic eggs are also known as **subitaneous** eggs (*subit* = G: sudden). Through the production of subitaneous eggs, the population size of a given rotifer species can double in as little as 15 hours. Typically, each female produces 4 to 40 amictic eggs per lifetime, and 20 to 40 or more generations of genetically identical amictic females can occur yearly.

Under certain conditions, however, a different type of monogonont female is produced.[4] These **mictic** females produce their eggs through meiosis, so that all

4. See *Topics for Further Discussion and Investigation,* no. 1.

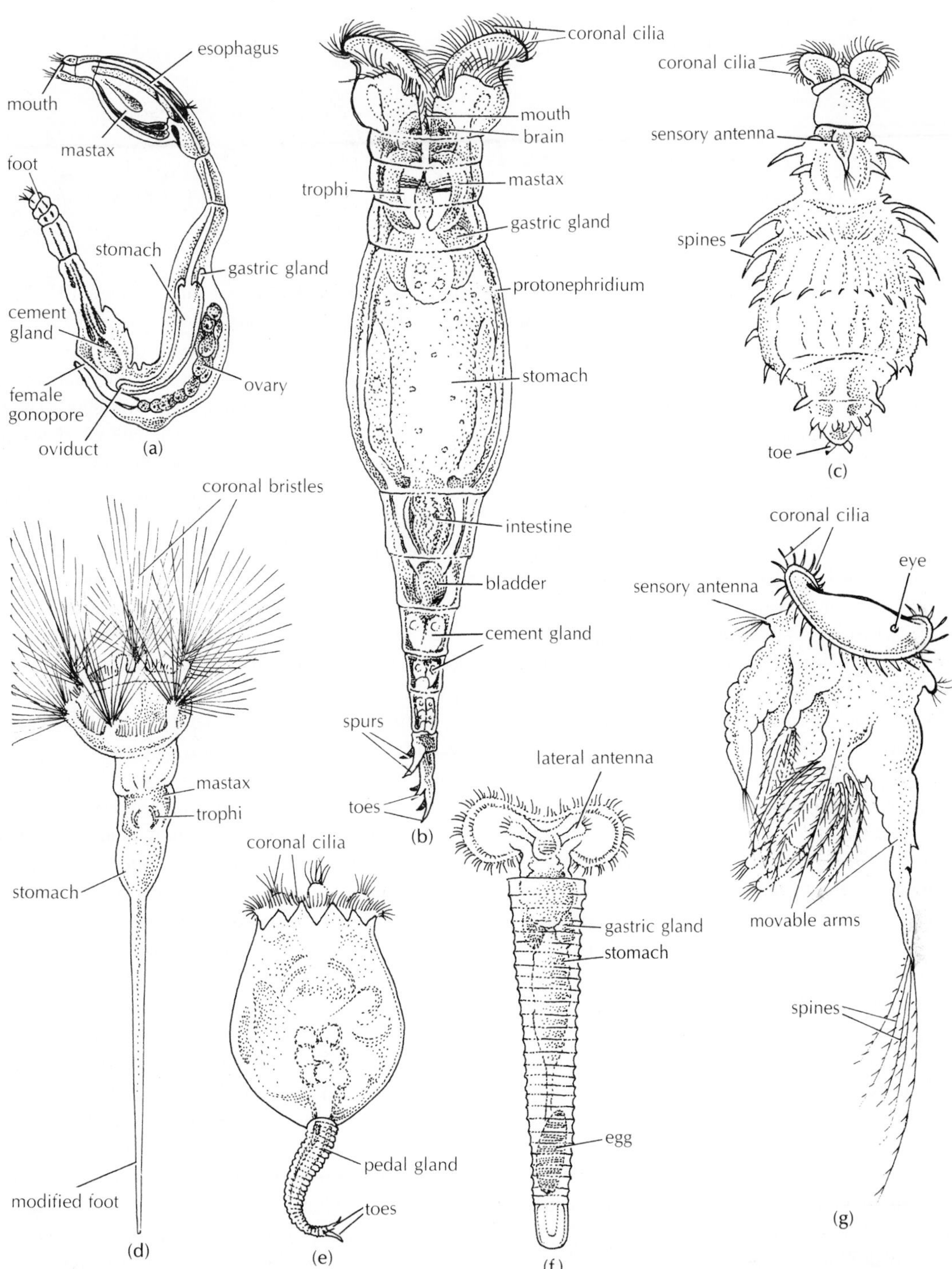

Figure 10.11

Representatives of the 3 rotifer classes. (a) Seisonidea (*Seison* sp.). All species are ectoparasitic on marine crustaceans and are gonochoristic. The illustrated individual is a female. Note the reduced ciliation associated with a parasitic lifestyle. (b) Bdelloidea (*Philodina roseola*). All species are free-living and motile suspension-feeders, and no males have ever been described; all reproduction is parthenogenetic. (c) Bdelloidea (*Macrotrachela multispinosus*). (d) Monogononta (*Collotheca* sp.). Note the highly modified corona, with 7 distinct lobes. The animal is sessile and secretes a gelatinous tube (not illustrated). (e) Monogononta (*Brachionus rubens*). This species is free-swimming and is commonly used as a food source for rearing larval fish. (f) Monogononta (*Limnius* sp., a sessile tube-dweller). (g) Monogononta (*Pedalia mira,* a free-swimming species). Members of this class show both sexual and asexual reproduction; the switch to sexual reproduction is triggered by alterations in the physical environment.

(a) From Meglitsch; after Plate. (b) From Hyman; after Hickernell. (c) From Hyman; after Murray. (d) After Hyman. (e) From Pennak; after Halbach. (f) From Meglitsch; after Edmondson. (g) From Hyman; after Hudson and Gosse.

RESEARCH FOCUS BOX 10.1

Adaptive Value of Rotifer Morphology

Gilbert, J. J., and R. S. Stemberger. 1984. *Asplanchna*-induced polymorphism in the rotifer *Keratella slacki. Limnol. Oceanogr.* 29:1309–16.

Certain aquatic invertebrates from several phyla exhibit **cyclomorphosis**—morphological changes in the subsequent generation induced by changing environmental factors, such as temperature, food level, predation pressure, or dissolved organic substances. The freshwater, suspension-feeding rotifer *Brachionus calyciflorus,* for example, exhibits such a change in response to some (uncharacterized) chemical released by a predaceous relative, rotifers in the genus *Asplanchna.* The chemical does not alter existing adult morphology, but it does affect the development of early-stage embryos. In the presence of *Asplanchna* spp., the offspring of *B. calyciflorus* develop much larger spines than usual, which probably protects them from predation by *Asplanchna.*

Do other suspension-feeding rotifers show similar responses to *Asplanchna*? Do the morphological changes actually reduce predation? And can the same morphological changes be induced by other predators, or are they specific responses to *Asplanchna*? These were the issues addressed by Gilbert and Stemberger (1984). They chose to work with the herbivorous rotifer *Keratella slacki* because it co-occurs with the predatory *Asplanchna* spp. and exhibits a wide range of body morphology in the field; indeed, field-collected individuals have longer spine lengths only when *Asplanchna* is present, suggesting a causal relationship worth examining. The questions are, is *Asplanchna* directly responsible for the morphological variation, and how does the variation benefit *K. slacki*?

Changes in body morphology of a species in the field over time can be explained in one of 3 ways: (1) individuals with certain morphologies survive better at some times than others; (2) individuals with certain genetically determined morphologies leave more offspring at some times than at others; or (3) the phenotype of most individuals is altered under some conditions without any underlying genetic change. These possibilities are readily evaluated for *K. slacki* (and many other rotifers) because it reproduces parthenogenetically and has a short generation time; a large, genetically identical population can be established rapidly from an individual female. In the experiments described, all individuals of *K. slacki* were clones derived from one individual.

To conduct their studies, Gilbert and Stemberger maintained high-density laboratory cultures of the predatory rotifer *Asplanchna girodi* and of a small, planktonic, predatory crustacean, *Tropocyclops prasinus.* Before each experiment, they passed the culture water through a glass-fiber filter to remove solid materials. Several adult females of *K. slacki* were placed into each of a number of small petri dishes in 5 ml of the *Asplanchna* filtrate, and they were fed on a small unicellular alga, *Cryptomonas* sp. As a control, additional populations of *K. slacki* were cultured with the same unicellular algae in filtrates of the protozoan *Paramecium aurelia,* which was used as a food source in culturing the *Asplanchna.* Other rotifer populations were all cultured in filtrates of the predatory crustacean. Under laboratory conditions, the rotifer populations increased rapidly. The researchers changed the food and water every few days for up to 9 days, and then they preserved all rotifers for measurement. The body and spine lengths of about 50 individuals from each treatment were measured, using a compound microscope equipped with a measuring scale (ocular micrometer).

In the presence of the predatory rotifer filtrate, the average body size in the population of *K. slacki* increased by about 15%, the right posterior spine length increased by an average of about 130%, and the lengths of the anterior spines increased by about 30%. Some individuals also developed conspicuous posterior spines on the left side (Focus Fig. 10.1). In contrast, filtrate from the predatory crustaceans or from the protozoan controls had no effect on the rotifer's body size or morphology (Focus Fig. 10.2). These data clearly indicate that *K. slacki* is not simply responding to higher levels of some general waste product, such as ammonia, since the crustaceans and protozoans were cultured at comparably high densities and excretory rates probably do not differ dramatically among these animals. Also, the response seems to be triggered by a specific water-soluble substance produced by *Asplanchna,* rather than by any physical contact, and it clearly reflects a developmental polymorphism rather than any genetic change in the population. As mentioned earlier, all the offspring of *K. slacki* in these experiments were produced parthenogenetically.

But is the response adaptively beneficial? To examine this, Gilbert and Stemberger placed an equal number of short- and long-spined *K. slacki* into each of several dishes of water, and then they added some *Asplanchna* predators to each dish. About 6 hours later, they counted the number of short- and long-spined *K. slacki* remaining in each dish. If short-spined individuals are more vulnerable to predation, fewer of these individuals should be left at the end of such an experiment. Indeed, this was the result obtained in each of the 3 separate studies conducted (Focus Table 10.1).

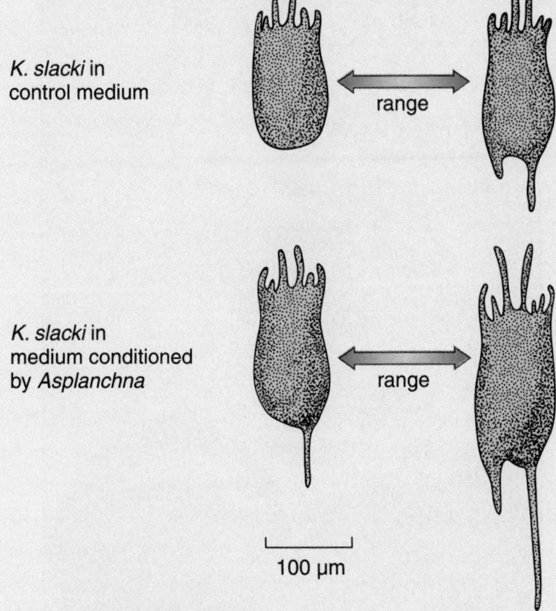

Focus Figure 10.1

Most common body types encountered in control cultures of *Keratella slacki* and those reared in *Asplanchna*-conditioned water. Animals in the experimental treatments have somewhat larger bodies and substantially longer spines.

Based on J. J. Gilbert and R. S. Stemberger. 1984. "*Asplanchna*-induced polymorphism in the rotifer *Keratella slacki*." *Limnology and Oceanography* 29:1309–16. 1984.

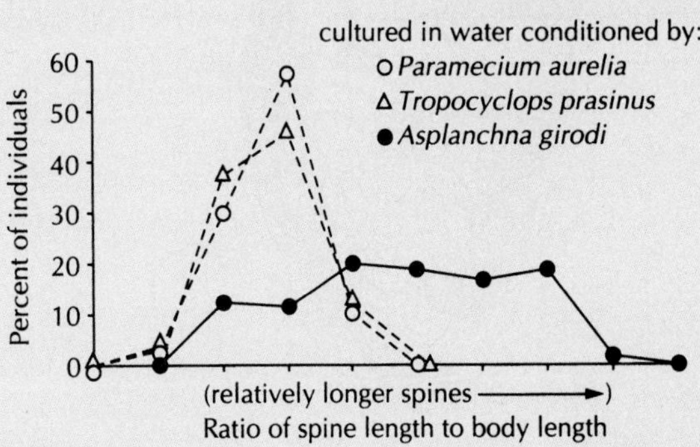

Focus Figure 10.2

The effect of water conditioned by a protozoan and 2 predators on the length of the right posterior spine in *Keratella slacki*.

Using a dissecting microscope to directly observe predator-prey interactions, the researchers confirmed these results: The predator-induced, large-spined morphs were captured and eaten significantly less often by *Asplanchna*, even though they were attacked just as often as the short-spined individuals.

As usual, the results of one scientific study trigger additional questions, leading to further studies. In particular, if having spines is such an advantage to *K. slacki*, why does it develop spines only in the presence of predaceous rotifers? Why are long spines and larger bodies not genetically programmed for all individuals? Gilbert and Stemberger have gone on to address this and related issues in subsequent papers. Can you envision how they might have designed their experiments?

Focus Table 10.1 The Influence of Body Size and Morphology on the Susceptibility of the Herbivous Rotifer *Keratella slacki* to Predation by *Asplanchna girodi*

Experiment	Percent Eaten Short-Spined	Long-Spined	Original Number of Each Type	Differences Statistically Significant?
1	40	10	10	No
2	62	0	8	Yes
3	50	20	18–20	Yes
All 3 combined	50	12	36–38	Yes

Note: The predaceous rotifers were allowed to feed for about 6 hours; the extent of predation on short- and long-spined rotifers was then assessed by the disappearance of individuals from containers and confirmed by finding remains of prey in predator guts or on the bottoms of the dishes. A statistically significant result is one that would occur by chance alone in fewer than 5 trials out of 100.

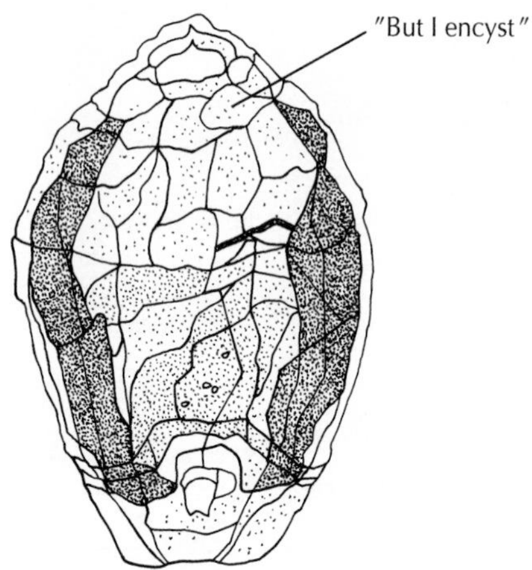

Figure 10.12
Philodina roseola, encysted. The adult is illustrated in Figure 10.11(b).
From Hyman; after Hickernell.

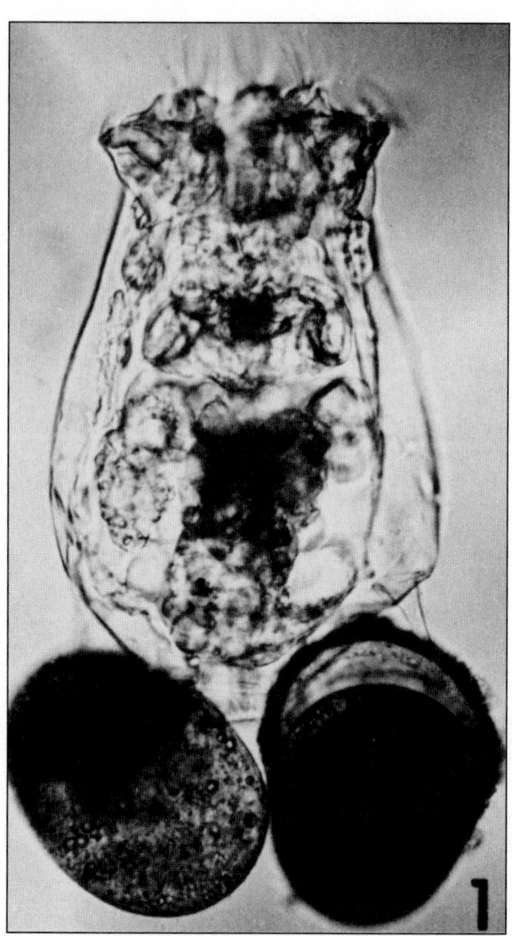

Figure 10.13
Photograph of female *Brachionus calyciflorus* with mictic resting eggs.
Courtesy John J. Gilbert. From Wurdak, Gilbert, and Jagels, 1978. *Trans. Amer. Microsc. Soc.* 97:49.

their eggs are haploid (Fig. 10.13). In the absence of fertilization, mictic eggs develop into haploid males. The males are usually smaller and morphologically dissimilar to the females of the same species, but they are always fast swimmers. The males are nonfeeding, lacking both mouth and anus. Their job must be done quickly, since they usually cannot survive for more than a few days. Typically, the males are ready to fertilize eggs within an hour of hatching! In at least one species, male rotifers can apparently use chemical cues to locate eggs containing females on the verge of hatching; the males then guard those eggs and mate with the females as soon as they emerge.[5] The fertilized haploid eggs form **resting eggs,** also known as **winter** or **diapause eggs** (Fig. 10.14). These resting eggs are highly resistant to a variety of physical and chemical stresses, permitting the developing embryos to withstand unfavorable conditions. During such periods, the embryos are in a state of developmental arrest—a state of **diapause** or **cryptobiosis.** Hatching occurs after conditions improve; some specimens have successfully hatched after 40 years of dormancy. Resting eggs always give rise to amictic females, which go on to reproduce by means of parthenogenesis (Fig. 10.15). Generally, only 1 or 2 mictic generations occur per year.[6]

Recently, researchers have discovered that a small percentage (< 0.5%) of females in a population may be characterized as neither mictic nor amictic. Instead, these females are **amphoteric;** a single female produces some haploid eggs by meiosis (which become males if unfertilized) and also produces some diploid eggs by mitosis.

It should be clear by now that most free-living rotifers encountered in the field, whether the rotifers are mobile or sessile, are females. Haplo-diploid sex determination is also discussed in Chapter 14 (pp. 341–419).

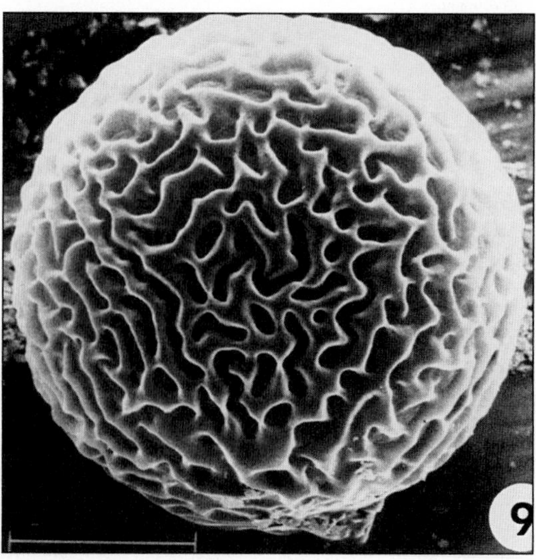

Figure 10.14
Scanning electron micrograph of resting egg of *Asplanchna intermedia.*
Courtesy John Gilbert. From Gilbert and Wurdack, 1978. *Trans. Amer. Microsc. Soc.* 97:330.

5. Schröder, T. 2003. *Proc. R. Soc.* London. 270:1965–70.

6. See *Topics for Further Discussion and Investigation,* no. 2.

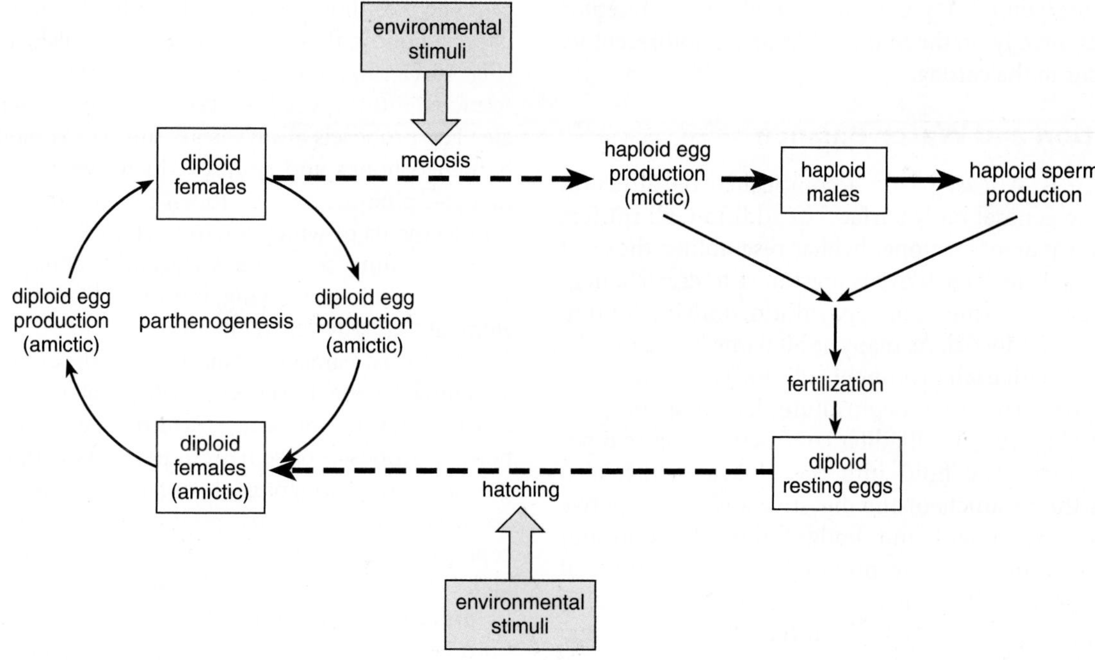

Figure 10.15

The life cycle of monogonont rotifers. In the absence of specific stimuli, reproduction is exclusively asexual, by means of parthenogenesis. Environmental factors that may influence reproduction include changes in food quality and quantity, changes in photoperiod, and changes in temperature. On the right side of the diagram, note that haploid eggs produce (haploid) males only in the absence of fertilization. Diploid eggs always produce diploid females, without additional fertilization.

Other Features of Rotifer Biology

Digestive System

Most rotifer species have a tubular digestive system with an anterior mouth and a posterior anus (Figs. 10.4 and 10.11). Cilia lining the inner surface of the gut move food through the digestive system. Digestion is largely extracellular and takes place in the stomach. **Gastric glands,** or **gastric caeca,** associated with the stomach, contribute digestive enzymes. Undigested wastes pass through a short intestine and discharge into a **cloaca** (Latin: a sewer), which, by definition, also receives the terminal ducts of the excretory and reproductive systems. The anus opens dorsally, near the junction of the trunk and the foot. Some bdelloid rotifers have no pronounced stomach cavity and no anus. Instead, the "stomach" is a continuous syncytial mass through which food is circulated. In such species, digestion is primarily intracellular. Except for members of the Seisonidea, male rotifers lack a functional digestive system.

Nervous and Sensory Systems

The rotifer brain consists of a bilobed mass of ganglia lying dorsal to the mastax (Fig. 10.16). Nerves extend throughout the body, connecting the brain to the musculature and organ systems and to a variety of sensory receptors. Sensory bristles and, usually, 3 antennae (2 lateral and one median-dorsal) serve as chemoreceptors and

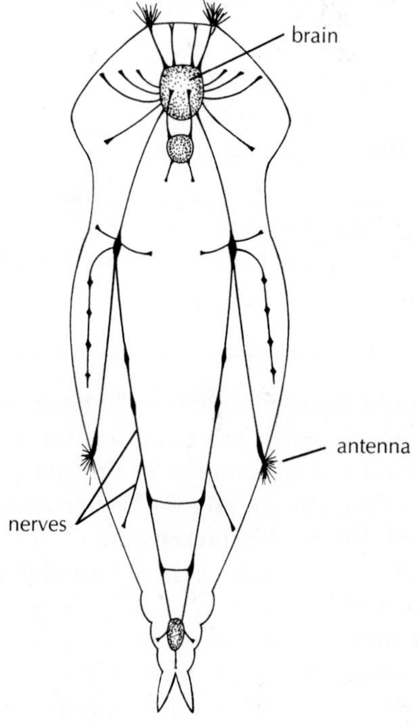

Figure 10.16

Rotifer nervous system.

From Pennak, *Fresh-Water Invertebrates of the United States*, 2d ed. Copyright © 1978 John Wiley & Sons, New York. Reprinted by permission of John Wiley & Sons, Inc.

mechanoreceptors. A pigmented, cuplike photoreceptor often lies directly on the brain. Additional photoreceptors may occur in the corona.

Excretion and Water Balance

Excretion is at least partly accomplished by diffusion across the general body surface. In addition, all rotifers contain a pair of protonephridia, resembling those of flatworms. Flagella activity is presumed to create a negative pressure within each nephridium, drawing fluid in from the pseudocoel. As many as 50 flame bulbs may be associated with each protonephridium. The 2 collecting tubules, one from each nephridium, lead to a common **bladder** (Fig. 10.11b). Bladder contractions, up to 6 per minute, expel the fluid into the **cloaca,** which also receives the products of the digestive and reproductive systems. The tissues and body fluids of freshwater rotifers contain a significantly higher concentration of dissolved materials than is found in the surrounding medium, so water continually diffuses into the animal across the permeable body surface. The major role of the protonephridia would thus seem to be the maintenance of water balance and body volume, rather than the removal of excretion products, at least in most species. The osmotic gradient that exists across the body wall of freshwater rotifers, and the accompanying diffusional influx of water, may aid the rotifer in maintaining body turgor.

Phylum Acanthocephala

Phylum Acantho • cephala
(G: spine- or thorn-head)
ah-can-thō-sef′-ah-lah

Defining Characteristics: 1) 1–2 large, acellular, collagenous sacs (ligament sacs) in the pseudocoel, supporting the gonads; 2) adults with proboscis containing intracellular hooks

The phylum Acanthocephala includes fewer than 1200 species, all of which are gut parasites in vertebrates. They are especially common in the small intestines of various fishes, particularly freshwater species, but also infect birds (including chickens and turkeys), mammals, and to a lesser extent, reptiles and amphibians. A single host may harbor hundreds of individual acanthocephalans. Arthropods are often required as intermediate hosts to complete the life cycle. Like most other aschelminths, acanthocephalans are typically cylindrical and generally small (a few millimeters to a few centimeters long), show a constant number of cells **(eutely),** lack any specialized respiratory structures, and exhibit a large pseudocoel.

The acanthocephalan body is divided into 3 distinct parts: a hollow, fluid-filled proboscis; a neck; and a trunk (Fig. 10.17). In all but a few species, the proboscis can be retracted into a specialized pouch, the **proboscis receptacle.** The proboscis always bears numerous biochemically hardened hooks and spines (which give the phylum its name—spine-head, or thorn-head), the sizes and arrangements of which are important in species identification. Calling someone an acanthocephalan makes a splendid insult—satisfying, but more likely to cause puzzlement than physical retaliation.

Intestinal parasites invariably require some means of attaching to the host's gut wall to avoid being swept downstream to the anus. Acanthocephalans use their hooked proboscis to remain in place. As with most other parasites, the nongonadal organ systems are greatly reduced; the capacious pseudocoel of the acanthocephalan trunk contains mostly gonad and associated glands (Fig. 10.17). Acanthocephalans are **gonochoristic** (= **dioecious),** with separate male and female sexes. In both sexes, the gonads are contained within 1 or 2 thin-walled **ligament sacs** that extend from the posterior end of the proboscis receptacle to near the genital pore, essentially subdividing the pseudocoel. Males bear conspicuous **cement glands** (Fig. 10.17a), whose cement hardens after copulation, securely plugging the female's vagina. Mature females have no well-defined ovary. Instead, the ovary fragments as it matures, forming numerous **ovarian balls** that float freely in the ligament sac fluid. Mature eggs enter the uterus by way of a posterior **uterine bell** (Fig. 10.17b) that somehow prevents immature eggs from leaving the ligament sac(s).

Following copulation and internal fertilization, the fertilized eggs develop within the female's pseudocoel to an advanced, differentiated state—the **acanthor** stage (Fig. 10.18). An individual female may release several hundred thousand acanthors daily, which soon exit the host with host fecal matter, encased in protective shells.

The acanthor can develop further only within certain invertebrate hosts; it cannot infect another vertebrate host directly. If the shelled embryo is eaten by the proper intermediate host—usually particular insects for terrestrial life cycles and particular crustaceans for aquatic life cycles—the acanthor emerges from its shell, bores through the gut tissue and into the blood sinus (hemocoel) of the arthropod, and continues to develop. Adulthood is reached only if the intermediate host is eaten by a suitable vertebrate, but this need not occur directly: The acanthocephalan can typically pass through one or more **transport hosts** before eventually arriving in the final, definitive host. For many acanthocephalan species, these transport hosts, in which the parasite cannot develop further, are essential links in the life history; the definitive host often feeds not on arthropods directly, but only on small fish, snails, or other transport hosts that themselves feed on arthropods (Fig. 10.18).

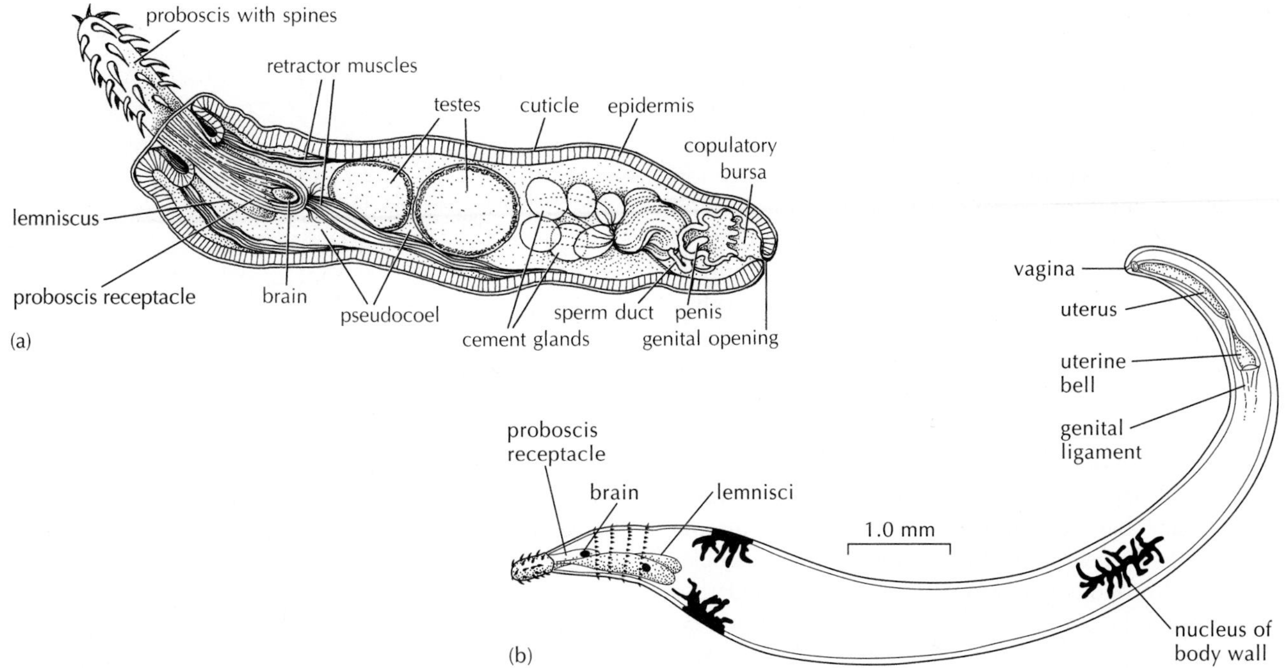

Figure 10.17

(a) A male acanthocephalan (*Acanthocephalus* sp.). The copulatory bursa is eversible and holds the female during copulation, using secretions from the cement glands. (b) Female of a different species, *Quadrigyrus nickolii.* The function of the paired lemnisci, found in both sexes, is not known. Note the brain near the posterior end of the proboscis receptacle.

(a) From Hyman; after Yamaguti.

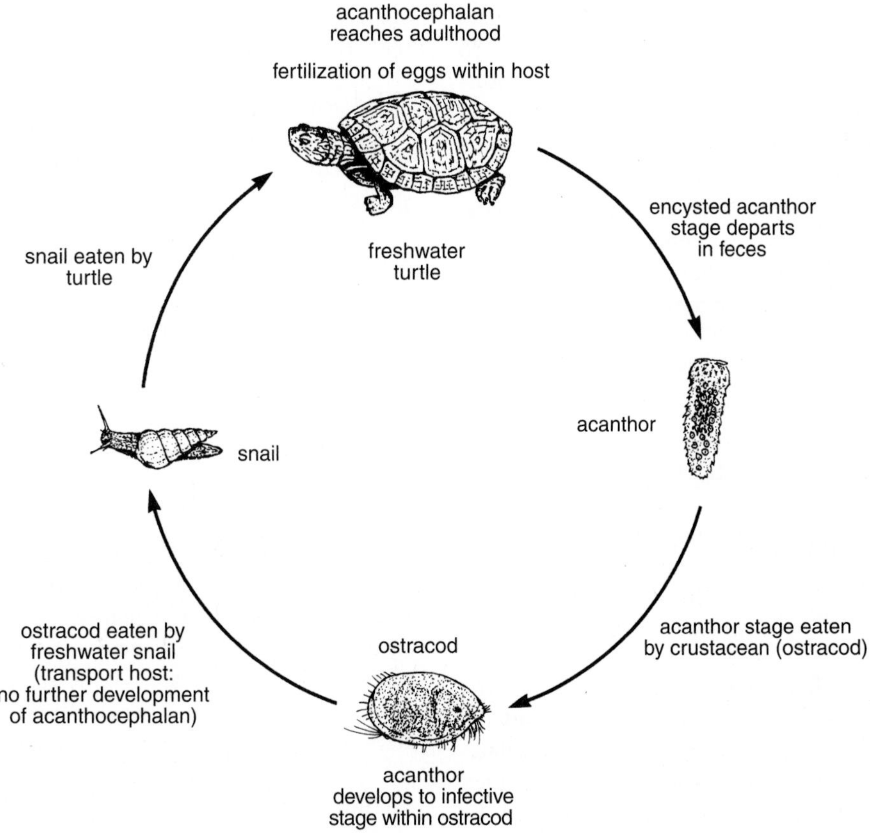

Figure 10.18

Life cycle of an acanthocephalan, *Neoechinorhynchus emydis.* The transport host is essential for completing the cycle in this species.

Throughout the life cycle, acanthocephalans must subsist on solubilized nutrients absorbed from the various hosts; acanthocephalans display no trace of a digestive tract at any stage of development. This is a common correlate of parasitic existence, along with the general reduction of sensory and many other organ systems.

The pronounced adaptations shown by acanthocephalans for the endoparasitic lifestyle leave few morphological clues about their evolutionary history. Fossil evidence suggests a close relationship to a small phylum of free-living worms called "priapulids" (Chapter 17); several species of fossilized priapulids recovered from the Burgess shale (formed about 540 million years ago) show apparent acanthocephalan morphologies. This suggests that acanthocephalans have descended from free-living, marine mud dwellers, possibly developing their present association with arthropods as an adaptation to intense predation by arthropods long ago: Parasitize or perish. Their present life histories would have evolved later, after the evolutionary appearance of vertebrates. On the other hand, and as reflected in their placement within this chapter, there is growing evidence that acanthocephalans are more closely related to rotifers. Although members of most acanthocephalan species lack a specialized excretory system, a system of protonephridia is found in some species, as in rotifers (and flatworms). Moreover, the acanthocephalan epidermis is syncytial and shows the same peculiar intracellular stiffening network of protein fibers found elsewhere only among the rotifers. Comparisons of 18S ribosomal RNA gene sequences support a close evolutionary relationship between Acanthocephala and Rotifera. Some biologists view the acanthocephalans as descendants of rotifers (probably bdelloid rotifers) and therefore place the 2 groups within a single phylum, the Syndermata. Other analyses present the Acanthocephala and Rotifera as sister groups, with the members having evolved in separate directions from a common ancestor that was neither a rotifer nor an acanthocephalan.[7] Over the next few years, molecular studies including more species and different gene sequences may resolve this issue.

Phylum Gnathostomulida

Phylum Gnatho • stomulida
(Greek: jaw mouth)
nath-ō-stom-ū-lē´-dah

The gnathostomulids (Fig. 10.19) resemble the turbellarian flatworms discussed in Chapter 8; indeed, they were originally described in 1956 as odd turbellarians. Since then, they have been provisionally associated with gastrotrichs (p. 459), annelids (Chapter 13), and the planula larvae of cnidarians (Chapter 6). Within the last 10 years or so, though, they have become convincingly allied with rotifers, through a combination of molecular analyses and ultrastructural studies.

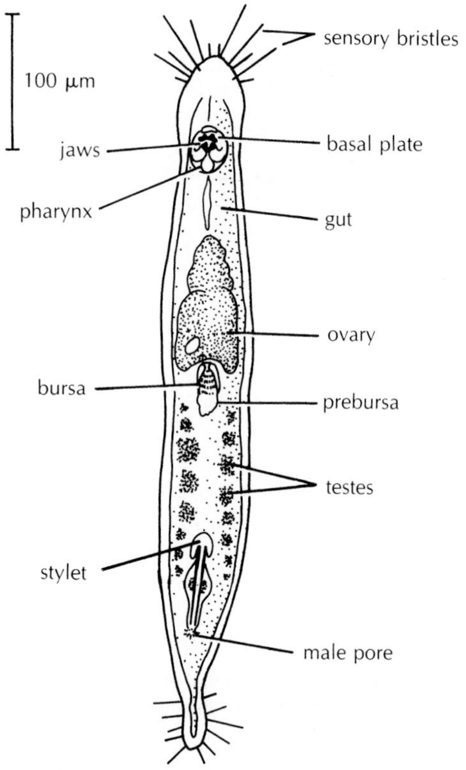

Figure 10.19
A gnathostomulid, *Gnathostomulida jenneri.* Most individuals are less than 1 mm long.
From Sterrer, 1972. *Syst. Zool.* 21:151.

Members of the approximately 80 gnathostomulid species described so far are all very small, soft-bodied, mostly interstitial worms living in the spaces between sand grains. Most individuals are less than 1 mm long, but worm concentrations as high as 6000 individuals per kilogram of sand have been reported, with the greatest worm aggregations occurring in deeper sediment layers high in hydrogen sulfide. Like turbellarian flatworms, gnathostomulids are acoelomate; are externally ciliated (but with only one cilium per cell, rather than the several per cell characterizing flatworms and most other spirally cleaving animals); are triploblastic; and have a mouth but no functional anus. Moreover, they lack specialized circulatory or respiratory systems, and their excretory organs are simple protonephridia. The gnathostomulids' most complex structures are the mouthparts, consisting of a pair of hardened jaws and an associated basal plate; the jaws, which are used to scrape bacteria and fungi from sand grains and other substrates, are now believed to have an origin in common with rotifer trophi.

Most gnathostomulid species are simultaneous hermaphrodites, and the eggs are fertilized internally and then released. As with other protostomes cleavage is spiral. However, gnathostomulid zygotes always develop directly into small worms: No free-swimming larvae are known.

Gnathostomulids are almost alone among spirally cleaving animals in having a monociliated epidermis, with a

7. See *Topics for Further Discussion and Investigation*, no. 5.

single cilium per cell; in this respect they more closely resemble the planula larvae of cnidarians, as alluded to earlier.

Phylum Micrognathozoa

In the year 2000, a new group of microscopic worms, the Micrognathozoa, was described from a single species *(Limnognathia maerski),* whose members are only about 140 μm long. No other species have yet been described. As illustrated (Fig. 10.20), these animals have a two-part head containing a highly complex series of jaws (Fig. 10.20b). The animals locomote by ciliary activity, using a double-row of ciliated cells. Near the animal's rear, on the ventral surface, there is an adhesive ciliary pad that secretes a sticky glue (Fig. 10.20a). The pad is quite unlike the adhesive toes of rotifers and may be unique to micrognathozoans. Little is yet known about the ecology, physiology, or behavior of these animals. So far only females have been found, suggesting that micrognathozoans reproduce by parthenogenesis, as do many rotifers.

Taxonomic Summary

Phylum Rotifera
- Class Seisonidea
- Class Bdelloidea
- Class Monogononta

Phylum Acanthocephala
Phylum Gnathostomulida
Phylum Micrognathozoa

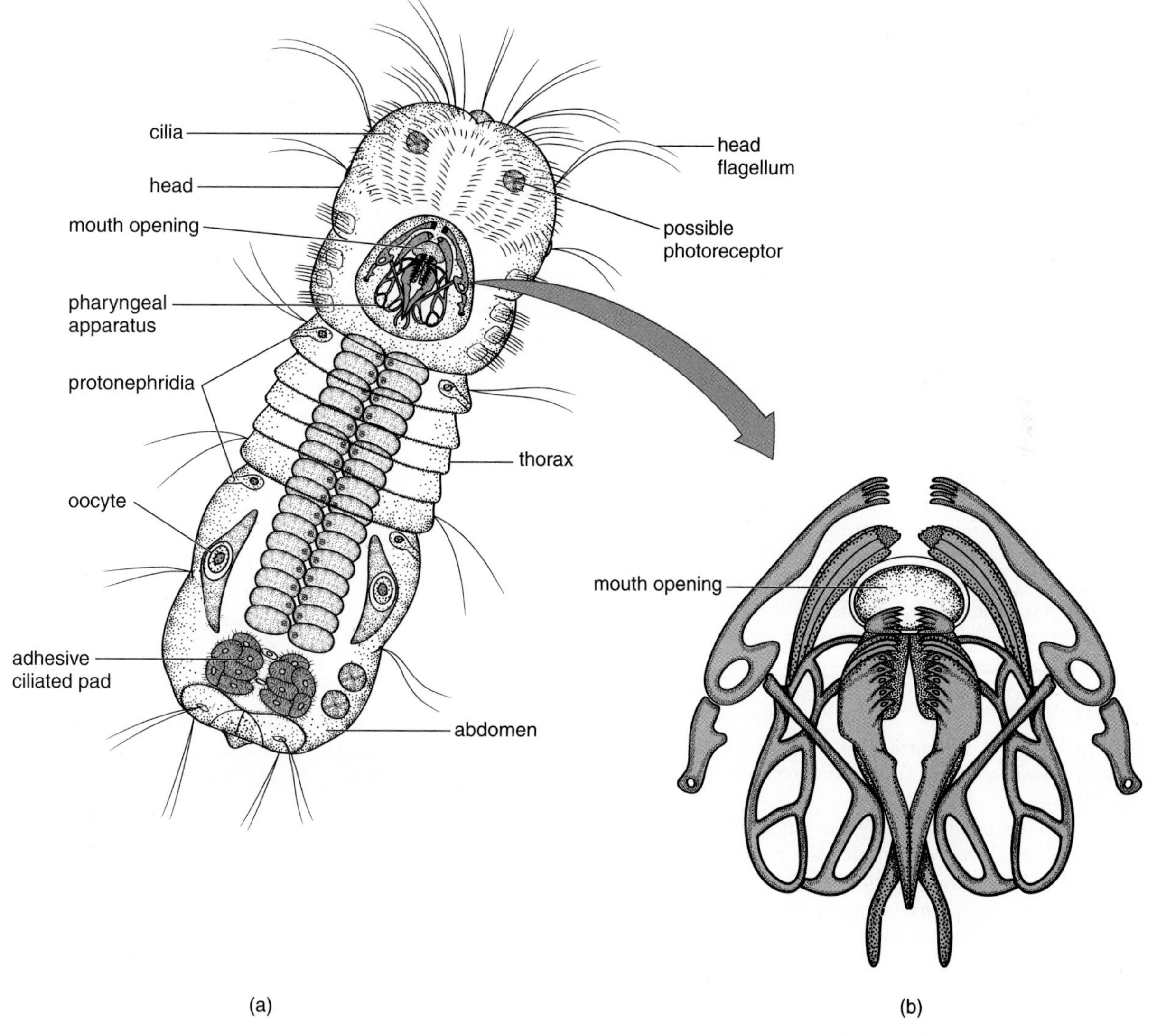

Figure 10.20
The micrognathozoan *Limnognathia maerski* (ventral view), the only described member of the phylum Micrognathozoa. The individual shown has a body length of approximately 120 μm. (a) Note the adhesive ciliary pad (10 cells), and the remarkably complex jaw structure in the pharynx, shown in greater detail in (b). Micrognathozoa: A new class with complicated jaws like those of Rotifera and Gnathostomulida.
From Kristensen, R.M., and P. Funch. 2000. *J. Morphol.* 246: 1–49.

Topics for Further Discussion and Investigation

1. Investigate the environmental and internal factors that appear to trigger the production of mictic females in monogonont rotifers.

Birky, C. W., Jr. 1964. Studies on the physiology and genetics of the rotifer, *Asplanchna*. I. Methods and physiology. *J. Exp. Zool.* 155:273.

Gallardo, W. G., A. Hagiwara, and T. W. Snell. 2000. Effect of juvenile hormone and serotonin (5-HT) on mixis induction of the rotifer *Brachionus plicatilus* Muller. *J. Exp. Marine Biol. Ecol.* 252:97.

Gilbert, J. J. 1963. Mictic-female production in the rotifer *Brachionus calyciflorus*. *J. Exp. Zool.* 153:113.

Gilbert, J. J. 2003. Specificity of crowding response that induces sexuality in the rotifer *Brachionus*. *Limnol. Oceanogr.* 48:1297.

Gilbert, J. J., and J. R. Litton, Jr. 1978. Sexual reproduction in the rotifer *Asplanchna girodi:* Effects of tocopherol and population density. *J. Exp. Zool.* 204:113.

Lubzens, E., and G. Minkoff. 1988. Influence of the age of algae fed to rotifers (*Brachionus plicatilis* O. F. Müller) on the expression of mixis in their progenies. *Oecologia (Berl.)* 75:430.

Snell, T. W., and E. M. Boyer. 1988. Thresholds for mictic female production in the rotifer *Brachionus plicatilis* (Müller). *J. Exp. Marine Biol. Ecol.* 124:73.

2. Discuss the physical and biological factors influencing the dynamics of natural rotifer populations.

Bosque, T., R. Hernández, R. Pérez, R. Todolí, and R. Oltra. 2001. Effects of salinity, temperature and food level on the demographic characteristics of the seawater rotifer *Synchaeta littoralis* Rousselet. *J. Exp. Marine Biol. Ecol.* 258:55.

Edmondson, W. T. 1945. Ecological studies of sessile Rotatoria. II. Dynamics of populations and social structures. *Ecol. Monog.* 15:141.

Gilbert, J. J. 1989. The effect of *Daphnia* interference on a natural rotifer and ciliate community: Short-term bottle experiments. *Limnol. Oceanogr.* 34:606.

Gilbert, J. J., and C. E. Williamson. 1978. Predator-prey behavior and its effect on rotifer survival in associations of *Mesocyclops edax, Asplanchna girodi, Polyarthra vulgaris,* and *Keratella cochlearis. Oecologia (Berl.)* 37:13.

King, C. E. 1972. Adaptation of rotifers to seasonal variation. *Ecology* 53:408.

Ricci, C., L. Vaghi, and M. L. Manzini. 1987. Desiccation of rotifers (*Macrotrachela quadricornifera*): Survival and reproduction. *Ecology* 68:1488.

Snell, T. W., R. Rico-Martinez, L. N. Kelly, and T. E. Battle. 1995. Identification of a sex pheromone from a rotifer. *Marine Biol.* 123:347.

Wallace, R. L., and W. T. Edmondson. 1986. Mechanism and adaptive significance of substrate selection by a sessile rotifer. *Ecology* 67:314.

3. In what respects does the life cycle of acanthocephalans resemble that of parasitic flatworms (phylum Platyhelminthes, Chapter 8)?

Nicholas, W. L. 1973. The biology of the Acanthocephala. *Adv. Parasitol.* 11:671.

Van Cleve, J., II. 1941. Relationships of the Acanthocephala. *Amer. Nat.* 75:31.

4. Investigate the influence of acanthocephalan infection on the crustacean host's tolerance of environmental pollutants.

Brown, A. F., and D. Pascoe. 1989. Parasitism and host sensitivity to cadmium: An acanthocephalan infection of the freshwater amphipod *Gammarus pulex*. *J. App. Ecol.* 26:473.

5. What is the evidence for and against the proposition that acanthocephalans are modified rotifers?

Garey, J. R., T. J. Near, M. R. Nonnemacher, and S. A. Nadler. 1996. Molecular evidence for Acanthocephala as a subtaxon of Rotifera. *J. Molec. Evol.* 43:287.

Herlyn, H., O. Piskurek, J. Schmitz, U. Ehlers, and H. Zischler. 2003. The syndermatan phylogeny and the evolution of acanthocephalan endoparasitism as inferred from 18S rDNA sequences. *Molec. Phylog. Evol.* 26:155–64.

Morris, S. C., and D. W. Crompton. 1982. The origins and evolution of the Acanthocephala. *Biol. Rev.* 57:85.

Near, T. J. 2002. Acanthocephalan phylogeny and the evolution of parasitism. *Integr. Comp. Biol.* 42:668.

Nielsen, C. 1995. *Animal Evolution: Interrelationships of the Living Phyla.* New York: Oxford University Press, pp. 248–53.

Sørensen, M. V., and G. Giribet. 2006. A modern approach to rotiferan phylogeny: Combining morphological and molecular data. *Molec. Phylog. Evol.* 40:61–72.

Welch, D. B. M. 2000. Evidence from a protein-coding gene that acanthocephalans are rotifers. *Invert. Biol.* 119:17.

6. How convincing is the evidence that bdelloid rotifers evolved millions of years ago from ancestors that lost the ability to reproduce sexually?

Welch, D. M., and M. Meselson. 2000. Evidence for the evolution of bdelloid rotifers without sexual reproduction or genetic exchange. *Science* 288:1211–15.

Welch, J. L. M., D. B. M. Welch, and M. Meselson. 2004. Cytogenetic evidence for asexual evolution of bdelloid rotifers. *Proc. Nat. Acad. Sci. U.S.A.* 101:1618–21.

Taxonomic Detail

Superphylum Gnathifera

The members of this group are united by molecular similarities and by their possessing jaws of unique structure.

Phylum Rotifera

The approximately 1850 species are divided among 3 classes.

Class Monogononta

This class contains at least 80% of all rotifer species. Three orders.

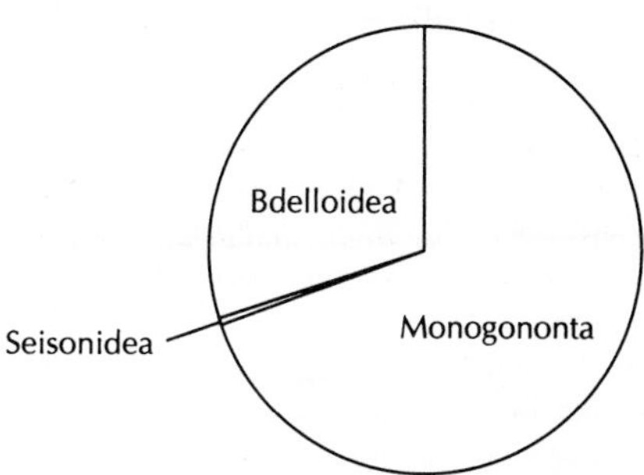

Order Ploima

Most rotifer species are contained in this order. Under most conditions, the populations consist of amictic females reproducing parthenogenetically, but mictic eggs are produced under certain environmental conditions. Most species live in freshwater, but a few species are marine. Fourteen families.

Family Brachionidae. *Brachionus, Keratella, Epiphanes, Lepadella.* Individuals are mostly planktonic, although the family also includes some benthic (bottom-dwelling) species, such as *Lepadella.*

Family Asplanchnidae. *Asplanchna.* All species are planktonic predators, often preying on each other. Individuals get as long as about 2 mm.

Family Lecanidae. *Lecane. Lecane* is one of the most species-rich of all rotifer genera. All species live in freshwater.

Family Synchaetidae. *Synchaeta, Polyarthra.* These rotifers are common in both marine and freshwater habitats. Most of the truly planktonic marine rotifers are found in this family.

Family Proalidae. *Proales.* A number of species are internal parasites of animals (living in such bizarre hosts as snail eggs or certain protozoans) or plants, while others live in symbiotic association with such diverse animals as colonial protozoans, hydrozoans, crustaceans, or insect larvae.

Order Flosculariaceae

Four families

Family Flosculariidae. *Floscularia.* This group includes some common benthic rotifers that build tubes on freshwater plants, using pellets formed from detritus and carefully cemented into place. Some species, including *Sinantherina,* form colonies in a common gelatinous mass, which may be attached to substrates or free-floating in the water.

Order Collothecaceae

One family

Family Collothecidae. *Collotheca, Stephanoceros.* Most species in this family are sessile and live attached to solid freshwater substrates, encased in gelatinous secretions. The anus empties anteriorly, on the neck, as an adaptation to living as a tube dweller.

Class Bdelloidea

Members of this group occupy a wide range of habitats, including damp soil, wet moss, hot springs, and Antarctic lakes, and a few species live in the sea. Bdelloid rotifers are extremely numerous at the bottoms of freshwater lakes and ponds worldwide. The life cycles seem entirely amictic; reproduction is exclusively by parthenogenesis, and males are unknown. Individuals can tolerate a remarkable range of temperatures and long periods of desiccation. About 370 species, distributed among four families.

Family Philodinidae. *Philodina, Rotaria, Zelinkiella.* Most species are free-living in freshwater and moist terrestrial environments, but a few species (e.g., *Embata*) are parasitic on the gills of certain crustaceans (phylum Arthropoda). *Zelinkiella* lives only in skin depressions on the bodies of sea cucumbers (Echinodermata: Holothuroidea).

Class Seisonidea

One family, containing only 2 species.

Family Seisonidae. *Seison.* These rotifers are all marine symbionts. Instead of a foot with toes, the animal's posterior end terminates in a leech-like adhesive disc. *Seison* lives exclusively on the gills of a European crustacean. The animals are typically 2–3 mm long (large, for rotifers), but they have a highly reduced corona. Males and females are equally abundant and equally well developed.

Phylum Acanthocephala

The approximately 1150 described species in this phylum are distributed among 3 classes.

Class Archiacanthocephala

The only acanthocephalans with specialized excretory systems are contained in this class. All species parasitize birds and mammals, with insects, centipedes, and millipedes serving as intermediate hosts. The class includes *Macracanthorhynchus hirudinaceus,* one of the largest acanthocephalans and the only one of economic significance; adults almost always parasitize swine, although human infestations have been reported occasionally. Females are 3 to 4 times larger than the males, and they reach lengths of about 70 cm. The larvae develop within developing scarabid beetles that ingest swine feces, and the swine become infected by eating the beetle grubs or adults. Four families.

Class Eoacanthocephala
Neoechinorhynchus (= *Neorhynchus*). Most species parasitize fish. Crustaceans serve most often as intermediate hosts. Three families.

Class Palaeacanthocephala
This class contains most acanthocephalan species. These acanthocephalans parasitize a wide range of vertebrate hosts: Definitive hosts include fish (including bass), amphibians, reptiles, birds, and mammals. Crustaceans usually serve as intermediate hosts. Some species parasitize turkeys and chickens, using terrestrial isopods as obligate intermediate hosts. Some species can alter host behavior in ways that increase the likelihood of capture by the definitive host. Thirteen families.

Phylum Gnathostomulida

Haplognathia, Gnathostomaria. The 80 or so species in this phylum are distributed among only 2 orders, based largely on differences in reproductive anatomy and sperm morphology. These tiny hermaphroditic worms (< about 4 mm) live interstitially, in the spaces between sand grains. The first members were discovered in 1956, when I was only 6 years old. All species are marine.

Order Filospermoidea
Haplognathia. Two families.

Order Bursovaginoidea
Agnathiella, Gnathostomula. This order contains about three-fourths of all gnathostomulid species. The reproductive anatomy is more complex for species in this order, and males have a stylet-bearing penis, which injects sperm through the body wall of females. Many members of this order can reverse the direction of ciliary beating, thereby swimming backward. Eight families, some of which contain a single species.

Phylum Micrognathozoa

This group currently contains but a single species, *Limnognathia maerski,* first described in 2000 from material collected from a cold spring in Greenland. Micrognathozoans probably have the most complex jaw system ever seen among microscopic animals.

General References about Rotifers and Other Gnathiferans

Harrison, F. W., and E. E. Ruppert, eds. 1991. *Microscopic Anatomy of Invertebrates, Volume 4: Aschelminthes.* New York: Wiley-Liss (Rotifers: 219–97; Acanthocephalans: 299–332).

Hyman, L. H. 1951. *The Invertebrates, Volume 3. Acanthocephala, Aschelminthes, and Entoprocta.* New York: McGraw-Hill.

Kristensen, R. M. 2002. An introduction to Loricifera, Cycliophora, and Micrognathozoa. *Integr. Comp. Biol.* 42:641–51.

Kristensen, R. M., and P. Funch. 2000. Micrognathozoa: A new class with complicated jaws like those of Rotifera and Gnathostomulida. *J. Morphol.* 246:1–49.

Morris, S. C., et al., eds. 1985. *The Origins and Relationships of Lower Invertebrates. Systematics Association, Special Volume 28.* Oxford: Clarendon Press.

Near, T. J. 2002. Acanthocephalan phylogeny and the evolution of parasitism. *Integr. Comp. Biol.* 42:668–77.

Nielsen, C. 2001. *Animal Evolution: Interrelationships of the Living Phyla,* 2nd ed. New York: Oxford University Press, 294–308.

Parker, S. P., ed. 1982. *Classification and Synopsis of Living Organisms,* vol. 1. New York: McGraw-Hill, 865–72, 933–40.

Ricci, C., and M. Caprioli. 2005. Anhydrobiosis in bdelloid species, populations and individuals. *Integr. Comp. Biol.* 45:759–63.

Roberts, L. S., J. Janovy, Jr., and P. Schmidt. 2004. *Foundations of Parasitology,* 7th ed. New York: McGraw-Hill.

Thorpe, J. H., and A. P. Covich, eds. 2001. *Ecology and Classification of North American Freshwater Invertebrates,* 2nd ed. New York: Academic Press, 187–248.

Wallace, R. L. 2002. Rotifers: Exquisite metazoans. *Integr. Comp. Biol.* 42:660–67.

Search the Web

1. www.ucmp.berkeley.edu/phyla/rotifera/rotifera.html

   This site is maintained by the University of California at Berkeley. It provides detailed information about rotifer biology, with photos and links to other relevant sites.

2. http://faculty.uml.edu/rhochberg/hochberglab/

   From Rick Hochberg's website (University of Massachusetts Lowell) you can access several videos, including one showing the rotifer mastax in action.

3. http://www.micrographia.com

   Choose "Specimen Galleries," then "Biological Specimens: Freshwater," and then "Rotifers" to see detailed information on rotifer biology along with exquisite photographs.

4. http://dmc.utep.edu/rotifer/

   This brings you to the Rotifer Systematic Database, sponsored by the Department of Biological Sciences, University of Texas at El Paso.

5. http://www.zmuc.dk/InverWeb/Dyr/Limnognathia/Limno_intro_UK.htm

   Learn about the new animal group, the Micrognathozoa, recently discovered on a field trip in Greenland.

11

The Nemertines

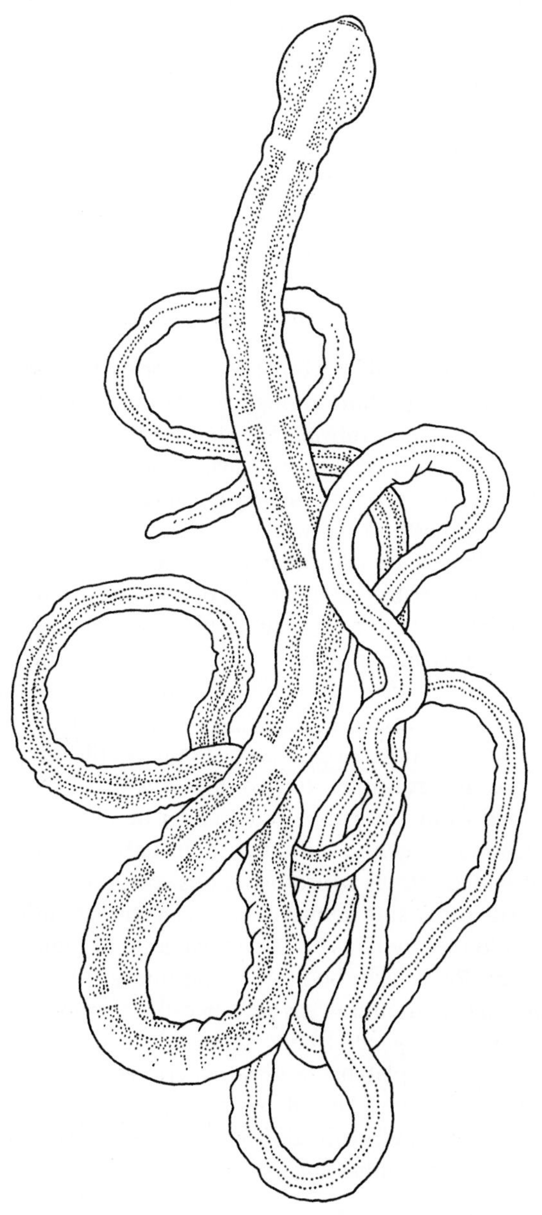

Introduction and General Characteristics

Phylum Nemertea (nem-er´-tē-ah)
(= phylum Rhynchocoela; rink´-ō-sēl´-ah)
(= phylum Nemertinea; nem-er-tē´-nē-ah)

Defining Characteristic:[1] Muscular eversible proboscis housed in a fluid-filled, schizocoelous cavity (the rhynchocoel)

The nemertines (also called "nemerteans") are a small group of elongated, unsegmented, soft-bodied worms. Most of the approximately 1150 described species in this phylum are marine. Nemertines are common inhabitants of marine shallow-water environments, crawling over solid substrates, burrowing into sediment, or lurking under stones, rocks, or mats of algae. Some other species (mostly in a single genus) are found in freshwater, and the members of several more small genera are terrestrial. The terrestrial species are most common in tropical, moist environments. The freshwater and terrestrial species most likely descended from marine ancestors. A variety of nemertine species live commensally with invertebrates from other phyla, particularly the Arthropoda and Mollusca, but only a few nemertine species are parasitic.

In several respects, the nemertines at least superficially resemble the free-living flatworms (phylum Platyhelminthes, class Turbellaria). Like the flatworms, nemertines are ciliated externally and secrete a mucus through which they progress. Small nemertines can move over substrates exclusively by means of ciliary beating. Larger individuals often generate waves of muscle contraction when moving over hard substrates or

1. Characteristics distinguishing the members of this phylum from members of other phyla.

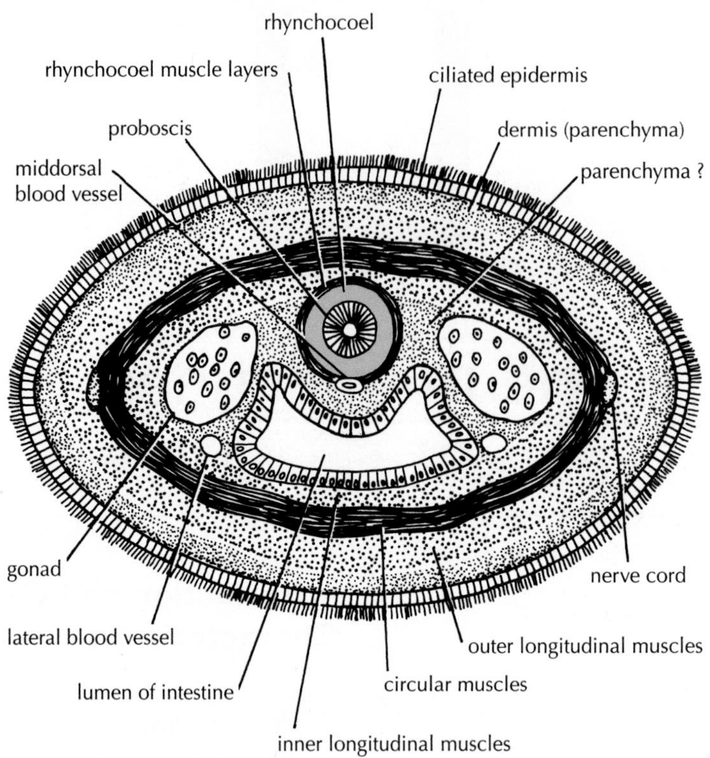

Figure 11.1
Diagrammatic cross section through the body of a nemertine worm. The functions of the proboscis and rhynchocoel are discussed later in the chapter.
After Harmer and Shipley.

through soft substrates. Members of a few species can swim, by generating relatively violent waves of muscular contraction.

As with the turbellarians, most nemertines are flattened dorsoventrally and possess circular, longitudinal, and dorsoventral muscles. In most nemertine species, the area lying between the outer body wall and the gut is occupied by extracellular matrix, but in at least a few species the space is apparently filled with mesodermally derived **parenchyma,** as in flatworms (Fig. 11.1). The nervous system of nemertines is also reminiscent of the turbellarian plan: The cerebral ganglia form a ring anteriorly and give rise to a ladder-like arrangement of longitudinal nerves with lateral connectives. Nemertines are equipped with chemoreceptors and mechanoreceptors, located in specialized pits and grooves on the body surface, and sensory bristles. Most species also possess pigmented photoreceptors ("eyes"), and a few species possess balance organs **(statocysts).** The number of eyes varies widely among species, from 0 to more than 80 per animal. As in the flatworms, nemertines generally have protonephridial excretory systems, and digestion is, as in the Platyhelminthes, largely intracellular. The larval stages also seem to share some key characteristics.

The body of most nemertine species is of considerable length, typically from several to 20 cm but in some cases as long as 30 meters. "Ribbon worm," the common name for members of this phylum, is especially appropriate for such elongated nemertines.

Nemertines differ from the flatworms in a number of respects, important enough to warrant placing these 2 groups in separate phyla. Indeed, molecular evidence and some developmental evidence suggests that flatworms and nemertines in fact have no recent ancestor in common and that nemertines are more closely related to coelomates than to flatworms (Research Focus Box 11.1). Although there are still serious differences of opinion on this issue, the weight of structural and molecular evidence currently locates nemertines as close relatives of annelids, molluscs, and some other coelomate animals within the "Lophotrochozoa" (see Figs. 2.10, 2.12c). Exactly how nemertines are related to other lophotrochozoans is still uncertain. There is, for example, some disagreement between data obtained from 18S rRNA genes and data obtained from mitochondrial genes,[2] although both data sets generally identify nemertines as protostomous coelomates.

Flatworms and nemertines differ conspicuously with respect to their systems for gas exchange, food capture, and digestion. Because the bodies of nemertines are flat and permeable, a fair amount of gas exchange likely occurs by diffusion across the general body surface. In addition, and in contrast to the platyhelminths, nemertines possess a true circulatory system. The blood circulates throughout the body through well-defined, contractile vessels (Fig. 11.2). Ultrastructural studies suggest that these vessels are highly

2. Turbeville, J. M., and D. M. Smith. 2007. *Molec. Phylog. Evol.* 43: 1056–65.

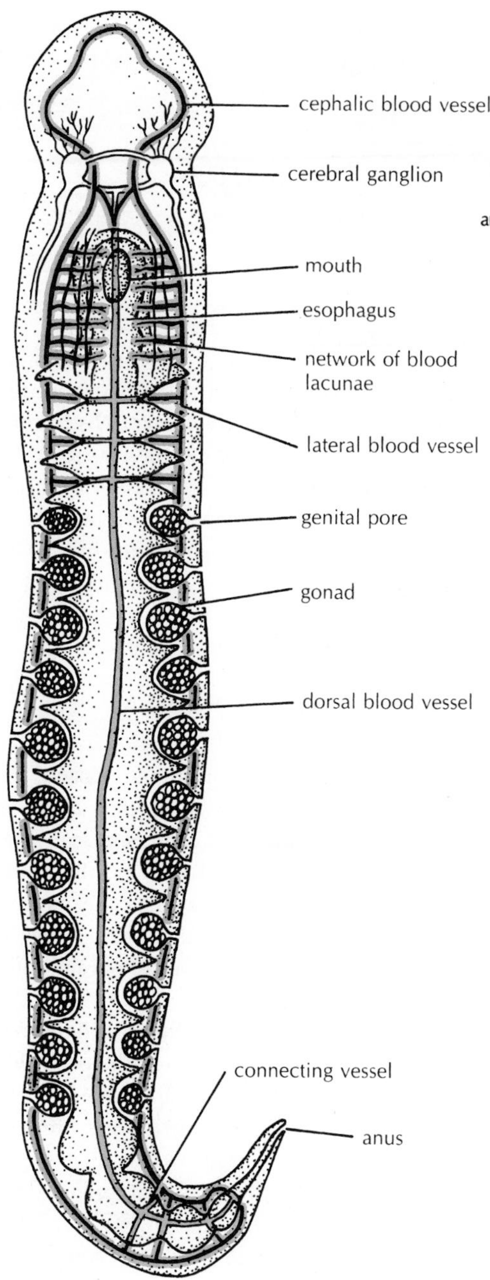

Figure 11.2
Schematic illustration of *Cerebratulus,* a common genus of ribbon worm. Note the well-developed circulatory system.
After Bayer and Owre; after Joubin.

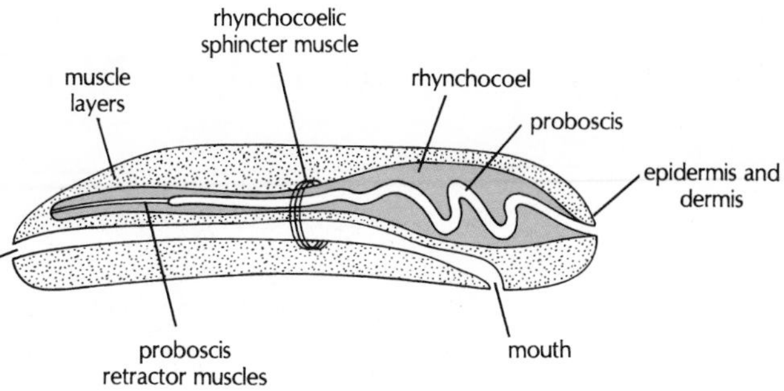

modified coelomic spaces,[3] providing more support for classifying nemertines as coelomates. There is no true heart, and the blood vessels lack one-way valves, so the blood does not circulate unidirectionally. Instead, it ebbs and flows erratically, propelled largely by muscular contractions associated with routine movements of the animal. A few species have hemoglobin in their blood, but most species lack any such oxygen-carrying blood pigment.

3. See *Topics for Further Discussion and Investigation,* no. 1, at the end of the chapter.

Figure 11.3
Diagrammatic longitudinal section through a nemertine, showing the tubular gut.
Modified from Turbeville and Ruppert, 1983, *Zoomorphol.* 103:103.

Unlike flatworms, ribbon worms have a one-way digestive tract, with an anterior mouth and a separate posterior anus (Fig. 11.3). The unidirectional flow of food material through this tubular gut permits an orderly digestive process: Digestive enzymes may be secreted in sequence as the food is moved through the gut by ciliary activity, and the animal can continue to eat without disrupting digestion. Because nutrients are absorbed only in a specialized posterior part of the gut, many body tissues are located far from the site of absorption. The nemertine circulatory system thus plays an important role in distributing nutrients throughout the body.

Most nemertines are meat eaters, with food preferences that are often highly specific. Nemertines seem to have a predilection for small annelids and crustaceans, and a given nemertine species often shows a preference for a particular species of prey.[4]

Unlike any flatworm, the ribbon worms possess a hollow, muscular **proboscis.** This structure is not homologous with the turbellarian pharynx, which is essentially a protrusible extension of the gut. In contrast, the nemertine proboscis is typically distinct from the digestive tract (Figs. 11.3 and 11.4b, c), floating in a separate, fluid-filled, tubular cavity called the **rhynchocoel** (Fig. 11.4a). The rhynchocoel forms during development as a split in the mesodermal tissue, and in this sense it is coelomic. Whether it is homologous with the coelomic cavities of annelids and other unquestionably coelomate protostomes is an open issue. The importance of the rhynchocoel as a defining characteristic of nemertines has led to an alternative name for the phylum, the Rhynchocoela (Greek: snout cavity).

The nemertine proboscis can be shot out with explosive force, something no turbellarian flatworm can do with its pharynx. What allows the nemertine proboscis to be extended with such rapidity?

The secret of proboscis functioning lies in the fluid-filled rhynchocoel. Although the nemertine proboscis is highly muscular, it is discharged not by its own muscular activity, but by the musculature of the tissue surrounding

4. See *Topics for Further Discussion and Investigation,* no. 2.

RESEARCH FOCUS BOX 11.1

Assessing Phylogenetic Relationships

Turbeville, J. M., K. G. Field, and R. A. Raff. 1992. Phylogenetic position of phylum Nemertini, inferred from 18S rRNA sequences: Molecular data as a test of morphological character homology. *Molec. Biol. Evol.* 9:235–49.

For many decades, the nemertine worms were closely allied with the platyhelminths as triploblastic acoelomates, the two being derived from a common ancestor. Two key characteristics that appear among nemertines but not flatworms—a fluid-filled rhynchocoel and a blood circulatory system—were believed to have evolved in nemertines after they diverged from their flatworm ancestors. Studies made by the senior author in the 1980s, however, confirmed suggestions made by German scientists many years earlier: (a) the rhynchocoel forms by a splitting of mesodermal tissue, as in the formation of coelomic cavities, and (b) the "blood vessels" have a continuous epithelial lining, musculature, and several other characteristics suggesting they are modified coelomic spaces.

But many biologists were not convinced that nemertines were true coelomates. At issue is the definition of coelom: Is it any cavity arising as a split in mesodermal tissue? Or must it specifically fill the space between the gut and body wall musculature? If we accept the latter, more restrictive definition, the nemertine rhynchocoel and blood vessels could be analogues, but not homologues, of the cavities found in annelids and other noncontroversial coelomates, with the structural resemblances among the cavities resulting from convergent evolution.

In this paper, James Turbeville and colleagues tested the hypothesis that nemertines are indeed closet coelomates, by comparing certain slow-evolving nucleotide sequences of 18S ribosomal RNA (rRNA) taken from the nemertine worm *Cerebratulus lacteus* with similar sequences documented previously from 9 other animals representing 6 other phyla. Three of the animals were flatworms (including one parasitic fluke species), 2 were annelids (including the common earthworm), 2 were other protostomous coelomates (1 chiton, 1 peanut worm—a sipunculan, see p. 314), and the last 2 were deuterostomous coelomates: a common sea star and a cephalochordate (a member of our own phylum, the Chordata, p. 539).

The researchers extracted all of the RNA from unfertilized nemertine eggs and then used 6 oligonucleotide primers to bind 6 particular regions of 18S rRNA. Using the enzyme reverse transcriptase, they then made DNA fragments complementary to the 6 primed regions of rRNA. The linear sequence of over 1000 nucleotide bases obtained from the 6 complementary DNA fragments (cDNA) was then compared, base by base, with those from the other 9 organisms. As discussed in Chapter 2, a greater degree of difference between sequences implies a longer period of independent evolution between the groups compared; the greater the similarities, the closer the evolutionary relationship between the groups being compared.

But how does one decide how close is close enough to implicate a closely shared evolutionary origin? How much difference does it take to establish that 2 groups are *not* closely related? And when differences in sequence patterns are detected, how does one tell which is the ancestral (plesiomorphic) situation and which is the more advanced (apomorphic) situation? In other words, how does one determine which groups are the ancestors and which are the descendants of other groups?

The answers to these questions are still controversial; as discussed in Chapter 2, establishing the polarity of evolutionary change is one of the key difficulties in phylogenetic analysis. To construct phylogenetic trees from their data, Turbeville et al. (1992) used 3 different computer techniques, including 2 techniques that provide a statistical indication of how likely it is that any given conclusion (i.e., phylogenetic tree) is in fact the correct conclusion for that particular data set. The analyses require comparison to comparable data from what is called an "outgroup," a phylum believed to be the sister group to all the other phyla being compared; this sister group is believed to share a common ancestor with all of the other phyla being studied. The outgroup provides a means of assessing the direction of evolutionary change in the gene sequences, with outgroup

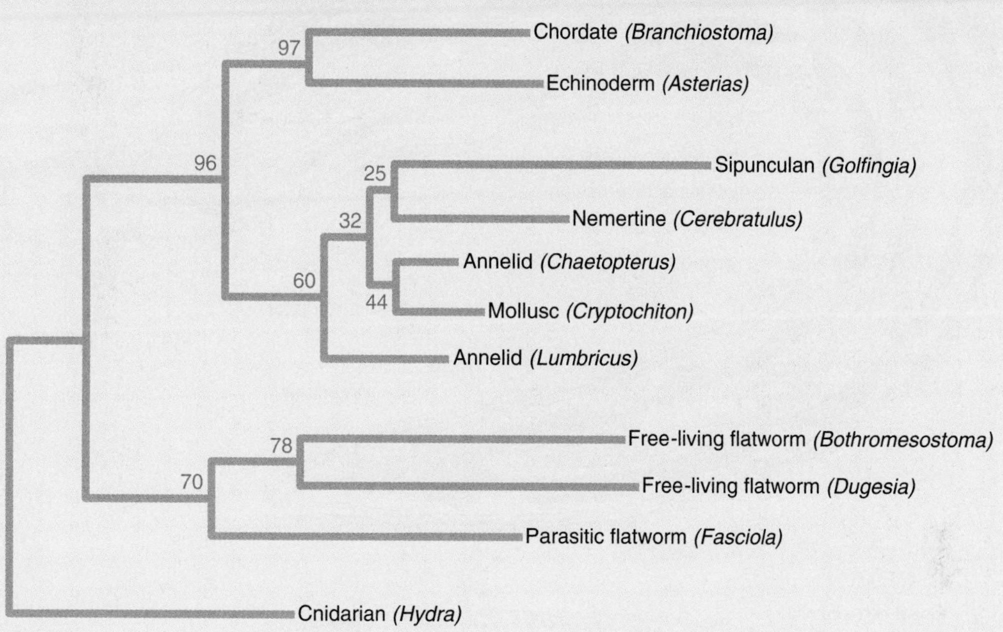

Focus Figure 11.1

One representative result of 18S rRNA nucleotide base sequence analysis. This particular analysis produces the phylogenetic tree requiring the smallest number of nucleotide base substitutions (874 to produce this particular tree). The number on each branch point was obtained by a statistical technique termed "bootstrapping," representing the degree of confidence in that particular branching pattern on a scale of 0–100.

Based on J. M. Turbeville et al. 1992. Phylogenetic position of phylum Nemertini, inferred from 18S rRNA sequences: Molecular data as a test of morphological character homology, in *Molecular Biology and Evolution* 9:235–49.

characteristics believed to be ancestral (plesiomorphic). For this study, the researchers chose a cnidarian (*Hydra*) for the outgroup comparison.

The results were clear and very similar for all 3 analyses. The phylogenetic tree resulting from one of the 3 analyses is shown in Focus Figure 11.1. It shows the 3 flatworm species clustering as a single cohesive acoelomate unit, the 2 deuterostome species clustering as a separate cohesive unit, and the nemertine species nestling comfortably with the protostomous coelomates. There is considerable uncertainty about exactly who is *most* closely related to the nemertine star of the show, but the grouping of nemertines with the protostomous coelomates is convincing. Do these results prove that nemertines are in fact coelomates? No, they do not, unless the traditional definition of "coelomic cavity" is altered. Could nemertines be degenerate coelomates?

How much do you think the results might differ if more representatives of each group, including more nemertine species, were included in the study, or if some of the included species were omitted from the analysis? How much do you think the results might differ if different RNA sequences were examined? Based on the ideas discussed in Chapter 2, what other factors can you think of that might alter the outcomes of these sorts of analyses?

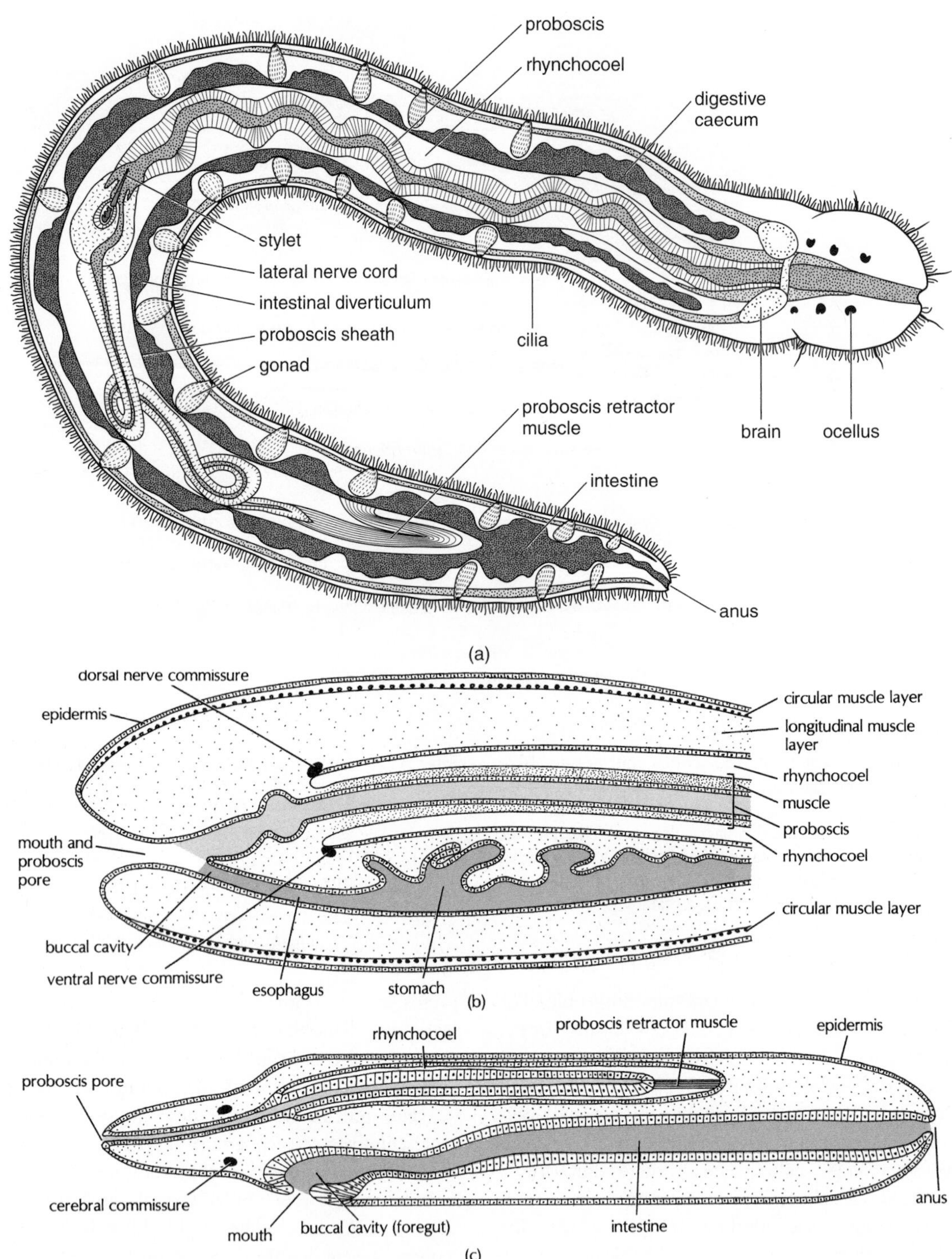

Figure 11.4

(a) Diagrammatic illustration of *Prostoma graecense,* a freshwater nemertine species. The intestine connects with various bulges (diverticulae) and sacs (caecae). (b) Diagrammatic illustration of *Nemertopsis* sp., anterior end in longitudinal section. Note that in this species the mouth opening has been lost so that there is a single opening for both the proboscis and the digestive system entrance. The nerve commissures connect to form a ring around the anterior portion of the rhynchocoel. Note the extensive folding of the stomach, which allows eversion of the foregut. (c) In *Cephalothrix bioculata,* the proboscis is ejected through an opening separate from the mouth.

(a) Based on Pennak, *Fresh-Water Invertebrates of the United States,* 2d ed. 1978. (b) After Bayre and Owre; after Burger. (c) From J. B. Jennings and Ray Gibson, 1969. *Biological Bulletin* 136:405. Reprinted by permission.

the rhynchocoel. When this musculature contracts, the pressure within the rhynchocoel rises. Even a small amount of contraction creates a substantial increase in pressure, since the rhynchocoel has a constant volume and the fluid within it is essentially incompressible; the fluid can neither escape nor be compressed. By default, hydrostatic pressure increases. The elevated pressure is relieved by relaxing the sphincter muscles surrounding the anterior end of the

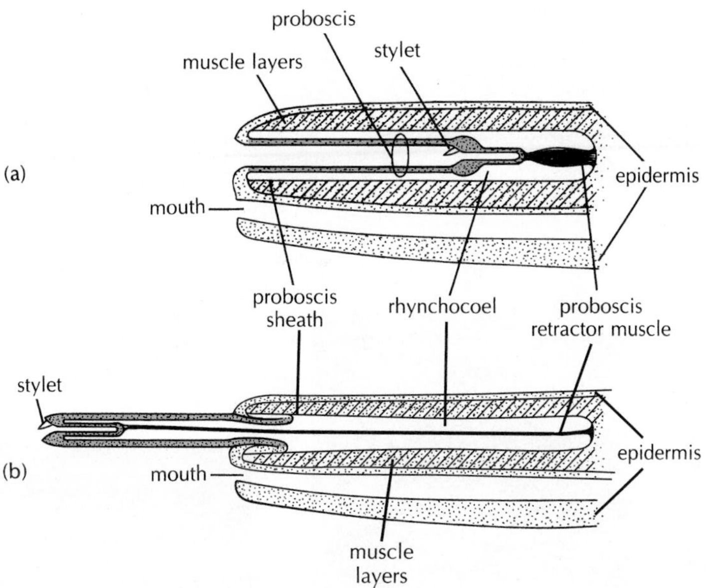

Figure 11.5
Diagrammatic illustration of the proboscis in (a) retracted and (b) extended positions. The digestive system has been omitted for clarity.

(a,b) After Harmer and Shipley; after Alexander.

rhynchocoel, allowing the proboscis to shoot out of the cavity with great speed and force (Fig. 11.5).

The workings of a proboscis can be visualized by picturing a rubber glove with one of the fingers pushed inside. If the wrist of the glove is tied off so that the air cannot escape, the glove becomes a constant-volume, fluid-filled cavity—a rhynchocoel, of sorts. The inverted finger is the proboscis in this rhynchocoel. If the glove is squeezed, the "proboscis" will be everted, turning inside out as it goes. If a hole is poked in the glove and the experiment repeated, the importance of the rhynchocoel's constant volume will be evident.

In most species, the proboscis is everted through a **proboscis pore** that is distinct from the mouth. In some species, however, the proboscis and digestive tract have come to share an opening, either because the mouth has been lost or because the proboscis pore has been lost.

Nemertines are carnivores to be reckoned with, even for rapidly moving prey. In one major group of species, the proboscis is armed with a piercing **stylet** (see Figs. 11.4a, 11.5, and 11.6d). A potent paralytic toxin found in the proboscis is often discharged into wounds made by the stylet. For those nemertine species possessing a stylet on the proboscis, prey may be harpooned directly. More commonly, the everted proboscis is first wound tightly around the prey, and the struggling animal is then stabbed repeatedly with the barb (Fig. 11.6). Species lacking a stylet capture prey by simply coiling the prehensile proboscis around the victim. A sticky mucus is generally secreted by the proboscis to help hold the prey.[5]

A **proboscis retractor muscle** runs from the tip of the everted proboscis to the inner wall of the rhynchocoel (Figs. 11.4a, c, and 11.5). Contraction of this muscle draws the proboscis back inside the rhynchocoel. For those species lacking stylets, captured food is usually transferred to the mouth as the proboscis is retracted. Nemertines with stylets first paralyze the prey and then retract the proboscis, completely losing contact with the stricken animal. The nemertine then moves to the prey to eat. In some species, prey are ingested whole, particularly if they are **vermiform** (worm shaped). In other species, the prey are typically too large or awkwardly shaped to be engulfed; in such cases, the nemertine inserts its foregut into the prey and, by peristaltic waves of muscular activity, pumps the juices of the victim into the gut of the victor.

Members of many nemertine species burrow into soft substrates, some using their body wall musculature and others using the proboscis, as follows. First, contraction of the body wall musculature elevates the internal pressure of the rhynchocoel. The proboscis is then everted into the sediment and allowed to widen at its tip, through a relaxation of the associated circular muscles. This distal dilation anchors the end of the proboscis in the sediment. The body of the worm can then be pulled downward, into the sediment, upon contraction of the proboscis retractor muscles and the longitudinal muscles of the body wall. Contracting the longitudinal muscles also forces additional fluid into the proboscis, under considerable pressure, causing additional dilation and anchoring. We will encounter similar burrowing mechanisms in several other animal phyla, phyla in which the coelom or blood sinuses subserve the same function as the rhynchocoel of nemertines.

Here, then, we see the principles of the hydrostatic skeleton (Chapter 5) in action. Advances in locomotion and food capture are achieved by exploiting the properties

5. See *Topics for Further Discussion and Investigation,* no. 2.

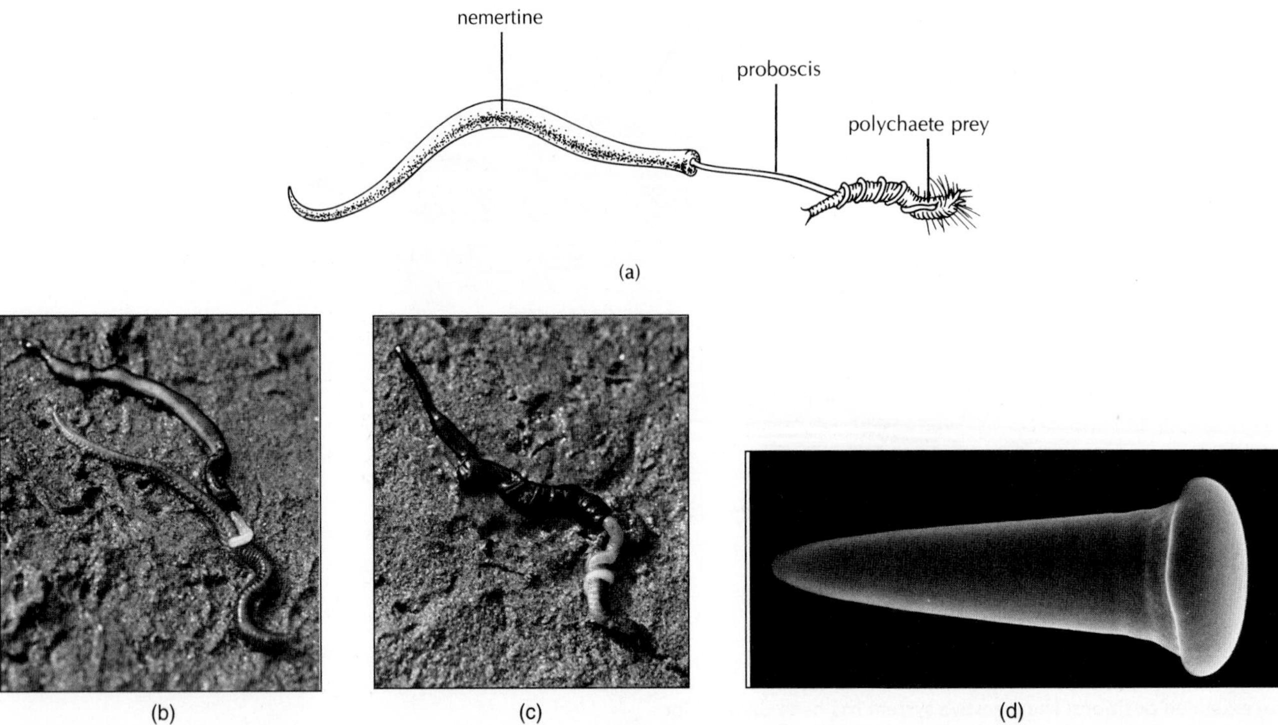

Figure 11.6
(a) Diagrammatic illustration of a nemertine (*Paranemertes peregrina*) subduing its prey. (b,c) *P. peregrina* attacking a polychaete worm (*Platynereis bicanaliculata*). (d) Scanning electron micrograph of the stylet of *Zygonemertes virescens;* the stylet is approximately 175 μm long.

(a) After MacGinitie and MacGinitie. (b,c) Courtesy of S.A. Stricker. From Stricker and Cloney. 1983. *J. Morphol.* 177:89. Reprinted with permission of Wiley-Liss, Inc., a subsidiary of John Wiley & Sons, Inc. (d) Courtesy of S.A. Stricker. From Stricker, 1983. *J. Morphol.* 175:153. Reprinted with permission of Wiley-Liss, Inc., a subsidiary of John Wiley & Sons, Inc.

of a constant-volume, fluid-filled cavity. Such cavities have apparently evolved at least 4 different times in metazoans: by enterocoely, by schizocoely, by the persistence of a blastocoel, and by the schizocoelous formation of a rhynchocoel. The selection pressures supporting the evolution of secondary body cavities obviously have been strong. Among the nemertines, the selective benefits would clearly seem to lie in mechanical advantages gained and in new lifestyles thereby made possible.

Other Features of Nemertine Biology

Classification

Nemertines are divided into 2 major classes: the Anopla and the Enopla. Members of the Anopla lack stylets; that is, the proboscis is unarmed. In addition, the mouth of anoplan nemertines is posterior to the brain. In contrast, the proboscis may be armed with a stylet among the Enopla—with all of the armed species belonging to a single order (the Hoplonemertea)—and the mouth is anterior to the brain.

Protection from Predators

How do you protect yourself if you're large and invitingly soft-bodied? Nemertines lack behavioral defenses and also lack shells, spines, and other protective hard parts. Burrowing is only a partial solution to the problem, as many predators search sediment for their prey. Nemertines seem to defend themselves chemically, but not entirely through their own efforts: In at least some nemertine species, bacterial symbionts (probably *Vibrio alginolyticus*) apparently synthesize tetrodotoxin (TTX)-like chemicals, powerful neurotoxins that probably provide an excellent defense against predators.

Reproduction and Development

Most nemertine species have separate sexes; that is, they are **gonochoristic (dioecious).** The few hermaphroditic species that have been described are **protandric;** that is, as it ages, each individual first becomes male and then becomes female. Following fertilization, cleavage is usually spiral and determinate, generally following the basic protostome pattern. Fertilization is usually external. In most species, the embryo develops into an elongated, ciliated, microscopic larval stage that resembles the juvenile worm, complete with rhynchocoel, proboscis, and digestive tract (which typically becomes functional only after metamorphosis). Within the order Heteronemertea, however, the embryos of many species differentiate into distinctive **pilidium** larvae. These are long-lived, ciliated, swimming, feeding individuals resembling football helmets with earflaps (Fig 11.7a, b). Only a portion of the pilidium gives rise

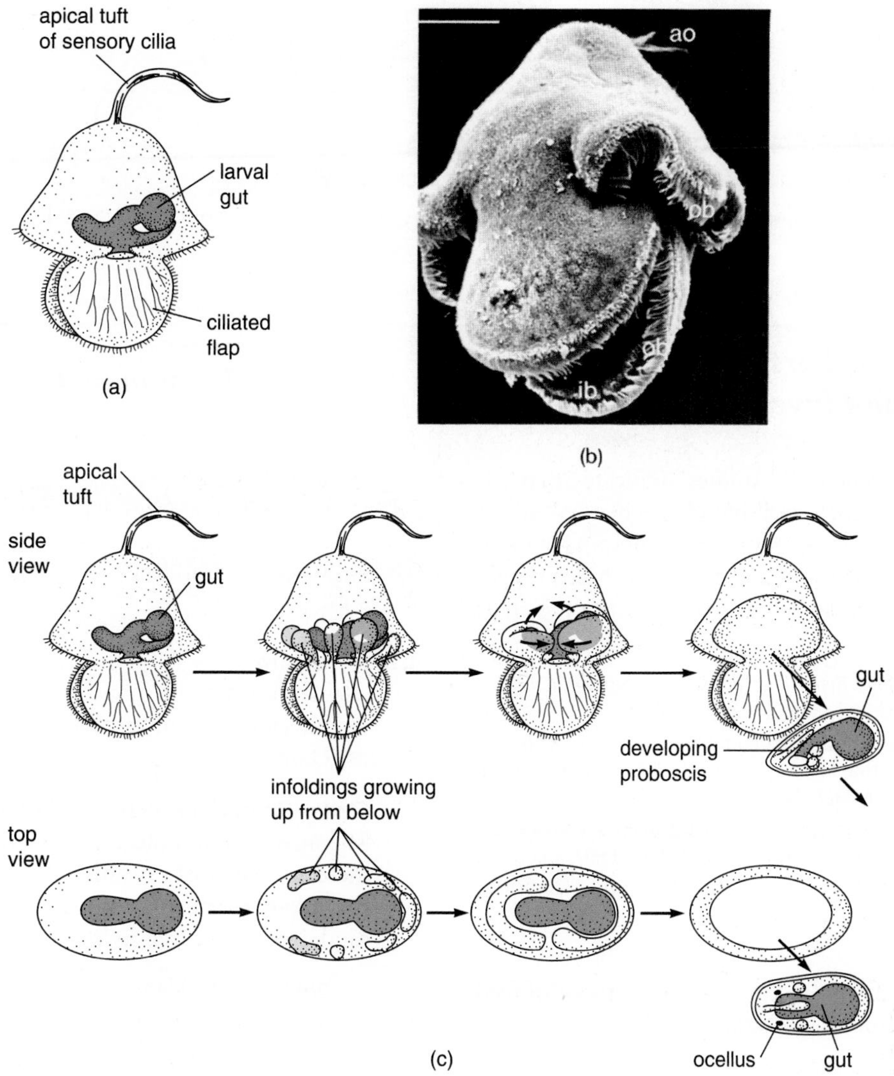

Figure 11.7

(a) The pilidium larva of a nemertine worm. (b) Scanning electron micrograph of an unidentified pilidium larva, about 400 μm long. ("ib" and "ob" = inner and outer ciliary bands, respectively; "ao" = apical tuft.) (c) Stages in the metamorphosis of the pilidium larva. Infoldings of the larval ectoderm form 7 imaginal discs (3 pairs and one posterior disc) that eventually fuse together, surrounding the future juvenile in a discrete sac. The head ectoderm, cerebral ganglion (brain), and proboscis are formed by the anterior pair of cephalic discs (light blue). At metamorphosis, the central mass detaches from the rest of the larva to take up residence on the sea bottom as a small nemertine. The other portion either continues swimming for a time, eventually starving to death in the absence of a digestive tract, or is eaten by the newly metamorphosed juvenile.
After Harmer and Shipley. (b) Courtesy of Claus Nielsen. From Nielsen. 1987. *Acta Zoologica* (Stockholm) 68:205–62. Permission of Royal Swedish Academy of Sciences. (c) From Hardy, *The Open Sea: Its Natural History.* Copyright © 1965 Houghton Mifflin Company, Boston, Massachusetts, and from Henry and Martindale, 1997. In S. F. Gilbert and A. M. Raunio, eds. *Embryology: Constructing the Organism.* Sinauer Associates, Inc., and from Henry and Martindale. In S. F. Gilbert and A. M. Raunio, eds. *Embryology: Constructing the Organism,* p. 160. Copyright © 1997 by Sinauer Associates, Inc.

to the juvenile, which develops within the larva from a series of **imaginal discs** that gradually fuse together (Fig. 11.7c). Such discrete masses of embryonic tissue, preprogrammed to form specific adult tissues and organs at metamorphosis, are also encountered in many insects (Chapter 14). At metamorphosis, what remains of the pilidium eventually swims off, in some species, and starves to death, since the juvenile takes the mouth and digestive system with it upon abandoning the larva (Fig. 11.7c). In other species, the juvenile worm ingests the larval tissues during or following metamorphosis. Recent embryological studies suggest that at least one primitive nemertine species has a *bona fide* trochophore larva in its development,[6] suggesting that the pilidium is derived from a trochophore ancestor.

Members of a number of nemertine species can also reproduce asexually, by fission or by fragmenting and then regenerating missing body parts. Most species, however, reproduce only sexually and lack the ability to regenerate.

6. Maslakova, S. A., M. Q. Martindale, and J. L. Norenberg. 2004. *Evol. Devel.* 6:219–26.

Taxonomic Summary

Phylum Nemertea (= Rhynchocoela)—the ribbon worms
Class Anopla—the unarmed nemertines (proboscis lacks a stylet)
Class Enopla—includes the armed nemertines (proboscis may be equipped with a stylet)

Topics for Further Discussion and Investigation

1. Recent morphological studies indicate that the nemertine rhynchocoel and blood circulatory system arise during embryonic development from splits in mesodermal tissue. What is the evidence on which this suggestion is based, and to what extent does it argue a coelomate ancestry for nemertines?

Nielsen, C. 2001. *Animal Evolution: Interrelationships of the Living Phyla,* 2nd ed. Oxford University Press, pp. 283–89.

Turbeville, J. M. 1986. An ultrastructural analysis of coelomogenesis in the hoplonemertine *Prosorhochmus americanus* and the polychaete *Magelona* sp. *J. Morphol.* 187:51.

Turbeville, J. M., and E. E. Ruppert. 1985. Comparative ultrastructure and the evolution of nemertines. *Amer. Zool.* 25:53.

Willmer, P. 1990. *Invertebrate Relationships.* New York: Cambridge University Press, 204–6.

2. Compare and contrast the different methods of feeding encountered among the Nemertea.

Fisher, F. M., Jr., and J. A. Oaks. 1978. Evidence for a nonintestinal nutritional mechanism in the rhynchocoelan, *Lineus ruber. Biol. Bull.* 154:213.

Jennings, J. B., and R. Gibson. 1969. Observations on the nutrition of 7 species of rhynchocoelan worms. *Biol. Bull.* 136:405.

McDermott, J. J. 1976. Observations on the food and feeding behavior of estuarine nemertean worms belonging to the order Hoplonemertea. *Biol. Bull.* 150:157.

Roe, P. 1976. Life history and predator-prey interactions of the nemertean *Paranemertes peregrina* Coe. *Biol. Bull.* 150:80.

Stricker, S. A., and R. A. Cloney. 1981. The stylet apparatus of the nemertean *Paranemertes peregrina:* Its ultrastructure and role in prey capture. *Zoomorphology* 97:205.

3. Discuss the limitations placed on shape changes in nemertines imposed by the structure of the body wall.

Clark, R. B., and J. B. Cowey. 1958. Factors controlling the change of shape of certain nemertean and turbellarian worms. *J. Exp. Biol.* 35:731.

Turbeville, J. M., and E. E. Ruppert. 1983. Epidermal muscles and peristaltic burrowing in *Carinoma tremaphoros* (Nemertini): Correlates of effective burrowing without segmentation. *Zoomorphology* 103:103.

4. How is a nemertine proboscis like a student studying?

Taxonomic Detail

Phylum Nemertea (= Rhynchocoela)

The approximately 1150 species are distributed among 2 classes.

Class Anopla
In these nemertines, the mouth and proboscis openings are separate, and the proboscis is usually unarmed. Two orders, 9 families.

Order Palaeonemertea (= Palaeonemertini)
The most recent molecular evidence indicates that these are the most primitive (basal) nemertines. All species are marine bottom-dwellers, living in sand or mud. None have free-living larvae. Four families.

Family Tubulanidae. *Tubulanus.* These nemertines may be striped or banded, and they are often brightly colored. The largest individuals are about 2 m long. Most species are restricted to shallow water, although some live as deep as 2500 m.

Order Heteronemertea
Most species are marine, although some are found in brackish or freshwaters. Pilidium (and related) larvae are found only among members of this order. Five families.

Family Baseodiscidae. *Baseodiscus.* Individuals are up to 2 m long and brightly colored, with conspicuous stripes or bands. Members are widely distributed in shallow waters of the Atlantic, Pacific, and Indian oceans and the Mediterranean Sea.

Family Lineidae. *Cerebratulus, Micrura, Lineus.* Most species are marine and free-living, although a few occur only in freshwater, and one (*Uchidana parasita*) lives only in the mantle cavity of marine bivalve molluscs. Members of the genus *Cerebratulus* can swim short distances by undulating. The family includes the largest known nemertines (*Lineus longissimus*), which can exceed 30 m in length.

Class Enopla

Two orders, 29 families. The proboscis of these nemertines typically exits through the mouth, rather than through a separate opening, and it is armed with one or more stylets in many species. All terrestrial nemertine species reside in this class.

Order Hoplonemertea (= Hoplonemertini)

Each member of this order possesses an armed, stylet-bearing proboscis. Studies of nervous and sensory systems have convinced some biologists that hoplonemertines gave rise to the vertebrates. Two suborders, 28 families.

Suborder Monostilifera

All species bear a proboscis armed with a single, central, piercing stylet and up to 12 accessory pouches housing replacement stylets. Members of this suborder occur in both marine and terrestrial habitats, and many feed suctorially on various arthropods, pumping out the body contents of their prey. Seven provisional families. The systematics of this group are particularly uncertain at present.

Family Carcinonemertidae. *Carcinonemertes.* The proboscis is not eversible. All species live on decapod crustaceans (crabs and spiny lobsters) as parasites or ectosymbionts. Members of the genus *Carcinonemertes* can apparently reach adulthood only on female hosts. Individuals normally live on the gill of the host crab, but they migrate to the host's abdomen to feed on developing embryos when the host is reproductively active. Members of at least some species can produce haploid larvae by parthenogenesis, although it is not yet clear if these larvae are viable or if they contribute to future population growth.

Family Cratenemertidae. *Nipponnemertes.* The members of some species in this family can swim, and individuals in at least one species are completely planktonic, spending their entire lives in the water.

Family Ototyphlonemertidae. *Ototyphlonemertes.* All individuals are small and live in the spaces between sand grains in marine shallow-water and intertidal habitats.

Family Prosorhochmidae. *Geonemertes, Gononemertes, Prosorhochmus.* Most species are marine and free-living, although members of the genus *Gononemertes* live only in association with sea squirts (ascidian urochordates), and members of several other genera live exclusively in terrestrial habitats (e.g., *Geonemertes:* Indo-Pacific and West Indies; *Leptonemertes:* European greenhouses, Azores and other N. Atlantic islands; *Argonemertes:* Australian species with up to 120 eyes!).

Family Tetrastemmatidae. *Tetrastemma, Prostoma.* Most members live in marine or brackish water, but some species (in the genus *Prostoma*) live only in freshwater. Species in this family are among the smallest nemertines, some attaining adult lengths of only 1–3 mm. A number of species live interstitially, in the spaces between sand grains.

Suborder Polystilifera

These nemertines are all marine and exhibit a proboscis armed with many small stylets. Twenty families.

Tribe Reptantia. All species are bottom-dwellers, some living at depths of several hundred meters.

Tribe Pelagica. *Nectonemertes, Pelagonemertes, Planktonemertes.* All species live free in the water, either swimming or floating, at depths of hundreds or even thousands of meters. About 10 families.

Order Bdellonemertea

Malacobdella. These strange nemertines are leech-like, with a posterior ventral sucker. All are commensal, living in the mantle cavities of certain marine bivalves and feeding on small food particles in suspension. The proboscis is unarmed. Molecular data suggest that these nemertines should be folded into the Monostilifera, as the order Malacobdellidae.[7] One family.

General References about the Nemertines

Gibson, R., J. Moore, and P. Sundberg, eds. 1993. *Advances in Nemertean Biology.* Kluwer Academic Publishers. Dordrecht, The Netherlands.

Harrison, F. W., and B. J. Bogitsh, eds. 1991. *Microscopic Anatomy of Invertebrates, Vol. 3: Platyhelminthes and Nemertinea.* New York: Wiley-Liss, 285–328.

Henry, J., and M. Q. Martindale. 1997. Nemerteans, the ribbon worms. In: Gilbert, S. F. and A. M. Raunio, eds. *Embryology: Constructing the Organism.* Sunderland, MA: Sinauer Associates, Inc.

Hyman, L. H. 1951. *The Invertebrates, Vol. 2. Platyhelminthes and Rhynchocoela: The Acoelomate Bilateria.* New York: McGraw-Hill.

McDermott, J. J., and P. Roe. 1985. Food, feeding behavior and feeding ecology of nemerteans. *Amer. Zool.* 25:113–25.

Norenberg, J. L., and P. Roe, eds. 1998. Fourth International Conference on Nemertean Biology. *Hydrobiologia* 365:1–310.

Parker, S. P., ed. 1982. *Classification and Synopsis of Living Organisms,* vol. 1. New York: McGraw-Hill, 823–46.

Roe, P., and J. L. Norenburg, eds. 1985. Comparative biology of nemertines. *Amer. Zool.* 25:3–151.

Thorpe, J. H., and A. P. Covich. 1991. *Ecology and Classification of North American Freshwater Invertebrates.* New York: Academic Press.

Turbeville, J. M. 2002. Progress in nemertean biology: Development and phylogeny. *Integr. Comp. Biol.* 42:692–703.

7. Thollesson, M., and J. L., Norenburg. 2003. *Proc. Roy. Soc.* London B 270: 407–15.

12

The Molluscs

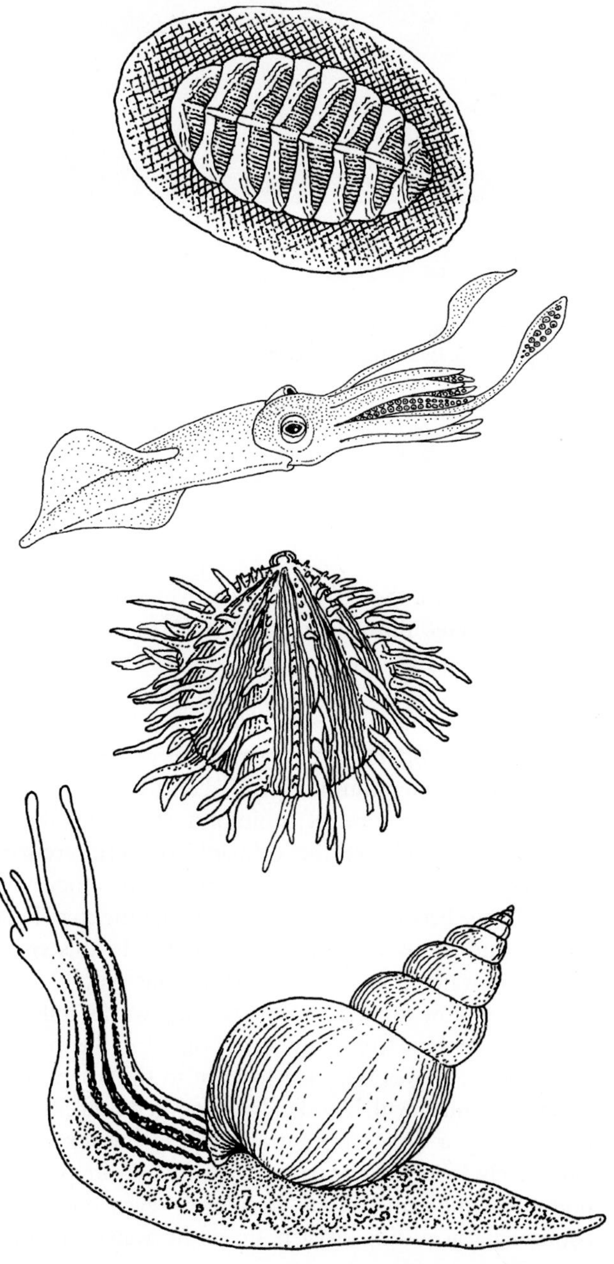

Introduction and General Characteristics

Phylum Mollusca
(Latin: soft)
mō-lusk´-ah

Defining Characteristics:[1] 1) Dorsal epithelium forming a mantle, which secretes calcareous spicules or one or more shells; 2) cuticular band of teeth (radula) in the esophagus, used for feeding (not present—lost?—in bivalves); 3) ventral body wall muscles develop into a locomotory or clinging foot

The Mollusca is an enormous phylum, including at least 50,000 and as many as 120,000 living species. These species are distributed among some extremely dissimilar-looking organisms, making the molluscan body plan probably the most malleable in the animal kingdom. Remarkably, clams, snails, and octopuses are all molluscs!

There really is no "typical" mollusc. Most, but not all, molluscs have shells consisting primarily of calcium carbonate set in a protein matrix. Organic material may comprise about 35% of the shell's dry weight in some gastropod species and up to about 70% of the dry weight in bivalves. The shells of most molluscs (including all gastropods and bivalves) have a thin, outer organic layer (the **periostracum**); a thin, innermost calcareous layer (the **nacreous layer**); and a thick, calcareous middle layer (the **prismatic layer**) (Fig. 12.1). Shell microstructure can differ dramatically among the members of different molluscan groups.

1. Characteristics distinguishing the members of this phylum from members of other phyla.

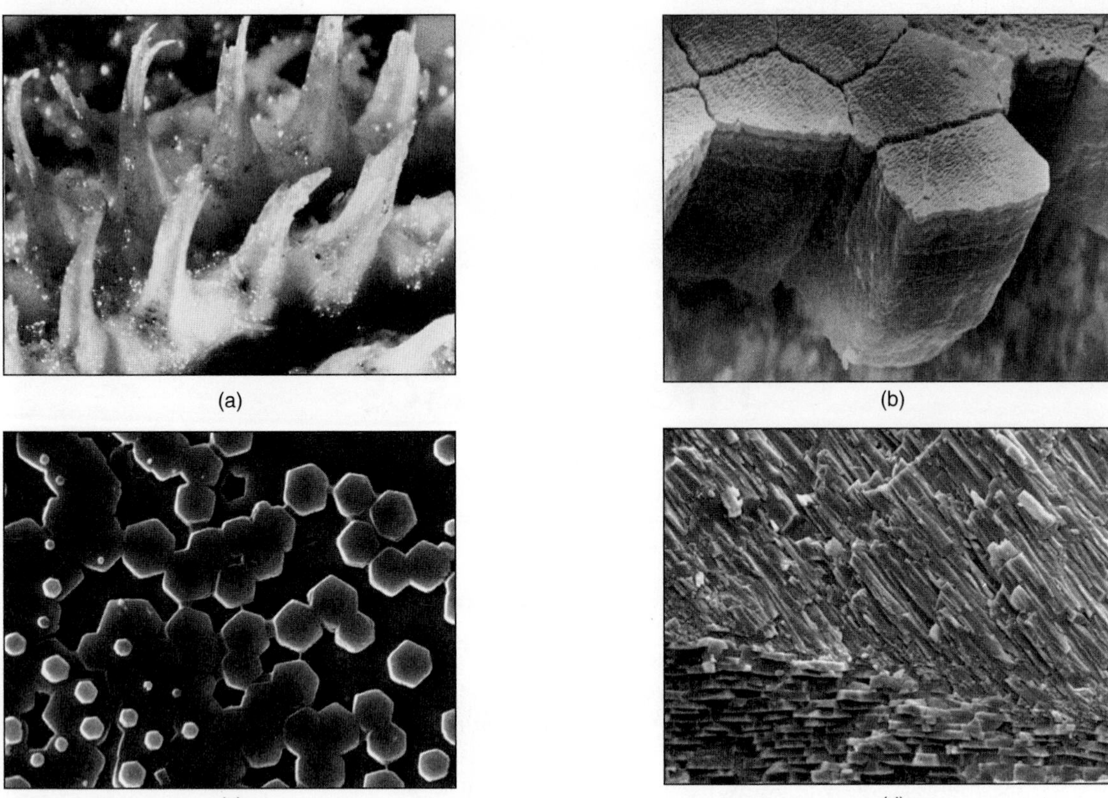

Figure 12.1

Molluscan shell structure. (a) Periostracal spines from the shell of the gastropod *Trichotropis cancellata.* (b) Tall calcite prisms from the prismatic layer of an oyster, *Crassostrea virginica* (Gmelin). The shell has been fractured and treated briefly with Chlorox to dissolve away the proteinaceous matrix (conchiolin) that normally occurs between individual prisms. (c) Individual calcareous tablets from the nacreous layer of a mussel shell, *Geukensia demissa* (Dillwyn). Each nacreous tablet is about 7 μm from side to side. (d) Fractured shell section of a deep-sea mussel, illustrating the outer prismatic layer (top) and calcareous tablets of the underlying nacreous layer.

(a) From D. J. Bottjer, Third North American Paleontological Convention, *Proceedings*, 1982, Vol. 1, pp. 51–56. (b) Courtesy of M. R. Carriker, from M. R. Carriker *et al.,* 1980. *Proc. Nat. Shellf. Assoc.* 70:139. (c) From R. A. Lutz and D. C. Rhoads, *Science* 198 (23 Dec 1977) pp. 1222–27, fig. 1. © AAAS. (d) Courtesy of R. A. Lutz.

Both the organic and inorganic components of the shell are secreted by specialized tissue known as the **mantle.** If a grain of sand, parasite, or other foreign particle becomes trapped between the mantle and the shell's inner surface, a pearl may form over a period of years. Natural pearl formation is a fairly rare event: Perhaps only one oyster in 1000 is likely to harbor a valuable pearl naturally. Humans increase the frequency of pearl production by surgically implanting pieces of shell (usually from freshwater bivalves) or plastic spheres between the shell and mantle of mature oysters, and then keeping the oysters alive for 5 to 7 years. Note that cultured pearls *form* in a perfectly normal manner; humans intervene only in getting the process started.

Although the mantle is a major molluscan characteristic, its role varies substantially in different molluscan groups. Similarly, the molluscan **foot** is also highly modified for a variety of functions in different groups.

Most molluscs have a characteristic cavity lying between the mantle and the viscera. This **mantle cavity** usually houses the comb-like molluscan gills, known as **ctenidia** (*ctenidi* = Greek: comb), and also generally serves as the exit site for the excretory, digestive, and reproductive systems. A **ctenidium** (the singular form of "ctenidia"), when present, may have a purely respiratory function or may also function in the collection and sorting of food particles. A chemoreceptor/tactile receptor known as the **osphradium** (*osphra* = G: a smell) is generally located adjacent to the ctenidium (Figs. 12.2 and 12.16).

With the exception of squid and other cephalopods, mollusc gills work on the principle of **countercurrent exchange,** a system that greatly increases the efficiency of gas exchange between blood flowing within the ctenidial filaments and the water flowing over them. In this system, blood and water flow in opposite directions (Fig. 12.3a). To understand how countercurrent exchange works, follow the path of *water flow* from point 1 to point 3 in Figure 12.3b. As the water moves from left to right in the diagram, it is always in contact with blood of lower oxygen concentration, maintaining a large concentration gradient for oxygen. Conversely, following the path of *blood flow* from point 3 to point 1 in the same figure, the blood—even though it is continually acquiring oxygen from the water—is always

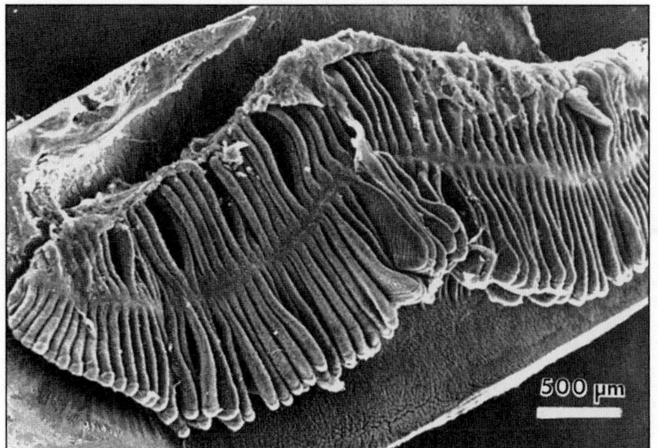

Figure 12.2

The osphradium (scanning electron micrograph) from the mantle cavity of the southern oyster drill, *Thais haemastoma canaliculata* (Mollusca: Gastropoda). The leaflets of the osphradium, which is 4–5 mm long in this species, hang down into the incurrent water stream within the mantle cavity; water thus contacts the osphradium before contacting the gill.

Courtesy of D.W. Garton and R.A. Roller. From Garton *et al.,* 1984. *Biological Bulletin* 167:310–21. Permission of *Biological Bulletin.*

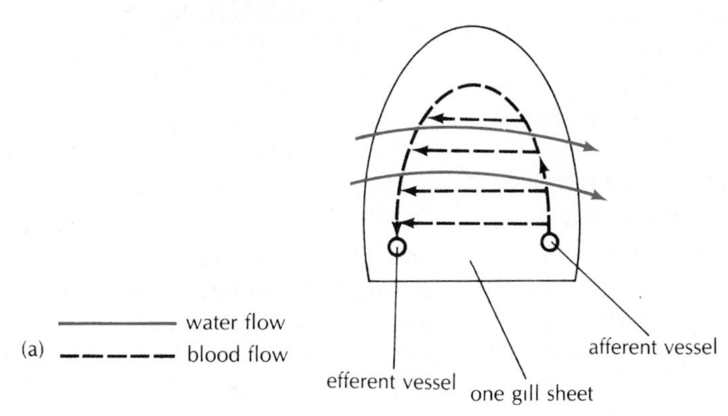

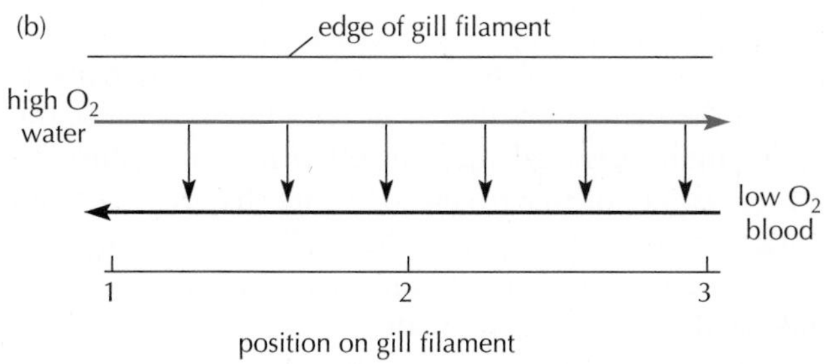

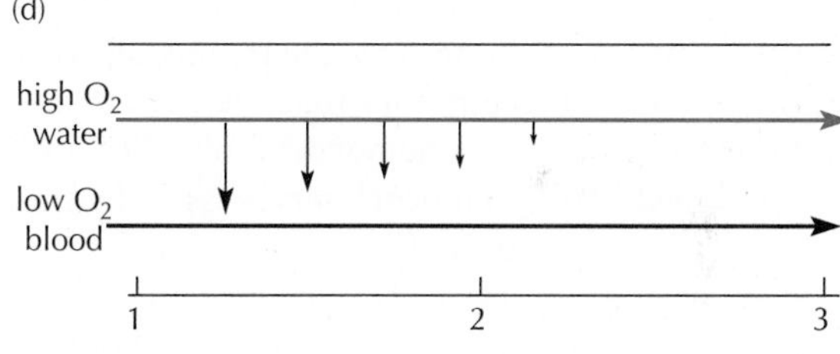

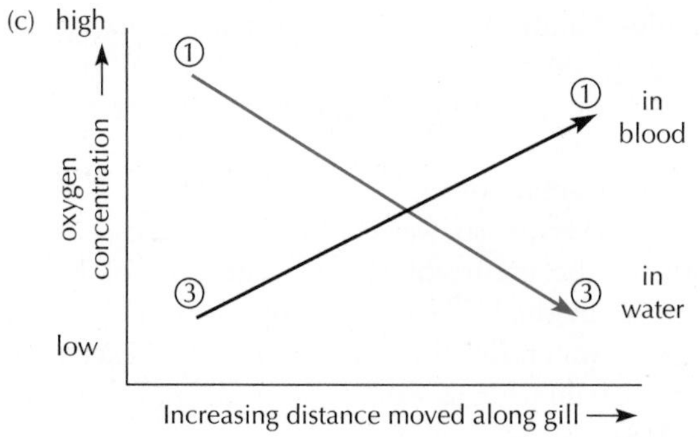

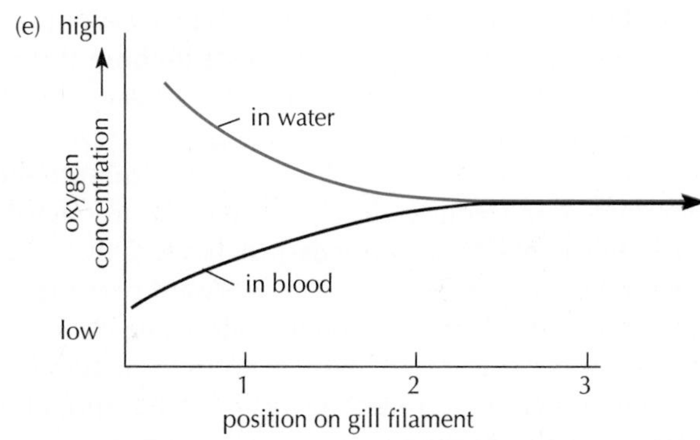

Figure 12.3

Countercurrent exchange in the molluscan gill. (a) The direction of water flow across the surface of each ctenidial leaflet is opposite to the direction of blood flow through the gill capillaries. (See Fig. 12.16b.) The afferent vessel carries deoxygenated blood to the ctenidial leaflets from the tissues. Oxygenated blood leaves the gill through the efferent vessel, carrying the blood to the heart and thence to the tissues. (b, c) Illustration of the principle of countercurrent exchange. Length of the vertical black arrows signifies magnitude of the concentration gradient for oxygen between water and blood. In (b), the countercurrent situation, equilibrium is never attained; oxygen diffuses along the entire gill-sheet surface, permitting a greater percentage of the oxygen in the water to be taken up by the blood (c). (d, e) Oxygen diffusion in a non-countercurrent situation. The oxygen concentration gradient diminishes rapidly (d), so that beyond point 2 on the gill very little further diffusion of oxygen takes place (e).

coming into contact with water of higher oxygen concentration as it moves from right to left in the figure. Thus, oxygen will diffuse from water to blood as the blood moves over the entire length of the gill filament (Fig. 12.3c).

In the alternative, hypothetical situation in which water and blood move in the same direction (Fig. 12.3d), the magnitude of the oxygen concentration gradient between the water and the blood continually decreases along the surface of the gill sheet as the water loses oxygen to the adjacent blood; the rate of gas exchange between water and blood will decrease correspondingly. By point 2 in Figure 12.3d, for example, the water has already given up much of its oxygen to the adjacent blood and is nearly at the same oxygen concentration as the blood next to it. Little further gas exchange between the 2 fluids will occur as the blood continues to move within the gill past point 2; in terms of oxygen exchange, the gill surface beyond point 2 is, in effect, wasted (Fig. 12.3e).

The molluscan coelom is very small, being restricted largely to the area surrounding the heart and gonads. Some zoologists have suggested that this cavity is in fact not homologous with the conspicuous coelomic cavities of annelids and other noncontroversial coelomates, and that molluscs evolved directly from acoelomate flatworm ancestors. Current molecular data, however, support an alternative view—that molluscs have descended instead from some coelomate ancestor and that the body cavity experienced a substantial reduction in size in the course of subsequent evolution. Either way, the molluscan "coelom" is small, and has no locomotory role. On the other hand, blood sinuses comprising a **hemocoel** ("blood cavity") are well developed. This hemocoel serves as a hydrostatic skeleton in the locomotion of some molluscs.

Many molluscs possess a feeding structure known as the **radula.** The radula consists of a firm ribbon, composed of chitin and protein, along which are found 2 rows of sharp, chitinous teeth (Fig. 12.4). The ribbon is produced from a **radular sac** and is underlain by a supportive cartilage-like structure called the **odontophore** (literally, G: tooth bearer). The odontophore-radular assembly, together with its complex musculature, is known as the **buccal mass** (*bucca* = L: cheek), or the **odontophore complex.** For feeding, the buccal mass is protracted so that the odontophore extends just beyond the mouth. The radular ribbon is then moved forward over the leading edge of the supporting odontophore and then pulled back. As each row of teeth passes back over the edge of the odontophore, the teeth automatically stand upright and rotate laterally, rasping food particles from the substrate and bringing them into the mouth as the radula is withdrawn. As old teeth are worn down or broken off at the anterior end of the radular ribbon, new teeth are continually being formed and added onto the ribbon's posterior end in the radular sac.

As might be inferred from the cautious wording in the preceding paragraphs, generalizations about molluscs are difficult to make.

Extant molluscs are distributed among 7 classes. Six of these 7 classes are represented by fossils formed some 450 million years ago, along with one additional class of molluscs, the Rostroconchia, whose clam-like members went extinct some 225 million years ago (Fig. 12.5). All told, there are at least 35,000 molluscan species known only as fossils. Only one class of molluscs, the Aplacophora, has apparently left no fossil record.[2]

Class Polyplacophora

Class Poly • placo • phora
(G: many plate bearing)
pol-ē-plah-koff′-or-ah

Defining Characteristic: Shell forms as a series of 7 to 8 separate plates

---

The 800 species in the class Polyplacophora are known as "chitons" (not to be confused with chitin, a polysaccharide!). Chitons are typically 3–10 cm (centimeters) long and generally found close to shore, particularly in the intertidal zone; they live only on hard substrata, especially rocks. A chiton's most distinctive external feature is its shell, which occurs as a series of overlapping and articulating plates (usually 8) covering the dorsal surface (Fig. 12.6a). These plates are partially or largely embedded in the mantle tissue that secretes them. Because the shell is multisectioned, the body can bend and conform to a wide variety of underlying substrate shapes. A chiton's thick lateral mantle is called the **girdle.** In most species, the girdle bears numerous calcareous spicules, secreted independently of the shell plates.

The mantle cavity of chitons takes the form of 2 lateral grooves, one on each side of the body (Fig. 12.6b). Up to about 80 **bipectinate** (L: double-combed, i.e., 2-branched) ctenidia hang down from the roof of each groove, dividing each elongated mantle cavity into incurrent and excurrent chambers. Water is drawn into the incurrent chamber by the action of the gill cilia. The flow of water is anterior to posterior, so waste products are discharged posteriorly in the excurrent stream. The flow of blood through the individual gill lamellae is opposite in direction to the flow of water, forming a **countercurrent exchange system** that facilitates gas exchange, as discussed earlier (pp. 216–218).

The foot extends along the animal's entire ventral surface and is completely covered by the overlying shell and girdle. Locomotion is accomplished by subtle waves of muscular activity called "pedal waves," as described later for

---

2. A fossil from Herefordshire, England, about 425 million years old, has been described as a possible aplacophoran (M. D. Sutton *et al.*, 2001. *Nature* 410: 461–63).

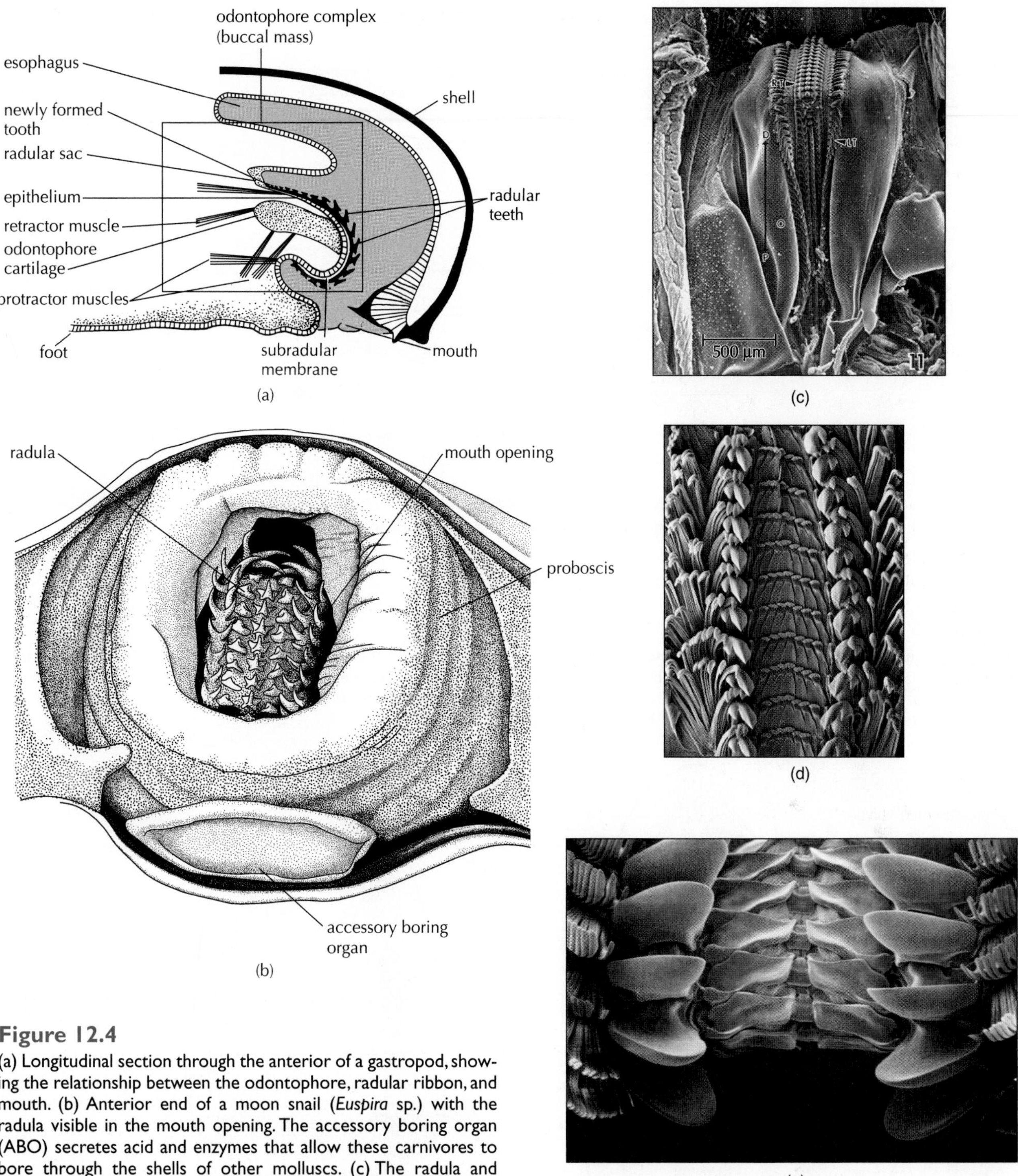

Figure 12.4

(a) Longitudinal section through the anterior of a gastropod, showing the relationship between the odontophore, radular ribbon, and mouth. (b) Anterior end of a moon snail (*Euspira* sp.) with the radula visible in the mouth opening. The accessory boring organ (ABO) secretes acid and enzymes that allow these carnivores to bore through the shells of other molluscs. (c) The radula and odontophore (scanning electron micrograph) projecting from the mouth of a marine snail, the southern oyster drill *Thais haemastoma canaliculata.* O = odontophore; RT and LT = radular teeth. (d, e) Scanning electron micrographs of radular teeth of 2 different snail species: (d) *Montfortula rugosa;* (e) *Nerita undata.*

(a) Modified after Runham and various sources. (b) From Hyman, 1967. *The Invertebrates,* Vol. 6. McGraw-Hill. (c) Courtesy of Roller *et al.,* 1984. *American Malacological Bulletin* 2:63–73. (d,e) Courtesy of C. S. Hickman, from Hickman, 1981. *Veliger* 23:189.

gastropods (p. 229). When disturbed, the chiton can press the girdle tightly against the substrate. By then lifting up the central portion of the foot (and the inner margin of the mantle tissue as well, if required), while retaining a tight seal against the substrate along the entire outer margin of the foot (and girdle), the chiton can generate a suction that holds the animal tightly against the substratum. Mucus secretion along the girdle helps to maintain the grip. This ability to cling tightly to the substrate is a particularly effective adaptation for life in areas of heavy wave action.

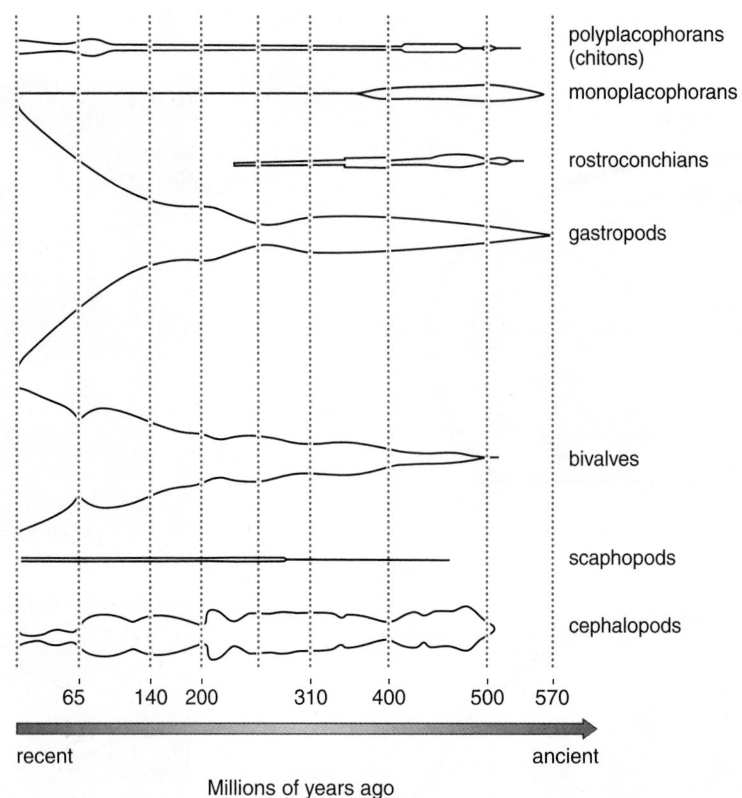

Figure 12.5

Molluscan species diversity as represented in the fossil record. The width of each bar is proportional to the number of species in each class. Members of the Rostroconchia superficially resembled the clams and related bivalves, except that they had hingeless, gaping shells. Bivalved molluscs and scaphopods may be descendants of rostroconchian ancestors.

Based on R. S. Boardman *et al.*, eds., *Fossil Invertebrates,* 1987.

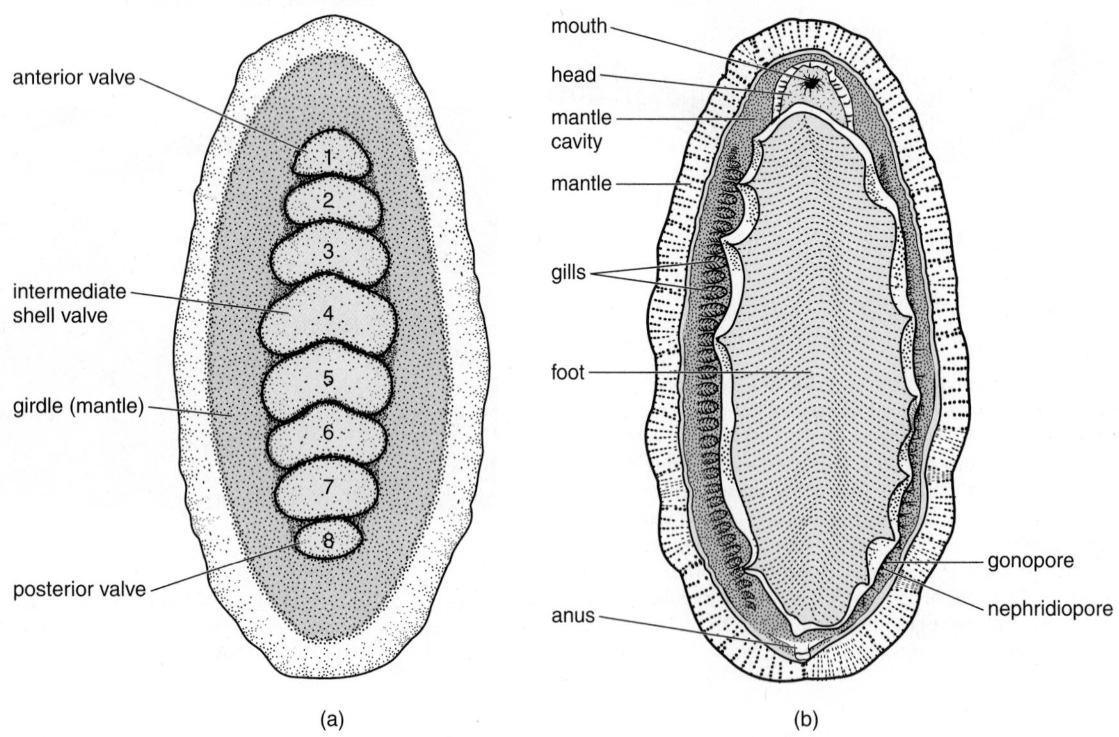

Figure 12.6a, b

The polyplacophoran *Katharina tunicata.* (a) Dorsal view. (b) Ventral view. All chitons are dorsoventrally flattened, with the foot forming a large suction cup for clinging to firm substrates. The shell is composed of 8 articulating plates, permitting the entire animal to curl and therefore conform to the topography of the underlying surface.

(a,b) Based on Beck/Braithwaite, *Invertebrate Zoology, Laboratory Workbook,* 3/e, 1968.

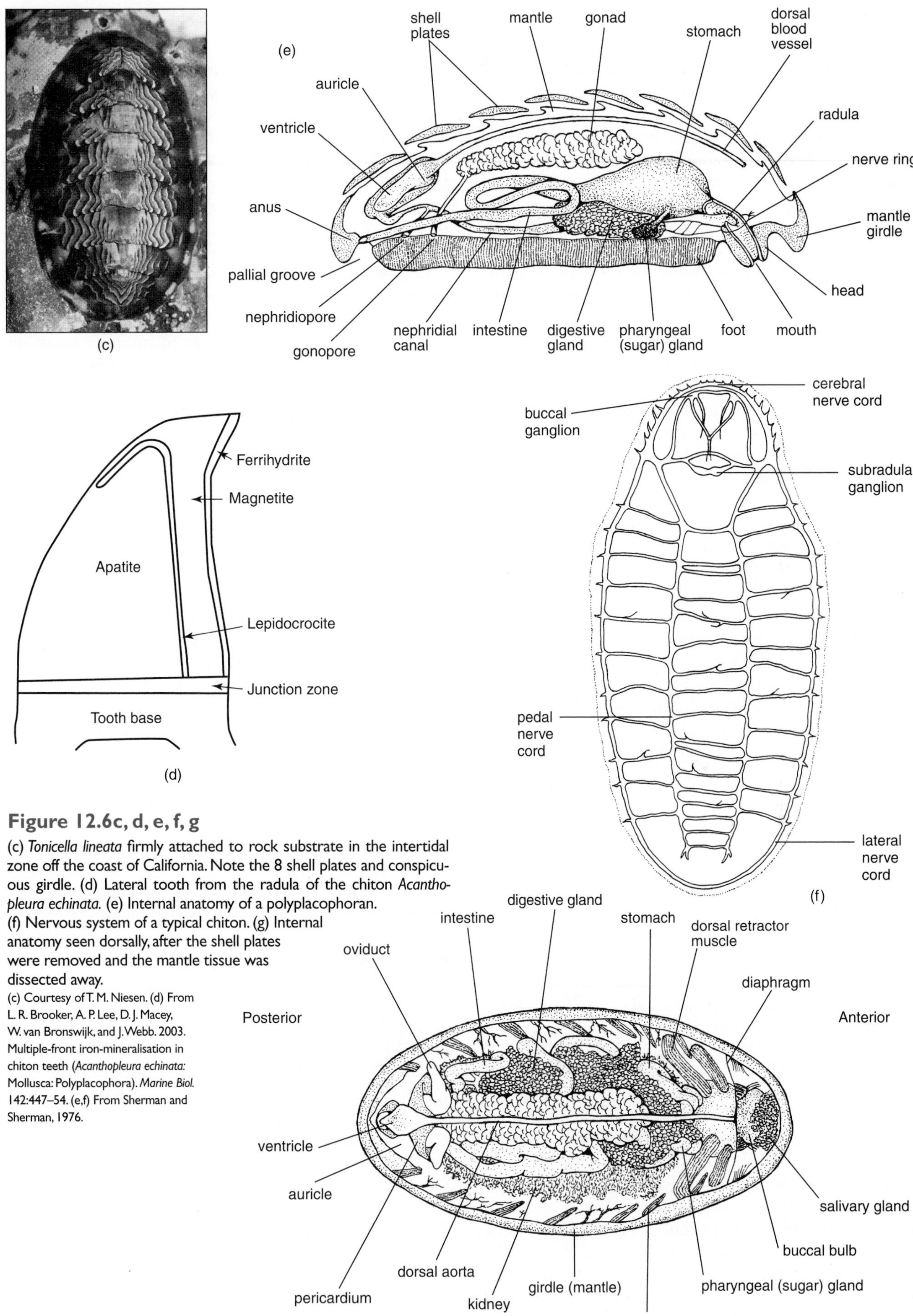

Figure 12.6c, d, e, f, g

(c) *Tonicella lineata* firmly attached to rock substrate in the intertidal zone off the coast of California. Note the 8 shell plates and conspicuous girdle. (d) Lateral tooth from the radula of the chiton *Acanthopleura echinata.* (e) Internal anatomy of a polyplacophoran. (f) Nervous system of a typical chiton. (g) Internal anatomy seen dorsally, after the shell plates were removed and the mantle tissue was dissected away.

(c) Courtesy of T. M. Niesen. (d) From L. R. Brooker, A. P. Lee, D. J. Macey, W. van Bronswijk, and J. Webb. 2003. Multiple-front iron-mineralisation in chiton teeth (*Acanthopleura echinata:* Mollusca: Polyplacophora). *Marine Biol.* 142:447–54. (e,f) From Sherman and Sherman, 1976.

The chiton nervous system is simple and ladder-like (Fig. 12.6f). Ganglia are lacking in many species, and only poorly developed in others. Sensory systems are also reduced: Adult chitons lack statocysts, tentacles, and eyes on the head. **Aesthetes**—abundant organs derived from mantle tissue and extending through holes in the shell plates—are thought to function, at least in some species, as light receptors. Recent ultrastructural studies, however, suggest that aesthetes may function primarily in secreting periostracum, replacing material that is naturally abraded away in the highly turbulent environment in which most chitons live.

Chitons have a linear digestive tract, with the mouth and anus at opposite ends of the body (Fig. 12.6b, e). Food particles, usually algae, are typically scraped from the substrate by a radula/odontophore complex, although a few species are carnivores. Many of the radular teeth are capped with iron-oxide (Fig. 12.6d). A pair of pharyngeal glands, often called **sugar glands,** release amylase-containing secretions into the stomach (Fig. 12.6e, g). As discussed later, most other herbivorous molluscs process food using something called a crystalline style (see p. 247).

Chitons have a fossil record extending back some 500 million years (Fig. 12.5). The evolutionary relationships between the Polyplacophora and other molluscan classes are unclear, although there is no reason to suspect that any other molluscs evolved directly from chiton ancestors. Chitons probably diverged from the main line of molluscan evolution early on, which is also likely to be the case for the worm-like aplacophoran molluscs to be described next.

Class Aplacophora

Class A • placo • phora
(G: not shell bearing)
ā-plak-off′-or-ah

Defining Characteristic: Cylindrical, vermiform body with the foot forming a narrow keel

---

Aplacophorans are worm-shaped (**vermiform**) molluscs (Fig. 12.7) found in all oceans, mostly in deep water. Most individuals are quite small—usually only a few millimeters and rarely more than a few centimeters long. Like the cephalopods, scaphopods, monoplacophorans, and chitons, aplacophorans are entirely marine. The body is unsegmented and bears numerous calcareous spines or scales embedded in an outer cuticle (Fig. 12.7a, c, f). The spines or scales are secreted by individual cells in the underlying epidermis; there is no true shell. At least some species possess a style sac (often complete with style and gastric shield), a small posterior mantle cavity with ctenidia, and a radula, although in most species the radula is apparently used for grasping rather than rasping. Aplacophorans have no conspicuous foot, although the members of one group (the Neomeniomorpha, or solenogastres) possess a nonmuscular, ciliated ridge located in a groove on the body's ventral surface and believed to be homologous with the foot of other molluscs. Solenogastres use the "foot" cilia to glide over the sediment along a mucous trail that they secrete as they go.

The Aplacophora is the only class of molluscs to have left no fossil record (see footnote, p. 218). About 320 extant aplacophoran species have been described to date, with many of the species known from only 1 or 2 specimens. Many aplacophorans burrow into or meander about on mud; other species live on cnidarians—primarily soft corals—on which they prey. A few species are interstitial, living in the spaces between sand grains and feeding on interstitial hydrozoans (phylum Cnidaria). The evolutionary relationships to other members of the phylum remain particularly obscure, in large part because aplacophoran biology has been so little studied.

For many years, aplacophorans were considered close relatives of the chitons and were placed with them in a single taxonomic group, the Amphineura. In the 1970s and 1980s, studies of aplacophoran spicule formation, radula structure, and anatomy argued against a close relationship with the chitons, but arguments in favor of such a relationship are again being made. In particular, the aplacophoran nervous system consists of paired cerebral ganglia, giving rise to 4 linear, ganglionated nerve cords—2 lateral, 2 ventral—interconnected in a ladder-like arrangement as in chitons (Fig. 12.7d). In addition, members of both groups form calcareous spicules in apparently identical fashion, through extracellular secretions from single cells. Molecular data that might help resolve the uncertainties have not yet been reported. In any event, it seems likely that aplacophorans are a primitive offshoot, diverging early from the main line of molluscan evolution before the advent of shell formation.

Class Monoplacophora

Class Mono • placo • phora
(G: one shell bearing)
mon-ō-plak-off′-or-ah

Defining Characteristics: 1) 3 to 6 pairs of ctenidia, 6 to 7 pairs of nephridia; 2) multiple (usually 8) pairs of foot (pedal) retractor muscles

---

Prior to 1952, the class Monoplacophora was known only from the fossil record. Actually, some representatives were collected in the 1890s but were categorized as gastropods and ignored. Since their rediscovery and recognition as a distinct class of molluscs in 1952, nearly 20 living species have been described, all marine and all collected from

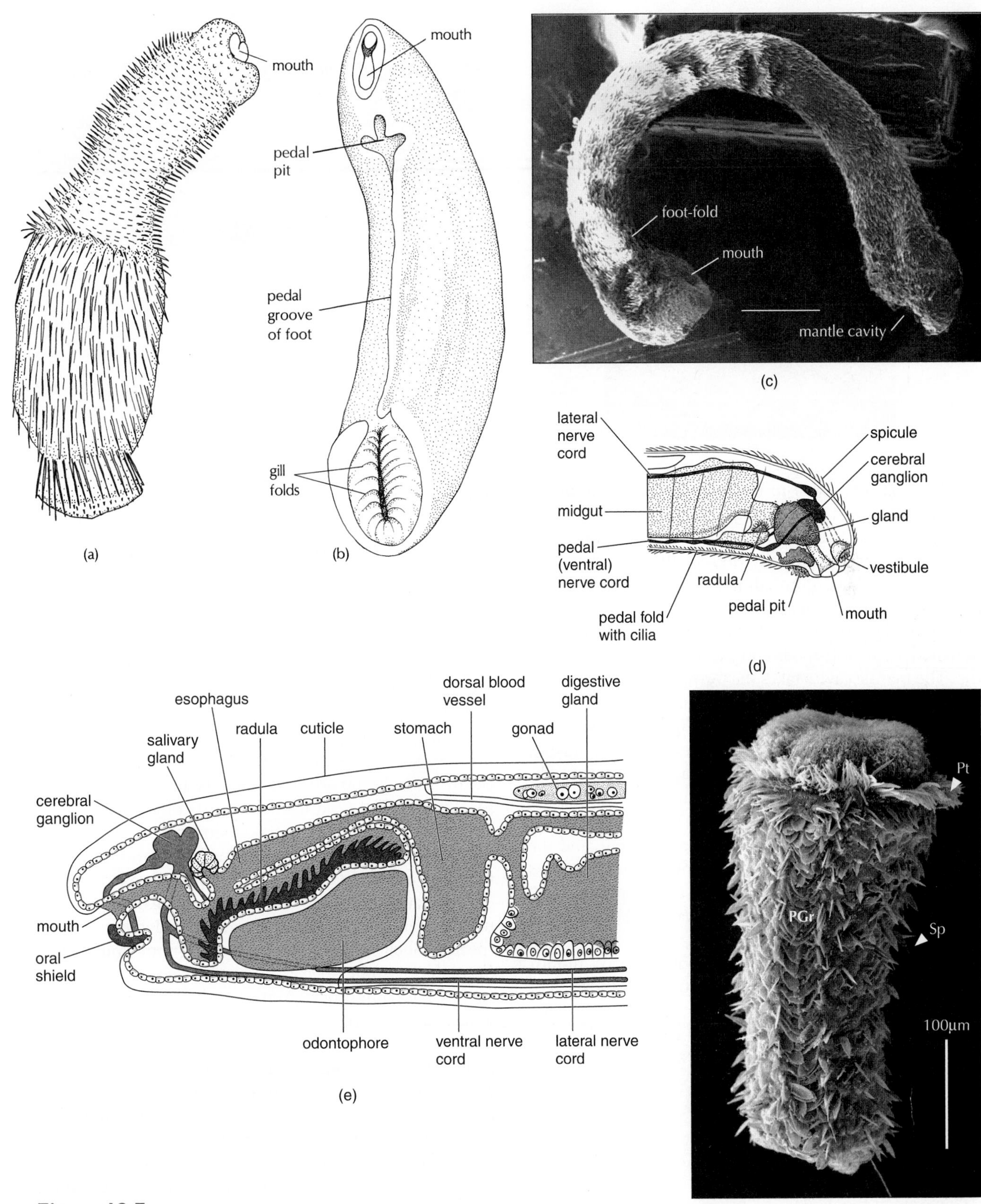

Figure 12.7

Members of the Aplacophora: (a) *Falcidens* sp. (b) *Neomenia carinata.* (c) Scanning electron micrograph of *Lyratoherpia* sp. (Scale bar = 0.5 mm.) (d) Anterior portion of an aplacophoran, showing the arrangement of nerve cords. (e) Internal anatomy, anterior end of *Limifossor talpoideus.* Note the well-developed radula and supporting odontophore. (f) Advanced larva (9–11 days old) of the aplacophoran *Epimenia babai* from Japan. Note the conspicuous spicules (Sp) and less conspicuous pedal groove (PGr). Pt = ring of larval locomotory cilia (prototroch).

(a) Courtesy of Dr. A. H. Scheltema. (b) From Hyman. (c) From Scheltema, A. H. *et al.* 1994. *Microscopic Anatomy of Invertebrates,* Vol. 5, New York: Wiley-Liss, 13–54. Photograph kindly provided by A. H. Scheltema. (d) Based on R. S. Boardman, *Fossil Invertebrates.* (e) Based on Harrison, *Microscopic Anatomy of Invertebrates,* Vol. 5. (f) Courtesy of Akiko Okusu. From Okusu, A. 2003. Embryogenesis and development of *Epimenia babai* (*Mollusca Neomeniomorpha*). *Biol. Bull.* 203:87–103.

(a)

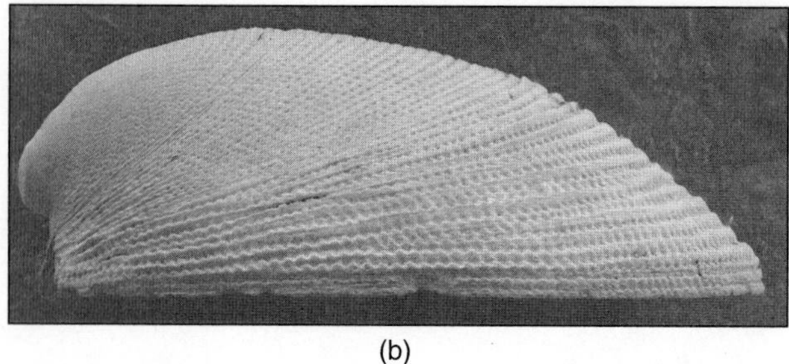

(b)

Figure 12.8

The shell of the monoplacophoran *Veleropilina reticulata* in (a) dorsal and (b) lateral views. The largest species found to date are only 1.6 mm long.

Courtesy of A. Warén. From A. Warén and S. Gofars, 1996. *Zoologica Scripta* 25:215–32 (figs. 3B, 4E).

depths of at least 2000 m (meters). A single, unhinged, cap-shaped shell is present (Fig. 12.8), as in many limpet-like gastropods. The shell of adult monoplacophorans is flattened rather than spirally wound, although the larval shell is spiral. Maximum adult shell lengths range from less than 1 mm in one species to about 37 mm in members of the largest species. The monoplacophoran foot is flattened, as in gastropods and polyplacophorans.

The mantle cavity takes the form of 2 lateral grooves, as in polyplacophorans, and 3, 5, or 6 pairs of gills hang down within the mantle grooves (Fig. 12.9a). Whether these gills are homologous with the typical molluscan ctenidium is, however, uncertain. In addition to the gills, the pedal retractor muscles, auricles and ventricles of the heart, gonads, and nephridia occur in multiple copies. Both a radula and a crystalline style are present, and the gut is linear, with the mouth being anterior and the anus posterior. As in the polyplacophorans and aplacophorans, the nervous system includes both lateral and pedal nerve cords (Fig. 12.9c).

Despite the small number of living species, monoplacophorans are well deserving of additional study: All of the molluscan groups remaining to be discussed—the scaphopods, gastropods, bivalves, and cephalopods—may have evolved from monoplacophoran ancestors.

Class Gastropoda[3]

Class Gastro • poda
(G: stomach foot)
gas-trop´-ō-dah

Defining Characteristics: 1) Visceral mass and nervous system become twisted 90–180° (exhibiting torsion) during embryonic development; 2) proteinaceous shield on the foot (operculum)

The Gastropoda is the largest molluscan class, containing more than half of all living mollusc species. Its snails and slugs are distributed among marine, freshwater, and terrestrial environments and occupy very diverse habitats, including rivers, lakes, trees, deserts, the marine intertidal zone, the plankton, and the deep sea. They exhibit a striking diversity of lifestyles, including suspension-feeding, carnivorous, herbivorous, deposit-feeding, and ectoparasitic species. The typical snail consists of a **visceral mass** (i.e., all of the internal organs) sitting atop a muscular **foot** (Fig. 12.10). The visceral mass is commonly protected by a univalved shell that is typically coiled, probably as an adaptation for efficient packaging of the visceral mass. Shell morphology differs considerably among species.[4] Shell size ranges from under 1 mm (millimeter) in adults of some species, to over 60 cm (centimeters) (nearly 2 feet) in others. In many other species, adults no longer have a shell.

For shelled species, the snail is attached to the inside of its shell by a **columellar muscle** (Fig. 12.10), which extends from within the animal's foot to the central axis of the shell; this central axis is known as the **columella** (Fig. 12.11b). The columellar muscle is important in most major body movements: protraction from the shell, retraction into the shell, twisting, raising the shell above the substratum, and lowering it back down.

The shell is typically carried so that it leans to the left side of the body. The shell axis is thus oblique to the long axis of the body, balancing the animal's center of mass over the foot. The shells of most gastropod species coil

3. Molluscan systematics are still unsettled, particularly for the gastropods and bivalves. Treatment of this edition is based largely on Beesley, P. L., G. J. B. Ross, and A. Wells (Eds.). 1998. *Mollusca: The Southern Synthesis. Fauna of Australia.* Vol. 5. CSIRO Publishing: Melbourne.

4. See *Topics for Further Discussion and Investigation,* no. 1, at the end of the chapter.

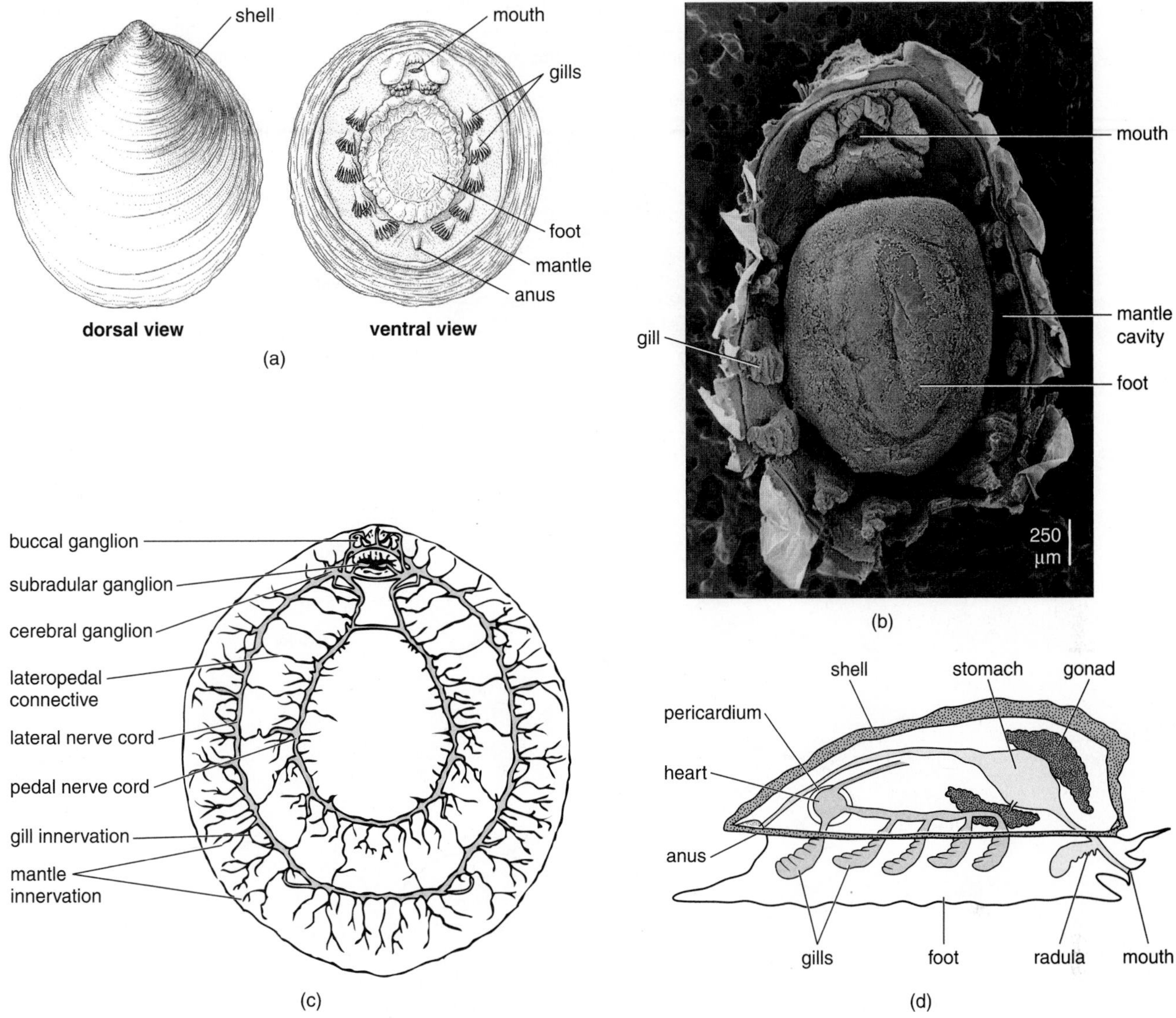

Figure 12.9

The monoplacophoran *Neopilina galatheae*. This animal was known only as a fossil until 1952, when it was dredged off the Pacific coast of Mexico from 5000 m deep. The shell is up to 37 mm in length. (a) Dorsal and ventral views. The adult shell is about 3 cm in length. (b) Ventral view of *Vema levinae*, about 2.5 mm long. Note the 6 pairs of gills. (c) Nervous system, dorsal view. (d) Internal anatomy, lateral view.

(a) After Lemche. (b) Courtesy of A. Warén. From A. Warén and S. Gofas. 1996. *Zoologica Scripta* 25:215–32. (c–d) Based on E. N. K. Clarkson, *Invertebrate Palaeontology & Evolution,* 2d ed. 1986.

clockwise, to the right (Figs. 12.11 and 12.12a). That is, the shells are "right-handed" or **"dextral"** (*dextro* = L: the right-hand side). Probably as a consequence of space limitations within the coiled shell, the ctenidium, osphradium, kidney (nephridium), and heart auricle on the right side of the body tend to be reduced or absent; only the primitive gastropods (i.e., those closest to the ancestral condition, including many archaeogastropod species) still exhibit paired structures (Fig. 12.13b, c, d). A relatively few snail species form shells that coil counterclockwise, to the left (Fig. 12.12b), and show a corresponding reduction or absence of the ctenidium, osphradium, kidney, and heart auricle on the left side of the body. Such species are said to be "left-handed," or **sinistral** in their coiling (*sinister* = L: the left-hand side). Note that it is possible for a dextrally coiled snail to produce a shell that appears to be "left-handed" (Fig. 12.12c, d); some gastropods exhibit such a pattern, particularly in the larval stage.

Many gastropod species possess, in addition to their shells, fairly elaborate behavioral or chemical defenses against predators. These adaptations commonly take one of the following forms: (1) the gastropod senses the presence of potential predators, either chemically or by touch, and initiates appropriate escape, avoidance, or deterrent behavior; (2) the gastropod chemically senses the presence

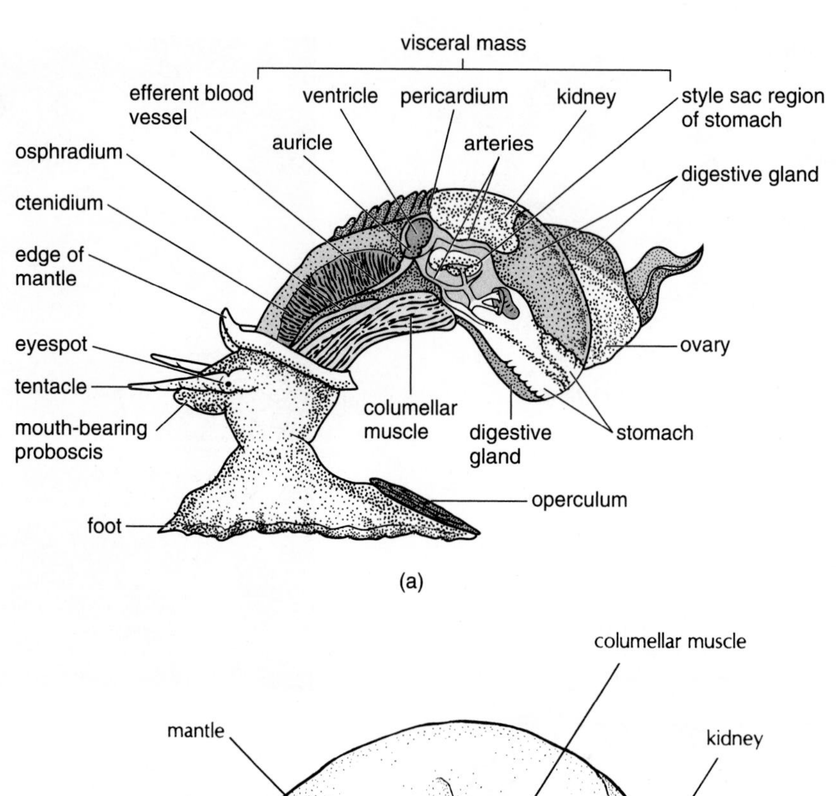

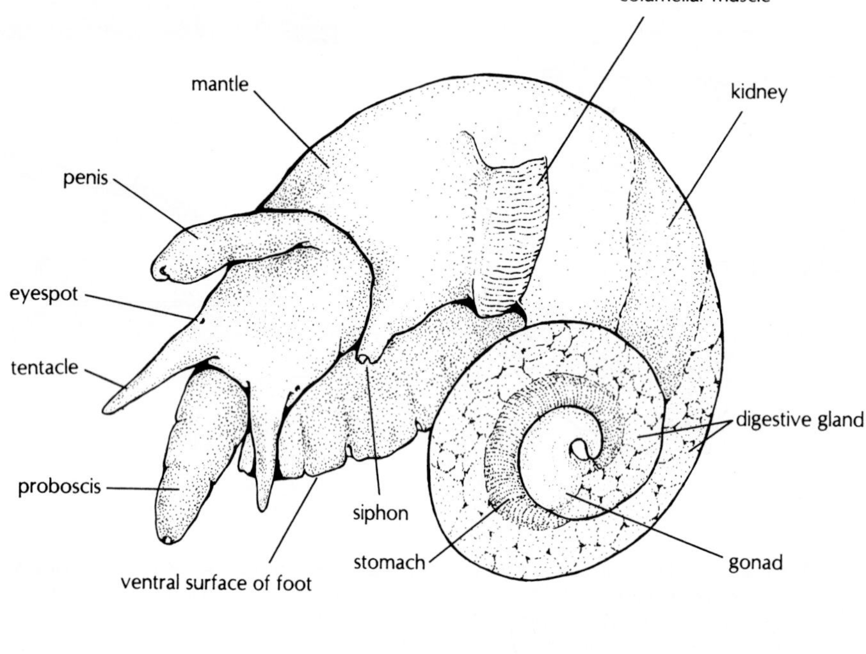

Figure 12.10

(a) Internal anatomy and gross morphology of the female periwinkle, *Littorina littorea,* removed from its shell. The animal occupies the entire inside of the shell, coiling around the columella as it grows. Blue structures are visible only by dissection. (b) The marine whelk *Busycon* sp., removed from its shell.

(a) Based on Fretter and Graham, *A Functional Anatomy of Invertebrates.* 1976.

of injured individuals of its own species (i.e., conspecific individuals) and initiates appropriate escape behavior; or (3) the gastropod accumulates noxious organic compounds in its tissues, thereby becoming distasteful to potential predators.[5]

5. See *Topics for Further Discussion and Investigation,* no. 5.

Of great importance in the evolutionary history and present-day biology of gastropods is the phenomenon of **torsion,** a counterclockwise 180° twisting of the head and foot relative to the shell, mantle, and the rest of the body (the **visceral mass**) during early development. As a consequence of torsion, the nervous and digestive systems become obviously twisted, and the mantle cavity moves from the rear of the animal to become positioned over the

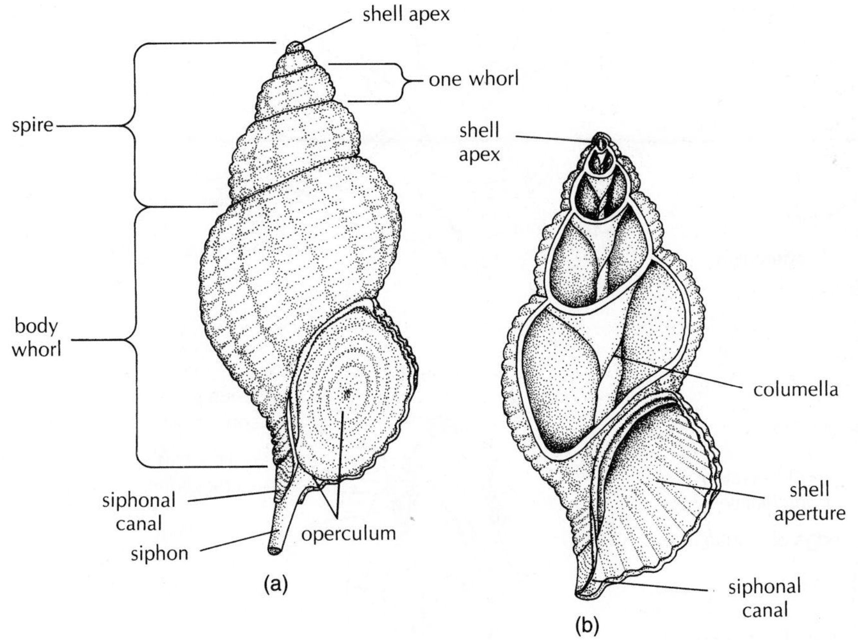

Figure 12.11

(a) Major external features of a gastropod shell. The siphon is a fold of mantle tissue through which water enters the mantle cavity. The operculum is a rigid, proteinaceous shield attached to the foot; when the animal fully withdraws into its shell, the operculum seals the aperture, protecting the animal from predators and physical stresses. (b) Longitudinal section through a typical gastropod shell. As the snail grows, its body coils around the columella.

(b) After Hyman; after Dakin.

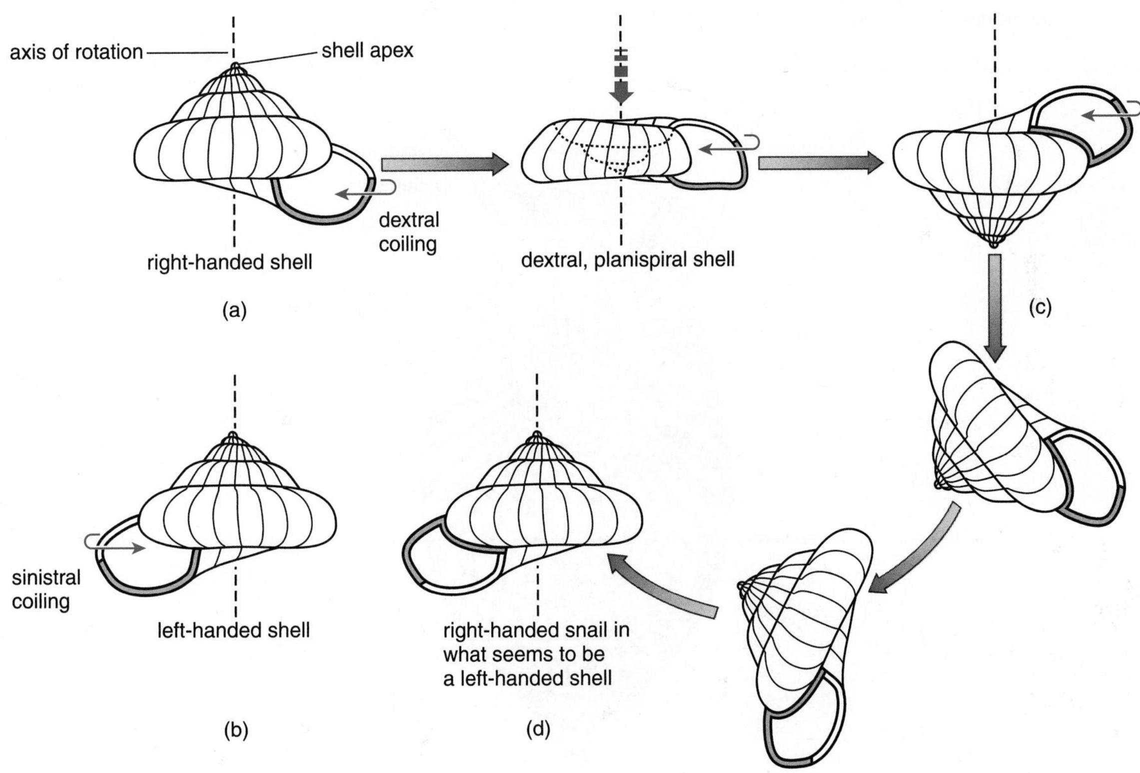

Figure 12.12

Major patterns of shell coiling among gastropods. Blue arrows indicate direction of shell coiling. Members of most species coil to the right and live in right-handed shells (a), although members of some species coil to the left and live in left-handed shells (b). As an intriguing complication, a right-handed snail can produce what appears to be a left-handed shell, by coiling up the central axis rather than down the central axis (c); only anatomical observations can confirm the true direction of body coiling. To visualize getting from (a) to (c), imagine pushing downward on the apex of the shell in (a), producing a planispiral shell, with all shell whorls in one plane, as an intermediate step. Continue pushing downward until you get to (c); note that the direction of shell coiling has not been altered in the process. Rotate shell to obtain (d).

Based on R. S. Boardman *et al.*, eds. *Fossil Invertebrates*. 1987.

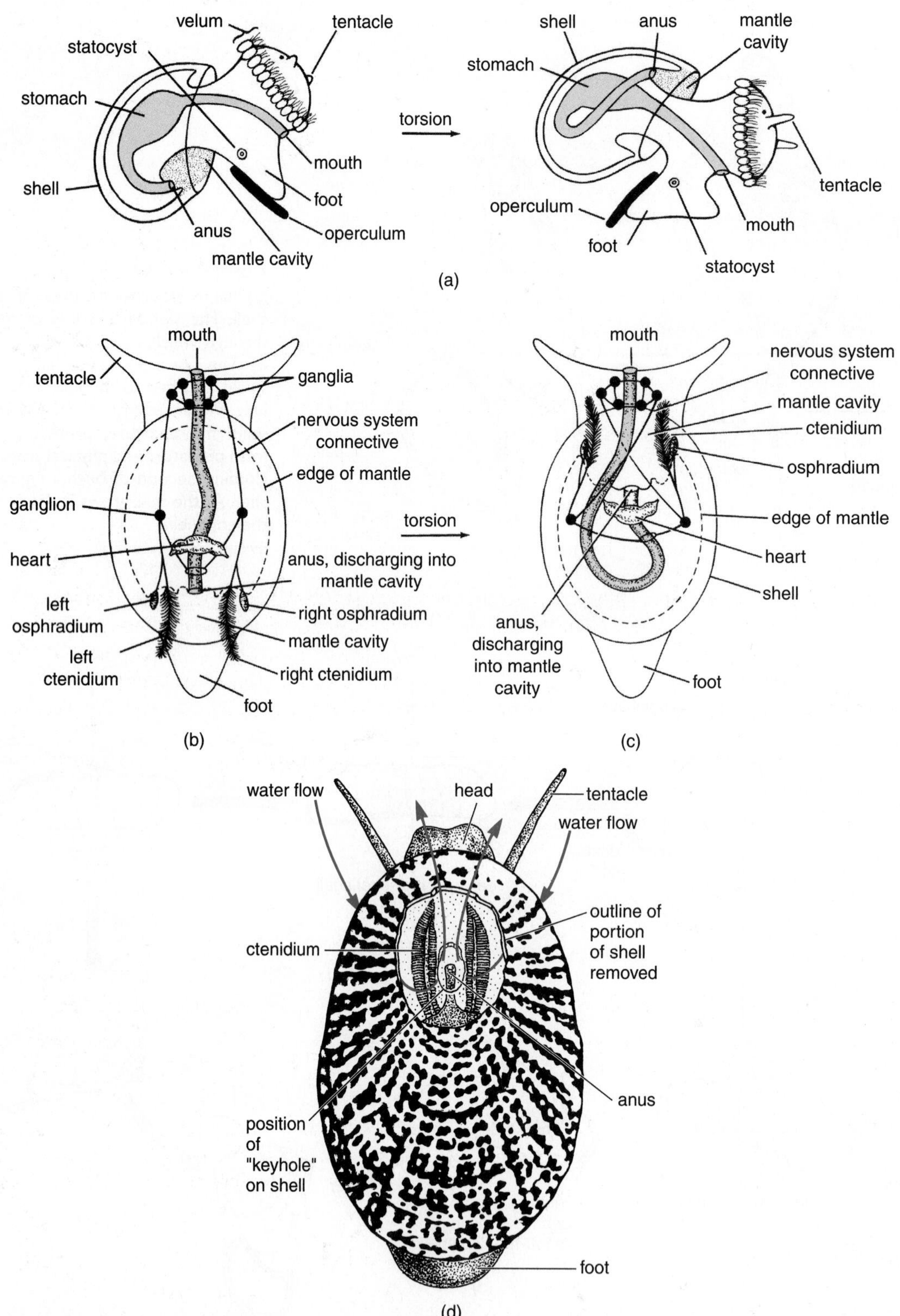

Figure 12.13

(a) Torsion in the free-swimming larva of a primitive prosobranch gastropod, *Patella* sp. Note that the mantle cavity is moved, along the right side of the animal, from the posterior to the anterior of the larva. Following torsion, the head and foot can be fully retracted into the mantle cavity and the aperture tightly sealed by the rigid operculum. (b, c) The consequences of torsion to the adult gastropod. (b) The untorted state of a hypothetical ancestral gastropod-like mollusc. (c) The rearrangement of internal anatomy following torsion. Note that the primitive gill has leaflets extending from both sides of the central axis. As will be discussed later, this is termed a bipectinate gill. (d) Pathway of water circulation through the mantle cavity of a primitive gastropod—a keyhole limpet, order Archaeogastropoda—with paired gills. Water enters on both sides of the head and leaves through a circular opening (the "keyhole") in the shell.

head (Fig. 12.13). This dramatic rearrangement of the internal and external anatomy is brought about largely through the particular orientation and asymmetric development of the retractor muscles attaching the head-foot of the developing embryo or larva to its shell. Because the retractor muscles on the body's right side develop before those on the left, the mantle cavity and associated organs apparently get pulled along the animal's right side toward the front.

Torsion may occur within a few hours or even minutes in some species. It has no direct relationship to shell coiling: The two are separate and independent processes.

The adaptive significance of torsion has been the subject of considerable speculation.[6] The controversy focuses largely on whether torsion benefits the larva or the adult, or both, and part of the difficulty in interpreting the "why" of torsion is that movement of the mantle cavity anteriorly would seem to be a mixed blessing at best. Certainly, through torsion, the ctenidia and osphradia come to be located at the front of the animal, in the direction of locomotion; but torsion also shifts the anus so that it discharges over the head, creating a potentially serious (and seemingly distasteful) sanitation problem. Moreover, the story of subsequent gastropod evolution clearly involves compensating for the results of embryonic or larval torsion; that is, the most advanced gastropods (opisthobranchs and pulmonates, known jointly as heterobranchs) exhibit a marked reduction in the degree to which torsion occurs during development. In some species, an apparent **detorsion** occurs subsequent to torsion, probably through a process of differential growth. The selective pressures responsible for the evolution of torsion, as with all "why did it happen?" evolutionary questions, are ultimately unknowable. Yet the phenomenon is too dramatic to be ignored (Research Focus Box 12.1). Torsion occurs in no other class of mollusc.

Small gastropod species may move largely through the action of cilia located on the ventral surface of the foot, but most species move by means of **pedal waves** of muscle contraction. Unlike peristaltic waves, pedal waves generally do not involve circular muscles or muscular contractions of great magnitude, and they are restricted to the animal's ventral surface.

The musculature of the foot is predominantly vertical (dorsoventral) and transverse (Fig. 12.14a). At the start of a pedal wave, the dorsoventral musculature contracts at the anterior portion of the foot. Apparently, the transverse muscles do not relax, which means that the foot cannot widen; instead, the foot is squeezed forward. A wave of contraction of the dorsoventral musculature then moves posteriorly, allowing the rest of the foot to catch up with the anterior (Fig. 12.14a). The edges of the foot are temporarily sealed against the substrate with mucus secreted by glands on the foot's ventral surface, so a small negative pressure (suction) is generated in the space between the substrate and the raised portion of the foot. The dorsoventral muscles are reextended when they relax, at least in part, by this negative external pressure; this small space thus acts as a hydrostatic skeleton, even though it is external to the body, allowing the musculature at the forward edge of the wave to antagonize that at the trailing edge. In this case, the hydrostatic skeleton operates through a temporary pressure decrease, essentially sucking the raised portion of the foot downward and extending the associated muscles, rather than through a temporary pressure increase.

The previous description applies to **retrograde waves** (*retro* = L: back; *grad* = L: step); the wave of muscular contraction travels in the direction opposite that in which the snail is moving. Pedal waves may also be **direct** (i.e., moving in the same direction as the animal) (Fig. 12.14b). The mechanical details of pedal wave formation may differ substantially among species.[7]

Gastropods play an important, although indirect, role in transmitting several major human diseases, with many species serving as obligate intermediate hosts in the life cycles of parasitic flatworms (phylum Platyhelminthes, class Trematoda; Chapter 8). Indeed, much research on the control of these flatworm parasites has focused on regulating the snail populations.

The Prosobranchs

Proso • branch
(G: anterior gill)
prō´-sō-brank

Defining Characteristic: Mantle cavity generally anterior, due to torsion

Gastropod systematics has been unsettled for many years. The previous system of 3 subclasses (Prosobranchia, Opisthobranchia, and Pulmonata) has given way to a more complex system that better reflects current understanding of the evolutionary relationships among the various groups. In particular, the Prosobranchia no longer exists as a formal taxonomic category.[8] It is still common practice to refer to all these species informally as "prosobranchs," however, and I adopt that approach here.

Most gastropod species are prosobranchs, and most prosobranch species are marine, although a small percentage live in fresh water or on land. At least 20,000 species

6. See *Topics for Further Discussion and Investigation,* no. 3.

7. See *Topics for Further Discussion and Investigation,* no. 16.

8. Prosobranch species are now assigned to either the subclass Eogastropoda, which contains the most primitive species (the true limpets, Order Patellogastropoda) or to the much larger subclass Orthogastropoda, which includes some archaeogastropods, all of the mesogastropods, and all of the neogastropods (see *Taxonomic Detail* at the end of this chapter).

RESEARCH FOCUS BOX 12.1

Gastropod Torsion

Pennington, J. T., and F. S. Chia. 1985. Gastropod torsion: A test of Garstang's hypothesis. *Biol. Bull.* 169:391–96.

In the late 1920s, the English zoologist Walter Garstang hypothesized that torsion arose as an adaptation for larval life rather than for adult life. Before torsion, he reasoned, the head and velum, the uniquely larval organ used for food collection and swimming, were more vulnerable to predation; because the untorted shell opened posteriorly, the head and velum were withdrawn last. Torsion brought the mantle cavity anteriorly, providing a space in which to rapidly retract these anterior structures when danger threatened. According to Garstang's hypothesis, torted individuals should therefore have survived better than their untorted counterparts, and the genes for larval torsion would quickly have become permanent residents in the life history.

Although Garstang's ideas have been much discussed and widely approved over the past 70 years, they were never directly tested until Pennington and Chia (1985) exposed abalone larvae (*Haliotis kamtschatkana*) to 7 different potential planktonic predators: crab larvae (the final larval stage—the megalopa), copepods, 2 ctenophore species, 2 species of hydromedusae, and one species of fish. Both pretorted and torted abalone larvae were tested, with 50 larvae put into each of 5 containers for each predator studied. Larvae that disappeared from test containers over 15-hour test periods were assumed to have been eaten; 5 control containers, each with 50 abalone larvae but without predators, were used to adjust for any losses due to the counting and handling errors inevitably associated with working with very small aquatic animals—these larvae are only a few hundred μm (micrometers) long. If Garstang's larval adaptation hypothesis is correct, there should be less predation on the abalone larvae after they undergo torsion.

The results did not support Garstang's hypothesis. Five of the 7 predators ate as many pretorted as torted larvae during the 15-hour experiment (Focus Fig. 12.1). Only the hydromedusae of *Aequorea victoria* showed a statistically significant preference for pretorted larvae. The crab larvae (species not determined) actually ate more torted than pretorted abalone larvae. On balance, there was no indication that torsion protected abalone larvae from predation.

In fairness to Garstang's hypothesis, it should be noted that Pennington and Chia did not test fully torted larvae; their "torted" larvae had only completed the first 90° of torsion. Even though each such larva could retract fully into its shell when provoked and could completely seal off the shell aperture with the operculum, fully (180°) torted larvae might be less vulnerable to attack. Also, larvae of other gastropod species might be better protected by torsion than are those of the abalone. In addition, the sorts of predators that may have brought about selection for torsion in ancestral gastropods perhaps no longer exist, or at least were not tested in this study. We cannot conclude that Garstang's ideas about the importance of predation on larvae in selecting for torsion are wrong; the limited experimental evidence now available can only suggest that torsion today does not effectively protect abalone larvae against most planktonic predators.

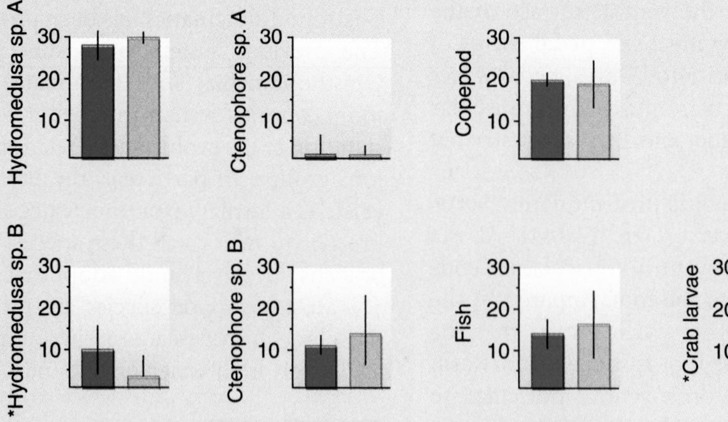

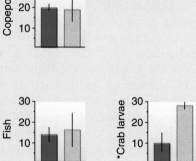

Focus Figure 12.1

Predation by 7 predators on pretorted (blue bars) and torted abalone larvae (*Haliotis kamtschatkana*). Data are mean number of larvae eaten (mean of 5 replicates, 50 larvae per replicate) within a 15-hour test period. Error bars show the standard deviation about the mean. An * indicates that predation on pretorted larvae was significantly different from predation on torted larvae for that predator ($P < 0.05$). Crab larvae = 5 species, mixed; copepod = *Epilabidocera longipedata;* fish = *Oncorhynchus gorbuscha;* hydromedusa sp. A = *Phialidium gregarium;* hydromedusa sp. B = *Aequorea victoria;* ctenophore sp. A = *Pleurobrachia bachei;* ctenophore sp. B = *Bolinopsis infundibulum.*

Based on Pennington, J. T., and F. S. Chia. 1985. "Gastropod torsion: a test of Garstang's hypothesis" in *Biological Bulletin* 169:391–96. 1985.

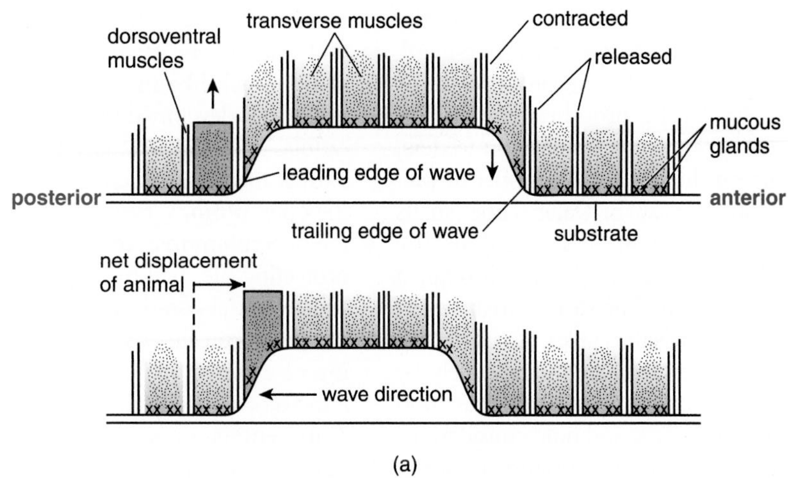

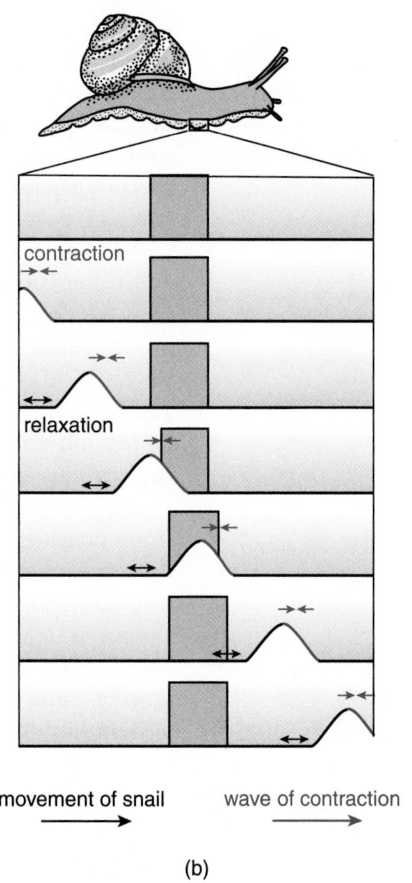

Figure 12.14

(a) Diagrammatic longitudinal section through the foot of a moving gastropod. A pedal wave is traversing the foot from anterior to posterior, while the animal advances in the opposite direction; this is called a retrograde pedal wave. Localized contraction of dorsoventral muscles raises a small portion of the foot away from the substrate, producing a small but measurable suction. This suction helps stretch the dorsoventral muscles when they relax, pulling the area at the rear of the wave back against the substrate. Note the forward progression of the blue region of the foot. (b) Direct waves in the locomotion of a terrestrial snail (Pulmonata). Direct waves move in the direction of locomotion, from posterior to anterior. Only the portion of the foot lifted away from the substrate moves forward (note the forward progression of the blue region of the foot). Many waves travel down the foot simultaneously, as shown in the snail at the left.

(a) Based on Jones and Trueman, in *Journal of Experimental Biology,* 52:201, 1970.
(b) Based on Lissman, in *Journal of Experimental Biology,* 21:58, 1945.

have been described, mostly contained within the Caenogastropoda. Prosobranchs are generally free-living and mobile, although some species have evolved sessile or even parasitic lifestyles. Free-living prosobranchs may be herbivores, deposit feeders, omnivores, suspension feeders, or carnivores, depending on the species. A number of carnivorous species—notably the warm-water cone snails (*Conus* spp.)—produce potent venoms (Fig. 12.15) that are injected into fish, molluscan, or annelid prey through a harpoon-like, hollow radular tooth. The various toxins exert their effects by binding to very specific classes of cell surface receptors and ion channels; they are being used widely by neurobiologists to study receptor and ion channel function in both invertebrates and vertebrates, and hold considerable promise for the treatment of pain, heart arrhythmia, clinical depression, and epilepsy. Unfortunately, many cone snail species seem close to extinction, due to severe habitat degradation and overzealous shell-collecting.[9]

An entirely different nutritional mode seems likely for a still-undescribed gastropod recently discovered in the deep sea. These snails contain symbiotic bacteria living in the gill tissue; the bacteria may provide the snail with nutrients (see p. 247).

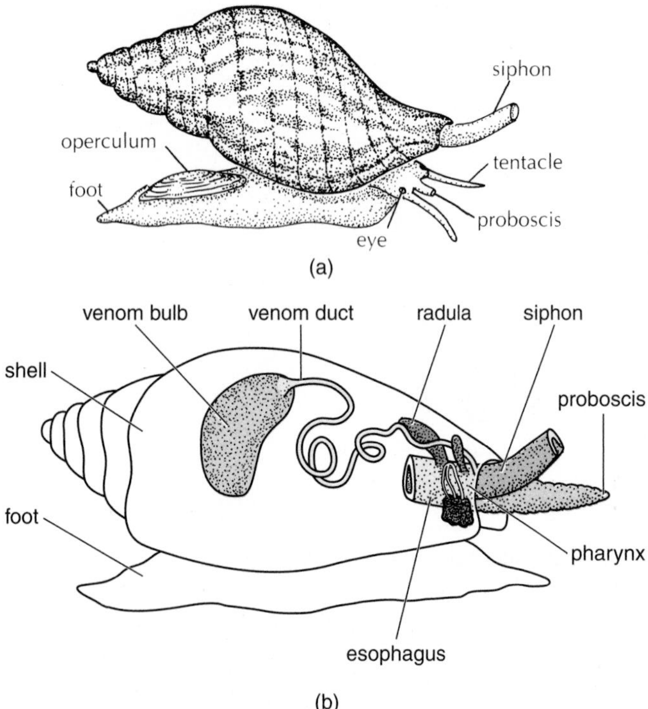

Figure 12.15
(a) Prosobranch gastropod (*Fasciolaria tulipa*) showing operculum and extended siphon. (b) Schematic diagram illustrating the venom apparatus in the carnivorous prosobranch gastropod *Conus* sp.
(a) After Niesen. (b) Based on F. E. Russell, in *Advances in Marine Biology*, 21:59, 1984.

Prosobranchs are the most primitive of gastropods; that is, other gastropods most likely evolved from prosobranch-like ancestors. Most prosobranch species possess a well-developed shell, mantle cavity, osphradium, and radula, and the foot usually bears a rigid disc of protein (sometimes strengthened with calcium carbonate) called the **operculum.** When the foot is withdrawn into the shell, the operculum may completely seal the shell aperture, thus protecting the snail from predators and from such physical stresses as dehydration and low salinity (Figs. 12.11, 12.15a).

The typical prosobranch gill is a **ctenidium,** consisting of a series of flattened, triangular sheets (filaments) lying one adjacent to the next (Fig. 12.16). Deoxygenated blood enters an afferent blood vessel from the animal's open system of blood sinuses (the **hemocoel**). Once distributed to the individual sheets of the ctenidium, the blood moves through the sheet, where it becomes oxygenated, and then on to the auricle of the heart through an efferent blood vessel. From the auricle, the blood is pumped into the single associated ventricle and is then distributed to the tissues through a single aorta leading to the blood sinuses of the hemocoel. Primitive prosobranch species (including the true limpets, order Patellogastropoda) possess a pair of auricles, a pair of efferent blood vessels, and a pair of ctenidia (Fig. 12.13d).

Water is drawn into the mantle cavity and across the gill sheets by the movements of gill cilia. In many prosobranch species, a portion of the mantle is drawn out into a cylindrical extension called the **siphon** (Figs. 12.15a and 12.16c); water is drawn through this siphon, by the action of the gill cilia, into the mantle cavity and across the **osphradium** (a chemical and tactile receptor organ—Figs. 12.2 and 12.16a, c). The snail moves the muscular siphon back and forth, sampling the water from different directions. In burrowing species, the siphon is extended through the substrate to the water above. The gastropod siphon is especially well developed in carnivores and scavengers—which often hunt their prey by chemical sensing—and generally reduced or absent in suspension feeders, herbivores, and deposit feeders.

In most gastropod species, water enters the mantle cavity at the left side of the head, passes over or between the gill filaments, and exits at the right side of the head (Fig. 12.16). In primitive archaeogastropod species with paired gills, water is necessarily drawn instead into the mantle cavity at both sides of the head; unidirectional water flow in such species is made possible by the presence of conspicuous lateral or posterior slits or circular openings in the shell, as in keyhole limpets (Fig. 12.13d) and abalone (Fig. 12.17e). In all species, water movement across the gill is unidirectional and counter to the direction of blood flow, so that the principle of countercurrent exchange always applies (pp. 216–218).

Much of prosobranch evolution is a story of changes in gill numbers (from 2 in the more primitive [ancestral] species, to one in the more advanced [derived] species), and changes in the orientation of gill filaments extending from

9. See *Topics for Further Discussion and Investigation*, no. 8.

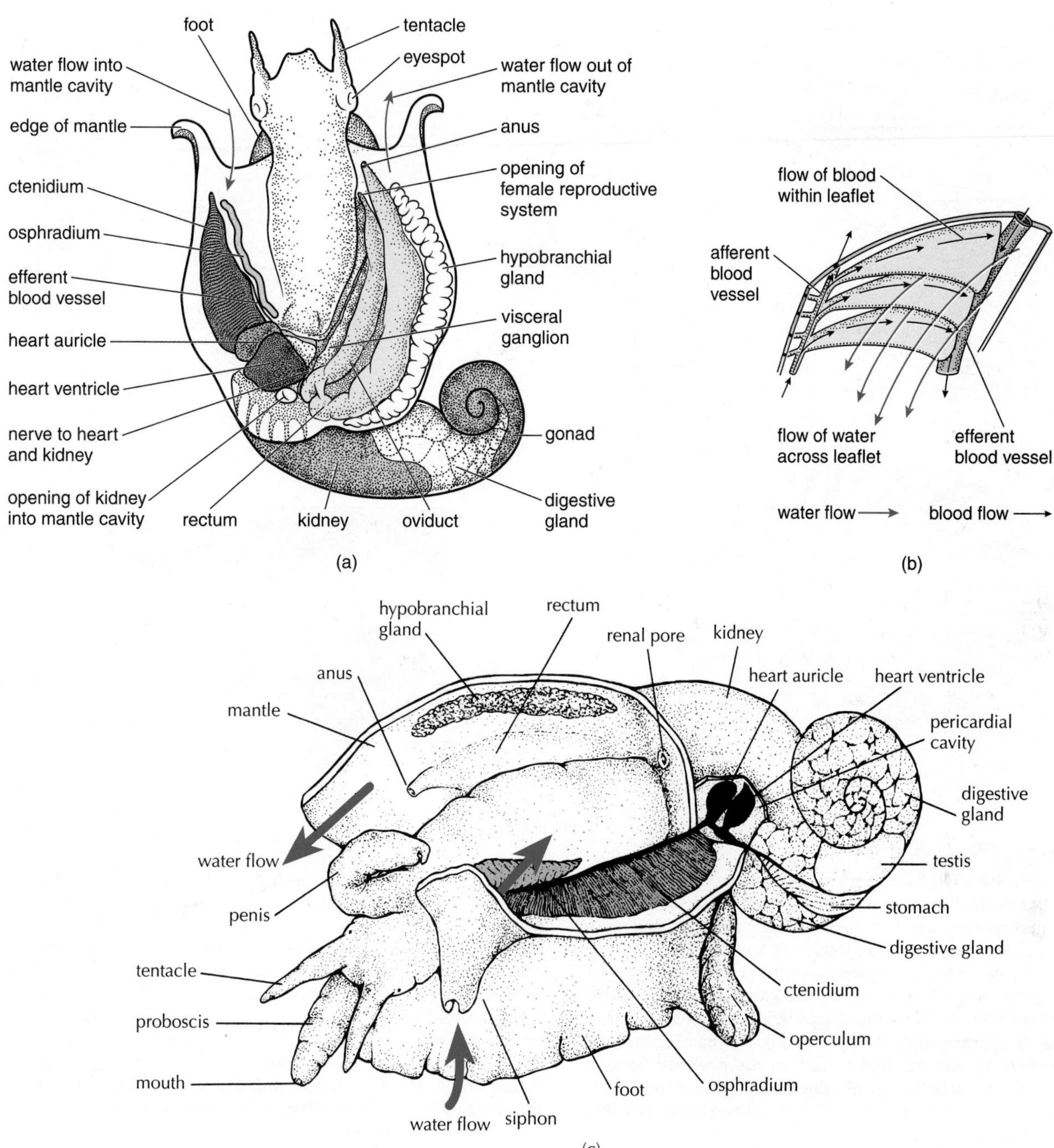

Figure 12.16

(a) The periwinkle *Littorina littorea,* removed from its shell and with its mantle cut middorsally to reveal the arrangement of organs within the mantle cavity. Note the major blood vessel (efferent blood vessel) that conducts blood from the gill to the auricle of the heart. From here, the blood moves to the ventricle and then to the tissues. The hypobranchial gland secretes mucus for binding particles carried by the ctenidial filaments. (b) Detail of ctenidium, showing direction of blood flow within individual gill sheets (black arrows) and water flow across the gill sheets (blue arrows). (c) The whelk *Busycon* sp., with mantle cavity exposed and the top of the pericardium dissected away.

(a, b) Based on Fretter and Graham, 1976. *A Functional Anatomy of Invertebrates.* (c) Modified from Brown, 1950. *Selected Invertebrate Types.*

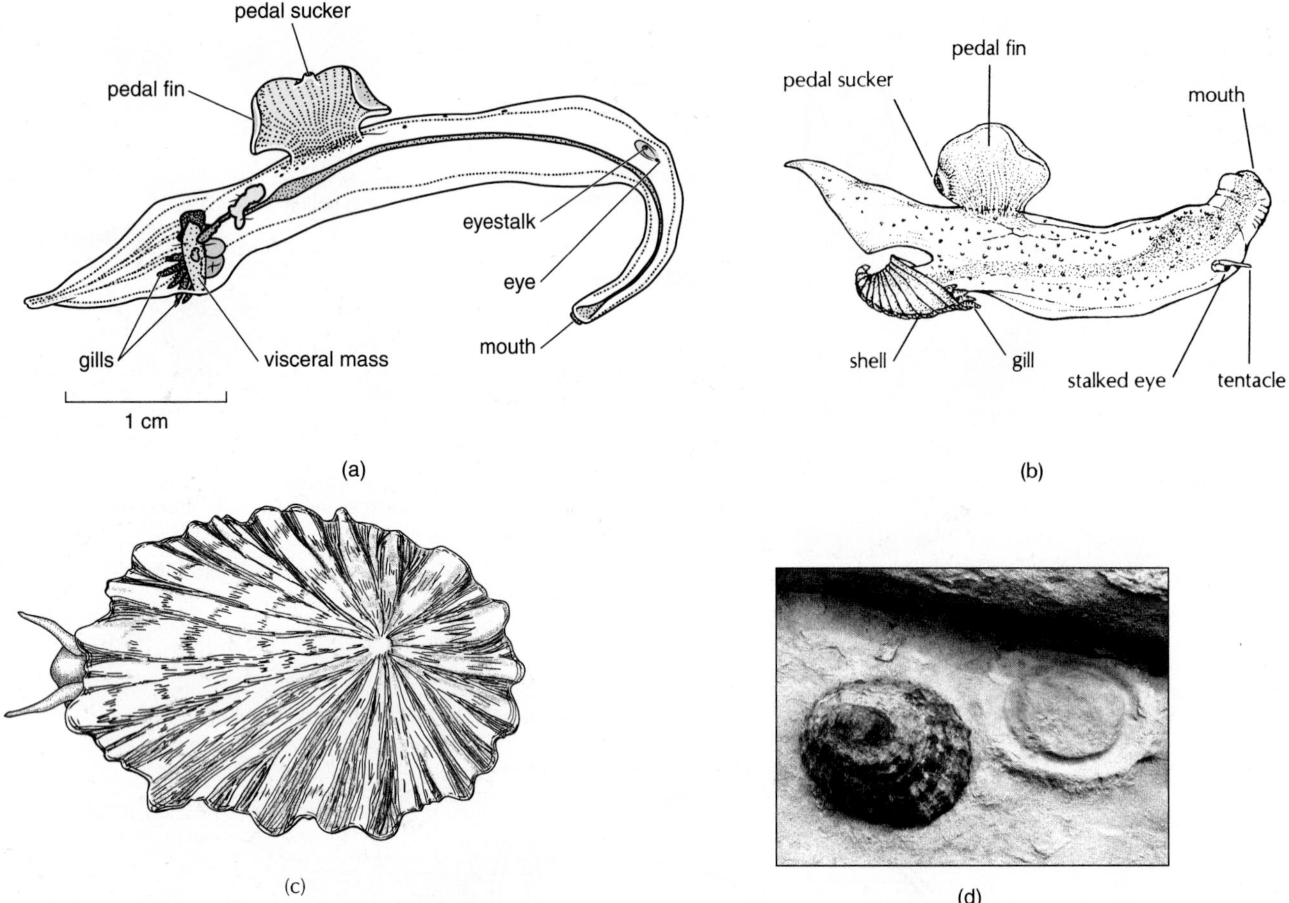

Figure 12.17a, b, c, d

Prosobranch diversity. (a) *Pterotrachea hippocampus,* a Hawaiian heteropod. The shell is lost at metamorphosis from the larval stage. The visceral mass contains the digestive tract, heart, nephridia, and much of the reproductive system. All heteropods are planktonic. (b) *Carinaria lamarcki,* a shelled heteropod. Note the small size of the shell relative to the rest of the animal. (c) The intertidal limpet *Collisella scabra* from California. Note the complete absence of shell coiling. The shell is approximately 2 cm long. (d) The patellogastropod limpet *Cellana* sp. and its "home scar" in the intertidal zone at Kaikoura, NZ. Intertidal limpets of many species graze actively while submerged and then return to their starting position before low tide, using chemical information in their mucous trails to find their way back home. The shell is about 3 cm long. (e) The abalone, *Haliotis* sp. Gill cilia draw water into the mantle cavity at the anterior end of the body. Water leaves through a series of posterior shell perforations. The openings of the digestive and excretory systems are located beneath one of these openings. (f) A worm-shell snail, *Vermicularia* sp. This is a sessile gastropod that lives attached to solid substrates, including rocks and other shells; the foot is reduced appropriately. As a juvenile, the animal produces a typical, spirally coiled shell, as observed near the apex of the adult shell. As the animal grows, however, the coiling becomes very loose, so that the whorls become disconnected. This corkscrew-shaped shell resembles that secreted by some species of sessile, polychaete worm. (g) A volute shell, *Scaphella* sp. All volutes are marine carnivores, feeding on other invertebrates. (h) The giant tun shell, *Tonna galea.* Note the conspicuous shell ridges, the low spire, and the large body whorl. The adult lacks an operculum. (i) The slipper shell snail, *Crepidula fornicata.* Adults live in stacks, as shown, with the large foot becoming little more than a strong suction device. Each individual eventually transforms from male to female, so that the oldest snails (at the base of the stack) are females. Females are typically 3 cm to 4 cm long.

(a) Based on R. Seapy, 1985, in *Malacologia,* 26:125–35. (c) Based on a photograph by Tom Niesen. (d) Photo by the author. (e) After Hyman. (g) After Hardy. (h) From Pimentel, after Abbott. (i) Modified from the drawings of W. R. Coe and others.

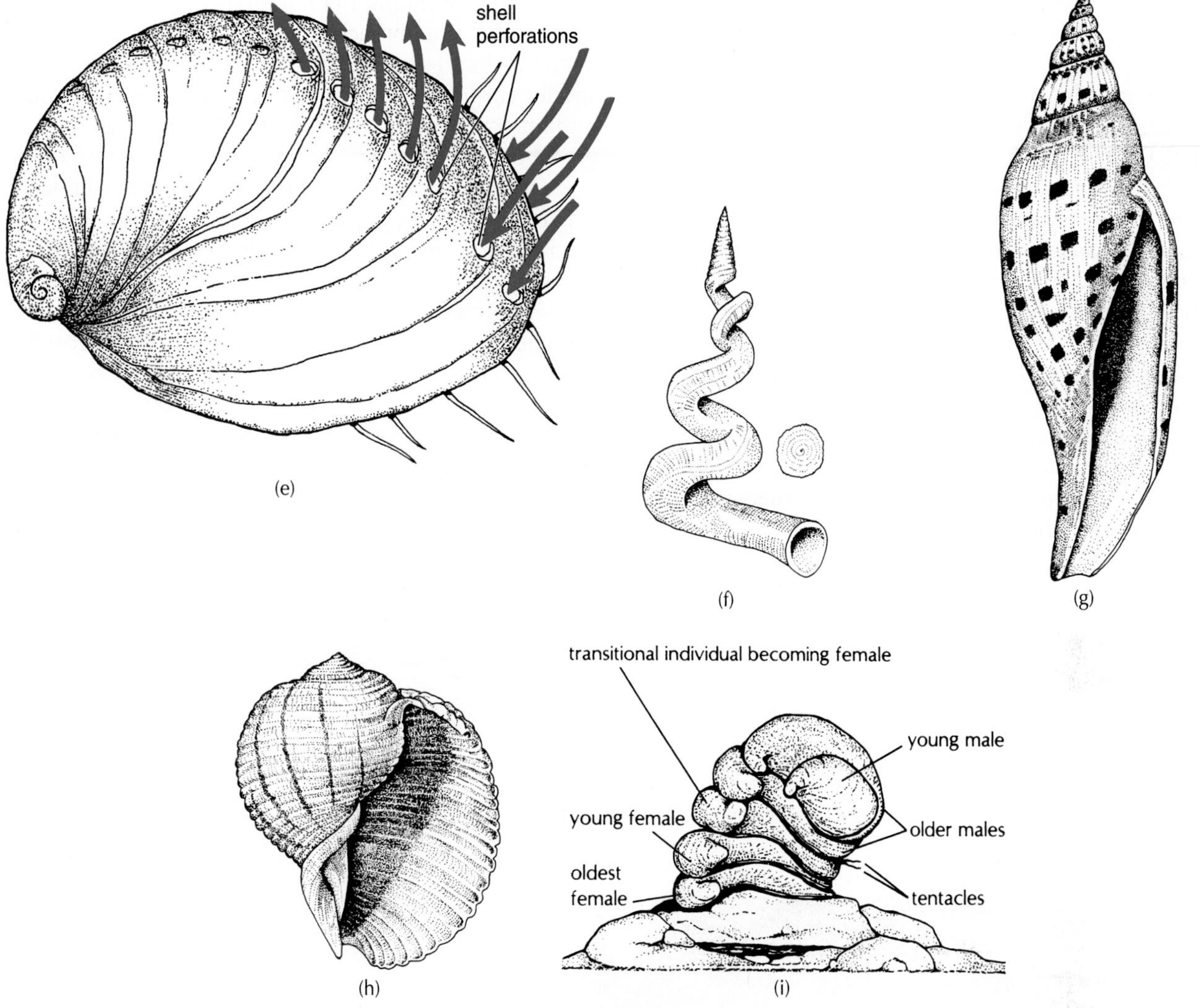

Figure 12.17e, f, g, h, i

the ctenidial axis. In the primitive (ancestral) **bipectinate** condition, gill filaments extend from both sides of the ctenidial axis, while in the relatively advanced (derived) **monopectinate** condition (Fig. 12.16), the filaments project from only one side (the "downstream" side) of the supporting axis. Regardless of the number of gills and the placement of gill filaments in different prosobranch species, however, the principle of countercurrent exchange applies to all. Indeed, the countercurrent principle applies to all ctenidia in the phylum Mollusca, except those of cephalopods.

Considerable anatomical and functional diversity exists among prosobranchs (Fig. 12.17). The members of one group, the **heteropods,** show especially striking modifications of the basic prosobranch body plan and lifestyle. The heteropods are planktonic, voracious carnivores, whose shell is reduced or absent and whose foot is a thin, undulating paddle that propels the animal through the water (Fig. 12.17a, b). Except for the viscera, the body is nearly transparent, an excellent adaptation for inconspicuous water travel. As illustrated in Figure 12.17a, b, heteropods swim upside down.

The Opisthobranchia[10]

The Opistho • branchia
(G: posterior gill)
ōpis-thō-brank´-ē-ah

Defining Characteristic: Mantle cavity lateral or posterior due to detorsion, or lost

Opisthobranchs, a group that includes the sea hares, sea slugs, and bubble shells, are almost all marine. Fewer than 2000 species have been described. The characteristics that distinguish adults of this group from those of the prosobranchs are (1) a trend toward reduction, internalization, or loss of the shell, (2) reduction or loss of the operculum

10. The Opisthobranchia is a valid taxonomic category, but there is disagreement about its taxonomic level. Opisthobranchs are now contained within the subclass Orthogastropoda, a group that also contains most prosobranch species and all pulmonate species.

(although it is present in all larvae), (3) limited torsion during embryogenesis, (4) reduction or loss of the mantle cavity, and (5) reduction or loss of the ctenidia. Most species that have lost the ctenidia have evolved other respiratory structures that are developmentally unrelated to the ancestral gill. For example, in many sea slugs (the nudibranchs—order Nudibranchia), gas exchange occurs across brightly colored dorsal projections called **cerata** (Fig. 12.18b, c), which also contain extensions of the digestive system. In at least one species, the cerata exhibit rhythmic muscular contractions, apparently serving to move blood through the hemocoelic sinuses for gas exchange.

Shell reduction or loss potentially increases vulnerability to predators, and it is reasonable to expect that pressures selecting for alternate means of defense have been quite strong. In particular, the cerata of many opisthobranch species house unfired defensive organelles (**nematocysts**) usurped from cnidarian prey; these **nematocysts** can then function in defending the nudibranch. Instead of cerata, many other nudibranchs possess feathery gills arising from the dorsal surface (Fig. 12.18a). Some opisthobranch species (including the sea hares, *Aplysia* spp., and many nudibranch species) are chemically defended against predation.

The opisthobranch head typically bears in addition to a pair of tentacles adjacent to the mouth as in prosobranchs, a second pair of tentacle-like structures located dorsally, called **rhinophores.** The rhinophores are believed to be chemosensory, making them analogous to the osphradium of prosobranch gastropods and to the osphradium of those opisthobranchs bearing a mantle cavity.

Opisthobranchs show various degrees of departure from the ancestral, prosobranch-like condition. Adult sea slugs, for example, have no mantle cavity, ctenidia, osphradium, shell, or operculum, and some species show no evidence of torsion as adults. Sea slug larvae, on the

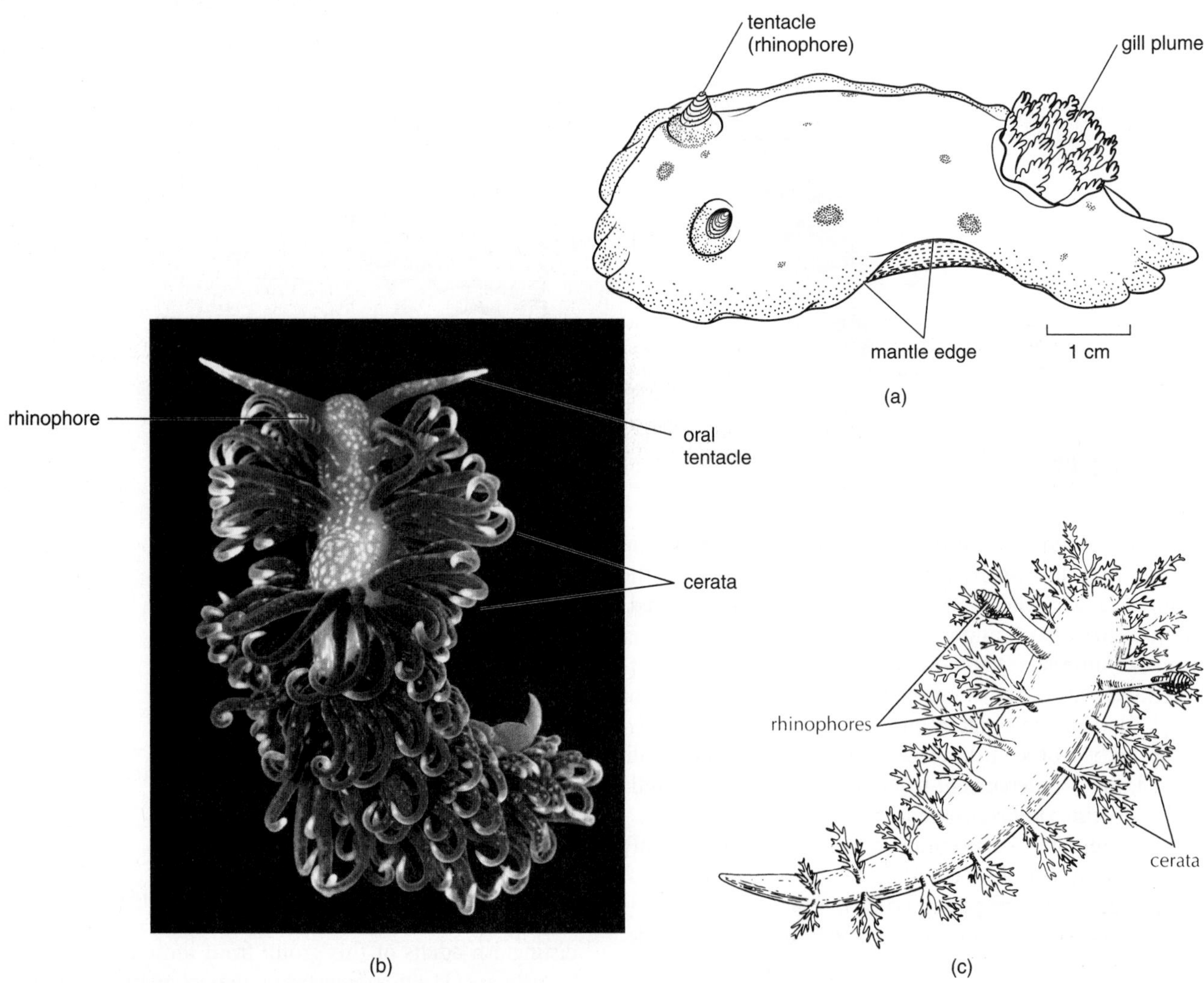

Figure 12.18

(a) A dorid nudibranch, *Dialula sandiegensis.* Note the gills arranged as a plume around the anus. (b) An eolid nudibranch, *Spurilla neapolitana.* The conspicuous dorsal projections are cerata, which serve for gas exchange and also contain outfoldings of the digestive system. (c) A dendronotid nudibranch, *Dendronotus arborescens.* Note the elaborate branching of the cerata. This and related species are capable of simple swimming by flexing the body from side to side.

(a) Based on Bayard H. McConnaughey and Robert Zottoli, *Introduction to Marine Biology*, 4th ed. (b) Copyright © L. S. Eyster. (c) After Kingsley.

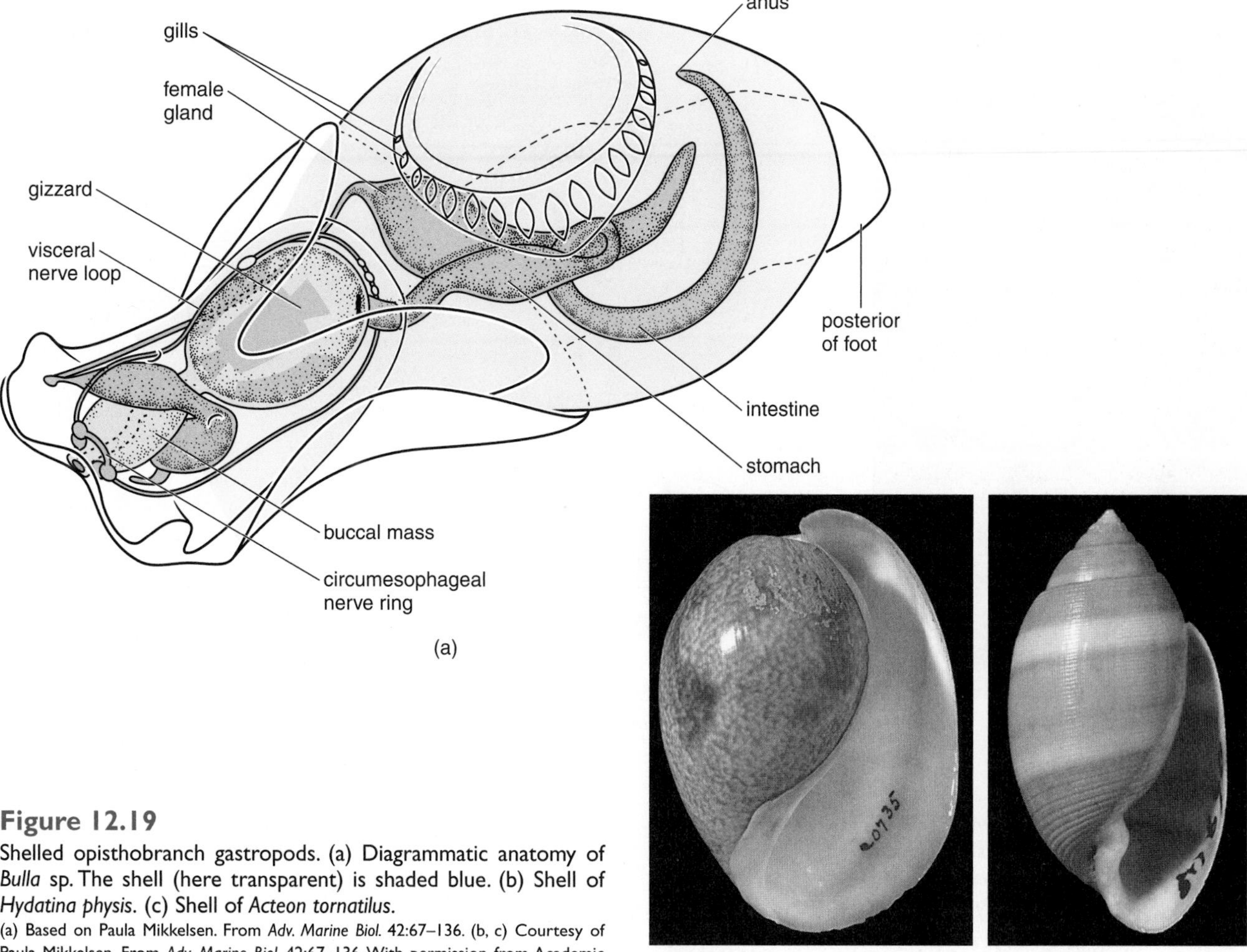

Figure 12.19
Shelled opisthobranch gastropods. (a) Diagrammatic anatomy of *Bulla* sp. The shell (here transparent) is shaded blue. (b) Shell of *Hydatina physis*. (c) Shell of *Acteon tornatilus*.
(a) Based on Paula Mikkelsen. From *Adv. Marine Biol.* 42:67–136. (b, c) Courtesy of Paula Mikkelsen. From *Adv. Marine Biol.* 42:67–136. With permission from Academic Press, NY.

other hand, have a pronounced mantle cavity, shell, and operculum, indicating a clear affinity with the prosobranchs. Another common opisthobranch group, the sea hares (order Anaspidea) have a mantle cavity (with gill and osphradium) as adults, and adults of most species also have a shell. However, the mantle cavity is very small and on the animal's right side, and the shell is reduced and internal. Clearly, the sea hares are closer to the ancestral, prosobranch-like condition than are the sea slugs.

Although shell reduction, internalization, or loss is exhibited by most opisthobranchs, a few species possess a conspicuous, external, spirally coiled shell (Fig. 12.19), with operculum. Moreover, the mantle cavity in such species (containing a gill and osphradium) is well developed, and the snails often show little sign of detorsion: In many species, the mantle cavity is still located anteriorly, and the nervous system still shows the fully twisted, "streptoneurous" condition (see Fig. 12.49c).

Although locomotion is generally by means of cilia and pedal waves along the ventral surface of the foot, some opisthobranchs, such as the sea hares, can swim in short spurts by flapping lateral folds of the foot called **parapodia.** In other members of this subclass, the entire foot is drawn out into 2 thin lobes, also called **parapodia,** which are used for swimming. These animals are known as pteropods ("wing-footed"), or sea butterflies (Fig. 12.20). Pteropods may or may not have shells, depending on the species, but all are permanent members of the plankton. Pteropods typically have no specialized respiratory organs, so gas exchange is accomplished across the general body surface.

The Pulmonata[11]

Pulmo • nata
(L: lung)
pul´-mō-not´-ah

Defining Characteristic: Mantle cavity highly vascularized and otherwise modified to form a lung

In contrast to the prosobranch and opisthobranch gastropods, few of the 17,000 pulmonate species are marine,

11. The Pulmonata is a valid taxonomic category, but there is disagreement about its taxonomic level. The group is contained within the subclasss Orthogastropoda, a group that also contains most prosobranch species and all opisthobranch species.

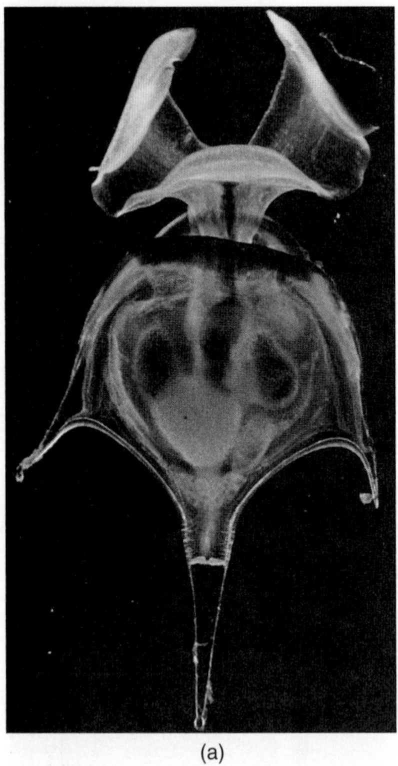

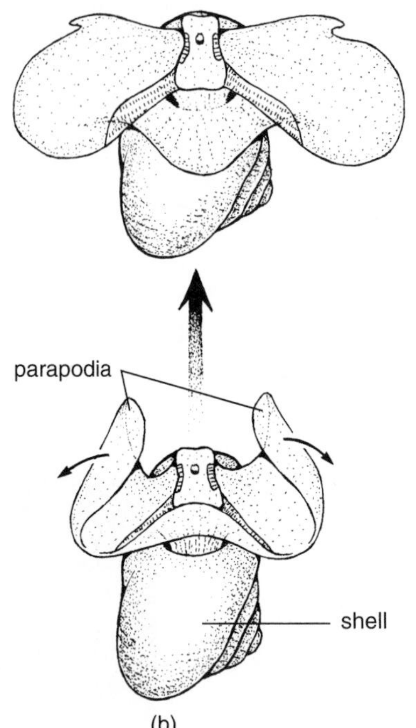

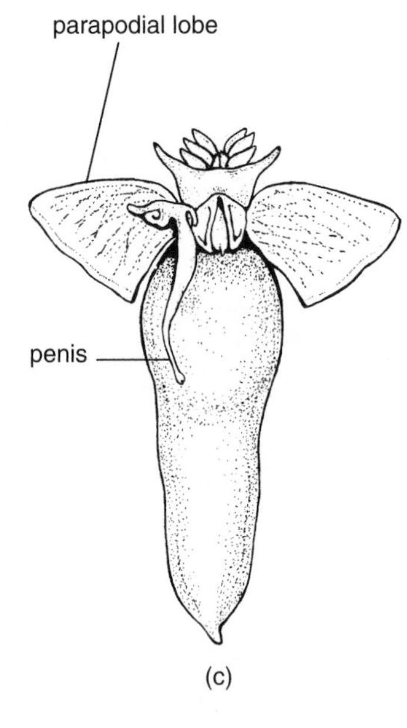

Figure 12.20
Pteropods. (a) *Cavolina* sp., with parapodia exposed. (b) *Spiratella* sp., swimming. The animal is shown executing its power stroke, in which the parapodia are swept downward forcefully, generating forward thrust. The muscular parapodia are elaborations of the ancestral foot. (c) A naked pteropod, *Clione limacina*. The animal loses its shell when the larva metamorphoses.
(a) Courtesy of R.W. Gilmer. (b) After Morton.

and those few species occur only intertidally and in estuaries. Most pulmonate species are found in terrestrial or freshwater environments; slugs and "escargot" (Fig. 12.21) are terrestrial members of this subclass.[12] A coiled shell is present in most pulmonate species, but the shell is reduced, internalized, or completely lost in others (the slugs) (Fig. 12.22c, d). Only a few species have an operculum on the foot. Most pulmonates possess a long radula, in keeping with their generally herbivorous diet, and the head commonly bears 2 pairs of tentacles. Torsion is limited to about 90°, so the nervous system is not so greatly twisted, and the mantle cavity opens on the right side of the body, as in many opisthobranchs.

A major feature distinguishing the pulmonates from other gastropods is that the mantle cavity is highly vascularized and functions as a lung (Fig. 12.21b). Downward movement of the floor of the mantle cavity increases the cavity's volume so that air, or in some cases water, is drawn into the mantle cavity for respiration. The fluid is then expelled by decreasing the volume of the mantle cavity. Air or water flows into and out of the lung through a single small opening called the **pneumostome** (*pneumo* = G: lung; *stoma* = G: mouth) (Figs. 12.21a and 12.22a, c).

12. See *Topics for Further Discussion and Investigation*, no. 10.

Although pulmonates lack ctenidia, a gill has secondarily evolved in some freshwater species. This gill takes the form of folds of mantle tissue near the pneumostome.

Class Bivalvia (= Pelecypoda)

Class Bi • valvia (= Pelecy • poda)
(L: two valved [G: hatchet foot])
bī-val′-vē-ah (pel-iss-ih-pō′-dah)

Defining Characteristics: 1) Two-valved shell; 2) body flattened laterally

The class Bivalvia contains about 15,000 contemporary species, including clams, scallops, mussels, and oysters. Major bivalve characteristics include (1) a hinged shell, the 2 sides (left and right "valves"—*valva* = L: a folding door) of which are closed by 1 or 2 **adductor muscles;** a springy **ligament** springs the shell valves apart when the adductor muscles relax; (2) lateral compression of the body and foot; (3) lack of cephalization: virtual absence of a head and associated sensory structures; (4) a spacious mantle cavity, relative to that found in other molluscan classes; (5) a sedentary lifestyle; and (6) the absence of a

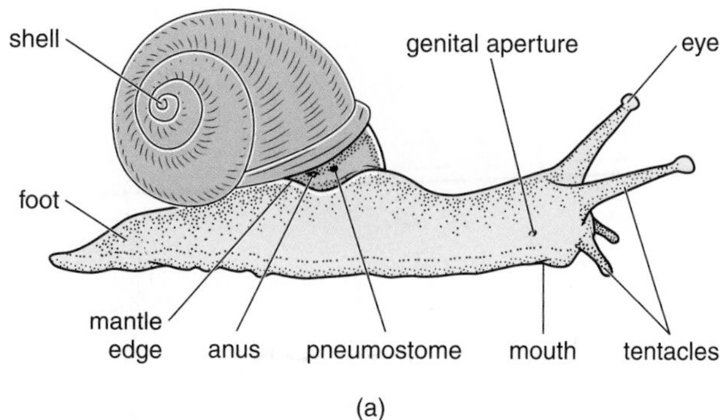

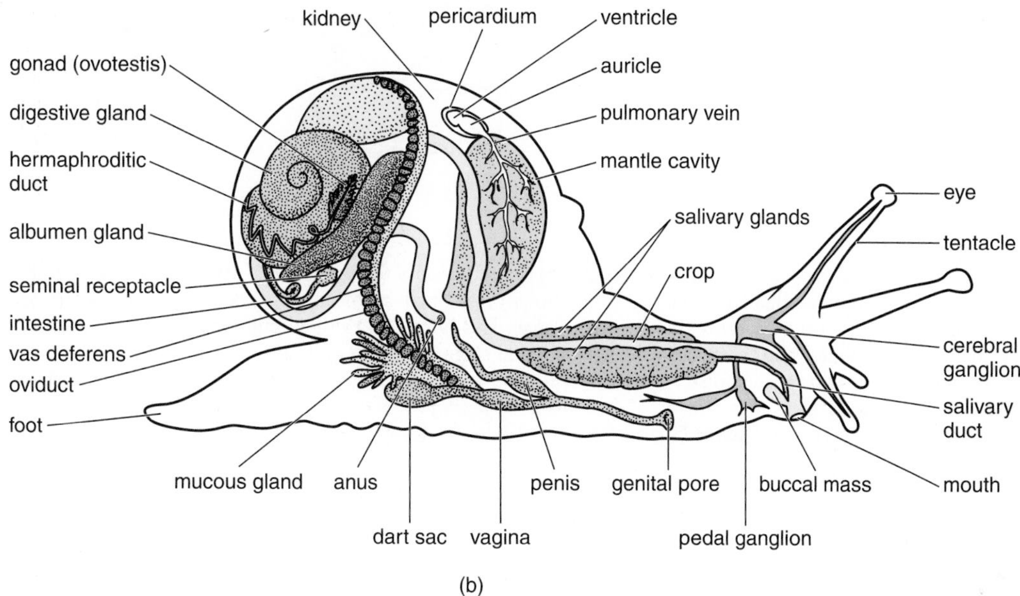

Figure 12.21
(a) External and (b) internal anatomy of the terrestrial pulmonate *Helix* sp. This is the animal eaten as "escargot."

(a) Based on Sherman/Sherman, *The Invertebrates: Function and Form,* 2/e, 1976, p. 236.
(b) Based on *Invertebrate Zoology,* 3d ed. by Joseph Engemann and Robert Hegner. 1981.

radula/odontophore complex. Bivalves are primarily marine, but about 10–15% of all species occur in freshwater. No bivalves are terrestrial. Members of most species are suspension-feeders, using their gill cilia to drive water through the mantle cavity and capture phytoplankton and other microscopic particles from the seawater.

The hinged portion of a bivalved shell is dorsal (Fig. 12.23). The shell valves, then, are on the animal's left and right sides. The shell opens ventrally. A conspicuous bulge in the shell is frequently seen on the dorsal surface, adjacent to the hinge. This bulge, termed the **umbo,** is comprised of the earliest shell material deposited by the animal. Distinct **growth lines** typically run parallel to the shell's outer margins, as illustrated in Figure 12.24. The foot projects ventrally and anteriorly, in the direction of movement, and the siphons, when present, project posteriorly.

For many years, bivalve classification was based mainly on gill structure. The Lamellibranchia, previously the largest of bivalve subclasses, is no longer a valid taxon. Its members have been distributed among 3 new subclasses (see *Taxonomic Detail,* pp. 286–290), based largely on difference in hinge characteristics. For simplicity, I will discuss these bivalves together as "lamellibranchs." The most primitive bivalve species are found within a 4th subclass, the Protobranchia, and the most bizarre bivalves are now placed within a 5th subclass, the Anomalodesmata.

Subclass Protobranchia

Subclass Proto • branchia
(G: first gill)
prō-tō-brank´-ē-ah

Defining Characteristics: 1) Gills small and resembling those of gastropods, functioning primarily as gas exchange surfaces; 2) food collected by long, thin, muscular extensions of tissue surrounding the mouth (palp proboscides)

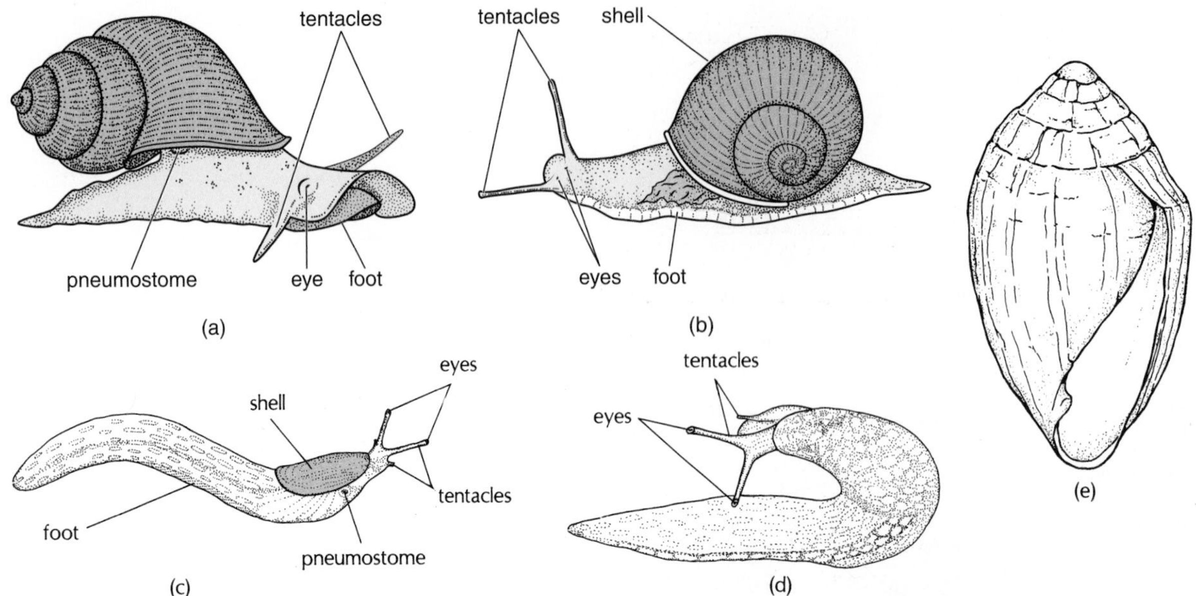

Figure 12.22

Pulmonate diversity. (a) *Lymnaea* sp., with conspicuous pneumostome. Members of this genus are common inhabitants of freshwater lakes and ponds throughout the world, and serve as intermediate hosts for a variety of trematodes and cestodes (Platyhelminthes). (b) *Helisoma* sp., another freshwater pulmonate. Note the planospiral shell, with all the whorls lying in a single plane. (c) A terrestrial slug, *Arion fuscus,* with a reduced, external shell. (d) *Limax flavus,* a terrestrial slug with a small internal shell (not shown). (e) Shell of *Melampus bidentatus,* a common salt-marsh snail. This is one of the few pulmonates with a free-living, marine larval stage in the life cycle.

(a, b) Based on Pennak, *Fresh-Water Invertebrates of the United States,* 2d ed. 1978. (c, d) After Kingsley.

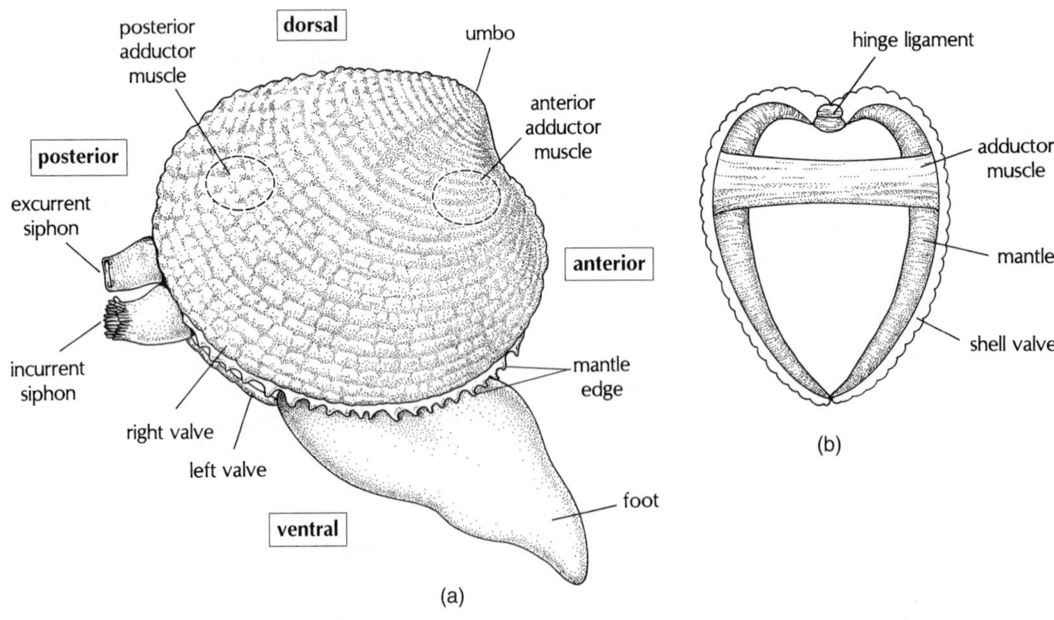

Figure 12.23

A bivalve, indicating the orientation of the body within the shell valves. Note that the hinge is located dorsally and the valves open ventrally. The siphons protrude posteriorly. Tentacles on the incurrent siphon function as chemo- and mechano receptors. (a) Lateral view. (b) Cross section through shell; siphons would face out of page.

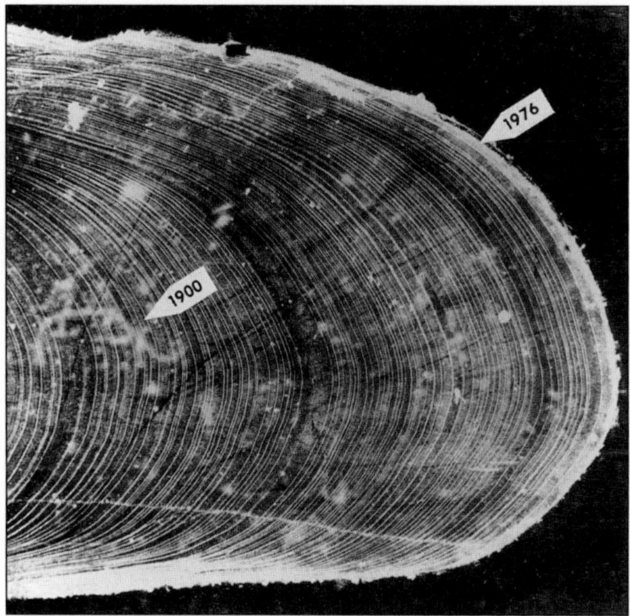

Figure 12.24
Growth lines in the shell of the bivalve *Arctica islandica*. The age of the shell can be determined by counting growth lines, since the patterns of shell growth are seasonal. The most recent growth lines are near the outer margin of the shell valve. This individual was estimated to be 149 years old.
Courtesy of Douglas S. Jones.

Members of the Protobranchia retain what seems to be a morphologically primitive state of bivalve organization. The group is entirely marine, and all species live in soft substrates (Fig. 12.25a). One pair of gills is present in the mantle cavity, with one gill on either side of the body (Fig. 12.25b, c). In many species, the 2 gills extend far enough posteriorly (beyond the foot) to join together by means of ciliary tufts on the gill filaments (Fig. 12.26a). Each gill consists of 2 parts, called **demibranchs** (Middle English: half-gills), extending from opposite sides of a central gill axis (Fig. 12.25b, c). Thus, the gills of protobranch bivalves are always **bipectinate** (L: double-combed). The many units making up each demibranch may be flat, rounded sheets (Fig. 12.25b), as in the typical gastropod ctenidium, or they may be finger-like structures (Fig. 12.25c) that have a more circular cross section; in both cases, each unit is called a **filament.** The gill filaments hang down into the mantle cavity, dividing it into an **incurrent** (ventral) chamber and an **excurrent** (dorsal) chamber. Water enters the mantle cavity ventrally, generally passes between gill filaments, and then exits dorsally (Fig. 12.25b, c). The protobranch gill functions primarily in gas exchange (as in most prosobranch gastropods), although it may also play some role in food collection by filtering unicellular algae from the water.

Gill cilia are confined to the area near the edges of the gill filaments, the **lateral cilia** always playing the primary role in driving water through the mantle cavity (Fig. 12.26a). The **frontal cilia** function primarily in cleaning the gill of sediment and debris.

Each gill filament is attached to adjacent filaments by circular patches of stiff, nonmotile cilia on the anterior and posterior surfaces of each sheet. The cilia of each sheet actually interlock with those of adjacent sheets, stabilizing the gill structure; ionic interactions between the ciliary surfaces apparently further strengthen the attachment between adjacent sheets. The ciliated discs thus fasten together adjacent filaments; that is, they form **interfilamental junctions**—stabilizing junctions between individual gill filaments (Fig. 12.26a). The ubiquitous fastening system known as Velcro®[13] works on a similar principle.

Although many species supplement their diets by suspension-feeding, most of the food collection in protobranch bivalves is accomplished not by the gills, but rather by **palp proboscides**—long, thin, muscular extensions of the tissue surrounding the mouth (Fig. 12.26b). The palp proboscides protrude between the shell valves and probe the surrounding mud substrate, entangling particles in mucus. The sediment-laden mucus is then transported into the mantle cavity by cilia along the ventral surface of the proboscides. Attached to the palp proboscides at their bases, within the mantle cavity, are flattened structures called **labial palps** (Fig. 12.26b); these have conspicuous ridges on their inner surfaces that sort particles by ciliary action, transporting small and nutritious particles to the mouth for ingestion and transferring large and less nutritious (and possibly toxic) particles to the margins of the labial palps, where they are ejected into the mantle cavity and expelled. This rejected material is termed **pseudofeces,** since it is material that has never been ingested. Protobranch bivalves' mode of feeding clearly restricts them to soft sediments. This type of feeding, in which sediment is taken in and the organic fraction is digested, is called **deposit feeding,** and is quite common among invertebrates. An individual protobranch may process more than 1.5 kg (kilograms) of sediment each year.

Although protobranchs are present in shallow water, they are much more common in deep water. Protobranch bivalves may account for about three-fourths of all bivalves in sediments sampled from depths of 1000 m (meters) or more.

The Lamellibranchs

Lamelli • branch
(G: plate gill)
lam-el′-ih-brank

Defining Characteristics: 1) Gills modified to collect suspended food particles, in addition to serving as gas exchange surfaces; 2) secretion of proteinaceous attachment material (usually in the form of threads) by a specialized gland (the byssus gland) in the foot

13. Velcro® is a registered trademark of Velcro U.S.A.

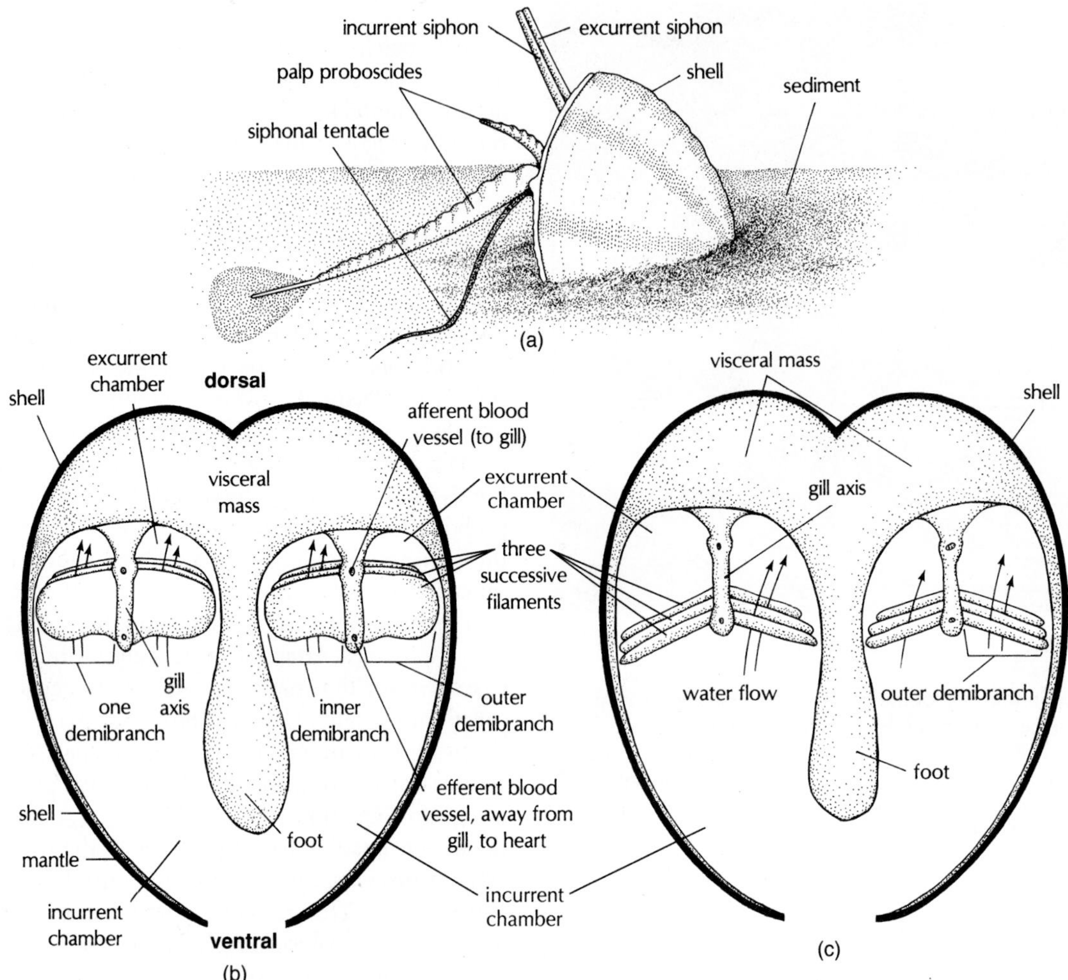

Figure 12.25

(a) A protobranch bivalve, *Yoldia limatula,* with its foot and the lower part of its shell buried in the sediment. (b, c) Two types of protobranch gill, with the more primitive (ancestral) gill type on the left. The anterior end of the animal projects into the page. Arrows indicate the path of water flow through the mantle cavity. The gills function primarily in gas exchange, not food collection.
(a) After Meglitsch. (b, c) After Russell-Hunter.

Most bivalves have lamellibranch characteristics. Although most lamellibranchs are marine, all freshwater bivalve species also belong to this subclass. Most freshwater species are contained within a single family, the Unionidae.

Both freshwater and marine lamellibranchs often play important ecological roles, especially in shallow water. Particularly in rivers, lakes, and estuaries, they often dominate the animal biomass. Commercially important for many years as food (e.g., oysters and scallops) and as sources of buttons and pearls, lamellibranchs have more recently become widely used to assess degrees of environmental pollution. Some recently introduced species, particularly the zebra mussel, *Dreissena polymorpha,* are exerting dramatic effects on freshwater ecosystem structure and operation.[14] Their substantial ecological impact (and their growing importance in monitoring aquatic pollution) is largely due to complex, evolutionary changes in gill function.

In bivalves of all subclasses, water typically enters and exits the mantle cavity posteriorly; the water generally enters through an **incurrent siphon,** passes dorsally between adjacent gill filaments, and then exits through a more dorsally located **excurrent siphon** (Figs. 12.27b, c and 12.32c). The siphons, where present, are tubular extensions of mantle tissue that often can be protruded far beyond the posterior shell margins, permitting the rest of the animal to live safely, deep within the surrounding substrate (Fig. 12.27c). As in the prosobranch gastropods and the protobranch bivalves, water currents are generated by the action of gill cilia—not by muscular activity of the siphons—and the gill is the primary gas exchange surface. But lamellibranch gills are, in addition, modified to collect food particles from the surrounding water. And it is largely the capacity of individuals to process water at high rates, in combination with what are often large numbers of individuals per m^2 of substratum, that accounts for their substantial ecological impact. How have lamellibranch gills been modified to take on the additional role of collecting food particles?

14. See *Topics for Further Discussion and Investigation,* no. 21.

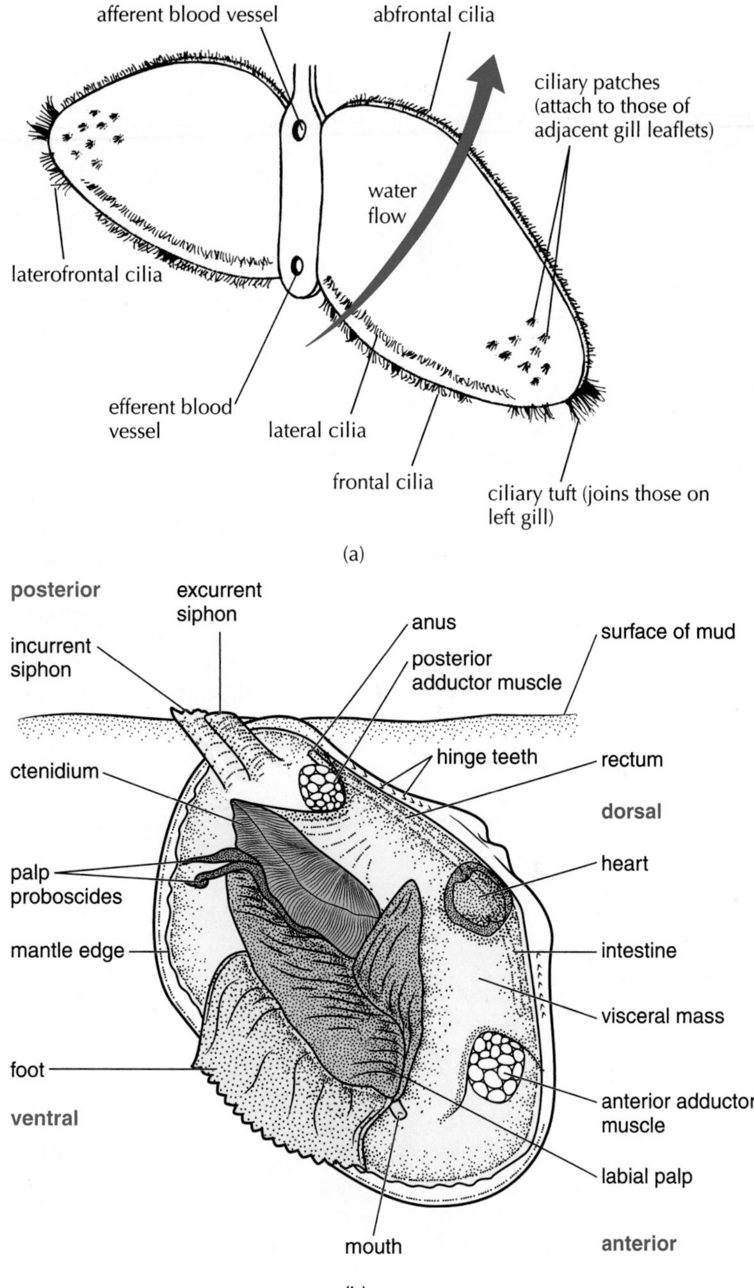

Figure 12.26

(a) Detail of right gill of the protobranch bivalve *Nucula* sp., showing ciliation on a single filament. (b) Internal anatomy of the protobranch bivalve *Yoldia eightsi,* showing the left side of the animal, including the left pair of palp lamellae and the left ctenidium. The left mantle tissue has been removed to reveal the underlying structures. Note the location of the mouth at the base of the labial palps.

(a) Based on Orton, 1912. *J. Marine Biol. Assoc.,* U.K. 9:444. (b) Based on J. Davenport, in *Proceedings of the Royal Society,* 232:431–42, 1988. With suggestions from J. Evan Ward.

The lamellibranch ctenidium is often much larger than that of protobranchs and is variously modified to provide an enormous surface area for collecting, sorting, and transporting suspended particles. Discussing the morphology of these ctenidia requires some unavoidably confusing terminology; the reader is urged to go through the next few pages slowly, with frequent reference to the appropriate illustrations. This discussion is limited to the basic terminology essential for laboratory observation and any discussion of bivalve evolution.

The individual filaments of the lamellibranch gill are thin and greatly elongated, typically bending to form a V shape so that the entire ctenidium is generally shaped like a W (Fig. 12.28). A ciliated ventral food groove lies

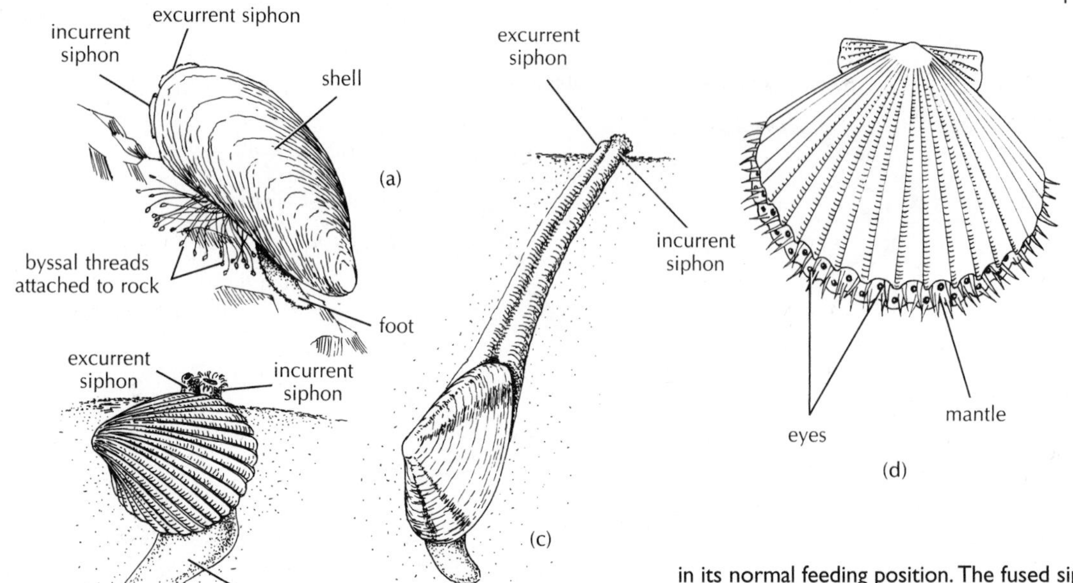

Figure 12.27

Lamellibranch diversity. (a) A blue mussel *Mytilus edulis* attached to a rock by proteinaceous threads. These byssal threads are secreted by a gland located at the base of the foot. (b) A cockle, *Cardium* sp., in its normal feeding position. (c) The soft-shell clam, *Mya arenaria,* in its normal feeding position. The fused siphons may extend more than 30 cm from the shell, permitting this clam to live well below the surface of the sediment. (d) A scallop, *Pecten* sp. Numerous light receptors are present along the edge of the mantle. The scallop can swim by forcefully contracting the well-developed posterior adductor muscle, ejecting water through openings near the shell hinge. Scallops lack siphons and have lost the anterior adductor muscle. It is the large, well-developed posterior adductor muscle that is served as "scallops" in restaurants.

(a–d) After Niesen.

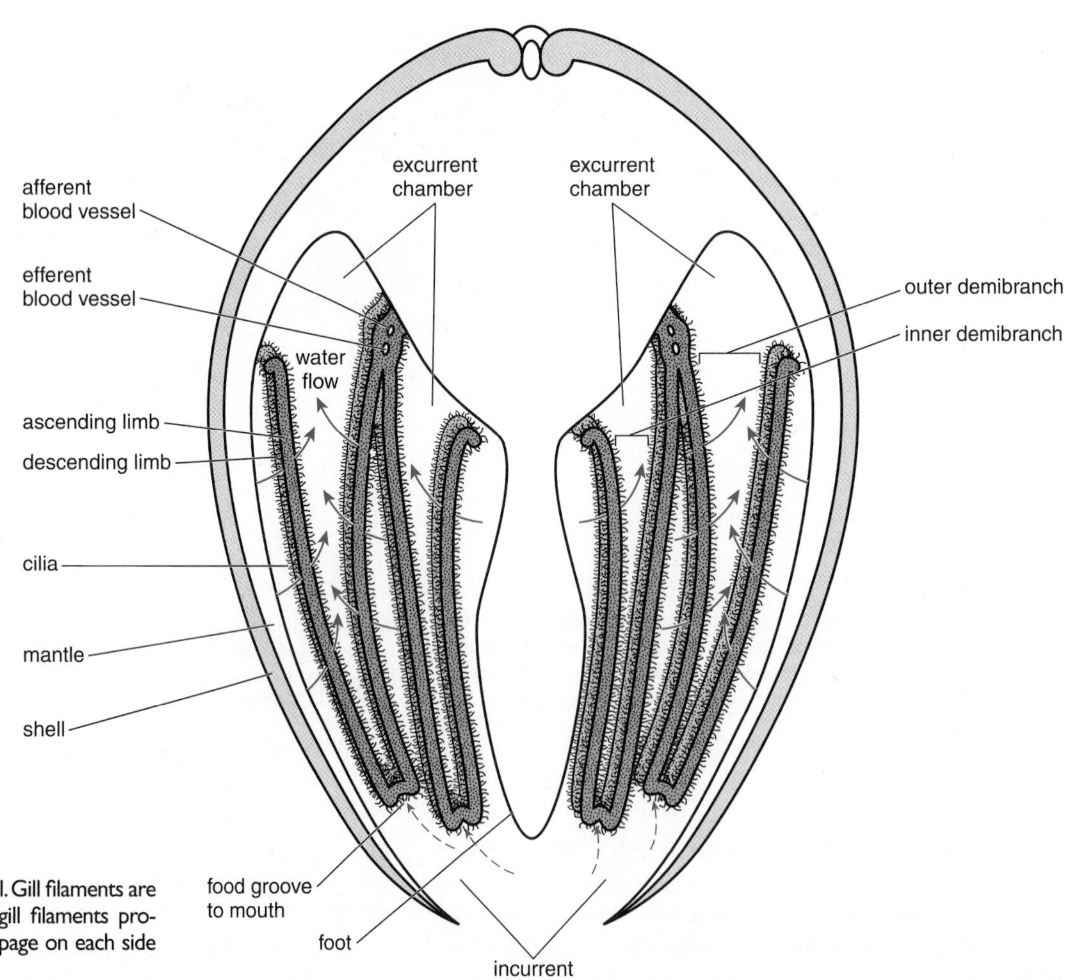

Figure 12.28

A simple lamellibranch gill. Gill filaments are shaded blue. Additional gill filaments project in a row out of the page on each side of the foot.

Based on W. D. Russell-Hunter. *A Life of Invertebrates.*

between the 2 limbs of each filament, at the base of each V (Figs. 12.28 and 12.29c), and the cilia within this groove pass food particles from one filament to the next, toward the mouth.

The 2 arms of each V are named in reference to the central axis from which each filament hangs. A **descending limb** descends from the central axis, and an **ascending limb** bends upward from the bottom of the descending limb (Fig. 12.29b). When 2 V-shaped filaments unite to form a W on each side of the foot (Figs. 12.28 and 12.29b), adjacent Ws in a gill are always attached to each other. Sometimes the Ws are attached by **interfilamental** (i.e., "between filaments") **ciliary junctions** (Fig. 12.29b), as in the protobranch gill. These lamellibranch gills—consisting of individual filaments linked together solely by ciliary disc junctions—are termed **filibranch gills.** In the most highly modified bivalve gills, called **eulamellibranch gills,** the junctions between adjacent filaments are made of tissue rather than cilia; these **interfilamental tissue junctions** completely and firmly attach adjacent filaments together (Fig. 12.29d). In both filibranch and eulamellibranch gills, the series of ascending and descending limbs form continuous sheets of tissue, or **lamellae.** Water passes through the gill lamellae—between adjacent gill filaments—before leaving the mantle cavity.

As with protobranch gills, lamellibranch gills are bipectinate, with one demibranch ("half-gill") extending from each side of the central gill axis; each demibranch typically consists of an ascending and a descending lamella, formed from the ascending and descending limbs of the numerous V-shaped filaments. The demibranch nearest the foot is termed the **inner demibranch;** that on the side nearest the mantle is termed the **outer demibranch** (Figs. 12.28 and 12.29b).

To summarize, on the left side of the foot and visceral mass is one gill. That gill consists of one inner and one outer demibranch, which are united dorsally at the gill axis. Each demibranch consists of many filaments; the filaments typically consist of descending and ascending limbs, so each demibranch consists of one descending and one ascending lamella. On the right side of the foot and visceral mass is another gill, the mirror image of the gill on the left side.

Commonly, the ascending and descending limbs of a single filament are connected by crosspieces of tissue, forming **interlamellar** (i.e., "between lamellae") **junctions** (Fig. 12.29b). In eulamellibranch gills, the interlamellar junctions between the ascending and descending limbs of each filament may be so extensive as to form a complete, solid sheet of tissue across the space between the 2 limbs. As shown in Figure 12.29d, the interlamellar and interfilamental junctions in such gills essentially turn pairs of adjacent filaments into lidless, rectangular boxes. Water must now pass between gill filaments, from the incurrent chamber of the mantle cavity to the excurrent chamber, through minute holes, called **ostia,** located in the sides of each box (Fig. 12.29d, e). In addition, the ascending lamellae of the inner demibranchs (e.g., the 2 demibranchs closest to the foot) often attach, at their tips, to the bivalve's foot, and the ascending lamellae of the 2 outer demibranchs often attach, at their tips, to the mantle (Figs. 12.29d and 12.30b). With the tips of the ascending lamellae firmly attached to other tissues of the bivalve, *all* water must pass between adjacent gill filaments before exiting the mantle cavity; this must increase the gill's filtering effectiveness considerably, since no particles can now escape the clutches of the food-gathering cilia.

The surfaces of lamellibranch gill filaments that face the incurrent chamber of the mantle cavity show complex patterns of ciliation (Fig. 12.29c, e). **Lateral cilia** along the sides of each filament create the water currents responsible for moving water into, and out of, the mantle cavity, as in protobranch bivalves; remember, even though water enters the mantle cavity through an incurrent siphon, the water is pulled in through the actions of these lateral gill cilia, not through any direct muscular action of the siphon itself. Even so, the rate of water flow through the bivalve mantle cavity can be impressive indeed: An American oyster moves 30–40 l (liters) of water past its gills each hour at 24°C. Dense mats of some freshwater bivalves filter up to 10 m^3 of water per m^2 of substratum per day. Individual bivalves can reduce water flow through the mantle cavity by contracting muscles to decrease siphon diameter, or by contracting smooth muscle within the gills themselves, or possibly by varying ciliary beat frequency.

Details of particle capture, transport, and sorting by bivalve gills are still being worked out.[15] The traditional model has food particles being captured by compound **laterofrontal** cilia, which then pass the captured particles to the nearby frontal cilia (Fig. 12.29c). More recent work indicates that particles are captured instead largely by hydrodynamic forces, without any physical contact between the particles and the laterofrontal cilia; in addition, some particles are intercepted directly by the gill filaments. In both cases, **frontal cilia** then move the captured food particles to specialized food grooves located at the ventral and dorsal margins of each demibranch (Figs. 12.28, 12.29c, e, and 12.31). Members of some bivalve species transport captured food particles primarily in the ventral food grooves, members of some other species transport captured food particles primarily in the dorsal food grooves, and the remaining bivalves transport food particles in both the ventral and dorsal food grooves (Fig. 12.31). In all cases, the particles are transported by gill cilia to the labial palps, where they are sorted according to particle size and nutritional value[15] and then carried to the mouth, as in protobranchs. In many species, particles are pre-sorted on the gills.

Periodically, the lamellibranch bivalve closes its valves forcefully, expelling water and unwanted particles **(pseudofeces)** from the mantle cavity through the incurrent siphon. Pseudofeces may be an important energy source for benthic deposit-feeders.

15. See *Topics for Further Discussion and Investigation,* no. 18.

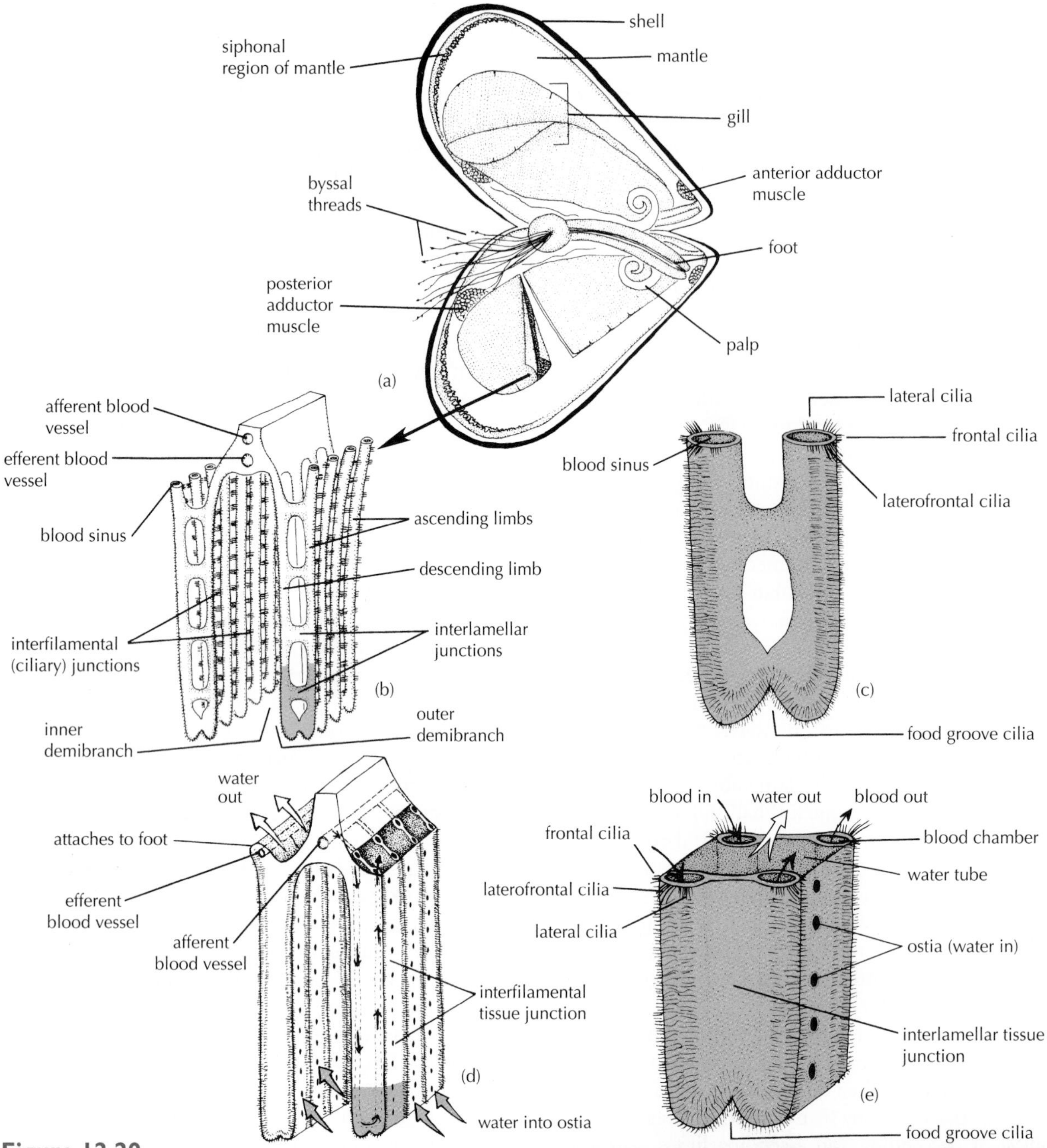

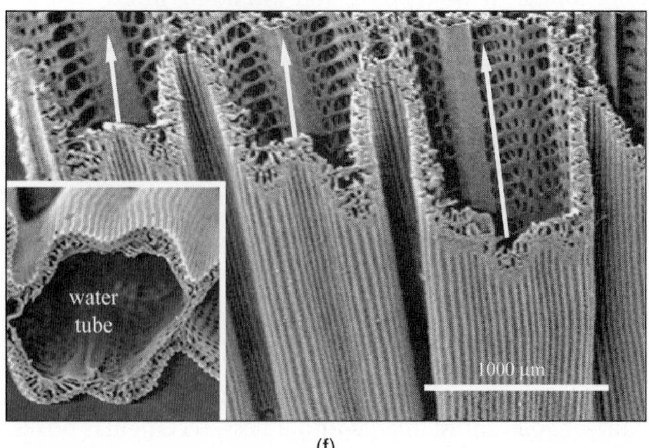

(f)

Figure 12.29

(a) The blue mussel, *Mytilus edulis,* with its shell opened to reveal the gills. (b) Filibranch gill of the blue mussel, *M. edulis.* In these relatively simple lamellibranch gills, adjacent filaments are locked together only by sporadic patches of adjoining cilia (interfilamental ciliary junctions). The ascending and descending lamellae of each demibranch are supported by thin crosspieces (interlamellar junctions). (c) Transverse section through one filibranch gill filament, showing pattern of ciliation. Lateral cilia generate water currents for feeding and gas exchange. Food particles are trapped by hydrodynamic forces or by laterofrontal cilia, which then transfer the food to frontal cilia. Frontal cilia move food particles along the gill filament to the food groove. If you remove one shell valve from a bivalve and lay the animal on its remaining valve, the frontal cilia project directly upward at you. (d) Diagrammatic representation of a eulamellibranch gill, such as that found in the hard-shell clam *Mercenaria mercenaria.* The interfilamental junctions between adjacent filaments are complete sheets of perforated tissue, rather than sporadic patches of interlocking cilia. The intrafilamental junction between the inner and outer lamellae of a single demibranch is now a solid sheet of tissue rather than a series of thin, separate bars. Black arrows represent blood flow within the gill filaments (e) Detail of one demibranch, showing basic structure and pattern of ciliation. All water enters through ostia in the interfilamental junctions and must flow past the blood vessels along the water tubes before exiting. Black arrows represent blood flow. (f) Water tubes in the gill of the clam *Mercenaria mercenaria* as seen with the scanning electron microscope (SEM). White arrows indicate the direction of water flow through the tubes after water has entered the gill through the ostia.

(f) Courtesy of S. Medler, from S. Medler and H. Silverman, 2001. Muscular alteration of gill geometry *in vitro:* Implications for bivalve pumping processes. *Biol. Bull.* 200:77–86.

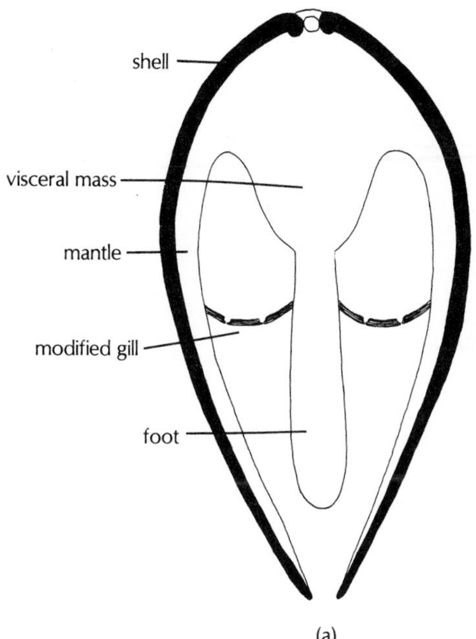

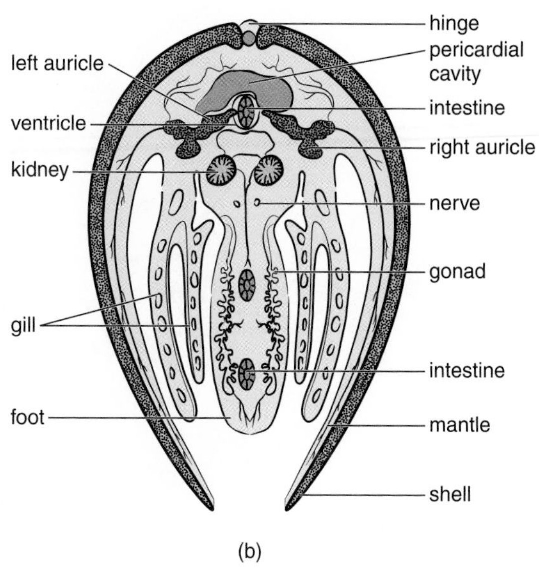

Figure 12.30
(a) Cross section through a septibranch bivalve, showing the highly modified gill. Compare with cross section through a freshwater eulamellibranch, the clam *Anodonta* sp., shown in (b). In (b), note that the heart is wrapped around the intestine, as in all bivalves.

(b) Based on Beck/Braithwaite, *Invertebrate Zoology,* Laboratory Workbook, 3/e, 1968.

After food particles are ingested at the mouth and passed down the esophagus, entangled in strings of mucus, they are drawn into the stomach, stirred, and, in part, digested by the action of a rotating translucent rod known as the **crystalline style.** The crystalline style is composed of structural protein and several digestive enzymes. One end of the rod lies in a **style sac,** a pouch of the intestine lined by cilia (Fig. 12.32b). The activity of these cilia causes the rod to rotate. The end of the style protruding into the stomach abrades against a chitinous **gastric shield** as the rod rotates, breaking food into smaller pieces. The abrasion also causes the rod to slowly degrade at the end, releasing digestive enzymes into the stomach. Additions to the crystalline style are made in the style sac. A morphologically and functionally similar crystalline style apparatus is also found in the stomachs of monoplacophorans and some suspension-feeding gastropod species. In contrast, protobranch bivalves possess a style sac containing mucus and digestive enzymes but do not have a crystalline style. As in most other molluscs, the bivalve stomach connects with larger **digestive glands (digestive diverticula),** which serve as the major sites of digestion and absorption.[16] Whether or not particles are sent to the digestive glands may depend on their nutritional value, so that the stomach may serve as an additional site of particle sorting.[15]

The nutrition of some lamellibranch bivalves differs considerably from what has just been described as the norm. In the early 1980s, bacteria were found living symbiotically in the gill tissue of a number of shallow- and deep-water bivalve species.[17] The bacteria-rich gills are unusually thick and typically more than 3 times heavier than the gills of bivalve species lacking the bacterial symbionts. The bacteria within the gills generate ATP and reducing power (NADH or NADPH) by oxidizing highly reduced chemical substrates (sulfides—HS^-—or methane) and use the energy so entrapped to fix carbon dioxide (CO_2) into organic compounds that then become available for the nutrition of the host bivalve. The bacteria, then, like plants, are autotrophs. In fact, to fix CO_2 into carbohydrates, they use the same enzymatic pathways of the Calvin cycle that plants use in photosynthesis, but the bacteria are **chemoautotrophs** rather than solar autotrophs, using chemical energy rather than light energy to fix carbon from CO_2. Bacterial chemosynthesis undoubtedly plays a role in the nutrition of the host, although the exact contribution has yet to be formally or fully quantified for any species. Many bivalves hosting gill bacteria have a greatly reduced digestive system, and in some species the digestive system has been lost completely; symbiotic bacteria surely contribute substantially to host nutrition in such animals.

The symbiosis between bivalves and sulfur-oxidizing bacteria is accompanied by certain morphological (Research Focus Box 12.2, p. 252) and physiological modifications on the part of the host, modifications that still are

16. See *Topics for Further Discussion and Investigation,* no. 9.

17. See *Topics for Further Discussion and Investigation,* no. 14.

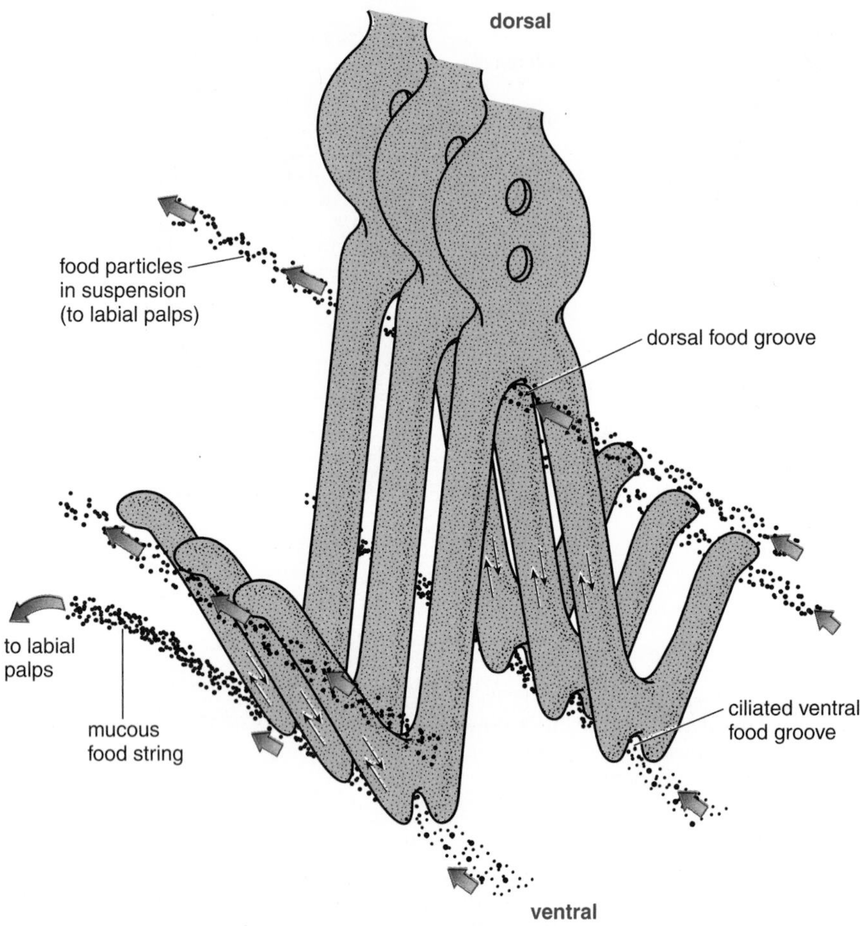

Figure 12.31

Capture and transport of food particles by lamellibranch bivalve gills. Particles entering the ventral food grooves are incorporated into a mucous string and transported by cilia from gill filament to gill filament toward the labial palps (Figs. 12.29a and 12.32a,c), where some or even most sorting of particles occurs before ingestion as in protobranchs. Particle transport by mucociliary mechanisms is indicated by solid arrows. Particles entering the dorsal food grooves, on the other hand, are captured by hydrodynamic mechanisms and carried to the labial palps in suspension (hollow arrows).

Based on J. E. Ward *et al.*, in *Limnol. Oceanogr.* 38:265–72, 1993.

being actively studied. The relationship with bacterial symbionts typically correlates with a reduction of the labial palps and digestive system, the loss of the outer gill demibranchs, and further modifications in gill morphology presumably associated with decreasing dependence on particulate food.

Physiological modifications are equally intriguing. HS^- is normally toxic to living tissues even at extremely low concentrations, disabling cytochrome-*c* oxidase, a key enzyme in the electron transport chain involved in ATP synthesis. Allowing symbiotic bacteria access to the sulfide needed for bacterial ATP production is thus a potentially risky proposition for the host. Some species have specialized blood proteins that bind HS^-, preventing free sulfide from contacting sulfide-sensitive enzymes of the electron transport system. This adaptation has the added benefit of protecting the sulfide from spontaneous oxidation while in transit to the bacteria; such spontaneous oxidation of sulfide would greatly reduce the sulfide's value as an energy source. Other species may detoxify HS^- by partially oxidizing it to thiosulfate, a nontoxic substance that can then be further processed by the bacteria. Finally, some of the host tissues may rely mostly on anaerobic metabolic pathways for ATP generation, reducing their dependency on the cytochrome-*c* oxidase system and thus decreasing their susceptibility to sulfide poisoning.

Bivalves exhibit a modest variety of lifestyles. Some lamellibranchs, such as mussels, live attached to hard substrates by means of proteinaceous secretions called **byssal threads.** A proteinaceous liquid is secreted by a byssal gland at the base of the foot, within the mantle cavity, and is quickly transported to the substrate along a groove in the foot. Shortly after contacting seawater, the secretion solidifies to form many thin—but very sturdy—anchoring threads (Figs. 12.27a and 12.29a). Some other bivalves (e.g., oysters, Fig. 12.32e) cement one shell valve permanently to a substrate. Biochemists have scrutinized these processes for many years; a sturdy glue that hardens quickly under water could have a bright commercial future as, for instance, a new dental adhesive. In most

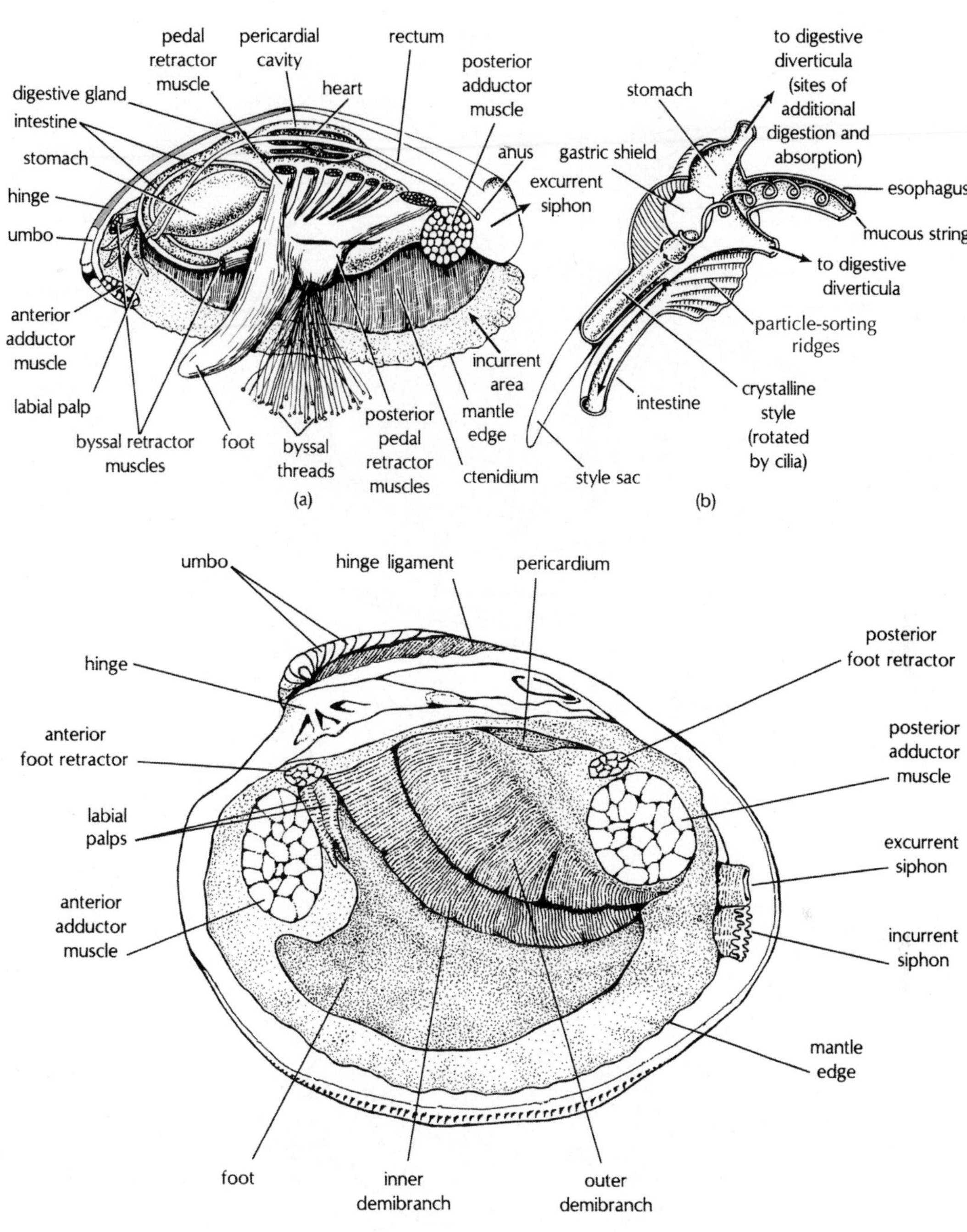

Figure 12.32a, b, c

(a) Internal anatomy of a lamellibranch bivalve, the marine mussel *Mytilus edulis*. The left gill has been removed to reveal underlying anatomy. Hinge shown in blue. (b) The digestive system of a suspension-feeding lamellibranch bivalve, such as the blue mussel or soft-shell clam. The crystalline style aids digestion; rotated by cilia in the style sac, the style releases digestive enzymes as the tip of the rod is abraded against the gastric shield. A crystalline style apparatus is encountered among some prosobranch gastropod species. (c) Internal anatomy of the hard-shell clam, *Mercenaria mercenaria*. The animal is lying on its right shell valve, with the left valve removed. Only the left gill is illustrated. The right gill lies beneath the foot in this drawing, and thus cannot be seen. (d) Same animal as in (c), but with the left gill removed and internal organs exposed by dissection. Note that the viscera extend well into the foot. (e) Internal structure of the American oyster, *Crassostrea virginica*, with the right shell valve removed. Adult oysters have no foot and lack an anterior adductor muscle. (*Continued on following page*.)

(a) After Turner. (b) After Morton. (c) After Kellogg. (d) Modified from several sources. (e) Modified from Kellogg and other sources.

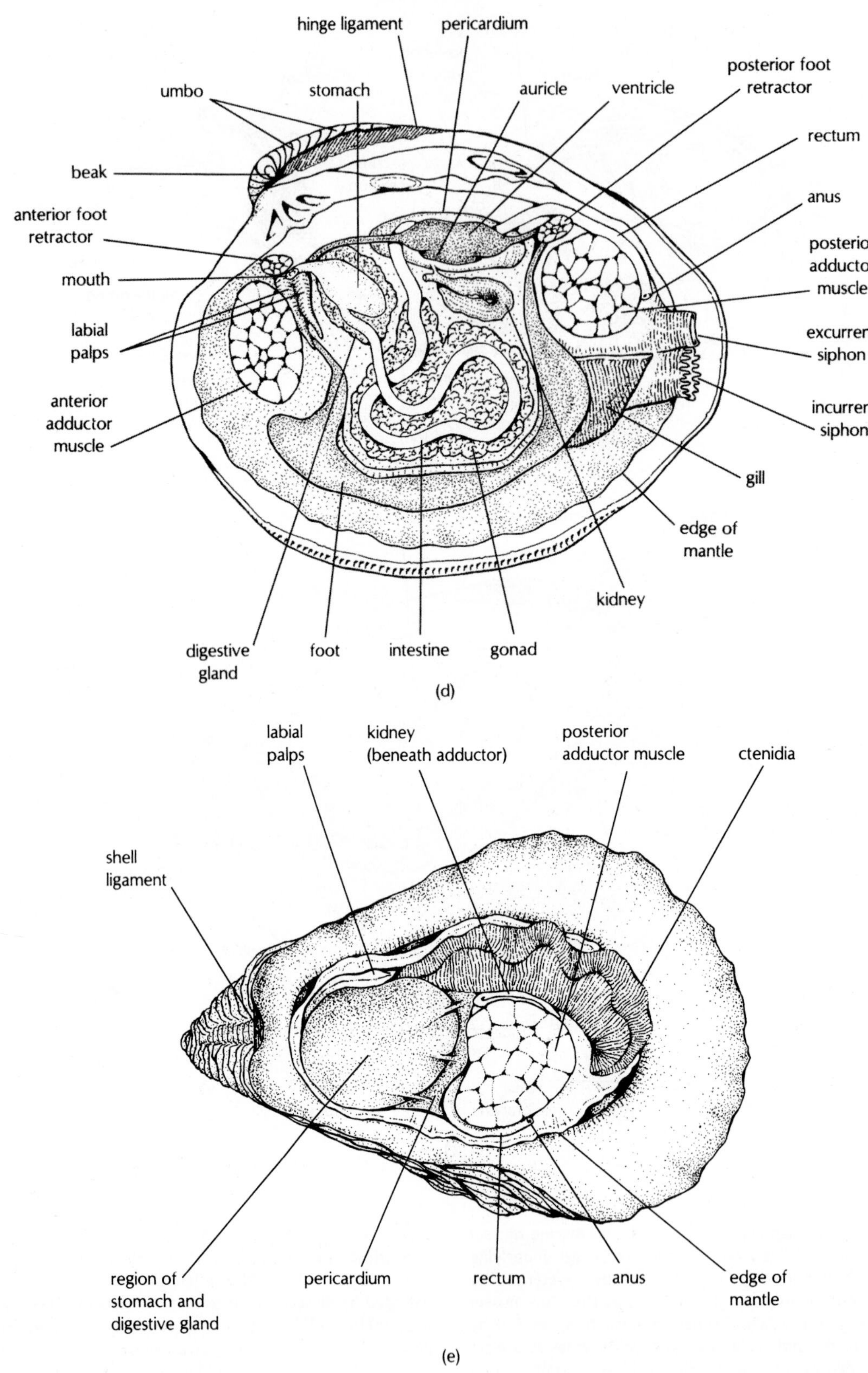
hinge ligament
pericardium
posterior foot retractor
umbo
stomach
auricle
ventricle
rectum
beak
anus
anterior foot retractor
posterior adductor muscle
mouth
excurrent siphon
labial palps
incurrent siphon
anterior adductor muscle
gill
edge of mantle
kidney
digestive gland
foot
intestine
gonad
(d)
labial palps
kidney (beneath adductor)
posterior adductor muscle
ctenidia
shell ligament
region of stomach and digestive gland
pericardium
rectum
anus
edge of mantle
(e)

Figure 12.32d, e

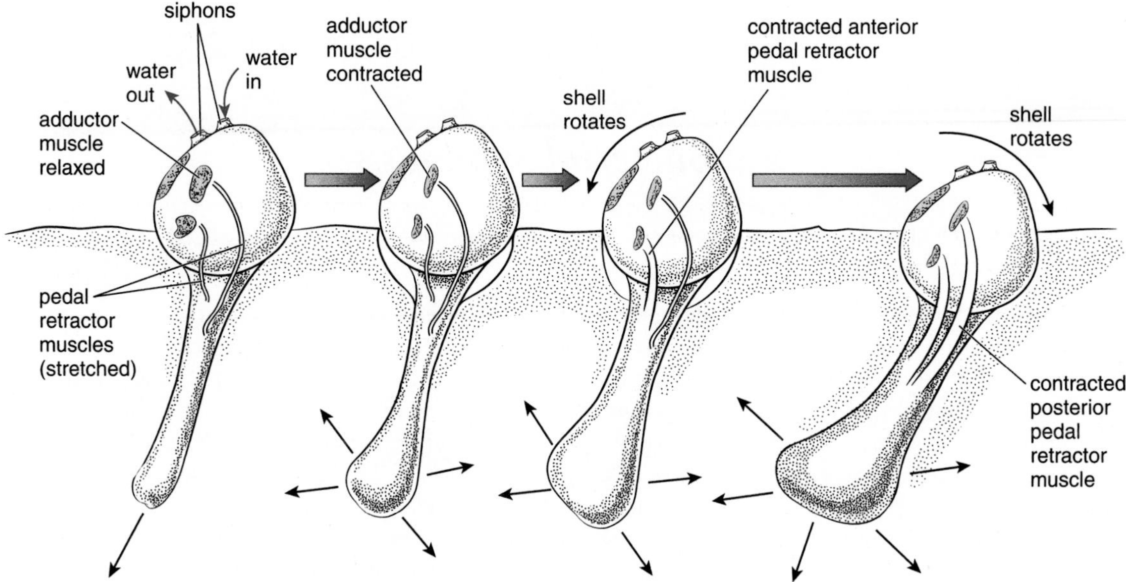

Figure 12.33

Burrowing by bivalves. The foot serves both to penetrate into the substrate and to anchor the animal as the body is pulled downward. See text for discussion.

Based on Jones and Trueman, in *Patella Locomotion*, 52:201, 1970.

lamellibranch species, only juveniles form byssal threads, the byssal gland becoming nonfunctional as the animals mature. No protobranch bivalves secrete a byssus.

A few lamellibranch species (some scallops, Fig. 12.27d) live unanchored atop the substrate and are capable of short bursts of "swimming," achieved by repeatedly clapping the shell valves together; rapid expulsion of water from the mantle cavity quickly displaces the animal, permitting escape from potential predators.

Most bivalves, however, avoid predators by living within a substrate—sediment, wood, or even rock—in burrows of their own making. Burrowing into a soft substrate, such as sand or mud, is accomplished by the muscular foot (Fig. 12.33). The foot is initially extended into the substratum by hydraulic or hydrostatic means, through contractions of the appropriate musculature in the foot. The **adductor muscles,** which attach the animal to the shell in all bivalve species, are relaxed at this time. This relaxation of the adductor muscles permits the release of energy that was stored in the compressed hinge ligament, so that the shell valves now move apart and press tightly against the surrounding substrate. This laterally directed force of the shell valves provides anchorage for the shell as the foot extends down into the substratum. Thus, downward extension of the foot need not eject the animal from the burrow. The adductor muscles then contract, drawing the shell valves toward each other. This action releases the shell anchor. Contracting the adductor muscles also pumps blood into the foot, which then dilates at the tip to form another anchor.

Abruptly closing the shell valves, in addition to swelling the foot, forces water out of the mantle cavity, blowing away and loosening some of the sediment adjacent to the opening between the shell valves. With the foot firmly anchored in the substrate, the shell can now be pulled downward by contracting the **pedal retractor muscles** that extend from the foot to the shell. The pedal retractor muscles do not contract simultaneously, but rather sequentially, so the bivalve shell "rocks" forward and then backward as it progresses, slicing into the substrate.

Many marine bivalve species burrow, as newly metamorphosed juveniles, into wood, shell, coral, and other hard substrata. The shell plays a major role in burrow formation by such species, actually carving into the substratum. As the animal grows, the body remains hidden within, and protected by, the substrate in which the burrow has been formed, with only the siphons protruding to the outside; the siphons of all burrowing species tend to be long, permitting the rest of the animal to live deep within the substrate. The damage done by wood-boring bivalves can be considerable, as any marina owner or boat owner knows all too well. Wood-boring bivalves have been found in all marine habitats, including the deep sea at several thousand meters.

A major group of wood-boring lamellibranch bivalves are misleadingly referred to as "shipworms" (Figs. 12.34 and 12.35). Filtering food particles from suspension in typical lamellibranch fashion may actually play a minor role in the feeding biology of these bivalves, since shipworms appear to meet many of their nutritional requirements with the aid of symbiotic bacteria living in high concentrations within a specialized portion of the shipworm stomach, called the **wood-storing caecum.** The bacteria (possibly assisted by cohabitating protozoans) fix nitrogen and digest the wood eaten by the shipworm.

RESEARCH FOCUS BOX 12.2

Bacterial Symbiosis

Distel, D. L., and H. Felbeck. 1987. Endosymbiosis in the lucinid clams *Lucinoma aequizonata, Lucinoma annulata,* and *Lucina floridana*: A reexamination of the functional morphology of the gills as bacteria-bearing organs. *Marine Biol.* 96:79–86.

The gill structure of typical bivalves—scallops, chowder clams, soft-shell clams, oysters, and mussels—has been well understood since the early 1900s. Members of the bivalve family Lucinidae, however, have unusually large gills that recently have been shown to harbor large numbers of symbiotic bacteria; the bacteria fix dissolved carbon dioxide (CO_2) into carbohydrates, using energy released through the oxidation of sulfides dissolved in seawater. The study by Distel and Felbeck (1987) asks, "How do lucinid gills differ from those of bivalves lacking endosymbiotic bacteria, and to what extent are the differences adaptively advantageous for the symbiosis?"

Distel and Felbeck collected representatives of 2 lucinid species (*Lucinoma aequizonata* and *L. annulata*) off the coast of California and one species (*Lucina floridana*) off the coast of Florida, and then they dissected out the gills. How are the bacteria distributed within the gill tissue? Study of the gill ultrastructure required that the gill tissue be killed, stabilized, and prepared for electron microscopy with appropriate chemicals. Even when chemically stabilized ("fixed"), the tissue is still too soft to be sliced ("sectioned") without being torn, crushed, or deformed. To stiffen the tissue, the researchers removed water by dehydrating the tissue in alcohol and then replaced the water with a plastic epoxy-like solution. Once the plastic hardened, the tissue samples were cut into sections using a microtome and glass or diamond knives. Thick sections (1–4 μm) were stained for light microscopy, and thin sections (less than 0.1 μm) were stained for transmission electron microscopy (TEM) to make certain cell components more visible.

Focus Figure 12.2b shows a chunk of tissue cut from a gill of one of the clams; these species have only one demibranch—the inner one—on each side of the foot/visceral mass system. When a gill is sectioned in the plane shown and stained appropriately, the light microscope reveals a normal gill structure at the surface: thin, parallel filaments bearing lateral, laterofrontal, and frontal cilia as in other bivalve gills (Fig. 12.29c, e). Deeper into the gill tissue, however, the filaments are unusual in being greatly swollen with large bacteria-containing cells called **bacteriocytes** (Focus Fig. 12.2c, d), each of which contains dozens of bacteria enclosed in membrane-bound vesicles. Cells in the bacteriocyte zone also house numerous dark-staining granules, termed "pigment granules."

After studying hundreds of adjacent sections of gill tissue, Distel and Felbeck constructed a 3-dimensional model of the gill (Focus Fig. 12.2e). As illustrated, the bacteria-containing cells are packed into stacks forming hollow cylinders; with this arrangement, each bacteriocyte has equal access to sulfide in the water flowing through the gill. Water flows into the gill tissue through these cylinders and moves past, but does not directly contact, the bacteriocytes; the bacteriocytes are separated from direct contact with seawater by a thin layer of epithelial cells. These epithelial cells bear numerous microvilli, greatly increasing the absorptive surface area available for sulfide uptake and for exchange of O_2 and CO_2. After passing through the hollow cylinders of the bacteriocytes and epithelial cells, water flows out of the gill into the space between adjacent lamellae of the demibranch, and then it exits the mantle cavity through the excurrent siphon.

Although Distel and Felbeck successfully documented the structure of the gill and the location of the bacteria within the gill, their work raised a number of interesting questions about the functioning of this symbiotic relationship between bivalve and bacterium:

1. Because the sulfide-rich water comes first into contact with the gill itself, there is no opportunity for other tissues to detoxify the sulfide before the water reaches the gill. What protects the outer layer of epithelial cells from sulfide poisoning? For that matter, how are other tissues in direct contact with the seawater, such as the mantle, protected from sulfide poisoning?
2. Why are the bacteriocytes not in direct contact with the seawater? What is the function of the thin layer of epithelial cells covering the bacteriocytes?
3. What is the function of the prominent pigment granules present in the bacteriocyte layer of the gill? Might they be involved in eliminating waste products or in digesting bacterial cells?
4. How does the host obtain nutrients from the bacteria? Are the bacteria digested by the bivalve? Or do the bacteria remain intact, releasing carbohydrates and other organic molecules to the host?
5. Where do the symbiotic bacteria come from? Are they continually taken up from the external environment? Distel and Felbeck consider this unlikely, since their electron microscope studies showed no evidence of

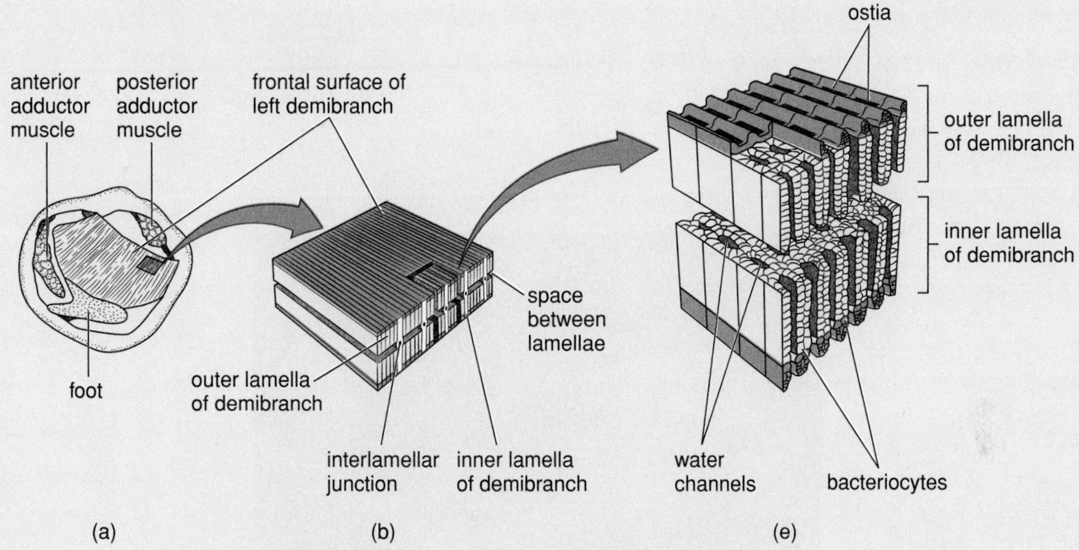

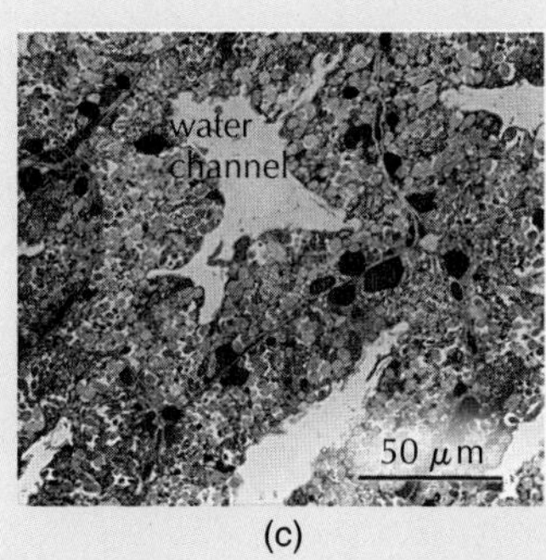

(c)

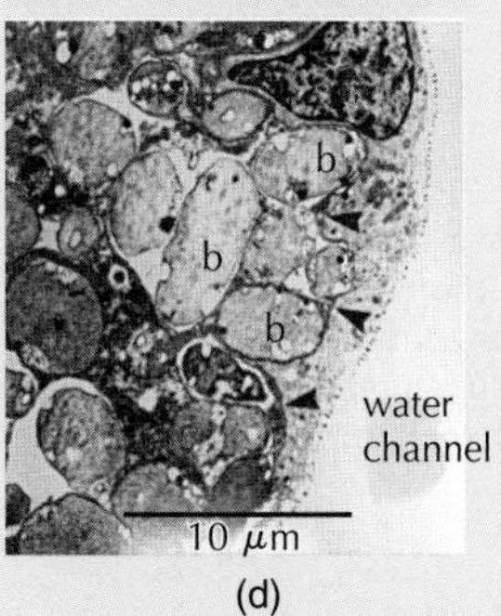

(d)

Focus Figure 12.2

(a) The deep-water bivalve *Lucinoma aequizonata* lying on its right side, with the left shell valve removed to expose the unusually thick gill. (b) Enlargement of square section shown in (a). The section has been rotated about 90° to show portions of the 2 lamellae composing the left demibranch. The shaded region is morphologically identical to the gills of other lamellibranch bivalves. (c) Transmission electron micrograph of sagittal section through the bacteria-laden portion of the gill lying between the 2 thin layers shaded blue in (b). Bacteria (labeled b) occur intracellularly in clusters around narrow water channels that extend perpendicular to the frontal surface of the gill. The bacteria are seen clearly at the higher magnification of (d). (e) Schematic reconstruction of gill structure based on electron micrographs, showing thick columns of bacteriocyte cells surrounding water channels.

(c, d) Courtesy © Springer-Verlag. (b, e) Based on D. L. Distel. From Distel and Felbeck, 1987. *Marine Biology* 96:79–86.

endocytotic activity in the gill tissue; the bacteria seem to be self-sustaining within the gill. But how does an individual gill first become "infected"? Are the bacteria passed directly from parent to fertilized egg? Do the bacteria first appear in the free-swimming larvae, or do they not appear until after metamorphosis?

How would you address each of these questions?

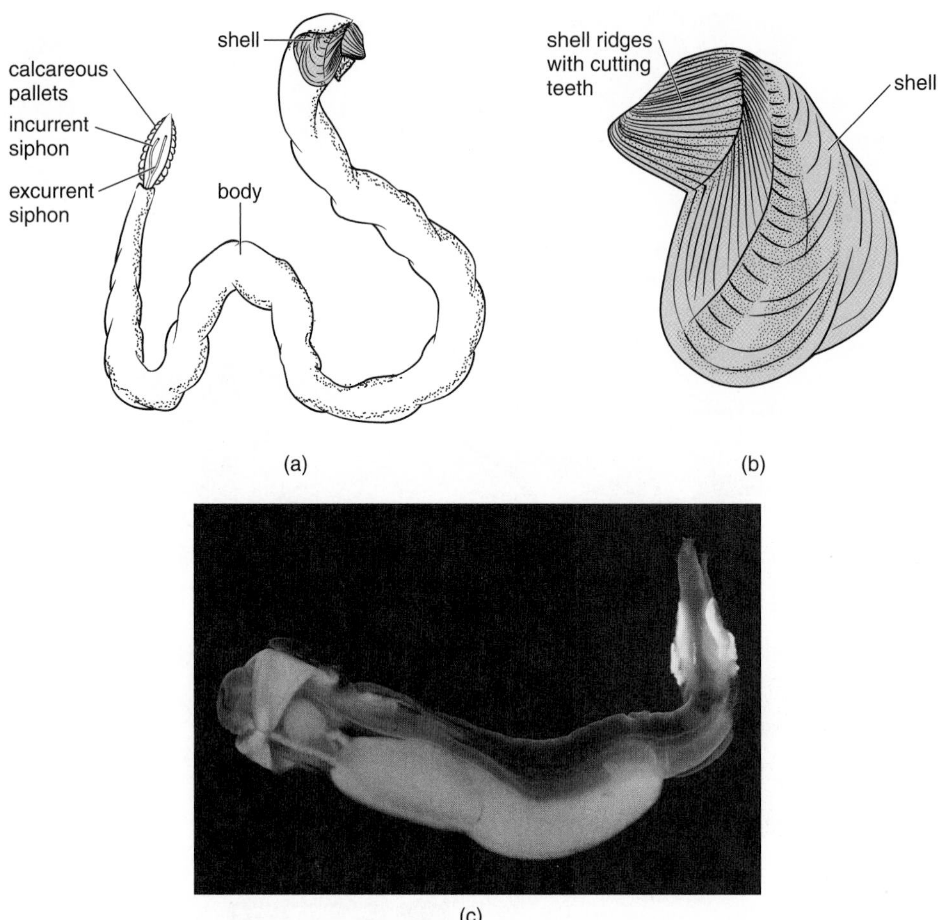

Figure 12.34

(a) External morphology of a shipworm removed from its wood burrow. The shell is used to excavate the burrow. The pallets seal off the opening of the burrow when the animal withdraws. The pallets are thus analogous to the operculum of gastropods. (b) Detail of the shell, showing the cutting teeth. (c) Photograph of the shipworm *Teredora malleolus.*

(a, b) Based on Ruth D. Turner, in *A Survey and Illustrated Catalogue of the Teredinidae,* 1966. (c) Courtesy of C. Bradford Calloway.

As noted earlier, lamellibranch species are now distributed among 3 subclasses (see *Taxonomic Detail,* pp. 286–290).

Subclass Anomalodesmata

Subclass Anomalo • desmata
(G: irregular ligament)
an-nō-mahl′-ō-dez-mah′-tah

Defining Characteristic: Hinge lacks true teeth (although secondary teeth are present in some species)

These unusual bivalves occur in marine and estuarine environments worldwide, from intertidal to abyssal depths. Some species have slightly modified lamellibranch gills and are either suspension-feeders or deposit-feeders. But perhaps the most bizarre members of the group have a highly modified ("septibranch") ctenidium that enables them to act as *bona fide* carnivores, feeding especially on polychaetes and crustaceans up to several mm long.

The septibranch ctenidium lacks filaments and instead forms a muscular septum (Fig. 12.30a). The septum divides the mantle cavity into ventral and dorsal chambers, as in other bivalves, but it is perforated by a series of ciliated openings. One-way valves regulate water flow through these openings in some species. The septum can be moved forcefully upward within the mantle cavity, drawing water in through the incurrent siphon and simultaneously expelling water through the excurrent siphon. These animals thus feed as organic vacuum cleaners, commonly sucking in small crustaceans and annelids. The stomach is lined with hardened chitin, serving to grind up ingested food. Labial palps, although present, are quite small and do not have any sorting function. A very reduced style sac is found but never a crystalline style. These truly are "anomalous" bivalves.

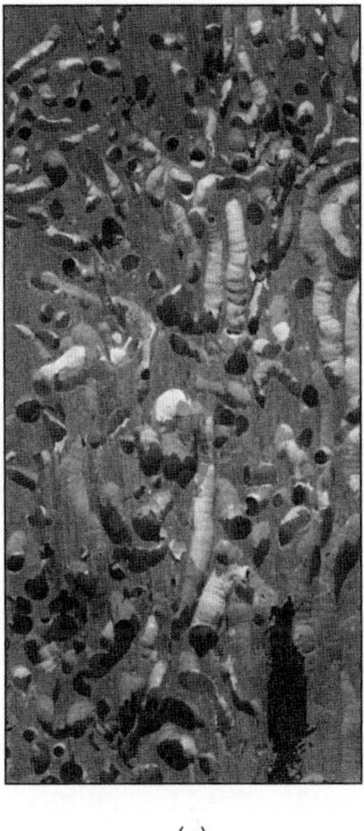
(a)

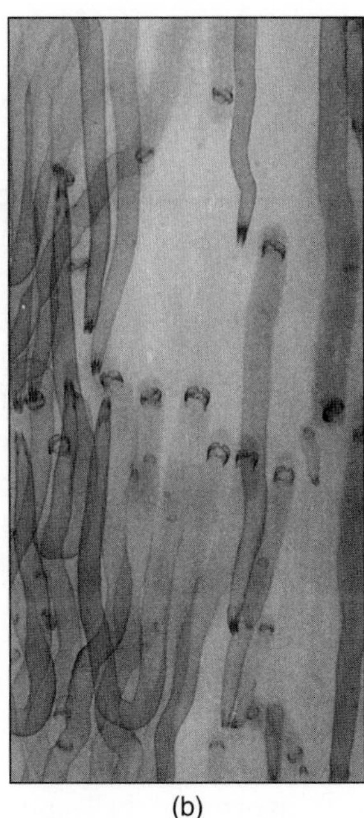
(b)

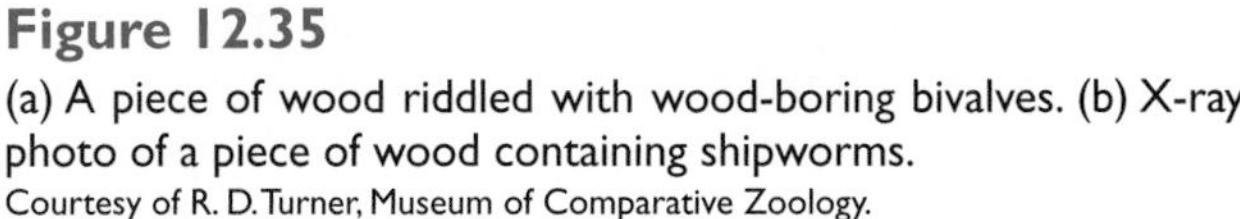

Figure 12.35
(a) A piece of wood riddled with wood-boring bivalves. (b) X-ray photo of a piece of wood containing shipworms.
Courtesy of R. D. Turner, Museum of Comparative Zoology.

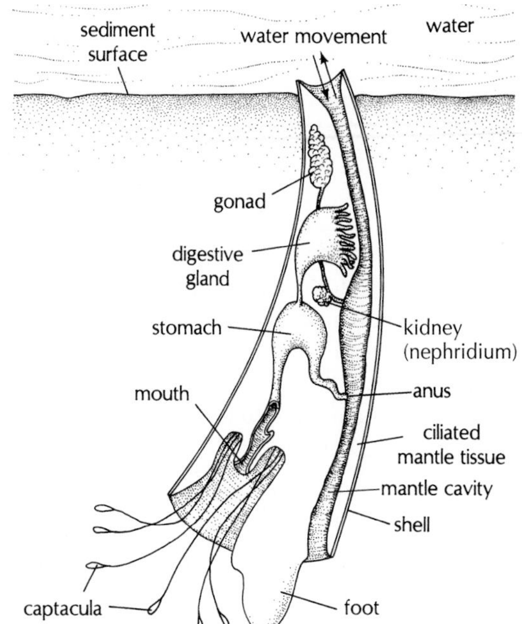

Figure 12.36
Dentalium sp., a scaphopod, in its normal feeding orientation. Food particles are captured by the captacula.
After Borradaile; after Naef.

Class Scaphopoda

Class Scapho • poda
(G: spade foot)
skaf-ō-pō´-dah

Defining Characteristics: 1) Tusk-shaped, conical shell, open at both ends; 2) development of anterior, threadlike, adhesive feeding tentacles

Of the 7 molluscan classes, 5 were well represented in the Cambrian fauna of some 550 million years ago. The morphological diversity of these fossils makes it clear that molluscan radiation had been going on for a considerable time. In comparison, the scaphopods are a young group, first appearing in the fossil record in the middle Ordovician—a mere 450 million years ago (Fig. 12.5). Present research suggests that the scaphopods and bivalves had the same ancestor: The now extinct class of molluscs known as "rostroconchs" (see p. 218).

The 300 to 400 species in the class Scaphopoda are all marine and live sedentary lives in sand or mud substrates, mostly in deep water. Scaphopods possess these "typical" molluscan features: foot, mantle tissue, mantle cavity, radula, and shell. The scaphopod shell is never spirally wound, but rather grows linearly as a hollow, curved tube; hence, the common names "tooth shell" and "tusk shell." The shell has an opening at each end (Fig. 12.36). Water enters at the narrower end, which protrudes above the substrate. Inflow of water is due to the action of ciliated cells restricted to ridges of mantle tissue. Periodically, water is expelled through this same opening by a sudden contraction of the foot musculature.

Unlike many other molluscs, scaphopods possess no ctenidia. They also lack a heart and circulatory system; instead, the blood circulates through the various large sinuses of the hemocoel as a consequence of the foot's rhythmic movements. Neither do any scaphopods possess an osphradium; specialized sensory receptors in the posterior region of the mantle cavity, where respiratory currents enter and leave, may provide the same water-sampling function provided by osphradia in other molluscs.

Burrowing into soft substratum is accomplished by the foot, essentially as described previously for the bivalves. Scaphopods capture small food particles, including foraminiferans (a group of shelled protozoans, see p. 65), from the surrounding sediment and water using specialized, thin tentacles known as **captacula** (Fig. 12.36). A typical individual possesses 100 to 200 such tentacles. Each captaculum terminates in a ciliated, sticky bulb; for feeding, the tentacles are extended by the creeping action of the cilia until the bulb contacts food. The food is then transported to the mouth by captacular cilia or, in the case of large food particles such as large foraminiferans and small bivalves, by muscular contractions of the tentacles themselves.

Class Cephalopoda

Class Cephalo • poda
(G: head foot)
sef′-ahl-ō-pō′-dah

Defining Characteristics: 1) Shell divided by septa, with chambers connected by the siphuncle: a vascularized strand of tissue contained within a tube of calcium carbonate (shell reduced or lost in many species); 2) closed circulatory system; 3) foot modified to form flexible arms and siphon; 4) ganglia fused to form a large brain encased in a cartilaginous cranium

It is truly wondrous that such animals as the squid and octopus belong to the same phylum as the clams, snails, scaphopods, and chitons. Unlike most molluscs, cephalopods are often fast-moving, active carnivores capable of remarkably complex behaviors. Members of this class are exclusively marine. The largest cephalopods—the giant squid, *Architeuthis*—weigh perhaps 1000 kg and reach lengths of up to 18 m, including the tentacles, which themselves can be 5 m or more long. Members of the smallest cephalopod species are less than 2 cm (centimeters) long, including the tentacles.

All cephalopods have a radula and ctenidia. A mantle cavity and foot are present as well, but they do not generally function in "typical" molluscan fashion. The head and associated sensory organs of cephalopod molluscs are extremely well developed. Cephalopods are the supreme testament to the impressive plasticity of the basic molluscan body plan.

Of the 600 or so extant cephalopod species, only the 5 or 6 species in the genus *Nautilus* possess a true external shell. The shell of *Nautilus* is spiral, but unlike that of gastropods it is divided by **septa** into a series of compartments (Figs. 12.37 and 12.38). The living animal is found only in the largest, outermost chamber. The septa are penetrated by the **siphuncle**—a calcified tube and its enclosed strand of vascularized tissue that spirals through the shell from the visceral mass, traversing all shell chambers. Liquid may be slowly transported to and from the shell chambers through the siphuncle, gas diffusing into or out of the chambers as the fluid volumes are altered.

This liquid transport is apparently made possible by enzymes in the siphuncular tissue; these enzymes actively concentrate solutes—probably ions—either inside or outside the siphuncular tissue, thereby establishing local osmotic gradients. If the osmotic concentration is higher within the siphuncular tissue, water will diffuse along the osmotic gradient from the liquid in the chamber—termed the **cameral fluid**—into the siphuncle and thence into the blood, to be discharged by the kidney. Gas probably diffuses from the blood—remember, the siphuncle is well vascularized—into the shell chamber as **cameral fluid** is removed.

As the gas content of each chamber is changed by this mechanism, the buoyancy of the shell, and therefore of the animal, also changes. In this way, the nautilus can maintain neutral buoyancy as it grows, exactly compensating for increased weight of shell and tissue through appropriate discharge of cameral fluid. Cameral fluid flux also may be involved in buoyancy regulation during the chambered nautilus's vertical migrations, in which the nautilus may ascend and then descend hundreds of meters each day.

The nautilus locomotes by jet propulsion, expelling water from the mantle cavity through a flexible, hollow tube called the **siphon** or **funnel** (Fig. 12.37a). Water is expelled through the contraction of the head retractor muscles. The muscular funnel, which is derived from the foot of the "typical" mollusc, may be turned in various ways to move the animal in different directions.

Other cephalopods also move by jet propulsion, but they expel a much greater volume of water and do so more forcefully. In contrast to the nautiloids, other cephalopods lack an external shell, and the mantle can thus play a major role in movement (Fig. 12.39a). The mantle tissue is thick and replete with both circular and radial musculature. Contracting the radial musculature of the mantle tissue while the circular muscle fibers are relaxed increases the volume of the mantle cavity. Water then enters the mantle cavity along the anterior mantle margin, often through one-way valves. (If you think of a squid or octopus as wearing a jacket [the mantle], the mantle cavity lies between the jacket and the body, and water enters the mantle cavity around the neck.) This influx of water causes the animal to be drawn forward slightly. When the radial muscles relax and the circular muscles then contract, the margins of the mantle tissue form a tight seal against the neck, and a large volume of water is forcefully expelled entirely through the flexible, hollow funnel. Since thrust = mass of expelled fluid × velocity, the expulsion of so much water through the funnel at great velocity enables these cephalopods to move at high speeds; for brief periods at least, squid probably can reach speeds of 5 m/sec (meters per second) to 10 m/sec. In such rapidly moving squid, the mantle cavity refills partly by radial muscle contractions, partly by elastic recoil of the compressed mantle, and partly (or largely) by pressure gradients set up along the sides of the body as the squid moves through the water.[18] Jet propulsion is used most commonly in escaping from predators and capturing prey. For more leisurely locomotion, many species use the arms (as in *Octopus*) or muscular lateral fins (as in squid and cuttlefish).

The mantle's active role in locomotion is incompatible with having an external shell. Most modern cephalopods that retain a shell (the cuttlefish, squid, and many octopuses) bear it internally (Fig. 12.40a, b). Cuttlefish, for example, have an internal chambered shell that is involved in buoyancy regulation, as in *Nautilus*.[19] The squid shell is

18. See *Topics for Further Discussion and Investigation,* no. 12.

19. See *Topics for Further Discussion and Investigation,* no. 2.

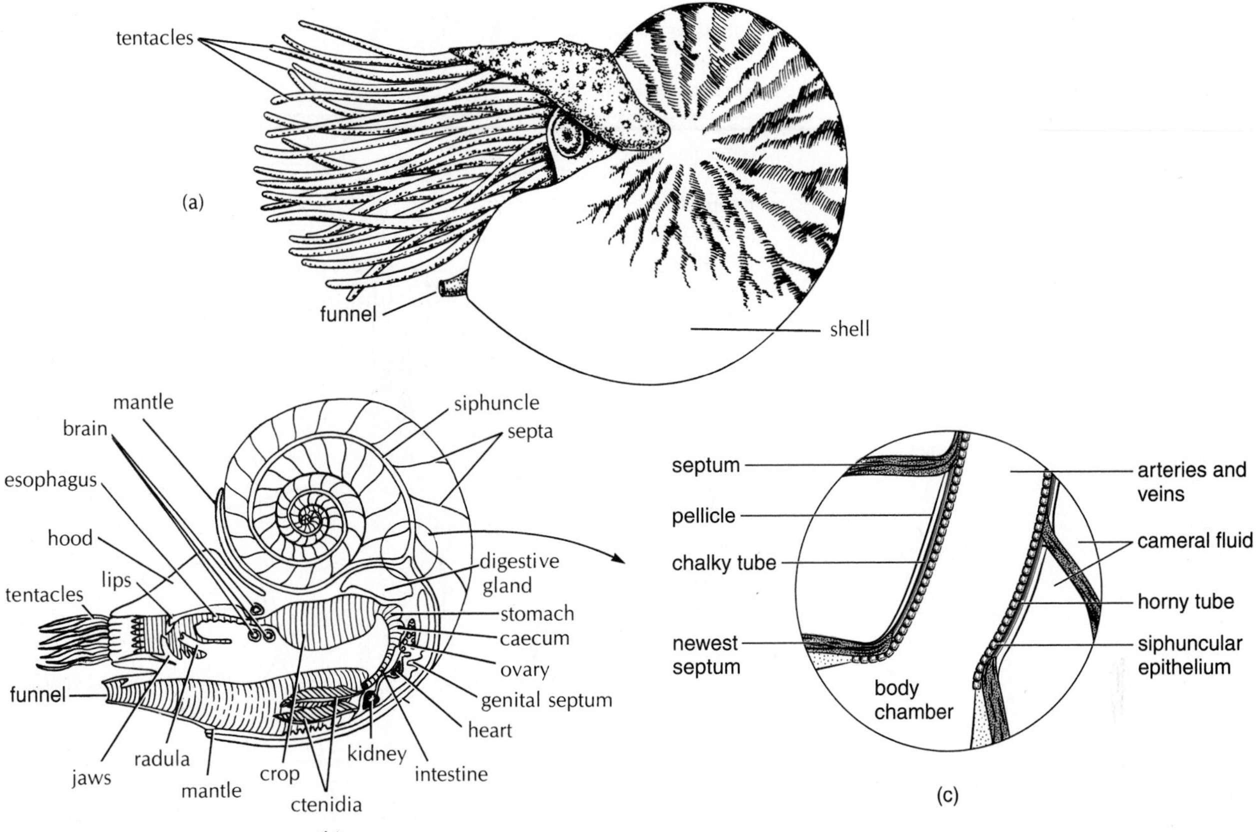

Figure 12.37

(a) *Nautilus* sp., in its normal swimming position. (b) Longitudinal section through the shell of the same individual, showing the septa isolating the different shell compartments, the siphuncle, and the internal anatomy. The entire body of the animal is found in the outermost chamber of the shell. Of the other shell chambers, all but the newest are filled with gas. (c) Detail of the newest and second most recent shell chambers.

(b) After Engemann and Hegner; after Borradaile and Potts. (c) Based on "The Buoyancy of the Chambered Nautilus" by Peter Ward, Lewis Greenwald, and Olive E. Greenwald, in *Scientific American*, October 1980.

also internal, but it is little more than a thin, stiff, proteinaceous sheet, called the **pen** (Fig. 12.39b). This nonchambered "shell" plays no role in buoyancy regulation; instead, squid compensate for their own body weight by accumulating high concentrations of ammonium ions in the coelomic fluid. In *Octopus*, no vestiges of a shell remain. The so-called paper nautilus (*Argonauta*) also lacks a true shell; it is much more closely related to the octopus than to *Nautilus*, despite its common name. When reproductively active, however, females produce a chamberless, spiral, fragile "shell," using glands on the arm—not the mantle tissue; the female paper nautilus lives in this shell while brooding her young.

Thus, of all living cephalopods, only the chambered nautilus forms a true external shell. Reduction or loss of the shell has clearly been a major theme in cephalopod evolution over the last several hundred million years; over 7000 species of shelled cephalopods are known from fossils (Fig. 12.40c), some of which have shells up to 4.5 m in diameter. As external shells became reduced or lost in evolution, cephalopods must have become increasingly vulnerable to predation. Such vulnerability probably led to selection for other means of protection. For example, the skin of most cephalopods contains several layers of tiny colored cells

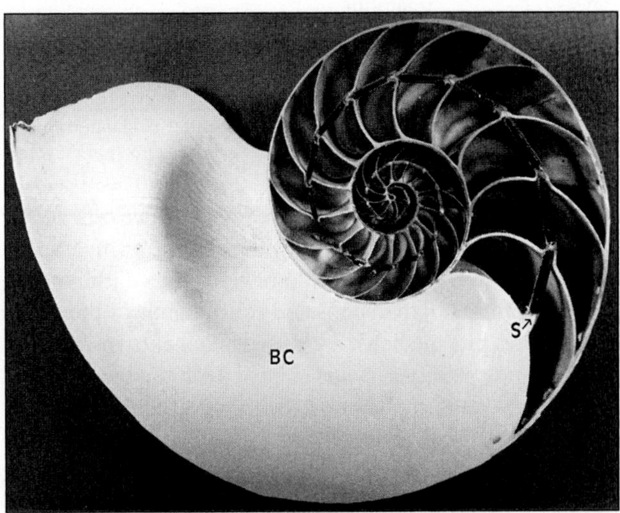

Figure 12.38

Section through the shell of *Nautilus*, showing the large body chamber (BC), siphuncle (S), and smaller chambers separated by septa. Holes have been drilled through the shell in several of the most recent chambers for experimental purposes; they are not natural.

Courtesy of L. Greenwald.

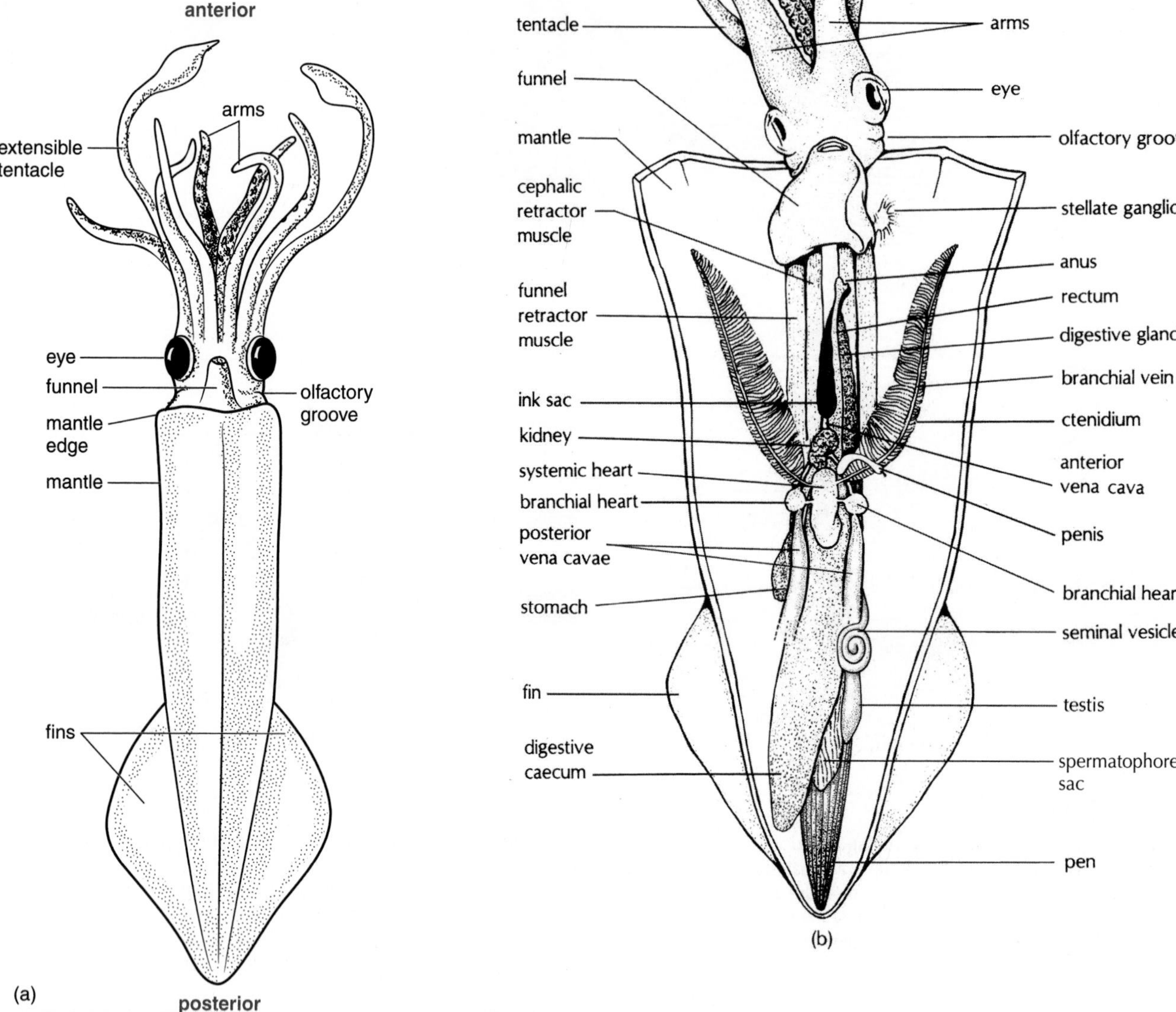

Figure 12.39a, b

(a) External anatomy of the squid *Loligo* sp. (b) The internal anatomy of a male squid (ventral surface facing up). Digestion is initiated in the stomach and completed in the large, associated caecum. (c) Dissected ripe female, showing major reproductive organs. The nidamental glands secrete a gelatinous coating around each egg.

(a) Based on Beck/Braithwaite, *Invertebrate Zoology, Laboratory Workbook,* 3/e, 1968.

called **chromatophores,** which overlay reflective cells called **iridocytes** (Fig. 12.41). Typically, the skin of a single individual contains hundreds of thousands, or even millions of chromatophores. Expansions and contractions of the chromatophores are mediated by muscle elements in the skin and are under direct nervous control from the brain. Thus, the coloration of cephalopod skin may change extremely quickly. Defensive, camouflaging, and courtship-related changes of color have been described.[20] Not surprisingly, the chambered *Nautilus* lacks chromatophores.

Many cephalopod species, particularly midwater and deep-water squid, possess numerous light organs, called **photophores.** These are distributed on the body in species-specific patterns, particularly on the ventral surface (Fig. 12.42). Light is produced in the photophores by biochemical reactions that parallel to a remarkable degree those demonstrated by many arthropods (including fireflies) and fishes. In at least 3 dozen squid species, the bioluminescence is produced by symbiotic bacteria living within the photophores. Although such **bioluminescence**—the biochemical production of light with minimal heat—may play a role in attracting or recognizing mates and in luring potential prey, it almost certainly also protects against predation. At night, or in very deep water where it is perpetually dark, bright coordinated flashes of light may startle would-be predators. At shallower depths, the photophores probably act instead to make the squid less visible when viewed from below.[20] Imagine looking up at an opaque object silhouetted against its surroundings during the day. Now imagine that object breaking up its silhouette by producing light of ambient intensity at many points along its ventral surface. Note that to achieve camouflage in this

20. See *Topics for Further Discussion and Investigation,* no. 5.

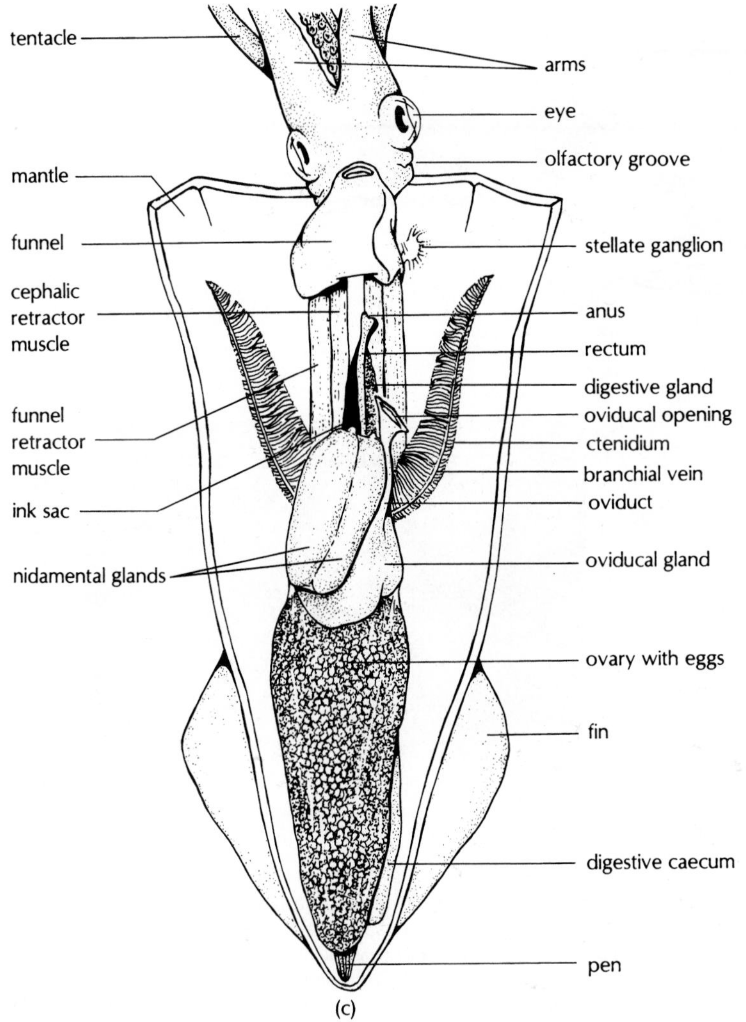

Figure 12.39c

manner the squid must be able to assess ambient light intensity and adjust the intensity of its own light production to match that intensity.

In addition, most cephalopods other than the chambered *Nautilus* have an **ink sac** associated with the digestive system (Fig. 12.39b, c). The dark-pigmented fluid secreted by the ink sac may be discharged deliberately through the anus, forming a cloud that presumably confuses potential predators and that also may act as a mild narcotic. The chambered *Nautilus* lacks an ink sac; its absence is consistent with the notion that the ink sac has been selected for in other species by the increased predation pressure that must have accompanied reduction and loss of the external shell.

The cephalopod mantle cavity generally contains a pair of ctenidia (Fig. 12.39b, c), but unlike the situation in other molluscs, blood and water do not flow continuously in opposite directions: There is no countercurrent exchange system. Moreover, the ctenidia of cephalopods are not ciliated. Water circulation is maintained by the continual emptying and refilling of the mantle cavity, accomplished through contraction of the mantle musculature.

Cephalopods are also unique among molluscs in having a completely closed circulatory system, in which blood flows entirely through a system of arteries, veins, and capillaries; the blood sinuses found in other molluscs are not present in cephalopods. In addition to a single **systemic heart,** which receives oxygenated blood from the gills and sends it back to the tissues, an accessory **(branchial)** heart is associated with each gill (Fig. 12.39b). The 2 branchial hearts increase blood pressure, helping to push blood through the gill capillaries. Concentrations of oxygen-binding blood pigments (hemocyanin, p. 266) are also unusually high in cephalopod blood. The cephalopod circulatory system is thus more efficient than that of other molluscs, supporting a far more active lifestyle.

In addition to forming the funnel, derivatives of the molluscan foot form the cephalopods' muscular **arms** and extensible **tentacles** (Fig. 12.43). The mouth thus lies in the center of the "foot." The total number of arms and tentacles is usually either 8 or 10, depending on the species, although some nautiloids possess 80–90. The tentacles of most cephalopods—all except those of the nautiloids—have small suction cups that are used for clinging to the substrate or to objects, including potential prey (Fig. 12.43a, b). The arms are also studded with receptors sensitive to touch and taste; all cephalopods tested to date have well-developed chemical sensory capabilities.

The degree of **cephalization** (concentration of sensory and nervous tissue at the anterior end of an animal) found among the Cephalopoda exceeds that found in any other invertebrate. Cephalopods possess a large, complex, highly differentiated brain (Fig. 12.44a). The brain of the common octopus, *Octopus vulgaris,* has over 10 distinct lobes. Studies of cephalopod behavior indicate a clear capability for memory and learning (Research Focus Box 12.3). In the basic experiments, an individual—usually an octopus—is trained, through appropriate rewards (food) and discouragements (mild electric shocks), to attack an object of a certain shape, color, or texture or to attack prey only if the particular object is or is not present. Once training has been completed, researchers test the individual at intervals to determine how long the animal remembers what it has learned. By changing an object's shape, texture, orientation, color, size, or weight, they can discover what differences the cephalopod can perceive: Can it tell the difference, for example, between a red circle and an otherwise identical white circle, or between a horizontal rectangle and a vertical rectangle, or between a heavy sphere and an identically shaped, but lighter sphere, or between a rough sphere and a smooth one, or between a scallop shell and a clam or mussel shell?

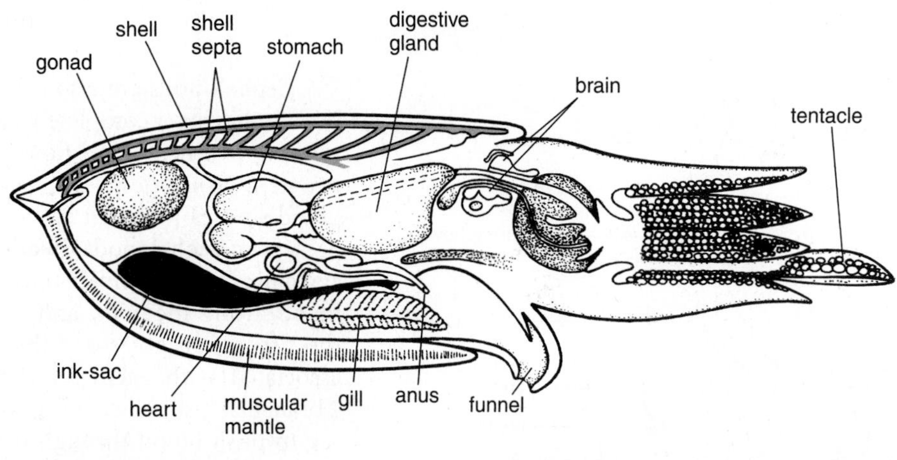

(a)

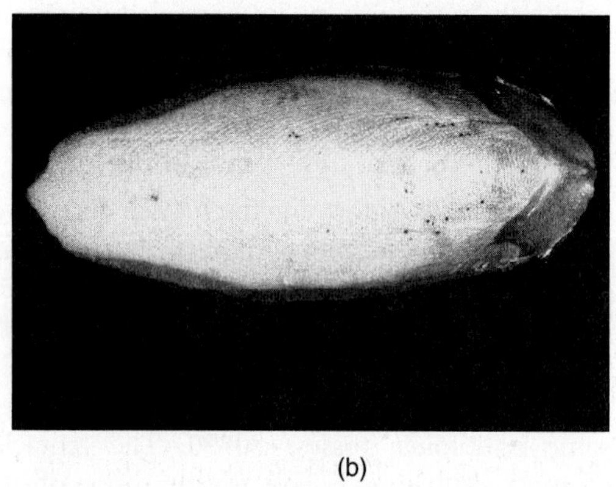

(b)

(c)

Figure 12.40

(a) Internal anatomy of the cuttlefish *Sepia* sp., showing location of internal shell (cuttlebone). (b) The internal shell of a cuttlefish. (c) Fossilized ammonite (about 5 cm in diameter), with the coiled shell typical of most cephalopods for several hundred million years. Shells of some species exceeded 4.5 m in diameter. Most shelled cephalopods were driven to extinction at the end of the Mesozoic, quite likely through competition with an increasingly successful evolutionary upstart: the bony fishes.

(a) Boardman *et al.,* 1987. *Fossil Invertebrates*. Palo Alto, Calif: Blackwell Scientific Publications. (b) Courtesy of Kerry M. Zimmerman. (c) Courtesy of H. Klinger, from Urreta and Klinger, 1986. *Annals of South African Museum* vol. 196 (specimen no. 11876). Sally Dove, photographer.

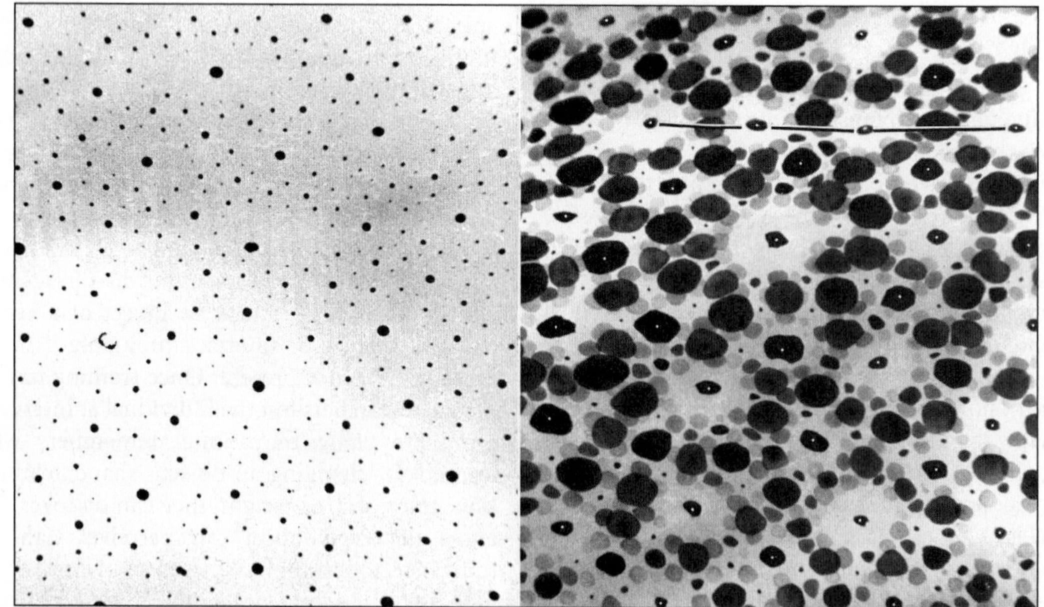

Figure 12.41

Chromatophores photographed from a living squid. The chromatophores are contracted at the left and expanded at the right.

Courtesy of R. Hanlon. 1982. *Malacologia* 23:89.

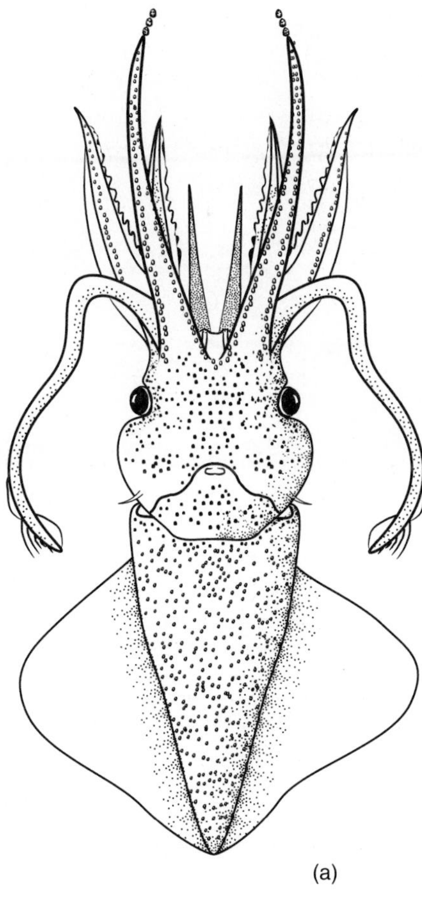
(a)

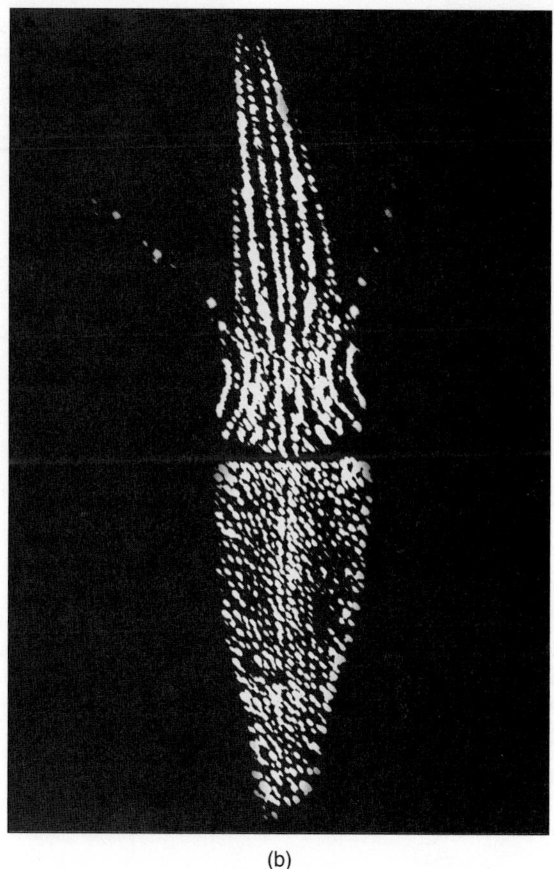
(b)

Figure 12.42

(a) Ventral surface of the squid *Abraliopsis* sp., showing its numerous photophores. During the day, the squid is found mainly 500–600 m below the ocean surface; it migrates into shallower water at night. (b) Photograph of bioluminescence, showing pattern of photophores on the squid *Abralia veranyi*.

(a) Based on R. E. Young, in *Science* 191:1046–48, 1976. (b) Courtesy of E. A. Widder. From Herring, P. J., *et al.*, 1992. *Marine Biol.* 112:293–98. © Springer-Verlag.

The results of such experiments clearly indicate that the octopus perceives shape, color intensity, and texture but, surprisingly, that it cannot distinguish between objects differing only in weight. Even more surprising, cephalopods show no behavioral response to sound except at very low frequencies. While direct electrical recordings clearly indicate that the statocysts are sensitive to sound stimuli, the information apparently is not suitably processed in the brain. Cephalopod deafness may be an adaptive response to millions of years of predation by toothed whales and dolphins, which may stun their prey with sound.[21] By tuning most sounds out, cephalopods might thus be immune to such a diabolical hunting mechanism; deaf cephalopods would have an edge over hearing cephalopods and so would be selected for over many generations, if we assume deafness to be a genetically determined trait.

Each cephalopod has 2 eyes. In *Nautilus* spp., the eyes are simple and function on the pinhole camera principle; there is no lens (Fig. 12.44c). The eyes form an image, but visual acuity is not great. In contrast, all other cephalopods have image-forming eyes that are incredibly similar to those of mammals (Fig. 12.44b). The eyes of these cephalopods and those of mammals are among the most beautiful examples of convergent evolution encountered among animals; the 2 groups of animals have independently evolved eyes that are amazingly similar in structure. Like the mammalian eye, the cephalopod eye possesses a cornea, lens, iris, diaphragm, and retina; both eyes are focusable and image-forming, although the process of image formation in mammals and cephalopods differs in detail. In most mammals, light is focused by altering the shape of the lens. In contrast, light is focused within the cephalopod eye by moving the lens toward or away from the retina. Vertebrate and cephalopod eyes also differ in how they develop: Whereas vertebrate eyes develop as outfolded extentions of the brain, cephalopod eyes develop through inpocketing of the ectoderm.

21. See *Topics for Further Discussion and Investigation*, no. 17.

(a)

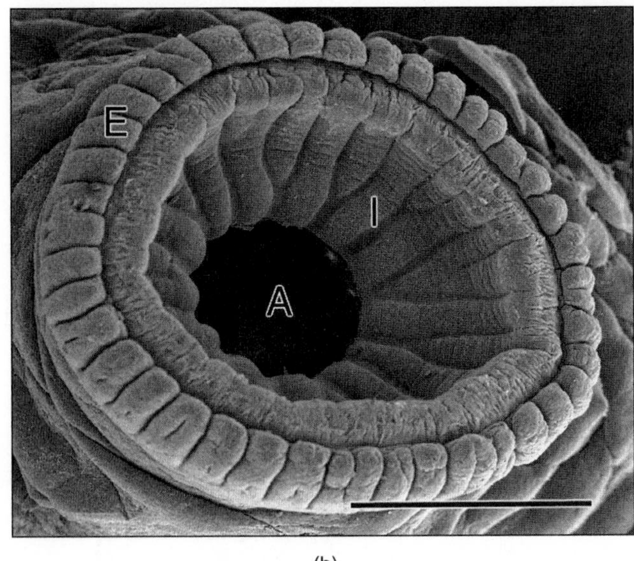

(b)

Figure 12.43a, b

(a) Suckers on the arms of *Octopus* sp. holding on to glass wall of an aquarium tank. (b) Detail of one sucker from *Octopus* sp. scale bar = 1 mm. I = sucker disc (infundibulum), A = acetabulum (opening into sucker cavity), E = epithelium. (c) The squid *Loligo pealei,* capturing prey with its suckered, extensible tentacles. (d) In this sequence, the crustacean prey was dropped in front of the squid at T_0 (time zero). The events shown occurred within 70 milliseconds (ms).

(a, b) Courtesy of Bill Kier. From Kier, W. M., and A. M. Smith. 2002. The structure and adhesive mechanism of *Octopus* suckers. *Integr. Comp. Biol.* 42: 1146–53. (c) Based on Kier, in *Journal of Morphology,* 172:179, 182. 1982. (d) Courtesy of W. M. Kier.

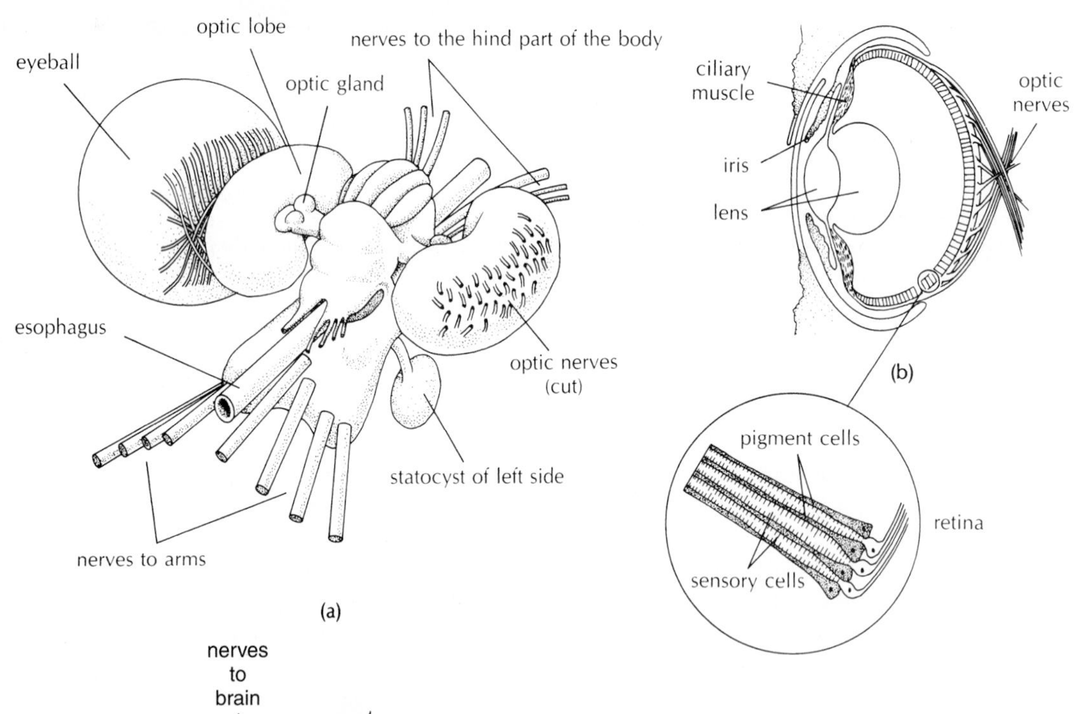

nerves to brain

pinhole

retina

(c)

Figure 12.44

(a) A cephalopod brain. Cephalopods possess the most complex brains found among invertebrates. (b) The eye of *Octopus,* demonstrating convergent evolution with the eye of vertebrates. Unlike the vertebrate eye, cephalopod eyes are focused by moving the lens back and forth, rather than by altering the shape of the lens. (c) Vertical section through the pinhole camera eye of the chambered *Nautilus.*

(c) From W. R. A. Muntz, 1991, *American Malacological Bulletin.* 9:69–74.

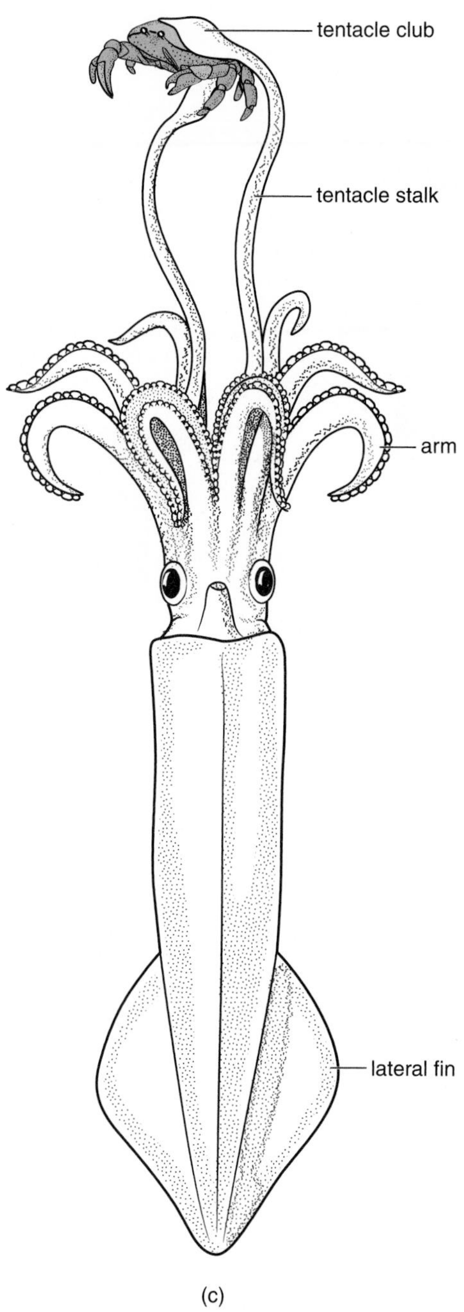

(c)

Figure 12.43c, d

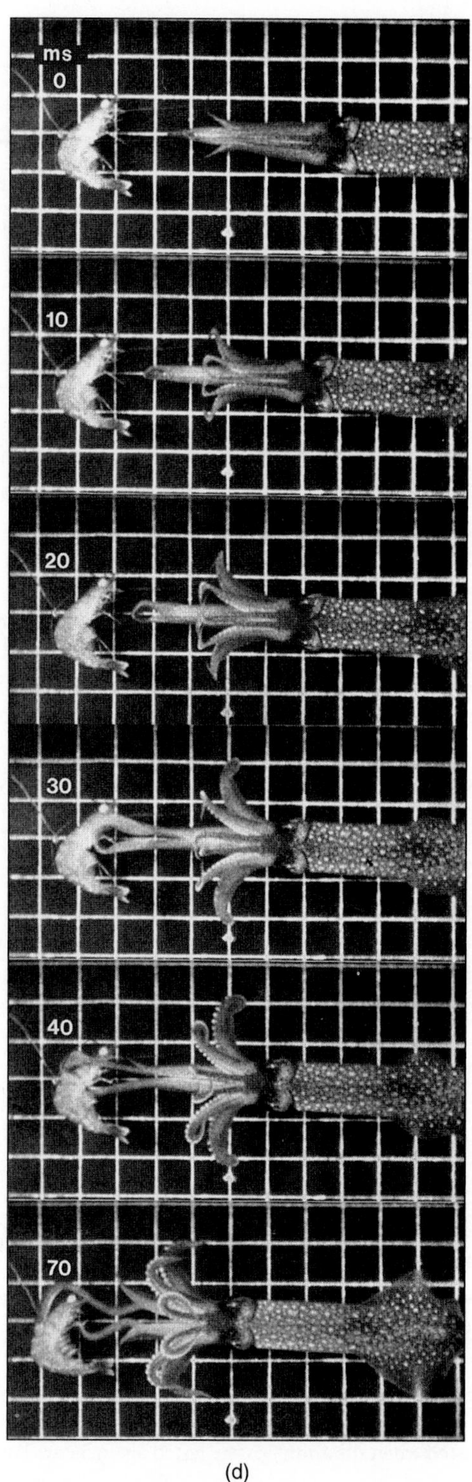

(d)

Clearly, the story of cephalopod evolution over the past several hundred million years has been one of developing adaptations for an active, carnivorous lifestyle: reduction and elimination of the shell; evolving alternate means of protection from predators; replacement of ciliary activity with muscular activity for locomotion and respiration; modification of the foot, mantle, and blood circulatory system; and extensive development of the head, sensory organs, brain, and nervous system. What might account for this dramatic evolution away from the standard molluscan body plan and lifestyle? What pressures could possibly have selected against a protective device as effective as a sturdy, external shell? Here is one attractive scenario: By the early Paleozoic, some 500 million years ago, the cephalopods, with their buoyant, chambered shells, were undoubtedly the most mobile and

RESEARCH FOCUS BOX 12.3

Cephalopod Behavior

Fiorito, G., and P. Scotto. 1992. Observational learning in *Octopus vulgaris. Science* 256:545–47.

How we learn and remember are 2 of the great mysteries of human biology. Biologists have long studied aspects of behavior in other animals, hoping to discover general principles of brain function that may apply equally to us, and to determine just how unique is our ability to learn, to remember, and to teach. Many such studies have been conducted on *Octopus vulgaris,* a species common in European coastal waters. Fiorito and Scotto (1992) are the first to ask whether an octopus can learn from other octopuses.

They performed their study with *Octopus vulgaris* collected from the Bay of Naples, Italy. First they trained each of 44 octopuses to grab either a red plastic ball or a white plastic ball, by mildly shocking the animal if it chose incorrectly and rewarding it with food if it chose correctly. After about 17 to 21 trials, the octopuses consistently would select balls of the color they were trained to prefer, even without additional reward or punishment. *Octopus vulgaris* is color-blind, so the animals must have been perceiving differences in brightness rather than in color *per se.* Then the researchers let each octopus demonstrate its acquired color preference to a freshly caught test octopus that had received no prior training. Each demonstrator octopus consistently chose either a red or a white ball 4 times (Focus Fig. 12.3a, c), while a test octopus looked on. Then each of the 44 test octopuses was allowed to choose between a red ball and a white ball, again without reward or punishment. Would the observer octopuses adopt the preferences of the demonstrators?

The results were truly remarkable. The 30 test octopuses exposed only to octopuses demonstrating a preference for red balls mostly chose red balls themselves (Focus Fig. 12.3d), even though they received no reward or punishment for choosing one ball over the other. Similarly, test octopuses exposed only to octopuses demonstrating a preference for white balls mostly chose white balls themselves (Focus Fig. 12.3b). Moreover, the preferences shown by observer octopuses persisted when the animals were tested again 5 days later. These results are all the more impressive in that the octopuses showed a clear preference for red balls for some reason; so choosing a white ball over a red ball went against their apparent preference for red.

It seems clear that individuals of *Octopus vulgaris* can learn from other individuals of the same species. In fact, they seem to learn more quickly by observing what other octopuses do than through human training; it took the octopuses at least 17 trials to learn by shock/reward training but only 4 trials to learn by example! We fully expect to find this sort of learning by copying among chimpanzees ("monkey see, monkey do") and other vertebrates, but to find it also among molluscs is indeed intriguing. One wonders about the extent to which octopuses can learn more complex behaviors by example, and about the things octopuses might actually be learning from each other in the field.

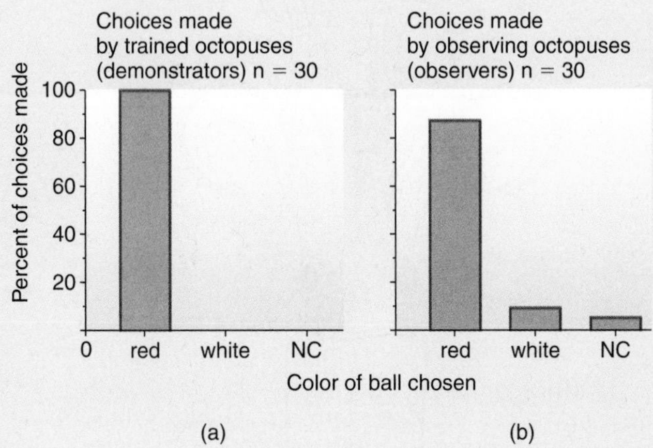

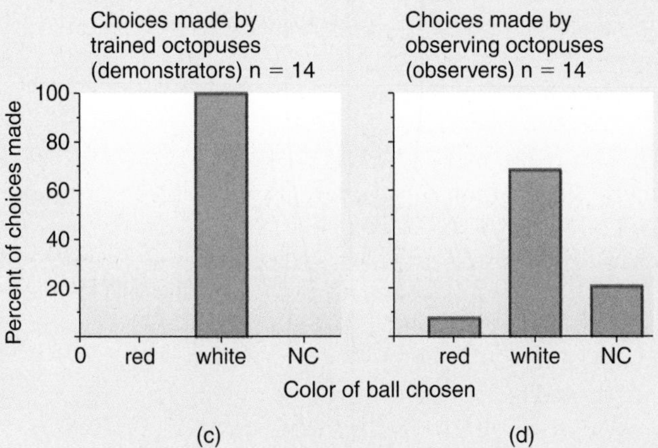

Focus Figure 12.3

The impact of training and observation on choices made by *Octopus vulgaris*. Demonstrators were trained to prefer either a white ball (a) or a red ball (c). Observers were then allowed to watch the choices made by demonstrators but received no training themselves. NC = no choice made. n = number of individuals participating in each experiment.

Based on Fiorito, G., and P. Scotto, 1992. Observational learning in *Octopus vulgaris* in *Science* 256:545–47.

successful of all marine predators. With the subsequent diversification and success of bony fishes must have come intense competitive pressures between fish and cephalopods. With the intriguing exception of the chambered nautilus, only those cephalopods that had or evolved a more fish-like morphology and lifestyle apparently met those pressures and persist today.[22]

Other Features of Molluscan Biology

Reproduction and Development

Although some gastropods are parthenogenetic, molluscs usually reproduce sexually. Most species are **gonochoristic** (i.e., sexes are separate), but there are many exceptions to this generalization. Some prosobranch gastropods, opisthobranch gastropods, and lamellibranch bivalves, for example, are **protandric hermaphrodites;** that is, the sex of a single individual changes from male to female with age. All pulmonates, and those opisthobranchs that are not protandric hermaphrodites, are **simultaneous hermaphrodites,** with a single individual producing both eggs and sperm simultaneously; the gonad of such an individual is termed an **ovotestis.** Simultaneous hermaphrodites often have reciprocal copulation, resulting in a mutual exchange of sperm (Fig 24.3b). There are no known hermaphroditic cephalopod species.

Generally, the molluscan genital ducts are associated with a portion, or a modified portion, of the excretory system. Fertilization of eggs is exclusively external in the Scaphopoda and probably in the Monoplacophora as well. Such external fertilization is quite common among the bivalves, chitons, and aplacophorans; less so among the gastropods; and nonexistent among the cephalopods. Terrestrial and freshwater molluscs (i.e., some gastropods and bivalves) fertilize only internally, as adaptations to stressful conditions that would otherwise be imposed upon the gametes and embryos by the environment (Chapter 1). Cephalopods show a particularly distinctive set of adaptations for achieving internal fertilization. In particular, one arm of the male is modified as a copulatory organ (Fig. 12.45). This sometimes highly modified arm, called a **hectocotylus** (*hecto* = G: 100; *cotylo* = G: sucker), transfers packets of sperm (**spermatophores,** Chapter 24, Fig. 24.9d, p. 564) to the female; in the paper nautilus, the hectocotylus detaches from the male and stays behind in the female's mantle cavity until sperm transfer is completed. Equally odd, in many terrestrial pulmonate gastropods, the genital organs produce specialized chitinous or calcareous "love darts" (Fig. 12.46) that are thrust into the partner during mating; while sperm are being transferred to the partner

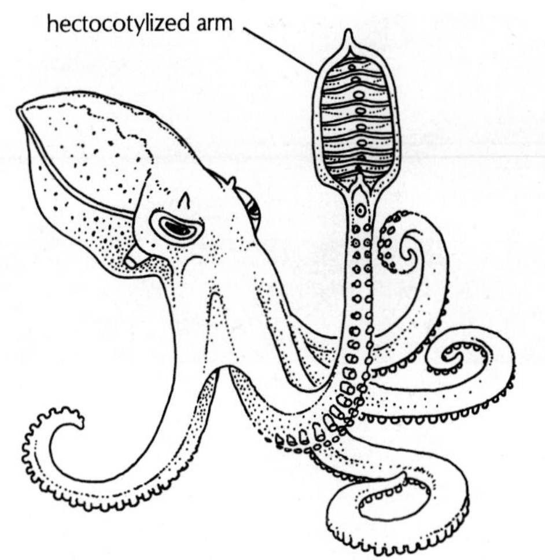

Figure 12.45
Octopus lentus male with hectocotylized arm. The arm is turned up to show where the spermatophores will be carried. The hectocotylus of most cephalopods is not so highly modified.
After Huxley; after Verrill.

through the penis, hormones delivered to the partner with the darts increase the likelihood of successful fertilization.

Not surprisingly, free-living larval stages are associated with the development of most species that fertilize their eggs externally in the surrounding seawater. The embryo passes through a conspicuous **trochophore** stage (Fig. 12.47a, b), resembling that of polychaete annelids (Fig. 13.11). It is presently unclear whether this similarity of larval stages indicates a close evolutionary relationship between annelids and molluscs, or whether the trochophore larva has instead arisen in the 2 phyla independently, by convergence. In any event, the prototroch of the trochophore stage of gastropods, bivalves, and scaphopods gradually becomes outfolded into a distinctive, ciliated organ known as the **velum.** A larva with a velum is called a **veliger** (Figs. 12.47d, e, i, j and 24.14e, f). The velum may be used for locomotion, food (phytoplankton) collection, and gas exchange and is lost upon metamorphosis to adult form. Veligers may spend hours, days, weeks, or months swimming in the plankton before metamorphosing. Among the shelled molluscs, metamorphosis often marks an abrupt change in shell morphology or ornamentation; in many species, the transition between larva and juvenile is thus clearly indicated on the shell (Fig. 12.48b). Among unshelled opisthobranchs, the larval shell is abandoned at metamorphosis.

Even when fertilization is internal, a dispersive larva often occurs at some point during development. Remarkably, considering the osmotic problems facing freshwater animals, a few freshwater bivalve species have free-living veliger larvae. The presence of such veligers in the life histories of zebra mussels (*Dreissena polymorpha*) has surely contributed to their rapid spread throughout freshwater ecosystems in North

22. See *Topics for Further Discussion and Investigation*, no. 19.

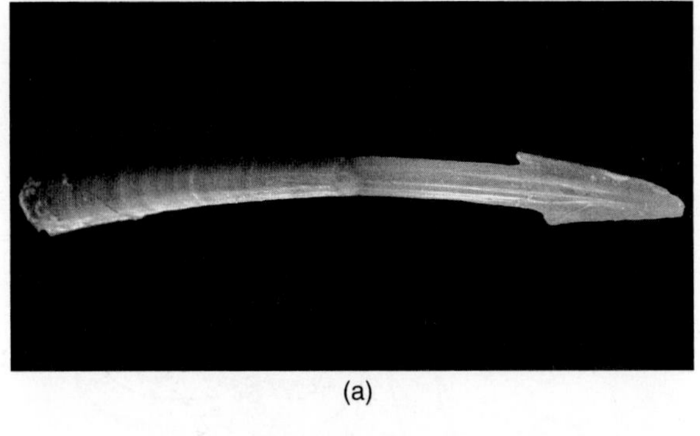

(a)

(b)

Figure 12.46
Love darts from the pulmonate gastropods *Monachoides vicinus* (a) and *Trichia hispida* (b). The darts shown are 5.1 mm (a) and 1.4 mm long (b). Courtesy of Joris M. Koene. From BioMed Central, Koene & Schulenburg 2005, *BMC Evol. Biol.* 5: article 25.

America over the past 10 years. But in most freshwater bivalve species the veliger has been greatly modified to form a **glochidium** larva (Fig. 12.47f), a microscopic, nonswimming, shelled larva that develops for weeks as an external parasite—almost always on fish—before dropping to the bottom to take up life as a benthic juvenile. Before their release from the parent mussel, glochidia develop within the parent's gills (Fig. 12.47g); once released, they can complete the life cycle only if they succeed in latching onto a passing, appropriate vertebrate host. Their dependence on particular species of host fish for completing the life cycle is likely contributing to the present-day extinction of many freshwater bivalve species; as many freshwater habitats are made unsuitable for host fish species, through pollution or the damming of rivers, for example, the bivalves' offspring cannot reach reproductive adulthood. In such cases, the present generation will be the last.[23]

In the most advanced mollusc species, particularly among the gastropods, the free-living larval stage is often suppressed. Embryos may develop to the juvenile stage entirely within a jelly mass, an egg capsule (Figs. 12.47h and 24.13), or a specialized brood chamber of the female. Most pulmonates develop to miniature snails within well-protected, calcified egg coverings, although some species retain a free-living veliger; obviously, none of the truly terrestrial gastropods have free-living larvae. Only among the cephalopods is a free-living, morphologically and ecologically distinct larval stage absent from the life cycle of all species. Cephalopod eggs develop within gelatinous masses, often attentively protected and ventilated by the mother until the young emerge as fully formed miniatures of the adult.

23. See *Topics for Further Discussion and Investigation,* no. 22.

Circulation, Blood Pigments, and Gas Exchange

All molluscs have a blood circulatory system. In cephalopods, the system is completely closed; all blood flow occurs through arteries, veins, and capillaries. In all other molluscs, the circulatory system is largely an open one, with blood moving through a series of large sinuses (the **hemocoel**) derived from the embryonic blastocoel (Fig. 12.49). In gastropods, the turgor of the tentacles and foot depends upon the amount of blood in the sinuses of these tissues. In most molluscs, including the apparently primitive (ancestral) monoplacophorans, the blood is pumped by a heart. However, in scaphopods, which lack a heart, muscular contractions of the foot have primary responsibility for moving the blood through the large sinuses.

The bloods of many mollusc species lack a specialized oxygen transport pigment. Where one is found, it is either hemoglobin, as in some bivalves and pulmonates, or hemocyanin, as in most pulmonates and all prosobranchs and cephalopods. **Hemocyanin** is a protein, like hemoglobin, except that the hemocyanin molecule contains copper rather than iron. The oxygen-binding capacity of hemocyanin is substantially less than that of hemoglobin.

Gas exchange may take place across gills housed within a mantle cavity (e.g., in cephalopods, bivalves, chitons, and prosobranch gastropods), across external gills (e.g., in some opisthobranch gastropods), or across other vascularized tissues (e.g., the cerata of opisthobranch gastropods and the tissue lining the mantle cavity of pulmonates and scaphopods).

Nervous System

The degree of development of the molluscan nervous system corresponds to the activity level of its possessor. Molluscan ganglia range from being nonexistent in many chitons to

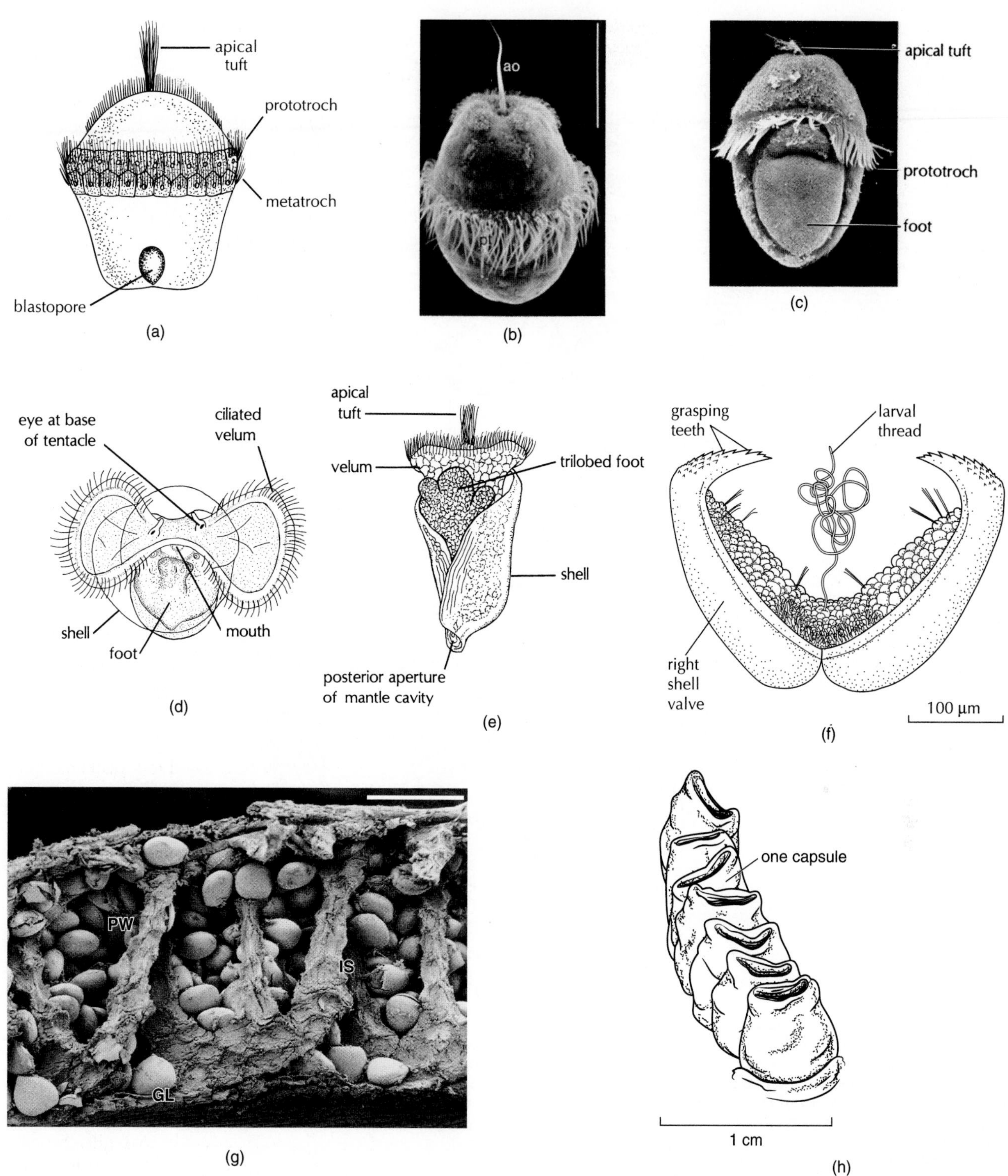

Figure 12.47

Molluscan development. (a) The trochophore larva of *Patella* sp., a primitive prosobranch gastropod. (b) Early trochophore larva of the chiton *Lepidochitona dentiens*. The larva is about 150 µm wide. ao = apical organ; pt = prototroch. (c) Advanced larva (7 days old) of the chiton *Lepidochitona hartweiqii;* note the pronounced developing foot. (d) A gastropod veliger larva in frontal view. (e) A scaphopod veliger. (f) Mature glochidium of the freshwater bivalve *Anodonta cyngea*. The larval thread possibly serves a sensory function, or for attachment to the fish host. (g) Glochidia developing within the parental gills of the Australian freshwater bivalve *Hyridella depressa*. GL = glochidium, PW = primary water canals, IS = interlamellar septa. Scale bar = 300 µm. (h) Egg capsules of the marine prosobranch gastropod *Conus abbreviatus*. Each capsule is about 1 cm high and contains numerous embryos. (i) Larva of the American oyster, *Crassostrea virginica,* in lateral view. (j) Veliger of the gastropod *Nassarius reticulatus*, showing internal anatomy in lateral view. Arrows in velar groove show movement of captured food particles toward mouth. (*Continued on the following page.*)

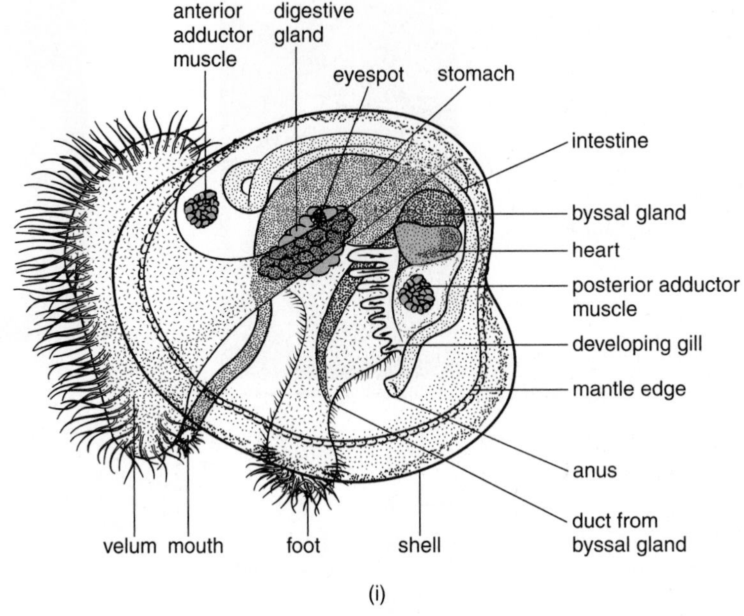

(i)

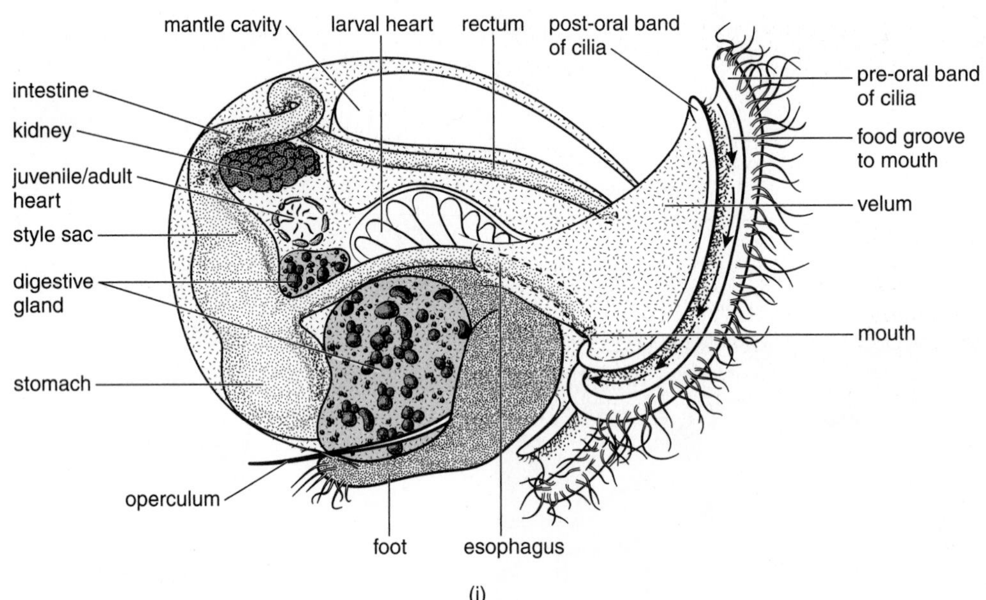

(j)

Figure 12.47 *Continued*

(a) From Hyman. (b) Courtesy of C. Nielsen. From Nielsen, 1987. *Acta Zoologica* (Stockholm) 68:205–62. Permission of Royal Swedish Academy of Sciences. (c) Courtesy of D. J. Eernisse. From D. J. Eernisse, 1988. *Biological Bulletin* 174:287–302. Permission of Biological Bulletin. (d) Courtesy Rudolph S. Scheltema. (e) From Giese and Pearse, *Reproduction of Marine Invertebrates*, Vol. 5 (after Lucaze-Duthiers, 1856). Copyright © 1979 Academic Press, Inc. Reprinted by permission. (f) From E. M. Wood, in *J. Zool.* London, 173:15–30. (g) Courtesy of M. Byrne. From S. D. Jupiter and M. Byrne, 1997. *Invert. Reprod. Devel.* 32:177–86. (h) Based on Kohn, in *Pacific Science*, 15:163–79, 1961. (i) Based on H. F. Prytherch, in *Ecological Monographs*, 4:56. 1934. (j) Based on Fretter and Montgomery, in *Journal of the Marine Biological Association of the United Kingdom*, 48:504, 1968.

being extremely well developed in the cephalopods. Among the cephalopods, the ganglia form a true brain (Fig. 12.44a). In other molluscs, the major ganglia are generally paired and are connected by nerve fibers, forming a ring through which the esophagus passes (Fig. 12.50a). Major nerve cords may run to the mantle, foot, gills and osphradium, viscera, and radula. The ganglia associated with innervation of these tissues are named, respectively, the pleural, pedal, parietal, visceral, and buccal ganglia. Details vary among members of the different classes.

In a number of molluscs, some neurons are specialized for very rapid impulse conduction. These unusually wide **giant fibers** are especially prominent among cephalopods, connecting the cerebral ganglia with the musculature of the

(a)

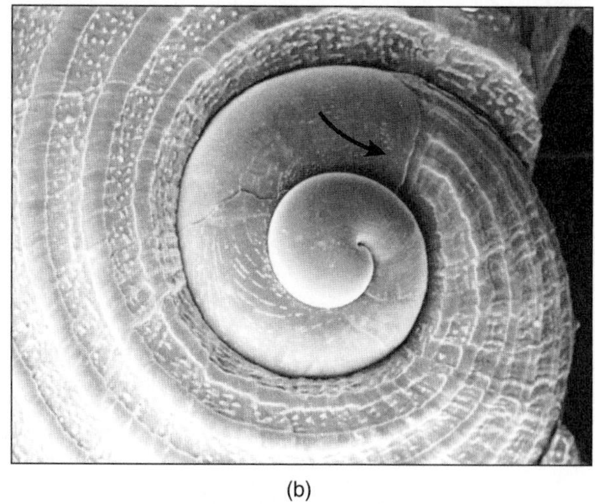

(b)

Figure 12.48

(a) The adult shell (scanning electron micrograph) of the prosobranch gastropod *Cyclostremiscus beauii* from Florida. The shell, 8 mm in diameter, is viewed from above; the aperture opens on the right. (b) High magnification of the shell apex, showing the transition between larval (protoconch) and juvenile shell (arrow); the larva metamorphosed when the shell was about 450 μm in diameter. At about 9 o'clock, there is another conspicuous line, which probably indicates an unsuccessful predatory attack on the larva; the larva survived, repaired its shell, and continued to grow.

(a,b) Courtesy of R. Bieler and P. M. Mikkelsen. From Bieler and Mikkelson. 1988. *The Nautilus* 102:1–29.

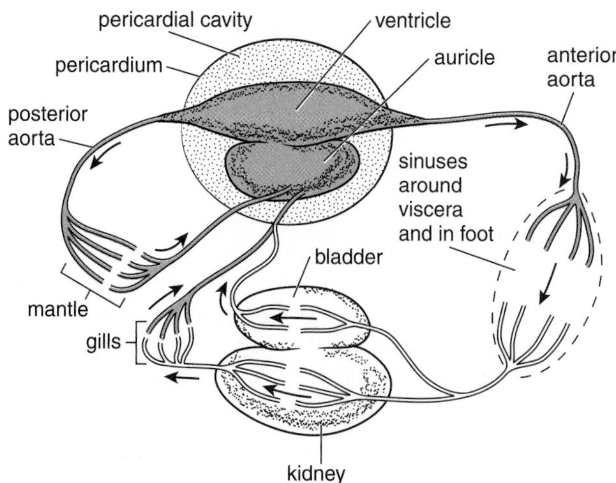

Figure 12.49

The pattern of blood circulation in the freshwater lamellibranch *Anodonta* sp. The shaded areas indicate the route of oxygenated blood.

Based on Cleveland P. Hickman, *Biology of the Invertebrates*, 2d ed. 1973.

mantle (Fig. 12.51). Giant fibers are extremely important in synchronizing contractions of the mantle musculature for jet propulsion, facilitating rapid escape responses. The fibers also have been important to humans; present knowledge of how nerve impulses are generated and transmitted comes largely from studies of cephalopod giant fibers.

Digestive System

Most molluscs have a complete digestive system with a separate mouth and anus. The mouth leads into a short esophagus, which, in turn, leads to a stomach. Associated with the stomach are one or more **digestive glands** or **digestive caeca.** Digestive enzymes are secreted into the lumen of these glands. Additional extracellular digestion takes place in the stomach. In cephalopods, digestion is entirely extracellular. In most other molluscs, the terminal stages of digestion are completed intracellularly, within the tissue of the digestive glands. The absorbed nutrients enter the blood circulatory system for

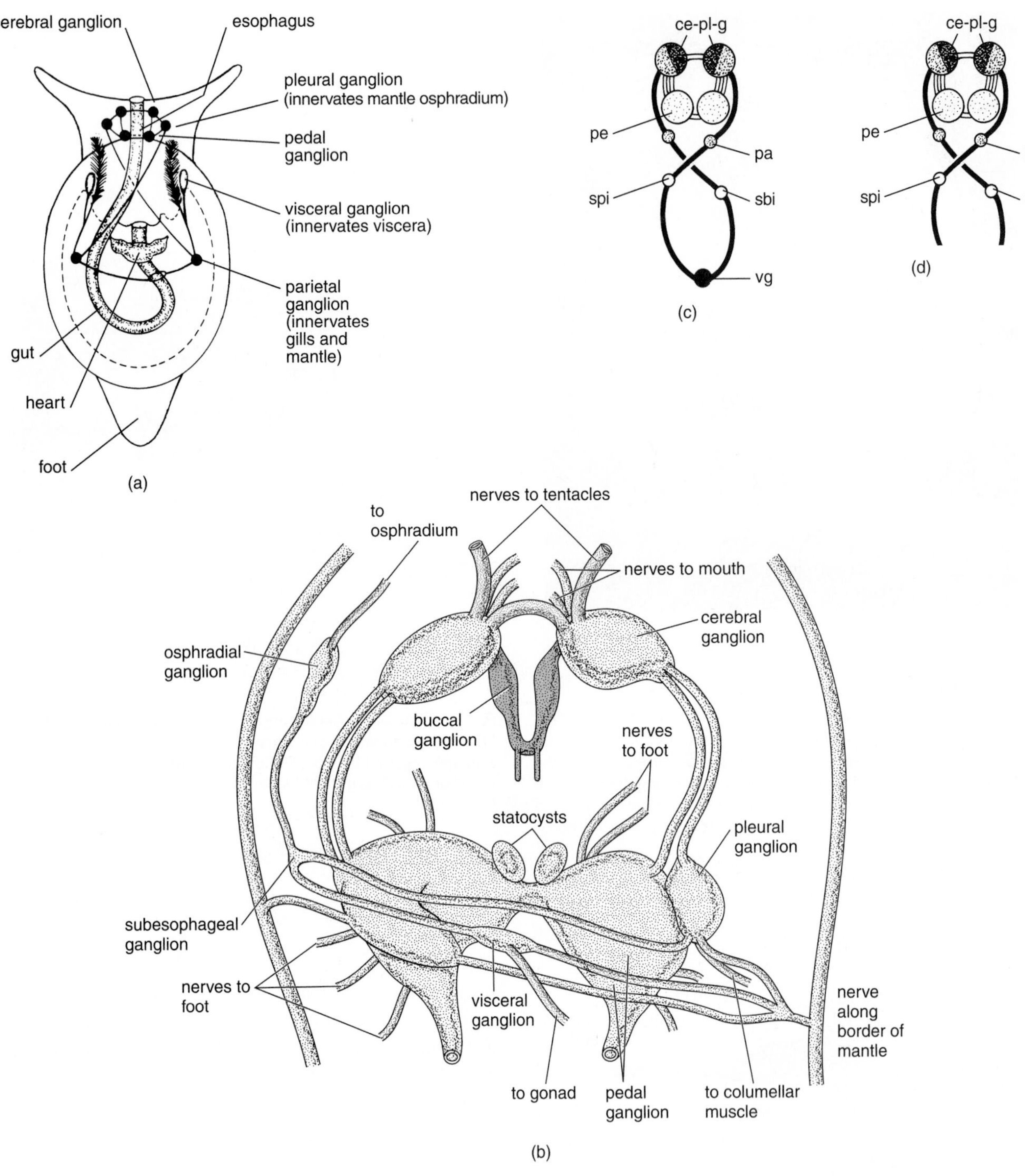

Figure 12.50

(a) Generalized nervous system of a prosobranch gastropod, demonstrating the effects of torsion. (b) Anterior portion of nervous system of the deep-sea limpet (class Gastropoda) *Cocculinella minutissima.* (c) Nervous system of a primitive opisthobranch gastropod, the bubble-shell snail *Acteon.* The members of this genus show no sign of detorsion; the nervous system still has the fully twisted, "streptoneurous" configuration (ce-pl-g = fused cerebral and plural ganglia; pa = parietal ganglion; pe = pedal ganglion; sbi = subintestinal ganglion; spi = supraintestinal ganglion; vg = visceral ganglion). (d) The nervous system of an advanced opisthobranch gastropod, the sea hare *Aplysia* sp. There is no longer any sign of torsion; the animal displays the "detorted" condition (pa-sbi-vg = parietal-subintestinal-visceral ganglia; ce = cerebral ganglion; pe = pedal ganglion; pl = pleural ganglion).

(b) Based on Haszprunar, in *Journal of Molluscan Studies,* 54:1–20, 1988. (c,d) From Louise Schmekel, in *The Mollusca,* Vol. 10, 1985. Copyright © 1985 Academic Press, Inc. Reprinted by permission.

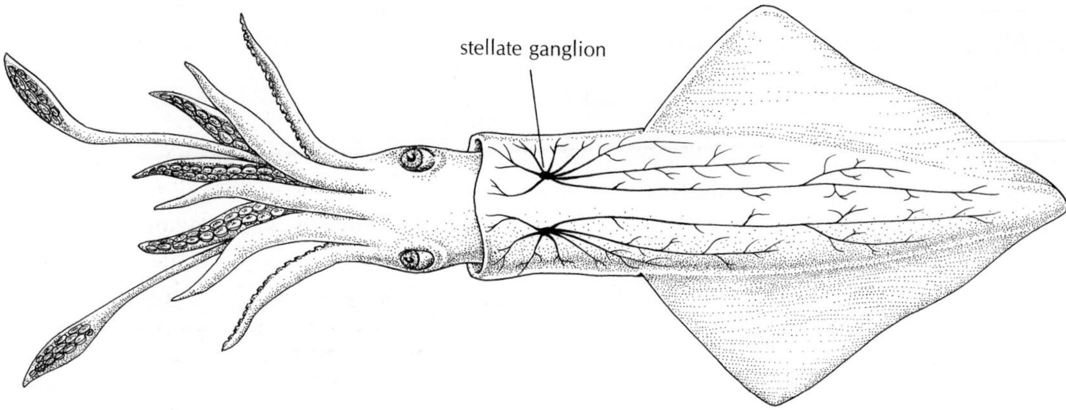

Figure 12.51
The nervous system of a cephalopod (squid), showing the system of giant fibers. Individual axons may be as large as 1 mm in diameter.

From Richard D. Keynes, "The Nerve Impulse and the Squid," illustrated by Bunji Tagawa in *Scientific American*, December 1958. Copyright © 1958 by Scientific American, Inc.

distribution throughout the body or are stored in the digestive glands for later use. Undigested wastes pass through an intestine and out through the anus.

Other aspects of food collection and processing have already been discussed where appropriate for each group.

Excretory System

Molluscan urine is typically formed as the coelomic fluid passes through one or more pairs of kidneys (sometimes called **metanephridia**). As in annelids (Chapter 13), coelomic fluid generally enters the kidney through a nephrostome. Recall that the molluscan coelom is little more than a small cavity **(pericardium)** surrounding the heart (*peri* = G: around; *cardio* = G: heart) (Fig. 12.49). Thus, the coelomic fluid appears to be largely a filtrate of blood, containing small waste molecules, such as ammonia, that are forced across the heart wall with the filtrate. The basal lamina itself, along with associated perforated cells called **podocytes,** perform the ultrafiltration as pressure forces fluid across the surface. Additional wastes are actively secreted into the coelomic fluid by glands lining the pericardium. As in annelids, the primary urine is further modified by selective resorption and secretion as it travels through the metanephridial tubules. The penultimate urine is then discharged into the mantle cavity through a **renal pore (nephridiopore)** and carried away by water currents.

Taxonomic Summary

Phylum Mollusca
- Class Polyplacophora—chitons
- Class Aplacophora
- Class Monoplacophora
- Class Gastropoda—snails and slugs
  - Subclass Eogastropoda
    - Order Patellogastropoda (true limpets)
  - Subclass Orthogastropoda (5 superorders)
    - Superorder Caenogastropoda—includes most "prosobranch" species
    - Superorder Heterobranchia
      - The Opisthobranchia
      - The Pulmonata
- Class Bivalvia (= Pelecypoda)—clams, mussels, oysters, and shipworms
  - Subclass Protobranchia
  - The lamellibranchs (3 subclasses)
  - Subclass Anomalodesmata—includes the septibranchs
- Class Scaphopoda—tooth shells, tusk shells
- Class Cephalopoda—squid, octopus, cuttlefish, and chambered nautilus

Topics for Further Discussion and Investigation

1. What is the functional significance of differences in shell morphology among gastropods and bivalves?

Appleton, R. D., and A. R. Palmer. 1988. Water-borne stimuli released by predatory crabs and damaged prey induce more predator-resistant shells in a marine gastropod. *Proc. Nat. Acad. Sci.* 85:4387.

Bottjer, D. J., and J. G. Carter. 1980. Functional and phylogenetic significance of projecting periostracal structures in the Bivalvia (Mollusca). *J. Paleontol.* 54:200.

Carefoot, T. H., and D. A. Donovan. 1995. Functional significance of varices in the muricid gastropod *Ceratostoma foliatum. Biol. Bull.* 189:59.

Conover, M. R. 1979. Effect of gastropod shell characteristics and hermit crabs on shell epifauna. *J. Exp. Marine Biol. Ecol.* 40:81.

Currey, J. D., and J. D. Taylor. 1974. The mechanical behavior of some mollusc hard tissues. *J. Zool., London* 173:395.

Krist, A. C., 2002. Crayfish induce a defensive shell shape in a freshwater snail. *Invert. Biol.* 121:235.

Leonard, G. H., M. D. Bertness, and P. O. Yund. 1999. Crab predation, waterborne cues, and inducible defenses in the blue mussel, *Mytilus edulis. Ecology* 80:1.

Palmer, A. R. 1977. Function of shell sculpture in marine gastropods: Hydrodynamic destabilization in *Ceratostoma foliatum. Science* 197:1293.

Palmer, A. R. 1979. Fish predation and the evolution of gastropod shell sculpture: Experimental and geographical evidence. *Evolution* 33:697.

Perry, D. M. 1985. Function of the shell spine in the predaceous rocky intertidal snail *Acanthina spirata* (Prosobranchia: Muricacea). *Marine Biol.* 88:51.

Schmitt, R. J. 1982. Consequences of dissimilar defenses against predation in a subtidal marine community. *Ecology* 63:1588.

Stanley, S. M. 1969. Bivalve mollusk burrowing aided by discordant shell ornamentation. *Science* 166:634.

Trussell, G. C., and M. O. Nicklin. 2002. Cue sensitivity, inducible defense, and trade-offs in a marine snail. *Ecology* 83:1635.

Vermeij, G. J. 1979. Shell architecture and causes of death of Micronesian reef snails. *Evolution* 33:686.

Vermeij, G. J., and A. P. Covich. 1978. Coevolution of freshwater gastropods and their predators. *Amer. Nat.* 112:833.

Vermeij, G. J., and J. D. Currey. 1980. Geographical variation in the strength of thaidid snail shells. *Biol. Bull.* 158:383.

Vogel, S. 1997. Squirt smugly, scallop! *Science* 385:21–22.

2. A number of cephalopod species retain shells as adults, either internally or externally. What role do these shells play in regulating buoyancy?

Denton, E. J. 1974. On buoyancy and the lives of modern and fossil cephalopods. *Proc. Royal Soc. London B* 185:273.

Denton, E. J., and J. B. Gilpin-Brown. 1961. The buoyancy of the cuttlefish *Sepia officinalis* (L.). *J. Marine Biol. Assoc. U.K.* 41:319.

Greenwald, L., C. B. Cook, and P. D. Ward. 1982. The structure of the chambered nautilus siphuncle: The siphuncular epithelium. *J. Morphol.* 172:5.

Greenwald, L., P. D. Ward, and O. E. Greenwald. 1980. Cameral liquid transport and buoyancy in chambered nautilus (*Nautilus macromphalus*). *Nature* 286:55.

Sherrard, K. M. 2000. Cuttlebone morphology limits habitat depth in eleven species of *Sepia* (Cephalopoda: Sepiidae). *Biol. Bull.* 198:404.

Ward, P. D., and L. Greenwald. 1981. Chamber refilling in *Nautilus. J. Marine Biol. Assoc. U.K.* 62:469.

3. Discuss the causes and adaptive significance of gastropod torsion.

Bondar, C. A., and L. R. Page. 2003. Development of asymmetry in the caenogastropods *Amphissa columbiana* and *Euspira lewisii. Invert. Biol.* 122:28.

Crofts, D. R. 1955. Muscle morphogenesis in primitive gastropods and its relation to torsion. *Proc. Zool. Soc. London* 125:711.

Fretter, V. 1969. Aspects of metamorphosis in prosobranch gastropods. *Proc. Malac. Soc. London* 38:375.

Ghiselin, M. T. 1966. The adaptive significance of gastropod torsion. *Evolution* 20:337.

Goodhart, C. B. 1987. Garstang's hypothesis and gastropod torsion. *J. Moll. Stud.* 53:33.

Hickman, C. S., and M. G. Hadfield. 2001. Larval muscle contraction fails to produce torsion in a trochoidean gastropod. *Biol. Bull.* 200:257.

Kriegstein, A. R. 1977. Stages in the post-hatching development of *Aplysia californica. J. Exp. Zool.* 199:275.

Pennington, J. T., and F.-S. Chia. 1985. Gastropod torsion: A test of Garstang's hypothesis. *Biol. Bull.* 169:391–96.

Thompson, T. E. 1967. Adaptive significance of gastropod torsion. *Malacologia* 5:423.

Underwood, A. J. 1972. Spawning, larval development and settlement behavior of *Gibbula cineraria* (Gastropoda: Prosobranchia) with a reappraisal of torsion in gastropods. *Marine Biol.* 17:341.

4. A number of molluscs demonstrate a symbiotic relationship with photosynthetic protists that is reminiscent of that between zooxanthellae and cnidarians. How are the photosynthetic symbionts acquired by the molluscs, and what benefits are obtained by the hosts?

Fitt, W. K., and R. K. Trench. 1981. Spawning, development, and acquisition of zooxanthellae by *Tridacna squamosa* (Mollusca, Bivalvia). *Biol. Bull.* 161:213.

Gallop, A., J. Bartrop, and D. C. Smith. 1980. The biology of chloroplast acquisition by *Elysia ciridis. Proc. Royal Soc. London B* 207:335.

Klumpp, D. W., B. L. Bayne, and A. J. S. Hawkins. 1992. Nutrition of the giant clam *Tricdacna gigas* (L.). I. Contribution of filter feeding and photosynthates to respiration and growth. *J. Exp. Marine Biol. Ecol.* 155:105.

Masuda, K., S. Miyachi, and T. Maruyama. 1994. Sensitivity of zooxanthellae and non-symbiotic microalgae to stimulation of photosynthate excretion by giant clam tissue homogenate. *Marine Biol.* 118:687.

Rees, T. A. V., W. K. Fitt, B. Baillie, and D. Yellowlees. 1993. A method for temporal measurement of hemolymph composition in the giant clam symbiosis and its application to glucose and glycerol levels during a diel cycle. *Limnol. Oceanogr.* 38:213.

Trench, R. K., D. S. Wethey, and J. W. Porter. 1981. Observations on the symbiosis with zooxanthellae among the Tridacnidae (Mollusca: Bivalvia). *Biol. Bull.* 161:180.

5. Investigate the behavioral and chemical defenses of gastropods, bivalves, or cephalopods against predation.

Alexander, J. E., Jr., and A. P. Covich. 1991. Predator avoidance by the freshwater snail *Physella virgata* in response to the crayfish *Procambarus simulans*. *Oecologia* 87:435.

Bullock, T. H. 1953. Predator recognition and escape responses of some intertidal gastropods in the presence of starfish. *Behavior* 5:130.

Cimino, G., S. De Rosa, S. De Stefano, and G. Sodano. 1982. The chemical defense of 4 Mediterranean nudibranchs. *Comp. Biochem. Physiol.* 73B:471.

Coelho, L., J. Prince, and T. G. Nolen. 1998. Processing of defensive pigment in *Aplysia californica:* Acquisition, modification and mobilization of the red algal pigment r-phycoerythrin by the digestive gland. *J. Exp. Biol.* 201:425.

Crowl, T. A., and A. P. Covich. 1990. Predator life-history shifts in a freshwater snail. *Science* 247:949.

Denny, M., and L. Miller. 2006. Jet propulsion in the cold: mechanics of swimming in the Antarctic scallop *Adamussium colbecki*. *J. Exp. Biol.* 209: 4503–14.

Dix, T. L., and P. V. Hamilton. 1993. Chemically mediated escape behavior in the marsh periwinkle *Littoraria irrorata* Say. *J. Exp. Marine Biol. Ecol.* 166:135.

Fainzilber, M., I. Napchi, D. Gordon, and D. Zlotkin. 1994. Marine warning via peptide toxin. *Nature* 369:192.

Feder, H. M. 1963. Gastropod defensive responses and their effectiveness in reducing predation by starfishes. *Ecology* 44:505.

Feifarek, B. P. 1987. Spines and epibionts as antipredator defenses in the thorny oyster *Spondylus americanus* Hermann. *J. Exp. Marine Biol. Ecol.* 105:39.

Ferguson, G. P., and J. B. Messenger. 1991. A countershading reflex in cephalopods. *Proc. Royal Soc. London B* 243:63.

Fishlyn, D. A., and D. W. Phillips. 1980. Chemical camouflaging and behavioral defenses against a predatory seastar by 3 species of gastropods from the surf grass *Phyllospadix* community. *Biol. Bull.* 158:34.

Frick, K. 2003. Response to nematocyst uptake by the nudibranch *Flabellin verrucosa* to the presence of various predators in the southern Gulf of Maine. *Biol. Bull.* 205: 367–76.

Garrity, S. D., and S. C. Levings. 1983. Homing to scars as a defense against predators in the pulmonate limpet *Siphonaria gigas* (Gastropoda). *Marine Biol.* 72:319.

Gillette, R., M. Saeki, and R.-C. Huang. 1991. Defense mechanisms in notaspid snails: Acid humor and evasiveness. *J. Exp. Biol.* 156:335.

Gilly, W. F., B. Hopkins, and G. O. Mackie. 1991. Development of giant motor axons and neural control of escape responses in squid embryos and hatchlings. *Biol. Bull.* 180:209.

Greenwood, P. G., and R. N. Mariscal. 1984. Immature nematocyst incorporation by the aeolid nudibranch *Spurilla neapolitana*. *Marine Biol.* 80:35.

Hanlon, R. T., M. J. Smale, and W. H. H. Sauer. 1994. An ethogram of body patterning behavior in the squid *Loligo vulgaris reynaudii* on spawning grounds in South Africa. *Biol. Bull.* 187:363.

Iken, K., C. Avila, A. Fontana, and M. Gavagnin. 2002. Chemical ecology and origin of defensive compounds in the Antarctic nudibranch *Austrodoris kerguelenensis* (Opisthobranchia: Gastropoda) *Marine Biol.* 141:101.

Kicklighter, C. E., S. Shabani, P. M. Johnson, and C. D. Derby. 2005. Sea hares use novel antipredatory chemical defenses. *Current Biol.* 15: 549–54.

Margolin, A. S. 1964. A running response of *Acmaea* to seastars. *Ecology* 45:191.

Marko, P. B., and A. R. Palmer. 1991. Responses of a rocky shore gastropod to the effluents of predatory and non-predatory crabs: Avoidance and attraction. *Biol. Bull.* 181:363.

Mäthger, L. M., and E. J. Denton. 2001. Reflective properties of iridophores and fluorescent "eyespots" in the loliginid squid *Allotheuthis subulata* and *Loligo vulgaris*. *J. Exp. Biol.* 204:2103.

Meyer, J. J., and J. E. Byers. 2005. As good as dead? Sublethal predation facilitates lethal predation on an intertidal clam. *Ecol. Lett.* 8: 160–66.

Packard, A., and G. D. Sanders. 1971. Body patterns of *Octopus vulgaris* and maturation of the response to disturbance. *Anim. Behav.* 19:780.

Parsons, S. W., and D. L. Macmillan. 1979. The escape responses of abalone (Mollusca, Prosobranchia, Haliotidae) to predatory gastropods. *Marine Behav. Physiol.* 6:65.

Penney, B. K. 2002. Lowered nutritional quality supplements nudibranch chemical defense. *Oecologia* 132:411.

Phillips, D. W. 1975. Distance chemoreception-triggered avoidance behavior of the limpets *Acmaea (Collisella) limatula* and *Acmaea (Notoacmaea) scutum* to the predatory starfish *Pisaster ochraceus*. *J. Exp. Zool.* 191:199.

Prior, D. J., A. M. Schneiderman, and S. I. Greene. 1979. Size-dependent variation in the evasive behaviour of the bivalve mollusc *Spisula solidissima*. *J. Exp. Biol.* 78:59.

Young, R. E., C. F. E. Roper, and J. F. Walters. 1979. Eyes and extraocular photoreceptors in midwater cephalopods and fishes: Their role in detecting downwelling light for counterillumination. *Marine Biol.* 51:371.

6. Discuss the role played by body coloration and bioluminescence in the mating and feeding behavior of cephalopods.

Chiao, C.-C., C. Chubb, and R. T. Hanlon. 2007. Interactive effects of size, contrast, intensity and configuration of background objects in evoking disruptive camouflage in cuttlefish. *Vision Res.* 47: 2223–35.

Jantzen, T. M., and J. N. Havenhand. 2003. Reproductive behavior in the squid *Sepioteuthis australis* from South Australia: Ethogram of reproductive body patterns. *Biol. Bull.* 204:290.

Johnsen, S., E. J. Balser, E. C. Fisher, and E. A. Widder. 1999. Bioluminescence in the deep-sea cirrate octopod *Stauroteuthis syrtensis* Verrill (Mollusca: Cephalopoda). *Biol. Bull.* 197:26.

7. Do molluscs show an immune response?

Adema, C. M., and E. S. Loker. 1997. Specificity and immunobiology of larval digenean-snail associations. In: B. Fried, and T. K. Graczyk (eds). *Advances in Trematode Biology.* New York: CRC Press, pp. 229–63.

Bayne, C. J., and T. P. Yoshino. 1989. Determinants of compatibility in mollusc-trematode parasitism. *Amer. Zool.* 29:399.

Cheng, T. C., K. H. Howland, and J. T. Sullivan. 1983. Enhanced reduction of T4D and T7 coliphage titres from *Biomphalaria glabrata* (Mollusca) hemolymph induced by previous homologous challenge. *Biol. Bull.* 164:418.

Hertel, L. A., S. A. Stricker, and E. S. Loker. 2000. Calcium dynamics of the gastropod *Biomphalaria glabrata:* Effects of digenetic trematodes and selected bioactive compounds. *Invert. Biol.* 119:27.

Tripp, M. R. 1960. Mechanisms of removal of injected microorganisms from the American oyster, *Crassostrea virginica* (Gmelin). *Biol. Bull.* 119:273.

8. Investigate adaptations for carnivorous behavior or parasitism in prosobranch gastropods.

Carriker, M. R., D. van Zandt, and T. J. Grant. 1978. Penetration of molluscan and nonmolluscan minerals by the boring gastropod *Urosalpinx cinerea*. *Biol. Bull.* 155:511.

Hermans, C. O., and R. A. Satterlie. 1992. Fast-strike feeding behavior in a pteropod mollusk, *Clione limacina* Phipps. *Biol. Bull.* 182:1.

Olivera, B. M. 2002. *Conus* venom peptides: Reflections from the biology of clades and species. *Annu. Rev. Ecol. Syst.* 33:25.

O'Sullivan, J. B., R. R. McConnaughey, and M. E. Huber. 1987. A blood-sucking snail: The Cooper's Nutmeg, *Cancellaria cooperi* Gabb, parasitizes the California electric ray, *Torpedo californica* Ayres. *Biol. Bull.* 172:362.

Perry, D. M. 1985. Function of the shell spine in the predaceous rocky intertidal snail *Acanthina spirata* (Prosobranchia: Muricacea). *Marine Biol.* 88:51.

Rittschoff, D., L. G. Williams, B. Brown, and M. R. Carriker. 1983. Chemical attraction of newly hatched oyster drills. *Biol. Bull.* 164:493.

Schulz, J. R., A. G. Norton, and W. F. Gilly. 2004. The projectile tooth of a fish-hunting cone snail: *Conus catus* injects venom into fish prey using a high-speed ballistic mechanism. *Biol. Bull.* 207: 77–79.

Stewart, J., and W. F. Gilly. 2005. Piscivorous behavior of a temperate cone snail, *Conus californicus*. *Biol. Bull.* 209: 146–53.

9. Investigate the cycles of digestive activity encountered among bivalved molluscs and prosobranch gastropods.

Curtis, L. A. 1980. Daily cycling of the crystalline style in the omnivorous, deposit-feeding estuarine snail *Ilyanassa obsoleta*. *Marine Biol.* 59:137.

Hawkins, A. J. S., B. L. Bayne, and K. R. Clarke. 1983. Coordinated rhythms of digestion, absorption and excretion in *Mytilus edulis* (Bivalvia: Mollusca). *Marine Biol.* 74:41.

Morton, J. E. 1956. The tidal rhythm and the action of the digestive system of the lamellibranch *Lasaea rubra*. *J. Marine Biol. Assoc. U.K.* 35:563.

Palmer, R. E. 1979. Histological and histochemical study of digestion in the bivalve *Arctica islandica* L. *Biol. Bull.* 156:115.

Robinson, W. E., and R. W. Langton. 1980. Digestion in a subtidal population of *Mercenaria mercenaria* (Bivalvia). *Marine Biol.* 58:173.

Yonge, C. M. 1926. Structure and physiology of the organs of feeding and digestion in *Ostrea edulis*. *J. Marine Biol. Assoc. U.K.* 14:295.

10. Investigate the adaptations of gastropods for terrestrial existence.

Boss, K. J. 1974. Oblomovism in the mollusca. *Trans. Amer. Microsc. Soc.* 93:460.

Sloan, W. C. 1964. The accumulation of nitrogenous compounds in terrestrial and aquatic eggs of prosobranch snails. *Biol. Bull.* 126:302.

Verderber, G. W., S. B. Cook, and C. B. Cook. 1983. The role of the home scar in reducing water loss during aerial exposure of the pulmonate limpet, *Siphonaria alternata* (Say). *Veliger* 25:235.

Wells, G. P. 1944. The water relations of snails and slugs. III. Factors determining activity in *Helix pomatia* L. *J. Exp. Biol.* 20:79.

Welsford, I. G., P. A. Banta, and D. J. Prior. 1990. Size-dependent responses to dehydration in the terrestrial slug, *Limax maximus* L: Locomotor activity and huddling behavior. *J. Exp. Zool.* 253:229.

11. Discuss the roles of mucus in the locomotion and feeding of gastropod molluscs.

Connor, V. M. 1987. The use of mucous trails by intertidal limpets to enhance food resources. *Biol. Bull.* 171:548.

Cook, A. 1992. The function of trail following in the pulmonate slug, *Limax pseudoflavus*. *Anim. Behav.* 43:813.

Davies, M. S., and P. Beckwith. 1999. Role of mucus trails and trail-following in the behaviour and nutrition of the periwinkle *Littorina littorea*. *Marine Ecol. Progr. Ser.* 179:247.

Kappner, I. S. M. Al-Moghrabi, and C. Richter. 2000. Mucus-net feeding by the vermetid gastropod *Dendropoma maxima* in coral reefs. *Marine Ecol. Progr. Ser.* 204:309.

Rice, S. H. 1986. An anti-predator chemical defense of the marine pulmonate gastropod *Trimusculus reticulatus*. *J. Exp. Marine Biol. Ecol.* 93:83.

Smith, A. M., T. J. Quick, and R. L. St. Peter. 1999. Differences in the composition of adhesive and non-adhesive mucus from the limpet *Lottia limatula*. *Biol. Bull.* 196:34.

12. Discuss the evidence indicating that radial muscle contractions alone do not completely account for refilling the cephalopod mantle cavity during jet propulsion.

Gosline, J. M., and R. E. Shadwick. 1983. The role of elastic energy storage in swimming: An analysis of mantle elasticity in escape jetting in the squid, *Loligo opalescens*. *Canadian J. Zool.* 61:1421.

Vogel, S. 1987. Flow-assisted mantle cavity refilling in jetting squid. *Biol. Bull.* 172:61.

Ward, D. V. 1972. Locomotor function of the squid mantle. *J. Zool., London* 167:437.

13. How humans learn and remember things is still one of the major unsolved biological mysteries. Much of what is currently understood about learning and memory and the biological basis of behavior comes from the study of invertebrates, including the gastropod molluscs. Describe the types of experiments that have been performed on gastropods and the extent to which each has increased our understanding of how animals learn.

Barnes, D. M. 1986. From genes to cognition. *Science* 231:1066.

Colebrook, E., and K. Lukowiak. 1988. Learning by the *Aplysia* model system: Lack of correlation between gill and gill motor neuron responses. *J. Exp. Biol.* 135:411.

Kandel, E. R., and J. H. Schwartz. 1982. Molecular biology of learning: Modulation of transmitter release. *Science* 218:433.

Lin, S. S., and I. B. Levitan. 1987. Concanavalin A alters synaptic specificity between cultured *Aplysia* neurons. *Science* 237:648.

Moffett, S., and K. Snyder. 1985. Behavioral recovery associated with the central nervous system regeneration in the snail *Melampus*. *J. Neurobiol.* 16:193.

14. Bacteria have been found in the gills of various marine bivalve species living in sediments rich in decomposing organic matter. What evidence suggests that these bacteria live in a symbiotic relationship with the bivalves, and what is the nature of that relationship?

Cavanaugh, C. M., R. R. Levering, J. S. Maki, R. Mitchell, and M. E. Lidstrom. 1987. Symbiosis of methylotrophic bacteria and deep-sea mussels. *Nature (London)* 325:346.

Childress, J. J., C. R. Fisher, J. M. Brooks, M. C. Kennicutt II, R. R. Bidigare, and A. E. Anderson. 1986. A methanotrophic marine molluscan (Bivalvia: Mytilidae) symbiosis: Mussels fueled by gas. *Science* 233:1306.

Hentschel, U., D. S. Millikan, C. Arndt, S. C. Cary, and H. Felbeck. 2000. Phenotypic variations in the gills of the symbiont-containing bivalve *Lucinoma aequizonata. Marine Biol.* 136:633.

Kádár, R. Bettencourt, V. Costa, R. S. Santos, A. Lobo-da-Cunha, and P. Dando. 2005. Experimentally induced endosymbiont loss and re-acquirement in the hydrothermal vent bivalve *Bathymodoilus azoricus. J. Exp. Mar. Biol. Ecol.* 318: 99–110.

Kochevar, R. E., J. J. Childress, C. R. Fisher, and E. Minnich. 1992. The methane mussel: Roles of symbiont and host in the metabolic utilization of methane. *Marine Biol.* 112:389.

Loomis, S. H., and M. Zinser. 2001. Isolation and identification of an ice-nucleating bacterium from the gills of the intertidal bivalve mollusc *Geukensia demissa. J. Exp. Marine Biol. Ecol.* 261:225.

Powell, M. A., and G. N. Somero. 1986. Adaptations to sulfide by hydrothermal vent animals: Sites and mechanisms of detoxification and metabolism. *Biol. Bull.* 171:274.

15. Discuss how molluscs have been used to detect and assess the impact of environmental pollutants.

Amiard, J. C., C. Amiard-Triquet, B. Berthet, and C. Métayer. 1986. Contribution to the ecotoxicological study of cadmium, lead, copper and zinc in the mussel *Mytilus edulis. Marine Biol.* 90:425.

Axiak, V., and J. J. George. 1987. Effects of exposure to petroleum hydrocarbons on the gill functions and ciliary activities of a marine bivalve. *Marine Biol.* 94:241.

Berger, B., and R. Dallinger. 1989. Accumulation of cadmium and copper by the terrestrial snail *Arianta arbustorum* L.: Kinetics and budgets. *Oecologia* 79:60.

Byrne, C. J., and J. A. Calder. 1977. Effect of the water-soluble fractions of crude, refined and waste oils on the embryonic and larval stages of the quahog clam *Mercenaria* sp. *Marine Biol.* 40:225.

Shi, H. H., C. J. Huang, S. X. Zhu, X. J. Yu, and W. Y. Xie. 2005. Generalized system of imposex and reproductive failure in female gastropods of coastal waters of mainland China. *Mar. Ecol. Progr. Ser.* 304: 179–89.

Stickle, W. B., S. D. Rice, and A. Moles. 1984. Bioenergetics and survival of the marine snail *Thais lima* during long-term oil exposure. *Marine Biol.* 80:281.

16. Compare the nature and role of hydrostatic skeletons in the locomotion of different gastropod species.

Bernard, F. R. 1968. The aquiferous system of *Polinices lewisi* (Gastropoda, Prosobranchiata). *J. Fish Res. Bd. Canada* 25:541.

Dale, B. 1973. Blood pressure and its hydraulic functions in *Helix pomatia* L. *J. Exp. Biol.* 59:477.

Jones, H. D. 1973. The mechanism of locomotion in *Agriolimax reticulatus* (Mollusca; Gastropoda). *J. Zool., London* 171:489.

Kier, W. M. 1988. The arrangement and function of molluscan muscle. In *The Mollusca, Vol. 11: Form and Function.* Trueman, E. R., and M. R. Clarke, eds. New York: Academic Press, 211–52.

Trueman, E. R., and A. C. Brown. 1976. Locomotion, pedal retraction and extension, and the hydraulic systems of *Bullia* (Gastropoda: Nassaridae). *J. Zool., London* 178:365.

Voltzow, J. 1986. Changes in pedal intramuscular pressure corresponding to behavior and locomotion in the marine gastropods *Busycon contrarium* and *Haliotis kamtschatkana. Canadian J. Zool.* 64:2288.

17. Discuss the hypothesis that deafness in cephalopods is an adaptive response to predation.

Hanlon, R. T., and B.-U. Budelmann. 1987. Why cephalopods are probably not "deaf." *Amer. Nat.* 129:312–17.

Moynihan, M. 1985. Why are cephalopods deaf? *Amer. Nat.* 125:465.

Packard, A., H. E. Karlsen, and O. Sand. 1990. Low frequency hearing in cephalopods. *J. Comp. Physiol.* A166:501.

18. How has the application of new technology altered our understanding of how lamellibranch bivalves collect and sort food particles?

Baker, S. M., J. S. Levinton, and J. E. Ward. 2000. Particle transport in the zebra mussel, *Dreissena polymorpha* (Pallas). *Biol. Bull.* 199:116.

Beninger, P. G., 2000. Limits and constraints: A comment on premises and methods in recent studies of particle capture mechanisms in bivalves. *Limnol. Oceangr.* 45:1196.

Brillant, M. G. S., and B. A. MacDonald. 2003. Postingestive sorting of living and heat-killed *Chlorella* within the sea scallop, *Placopecten magellanicus* (Gmelin). *J. Exp. Marine Biol. Ecol.* 290:81.

Levinton, J. S., J. E. Ward, and S. E. Shumway. 2002. Feeding responses of the bivalves *Crassostrea gigas* and *Mytilus trossulus* to chemical composition of fresh and aged kelp detritus. *Marine Biol.* 141:367.

Medler, S., and H. Silverman. 2001. Muscular alteration of gill geometry *in vitro:* Implications for bivalve pumping processes. *Biol. Bull.* 200:77.

Milke, L. M., and J. E. Ward. 2003. Influence of diet on pre-ingestive particle processing in bivalves II. Residence time in the pallial cavity and handling time on the labial palps. *J. Exp. Marine Biol. Ecol.* 293:151.

Nielsen, N. F., P. S. Larsen, H. U. Riisgård, and C. B. Jørgensen. 1993. Fluid motion and particle retention in the gill of *Mytilus edulis:* Video recordings and numerical modelling. *Marine Biol.* 116:61.

Richoux, N. B., and R. J. Thompson. 2001. Regulation of particle transport within the ventral groove of the mussel (*Mytilus edulis*) gill in response to environmental conditions. *J. Exp. Marine Biol. Ecol.* 260:199.

Silverman, H., J. W. Lynn, P. G. Beninger, and T. H. Dietz. 1999. The role of latero-frontal cirri in particle capture by the gills of *Mytilus edulis. Biol. Bull.* 197:368.

Silverman, H., J. W. Lynn, and T. H. Dietz. 2000. In vitro studies of particle capture and transport in suspension-feeding bivalves. *Limnol. Oceanogr.* 45:1199.

Ward, J. E., and J. S. Levinton. 1997. Site of particle selection in a bivalve mollusc. *Nature* 390:131.

Ward, J. E., and S. E. Shumway. 2004. Separating the grain from the chaff: Particle selection in suspension- and deposit-feeding bivalves. *J. Exp. Mar. Biol. Ecol.* 300: 83–130.

19. Evaluate the hypothesis that shell reduction and loss among cephalopods was driven by competition with bony fishes. What additional evidence could be sought that might argue for or against the hypothesis?

Packard, A. 1972. Cephalopods and fish: The limits of convergence. *Biol. Rev.* 47:241.

20. By cladistic methods (Chapter 2), animal groups are defined by the shared possession of characters not present in the immediate ancestor to those groups. These shared derived characters (synapomorphies) are essentially what I have been calling Defining Characteristics throughout this book. Explain why each of the following is *not* a valid synapomorphy for the Cephalopoda: ink sac; chromatophores; loss of shell; development of a spiral shell.

21. Zebra mussels (*Dreissena polymorpha*) were first reported in the Great Lakes in 1986, probably transported here from Europe in ship ballast water. They have since spread rapidly through major waterways south to Louisiana and west to Oklahoma, and are ultimately expected to colonize three-fourths of the freshwater rivers and lakes in the United States. They are also spreading in Canadian rivers and lakes. What features make zebra mussels such successful invaders, and how has the invasion altered the ecological relationships in U.S. lakes, rivers, and streams?

Berkman, P. A., M. A. Haltuch, E. Tichich, D. W. Garton, G. W. Kennedy, J. E. Gannon, S. D. Mackey, J. A. Fuller, and D. L. Liebenthal. 1998. Zebra mussels invade Lake Erie muds. *Nature* 393:27.

Kolar, C. S., A. H. Fullerton, K. M. Martin, and G. A. Lamberti. 2002. Interactions among zebra mussel shells, invertebrate prey, and Eurasian ruffe or yellow perch. *J. Great Lakes Res.* 28:664.

MacIsaac, H. J. 1996. Potential abiotic and biotic impacts of zebra mussels on the inland waters of North America. *Amer. Zool.* 36:287.

Strayer, D. L., N. F. Caraco, J. J. Cole, S. Findlay, and M. L. Pace. 1998. Transformation of freshwater ecosystems by bivalves. *BioScience* 49:19.

22. What factors are contributing to the extinction of freshwater, marine, and terrestrial molluscs, and what makes some species more vulnerable than others?

Bogan, A. E. 1993. Freshwater bivalve extinctions (Mollusca: Unionoida): A search for causes. *Amer. Zool.* 33:599.

Carlton, J. T., G. J. Vermeij, D. R. Lindberg, D. A. Carlton, and E. C. Dudley. 1991. The first historical extinction of a marine invertebrate in an ocean basin: The demise of the eelgrass limpet *Lottia alveus*. *Biol. Bull.* 180:72.

Hadfield, M. G., S. E. Miller, and A. H. Carwile. 1993. The decimation of endemic Hawai'ian tree snails by alien predators. *Amer. Zool.* 33:610.

Lorenz, S. 2003. E. U. shifts endocrine disrupter research into overdrive. *Science* 300:1069.

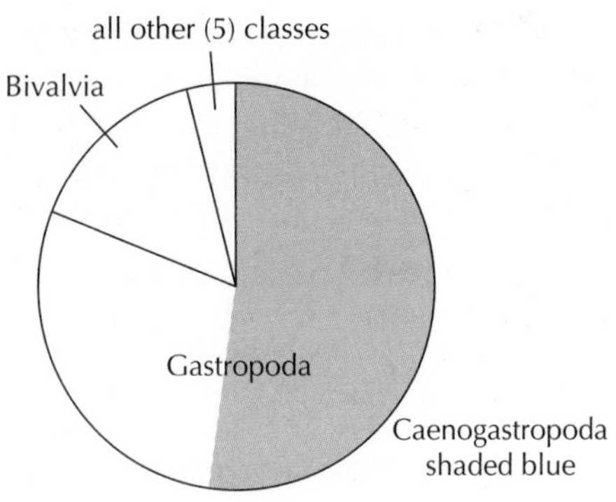

Taxonomic Detail

Phylum Mollusca[24]

Subphylum Aculifera

Calcareous spicules are formed within individual cells in the mantle tissue, or in specialized regions of the mantle. Two classes.

Class Polyplacophora

The approximately 500 species are divided among thirteen families. Chitons lack copulatory organs and fertilization is usually external, although a few species brood embryos to the trochophore stage.

Family Lepidopleuridae. *Leptochiton, Lepidopleurus.* Unlike most chitons, which occur in shallow water or intertidally, most species in this family are collected only in deep water, down to 7000 m depth.

Family Ischnochitonidae (= Bathychitonidae). *Callochiton, Chaetopleura, Ischnochiton, Lepidochitona, Tonicella.* These are mostly shallow-water species living on stones and in oyster beds. This family contains up to 40% of all chiton species.

Family Mopaliidae. *Mopalia*—the hairy, or mossy, chiton, up to 13 cm long; *Katharina.* The girdle of *Mopalia* spp. is studded with conspicuous spines (the "hairy girdle" syndrome). Members of one genus, *Placiphorella,* are carnivores, preying on polychaetes and crustaceans.

24. Molluscan systematics are still unsettled, particularly for the gastropods and bivalves. My treatment here is based largely on Beesley, P. L., G. J. B. Ross, and A. Wells (Eds.). 1998. *Mollusca: The Southern Synthesis. Fauna of Australia.* CSIRO Publishing: Melbourne. Numbers of species given are minimum estimates.

Family Chitonidae. *Chiton, Acanthopleura.* About 20% of all chiton species are members of this family. Bodies can grow to 20 cm.

Family Acanthochitonidae. *Cryptochiton*—The largest of the chitons, up to 36 cm long. Shell valves are completely covered by the extensive girdle. *Cryptoplax*—a strange, worm-like chiton with a highly extensible body; it lives in rock crevices in the Indo-Pacific. About 20% of all chiton species are placed in this very diverse family.

Family Cryptoplacidae. *Cryptoplax.* These are large (up to 15 cm long), worm-shaped chitons with small shell valves spaced far apart from each other in adults.

Class Aplacophora

Approximately 320 species.

Subclass Neomeniomorpha (= Solenogastres)

Members all possess a slender foot, seated in a narrow groove that runs along the ventral surface from just posterior to the mouth to the posterior mantle cavity. Many species lack a radula, and the rest possess a highly simplified radula. None have true ctenidia. All species are hermaphroditic. Members of this subclass either crawl over the sediment or live among cnidarians, on which they feed. All are marine. The twenty-one families contain about 70% of all aplacophoran species. Nearly 45% come from Antarctic waters.

Family Neomeniidae. *Neomenia.*

Subclass Chaetodermomorpha (= Caudofoveata)

Members possess neither a foot nor a ventral groove, but do have a distinctive cuticular shield around the mouth. All species possess a radula and a mantle cavity housing one pair of bipectinate ctenidia. All species are marine and burrow through muddy substrate. Three families.

Family Chaetodermatidae. *Chaetoderma, Falcidens.*

Subphylum Conchifera

The mantle tissue secretes one or more calcareous shells, but no spicules. Five classes.

Class Monoplacophora

The 19 described species are all in one family (Neopilinidae). *Micropilina, Neopilina, Rokopella, Vema.* The species with the largest shell is *Neopilina galatheae* (37 mm), that with the smallest shell (< 1 mm) is the species *Micropilina arntzi.*

Class Gastropoda[25]

This class contains at least 40,000 described recent species. It may contain nearly 200,000 species, including many that are not yet described.

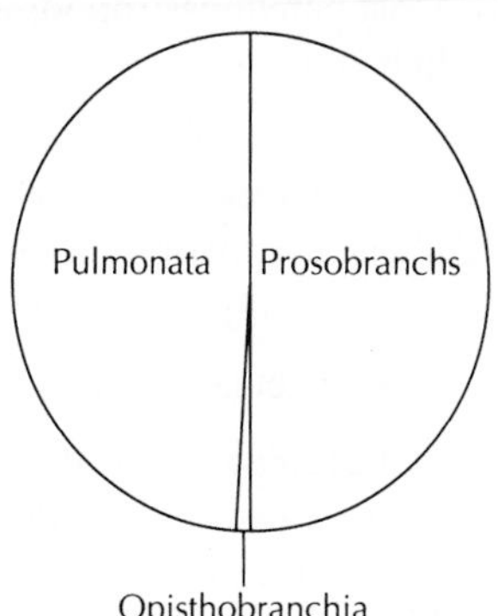

The prosobranchs

This group contains at least 20,000 species, distributed among over one hundred forty families.

The archaeogastropods

These snails possess many primitive prosobranch characteristics, with many species bearing a pair of hypobranchial glands, a pair of osphradia, a pair of auricles, a pair of kidneys, and a pair of ctenidia (always bipectinate). Ganglia are paired, rarely fused, and typically separated by long commissures. The radula has many teeth. Most archaeogastropod species are marine herbivores, including about 50 species of abalone; a few terrestrial species and a few freshwater species are also known. Archaeogastropods typically fertilize externally and are unique among gastropods in commonly developing as free-swimming trochophore larvae. Approximately 5000 species, distributed among twenty-six families.

Subclass Eogastropoda. This group contains the limpets and their presumed ancestors. The Eogastropoda appears to be the sister group to all other gastropods.

Order Patellogastropoda. The true limpets, all marine, less than 1% of named gastropod species. The radula differs in form and function from that of all other gastropods. They are the most primitive

25. Under the currently accepted taxonomic arrangement, the Archaeogastropoda and Mesogastropoda are no longer formally recognized taxa, although the terms archaeogastropod and mesogastropod are still widely used to refer informally to different grades of organization. I have used shading in this *Taxonomic Detail* to indicate how the archaeogastropods and mesogastropods are now distributed among the modern taxonomic groups (the archaeogastropods are shaded blue, the mesogastropods are not shaded). The Neogastropoda (a different shade of blue) still exist as a formal grouping, but as an infraorder within the superorder Caenogastropoda (pp. 271, 281).

group of prosobranchs, and the most primitive group of gastropods. Their shells are flattened and with no sign of coiling, although their ancestors probably had coiled shells. Most patellogastropods are herbivores, using their radulae as rasps. All fertilize their eggs externally. Five families.

Family Patellidae. *Patella, Scutellastra.* Members are distributed throughout the world, especially in temperate seas.

Family Acmaeidae. *Acmaea.*

Family Lottiidae. *Lottia (= Collisella), Notoacmea, Patelloida, Tectura.* This is the most diverse and species-rich family of patellid gastropods.

Family Nacellidae. *Cellana, Nacella.*

Subclass Orthogastropoda. This subclass contains all other gastropods: the remaining prosobranchs, the opisthobranchs, and the pulmonates.

Superorder Cocculiniformia. Seven families, with members living at depths of 30 m to over 9000 m.

Family Cocculinidae. *Cocculina.* Deep-sea limpets, about 130 species, collected from depths of 30 to over 3700 m. Most species eat wood, but some species feed on such restricted substrates as cephalopod beaks, polychaete tubes, and the bones of dead fish and whales.

Family Choristellidae. *Choristella.* Unlike other members of this superfamily, these deepwater snails all have coiled shells.

Superorder Vestigastropoda. Nine families.

Superfamily Pleurotomarioidea. Four families.

Family Haliotidae. *Haliotis*—the abalone. Shallow-water herbivorous snails—about 50 species—living on solid substrates, to which they can cling very tightly. The tissue is commonly eaten by humans, and the broad shells are commonly used by humans as ornaments.

Family Neomphalidae. *Neomphalus.* A deep-sea archaeogastropod recently collected at depths of 2400 m in the Galápagos Rift of the Pacific Ocean. Considered a living fossil because of its many primitive characteristics.

Superfamily Fissurellacea. One family (Fissurellidae). *Fissurella, Emarginula, Emarginella, Diodora, Puncturella*—keyhole limpets. So named because the shell bears a conspicuous hole at the top, through which water exits after passing over the gills. The operculum is lost at metamorphosis. All marine.

Superfamily Trochoidea. Eight families.

Family Trochidae. *Gibbulla, Margarites, Calliostoma, Cittarium, Tegula, Austrocochlea, Stomatella, Umbonium, Bankivia, Monodonta*—top shells. All species are marine, and most are herbivores.

Family Turbinidae. *Astraea*—star shells. *Turbo*—turban shells. All species are marine, and most are herbivores. The operculum is calcified.

Superorder Neritopsina

Superfamily Neritoidea. Six families, about 300 species.

Family Neritidae. *Nerita, Smaragdia, Neritina.* Among the most advanced of the archaeogastropods, reflected by the loss of the right kidney and the right ctenidium and by unusually complex reproductive anatomy and biology. Most species are marine, but some occupy estuarine, freshwater, or terrestrial habitats.

Family Hydrocenidae. *Hydrocena, Georissa.* Terrestrial snails with mantle cavity forming a lung; no ctenidia.

Family Helicinidae. *Helicina, Alcadia.* Terrestrial snails living on the ground or, more rarely, in trees.

The mesogastropods (= taenioglossans)
The right ctenidium has been lost, and the remaining (left) ctenidium is always monopectinate; some terrestrial species have lost both ctenidia. Ganglia tend to be fused. The head typically bears a protrusible proboscis and one pair of cephalic tentacles, with a simple eye at the base of each. The right kidney is lost or modified for reproduction. The heart has a single auricle. The left hypobranchial gland is lost. The radula usually has 7 teeth in each row (the "taenioglossate" condition). All species have internal fertilization, and most produce free-living veliger larvae. In some species, the mantle edge forms an incurrent siphon. Most mesogastropods are free-living and

marine, but some species are terrestrial, freshwater, or parasitic. Approximately 10,000 species, distributed among 95 families.

Superorder Caenogastropoda. This group probably contains more than half of all gastropod species (see p. 276).

Order Architaenioglossa.[26] Operculum-bearing land snails lacking gills of any kind: The mantle cavity is modified for respiration. Five families.

Family Cyclophoridae. *Leptopoma.*

Family Pupinidae. *Pupina.*

Family Viviparidae. *Viviparus, Notopala.* Freshwater suspension feeders. Right tentacle of male modified for copulation.

Family Ampullariidae (= Pilidae). *Asolene, Pomacea, Marisa*—apple snails. Large (to 6 cm) tropical freshwater snails, commonly sold for home aquaria. The mantle cavity has both a lung and a gill. Apple snails prey on the eggs and juveniles of other gastropods, and are commonly used to reduce populations of gastropods that serve as intermediate hosts in parasitic trematode life cycles.

Order Sorbeoconcha

Superfamily Cerithioidea. 15–25 families.

Family Cerithiidae. *Cerithium, Bittium, Litiopa, Batillaria.* All are marine, mostly in shallow water, and most eat detritus.

Family Pleuroceridae. *Pleurocera, Pachychilus, Leptoxis.* This is a group of widely distributed freshwater gastropods, common in streams, ponds, and lakes.

Family Turritellidae. *Turritella, Vermicularia.* Suspension-feeding marine species.

Family Potamididae. *Cerithidea, Terebralia.* Abundant in mangroves and estuarine mudflats in tropical and subtropical areas.

Family Thiaridae. *Melanoides, Thiara.* Tropical freshwater and brackish water snails, often serving as intermediate hosts in parasitic trematode life cycles.

Infraorder Littorinimorpha
~Fifty families.

Superfamily Littorinoidea

Family Littorinidae. *Littorina, Lacuna, Bembicium, Melarhaphe, Tectarius*—the true periwinkles. All are marine herbivores, typically smaller than about 2.5 cm. Most live intertidally; some species are found only above high tide.

Family Pomatiasidae. *Pomatias.* Terrestrial snails living among fallen leaves and moss. Possess a unique form of bipedal, walking locomotion, in which the 2 sides of the foot move forward in alternation.

Superfamily Rissooidea

Family Bithyniidae. *Bithynia.* Freshwater suspension feeders that serve as intermediate hosts in the life cycle of the trematode (fluke) intestinal parasite *Opisthorchis tenuicollis.*

Family Hydrobiidae. *Hydrobia.* Small (approximately 6 mm high), mostly freshwater prosobranchs; some brackish species and a few terrestrial species. Most hatch as small juvenile snails, but one species (*H. ulvae*) has free-living veliger larvae.

Family Caecidae. *Caecum.* Very small (typically a few millimeters long) marine species with tubular shells.

Superfamily Stromboidea. Three families.

Family Strombidae. *Strombus*—conchs. Marine. Economically important as a food source in some tropical areas.

Superfamily Calyptraeacea. Four families.

Family Calyptraeoidea. *Calyptraea*—cup-and-saucer shells; *Crepidula*—slipper shell snails.

26. Some workers consider these to be archaeogastropods.

Limpet-like marine herbivores. Many species are suspension feeders, collecting food on the ciliated gill. Some species live in communal stacks, with females at the bottom and males at the top. All species are sequential hermaphrodites.

Superfamily Veretoidea

Family Vermetidae. *Vermetus, Serpulorbis, Petaloconchus, Dendropoma.* Suspension-feeding, immobile marine snails living in loosely coiled or uncoiled shells resembling those made by many sedentary polychaete worms. The shells are permanently cemented to rocks.

Superfamily Cypraeoidea. Six families.

Family Cypraeidae. *Cypraea*—the cowries. All marine, mostly in shallow, tropical seas. The shells are very smooth and glossy, usually about 4.0–7.5 cm long. Juveniles and adults lack an operculum.

Family Ovulidae. *Cyphoma*—flamingo tongue snails; *Ovula, Simnia.* Mostly tropical marine carnivores, feeding on colonial cnidarians.

Superfamily Naticoidea. One family (Naticidae). *Natica, Polinices, Lunatia*—moon snails. Marine, usually sand- or mud-dwelling predators on bivalves and other gastropods; they bore circular holes through prey shells using glandular secretions and radular activity and then insert a highly protrusible proboscis for feeding. Naticids have an extensive, water-filled cavity in the foot; the water must be expelled from pores at the rear of the foot before the foot can be fully retracted into the shell.

Superfamily Tonnoidea. Eight families.

Family Tonnidae. *Tonna*—tun shells. Marine, typically sand-dwelling carnivores living to depths greater than 5000 m. Prey on a variety of invertebrates and fish by injecting sulfuric acid and paralytic secretions.

Family Ranellidae (= Cymatiidae). *Cymatium, Fusitriton, Charonia*—the tritons. Marine, primarily warm-water snails. Prey on molluscs and holothurians, incapacitating prey with sulfuric acid and often ingesting them whole. Shells up to 50 cm long.

Family Bursidae. *Bursa*—frog shells. Mostly shallow-water marine carnivores feeding on various worms, which they paralyze with an acidic secretion and then ingest whole through a greatly expandable proboscis and esophagus.

Superfamily Carinariodae—The Heteropods. Three families. All are pelagic carnivores.

Family Atlantidae. *Atlanta.* Heteropods with small, thin, fragile, coiled shells. Animal can pull completely into shell. Head has large eyes and proboscis. Apparently feeds preferentially on other planktonic gastropods (pteropods).

Family Carinariidae. *Carinaria.* Heteropods with thin, flattened shell too small to contain entire animal. Proboscis extremely extensible. Body up to 0.5 m long.

Family Pterotracheidae. *Pterotrachea.* Heteropods lacking shells, mantle, mantle cavity, and proboscis. Body up to 20 cm long, transparent and cylindrical.

Superfamily Janthinoidea. Three families.

Family Epitoniidae. *Epitonium*—wentletraps. Typically ectoparasitic on anthozoans.

Family Janthinidae. *Janthina, Recluzia*—violet snails. All are pelagic, drifting attached to a float composed of secreted mucus and air. All are carnivores, feeding on planktonic hydrozoans (siphonophores), and all species are sequential hermaphrodites, with each individual being first male and later female.

Superfamily Eulimoidea. Six families.

Family Eulimidae. *Eulima, Melanella, Stilifer.* Marine snails, mostly ectoparasitic on echinoderms; body fluids are pumped in through a long proboscis. Females are much larger than males.

Family Entoconchidae. *Entoconcha, Thyonicola.* All species are internal parasites of sea cucumbers (holothurians). Adult females are typically shell-less and vermiform, and they may attain lengths of up to 1.3 m. Males are microscopic (they are "dwarf" males) and usually embedded in tissues of the female as little more than testicular sacs. Some workers include all members of this family within the Eulimidae.

Superfamily Triphoroidea

Family Triphoridae. *Triphora, Inella.* Marine snails, primarily with sinistrally coiled shells. Adults generally feed on sponges.

The Neogastropods—Infraorder Neogastropoda

These are the most highly evolved gastropods. All species are marine, and most are carnivores. Like the mesogastropods, the neogastropods have a single, monopectinate ctenidium, a single kidney, and a heart with one auricle. The radula, however, has no more than 3 teeth per row (the "stenoglossate" condition), and the osphradium is especially well developed and bipectinate. Fertilization is always internal. Many species develop to the juvenile stage without having a free-living veliger. The mantle edge always forms an incurrent siphon. Approximately 5000 species, distributed among 21 families.

Superfamily Muricoidea. Seventeen families.

Family Muricidae. *Urosalpinx*—the oyster drill; *Thais, Nucella*—dogwinkles; *Concholepas*—the economically important, highly edible South American "loco;" *Drupa, Murex, Ocenebra.* Shallow-water marine predators that bore holes through shells of barnacles, gastropods, and bivalves, using a specialized gland (the accessory boring organ, ABO) on the foot, in conjunction with radular rasping. The ABO secretes a shell-dissolving mixture of acid and enzymes. Muricids typically lack free-living veliger larvae in the life cycle; most emerge as juveniles from egg capsules cemented to hard substrates. Phoenicians and Romans dyed their ceremonial robes purple using a substance secreted by many species of *Murex.* In fact, that's how the Phoenicians got their name (*phoenix* = G: reddish purple).

Family Buccinidae. *Buccinum, Neptunea, Colus*—the whelks. All marine. A very large family of mostly carnivorous species.

Family Columbellidae. *Columbella, Anachis*—dove shells. All marine. Mostly small snails, less than 0.5 cm high. Carnivorous and herbivorous species.

Family Nassariidae. *Nassarius, Ilyanassa*—mudsnails, basket shells. All are marine and include carnivores, scavengers, and herbivores. Live mostly on mud or sand in marine habitats; some brackish, some freshwater species.

Family Melongenidae. *Busycon*—whelks; *Melongena*—crown conchs. Large marine snails, up to 60 cm long. Mostly carnivores and scavengers in shallow water; most species are tropical.

Family Fasciolariidae. *Fasciolaria*—tulip shells; *Pleuroplaca.* Large marine carnivores, up to 60 cm long.

Family Olividae. *Oliva, Olivella*—olive shells. These sand-dwelling carnivores are common in all tropical and subtropical seas.

Superfamily Conoidea. Three families.

Family Conidae. *Conus*—cone shells. Marine carnivores, about 500 species, mostly tropical on coral reefs. Highly toxic venom is produced by poison glands and injected into prey through the sharp lateral teeth of the radula.

Superorder Heterobranchia (= Euthyneura). This newly erected group contains the opisthobranch and pulmonate snails, as well as some species (sundials and pyramidellids—so-called "lower heterobranchs") previously thought to be prosobranchs.

Superfamily Architectonicoidea. Two families.

Family Architectonicidae. *Architectonica, Philippia*—sundial shells. Warm-water marine species with conical or disc-shaped shells, living as deep as 2000 m. Typically feed on cnidarians.

Superfamily Pyramidelloidea. Five families.

Family Pyramidellidae. *Boonea, Odostomia, Pyramidella.* All 6000 species are marine predators or ectoparasites on other invertebrates, including other molluscs; they pierce their host with a sharp stylet and then pump in body fluids through the proboscis. Some workers consider these snails to be members of the subclass Opisthobranchia.

The Opisthobranchia[27]

Approximately 2000 heterobranch species, mostly marine, distributed among over 120 families.

Order Cephalaspidea. *Bulla, Haminoea, Hydatina, Retusa, Runcina*—the bubble-shell snails. All members have shells, which may be either external, as in most prosobranchs, or hidden internally. As with the prosobranchs, the nervous system is fully torted (i.e., displays streptoneury). Cephalaspids are typically marine carnivores and ingest prey whole, crushing them using hard, calcareous plates in the gizzard. All haminoeids, however, are herbivorous. 31 families.

27. The Opisthobranchia is a valid taxonomic category, but there is disagreement about its taxonomic level. The group is contained within the subclass Orthogastropoda.

Order Acochlidioidea. Seven families.

Family Acochlidiidae. *Acochlidium, Microhedyle.* Mostly marine, interstitial, vermiform snails living in the spaces between sand grains. No mantle cavity, ctenidia, or shell. A few species live in freshwater in Indonesia, Palau, and the West Indies. Most species only 2–5 mm long. All individuals are hermaphroditic, and the penis is in some species armed with a sharp stylet that injects sperm into a mate by hypodermic impregnation.

Order Rhodopemorpha. Two families.

Family Rhodopidae. *Rhodope.* A bizarre, very small (less than 0.4 cm long), shell-less, vermiform snail. These snails lack tentacles, mantle cavity, gills of any kind, and heart; there is not even any trace of a pericardium. Calcareous spicules are embedded in the flesh. Members of this family are exclusively interstitial, living in the spaces between sand grains in both the Atlantic Ocean and Mediterranean Sea. There is some support for relating them more closely to pulmonates.

Order Sacoglossa (= Ascoglossa). The sacoglossans, or ascoglossans. 12 families.

Family Elysiidae. *Elysia.* Members of this family lose the shell at metamorphosis and lack cerata. The foot usually has lateral folds, called "parapodia," which may be folded over the dorsal surface. The mantle cavity, associated ctenidium, and osphradium have been lost. The anus opens on the animal's right side. All are hermaphroditic herbivores, and most adults are less than 1 cm long. At least some species contain unicellular algae living symbiotically in the tissues of the snail, giving the animal a distinctly green color.

Family Juliidae. *Berthelinia, Julia.* These opisthobranchs form a 2-valved shell, superficially resembling that of the bivalved molluscs; however, the remains of a coiled larval shell, like those of other gastropods, can be found on the left shell valve. Mantle cavity, osphradium, and gill are present; cerata are lacking. Species are typically green due to symbiotic algae in the tissues, smaller than 1 cm, and tropical, feeding on a particular genus of macroalgae (*Caulerpa*).

Order Anaspidea (= Aplysiacea). Two families.

Family Aplysiidae. *Aplysia*—the sea hares. Large, herbivorous animals, up to 75 cm long and weighing up to 16 kg. Rhinophores resemble rabbit ears. A thin, internal shell is covered by the mantle. A mantle cavity (containing one ctenidium) is present, but it has shifted to the animal's right side as a result of detorsion. All species are simultaneous hermaphrodites. Large, lateral projections called "parapodia" extend from the foot; by flapping these, the animal can swim. Research on *Aplysia* has led to new understanding of the biochemical basis of learning and memory. The sea hare's neurons are 100 to 1000 times wider than those of humans.

Order Notaspidea. Three families.

Family Pleurobranchidae. *Pleurobranchus, Berthella, Pleurobranchaea.* The shell has been lost in many species; if present, the shell is internal. Individuals retain a gill. Highly acidic secretions are produced by glands opening into the pharynx and by glands distributed on the mantle. Members of this family are predators, especially on sponges and ascidians. Some species can swim.

Order Thecosomata. The shelled pteropods. All are marine. All use a modified foot (the parapodia) to swim, and many secrete an external, mucous web for collecting food. Five families.

Family Limacinidae (= Spiratellidae). *Limacina.* Pteropods living in small (less than 1 cm), sinistrally coiled shells; mantle cavity, operculum, and osphradium are present. All species are protandric hermaphrodites.

Family Cavoliniidae (= Cuvieriidae). *Cavolina, Clio.* Pteropods living in uncoiled shells up to 5 cm long. Shells may be bottle shaped, bulbous and shield-like, or conical. The operculum has been lost, but the mantle cavity, osphradium, and ctenidium are retained. After death, empty shells often form an important component of temperate and warm-water sediments.

Family Cymbulidae. *Gleba, Corolla, Cymbulia.* The typical gastropod spiral shell is discarded at metamorphosis, and the adult then secretes a transparent, internal, gelatinous "shell." All species are hermaphroditic.

Order Gymnosomata. The unshelled pteropods. All are marine. Seven families.

Family Clionidae. *Clione.* Unshelled pteropods, also lacking mantle cavity and gills; an osphradium is present. Typically less than 4 cm long. All are hermaphroditic. *Clione limacina* exhibits highly specialized feeding appendages to capture the shelled pteropods on which it exclusively feeds.

Order Nudibranchia. The nudibranchs, with 68 families containing 40–50% of all opisthobranch species. The shells are discarded at metamorphosis, and adults lack a mantle cavity and ctenidia. Cerata typically serve as secondary gills. All are marine hermaphrodites.

Suborder Doridina. The dorid nudibranchs. Adults lack a shell, ctenidia, and mantle cavity. All members have evolved secondary gills, arranged in a circular tuft around the anus; true cerata, containing extensions of the digestive system, are lacking. All are marine hermaphrodites. Most are carnivores, feeding most commonly on sponges. Twelve families.

Family Polyceridae (= Polyceratidae). *Polycera.*

Family Hexabranchidae. *Hexabranchus.* A widespread tropical group containing snails up to 30 cm long and weighing 350 g (grams). The members of this order can swim as well as crawl.

Family Dorididae. *Doris, Austrodoris, Rostanga.*

Family Archidorididae. *Archidoris.* Plump, particularly soft-bodied snails, usually less than 10 cm long.

Family Discodorididae (= Diaululidae). *Discodoris, Diaulula.*

Suborder Dendronotina. All dendronotid species are marine and hermaphroditic, and most have true cerata. The cerata are highly branched in many species. Ten families.

Family Tritoniidae (= Duvaucellidae). *Tritonia.* Most are carnivores, feeding on soft corals (alcyonarians).

Family Dendronotidae. *Dendronotus.* These snails typically feed on hydroids.

Family Tethydidae. *Melibe, Tethys.* Up to 30 cm long. The cerata are unusually wide and flattened, and the head bears a conspicuous oral hood, fringed with tentacles. The hood may be as wide as 15 cm when fully expanded and is used to capture active prey, including crustaceans and fish.

Family Dotoidae. *Doto, Tenellia.*

Suborder Arminina. Another group of shell-less marine hermaphrodites. Six families.

Family Arminidae. *Armina.* Smallish slugs, typically less than 5 cm long. Rather than cerata, secondary gills are found under the surface of the unusually thick mantle. These shallow-water carnivores feed at night, often on bioluminescent sea pansies (colonial anthozoans); they themselves may become bioluminescent as a consequence of their feeding activities.

Suborder Aeolidina. Slugs with numerous cerata on the dorsal surface. Most species feed on cnidarians in shallow waters, appropriating their prey's nematocysts for their own defense. Seventeen families.

Family Flabellinidae. *Coryphella, Flabellina.* Most species feed on hydrozoans.

Family Pseudovermidae. *Pseudovermis.* Small slugs, less than 0.6 cm long, adapted for interstitial life in the spaces between sand grains. Cephalic tentacles and rhinophores are absent, and the cerata are very short and positioned laterally.

Family Tergipedidae (= Cuthonidae). *Cuthona, Tenellia, Cratena, Phestilla.* Individuals are less than 2.5 cm long. Some species feed on hydroids.

Family Glaucidae. *Glaucus, Hermissenda, Phidiana.* All are carnivores, feeding on cnidarians, annelids, and molluscs. Species in—and closely related to—the genus *Glaucus* are pelagic, feeding on siphonophores; by usurping their prey's nematocysts, these slugs themselves become dangerous, even to humans. The penis is often armed with spines.

Family Aeolidiidae. *Aeolidia, Spurilla.*

The Pulmonata[28]

Another major group of heterobranch species. Members of most species possess a shell, which in slugs is often enclosed within the mantle and therefore not readily seen. The mantle cavity lacks a ctenidium, but sometimes it houses secondarily evolved gills. All species are hermaphroditic, and most are oviparous; some species protect their fertilized eggs with calcareous shells. Approximately 70 families containing about 17,000 species.

Order Eupulmonato

Suborder Actophila

Family Ellobiidae. *Melampus, Ovatella, Ellobium.* Shallow-water, or

28. The Pulmonata is a valid taxonomic category, but there is disagreement about its taxonomic level. The group is contained within the subclass Orthogastropoda, superorder Heterobranchia.

intertidal, typically marine or estuarine pulmonates (a few species are terrestrial), with spirally coiled external shells; an operculum and osphradium are lacking. All are hermaphroditic, either sequentially or simultaneously. A few species release free-living planktonic veliger larvae closely resembling those produced by prosobranch and opisthobranch gastropods. The penis often bears a sharp stylet.

Suborder Stylommatophora. Individuals of most species have spiral shells that, in slugs, may be completely enveloped by the mantle. An operculum is never present. All individuals possess 2 pairs of tentacles, with the upper ones each bearing an eye at the tip. Over 50 families, with about 15,000 species.

Family Achatinellidae. *Achatinella.* These are tree-dwelling terrestrial snails found primarily on Pacific islands, including Hawaii, but also in Japan, New Zealand, and Australia.

Family Pupillidae. *Pupilla, Orcula.* Small, terrestrial snails, usually less than 1 cm long. This large group contains nearly 500 species.

Family Clausiliidae. *Clausila, Papillifera, Vestia*—door snails. Exclusively terrestrial, ovoviviparous snails distributed among more than 200 genera in Asia, South America, Europe, and Asia Minor. All members possess a unique, morphologically and functionally complex mechanism (the "clausilium") for sealing the aperture after withdrawing the body into the shell.

Family Succineidae. The amber snails. All are terrestrial, with very thin, fragile shells smaller than 2 cm.

Family Athoracophoridae. *Triboniophorus.* Slug-like terrestrial snails, with the shell reduced to a mass of calcified pieces embedded in the integument. The unusual respiratory system resembles the tracheal system of insects. The snails usually live in trees and bushes, mostly in Australia and New Zealand.

Family Achatinidae. *Achatina.* All are terrestrial, with shells as large as 23 cm; these are the largest of all terrestrial pulmonates. The African species estivate during the dry season, forming a particularly strong epiphragm. *A. fulica* is a major agricultural pest and is commonly eaten by humans in some parts of the world; the shells are used as utensils and as decorations.

Family Streptaxidae. A large group of tropical, terrestrial snails, with perhaps 500 species. They live in leaf litter and under logs and can withstand long periods of drought by estivating. All are carnivorous on other snails and on annelids.

Family Limacidae. *Deroceras, Limax.* Slug-like terrestrial snails, with the shell reduced to a flat plate and mostly enveloped by the mantle. These are major agricultural pests, particularly in Europe and Africa. Individuals can self-fertilize.

Family Helicidae. *Helix, Cepaea.* These terrestrial, distinctly edible snails ("escargot") have an external, spiral shell. They overwinter by burrowing in soil, covering the burrow with a substantial epiphragm. A specialized dart sac is associated with the vagina of each individual. During mating, each partner injects a calcareous dart into the flesh of its mate, as happens in members of several other pulmonate groups and in some opisthobranchs.

Order Basommatophora. These are smallish snails (usually less than 10 cm), with eyes usually located at the base of the tentacles. Fifteen families, containing about 1000 species.

Family Siphonariidae. *Siphonaria.* The shell is cap shaped, with an irregular bulge on the right side, and the mantle cavity houses a secondary gill. Some species release free-living veliger larvae. All species are marine. Many are tropical, living in the high intertidal zone.

Family Amphibolidae. *Amphibola.* Snails in this family are unique among pulmonates in having an operculum as adults. Although gills of any sort are lacking, amphibolids do possess an osphradium in the mantle cavity. Species are estuarine and produce a distinct veliger, but the veliger apparently emerges as a swimming, feeding larva in only one species (*Amphibola crenata,* from New Zealand).

Family Lymnaeidae. *Lymnaea.* Pond snails, with a spiral, external shell. All species live in freshwater; they commonly serve as intermediate hosts in the life histories of various parasitic trematode flatworm species.

Family Physidae. *Physa.* These freshwater hermaphrodites can self-fertilize. The shell is always sinistral.

Family Planorbidae. *Biomphalaria, Bulinus, Planorbis, Helisoma*—ram's horn snails. All are freshwater snails, often with a planispiral shell, which coils sinistrally. Members of this family are extremely important as intermediate hosts in the transmission of schistosomiasis, a devastating disease caused by trematode flatworms in the genus *Schistosoma.* The blood of these snails is often rich in hemoglobin, allowing them to inhabit environments—including polluted areas—with very low dissolved oxygen concentrations. Populations are difficult to control, in part because these snails can self-fertilize.

Order Systellommatophora.
Onchidium, Smeagol, Vaginulus. These slugs usually have no trace of a shell, internally or externally. The pulmonary cavity, if present, is always located posteriorly. The head bears 2 pairs of tentacles, with the upper pair bearing the eyes as in basommatophorans. Most species are terrestrial, although a few are amphibious, living partly in air and partly in the sea. Some species are herbivores; others feed on other pulmonates. Four families.

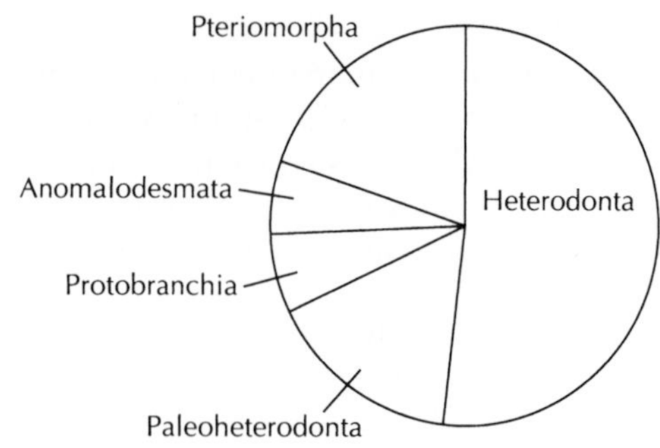

Class Bivalvia[29]
Approximately 7,650 species are distributed among over ninety families.

Subclass Protobranchia (= Paleotaxodonta = Cryptodonta)
Seven families, all marine, with approximately 500 species.

Family Nuculidae. *Nucula*—nut shells. These are deposit feeders living in sandy sediments, with shells typically only 2–3 cm long. These bivalves lack siphons, and water enters the mantle cavity anteriorly.

Family Nuculanidae. *Yoldia.* These small protobranchs (less than 7 cm long) live mostly in deep-water sediments. They possess both incurrent and excurrent siphons posteriorly.

Family Solemyidae. *Solemya.* Solemyids burrow in sand and mud and occur over an extremely wide depth range. The digestive system is substantially reduced, or even absent, in some species; nutrients may be obtained through activities of symbiotic sulfur bacteria living in the gills. These clams lack siphons, and water enters the mantle cavity anteriorly. Solemyids can swim for up to a minute at a time by forcefully ejecting water from the excurrent siphon.

29. In this classification scheme, members of the Lamellibranchia are now distributed among the following 4 subclasses: Pteriomorphia, Paleoheterodonta, Heterodonta, and Anomalodesmata. The septibranch bivalves are now treated as 2 families (Poromyidae and Cuspidariidae) within the subclass Anomalodesmata.

Subclass Pteriomorphia

Most species are epifaunal, attaching to solid substrates with byssal threads, and lack a true incurrent siphon. Twenty-four families with approximately 1500 species.

Family Mytilidae. *Modiolus, Bathymodiolus, Mytilus, Lithophaga*—mussels. Most species live attached by byssal threads to solid substrates in marine and estuarine habitats; a few species live in freshwater. A few of the marine species (*Lithophaga* spp.) bore into calcareous substrate (including coral) or live commensally with sea squirts (ascidians). *Mytilus edulis* has become a major pollution bioindicator species. *Bathymodiolus azoricus* dominates the fauna on many hydrothermal vents along the mid-Atlantic ridge. Although these bivalves can ingest particulate material, they also derive nutrients from symbiotic sulfur oxidizing bacteria and can take up dissolved amino acids directly from seawater.

Family Pinnidae. *Pinna*—pen shells. The shells are thin and fragile, and may be as long as 1 m. The posterior adductor muscles are much larger than the anterior adductor muscles. The animals live in shallow tropical seas, partly buried in sediment and attached to the underlying solid substrate by silky byssal threads.

Family Ostreidae. *Ostrea, Crassostrea*—oysters, of considerable commercial importance. These bivalves lie on the left valve, which may be firmly cemented to the substrate. Adults lack a foot and do not secrete byssal threads. There is no anterior adductor muscle, and the shells lack a nacreous layer. Individuals alter their sex every few years throughout life. Each female may produce more than one million eggs yearly.

Family Pectinidae. *Chlamys, Pecten, Aequipecten, Argopecten, Placopecten*—scallops. Many species can swim by snapping the 2 shell valves closed, but a few species cannot swim and lie on the substrate or live attached to firm substrate by byssal threads. The anterior adductor muscle is lacking. The posterior adductor muscle, however, is large and is the only part of the scallop that humans eat.

Family Anomiidae. *Anomia*—jingle shells. The shells are round or oval and decidedly shiny. Jingle shells live attached to solid substrates; a chitinous, often calcareous plug of byssal material reaches the substrate through a hole in the right shell valve. There is no anterior adductor muscle, and the posterior adductor muscle is much reduced.

Subclass Paleoheterodonta ("ancient and differently toothed")

Eight families, with approximately 1000 species.

Family Unionidae. *Lampsilis, Ligumia, Medionidus, Villosa, Unio, Anodonta, Pyganodon.* All members of this family (over 300 species in North America) live in freshwater. Females brood embryos in the gills and release glochidia, which continue their development parasitically on fish. The adults are free-living, with a particularly well-developed periostracum and 2 strong adductor muscles. Hundreds of species are currently imperiled or already extinct.

Subclass Heterodonta ("differently toothed")

Forty-two families, with approximately 4000 species. The adult foot usually lacks a byssal gland.

Order Veneroida. This group holds over one-third of all current bivalve species.

Family Lucinidae. *Lucina, Lucinoma.* These clams are common in sulfide-rich sediments and build (with the foot) fairly complex tunnels that extend deep into the sediment below the animal. The long, worm-like foot of these suspension-feeding bivalves typically builds an anterior feeding tube and lines it with mucus. The position of this inhalant tube is changed frequently. All lucinids studied so far have chemoautotrophic bacteria living symbiotically in the gills; the bacteria form carbohydrates from CO_2, using energy obtained in sulfide oxidation. The gills are unusually thick and have only the inner demibranch.

Family Thyasiridae. *Thyasira.* Like their close relatives, the lucinids, these bivalves commonly have enlarged gills harboring symbiotic chemoautotrophic bacteria, and build a mucus-lined anterior feeding tube with the foot. Unlike that of the lucinids, the feeding tube is fixed in position and the gills are complete, bearing both the inner and outer demibranchs. Populations as dense as 4000 clams/m^2 have been described in the North Atlantic Ocean, in sediments rich in organic matter. The foot constructs an extensive network of tunnels that extend deep into the sediment below the clams; these tunnels are more complex than those of the lucinids.

Family Galeommatidae. *Lasaea, Montacuta.* A group of small (typically less than 2 cm), actively crawling bivalves that are commonly commensal on other marine invertebrates, especially anemones, annelids, echinoderms, crustaceans, and other bivalves. Some species of *Entovalva* live only within the guts of holothurians. One deep-water species (*Mysella verrilli*) seems to be a suctorial ectoparasite on cnidarians. Individuals in this family are often hermaphroditic and typically brood embryos in the mantle cavity before releasing offspring as free-living veliger larvae. In some species, males live within the female's mantle cavity.

Family Carditidae. *Cardita*—little heart shells. These shallow-water suspension feeders attach to firm substrates by byssal threads. The blood contains hemoglobin. The sexes are separate, and females brood embryos in the gill.

Family Cardiidae. *Cardium, Laevicardium*—cockles. This group contains about 200 species of shallow-water suspension feeders, generally living in sandy substrates. The foot is highly muscular and used for burrowing, jumping, and even swimming, although the swimming is hardly graceful. The periostracum is poorly developed.

Family Tridacnidae. *Tridacna, Hippopus*—giant clams. Individuals can weigh up to 180 kg. Most species have a very small foot and live attached to substrates by a huge byssus. The mantle is replete with unicellular algae (zooxanthellae) living symbiotically in the tissue. No adult lacking zooxanthellae has ever been reported. By living in the mantle tissue of the bivalve, zooxanthellae gain protection from UV irradiation. All 6 tridacnid species live in the tropical Indo-Pacific, in shallow water.

Family Mactridae. *Mactra, Spisula*—the surf clams; *Mulinia, Rangia.* Most species are marine, although a few live only in freshwater. Marine species burrow in shallow-water sediments, using a large foot that lacks a byssus. Several species are commercially important foods.

Family Pharidae (= Cultellidae). *Cultellus, Ensis*—razor clams. Rapidly burrowing suspension feeders in marine and estuarine habitats.

Family Tellinidae. *Tellina, Macoma.* These are all marine and primarily deposit feeders, living in sand or mud. The outer demibranch is very small and lacks all or most of the ascending lamellae.

Family Donacidae. *Donax.* Burrowing suspension feeders, all marine. At least some species migrate up and down sandy beaches by leaping out of the sand (apparently cued by the sound made by the collapsing of especially large waves) and being carried by the incoming or outgoing tide.

Family Arcticidae. *Arctica islandica.* These clams live in fairly deep water off the New England coast. They are commercially important as chowder clams; the United States harvests about 20,000 metric tons of clams yearly.

Family Corbiculidae. *Corbicula.* Burrowing suspension feeders found in estuaries and in freshwater. About 100 species, one of which was introduced to the United States from Asia probably in the 1930s and is now a major economic nuisance throughout much of the United States. Populations can exceed 1000 individuals per m^2. Most species release free-swimming veliger larvae. The introduced species, *C. fluminea,* can self-fertilize, broods embryos in the gill, and releases small juvenile bivalves that can be carried considerable distances downstream by currents.

Family Dreissenidae. *Dreissena*—zebra mussels. These small (< 5 cm) European and Asian bivalves recently have invaded the eastern United States and Canada, probably carried here in ship ballast water in 1986. They thrive in fresh and salty water, and they are rapidly becoming major economic pests; they compete for suspended food with commercially important fish and native bivalve species and clog the water pipes of power plants, boats, and industrial cooling systems. Adults attain densities exceeding 50,000 individuals per m^2, reproduce prolifically (fertilizing eggs externally and releasing free-swimming veliger larvae), and form thick mats of byssal threads that make them extremely difficult to dislodge from clogged pipes. No other North American freshwater bivalves form byssal threads as adults.

Family Sphaeriidae (= Pisidiidae). *Pisidium, Sphaerium*—fingernail clams. Freshwater suspension feeders. The shells are usually smaller than about 0.5 cm. Some species live out of the water in moist leaf litter along the shores of ponds and streams, so they are, in a sense, terrestrial.

Family Vesicomyidae. *Calyptogena, Vesicomya.* The approximately 50 species of clams in this family all live in sulfide-rich habitats, such as near hydrothermal vents and cold-water sulfide seeps, and all harbor chemoautotrophic endosymbiotic bacteria in the gills.

Family Veneridae. *Mercenaria* (= *Venus*)—quahogs (pronounced kwō-hogs), or "hard-shell clams"; *Gemma, Tapes.* A large group of some 500 suspension-feeding species, all of which are marine. The northern quahog, *Mercenaria mercenaria,* is used in chowders and eaten on the half-shell as "cherry-stone" clams.

Family Petricolidae. *Petricola, Mysia.* These clams typically bore into a variety of substrates, such as mud, chalk, and coral. All are marine.

Order Myoida. Six families

Family Myidae. *Mya*—soft-shell clams, steamers. Most species are burrowing suspension feeders. The siphons are fused and covered by a periostracal sheath.

Family Hiatellidae. *Panopea*—the geoduck (pronounced "gooey-duck"). Geoducks are large clams found along the U.S. and Canadian Pacific coasts. Shells grow to lengths of 20 cm, with siphon lengths exceeding 75 cm, permitting the animals to live far beneath the sediment's surface.

Family Pholadidae. *Martesia, Xylophaga, Zirphaea*—the piddocks. These marine bivalves bore into hard substrates including hard clay, peat, shale, shells, and wood. The siphons protrude outside the substrate, for suspension feeding. These animals do considerable damage to wooden boats, docks, and pilings.

Family Teredinidae. *Teredo, Bankia*—the shipworms. These bivalves mostly bore into wood. They line their burrows with calcium deposits. Their shell is quite small—typically about 0.4 cm in an individual that is 6–7 cm long—and most of the animal protrudes posteriorly from the shell as a long "worm." The animal lives hidden inside the wood except for the siphons, which protrude for feeding. Unlike pholads (see p. 289), teredinids possess calcareous pallets, with which they seal the burrow after retracting the siphons. Most species are marine or estuarine. Some are gonochoristic; others are hermaphroditic. Young are brooded in the gills of some species. These bivalves cause considerable economic damage to wooden ships and pilings, which they enter during larval metamorphosis.

Subclass Anomalodesmata

This class includes all "septibranch" species. Thirteen families with approximately 450 species.

Family Pandoridae. *Pandora.* A small group (about 25 species) of marine, burrowing suspension feeders living in shallow-water sediments. The inner demibranchs of the ctenidia are fully developed, but the outer demibranchs are greatly reduced. All species are hermaphroditic.

Family Poromyidae. *Poromya.* These "septibranch" species are carnivores—especially on annelids—and most live in the deep sea. The ctenidia are substantially modified. All species are marine and hermaphroditic.

Family Cuspidariidae. *Cuspidaria.* These "septibranch" animals are highly modified carnivores, living in deep-sea sediments. The foot is small but byssate; the labial palps may be missing entirely; the mantle cavity is divided into 2 chambers by a muscular, perforate septum; true ctenidia are absent. All species are marine, and individuals eat mostly crustaceans and annelids.

Family Clavagellidae. *Clavagella*—watering pot shells. This family contains some of the oddest and least understood bivalves. Adults live within long (perhaps 12–30 cm) calaceous tubes that are perforated at the anterior end as in a garden watering pot. At least one shell valve fuses with the tube. Nothing is known about the life history.

Class Scaphopoda

The approximately 350 species are distributed among 10 families.

Family Dentaliidae. *Dentalium.* The shells are up to 15 cm long, which is as large as any scaphopod gets. Species in this group are very widespread, with some living in shallow water and others at great depths, in all the major oceans: Atlantic, Pacific, Indian, and Arctic.

Class Cephalopoda

The approximately 600 species are divided among 45 families. All species are marine, gonochoristic, and carnivorous.

Subclass Nautiloidea

Nautiloids are the only cephalopods with true external shells secreted by the mantle. Although this group once contained thousands of species distributed among many families, only 6 species have escaped extinction, and these species are so closely related as to be placed in a single family.

Family Nautilidae. *Nautilus*—the chambered nautilus. The external, calcareous shell is coiled and divided internally by transverse septa; the animal lives only in the outermost chamber. Shells of adults can reach about 27 cm across. Nautiloids have simple lens-less eyes, 80 to 90 tentacles, 2 pairs of ctenidia, and 2 pairs of osphradia; none have an ink sac. All 6 species occur in the Indo-Pacific, typically living at depths between 50 and 500–600 m.

Subclass Coleoidea (= Dibranchiata)

Most species possess internal shells that are completely surrounded by mantle tissue. Four orders, 44 families.

Order Sepioidea. All members bear 8 arms and 2 tentacles. Five families.

Family Spirulidae. *Spirula*—the ram's horn. The internal shell is spiralled and calcareous, and functions in buoyancy regulation. These animals are pelagic in deep water (200–600 m), and they have an ink sac and, posteriorly, a specialized external bioluminescent organ; they lack a radula.

Family Idiosepiidae. *Idiosepius.* These tiny cephalopods rarely exceed lengths of 1.5 cm, and they lack even an internal shell.

Family Sepiidae. *Sepia*—the cuttlefish. The lightweight, calcareous shell ("cuttlebone") is internal and functions in buoyancy regulation. The tentacles can be completely retracted into a special pocket.

Order Teuthoidea (= Decapoda). The squids. Shells are always internal and uncalcified. The head is surrounded by 8 arms and 2 tentacles, and it bears advanced eyes with lenses. All members have a well-developed radula. Squid are commercially important, some 200 million metric tons of certain species being caught for food each year. Twenty-five families.

Family Loliginidae. *Loligo*—the common Atlantic squid; *Sepioteuthis, Lolliguncula. Loligo* squid attain lengths of about 50 cm and tend to live in large aggregations.

Family Ommastrephidae. *Illex, Todarodes*—arrow squids. These squid support enormous N. Atlantic and Japanese fisheries, respectively.

Family Lycoteuthidae. *Lycoteuthis.* These are small (less than 10 cm long), deep-water squid, found as deep as 3000 m. They possess bioluminescent organs in species-specific patterns. Particularly patriotic species emit red, blue, and white light in different regions of the body.

Family Architeuthidae. *Architeuthis*—giant squids; Mesonychoteuthis—colossal squids. These are the largest of all invertebrates, attaining lengths of 20 m, including the tentacles, and weights exceeding 1 ton. The eyes are up to 20 cm in diameter, making them the largest animal eyes on record. These squid have no bioluminescent organs and live at depths of 500–1000 m. They are heavily preyed upon by sperm whales.

Family Cranchiidae. *Galiteuthis.*

Order Vampyromorpha. One family (Vampyroteuthidae). *Vampyroteuthis*—vampire squids. These deep-water (300–3000 m), dark-bodied squids have 8 normal arms plus a pair of highly modified, thin, elongated, tendril-like arms. Conspicuous sheets of tissue occur between arms, forming a vampire-like cloak. Species bear well-developed bioluminescent organs, a radula, and large, red eyes. The shell is internal, uncalcified, and nearly transparent.

Order Octopoda. These cephalopods possess 8 arms and no tentacles. The shell, if present, is reduced to small cartilaginous support structures. Twelve families, containing at least 600 species.

Family Cirroteuthidae. *Cirrothauma.* These octopods occur as deep as 4000 m, and at least one species is unique among cephalopods in apparently being blind. They lack a radula and ink sac, and some species have a gelatinous appearance, looking more like large jellyfish than cephalopods.

Family Octopodidae. *Octopus*—the octopuses (at least 250 species). These shallow-water cephalopods are up to 3 m long and 2.5 kg in weight. They are bottom dwellers, not swimmers, and individuals tend to live by themselves in small, protective caves. Octopodids can bore holes in mollusc shells, and can kill other prey using a toxic salivary gland secretion. The well-developed brain is surrounded by a cartilaginous skull. Octopuses have complex behavior; they can learn and remember. About 150 new octopus species have been identified since 1995.

Family Argonautidae. *Argonauta*—the paper nautilus. These octopods are exclusively pelagic. The females are up to 30 cm long and secrete a large, but paper-thin, chamberless shell in which the female lives and broods her developing embryos; the shell is produced by specialized glands on several of the tentacles, and is not homologous with other molluscan shells. The male is only about 1.5 cm long and shell-less. During mating, the disproportionately long hectocotyl arm of the male detaches completely after entering the female's mantle cavity.

Family Opisthoteuthidae. *Opisthoteuthis.* These deep-water animals are semi-gelatinous in consistency, possess a cartilaginous internal shell, and lack a radula. They are found in tropical and temperate seas, mostly at depths of 100–1000 m.

Family Ocythoidae. *Ocythoe.* These Atlantic, Pacific, and Mediterranean pelagic octopods are sexually dimorphic, with dwarf males (about 3–4 cm long) living in abandoned salp tests; females are up to 30 cm long and free-living. Females of *Ocythoe tuberculata* are unique among cephalopods in having a fish-like swim bladder for buoyancy regulation.

General References about the Molluscs

Ballarini, R., and A. H. Heuer. 2007. Secrets in the shell. *Amer. Sci.* 95: 422–29.

Boardman, R. S., A. H. Cheetham, and A. J. Rowell, eds. 1987. *Fossil Invertebrates.* Palo Alto, Calif: Blackwell Scientific, 270–435.

Fretter, V., and A. Graham. 1994. *British Prosobranch Molluscs,* 2d ed. London: The Ray Society.

Hanlon, R., and J. B. Messenger. 1996. *Cephalopod Behaviour.* New York: Cambridge University Press.

Harrison, F. W., ed. 1992. *Microscopic Anatomy of Invertebrates, Vols. 5 and 6: Molluscs.* New York: Wiley-Liss.

Hyman, L. H. 1967. *The Invertebrates, Vol. 6. Mollusca I.* New York: McGraw-Hill.

Kat, P. W. 1984. Parasitism and the Unionacea (Bivalvia). *Biol. Rev.* 59:189–207.

Lalli, C. M., and R. W. Gilmer. 1989. *Pelagic Snails: The Biology of Holoplanktonic Gastropod Mollusks.* Stanford, Calif.: Stanford Univ. Press.

Lydeard, C., et al. 2004. The global decline of nonmarine mollusks. *BioScience* 54: 321–30.

Morton, J. E., 1979. *Molluscs,* 5th ed. London: Hutchinson Univ. Press.

Parker, S. P., ed. 1982. *Classification and Synopsis of Living Organisms,* vol. 2. New York: McGraw-Hill, 946–1166.

Ponder, W. F., and D. R. Lindberg, eds. 2008. *Phylogeny and Evolution of the Mollusca.* Berkeley: University of California Press.

Pörtner, H. O., R. K. O'Dor, and D. L. Macmillan, eds. 1994. *Physiology of Cephalopod Molluscs: Lifestyle and Performance Adaptations.* Switzerland: Gordon & Braech, Publishers.

Reynolds, P. D. 2002. The Scaphopoda. *Adv. Mar. Biol.* 42: 137–236.

Southward, A. J., P. A. Tyler, C. M. Young, and L. A. Fuiman. 2002. *Advances in Marine Biology: Molluscan Radiation—Lesser-known Branches.* New York: Academic Press. This volume contains chapters on the protobranch bivalves (by J. D. Zardus, pp. 2–65), shelled opisthobranch gastropods (by P. M. Mikkelsen, pp. 69–136), scaphopods (by P. D. Reynolds, pp. 139–236), and pleurotomariodean gastropods (by M. G. Harasewych, pp. 238–94).

Thorpe, J. H., and A. P. Covich, eds. 2001. *Ecology and Classification of North American Freshwater Invertebrates,* 2nd ed. New York: Academic Press.

Vermeij, G. 1993. *A Natural History of Shells.* Princeton N.J.: Princeton Univ. Press.

Ward, P. D. 1987. *The Natural History of* Nautilus. Boston: Allen & Unwin.

Wilbur, K. M., ed. 1983–1988. *The Mollusca.* (Vol. 1: Metabolic Biochemistry; Vol. 2: Environmental Biochemistry and Physiology; Vol. 3: Development; Vols. 4 and 5: Physiology; Vol. 6: Ecology; Vol. 7: Reproduction; Vols. 8 and 9: Neurobiology and Behavior; Vol. 10: Evolution; Vol. 11: Form and Function; Vol. 12: Paleontology and Neontology of Cephalopods.) New York: Academic Press.

Search the Web

1. www.nrcc.utmb.edu/

   This site is sponsored by the Natural Resource Center for Cephalopods, University of Texas at Galveston. It includes numerous color images, along with current information about cephalopod research.

2. www.mbl.edu/

   Select "Marine Organisms" under Quick Links and then "Marine Organisms Database" and then "Mollusca." This site, operated by the Marine Biological Laboratory at Woods Hole, MA, includes identification keys to the common molluscs found near Woods Hole. It also includes a section explaining terminology relevant to the group.

3. Two excellent sites about zebra mussels (*Dreissena polymorpha*):

   http://nas.er.usgs.gov/taxgroup/mollusks/zebramussel/

   This site is part of the National Zebra Mussel Information Network.

   www.anr.state.vt.us/dec/waterq/lakes/htm/ans/lp_zebra.htm

   This is part of the Vermont Agency of Natural Resources World Wide Web site.

4. http://tolweb.org/tree/eukaryotes/animals/mollusca/cephalopoda/cephalopoda.html

   This site (part of the Tree of Life project) was prepared by R. Young, M. Vecchione, and K. M. Mangold, at the University of Hawaii. It contains extensive information about cephalopod systematics, with accompanying photographs.

5. www.slugsite.tierranet.com

   This site, maintained by Steve Long, contains wonderful photographs of nudibranchs, along with information about the animals.

6. http://www.conchologistsofamerica.org/conchology

   This is an excellent introduction to molluscan systematics and diversity, written and maintained by Dr. Gary Rosenberg.

7. http://shells.tricity.wsu.edu

   Photographs and taxonomic information on the various molluscan groups, provided by the Washington State Universities Natural History Museum.

8. www.whoi.edu/science/B/aplacophora/defchaetneo.html

   This site provides several color photographs of aplacophoran molluscs, courtesy of Robert Robertson and Amelie Scheltema.

9. http://www.thecephalopodpage.org

   This is the personal Web page of James B. Wood, Bermuda Institute of Ocean Sciences. It includes a list of species with accompanying color images, along with general information and links to other sites.

10. http://www.cephbase.utmb.edu

    This is another excellent source of information about cephalopods, including over 1500 images and many videos, and links to other relevant Web sites. It is maintained by J. Wood and C. L. Day, under the guidance of R. K. O'Dor.

13

The Annelids

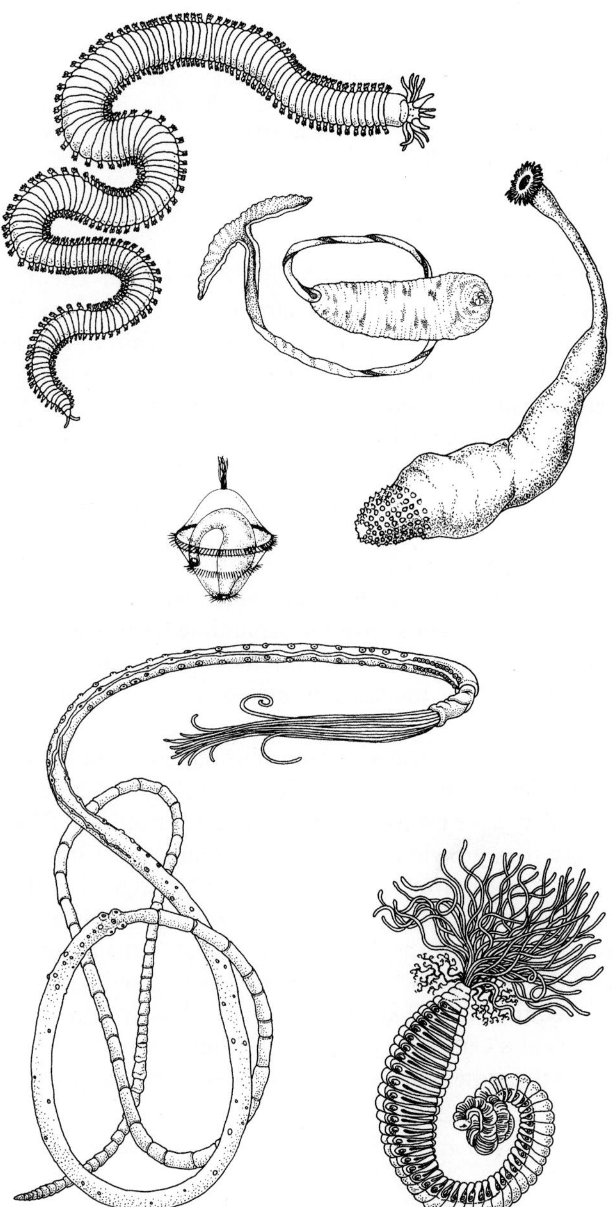

Introduction

The phylum Annelida has become a more diverse group in the past 15 years; it now includes 3 groups of animals formerly given status as separate phyla: pogonophorans (now called siboglinids), echiurans, and at least provisionally, the sipunculans. Morphological, developmental, and/or gene-sequence data suggest that all of these protostomes belong together within a single phylum.

Phylum Annelida

Phylum Annelida
(annulus = Latin: ring)
an-el-ē´-dah

Defining Characteristic:[1] One or more pairs of chitinous setae

General Annelid Characteristics

There are at least 12,500 described annelid species. All adult annelids other than sipunculans possess at least one pair of chitinous bristles, called **setae,** or **chaetae,** and all are **vermiform** (worm-shaped); that is, these animals, like those in a number of other phyla, are soft-bodied, reasonably circular in cross section, and longer than they are wide. Unlike those of most other vermiform animals, however, the bodies of most annelids consist of a series of repeating segments. This serial repetition of segments and organ systems (skin, musculature, nervous system, circulatory system, reproductive system, and excretory system) is known as **metamerism,** or metameric segmentation (Fig. 13.1).

1. Characteristics distinguishing the members of this phylum from members of other phyla.

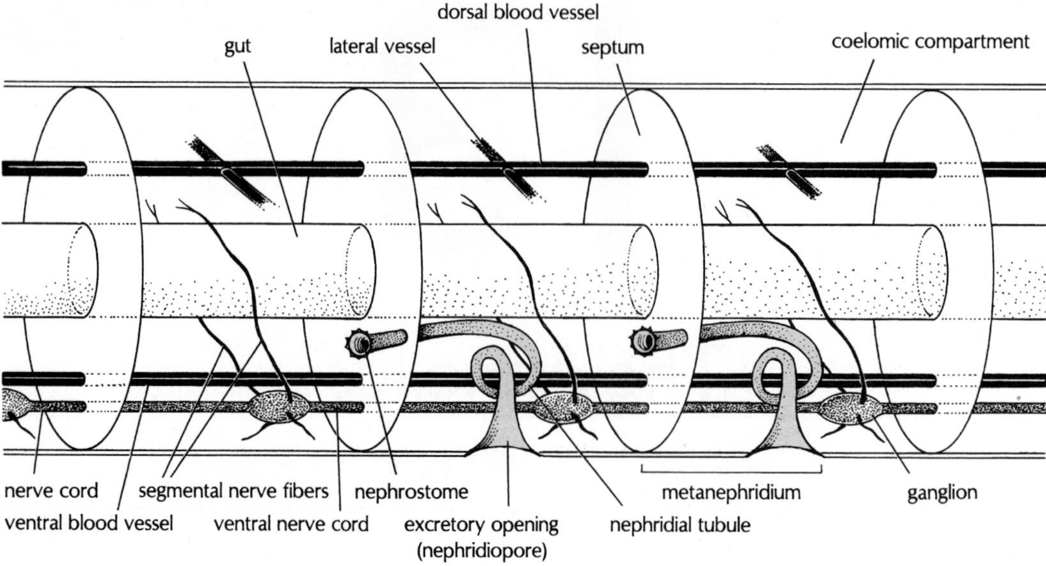

Figure 13.1

Schematic illustration of metameric organization in annelids. The annelid body consists of a linear series of segments, separated from each other by transverse, mesodermally derived septa. Much of the internal anatomy, including excretory, nervous, coelomic, and muscle systems, is segmentally arranged. Body wall musculature has been omitted for clarity. Note the ventral nerve cord; the vertebrate nerve cord is located dorsally. Also note that the coelomic fluid collected from one segment is discharged through an opening in an adjacent segment.

The annelid outer body wall is generally flexible and can play an active role in locomotion. Moreover, the thin body wall can serve as a general surface for gas exchange, provided that it is kept moist. Even when the epidermis secretes a protective cuticle, the cuticle remains permeable to both water and gases. For this reason, annelids are restricted to moist environments.

The individual annelid segments are generally separated from each other to a large degree by **septa,** which are thin sheets of mesodermally derived tissue (**peritoneum**) that essentially isolate the coelomic fluid in one segment from that in adjacent segments (Fig. 13.1). This allows for localized deformation of the outer body wall, brought about by contractions of circular and longitudinal musculature within a single segment. Thus, muscle contractions in any one segment do not alter the hydrostatic pressure in other parts of the animal.

Although some wastes are excreted across the general body surface, excretion generally occurs by means of structures called **nephridia** ("little kidneys"). Most annelid segments contain 2 nephridia, each of which is open at both ends. This type of nephridium is called a **metanephridium.**

Coelomic fluid is drawn into the nephridium at the **nephrostome** (kidney-mouth; *stoma* = Greek: mouth) by the action of cilia (Fig. 13.2). As the fluid passes through the convoluted tubule of the nephridium, some substances (including salts, amino acids, and water) may be selectively resorbed, and other substances (including metabolic waste products) may be actively secreted into the lumen of the tubule. The final urine emerging from the nephridiopore is thus quite different in chemical composition from the primary urine entering at the nephrostome. In addition to providing an outlet for metabolic waste products, nephridia may be used to regulate the water content of the coelomic fluid.[2]

In many annelid species, ducts leading from the gonadal tissue merge with the nephridial tubule. Thus, the nephridium generally plays a role in discharging gametes as well as urine.

Annelid systematics are presently in considerable disarray, making it difficult to generalize about annelid characteristics. Annelids have for some time been distributed among 3 major groups: Polychaeta, Oligochaeta, and Hirudinea, with the last 2 groups combined to form the class Clitellata. As discussed later in this chapter, however, there is growing evidence from gene sequence studies that Polychaeta is not monophyletic, and that in fact the clitellates may have evolved from polychaete ancestors.[3] There is also now growing evidence that echiuran and sipunculan worms, 2 groups that were formerly assigned to their own phyla, are in fact polychaetes, although their precise relationship to other polychaete groups remains uncertain. Since no new system has yet been agreed upon, for this edition I have retained the two class system (Polychaeta and Clitellata) from the previous edition, while trying at the same time to acknowledge echiurans as probable

2. See *Topics for Further Discussion and Investigation,* no. 4, at the end of the chapter.

3. See *Topics for Further Discussion and Investigation,* no. 13.

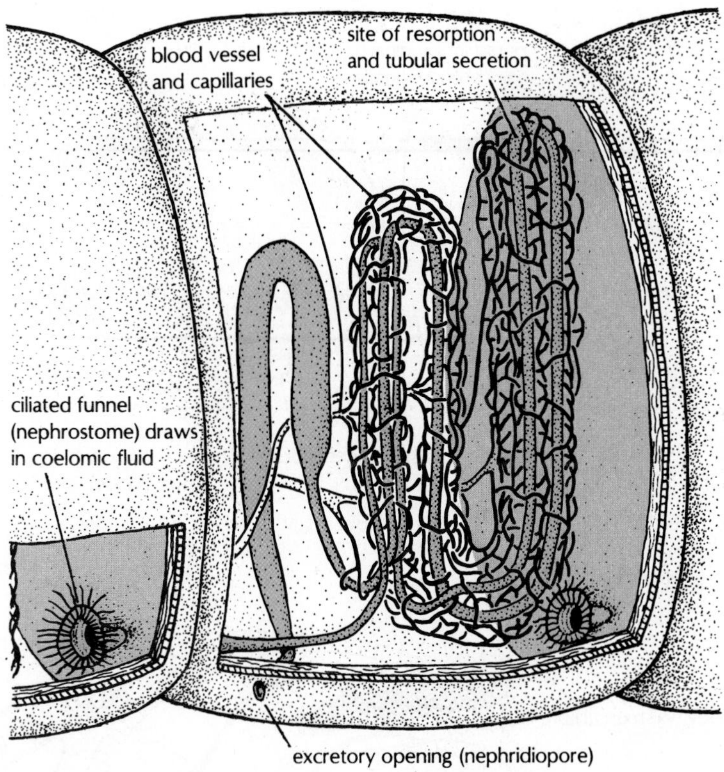

Figure 13.2
Diagrammatic representation of a typical metanephridium. The chemical composition of the primary urine drawn through the nephrostome is altered by selective resorption and secretion as fluid moves through the nephridial tubules. The final urine is discharged through the nephridiopore.

polychaetes and sipunculans as possible annelids (and possible polychaete annelids). If echiurans and sipunculans and clitellates are all polychaetes, then the terms Polychaeta and Annelida are essentially equivalent. The current turmoil is nicely summarized in the following quote from a 2007 paper by V. Rousset et al. (*Cladistics* 23: p. 54): "This study was intended to be the most ambitious attempt yet to resolve annelid relationships. Still, overall resolution remains discouraging: rarely (have) so many taxa . . . been sequenced for so many nucleotides with such sparing results."

Class Polychaeta

Class Poly • chaeta
(G: many setae)
pahl′-ē-kē′-tah

Defining Characteristic:[4] ? paired lateral outfoldings of the body wall (parapodia) ?

4. Characteristics distinguishing the members of this class from the members of other classes within the phylum.

Approximately 70% of all annelid species are placed in the class Polychaeta. Nearly all polychaetes live in salt water. Polychaetes generally possess at least one pair of eyes and at least one pair of sensory appendages **(tentacles)** on the anteriormost part of the body (the **prostomium**). Generally, the body wall is extended laterally into a series of thin, flattened outgrowths called **parapodia** (Figs. 13.3 and 13.4). Parapodial morphology differs significantly among species and therefore plays an important role in polychaete identification. These outfoldings increase the animal's exposed surface area and, because parapodia are highly vascularized (Fig. 13.3b), they function in gas exchange between the worm and its environment. Parapodia also have a locomotory function in many species, being stiffened by the presence of chitinous support rods called **acicula** (Fig. 13.3a). In addition, siliceous, chitinous, or, more rarely, calcareous bristles called **setae** protrude from each parapodium; setal morphology also differs substantially among polychaete species. The body is covered by a series of overlapping protective plates **(elytra)** in some species (Figs. 13.3a and 13.4b).

The septa present between most segments in polychaete worms enable the hydrostatic skeletal system to function independently in each segment. Localized contractions in one part of the body result in localized deformations without interfering with the musculature in the worm's other segments. Setae form temporary attachment sites and

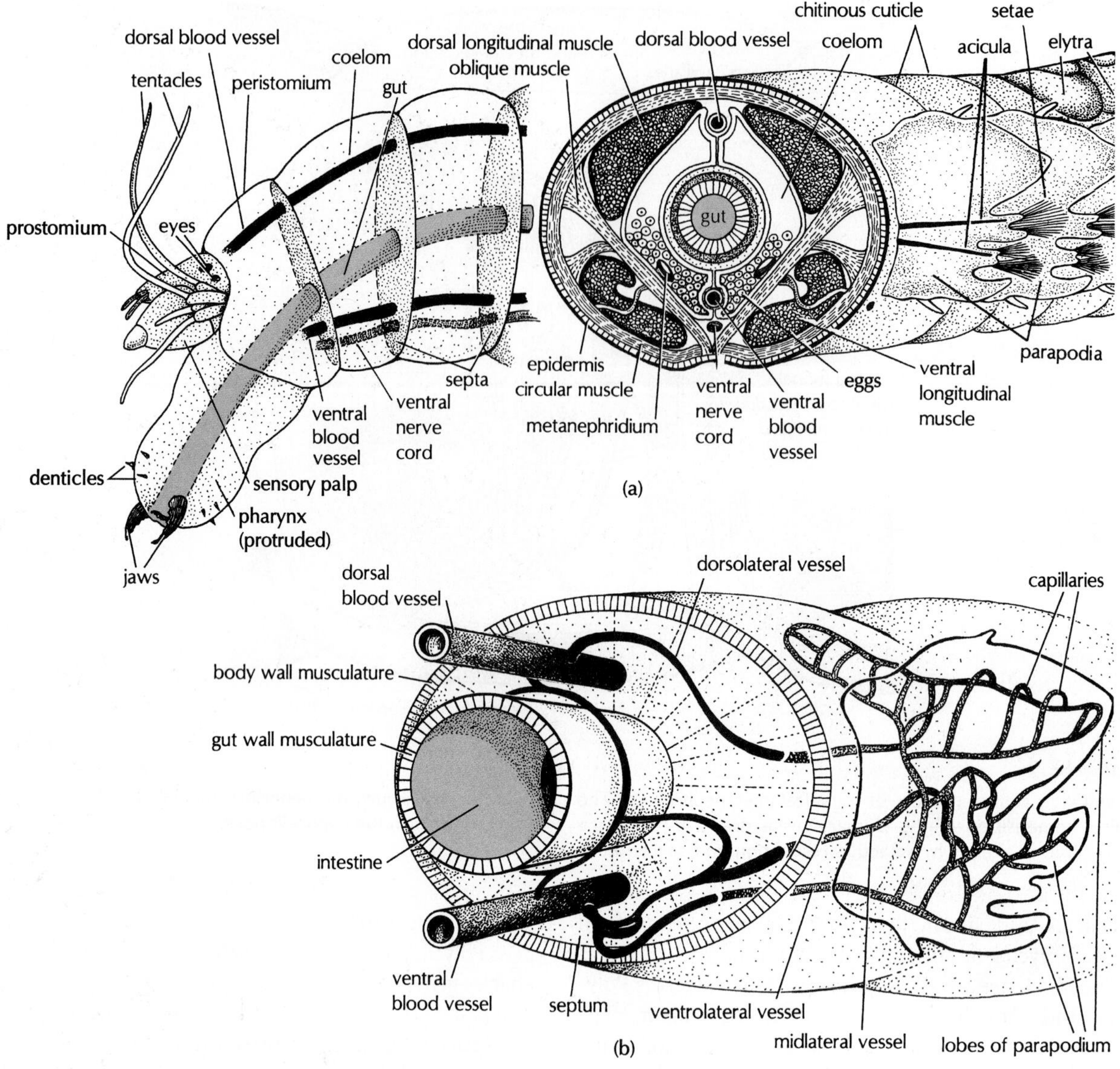

Figure 13.3
(a) Hypothetical polychaete worm, a composite of several species, showing typical major features. (b) Detail of a parapodium, showing the high degree of vascularization.

prevent backsliding during locomotion on or within the substrate or burrow. Septa between anterior adjacent segments are often absent or incomplete (**perforate**) in active, burrowing forms, enabling a greater volume of coelomic fluid to be utilized. This, in turn, permits greater shape changes to be associated with extending and anchoring the penetration organ (the everted pharynx, termed the **proboscis**) into the sediment.[5]

A cross section of a polychaete worm reveals a secreted, nonliving cuticle and a layer of circular muscles underlying the epidermis (Fig. 13.3a). Beneath the layer of circular muscles lies a layer of longitudinal muscle fibers. Oblique muscles are often found as well, serving to maintain body turgor and to operate the parapodia, which can be used as oars for locomotion through the water (i.e., swimming), over surfaces, or within burrows.

Many polychaetes form burrows in the sediment, largely by the mutual antagonism of the longitudinal and circular muscles through the hydrostatic skeleton of each coelomic compartment. In some burrowing polychaetes, the proboscis is everted into the sediment (Fig. 13.5a), penetrating and pushing aside the sand. The segments just behind the proboscis flange markedly as the circular muscles in those segments contract (Fig. 13.5b), preventing the worm from backsliding. Protrusion of setae also deters backsliding. The longitudinal muscles then contract in the

5. See *Topics for Further Discussion and Investigation,* no. 2.

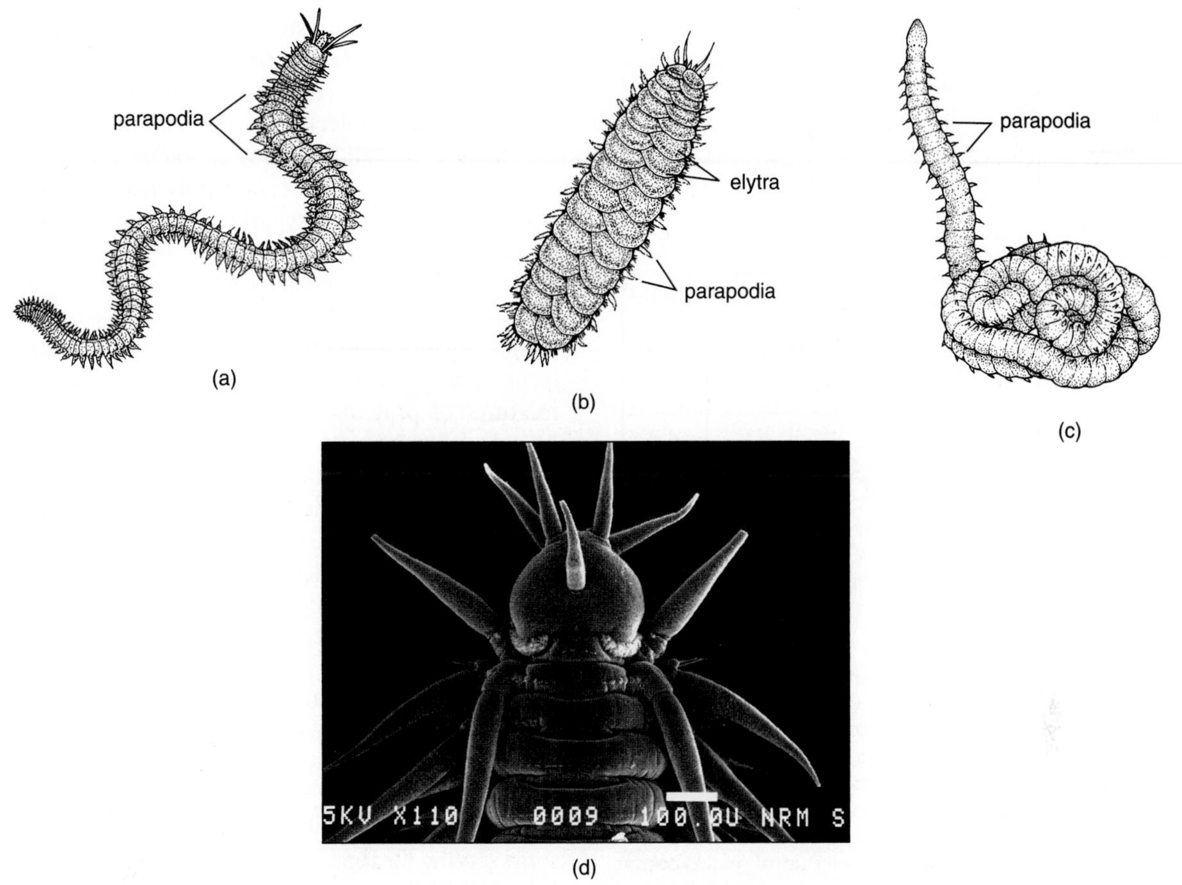

Figure 13.4

Representative errant polychaetes. (a) *Nereis virens.* (b) *Harmothoe imbricata.* (c) *Arabella iricolor.* (d) Anterior end of the phyllodocid polychaete *Sige fusigera.*

(a,b) Modified from McConnaughey and Zottoli, *Introduction to Marine Biology,* 4th ed., 1983. (c) Modified from Ruppert and Fox, *Seashore Animals of the Southeast United States.* South Carolina Press, Columbia, S.C., 1988. (d) Courtesy of Fredrik Pleijel. From Pleijel, F. 1990. *Zool. J. Linnean Soc.* 98:161–84.

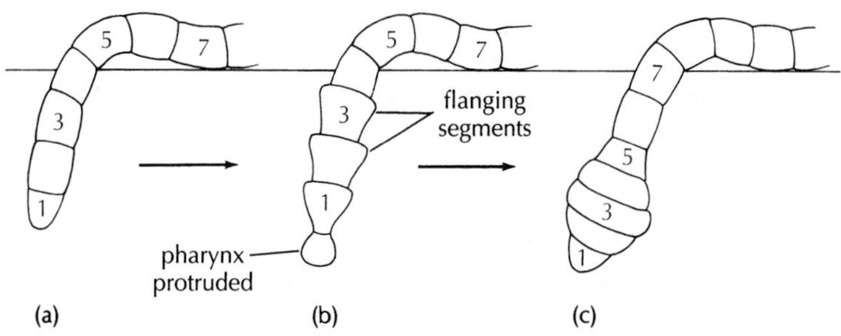

Figure 13.5

Sequence of movements involved in burrowing by the marine lugworm *Arenicola marina.* (a) The worm has already burrowed part way into the sediment. (b) The pharynx is protruded anteriorly, forming a proboscis. Much of the sediment is ingested; the rest is forced aside anteriorly. Appropriate muscle contractions cause adjacent segments to flange, preventing the worm from being pushed backward, out of the burrow, by the forward thrust of the pharynx. (c) Sequential contraction of the longitudinal musculature of each segment draws the worm downward. The increased width of the anterior segments, reflecting relaxation of circular muscles, anchors the worm as the more posterior segments are drawn forward.

After Trueman.

anterior segments, while the circular muscles in these segments relax. This dilates the worm's anterior segments (Fig. 13.5c). The anterior end is thus firmly anchored in the substratum, and the more posterior segments can then be drawn forward by contracting the longitudinal muscles. The worm is then ready to repeat the cycle. Note that burrow formation is accomplished by exploiting the properties of a compartmentalized hydrostatic skeleton (Chapter 5).

Polychaetes may be divided into 2 general groups. One group includes the generally active, mobile species

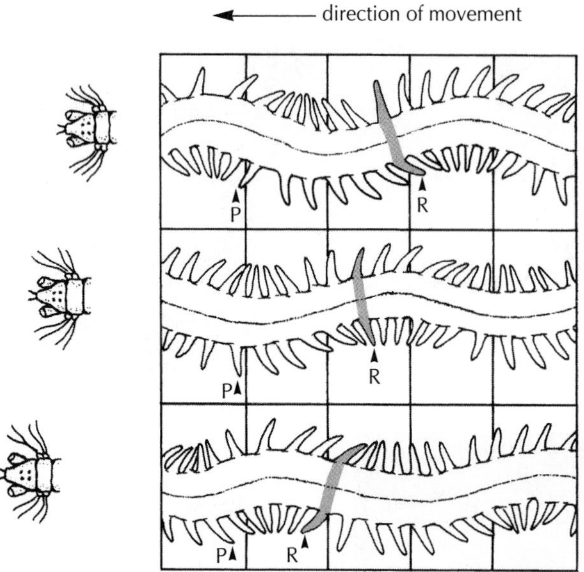

Figure 13.6

The errant polychaete *Nereis virens*, crawling rapidly over the substrate. *P* indicates a parapodium executing its power stroke. Its tip is applied to the substrate, with setae protruded for added stability; note that the tip remains stationary as the body is levered forward. *R* indicates a parapodium executing its recovery stroke. Its setae are retracted as the parapodium is lifted up and swung forward above the substrate; the parapodium is then lowered to contact the substrate, ready to perform a power stroke. Simultaneously, waves of longitudinal muscle contraction pass posteriorly; the longitudinal muscles on each side of the body contract out of phase, producing a rapid wriggling motion.

Source: J. Gray, 1939, "Studies in animal locomotion." VIII. *Nereis diversicolor* in *Journal of Experimental Biology*, 16:9–17, Plate I.

(i.e., the **errant** species, which were formerly placed in a separate subclass, the Errantia). Circular muscles typically play a minor role in the locomotion of these species; instead, the polychaetes are levered forward by the action of parapodia, which are operated in a complex pattern as oars. The acicula play an important role as stiffening elements during this activity, preventing the thin parapodial tissue from collapsing as thrust is applied against the substratum. Errant polychaetes typically augment parapodial movements with carefully coordinated contractions of the longitudinal body wall musculature, generating sinusoidal waves of activity; here, the longitudinal muscles on one side of the body antagonize those on the opposite side, resulting in rapid, eel-like movement over the substratum (Fig. 13.6) or, in some species, swimming.

Not all errant polychaetes are surface dwellers: Members of some "errant" species form simple or complex burrows within substrates, but these polychaetes nevertheless typically have well-developed parapodia, complete with acicula and setae, and a well-developed head equipped with a toothed or jawed protrusible pharynx. Most errant species are carnivorous, but the group also includes suspension feeders, detritus feeders, and omnivores.

A number of burrowing polychaete species lack parapodia, but it is unclear whether this represents a secondary loss of parapodia or the ancestral polychaete condition. If the ancestral polychaetes had parapodia, then "presence of parapodia" suffices to define the Polychaeta as a separate class; if parapodia evolved secondarily from burrowing polychaetes that had no parapodia, then the class has no valid defining characteristics.

In contrast to these errant polychaetes, other species (formerly placed in a separate subclass, the Sedentaria) typically spend their entire lives in simple burrows in the sediment or in rigid, protective tubes—that is, the animals live sedentary lives.[6] Their tubes vary in construction from simple organic secretions mixed with sand grains or mud to tubes composed of calcium carbonate and/or complex mixtures of proteins and polysaccharides. Other species actively bore into calcareous substrates and live in the resulting burrows. The parapodia tend to be greatly reduced, highly modified, or absent among the sedentary polychaetes (Fig. 13.7). Not surprisingly, acicula are absent as well, confirming their locomotory function in errant species. Most sedentary species also lack a protrusible pharynx (proboscis). Water is moved through the tubes of sedentary species for respiration, feeding, and/or waste removal by ciliary action, rhythmic movements of modified parapodia, or waves of muscle contraction passing from one end of the worm to the other (Fig. 13.8). Many sedentary polychaete species display an abundance of thread-like or feathery appendages at the anterior end, some serving for food capture and others being highly vascularized and serving as gills for gas exchange (Figs. 13.7 and 13.9a). All polychaetes in this ecological grouping are either suspension feeders or deposit feeders.

Species in a number of polychaete families have **protonephridia** similar to those found among the Platyhelminthes (p. 150) and several other invertebrate groups. Thus, in these polychaete species, coelomic fluid is probably ultrafiltered as it is drawn across the terminal meshwork into the nephridial tubule by ciliary beating. In other respects, protonephridia and metanephridia are believed to function similarly. All other polychaetes—indeed, all other annelids except, perhaps, for some siboglinid species—have metanephridia. Intriguingly, polychaetes of most species have protonephridia in their larval stages, even when the adults have metanephridia. Protonephridia have also been described from the developmental stages of some clitellates. This evidence suggests that annelids have evolved from ancestors with protonephridia.

Feeding habits are highly diverse among polychaetes and include predation, scavenging, suspension feeding, and deposit feeding.

Polychaete Reproduction

Reproduction is exclusively sexual in most polychaete species, and most polychaetes are **gonochoristic** (i.e., they have separate sexes). Gametes are produced by peritoneal tissue, rather than in distinct gonads, and are then released

6. See *Topics for Further Discussion and Investigation*, no. 1.

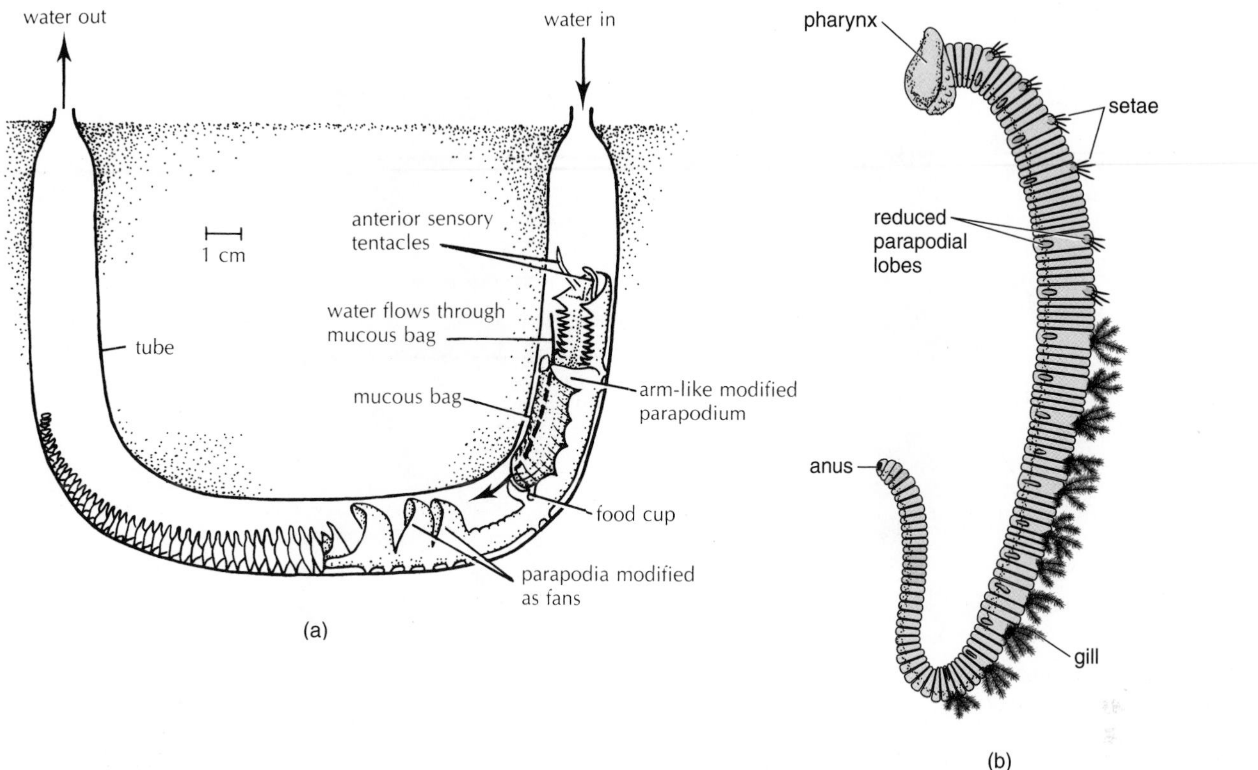

Figure 13.7

Representative sedentary polychaetes, showing a variety of morphological adaptations for a nonmotile existence. Most animals are drawn removed from their tubes or burrows. (a) *Chaetopterus variopedatus.* Feeding currents are generated by rhythmic movements of fanlike parapodia. One pair of anterior parapodia is modified for holding open a mucous bag, which filters out food particles as water moves through the tube. The bag is held posteriorly by a ciliated food cup, which continuously gathers the food-laden mucus into a ball. Periodically, the ball is released from the rest of the net and passed forward along a ciliated tract to the mouth for ingestion. (b) *Arenicola marina,* the lugworm. This worm is a deposit feeder and lives burrowed in sand. For gas exchange, water is pumped through the burrow by peristaltic waves of muscle contraction; parapodia play no role in locomotion or water flow and are greatly reduced in size in most segments. In a number of segments, a small portion of the parapodial lobes has become modified to form gills—clusters of thin, vascularized, branching filaments. (*Continued on following page.*)

into the associated coelomic compartments, where they mature. At least 6 adjacent segments of a given individual are involved in gamete production, and in some species gametes are produced within nearly all segments.

In some polychaete families, even many tube-dwelling or otherwise sedentary species undergo **epitoky,** a marked morphological transformation in preparation for reproductive activity. The result of this transformation is an **epitoke:** a sexually mature being (male or female) that is highly specialized for swimming and sexual reproduction (*tokus* = G: birth). In many species, epitoky involves asexual budding (Fig. 13.10a). One or more new, reproductive modules (epitokes) are budded, one segment at a time, from the posterior portion of the original animal (the **atoke**).[7] These epitokes subsequently detach from the atoke and swim off to commingle with other epitokes and discharge their eggs or sperm. The atoke remains safely behind in its burrow or tube. Note that the epitokes are genetically identical with the atokes that produced them.

In some other epitokous polychaete species, epitoky involves the remodeling of preexisting structures, rather than the budding off of new segments. The original worm thus becomes the epitoke, through adaptations for gamete production and storage, and for active swimming (Fig. 13.10c). In these species, the entire worm swims off to reproduce. Gamete maturation and epitoke formation have been shown to be under environmental and hormonal control in a number of polychaete species.

Polychaetes usually fertilize their eggs externally, in the surrounding water. Typically, the free-living embryo soon develops a digestive system and 2 rings of cilia (Fig. 13.11). These rings are the **prototroch** (G: first wheel), located around the equator of the animal, anterior to the mouth, and the **telotroch** (G: tail wheel), located posteriorly on what will become the terminal portion of the adult, the **pygidium.** The prototroch is the larva's major locomotory organ. Because of the ciliated rings, the larva is called a **trochophore,** meaning "wheel-bearer" (Fig. 13.11a,b). A third band of cilia, the **metatroch** ("between wheel"), later forms between the prototroch and the telotroch.

7. See *Topics for Further Discussion and Investigation,* nos. 10, 12.

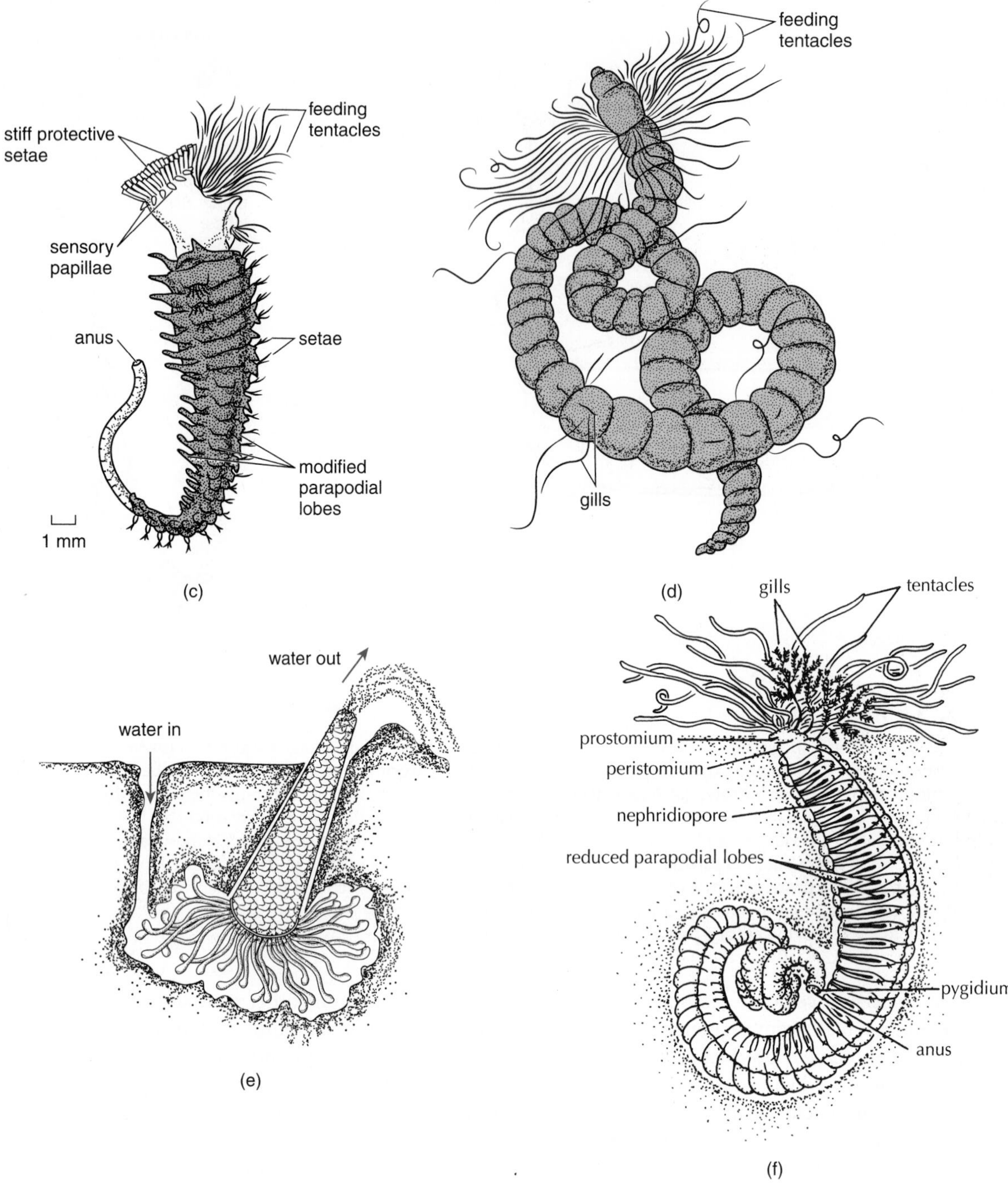

Figure 13.7 *Continued*

(c) *Sabellaria alveolata,* in lateral view. Members of this species build tubes of sand grains and mucus and attach them to a variety of solid substrates, including the tubes of neighboring individuals. Segments have fused anteriorly to form a densely bristled surface that occludes the opening of the tube when the animal pulls inside. (d) *Cirratulus cirratus.* These animals live within muddy substrates. Although the parapodia are much reduced, a portion of each parapodium has become greatly elongated in some segments, forming gills for gas exchange. (e) *Pectinaria belgica,* in its conical tube of mucus and cemented sand grains. The worm lives within this tube, head down in the sediment. Members of the genus *Pectinaria* (= *Cistenides*) are selective deposit feeders, ingesting particles high in organic content. The undigestible fraction is expelled at the posterior, elevated end of the tube, as illustrated. The long tentacles, seen anteriorly, are involved in food collection. (f) *Amphitrite ornata,* a species living in shallow water, in burrows of sand or mud lined with mucus. The first 3 segments each bear a pair of highly branched gills for gas exchange. Numerous grooved tentacles project from the prostomium. When the animal is feeding, these tentacles lie against the substrate, trapping food particles. Food is moved to the mouth by cilia lining the tentacular grooves or by shortening the tentacles themselves and bringing the food directly to the mouth. Note the absence of conspicuous parapodia; water is moved through the burrow by peristalsis of the body wall musculature.

(b) Based on Frank Brown, *Selected Invertebrate Types.* (c,e) Based on Fretter and Graham, *A Functional Anatomy of Invertebrates.* (d) Based on Bayard H. McConnaughey and Robert Zottoli, *Introduction to Marine Biology,* 4th ed. 1983; after W. C. McIntosh, 1915, *British Marine Annelids,* Vol. III, The Royal Society. (f) After Brown; after Barnes.

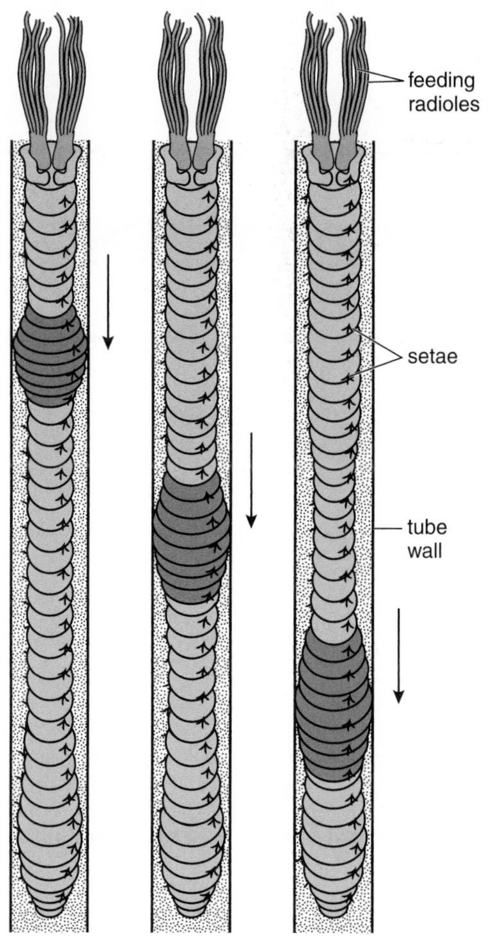

Figure 13.8
The tube-dwelling sabellid polychaete *Eudistylia vancouveri* irrigating its tube with a peristaltic wave of muscle contraction.
Based on A. Giangrande, "Irrigation of *Eudistylia* tube," in *Journal of the Marine Biological Association of the U.K.* 71:27–35. 1991.

(a)

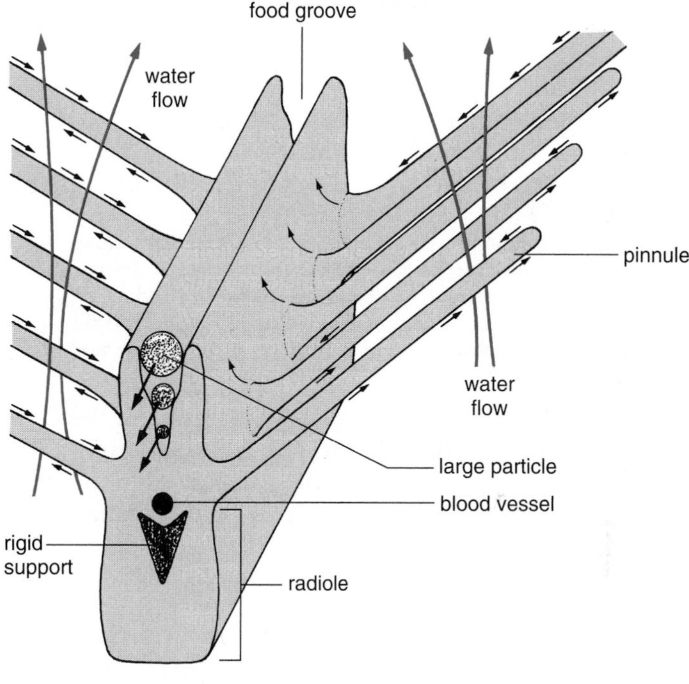

(b)

Figure 13.9
(a) The sedentary polychaete *Sabella pavonia.* The stiff ciliated filaments (called "radioles") form a fan surrounding the mouth. Water currents pass between the tentacles, and captured food particles are conducted to the mouth area for sorting and subsequent ingestion or rejection. When the worm is startled, the tentacles are rapidly withdrawn within the protective tube. (b) Detail of a single filament (radiole) seen in cross section. Food particles are captured by the ciliated *pinnules* and transported to a ciliated food groove, which runs the length of each filament to the animal's mouth. Blue arrows show direction of water flow; small black arrows show path taken by captured food particles along the pinnules to the major food groove of the radiole. Particles of different sizes are transported in different regions of the food groove as shown, for ingestion, tube-building, or rejection.
(a) © D. P. Wilson/Frank Lane Picture Agency Ltd. (b) From W. D. Russell-Hunter, *A Life of Invertebrates.* 1979. W. D. Russell-Hunter.

Trochophore larvae are also produced by echiurans and sipunculans, but *not* by any members of the Clitellata. However, they *are* produced by many mollusc species! What trochophores tell us about the evolutionary relationship between annelids and molluscs is hotly debated—did trochophore larvae evolve once, or did they evolve independently through convergence in the different groups?

As the polychaete trochophore larva swims and, generally, feeds in the water, body segments are repeatedly budded from its posterior region, just anterior to the pygidium. Each new segment bears a ring of locomotory cilia, permitting the larva to continue swimming despite the increasing body weight (Fig. 13.11c). The larva becomes more like the adult as it continues to develop (Fig. 13.11d, e), and eventually it metamorphoses to adult form and habitat, often after many weeks of dispersing in the plankton. Some polychaete species produce a nonfeeding, short-lived trochophore stage that subsists on stored yolk reserves and metamorphoses after swimming for only a few hours or, in some cases, only a few minutes.

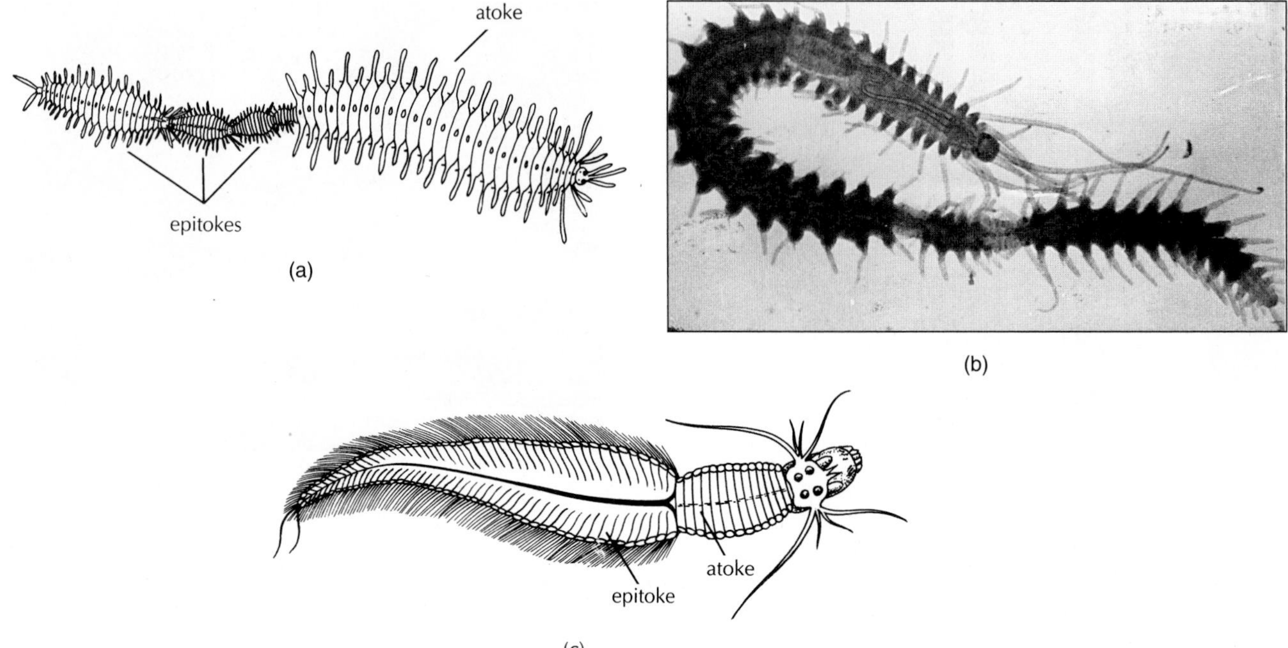

Figure 13.10
(a) Epitoke formation by asexual budding in *Autolytus*. Each atoke produces either male or female epitokes. (b) *Autolytus* sp. Female atoke with several female epitokes. (c) Epitoke formation via remodeling of original structures. Note that in (a) and (b) the original worm becomes the atoke, left behind in the burrow, and does not participate directly in reproduction, while in (c) a single individual is both epitoke and atoke and participates directly in reproduction.
(a) After Harmer and Shipley; after Malaquin. (b) Courtesy of K. L. Schiedges. (c) After Fauvel.

Some species lack any free-living larval stage in the life history. Instead, the embryos may develop within a gelatinous egg mass anchored in the sediment or to the inside surface of the female's tube. In some species, embryos may be directly protected by the parent, developing to a larval or juvenile stage within specialized adult brooding chambers.

Some species also reproduce asexually, by fragmenting and subsequently regenerating missing parts. All of the fragments regenerating from a single source individual are genetically identical, and so are not truly "individuals." Each regenerated fragment is best thought of as a **ramet** of the original genotype (*ramus* = L: a branch).

Family Siboglinidae (formerly the Pogonophora)

General Characteristics

Defining Characteristics: 1) Gut tissue (endoderm) forms an organ (the trophosome) that becomes filled with chemosynthetic bacteria; 2) segmentation confined to small rear portion of animal (the opisthosoma)

The siboglinids are a small but especially intriguing group of tube-dwelling polychaetes distributed throughout the world's oceans. All 150 described species are marine. Several thousands of individuals per square meter have been reported from some areas of the ocean bottom. Yet, siboglinids were unknown before the twentieth century, in part because there are no shallow-water species: A few species can be collected at depths of 20–25 m (meters), but most siboglinids live much deeper, at depths of hundreds or thousands of meters. Until recently, the animals were collected mostly by dredging, a technique that generally mutilates these often fragile organisms.

Siboglinids have had a colorful taxonomic history, with a variety of different names and a variety of presumed affiliations. For most of the past 20 years they were placed in their own phylum—the Pogonophora (G: beard bearer)—and at one point were divided between 2 phyla. Now that their status as derived polychaetes is almost universally acknowledged, their name reverts back to that of the original description. Their exact taxonomic status within the Polychaeta is still uncertain. With the apparent demotion of siboglinids from representatives of separate phyla to status as simply one of many polychaete families, why do I give them such prominence in this book? As you read what follows, I hope you will agree that relegating them to a few sentences in the *Taxonomic Detail* at the end of this chapter would have been a pedagogical tragedy.

Most of the siboglinid body bears little hint of annelid affinity, and in fact, siboglinids were originally declared to be deuterostomes, not protostomes. The anteriormost region of the body typically bears a **cephalic**

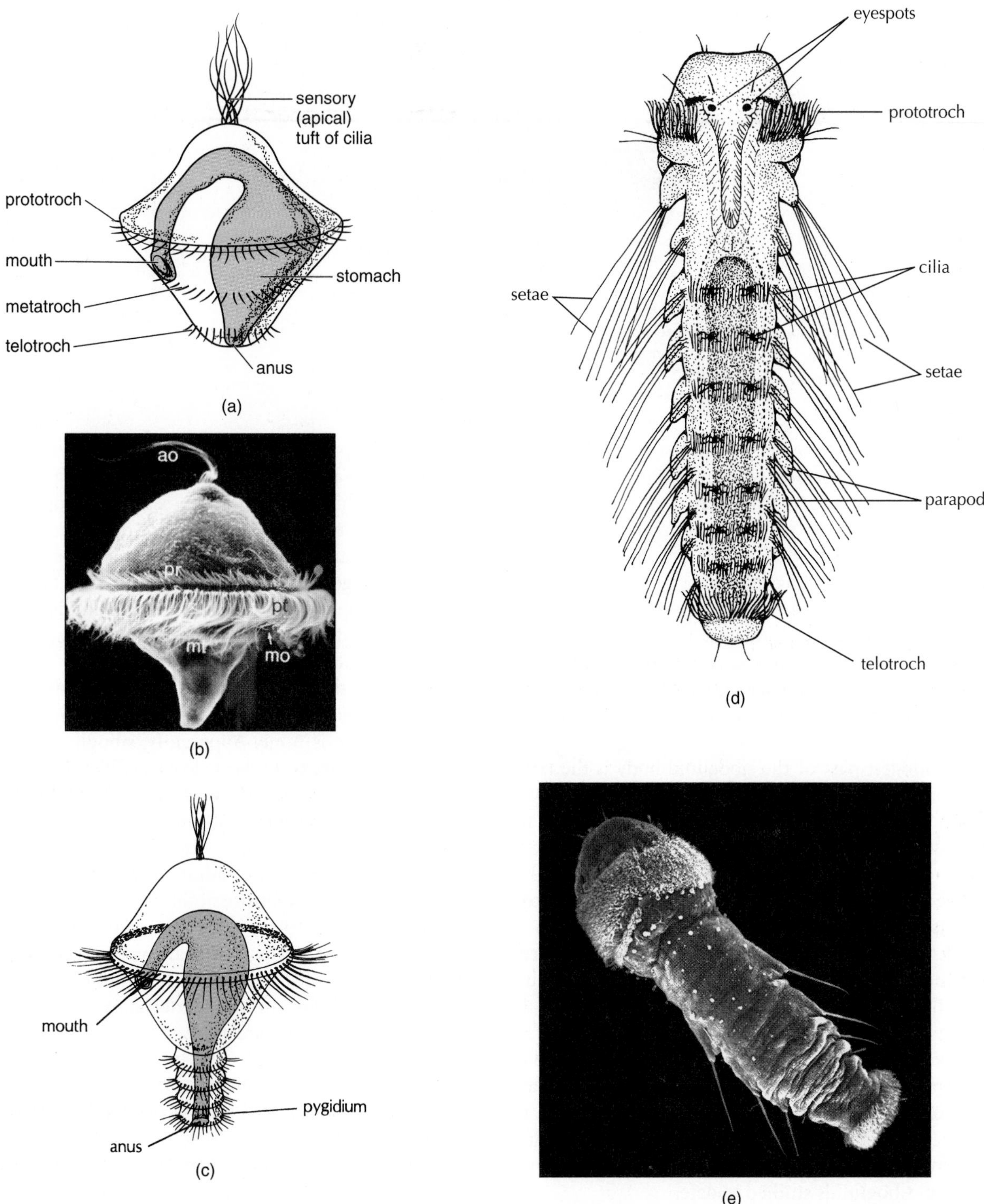

Figure 13.11

(a) Polychaete trochophore larva. The larva feeds on unicellular algae collected by the ciliary bands. (b) Scanning electron micrograph of the trochophore larva of *Serpula vermicularis*. The larva is about 110 μm wide. (ao = apical tuft; mo = mouth opening; mt = metatrochal cilia; pr = pretrochal cilia; pt = prototrochal cilia.) (c) Advanced trochophore larva, showing additional ciliated segments budded posteriorly from the initial trochophore. Parapodia will soon develop from each of these segments. New segments are added just anterior to the pygidium, which bears the anus. Thus, the youngest segments are located posteriorly, and the oldest segments of an annelid are anterior. (d) Advanced larva, about 600 μm long, of the polychaete *Polydora ciliata* in dorsal view. (e) Scanning electron micrograph of an advanced larva of the polychaete *Eupolymnia nebulosa,* viewed dorsally. The animal is about 13 days old. Note the conspicuous anterior prototroch and posterior telotroch.

(a) Based on Purves and Orians, *Life: The Science of Biology,* 2d ed. 1983. (b) Courtesy of Claus Nielsen, from Nielsen, 1987. *Acta Zoolog.* (Stockholm) 68:205–62. Permission of the Royal Swedish Academy of Sciences. (c) From Hardy, *The Open Sea: Its Natural History,* 1965. Boston: Houghton Mifflin Company. (d) From D. P. Wilson, in *Journal of Marine Biology Association,* U. K. 15:567, 1928. (e) Courtesy of Michel Bhaud. From Bhaud and Grémare, 1988. *Zoologica Scripta* 17:347–56. Courtesy of Pergamon Press.

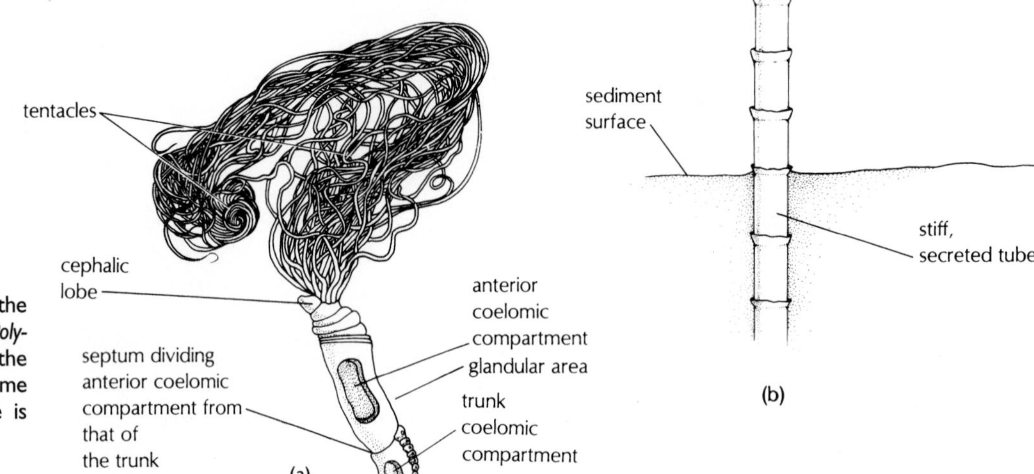

Figure 13.12
(a) Anterior end of the siboglinid polychaete *Polybrachia* sp. (b) A portion of the tube secreted by the same animal. Much of the tube is anchored in soft sediment.
From Hyman; after Ivanov.

lobe; a "beard," consisting of one to many thousands of ciliated tentacles; and a glandular area that secretes a chitinous tube, within which the animal spends its life (Fig. 13.12). All siboglinids live in chitinous tubes. Although sedentary, siboglinids are free to move up and down within their tubes, which always exceed the length of the animal's body. Each tentacle is serviced by 2 blood vessels, so that the tentacles form the primary gas exchange surface. The animal's anterior section contains a single small coelomic cavity that extends into each of the tentacles.

The longest part of the siboglinid body is the **trunk** (Fig. 13.13), which contains a pair of uninterrupted coelomic cavities. The body wall of the trunk contains both outer circular and inner longitudinal muscles. Externally, the trunk is often marked by large numbers of **papillae** (small bumps), 2 regions of ciliation, and 2 conspicuous rings of setae about halfway down the body (Fig. 13.13). The trunk is unsegmented, and the body cavity of the trunk is not septate. The major organs found within this coelomic cavity are the gonads and a multilobed structure called the **trophosome** (Fig. 13.14a). The trophosome of all species examined so far contains many closely packed bacteria (Fig. 13.14b), which likely play a major role in siboglinid nutrition, as discussed later in the chapter (*tropho* = G: nourish; *soma* = G: body). Presumed excretory organs have been described from the anterior ends of all siboglinids studied to date.

The annelid affinities of the group were first suggested only about 35 years ago, when the first intact specimens were collected. At this time, a third, more posterior body region was discovered, the **opisthosoma** (*opistho* = G: hind; *soma* = G: body). Discovering the opisthosoma must have been terribly exciting, and surprising: the opisthosoma shows the same sort of conspicuous segmentation found among polychaetes and clitellates (Fig. 13.13), with each of the approximately 6 to 25 segments containing a coelomic compartment that is isolated from adjacent compartments by muscular septa. Each segment bears chitinous setae. The opisthosoma is used for digging in sediment by most species, or for anchoring the animal within its chitinous tube. The segmented opisthosoma, along with developmental evidence described in the following text and DNA base sequence comparisons from 2 different genes, one mitochondrial and one nuclear, strongly support the inclusion of siboglinids within the Annelida, as do comparisons of hemoglobin structure and the placement and morphology of setae.

Perhaps the single most fascinating component of the biology studied to date relates to the siboglinid digestive system: There is none. Apparently, siboglinids are polychaetes that have secondarily lost the digestive tract. Adult siboglinids lack a mouth, an anus, and anything resembling a digestive tract, although a complete digestive system appears for a short time in the development of some species. How do adults meet their nutritional requirements? Several possibilities exist, and some evidence supports a role for each. For example, the tentacles at the animal's anterior end bear many surface microvilli and secretory cells. Food particles might be trapped by the tentacles and digested externally. The solubilized nutrients could then be absorbed across the microvillous surface of the tentacles. The evidence for this mode of feeding is strictly morphological, and it currently seems unlikely that particle feeding plays a substantial role in siboglinid nutrition.

On the other hand, considerable experimental evidence indicates that siboglinids, like many other soft-bodied invertebrates, can take up dissolved organic matter (DOM) from seawater, even at the low concentrations found naturally. Basically, the experiments consist of demonstrating that radioactively labeled amino acids, carbohydrates, and other organic molecules are accumulated by siboglinids against remarkable concentration gradients. Some compounds can be taken up from concentrations as low as 1 μM (micromolar, i.e., 1×10^{-6} moles per liter). Uptake is mediated through an active, energy-requiring process. Calculations based on respiration rates and rates of DOM uptake suggest that for some, but not all, species, the uptake of dissolved organic

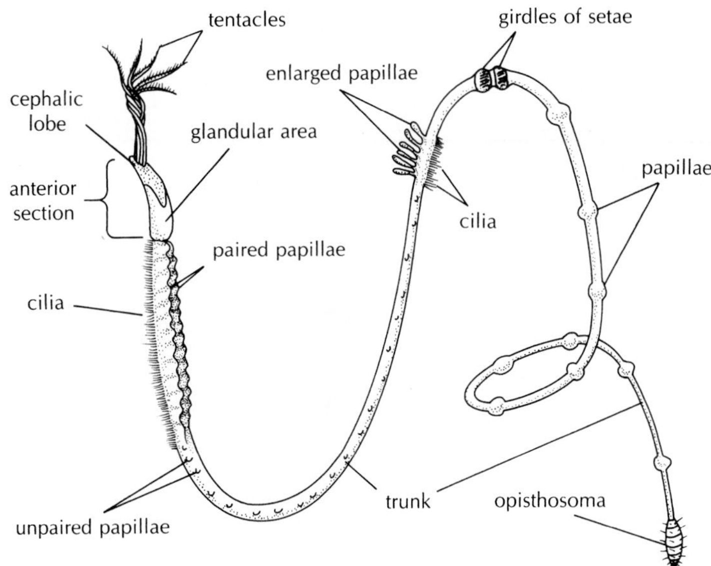

Figure 13.13

Intact siboglinid removed from its tube. Note the 3 conspicuous body divisions: cephalic lobe, with tentacles; trunk; and segmented opisthosoma.

From George and Southward, in *Journal of the Marine Biological Association,* 53:403, 1973. Copyright 1973 Marine Biological Association, U.K. Reprinted by permission

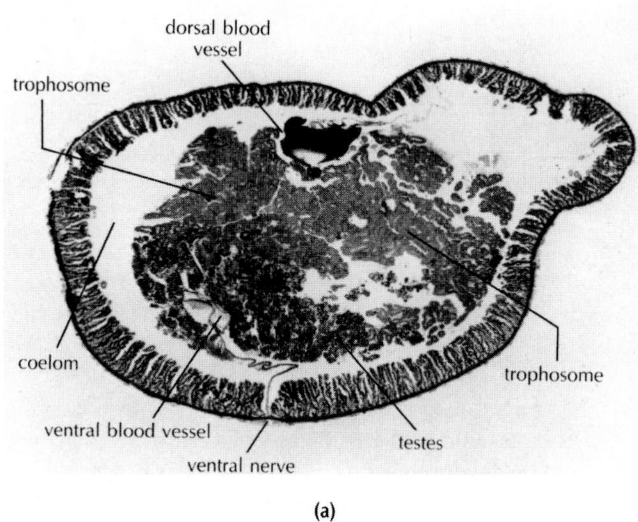

(b)

Figure 13.14

(a) Transverse section through the trunk of *Riftia pachyptila* Jones, showing location of the trophosome. In fact, most of the coelomic space is occupied by the bacteria-laden trophosome. (b) Scanning electron micrograph of the trophosome, showing many spherical, symbiotic, sulfide-reducing bacteria. The bacteria are globular, and 3–5 μm wide.

(a) Courtesy of M. L. Jones. From Jones, *Science* 213 (7 Jul 1981), p. 333. (b) From C. M. Cavanaugh *et al. Science*, Vol. 213 (7 Jul, 1981), pp. 340–41, fig. 1, © AAAS.

molecules from the surrounding water may be sufficient to meet all of a siboglinid's metabolic maintenance requirements.

Most siboglinids—specifically the frenulates—are only about 6–36 cm (centimeters) long and typically less than a millimeter wide. These species thus resemble long, thin threads. The ratio of surface area to volume in such animals is very high, increasing the likelihood that the general body surface can play a major role in taking up dissolved nutrients from seawater. The tubes of these perviate species are thin-walled and open at both ends, and some evidence indicates that the tubes are permeable to DOM. Moreover, all perviate siboglinids are **infaunal;** that is, they live with most of the body (and tube) implanted into soft, muddy substrates, in which dissolved nutrients accumulate to high concentrations relative to the concentrations found in seawater. DOM may well make a major contribution to the nutritional biology of these siboglinids.[8]

However, DOM uptake likely plays a less important nutritional role in the more recently discovered siboglinid group known as the **vestimentiferans.** About two dozen

8. See *Topics for Further Discussion and Investigation,* no. 15.

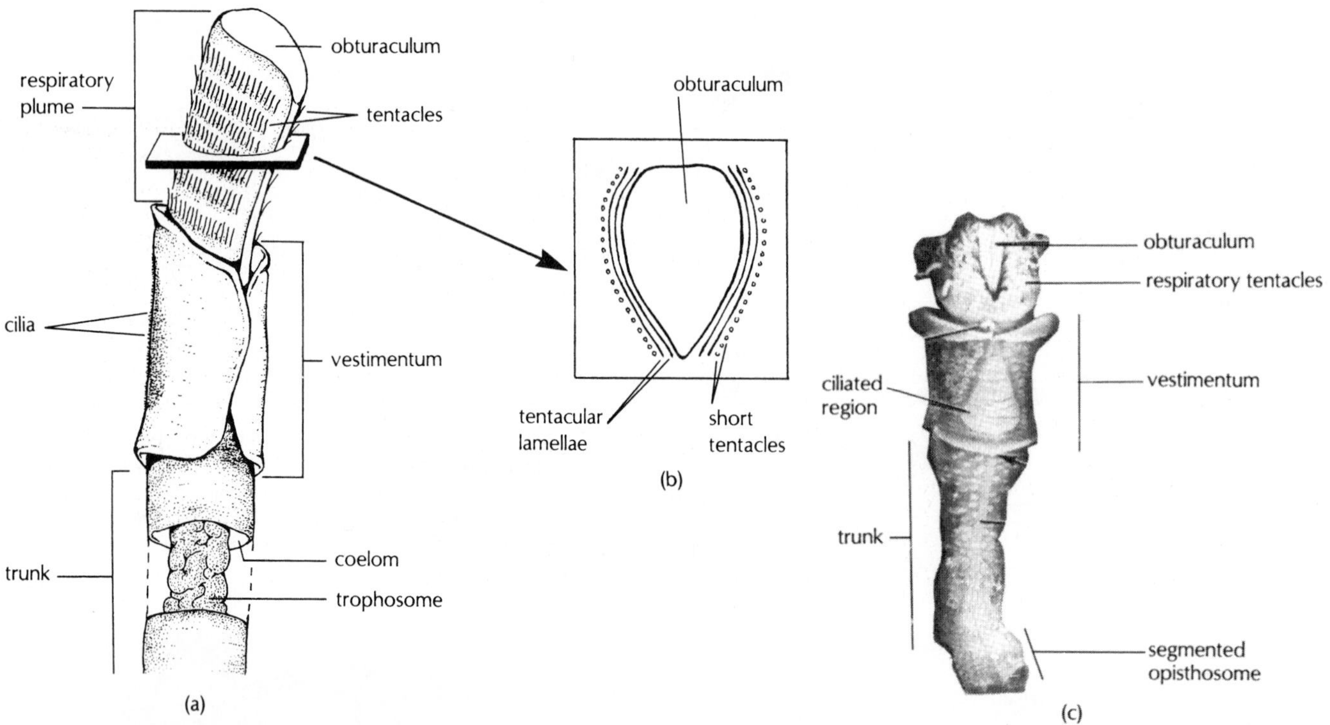

Figure 13.15

(a) Anterior portion of the hot-vent siboglinid *Riftia pachyptila*. The tissue below the vestimentum has been cut open to reveal a portion of the extensive trophosome within the animal's trunk. (b) Cross section through the branchial (tentacular) plume of *Riftia pachyptila,* showing the lamellae formed from the partial fusion of hundreds of individual tentacles. (c) Scanning electron micrograph of a juvenile tube worm, *Ridgeia* sp. This juvenile was only a few millimeters long.

(a) Modified from Southward and Southward, *Animal Energetics*, Vol. 2, pp. 201–28, Academic Press, 1987; and G. N. Somero, *Oceanus*, 27:69, 1984. (b) From Southward and Southward, 1987. (c) Courtesy of M. L. Jones. From M. L. Jones and S. L. Gardiner, 1989. *Biological Bulletin* 177:254–76. Permission of *Biological Bulletin*.

vestimentiferan species have been described to date, all placed in a separate subgroup, the Vestimentifera.

Vestimentiferans are huge, with bodies up to 2 m long and 25–40 mm wide. Moreover, they are morphologically dissimilar to other siboglinids in possessing a conspicuous, long collar called the **vestimentum,** just posterior to the plume of tentacles (the **respiratory,** or **branchial, plume,** Fig. 13.15), and a large, solid, flat-topped anterior structure (the **obturaculum,** Fig. 13.15) that supports the respiratory plume and plugs the opening of the tube when its resident withdraws.

The animal's excretory pores open dorsally on the vestimentum.

The tentacles of these vestimentiferan species are very numerous (thousands or even several hundred thousand per individual) and fused for much or all of their length, forming flattened sheets, or lamellae, encircling the obturaculum (Fig. 13.15b). Moreover, the vestimentiferans lack trunk setae, have a far more extensive trophosome, and have opisthosomal segments with paired coelomic cavities; the perviate siboglinids possess only one coelomic compartment per segment.

Despite their substantial anatomical differences as adults, however, gene-sequence comparisons (18S rDNA and a mitochondrial cytochrome oxidase sequence) support the placement of frenulates and vestimentiferans within a single taxonomic group. Indeed, all siboglinids seem to develop in much the same way initially and look very much alike as juveniles. In particular, juveniles of all species examined to date show paired opisthosomal coelomic compartments within each segment; the mesodermal partitions between compartments are simply retained to adulthood in vestimentiferans and lost during further development in the perviate species.

Vestimentiferans were first discovered in 1977 while researchers aboard the deep-sea submersible *Alvin* were exploring a major ocean-spreading center near the Galápagos Islands, in the Pacific Ocean off the coast of Ecuador. The ocean-bottom rifts of this area are regions where new crustal material is being generated from deepz within the earth. The rifts are characterized by scattered **hydrothermal vents,** small openings in the new crust that emit hot seawater rich in reduced inorganic compounds such as hydrogen sulfide and methane. The hydrothermal vent species *Riftia pachyptila* and *Ridgeia* spp. live upright in blind-ended tubes attached to solid rock, directly in the path of this outflow (Fig. 13.16), bathed in water cooled by mixing to about 10°–15°C. These animals live at considerable

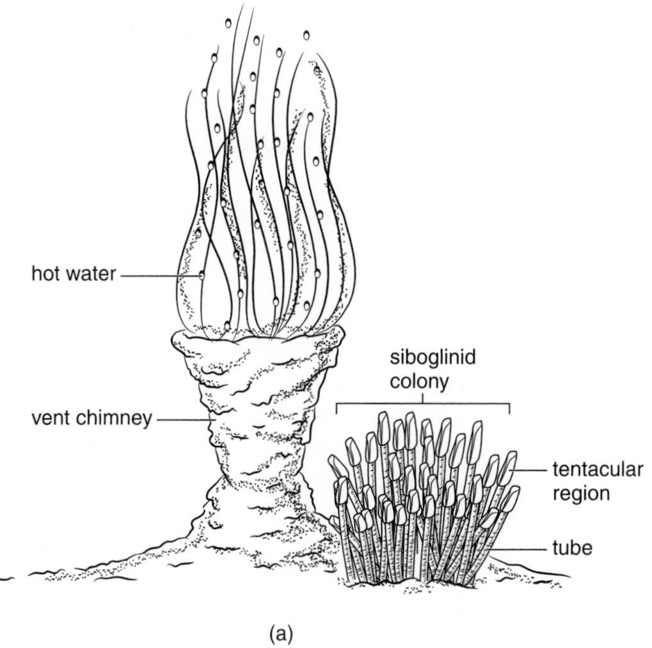

Figure 13.16
(a) Animal life thrives in the warm, sulfide-rich water near hydrothermal vents. These siboglinids (*Riftia pachyptila*) were found at a depth of about 2500 m by a team of scientists aboard the research submersible *Alvin* in 1977. The blind-ended tubes of these worms approach 3 m in length and are white, cylindrical, and flexible. (b) Dense cluster of the tube worms *Riftia pachyptila* congregated near a hydrothermal vent in the Galápagos, at a depth of over 2500 m.
(a) Based on *Science at Sea: Tales of an Old Ocean* by Van Andel. 1981. (b) Courtesy of H. W. Jannasch. From Jannasch, 1989. *American Society for Microbiology News* 55:413–16.

densities and form a conspicuous component of a thriving undersea community that would seem more likely in a tropical environment than in the ocean depths (Fig. 13.16b).

Similar communities were later discovered in cold water (only about 2°–4°C) at depths of about 600–3000 m in the Gulf of Mexico and off the coasts of Oregon and Louisiana. These so-called "seep" communities, which include large populations of approximately meter-long siboglinids, are restricted to well-defined areas where hypersaline sulfide- and methane-containing water seeps up through the sediment. Vestimentiferan communities have also been found around the decomposing bodies of dead whales, another site rich in hydrogen sulfide.

The surface-area-to-volume ratio of these very large "vent," "seep," and "whale fall" vestimentiferans is considerably less than that of the smaller, thinner perviate species, making it unlikely that the larger siboglinids could take up DOM rapidly enough to satisfy their more substantial metabolic requirements. Moreover, the thick tube walls surrounding the vestimentiferans probably act as barriers to DOM uptake across much of the body surface. Although DOM uptake has not been discounted in these species, biologists began looking for an alternative food source for these animals. What they discovered explains very clearly why these thriving deep-water communities are uniquely associated with waters rich in reduced inorganic substrates such as hydrogen sulfide and methane.

The trophosome of these "hot-vent" and "cold-seep" siboglinids contains vast numbers of bacteria (Fig. 13.14), at concentrations up to 10^{10} (10 billion) bacteria per gram (wet weight) of trophosome tissue. Biochemical evidence indicates that these bacteria obtain energy by oxidizing the hydrogen sulfide or methane in the surrounding seawater (Fig. 13.17). Since the seawater is outside the animal and the bacteria are inside the animal, the animal must provide the methane or sulfides to the bacteria. The sulfides and methane apparently diffuse into the siboglinid's blood across the tentacular plume and are then delivered to the trophosome by hemoglobin transport proteins in the circulatory system along with the oxygen needed to oxidize these reduced compounds. The bacteria use the energy they obtain from oxidizing these substrates in the trophosome to generate ATP, which in turn is used to synthesize organic molecules from the carbon in carbon dioxide (CO_2) as in plant photosynthesis:

$$CO_2 + 4\,H_2S + O_2 \rightarrow [CH_2O]_n + 4\,S + 3\,H_2O$$

The bacteria use these carbohydrates for their own growth and reproduction. The growing, internal bacterial population then can be digested by the host siboglinid or can feed the host indirectly by liberating soluble carbohydrates to host tissues.[9]

9. See *Topics for Further Discussion and Investigation*, no. 16.

RESEARCH FOCUS BOX 13.1

Living with Sulfides

Powell, M. A., and G. N. Somero. 1986. Adaptations to sulfide by hydrothermal vent animals: Sites and mechanisms of detoxification and metabolism. *Biol. Bull.* 171:274–90.

Much remains to be learned about the symbiotic relationship between sulfur-oxidizing bacteria and their animal hosts. For one thing, how do siboglinids withstand exposure to sulfide? All animals rely on the sulfide-sensitive enzyme cytochrome-*c* oxidase for aerobic ATP generation. This protein, located in the inner membrane of mitochondria, transfers electrons from cytochrome *c* to oxygen, the final step in the electron transport chain; cytochrome-*c* oxidase is typically poisoned at even very low sulfide levels, levels substantially lower than those found in the water surrounding hydrothermal vents and cold-water seeps. How, then, are siboglinids able to survive constant exposure to these high concentrations of hydrogen sulfide (H_2S)? Why are their cytochrome-*c* oxidase systems not disabled? Either (1) the siboglinid cytochrome-*c* oxidase somehow tolerates higher sulfide concentrations than do comparable enzyme systems of animals living in sulfide-free environments, or (2) the sulfide is somehow detoxified during its transport through the siboglinid to the bacteria living in the trophosome. Research by Powell and Somero (1986) indicates that the siboglinid survives by detoxifying sulfide and that there are 2 likely detoxification sites and mechanisms.

To examine the sensitivity of siboglinid cytochrome-*c* oxidase to sulfides, Powell and Somero collected—by submarine—specimens of the hot-vent species *Riftia pachyptila* from a depth of 2600 m and then homogenized the branchial plume of tentacular tissue. This is the tissue most directly exposed to the sulfides in seawater. They then ascertained the ability of siboglinid cytochrome-*c* oxidase from the plume to catalyze the transfer of electrons from cytochrome *c* to oxygen in the absence of sulfides; these data served as controls for the experiment.

Different amounts of sulfide were then added to other samples of the same homogenate, and cytochrome-*c* oxidase activity was again measured by monitoring the disappearance of substrate (reduced cytochrome *c*) from the mixture. Reaction temperatures were held constant at 20°C in all experiments, so that any differences in reaction rates could not be due to temperature effects. Cytochrome-*c* oxidase activity in *Riftia pachyptila* is obviously very sensitive to sulfide (Focus Fig. 13.1).

The inhibitory effects of sulfides on cytochrome-*c* oxidase activity were dramatically counteracted by adding siboglinid blood to the plume homogenates (Focus Fig. 13.2). The blood apparently contains high concentrations of a protein (hemoglobin, as it turns out) that binds tightly to sulfide molecules, so the sulfide is essentially inactivated during transport to the trophosome. As long as blood concentrations of *free* sulfide remain low, aerobic respiration of siboglinid tissues is not threatened. The problem of how the sulfide becomes unbound in the trophosome was not studied.

Another way to detoxify sulfides is to oxidize them. Such oxidation by the siboglinid would make the molecules far less useful as an energy source for the bacteria, but it would protect the siboglinid tissues. Powell and Somero homogenized various siboglinid tissues and used standard biochemical techniques to assess the ability of the purified homogenates to oxidize sulfur. The highest activity (per gram of animal tissue) was, of course, found in the trophosome, reflecting sulfide oxidation by the dense bacterial aggregations in that tissue. However, respectable

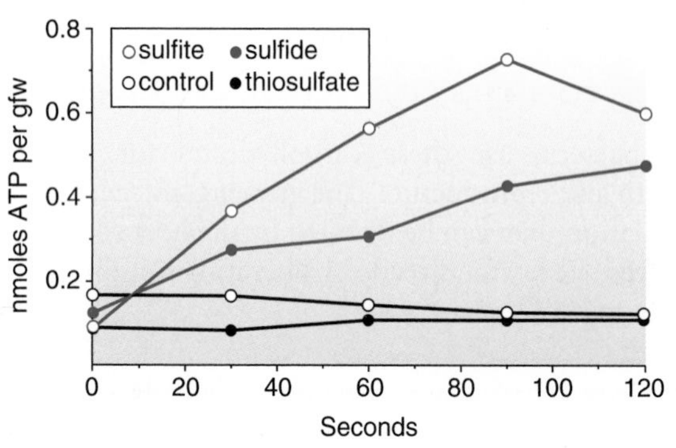

Figure 13.17

The influence of sulfides and related compounds on ATP production by homogenized, bacteria-laden trophosome tissues obtained from the hot-vent siboglinid *Riftia pachyptila*. Each point represents the mean result from 2 replicates. ATP concentration was determined by adding known quantities of luciferin and firefly luciferase, which react to produce light only if ATP is present, and measuring the intensity of the light produced. The experiment was conducted at 20°C. Results are expressed as nano-moles (10^{-9} moles) of ATP produced per gram fresh weight (gfw) of trophosome tissue. Both sulfide and sulfite additions dramatically increased ATP production above control levels within only 1 or 2 minutes. The ATP would presumably then be used by the sulfur-oxidizing bacteria to fix carbon from CO_2 into carbohydrates.

Based on M. A. Powell and G. N. Somero, "Adaptations to sulfide by hydrothermal vent animals: sites and mechanisms of detoxification and metabolism," in *Biological Bulletin*, 171:274–90, 1986.

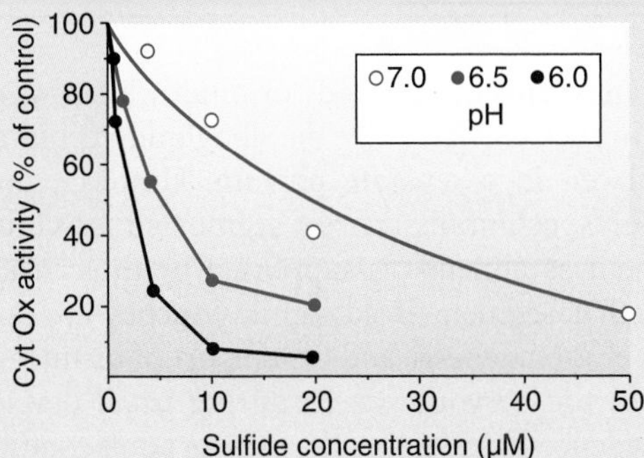

Focus Figure 13.1

The influence of sulfide on cytochrome-c oxidase activity in tissues of the hot-vent siboglinid *Riftia pachyptila*. The assays were carried out at 20°C, using tentacular plume samples. Cytochrome-c oxidase activity was determined using a spectrophotometer to monitor the rate of disappearance of substrate (reduced cytochrome c) from the incubation medium. Even low sulfide levels substantially inhibited cytochrome-c oxidase activity, particularly under slightly acidic conditions.

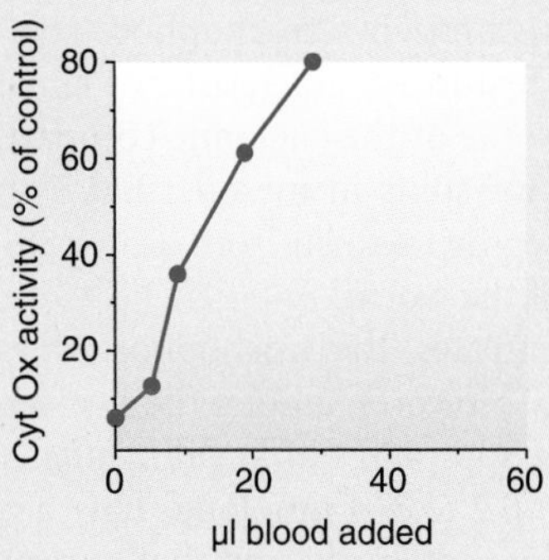

Focus Figure 13.2

The ability of siboglinid blood to protect mitochondrial cytochrome-c oxidase activity from sulfide poisoning in plume tissue of the vent species *Riftia pachyptila*. The experiment was conducted as in Focus Figure 13.1, except that different amounts of blood were added to the reaction mixture as shown. Without blood (0 μl [microliters] blood added), sulfides suppressed cytochrome-c oxidase activity dramatically, to less than 10% of its activity in the absence of sulfides. This sulfide inhibition is clearly removed by some component of siboglinid blood, now known to be the extracellular hemoglobin itself. Adding as little as 20 μl of siboglinid blood substantially increased the level of cytochrome-c oxidase activity despite the presence of sulfides.

Focus Figures 13.1 and 13.2 based on M. A. Powell and G. N. Somero, "Adaptations to sulfide by hydrothermal vent animals: sites and mechanisms of detoxification and metabolism," in *Biological Bulletin*, 171:274–90, 1986.

levels of sulfide oxidase activity were also found in the siboglinid body wall, suggesting that enzymes in the outer body wall oxidize sulfide as it diffuses inward across the general body surface and thereby protect body wall tissues from sulfide poisoning. The lowest sulfide oxidase concentrations were found in plume tissue; this finding is consistent with the plume → circulatory system → trophosome model of how the bacterial symbionts gain access to unoxidized sulfide. But how are the respiratory activities of the plume protected, since the plume tissues do not seem to detoxify sulfides (at least not by oxidation)? And do the siboglinid body wall tissues obtain energy from sulfide oxidation? These are among the many questions that have been investigated since this paper was published.

In either case, these deep-sea siboglinids are participating in a novel food chain, one that does not depend on light as an energy source. This food chain begins with **chemosynthesis** rather than photosynthesis: The energy source for fixing carbon dioxide into organic molecules is chemical rather than light. Similar chemosynthetic bacteria have now been found in the tissues of the mud-dwelling frenulate siboglinids, as well as in the tissues of several mollusc species and at least one polychaete annelid species. A symbiotic, chemotrophic mode of nutrition is turning out to be common among animals living in water and sediments high in hydrogen sulfide or methane, regardless of depth or ambient temperature; how the animals themselves withstand exposure to such high concentrations of these toxic chemicals is still being studied (Research Focus Box 13.1).

Methane-oxidizing bacteria also are abundant and active on the exterior surface of vestimentiferan tubes, possibly serving as a major food source for vent gastropods and other grazing invertebrates.

Siboglinid Reproduction and Development

Most siboglinids are **gonochoristic,** with male and female gonads located in the trunks of separate individuals. Details of fertilization and early development are known from only a few species, and even then knowledge is limited. In 1997, eggs from 2 vestimentiferan species were fertilized in the laboratory for the first time, with sperm removed artificially from other individuals. The embryos showed clear-cut spiral cleavage, formed polar lobes (Chapter 2), and developed into trochophore larvae, as in other polychaetes. Trochophore-like ciliation patterns had previously been reported from frenulates found developing within parental tubes. So, all siboglinids studied to date pass through a trochophore stage. They also display 2 other features characteristic of annelid development: At least some of the coelomic compartments form by schizocoely (as splits in mesodermal tissue), and the opisthosomal segments form from a discrete zone at the posterior end of the worm.

Among frenulates, the trochophore does not ingest particulate food. However, among the hot-vent vestimentiferans *Ridgeia* spp. and *Riftia pachyptila,* the smallest juveniles (about 0.2 to 3–4 mm long) have a conspicuous, ventral appendage anteriorly, and that appendage bears a terminal mouth. The mouth leads into a ciliated gut that terminates in an anus; that is, the early vestimentiferan juvenile has a complete digestive tract! The gutless condition of the older juveniles and adults thus reflects a secondary degradation and loss during later development, supporting the contention that siboglinids have developed from feeding polychaete ancestors. One developmental observation that does not fit this scenario is a report that mesoderm forms in frenulate siboglinids through enterocoely, a deuterostome trait that is otherwise unknown among protostomes. More embryological studies are needed to more clearly define the development of coelomic compartments and the origins of mesodermal tissue.

Although free-living siboglinid larvae have never been collected in plankton samples, current thinking suggests that at least the vestimentiferans release such larvae, enabling each generation to disperse to new vents and seeps. Consistent with the idea of larval dispersal, colonization of new sites is fairly rapid: One desolate area visited by submarine in 1991 supported a thriving colony of one vestimentiferan species 11 months later and a thriving colony of a second species (*R. pachyptila*) when revisited 21 months after that. Moreover, to reach the size observed during intervening periods, the worms must have elongated their tubes at remarkably rapid rates: at least 30 $cm \cdot yr^{-1}$ (centimeters per year) for the first species and 85 $cm \cdot yr^{-1}$ for the second. At least some of the frenulate siboglinids probably also release larvae—the larvae of one species have been kept swimming in laboratory experiments for several days in the absence of sediment—but the data are insufficient for any generalizations.

The Echiurans

Echiura
(*echis* = G: serpent-like)
ek-ē-your′-ah

Defining Characteristic: Muscular organs (anal sacs) outpocketing from the rectum into the coelomic space, bearing numerous funnels that discharge coelomic fluid (and wastes?) through the anus

Echiuran worms live in sandy or muddy burrows or, more rarely, in rock crevices. Like the siboglinids, echiurans were long placed in a separate phylum. However, unlike the siboglinids, echiurans are not segmented as adults. Their placement as annelids is supported by molecular data, a report that segmented coelomic pouches appear briefly during embryogenesis, and by a report of distinct segmentation in the nervous system during larval development. This evidence suggests that the present adult condition represents a secondary loss of segmentation from a segmented ancestor. This is not a particularly far-fetched possibility, since nearly all leech species show just such a loss of septa.

About 160 echiuran species have been described, mostly from shallow water. The length of the cylindrical, sausage-like body varies considerably among species, from several millimeters to more than 8 cm. Perhaps the echiurans' most conspicuous feature is an anterior cephalic projection commonly called the **proboscis** (Fig. 13.18), which contains the brain. The proboscis is muscular and quite mobile—in some species, it can be extended more than 25 times the length of the body proper (200 cm in an 8-cm individual).

The proboscis is ciliated on the ventral surface and serves as the organ of food collection in all echiuran species. Although the length of the proboscis can be altered greatly by its owner, the proboscis can never be retracted into the body. The edges of the proboscis curl to form a gutter; mud and detritus, trapped by mucous secretions of the proboscis, are moved by cilia posteriorly toward the mouth, which is located at the base of the proboscis near the junction of the proboscis and the body proper (the **trunk**). The digestive tract is contained within the trunk and is very long and convoluted, with the anus opening posteriorly.

Although most echiurans are deposit feeders, all members of the genus *Urechis* (Fig. 13.19) are suspension feeders, straining out food particles as water passes through a mucous net that the animals secrete across the lumen of their burrows. Water flow through the U-shaped burrow characteristic of this genus is maintained by peristaltic waves of muscular contraction moving down the body (Fig. 13.20).[10] The adjoining sediments are often rich in hydrogen sulfide, a metabolic poison. The coelomic fluid of *U. caupo* has recently been found to contain millimolar (mM) concentrations of an iron-containing compound that oxidizes this sulfide, thereby detoxifying it.

10. See *Topics for Further Discussion and Investigation*, no. 20.

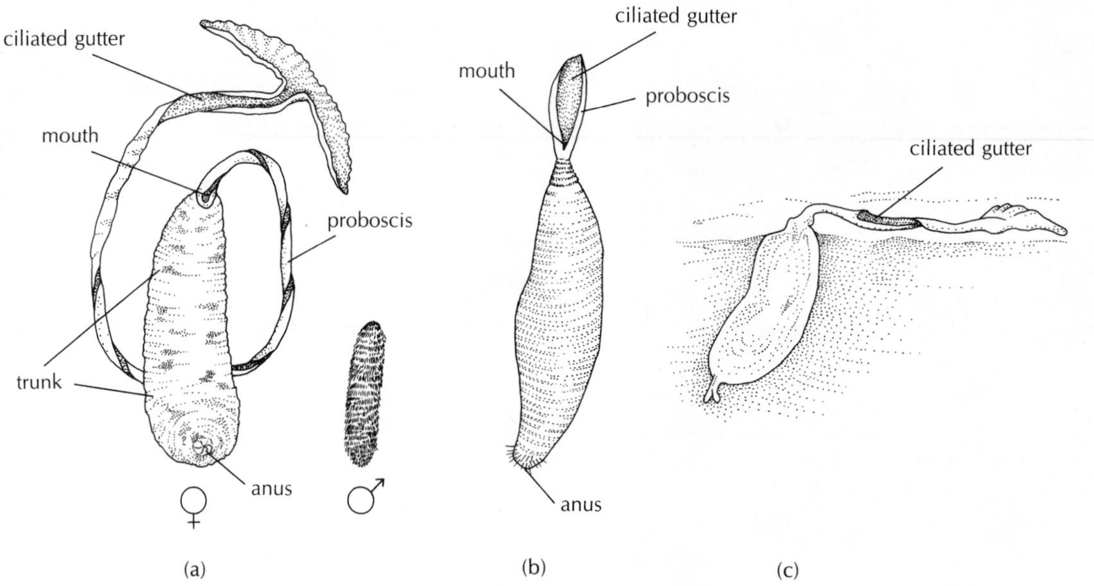

Figure 13.18

Representative echiurans. (a) *Bonellia viridis* (female). The trunk is about 2.5 cm in length, and the proboscis may exceed 1 m when fully extended. Note the ciliated gutter of the proboscis, leading to the mouth. The nondescript male has a total body length of only a few millimeters. (b) *Echiurus echiurus,* in ventral view. Males and females are similar in shape. The trunk length may reach 15 cm, while the proboscis extends about 2–4 cm. Both species live in soft sediments, in burrows. (c) Typical echiuran in feeding position.

(b) After Harmer and Shipley. After Zenkevitch; modified from Grasse, *Traité de Zoologie,* vol. 5.

Aside from the **papillae** (small external bumps) found in some species, the only conspicuous projections from the echiuran trunk are a single pair of large, chitinous, annelid-like setae located just posterior to the proboscis. Some echiuran species bear additional setae in the form of a ring just anterior to the anus. The setae probably aid in burrowing.

The trunk contains a single, uninterrupted, large coelomic space that extends into the proboscis (Fig. 13.21). There is no metameric segmentation of the coelomic space, nor of the ventral nerve cord that runs through it. In addition to containing the digestive system, the coelomic space houses from one to many hundreds of metanephridia, as found in other annelids; the members of most species possess 1 to 5 pairs of nephridia. Coelomic fluid is presumably drawn into each nephrostome by ciliary action, and a final urine is excreted through nephridiopores opening to the surrounding seawater.

The one additional organ within the echiuran coelom is also believed to serve an excretory role. A pair of **anal sacs** outpocket from the rectum (Fig. 13.21). The sacs are muscular, and their surface is dotted with many thousands of ciliated funnels resembling nephrostomes. Like metanephridia, these funnels collect coelomic fluid, but there is no indication that the collected fluid is then modified by either secretion or resorption. Moreover, the fluid is discharged to the outside through the anus, rather than through a nephridiopore. Discharge is accomplished by periodic muscular contractions of the sacs; one-way valves at the base of the funnels prevent fluid from moving back into the coelom.

Echiurans lack specialized respiratory organs; diffusion across the general body surface apparently satisfies requirements for gas exchange in most species.[11]

Similarly, echiurans lack distinct gonads. Instead, gametes are produced by the peritoneal lining of the coelom and are released into the coelomic cavity as in many other polychaetes. The gametes of echiurans leave the body by passing through the metanephridia, exiting at the nephridiopores. Gamete formation and release are by essentially identical methods among sipunculans.

Echiuran sex lives appear to be rather unexciting, with only a few exceptions (as discussed in the next paragraph). Gametes are always liberated through nephridiopores into the surrounding seawater, where fertilization occurs. Development is typically protostomous, culminating in the production of a **trochophore larva** (Fig. 13.22), as among the polychaetes and siboglinids.

All echiuran species are **gonochoristic (dioecious)** (one sex per individual), and males and females are generally similar in appearance. However, *Bonellia viridis* and a few close relatives are sexually **dimorphic** (characterized by 2 body forms) (Fig. 13.18a). Compared with the females, the males of *B. viridis* are much smaller, lack a proboscis and a specialized circulatory system, and

11. See *Topics for Further Discussion and Investigation,* no. 21.

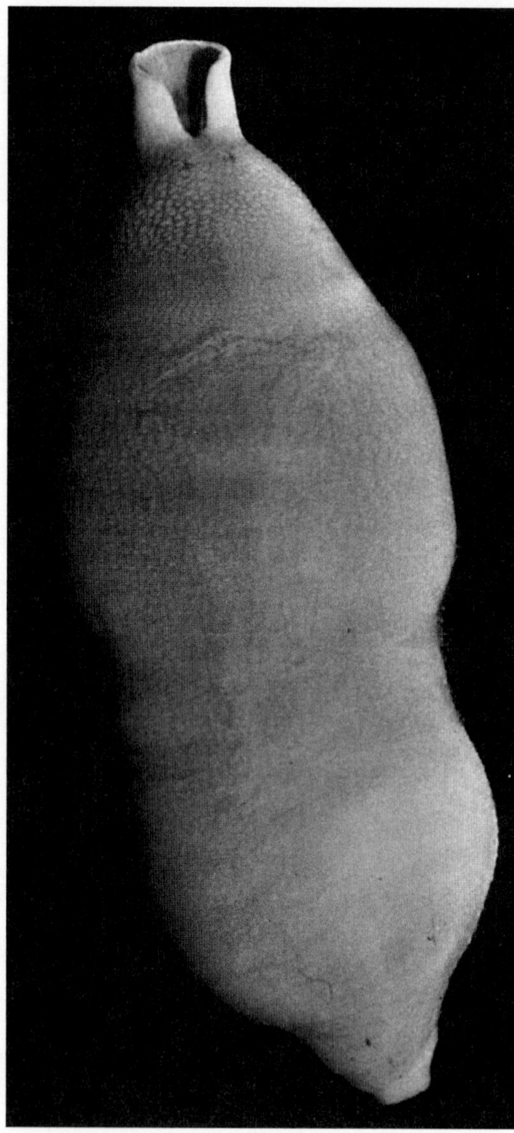

Figure 13.19
Photograph of *Urechis caupo* (about 8 cm long), removed from its U-shaped burrow. The conspicuous constriction of the body wall represents a peristaltic contraction. These waves of muscle contraction drive water through the burrow for food collection, gas exchange, and waste removal, and circulate the blood, which lies free in the body cavity; blood vessels are lacking in this species. To facilitate gas exchange, water is periodically taken into the thin-walled intestine through the anus and expelled through this same opening.
Courtesy of C. Bradford Calloway and Brent D. Opell.

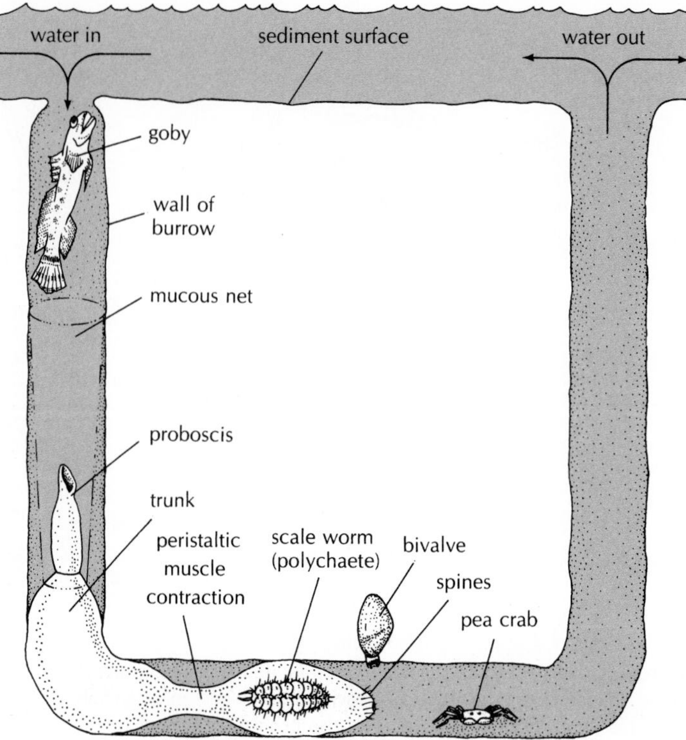

Figure 13.20
Urechis caupo, the innkeeper worm, driving water through its U-shaped burrow by waves of muscle contraction (peristaltic waves). The adults reach approximately 50 cm in length. The burrows of *U. caupo* are commonly shared with various symbionts, most notably a polychaete worm (*Hesperonoe adventor*), several species of pea crab (especially *Pinnixa franciscana* and *P. schmitti*), a small bivalve (*Cryptomya californica*), and the goby fish (*Clevelandia ios*).
From Pimentel; after Fisher and MacGinitie.

possess a degenerate digestive tract. These males are not free-living; rather, they dwell within the body of the female—often in the nephridia, which take on the function of uteri. A single female commonly has 20 males living within her body at any one time, awaiting their opportunity to fertilize eggs. Perhaps the most bizarre aspect of this peculiar life history is that the sex of most embryos is not determined at fertilization. Instead, for the majority of individuals, maleness is induced only following contact of the larva with the proboscis of an adult female. In the absence of such contact, the larva generally becomes a female.[12]

The Sipunculans

The Sipuncula
(*siphunculus* = G: little tube)
sī-puhn´-kū-lah

Defining Characteristics: 1) Anterior part of body forms an eversible and fully retractable introvert, with the mouth at its end; 2) multicellular bodies (urns) in the coelomic fluid, specialized for accumulating particulate wastes; 3) anterior tentacles connected to a series of muscular sacs (compensatory sacs) that pump fluid into the tentacles and store fluid when the tentacles retract

12. See *Topics for Further Discussion and Interpretation,* no. 18.

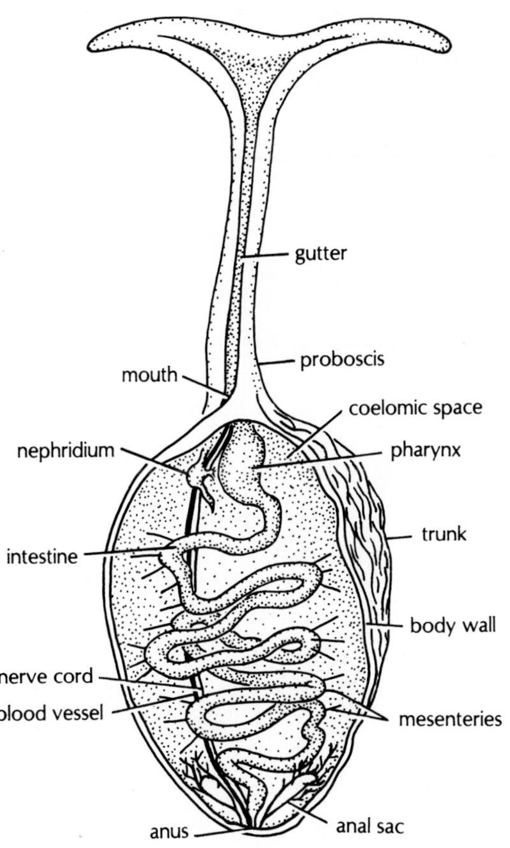

Figure 13.21
Internal anatomy of the echiuran *Bonellia viridis*. Note the capacious coelomic cavity.
After Harmer and Shipley.

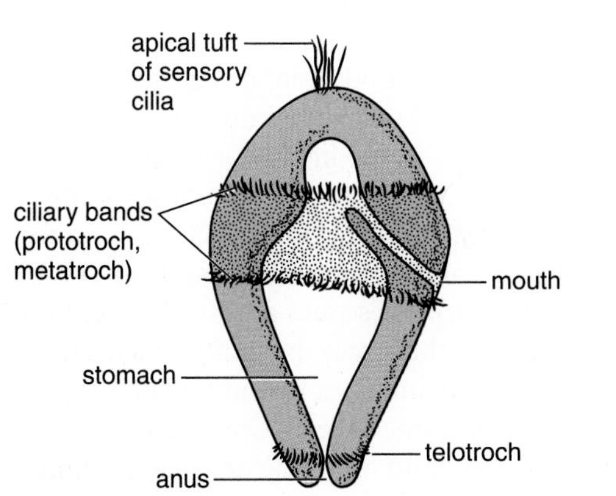

Figure 13.22
A typical echiuran trochophore larva with 3 conspicuous bands of cilia. Compare with Figure 13.11a.
Based on Gosta Jägersten, *Evolution of the Metazoan Life Cycle*. 1972.

Like the other worms discussed in this chapter, sipunculans have a body wall composed, in sequence, of an outer cuticle, an epidermis, a layer of circular muscles, a layer of longitudinal muscles, and a peritoneum lining the coelom (Fig. 13.23a); show spiral cleavage, mesoderm formation from the 4d cell (Chapter 2), and schizocoelous coelom formation (Chapter 2); and generally produce ciliated trochophore larvae (Table 13.1). However, sipunculans lack setae and show no trace of segmentation at any point in development. Some developmental details suggest a close relationship with molluscs, but recent gene sequence studies suggest clear annelid affinities. Some molecular studies suggest that sipunculans are in fact polychaetes.[13] If sipunculans really are polychaete annelids, then they have apparently lost all traces of segmentation and parapodia from an ancestor that possessed those features. There is still uncertainty here, and it may turn out, for example, that sipunculans are not annelids but rather the sister group to annelids. One piece of morphological evidence supporting a linkage with polychaetes, however, is that the introvert of some sipunculan species bears a pair of presumed chemoreceptors, called **nuchal organs,** resembling those found elsewhere only among acknowledged polychaetes.

Like most echiurans, sipunculans are all marine and found mostly in shallow waters, although some species in both phyla are found at depths of 7,000 m to 10,000 m. Most species of both phyla are deposit-detritus feeders and have separate sexes. Neither echiurans nor sipunculans secrete tubes, but both may form burrows. Sipunculans are commonly found in burrows in shallow-water muddy or sandy sediments[14] and in empty mollusc shells or polychaete tubes. One very small species is known to inhabit the calcareous shells built by foraminiferans (protozoa). Some other species are found in rock crevices, and a few species can bore into the calcareous substrates of coral reefs. There are about 2.5 times as many sipunculan species as there are echiuran species. Most of the 350 sipunculan species are only a few millimeters long, although some species attain lengths in excess of 1 m.

The sipunculan body consists of a plump, unsegmented trunk and an anterior **introvert.** Unlike the echiuran proboscis, the sipunculan introvert is fully retractable into the body (*intro* = L: within; *verta* = L: turn; i.e., "turn within"), and the mouth opens at the end of the introvert (Fig. 13.24). Contracting the body wall musculature causes the introvert to be thrust out of the body (Fig. 13.24b), while contracting well-developed introvert retractor muscles draws the anterior end of the animal back inside the trunk (Fig. 13.24a). The introvert commonly bears numerous mucus-covered tentacles

13. For example Hall, K. A., P. A. Hutchings, and D. J. Colgan. 2004. *J. Mar. Biol. Ass. U.K.* 84: 949–60; Struck, T. H., et al., 2007. *BMC Evol. Biol.* 7: article 57.

14. See *Topics for Further Discussion and Investigation,* no. 19.

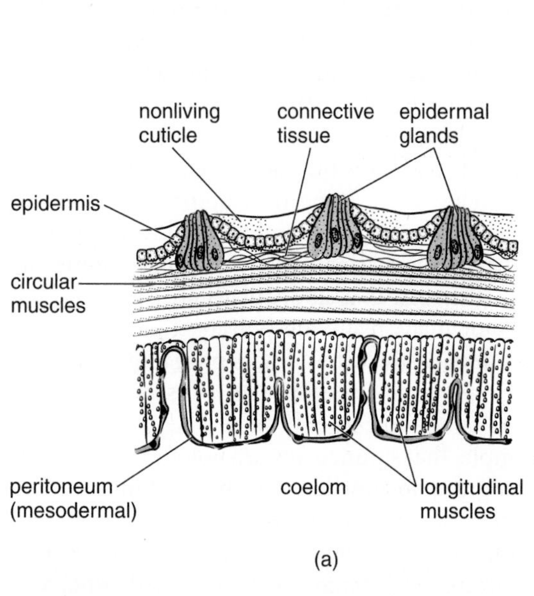

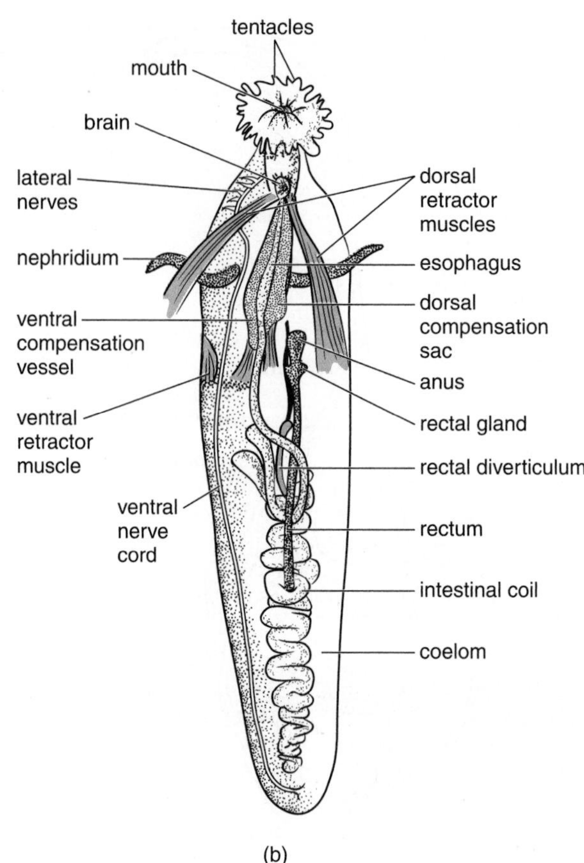

Figure 13.23

(a) Cross section through the body wall of the sipunculan *Phascolosoma gouldi*, illustrating the typical wormlike arrangement of circular and longitudinal musculature. (b) Internal anatomy of *Sipunculus nudus*. Note that the anus terminates anteriorly rather than posteriorly; that is, the gut is U-shaped. Also note the spacious coelomic cavity.

(a) Based on Frank Brown, *Selected Invertebrate Types*. (b) Based on *Invertebrate Zoology*, 3rd ed., by Paul A. Meglitsch.

surrounding or adjacent to the terminal mouth. These tentacles trap particles from the surrounding water or are pressed into the substrate, trapping mud and detritus. The entire introvert may then be withdrawn into the body and the captured particulate material ingested, or the particles may be moved to the mouth by ciliary tracts on the tentacles. Thus, most sipunculans, like echiurans, are deposit feeders. During burrow formation, sipunculans may ingest mud directly.

Unlike the typical annelid digestive tract, that of sipunculans is U-shaped, with the anus opening at about the midpoint of the trunck (Fig. 13.23b). This would seem to be an adaptation for a sedentary existence in burrows having only a single opening to the outside; solid wastes are discharged near the mouth of the burrow.

In addition to the nuchal organs mentioned earlier, the sipunculan introvert bears surface sensory cells. Sipunculans may also have several pairs of light-sensitive ocelli within the brain. Other than the tentacles borne on the introvert, the sipunculan body lacks substantial projections; it bears neither setae nor body appendages (Fig. 13.24).

The sipunculan body contains an enormous coelomic cavity (Fig. 13.23b) that shows no sign of segmentation or septa at any stage in the life history. Surprisingly, the hollow tentacles of the introvert are not associated with the coelomic cavity. Instead, the tentacles are connected to a series of sacs, called **contractile vessels** or **compensatory sacs** (Fig. 13.23b). These sacs are attached to the surface of the esophagus; they may be quite extensive, ramifying throughout the trunk coelom and forming a type of circulatory system in some species. Contraction of the sac musculature drives fluid into the tentacles, bringing about their extension. Muscular retraction of the tentacles drives fluid back into the sacs. A similar system works in the locomotion of sea urchins and sea stars and in operating the tentacles of sea cucumbers (phylum Echinodermata), as will be discussed in Chapter 21.

Table 13.1 Comparison of Features Encountered among Polychaetes, Clitellates, Echiurans, and Sipunculans.

Characteristic	Polychaeta and Clitellata	Echiurans	Sipunculans
Body wall musculature	Outer: circular; inner: longitudinal	Outer: circular; inner: longitudinal	Outer: circular; inner: longitudinal
Excretory system	Metanephridia (one to many pairs)	Metanephridia (one to many pairs) plus anal sacs	Metanephridia (one pair) plus urns
Nervous system	Brain with ventral nerve cord	Brain with ventral nerve cord	Brain with ventral nerve cord
Anteriormost feature	Prostomium (cephalic lobe in siboglinids)	Proboscis (not retractable)	Introvert (fully retractable)
Blood circulatory system	Dorsal and ventral blood vessels	Dorsal and ventral blood vessels	None
Oxygen-binding pigments	Hemoglobin, hemerythrin, chlorocruorin	Hemoglobin	Hemerythrin
Setae	Present	Present	None
Metamerism	Present throughout life	Only during development	None
Coelom formation	Schizocoely	Schizocoely	Schizocoely
Cleavage pattern	Spiral, determinate	Spiral, determinate	Spiral, determinate
Larval form	Trochophore	Trochophore	Trochophore, pelagosphera
Gamete production	May arise from peritoneum and mature in coelom; may exit through nephridiopores	Arise from peritoneum; mature in coelom; exit through nephridiopores	Arise from peritoneum; mature in coelom; exit through nephridiopores
Digestive tract	Linear (i.e., mouth and anus at opposite ends of body)	Linear	U-shaped

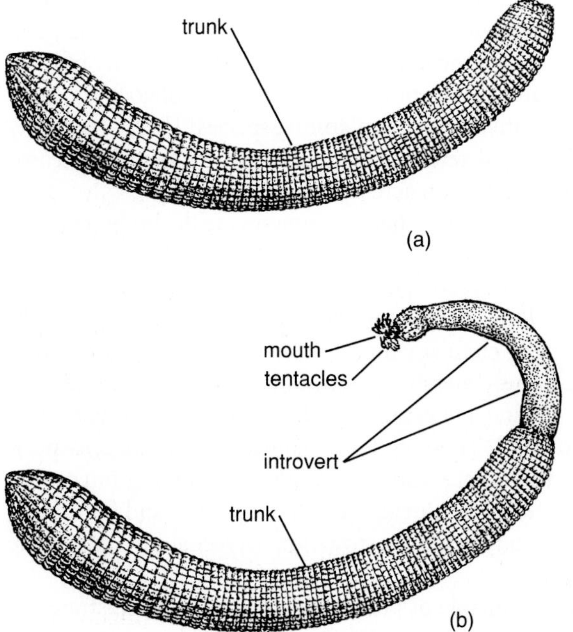

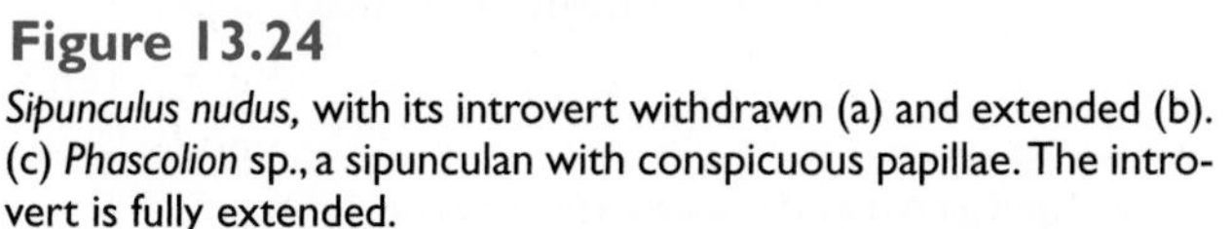

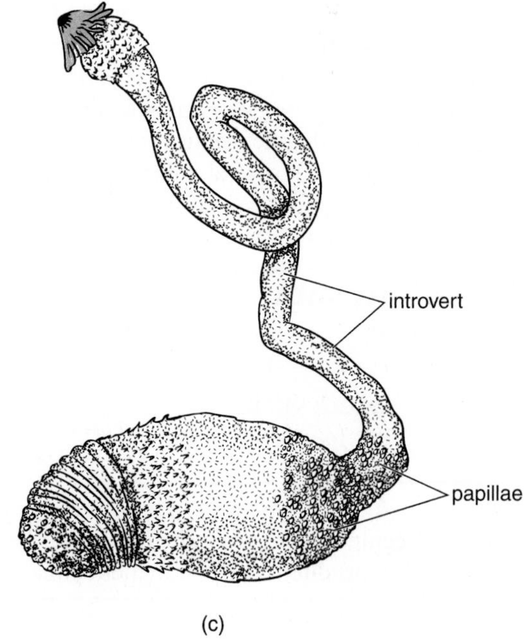

Figure 13.24

Sipunculus nudus, with its introvert withdrawn (a) and extended (b). (c) *Phascolion* sp., a sipunculan with conspicuous papillae. The introvert is fully extended.

(c) Based on R. I. Smith, *Key to Marine Invertebrates of the Woods Hole Region*. 1964.

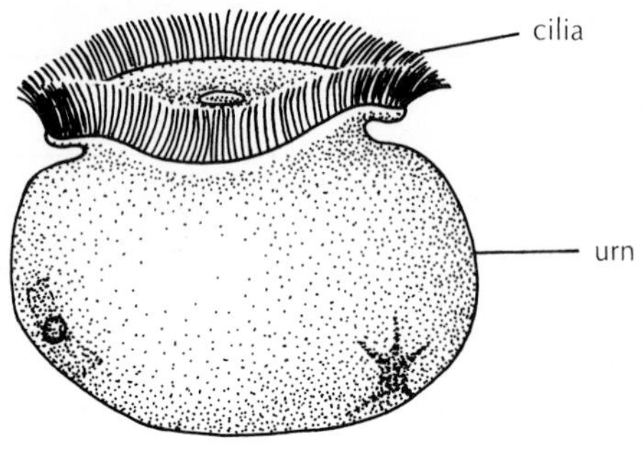

Figure 13.25
An urn cell recovered from the coelomic fluid of *Sipunculus nudus.*
From Hyman; after Selensky.

In contrast to other annelids, sipunculans lack a circulatory system with heart and blood vessels. However, cells in the coelomic fluid contain oxygen-binding pigment. Gas exchange takes place via the tentacles, introvert, or body wall, with coelomic fluid serving as a circulatory medium.

Excretion by sipunculans is accomplished largely by metanephridia, as in other annelids. However, most sipunculan species bear only a single pair of these organs (Fig. 13.23b). The nephridial system is supplemented by a peculiar system of **urns.** These urns are clusters of cells that arise and detach from the peritoneal lining of the coelom (Fig. 13.25). Floating freely in the coelomic fluid, urns collect solid wastes and eventually deposit these wastes in the body wall of the animal or exit the animal through the nephridial system.

Development of a trochophore larva, as also found among molluscs, has already been mentioned (Fig. 13.26a). In some sipunculans, the trochophore develops further to form a **pelagosphera** larva (Fig. 13.26b), which at one time was thought to be a free-swimming adult sipunculan and placed in the now-defunct genus *Pelagosphaera.* These larvae are typically large (several millimeters long) and long-lived: They can disperse over great distances.

Class Clitellata

Class Clitell • ata
(clitell = L: a pack saddle)

Defining Characteristics: 1) Pronounced cylindrical glandular region of the body (clitellum) that plays important roles in reproduction; 2) permanent gonads

This class includes 2 major groups of worms, the oligochaetes and the leeches, which are placed in separate subclasses. Both oligochaetes and leeches are **hermaphroditic** (both male and female reproductive apparatus are contained within a single individual) and demonstrate a specialized region of the epidermis known as the **clitellum** (Figs. 13.27a and 13.30). In both classes, the clitellum secretes a cocoon within which embryos develop. It also secretes a mucus that assists in transfering sperm between individuals, and produces albumen, which serves as a food resource for embryos as they develop within the cocoon.

More than 85% of clitellate species are oligochaetes.

Subclass Oligochaeta

Subclass Oligo • chaeta
(G: few setae)
ō-lig´-ō-kē´-tah

Approximately 3500 oligochaete species have been described. In contrast to the Polychaeta, only about 6.5% of oligochaete species are marine; most are found in freshwater or terrestrial habitats. The common earthworm, *Lumbricus terrestris,* is a familiar example of this subclass (Fig. 13.27a); this species has become widely used as a biomonitor of pollution stress, and has gained considerable attention for its potential role in transferring pollutants to birds and other terrestrial vertebrates[15] (see Research Focus Box 13.2). Earthworms are also well known for their essential roles in aerating and draining soil; indeed, Charles Darwin calculated that in each acre of land, earthworms bring about 18 tons of soil to the surface each year.

Oligochaetes are more streamlined in appearance than most polychaetes. In particular, parapodia are lacking and the anteriormost region of the body, the **prostomium,** lacks conspicuous sensory structures, such as eyes and tentacles. Oligochaetes do have setae, but the setae are less densely distributed along the body. In contrast to the diversity of body plans found among polychaete worms, the oligochaete body plan is relatively invariant among species. There are usually no specialized respiratory organs; gas exchange is accomplished by diffusion across a moist body wall.

As in most polychaetes, septa divide the body coelomic cavity into a series of semi-isolated compartments. The musculature is arranged as in the polychaetes, with an inner layer of longitudinal muscles overlain by a layer of well-developed circular muscles (Fig. 13.27b). To move, oligochaetes generate a continuous series of localized contractions and relaxations of the circular and longitudinal musculature. These cycles of contraction, known as **peristaltic waves,** are also commonly used among burrowing polychaetes, both during burrow formation and for driving water through the burrow for oxygenation and waste removal.

The principle of peristaltic locomotion is illustrated in Figure 13.28. In Figure 13.28a, the musculature is partially relaxed, and the worm is uniform in diameter. A portion of

15. See *Topics for Further Discussion and Investigation,* no. 14.

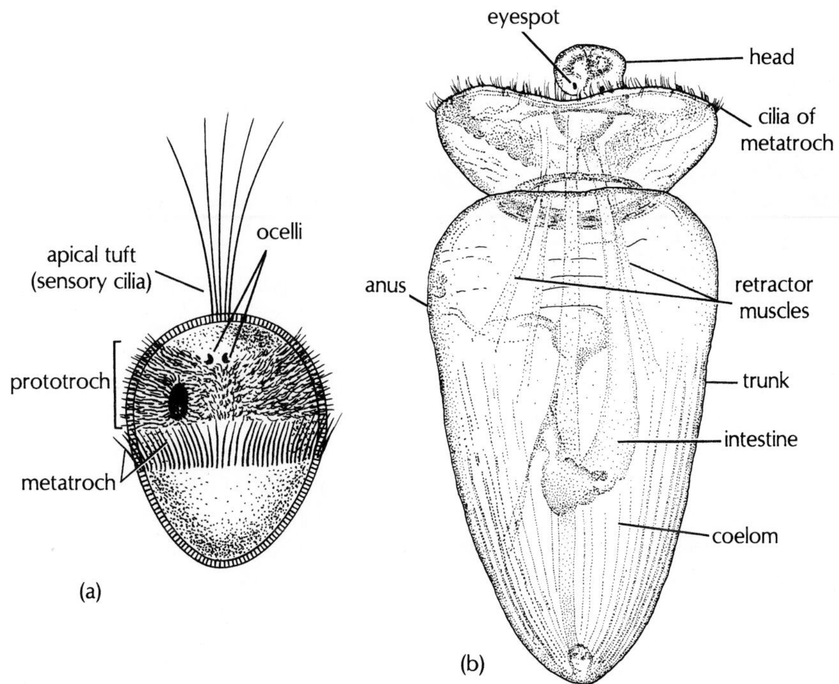

Figure 13.26

(a) Trochophore larva of *Golfingia* sp. Compare with Figures 13.11a and 13.30. (b) Pelagosphera larva of *Sipunculus polymyotus.* The trochophore larva is always a nonfeeding stage in sipunculans, whereas the pelagosphera larva may feed on microscopic algae in the plankton.

(a) From Hyman; after Gerould. (b) Courtesy Dr. Rudolph S. Scheltema.

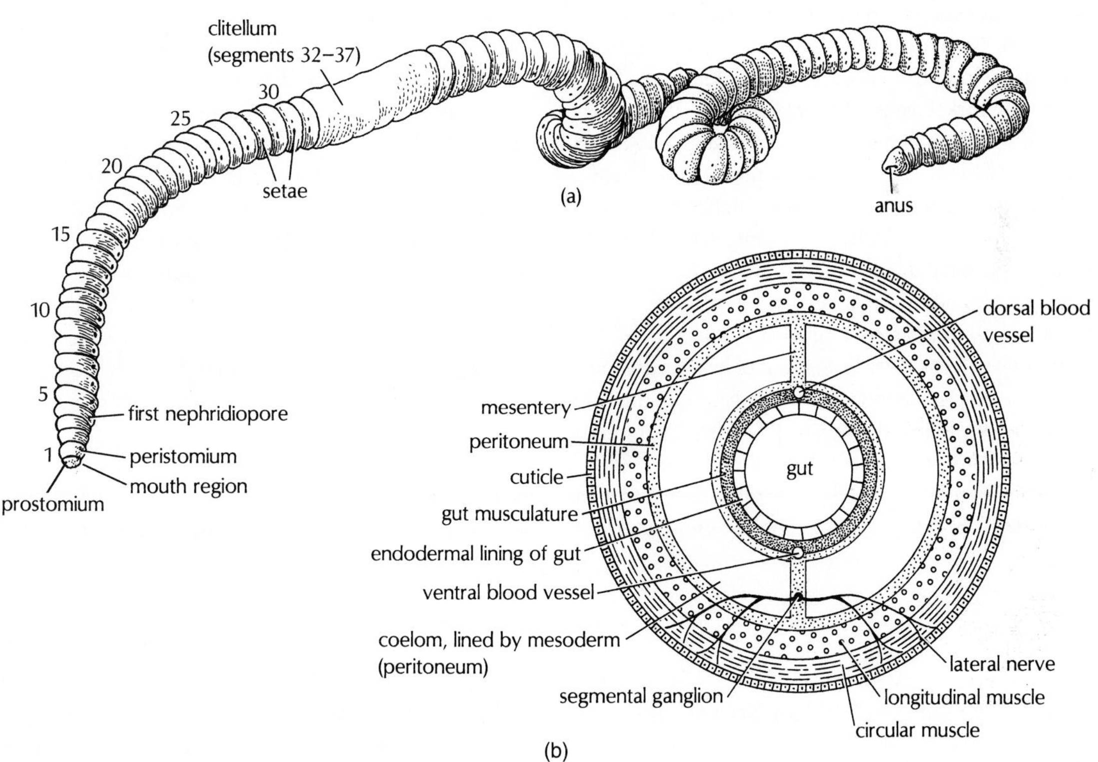

Figure 13.27

(a) External morphology of the terrestrial oligochaete *Lumbricus terrestris.* Note the absence of anterior and lateral appendages. The clitellum is a glandular region characteristic of both oligochaetes and leeches. It produces a mucus that aids copulation and encloses fertilized eggs in a protective cocoon. (b) Oligochaete in diagrammatic cross section, showing the arrangement of muscle layers of the body wall and gut.

(a) After Sherman and Sherman. (b) After Russell-Hunter.

RESEARCH FOCUS BOX 13.2

Unexpected Responses to Pollution

Klerks, P. L., and J. S. Levinton. 1989. Rapid evolution of metal resistance in a benthic oligochaete inhabiting a metal-polluted site. *Biol. Bull.* 176: 135–41.

As human populations continue to grow—some 6 billon people now occupy the planet, about twice the number present in 1960, and 20% more than were here only 17 years ago—our impact on the environment increases dramatically. One product of our increasing numbers and increasingly sophisticated activites is increased pollution, which has in turn created many new research opportunities for environmental researchers. One of their goals is to determine the concentrations at which particular pollutants become toxic to organisms. But what if populations can gradually become more tolerant to pollutants? After all, pollution, through its direct effects on survival and reproductive success, must exert strong selective pressure on populations, so that individuals in future generations might be able to withstand greater and greater degrees of environmental insult. If pollution tolerance can evolve, then short-term laboratory studies will not accurately predict long-term consequences.

Klerks and Levinton (1989) studied a population of oligochaetes from Foundry Cove, a freshwater bay on the Hudson River in New York. Over nearly 20 years, until 1979, a battery factory in the area discharged large quantities—about 53 tons!—of cadmium (Cd) and other heavy metals into the cove, resulting in concentrations of more than 10,000 μg Cd per gram of sediment (μg Cd/g); deposit-feeding animals should be particularly impacted by such conditions. Yet, the cove supports large populations of the deposit-feeding oligochaete *Limnodrilus hoffmeisteri* (family Tubificidae . . . see p. 336). Are members of this species unusually tolerant of heavy metal pollution? Or has the population of *L. hoffmeisteri* at this site evolved an unusually high tolerance, and if so, how many generations did it take to evolve such increased resistance?

Klerks and Levinton found that worms collected from a control site were dramatically intolerant of sediments from the polluted Foundry Cove site (Focus Fig. 13.3). While worms from both sites survived well in uncontaminated sediments during the 28-day study (Focus Fig. 13.3, left pair of bars), only worms taken from the contaminated control site survived a 28-day exposure to the contaminated sediments; no worms collected from the uncontaminated control site survived such exposure (Focus Fig. 13.3, right pair of bars). But these results don't necessarily mean that the greater resistance of Foundry Cove worms resulted from evolution; the greater resistance of Foundry Cove worms could instead reflect the effects of some biochemical detoxification mechanism that is called to action within a generation, as individual worms acclimate to the contaminated conditions. Therefore, the authors reared offspring from Foundry Cove adults to a second generation in clean water and uncontaminated sediment and then tested the tolerance of the offspring to the same pollutants. When exposed to concentrations of 20,000 and 30,000 μg Cd/g sediment (Focus Fig. 13.4), offspring of Foundry Cove worms survived about as well as their parents had (i.e., the mean survival did not differ significantly for the 2 groups), even though the offspring had never previously experienced conditions of elevated cadmium. In contrast, worms taken from the control site did not survive nearly as well at these concentrations (Focus Fig. 13.4, x-symbols). Cadmium at concentrations of at least 60,000 μg/g devastated all experimental populations (Focus Fig. 13.4), but some Foundry Cove worms were still alive at the end of the 28-day study while no worms from the control populations survived; again, Foundry Cove offspring did about as well as their parents.

Clearly, the increased tolerance to elevated Cd has a genetic basis, and has resulted from selection. In additional laboratory studies using offspring from control site adults,

the body wall in each segment is in contact with the substrate. In Figure 13.28b, the circular musculature of segment one has contracted. Since the adjacent segments are pushing laterally against the substrate, the anteriormost region of the body (the prostomium) moves forward, pushing off against segment 2 and other, more posterior segments. The setae are protruded in the segments behind segment 1, helping to prevent slippage. The circular muscles in segments 2, 3, and 4 then contract in succession, resulting in anterior extension of the body (and retraction of setae), as illustrated in Figure 13.28c. At about this time, a wave of contraction of the longitudinal musculature of each segment begins at the animal's anterior end (Fig. 13.28d), and this wave also passes from anterior to posterior.

As the peristaltic wave of circular muscle contraction passes along the body of the worm, the adjacent uncontracted segments serve as temporary anchors, so that the contraction of circular muscles generates a forward thrust. As the longitudinal musculature subsequently contracts, the anterior segments become fat and anchor the animal. The

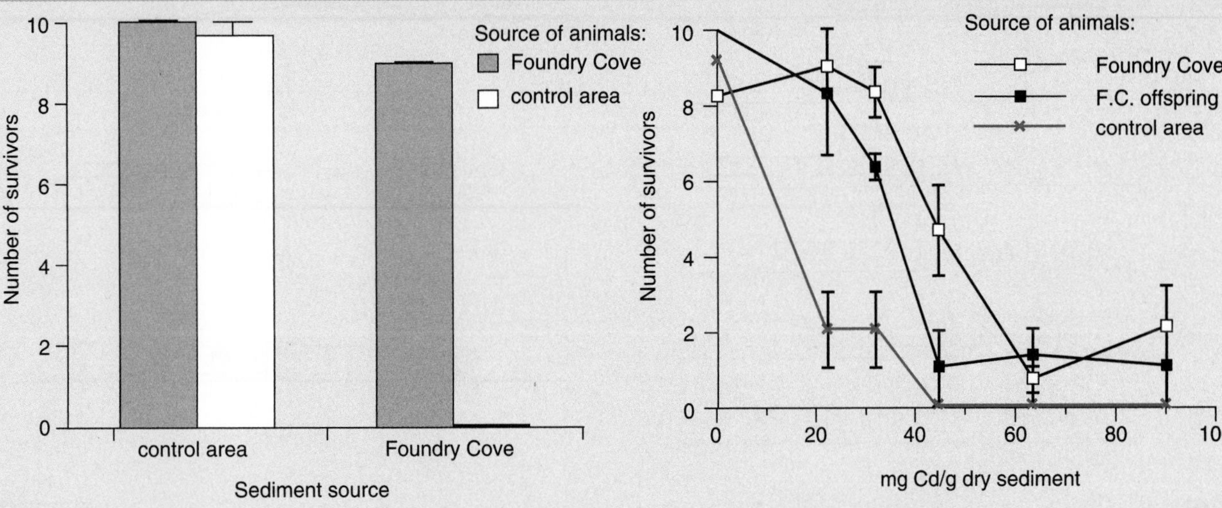

Focus Figure 13.3

Effect of cadmium exposure on the survival of worms (*Limnodrilus hoffmeisteri*) after 28 days of exposure to sediments (1 cm deep in glass beakers) from either the contaminated Foundry Cove site on the Hudson River, NY, or a relatively unpolluted control area about 2 km away. Cadmium concentration at the Foundry Cove site was about 7000 μg Cd g^{-1} sediment. The animals were fed on ground fishfood flakes during the study. Each bar represents the mean (+1 Standard Error of the mean) of 3 replicates with 10 worms per replicate.

Focus Figure 13.4

Effect of cadmium exposure on the survival of worms collected from the contaminated site (Foundry Cove, NY) and exposed for 28 days to sediment with the different cadmium concentrations indicated. Control worms came from a relatively unpolluted site about 2 km away. Solid symbols represent data for worms that were collected from the polluted site but then reared for 2 generations under unpolluted conditions in the laboratory before testing. Worms were fed ground fishflakes during the study. Each point represents the mean survival (+1 S.E.) of 3 replicates, 10 worms per replicate.

the authors found that worms with dramatically increased cadmium tolerance could be obtained after only 3 generations of selection. Subsequent work* showed that tolerant worms produce high concentrations of a metal-binding protein in response to high environmental Cd concentrations. The protein binds to the Cd; this detoxifies the metal but also results in high intracellular Cd concentrations.

*Wallace, W. G., G. R. Lopez, and J. S. Levinton. 1998. *Marine Ecol. Progr. Ser.* 172: 225–37.

Although it seems clear that populations of *L. hoffmeisteri* can evolve increased resistance to at least some pollutants, a few aspects of the data are perhaps puzzling: At 40,000 μg/g sediment, Foundry Cove offspring survived significantly less well than their parents (Focus Fig. 13.4). How might you explain that result, if tolerance is genetically determined? But the biggest puzzle concerns the implications of the data. Do you think that the results of this study are cause for rejoicing, or cause for alarm? Can you think of any negative consequences that might accompany increased pollution resistance by invertebrates?

adjacent thin segments can then be pulled forward. Note that the muscular waves pass in one direction, but the animal moves in the opposite direction (Figs. 13.28 and 13.29). Such waves are known as **retrograde,** as opposed to **direct.**[16]

Peristaltic waves are especially effective in burrow formation and in movement within a burrow or tube. When the longitudinal musculature is contracted, thickening the worm's body, the entire surface area of the segments becomes tightly pressed against the walls of the burrow, greatly increasing the magnitude of the pushing and pulling forces that can be generated without causing the worm to slip within the burrow. Protrusion of setae by the worm, at appropriate times in the cycles of muscular contraction and relaxation, aids in anchoring these portions of the worm and in preventing backward slippage when thrusting and pulling forces are generated (Fig. 13.29).

16. See *Topics for Further Discussion and Investigation,* no. 3.

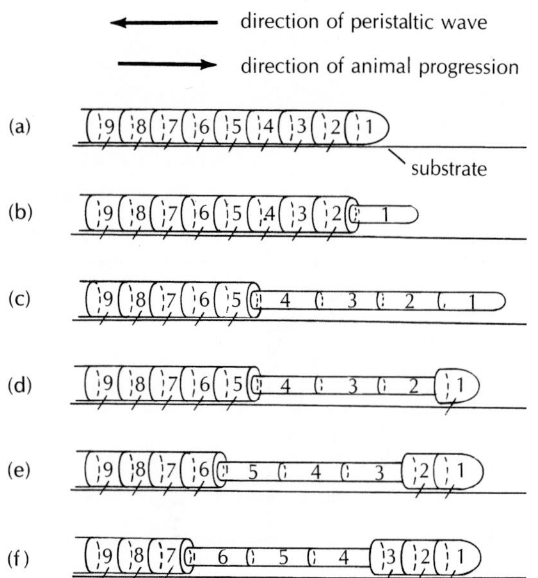

Figure 13.28
Diagrammatic illustration of peristaltic locomotion in oligochaetes. (a–c) A wave of contraction of the circular muscles begins passing down the body, resulting in forward extension of the worm. As the wave continues to travel down the worm's body, from segment to segment, a consolidation phase is initiated anteriorly; that is, a wave of longitudinal muscle contractions moves posteriorly (d–f), shortening and thickening each segment in preparation for the next extension phase.

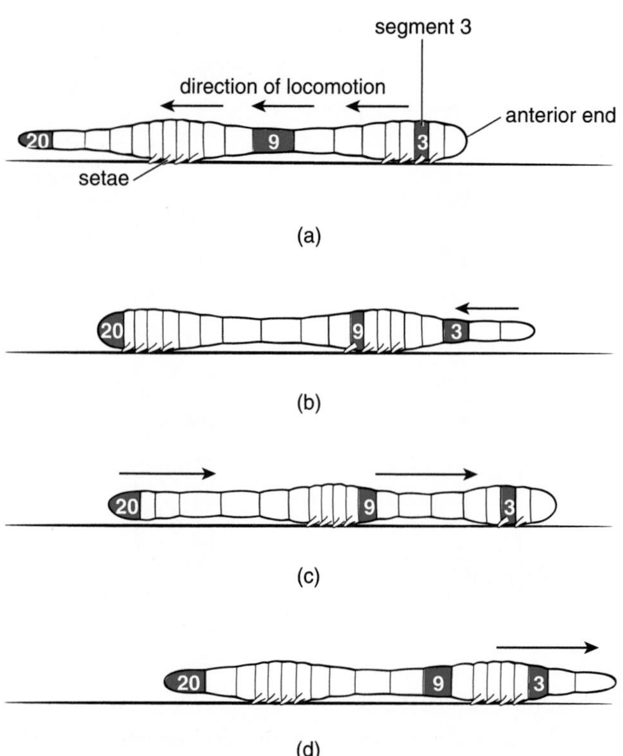

Figure 13.29
Locomotion of an earthworm, based upon movie footage. (a) The circular muscles of the first 5 to 6 anterior-most segments are relaxed, and the longitudinal muscles of these segments are fully contracted. In (b), a wave of circular muscle contraction has been initiated anteriorly; segment 3 and its immediate neighbors have become longer and thinner. Segment 3 has also moved forward, in part due to the contraction of its own circular musculature (and relaxation of its longitudinal musculature), and in part due to the forward thrust generated by the muscle contractions of segments 4 and 5. At the same time, the wave of longitudinal muscle contraction has passed posteriorly, encompassing segment 9. Note that segments 6–9 are prevented from backsliding by the protrusion of setae as the anterior segments are thrust forward. A second wave of longitudinal muscle contraction has passed from segment 15 to the terminal segments of the worm; consequently, segment 20 has been pulled forward. In (c) and (d), waves of longitudinal and then circular muscle contraction continue to be generated anteriorly, thrusting and pulling the segments of the worm forward.
Based on Purves and Orians, *Life: The Science of Biology*, 2d ed. 1983.

Oligochaete Reproduction

Unlike polychaetes, all oligochaete species are hermaphroditic, and generally only a few segments of each individual produce gametes. Moreover, gametes are produced within distinct testes and ovaries, rather than from the peritoneum lining the coelomic cavity. Sperm are generally exchanged simultaneously between 2 mating individuals (Fig. 13.30a); the sperm are stored for later use in specialized organs called **spermathecae** (G: sperm boxes). Eventually, eggs and sperm are extruded from separate openings into a complex cocoon secreted by the **clitellum,** so that fertilization occurs externally. The embryos develop in and feed upon a nutritive fluid found within the cocoon. When development is completed, miniature worms emerge from the cocoons; oligochaetes lack free-living larval stages, even in marine environments.

Although asexual reproduction does occur in some polychaete groups, it is more common among oligochaetes, particularly in freshwater species. The process of asexual reproduction involves the transverse (i.e., crosswise) division of the "adult" into a number of separate sections and the subsequent regeneration of each section into a complete individual. In addition, quite a few oligochaete species, especially those living in terrestrial environments, are **parthenogenetic;** that is, eggs may develop normally in the absence of fertilization.

Subclass Hirudinea

Subclass Hirudinea
(G: leech)
here-oo-din´-ē-ah

Defining Characteristic: Posterior sucker

The subclass Hirudinea, which includes the leeches, has approximately 500 to 630 described species, which are believed to have evolved from oligochaete stock. Reflecting

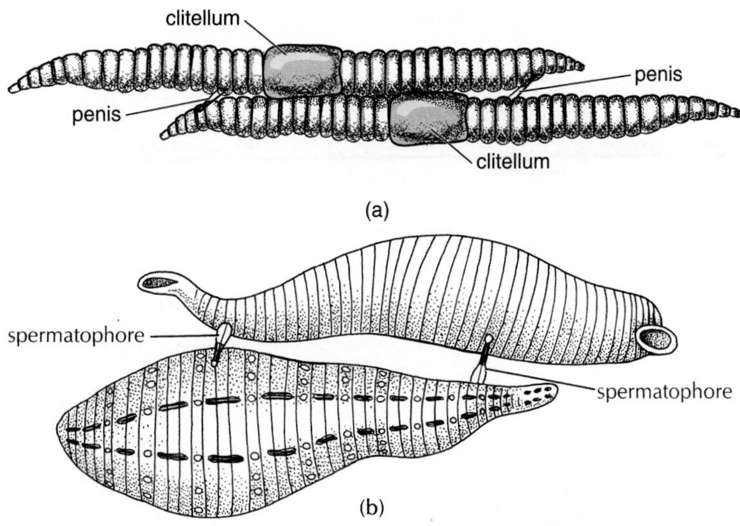

Figure 13.30

All clitellate annelids are simultaneous hermaphrodites. (a) Copulating oligochaetes, *Pheretima communissima.* (b) Copulating leeches, *Glossiphonia* sp. The members of this family generally lack a penis; spermatophores are forcefully injected into the partner, and the sperm migrate to the reproductive tract for storage and subsequent fertilization of eggs. Fertilization is external among oligochaetes, and internal among leeches. In both oligochaetes and leeches, the fertilized eggs are incubated in a gelatinous cocoon secreted by the clitellum.

(a) Based on Fretter and Graham, *A Functional Anatomy of Invertebrates.* 1976. (b) After Meglitsch; after Brumpt.

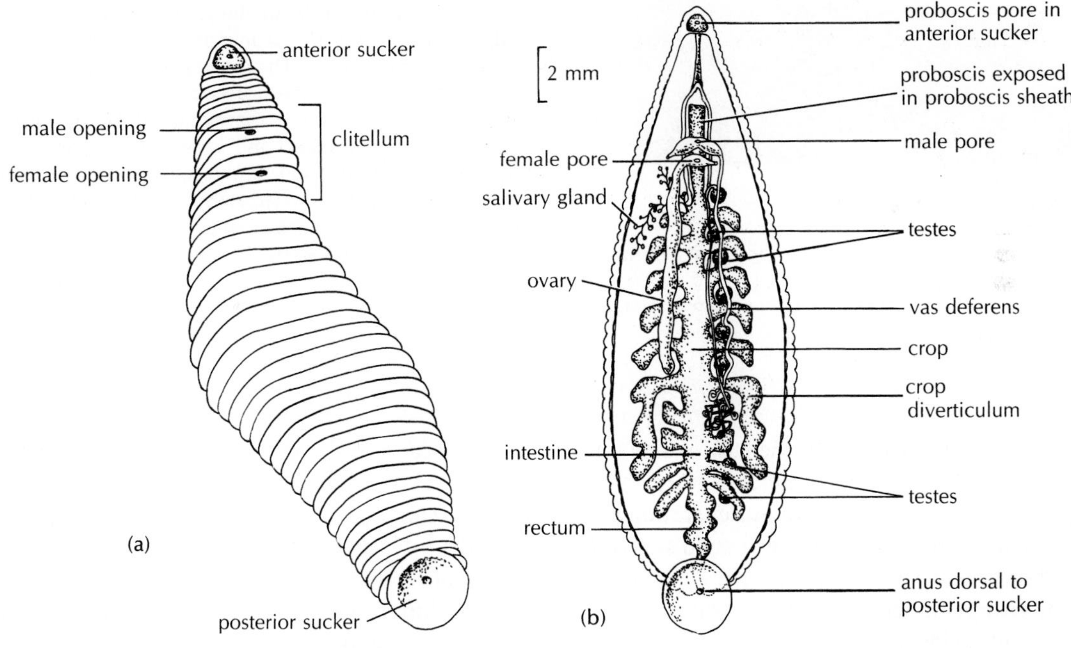

Figure 13.31

(a) External anatomy of a leech. (b) Internal anatomy of *Glossiphonia complanata.*

(a) After Pimentel. (b) Mann (after Harding), *Leeches.* Oxford, England: Pergamon Press, 1962.

their presumed oligochaete ancestry, most leeches occupy freshwater or terrestrial habitats; only a small proportion of species is marine. Like the oligochaetes, the basic body plan of the Hirudinea varies little among species. Conspicuously absent are parapodia, other specialized respiratory appendages, and head appendages, as in oligochaetes (Figs. 13.30b and 13.31).

In contrast to oligochaetes, most leeches lack setae. Moreover, the body is generally not separated into compartments by septa, and the continuous coelomic space is largely filled with connective tissue, or **mesenchyme** (Fig. 13.32). The remaining channels and sinuses within this tissue serve a blood transport function in most leech species; thus, the circulatory medium is the coelomic

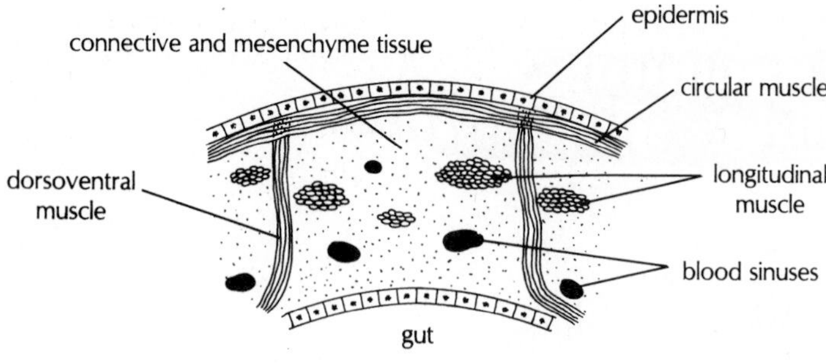

Figure 13.32
A leech in cross section.
After Mann.

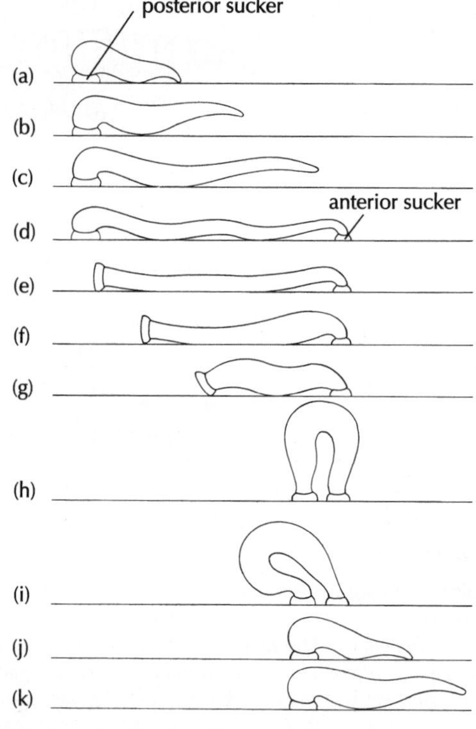

Figure 13.34
Leech locomotion. In (a–d), the posterior sucker is affixed to the substrate while the circular muscles of the leech contract, extending the animal forward. The longitudinal muscles are stretched in the process. In (e–g), the posterior sucker is released while the anterior sucker is applied. Contraction of the longitudinal musculature shortens the animal and reextends the circular muscles. In (h), selective contraction of the longitudinal musculature places the posterior sucker immediately behind the anterior sucker, and the entire sequence is repeated (i–k).
After Gray, Lissman, and Pumphrey.

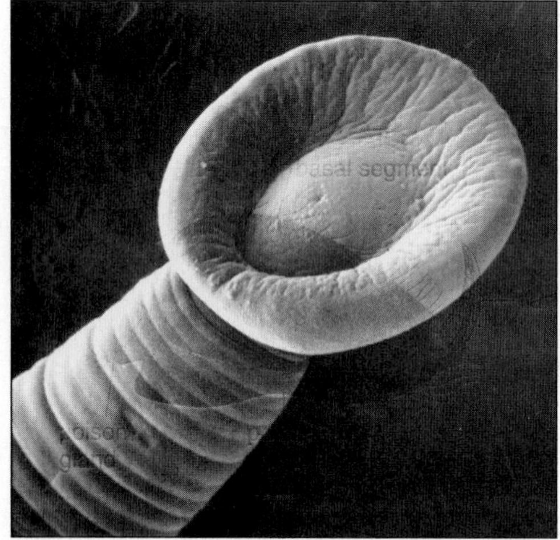

Figure 13.33
Scanning electron micrograph showing anterior end of a freshwater leech, *Johanssonia arctica*.
Courtesy of Dr. R.A. Khan. From Khan and Emerson, 1981. *Amer. Microsc. Soc. Trans.* 100:51.

fluid. Not surprisingly, considering the lack of septa, the locomotion of leeches differs considerably from that of other annelids. Leeches do not move by generating peristaltic waves; instead, they locomote over solid substrates by using suckers as temporary anchors. These suckers are formed from groups of segments at the body's anterior and posterior ends (Figs. 13.31 and 13.33). With the posterior sucker attached to the substrate, the worm extends itself forward by contracting the circular muscles of the body wall (Fig. 13.34a–c). The anterior sucker is then applied to the substrate, the circular muscles are relaxed, and the longitudinal muscles are contracted. The leech becomes shorter and, as the circular muscles are stretched, fatter (Fig. 13.34d–g). By further contracting the longitudinal musculature on the ventral surface, the leech arches its body, bringing the posterior and anterior suckers in close proximity (Fig. 13.34h). The posterior sucker is then applied to the substrate, the anterior sucker is released, and the cycle can be repeated. The mouth is included within the anterior sucker.

Leeches are generally ectoparasitic, feeding either on the blood of other invertebrates or, more commonly, on the blood of vertebrates. Most parasitic species possess 3 toothed jaws within the mouth; the leech uses these jaws to make an incision in the host. Other species have a protrusible proboscis, through which blood is removed from the host.

The blood-sucking habits of leeches have been long employed in the practice of medicine, for bloodletting through much of the nineteenth century and, more recently, to alleviate fluid pressure following damage to vascular tissue (e.g., following a snakebite or the reattachment of a severed finger or ear). When a finger is surgically reattached, the arteries supplying blood can be rejoined to those in the finger, but the veins, because of their smaller size, generally cannot. Thus, blood flows into the reattached finger but cannot escape. An application of

Hirudo medicinalis, the medicinal leech, helps in 2 ways: The leech (1) drains off the excess blood from the finger and (2) injects an anticoagulant—hirudin—that keeps the blood flowing out from the bite-wound long after the sated leech has dropped off. Eventually, the patient's venous system reestablishes itself in the reattached appendage, and the leech's services are no longer needed.

Leeches are also attracting the interest of modern drug companies as sources of such useful substances as local anesthetics (which enable leeches to bite unnoticed by the host) and certain antibiotics. Very recently, an Amazonian leech has been found to produce a protein that dissolves blood clots; the hirudin produced by *H. medicinalis* only prevents such clots from forming. If the genes coding for these substances can be isolated, commercial mass production of these substances should become possible. Finally, leeches are becoming leading research models in the study of nervous system development and function.

Less than 100 years ago, selling leeches to the medical establishment was a booming business, and perhaps it will become so again. As described by the late Reverend J. G. Wood (*Animal Creation,* 1885), the collection of leeches was a colorful enterprise:

> The Leech-gatherers take them in various ways. The simplest and most successful method is to wade into the water and pick off the leeches as fast as they settle on the bare legs. This plan, however, is by no means calculated to improve the health of the Leech-gatherer, who becomes thin, pale, and almost spectre-like, from the constant drain of blood, and seems to be a fit companion for the old worn-out horses and cattle that are occasionally driven into the Leech-ponds in order to feed these bloodthirsty annelids.

Today, leeches are reared commercially, particularly in England.

The nonparasitic leeches, about 25% of all known species, are predators upon other invertebrate animals.

Leech Reproduction

As in the Oligochaeta, adult leeches are simultaneous hermaphrodites and a free-living larval stage is absent from the life cycle. Only a few segments of each individual are directly involved in **gametogenesis** (gamete production). Whereas oligochaetes fertilize their eggs externally, fertilization among leeches occurs internally, either through copulation or, in species lacking a penis, by jabbing packets of sperm **(spermatophores)** into the partner's body or through penetration of the body wall by the spermatophores themselves. Sperm exchange is mutual between two mating individuals (Fig. 13.30b), and the fertilized eggs of each individual generally develop within external cocoons. The **clitellum** generally functions in producing the cocoon and a nutritive fluid, as in the oligochaetes. A miniature leech eventually emerges from the cocoon.

In contrast to the polychaetes and oligochaetes, no members of the Hirudinea exhibit asexual reproduction.

Other Features of Annelid Biology

Digestive System

The typical gut is linear and unsegmented, with a mouth opening on the peristomium and an anus opening at the posterior end of the animal (pygidium). Food is moved through the gut by cilia and/or by muscular contractions. Digestion is primarily extracellular, although some species show an intracellular component as well.[17]

Class Polychaeta

There is no gut in siboglinid adults, and that of echiurans is long and highly convoluted, as mentioned earlier. The digestive tract of other polychaetes is typically divided into a pharynx, esophagus, intestine, and rectum (Fig. 13.35a). In some species, evaginations of the gut form blind-ending **digestive glands** or **digestive caeca** (*caec* = L: blind), increasing the amount of surface area available for digestion and absorption.

Class Clitellata

Subclass Oligochaeta

Oligochaetes generally show more modification of the basic arrangement just described for polychaetes. The esophagus may be modified to form a **crop,** for food storage, and/or a **gizzard,** a highly muscular structure lined with hardened cuticle, for grinding food (Fig. 13.35b). **Calciferous glands** are associated with the esophagus; these may function in regulating blood pH by controlling the concentration of carbonate ions.

The intestine of many terrestrial oligochaete species is thrown into a ridge or fold **(typhlosole),** which increases the gut's effective surface area (Fig. 13.36).

Associated with the intestine (and dorsal blood vessel) of oligochaetes is a characteristic yellow tissue called **chloragogen.** Chloragogen cells play major roles in protein, carbohydrate, and lipid metabolism of oligochaetes.

Subclass Hirudinea

The mouth of a leech opens into a muscular, pumping pharynx. **Salivary glands** associated with the pharynx (Fig. 13.31b) secrete **hirudin,** an anticoagulant. A crop and digestive glands are found in some species.

---

17. See *Topics for Further Discussion and Investigation,* no. 8.

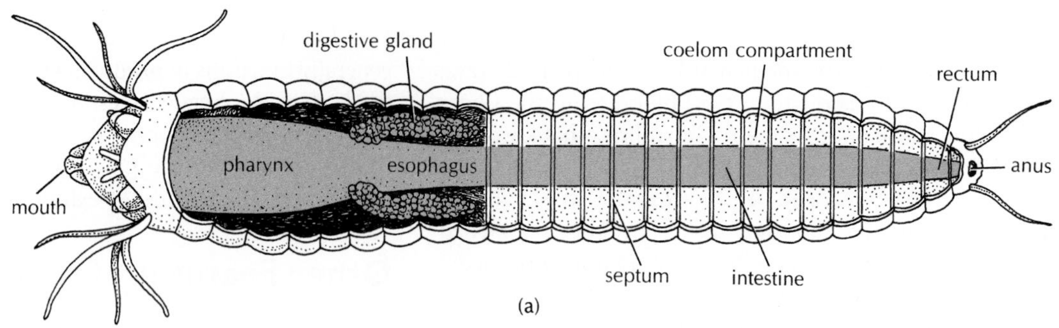

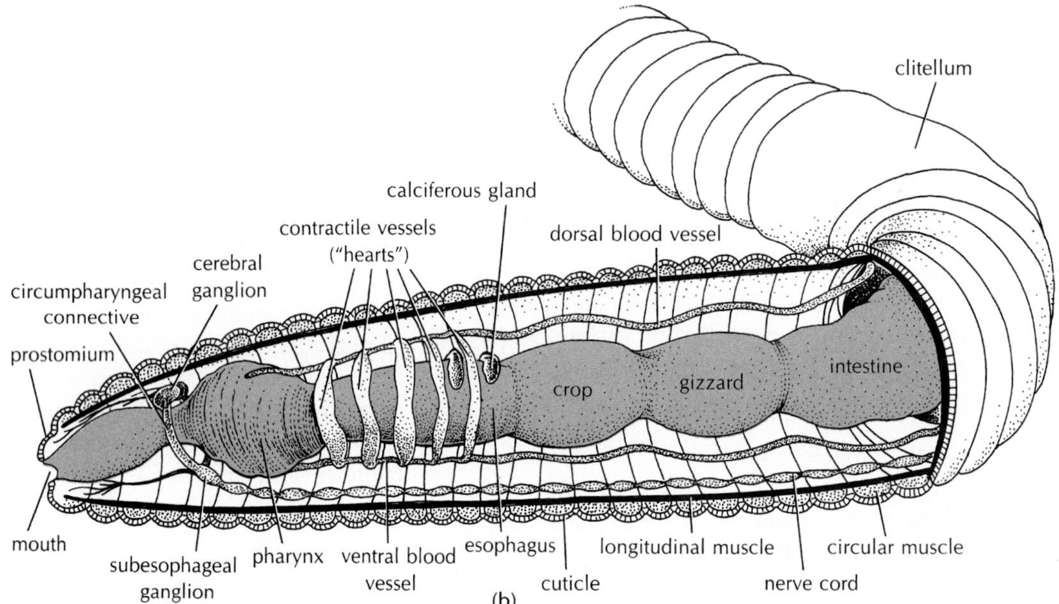

Figure 13.35
(a) Digestive system of the errant polychaete *Nereis virens*. (b) Digestive system of an oligochaete, *Lumbricus terrestris*. Elements of the nervous and circulatory systems are also shown.

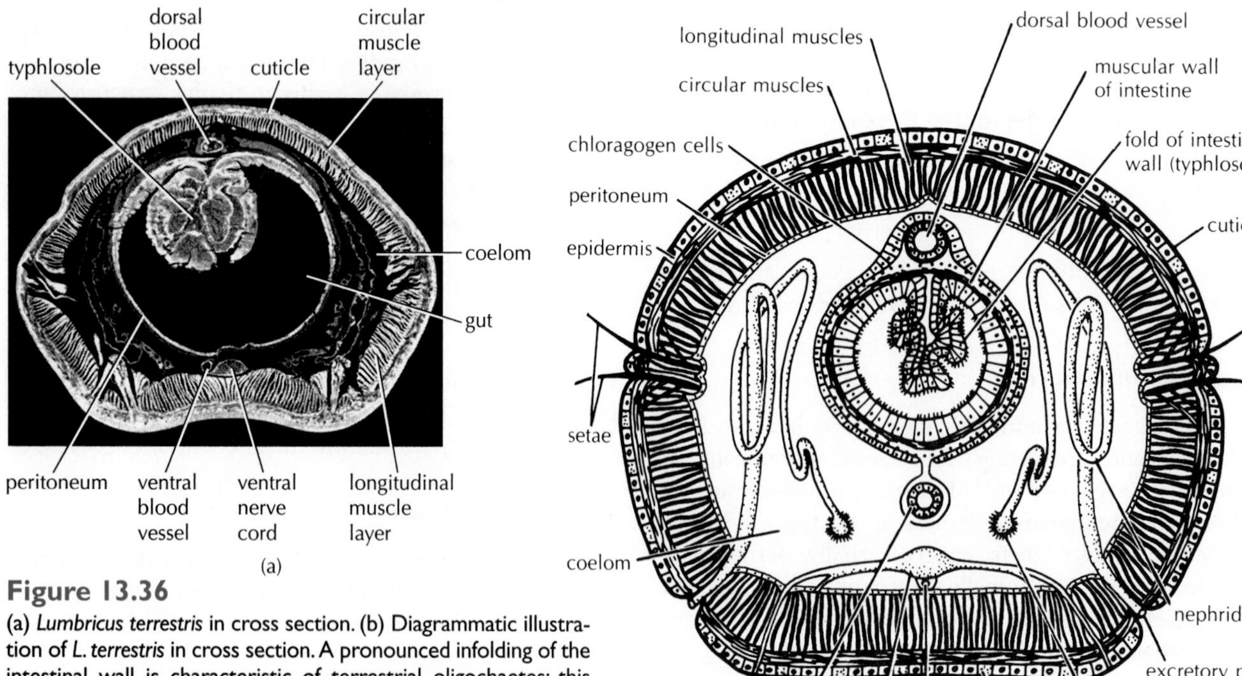

Figure 13.36
(a) *Lumbricus terrestris* in cross section. (b) Diagrammatic illustration of *L. terrestris* in cross section. A pronounced infolding of the intestinal wall is characteristic of terrestrial oligochaetes; this typhlosole increases the gut surface area available for absorbing nutrients. Note the chloragogen tissue surrounding the intestine. This tissue, found in oligochaetes and polychaetes, plays a role in excretion, hemoglobin synthesis, and basic metabolism.
(a) © L. S. Eyster. (b) After Buchsbaum and other sources.

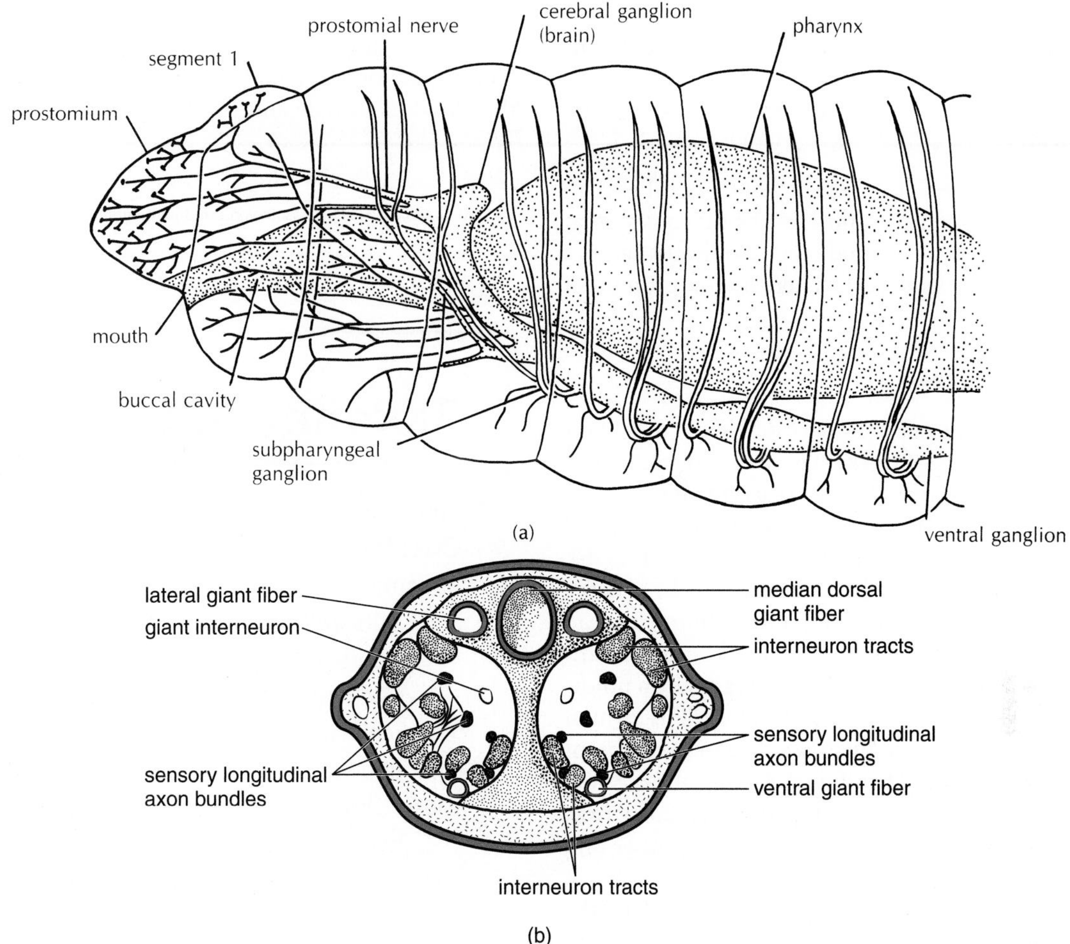

Figure 13.37

(a) Nervous system of *Lumbricus terrestris,* anterior end of animal. (b) Diagrammatic illustration of ventral nerve cord in cross section, showing a complex array of nerve fibers, including giant fibers. The giant fibers mediate rapid motor responses. The interneuron tracts are primarily involved in coordinating the activities of the various body segments. The giant fibers conduct nerve impulses up to about 1600 times faster than the smaller nerve fibers.

(a) Based on Barnes; after Hess; from Avel. (b) Based on P. J. Mill in *Comparative Biochemistry and Physiology,* 73A:641, 1982.

Nervous System and Sense Organs

Classes Polychaeta and Clitellata

Polychaetes and clitellates have a mass of **ganglia** (aggregations of nerve tissue) forming a brain (Fig. 13.37a). A solid ventral nerve cord (or a pair of nerve cords in the primitive condition) passes from the anterior to the posterior end of each individual. For segmented species, swellings of the cord in each segment form segmental ganglia. A variety of sense organs, including touch receptors, statocysts, light receptors, vibration receptors, and chemoreceptors, are distributed along the length of the body. In most polychaete species, the head bears a pair of ciliated depressions or slits, called **nuchal organs,** which are believed to be chemosensory. In some species the nuchal organs are eversible, while in some other species the organs are small and internal, and can be found only by histological sectioning. Nuchal organs are also found among the sipunculans, as discussed earlier. Oligochaetes, leeches, and echiurans lack nuchal organs.

Among polychaetes and clitellates, the various receptors are either connected to the ventral nerve cord by means of segmental nerves or (as in the case of the eyes and nuchal organs) connected directly to the brain by other nerves. The eyes of most species are not image forming, but they are sensitive to light and to changes in light intensity. Lateral nerves extending from the ventral nerve cord also innervate the digestive tract and the parapodial and body wall musculature of each segment.

Generally, a few of the nerves in the ventral cord are of considerably greater diameter than the others (Fig. 13.37b). These "giant" fibers may conduct nerve impulses 20 to more than 1000 times faster than other fibers, making possible the nearly simultaneous contraction of appropriate musculature throughout the worm's body. These giant fibers permit a very rapid, coordinated response to potential predators.[18]

The Echiurans

Echiurans have no specialized sensory systems, other than sensory cells studding the proboscis. The nervous system includes a nerve ring around the esophagus and a conspicuous ventral nerve cord. There are no prominent ganglia.

Circulatory System

Except for sipunculans, which have neither a heart nor blood vessels, annelids generally have a closed circulatory system consisting of a dorsal vessel (carrying blood anteriorly), a ventral vessel (carrying blood posteriorly), and capillaries connecting the two (Figs. 13.3b and 13.35b). Only siboglinid polychaetes have what might be called a specialized heart. In other annelids, circulation is instead maintained by contractions of the blood vessels themselves, especially of the dorsal vessel. Valves assure a unidirectional flow of blood throughout the body. This form of blood circulatory system is much reduced or even absent among members of the Hirudinea. Among the leeches, coelomic fluid assumes all or part of the circulatory role, reaching the tissues via contractile coelomic sinuses and channels.

Oxygen-carrying blood pigments are found in the circulatory fluid of most polychaete, oligochaete, and leech species. Blood pigments are absent from the echiuran circulatory system, but present in the coelomic fluid, and present also in the coelomic fluid of sipunculans. Several chemically and functionally distinct pigments are found among annelids. The blood of most species (and the coelomic fluid of echiurans) contains **hemoglobin,** in which iron atoms serve as the binding sites for oxygen. Hemoglobin may occur in corpuscles or may be found in solution in the blood fluid, depending upon the species examined. Hemoglobin-containing blood is especially characteristic of leeches, oligochaetes, and siboglinids, although it is found as well in the blood of some other polychaete species. In addition to hemoglobin, 2 other blood pigments have been encountered among the Polychaeta. **Chlorocruorin,** another iron-containing pigment, is found in solution in the blood of several polychaete species. This pigment is chemically quite similar to hemoglobin, but it has a greenish coloration. Yet a third iron-containing pigment, **hemerythrin,** is found in at least one acknowledged polychaete species and also among sipunculans. Hemerythrin is structurally quite dissimilar to the other 2 pigments, and it is always contained within cells. More than one type of blood pigment (including several structurally and functionally different types of hemoglobin) may occur simultaneously in the blood of a single annelid.[19]

---

18. See *Topics for Further Discussion and Investigation,* no. 9.

19. See *Topics for Further Discussion and Investigation,* no. 5.

Taxonomic Summary

Phylum Annelida
- Class Polychaeta
  - Family Siboglinidae
    - Subfamily Frenulata
    - Subfamily Vestimentifera (or Obturata)—the vestimentiferans
- The Echiurans
- The Sipunculans
- Class Clitellata
  - Subclass Oligochaeta
  - Subclass Hirudinea—the leeches

Topics for Further Discussion and Investigation

1. Explore the morphological, behavioral, and physiological adaptations associated with a sedentary existence among polychaetes.

Barnes, R. D. 1965. Tube-building and feeding in chaetopterid polychaetes. *Biol. Bull.* 129:217.

Brown, S. C., and J. S. Rosen. 1978. Tube-cleaning behavior in the polychaete annelid *Chaetopterus variopedatus* (Renier). *Anim. Behav.* 26:160.

Chughtai, I., and E. W. Knight-Jones. 1988. Burrowing into limestone by sabellid polychaetes. *Zool. Scripta* 17:231.

Dauer, D. M. 1985. Functional morphology and feeding behavior of *Paraprionospio pinnata* (Polychaeta: Spionidae). *Marine Biol.* 85:143–51.

Dubois, S., L. Barillé, B. Cognie, and P. G. Beninger. 2005. Particle capture and processing mechanisms in *Sabellaria alveolata* (Polychaeta: Sabellariidae). *Mar. Ecol. Progr. Ser.* 301: 159–71.

Flood, P. R., and A. Fiala-Médioni. 1982. Structure of the mucous feeding filter of *Chaetopterus variopedatus* (Polychaeta). *Marine Biol.* 72:27.

Grelon, D., M. Morineaux, G. Desrosiers, and S. K. Juniper. 2006. Feeding and territorial behavior of *Paralvinella sulfincola*, a polychaete worm at deep-sea hydrothermal vents of the Northeast Pacific Ocean. *J. Exp. Mar. Biol. Ecol.* 329: 174–86.

Hedley, R. H. 1956. Studies of serpulid tube formation. I. The secretion of the calcareous and organic components of the tube of *Pomatoceros triqueter. Q. J. Microsc. Sci.* 97:411.

Hentschel, B. T. 1996. Ontogenetic changes in particle-size selection by deposit-feeding spionid polychaetes: The influence of palp size on particle contact. *J. Exp. Marine Biol. Ecol.* 206:1.

Hoffmann, R. J., and C. P. Mangum. 1972. Passive ventilation in benthic animals? *Science* 176:1356.

Mahon, H. K., and D. M. Dauer. 2005. Organic coatings and ontogenetic particle selection in *Streblospio benedicti* Webster (Spionidae: Polychaeta). *J. Exp. Mar. Biol. Ecol.* 323: 84–92.

Mattila, J. 1997. The importance of shelter, disturbance and prey interactions for predation rates of tube-building polychaetes (*Pygospio elegans* [Claparéde]) and free-living tubicifid oligochaetes. *J. Exp. Marine Biol. Ecol.* 218:215.

Merz, R. A. 1984. Self-generated *versus* environmentally produced feeding currents: A comparison for the sabellid polychaete *Eudistylia vancouveri. Biol. Bull.* 167:200.

Shimeta, J., and M. A. R. Koehl. 1998. Mechanisms of particle selection by tentaculate suspension feeders during encounter, retention, and handling. *J. Exp. Marine Biol. Ecol.* 209:47.

Strathmann, R. R., R. A. Cameron, and M. F. Strathmann. 1984. *Spirobranchus giganteus* (Pallas) breaks a rule for suspension feeders. *J. Exp. Marine Biol. Ecol.* 79:245.

Völkel, S., and M. K. Grieshaber. 1992. Mechanisms of sulphide tolerance in the peanut worm, *Sipunculus nudus* (Sipunculidae), and in the lugworm, *Arenicola marina* (Polychaeta). *J. Comp. Physiol. B* 162:469.

Whitlatch, R. B., and J. R. Weinberg. 1982. Factors influencing particle selection and feeding rate in the polychaete *Cistenides (Pectinaria) gouldii. Marine Biol.* 71:33.

2. What mechanical advantages are gained and lost through the elimination of septa among annelids?

Chapman, G. 1958. The hydrostatic skeleton in the invertebrates. *Biol. Rev.* 33:338.

Chapman, G., and G. E. Newell. 1947. The role of the body fluid in relation to movement in soft-bodied invertebrates. I. The burrowing of *Arenicola. Proc. Royal Soc.* London *B* 134:431.

Elder, H. Y. 1973. Direct peristaltic progression and the functional significance of the dermal connective tissues during burrowing in the polychaete *Polyphysia crassa* (Oersted). *J. Exp. Biol.* 58:637.

Gray, J., H. W. Lissmann, and R. J. Pumphrey. 1938. The mechanism of locomotion in the leech (*Hirudo medicinalis* Ray). *J. Exp. Biol.* 15:408.

Trueman, E. R. 1966. The mechanism of burrowing in the polychaete worm, *Arenicola marina* (L.). *Biol. Bull.* 131:369.

3. Compare the locomotion of errant polychaetes with that of oligochaetes.

Gray, J. 1939. Studies in animal locomotion. VIII. *Nereis diversicolor. J. Exp. Biol.* 16:9.

Merz, R. A., and D. R. Edwards. 1998. Jointed setae—Their role in locomotion and gait transitions in polychaete worms. *J. Exp. Marine Biol. Ecol.* 228:273.

Quillin, K. J. 1998. Ontogenetic scaling of hydrostatic skeletons: Geometric, static stress and dynamic stress scaling of the earthworm *Lumbricus terrestris. J. Exp. Biol.* 201:1871.

Seymour, M. K. 1969. Locomotion and coelomic pressure in *Lumbricus terrestris. J. Exp. Biol.* 51:47.

4. What differences in nephridial structure and function might you expect to see among marine, freshwater, and terrestrial annelids?

5. A variety of blood pigments are found within the Annelida. All blood pigments must have one key feature to be functional: They must be able to combine reversibly with oxygen. The amount of oxygen potentially carried per milliliter of blood, and the conditions under which a blood will become saturated with oxygen, are determined by the abundance and properties of the particular blood pigment present. In addition, blood pigments differ in how readily they give up oxygen under any given set of environmental conditions. What are the structural and functional differences among the various annelid blood pigments, and how do these differences relate to the environments in which the different species occur?

Baldwin, E. 1964. *An Introduction to Comparative Biochemistry.* New York: Cambridge University Press, 88–106.

Dorgan, K. M., P. A. Jumars, B. Johnson, B. P. Boudreau, and E. Landis. 2005. Burrow extension by crack propagation. *Nature* 433: 475.

Kayar, S. R. 1981. Oxygen uptake in *Sabella melanostigma* (Polychaeta: Sabellidae): The role of chlorocruorin. *Comp. Biochem. Physiol.* 69A:487.

Mangum, C. P. 1985. Oxygen transport in invertebrates. *Amer. J. Physiol.* 248:R505.

Mangum, C. P., J. M. Colacino, and J. P. Grassle. 1992. Red blood cell oxygen binding in Capitellid polychaetes. *Biol. Bull.* 182:129.

Toulmond, A., F. E. I. Slitine, J. de Frescheville, and C. Jouin. 1990. Extracellular hemoglobins of hydrothermal vent annelids: Structural and functional characteristics in three alvinellid species. *Biol. Bull.* 179:366.

Wood, S. C., ed. 1980. Respiratory pigments. *Amer. Zool.* 20:3.

6. What is the influence of deposit-feeding polychaetes on the distribution and abundance of other annelids in the community?

Wilson, W. H., Jr. 1981. Sediment-mediated interactions in a densely populated infaunal assemblage: The effects of the polychaete *Abarenicola pacifica. J. Marine Res.* 39:735.

Woodin, S. A. 1985. Effects of defecation by arenicolid polychaete adults on spionid polychaete juveniles in field experiments: Selective settlement or differential mortality? *J. Exp. Marine Biol. Ecol.* 87:119.

7. Do annelids show an immune response? What is the evidence?

Anderson, R. S. 1980. Hemolysins and hemagglutinins in the coelomic fluid of a polychaete annelid, *Glycera dibranchiata*. *Biol. Bull.* 159:259.

Chain, B. M., and R. S. Anderson. 1983. Antibacterial activity of the coelomic fluid of the polychaete, *Glycera dibranchiata*. I. The kinetics of the bactericidal reaction. *Biol. Bull.* 164:28.

Cooper, E. L., and P. Roch. 1984. Earthworm leukocyte interactions during early stages of graft rejection. *J. Exp. Zool.* 232:67.

Çotuk, A., and R. P. Dales. 1984. The effect of the coelomic fluid of the earthworm *Eisenia foetida* Sav. on certain bacteria and the role of the coelomocytes in internal defense. *Comp. Biochem. Physiol.* 78A:271.

Dhainaut A., and P. Scaps. 2001. Immune defense and biological responses induced by toxics in Annelida. *Can. J. Zool.* 79: 233–53.

Fitzgerald, S. W., and N. A. Ratcliffe. 1989. In vivo cellular reactions and clearance of bacteria from the coelomic fluid of the marine annelid, *Arenicola marina* L. (Polychaeta). *J. Exp. Zool.* 249:293.

8. More than 99% of all the organic matter in the ocean is in dissolved form, rather than particulate (Hedges, J. I. 1987. Organic matter in sea water. *Nature* [London] 330:205). What is the potential role of this dissolved organic material in meeting the nutritional requirements of aquatic annelids?

Ahearn, G. A., and S. J. Townsley. 1975. Transport of exogenous D-glucose by the integument of a polychaete worm (*Nereis diversicolor* Müller). *J. Exp. Biol.* 62:243.

Manahan, D. T. 1990. Adaptations by invertebrate larvae for nutrient acquisition from seawater. *Amer. Zool.* 30:147.

Stephens, G. C. 1975. Uptake of naturally occurring primary amines by marine annelids. *Biol. Bull.* 149:397.

Taylor, A. G. 1969. The direct uptake of amino acids and other small molecules from seawater by *Nereis virens* Sars. *Comp. Biochem. Physiol.* 29:243.

9. Investigate the role of giant fibers in mediating escape responses by aquatic oligochaetes.

Drewes, C. D., and C. R. Fourtner. 1989. Hindsight and rapid escape in a freshwater oligochaete. *Biol. Bull.* 177:363.

Zoran, M. J., and C. D. Drewes. 1987. Rapid escape reflexes in aquatic oligochaetes: Variations in design and function of evolutionarily conserved giant fiber systems. *J. Comp. Physiol. (A)* 161:729.

10. In what respects is epitoke formation similar to the strobilation of polyps in the life cycle of scyphozoan jellyfish (p. 108)?

11. Where does a polychaete go when its feet hurt?

12. Compare and contrast the phenomenon of epitoky among polychaetes with the life cycle of marine hydrozoans (p. 112).

13. The evolutionary origins of annelids, and the evolutionary relationships among the various annelid groups, are far from certain. For example, does the absence of parapodia and nuchal organs reflect the original clitellate condition, or a secondary loss? Some regard the primitive annelid as polychaete-like, with well-developed parapodia; oligochaetes would then have evolved through secondary reduction and loss of parapodia. Others place the streamlined oligochaetes, or even the leeches, in the ancestral position. And some have suggested that polychaetes and oligochaetes evolved independently from early coelomate, segmented ancestors. What is the evidence in favor of each alternative?

Brinkhurst, R. O. 1982. Evolution in the Annelida. *Canadian J. Zool.* 60:1043.

Clark, R. B. 1964. *Dynamics in Metazoan Evolution.* London: Oxford Univ. Press.

McHugh, D. 1997. Molecular evidence that echiurans and pogonophorans are derived annelids. *Proc. Natl. Acad. Sci. USA* 94:8006.

Nielsen, C. 1995. *Animal Evolution: Interrelationships of the Animal Phyla.* New York: Oxford Univ. Press.

Rousset, V., F. Pleijel, G. W. Rouse, C. Erséus, and M. E. Siddall. 2007. A molecular phylogeny of annelids. *Cladistics* 23: 41–63.

Siddall, M. E., K. Fitzhugh, and K. A. Coates. 1998. Problems determining the phylogenetic position of echiurans and pogonophorans with limited data. *Cladistics* 14:401.

Westheide, W. 1997. The direction of evolution within the Polychaeta. *J. Nat. Hist.* 31:1.

Willmer, P. 1990. *Invertebrate Relationships: Patterns in Animal Evolution.* Cambridge: Cambridge Univ. Press.

Winnepenninckx, B., T. Backeljau, and R. De Wachter. 1995. Phylogeny of protostome worms derived from 18S rRNA sequences. *Molec. Biol. Evol.* 12:641.

14. Ecological risk assessment is widely used to manage waste disposal, waste site cleanup, and other complex environmental problems. As deposit feeders and important components of many terrestrial food webs, earthworms may play major roles in transferring various pollutants from soil to vertebrates. What are the benefits and complications associated with using earthworms to monitor soil pollutant concentrations and to assess the potential risks associated with those concentrations?

Menzie, C. A., D. E. Burmaster, J. S. Freshman, and C. A. Callahan. 1992. Assessment of methods for estimating ecological risk in the terrestrial component: A case study at the Baird & McGuire superfund site in Holbrook, Massachusetts. *Environ. Toxicol. Chem.* 11:245.

Morgan, J. E., and A. J. Morgan. 1993. Seasonal changes in the tissue-metal (Cd, Zn, and Pb) concentrations in two ecophysiologically dissimilar earthworm species: Pollution-monitoring implications. *Environ. Poll.* 82:1.

Vandecasteele, B., J. Samyn, P. Quataert, B. Muys, and F. M. G. Tack. 2004. Earthworm biomass as additional information for risk assessment of heavy metal biomagnification: A case study for dredged sediment-derived soils and polluted floodplain soils. *Environ. Poll.* 129: 363–75.

15. Discuss the evidence indicating that the uptake of dissolved organic material (DOM) from the surrounding seawater plays a role in siboglinid ("pogonophoran") nutrition.

Southward, A. J., and E. C. Southward. 1982. The role of dissolved organic matter in the nutrition of deep-sea benthos. *Amer. Zool.* 22:647.

Southward, A. J., and E. C. Southward. 1987. Pogonophora. *Animal Energetics,* vol. 2. Pandian, T. J., and F. J. Vernberg, eds. New York: Academic Press, 201–28.

16. Discuss the evidence indicating that sulfur- and methane-oxidizing bacteria play a major role in meeting the nutritional needs of siboglinids ("pogonophorans") and other annelids.

Belkin, S., D. C. Nelson, and H. W. Jannasch. 1986. Symbiotic assimilation of CO_2 in two hydrothermal vent animals, the mussel *Bathymodiolus thermophilus,* and the tube worm, *Riftia pachyptila. Biol. Bull.* 170:110.

Cavanaugh, C. M., S. L. Gardiner, M. L. Jones, H. W. Jannasch, and J. B. Waterbury. 1981. Prokaryotic cells in the hydrothermal vent tube worm *Riftia pachyptila* Jones: Possible chemoautotrophic symbionts. *Science* 213:340.

Felbeck, H. 1981. Chemoautotrophic potential of the hydrothermal vent tube worm, *Riftia pachyptila* Jones (Vestimentifera). *Science* 213:336.

Giere, O., N. M. Conway, G. Gastrock, and C. Schmidt. 1991. Regulation of gutless annelid ecology by endosymbiotic bacteria. *Mar. Ecol. Progr. Ser.* 68: 287–99.

Southward, A. J., E. C. Southward, P. R. Dando, G. H. Rau, H. Felbeck, and H. Flügel. 1981. Bacterial symbionts and low $^{13}C/^{12}C$ ratios in tissues of Pogonophora indicate unusual nutrition and metabolism. *Nature (London)* 293:616.

Southward, A. J., E. C. Southward, P. R. Dando, R. L. Barrett, and R. Ling. 1986. Chemoautotrophic function of bacterial symbionts in small pogonophora. *J. Marine Biol. Assoc. U.K.* 66:415.

17. Evidence to date suggests that siboglinid eggs are uncontaminated by any bacteria when they leave the oviduct. When in development, and by what mechanism, do siboglinids acquire their endosymbiotic bacteria?

Distel, D. L., D. J. Lane, G. J. Olsen, S. J. Giovannoni, B. Pace, N. R. Pace, D. A. Stahl, and H. Felbeck. 1988. Sulfur-oxidizing bacterial endosymbionts: Analysis of phylogeny and specificity by 16S rRNA sequences. *J. Bacteriol.* 170:2506.

Jones, M. L., and S. L. Gardiner. 1989. On the early development of the vestimentiferan tube worm *Ridgeia* sp. and observations on the nervous system and trophosome of *Ridgeia* sp. and *Riftia pachyptila. Biol. Bull.* 177:254.

Southward, E. C. 1988. Development of the gut and segmentation of newly settled stages of *Ridgeia* (Vestimentifera): Implications for relationship between Vestimentifera and Pogonophora. *J. Marine Biol. Assoc. U.K.* 68:465.

18. A green pigment called "bonellin" has been isolated from the body wall of the echiuran *Bonellia viridis.* Discuss the evidence implicating bonellin as the factor that causes metamorphosing larvae to become males.

Aguis, L. 1979. Larval settlement in the echiuran worm *Bonellia viridis:* Settlement on both the adult proboscis and body trunk. *Marine Biol.* 53:125.

Jaccarini, V., L. Aguis, P. J. Schembri, and M. Rizzo. 1983. Sex determination and larval sexual interaction in *Bonellia viridis* Rolando (Echiura: Bonellidae). *J. Exp. Marine Biol. Ecol.* 66:25.

19. Compare and contrast burrow formation by sipunculans (or echiurans) and polychaetes.

Trueman, E. R. 1966. The mechanism of burrowing in the polychaete worm, *Arenicola marina* (L.). *Biol. Bull.* 131:369.

Trueman, E. R., and R. L. Foster-Smith. 1976. The mechanism of burrowing of *Sipunculus nudus. J. Zool. (London)* 179:373.

Wilson, C. B. 1900. Our North American echiurids. A contribution to the habits and geographic range of the group. *Biol. Bull.* 1:163.

20. Compare and contrast the feeding mechanism of the echiuran *Urechis caupo* with that of the polychaete *Chaetopterus variopedatus.*

Barnes, R. D. 1965. Tube-building and feeding in chaetopterid polychaetes. *Biol. Bull.* 129:217.

MacGinitie, G. E. 1939. The method of feeding of *Chaetopterus. Biol. Bull.* 77:115.

MacGinitie, G. E. 1945. The size of the mesh openings in mucous feeding nets of marine animals. *Biol. Bull.* 88:107.

21. The innkeeper worm, *Urechis caupo* (Echiura), supplements gas exchange across its body surface by pumping water in and out of the cloaca. Discuss the mechanism by which this movement of water is accomplished.

Wolcott, T. G. 1981. Inhaling without ribs: The problem of suction in soft-bodied invertebrates. *Biol. Bull.* 160:189.

22. Many annelid species (including sipunculans) are exposed to substantial concentrations of the metabolic poison hydrogen sulfide. Discuss the morphological and physiological adaptations to sulfide exposure exhibited by these animals.

Arp, A. J., and C. R. Fisher (eds). 1995. Life with sulfide. *Amer. Zool.* 35:81–185.

Childress, J. J., C. R. Fisher, J. A. Favuzzi, R. E. Kochevar, N. K. Sanders, and A. M. Alayse. 1991. Sulfide-driven autotrophic balance in the bacterial symbiont-containing hydrothermal vent tubeworm, *Riftia pachyptila* Jones. *Biol. Bull.* 180:135.

Menon, J., and A. J. Arp. 1998. Ultrastructural evidence of detoxification in the alimentary canal of *Urechis caupo. Invert. Biol.* 117:307.

Ruppert, E. E., and M. E. Rice. 1995. Functional organization of dermal coelomic canals in *Sipunculus nudus* (Sipuncula) with a discussion of respiratory designs in sipunculans. *Invert. Biol.* 114:51.

Völkel, S., and M. K. Grieshaber. 1992. Mechanisms of sulfide tolerance in the peanut worm, *Sipunculus nudus* (Sipunculidae) and in the lugworm, *Arenicola marina* (Polychaeta). *J. Comp. Physiol. B* 162:469.

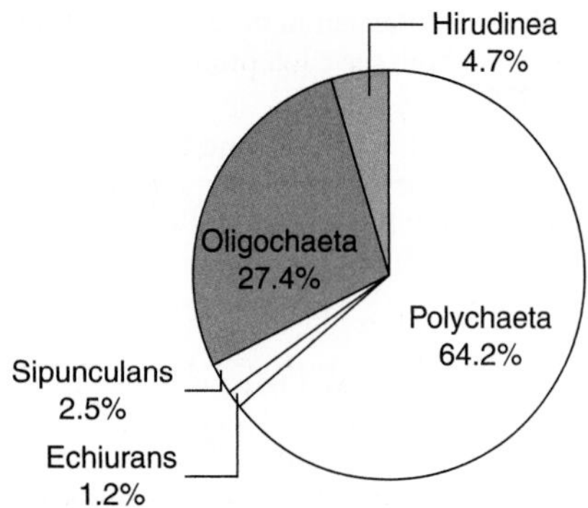

Taxonomic Detail

Phylum Annelida

This phylum has at least 12,500 species distributed among 2 classes. The 2 subclasses of the Clitellata are shaded blue. The presentation below is based largely on the work of G. W. Rouse and F. Pleijel, 2001 (*Polychaetes*, Oxford University Press). It will likely undergo dramatic change in the future.

Class Polychaeta
"The systematic organization of the polychaetous annelids is one of the most unsatisfactorily resolved problems in invertebrate phylogeny." W. Westheide, 1997. *J. Nat. Hist.* 31:1–15.

The approximately 8000 species are distributed among 25 orders. Most species are marine or estuarine.

Order Phyllodocida
Twenty-seven families.

Superfamily Phyllodocidacea. Eight families, many of which contain small numbers of exclusively pelagic, swimming worms. Most phyllodocid species are benthic. All members are marine.

Family Phyllodocidae. *Eulalia, Notophyllum, Phyllodoce, Eteone, Sige.* Adults are benthic, and only a few species burrow; most are active carnivores, crawling over hard substrates. There are over 300 described species. When sexually mature, the entire worm may become epitokous and swim up into the water to reproduce. Individuals are composed of up to 700 segments.

Family Tomopteridae. *Tomopteris.* These small (fewer than 40 segments), transparent worms live their entire lives swimming in open-ocean waters. All are voracious carnivores.

Family Typhloscolecidae. *Typhloscolex.* Most of the 15 species are free-swimming in the deep sea.

Superfamily Glyceracea. Three families, all marine.

Family Glyceridae. *Glycera*—bloodworms. All species use the muscular proboscis to burrow forcefully through sediment. Glycerids are notable carnivores, subduing prey with potent toxins discharged through oral fangs. Morphological alterations occurring during sexual maturation (epitoke formation) include degeneration of the proboscis, gut, and body wall musculature and elongation of the parapodia and setae. Fertilization is external, with gametes departing through the mouth. Adults die after mating. *Glycera dibranchiata* is a commercially important cold-water bait worm, and one that can inflict a nasty bite on its collector. Like some other polychaetes, glycerids possess blind-ended protonephridia rather than metanephridia.

Superfamily Nereididacea. Six families.

Family Hesionidae. *Hesionides, Leocrates, Podarke.* Most hesionids are surface-dwelling scavengers, although some species are interstitial, living between sand grains. Many species are commensal with other invertebrates, including echinoderms and other polychaetes. One species has recently been reported from freshwater sandy beaches, but all other species are marine.

Family Antonbruuniidae. This family contains a single species (*Antonbruunia viridis*) restricted to the waters off Mozambique. These worms live exclusively within the mantle cavities of one bivalve species (*Lucina fosteri*), with each host housing only one male worm and one female worm.

Family Syllidae. *Syllis, Odontosyllis, Exogone, Autolytus.* Some species are interstitial, and all are active carnivores. One species (*Calamyzas amphictenicola*) lives ectoparasitically on certain other polychaetes. Syllids show a variety of complex reproductive adaptations. In some species, the entire animal becomes an epitoke, whereas in others, epitokes bud off posteriorly as stolons. The epitokes often swarm and may do so at marvelously precise times in the lunar cycle.

Family Nereididae. *Nereis, Platynereis.* Most species live in tubes or burrows, while some are commensal with hermit crabs or certain other polychaetes. A few species live in freshwater. The marine sandworm *Nereis virens* is an economically important bait worm along the northern New England coast. Some species in this family form extremely modified epitokes, while others do not form epitokes at all. Animals often die after spawning.

Superfamily Nephtyidacea. Two families.

Family Nephtyidae. *Nephtys.* Most species are shallow-water, marine carnivores that use a muscular proboscis to burrow through sediment. The group contains some interstitial species, some deep-water species, and some freshwater species. Members of this family are unlike other polychaetes and most other animals in that the striated muscle fibers commonly contain high concentrations of intracellular calcium phosphate granules; the granules possibly strengthen the muscle fibers and aid the worms in their vigorous burrowing activities. Like some other polychaetes, nepthyids possess blind-ended protonephridia rather than metanephridia.

Superfamily Aphroditacea. All members are marine and have characteristic dorsal scales (elytra). Six families.

Family Aphroditidae. *Aphrodita*—the sea mouse. The common name refers to the often thick, hair-like dorsal surface, made of numerous fine setae. These bottom-dwelling polychaetes spend most of their time partly buried in fine muds, using thick, tubular, highly muscular parapodia to walk over the substrate. The largest individuals are about 22 cm long, but none have more than 60 segments.

Family Polynoidae. *Harmothoe, Hermenia, Lepidonotus.* These are mostly shallow-water marine carnivores, although some species are commensal with other invertebrates. This is a large family, containing over 600 species. The largest individuals, which live in the Antarctic, grow as long as 19 cm.

Family Pholoidae. *Pholoe.* These widely distributed worms live in muddy sediments or crawl under rocks.

Family Sigalionidae. *Sigalion.* Most species are active, burrowing carnivores, composed of up to 300 segments.

Family Polyodontidae. *Polyodontes*—the sea wolf. These tube-dwelling, fierce predators can become over 0.5 m long.

Family Pisionidae. *Pisionella.* These small, carnivorous worms are highly modified for interstitial life between sand grains.

Order Spintherida

Spinther. The 9 species in this order all live as ectoparasites on sponges.

Order Eunicida

Nine families.

Family Onuphidae. *Diopatra, Kinbergonuphis, Onuphis.* These polychaetes live in characteristic parchment-like tubes or, less commonly, in burrows. They are predators or scavengers and are common over a wide depth range. Some species "farm" algae growing on their tubes. The largest individuals may reach 200 cm in length.

Family Eunicidae. *Eunice, Palola, Marphysa.* This family contains some of the largest of all polychaetes, with the members of some species exceeding 2 m in length and possessing more than 600 segments. All are predators, mostly in shallow waters, and most live in crevices and burrows in hard substrates. The reproductive behavior of palolo worms has been documented especially well; epitokes break free from the parent and swarm at very predictable times in the lunar cycle. In some places (e. g., Samoa) the epitokes are harvested and eaten.

Family Lumbrinereidae. *Lumbrinereis, Lumbrinerides.* These marine, free-living annelids resemble earthworms in appearance. Most feeding types, including deposit feeders and carnivores, are represented within the group.

Family Arabellidae. *Arabella.* All species are free-living and predaceous as adults, but some are parasitic in other invertebrates as juveniles. Arabellids resemble earthworms; parapodia are inconspicuous and anterior appendages are absent, as adaptations to burrowing.

Family Histriobdellidae. *Histriobdella.* This group of small (usually less than 2 mm long) worms contains only 6 known species. All are symbiotic with certain crustaceans (including lobsters and crayfish) and highly modified for this association: The anterior and posterior appendages form flattened, adhesive discs, with which the worms walk about and attach within the host's gill chamber. Setae are absent. The worms feed on bacteria and other microorganisms coating the walls of the crustacean gill chamber. The group includes both marine and freshwater species.

Order Spionida

Five families.

Family Spionidae. *Polydora, Scolelepis, Spio, Streblospio.* Most species burrow in sediments, but species in the genus *Polydora* burrow into calcareous substrates, such as bivalve and snail shells; members of some species prey on embryos brooded by hermit crabs occupying these shells. Some species secrete their own tubes. The group includes both deposit-feeding and suspension-feeding species, and members of some species can alternate between the 2 feeding modes as conditions warrant. Over 300 species have been described, including a few freshwater species.

Order Chaetopterida

One family (Chaetopteridae). *Chaetopterus, Phyllochaetopterus.* These sedentary polychaetes are entirely marine and live in secreted, leathery tubes. The body is always divided into 3 distinct tagmata. The worms capture suspended food particles with mucous nets, with most species drawing water through the tube and across the net by ciliary beating. Individuals of *Chaetopterus variopedatus* pump water through their tubes by the synchronized waving of 3 highly modified central notopodia. This species also differs from other family members in producing a remarkably bioluminescent mucus when disturbed.

Order Magelonida

One family (Magelonidae). *Magelona.* These worms are all detrital feeders and are common in shallow coastal waters. Among the annelids, only worms in this family possess the blood pigment hemerythrin.

Order Psammodrilida

Psammodrilus. One family (Psammodrilidae). Both species in this small order are marine and highly modified for interstitial life. They are small (less than about 9 mm) and narrow, and are streamlined by having very reduced parapodia. The animals are widespread in cooler waters of the Northern Hemisphere, feeding mostly on benthic diatoms.

Order Cirratulida

Three families.

Family Cirratulidae. *Cirratulus.* Most species form burrows in mud, collecting sediment with thin tentacles that creep actively over the substrate surface. Numerous filamentous gills project anteriorly, lying on the substrate. Species in the genus *Dodecaceria* instead burrow into bivalve mollusc shells and other calcareous substrates. Reproductive patterns among the members of this family are complex and include asexual reproduction by fragmentation, with the subsequent regeneration of missing parts.

Order Flabelligerida

Three families.

Family Flabelligeridae. *Flabelligera.* These deposit-feeding marine worms live mostly under stones and in shallow burrows, collecting sediment with their palps and tentacles. One species (*F. commensalis*) lives exclusively among the spines of sea urchins, feasting on their fecal products.

Order Opheliida

Two families.

Family Opheliidae. *Armandia, Ophelia.* All members are burrowing deposit feeders, typically restricted to sediments of specific particle sizes. Most species can swim for short periods by undulating the entire body, and some become completely free-swimming as epitokes.

Order Capitellida

Three families.

Family Capitellidae. *Capitella, Mediomastus, Notomastus.* In general morphology, capitellids resemble thin earthworms. The body is thread-like, with inconspicuous parapodia and no anterior appendages. Some species are opportunistic, having very short generation times (less than 30 days) and being capable of living under environmental conditions not tolerated by most other species; as such, they are excellent indicators of pollution and other environmental disturbances. A few species occur in freshwater, but most are marine or estuarine. All capitellids are deposit feeders, ingesting sediment and digesting the organic components.

Family Maldanidae. *Clymenella*—bamboo worms. In external appearance, these worms resemble a stick of bamboo. They live head down in the sediment, in vertical tubes of their own manufacture.

Family Arenicolidae. *Arenicola, Abarenicola*—lugworms. Lugworms are sluggish, deposit-feeding worms common in intertidal areas of estuaries and bays. They live head down in their burrows, defecating conspicuous, coiled fecal castings on the sand at the burrow opening. They commonly deposit their eggs in enormous gelatinous cylinders that sometimes approach 2 m in length. The worms themselves may grow to about 1 m long. The family is fairly small, containing only about 30 described species, but representatives are widespread.

Order Oweniida

One family (Oweniidae). *Owenia.* These worms live in tubes in sand or mud bottoms. Individuals of most species have ciliated tentacles for suspension feeding, while members of some species subsist instead on ingested sediment.

Order Terebellida

Six families.

Family Pectinariidae (= Amphictenidae). *Cistena* (= *Pectinaria, Cistenides*). These worms are active deposit feeders, living head down in conical tubes open at both ends. The tubes are commonly constructed of sand grains, selected and cemented in place with remarkable precision.

Family Sabellariidae. *Sabellaria, Phragmatopoma.* These suspension-feeding worms typically live in tubes of cemented sand and shell bits, often in massive, reef-forming aggregations; these reefs can extend along hundreds of kilometers of coastline. The body is divided into 4 distinct tagmata.

Family Ampharetidae. *Melinna.* This group of some 230 deposit-feeding marine polychaetes is particularly common in muddy substrates in shallow water. The animals live protected within tubes, collecting sediment with thin, active, feeding tentacles.

Family Terebellidae. *Amphitrite, Eupolymnia, Polycirrus.* Most species in this family are tube dwellers, but the structure and composition of the tubes vary widely among species. All are specialized deposit feeders. Nearly 400 species have been described, some of them reaching nearly 0.33 m in length; the tentacles of such large individuals can extend more than a meter from the tube or burrow.

Family Alvinellidae. *Alvinella, Paralvinella.* These large, exclusively deep-water worms, commonly called Pompeii worms, or palm worms, can comprise a major part of the biomass found at hydrothermal vents.

Order Sabellida

The species in this order are known as feather-duster worms or fanworms. Four families.

Family Sabellidae. *Sabella, Schizobranchia, Myxicola, Eudistylia, Fabricia, Manayunkia.* The 300 fanworm species in this group are mostly tube-dwelling suspension feeders. The worm's posterior end is at the bottom of the blind-ended tube or burrow; ciliated tracts carry fecal wastes upward for expulsion. Many species produce a freestanding tube elevated above the substrate (Fig. 13.9a). A number of other species burrow into coral, rock, or other calcareous substrates, and others burrow into sediment; in most such cases, the worms secrete mucous tubes within the burrows. In addition to reproducing sexually, many species routinely reproduce asexually by fission.

Family Caobangidae. *Caobangia.* The half-dozen or so species in this small family burrow into the shells of freshwater gastropods and bivalves in Southeast Asia. As an adaptation to living in blind-ended burrows in the shells, the digestive tract has become U-shaped, with the anus opening anteriorly. The worms are hermaphroditic suspension feeders, with internal fertilization. Despite their freshwater habitat, they release a swimming larva, which eventually metamorphoses on a suitable shell.

Family Serpulidae. *Hydroides, Serpula; Spirobranchus*—the tropical "Christmas-tree worm." All species live in permanent, calcareous tubes, usually attached to rocks, shells, ship bottoms, or other solid substrates. Elaborate, ciliated filaments are projected from the tube opening for suspension feeding; these are retracted rapidly if danger threatens. Although most of the 350 serpulid species are marine, at least one (*Ficopomatus* [= *Mercierella*] *enigmaticus*) occurs in freshwater.

Family Spirorbidae. *Spirorbis.* All 170 species live in permanent calcareous tubes cemented to algae, shells, rocks, and other solid objects. The tubes are typically small (a few millimeters or less in diameter) and coiled. The short-lived larvae have been widely used in studies of habitat selection. All species are marine.

Order Protodrilida

Protodrilus, Saccocirrus. A small group (fewer than 50 species) of common interstitial polychaetes. Many species lack parapodia, and a few live in freshwater; most are marine or estuarine. Two families.

Order Myzostomida

Myzostomum. The 170 or so species in this order are exclusively parasitic in or on sea stars (asteroids), brittle stars (ophiuroids), and especially, sea lilies and feather stars (crinoids). Some molecular data place these worms outside the Annelida, but other data (mitochondrial gene order data in particular) support their position here. They have at times been thought to be trematode flatworms, crustaceans, crustacean relatives, or bryozoans. They have no coelom, the body cavity instead being filled with parenchymal cells. Seven families.

The Archiannelids

Troglochaetus (freshwater), *Polygordius, Dinophilus, Protodrillus.* Formerly placed in a separate class of mostly marine annelids, these segmented worms are now considered to be highly modified polychaetes and are distributed among several polychaete families (including the Protodrilida) accordingly. Archiannelids are mostly small and highly modified for life between sand grains. Individuals of many species retain certain larval characteristics, including external cilary bands.

The "pogonophorans" (= Family Siboglinidae)

This group contains about 160 species. Their relationships with other polychaetes are not yet clear, although one analysis suggests an affiliation with the order Sabellida.

Subfamily Frenulata (= Perviata)

Siboglinum, Oligobrachia, Polybrachia. Small, thin siboglinids that lack a vestimentum and obturaculum. About 130 species distributed among 5 families.

Subfamily Vestimentifera (= Obturata)

Riftia, Ridgeia, Lamellibrachia, Tevnia. The vestimentiferans. Large, thick-bodied siboglinids up to 2 m long, bearing both a vestimentum and obturaculum anteriorly. About 25 species distributed among 5 families.

Subfamily Monilifera

Sclerolinum. Moniliferans (7 described species) may be the sister group to vestimentiferans. Members lack a vestimentum and show rings of chaetae (uncini) in the opisthosoma. Moniliferans are found on reduced sediments, but sometimes also among decaying organic materials such as wood and rope made from natural fibers.

Class Clitellata

Subclass Oligochaeta

The approximately 3500 species are distributed among 3 orders.

Order Lumbriculida

One family (Lumbriculidae). *Lumbriculus.* These worms live in freshwater or are semiterrestrial.

Order Haplotaxida

Suborder Tubificina

Six families.

Family Tubificidae. *Limnodrilus, Clitellio, Tubifex*—sludge-worms. This group contains numerous marine species as well as many freshwater species; recently, some species have even been found living in the deep sea. Members of this family typically thrive in heavily polluted water, as the common name implies. Individual worms are thread-like and rarely longer than a few centimeters. They live head down in sedimentary tubes, waving their posteriors about in the water, presumably to facilitate gas exchange. Asexual reproduction by fission is common.

Family Naididae. *Branchiodrilus, Chaetogaster, Dero, Nais, Pristina.* The members of this widespread family often resemble some polychaetes in possessing eyes and gills and in being able to swim. Asexual reproduction by fission is extremely common and may be the primary or sole form of reproduction in some species. Most of the approximately 100 species are free-living, but species of *Chaetogaster* are either carnivorous or symbiotic in the mantle cavity or tissues of some freshwater gastropods and bivalves. A few species live in estuaries or in the ocean.

Suborder Lumbricina

This large group of well over 3000 species includes the earthworms and their freshwater counterparts. Earthworm feeding and burrowing activities bring many tons of soil to the earth's surface per acre each year. The first detailed study of earthworm activity was published by Charles Darwin in 1881. About 14 families.

Family Lumbricidae. *Lumbricus.* This group of some 300 species includes the most widespread and familiar earthworms.

Family Ailoscolecidae. *Komarekiona.* These earthworms are especially common in North Carolina, Tennessee, and Indiana.

Family Glossoscolecidae. *Pontoscolex.* These are the most commonly encountered earthworms in South America and in the Caribbean. Although found mostly in forests, one species lives primarily in coastal beach sand. Members of this family can reach 2 m in length. These species are apparently eaten by some South American forest peoples.

Family Megascolecidae. This family of earthworms holds over 1000 species, mostly in Asia and Australia. The family includes the largest known annelids, members of *Megascolides australis,* which may grow to lengths exceeding 3 m; the diameters are correspondingly great—about 3 cm or more. These worms, listed as one of the world's dozen most endangered animal species, construct permanent burrows as deep as 2 m underground.

Family Eudrilidae. *Eudrilus.* This family of some 500 African earthworm species includes the most structurally complex of all oligochaetes.

Subclass Hirudinea

The approximately 630 species are distributed among 4 orders.

Order Rhynchobdellae

These species lack jaws, obtaining the host's blood through a rigid, muscular proboscis instead. All are aquatic parasites or predators in both freshwater and marine habitats. They feed on hosts as diverse as elephants and small invertebrates. There is no penis; sperm are exchanged by transferring sperm sacs (spermatophores), which somehow penetrate the partner's body wall. Three families.

Family Glossiphoniidae. *Glossiphonia, Placobdella.* This group includes parasites and predators, all restricted to freshwater. The bodies are broad and leaf shaped.

Family Piscicolidae. *Calliobdella, Myzobdella, Piscicola, Pontobdella.* Most marine species are included in this family, although some freshwater species are included as well. All species are parasitic on fish or other aquatic vertebrates.

Family Ozobranchidae. *Ozobranchus.* This family includes both marine and freshwater species, which are mostly parasitic on turtles and crocodiles.

Order Arhynchobdellae

Fertilization is accomplished by inserting a penis into the vagina of another individual. Nine families.

Family Hirudinidae. *Macrobdella; Hirudo medicinalis*—the medicinal leech. The more than 80 species in this group are mostly freshwater bloodsuckers with 3 well-developed, toothed jaws. The largest known leeches (*Haemopis* spp.) occur in this family, reaching lengths of 45 cm.

Family Haemadipsidae. *Haemadipsa.* These so-called "land leeches" are all terrestrial bloodsuckers, typically living in leaf litter or in bushes and trees. Most species are restricted to Australia and the Orient. Like hirudinids, these leeches possess toothed jaws.

Family Erpobdellidae. *Erpobdella.* These "worm leeches" are cylindrically shaped, lack jaws, and for the most part, lack teeth. All are predators, living in terrestrial and freshwater habitats worldwide.

Order Branchiobdellida

Cambarincola, Stephanodrilus, Xironodrilus. The 130 or so species in this group live primarily as ectocommensals on freshwater crustaceans. About 10% of species live as commensals or parasites in their hosts' gill chambers. Morphologically, these worms are intermediate between the true leeches and the oligochaetes; they attach to their host by means of anterior and posterior suckers. A single freshwater crayfish may be infected with several hundred worms simultaneously. Many workers consider branchiobdellids to be a specialized family of the Oligochaeta.

Order Acanthodbellida

The order contains one species in a single genus: *Acanthobdella.* Uniquely among leeches, these animals retain a compartmented coelom, and setae, in the first 5 anterior segments, and lack an anterior sucker. Adults parasitize salmonid fish.

The Echiurans

The approximately 160 species are distributed among 3 orders. All echiurans are marine. Some molecular studies place echiurans as the sister group to the polychaete family Capitellidae (p. 334).

Order Echiuroidea

This order contains nearly 95% of all echiuran species. Two families.

Family Bonelliidae. *Bonellia.* The various members of this family occur over a remarkable depth range, from the intertidal zone to the abyss. All species show pronounced sexual dimorphism, with females typically being at least 20 times larger than males. The males live symbiotically on or in the females.

Family Echiuridae. *Echiuris.* This is the largest echiurid family, holding nearly 60% of all species. Most species occur in shallow water and typically show little or no sexual dimorphism.

Order Xenopneusta

Urechis. The members of this well-known genus are atypical echiurans in having an unusually short proboscis, an open circulatory system, and an enlarged cloacal region that serves in gas exchange. One family, with only 4 species.

Order Heteromyota

Ikeda taenioides. This small order contains a single family, which itself contains a single species found only around Japan. The proboscis can be over 1 m long, more than 3 times longer than the rest of the animal. Unlike other echiurans, the members of this species have hundreds of unpaired nephridia in the coelomic cavity. The arrangement of the body wall musculature also departs from the standard echiuran plan.

The Sipunculans

The approximately 350 species of sipunculans (peanut worms) are distributed among four families. All species are marine.

Family Golfingiidae. *Golfingia, Themiste, Phascolion.* This family contains nearly half of all sipunculan species. *Golfingia minuta* is unusual in being the only known hermaphroditic sipunculan. Some species in the genus *Themiste* bore into coral and other calcareous substrates. Various species in the genus *Phascolion* commonly inhabit structures made by other animals, including the tubes of polychaete worms and the empty shells of dead gastropods.

Family Phascolosomatidae. *Phascolosoma.* Many of these species bore into rock.

Family Sipunculidae. *Sipunculus.* Most individuals in this family are large and burrow in sand, ingesting sediment as they travel. Some species can swim short distances by flexing the body rapidly.

Family Aspidosiphonidae. This group of nearly 60 described species is commonly encountered in coral reef rubble.

General References about the Annelids

Clark, R. B. 1964. *Dynamics in Metazoan Evolution: The Origin of the Coelom and Segments.* London: Oxford Univ. Press.

Colapinto, J. 2005. Blocksuckers: How the leech made a comeback. *The New Yorker,* July 25: 72–81.

Cuttler, E. B. 1994. *The Sipuncula: Their Systematics, Biology, and Evolution.* Ithaca, N.Y.: Cornell Univ. Press.

Dales, R. P. 1967. *Annelids,* 2d ed. London: Hutchinson University Library.

Fauchald, K., and P. A. Jumars. 1979. The diet of worms: A study of polychaete feeding guilds. *Oceanogr. Marine Biol. Ann. Rev.* 17:193–284.

Gage, J. D., and P. A. Tyler. 1991. *Deep-Sea Biology: A Natural History of Organisms at the Deep-Sea Floor.* New York: Cambridge Univ. Press.

Gilbert, S. F., and A. M. Raunio. 1997. *Embryology: Constructing the Organism.* Sunderland, MA: Sinauer Assoc., Inc., pp. 167–88 (echiurans) and pp. 219–35 (other annelids).

Halanych, K. M., T. G. Dahlgren, and D. McHugh. 2002. Unsegmented annelids? Possible origins of four lophotrochozoan worm taxa. *Integ. Comp. Biol.* 452:678–84.

Halanych, K. M., and A. M. Janosik. 2006. A review of molecular markers used for Annelid phylogenetics. *Integr. Comp. Biol.* 46: 533–43.

Harrison, F. W., and S. L. Gardiner, eds. 1992. *Microscopic Anatomy of Invertebrates,* vol. 7 (polychaetes and oligochaetes), vol. 12 (pp. 327–460, pogonophorans). New York: Wiley-Liss.

Hyman, L. H. 1959. *The Invertebrates, Vol. 5: Smaller Coelomate Groups.* New York: McGraw-Hill (pogonophorans and echiurans).

Ivanov, A. V. 1963. *Pogonophora.* New York: Academic Press.

McHugh, D. 2000. Molecular phylogeny of the Annelida. *Canadian J. Zool.* 78:1873–84.

Parker, S. P., ed. 1982. *Classification and Synopsis of Living Organisms,* vol. 2. New York: McGraw-Hill, pp. 1–43 (polychaetes and clitellates), and pp. 65–66 (echiurans).

Piger, J. F. 1997. In: Gilbert, S. B., and A. M. Raunio. *Embryology: Constructing the Organism.* Sunderland, MA: Sinauer Assoc., Inc., pp. 167–78.

Rice, M. E., and M. Todorovic. 1975. *Proceedings of the International Symposium on the Biology of the Sipuncula and Echiura.* Washington, D.C.: National Museum of Natural History, Smithsonian Institution.

Rouse, G., and F. Pleijel (eds.). 2006. *Reproductive Biology, and Phylogeny of Annelida.* Enfield, NH: Science Publishers, 675 pp.

Rouse, G. W. 2001. A cladistic analysis of Siboglinidae Caullery, 1914 (Polychaeta, Annelida): formerly the phyla Pogonophora and Vestimentifera. *Zool. J. Linn. Soc.* 132:55–80.

Rouse, G. W., and F. Pleijel. 2001. *Polychaetes.* New York: Oxford University Press.

Rousset, V., F. Pleijel, G. W. Rouse, C. Erséus, and M. E. Siddall. 2007. A molecular phylogeny of annelids. *Cladistics* 23: 41–63.

Struck, T. H., N. Schult, T. Kusen, E. Hickman, C. Bleidorn, D. McHugh, and K. M. Halanych. 2007. Annelid phylogeny and the status of Sipuncula and Echiura. *BMC Evol. Biol.* 7: 57.

Thorpe, J. H., and A. P. Covich, eds. 2001. *Ecology and Classification of North American Freshwater Invertebrates,* 2nd ed. New York: Academic Press.

Search the Web

1. Each of the following addresses contains at least one excellent color image of a representative species. The Pogonophora page illustrates a cluster of vestimentiferans. The sites are maintained by the University of California at Berkeley.

   www.ucmp.berkeley.edu/annelida/polyintro.html
   www.ucmp.berkeley.edu/annelida/pogonophora.html
   www.ucmp.berkeley.edu/annelida/echiura.html
   www.ucmp.berkeley.edu/sipuncula/sipuncula.html

2. www.mbl.edu/

   This site is operated by the Marine Biological Laboratory at Woods Hole, MA. Choose "Biological Bulletin" under Quick Links, and then "Other

Biological Bulletin Publications." Next select "Keys to Marine Invertebrates of the Woods Hole Region," and then choose "Annelida." This brings up a taxonomic key to polychaete families represented near Woods Hole, and a section explaining terminology relevant to the group.

3. http://biodiversity.uno.edu/~worms/annelid.html

   This brings you to the Annelid Worm Biodiversity Resources web page, maintained by Geoffrey B. Read. It includes geographic faunal lists of annelid species and extensive discussion of annelid phylogeny. It also includes links to many other sites concerning the biology of polychaetes, clitellates, and sipunculans.

4. http://tolweb.org/Annelida/2486

   This brings you to the Tree of Life site at the University of Arizona, Tucson. The annelid entry is provided by Greg W. Rouse, Fredrik Pleijel, and Damhnait McHugh.

5. http://biodidac.bio.uottawa.ca

   Choose "Organismal Biology," "Animalia," and then "Annelida" for photographs and drawings, including sectioned material.

14

The Arthropods

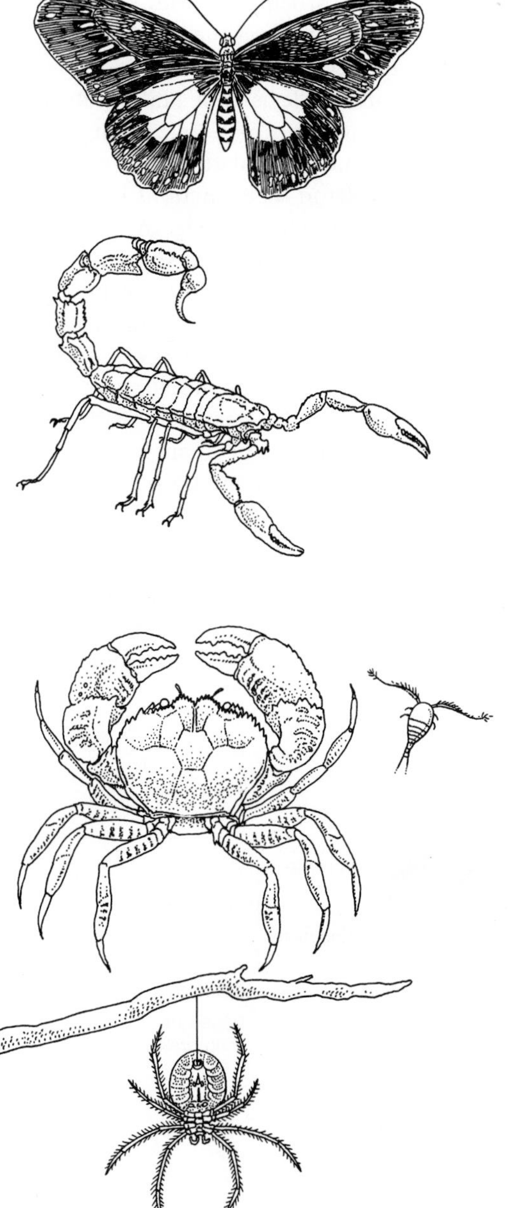

Phylum Arthro • poda
(Greek: jointed foot)
ar-throp´-ō-dah

Defining Characteristics:[1] 1) Epidermis produces a segmented, jointed, and hardened (sclerotized) chitinous exoskeleton, with intrinsic musculature between individual joints of appendages; 2) complete loss of motile cilia in adult and larval stages

Introduction and General Characteristics

At least 75% of all animal species described to date belong to the phylum Arthropoda, making the arthropod body plan by far the best represented in the animal kingdom. Insects, spiders, scorpions, pseudoscorpions, centipedes, crabs, lobsters, brine shrimp, copepods, and barnacles are all arthropods. Like annelids, arthropods are basically metameric, with new segments arising during development from a specific budding zone at the rear of the animal. In most modern members of the phylum, however, the underlying metameric, serial repetition of like segments is masked by the fusion and modification of different regions of the body for highly specialized functions. This specialization of groups of segments, known as **tagmatization,** is also seen in some polychaete annelids, but it reaches its greatest extent in the Arthropoda. Two of the major arthropod groups (Insecta and Crustacea) have 3 distinct tagmata: head, thorax, and abdomen.

Arthropods are unusual in lacking cilia, even in the larval stages.

1. Characteristics distinguishing the members of this phylum from members of other phyla.

The Exoskeleton

Arthropods have one conspicuous feature in common with molluscs: Individuals of both groups generally have a hard, external, protective covering—but here the similarity ends. The external coverings in the 2 phyla are produced by entirely different mechanisms, differ greatly in chemical composition, have distinctly different physical properties, and perform largely different functions. Whereas the molluscan shell functions largely to protect the soft parts within, the arthropod integument functions additionally as a locomotory skeleton.

The arthropod exoskeleton (Fig. 14.1) is secreted by epidermal cells. Its outermost layer (the **epicuticle**) is generally waxy, being composed of a firm lipoprotein layer underlain by layers of lipid. The cuticle is thus water-impermeable, so that the outer body surface cannot serve for gas exchange. On the other hand, the waterproof cuticle makes arthropods unusually resistant to water loss by dehydration. The epicuticle is quite thin, perhaps only 3% of the exoskeleton's total thickness. The bulk of the exoskeleton is made up of the **procuticle,** composed largely of the polysaccharide **chitin** in association with a number of proteins.

Commercial interest in chitin has slowly increased in the past several years because chitin is strong, nonallergenic, and biodegradable. Chitin can, for example, be solubilized and then re-formed into fibers, which can then be used in making fabrics and surgical sutures. It might also be used to make biodegradable capsules that could be implanted in the body, where they would gradually release therapeutic drugs over long time periods. Since chitin can be produced as a clear film, it might someday be used to manufacture a substitute for plastic wrap. Moreover, chitin and its derivatives bind readily to numerous inorganic and organic compounds, including fats, but they are themselves not digestible by vertebrates. As a food additive, chitin thus might reduce caloric and cholesterol uptake. Its binding abilities also make it a good candidate for removing toxic organic and inorganic compounds from drinking water and during sewage treatment.

Among arthropods, chitin functions in protection, support, and movement, providing a rigid skeletal system. The arthropod procuticle is strengthened by various hardening elements. Among crustaceans, this hardening is partly achieved through the deposition of calcium carbonate in some procuticle layers. Hardening is also accomplished by "tanning" the procuticle's protein component. The tanning process, also called **sclerotization,** involves the formation of cross-linkages between protein chains; this contributes to cuticle hardening in all arthropods. In the insects, the cuticle is hardened entirely through tanning. A similar process commonly occurs within other phyla as well: Sclerotization is involved in forming the hinges and byssal threads of bivalves; strengthening the periostracum of the molluscan shell; and hardening the molluscan radula, the setae of polychaetes, the jaws of some annelids, the egg coverings of some platyhelminths, and the trophi of rotifers.

The arthropod procuticle varies in thickness and is not hardened uniformly over the entire body, and therein lies its major functional significance. In many regions of the body the procuticle is thin and flexible in certain directions, forming joints (Fig. 14.2). Through the presence of appropriate musculature, the arthropods thus have a jointed skeleton that functions in much the same way as does the vertebrate skeleton; pairs of muscles antagonize each other through a system of rigid levers. Some arthropod joints—such as wing joints and those involved in jumping—contain a substance called "animal rubber," or resilin, which stores energy upon compression and releases it efficiently. The development of a jointed, flexible exoskeleton is the essence of arthropod success, and as discussed later in the chapter, has opened up a lifestyle accessible only to members of this group: flight.

The Hemocoel

The coelom can play no major role in the locomotion of animals encased in a suit of rigid plates, and the arthropod coelom is greatly reduced accordingly. The main

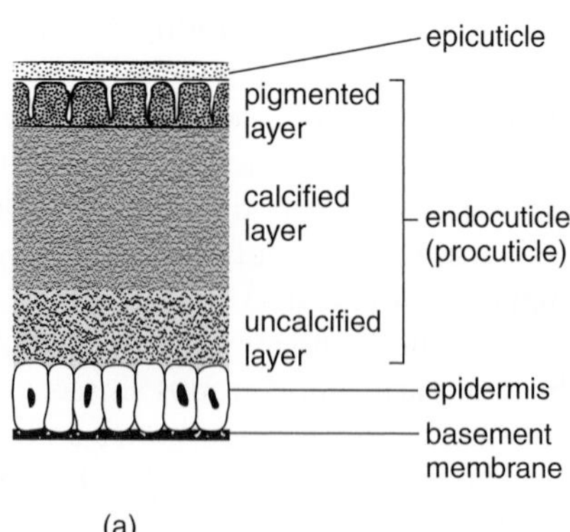

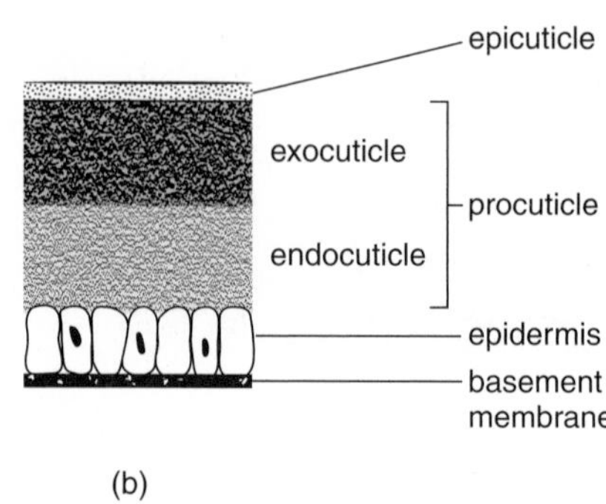

Figure 14.1

The cuticle of (a) crustaceans and (b) insects. The cuticles of both groups of animals are secreted by the underlying epidermis.
Based on W. D. Russell-Hunter, *A Life of Invertebrates.* 1979.

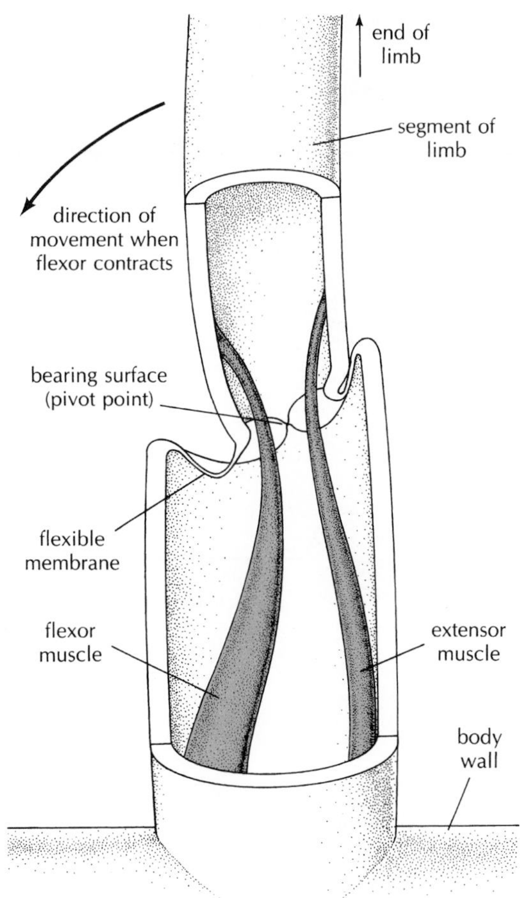

Figure 14.2
The articulation of a crustacean limb. The procuticle is hardened everywhere except at the joints, as indicated.
After Russell-Hunter.

body cavity is instead a **hemocoel,** part of the blood circulatory system, as in the Mollusca. Spiders and other arachnids extend their legs by increasing the blood pressure within the hemocoel, but otherwise, fluids and body cavities typically have little to do with arthropod locomotion.

Molting

In contrast to shell growth among molluscs, an arthropod's outer protective covering is not added gradually at a growing edge. Instead, it is secreted over all regions of the body simultaneously. Once the hardening process is completed, the arthropod is literally encased in its armor, except where the armor is pierced by sensory hairs and gland openings. Major regions of the foregut and hindgut are also lined with cuticle. To increase in body size, the arthropod must shed the cuticle—including that lining the gut—grow larger, and then harden a new cuticle to fit the larger (and, in certain instances, morphologically altered) body. The old cuticle is partially degraded by enzymatic secretions and is split prior to its removal. Among crustaceans, formation of the new exoskeleton requires a substance called cryptocyanin, a molecule that apparently evolved from hemocyanin, a copper-based protein involved in oxygen transport.[2]

The old cuticle is split by uptake of water or air and by increased blood pressure, all of which cause the body to swell. The process of removing the existing exoskeleton is called **ecdysis,** from the Greek word meaning "an escape" or "a slipping out of." In practice, the new cuticle is actually secreted before the old one is shed, which may partially explain why arthropods do not become totally nonfunctional during the molting process. Soft-bodied crabs—and possibly other molting arthropods as well—rely on high internal blood pressure in the hemocoel to maintain locomotory function; thus, the hemocoel acts as an internal hydrostatic skeleton until the new exoskeleton hardens (see Chapter 5).[3] Of course, potential collapse of the body during molting is more of a problem in air than in water, since air is relatively unsupportive. This may help to explain why terrestrial arthropods are smaller than aquatic species. The cuticle does not harden until morphological alterations, if any, and the increase in body size have taken place. Thus, the time between ecdysis and hardening of the new cuticle is a period of increased vulnerability to predators; most arthropods seek protective shelter during ecdysis.

Although increases in size are discontinuous in arthropods, growth of tissue (**biomass**) is a continuous process. The number of cells in the epidermis, for example, increases continuously in many arthropods, the additional tissue often becoming folded or pleated until the old cuticle is shed and the increase in body size can take place.

The processes of ecdysis and formation of the new exoskeleton are under both neural and hormonal control. Basically, one gland (the **Y-organ,** located in the head of crustaceans, for example, or the **prothoracic glands,** located in the thorax of insects) produces **ecdysteroid hormones** that stimulate molting. Among insects, ecdysteroid production is triggered by the brain's production of another hormone, which activates the prothoracic glands. In contrast, among crustaceans, ecdysteroid production is inhibited between molts by a second hormone, which is produced by a neurosecretory complex (the **X-organ**) located in the eyestalks. When the X-organ ceases effective production of its hormone, Y-organ activity is no longer inhibited and ecdysone may be produced. Ecdysis cannot occur until the X-organ stops producing its inhibitory hormone (Fig. 14.3); surgical removal of crustacean eyestalks results in premature ecdysis. A number of other important arthropod functions are also known to be under neurohormonal control, including regulation of the reproductive cycle, regulation of body fluid osmotic concentration, migration of light-screening pigments in the eye, and movement of pigment granules within **chromatophore** cells, leading to gradual changes in body color.

2. Terwilliger, N. B., M. C. Ryan, and D. Towle. 2005. *J. Exp. Biol.* 208: 2467–74.

3. Taylor, J. R., A., and W. M. Kier. 2003. *Science* 301: 209–10.

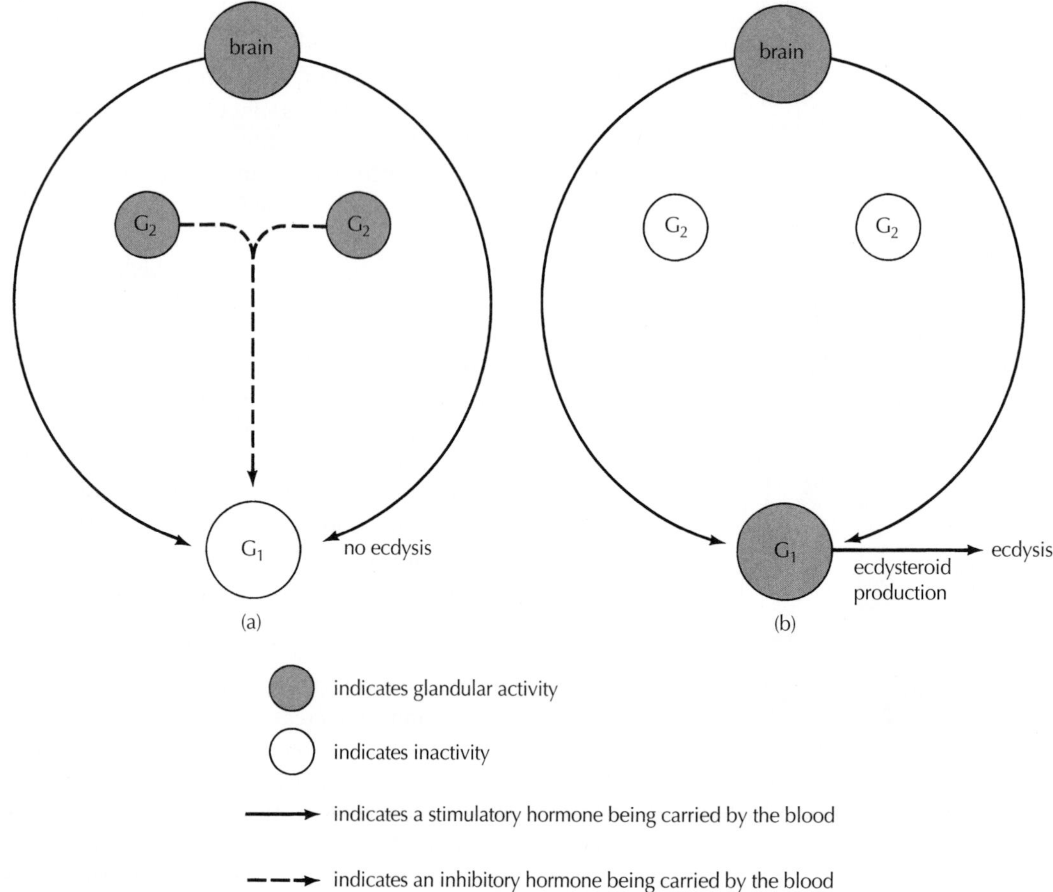

Figure 14.3
Schematic diagram of a likely mechanism regulating ecdysis in crustaceans. Glands associated with molting are indicated by G_1 and G_2. (a) The hormone produced by G_2 inhibits ecdysteroid production by G_1, and no molting occurs. (b) Gland G_2 is inactive; the brain hormone stimulates ecdysteroid production by G_1, and the animal molts.

Nerves and Muscles

The arthropod nervous system merits special mention, since it is operationally quite different from the nervous systems of both vertebrates and other invertebrates. In vertebrates, each muscle fiber is innervated by a single neuron. The strength of muscle contraction depends upon the number of fibers contracting, and the number of fibers contracting in a given muscle depends upon the number of axons fired. In arthropod muscle, in contrast, the strength of contraction depends upon the rate at which nerve impulses are delivered to the fibers. Moreover, a single muscle fiber may be innervated by as many as 5 different types of neurons (Fig. 14.4). The type of contraction (fast but brief *versus* slow and sustained) depends in part upon the source of the stimulation to the muscle. In addition, some of the neurons are inhibitory; action potentials delivered to such inhibitory neurons can alter the outcome of signals delivered down other axons to the same muscle fiber. Another complication is that arthropods have several physiological and functional types of muscle fiber; that is, the rate of contraction is partly a property of the individual muscle fiber. Fine control of arthropod movement therefore depends upon the types of muscle fiber stimulated and the interaction of several types of neurons terminating on a single muscle fiber. Finally, a single arthropod neuron may innervate a large number of muscle fibers so that a given muscle may be innervated by very few neurons (2 to 3, in some cases). In contrast, a given vertebrate muscle may be innervated by hundreds of millions of neurons.

The musculature of arthropods also differs significantly from that of other invertebrate groups: Arthropod muscle is entirely striated, whereas most other invertebrates possess primarily (or entirely) smooth muscle. The functional ramifications of this difference are considerable, in that striated muscle can contract far more quickly than smooth muscle (Table 14.1). Without striated muscle, arthropods would never have achieved flight.

The Circulatory System

The arthropod circulatory system also is of interest in that although the blood leaves the heart through closed vessels in most species, it enters the heart directly from the hemocoel through perforations, called **ostia,** in the heart wall (Fig. 14.5). The circulatory system is thus open, with the oxygenated blood moving through a series of sinuses

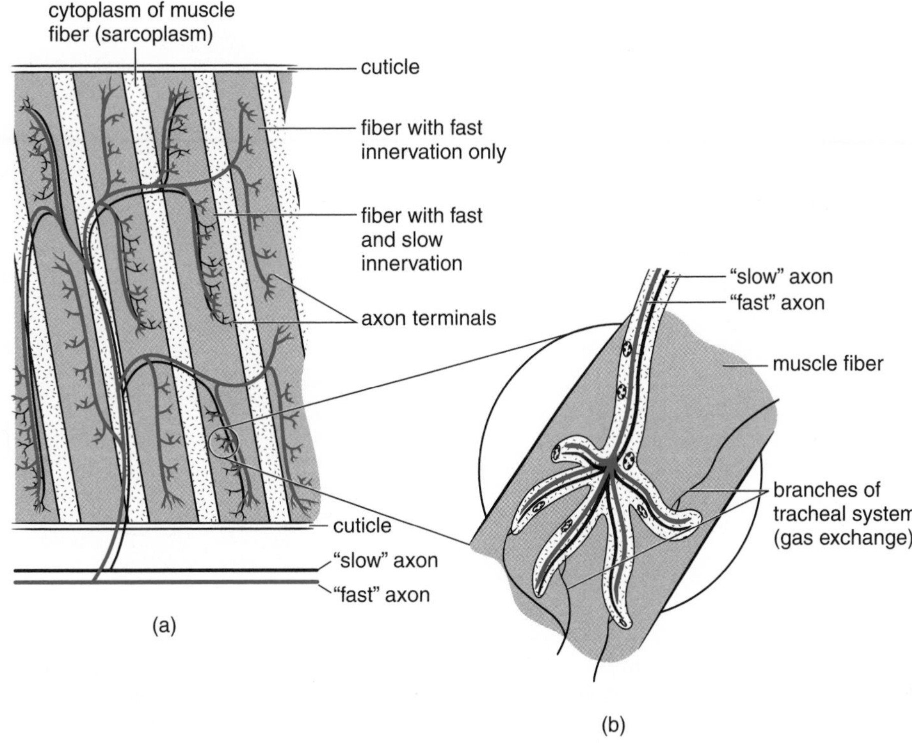

Figure 14.4

(a) Innervation of a typical arthropod muscle. Note that only some of the muscle fibers receive innervation from the "slow" axon. (b) Neuromuscular junction (insert), in detail.

Based on G. Hoyle, in M. Rockstein, ed., *The Physiology of the Insecta*, Vol. 4. 1974.

Table 14.1 Contraction Times of Invertebrate Muscle

Source	Contraction Time (seconds)
Anthozoa	
Sphincter muscle	5.0
Circular muscle	60.0–180.0
Scyphozoa	0.5–1.0
Annelida	
Earthworm circular muscle	0.3–0.5
Bivalvia	
Anterior byssus retractor muscle	1.0
Gastropoda	
Tentacle retractor muscle	2.5
Arthropoda	
Limulus abdominal muscle	0.195
Insect flight muscle	0.025

From various sources.

and finally being drawn back into the heart through the ostia as the heart expands. A "heart with ostia" is one of the diagnostic features of the Arthropoda, although gas exchange is achieved by radically different means in a number of arthropod groups.

Arthropod Visual Systems

Arthropod visual systems take one of 2 forms: ocelli, or compound eyes.

An **ocellus** is simply a small cup with a light-sensitive surface backed by light-absorbing pigment. Such simple photoreceptors are found in many phyla, including the Platyhelminthes, Annelida, Mollusca, and Arthropoda. The cup is often covered by a lens. As in all known visual systems, the photosensitive pigment of the ocellus is a vitamin A derivative in combination with a protein. Stimulation by light causes a chemical change in the photoreceptor pigment, generating action potentials that are then carried by nerve fibers to be interpreted elsewhere. Ocelli are generally not image forming.

On the other hand, compound eyes *can* form images. Such eyes are especially common and well-studied among insects and crustaceans, and may be present in addition to ocelli. Some species of polychaete annelid and bivalve mollusc have independently evolved compound eyes that operate on principles remarkably similar to those demonstrated by insect and crustacean eyes.

For any eye to form an image, light must first be focused on the receptor surface. The animal must then be able to examine each component of the scene independently and monitor how light intensity varies across the image formed. The animal also must possess a nervous system sufficiently sophisticated to reconstruct the image detected by the sensory system.

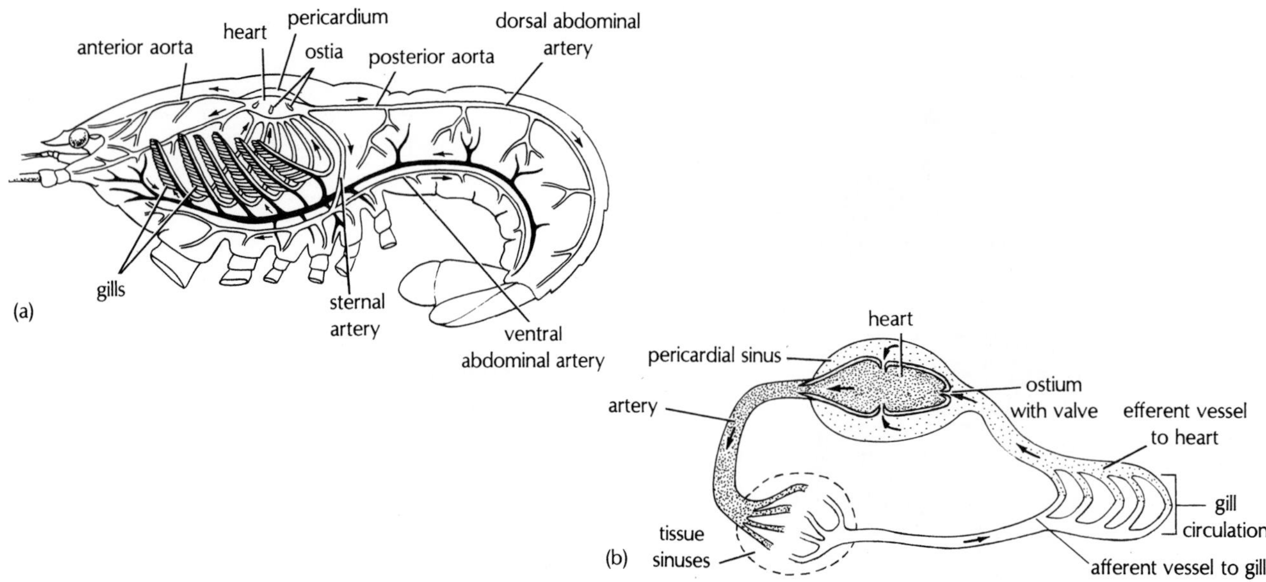

Figure 14.5

(a) Circulatory system of a typical arthropod, demonstrating the heart with ostia. The animal illustrated is a lobster. (b) Diagrammatic illustration of the blood circulatory pattern. Oxygenated blood is transported through the unshaded vessels to hemocoelic channels, in which gas exchange between blood and tissues occurs. Deoxygenated blood (shaded vessels in [a]) collects in a series of ventral venous sinuses. From here, the blood moves to the gills, is oxygenated, and returns to the pericardial sinus surrounding the heart.
(a) After Engemann and Hegner; after Gegenbauer. (b) From Cleveland P. Hickman, *Biology of the Invertebrates,* 2d ed. Copyright © 1973. The C. V. Mosby Company, St. Louis, Missouri. Reprinted by permission.

In the human camera-type eye, light enters through a single lens and is focused on the retina at the rear, in somewhat the same way that a motion picture image is focused on a screen. The components of the inverted image formed at the back of a human eye are sampled by many millions of closely packed receptor cells in the retina—the rods and cones. Nerve impulses from the individual receptor cells are then integrated and interpreted by the brain. A compound eye works in much the same way, except that (1) there are many lenses, (2) the focus of each lens cannot be varied, and (3) there are many fewer receptor cells to sample the image, which, by the way, is upright rather than inverted.

The compound eye is composed of many individual units called **ommatidia** (*ommato* = G: eye; *ium* = G: little) (hence, the term "compound" eye). The compound eyes of some insect species contain many thousands of ommatidia, each oriented in a slightly different direction from the others as a result of the eye's convex shape (Fig. 14.6). The visual field of such a multifaceted, convex eye is very wide, as anyone who has tried to surprise a fly will know.

The simplest type of compound eye is also the most common, found in insects as diverse as bees, ants, and cockroaches and in such noninsects as horseshoe crabs, true crabs (Infraorder Brachyura), myodocarpid ostracods, and isopods. Remarkably, recent molecular evidence suggests an independent evolutionary origin for the compound eyes of myodocarpid ostracods.[4] In contrast to how human eyes work, compound eyes break up the image before it reaches the retina, so that each ommatidium samples only a small part of the complete image. The horror movies thus have it wrong: An insect does not see thousands of identical, complete images at once, but rather only a single, fairly coarse-grained picture at a time. Each ommatidium consists of (1) a fixed-focus lens

Figure 14.6

Compound eye of a fly, containing hundreds of ommatidia. Note the eye's convex shape; no 2 ommatidia are oriented in precisely the same direction.
©Thomas Eisner.

4. Oakley, T. H., and C. W. Cunningham. 2002. *Proc. Natl. Acad. Sci. USA* 99: 1426–30.

(the cornea), which has such great depth-of-field that objects from 1 mm to several meters away are in focus at the receptor; (2) an underlying gelatinous **crystalline cone,** which serves as a lens in some insects and in most crustaceans; (3) a series of up to 8 cylindrical bodies (the photoreceptors) called **retinular cells,** each containing light-sensitive pigment; (4) cylindrical cells (collars) containing **shielding pigment,** optically isolating every ommatidium from surrounding ommatidia; and (5) at the basal end, a **neural cartridge,** a cluster of neurons receiving the information carried by the retinular cells and sending action potentials to the optic ganglia for processing (Fig. 14.7a). Insect and crustacean eyes are sensitive to the polarization of light. Polarized light is used by many insects as a navigational cue during flight and by at least some butterfly species for mate recognition. Their sensitivity to ultraviolet light (UV) also permits insects to see patterns (on flowers and other objects) that are invisible to humans. Some insects, especially those that fly in the evening and before dawn, are also sensitive to light of infrared wavelengths.

The light-sensitive pigment of the retinular cells is contained within tens of thousands of **rhabdomeres,** which are fine, microvillar outfoldings of the retinular cell walls. The rhabdomeres within each ommatidium form a discrete, ordered association called a **rhabdom** (Fig. 14.7a–c).

In the "closed" or "fused" rhabdoms found in most insects, including bees and many mosquitoes (Fig. 14.7b, c), there is no space between adjacent rhabdomeres and the rhabdom functions as a single functional unit. The fused rhabdom, in other words, is not really a single structure, but rather a central entity formed cooperatively by microvilli of the participating retinular cells. The rhabdom records the light intensity at the *center* of the image that falls at its tip; it does not record the entire image. The tip of one rhabdom is essentially analogous to a single rod in our eyes. The rest of the cylindrical rhabdom acts as a light guide, down which this small component of the image travels to the neural cartridge at the base. The brain then reconstructs the complete image from all the signals input from the dozens, hundreds, or many thousands of individual ommatidia.

Less commonly, the rhabdomeres inside each ommatidium are physically separated from each other (Fig. 14.7e) so that each rhabdomere acts as a separate light guide, thus increasing visual sensitivity to motion. Such rhabdoms are seen, for example, in fruitflies and houseflies. Remarkably, the development of open or closed rhabdoms is under the control of a single gene, and thus a single structural protein, facilitating the evolution from closed to open systems.[5]

In many species, particularly those active at night, each ommatidium is functionally isolated from its neighbors by permanent shielding pigment (Fig. 14.7a) or by reflective tracheoles, so that visual acuity is high. This basic compound eye is called an **apposition eye,** because the lens is directly apposed to the receiving rhabdom. Because each lens is very small, each rhabdom receives only a small amount of light; apposition eyes therefore work best at fairly high light intensities. For a compound eye to work well at low light intensities, each neural cartridge must receive light from more than one ommatidium. If the screening pigment between adjacent ommatidia is lacking, many facets can combine, or superimpose, the light they receive into a single image on the retina. This type of eye is called a **superposition eye.** In such an eye, each ommatidium has a large space between the distal end of the crystalline cone and the rhabdom; without the shielding pigment in the way, light from a single point in the visual field can be received by many lenses and focused onto a single rhabdom, producing a signal of substantially greater intensity than that received through a single lens (Fig. 14.8a). Superposition eyes are especially common among insects and crustaceans active at night.

The superposition eye adapts easily, with the help of pigment-containing collars flanking each rhabdom, to different light conditions; as light intensity increases, dense pigment granules migrate down each collar, so the collar acts as an iris, blocking the light received from adjacent ommatidia (Fig. 14.8b). In bright light, then, the superposition eye functions essentially as an apposition eye. If the light intensity is reduced, the pigment granules migrate out of the way so that light rays received by the lens of one ommatidium can again cross over to adjacent ommatidia, increasing the strength of the signal received. Because pigment migration is under hormonal control, it takes some time to dark-adapt or light-adapt a superposition eye.

The sharpness of the image formed by a compound eye depends upon a number of factors: (1) the extent to which the light impinging on the rhabdomeres of a single ommatidium enters along a pathway parallel to the optic axis (i.e., the long axis) of that ommatidium (increased resolution); (2) the extent to which light from adjacent ommatidia impinges upon the receptor pigment of an ommatidium (decreased resolution); (3) the amount of difference in direction in which adjacent ommatidia are oriented (decreased angle gives increased resolution); (4) the number of ommatidia per eye (increased number gives increased resolving potential); and (5) the complexity of the information center (i.e., brain) receiving and processing the impulses sent from the ommatidia. Even the largest and most sophisticated of compound eyes must produce a rather coarse-grained image. The image that humans see is synthesized from light-intensity data collected by millions of individual rods and cones; the image seen by a compound eye results from the collaborative efforts of as few as about a half-dozen and as many as several thousands of individual neural cartridges. In addition to image-forming eyes on the head, many insects,

5. Zelhof, A. C., R. W. Hardy, A. Becker, and C. S. Zucker. 2006. *Nature* 443: 696–99.

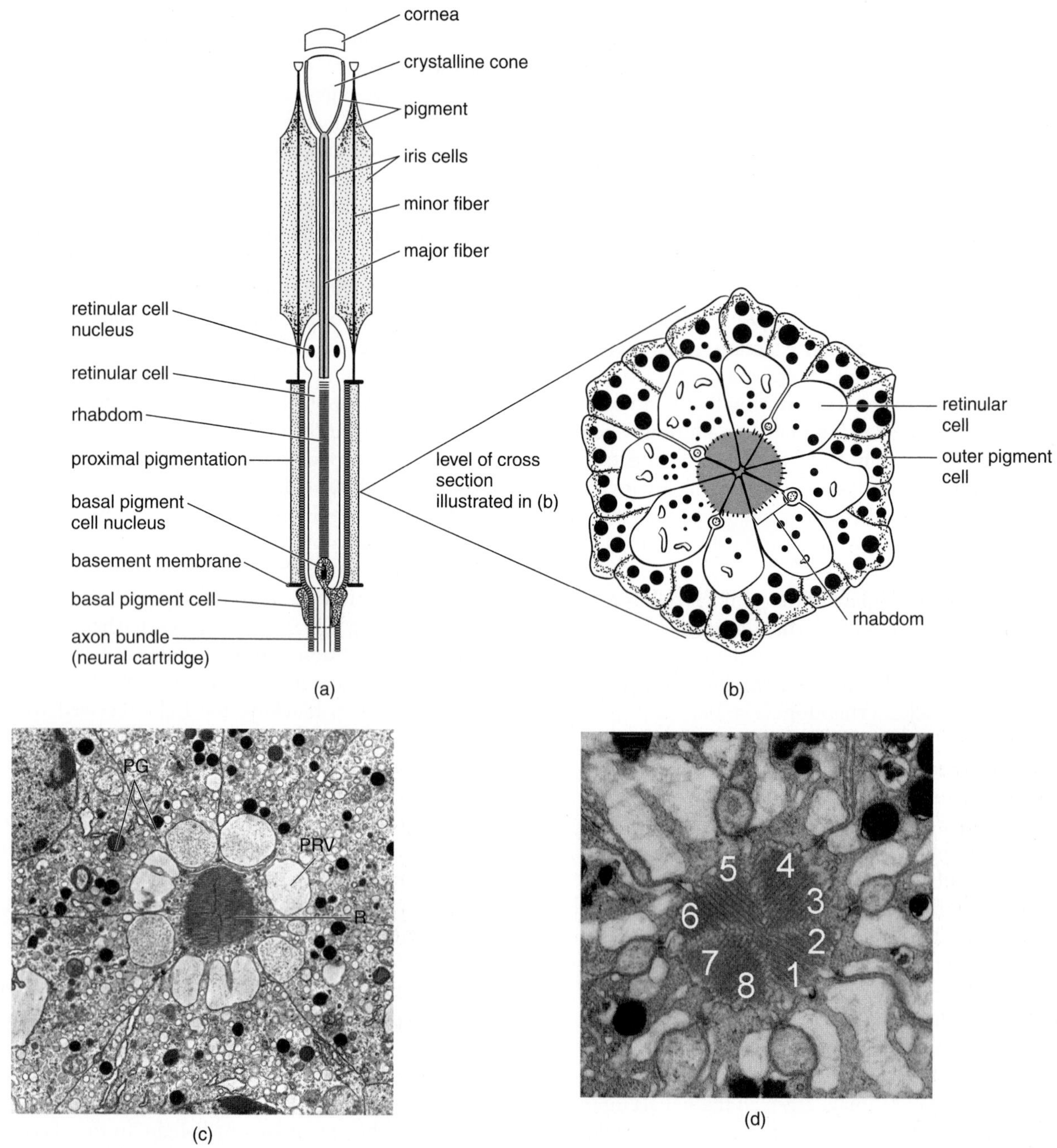

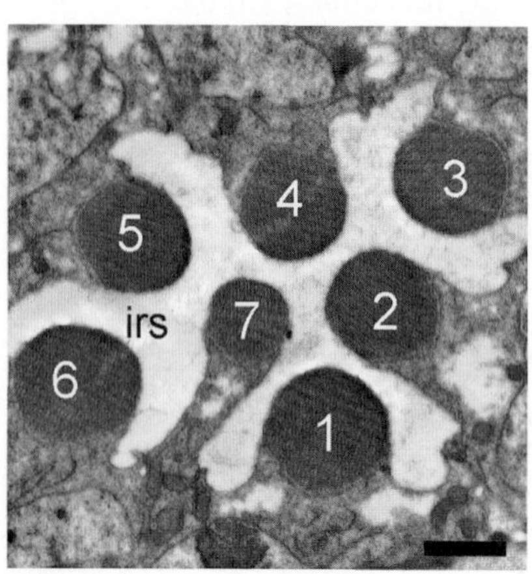

(e)

Figure 14.7

(a) Structure of a single ommatidium in a butterfly eye. The light-sensitive pigment is contained in the rhabdom. (b) Cross section through the rhabdom region of an ommatidium (at approximate level indicated in [a]), showing the relationship between the rhabdom and the reticular cells contributing to it. (c) Transmission electron micrograph of a crab ommatidium in cross section. PG = pigment granules in retinular cells; R = rhabdom; PRV = perirhabdomal vacuole. (d) Cross section through the closed ("fused") ommatidium of the honeybee *Apis mellifera* and (e) through the open ommatidium of the fruitfly *Drosophila melanogaster*. A single mutation can turn an eye like "b," "c," or "d" into one like "e."

(a, b) Based on S. L. Swihart, in *Journal of Insect Physiology,* 15:37. 1969. (c) Courtesy of K. Arikawa, from Arikawa et al., 1987. Daily changes of structure, function and rhodopsin content in the compound eye of the crab *Hemigrapsus sanguinensis J. Comp. Physiol. A* 161:161–74. (d, e) Courtesy of A. Zelhof, from Zelhof et al., 2006. Transforming the architecture of compound eyes. *Nature* 443: 696–99.

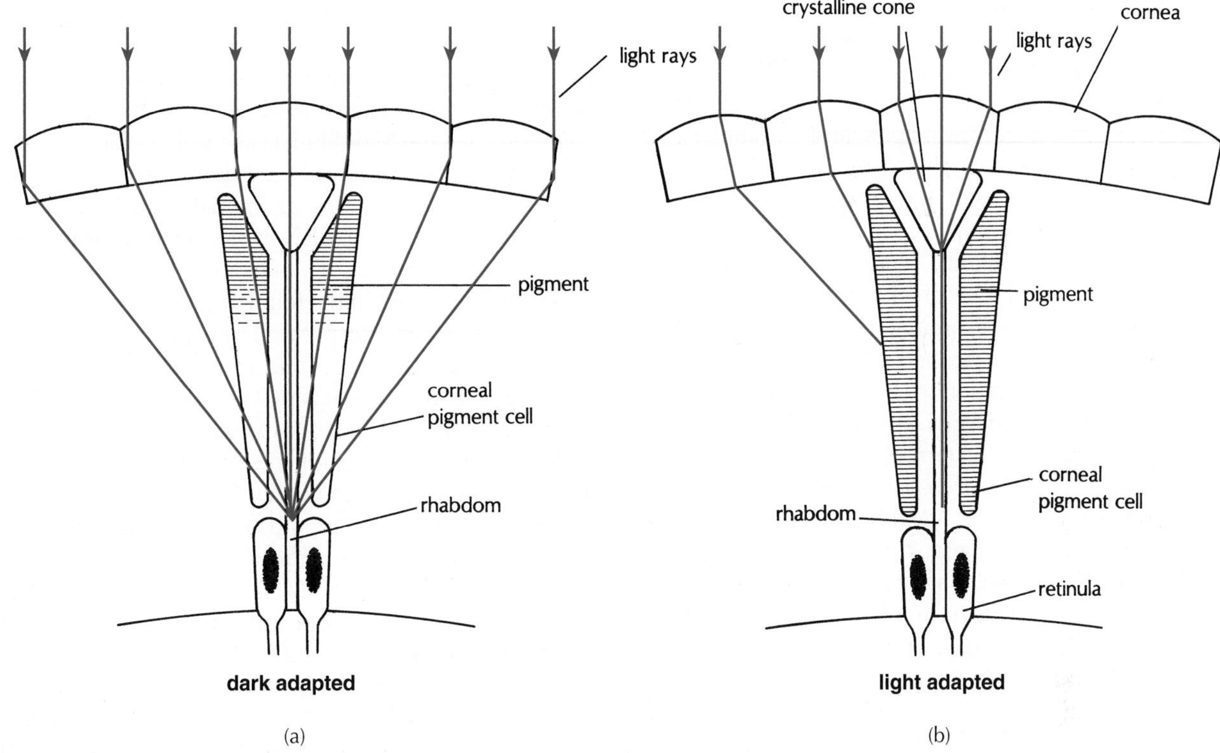

Figure 14.8

A superposition eye in the (a) dark-adapted and (b) light-adapted conditions. Note that in the dark-adapted condition, light entering through the lenses of several adjacent ommatidia impinges upon a single rhabdom. Migration of pigment within the pigment collars prevents this from happening in the light-adapted eye, improving visual acuity and directional sensitivity. For better clarity, the paths of reflected and refracted light are shown separately in (a) and (b), respectively.

crustaceans, spiders, and other arthropods have photosensitive neurons distributed elsewhere on the body (Fig. 14.10).

Arthropod Reproduction

Sexual reproduction is the rule among arthropods. Fertilization is internal in most species, but it is external in some. Most species are **gonochoristic (= dioecious)** (i.e., they have separate sexes), although some species—especially sedentary and parasitic ones—are hermaphroditic (Fig. 24.3a), and some free-living species even exhibit various degrees of asexual reproduction. **Parthenogenesis**—that is, production of offspring from unfertilized eggs—is commonly encountered among the Insecta and Branchiopoda and in some freshwater copepods. Indeed, males have never been found in some species of these groups. In both terrestrial and aquatic species, individuals often locate potential mates through pheromone production and chemical sensing.

Although internal fertilization is sporadically distributed among marine arthropods, it is the rule among terrestrial species. Indeed, internal fertilization in marine forms must have been a prerequisite, a preadaptation, for the invasion of land. Many arthropods, including insects, scorpions, mites, and copepods, transfer sperm to females indirectly, often by means of specialized containers **(spermatophores)** that are passed to the female in a variety of ways, as described in Chapter 24 (pp. 562–565; Figs. 24.9, 24.10, and 24.11b).

Further details of arthropod development are discussed later in this chapter, for insects (p. 370) and crustaceans (p. 389).

Classification and Evolutionary Relationships

In the classification scheme adopted in this text, there are 6 classes, 80 orders, and about 2,400 families contained within the phylum Arthropoda. Arachnids (mites, ticks, and spiders) alone are distributed among nearly 550 families. Despite the large number of major and minor taxonomic groupings found within the Arthropoda, however, most adult arthropods show only minor deviations from the general body plan. Taxonomic distinctions among arthropods depend largely on the number, distribution, embryological origin, form, and function of appendages.

Since this book focuses on the basic vocabulary and grammar of invertebrate zoology, rather than on the diversity of form and function encountered within each group, some of the arthropod classes and subclasses have been omitted from the discussion that follows. Selecting the "essential" arthropod groups has not been easy. I have included those groups whose representatives are generally familiar to the average person and those groups having major ecological and/or evolutionary significance. The chosen groups are also those most likely to be encountered in the field and in the research literature. Together, they illustrate the major principles and vocabulary of arthropod biology and architecture. The *Taxonomic Detail* section at the end of the chapter includes basic information about the groups I have had to omit here.

Three groups (Arachnida, Insecta, and Crustacea) include well in excess of 95% of all arthropod species. The phylogenetic relationships among these 3 groups, and among these and other arthropod groups, have long been controversial; the classification scheme adopted here is only one of several that have been proposed. The nature of the relationship between the terrestrial arthropods and the other major groups (crustaceans, horseshoe crabs and spiders, and the extinct groups, including trilobites) is especially obscure. Some zoologists have argued persuasively, based on detailed anatomical studies, that members of the terrestrial classes Chilopoda (centipedes), Diplopoda (millipedes), Symphyla, Pauropoda, and Insecta evolved independently of the Crustacea from entirely different ancestors, and thus place the terrestrial arthropods within a separate phylum, the Uniramia. Others have argued, also persuasively, that even the crustaceans and horseshoe crabs have no common ancestor. On the other hand, recent analyses of morphological data, new anatomical evidence from fossilized insects, new molecular analyses, and a growing understanding of how limb development is genetically controlled all support the Arthropoda as a monophyletic grouping.[6] I therefore retain the Arthropoda in this edition as a *bona fide* phylum, although the Uniramia is now abandoned as a valid taxonomic category. The disposition of the former uniramian arthropod groups under such a scheme remains particularly contentious, however. Some workers believe that insects, centipedes, millipedes, and a few other small terrestrial groups evolved from a common ancestor, and therefore retain insects and myriapods as sister groups (as the Atelocerata). However, accumulating molecular evidence suggests that the centipedes and millipedes (myriapods) are more closely related to horseshoe crabs and spiders than to insects and crustaceans, and that the insects have evolved from *within* the Crustacea, creating a new group that contains both insects (and other 6-legged arthropods) and crustaceans: the Pancrustacea (Fig 14.42). Here I retain the Crustacea and Insecta as distinct classes, something that will certainly change in the next edition if further molecular work supports the origins of insects from crustacean ancestors. It is certainly interesting to think of insects as modified crustaceans that have invaded the land.

In addition, some molecular data suggest that myriapods should be placed together with chelicerates (to form the new group Paradoxopoda, highlighting the fact that such a relationship is not expected from morphological comparisons) (Fig. 14.42). A recent (2006),[7] particularly detailed molecular analysis based on nearly complete 28S and 18S rRNA gene sequences from over 80 species (see p. 396) has failed to resolve the issue. For this edition, I have retained the Mandibulata as a valid category.

Because determining shared derived (or lost) characters for particular groups of animals usually requires a good understanding of their evolutionary history, I have had to omit the *Defining Characteristics* section for some of the groups discussed in the following text.

As implied in the previous paragraphs, the evolutionary origins of arthropods are unclear. Compare, for example, the various parts of Figure 2.12. Over the past several decades, evidence has been mustered in support of the following: evolution of arthropods from one or more annelid-like ancestors; evolution of annelids from arthropod-like ancestors; and independent evolution of annelids and arthropods from separate onycophoran-like ancestors (Chapter 15). The issue is contentious. Although arthropods have an interesting fossil record extending back some 550 million years, the fossils have to date been of little use in resolving these controversies. Of particular concern is the issue of how many times metameric segmentation evolved. If annelids and arthropods are closely related, then metameric segmentation probably evolved only once. But if annelids and arthropods arose independently from different ancestors, then metamerism in the 2 groups must reflect evolutionary convergence. Some recent analyses do support a close link between annelids and arthropods, but most do not. Instead, annelids seem more closely related to molluscs, flatworms, and other members of the Lophotrochozoa (see Fig. 2.12c) while arthropods seem more closely related to nematodes and other molting animals placed in the major new taxonomic grouping Ecdysozoa (see Fig. 2.12c,d).

Subphylum Trilobitomorpha

Subphylum Trilobitomorpha
(Latin: Three-lobed form)
try-lō-bite′-oh-morph′-ah

Defining Characteristic:[8] 2 anterior-posterior furrows divide the body into 3 regions (2 lateral, one central)

6. See *Topics for Further Discussion and Investigation*, nos. 1 and 22, at the end of the chapter.

7. Taxonomic treatment of the arthropod groups in this edition is largely based on Mallatt and Giribet. 2006. *Molec. Phylog. Evol.* 40: 772–94; Grimaldi and Engel. 2005. *Evolution of the Insects*. Cambridge University Press; http://animaldiversity. ummz.umich.edu/site/index.html; http://www.tolweb.org/tree

8. Characteristics distinguishing the members of this subphylum from members of other subphyla.

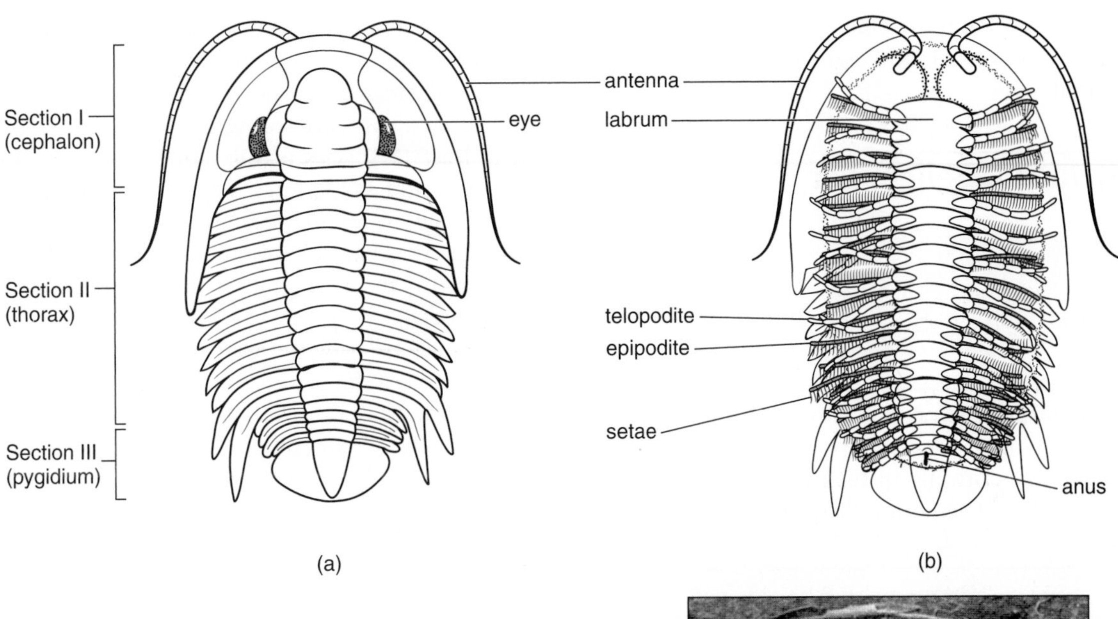

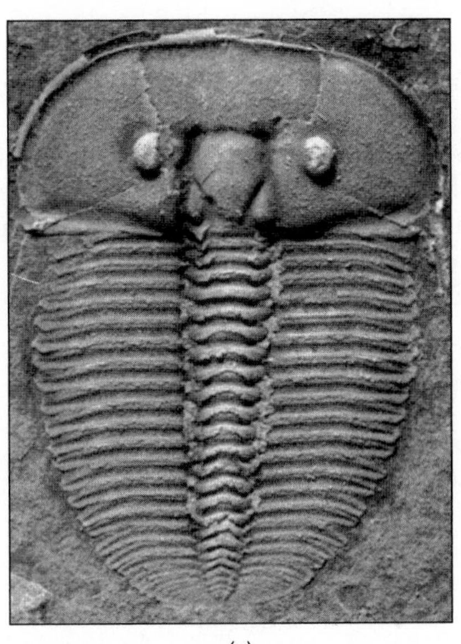

Figure 14.9
Trilobite as seen in (a) dorsal view and (b) ventral view. Note the pair of biramous appendages associated with each segment; the appendages are very uniform in structure all along the body. For a given appendage, the epipodite is the branch that bears the long filaments, the function of which remains uncertain. (c) Dorsal view of the tribolite *Aulacopleura konincki,* obtained from Silurian deposits (about 430 million years old) in the Czech Republic. This individual is about 2 cm long. Note the large compound eyes and the sharp break between the head (cephalon) and thorax.
(a,b) Based on Beck/Braithwaite, *Invertebrate Zoology, Laboratory Workbook,* 3/e, 1968. (c) Courtesy of Nigel Hughes. From Hughes, N. C. 2003. Trilobite tagmosis and body patterning from morphological and developmental perspectives. *Integr. Comp. Biol.* 43: 185–206.

Class Trilobita

Although approximately 4000 species have been described from the fossil record, the group has no living representatives. The trilobites were especially common about 500 million years ago, and highly diverse—distributed among about 5000 genera—but they were all extinct by about 225 million years later, remains near the close of the catastrophic end-Permian mass extinction defining the close of the Paleozoic era. Morphological differences apparent among the fossilized remains imply that a significant ecological diversity existed within the group at one time, varying from burrowing trilobites to walking and swimming forms.

The trilobite body was flattened dorsoventrally and divided into 3 sections (Fig. 14.9). Sections I and III were covered by a continuous unjointed sheet of exoskeleton (a **carapace**), so the underlying metameric segmentation was not visible when viewed from the dorsal surface. A pair of **compound eyes,** each composed of many **ommatidia,** was found laterally on the first body section. The median portion of the dorsal surface was partially divided from the lateral portions by 2 anterior-posterior furrows, giving the body a "3-lobed" look, acknowledged by the name of the group.

Adjacent to the mouth, on the ventral surface of one of the segments of body region I, was a chitinous lip, the **labrum,** and each body segment posterior to the mouth bore a pair of 2-branched (**biramous**) appendages (Fig. 14.9b). The innermost branch was devoid of long setae and presumably functioned in walking. The segments of the outer branch bore long filaments, which may have been gill filaments or simply may have been setae used for swimming, filtering food, or digging in loose substrate. This serial repetition of identical biramous appendages along the entire body length was clearly the primitive arthropod condition. More advanced groups show increasing specialization of appendages for specific tasks. This specialization has often involved the reduction or complete loss of one of the 2 branches of each primitive biramous limb.

Subphylum Chelicerata

Subphylum Chelicerata
(Greek: claw)
chel-iss-er-ah´-tah

Defining Characteristics: 1) Absence of antennae; 2) body divided into 2 distinct portions (the prosoma and the opisthosoma), with n.o distinct head; 3) first pair of appendages (the chelicerae) on the prosoma are adapted for feeding

---

Chelicerates are the only arthropods without antennae. Indeed, the first anterior segment bears no appendages at all. The second anterior segment bears a pair of clawed appendages **(chelicerae),** adjacent to the mouth, for grabbing and shredding food. Members of the Chelicerata also lack mandibles, appendages found adjacent to the mouth in many other arthropod groups and used for chewing and grinding food during ingestion.

Class Merostomata

Class Merostomata
mer´-oh-stō-mah´-tah

Defining Characteristics: 1) Appendages on the opisthosoma are flattened and modified for gas exchange as "book gills;" 2) terminal portion of body (telson) drawn out into an elongated spike

---

The class Merostomata is composed primarily of extinct species. Only 4 species are currently living, including the so-called horseshoe crab, *Limulus polyphemus* (Fig 14.10). Despite the common name, these animals are not true crabs; the true crabs are members of the Crustacea (p. 373). Horseshoe crabs burrow through surface layers of mud, ingesting smaller animals that they come across and oxygenating the sediment in the process. All members of the Merostomata are marine. Curiously, living representatives are found only in the waters of eastern North America, Southeast Asia, and Indonesia. Despite the small number of horseshoe crab species and their limited geographical distributions, humans have profited considerably from studying their biology. Much of what is understood about the basic principles of vision is based on studies of horseshoe crab eyes. Moreover, components of horseshoe crab blood are routinely used to test injectable pharmaceutical solutions for contamination by bacterial endotoxins and to screen for gonorrhea, spinal meningitis and several other diseases. Their embryos, deposited in vast numbers in shallow coastal waters each spring, provide a critical fuel source for at least 11 species of migrating shorebirds, including red knots and sanderlings. Egg production has dropped dramatically in recent years, as adult horseshoe crabs are widely used as bait for eels and conch, as well as for their blood; thus, many migrating bird populations are also under threat.

Characteristically, the head and thorax of merostomates are fused into a single functional unit, the **prosoma** (*pro* = G: forward; *soma* = G: body) or **cephalothorax,** and are covered with a single unjointed sheet of exoskeleton, the **carapace** (Fig. 14.10a). Patterns of *hox* gene expression during development suggest that the prosoma of horseshoe crabs and other chelicerates corresponds to the head of other arthropods. A pair of compound eyes is present laterally, on the dorsal surface of the prosoma. No other living chelicerate has compound eyes.

The first pair of appendages found ventrally on the prosoma are the **chelicerae** (*cheli* = G: claw) (Fig. 14.10b). These are followed by 5 pairs of similar appendages, the **walking legs,** all but the last of which bear claws. The first pair of walking legs (i.e., the appendages on the third anterior segment) are called **pedipalps,** but among females they are morphologically and functionally similar to the other walking legs. In males, the first pair is modified for grasping the female during mating. In both sexes, the first 4 pairs of walking legs, including the pedipalps, are each modified near the base to form a toothed food-grinding surface called a **gnathobase.** The fifth pair of walking legs is slightly modified for cleaning the gills, and for removing mud during burrowing. A pair of small, hairy appendages **(chilaria)** are found on the last segment of the prosoma; these may be involved in crushing food or simply moving it anteriorly prior to ingestion.

The abdomen, or **opisthosoma** (*opistho* = G: behind; *soma* = G: body), bears 6 pairs of appendages. The first pair is modified for reproduction, and the subsequent 5 pairs are modified to serve as gills. The underside of each gill flap bears approximately 150 leaf-like gas exchange surfaces, called **book gills,** through which blood circulates.

Class Arachnida

Class Arachnida
(*arachni* = G: a spider)
ah-rak´-nid-ah

Although the earliest members of the class Arachnida were undoubtedly marine, the more than 70,000 living arachnid species so far described are primarily terrestrial (Fig. 14.11). Moreover, the relatively few living aquatic species are clearly derived from terrestrial forms.

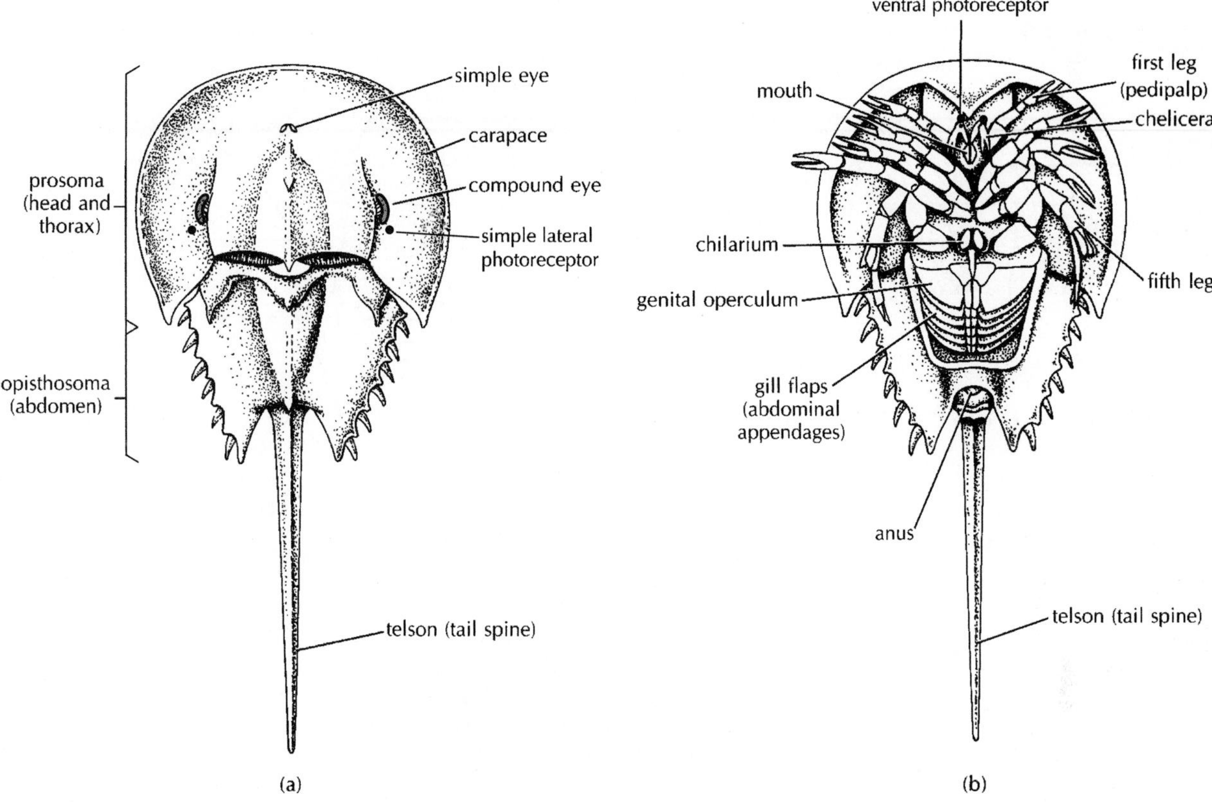

Figure 14.10

Limulus polyphemus, the horseshoe crab. (a) Dorsal view, showing major body divisions. In addition to the 6 photoreceptors illustrated, horseshoe crabs also have 2 ventral photoreceptors anterior to the mouth, and a series of photoreceptors along the telson. (b) Ventral view, showing appendages. Note that the abdominal appendages are modified as flattened sheets for gas exchange.

In the course of their evolution, the arachnids obviously have left the sea. This class includes many familiar but generally unpopular organisms, including spiders, mites, ticks, and scorpions. Nearly half of all arachnid species are spiders, and most of the remaining species—all but about 9,000—are mites and ticks. Spiders are major insect eaters and are increasingly used to control insect populations.

As with members of the Merostomata, the head and thorax of arachnids are fused to form a **prosoma,** which is covered by a carapace (Fig. 14.12). As mentioned earlier, recent studies of *hox* gene expression during development suggest that the head of millipedes, centipedes, insects, and crustaceans has a common evolutionary origin with the chelicerate prosoma. From 0 to 4 pairs of eyes are found on the prosoma, with 4 pairs being most common. The anteriormost pair of appendages borne by the prosoma are **chelicerae,** which generally tear apart food prior to ingestion. The next pair of appendages are the **pedipalps,** which are variously modified for grabbing, killing, or reproducing and in some species may have a sensory function as well. The basal segment of each pedipalp forms a **maxilla** (endite), which, like the chelicerae, aids in the mechanical preparation of food. The pedipalps are followed by 4 pairs of **walking legs.**

The arachnid **abdomen,** or **opisthosoma,** is generally distinct from the prosoma; in some arachnids, including the spiders, the 2 divisions are connected by a narrow stalk called a **pedicel,** which increases the abdomen's range of movement, facilitating the precise placement of silk threads in web building and prey capture. In a few arachnid groups, notably the ticks and mites, the prosoma and opisthosoma have fused together, and the entire dorsal surface is covered by a single carapace (Fig. 14.11c–e).

Respiration in the more primitive arachnid forms is by means of pairs of modified, internalized book gills, now known as **book lungs.** These flattened respiratory surfaces in the abdomen are connected to the outside by means of openings called **spiracles.** The spiracles of some species can be closed between "breaths," to limit water loss. In many species with small bodies, the spiracles may lead into a system of tubules known as **tracheae.** The tracheae form a system of branching tubules that ultimately terminate directly on the tissues. Gas exchange therefore occurs without use of the blood circulatory system. In some arachnid species, both book lungs and tracheae are present.

Some arachnids—the spiders (order Araneae)—bear up to 4 pairs of small abdominal appendages called **spinnerets** (Fig. 14.12b–d). These appendages are located ventrally and posteriorly, near the anus, and bear spigots

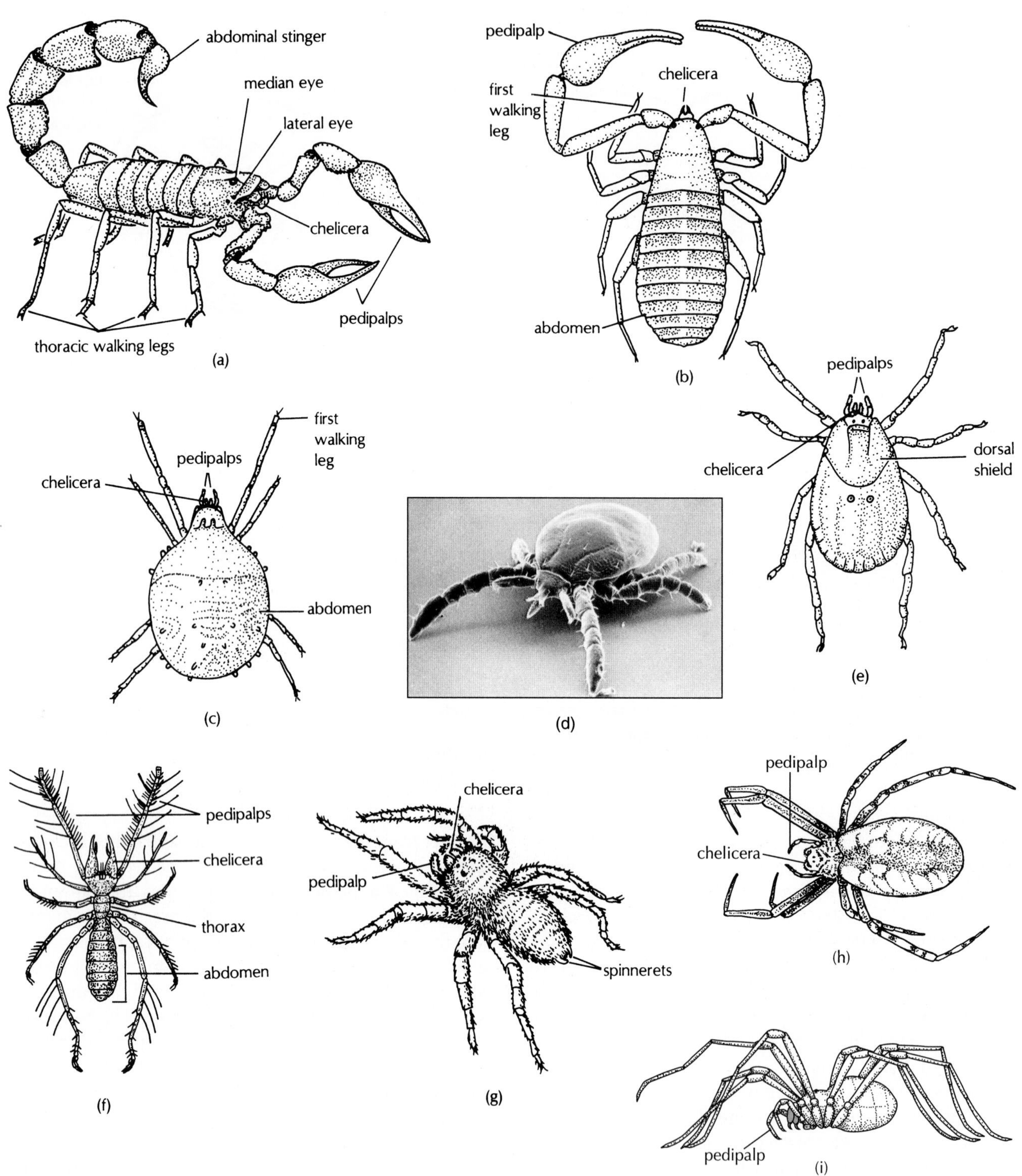

Figure 14.11

Arachnid diversity. (a) Scorpion. (b) Pseudoscorpion. (c) Red spider mite. (d) Scanning electron micrograph of the black-legged tick *Ixodes scapularis* (formerly *I. dammini*), the tick that transmits Lyme disease to humans throughout most of the United States and parts of Canada. The vector for Lyme disease in the western United States and western Canada is the western black-legged tick, *I. pacificus*. (e) *Dermacentor andersoni,* the tick that transmits Rocky Mountain spotted fever. (f) A solpugid, *Galeodes dastuguei.* The enormous chelicerae are used to tear apart prey. (g) A jumping spider (family Salticidae). (h) A common garden spider, *Argiope* sp., which forms an orb web like those illustrated in Figure 14.12h. Most of the orb-weaving spiders are found in a single family (the Araneidae). (i) An opilionid (daddy long-legs); chelicerae shaded blue.

(d) Courtesy of A. Spielman, Dept. of Tropical Public Health, Harvard School of Public Health.

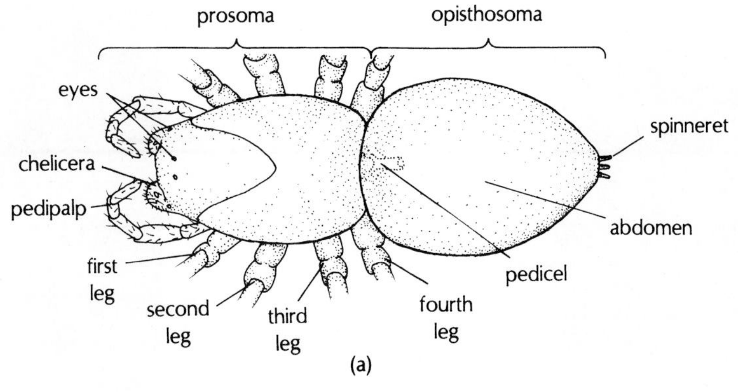

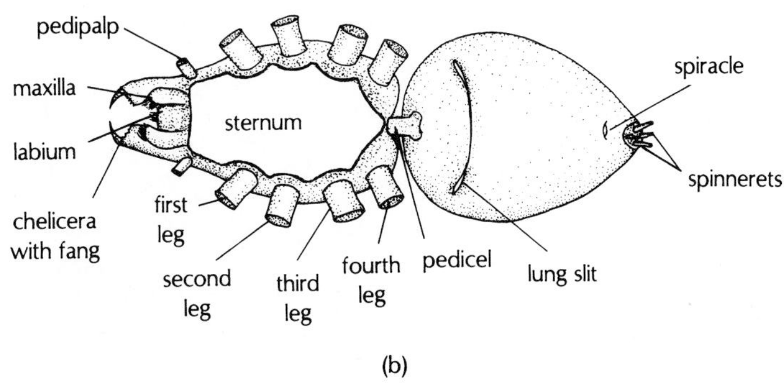

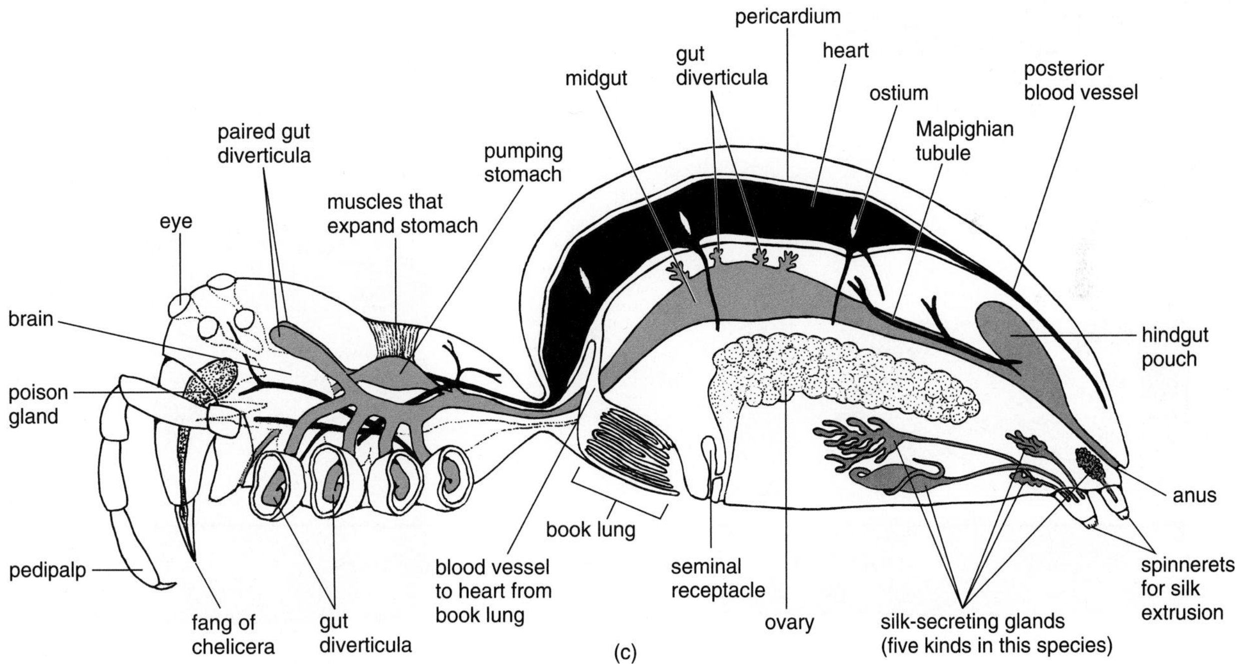

Figure 14.12

(a) Typical spider, in dorsal view. (b) Diagrammatic ventral view of spider. The legs have been removed for clarity. (c) Internal anatomy of a typical female web-spinning spider, seen in lateral view. Each silk-secreting gland secretes a different kind of silk, each specialized for a different function. (*Continued on following page.*)

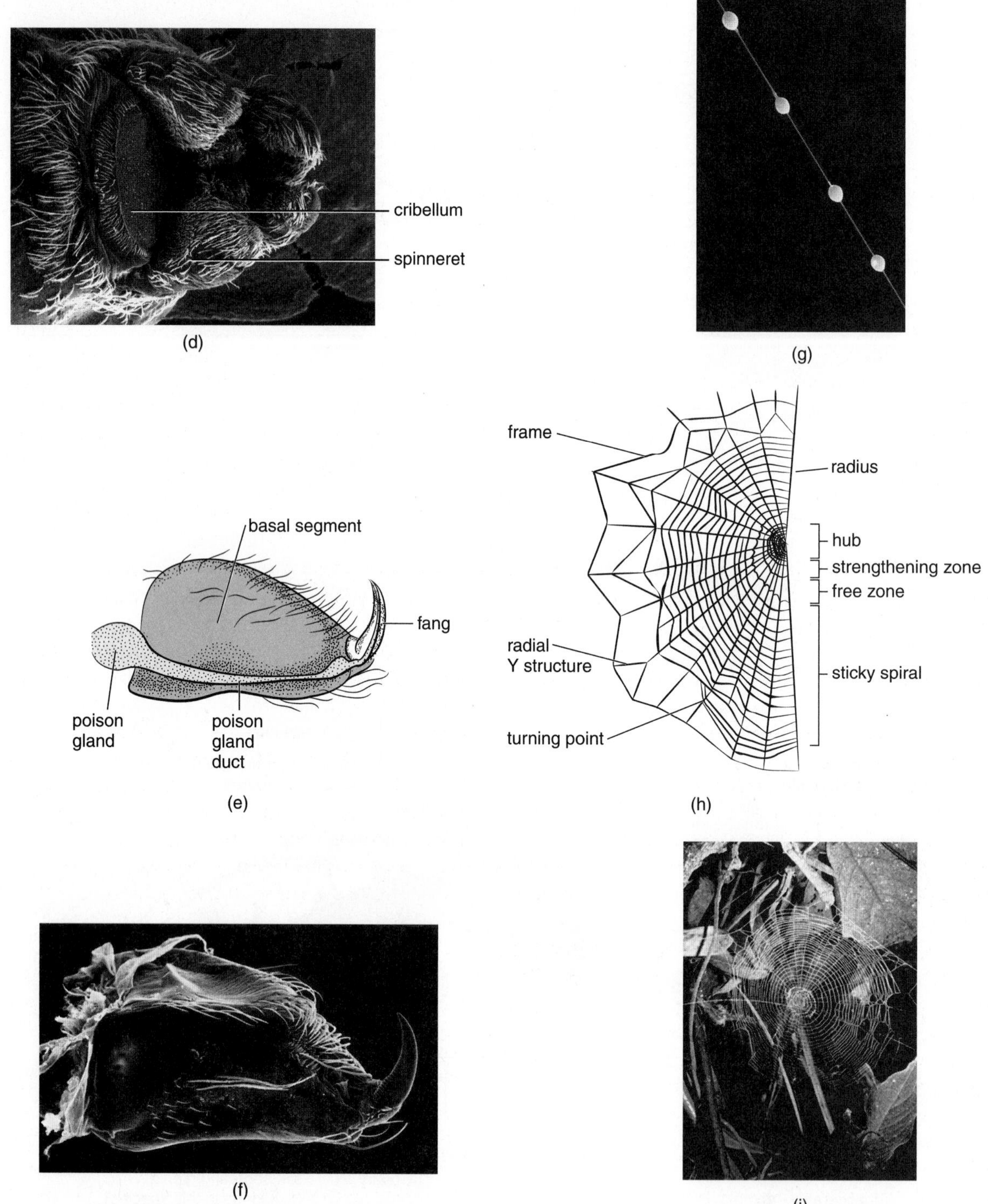

Figure 14.12 *Continued*

(d) Ventral view showing posterior portion of spider abdomen; note the 3 pairs of jointed, flexible spinnerets and the platelike cribellum, which bears additional spigots. Silk is extruded through the lumen of the hollow bristles. (e) Detail of an arachnid chelicera, consisting of a basal segment and a fang (seen here in lateral view). (f) Scanning electron micrograph of the chelicera of *Zosis geniculatus*. This species belongs to one of the few families lacking poison glands; the fang is used primarily for grooming rather than prey capture in this species. Prey are subdued by quickly wrapping them in silk. (g) A silk strand with adhesive droplets, from the spiral of an orb web built by *Mangora* sp. This silk can be stretched to nearly 3 times its resting length before it breaks. (h) Structure of an orb web. (i) Photograph of an orb web.

(a) After Sherman and Sherman. (b) After the Kastons. (c) Based on L. A. Borradalle and F.A. Potts, *The Invertebrates,* 2d ed. (d,f,g,i) Courtesy of Brent Opell. (e,h) Based on Foelix, *Biology of Spiders.*

connecting to internal abdominal glands that secrete silk proteins (Fig. 14.12c). These proteins are extruded through the spinnerets to produce silk, which may be used to form safety lines during climbing; egg sacs that protect developing embryos; fine threads for the aerial dispersal of newly emerged young; air-tapping diving bells for underwater foraging; and webs for trapping prey, building homes, or mating.[9] Humans also have put arachnid silk to good use, notably as crosshairs in optical equipment. (The silk used for over 4000 years to make silk clothing comes not from arachnids but from the cocoons of the silkworm *Bombyx mori* and some other lepidopterans, the order of insects including the butterflies and moths.) Spiders have been producing silk for webs for at least 130 million years.

An individual spider may contain 7 or more different silk glands that produce different, biochemically distinct forms of silk for different uses (Fig. 14.12c). Some nonarachnids (centipedes and larval insects) also produce silk, but only one kind.

The mites and ticks are contained in a separate arachnid order, the Acari, a group containing many species of economic and medical importance despite their small physical size. Few mites exceed 1 mm in length, and members of some species are only 100 μm long. The largest individuals are the ticks, which can reach 5–6 mm (and considerably more after feeding). The group is tremendously diverse and includes omnivores, carnivores, herbivores, fungivores, and parasites. Like other arachnids, mites and ticks feed exclusively on fluids, which they suck in through a muscular pharynx, and some species produce silk. Acarines occupy diverse habitats, such as wood, moss, ant colonies, bird nests, bat guano, water-filled tree holes, and decomposing vertebrates. Members live in freshwater, marine, and terrestrial habitats, with some species found 30 cm to 50 cm deep in desert sand dunes. Species are parasitic, as larvae or as adults, on a tremendous variety of hosts, including birds, lizards, humans (in hair follicles), mosquitoes, bears, frogs, butterflies, scorpions, spiders, dipterans (flies), beetles, chitons, slugs, grasshoppers, sea urchins, and various crustaceans. Some species are severe agricultural pests—directly or as vectors of plant viruses—on cranberry bushes, tobacco and tea plants, fruits, vegetables, flowers, and grasses (such as wheat, oats, and corn), while others greatly lower the quality of sheep wool. Many ticks transmit to humans a variety of diseases, including Rocky Mountain spotted fever (Fig. 14.11e), Q fever, Lyme disease (Fig. 14.11d), and encephalitis. Finally, many people develop allergies to the feces or exoskeletons of mites living in household dust.

Members of other arachnid orders, including the pseudoscorpions, solfugids, and opiliones (daddy longlegs), are described briefly in the *Taxonomic Detail* at the end of this chapter.

9. See *Topics for Further Discussion and Investigation*, no. 5.

Class Pycnogonida (= Pantopoda)

Class Pycno • gonida (= Panto • poda)
(G: thick knees [G: all leg])
pik-nō-gon′-id-ah

Defining Characteristics: 1) Body not divided into distinct regions (tagmata); 2) unique proboscis at the anterior end, with an opening at its tip; 3) variable numbers of walking legs among species

Zoologist Paul Meglitsch once wrote, "Pycnogonids are queer creatures, with queer habits," and that statement seems a good introduction to this group of some 1160 chelicerate species. Pycnogonids are known as sea spiders, since all species are marine and bear conspicuously long legs; the legs are typically about 3 times the length of the body and may be nearly 16 times longer than the body in some species. Many species have bodies only a few millimeters long, while a few deep-water and antarctic pycnogonids may exceed 10 cm in body length. Most of the body is prosoma; the abdomen (opisthosoma) is reduced to a short stump (Fig. 14.13a). Pycnogonids are found in all of the world's oceans, and have a fossil record extending back at least 425 million years. Possibly they are basal arthropods, from which all other arthropod groups evolved.

Unlike the true spiders, sea spiders lack specialized respiratory or excretory systems. They do, however, have a complete digestive system, with a sucking mouth that opens at the tip of an often greatly elongated proboscis. The digestive system extends well into the legs, as do the gonads.

Like arachnids, most pycnogonid species have 4 pairs of walking legs posterior to the pair of chelicerae and pair of palps (Fig. 14.13), although some species have 5 or 6 pairs of walking legs. In addition, the head bears a posterior pair of **ovigers,** which are used by both sexes to groom the other legs and the trunk and by males to carry the eggs after they are fertilized. Unlike most other arthropods, juvenile pycnogonids increase in size not only when they molt, but also during intermolt periods; the thin flexible membranes at the joints apparently stretch as the animal's tissue mass increases.

The adults are mostly free-living, although almost comically slow moving. However, the larva, if it leaves the egg mass before completing development, apparently grows as a parasite, particularly of cnidarians (such as hydroid polyps, anemones, and jellyfish). Many juvenile pycnogonids, and some adults, are parasitic or commensal in or on various marine invertebrates, including gastropods, bivalves, echinoderms, and jellyfish. Most adults are carnivores, feeding on prey that move even more slowly than they do: Bryozoans (moss animals), colonial hydrozoans, and sponges seem to be favored foods. Remarkably, the juveniles and adults of one well-studied species (*Pycnogonum litorale*) can withstand starvation for at least several months.

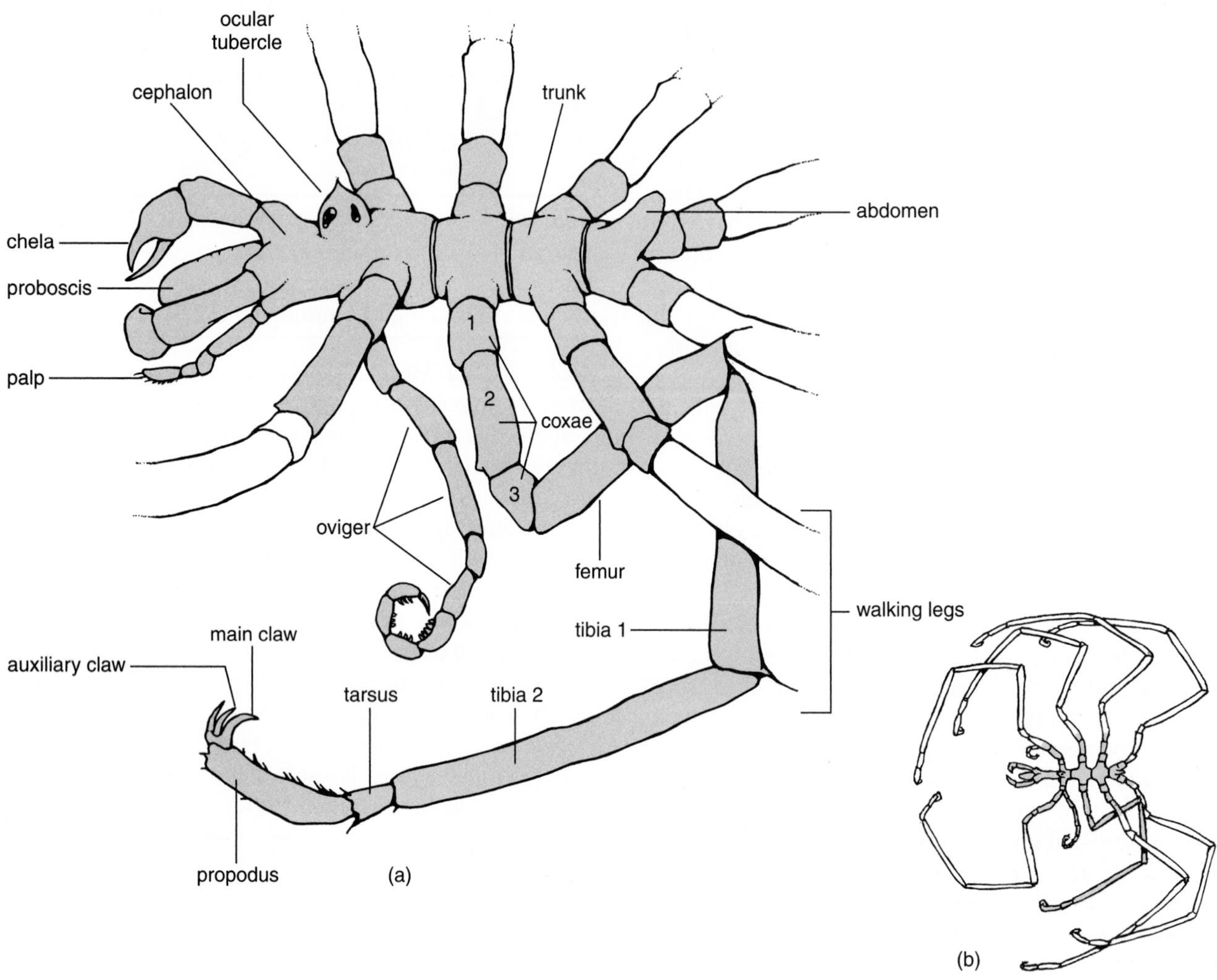

Figure 14.13

(a) Diagrammatic detailed representation of a male pycnogonid. The oviger is a leg modified for carrying an egg mass contributed by the female. (b) Entire pycnogonid in dorsal view, showing length of legs relative to body length.

Source: After Hamber and Arnaud, 1987, *Advances in Marine Biology*, Vol. 24, Academic Press; after Child, 1979.

Subphylum Mandibulata

Subphylum Mandibulata
(*mandibul* = L: a jaw)
man-dib´-ū-lah´-tah

Defining Characteristics: 1) Appendages on the third head segment are modified as mandibles, for chewing or grinding food; 2) retinula of compound eyes contains 8 cells

All members of the subphylum Mandibulata bear appendages (mandibles) on the third head segment that are modified for feeding. Mandibulates possess both uniramous and biramous appendages. Much molecular and morphological evidence currently supports this subphylum as a monophyletic group; support weakens when representatives of extinct taxa are included in the analyses. In its present configuration, the Mandibulata includes 3 classes (Myriapoda, Insecta, and Crustacea).

Class Myriapoda

Class Myria • poda
(G: many feet)
meer´-ē-ah-pō´-dah

Orders Chilopoda and Diplopoda

The Chilopoda (chil´-ō-pō´-dah) and Diplopoda (dip´-lō-pō´-dah) contain the centipedes ("one hundred feet") and millipedes ("one thousand feet"), respectively. Because of their many legs (Fig. 14.14), the members of both groups

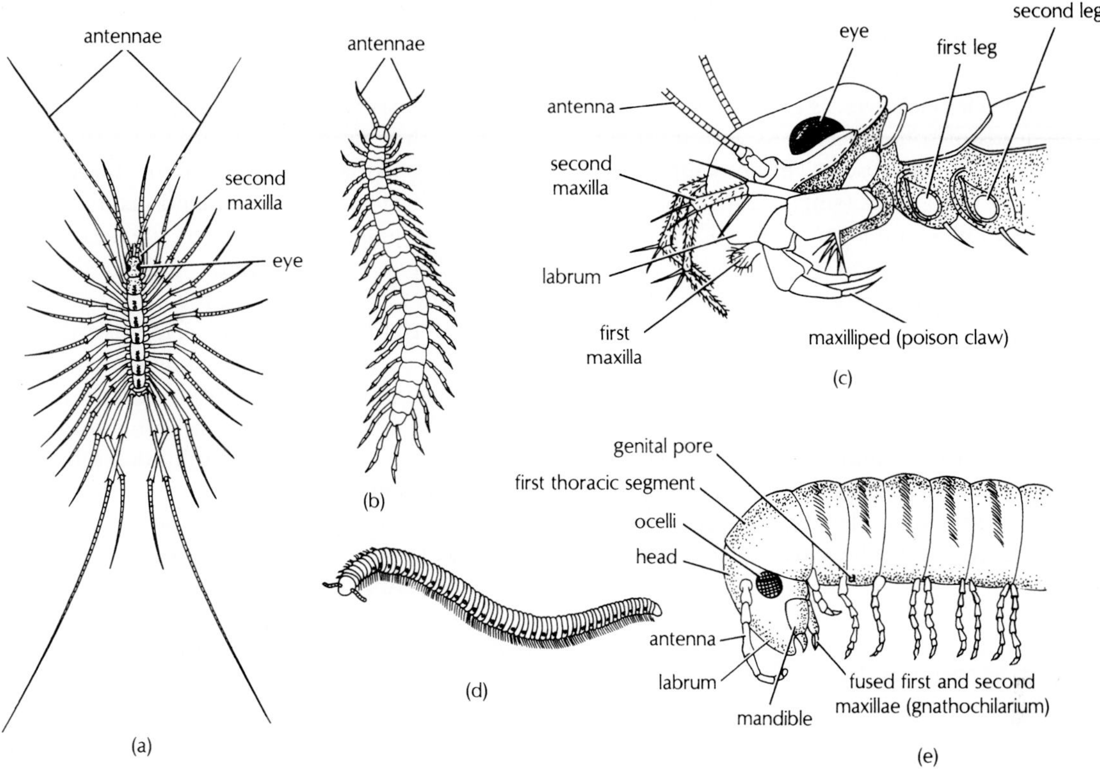

Figure 14.14

(a) A long-legged centipede, *Scutigera coleoptrata,* capable of especially rapid locomotion. Members of this species are seldom greater than 3 cm long; they are often encountered in moist areas (e.g., bathrooms) in buildings. (b) *Scolopendra* sp., a warm-climate centipede that may grow to a length of 25 cm. (c) Detail of centipede head, *Scutigera coleoptrata,* in lateral view. Note the large eye—actually a dense and organized cluster of ocelli—and the conspicuous poison claw. A poison-secreting gland is housed within the maxilliped. (d) Millipede, in dorsolateral view. (e) Detail of millipede head. Note that 2 pairs of legs are borne by each diplosegment of the abdomen.

(a) After Pimentel. (b,d) After Huxley. (c) After Snodgrass. (e) Sherman/Sherman, *The Invertebrates: Function and Form,* 2/e, © 1976, pp. 111, 112, 236, 172, 169, 45, 9, 15. Reprinted by permission of Prentice Hall, Upper Saddle River, New Jersey.

are referred to as myriapods ("many feet"). All of their appendages are uniramous (single-branched).

Chilopods are generally fast-moving carnivores, living in soil, in humus, under logs, and occasionally, in people's homes. Although most of the 3000 species are terrestrial, some are marine. The body is covered by a cuticle, but the cuticle is unwaxed. Moreover, respiration is accomplished by tracheae, but the spiracles cannot be closed. Thus, most centipedes are restricted to moist environments (or moist microenvironments) because of difficulty in restricting water loss. Many species conserve water by being nocturnal—that is, by avoiding the heat of the day and becoming active only at night.

The chilopod head bears a single pair of antennae, a pair of mandibles for chewing, a pair of first and second **maxillae,** and a pair of **maxillipeds.** The maxillipeds are modified for subduing prey (*Chilopoda* = G: jaw foot); they contain poison glands and resemble fangs. Chilopods often lack eyes; when present, the eyes generally are simple light receptors called **ocelli.**

The chilopod head is followed by 15 or more leg-bearing segments. Some species have **repugnatorial glands** on the ventral surface of each segment or on some of the legs themselves. These glands discourage predation by producing an adhesive ejaculate. A number of species produce silk from silk glands. Although most centipede species are long-legged runners, some species are adapted for burrowing through soil. In these species, the legs are reduced and thrust is generated by exploiting the properties of a hydrostatic skeleton, earthworm-style.

There are approximately 10,000 millipede species, over 3 times the number of known centipede species. In contrast to the chilopods, the diplopods are primarily slow-moving deposit feeders that plow through soil and decaying organic material. Some carnivorous species also exist. Pairs of segments have become fused in the millipedes, so each new segment (a diplosegment) bears 2 pairs of legs (*Diplopoda* = G: double foot) (Fig. 14.14e), as well as 2 pairs of spiracles and ventral ganglia. In many species, the integument (body covering) is impregnated with calcium salts, as in crustaceans. The covering of millipedes is therefore more protective against abrasion and predation than that of the centipedes. As in the centipedes, however, the cuticle is not waxy. Although many diplopod species lack eyes, as many as 80 ocelli are found on the heads of some species. As with most centipedes, compound eyes are absent. The head appendages consist of a pair of uniramous (single-branched) antennae, a pair of mandibles, and a pair of maxillae. Distinct second maxillae are lacking among millipedes. Instead, the first and second maxillae on each side are fused to form a single appendage (the gnathochilarium). Most species have an abundance of repugnatorial glands, which eject a variety of toxic, repellent secretions.

Both chilopods and diplopods are generally small animals, often only a few millimeters or at most 1 cm long, although some tropical species in both groups have been reported to attain lengths of nearly 30 cm. Myriapods are believed to be close relatives of the insects.

Superclass Hexapoda

(= *Hexa • poda*)
(G: six-footed)

Hexapods are 6-legged arthropods. Most species are insects, although a few groups of primitively wingless hexapods are contained in a separate grouping, the Entognatha (see *Taxonomic Detail,* p. 405). Molecular data suggest that these hexapods, including the common "springtails," branched off from a common ancestor before the evolution of insects, representing an independent invasion of land, and of the 6-legged condition. One group of primitively wingless hexapods, the silverfish, are true insects.[10]

Class Insecta

Class Insecta
(*insecti* = L: an insect)

Defining Characteristics: 1) Fusion of one pair of head appendages (the second maxillae) to form a lower lip (the labium); 2) loss of abdominal appendages

Insects have been reported from nearly every habitat except the deep sea. Although most species are terrestrial, many species live, either as adults or as larvae, in freshwater or in saltwater marshes (Fig. 14.15j, k). A few species (e.g., the ocean striders, all in the genus *Halobates*) live on the surface waters of the open ocean, although adult insects have otherwise been surprisingly unsuccessful in colonizing the ocean. Nearly 1 million species of insects have been described so far, with at least 4 times that number probably awaiting description. Something like another 95 million insect species now exist only as fossils. This tremendous number of species is in large part attributable to the feeding specializations, dispersal capabilities, and predator-avoidance possibilities associated with the evolution of flight. No other invertebrates, and relatively few vertebrate species, have evolved this capability. Indeed, when insects first evolved flight, they achieved access to a lifestyle previously unexploited by any other organism. Their adaptive radiation was thus unhindered by competition from other animal groups.

Insects are among the best studied of invertebrates, in large part because of their omnipresent impact on humans. Most flowering plants (angiosperms), including many species of agricultural importance, depend upon insects for pollination. Indeed, the present diversity of insect species may in large part owe to the proliferation and diversification of flowering plants over the past 100 million years or so, and to the complexity of their associations with insects. In any event, the lives of many insects and angiosperms are now inexorably linked, and the study of the mutual interdependence of plants and insects, and the evolution of those interactions, is a burgeoning field. Now for the bad news: The rapid rate at which many angiosperm communities are being destroyed likely will cause the extinction of many insect species around the world over the next 20 years.

Insects are also widely studied because they are significant vectors of human disease (e.g., malaria, bubonic plague, typhoid fever, yellow fever)[11] and are major threats to agriculture, both as predators (especially as larvae) and as vectors of plant diseases. A few insect species (e.g., bees, wasps, and some beetles) produce secretions that are toxic to humans and other animals. At least some tropical frog species obtain their toxins (used by humans to make their arrow tips poisonous) by ingesting large numbers of ants. Alternatively, other insect products (e.g., silk, honey, and beeswax) are of commercial importance; beeswax is a key ingredient in candles, cosmetics, facial creams, adhesives, crayons, inks, ski waxes, chewing gums, and waterproofing materials. Some insects are also being used to control the population sizes of other (pest) insect species. The most useful biological control agents are **parasitoids,** which develop inside developing embryos or larvae produced by other insect species, slowly devouring the hosts' tissues from the inside out and killing the host in the process. Parasitoids are especially common among

10. Thus, the former category containing all primitively wingless hexapods, the Apterygota, has been abandoned.

11. See Research Focus Box 14.1, p. 362.

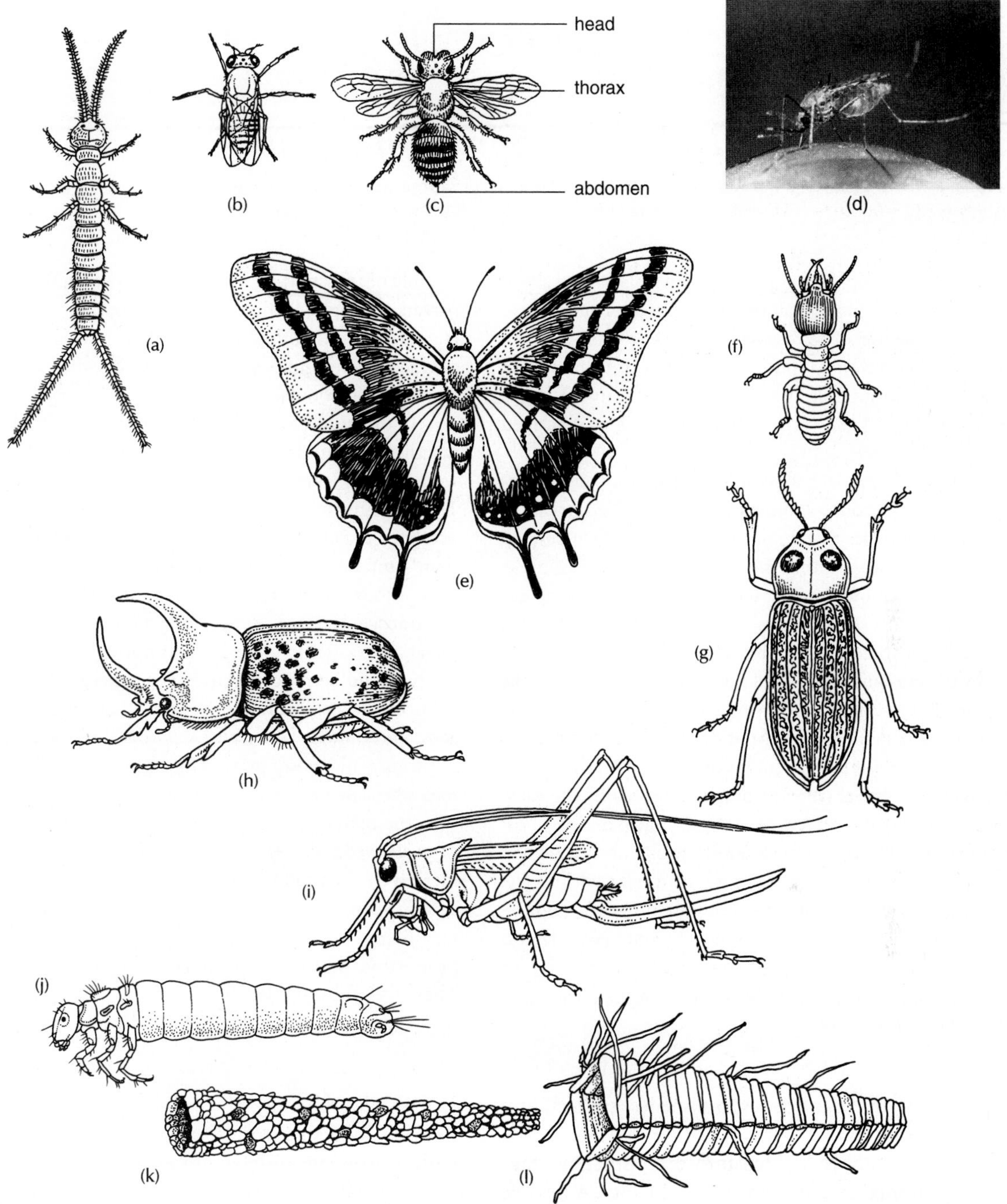

Figure 14.15

Insect diversity. (a) An apterygote, *Campodea staphylinus*. These, like the related silverfish, bristletails, and springtails, are primitive, wingless animals descended from wingless ancestors. (b) Fruit fly, *Drosophila* sp. (c) Leaf-cutter bee. (d) The mosquito *Anopheles gambiae* feeding on a mammalian host. This is one of the 3 main mosquito species that transmit the agent of malaria, a parasitic protozoan, throughout tropical Africa. (e) Butterfly. (f) Termite soldier. (g,h) Beetles. (i) Grasshopper. (j) Caddisfly larva. (k,l) Protective cases made by the larvae of 2 caddisfly species.

(a) After Huxley. (b,i) After Pimentel. (d) Courtesy of F. H. Collins. From Collins, F. H., and N. J. Besansky. 1994. *Science* 264:1874–75. Copyright © 1993. (f) After Romoser. (j,k,l) From McCafferty, *Aquatic Entomology*. Copyright © 1983 Jones and Bartlett Publishers. Reprinted by permission.

the true flies (order Diptera) and wasps (order Hymenoptera). Parasitoids are not usually parasitic as adults; the free-living adult females' major tasks are to mate and then to locate the ill-fated host for its offspring (Fig. 14.16). Perhaps 20% of all insect species are parasitoids, so the number of potential biological control agents is very large. Finally, the life histories, social interactions,[12] and division of labor seen among some insect groups are magnificently complex and have long occupied the attention of animal behaviorists.

12. See *Topics for Further Discussion and Investigation*, no. 9.

RESEARCH FOCUS BOX 14.1

Halting the Spread of Malaria

Marrelli, M. T., L. Chaoyang, J. L. Rasgon, and M. Jacobs-Lorena. 2007. Transgenic malaria-resistant mosquitoes have a fitness advantage when feeding on *Plasmodium*-infected blood. *Proc. Natl. Acad. Sci. USA* 104: 5580–83.

Several million people die every year from malaria, and about 3 billion people—nearly half of the human population—are at risk of infection. The disease is caused by protozoans in the genus *Plasmodium,* and is transmitted from person to person only through the bites of mosquitoes in the genus *Anopheles* (Fig. 14.15d). The malarial life cycle is described in Chapter 3 (p. 59).

In a recent approach to reducing transmission of the parasite from one person to another, biologists have been trying to engineer mosquitoes that do not allow the parasite to develop to the infective stage within the mosquito host. For example, once inside the mosquito host the malarial "eggs" become fertilized and then must move to the mosquito's stomach ("midgut") to become infective. Preventing such migration within the mosquito host would prevent transmission of the disease. In a previous study (Ito et al., 2002),* researchers showed that engineering mosquitoes to express the peptide SM1 in the midgut epithelium prevents such parasite migration, so that malarial parasites that are ingested by the mosquito with a blood meal will not get transmitted to a new host because they will not be able to develop to an infective stage.

If such engineered mosquitoes were released into the wild and that trait then became common in a natual population, mosquitoes in that area should become less and less capable of transmitting the disease. However, the trait won't spread throughout the mosquito population unless the engineered mosquitoes are fitter than normal mosquitoes; engineered mosquitoes that are less fit will be eliminated from natural populations by selection.

In this study, Marrelli et al. (2007) sought to determine whether mosquitoes engineered to express the SM1 peptide were more or less fit than normal ("wild-type") mosquitoes. They conducted their experiments using *Plasmodium berghei,* a malaria parasite species that infects rodents rather than humans, and two types of mosquitoes: transgenic mosquitoes engineered to express the SM1 peptide, and nontransgenic ("wild-type") mosquitoes that allowed the malaria parasites to undergo their normal migration within the host and become infective. The experiments were done using cages. Into each cage the researchers placed 250 transgenic mosquitoes (*Anopheles stephensi*) of one sex and 250 nontransgenic mosquitoes of the opposite sex. The mosquitoes were allowed to feed only on mice that were infected with the malaria parasite. After 4 to 6 days the mosquitoes were allowed to lay eggs, and 250 offspring from each treatment were then randomly selected and placed into another cage for continued rearing. The researchers thus reared the mosquitoes through 13 generations, and determined the proportion of transgenic individuals in each generation. They also sub-sampled females from the transgenic and wild-type populations after the females had been allowed to

*Ito, J., A. Ghosh, L. A. Moreira, E. A. Wimmer, and M. Jacobs-Lorena. 2002. *Nature* 417: 452–54.

Please note that centipedes, millipedes, spiders, and mites are *not* insects.

The insect body is divided into 3 conspicuous tagmata: head, thorax, and abdomen (Fig. 14.15c). A flexible joint separates the head and thorax. A pair of head appendages (the second maxillae) are fused to form a lower lip, the **labium.** This also occurs among a small group of myriapods (the Symphyla)—a convergence?

Two pairs of wings are generally carried dorsally on the thorax. The wings are outfoldings of the thoracic integument and consist of 2 thin, chitinous sheets. Among true flies (dipterans), the 2 hind wings have become modified into small club-shaped organs (**halteres**) that measure angular velocity, relaying to the fly information about its rotation in space. In many insect groups, some species have reverted to a wingless condition, and there is now some evidence that wings may have been secondarily reacquired in some of those lineages (see Chapter 2, pp. 23–25). In addition, the thorax generally bears 3 pairs of legs directed ventrally.

As with the myriapods, all insect appendages are uniramous. Evidence from fossilized insects in combination with a growing understanding of the genetic control of limb formation in arthropod development suggests that insect (and myriapod) appendages are only secondarily single-branched; that is, that they evolved from multibranched ancestral appendages. Insect legs are modified for walking, jumping, swimming, digging, or grasping, and are generally studded with a variety of sensory receptors, including receptors for taste, smell, and touch. Such receptors are also found on the mouthparts and elsewhere on the body. Current studies of how insect and crustacean limb and body movements are controlled may lead to new robot designs.

Many insects have organs of hearing, called **tympanal organs.** Typically, auditory sensory cells attach to a thin, external, vibrating membrane (the **tympanum**) associated with the tracheal air system. These "ears" respond to a wide range of sounds, including feed on infected mice, and determined the number of eggs produced by each of these females.

The results were striking. With each new generation, the proportion of transgenic mosquitoes in the caged population increased, rising from the initial 50% to about 70% after 12 to 13 generations (Focus Figure 14.1). The fitness of the engineered mosquitoes feeding on infected mice was thus higher than that of the wild-type individuals, mediated largely through an effect on fecundity: Infection by the malarial parasite reduced mosquito fecundity drastically in wild-type mosquitoes, probably because parasite development robs the mosquito of nutrients it would otherwise put into its own eggs; the fecundity of engineered mosquitoes was reduced far less, so that engineered mosquitoes feeding on infected mice deposited, on average, about 43% more eggs than wild-type mosquitoes feeding on infected mice. The engineered mosquitoes also showed higher survival in the laboratory. All in all, engineered mosquitoes were found to be more fit than wild-type mosquitoes when feeding on infected hosts.

So these results are encouraging, assuming that the results from this model system resemble those that would be obtained for humans and human malaria. Another strategy being investigated is to use the intracellular bacterium *Wolbachia* to help transmit engineered genes throughout natural populations.** These bacteria are found in at least 20% of all insect species (and interestingly enough also in filarial nematodes—see p. 443), and are readily spread through insect populations by direct transmission from mothers to their offspring.

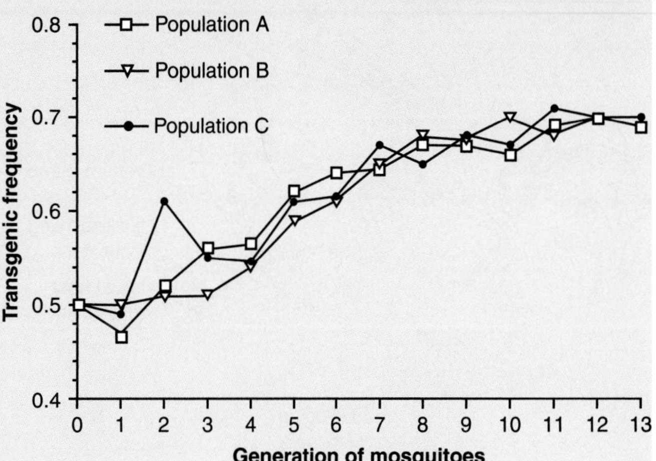

Focus Figure 14.1

Change in frequency of genetically engineered mosquitoes over time. Initial status of Population A: 250 transgenic virgin females and 250 wild-type males. Initial status of Population B: 250 transgenic males and 250 wild-type females. Initial status of Population C: 250 transgenic virgin females and 250 wild-type males from a different genetic line. The mosquitoes (*Anopheles stephensi*) were engineered to express a gene (SM1) that prevents early developmental stages of the rodent malaria parasite *Plasmodium berghei* from becoming infective to its rodent host.

Modified from Marrelli, M. T., L. Chaoyang, J. L. Rasgon, and M. Jacobs-Lorena. 2007. Transgenic malaria-resistant mosquitoes have a fitness advantage when feeding on *Plasmodium*-infected blood. *Proc. Natl. Acad. Sci. USA* 104: 5580–83.

**Xi, Z., C. C. H. Khoo, and S. L. Dobson. 2005. *Science* 310: 326–28.

those of very high frequency, such as those emitted by predatory bats; thus, many insects use their tympanal organs to provide early warning signals. Many insects also use high frequency sounds to communicate with each other.

The main light receptors are a pair of **compound eyes,** but 3 single-unit eyes (**ocelli**) are usually present on the head as well.

The insect head also bears 4 pairs of appendages: one pair of antennae (which are always uniramous or single-branched) and 3 pairs of mouthparts (Fig. 14.17). In sequence, the mouthparts are the **mandibles,** the **maxillae,** and finally, a pair of **second maxillae** that have fused to form a single appendage called the **labium.** The mandibles are shielded anteriorly by a downward extension of the head called the **labrum.** The precise morphology of these mouthparts varies considerably according to the insect's feeding biology (Fig. 14.18). The abdomen lacks appendages, except for a pair of sensory **cerci** borne on the last abdominal segment. The abdomen also may house receptors that monitor the degree to which the body wall is stretched during feeding.

Gas Exchange and Water Conservation

In keeping with a largely terrestrial lifestyle, an insect's gas exchange surfaces have been internalized. Gas exchange is accomplished in almost all insect species by means of a **tracheal system** (Fig. 14.19); in the few groups in which trachea are lacking (e.g., the primitive "springtails"), gases exchange by diffusion across the cuticle. Although resembling the tracheal system found in more advanced arachnids, the insect tracheal system is thought to have been independently evolved; that is, the tracheal systems in the 2 groups are probably convergent, evolving independently in different ancestors. One or 2 pairs of openings (**spiracles**) into the tracheal system are on the thorax, and additional pairs of spiracles are generally located on each of the

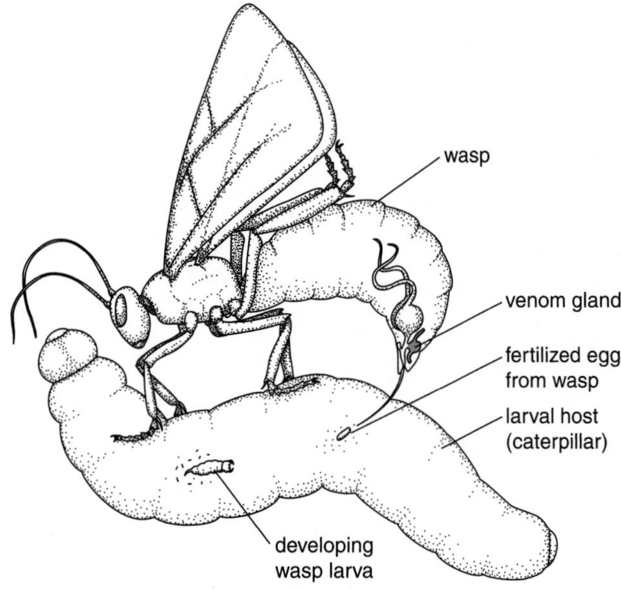

Figure 14.16
Female parasitoid (wasp) ovipositing into the larva of another insect species. Along with fertilized eggs, females typically inject a paralytic venom and sometimes other materials that prevent the host larva from completing its metamorphosis.
Modified from M. R. Strand and J. J. Obrycki. 1996. *BioScience* 46:422–29.

abdominal segments. The spiracles of most species can be closed, deterring evaporative loss of water. The tracheae are lined by cuticle, which is shed and resecreted by the underlying epidermis each time the insect molts, and the tracheal tubules are kept from collapsing by means of chitinous rings embedded in the walls. The tracheae branch to form a network of smaller tubules called **tracheoles,** which are less than 1 μm in diameter. These branch again and terminate directly on the insect's tissues. Thus, gas is exchanged between the tissues and the environment without the involvement of the blood circulatory system. High-definition X-ray observations show[13] that at least some insects contract and relax the tracheal walls, forcing air in and out in regular pulses: that is, many insects breathe. Some insect species lack tracheae, either as adults or during development. Gas exchange in such animals must occur across general body surfaces. Such surfaces obviously cannot be waxy, and the animals are thus restricted to moist habitats.

Water conservation is another correlate of a terrestrial lifestyle.[14] Uric acid is the primary end product of protein metabolism among insects; this nontoxic nitrogenous compound is excreted in nearly dry, solid form. The major excretory organs are long, slender, blind-ending tubes called **Malpighian tubules,** which, unlike nephridia, empty into the digestive tract (Fig. 14.20). Up to 250 pairs are found in the insect hemocoel. Waste products, notably a soluble derivative of uric acid, are actively transported

13. Westneat, M.W., *et al.* 2003. *Science* 299: 558–60.

14. See *Topics for Further Discussion and Investigation,* no. 2.

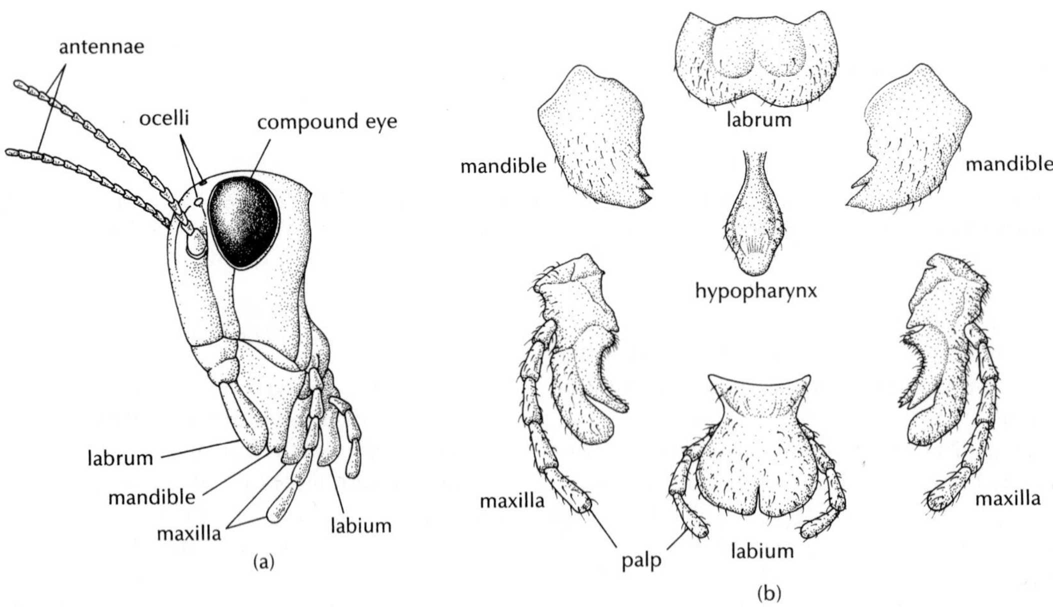

Figure 14.17
(a) Insect head, showing the various appendages and eyes. (b) Detail of appendages, drawn in proper orientation viewed anteriorly. The morphology of the mouthparts differs widely among species and correlates with feeding biology. In butterflies and moths, for example, the maxillae are greatly elongated for taking up nectar.
(a) After Snodgrass. (b) After James and Harwood.

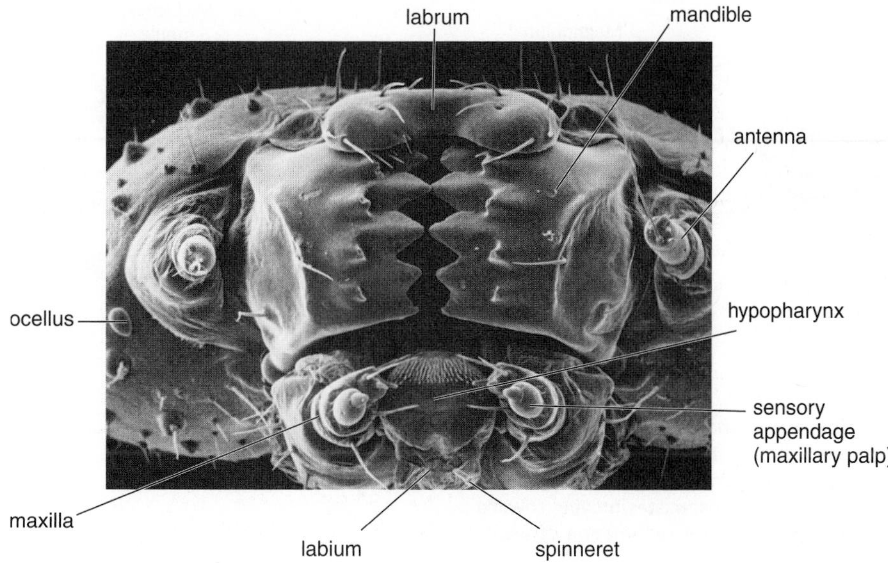

Figure 14.18
Scanning electron micrograph of the head of an insect, the tobacco hornworm *Manduca sexta* (caterpillar stage).
Courtesy of Nancy Milburn.

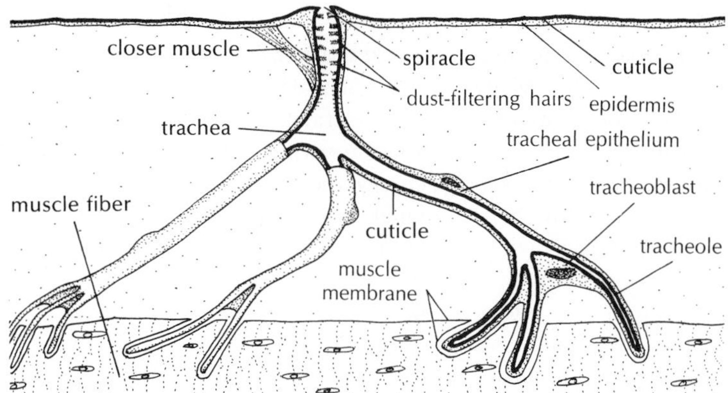

Figure 14.19
The insect tracheal system. The spiracles of terrestrial insects can generally be closed to regulate water loss by contracting appropriate musculature. The finest tubes of the tracheal system, the tracheoles, develop from the tracheoblast cell.
After Chapman; after Meglitsch.

from the blood into the distal, blind-ending portion of the Malpighian tubules. Increased acidity in the proximal portion of the tubules causes the uric acid to precipitate out of solution. Most of the water contained in the urine is then resorbed during its passage through the rectum.

Insect Flight

One feature that sets the insects apart from all other invertebrates, and most vertebrates as well, is the ability to fly, an innovation that first evolved some 300 to 400 million years ago.

The major characteristics that make flight possible in the Insecta follow:

1. abundance of striated muscle specialized for rapid, strong contractions;
2. muscle antagonism by means of a lightweight, jointed skeleton, permitting a great amount of movement to be generated from relatively short changes in muscle length;
3. small body size;
4. water-impermeable outer body covering, preventing dehydration;
5. efficient systems for gas exchange, nutrient storage, and distribution of nutrients to the musculature; and
6. highly developed nervous and sensory systems for steering, navigating, and sensing wind direction.

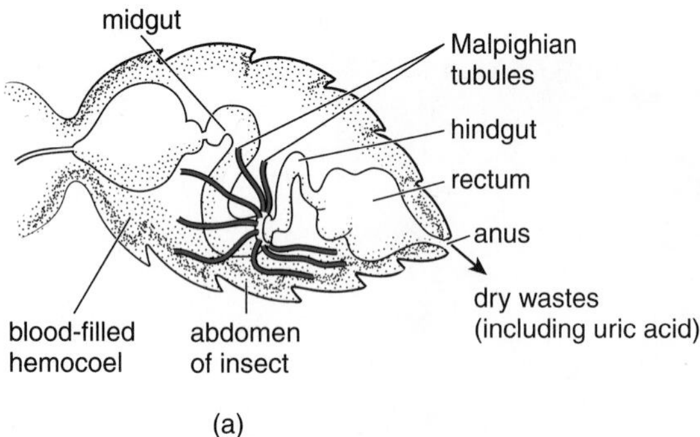

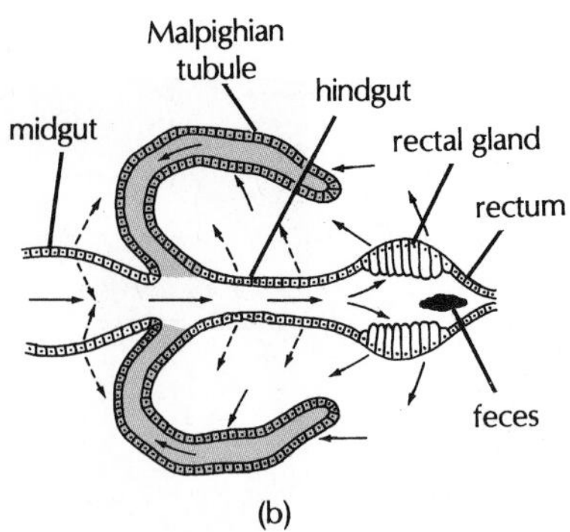

Figure 14.20

(a) Malpighian tubules in the insect abdomen. (b) Diagrammatic illustration of the relationship between the Malpighian tubules and the posterior portion of the digestive tract. Fluid moves from the hemocoel into the tubules, where it joins wastes moving toward the anus. The arrows indicate the extensive reclamation of water that occurs in the hindgut and rectum.

(a) Based on Purves and Orians, *Life: The Science of Biology,* 2d ed. 1983. (b) After Wilson; after Wigglesworth.

The insect wing is a lateral outfolding of the body wall. These outfoldings are very thin and lightweight, and are structurally supported by a characteristic network of veins connecting with the blood circulatory system, by tracheal tubules, and by rows of pleats radiating outward from the base of the wings to the tips.

No one knows how insect flight evolved, although various possibilities have been much debated.[15] Many flying insects use their wings to regulate body temperature, either by varying the amount of beating activity or the amount of wing exposed to sun and wind. Accordingly, some researchers have argued that the selective benefits associated with the evolution of the first insect wings likely had more to do with temperature regulation than with flight. Others believe that wings evolved from lateral outgrowths originally used to stabilize ancestral forms during jumping. Still others argue that insect wings probably evolved directly from the gills of aquatic mayfly-like developmental stages, which currently use their gills for both gas exchange and locomotion; such gills may have been used first in skimming across the water surface, as in modern stonefly nymphs, as a prelude to flapping flight. This hypothesis has recently gained additional support from similarities in the patterns of gene expression documented during the development of gill-like appendages in crustaceans and of wings in insects. Whatever their origins, the evolution of wings opened up to insects a lifestyle that is virtually inaccessible to most other animals.

Flight requires the generation of both lift and thrust, which flying insects achieve through a combination of body shape, wing morphology, and highly complex wing behavior.[16] Thrust is something we have an intuitive feel for; it is the force we exert in one direction that creates motion in the opposite direction (every action produces an equal and opposite reaction). The generation of lift is more mysterious. The secret is contained in the equation that follows, modified from the original of Daniel Bernoulli (1700–1782). When dealing with flight, the equation pertains to air moving across a solid surface, such as a wing:

$$\frac{1}{2} dv^2 + p + dgh = \text{a constant}$$

where d = density of the air; v^2 = the square of the air velocity relative to the wing; p = air pressure at the wing's surface; g = the gravitational constant; and h = the height of the air above the wing surface. The term dgh is related to potential energy and the $\frac{1}{2}\, dv^2$ term is related to the expression for kinetic energy. When we fly in an airplane, the principles embodied within this equation are what keep us aloft. In this situation, the air we are talking about does not change density and is always the same height above the wing, so d and dgh are constants. The equation states, then, that if the velocity of the air moving across the wing increases, the pressure above the wing surface must decrease; this must occur if the sum total of the 3 expressions on the left side of the equation is to remain constant. This pressure decrease above the wing produces lift; that is, the pressure below exceeds the pressure above, and the body rises. You can prove to yourself that lift is generated by differential air flow over the upper and lower surfaces of an object by blowing along the length of a strip of paper. Try a strip about 1 inch wide and 6–8 inches long, holding one end of the paper just below your mouth. If you blow hard enough along the length of the strip, the end of the paper will rise.

Obviously, an insect does not blow over its wings to generate lift. Neither does the airplane propeller or jet engine function by blowing air over the wings' upper surfaces. Instead, the beating of an insect wing, like the spinning of a propeller, moves the wing through the air. Wing beat frequencies vary from less than 10 to more than 1000 beats per second in different insect species.

While flying, an insect is not simply moving its wings up and down. Instead, the wings are continually being brought either forward or backward and bent or twisted

15. See *Topics for Further Discussion and Investigation,* no. 19.

16. See *Topics for Further Discussion and Investigation,* no. 6.

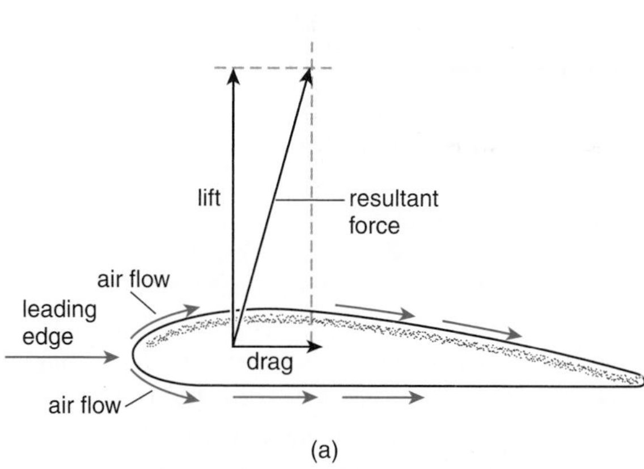

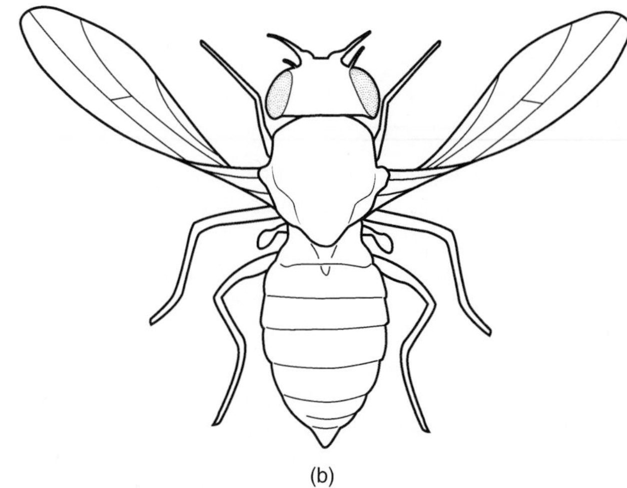

Figure 14.21
(a) The principle of lift generation for an airfoil. Because the upper surface is convex, air moves more quickly over this surface than over the lower surface: The air must move a greater distance in the same amount of time. By Bernoulli's principle, this lowers the relative pressure above the surface, creating net lift as illustrated.

(b) The wings of the fly *Drosophila melanogaster* beginning to move back upwards. The bending of the wing causes a short-lived but substantial increase in lift.

(a) Based on *Life in Moving Fluids: The Physical Biology of Flow* by Steven Vogel 1981. (b) From R. Wootton. 1999. How flies fly. *Nature* 400: 112–13.

(Fig. 14.21b). The net effect is that the air moving over the wing's upper surface has to travel farther (and thus faster) to reach the back of the wing than does the air moving over the lower surface, producing lift. When lift exceeds the insect's weight, the insect rises. The principle is most easily illustrated with the forward movement of a simple shape termed an **airfoil** (Fig. 14.21a). Since the entire airfoil moves a given distance per unit of time, the shape of the airfoil ensures that air moves faster across the upper surface than across the lower surface, generating lift according to Bernoulli's principle. By varying the angle at which the wing moves through the air (the "angle of attack"), the insect can alter the direction of the net forces generated, producing movement, via thrust, in the direction of choice.

Several other factors, including the shape of the insect's body and what are called "non-steady-state" flight dynamics, may also contribute to lift. Bernoulli's equation, discussed above, applies only to "steady-state" flight, in which lift is generated from a steady flow of air over the upper wing and body surfaces, as is the case for standard aircraft. But because insects generate propulsive force by flapping their wings, they typically also exhibit "non-steady-state" flight dynamics, with additional lift being generated by small whirlpools of air (vortices) that swirl around the long axis and tips of the wings; these vortices are often shaped, at least in part, by "clap and fling" motions, in which the wings "clap" together at the end of the upstroke and then "fling" air into a lift-generating vortex. Very small insects can't induce lift at all by the steady-state mechanism previously described because they operate at very low Reynolds numbers (see p. 5); they rely, therefore, upon unsteady mechanisms exclusively, while many larger insects use non-steady-state mechanisms only as a supplement, possibly decreasing the energetic cost of flying.

Most insects have 2 pairs of wings, although many species have only a single pair. Beetle species, with 2 pairs of wings, typically use only one pair for flying, holding the hardened, protective first pair out of the way while flapping the second pair.

Insect species differ considerably with respect to wing morphology and the manner in which the wings are operated by the thoracic musculature. Much of the modification in wing structure and function encountered among different insect groups seems to reflect selection for increased energy efficiency and increased fine directional control during flight. Many insects can hover and even fly backward, a great advantage for mating and egg laying "on the wing." Other morphological modifications serve to protect the insect body or the wings themselves.

In many of the faster-flying insects, the flight muscle is highly specialized. In these species, the muscle fibers are capable of contracting many times following stimulation by a single nerve impulse. This type of flight is termed **asynchronous flight,** since wing-beat frequency does not correspond to the frequency of nerve impulse generation. The wing beat frequencies of over 1000 beats per second that have been recorded in mosquitoes are made possible by this unique insect invention.

Flight, of course, implies mental activity as well as physical complexity: Flying insects must, for example, be able to orient to their surroundings and to adjust their speed and direction for smooth landings, and social insects (see p. 372) must be able to communicate information about their flights to nestmates (Research Focus Box 14.2). Detailed studies of insect flight behavior may lead to the design of more efficient computerized navigational systems.

RESEARCH FOCUS BOX 14.2

Honeybee Navigation

Srinivasan, M. V., S. Zhang, M. Altwein, and J. Tautz. 2000. Honeybee navigation: Nature and calibration of the "odometer." *Science* 287: 851–53.

In the 1930s and 1940s, Karl von Frisch showed that honeybees, upon locating a distant source of nectar, inform their nestmates of its location and its distance from the hive by repeatedly performing a "waggle dance" back at the nest. Direction is indicated by the dancing bees' orientation relative to the current position of the sun (Focus Fig. 14.2), while distance is communicated by how long each iteration of the dance lasts. How do bees accurately judge the distance they have flown from the food source to the hive? They might estimate distance from their flight time, or from the amount of energy expended during flight, but both characteristics would surely vary with wind conditions.

Another, perhaps more attractive possibility (suggested by von Frisch) is that honeybee scouts estimate distance flown from the amount of visual information that they process *en route:* that is, from the amount of what researchers call "visual flow." To test this longstanding hypothesis that honeybees estimate distance from the amount of visual information perceived during flight, Srinivasan et al. (2000) first marked some bees, allowed them to forage at outdoor feeding stations located 60 to 350 m* from the hive, and then timed the bees' dances after their return to the hive. Observers at the feeding stations noted the arrival of marked bees, so that they knew which feeding stations each of the marked bees had visited, and thus they also knew the actual distance flown by each bee from the feeding station back to the hive. Thus, the researchers could now determine the relationship between distance flown and duration of the waggle dance enacted (Focus Fig. 14.3). Note that r^2 is very close to 1.0, which means that nearly all (> 99% in this case, because $r^2 = 0.998$) of the variation in waggle dance duration was explained by variation in distance flown. Also note that the slope of the line is 1.88, which in this case means that the dance duration increased by 1.88 msec** for each additional meter flown. Now the researchers were in an excellent

*meters

**milliseconds (1/1000th of a second)

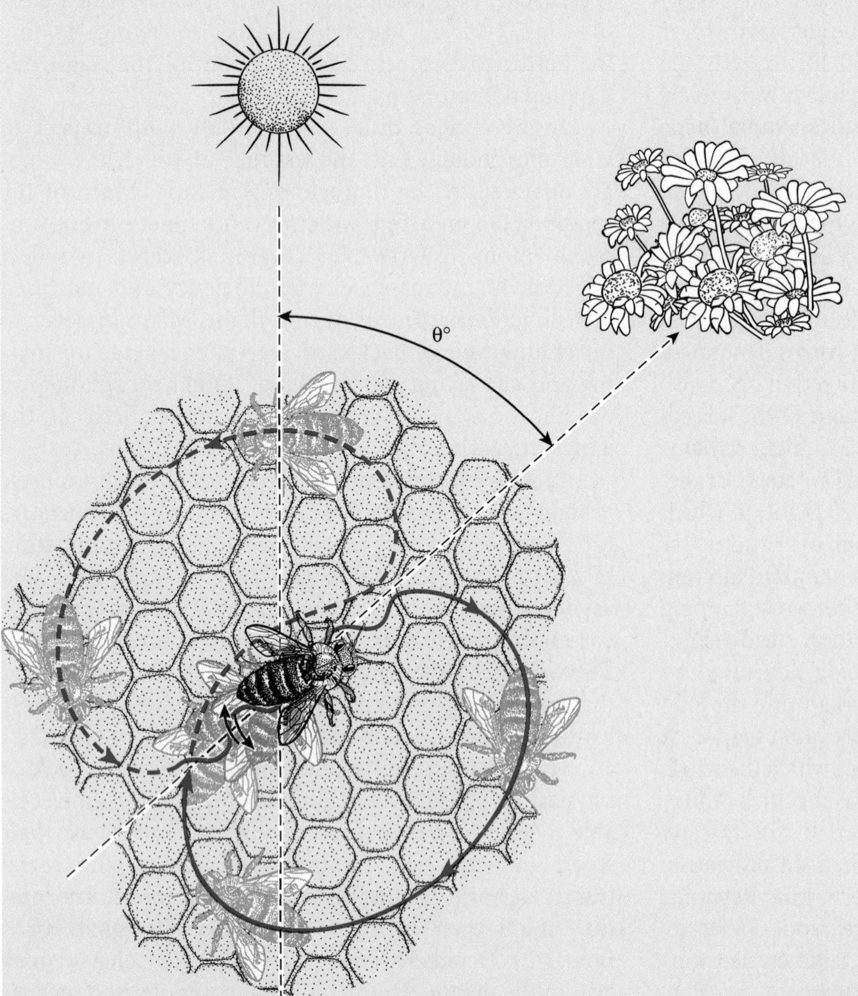

Focus Figure 14.2
The waggle dance of honeybees. Bees returning from a distant food source walk a short distance in a straight line, waggling their bodies rapidly from side to side as they walk, and then walk in a semicircle alternately to the right and to the left. The angle of the "waggle" portion of the dance (wiggly line) conveys information about the location of the food source relative to the position of the sun.
Based on drawings in K. von Frisch. *Dance Language and Orientation of Bees*. 1967.

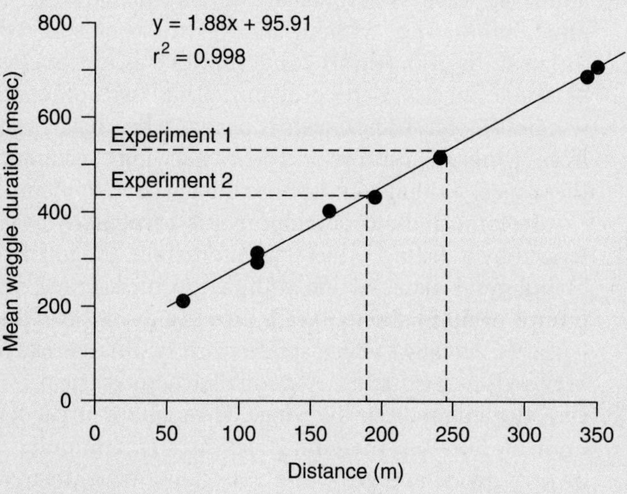

Focus Figure 14.3

Relationship between distance flown and mean duration of waggle dances by honeybees. Feeding stations were placed at different distances from a hive, from 60 to 350 m, and the activities of marked bees were recorded. Each point represents the mean duration of 65–345 waggle phases during 7–23 dances performed by 3–10 bees.

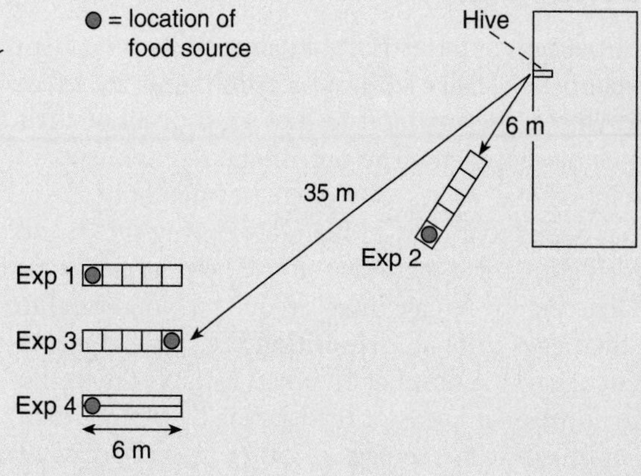

Focus Figure 14.4

Summary of the 4 experiments, showing pattern on the tunnel walls, placement of the food source (circle), and distance of the tunnel from the beehive. See text for details.

position to play Fool the Bee: The plan was to alter the amount of visual information perceived by marked, flying bees without altering the actual distance that they flew, and then see from their dances how far the bees *thought* they had flown. The scientists conducted 4 experiments (Focus Fig. 14.4).

Here the experiments get really ingenious. The researchers constructed a tunnel (11 cm wide, 20 cm high, and 6.4 m long) with only one opening, and placed it with its entrance 35 m from the beehive (Focus Fig. 14.4, Experiment 1). A sugar solution (food source) was placed inside the tunnel, at the closed end. The tunnel walls were lined with a very busy design to increase the amount of visual information that bees would see as they flew. Bees returning from the feeding station in this first experiment waggled for an average of about 529 msec**, which indicates, from Focus Figure 14.3 (Experiment 1), a perceived flight distance of 230 m even though the bees had really flown only 35 m + 6 m = 41 m. Even when the tunnel was later moved so that its entrance was only 6 m from the hive (Focus Fig. 14.4, Experiment 2), the returning bees waggled for 441 msec, equivalent to an outdoor distance of 184 m (Focus Fig. 14.3, Experiment 2). Apparently, the visual input received by bees flying just the length of the 6 m tunnel was perceived by the flyers as a distance of more than 200 m in the first instance, and more than 150 m in the second!

If the researchers placed the feeding solution at the entrance to the tunnel (Focus Fig. 14.4, Experiment 3), so that foraging bees could feed without entering it, most of the returning bees didn't do a waggle dance at all (Focus Fig. 14.5, Experiment 3)—they did a different sort of dance (a round dance), one they do only when they fly less than 50 m. Here, the bees gave—back at the hive—an accurate report of the distance traveled, as they also did if the researchers placed the sugar solution back at the end of the tunnel but lined the walls of the same tunnel with stripes oriented in the direction of flight (Focus Figs. 14.4, 14.5, Experiment 4), thus reducing the amount of visual stimulation offered to flying bees. Thus, there was nothing about the sight or smell of the tunnel itself that was confusing the bees about the distance flown.

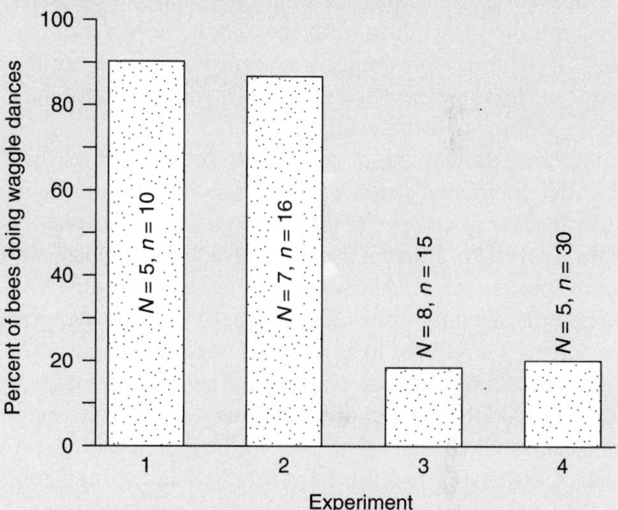

Focus Figure 14.5

Proportion of marked honeybees doing waggle dances after returning from a food source. N = number of bees dancing, n = number of dances analyzed. Bees that did not do waggle dances did dances associated with food sources located within 50 m of a hive. For experiments 1 and 2, bees had to fly through a 6 m tunnel lined with a visually busy pattern in returning from the food source. The tunnel was much closer to the hive in Experiment 2 than in Experiment 1 (see Focus Fig. 14.4). In the other 2 experiments, the tunnel walls were either covered with stripes running lengthwise (Experiment 4) or bees did not have to enter the tunnel to reach the feeding solution (Experiment 3) (Focus Fig. 14.4). Note that about 90% of all dances were waggle dances in the first 2 experiments, and that hardly any of the dances were waggle dances in the second 2 experiments.

It seems that honeybees record and report not absolute distance, but rather the amount of visual stimulation perceived during the flight back to the hive from a food source. Can you think of any alternative explanations for the data reported? If the authors are correct, the calibration curve shown in Focus Figure 14.3 should be different for different natural environments, depending on how much visual stimulation each environment offered. How would you test that hypothesis?

Insect Development

The fertilized eggs of terrestrial animals require some form of protection, especially from desiccation, and are often provided with sufficient food to fuel most or all of their prejuvenile development. The nutritional requirements of the developmental stages can often be met only if the female has access to a high-protein diet during the period of egg formation, or **oogenesis** (*oo* = G: egg; *genesis* = G: birth). Hence, many female insects require a blood meal to mature their eggs prior to **oviposition** (i.e., discharge and placement of eggs). A number of insects, notably the wasps, meet the nutritional needs of their larvae by placing their eggs in or adjacent to the eggs of other insect species or within the bodies of other adult insects, which are then devoured by the developing young from the inside out.[17] Some insects deposit their eggs in plants, which respond by forming protective galls. The eggs are inserted into these various substrates through a long tube, called an **ovipositor,** typically protruding from the abdomen. Harvestmen (class Arachnida) are similarly equipped, and for a similar purpose. In some insects (e.g., bees), the ovipositor has been modified to form a stinger.

During development, insects pass through several distinct developmental stages, called **instars.** This is conspicuously true for insect species that undergo a **metamorphosis** from a larval to a distinctly different adult body plan. In some species, this transition is gradual, and the different instars are called **nymphs** (Fig. 14.22a, b). Aquatic nymphs are sometimes referred to as **naiads.** Dragonflies, grasshoppers, and cockroaches, for example, develop in this manner and are said to be **hemimetabolous** (*hemi* = G: half; *metabolo* = G: change) (Fig. 14.23a). In most other insect species, the change to adult form is radical and abrupt, and termed **holometabolous** (*holo* = G: whole; *metabolo* = G: change). The feeding, immature stages are termed **larvae** (Fig. 14.22c, d). After passing through several larval instars of ever-increasing size, a morphologically distinct, nonfeeding pupal stage is formed. The **pupa** then undergoes extensive internal and external reorganization to form the adult morph. Butterflies provide what is probably the most familiar example of holometabolous development (Fig. 14.23b). Wasps and ants are other noteworthy examples, with the adults laboriously tending the helpless larvae and pupae. Holometabolous development characterizes about 88% of all extant insect genera, as opposed to less than 50% of insect genera known from 250-million-year-old fossils; life histories with ecologically distinct larval stages and a dramatic transition to adulthood clearly have had a selective advantage over those exhibiting more gradual development to the adult stage.

Adults of wingless species, such as bristletails, do not exhibit a pronounced metamorphosis as they develop. Instead, immatures simply get larger with each succeeding molt, and the body plan resembles that of the final adult at each stage. Such development is termed **ametabolous** (i.e., without change) and seems to reflect the primitive (ancestral) condition; that is, the ancestors of these species were probably wingless. Some other wingless insect species, such as fleas and lice, have evolved from winged ancestors and lost the wings secondarily; these species do metamorphose during development.

Holometabolous development is particularly characterized by a distinctly insect phenomenon, the formation of **imaginal discs.** At the completion of cleavage, small groups of up to about 30 cells give rise to discrete "discs" (spheres, actually), which are destined to differentiate into very well-defined adult epidermal structures, such as the eyes, the antennae, or the wings. Throughout larval development, however, these imaginal discs remain quiescent, or they divide and grow at a very slow rate compared to the rest of the larva. The cells of one disc are distinguishable from those of other discs found elsewhere in the body only by their position and a small number of structural details. Nevertheless, the eventual fate of the cells in a disc is fixed, and at metamorphosis a particular disc always will give rise to the same structure, even if surgically transplanted elsewhere in the body. The orientation of the structure within the body may be incorrect following such disc transplantation, but the structure itself will be perfectly formed. Imaginal discs long have been utilized by developmental biologists to probe the manner in which the expression of genes in individual cells is controlled.

Molting and metamorphosis during insect development are under complex environmental and hormonal control. The molting process—in particular, the resorption of some of the old cuticle and the development of new cuticle—is triggered by a steroid called **ecdysone,** but ecdysone production is itself regulated in a surprisingly complex manner. Neurosecretory cells in the brain secrete a **prothoracicotropic hormone (PTTH),** a polypeptide that activates (*tropic* = G: stimulate) a pair of glands in the anterior portion of the thorax, the **prothoracic glands (PG)** (Fig. 14.24). The PG, in turn, typically secrete an ecdysone precursor that is rapidly converted to ecdysone, the molting hormone, by particular enzymes in the hemolymph or target tissues; the ecdysone in turn is converted to a final, active form (20-hydroxyecdysone) by cells in target tissues, triggering the molting process. The PG thus seem to perform the same role as the Y-organ in crustaceans. Adult insects do not molt, and the prothoracic glands degrade during pupal development or shortly after metamorphosis to adulthood is completed.

The extent to which morphological differentiation occurs during molting depends upon whether another hormone, **juvenile hormone (JH),** is present in the blood at certain critical periods **(gates).** JH, a sesquiterpene lipid, is produced by yet another gland, the paired **corpora allata,** located just behind the brain. Experiments have demonstrated that high quantities of JH in the blood generally inhibit differentiation. In particular, the switch from larval to pupal development is inhibited

17. See *Topics for Further Discussion and Investigation,* no. 4.

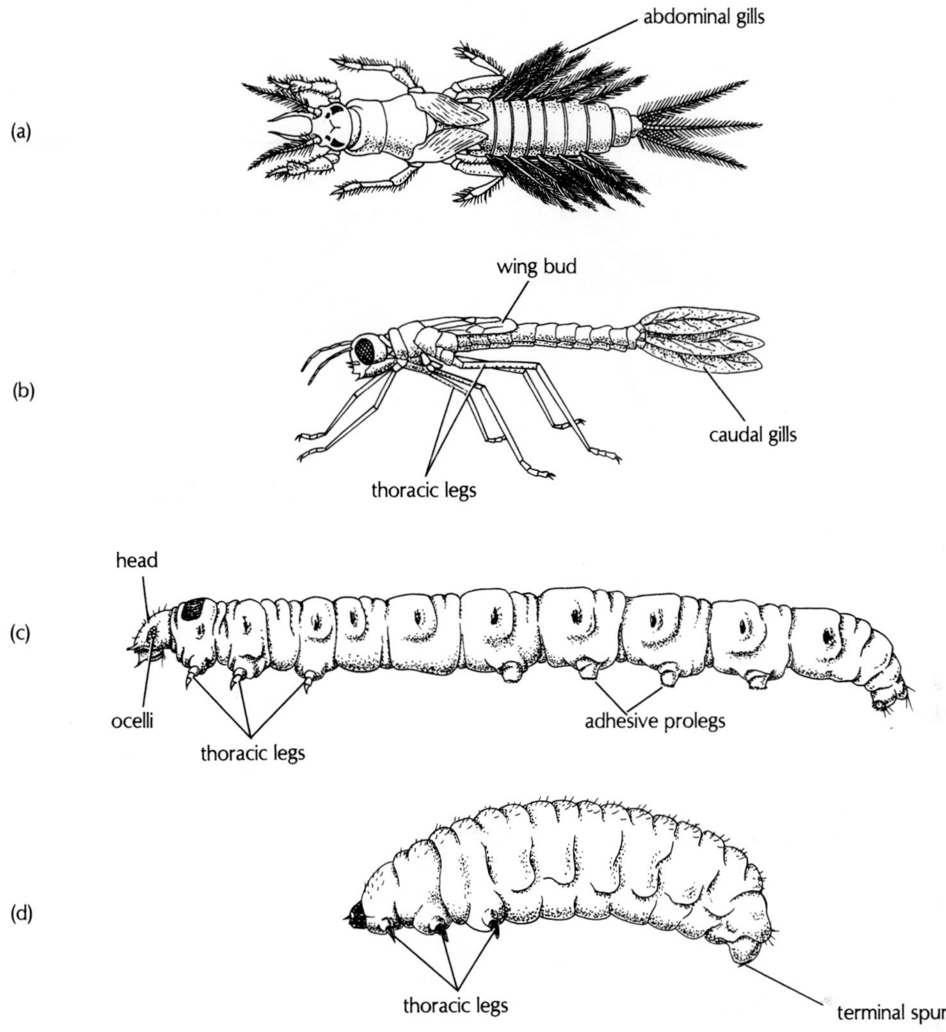

Figure 14.22
Insect development. The nymphs show a gradual transition to the adult form, exhibiting hemimetabolous development. In contrast, the larval stages of moths, beetles, and related species show little resemblance to the adult; development to adulthood is radical and abrupt, and is termed holometabolous. (a) Mayfly nymph, *Ephemera varia.* (b) Damselfly nymph. (c) Moth larva, *Bellura* sp. (d) Beetle larva *Donacia* sp.
(a) After Pennak: after Needham. (b) After Sherman and Sherman. (c,d) After McCafferty.

by JH; insects metamorphose only after the corpora allata cease JH production and all circulating JH has been destroyed by specific enzymes in the hemolymph. The normal sequence of insect development depends upon pulses of JH secretion being critically timed to coincide with gates of sensitivity of target tissues to the hormone. JH also plays other important roles during development—for example, in determining caste in social insects and in stimulating yolk deposition during oogenesis in females and, at least in some species, sperm development in males. A number of intriguing behavioral patterns also are known to be under hormonal control. Curiously, peptide and protein hormones first identified as endocrine products in humans have recently been found in the insect nervous system; at least some of these hormones seem to control aspects of insect physiology and behavior. Recent studies indicate that juvenile hormone precursors (including methyl farnesoate and farnesoic acid) are synthesized by crustacean mandibular organs, which may be homologous with insect corpora allata; the role played by these compounds in crustacean biology is still unclear, although it seems they may regulate aspects of female gametogenesis.

A number of insects enter into a resting state (**diapause**) at some point in their development, as an adaptation for withstanding adverse conditions, such as cold winters. Entrance into the diapause state is under hormonal control (JH secretion again plays a role here) and is often triggered (and later released) by changing day length (**photoperiod**) and/or changing temperature. The stage at which diapause occurs is species dependent: In some species, an early embryonic stage enters diapause; in other species, either a larval instar, the pupal instar, or even the adult typically enters diapause. Diapause is also commonly encountered among some crustaceans (copepods and branchiopods).

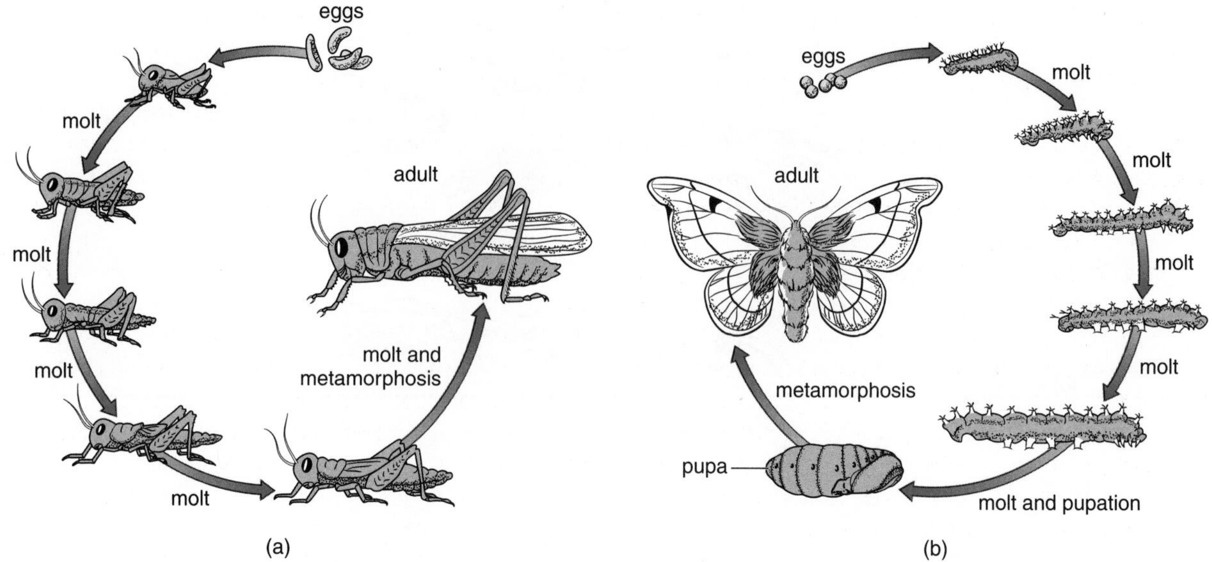

Figure 14.23

(a) Hemimetabolous development of a grasshopper.
(b) Holometabolous development in the silkworm moth.
Based on *General Endocrinology*, by C. Donnell Turner, 1948.

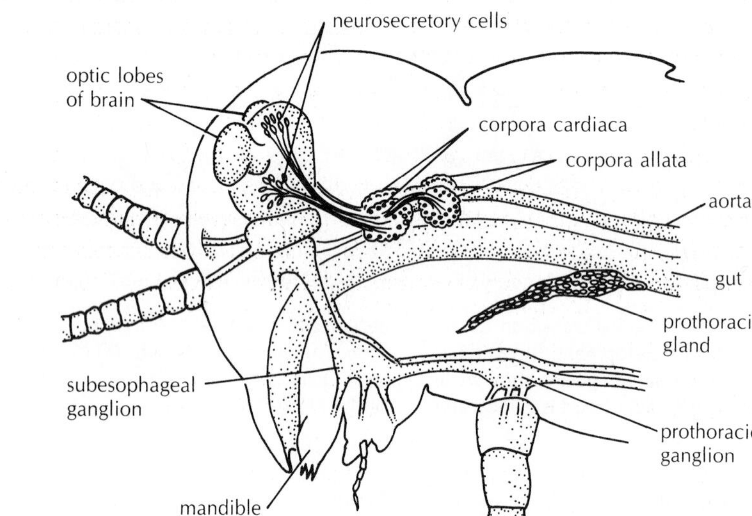

Figure 14.24

Anterior end of an insect, showing the location of the brain hormone, juvenile hormone, and ecdysone secretory centers. Neurosecretory cells in the brain secrete a brain hormone (PTTH) that stimulates the prothoracic glands to secrete ecdysone. The corpora allata secrete juvenile hormone. The corpora cardiaca innervate the corpora allata and are also neurosecretory. One of their major roles is the regulation of heartbeat rate.

Insect Social Systems

The study of insect social systems is an active and fascinating field, encompassing many important issues in behavioral ecology and evolutionary theory. Truly social insects (**eusocial** species) include many hymenopterans (all ant species, some bee species, some wasp species—all in the insect order Hymenoptera) and all termites (order Isoptera). By definition, eusocial insects form colonies composed of more or less sterile workers and one or more reproductive queens; exhibit multiple generations within a colony, so the queen is protected and cared for by her offspring; and cooperate in the care of developing embryos and larvae. A few arachnid species exhibit high degrees of social development, and one marine crustacean species (a snapping shrimp that lives in the internal water canal system of some tropical sponges) is eusocial, but among invertebrates eusociality is otherwise unknown outside the Insecta.

Ants evolved some 100 million years ago from nonsocial wasps, but did not become common until about 45 million years ago, based on fossil evidence. All nearly 10,000 described species are eusocial, which has surely contributed to ant success. Accounting for only about 2% of all insect species, ants make up about half of all insect biomass. A new ant colony is typically formed by a winged female shortly after she mates. The female then discards her wings, builds a nest, and produces many daughters over a number of years, all from sperm stored from matings accomplished during her one brief nuptial flight. The daughters form a worker caste, whose job it is to care for the queen, maintain the nest, care for embryos and larvae, defend the nest, and forage for food. In many ant species, some workers develop into soldiers, morphologically and behaviorally specialized for defense and aggression. Workers communicate through a complex system of mechanical and chemical cues (Fig. 14.25). Not only are all workers female, but in most colonies they are in fact all sisters, and the larvae they care for are also their sisters. Males are produced only after the colony has achieved a substantial size, which often takes several years. The queen produces males deliberately, by releasing eggs without fertilizing them. Males are thus all haploid, developing parthenogenetically without the aid of sperm. Males are winged, and do no work; they are eventually chased out of the colony by workers. Meanwhile, perhaps by altering the quality of food provided to certain diploid embryos, hormonal titres are altered such that these particular female embryos develop into winged reproductives—future queens—which leave the colony for a short period of frenzied mating. Males die shortly after mating, while the mated females excavate new nests and initiate new colonies.

Note that the queen's female offspring obtain half of their mother's alleles but all of their father's alleles, since the father is haploid and therefore passes along everything he has to each daughter in turn. In consequence, the workers share more alleles with their sisters than they would with their own daughters, if they had any; they therefore do more to propagate their own genotypes by taking care of their sisters than by having offspring of their own!

Ants exhibit a considerable diversity of lifestyles, the scope of which I can only hint at here. Members of most ant species prey on other insects (especially termites) and arachnids, but some feed largely or exclusively on excretions produced by other insects, including aphids and butterfly caterpillars. Members of some ant species (the leaf-cutter ants) actively cultivate fungal gardens within their nests for food. There are even slave-making ant species: Incapable of feeding themselves or their nestmates, the workers of such species capture larvae or pupae from the nests of other ants; captured workers tend to the needs of the captors with apparent indifference.

Only a small proportion of bee and wasp species exhibit social development rivaling that encountered among ants; members of most bee and wasp species live as solitary individuals. Among the wasps, paper wasps, yellow jackets, and hornets show the greatest degree of social organization. Among bees, honeybees show particularly advanced social behavior. In both groups, workers are always diploid females, as in ants; they play many roles in the colony—caring for the queen and her offspring, scouting out suitable nest sites, and controlling nest temperatures, for example—and those roles typically change as workers age (and exhibit altered hormonal concentrations). As with ants, males are always haploid, do no useful work in the colony, and are eventually expelled. Queens mate for only a brief period of time, and they store enough sperm to last for life.

Termites (order Isoptera), derived from some cockroach-like ancestor some 150 to 200 million years ago, have evolved eusociality independently of the hymenopterans. Unlike hymenopteran queens, termite queens cannot control the sex of their offspring, so termite workers may be male or female. In both cases, workers are always sterile, and they are usually blind. Advanced termite species produce genuine worker, soldier, and reproductive castes (Fig. 14.26). In contrast, workers in more primitive species are made up of developmentally arrested immature stages called **pseudergates** ("false workers"). Pseudergates can remain workers for life, metamorphose into winged reproductives and found new colonies, or metamorphose into replacement reproductives or soldiers within the same colony. Soldiers are usually blind and bear huge, heavily sclerotized mandibles (Fig. 14.26c), which they use to defend the colony against intruders. Soldiers also employ a variety of chemical defenses. The fate of workers is regulated by a complex system of pheromone production by both queens and soldiers, which, in turn, regulates hormonal concentrations within the pseudergates.

Class Crustacea

Class Crusta • cea
(L: a crust)
kruss-tā´-shuh

Defining Characteristics: 1) Head bears 5 pairs of appendages, including 2 pairs of antennae; 2) development includes a triangular larval form (the nauplius) bearing 3 pairs of appendages and a single medial eye (even species that hatch at a later stage of development pass through this naupliar stage)

Most of the approximately 42,000 crustacean species are divided among 6 major subclasses.

(a)

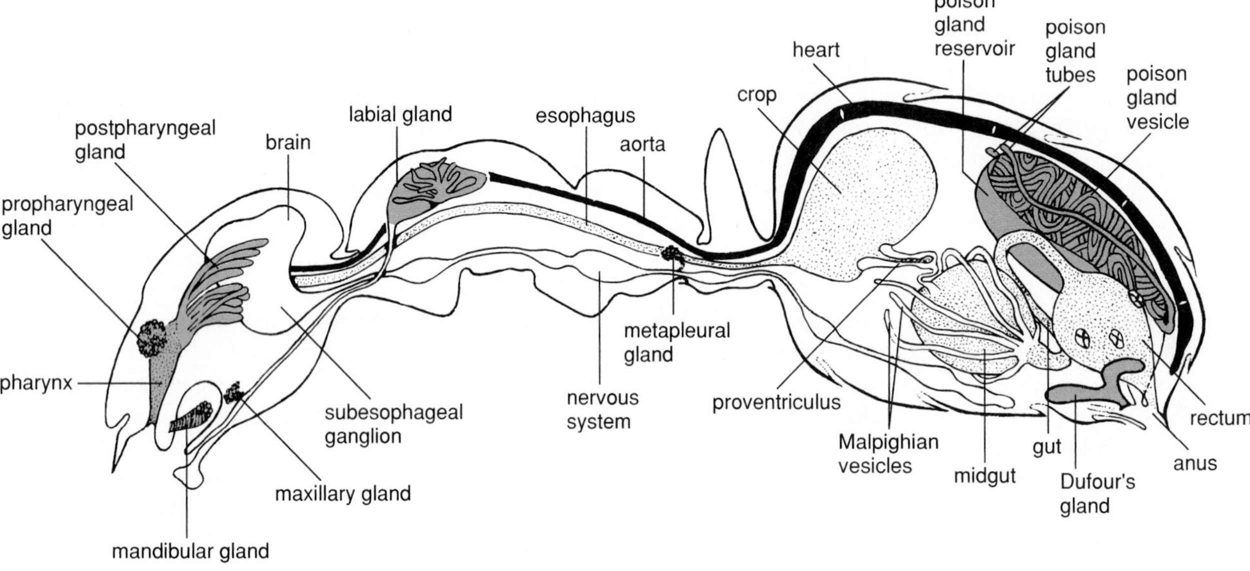

(b)

Figure 14.25

(a) A worker ant in the genus *Formica.* (b) Sagittal section through a *Formica* worker showing internal anatomy, including the numerous exocrine glands (in blue), which release their products outside the body (*exo* = G: outside). Dufour's gland secretions are mostly involved in alarm signalling and recruiting nestmates for foraging, attack, or defense; the poison gland produces formic acid for trail marking or venoms used in predation or defense; the mandibular glands produce defense and alarm pheromones; the metapleural glands secrete antibiotics, protecting the body surface and the nest itself against microbial infection.

Subclass Malacostraca

Subclass Malaco • straca
(G: soft shell)
mal′-ack-ō-strak′-ah

Defining Characteristics: 1) Thorax with 8 segments, abdomen with 6 to 7 segments plus a telson; 2) appendages on the sixth abdominal segment are flattened to form uropods

The subclass Malacostraca contains nearly 60% of all described crustacean species, including decapods, euphausiids, stomatopods, isopods, and amphipods. The most familiar malacostracans, such as the crabs, hermit crabs, shrimp, and lobsters, are decapods. The basic malacostracan body is tripartite, consisting of a **head, thorax,** and **abdomen** (Fig. 14.27a). Whereas the insect head and thorax are separated by a flexible joint, the crustacean head and thorax are almost always rigidly fused. The head and thorax may be covered by a **carapace,** extending posteriorly from the head, and therefore may function as a single unit, the **cephalothorax.** In some species, the carapace bears a prominent anterior projection called the **rostrum.** Large, stalked **compound eyes** are conspicuous, as are 2 pairs of head appendages, the first and second **antennae.** Often in the zoological literature, the first pair of antennae are known as the "antennules," and the second pair are simply referred to as the "antennae." Insects (and myriapods) bear only a

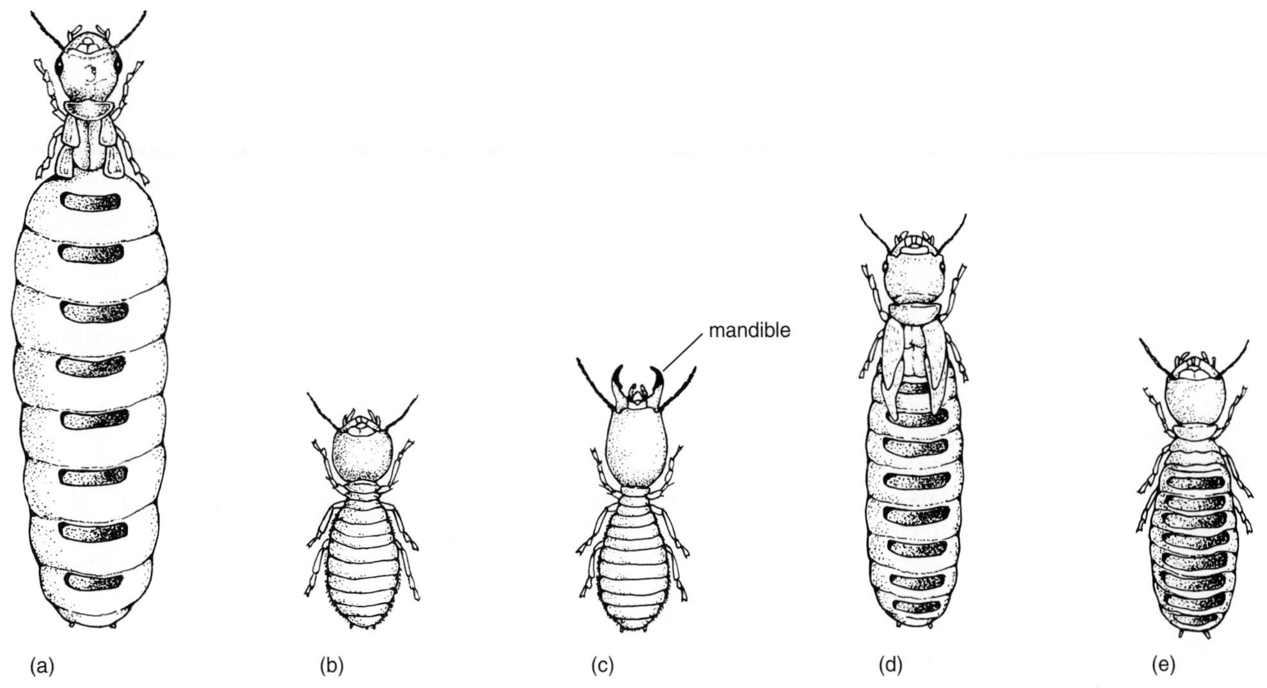

Figure 14.26

Female caste members of the termite *Ameritermes hastatus,* all drawn to the same scale. (a) Queen. (b) Worker. (c) Soldier. (d) Secondary queen. (e) Tertiary queen. Supplemental queens (d,e) replace the original queen when she dies. Note the large mandibles of the soldier (c).

Reprinted by permission of the publishers from *The Insect Societies* by Edward O. Wilson (after Atkins, 1978, after Skaife, 1954), Cambridge, Mass.: The Belknap Press of Harvard University Press. Copyright © 1971 by the President and Fellows of Harvard College.

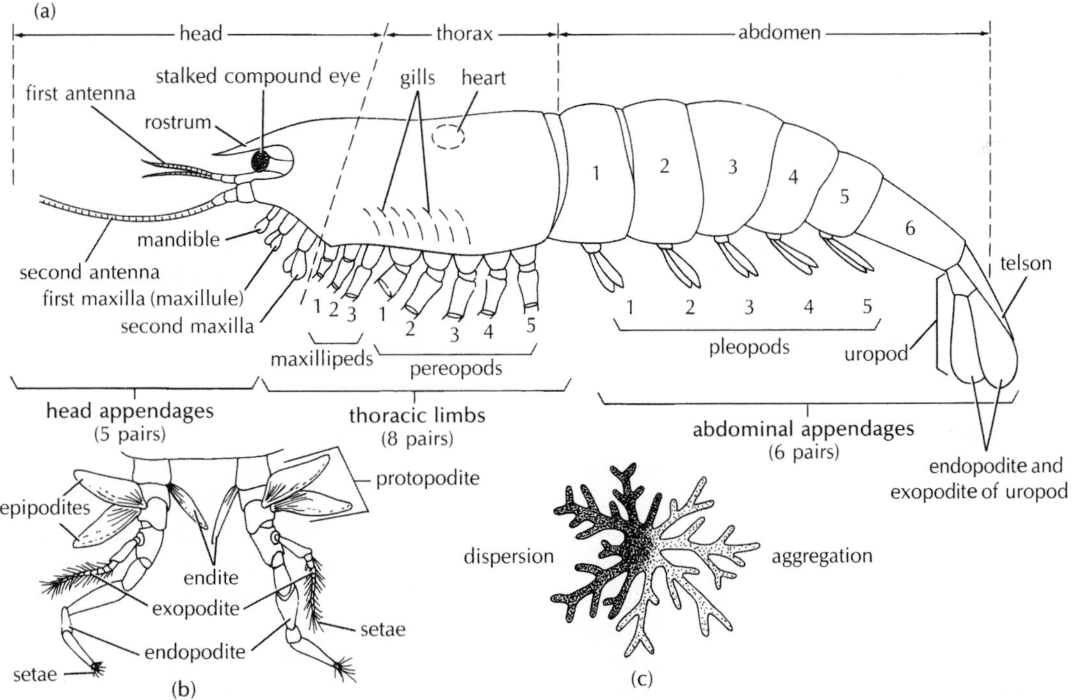

Figure 14.27

(a) General external anatomy of a crustacean, showing the head, thorax, abdomen, and the bases of associated appendages. The animal illustrated is a decapod. (b) Illustration of biramous appendages. (c) Patterns of pigment dispersion in crustacean chromatophores. In the dispersed configuration, the cuticle becomes dark; pigment aggregation within the chromatophores causes the cuticle to become lighter in color. Movement of pigment within the chromatophores is under hormonal control in response to changing light intensities.

(a) Modified after Russell-Hunter and other sources. (c) From Weber, 1983. *American Zoologist.* Thousand Oaks, California, American Society of Zoologists.

single pair of antennae, apparently a secondary loss. In malacostracans, both pairs of antennae are primarily sensory. In other crustacean groups, the second antennae also may play roles in feeding, locomotion, and mating. In addition to the 2 pairs of antennae, the malacostracan head bears 3 pairs of smaller appendages that are involved in feeding or in generating respiratory currents. These appendages are, in sequence beginning from the mouth and moving posteriorly, the **mandibles** (which crush food) and the first and second **maxillae** (which generate water currents and manipulate food). The next 8 segments of the cephalothorax are thoracic segments, commonly bearing in sequence the first, second, and third **maxillipeds** (for food manipulation), and 5 pairs of thoracic **walking legs,** commonly known as **pereopods.** The first 1–3 pairs of pereopods may be chelate (claw bearing), in which case they also function in feeding and in defense.

Each of the 6 abdominal segments bears a pair of appendages as well. The first 5 pairs of abdominal appendages are referred to as **pleopods;** these function primarily in swimming or generating respiratory currents and, in females of some taxa, in the brooding of eggs and developing young. The last pair of abdominal appendages are the **uropods.** These flat appendages lie on either side of the telson, forming a tail (Fig. 14.27a).

Malacostracan appendages are generally biramous; that is, they have 2 branches (*bi* = L: two; *rami* = L: branch). The portion of the limb proximal to the branch point is the protopodite (*proto* = G: first). The inner and outer branches are the endopodite and exopodite, respectively (Fig. 14.27b). The exopodite is often less well developed than is the endopodite. Frequently, lateral protuberances extend from the protopodites themselves; epipodites, for example, commonly function as gills or as gill-cleaners. Some crustacean appendages no longer have both the endopodite and exopodite and are, therefore, secondarily uniramous (one-branched). The first antennae and the maxillae of lobsters, for example, are biramous, whereas the second antennae and the thoracic appendages (pereopods) are uniramous. The abdominal appendages are biramous, a uniquely malacostracan characteristic.

The body surface of many malacostracan species is covered with **chromatophores** (Fig. 14.27c), which are highly branched cells containing pigment granules. The pigments come in a variety of colors, including red, black, yellow, and blue. More than one pigment may be found within a single chromatophore, and pigment distribution differs among the chromatophores of a single animal. By varying the pigment distribution in the different chromatophores over time, the animal can alter its body color considerably.[18] The migration of pigment granules within chromatophores is under hormonal control.[19] The hormones are manufactured by the so-called X-organ located in the eyestalks and are transported a short distance to the **sinus gland** for storage. From here, the hormones are transported as needed through the bloodstream. Because chromatophore operation is under hormonal control rather than under direct nervous control, arthropod color changes never occur as rapidly as do those of cephalopods.

The description of a typical malacostracan just given applies best to members of the order Decapoda, which includes about 10,000 species and is the largest of the malacostracan orders. The term *decapod* (*deca* = G: ten; *pod* = G: foot) refers to the fact that members of the order Decapoda have only 10 thoracic legs (5 pairs); the first 3 pairs of thoracic appendages (the **maxillipeds**) are modified for feeding. The term *decapod* is also useful in recalling that the abdomen and thorax together bear a total of 10 pairs of leg-like appendages. Decapods are probably the best-known crustaceans and include lobsters, crayfish, hermit crabs, true crabs, and shrimp (Fig. 14.28a–f).

One of the smallest malacostracan orders is the order Euphausiacea (pronounced ū-fow′-zē-ā′-sē-ah), whose members are more commonly known as "krill." Only about 85 species have been described. Yet, the euphausiids' commercial and ecological importance far exceeds their limited species diversity. The annual production of euphausiid biomass in Antarctic waters alone is estimated to at least equal the current world harvest of all other marine animals combined, some 99 million metric tons per year. Some countries, notably Japan and the former Soviet Union, are now exploiting these animals for human use. An international regulatory agency has limited the amount of krill that can be captured yearly to 1.5 million tons, more than 3 times the amount currently harvested. The ecological impact of increased krill exploitation is uncertain: Euphausiids constitute the primary diet of seals and of many baleen whale and seabird species; some whales consume perhaps 4 tons of krill each day.

Krill have the general appearance of decapod shrimp, except that they have 8 pairs of thoracic walking legs rather than 5 pairs (Fig. 14.28g); none of the thoracic legs are specialized as maxillipeds, although the 8th and sometimes the 7th pairs are reduced in size in some species. Also, the thoracic appendages of euphausiids bear conspicuous, feathery gills (Fig. 14.28g). Euphausiids grow to lengths of about 6 cm. They are remarkably tolerant of starvation—individuals have survived more than 200 days without food in the laboratory—and starved euphausiids actually decrease their size at each molt, a possible adaptation to the low productivity of Antarctic winters.

As with many shrimps and copepods (to be discussed shortly), and a variety of animals sporadically distributed among most other phyla, euphausiids typically can produce

18. See *Topics for Further Discussion and Investigation,* no. 21.

19. See *Topics for Further Discussion and Investigation,* no. 12.

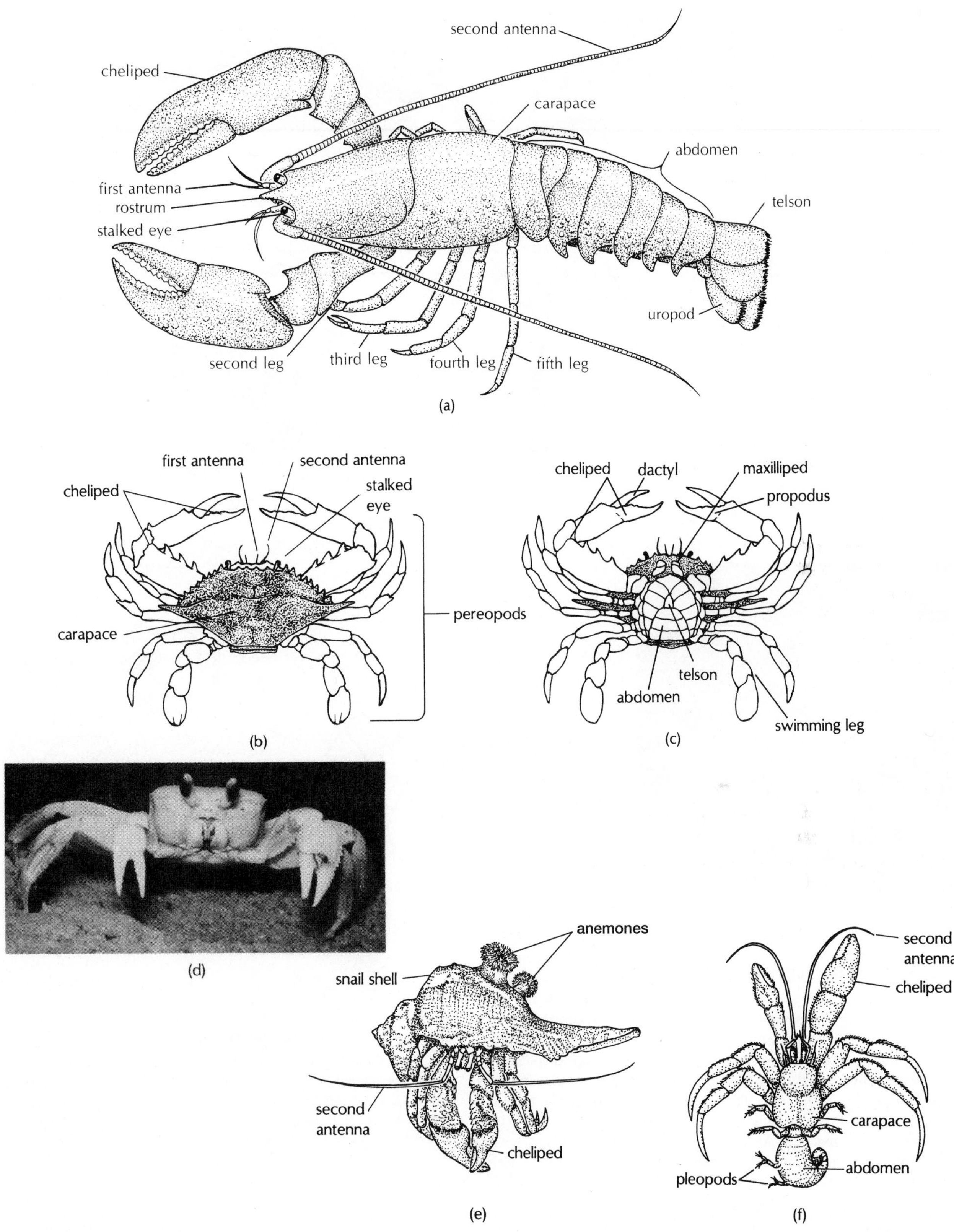

Figure 14.28

Malacostracan diversity. (a) A lobster, *Homarus americanus*. The large appendage (cheliped) is the equivalent of the first thoracic leg (pereopod) illustrated in Fig. 14.27a. (b) Dorsal view of the blue crab. *Callinectes sapidus*. The fifth pair of thoracic legs have become modified into flattened paddles for swimming. (c) Ventral view of same individual in (b), showing the abdomen (♀). The abdomen can be pulled away from the body to reveal abdominal appendages (pleopods) specialized for mating in males or carrying eggs in females. (d) Photograph of a marine crab. (e) Hermit crab in a gastropod shell. Several dozen hermit crab species deliberately position sea anemones on their shells, gaining added protection from predators and providing the anemones with a firm, mobile substrate in return. (f) Same individual in (e), removed from the shell; note the soft abdomen, which is not tucked up under the body. (*Continued on following page.*)

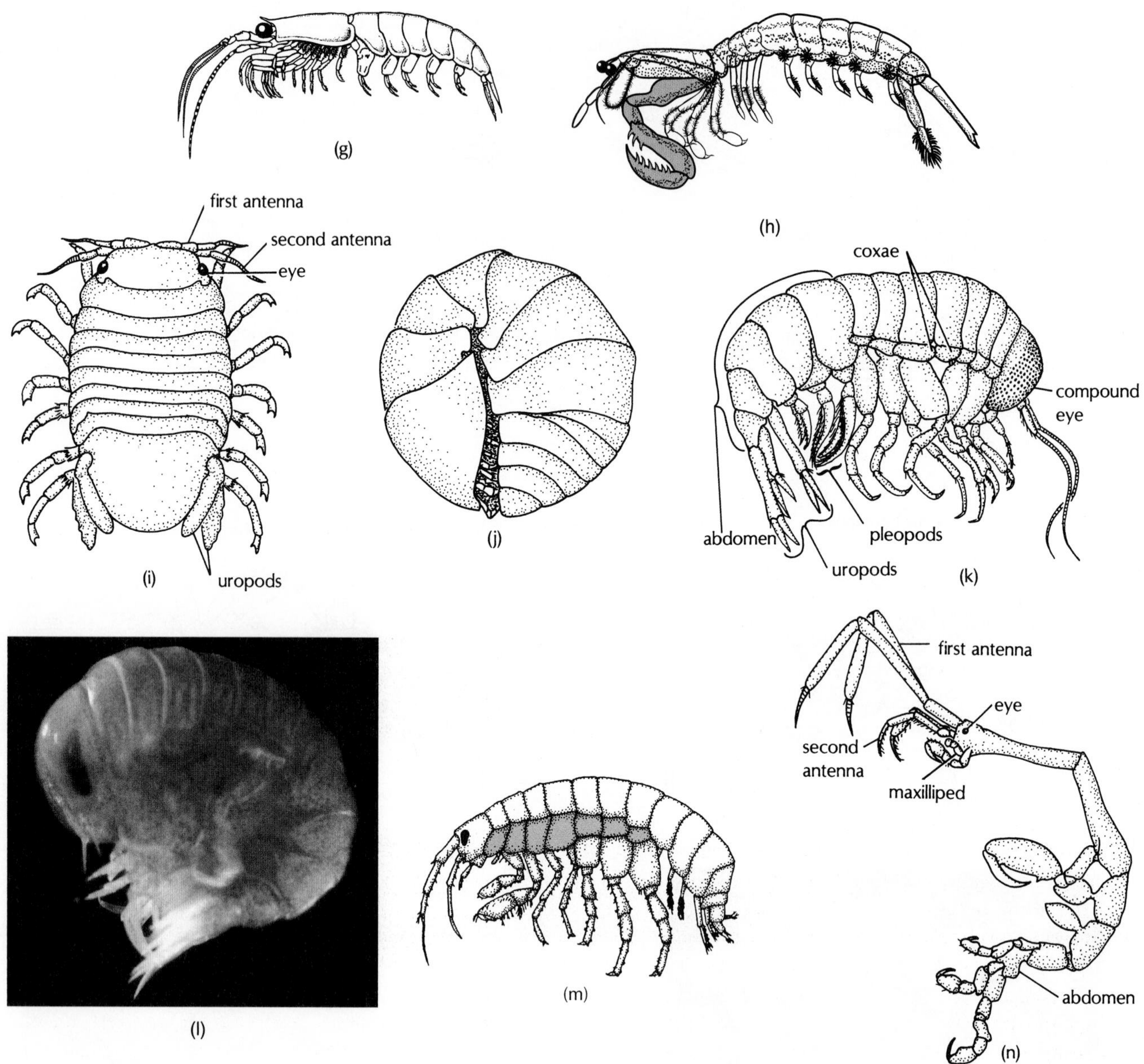

Figure 14.28 *Continued*

(g) A euphausiid. (h) Lateral view of a stomatopod. (i) A marine isopod, *Sphaeroma quadridentatum.* Note the dorsoventral flattening of the body, the uniramous antennae, and the absence of a carapace. The eyes are not stalked. (j) Terrestrial isopod (pillbug), curled to form a ball. (k) The amphipod *Hyperia gaudichaudii* in side view. Note the large, unstalked compound eyes (only one is shown, covering nearly one entire side of the head), and the absence of a carapace. (l) An unidentified hyperiid amphipod, seen in lateral view. This individual was about 0.5 cm long. Note the lack of carapace and the enormous compound eye (eye on other side of head not visible). (m) A generalized gammarid amphipod, with coxal plates shaded blue. (n) The unusual amphipod *Caprella equilibra,* an animal highly modified for clinging to algae, hydroids, and other substrates. Caprellid amphipods are from 1–32 mm long.

(c) After D. Krauss. (d) Courtesy of W. Lang. (h) Based on: *Comparative Morphology of Recent Crustacea* by McLaughlin, 1980. (i) After Harger. (j) After Pimental. (k) After Stebbing. (l) Photo by J. A. Pechenik. (n) After Light.

bioluminescence: light produced chemically, without heat. The light-producing organs (**photophores**) of some euphausiid species are among the most complex known (Fig. 14.29). The photophores are distributed on the body in species-specific patterns and may therefore function in species and mate recognition. Photophores also may protect against predation in surface water, by breaking up the silhouette of a euphausiid when seen by a predator from below.

Stomatopods (order Stomatopoda) are another interesting group of shrimp-like malacostracans, with about 350 species. Stomatopods are rather large (up to about 35 cm long), bottom-dwelling, violent carnivores. Unlike those of the euphausiids, the abdominal appendages of stomatopods bear conspicuous gills (Fig. 14.28h). Stomatopods resemble flattened shrimp but, not being decapods, have 8 pairs of thoracic legs like euphausiids. The second pair are large and extremely powerful,

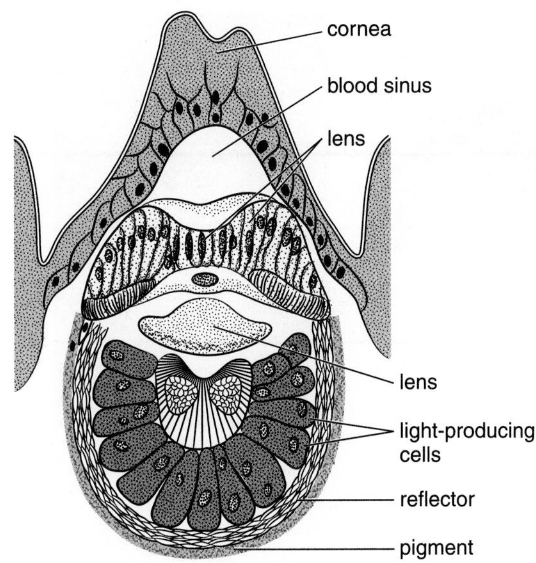

Figure 14.29

A complex photophore of the euphausiid *Meganyctiphanes norvegica*. Such organs produce light intracellularly and are distributed in species-specific patterns on the appendages and ventrally on the thorax and abdomen.

Based on J. A. Colin Nichol, in *The Biology of Marine Animals,* 2d ed. 1967.

modified for smashing into hard-shelled prey, such as bivalves, gastropods, and crabs, or for spearing fish and other soft-bodied prey. Stomatopods deal with their prey with impressive speed; the specialized raptorial legs have been reported to strike their victims with speeds of up to 1000 cm per second. In this respect, stomatopods resemble the terrestrial praying mantids and thus are commonly known as "mantis shrimp." The stomatopods are unique among crustaceans in having a jointed head, enabling the anterior and posterior portions to move independently and adding to the praying mantids appearance. They also possess particularly amazing compound eyes, with probably the most complex set of color receptors known.

Most stomatopods are tropical, but some species do very well in temperate areas. They typically live in burrows—in rock, coral, or mud—in shallow water.

Two of the largest orders of nondecapod malacostracan crustaceans are the Isopoda and Amphipoda. The order Isopoda contains at least 10,000 species, about as many species as described in the Decapoda. Most isopods are marine, although both freshwater and terrestrial species (including the familiar "pillbugs" or "sow bugs") occur. Unlike the decapods, isopods have no carapace. Moreover, they have only a single pair of maxillipeds, in contrast to the 3 pairs found in decapods, and they have uniramous first antennae, as opposed to the decapods' biramous first antennae. Isopods tend to be small, about 0.5–3.0 cm in length. Compound eyes, if present at all, are not on movable stalks. Isopods are characteristically flattened dorsoventrally (Fig. 14.28i) and accomplish gas exchange by means of flattened pleopods. Thus, respiratory appendages are associated with the abdomen. A number of terrestrial isopod species possess a system of tracheae. This is another example of convergent evolution, in which 2 or more groups of animals have independently evolved similar adaptations in response to similar selective pressures. In this case, as in the insects and arachnids, selection has favored internalization of the respiratory system as a means of deterring water loss in a terrestrial environment.

In contrast to the isopods, members of the order Amphipoda tend to be flattened laterally (Fig. 14.28k–m). Again, no carapace is present, individuals possess only a single pair of maxillipeds, and the compound eyes are sessile. About 6000 amphipod species have been described, mostly from salt water. However, many species occupy freshwater habitats, and some others are terrestrial. In contrast to the isopods, amphipod gills are found on the thorax, attached to the pereopods. In many species, a portion of the protopodite (the **coxa**) of each of several pairs of anterior appendages is elaborated into a large, flattened sheet (Fig. 14.28m), contributing significantly to the flattened appearance of the amphipod body.

Subclass Branchiopoda

Subclass Branchio • poda
(G: gill foot)
brank-ē-ō-pō´-dah

Branchiopods are a diverse group of small, primarily freshwater crustaceans. On each thoracic appendage, the **coxa** (one of the basal segments of the typical crustacean appendage) is modified to form a large, flattened paddle; this paddle functions in gas exchange and locomotion, giving rise to the name of the class (*branchio* = G: a gill; *pod* = G: foot). Most species are filter feeders, although a few are carnivorous. The bodies of most branchiopods are at least partially enclosed in a bivalved carapace (Fig. 14.30a–c). However, some species (brine shrimp, and the freshwater fairy shrimp) completely lack a carapace (Fig. 14.30d). The well-known brine shrimp (*Artemia salina*) are found in waters whose salinities range from about 0.1 to 10.0 times the salt concentration of open-ocean seawater. The fertilized eggs of this species are commonly marketed as "sea monkeys." Fairy shrimp are also restricted to harsh environments (temporary ponds). Presumably, brine shrimp and fairy shrimp thrive in these stressful habitats because their major predators do not. Both groups produce resting eggs that resist extreme temperature and desiccation.

Of the approximately 800 branchiopod species so far described, at least 50% are contained within the order Cladocera, the water fleas. Cladocerans, including the familiar genus *Daphnia* (Fig. 14.30a, b), dominate the zooplankton of freshwater lakes, although a few species are marine. All species are microscopic, with most of the body contained within a bivalved carapace. Protruding from this carapace is the head, bearing a pair of large, biramous second antennae that are used to propel the

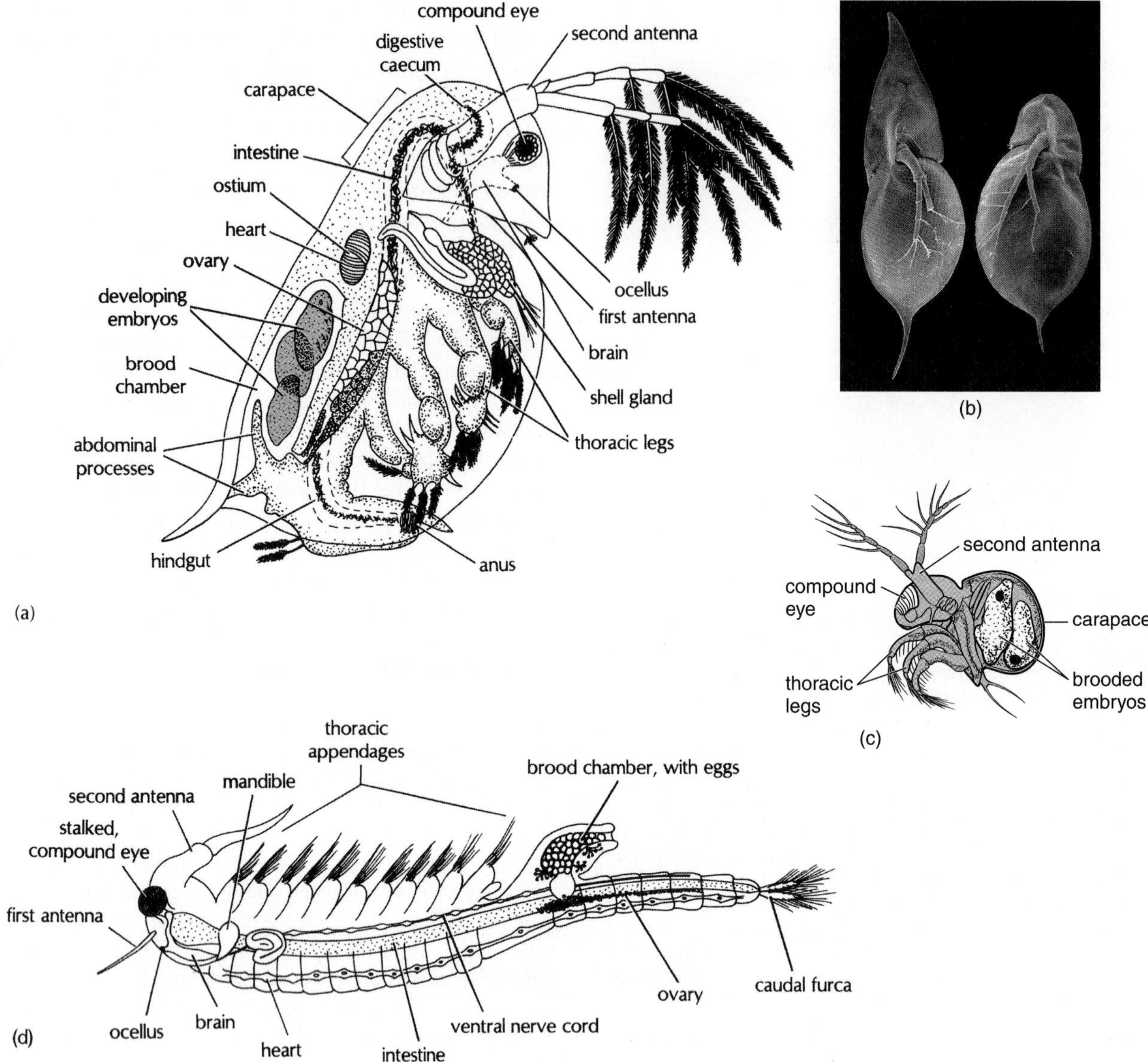

Figure 14.30

Branchiopod diversity. (a) *Daphnia pulex* (female), the freshwater water flea. Except for the head, most of the body lies within a laterally flattened, bivalve-like carapace. The long, biramous second antennae are the primary locomotory appendages of cladocerans. The compound eyes are paired and sessile (not stalked). (b) Phenotypic plasticity in the cladoceran *Daphnia cucullata*. The animals were raised in the presence (left) or absence (right) of predators, which had remarkable effects on development, particularly on head ("helmet") size. Individuals with the larger helmets were found to be much less vulnerable to predators in controlled experiments. One more Cladoceran: (c) *Podon intermedius*. Cladocerans range from a few hundred microns to nearly 2 cm in length. (d) *Artemia salina,* the brine shrimp, in normal swimming orientation. Brine shrimp may attain lengths of about 1 cm, and possess a pair of stalked, compound eyes.

(a) After Pennak, after Claus. (b) Courtesy of A. Agrawal, from Agrawal et al., 1999. Transgenerational induction of defences in animals and plants. *Nature* 401: 60–63. (c) Based on G. Hutchinson, *A Treatise on Limnology*, Vol. 2. 1967. (d) After Brown; after Lochhead.

animal through the water. The first antennae and the second maxillae are much reduced. The thoracic region bears 5 to 6 pairs of appendages, which generate feeding and respiratory currents; food particles are filtered from the water by fine setae on these appendages. There are no abdominal appendages. The head bears, in addition to the second antennae, a single, huge, compound eye, formed by the fusion of the ancestral compound eyes from each side of the head. The eye is not stalked but can be rotated in various directions by associated musculature. A number of cladoceran species show seasonal alterations in morphology **(cyclomorphosis),** apparently as an adaptation against the seasonal appearance of certain size-specific predators (see also Research Focus Box 10.1, describing the same phenomenon among freshwater rotifers).

Members of another branchiopod group (the tadpole shrimp) completely lack second antennae; only a vestige remains. Clearly, the branchiopod body plan is rather plastic. Recent molecular data indicate that branchiopods share a common ancestor with hexapods (Fig. 14.42), suggesting that insects evolved from crustacean ancestors.

Subclass Ostracoda

Subclass Ostra • coda
(G: a shell)
os-truh-cō´-dah

Defining Characteristics: 1) Head and body are enclosed in a bivalved carapace, which lacks concentric growth rings; 2) trunk of body possesses no more than 2 pairs of limbs

---

Ostracods are small (rarely more than a few millimeters across) but widespread crustaceans, common in both marine and freshwater habitats. Remarkably, a few species are terrestrial. Ostracods are mostly head; the rest of the body is greatly reduced, bearing at most 2 pairs of appendages (fewer than in any other crustacean), and showing no sign of segmentation externally. The body is completely encased within a partially calcified, bivalved carapace (Fig. 14.31). Most of the approximately 6650 species are free-living, although some ostracods are commensal with other crustaceans or with certain echinoderms; there are no parasitic ostracods. Another 10,000 ostracod species are known only as fossils, with an extensive record extending back some 530 million years.

A few extant species are completely planktonic, but most ostracods leave the bottom only periodically, using their first or second antennae, or sometimes both pairs of antennae, for swimming. A number of species are completely benthic bottom-dwellers, in some cases burrowing within the sediment. Ostracods include carnivorous, suspension-feeding, scavenging, and herbivorous species.

Subclass Copepoda

Subclass Cope • poda
(G: oar foot)
cō-peh-pō´-dah

Defining Characteristics: 1) Thorax with 6 segments, abdomen with 5 segments; 2) first segment of thorax fused to head; 3) loss of all abdominal appendages; 4) most species bear a single, "naupliar" eye

---

Most of the approximately 8500 species in the class Copepoda are marine and feed on unicellular, free-floating, photosynthesizing protists called **phytoplankton** (*phyto* = G: plant; *plankton* = G: that which is forced to wander). Copepods are invariably small, usually less than 1–2 mm long. Some copepod species occur in freshwater lakes and ponds, while terrestrial species live in soil or in moist surface films in humid environments. Perhaps two-thirds of

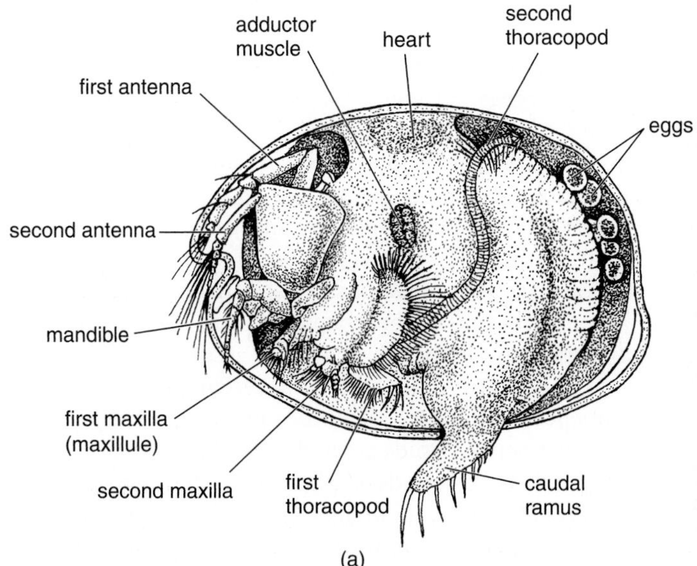

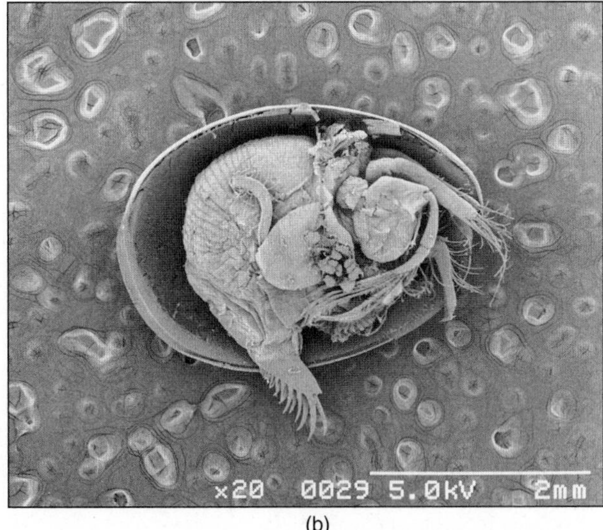

Figure 14.31
(a) A marine ostracod seen in lateral view, with the left valve removed to show internal structure. (b) Scanning electron micrograph of the ostracod *Vargula hilgendorfi,* with one valve removed. (magnification = 20×)
Courtesy of J. Vannier. From Abe, K., and J. Vannier, 1995. *Marine Biol.* 124:51–58.

all copepod species are planktonic in the ocean. Because of their small size, such copepods are, to a large extent, at the mercy of currents. Together with other animals of limited locomotory capability (relative to movement of water currents), these copepods form a major component of the **zooplankton** (*zoo* = G: animal). Most other copepod species are specialized for life in or on substrates, forming a major component of the **meiobenthos** (i.e., the community of small animals living in association with sediment). Locomotion of planktonic copepods is accomplished primarily by the actions of a pair of biramous second antennae, while benthic species use their thoracic appendages to walk over surfaces.

Copepods are among the most abundant animals on earth and are also among the most important herbivores of the ocean. In part, they collect phytoplankton through the activities of the first and second maxillae (Fig. 14.32a), although the details of food capture are complex and incompletely understood (Research Focus Box 14.3).[20] Copepods are at the base of the oceanic food chain in another respect as well: They are a major food source for primary carnivores, including both larval and juvenile fish of commercial importance.

Most free-living copepods have a single, median eye on the head (Fig. 14.32a, b, d). The eye usually consists of 3 lens-bearing ocelli; 2 of the units look forward and upward, and the third ocellus is directed downward. Yet, exceptions do exist; some species have a pair of eyes, placed laterally and with conspicuous lenses (Fig. 14.32c), while other species lack eyes entirely. Compound eyes are never encountered among the copepods.

In contrast to many other crustaceans, copepods lack gills and abdominal appendages. However, the other tagmata (head and thorax) do bear appendages (Fig. 14.33). The structure and function of copepod appendages vary substantially with species and lifestyle and, often, with sex as well. One or both first antennae may be hinged in the male (Fig. 14.32a[3]), functioning to capture females for mating, in addition to performing the standard sensory function. In addition, one or both of the male's fifth thoracic appendages may terminate in a claw (Fig. 14.32a[2]), which is used to hold the female during mating (Fig. 24.11a).

Although a free-living, suspension-feeding existence is the norm for planktonic copepods, a number of planktonic species are carnivorous, and many benthic species, living on sediment or in the spaces between sand grains, scrape off food particles from solid substrates. Still other copepods, about 25% of all described species, parasitize a variety of vertebrates and invertebrates. As might be expected, the head appendages and other body parts are modified, sometimes extravagantly, as reflections of these different lifestyles (Fig. 14.34).

20. See *Topics for Further Discussion and Investigation*, no. 11.

Subclass Pentastomida

Subclass Penta • stomida
(G: five mouths)
pen-tah-stō´-mid-ah

Defining Characteristics: 1) All species are parasitic in the nasal passages of vertebrate hosts; 2) body bears only 2 pairs of appendages, with claws

The taxonomic affinities of pentastomids ("tongue worms") have long been uncertain. All pentastomids are internal parasites of vertebrates, and most characteristics of potential diagnostic utility have been lost; most aspects of adult external morphology—and of the sensory, digestive, excretory, and reproductive systems—have become highly modified as adaptations for an exclusively endoparasitic existence (Fig. 14.35). The larval stages of pentastomids also are parasitic, with an associated disappearance or extreme reduction of virtually all potentially revealing ontogenetic characteristics. The fertilized egg exhibits spiral cleavage, so at least the protostome affiliation of the group is clear. But a consensus for placing pentastomids among the Crustacea—and even among the Arthropoda—has emerged only recently; for many years, the Pentastomida was considered a separate phylum.

In support of their affiliation with arthropods, pentastomids show such arthropod characteristics as an exoskeleton that is periodically molted, striated musculature arranged metamerically, a spacious hemocoel, and larval stages often possessing 3 pairs of appendages. However, pentastomids also show some decidedly unarthropod characteristics: The larval legs are not jointed, and the adult is legless; the only adult appendages are 4 pairs of anterior, chitinous hooks. Moreover, the pentastomid cuticle is not chitinous. But the weight of available evidence argues strongly for pentastomid membership within the Arthropoda. Perhaps the most compelling evidence of their arthropod affinity is found in the unusual morphology of their sperm, the characteristics of which suggest that pentastomids are most closely related to a group of marine crustaceans, the Branchiura, which are external parasites (**ectoparasites**) of fish (see *Taxonomic Detail* at the end of this chapter, p. 416). Detailed studies of cuticle ultrastructure that have accumulated over the past 15 years or so, together with recent analyses of DNA sequences coding for 18S ribosomal RNA support the idea that pentastomids are highly modified crustaceans.

Adult pentastomids dwell mainly in the lungs and nasal passages of reptiles, although some species instead specialize in the same areas within amphibians, birds, and even some mammals, including dogs and horses. All species feast on the host's blood, clinging to host tissues with sharp anterior hooks. Variations in hook morphology provide one of the few useful criteria for classifying pentastomid species. The parasites are gonochoristic (i.e., there are separate males and

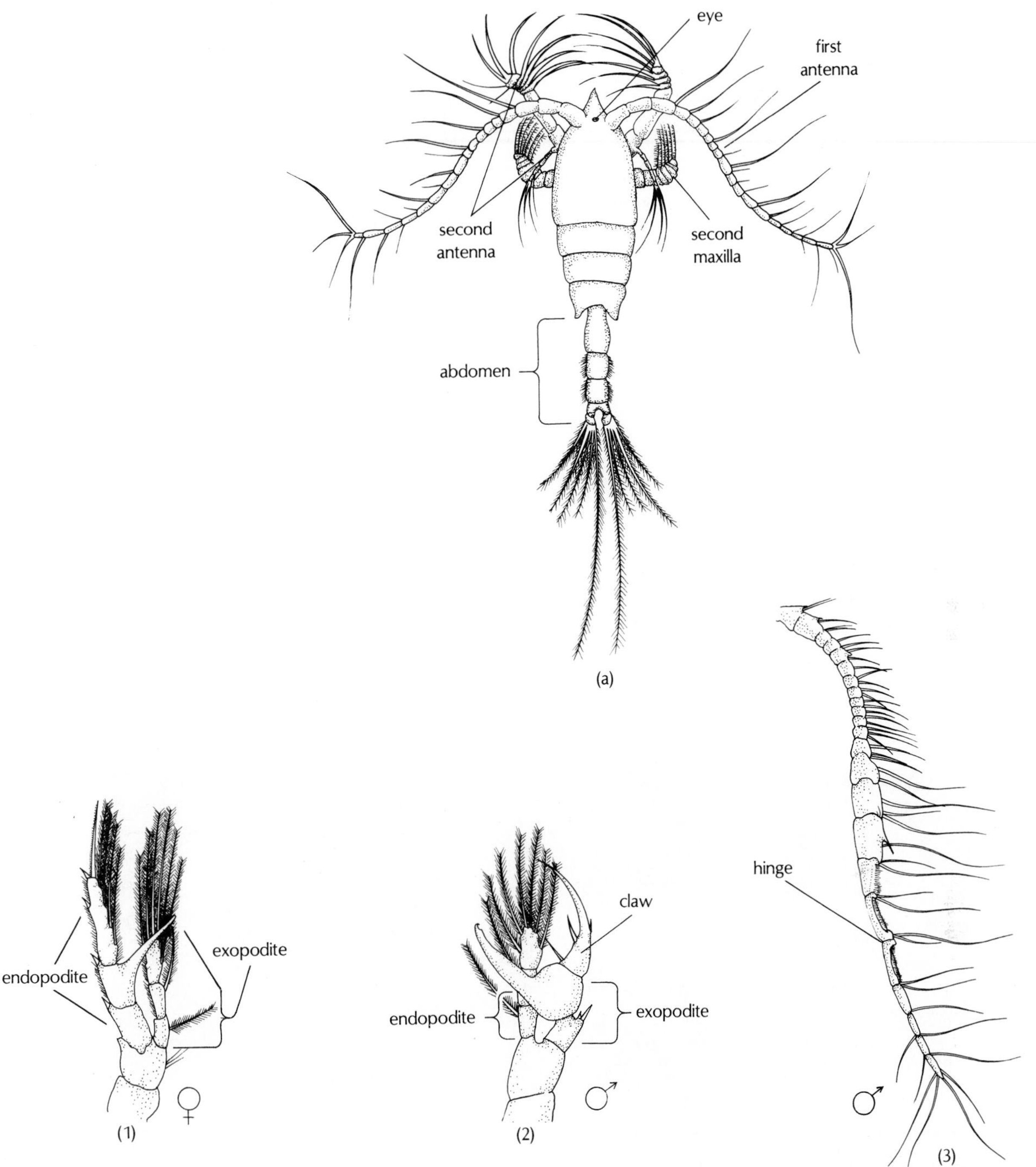

Figure 14.32
Diversity among the free-living copepods. (a) *Euchaeta prestandreae,* a typical, actively swimming copepod. Copepod appendages: The fifth right thoracic legs of a female (1) and a male (2) copepod, *Centropages typicus,* illustrating the clawed appendage of the male. The right first antenna of the male (3) illustrates the hinge. (*Continued on following page.*)

females), and embryogenesis occurs within a shelled capsule. The nascent, encysted parasite usually leaves the host in nasal secretions and saliva. It develops no further unless eaten by the proper intermediate host, usually another vertebrate. Some insects act as obligate intermediate hosts for a few species. If taken inside the correct intermediate host, the larvae emerge, tear through the gut wall, and develop further within species-specific tissues, undergoing up to 9 molts before reaching a stage infective to the final, definitive host. As with other internal parasites, pentastomids cannot reach sexual maturity in the intermediate host; that is, after all, why these hosts are termed "intermediate." If the intermediate host is ingested by the right definitive host, the juvenile migrates back up to the pharynx and thence to the desired location in the respiratory system, using its piercing claws.

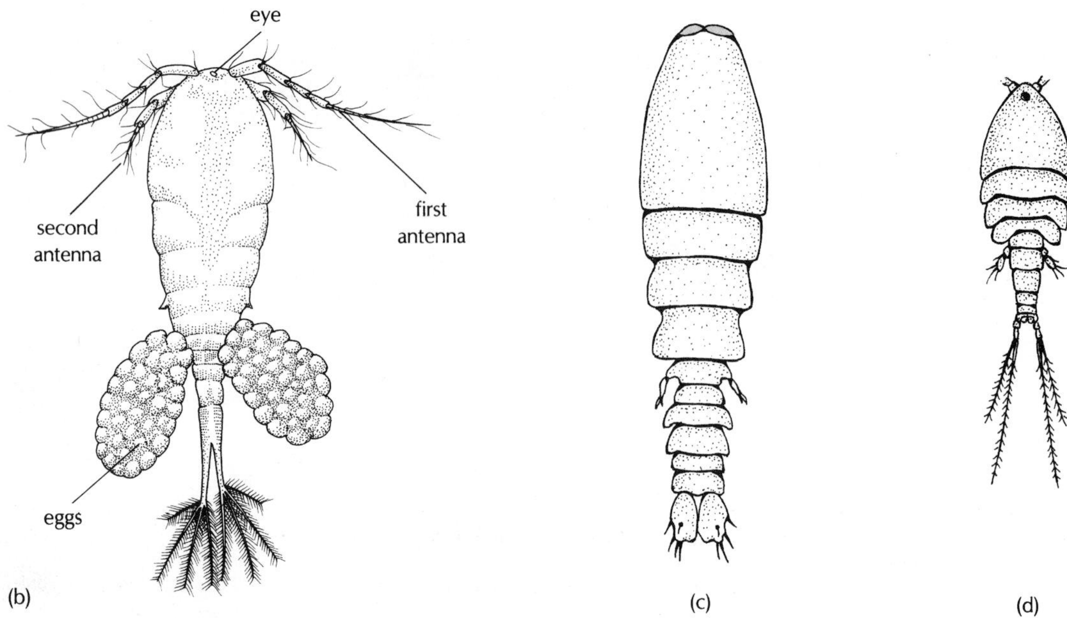

Figure 14.32 *Continued*

(b) Free-living copepod carrying 2 clusters of fertilized eggs on the abdomen. (c) A harpacticoid copepod, *Tisbe furcata.* (d) A cyclopoid copepod, *Sapphirina angusta.*

(a) After McConnaughey and Zottoli; after Brady, 1883. *Challenger Reports,* Vol. 8. From C. B. Wilson, 1932, "The Copepods of the Woods Hole Region." *Smithsonian Institution Bulletin* 158. (b) From Pimentel. (c,d) Reprinted from C. B. Wilson, "The Copepods of the Woods Hole Region," 1932, from the Bureau of American Ethnology. Washington, DC: Smithsonian Institution Press, pages 197, 343, 354, by permission of the publisher. Copyright 1932.

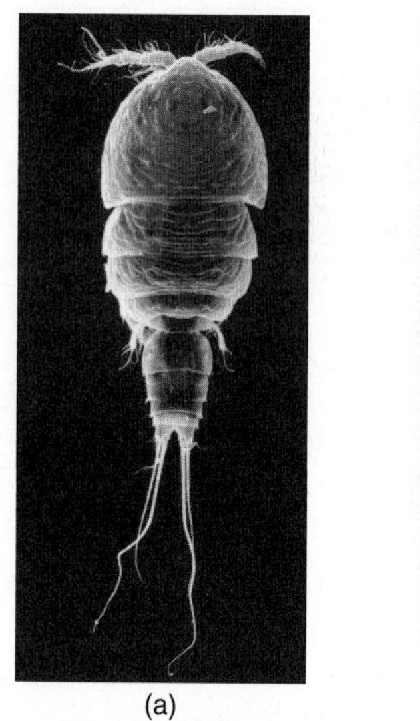

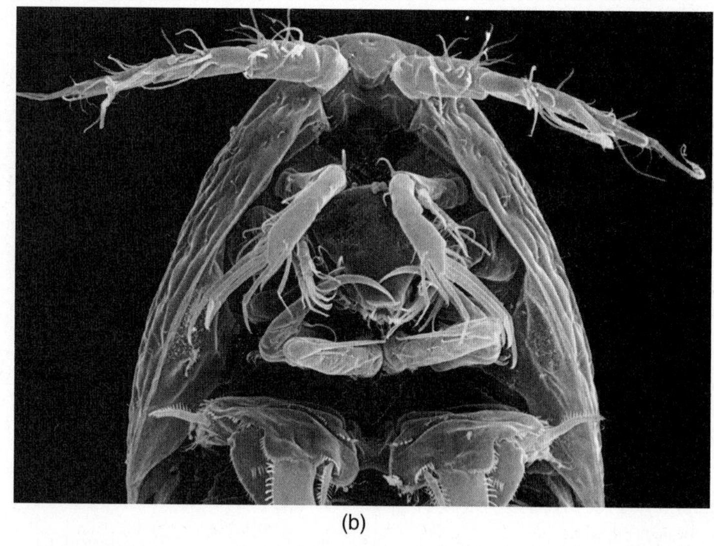

Figure 14.33

(a) Scanning electron micrograph of a free-living cyclopoid copepod in dorsal view. The total length is about 1 mm. Note the point of articulation between the fourth and fifth thoracic segments, and the short first antennae. (b) Ventral view of same animal, showing head appendages.

(a,b) Courtesy of C. Bradford Calloway.

In at least one pentastomid species, the intermediate host can be bypassed. In this single documented case, the larvae can develop and become infective within the definitive host (Norwegian reindeer) and then, in female hosts, penetrate placental tissue to infect the next generation of reindeer directly. This finding provides one of the few examples of likely recent evolution within a parasite life cycle; perhaps the intermediate host will eventually be eliminated from the life cycle of this species. How this pentastomid species is able to develop to adulthood entirely within the definitive host when an intermediate host is absolutely required for other parasitic species in this phylum (and most other phyla) is unknown.

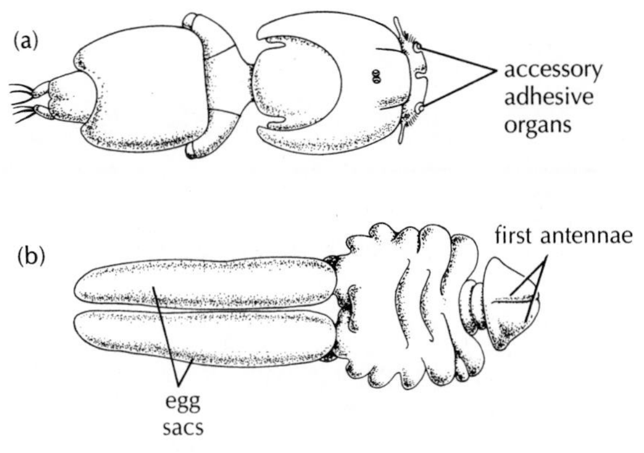

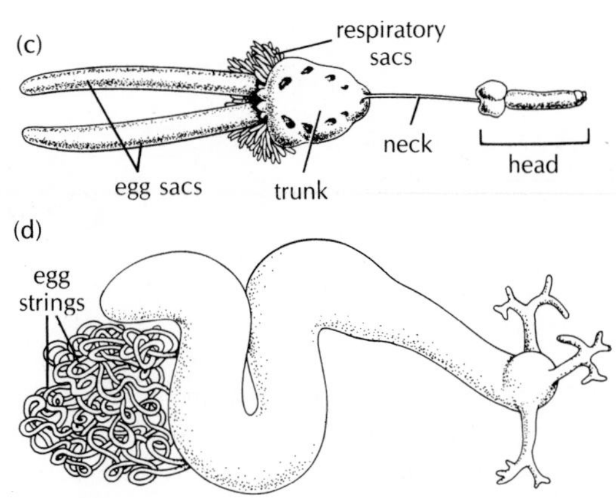

Figure 14.34

Parasitic copepods. (a) *Caligus curtus,* female, found on the outside surface of various marine fish, including cod, pollock, and halibut. Total length is 8–12 mm. (b) *Chondracanthus nodosus,* female, taken from the gills of redfish. The body is about 7 mm long. (c) *Lophoura* (= *Rebelula*) *bouvieri,* female, taken from the flesh of a marine fish. The egg sacs alone are 30–40 mm long. (d) *Lernaeocera branchialis,* a parasite in the gills of flounder. The body of the copepod may reach 40 mm in length, while the egg strings may be several hundred mm long.

Reprinted from C. B. Wilson, "The Copepods of the Woods Hole Region," 1932, from the Bureau of American Ethnology. Washington, DC: Smithsonian Institution Press, pages 197, 343, 354, by permission of the publisher. Copyright 1932.

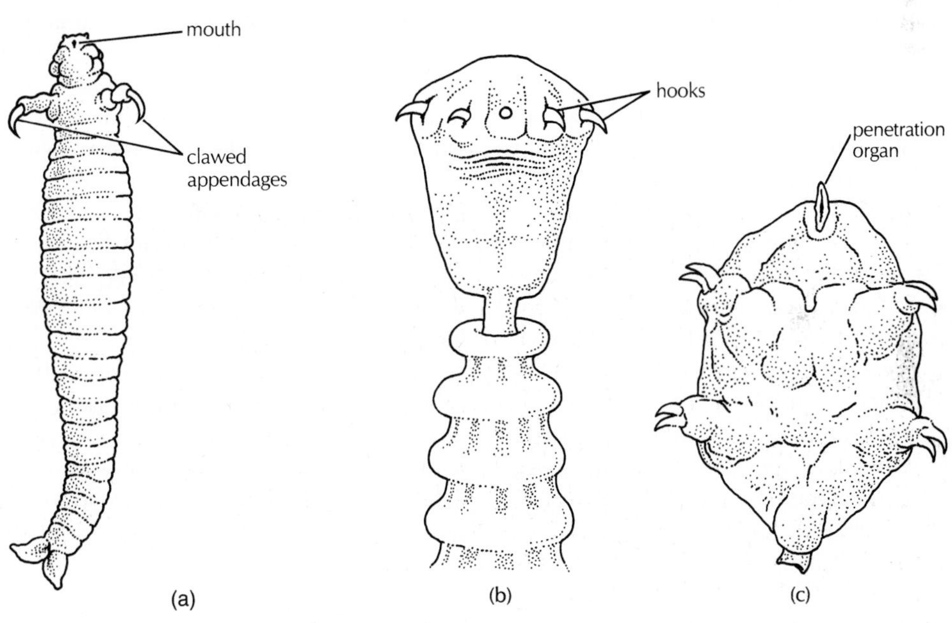

Figure 14.35

Representative pentastomids. The adults are parasitic in the nasal passages of snakes and lizards, while cockroaches, bats, raccoons, muskrats, and armadillos commonly serve as intermediate hosts for the larval stages. (a) *Raillietiella mabuiae,* entire individual. (b) Anterior end of *Armillifer annulatus.* (c) Intermediate larval stage of *Porocephalus crotali.*

(a,b) From Baer, 1951. *Ecology of Animal Parasites.* Champaign: University of Illinois Press. (c) From E. R. Noble and G. A. Noble, *Parasitology: The Biology of Animal Parasites,* 5th ed. Philadelphia: Lea & Febiger, 1982. After Penn. Courtesy of *The Journal of Parasitology.*

Each pentastomid species is highly specific for particular hosts, but the host is never seriously harmed. People occasionally become infected by eating intermediate stages in uncooked meat, but the parasites cannot reach adulthood in humans and die within a few weeks, after causing only minor irritation.

Since all extant pentastomids parasitize vertebrates, one might expect that pentastomids evolved (presumably from crustacean ancestors) only after the evolution of vertebrates. However, what appear to be fully convincing pentastomids were described in 1994 from Cambrian rocks formed some 500 million years ago, well before the advent of vertebrates, or even of terrestrial arthropods. One can but wonder about the original pentastomid hosts—they must have been aquatic and invertebrate—and about why modern pentastomids parasitize only vertebrates.

RESEARCH FOCUS BOX 14.3

Copepod Feeding

Cowles, T. J., R. J. Olson, and S. W. Chisholm. 1988. Food selection by copepods: Discrimination on the basis of food quality. *Marine Biol.* 100:41.

Over the past 40–50 years, the feeding biology of herbivorous marine copepods has been the subject of a growing number of studies, in recognition of the great role these small planktonic crustaceans play in marine food chains that lead to commercially important fish species. Until about 1970, most biologists believed that copepods passively collected food particles from the surrounding water, using the second antennae to generate feeding currents and setae of the first and second maxillae as sieves. This mechanism implied that copepods indiscriminately ingest all particles large enough to be captured by the "sieve." However, research in the past 20 years has indicated that both marine and freshwater planktonic crustaceans deliberately can accept some algal cells and reject others. *How* planktonic crustaceans make these discriminations among food particles is not yet clear. Cowles, Olson, and Chisholm (1988) have demonstrated that at least one copepod species can distinguish between foods on the apparent basis of differences in nutritional value.

To study whether herbivorous copepods (*Acartia tonsa*—about 500 μm long) can recognize differences in nutritional value requires using 2 diets identical in all physical characteristics and apparently differing only in intracellular chemistry. The researchers achieved this by growing the marine diatom *Thalassiosira weissflogii* (about 14 μm in diameter) at 2 different nitrogen concentrations (provided as nitrate). The resulting cells were identical in cell size and in average carbon content, but they differed in individual nitrogen content, chlorophyll content, protein content, and amino acid content; the nutrient-deprived cells were lower in each of these biochemical measures (Focus Table 14.1).

After collecting copepods with plankton nets and acclimating them to laboratory conditions for several days, the researchers determined copepod feeding rates (cells ingested per 5 hours) on each of the 2 chemically distinct algal diets. If the initial number of algal cells per milliliter of seawater in the test container, the final number of cells per milliliter after 5 hours of feeding, the volume of seawater in the container, and the number of copepods put into the container are all known, then feeding rates are easily determined; if cells disappear, it can only be through ingestion. In 5 of the 8 experiments reported in the paper by Cowles et al. (1988), copepods fed substantially more quickly on the cells grown in the high-nutrient medium (Focus Fig. 14.6). These high-nutrient cells thus seemed to somehow stimulate the copepods to feed more actively. Other experiments reported in this paper indicated that the stimulatory agent is a chemical released by the diatom. In all 8 experiments, copepods ingested protein at a significantly faster rate when feeding on the high-nutrient cells, even when eating comparable numbers of cells per hour (Focus Fig. 14.7); this is possible due to the higher protein content of the cells grown in the high-nutrient medium.

What would happen if the copepods were offered a choice of diets? Could they actually distinguish between cells differing in nutrient composition and *preferentially* choose to eat the most nutritionally valuable ones? How would you design such an experiment? If the 2 types of cells are physically identical, how could researchers determine how many of each type the copepods ate?

Focus Table 14.1 Chemical Composition of the Diatom *Thalassiosira weissflogii* Cultured at 2 Nutrient Levels

Nutrient Level	Cell Diameter (μm)	Carbon (pg cell^{-1})	Nitrogen (pg cell^{-1})	C:N	Chl *a* (pg cell^{-1})	Protein (pg cell^{-1})	Free Amino Acids (pg cell^{-1})
Low	14.1 ± 2.3	202.4 ± 29.8	11.8 ± 2.6	17.2 ± 3.7	0.65 ± 0.5	41.2 ± 9.0	735.5
High	14.0 ± 2.1	179.2 ± 25.0	16.5 ± 3.9*	10.9 ± 2.0*	1.85 ± 0.38*	93.6 ± 30.0*	1537.6

Note: Values are means ±2 SE. n = 12, except for protein, where n = 7, and free amino acids, where n = 1. pg = picograms.

*Significant difference ($p < 0.05$) between cell types, paired comparison t-test (df = $2n - 2$).

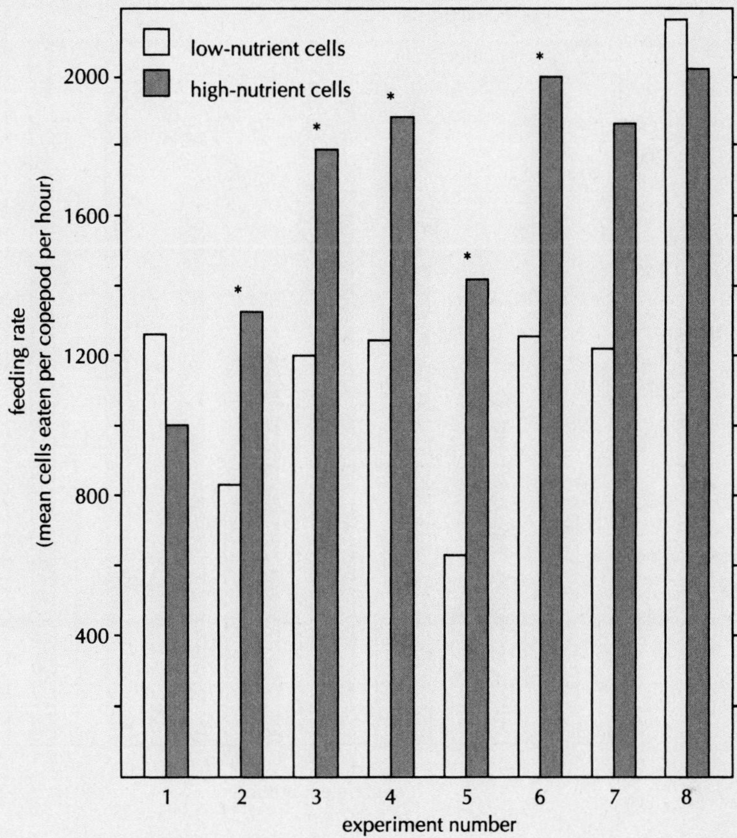

Focus Figure 14.6

The rate at which the copepod *Acartia tonsa* fed on cells of the diatom *Thalassiosira weissflogii* when those cells were cultured either under high-nutrient or low-nutrient conditions. The height of each bar represents the mean of 3 determinations. Asterisks indicate significant differences between mean feeding rates.

It turns out that when cells of the 2 nutrient types are exposed to light and then darkness, they fluoresce at markedly different wavelengths. If a sample of the feeding mixture is placed—before and after the 5-hour feeding study—into a complex machine called a "flow cytometer," the machine will examine each cell's fluorescence pattern and then sort the sample cell by cell into a high-nutrient or low-nutrient category. The flow cytometer thus makes it possible to distinguish between the cells, even though they look identical. In this experiment, many more high-nutrient cells disappeared from the medium during feeding than would have been predicted from their abundance in the mixture at the start of the study; these nutritionally superior cells seem to have been eaten preferentially, certainly not in direct proportion to their abundance (Focus Fig. 14.8).

How can a copepod possibly tell the difference between one algal cell and the next if the cells are morphologically indistinguishable? The researchers suggest that either (1) the copepod recognizes a nutritionally more valuable cell from a distance and preferentially captures such a cell, or (2) the copepod captures both cell types at equal rates but then preferentially *ingests* the high-nutrient cells, rejecting the others. In both cases, the copepod is presumed capable of perceiving a subtle chemical gradient surrounding each cell and behaving accordingly. Certainly, diatoms and other unicellular algae are known to produce a variety of chemical exudates. But can a copepod really be so sensitive to what are likely very subtle differences in chemical gradients surrounding individual algal cells? Can you think of any alternative explanations for the data?

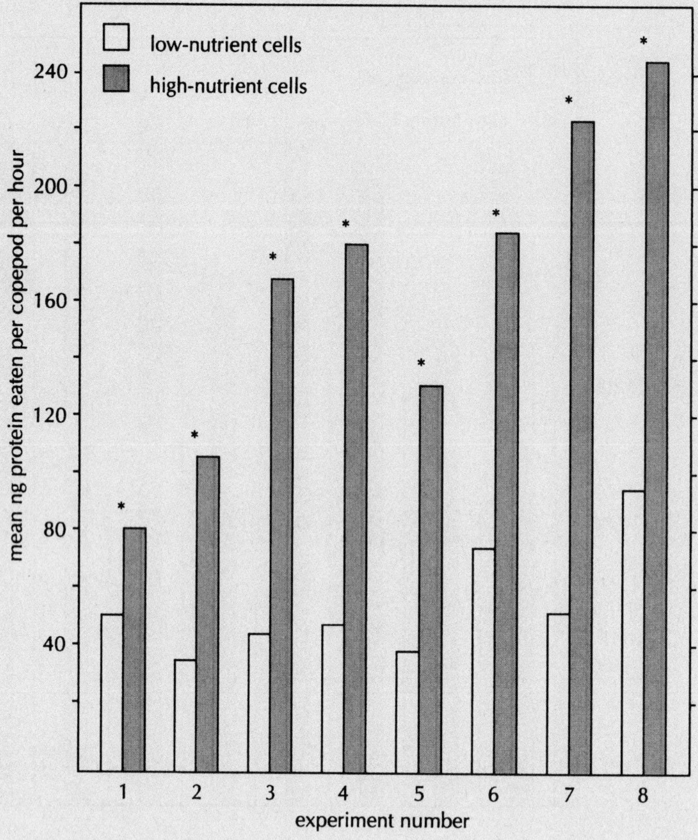

Focus Figure 14.7

The rate at which the copepod *Acartia tonsa* ingested protein when fed diatoms (*Thalassiosira weissflogii*) reared under either high-nutrient or low-nutrient conditions. The height of each bar represents the mean of 3 replicates. Asterisks indicate significant differences between mean protein ingestion rates. (ng = nanograms.)

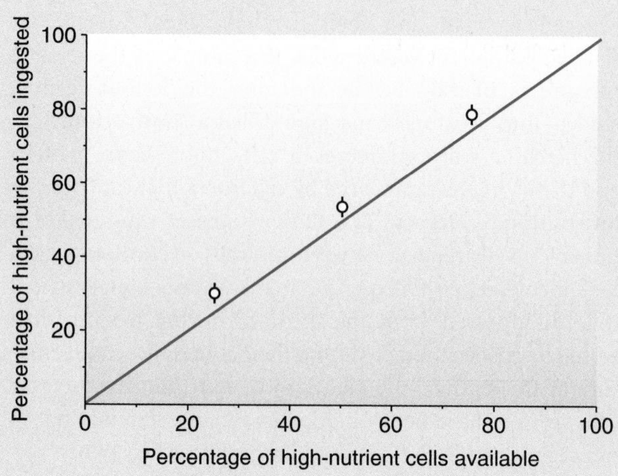

Focus Figure 14.8

Preference for high-nutrient algal cells by copepods (*Acartia tonsa*) fed mixtures of high- and low-nutrient cells. If copepods were eating cells in proportion to their abundance, data points would fall on the dashed line. Points above the dashed line indicate a preference for high-nutrient cells. Vertical bars attached to each point indicate one standard deviation about the mean. Each point represents the mean of 3 replicates.

Based on T. J. Cowles, R. J. Olson, and S. W. Chisholm, "Food Selection by Copepods: Discrimination on the Basis of Food Quality" in *Marine Biology*, 100:41–49, 1988.

Subclass Cirripedia

Subclass Cirri • pedia
(L: hairy foot)
sēr-i-pē´-dē-ah

Defining Characteristics: 1) All species highly modified for attachment to hard substrates, including the outer surfaces of other animals, or for parasitic life; 2) thoracic limbs modified as filtering cirri; 3) no abdomen

Cirripedes are more commonly billed as the barnacles. The approximately 1000 species in this subclass are exclusively marine and show a greater departure from the basic crustacean body plan than the members of any other subclass. Unlike most other crustaceans, all cirripedes are exclusively sedentary organisms, permanently affixed to, or burrowed into, living (including whales) or nonliving substrates. Barnacles attached to moving or floating substrata often are conspicuously stalked (Fig. 14.36c). Other species may be cemented directly to the substratum at the "basis." In keeping with a nonmotile existence, the head is greatly reduced, the first antennae are much reduced, and the second antennae are absent. Most species live within a thick calcium carbonate protective shell (Fig. 14.36), which they secrete. Because of this shell, barnacles were classified as molluscs until about 150 years ago. Indeed, the barnacle shell is secreted by the "mantle" tissue, and the space between the inner wall of the shell and the animal itself is called the "mantle cavity," as in molluscs.

Barnacles have attracted a reasonable amount of attention as foulers of ship bottoms; even a moderate encrustation of barnacles can greatly reduce ship speed and fuel efficiency. The happy consequence of barnacle encrustation is that research on the biology of cirripedes has not been difficult to justify.

The shell of barnacles is composed of numerous plates, including the **carina, rostrum, scuta,** and **terga** (Fig. 14.36a–c). The rostrum represents the side of the shell at which the body of the barnacle attaches to the mantle. Certain of the scuta and the terga are movable. Thus, the scuta and terga can occlude the opening at the top of the shell when the animal withdraws, and they can move apart when the feeding appendages are to be protruded. The feeding appendages, called **cirri,** are modified thoracic appendages (Fig. 14.36a); these are used by free-living barnacles to filter food particles from the water. Recent studies indicate that members of at least some species feed actively when water velocities moving past the barnacles are low but switch to passive filtering when water velocities exceed a particular threshold (Fig. 14.37).

Barnacles lack abdominal segments, gills, and a heart; the blood circulates through sinuses entirely by movements of the body. The circulatory system is open, as it is in all arthropods.

A number of cirripedes live in burrows within calcareous substrates (e.g., shell, coral) or as parasites within the bodies of other animals, including other crustaceans. As might be expected, such specializations in habitat and lifestyle are reflected by major morphological modifications, particularly by the presence of root-like absorptive structures (Fig. 14.36d,e).[21] The life cycles of parasitic barnacles are often bizarre and certainly unique among crustaceans, as discussed in the next section.

Crustacean Development

Marine species often have a free-living larval stage in the life history (Fig. 14.38). A free-living **nauplius** larva is typical of several diverse groups of crustaceans, including the copepods, ostracods, branchiopods, euphausiids, and cirripedes. Although the adult body plan of barnacles has become highly modified from the basic crustacean pattern, the typical nauplius larva remains unchanged; even highly modified parasitic species produce typical nauplii. In all species, the nauplius has a characteristic triangular shape, a good crustacean carapace, and a median eye composed of 3 ocelli (Fig. 14.38a, b). Periodically, the nauplius larva molts and subsequently adds or modifies appendages and gets larger. After going through several naupliar stages, the nauplius metamorphoses into a morphologically distinct larval stage. Among the Copepoda, the final naupliar stage undergoes a transition to a **copepodite** form, and then the individual goes through 5 copepodite stages before the final adult body plan is attained. Among the barnacles, the larva proceeds through several feeding naupliar stages and then metamorphoses into a remarkable nonfeeding **cypris** stage (Figs. 14.38c and 24.14a). The cypris larva, or **cyprid,** is housed in a bivalved, noncalcified carapace. The thoracic appendages with which the larva swims become the filtering appendages of the adult barnacle. A pair of sensory antennae are conspicuous at the animal's anterior end. When the cyprid of nonparasitic species locates a suitable substrate, it secretes a glue from anterior cement glands and the animal then becomes permanently attached to the substrate by its head. Dramatic internal reorganization and production of the adult body enclosure quickly ensue.

Parasitic barnacles also have free-living, perfectly normal cyprids as the terminal larval stage, but their life cycles are often strikingly modified from the moment of substrate selection. A short summary will illustrate just how modified the life cycle can become. Female cyprids of several well-studied species (e.g., members of the genus *Sacculina*) attach to a prospective crustacean host and metamorphose into another form (the **kentrogon**), specialized for piercing through the host cuticle. The entire kentrogon does not enter the host, however; rather, it injects only a part of itself through the cuticle—a microscopic, mobile, multicellular, worm-shaped mass that eventually ruptures to release a number of large,

21. See *Topics for Further Discussion and Investigation,* no. 4.

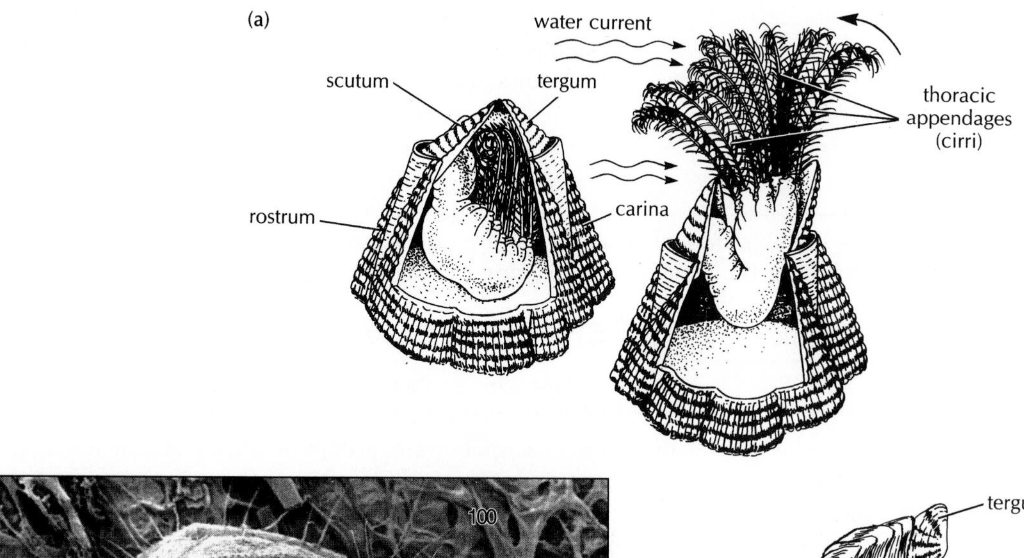

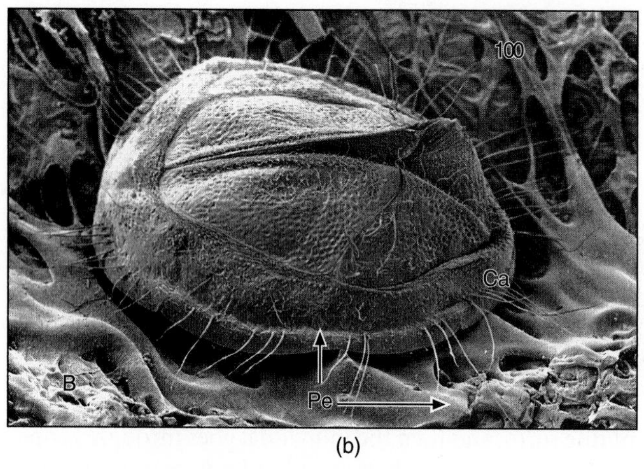

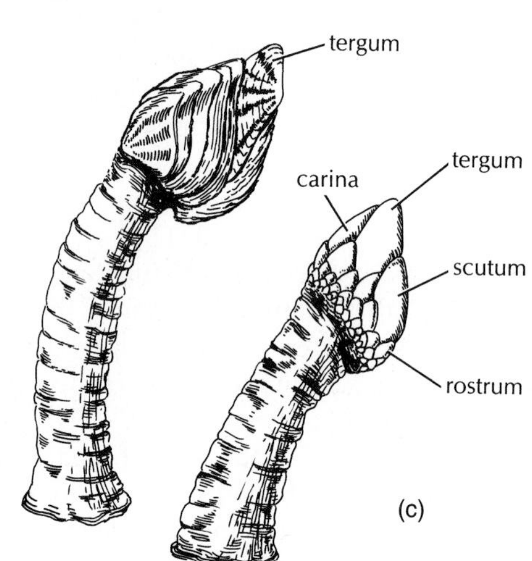

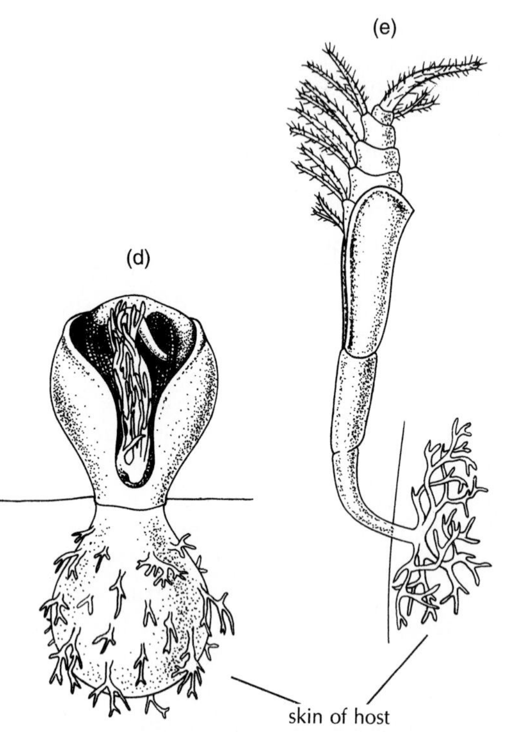

Figure 14.36

(a) The free-living barnacle *Balanus* sp., an "acorn barnacle." The scutum and tergum of the barnacle shell can move apart, allowing the animal to extend its cirri in the water currents and filter out food particles. The scuta and terga form a type of operculum, protecting the animal from predators, salinity stress, and desiccation stress once the thoracic appendages have been withdrawn. (b) Juvenile of the barnacle *Balanus improvisus* 4–6 hours after the free-swimming larval stage attached to the surface and shed its larval cuticle. The plates of the adult shell are already conspicuous. This individual is about 680 μm (micrometers) in longest dimension. Ca = carina, Pe = rudimentary peduncle. (c) Stalked gooseneck barnacles; *Lepas* sp. on the left and *Pollicipes* sp. on the right. Barnacles encrust a variety of substrates, including rocks, ship bottoms, turtle shells, and whales. (d,e) Parasitic barnacles, shown in the skin of a shark (d) and an annelid (e).

(b) From H. Glenner and J. T. Høeg. 1993. *Marine Biology.* 117:431–39. Springer-Verlag, New York. (c) After Pimentel. (d,e) From Baer, 1951. *Ecology of Animal Parasites.* Champaign: University of Illinois Press.

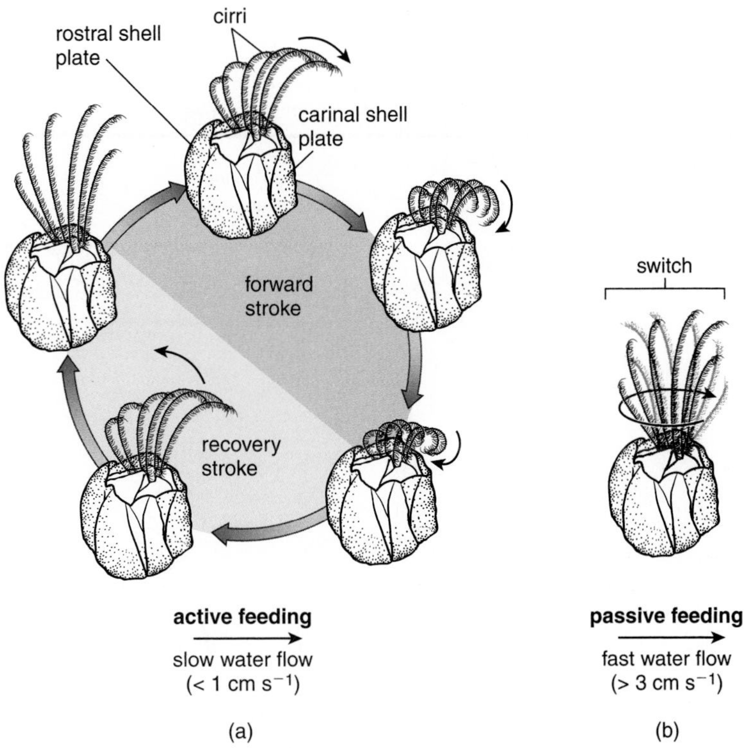

Figure 14.37

Feeding by the common barnacle *Semibalanus balanoides.* (a) Active capture of food particles by cirri when water is flowing slowly. (b) Under conditions of faster water flow, the cirri reverse their orientation and filter particles from the water passively. Barnacles switched to passive feeding when mean water velocity exceeded 3.10 cm • sec^{-1} (centimeters per second).
(a) Based on C. C. Trager et al., in *Marine Biology,* 105:117–27, 1990.

mobile cells. At least one of these cells then develops extensively within the host. Eventually, a large, specialized reproductive mass, containing the ovary, emerges on the host's lower abdominal surface (Fig. 14.39a). Once a small aperture develops on this external female gonad, the gonad becomes attractive to male cyprids, which walk into the opening and attach to the female. Within an hour, a spiny, amorphous mass of tissue is discharged through one of the male cyprid's antennae. This small mass of tissue, which is best thought of as a juvenile male, somehow migrates within the female mantle cavity to one of 2 receptacle ducts and eventually implants at the distal end of a waiting receptacle chamber (Fig. 14.39b). The dwarf male then begins differentiating gonadal tissue, and little else. Only one male is apparently ever found in a single receptacle, so each female supports and nourishes up to 2 males within her reproductive tract for their entire lives. The female gonad does not mature unless at least one of the 2 receptacle chambers houses a male. Eggs can then be fertilized and free-living nauplii released into the surrounding water.

Not all crustaceans produce free-living nauplius larvae, although all pass through a naupliar stage as they develop. Decapod crabs and shrimp, for instance, usually brood their embryos (externally, beneath the abdomen) to a more advanced stage, releasing **zoea** larvae characterized by a pair of large compound eyes and a spiny carapace (Figs. 14.38d and 24.14b). Crabs typically go through a number of zoeal stages before metamorphosing into a **megalopa** stage, which looks much like the adult except that the abdomen is not tucked up under the thorax (Fig. 14.38e). Other crustacean groups produce other types of larval stages, always recognizable as being arthropod. Freshwater crustaceans often lack free-living larval stages, probably for reasons discussed in Chapter 1; freshwater copepods, however, often have free-swimming naupliar stages in the life history.

As with the branchiopods discussed earlier (p. 379), many marine and freshwater copepods release "resting" or "dormant" eggs at certain times of year. These accumulate in sediments at up to 1 million eggs per m^2. They can remain viable for up to several decades before releasing nauplius larvae, potentially contributing to population growth many years after their inital fertilization.

Isopods, amphipods, and other peracarid malacostracans (see *Taxonomic Detail,* p. 412) lack larval stages. Young emerge as miniatures of the adult, after a period of protection in a ventral brood chamber (**marsupium**) formed by flat, overlapping extensions of certain thoracic appendages on the female; the ventral surface of the female's thorax forms the roof of the chamber. The lack of free-living larval stages in aquatic species may be a preadaptation for a terrestrial existence. Not surprisingly,

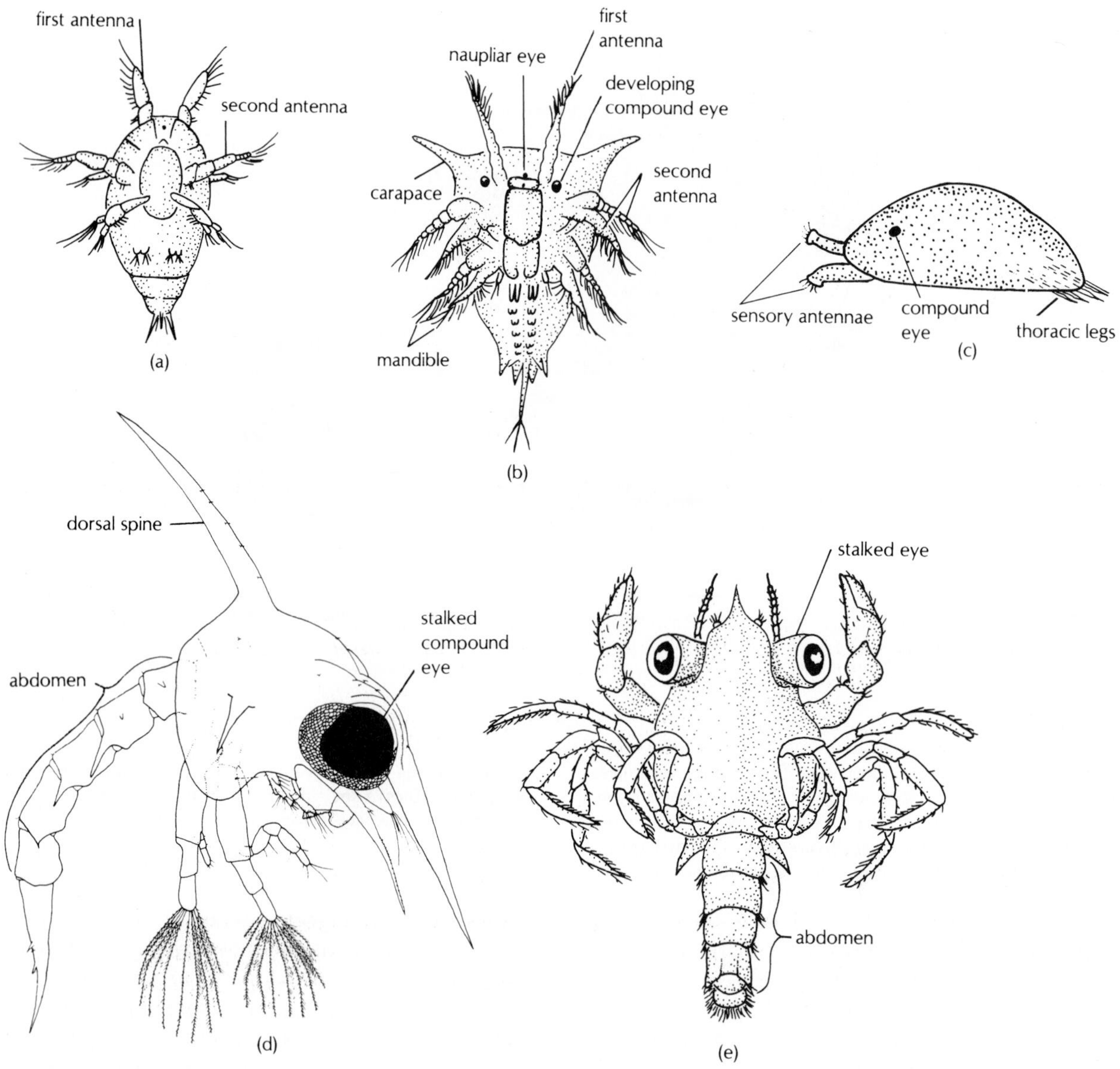

Figure 14.38
Crustacean larval stages. (a) Copepod nauplius. (b) Barnacle nauplius. (c) Barnacle cyprid. (d) Decapod zoea: *Portunus sayi,* stage IV. (e) Decapod megalopa larva.
(b) After Korschelt. (d) Courtesy of I. P. Williams. (e) After Hardy.

isopods and amphipods are the only groups of malacostracans to have accomplished major radiations into terrestrial or semiterrestrial habitats.

Other Features of Arthropod Biology

Digestion

The arthropod gut is divisible into 3 areas: foregut, midgut, and hindgut. All free-living species exhibit a distinct and separate mouth and anus, and in all species, food must be moved through the digestive tract by muscular activity rather than ciliary activity, since the lumen of the foregut and hindgut is lined with cuticle. Digestion is generally extracellular. Nutrients are distributed to the tissues through the hemal system.

Class Merostomata

Members of the class Merostomata are carnivorous. The mouth leads into an esophagus and thence into a **gizzard.** Both the esophagus and the gizzard are lined

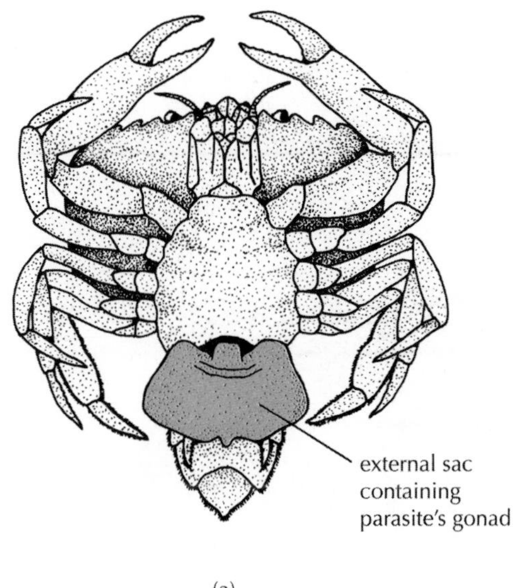

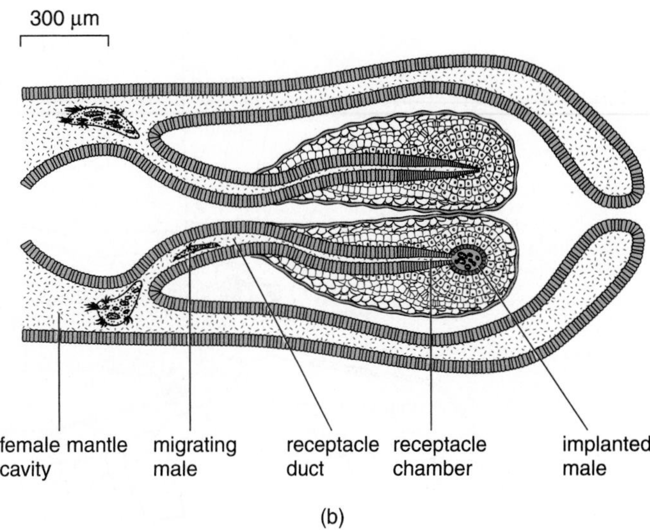

Figure 14.39

(a) External gonad of the rhizocephalan *Sacculina carcini,* parasitic on (and in) the green crab *Carcinus maenas*. The crab is shown in ventral view. Internally, the parasite ramifies throughout the crab's tissues. (b) Diagrammatic cross section through a portion of the external gonad of a female rhizocephalan barnacle, *Sacculina carcini.* As shown, the female has 2 narrow receptacle ducts opening into the mantle chamber. A dwarf male, expelled through one antenna of a cypris larva, is shown entering the lower of the 2 ducts. The upper duct already houses one male at its distal end; another male, shown about to enter this upper duct, will never reach the end: The passageway is blocked by the cuticle shed previously by the first male to enter.

(a) From K. Rhode, 1982. *Ecology of Marine Parasites.* New York: Queensland Press; after Boas. (b) Based on J. T. Høeg, *Philosophical Transactions of the Royal Society,* 317:47–63, 1987.

with cuticle, which is molted periodically along with the rest of the exoskeleton. The gizzard is equipped with chitinous teeth for grinding ingested food. A valve prevents passage of undigestible material from the gizzard into the stomach. A pair of elongated pouches (**hepatic,** or digestive, **ceca**) extend laterally from the stomach. Most food digestion and nutrient absorption occur in these hepatic ceca. Wastes pass through the rectum and out an anus, located ventrally at the base of the caudal spine.

Class Arachnida

Most arachnid species are carnivores. Because arachnids lack mandibles, the chelicerae must tear and grind the food prior to ingestion. In addition, glands associated with the oral cavity release enzymes into the prey, and the food is thus predigested externally. Digestive enzymes contributed by glands in the chelicerae or pedipalps may also participate in this process. After sufficient time has elapsed, the solubilized tissue is moved into the foregut through a muscular pump and then travels from there into the stomach; with few exceptions, arachnids do not ingest food in particulate form. Digestion and absorption take place primarily in the extensive tubular outgrowths of the stomach wall, the digestive diverticula. Undigested material and wastes pass through an intestine and exit from a posterior anus.

Class Myriapoda (Orders Chilopoda and Diplopoda)

Most species of chilopod (centipede) are predaceous carnivores on other invertebrates and on some smaller vertebrates, using their modified maxillipeds to hold and poison prey. The digestive tract is usually a straight tube. Diplopods (millipedes), on the other hand, are usually herbivores, feeding on both living and, especially, dead or decaying plant material or the juices of living plants; some carnivorous species also exist. A **peritrophic membrane** lines the midgut of millipedes, presumably to protect against abrasion. Food becomes enclosed by this membrane as it moves through the gut, and new peritrophic membrane is then secreted. As in the centipedes, the millipede gut is essentially a linear tube.

Class Insecta

Insects feed on a variety of nutritive sources, including plant and animal tissues and fluids. The esophagus is highly muscularized and serves as a pump, moving food into a crop for storage and/or preliminary digestion. A **proventriculus** is generally present, functioning as a valve to regulate the passage of food and, in some species, to grind ingested food. Digestion and absorption occur in the midgut and its associated gastric ceca. Symbiotic bacteria and protozoans harbored in the gastric ceca of many species participate in the digestive process. The walls of the midgut are often lined by peritrophic membrane, as

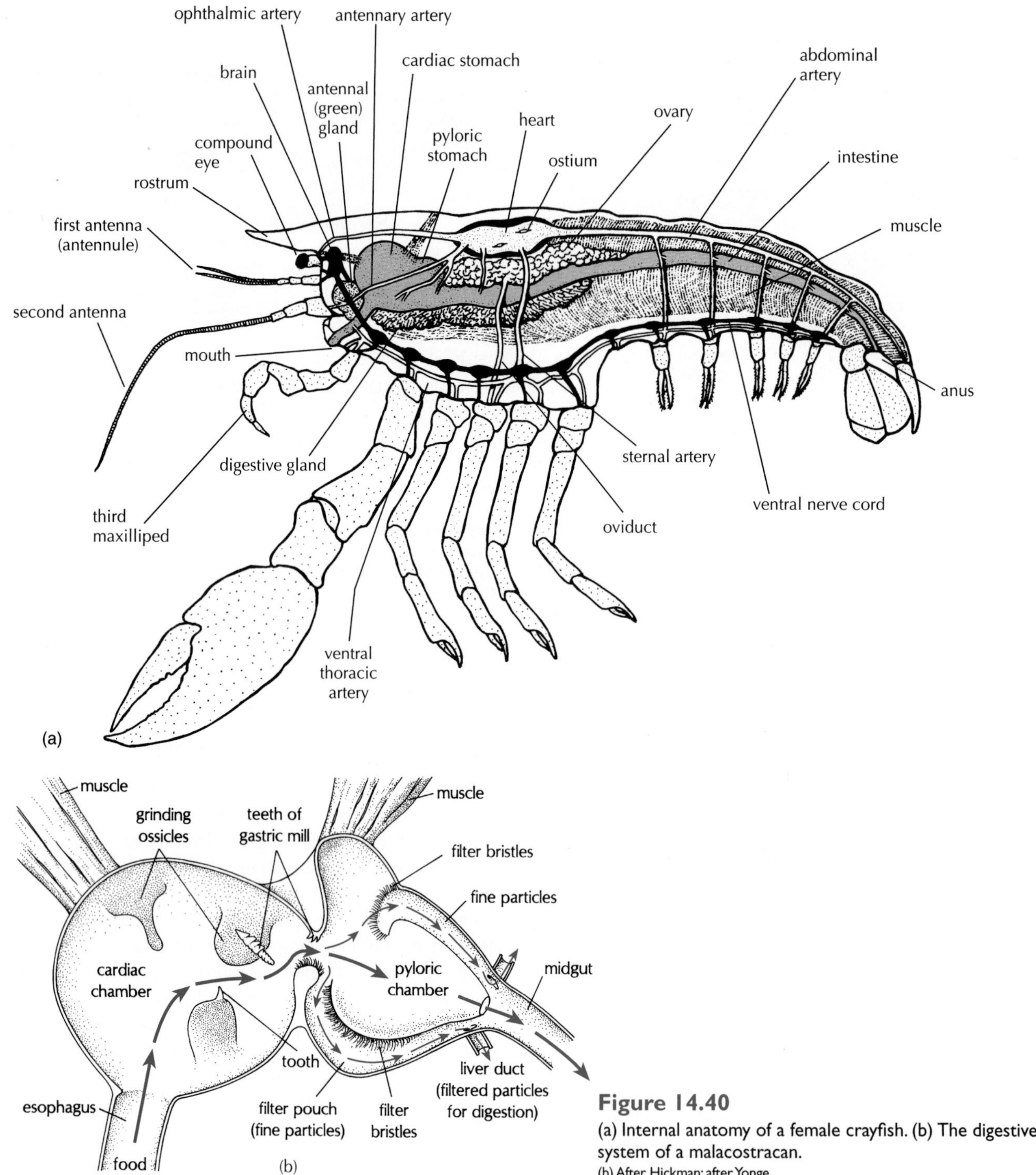

Figure 14.40
(a) Internal anatomy of a female crayfish. (b) The digestive system of a malacostracan.
(b) After Hickman; after Yonge.

in the Diplopoda. This membrane is discarded and renewed periodically. Before wastes reach the anus, most of the water is resorbed from the fecal material by the rectal glands of the hindgut (Fig. 14.20).

Class Crustacea

In the crustaceans, food is ground and strained in a muscular foregut consisting of a large **cardiac stomach** and a smaller **pyloric stomach** (Fig. 14.40). The food is ground by the chitinous, toothed ridges of a **gastric mill.** Stiff setae often prevent food particles from passing farther until they have reached the proper consistency. Food passes on to the midgut, which is associated with an extensive array of digestive ceca in which the food is digested, absorbed, and stored. The digestive tubules often form a distinct organ, the **hepatopancreas,** or liver.

Excretion

Class Merostomata

Members of the Merostomata bear 4 pairs of excretory **coxal glands** adjacent to the gizzard. The glands empty into a common chamber, which then leads through a coiled tubule and into a bladder. Salts and amino acids may be selectively resorbed as they pass through the system, the final urine being discharged through pores at the base of the last pair of walking legs.

Class Arachnida

Coxal glands are also common among arachnids. Here they are spherical sacs resembling annelid nephridia. Wastes are collected from the surrounding blood of the hemocoel and discharged through pores on from one to several pairs of appendages. Recent evidence suggests that the coxal glands may also function in the release of pheromones.

Some arachnid species have Malpighian tubules instead of, or in addition to, the coxal glands. In some of these species, however, the Malpighian tubules seem to function in silk production rather than excretion. The excretory role of Malpighian tubules in insects has already been discussed (p. 364). Arachnids possess in addition to the coxal glands and Malpighian tubules, a number of strategically placed cells called **nephrocytes,** which phagocytize waste particles. The major waste product of protein metabolism among arachnids is guanine, a purine biochemically related to uric acid.

Orders Chilopoda and Diplopoda

The major excretory structures in the Chilopoda and Diplopoda are Malpighian tubules. Although some uric acid is produced, the major waste product of centipedes is ammonia.

Class Insecta

The major excretory structures of insects are Malpighian tubules (Fig. 14.41a), as discussed earlier (p. 364). Several other mechanisms also may be involved in insect waste elimination. In particular, evidence indicates that some wastes are incorporated into the cuticle, to be shed at ecdysis. Many insects also possess nephrocytes, as found in arachnids.

Insects eliminate a significant fraction of their nitrogenous wastes as water-insoluble uric acid and related compounds. These compounds are eliminated from the body in nearly dry form, as most of the water in the urine is resorbed in transit through the rectum. This is an outstanding physiological adaptation for terrestrial existence.

Class Crustacea

In crustaceans, nitrogenous wastes are generally removed by diffusion across the gills—for those species that have gills. Most crustaceans release ammonia, although some urea and uric acid are also produced. The so-called excretory organs may be more involved with elimination of water than with discharge of nitrogenous wastes in

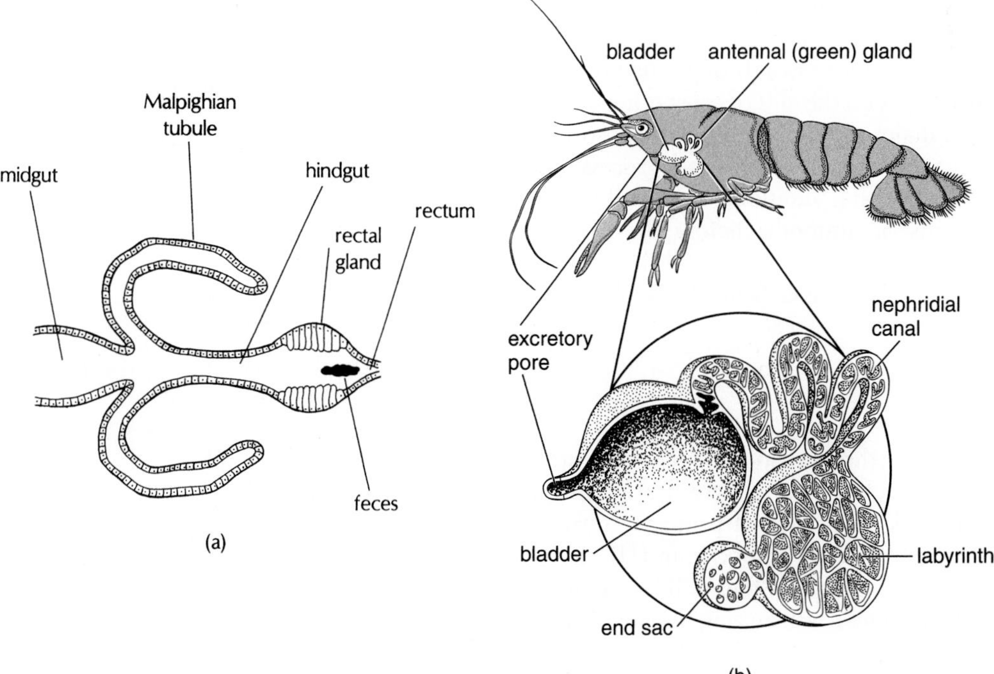

Figure 14.41
(a) Malpighian tubule system from an insect. (b) Antennal gland of the crayfish. In the enlargement, the gland has been "untangled" for clarity.

(b) Based on Purves and Orians, *Life: The Science of Biology,* 2d ed. 1983.

Taxonomic Summary[22]

Phylum Arthropoda
- Subphylum Trilobitomorpha
  - Class Trilobita—the trilobites
- Subphylum Chelicerata
  - Class Merostomata—horseshoe crabs
  - Class Arachnida—spiders, mites, ticks, scorpions
  - Class Pycnogonida (= Pantopoda)—sea spiders
- Subphylum Mandibulata
  - Class Myriapoda
    - Order Chilopoda—centipedes
    - Order Diplopoda—millipedes
  - Superclass Hexapoda
  - Class Entognatha
  - Class Insecta (= Ectognatha)

Class Crustacea
- Subclass Malacostraca
  - Order Isopoda—pillbugs, woodlice
  - Order Amphipoda—sand fleas
  - Order Euphausiacea—euphausiids (krill)
  - Order Stomatopoda—stomatopods
  - Order Decapoda—crabs, lobsters, shrimp, hermit crabs
- Subclass Branchiopoda—brine (fairy) shrimp, clam shrimp, water fleas
- Subclass Ostracoda—the ostracods
- Subclass Copepoda—the copepods
- Subclass Pentastomida
- Subclass Cirripedia—the barnacles

most species. In some species, these paired "excretory" organs are called **antennal glands** or **green glands** because of their location near the base of the antennae (Fig. 14.41) and their color. In other species, the organs are found near the maxillary segments and are termed **maxillary glands.** These excretory organs are structurally similar to the coxal glands of chelicerates. Fluid collects within the tubules from the surrounding blood of the hemocoel. Among freshwater crayfish, this primary urine is modified substantially by selective reabsorption and secretion as it moves through the excretory system. Specialized cells on the gills of most other species generally play the major role in expelling and absorbing salts. At least some terrestrial crab species reprocess their urine at the gills, resorbing additional ions and adding substantial amounts of ammonia before the urine is finally discharged; the urine reaches the gills by trickling downward into the gill chamber from the nephropores (G: kidney openings).

Blood Pigments

The bloods of many arthropod species lack respiratory pigments. This is particularly true of terrestrial species, in which gas exchange is accomplished primarily through tracheae. In other species, both hemocyanin (HCy) and, more rarely, hemoglobin (Hb) are found. Both pigments are sporadically distributed, even among the members of a given class.

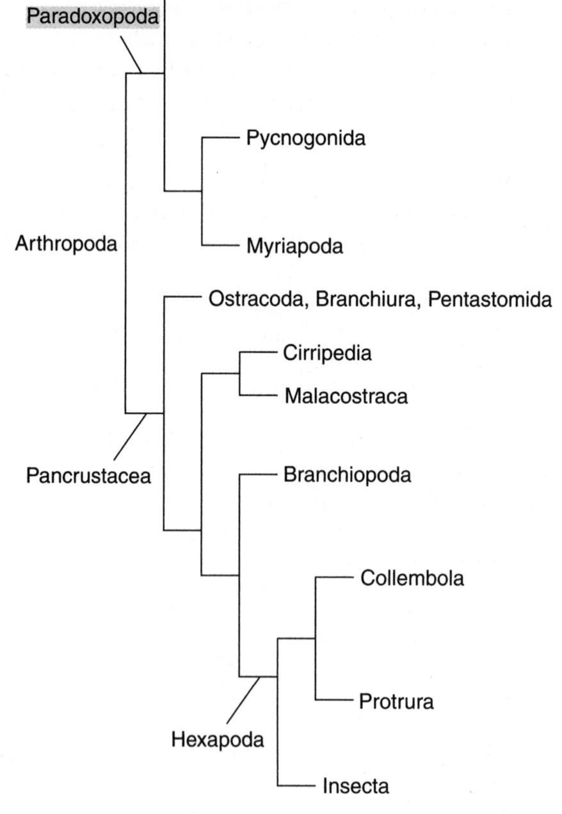

Figure 14.42
A recent view of arthropod relationships. At present, there is approximately equal support for the Paradoxopoda (an unexpectedly close relationship between myriapods and chelicerates) and the Mandibulata (as retained in this text, linking the myriapods to crustaceans and insects).
Simplified from Mallat, J. and G. Giribet. 2006. *Molec. Phylog. Evol.* 40: 772–794.

22. See Figure 14.42 for a competing view, in which myriapods are more closely allied with chelicerates than with crustaceans.

Topics for Further Discussion and Investigation

1. Arthropod affinities and origins are far from certain, even at the highest taxonomic levels. Arthropods may have derived from annelid-like ancestors (the Articulata hypothesis). However, embryological studies and studies of limb structure and function in various arthropod and polychaete species have long suggested that insects, millipedes, and centipedes have had a very different evolutionary origin from that of other arthropod groups, such as the Crustacea and Arachnida. What is the evidence in favor of and against an independent origin for these arthropods, and what difficulties does each scheme present in terms of the numbers of different arthropod characteristics that must be presumed to have evolved independently through convergent evolution?

Aguinaldo, A. M. A., J. M. Turbeville, L. S. Linford, M. C. Rivera, J. R. Garey, R. A. Raff, and J. A. Lake. 1997. Evidence for a clade of nematodes, arthropods and other molting animals. *Nature* 387:489.

Ballard, J. W. O., G. J. Olsen, D. P. Faith, W. A. Odgers, D. M. Rowell, and P. W. Atkinson. 1992. Evidence from 12S ribosomal RNA sequences that onychophorans are modified arthropods. *Science* 258:1345.

Boore, J. L., T. M. Collins, D. Stanton, L. L. Daehler, and W. M. Brown. 1995. Deducing the pattern of arthropod phylogeny from mitochondrial DNA rearrangements. *Nature* 376:163.

Cisne, J. L. 1974. Trilobites and the evolution of arthropods. *Science* 146:13.

Emerson, M. J., and F. R. Schram. 1990. The origin of crustacean biramous appendages and the evolution of Arthropoda. *Science* 250:667–69.

Evans, H. E. 1959. Some comments on the evolution of the Arthropoda. *Evolution* 13:147.

Mallatt, J., and G. Giribet. 2006. Further use of nearly complete 28S and 18S rRNA genes to classify Ecdysozoa: 37 more arthropods and a kinorhynch. *Molec. Phylog. Evol.* 40: 772–94.

Mayer, G. 2006. Orgin and differentiation of nephridia in the Onychophora provide no support for the Articulata. *Zoomorphol.* 125: 1–12.

Moore, J., and P. Willmer. 1997. Convergent evolution in invertebrates. *Biol. Rev.* 72:1.

Nardi, F., G. Spinsanti, J. L. Boore, A. Carapelli, R. Dallai, and F. Frati. 2003. Hexapod origins: Monophyletic or paraphyletic? *Science* 299: 1887–89.

Panganiban, G., S. M. Irvine, C. Lowe, H. Roehl, L. S. Corley, B. Sherbon, J. K. Grenier, J. F. Fallon, J. Kimble, M. Walker, G. A. Wray, B. J. Swalla, M. Q. Martindale, and S. B. Carroll. 1997. The origin and evolution of animal appendages. *Proc. Natl. Acad. Sci. USA* 94:5162.

Regier, J. C., J. W. Schultz, and R. E. Kambic. 2005. Pancrustacean phylogeny: Hexapods are terrestrial crustaceans and maxillopods are not monophyletic. *Proc. Royal Soc. London B* 272: 395–401.

Schmidt-Rhaesa, A., T. Bartolomaeus, C. Lemburg, U. Ehlers, and J. R. Garey. 1998. The position of the Arthropoda in the phylogenetic system. *J. Morphol.* 238:263.

Shear, W. A. 1992. End of the "Uniramia" taxon. *Nature* 359:477.

Shubin, N., C. Tabin, and S. Carroll. 1997. Fossils, genes and the evolution of animal limbs. *Science* 388:639.

Tiegs, O. W., and S. M. Manton. 1958. The evolution of the Arthropoda. *Biol. Rev.* 33:255.

2. Living in the terrestrial environment poses many difficulties for invertebrates. Not least among these is that the air is dry and that temperatures fluctuate considerably, even on a day-to-day or hour-to-hour basis. Investigate some of the physiological and behavioral adaptations that enable terrestrial and semiterrestrial arthropods to tolerate dehydrating conditions and marked shifts in environmental temperature.

a. References concerning temperature regulation:

Dorsett, D. A. 1962. Preparation for flight by hawkmoths. *J. Exp. Biol.* 39:579.

Duman, J. G. 1977. Environmental effects on antifreeze levels in larvae of the darkling beetle, *Meracantha contracta. J. Exp. Zool.* 201:333.

Edney, E. B. 1953. The temperature of woodlice in the sun. *J. Exp. Biol.* 30:331.

Heinrich, B. 1974. Thermoregulation in endothermic insects. *Science* 185:747.

Heinrich, B., and H. Esch. 1994. Thermoregulation in bees. *Amer. Sci.* 82:164.

Kingsolver, J. G. 1987. Predation, thermoregulation, and wing color in pierid butterflies. *Oecologia* (*Berlin*) 73:301.

Kugal, O., B. Heinrich, and J. G. Duman. 1988. Behavioural thermoregulation in the freeze-tolerant arctic caterpillar, *Gynaephora groenlandica J. Exp. Biol.* 138:181.

Morgan, K. R. 1985. Body temperature regulation and terrestrial activity in the ectothermic beetle *Cicindela tranquebarica. Physiol. Zool.* 58:29.

Southwick, E. E., and R. A. Moritz. 1987. Social control of air ventilation in colonies of honey bees, *Apis mellifera. J. Insect Physiol.* 33:623.

Vogt, F. D. 1986. Thermoregulation in bumblebee colonies. I. Thermoregulatory versus brood-maintenance behaviors during acute changes in ambient temperature. *Physiol. Zool.* 59:55.

Wilkens, J. L., and M. Fingerman. 1965. Heat tolerance and temperature relationships of the fiddler crab, *Uca pugilator,* with reference to its body coloration. *Biol. Bull.* 128:133.

Willmer, P. G. 1982. Thermoregulatory mechanisms in Sarcophaga. *Oecologia* (*Berlin*) 53:382.

b. References concerning water balance:

Cohen, A. C., R. B. March, and J. D. Pinto. 1981. Water relations of the desert blister beetle *Cysteodemus armatus* (Leconte) (Coleoptera: Meloidae). *Physiol. Zool.* 54:179.

Combs, C. A., N. Alford, A. Boynton, M. Dvornak, and R. P. Henry. 1992. Behavioral regulation of hemolymph osmolarity through selective drinking in land crabs, *Birgus latro* and *Gecarcoidea lalandii. Biol. Bull.* 182:416.

Dresel, E. I. B., and V. Moyle. 1950. Nitrogenous excretion in amphipods and isopods. *J. Exp. Biol.* 27:210.

Edney, E. B. 1966. Absorption of water vapour from unsaturated air by *Arenivaga* sp. (Polyphgidae, Dictyoptera). *Comp. Biochem. Physiol.* 19:387.

Hamilton, W. J., III, and M. K. Seely. 1976. Fog basking by the Nambi desert beetle, *Onymacris unguicularis. Nature* (*London*) 262:284.

Noble-Nesbitt, J. 1970. Water uptake from subsaturated atmospheres: Its site in insects. *Nature* (*London*) 225:753.

Schubart, C. D., R. Diesel, and S. B. Hedges. 1998. Rapid evolution to terrestrial life in Jamaican crabs. *Nature* 393:363.

Seely, M. K., and W. J. Hamilton III. 1976. Fog catchment sand trenches constructed by tenebrionid beetles, *Lepidochora,* from the Namib Desert. *Science* 193:484.

Standing, J. D., and D. D. Beatty. 1978. Humidity behaviour and reception in the sphaeromatid isopod *Gnorimosphaeroma oregonensis* (Dana). *Canadian J. Zool.* 56:2004.

Willmer, P. G., M. Baylis, and C. L. Simpson. 1989. The roles of colour change and behavior in the hygrothermal balance of a littoral isopod, *Ligia oceanica. Oecologia* (*Berlin*) 78:349.

Wolcott, D. L. 1991. Nitrogen excretion is enhanced during urine recycling in 2 species of terrestrial crab. *J. Exp. Zool.* 259:181.

3. Invertebrates are basically poikilothermic; that is, in the absence of behavioral modification (including flight), body temperatures generally follow those of the surrounding air or water rather closely. This poses a particular problem for terrestrial and shallow-water invertebrates when temperatures fall below freezing. Investigate the mechanisms used by arthropods to prevent freezing of tissues during periods of cold weather.

Horwarth, K. L., and J. G. Duman. 1982. Involvement of the circadian system in photoperiodic regulation of insect antifreeze proteins. *J. Exp. Zool.* 219:267.

van der Laak, S. 1982. Physiological adaptations to low temperature in freezing-tolerant *Phylodecta laticollis* beetles. *Comp. Biochem. Physiol.* 73A:613.

Lee, R. E., Jr. 1989. Insect cold-hardiness: To freeze or not to freeze? *BioScience* 39:308.

Lee, R. E., Jr., J. M. Strong-Gunderson, M. R. Lee, K. S. Grove, and T. J. Riga. 1991. Isolation of ice nucleating active bacteria from insects. *J. Exp. Zool.* 257:124.

Tursman, D., J. G. Duman, and C. A. Knight. 1994. Freeze tolerance adaptations in the centipede *Lithobius forficatus. J. Exp. Zool.* 268:347.

Williams, J. B., J. D. Shorthouse, and R. E. Lee, Jr. 2002. Extreme resistance to desiccation and microclimate-related differences in cold-hardiness of gall wasps (Hymenoptera: Cynipidae) overwintering on roses in southern Canada. *J. Exp. Biol.* 205:2115.

Yingst, D. R. 1978. The freezing resistance of the arctic subtidal isopod, *Mesidotea entomon. J. Comp. Physiol.* 125:165.

4. Symbiotic relationships (including parasitism) are commonly encountered among the Crustacea, Insecta, and Arachnida. For one of these groups, discuss the morphological, behavioral, and/or physiological adaptations for the symbiotic lifestyle.

A general treatment can be found in any parasitology text, such as Noble, E. R., and G. A. Noble. 1982. *Parasitology: The Biology of Animal Parasites,* 5th ed. Philadelphia: Lea & Febiger. Or Cheng, T. C. 1986. *General Parasitology,* 2d ed. New York: Academic Press.

Christensen, A. M., and J. J. McDermott. 1958. Life-history and biology of the oyster crab, *Pinnotheres ostreum* Say. *Biol. Bull.* 114:146.

Day, J. H. 1935. The life-history of *Sacculina. Q. J. Microsc. Sci.* 77:549.

DeVries, P. J. 1990. Enhancement of symbioses between butterfly caterpillars and ants by vibrational communication. *Science* 248:1104.

Gotto, R. V. 1979. The association of copepods with marine invertebrates. *Adv. Marine Biol.* 16:1.

Høeg, J. T. 1987. Male cypris metamorphosis and a new male larval form, the trichogon, in the parasitic barnacle *Sacculina carcini* (Crustacea: Cirripedia: Rhizocephala). *Phil. Trans. Royal Soc. London* (*B*) 317:47.

Letourneau, D. K. 1990. Code of ant-plant mutualism broken by parasite. *Science* 248:215.

Lewis, W. J., and J. H. Tomlinson. 1988. Host detection by chemically mediated associative learning in a parasitic wasp. *Nature* 331:257.

McClintock, J. B., and J. Janssen. 1990. Chemical defense in a pelagic antarctic amphipod via pteropod abduction: A novel symbiosis. *Nature* (*London*) 346:462.

Mitchell, R. 1968. Site selection by larval water mites parasitic on the damselfly *Cercion hieroglyphicum* Brauer. *Ecology* 49:40.

Moyse, J. 1983. *Isadascus bassindalei* gen. nov., sp. nov. (Ascothoracica: Crustacea) from northeast Atlantic with a note on the origin of barnacles. *J. Mar. Biol. Assoc. U.K.* 63:161.

Price, P. W. 1972. Parasitoids utilizing the same host: Adaptive nature of differences in size and form. *Ecology* 53:190.

Salt, G. 1968. The resistance of insect parasitoids to the defense reactions of their hosts. *Biol. Rev.* 43:200.

Takasu, K., and Y. Hirose. 1991. The parasitoid *Ooencyrtus nezarae* (Hymenoptera: Encyrtidae) prefers hosts parasitized by conspecifics over unparasitized hosts. *Oecologia* 87:319.

5. Investigate form and function in spider webs.

Craig, C. L. 1986. Orb-web visibility: The influence of insect flight behaviour and visual physiology on the evolution of web designs within the Araneoidea. *Anim. Behav.* 34:54.

Denny, M. 1976. The physical properties of spiders' silk and their role in the design of orb webs. *J. Exp. Biol.* 65:483.

Gosline, J. M., P. A. Guerette, C. S. Ortlepp, and K. N. Savage. 1999. The mechanical design of spider silks: From fibroin sequences to mechanical function. *J. Exp. Biol.* 202:3295.

Hyoung-Joon, J., and D. L. Kaplan. 2003. Mechanism of silk processing in insects and spiders. *Nature* 424:1057.

Köhler, T., and F. Vollrath. 1995. Thread biomechanics in the two orb-weaving spiders *Araneus diadematus* (Araneae, Araneidae) and *Uloborus walckenaerius* (Araneae, Uloboridae). *J. Exp. Zool.* 271:1.

Krink, K., and F. Vollrath. 1998. Emergent properties in the behaviour of a virtual spider robot. *Proc. Royal Soc. London* (*B*) 265:2051.

Nentwig, W. 1983. Why do only certain insects escape from a spider's web? *Oecologia* (*Berlin*) 53:412.

Opell, B. D. 1990. Material investment and prey capture potential of reduced spider webs. *Behav. Ecol. Sociobiol.* 26:375.

Opell, B. D. 1996. Functional similarities of spider webs with diverse architectures. *Amer. Nat.* 148:630.

Palmer, J. M., F. A. Coyle, and F. W. Harrison. 1982. Structure and cytochemistry of the silk glands of the mygalomorph spider *Antrodiaetus unicolor* (Araneae, Antrodiaetidae). *J. Morphol.* 174:269.

Popock, R. I. 1895. Some suggestions on the origin and evolution of web spinning in spiders. *Nature* (*London*) 51:417.

Rypstra, A. L. 1982. Building a better insect trap: An experimental investigation of prey capture in a variety of spider webs. *Oecologia* (*Berlin*) 52:31.

Seibt, U., and W. Wickler. 1990. The protective function of the compact silk nest of social *Stegodyphus* spiders (Araneae, Eresidae). *Oecologia* (*Berlin*) 82:317.

Vollrath, F., and D. P. Knight. 2001. Liquid crystalline spinning of spider silk. *Nature* 410: 541.

6. How do insects fly?

Alexander, R. M. 1995. Springs for wings. *Science* 268:50.

Betts, C. R. 1986. Functioning of the wings and axillary sclerites of Heteroptera during flight. *J. Zool., London (B)* 1:283.

Boettiger, E. G., and E. Furshpan. 1952. The mechanics of flight movements in Diptera. *Biol. Bull.* 102:200.

Chan, W. P., F. Prete, and M. H. Dickinson. 1998. Visual input to the efferent control system of a fly's "gyroscope". *Science* 280:289.

Dickinson, M. H., F.-O. Lehmann, and S. P. Sane. 1999. Wing rotation and the aerodynamic basis of insect flight. *Science* 284: 1954.

Froy, O., A. L. Gotter, A. L. Casselman, and S. M. Reppert. 2003. Illuminating the circadian clock in monarch butterfly migration. *Science* 300: 1303–5.

Hedenström, A. 1997. Aerodynamics and insect flight. *TREE* 12:174.

Hengstenberg, R. 1998. Controlling the fly's gyroscopes. *Nature* 392:757.

Lehmann, F.-O., and S. Pick. 2007. The aerodynamic benefit of wing–wing interaction depends in flapping insect wings. *J. Exp. Biol.* 210: 1362–77.

Marden, J. H. 1987. Maximum lift production during take-off in flying animals. *J. Exp. Biol.* 130: 235–58.

Mouritsen, H., and B. J. Frost. 2002. Virtual migration in tethered flying monarch butterflies reveals their orientation mechanisms. *Proc. Natl. Acad. Sci.* 99: 10162.

Srinivasan, M. V., S. Zhang, and J. S. Chahl. 2001. Landing strategies in honeybees, and possible applications to autonomous airborne vehicles. *Biol. Bull.* 200: 216.

Usherwood, J. R., and C. P. Ellington. 2002. The aerodynamics of revolving wings. I. Model hawkmoth wings. *J. Exp. Biol.* 205: 1547.

Weis-Fogh, T. 1973. Quick estimates of flight fitness in hovering animals, including novel mechanisms for lift production. *J. Exp. Biol.* 59: 169–230.

Weis-Fogh, T. 1975. Unusual mechanisms for the generation of lift in flying animals. *Sci. Amer.* 233:80.

7. Hermit crabs are an active group of marine decapod crustaceans whose abdominal exoskeleton is too thin to provide protection from predators. Instead, the crabs protect their soft parts by taking up residence inside empty gastropod shells. How do hermit crabs choose their shells, and what selective pressures probably have molded these choices?

Abrams, P. A. 1987. An analysis of competitive interactions between three hermit crab species. *Oecologia (Berlin)* 72:233.

Elwood, R. W., K. E. Wood, M. B. Gallagher, and J. T. A. Dick. 1998. Probing motivational state during agonistic encounters in animals. *Nature* 393:66.

Fotheringham, N. 1976. Population consequences of shell utilization by hermit crabs. *Ecology* 57:570.

Gherardi, F. 1996. Non-conventional hermit crabs: Pros and cons of a sessile, tube-dwelling life in *Discorsopagurus schmitti* (Stevens). *J. Exp. Marine Biol. Ecol.* 202:119.

Gherardi, F., and J. Atema. 2005. Effects of chemical context on shell investigation behavior in hermit crabs. *J. Exp. Marine Biol. Ecol.* 320: 1–7.

Hazlett, B. A. 1981. The behavioral ecology of hermit crabs. *Ann. Rev. Ecol. Syst.* 12:1.

McClintock, T. S. 1985. Effects of shell condition and size upon the shell choice behavior of a hermit crab. *J. Exp. Marine Biol. Ecol.* 88:271.

Mercando, N. A., and C. F. Lytle. 1980. Specificity in the association between *Hydractinia echinata* and sympatric species of hermit crabs. *Biol. Bull.* 159:337.

Pechenik, J. A., J. Hsieh, S. Owara, S. Untersee, D. Marshall, and W. Li. 2001. Factors selecting for avoidance of drilled shells by the hermit crab *Pagurus longicarpus*. *J. Exp. Marine Biol. Ecol.* 262: 75.

Vance, R. 1972. Competition and mechanism of coexistence in three sympatric species of intertidal hermit crabs. *Ecology* 53:1062.

Wilber, T. P., Jr. 1990. Influence of size, species, and damage on shell selection by the hermit crab *Pagurus longicarpus*. *Marine Biol.* 104:31.

8. Through the process of natural selection, many arthropods have developed remarkable abilities to disguise themselves, either by blending in with their surroundings or by imitating other species. Investigate the adaptive value of mimicry and camouflage among the Arthropoda.

Blest, A. D. 1963. Longevity, palatability, and natural selection in five species of New World Saturuiid moth. *Nature (London)* 197:1183.

Brakefield, P. M., J. Gates, D. Keys, F. Kesbeke, P. J. Wijngaarden, A. Monteiro, V. French, and S. B. Carroll. 1996. Development, plasticity and evolution of butterfly eyespot patterns. *Nature* 384:236.

Eisner, T., K. Hicks, M. Eisner, and D. S. Robson. 1978. "Wolf-in-sheep's-clothing" strategy of a predaceous insect larva. *Science* 199:790.

Finday, R., M. R. Young, and J. A. Finday. 1983. Orientation behaviour in the Grayling butterfly: Thermoregulation or crypsis? *Ecol. Entomol.* 8:145.

Greene, E. 1989. A diet-induced developmental polymorphism in a caterpillar. *Science* 243:643.

Greene, E., L. J. Orsak, and D. W. Whitman. 1987. A tephritid fly mimics the territorial displays of its jumping spider predators. *Science* 236:310.

Kettlewell, H. B. D. 1956. Further selection experiments on industrial melanism in the Lepidoptera. *Heredity* 10:287.

Körner, H. K. 1982. Countershading by physiological colour change in the fish louse *Anilocra physodes* L. (Crustacea: Isopoda). *Oecologia (Berlin)* 55:248.

Mather, M. H., and B. D. Roitberg. 1987. A sheep in wolf's clothing: Tephritid flies mimic spider predators. *Science* 236:308.

Platt, A., R. Coppinger, and L. Brower. 1971. Demonstration of the selective advantage of mimetic *Limentis* butterflies presented to caged avian predators. *Evolution* 25:692.

9. Much communication among arthropods is accomplished by means of chemicals. Investigate the mechanism and/or the adaptive significance of chemical communication among the Arthropoda.

Bagøien, E., and T. Kiørboe. 2005. Blind dating—mate finding in planktonic copepods. I. Tracking the pheromone trail of *Centropages typicus*. *Marine Ecol. Progr. Ser.* 300: 105–15.

Boeckh, J., H. Sass, and D. R. A. Wharton. 1970. Antennal receptors: Reactions to female sex attractant in *Periplaneta americana*. *Science* 168:589.

Dahl, E., H. Emanuelsson, and C. von Mecklenburg. 1970. Pheromone transport and reception in an amphipod. *Science* 170:739.

Fitzgerald. T. D. 1976. Trail marking by larvae of the eastern tent caterpillar. *Science* 194:961.

Lewis, W. J., and J. H. Tumlinson. 1988. Host detection by chemically mediated associative learning in a parasitic wasp. *Nature* 331:257.

Linn, C. E., Jr., M. G. Campbell, and W. L. Roelofs. 1987. Pheromone components and active spaces: What do moths smell and where do they smell it? *Science* 237:650.

Mafra-Neto, A., and R. T. Cardé. 1994. Fine-scale structure of pheromone plumes modulates upwind orientation of flying moths. *Nature* 369:142.

McAllister, M. K., and B. D. Roitberg. 1987. Adaptive suicidal behaviour in pea aphids. *Nature* (*London*) 328:797.

Myers, J., and L. P. Brower. 1969. A behavioural analysis of the courtship pheromone receptors of the queen butterfly, *Danaus gilippus berenice*. *J. Insect Physiol.* 15:2117.

Nault, L. R., M. E. Montgomery, and W. S. Bowers. 1976. Ant-aphid association: Role of aphid alarm pheromone. *Science* 192:1349.

Pierce, N. E., and S. Eastseal. 1986. The selective advantage of attendant ants for the larvae of a Lycaenid butterfly, *Glaucopsyche lygdamus*. *J. Anim. Ecol.* 55:451.

Price, P. W. 1970. Trail odors: Recognition by insects parasitic on cocoons. *Science* 170:546.

Rust, M. K., T. Burk, and W. J. Bell. 1976. Pheromone-stimulated locomotory and orientation responses in the American cockroach *Periplaneta americana*. *Anim. Behav.* 24:52.

Ruther, J., A. Reinecke, T. Tolasch, and M. Hilker. 2001. Make love not war: A common arthropod defence compound as sex pheromone in the forest cockchafer *Melolontha hippocastani*. *Oecologia* 128: 44.

Schneider, D. 1969. Insect olfaction: Deciphering system for chemical messages. *Science* 163:1031.

Tsuda, A., and C. B. Miller. 1998. Mate-finding behaviour in *Calanus marshallae* Frost. *Phil. Trans. Royal Soc. London B* 353: 713–20.

10. It is becoming clear that all invertebrates possess some form of immune response; that is, they have the ability to distinguish self from nonself at the cellular level. Investigate the ability of arthropods to make this distinction.

Berg, R., I. Schuchmann-Feddersen, and O. Schmidt. 1988. Bacterial infection induces a moth (*Ephestia kuhniella*) protein which has antigenic similarity to virus-like particle proteins of a parasitoid wasp (*Venturia canescens*). *J. Insect Physiol.* 34:473.

Brehélin, M., and J. A. Hoffmann. 1980. Phagocytosis of inert particles in *Locusta migratoria* and *Galleria mellonella:* A study of ultrastructure and clearance. *J. Insect Physiol.* 26:103.

Briggs, J. D. 1958. Humoral immunity in lepidopterous larvae. *J. Exp. Zool.* 138:155.

Chisholm, J. R. S., and V. J. Smith. 1992. Antibacterial activity in the haemocytes of the shore crab, *Carcinus maenas*. *J. Marine Biol. Assoc. U.K.* 72:529.

Dunn, P. E. 1990. Humoral immunity in insects. *BioScience* 40:738.

Edson, K. M., S. B. Vinson, D. B. Stoltz, and M. D. Summers. 1981. Virus in a parasitoid wasp: Suppression of the cellular immune response in the parasitoid's host. *Science* 211:582.

Ennesser, C. A., and A. J. Nappi. 1984. Ultrastructural study of the encapsulation response of the American cockroach, *Periplaneta americana*. *J. Ultrastr. Res.* 87:31.

Kurtz, J., and K. Franz. 2003. Evidence for memory in invertebrate immunity. *Nature* 425: 37–38.

Ratcliffe, N. A., C. Leonard, and A. F. Rowley. 1984. Prophenoloxidase activation: Nonself recognition and cell cooperation in insect immunity. *Science* 226:557.

Salt, G. 1968. The resistance of insect parasitoids to the defense reactions of their hosts. *Biol. Rev.* 43:200.

Sloan, B., C. Yocum, and L. W. Clem. 1975. Recognition of self from non-self in crustaceans. *Nature* (*London*) 258:521.

Watson, F. L., et al. 2005. Extensive diversity of Ig-superfamily proteins in the immune system of insects. *Science* 309: 1874–78.

White, K. N., and N. A. Ratcliffe. 1982. The segregation and elimination of radio- and fluorescent-labelled marine bacteria from the haemolymph of the shore crab, *Carcinas maenas*. *J. Mar. Biol. Assoc. U.K.* 62:819.

11. Small planktonic crustaceans, such as copepods and cladocerans, are near the base of the food chain in aquatic ecosystems; that is, the growth rate and reproductive potential of many fish depend largely on the size of the herbivorous zooplankton population available as food. In turn, the size of the herbivorous zooplankton population depends on the amount of phytoplankton available, and the rate at which the phytoplankton can be captured, ingested, and converted into new zooplankton biomass and offspring. Consequently, considerable effort has gone into studying zooplankton feeding biology. What factors are involved in the collection of phytoplankton by crustacean zooplankton?

Anraku, M., and M. Omori. 1963. Preliminary survey of the relationship between the feeding habit and structure of the mouth parts of marine copepods. *Limnol. Oceanogr.* 8:116.

Butler, N. M., C. A. Suttle, and W. E. Neill. 1989. Discrimination by freshwater zooplankton between single algal cells differing in nutritional status. *Oecologia* (*Berlin*) 78:368.

Costello, J. H., J. R. Strickler, C. Marvasé, G. Trager, R. Zeller, and A. J. Freise. 1990. Grazing in a turbulent environment: Behavioral response of a calanoid copepod, *Centropages hamatus*. *Proc. Nat. Acad. Sci. USA* 87:1648.

DeMott, W. R. 1989. Optimal foraging theory as a predictor of chemically mediated food selection by suspension feeding copepods. *Limnol. Oceanogr.* 34:140.

Gerritsen, J., and K. G. Porter. 1982. The role of surface chemistry in filter feeding by zooplankton. *Science* 216:1225.

Gophen, M., and W. Geller. 1984. Filter mesh size and food particle uptake by *Daphnia*. *Oecologia* (*Berlin*) 64:408.

Hamner, W. M., P. P. Hamner, S. W. Strand, and R. W. Gilmer. 1983. Behavior of antarctic krill, *Euphausia superba:* Chemoreception, feeding, schooling, and molting. *Science* 220:433.

Huntley, M. E., K.-G. Barthel, and J. L. Star. 1983. Particle rejection by *Calanus pacificus:* Discrimination between similarly sized particles. *Marine Biol.* 74:151.

Poulet, S. A., and P. Marsot. 1978. Chemosensory grazing by marine calanoid copepods (Arthropoda: Crustacea). *Science* 200:1403.

Price, H. J., G.-A. Paffenhöfer, and J. R. Strickler. 1983. Modes of cell capture in calanoid copepods. *Limnol. Oceanogr.* 28:116.

Richman, S., and J. N. Rogers. 1969. The feeding of *Calanus helgolandicus* on synchronously growing populations of the marine diatom *Ditylum brightwelli*. *Limnol. Oceanogr.* 14:701.

Turner, J. T., P. A. Tester, and J. R. Strickler. 1993. Zooplankton feeding ecology: A cinematographic study of animal-to-animal variability in the feeding behavior of *Calanus finmarchicus. Limnol. Oceanogr.* 38:255.

12. What is the evidence indicating that color changes in arthropods are regulated by hormones?

Brown, F. A., Jr. 1935. Control of pigment migration within the chromatophores of *Palaemonetes vulgaris. J. Exp. Zool.* 71:1.

Brown, F. A., Jr. 1940. The crustacean sinus gland and chromatophore activation. *Physiol. Zool.* 13:343.

Brown, F. A., Jr., and H. E. Ederstrom. 1940. Dual control of certain black chromatophores of *Crago. J. Exp. Zool.* 85:53.

Fingerman, M. 1969. Cellular aspects of the control of physiological color changes in crustaceans. *Amer. Zool.* 9:443.

McNamara, J.C., and M. R. Ribeiro. 2000. The calcium dependence of pigment translocation in freshwater shrimp red ovarian chromatophores. *Biol. Bull.* 198: 357.

McWhinnie, M. A., and H. M. Sweeney. 1955. The demonstration of two chromatophorotropically active substances in the land isopod, *Trachelipus rathkei. Biol. Bull.* 108:160.

Pérez-González, M. D. 1957. Evidence for hormone-containing granules in sinus glands of the fiddler crab *Uca pugilator. Biol. Bull.* 113:426.

13. What are the roles of predation and competition in regulating the sizes of arthropod populations?

Bertness, M. D. 1989. Intraspecific competition and facilitation in a northern acorn barnacle population. *Ecology* 70:257.

Chew, F. S. 1981. Coexistence and local extinction in two pierid butterflies. *Amer. Nat.* 118:655.

El-Dessouki, S. A. 1970. Intraspecific competition between larvae of *Sitona* sp. (Coleoptera, Curculionidae). *Oecologia (Berlin)* 6:106.

Frank, J. H. 1967. The insect predators of the pupal stage of the winter moth, *Operophtera brumata* (L.) (Lepidoptera: Hydiomenidae). *J. Anim. Ecol.* 36:375.

Gotelli, N. J. 1996. Ant community structure: Effects of predatory ant lions. *Ecology* 77:630.

MacKay, W. P. 1982. The effect of predation of western widow spiders (Araneae: Theridiidae) on harvester ants (Hymenoptera: Formicidae). *Oecologia (Berlin)* 53:406.

Moore, N. W. 1964. Intra- and interspecific competition among dragonflies (Odonata). *J. Anim. Ecol.* 33:49.

Wise, D. H. 1983. Competitive mechanisms in a food-limited species: Relative importance of interference and exploitative interactions among labyrinth spiders (Araneae: Araneidae). *Oecologia (Berlin)* 58:1.

14. Based upon lectures and your reading in this text, what factors contribute to making arthropods the most abundant animals (with the possible exception of the nematodes) on the earth?

15. Based upon lectures and your reading in this text, what functions does the blood of insects serve, now that gas exchange is achieved by means of tracheae rather than by a circulatory system?

16. Growing evidence indicates that the morphology of some aquatic arthropods is influenced by predation. Discuss the evidence supporting this statement, and the adaptive value and costs of the response.

Barry, M. J. 1994. The costs of crest induction for *Daphnia carinata. Oecologia* 97:278.

Barry, M. J. 2000. Inducible defences in *Daphnia:* Responses to two closely related predatory species. *Oecologia* 124: 396.

Black, A. R. 1993. Predator-induced phenotypic plasticity in *Daphnia pulex:* Life history and morphological responses to *Notonecta* and *Chaoborus. Limnol. Oceanogr.* 38:986.

LaForsch, C., and R. Tollrian. 2004. Inducible defenses in multipredator environments: Cyclomorphosis in *Daphnia cucullata. Ecology* 85: 2302–11.

Lively, C. M. 1986. Predator-induced shell dimorphism in the acorn barnacle *Chthamalus anisopoma. Evolution* 40:232.

Morgan, S. G. 1989. Adaptive significance of spination in estuarine crab zoeae. *Ecology* 70:464.

O'Brien, W. J., J. D. Kettle, and H. P. Riessen. 1979. Helmets and invisible armor: Structures reducing predation from tactile and visual planktivores. *Ecology* 60:287.

Stenson, J. A. E. 1987. Variation in capsule size of *Holopedium gibberum* (Zaddach): A response to invertebrate predation. *Ecology* 68:928.

Walls, M., and M. Ketola. 1989. Effects of predator-induced spines on individual fitness in *Daphnia pulex. Limnol. Oceanogr.* 34:390.

Zaret, T. M. 1969. Predation-balanced polymorphism of *Ceriodaphnia cornuta* Sars. *Limnol. Oceanogr.* 14:301.

17. Many plants have evolved chemical defenses against herbivore feeding. Discuss the manner in which at least some insect species circumvent these defenses or even appropriate the plant chemicals for their own uses.

Conner, W. E., R. Boada, F. C. Schroeder, A. González, J. Meinwald, and T. Eisner. 2000. Chemical defense: Bestowal of a nuptial alkaloidal garment by a male moth on its mate. *Proc. Nat. Acad. Sci.* 97:14406.

Dussourd, D. E., and T. Eisner. 1987. Vein-cutting behavior: Insect counterploy to the latex defense of plants. *Science* 237:898.

Eisner, T., J. S. Johanessee, J. Carrel, L. B. Hendry, and J. Meinwald. 1974. Defensive use by an insect of a plant resin. *Science* 184:996.

Kearsley, M. J. C., and T. G. Whitham. 1992. Guns and butter: A no cost defense against predation for *Chrysomela confluens. Oecologia* 92:556.

Peterson, S. C., N. D. Johnson, and J. L. LeGuyader. 1987. Defensive regurgitation of allelochemicals derived from host cyanogenesis by eastern tent caterpillars. *Ecology* 68:1268.

18. Discuss the evidence that bees and ants use memory in navigating to and from their hives and nests.

Gould, J. L. 1986. The locale map of honeybees: Do insects have cognitive maps? *Science* 232:861.

Judd, S. P. D., and T. S. Collett. 1998. Multiple stored views and landmark guidance in ants. *Nature* 392: 710.

Menzel, R., R. Brandt, A. Gumbert, B. Komischke, and J. Kunze. 2000. Two spatial memories for honeybee navigation. *Proc. R. Soc. London* B 267: 961.

Zhang, S., S. Schwarz, H. Zhu, and J. Tautz. 2006. Honeybee memory: A honeybee knows what to do and when. *J. Exp. Biol.* 209: 4420–28.

19. Most biologists agree that insect wings originally must have evolved in response to pressures not associated with flight. Only after the basic wing structure evolved could the further evolution of wings as organs of flight become possible. Compare and contrast the different hypotheses accounting for the origin of insect wings. Is the evidence supporting any one hypothesis more compelling than that supporting the other hypotheses?

Averof, M., and S. M. Cohen. 1997. Evolutionary origin of insect wings from ancestral gills. *Nature* 385:627.

Douglas, M. W. 1981. Thermoregulatory significance of thoracic lobes in the evolution of insect wings. *Science* 211:84.

Heinrich, B. 1993. *The Hot-blooded Insects: Strategies and Mechanisms of Thermoregulation.* Cambridge, Mass.: Harvard Univ. Press, 104–13.

Kingsolver, J. G., and M. A. R. Koehl. 1985. Aerodynamics, thermoregulation, and the evolution of insect wings: Differential scaling and evolutionary change. *Evolution* 39:488.

Kukalova-Peck, J. 1978. Origin and evolution of insect wings and their relation to metamorphosis, as documented by the fossil record. *J. Morphol.* 156:53.

Marden, J. H., and M. G. Kramer. 1994. Surface-skimming stoneflies: A possible intermediate stage in insect flight evolution. *Science* 266:427.

Thomas, A. L. R., and R. Å. Norberg. 1996. Skimming the surface—The origin of flight in insects? *TREE* 11:187 (see also *TREE* 11:471 for rebuttals).

Will, K. W. 1995. Plecopteran surface-skimming and insect flight evolution. *Science* 270:1684.

Yanoviak, S. P., R. Dudley, and M. Kaspari. 2005. Directed aerial descent in canopy ants. *Nature* 433: 624–26.

20. Many small aquatic crustaceans make extensive daily vertical migrations, often ascending dozens of meters (or more), only to return to greater depths a few hours later. Discuss the adaptive value of this behavior. What are the advantages and limitations of each hypothesis you discuss?

Dini, M. L., and S. R. Carpenter. 1991. The effect of whole-lake fish community manipulations on *Daphnia* migratory behavior. *Limnol. Oceanogr.* 36:370.

Enright, J. T. 1977. Diurnal vertical migration: Adaptive significance and timing. Part I. Selective advantage: A metabolic model. *Limnol. Oceanogr.* 22:856.

Gliwicz, M. Z. 1986. Predation and the evolution of vertical migration in zooplankton. *Nature (London)* 320:746.

McLaren, I. A. 1974. Demographic strategy of vertical migration by a marine copepod. *Amer. Nat.* 108:91.

Neill, W. E. 1990. Induced vertical migration in copepods as a defense against invertebrate predation. *Nature* 345:524.

Ohman, M. D., B. W. Frost, and E. B. Cohen. 1983. Reverse vertical migration: An escape from invertebrate predators. *Science* 220:1404.

Orcutt, J. D., Jr., and K. G. Porter. 1983. Diel vertical migration by zooplankton: Constant and fluctuating temperature effects on life history parameters of *Daphnia. Limnol. Oceanogr.* 28:720.

21. Chromatophores are widespread among both crustaceans and cephalopods. Based upon your readings in this book, argue either that this probably reflects evolution from a common ancestor, or that it reflects the independent evolution of color change systems in the 2 groups.

22. The past few years have seen the publication of increasing numbers of studies exploring the genetic and molecular bases of morphological shifts in developmental patterns among insects and other arthropods. How have these so far contributed to or altered our understanding of insect biology?

Averoff, M. 1998. Origin of the spider's head. *Nature* 395:436.

Brakefield, P. M., J. Gates, D. Keys, F. Kesbeke, P. J. Wijngaarden, A. Monteiro, V. French, and S. B. Carroll. 1996. Development, plasticity and evolution of butterfly eyespot patterns. *Nature* 384:236.

Goodisman, M. A. D., J. Isoe, D. E. Wheeler, and M. A. Wells. 2005. Evolution of insect metamorphosis: A microarray-based study of larval and adult gene expression in the ant *Camponotus festinatus. Evolution* 59: 858–70.

Konopova, B., and M. Jindra. 2007. Juvenile hormone resistance gene *Methoprene-tolerant* controls entry into metamorphosis in the beetle *Tribolium castaneum. Proc. Natl. Acad. Sci. USA* 104: 10488–93.

Mahfooz, N. S., H. Li, and A. Popadić. 2004. Differential expression patterns of the *hox* gene are associated with differential growth of insect hind legs. *Proc. Natl. Acad. Sci. USA* 101: 4877–82.

Nijhout, H. F. 1994. Developmental perspectives on evolution of butterfly mimicry. *BioScience* 44:148.

Shubin, N., C. Tabin, and S. Carroll. 1997. Fossils, genes and the evolution of animal limbs. *Science* 388:639.

Zelhof, A. C., R. W. Hardy, A. Becker, and C. S. Zuker. 2006. Transforming the architecture of compound eyes. *Nature* 443: 696–99.

Taxonomic Detail

Phylum Arthropoda

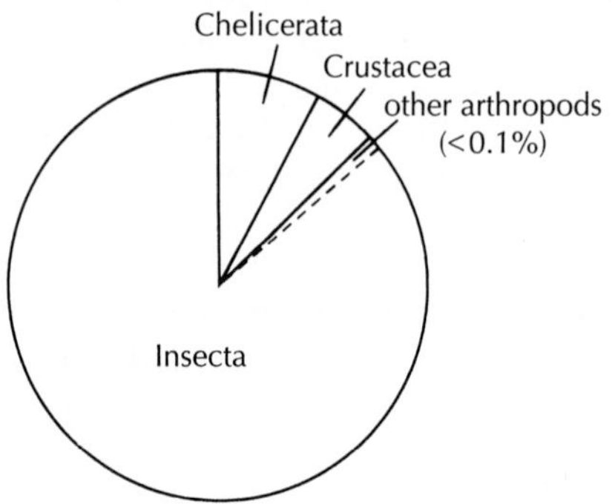

"Arthropods rule the world, at least among multicellular animals, and we'd like to think that we understand the basic outlines of their evolutionary relationships. Unfortunately, we don't." Richard H. Thomas. 2003. *Science* 299: 1854.

This phylum probably contains up to 5 million species, of which about 1 million have been described so far.

Subphylum Chelicerata

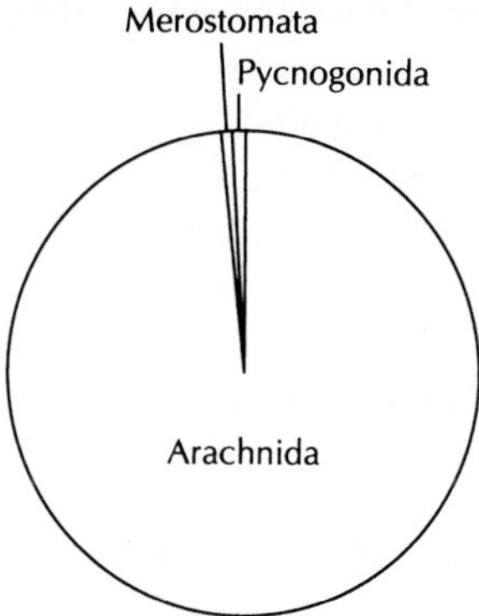

The approximately 75,000 species in this subphylum are divided among 3 classes.

Class Merostomata

Only 4 living species.

Order Xiphosura

Limulus—the horseshoe crabs. All species are marine bottom-dwellers, growing up to 60 cm long, with a life span of up to 19 years.

Order Eurypterida

This group (the "water scorpions") went extinct some 230 million years ago, and was maximally abundant about 350 to 450 million years ago. Originating in the sea, where members probably both swam and walked along the bottom, eurypterids later invaded estuarine and freshwater environments, and some even may have been amphibious, spending part of their time on land. Most preserved individuals are less than 20 cm long, but members of some species attained lengths exceeding 2 m, making them the largest arthropods that ever lived. Many biologists believe that arachnids evolved from eurypterid ancestors.

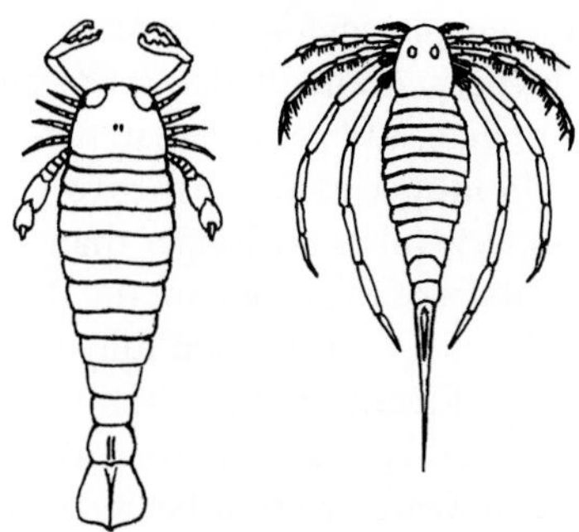

From E. N. K. Clarkson, *Invertebrate Paleontology & Evolution,* 2d ed. Copyright © 1986. Chapman & Hall, Div. International Thomson Publishing Services Ltd., U. K. With kind permission from Kluwer Academic Publishers.

Class Arachnida

The approximately 74,000 species are divided among 11 orders and about 550 families.

Order Scorpiones

Diplocentrus, Centruroides—the scorpions. This group contains the largest of all arachnids, with individuals of some species reaching lengths of 18 cm. All species are terrestrial and carnivorous. There are 1,500 species divided among 9 families.

Order Uropygi

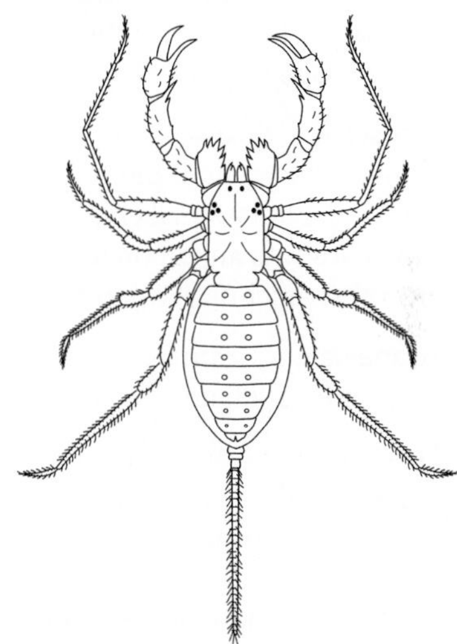

Mastigoproctu—the whip scorpions. Eighty-five species, one family.

Order Amblypygi

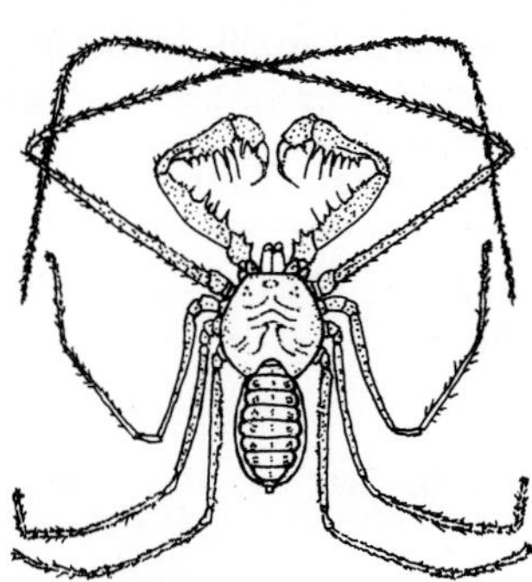

The tail-less whip scorpions.

Order Araneae

The spiders. This group contains about half of all arachnid species. All are predators, primarily on insects. About 36,000 described species (with many more species yet to be described) in 3 suborders (1 minor, omitted here); about 90 families.

Suborder Orthognatha

Bothriocyrtum, Actinopus—trapdoor spiders; *Aphonopelma, Acanthoscurria*—tarantulas. Over 1200 species divided among 11 families. Mostly tropical.

Suborder Labidognatha

Seventy-four families, about 35,000 species.

Family Symphytognathidae. This family contains the smallest of all known spider species (about 0.3 mm in body length). The group occurs only in New Zealand, Australia, and the neotropics.

Family Argyronetidae. *Argyronecta*—water spiders. All live submerged in freshwater, coming to the surface periodically to capture air on fine abdominal hairs. They then submerge and brush the air into a dome-shaped underwater web for storage.

Family Lycosidae. *Lycosa*—wolf spiders. Most species hunt on the ground, and only a few species make webs. Males follow a silken thread to locate the female, which leads to an elaborate courtship ritual as in many other arachnids. Up to 3000 species.

Family Araneidae. *Araneus, Argiope*—orb weavers. Between 3000 and 4000 species.

Family Theridiidae. Comb-footed spiders. About 2000 species. Females in the genus *Latrodectus* (the black widow spider) produce a venom that is highly toxic to humans (usually not deadly) and other vertebrates. Females have bodies as long as 1.5 cm and total lengths (including legs) of up to 5 cm.

Family Salticidae. Jumping spiders. This is one of the largest of all spider families, containing nearly 5000 species. Jumping spiders have the most highly developed of all arachnid eyes and use them to pounce on prey from a considerable distance. Salticids use silk primarily for nest building rather than for prey capture.

Family Thomisidae. Crab spiders. About 1500 species.

Order Ricinulei

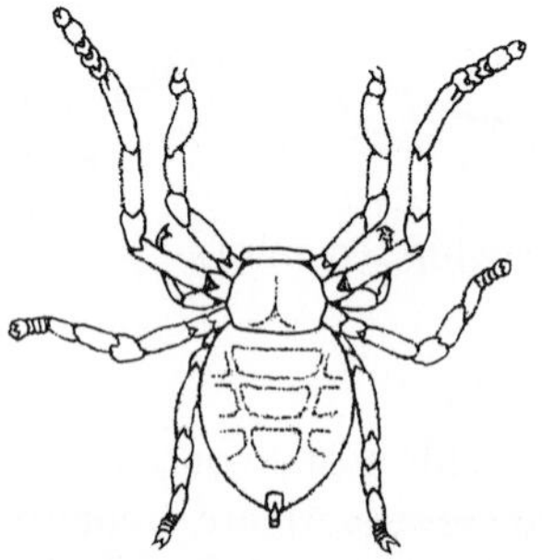

The 33 species in this order form a relatively inconspicuous group of tropical and subtropical eyeless arachnids. One family.

Order Pseudoscorpiones

Chelifer. The 2000 species in this group resemble the true scorpions, but they lack the long tail and stinger, and most are only 1–7 mm long. Most species are terrestrial, but some live intertidally, among seaweed stranded at high tide. All feed on the body juices of other invertebrates. Many hitch rides on insects, a traveling behavior termed "phoresy." Most species manufacture silken nests, spun from the tips of the chelicerae, and many indulge in elaborate courtship rituals. Twenty-two families.

Order Solifugae (= Solpugida)

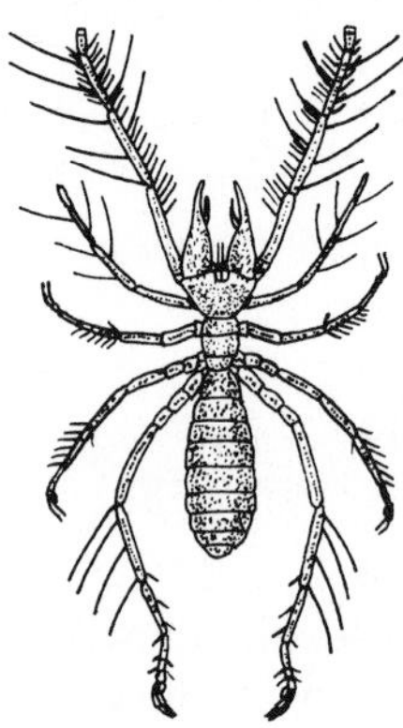

The solifugids (or solpugids)—wind scorpions (because they run "like the wind"). Individuals can be up to 7 cm long. Unlike other arachnids, these bear huge chelate chelicerae that point forward. The more than 900 species are found mostly in tropical and subtropical areas other than Australia and New Zealand. All are fast-moving predators. Members of many species eat termites. Twelve families.

Order Opiliones

Leiobunum—the harvestmen, "daddy longlegs." Between 4500 and 5000 known species. Individuals range from less than a millimeter to nearly 22 mm in body length. Legs often very long and thin. These arachnids prey on other arthropods and on gastropods, or scavenge on fruit, vegetables, and decomposing animals. Twenty-eight families.

Order Acari

Oppia, Dermatophagoides—the mites; *Dermacentor, Ixodes*—the ticks. About 30,000 species have been described to date, and some researchers estimate that another million species await discovery and description. Few mites are longer than 1 mm. The largest members of the order are the ticks, which can reach lengths of 5–6 mm; when fully fed, they can reach lengths of nearly 30 mm. Many species are major agricultural pests, and others transmit debilitating or deadly viral or bacterial diseases to humans and livestock. Lyme disease, a bacterial infection transmitted by the deer tick, now infects nearly 13,000 people in the United States each year, and is similarly widespread in Europe and Asia. Other mite and tick species help to decompose and recycle organic material. Nearly four hundred families.

Class Pycnogonida (= Pantopoda)

Nymphon—the sea spiders. All 1000 or so species are marine, living from shallow water to depths of 6800 m. Most are external parasites or sucking predators. If the nauplius-like larva leaves the egg before development is completed, it grows as a parasite of cnidarians. Eight to nine families.

Subphylum Mandibulata[23]

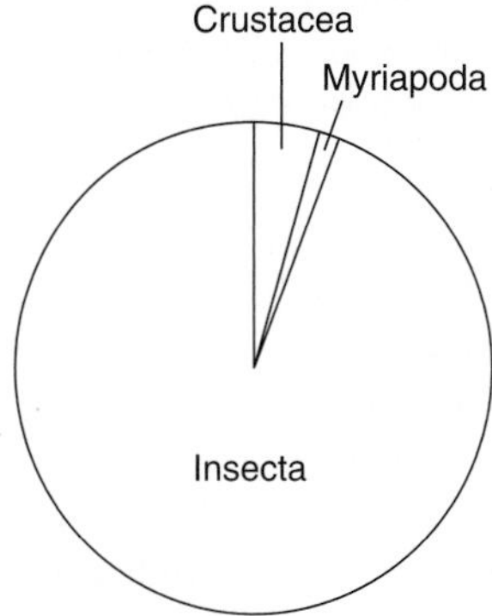

Class Myriapoda

Over 13,000 species have been described, distributed among 4 orders.

Order Chilopoda

The centipedes. All approximately 3000 species are predators: The first pair of trunk appendages are modified into a pair of poison fangs, which are used to paralyze prey before consumption. Most species are tropical, although many are temperate. *Scutigera,* the common household centipede, is unusual among centipedes in possessing large, multifaceted (compound) eyes, apparently independently evolved from those of insects. The smallest centipedes are about 4 mm long, while members of the largest species, *Scolopendra gigantea,* reach lengths of nearly 30 cm. A small number of species (e.g., *Hydroschendyla submarina*) are marine. Twenty families.

Order Diplopoda

The millipedes. Nearly 10,000 species have been described to date. Each millipede segment typically bears a pair of poison glands that secrete a volatile and irritating, or even deadly, liquid; among a variety of other organic constituents, the secretions usually contain hydrogen cyanide. At least one species can spray organisms a meter away. Most species are scavengers, relatively few are carnivores, and a few others are vegetarians. Approximately 120 families.

Order Symphyla

Scutigerella, Symphella. These small (1–8 mm long) terrestrial arthropods live in damp habitats, such as beneath rotting wood, or in moist soil or leaf litter. Individuals usually have 10 to 12 pairs of legs. Most of the 160 species are vegetarians, and at least one species is an agricultural pest. Two families.

Order Pauropoda

These are extremely small (usually less than 1.5 mm long) terrestrial arthropods that live in leaf litter and soil in forests. Each individual has 9 pairs of legs. Most species are fast-moving and feed on fungi. Fewer than 500 species have been described. Five families.

Superclass Hexapoda

Class Entognatha

Primitively wingless hexapods whose mouthpart appendages are recessed within a special pouch on the head.

Order Collembola

Springtails. A group of small hexapods, no more than several millimeters long, possessing a characteristic abdominal jumping organ. Unlike any other hexapods, collembolids possess only 6 abdominal segments. Springtails are common in various freshwater, coastal, marine, and terrestrial habitats. Their fossil record extends back approximately 400 million years. Five families, containing nearly 4000 described species.

Order Protura

A group of small (less than 2 mm), wingless, eyeless creatures living in leaf litter and decomposing vegetation. 250 species.

Order Diplura

Small (less than 4 mm), white, wingless and eyeless herbivores and predators lacking Malpighian tubules. 650 species.

Class Insecta (= Ectognatha)

At least 1 million species have been described to date, and probably several times that number await discovery. Insects are found in every known terrestrial and freshwater habitat; they even occur in the Antarctic, living as ectoparasites on seals and seabirds.

Order Archaeognatha

Bristletails. These are the most primitive living insects; a recently discovered fossil is nearly 400 million years old. The approximately 400, mostly nocturnal species in this group were formerly

23. See Figure 14.42 for an alternative view, in which this subphylum is abandoned.

united with silverfish in the now-abandoned order Thysanura. Abdominal flexing periodically propels the animals into the air. Four families.

Order Zygentoma

Silverfish. The approximately 300 species in this group were formerly united with bristletails in the now-abandoned order Thysanura. Silverfish are very fast runners. Some live exclusively in nests of ants and termites. Five families.

Subclass Pterygota

These are the winged (*pteron* = G: wing) insects. Although in many species the wings have been secondarily lost, all members of this group are descended from winged ancestors. Twenty-seven orders.

Order Mantophasmatodea

This order was first erected in 2001—the first new insect order in nearly 100 years—to accommodate certain fossilized insects, but living representatives have now been found in South Africa. These predators (common name: gladiators!) resemble a cross between a stick insect, a mantid, and a grasshopper. One family.

Order Ephemeroptera

Ephemera, Ephemerella—mayflies. Over 2000 species have been described from all over the world. Adults are nonfeeding and short-lived: Few live longer than a few days, and some live only a few hours. Fertilized eggs are deposited exclusively into freshwater, where the larvae (nymphs) develop for up to several years, passing through as many as 55 distinct stages (instars). Ephemeropterans are the only insects to have wings before adulthood: The penultimate nymphal stage (the "subimago"), which lasts only a few days at most, is also winged. Individuals of most species bear 2 pairs of wings. Like members of the following order, these insects cannot fold the wings flat against the body when at rest, believed to be a primitive condition. Nineteen families.

Order Odonata

The damselflies (suborder Zygoptera) and dragonflies (suborder Anisoptera), about 5200 described species. These insects are winged (2 pairs), but, as with the Ephemeroptera, the wings cannot be folded along the abdomen when the flies are resting. The gilled nymphs develop in freshwater, where they are major predators and also key food resources for fish. The largest dragonflies *look* intimidating—they are about 10 cm long—but they do not bite or sting. Adults eat only other insects. Twenty-five families.

Order Blattaria

Periplaneta, Blatella, Blatta, Blaberus, Blatteria, Cryptocercus—cockroaches. Most of the over 4000 species in this order are tropical; a few species are house pests. Although many species are omnivorous, others specialize on such diets as wood and live in such unlikely habitats as deserts, caves, or ant nests. Some species (members of the primitive family Cryptocercidae) harbor in their hindguts protozoan symbionts closely related to those inhabiting the hindguts of termites. Many species lack wings. One wingless Australian species, *Macroparesthia rhinoceros,* weighs 20 gm despite being only 6.5 cm long. Another large cockroach species, *Blaberus giganteus,* found in Mexico and South America, grows to over 6 cm in body length. Five families.

Order Mantodea

Mantids, or "praying" mantids. All 2000 species prey on other insects. The nymphs resemble small adults and develop without a pronounced metamorphosis. Adults generally possess 2 pairs of wings, although females may lack both pairs. Eight families.

Order Isoptera

Termites ("white ants"). These insects are believed to have evolved from primitive wood-eating cockroaches. Like those cockroaches, termites house symbiotic protozoa or bacteria that digest cellulose and release the nutrients to the insect host. Some termite species acquire their cellulose by eating fungi, which they farm. A single termite colony may contain over 1 million individuals in belowground colonies or in aboveground earthen colonies as large as a house. Termites are remarkably destructive of wooden structures (homes, trees), but are extremely important in nutrient and energy recycling, especially in tropical areas. All species are eusocial, and individual termites fall into certain castes that define their social rank and life's work: workers, males, soldiers, and queens. The approximately 2100 described termite species are distributed among 6 families.

Order Grylloblattaria

The 17 species of wingless omnivores contained in this order are found only in cold environments, including glaciers and ice caves. One family.

Order Orthoptera

Crickets, katydids, grasshoppers, locusts, monkey hoppers (referring to their agility in trees and shrubs). This is a large group of some 20,000 species, some members of which grow to over 11 cm long, with wingspans exceeding 22 cm. Many species "stridulate," producing a species-specific song by rubbing specialized portions of their wings (*not* the hind legs) together. Members of 1 genus, *Mecopoda,* are commonly caged as long-lived pets in China and Japan. Sixty-one families.

Order Phasmida (= Phasmatoptera)

Carausius, Diapheromera—walking sticks (stick insects); *Phyllium*—walking leaves (leaf insects). The approximately 2500 species are infamous for their ability to almost perfectly mimic the stems and leaves of the plants on which they feed. Most species can alter their body coloration with each molt to better match the color pattern of their surroundings. The largest stick insects reach about 33 cm in length. Eleven families.

Order Dermaptera

Earwigs. Most earwigs (about 99% of the 2000 described species) are free-living herbivores or carnivores, but about 20 species are exclusively parasitic or commensal on bats and rodents. Most species are tropical, and many lack wings. Eleven families.

Order Embiidina

Embiids. This group contains about 2000 species of tropical or near-tropical silk-spinning insects that live mostly in narrow galleries lined with silk, in soil, in wood, or in leaf litter. Males usually have wings, but the females are always wingless. Embiids primarily eat decaying plant matter. Eight families.

Order Plecoptera

Stoneflies. These insects (about 1800 species) are distributed worldwide, except for Antarctica. The adult life span is typically short—just long enough for mating and egg laying; most of the life cycle is spent in immaturity, usually in freshwater. Stonefly nymphs closely resemble those of mayflies (order Ephemeroptera), but have only 2 tails instead of 3 and bear gills on the thorax instead of on the abdomen. Fifteen families.

Order Psocoptera

Book lice, bark lice. Despite the connotation of their common name, these insects are not parasites. Most feed on algae, mold, lichens, pollen, or dead insects. Individuals are small, usually 1–6 mm long. Most of the nearly 2600 species are found in leaf litter, under bark, on leaves, under stones, and in caves, particularly in the tropics. Some species are annoying to humans, thriving on stored foods in pantries and cupboards. Thirty-seven families.

Order Anoplura

Sucking lice; crab lice (transmitted venereally among humans); *Pediculus*—human body lice. These 520 species of blood-sucking ectoparasites are small, wingless insects that never exceed about 4 mm in length. The abdomen swells extensively to accommodate a large volume of blood. All parasitize mammals, including such diverse animals as aardvarks, camels, monkeys, llamas, seals, ungulates, and humans. Livestock infestations can be extremely debilitating. Human head and body lice transmit typhus. Fifteen families.

Order Mallophaga

Chewing lice, biting lice. All are small (less than 5–6 mm), wingless parasites of birds and mammals. About 2500 species have been described. Eleven families.

Order Thysanoptera

Thrips. All 5300 species are small (rarely larger than 5.0 mm, with some as small as 0.5 mm). Some species are winged; some are not. Most species feed on various parts of plants and may transmit diseases among the plants on which they feed. Five families.

Order Hemiptera

The true bugs. Many of the 50,000 described species in this large group are major agricultural pests or transmitters of diseases; many other species are beneficial. Most species feed on various portions of plants, but some prey on other arthropods, and a few are ectoparasites of vertebrates. This order also contains the only open-ocean insects, 5 species in the genus *Halobates*. Seventy-four families.

Order Homoptera

This large order (some 35,000 species have been described to date) includes the cicadas, aphids, mealy bugs, spittle bugs, jumping plant lice, and leaf hoppers. All species feed on plants and often require specific plant hosts. The group includes many agricultural pest species. Fifty-five families.

Family Aphididae. Aphids. Many of the 3500 aphid species ("plant lice") are severe agricultural pests and vectors of serious plant diseases. The life cycle usually involves several asexual generations: Females hatch in the spring and deposit eggs, which develop parthenogenetically into more females, which continue to reproduce parthenogenetically for several more generations. The last generation of the summer develops into males and females; these mate and deposit fertilized eggs that overwinter (lay dormant) until the following spring.

Family Cicadidae. Cicadas. These large insects have 2 pairs of wings. The male usually has sound-producing organs at the base of the abdomen, although some species produce sound using the wings. Eggs are deposited in trees; nymphs and adults feed on tree sap.

Family Cercopidae. Spittlebugs (= froghoppers). About 23,000 species have been described to date. Adults are herbivorous and usually require particular host species. Females lay eggs in plant tissue, and the developing nymphs usually produce a conspicuous, foamy white mass, which protects them from predation and desiccation. Nymphs of other species instead secrete a calcareous tubular house on the host plant. Some species stunt or otherwise damage pine trees and clover.

Superorder Holometabola
The 9 orders in this group include most insect families and most insect species. All exhibit complete, dramatic metamorphosis.

Order Neuroptera
Dobson flies, lacewings, ant lions, and snake flies (so-named from the snake-like movements and shape of the prothorax). This is a small (about 5100 species) but diverse and widespread group of rather primitive holometabolous insects. The larvae are aquatic and secrete silken cocoons from the Malpighian tubules. Twenty-one families.

Order Coleoptera
The beetles. This is the largest of all insect orders, including over 360,000 described species. Nearly 70% of these species are contained within only 7 families. Some families contain fewer than a dozen species, but most contain hundreds or thousands, and a few contain more than 30,000 species each. In addition to the families about to be discussed, coleopterans include whirligig beetles, ladybugs (ladybirds), click beetles (which click while jumping in the air), Japanese beetles, and waterpenny beetles. Most beetles have 2 pairs of wings, with the front pair serving only as a protective sheath for the rear pair, which are used for flying. Many species produce sounds in various ways, including stridulation (rubbing various specialized body parts together). One hundred fifty-three families.

Family Carabidae. This group of some 30,000 species includes the colorful tiger beetles and the remarkable bombardier beetles, which explosively discharge a severe irritant as a potent defense measure. Most species are carnivores, even as larvae; the larvae digest their prey before ingestion and then slurp up the liquid food.

Family Ptiliidae. This is a small group of about 430 species. Many feed on fungal spores. A few highly specialized species live only in ant colonies, feasting on the excretory products of ant larvae.

Family Staphylinidae. Rove beetles. The approximately 30,000 species in this family live mostly among leaf litter. Many species live in ant or termite nests and feed on fungal spores and hyphae, but most are carnivorous.

Family Scarabaeidae. Dung beetles, scarab beetles. This group contains about 25,000 species. Members of most species eat dung, although some feed on fungi, flowers, and grasses. Some species are serious pests on golf courses, ruining the greens. Larvae often destroy crops by eating the roots.

Family Buprestidae. Jewel beetles (= metallic beetles). Adults often have a distinct metallic coloration. The larvae often do serious damage to shrubs and trees, especially fruit trees. About 15,000 species have been described.

Family Lampyridae. *Photinus, Photurus.* The fireflies. The 2000 species are characterized by specialized bioluminescent organs at the tip of the abdomen that produce a light signal that attracts mates. The larvae are ground-dwelling predators of other terrestrial invertebrates, including snails, slugs, caterpillars, and earthworms.

Family Dermestidae. These are small beetles (1–12 mm long) with a wide range of tastes, feeding on such diverse foods as pollen, nectar, carpets, upholstery, grains, dead and decaying vertebrates, and dead insects. Members of the genus *Dermestes* are routinely used to help clean vertebrate skeletons for display or study. Other species destroy prized insect collections. The larvae are especially damaging. The group contains about 850 species.

Family Tenebrionidae. This group includes some 18,000 species, many wingless, which feed mostly on plant material. Several genera, including *Tribolium* and *Tenebrio,* commonly infest stored foods. *Tribolium* (the flour beetle), in particular, has been used widely in ecological studies of population growth.

Family Cerambycidae. Timber beetles. The larvae of these 35,000 species bore into plant tissue, living or dead. Adults mostly feed on pollen and nectar and are therefore often seen on flowers.

Family Curculionidae. The weevils. This large group of some 50,000 described species includes major agricultural pests (e.g., rice weevils, cotton-boll weevils). Most species feed on various parts of flowering plants.

Family Chrysomelidae. Leaf beetles. This group contains about 35,000 species. All adults and many larvae feed on plant leaves. The larvae of some species feed underground, on plant roots. Many species are agricultural pests.

Order Strepsiptera
This group of holometabolous insects contains nearly 400 species. The females all are wingless, often legless, endoparasites of other insects, including bees, wasps, thysanurans, and cockroaches. Females spend their entire adult lives within the host's body, often with only the head protruding between a pair of adjacent host abdominal segments. Males are winged (although the front pair of wings are greatly reduced in size) and free-living, and they soon locate a female and fertilize her eggs. Hundreds or thousands of very active 6-legged larvae, less than

300 μm long, escape from the parental host. The larvae must then locate and bore into the larval stage of the host insect; larvae that mature as males then leave the host and fly off, while those that mature as females remain forever within the host. Eight families.

Order Mecoptera

Scorpion flies, snow fleas. These common forest insects feed on nectar or eat other insects. The abdomen of males ends in an upward, pointed curve, resembling a scorpion's stinger; nevertheless, the flies do not sting. Eight to nine families. The group contains about 500 species.

Order Siphonaptera

Fleas, jiggers. Approximately 2000 species of these wingless, holometabolous, biting, and blood-sucking insects have been described. Adults are parasitic—usually ectoparasitic—on warm-blooded animals, usually mammals (especially rodents). The larvae are typically not parasitic and pupate within silken cocoons. Because adults frequently jump from one host to another, fleas are excellent vehicles for transferring diseases among hosts. In particular, fleas are vectors for bubonic plague (the black death). Certain flea species are also obligate intermediate hosts for the common tapeworm of dogs and cats. Fleas lack compound eyes. Fifteen families.

Order Diptera

The true flies. This immense group of some 125,000 to 150,000 species contains such beloved insects as mosquitoes, gnats, black- and greenflies, no-see-ums, botflies, fruit flies, dung flies, and houseflies. Unlike other so-called flies, adult dipterans exhibit a posterior pair of club-shaped reduced wings (halteres, used for balancing during flight) and only one pair of flying wings. The members of this group occur worldwide and some can breed successfully in such unlikely places as oil seeps, hot springs, and the seafloor. The larvae show a terrific diversity of feeding patterns, including leaf-mining, predation, detritus feeding, and ecto- and endoparasitism. The larvae of many species lack legs and are known as "maggots." Many species transmit diseases, such as malaria, typhoid, yellow fever, and dysentery, and many other species are important agricultural pests. On the other hand, many other dipteran species eat or parasitize various insect pests, pollinate flowers, or destroy certain weeds; these dipterans thus are undeniably beneficial. One hundred sixty-two families.

Family Chironomidae. Midges. Approximately 5,000 species of ubiquitous, nonbiting, flying insects. A number of species are found in coastal marine habitats. The larvae are generally aquatic.

Family Tipulidae. Crane flies. With over 13,000 described species, this is the largest dipteran family.

Family Chaoboridae. Phantom midges (e.g., *Chaoborus*). The larvae are aquatic and commonly prey on larval mosquitoes, serving as natural control agents. Only about 75 species have been described.

Family Culicidae. Mosquitoes (e.g., *Culex, Anopheles, Aedes*). The female adult proboscis is modified for piercing; females require a blood meal before egg laying. Mosquitoes play major roles in transmitting such devastating diseases as malaria, yellow fever, and filariasis. The larvae are aquatic. About 3000 species have been described.

Family Simuliidae. Blackflies (buffalo gnats). Females are blood-sucking parasites that can inflict a memorable bite. A swarm of adults can kill livestock and even humans. One species is essential in the transmission of river blindness (Africa and Central America). The larvae develop in streams.

Family Tabanidae. Deerflies, horseflies. The females are bloodsuckers and often large-bodied ones. The larvae are aquatic predators on various other aquatic invertebrates. Deerflies transmit anthrax. More than 3000 species have been described.

Family Tephritidae. Fruit flies. Over 4000 species are known. Larvae mostly feed on fruits, such as apples and cherries, making the larvae major agricultural pests.

Family Drosophilidae. Vinegar flies. The best-known genus in this group of some 1500 species is *Drosophila,* widely used in evolutionary, genetic, and developmental studies.

Family Muscidae. This group includes houseflies (*Musca domestica*), which transmit typhoid, anthrax, dysentery, and conjunctivitis; the cattle face fly (*M. autumnalis*), which is commonly seen clustering around the heads of cattle; and the tsetse fly (*Glossina,* sometimes placed in a separate family, the Glossinidae), which transmits sleeping sickness and other similar diseases caused by trypanosomes. Larvae generally feed on decaying animal and plant matter or on feces.

Family Calliphoridae. Blowflies (the "housefly" of the western and southwestern United States and the "green bottle fly" of Southern Canada); screwworm. Most larvae (maggots) develop on decaying animals. Some species preferentially deposit their eggs on open sores of living animals, rather than on the carcasses of dead ones.

Family Oestridae. Botflies. Most of the 65 described species resemble bees. The larvae are endoparasites of mammals, including sheep, cattle, and other livestock. Horse botflies (about 45 species) are sometimes placed in another family.

Family Bombylliidae. Bee flies. Many of the approximately 4000 species closely resemble bees or wasps. Although adults generally feed on nectar, the larvae are always parasitic on developmental stages of other insects or on other insect parasites; as such, bee-fly larvae control many insect pest populations, including locusts and tsetse flies.

Order Trichoptera

Caddisflies. About 7000 species have been described to date. Adults resemble small moths but feed exclusively on liquids. Some species never get longer than about 2 mm. The larvae and pupae are generally aquatic (mostly in freshwater, although some species develop in salt marshes), but the developmental stages of some species are fully terrestrial. Larvae typically feed on algae, fungi, and bacteria. Approximately 40 families.

Order Lepidoptera

Moths and butterflies. This enormous group of insects contains nearly 160,000 described species. The females of some species are wingless, and many species are strictly nocturnal (active only at night). The larvae typically feed on plants (leaves, stems) or plant products (fruits, seeds), but some prey on other insects; one species (*Hyposmucoma molluscivora*) eats terrestrial gastropods! There are 137 families (with some families containing fewer than a dozen species each).

Family Noctuidae. Cutworms, armyworms. This is the largest lepidopteran family, with about 25,000 described species. The larvae of many species are major agricultural pests, feeding on plants and fruits.

Family Cyclotornidae. This group of only 5 Australian species is interesting despite its small size. The larvae are initially ectoparasitic on ants. As the larvae get older, they drop off the host back at the nest, where they then provide the ants with nectar and feed on the ants' larvae.

Family Pieridae. The 2000 butterfly species in this group often show highly specific feeding preferences, and several species have been much studied by ecologists interested in plant-insect coevolution. Some species are pests, particularly those feeding on legumes and crucifers.

Family Danaidae. *Danaus plexippus* (the monarch butterfly). The caterpillars in this group of about 150 species all feed on various milkweed species. Monarch butterflies migrate about 3500 km each fall (from northeastern North America to Mexico), using a time-compensated sun compass.

Family Pyralidae. This major (about 20,000 species) group of moths contains many agricultural pest species.

Family Bombycidae. This group of only about 100 Asiatic moth species includes one of the most famous of all lepidopteran species, the well-known silkworm *Bombyx mori*. The silkworm, in addition to its long-standing commercial importance, has played a major role in the development of molecular biology; the first messenger RNA molecules to be isolated in large quantities from any eukaryote were the mRNAs coding for the silk protein of *B. mori*.

Family Saturniidae. *Hyalophora cecropia*—giant silk moths. This group of about 1000 species includes the largest of all moths, with wingspreads up to 25 cm. Some species are economic pests on various trees, while others produce a commercially valuable silk. Females are well known for using pheromones to attract mates from long distances.

Family Sphingidae. Sphinx moths. This group of some 850 species includes the well-studied tobacco hornworm caterpillar *Manduca sexta*, a serious pest of tobacco and tomato plants.

Order Hymenoptera

This group of some 130,000 holometabolous insect species includes the familiar ants, bees, and wasps. Many species form functionally complex societies. Ninety-nine families.

Suborder Symphyta

The sawflies. The caterpillars (larvae) of all species feed on terrestrial plant tissues and often specialize on particular plant species or groups. Adults of some species prey on other insects. Fourteen families.

Suborder Apocrita

Wasps, ants, and bees. Adults are often nectar feeders, although members of some species suck the body juices from other arthropods. The larvae are usually legless and blind. Many of these white grubs and maggots feed in or on the bodies of a host arthropod or its larvae; others develop within plant galls, fruits, or seeds. Seventy-five families.

Family Ichneumonidae. These wasps are mostly parasitoids, living freely as adults but developing at the expense of an arthropod host, usually an insect but sometimes a spider, bee, or pseudoscorpion; the larva feeds on the host and eventually kills it. At least 15,000 species have been described to date, although perhaps 3 times as many species await discovery. Only about 5% of species are eusocial, and only a few species sting people.

Family Formicidae. *Formica, Myrmica, Solenopsis*—the ants. Adults typically feed on fungi or nectar, or they prey on other terrestrial arthropods. In tropical rain forests, as many as 72 ant species have been found living in a single tree. At least 9500 ant species are known, each of which forms complex social groups of dozens to thousands of cooperating individuals per colony. Most ant colonies include members of at least 3 distinct castes: workers (wingless, sterile females), males, and queens. Probably another 20,000 ant species remain to be described, primarily from tropical rain forests. At least some tropical frog species derive their toxins from ingesting large quantities of ants.

Family Apidae. *Apis*—honeybees; *Bombus*—bumblebees. This is one of 8 bee families, the entire group encompassing perhaps 20,000 species. Not all bee species bear stingers. Only about 5% of the species, including the honeybees, are eusocial. Unlike most other bees, the members of this family do not dig burrows, but rather nest in cells of wax or resin, sometimes supplementing the structure with other materials, such as bark, mud, or even vertebrate feces. Bees are major flower pollinators, and many flowers, including orchids, depend on bees for pollination; both adults and larvae subsist on nectar and pollen. Honeybees pollinate about $10 billion worth of crops in the United States each year. Recent unintended introductions of 2 species of parasitic mite are decimating U.S. honeybee populations, either by plugging up the tracheae and suffocating the bees or by feeding directly on the bees' hemolymph; thus, it has been illegal to import U. S. honeybees into Canada since 1987. The so-called "killer bees" now working their way northward from Central and South America are unusually aggressive honeybees, but they may produce more honey than most other bees. Honeybees are well known for their complex dances, through which workers communicate the location of desirable flowers.

Class Crustacea

This class contains approximately 45,000 species.

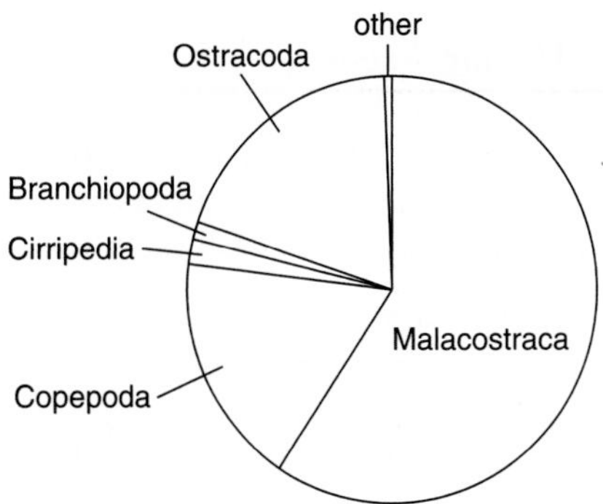

Subclass Cephalocarida

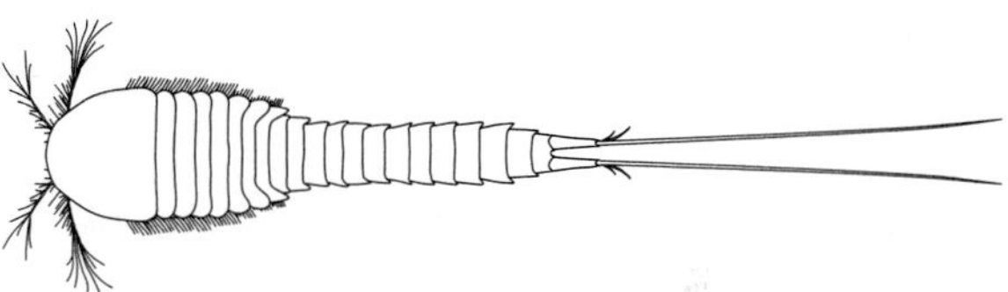

Hutchinsoniella. The 9 species in this group are thought to be among the most primitive of living crustaceans. All are marine bottom-dwellers living in soft sediments intertidally to depths exceeding 1500 m. No species exceeds 3.7 mm in length. All are suspension feeders, collecting food particles using spines on the legs and circulating water through the legs by the rhythmic beating of the limbs.

Subclass Malacostraca

The over 20,000 described species are distributed among 12 orders.

Superorder Syncarida

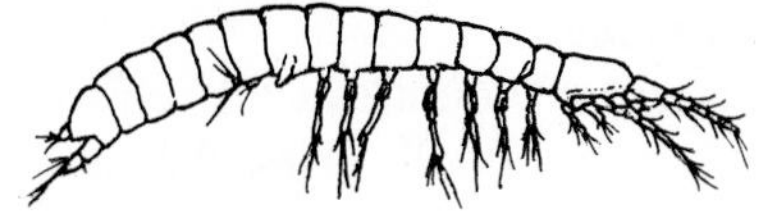

Anaspides, Bathynella—the syncarids. The 150 species contained in this group are mostly freshwater, elongated crustaceans lacking a carapace. A number of species are interstitial, living in the spaces between sand grains. A few other species live only in burrows made by freshwater crayfish. Six families.

Superorder Hoplocarida

Order Stomatopoda

Squilla, Pseudosquilla, Gonodactylus—mantis shrimp. The 350 stomatopod species are active, hard-hitting predators living in crevices and holes in hard substrate or in extensive burrows of their own making in soft substrates. Some species are fished commercially. Twelve families.

Superorder Peracarida

The over 11,000 described species are distributed among 7 orders.

Order Thermosbaenacea

Thermosbaena. Small (about 3 mm long) crustaceans living only in saline hot springs, at water temperatures of up to 43°C. Other members of the order live in caves, in cold, freshwater. Two families.

Order Mysidacea

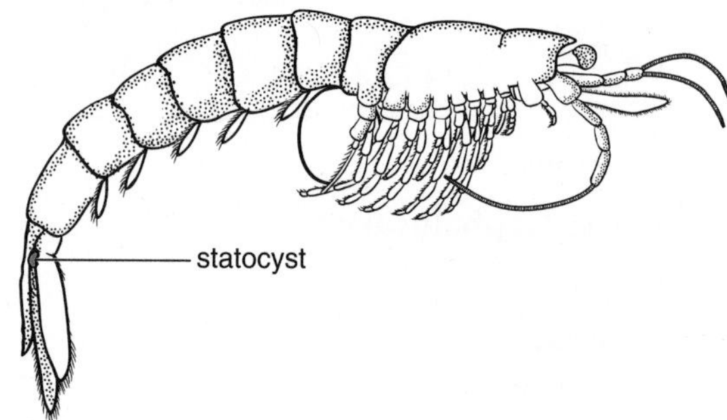

Based on *Comparative Morphology of Recent Crustacea* by McLaughlin. 1980.

Mysis—mysid shrimp, "opossum shrimp." Most of the nearly 1000 shrimp-like species in this group are marine or estuarine, but several dozen inhabit freshwater environments, including lakes, wells, and caves. Bodies are typically 5–25 mm long, and mature females have a distinctive brood pouch ventrally. Many species are heavily preyed on by various fish. Six families.

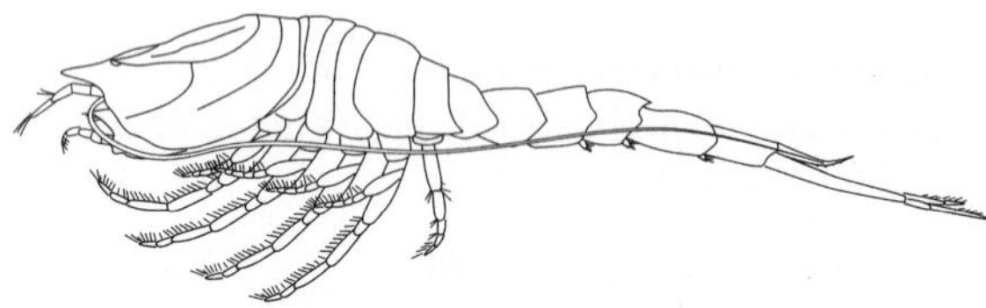

Order Cumacea

The cumaceans. These are exclusively marine, living in soft substrates, usually at less than about 200 m depth. The almost 1000 species are distributed among 8 families.

Order Tanaidacea

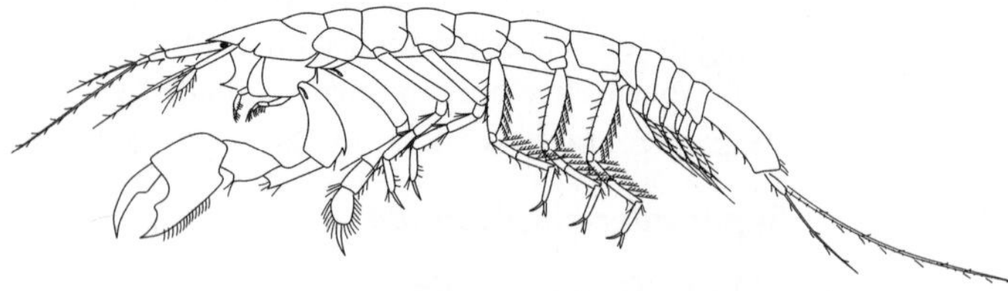

The tanaids. The 550 species are all marine, occurring from shallow to abyssal depths. They are among the most commonly encountered malacostracans in the deep sea. Sixteen families.

Order Isopoda

Caecidotea (= *Asellus*), *Ligia, Idotea, Paracerceis, Sphaeroma*—the isopods; woodlice; pillbugs. This large group of some 10,000 species occurs in marine, estuarine, freshwater, interstitial, and terrestrial habitats. Some species are blind cave dwellers. Others parasitize fish, cephalopods, or other crustaceans, including other isopods. *Bathynomus giganteus,* a deep-sea species, attains a length of 42 cm (nearly 1.5 ft) and a width of about 15 cm. Most species are considerably smaller. *Limnoria,* although less than 0.5 cm long, is a very destructive, marine, wood-boring isopod. In some, males are greatly reduced and live in a pouch within the female or attached externally to her antennae. Isopods are among the most common deep-sea malacostracans and are the dominant terrestrial malacostracans. One hundred families.

Order Amphipoda

Amphipods dominate the small (1–50 mm) malacostracan fauna in freshwater and in shallow coastal waters in temperate regions. Some of the approximately 6000 species are common in underground streams and in caves. About one hundred twenty families.

Suborder Gammaridea

Ampelisca, Corophium, Gammarus; Orchestia, Talorchestia (beachhoppers and sand fleas)—the gammarid amphipods. The over 4700 species are distributed among some 91 families. Most individuals are only 1–15 mm long, although one abyssal species reaches a length of 25 cm. Nearly one-third of the species live in freshwater. Another 3350 species are marine, mostly living intertidally or in shallow water, and a few species are terrestrial. Gammarid amphipods are major food sources for many fish species and for some marine mammals. Amphipod concentrations can reach 73,000 individuals per m^2. Gammarids exhibit a variety of lifestyles. Various species are herbivores, carnivores, deposit feeders, commensals, or external parasites (of invertebrates and fishes). Some species display complex mating behavior.

Suborder Hyperiidea

Hyperia, Phronima—the hyperiid amphipods. All 300 or so species are marine and planktonic and occur in all ocean waters. Despite the relatively small number of species, hyperiid amphipods are important in marine food chains: as key food items for marine mammals, seabirds, and large fish and as carnivores of other invertebrates. These amphipods are typically associated with gelatinous animals, such as salps, jellyfish, and siphonophores. They often exhibit extensive diurnal vertical migrations (presumably with their hosts), often traveling over 1000 m vertically in a 24-hour period. *Phronima* can apparently kill the gelatinous host and live inside the hollowed-out test, propelling it by beating its pleopods.

exclusively parasitic on, or commensal with, echinoderms, bivalves, or tube-dwelling polychaetes. About 225 species.

Family Hapalocarcinidae. The coral gall crabs. This small group (27 species) is exclusively tropical, and its members are all tiny; individuals typically measure only a few millimeters across the carapace. The female lives within a lump of coral tissue produced by hard corals as the coral slowly grows around and eventually encloses the patient crab, which then lives protected but imprisoned.

Family Ocypodidae. *Uca*—the fiddler crabs. The several hundred species in this group are mostly semiterrestrial, living intertidally in estuaries and salt marshes.

Subclass Branchiopoda

This remarkably varied group of crustaceans includes the water fleas (*Daphnia*), brine shrimp (*Artemia*), and about 800 other aquatic crustacean species.

Infraclass Sarsostraca

Order Anostraca

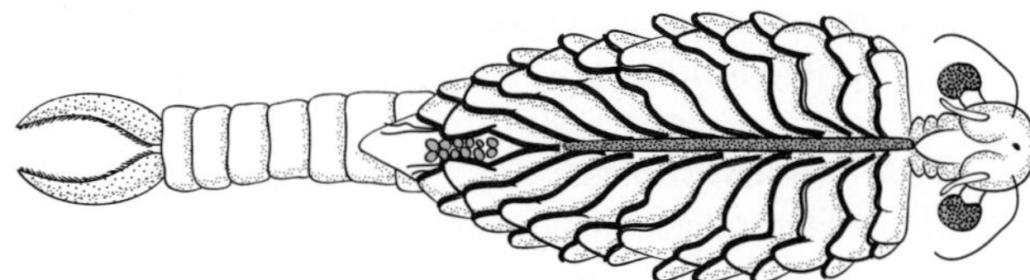

Based on R. Pennak, *Freshwater Invertebrates of the U.S.*, 3d ed. 1978.

Branchinecta; Eubranchipus; Streptocephalus—the fairy shrimp; *Artemia*—the brine shrimp. This group contains about 200 species, which grow to 10 cm in length. These animals live in stressful habitats, either temporary freshwater pools and ponds or under hypersaline conditions. Some groups, such as members of the genus *Artemia,* are found worldwide except for the poles, while others occur mainly, or only, at the poles. Members lack a carapace, and as many as 19 of the body segments bear swimming appendages. Eight families.

Infraclass Phyllopoda

Most of the approximately 600 species in this group live in freshwater, with only a few species living in the sea. Three orders.

Superorder Diplostraca

Order Cladocera

Daphnia, Bosmina, Polyphemus, Moina, Podon, Evadne—water fleas. Individuals are usually smaller than 3 mm. This order contains about 400 species. Nine families.

Suborder Anomopoda
Bosmina, Daphnia, Macrothrix

Suborder Ctenopoda
Sida

Suborder Eucladocera (= Onychopoda)
Evadne, Moina, Podon, Polyphemus

Suborder Haplopoda

Order Conchostraca

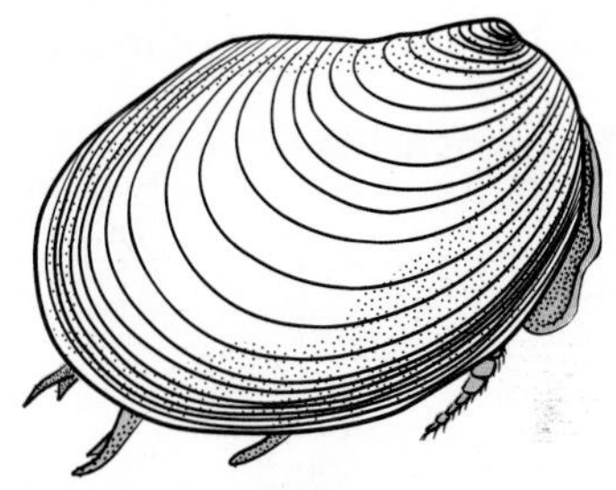

Based on R. Pennak, *Freshwater Invertebrates of the U.S.*, 3d ed. 1978.

Clam shrimp. This group contains 3 suborders: Cyclestherida, Leavicaudata, and Spinicaudata. In some treatments the name Conchostraca has been abandoned, and the suborders are each raised to the level of order. Clam shrimp are believed to have given rise to the cladocerans. The fewer than 200 species are common in temporary freshwater ponds, and the eggs withstand considerable dehydration. As with cladocerans, embryos are brooded within the female carapace, which completely encloses the adult conchostracan body. Individuals can grow to lengths of 2 cm. Five families.

Order Notostraca

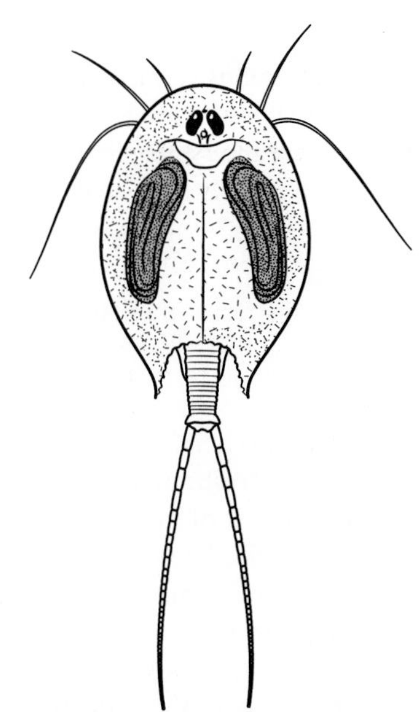

Based on *Comparative Morphology of Recent Crustacea* by McLaughlin. 1980.

Triops—tadpole shrimp. The 9 species in this order generally inhabit temporary lakes. The eggs withstand considerable dehydration. One family.

Subclass Ostracoda

Cypridina, Cypris, Gigantocypris, Pontocypris—seed shrimp. Most of the 6650 ostracod species are only a few millimeters long, but females of *Gigantocypris* can exceed 3 cm. Most species are marine, estuarine, or freshwater, but a few occur in moist terrestrial habitats. Ostracods exhibit a variety of lifestyles, but none are parasitic. Marine species are found from shallow to abyssal depths. Ostracods are mostly divided among the 2 major orders, Myodocopa and Podocopa. Forty-three families.

Subclass Mystacocarida

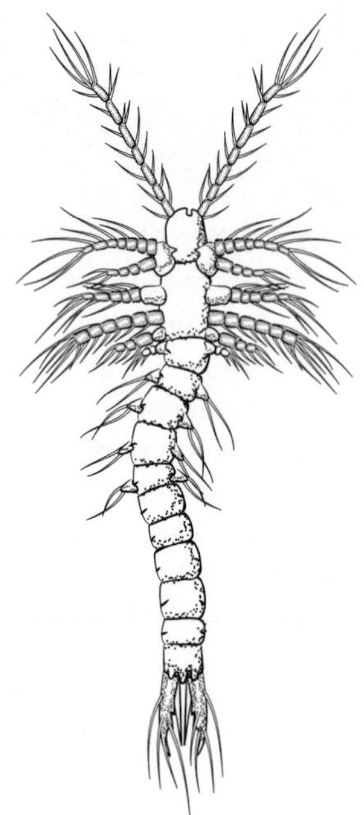

Base on *Comparative Morphology of Recent Crustacea* by McLaughlin. 1980.

The 8 known species are exclusively marine and very small: Most are less than 0.5 mm long, and none are larger than 1 mm. All species are interstitial, living in the spaces between sand grains intertidally or in shallow water.

Subclass Copepoda

The approximately 8500 species are distributed among 6 orders.

Order Calanoida

Calanus, Euchaeta, Eurytemora, Centropages, Diaptomus (occurs in freshwater only), *Candacia, Bathycalanus* (the largest of free-living copepods, up to 17 mm long), *Acartia.* These animals are mostly marine, and in fact they dominate the marine plankton; they are extremely important in marine food chains leading to many important fish species. Copepods occur in all ocean waters, from the surface to depths exceeding 5000 m. Many planktonic species perform extensive vertical migrations, swimming to the surface at night and moving into deeper water at dawn. The members of one family (Diaptomidae) are found exclusively in freshwater. The order contains over 1900 species, distributed among 37 families.

Order Harpacticoida

Euterpina, Psammis, Tisbe. These small copepods (less than 2.5 mm long) are found in marine, estuarine, and freshwaters. Most species live in or on the bottom, many living interstitially in the spaces between sand grains. Some freshwater species live in damp moss. A few species are commensal in the baleen of certain whale species, and a few spend their lives swimming in the plankton. Most species occur only in shallow water, but a few are restricted to the deep sea. The order contains 34 families, housing about 2250 species.

Order Cyclopoida

These copepods are found in marine and freshwater habitats. Most species are free-living, but some are commensal, and others are parasitic. Members of the genus *Mytilicola* are intestinal parasites of bivalves. Others are ectoparasites on gill filaments of fish. Some species are intermediate hosts for the parasitic nematode *Dracunculus medinensis* (p. 442). The over 3000 species are distributed among 12–16 families.

Order Monstrilloida

Monstrilla. All are marine, bizarre, often highly degenerate crustaceans, and all are parasites, as larvae, on various invertebrates. The adults are free-living but nonfeeding, lacking mouthparts or gut; the adult stage of the life cycle is therefore short. Adults also lack second antennae. The approximately 80 species are distributed among only 2 families.

Order Siphonostomatoida

Caligus, Salmincola (the salmon gill maggot). Although the nauplius larvae are typical and free-living, the adults are exclusively parasites of marine fish, commonly affixed to the gills. Eleven families.

Order Poecilostomatoida

Bomolochus, Ergasilus, Tucca. Although the nauplii are free-living, the adults are parasitic on fish (mostly marine) and marine invertebrates. Often the antennae end in claws, for attaching to the host.

Subclass Branchiura

Argulus. All 125 species are ectoparasitic on freshwater or marine fish, even in the larval stages. The animals are typically smaller than 2 cm. One family.

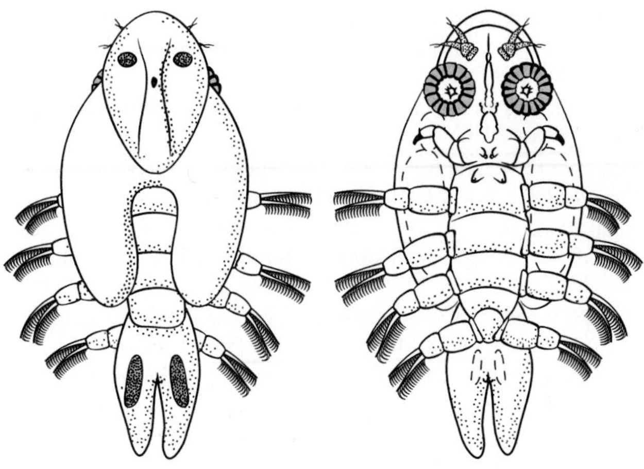

Based on L. Margolis and Z. Kabata, "Guide to the Parasites of Fishes of Canada, Part II: Crustacea," in *Fisheries & Oceans*. 1988.

Subclass Pentastomida

The approximately 100 species are distributed among 2 orders and 7 families.

Order Cephalobaenida

Cephalobaena, Reighardia. The life cycles of the species in this order are largely undocumented. This order contains one of the few pentastomid species living in a nonreptilian host; members of the genus *Reighardia* parasitize only birds, including gulls. Two families.

Order Porocephalida

This order includes the largest of all pentastomids, members of one species reaching lengths of 9 cm. Five families with about 65 species, most of which parasitize reptiles.

Family Linguatulidae. *Lingulata.* This family is unusual in that all its members parasitize mammals rather than reptiles. Developmental stages occur in the glandular tissues of rabbits, horses, cows, sheep, and pigs, while adulthood is reached only in canines. Developmental stages can persist for several weeks in humans, who become infected by eating raw glands from slaughtered mammalian hosts. In humans, the nymphs migrate out of the gut and take up residence in the nasopharynx, or sometimes the eye, before dying a few weeks later.

Subclass Tantulocarida

A recently recognized group of only 5 known species, all ectoparasitic on deep-water crustaceans. The members resemble copepods, but lack thoracic legs.

Subclass Remipedia

This group was first recognized in 1983. The 8 species described so far are restricted to tropical, underwater marine caves. The long body with its abundant lateral appendages resembles that of a polychaete. Remipedians

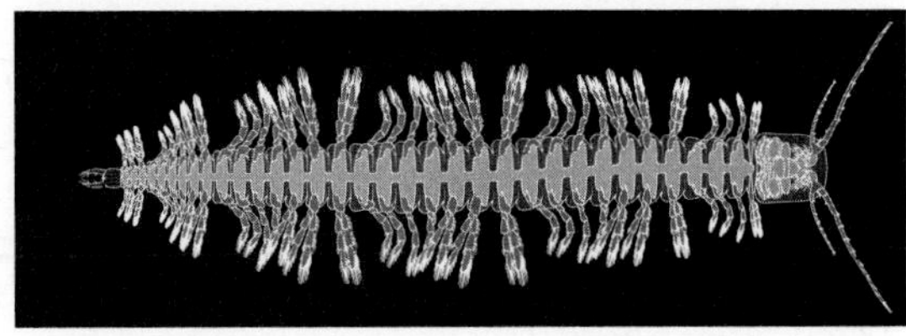

may be the closest of all animals to the ancestral crustacean condition.

Subclass Cirripedia

All nearly 1000 species are entirely sedentary as adults and are found only in marine and estuarine waters. The approximately 25 families are distributed among 4 orders. This class includes the barnacles.

Order Acrothoracica

Trypetesa. Most species lack calcareous protective plates, burrowing instead into mollusc shells, coral skeletons, or limestone. The females are fully developed, but the males are little more than inconspicuous sacs of sperm living adjacent to females. A few species are found at depths of 600–1000 m, but most are found in shallow water. Three families.

Order Ascothoracica

Ulophysema. All 30 species are marine ecto- or endoparasites of cnidarians or echinoderms. The body is enclosed within a bivalved carapace or sac. These animals occur over a wide depth range, from shallow water to abyssal depths.

Order Thoracica

Balanus, Lepas, Chthamalus, Verruca, Elminius Pollicipes—the true barnacles. Most cirripedes are placed in this order. All of the approximately 800 species are marine; species in this group are found from the intertidal zone to the deepest reaches of the sea. Individuals of most species are enclosed within a complex of calcareous plates. A wide variety of lifestyles is exhibited: Most species are sessile suspension feeders, but others are parasitic in sharks, polychaetes, or corals, and a few are pelagic. Some species live attached to the outsides of fish, jellyfish, turtles, or whales, and others are commensal with anthozoans, crustaceans, or sponges or in the mantle cavity of certain bivalves. This order contains the oldest known cirripedes, with a fossil record extending back over 400 million years. Seventeen families.

Order Rhizocephala

Sacculina, Lernaeodiscus, Loxothylacus. Of the 230 species in this order, all but 3 or 4 are exclusively marine; the few misfits occur only in freshwater.

Regardless of habitat, all species are internal parasites of other crustaceans, especially of crabs and other decapods. A sac containing the female gonad of the parasite protrudes conspicuously from the host, usually ventrally, while the large absorptive portion of the parasite remains inside. Males are small and inconspicuous, little more than masses of sperm cells. Species occur from shallow waters to depths exceeding 4000 m. Seven families.

General References about the Arthropods

Ali, M. F., and E. D. Morgan. 1990. Chemical communication in insect communities: A guide to insect pheromones with special emphasis on social insects. *Biol. Rev.* 65:227–47.

Anderson, D.T. 1994. *Barnacles: Structure, Function, Development and Evolution.* New York: Chapman & Hall.

Arnett, R. H., Jr. 2000. *American Insects: A Handbook of the Insects of America North of Mexico,* 2d ed. CRC Press.

Bauer, R. T. 2004. *Remarkable Shrimps: Adaptations and Natural History of the Carideans.* University of Oklahoma Press.

Bliss, D. E., ed. 1982–1987. *The Biology of Crustacea,* vols. 1–9. New York: Academic Press.

Chapman, R. F. 1998. *The Insects: Structure and Function,* 4th ed. New York: Cambridge Univ. Press.

Coddington, J. A., and R. K. Colwell. 2001. Arachnids. In *Encyclopedia of Biodiversity,* ed. S. A. Levin, pp. 199–218. San Diego, CA: Academic Press.

Coddington, J. A., and H. W. Levi. 1991. Systematics and evolution of spiders (Araneae). *Ann. Rev. Ecol. Syst.* 22:565–92.

Davies, R. G. 1988. *Outlines of Entomology,* 7th ed. London: Chapman & Hall.

Drosopoulos, S., and Michael F. Claridge (eds). 2006. *Insect Sounds and Communication: Physiology, Behaviour, Ecology, and Evolution.* Boca Raton, FL: Taylor & Francis.

Dudley, R. 1999. *The Biomechanics of Insect Flight.* Princeton University Press, 476 pp.

Eisner, T. 2005. *For Love of Insects.* Belknap Press.

Foelix, R. F. 1996. *Biology of Spiders,* 2d ed. New York: Oxford Univ. Press.

Gaskett, A. C. 2007. Spider sex pheromones: Emission, reception, structures, and functions. *Biol. Rev.*: 26–48.

Gilbert, S. F., and A. M. Raunio, eds. 1997. *Embryology: Constructing the Organism.* Sinauer Associates Inc., Publishers, Sunderland, MA, pp. 237–78.

Grimaldi, D., and M. S. Engel. 2005. *Evolution of the Insects.* New York: Cambridge University Press, 755 pp.

Harrison, F. W., and R. F. Foelix, eds. 1999. *Microscopic Anatomy of the Invertebrates, Vol. 8: Chelicerate Arthropoda.* New York: Wiley-Liss.

Harrison, F. W., and A. G. Humes, eds. 1992. *Microscopic Anatomy of the Invertebrates, Vol. 10: Decapod Crustacea.* New York: Wiley-Liss.

Harrison, F. W., A. G. Humes, and E. E. Ruppert, eds. 1992. *Microscopic Anatomy of the Invertebrates, Vol. 9: Crustacea.* New York: Wiley-Liss.

Harrison, F. W., and M. Locke, eds. 1998. *Microscopic Anatomy of the Invertebrates, Vol. 11: Insecta.* New York: Wiley-Liss.

Harrison, F. W., and M. E. Rice, eds. 1993. *Microscopic Anatomy of the Invertebrates, Vol. 12: Onychophora, Chilopoda, and Lesser Protostomata.* New York: Wiley-Liss.

Hölldobler, B., and E. O. Wilson. 1990. *The Ants.* Cambridge, Mass.: Harvard Univ. Press.

McCravy, K. W., and F. R. Prete. 2007. Hundred-legged hunters. In *Predator,* ed. F. R. Prete. Chicago: Univ. Chicago Press.

Papaj, D. R., and A. C. Lewis. 1993. *Insect Learning: Ecological and Evolutionary Perspectives.* New York: Routledge, Chapman and Hall, Inc.

Phillips, B. (ed.). 2006. *Lobsters—Biology, Management, Aquaculture and Fisheries.* Oxford, UK: Blackwell Publishers.

Polis, G. A., ed. 1990. *The Biology of Scorpions.* Stanford, Calif.: Stanford Univ. Press.

Punzo, F. 2007. *Spiders—Biology, Ecology, Natural History and Behavior.* Boston: Brill.

Savory, T. 1977. *Arachnida,* 2d ed. New York: Academic Press.

Schram, F. R. 1986. *Crustacea.* New York: Oxford Univ. Press.

Schultz, J. W. 1987. The origin of the spinning apparatus in spiders. *Biol. Rev.* 62:89–113.

Shear, W. A. 2000. Millipeds. *Amer. Sci.* 87: 232–39.

Shuster, C. N., Jr., R. B. Barlow, and H. J. Brockmann (eds.). 2003. *The American Horseshoe Crab.* Harvard University Press.

Sutherland, S. K., and J. Tibballs. 2001. *Australian Animal Toxins.* Oxford University Press.

Taylor, P. D., and D. N. Lewis. 2005. *Fossil Invertebrates.* Harvard University Press.

Triplehorn, C. A., and N. H. Johnson. 2005. *Borror and DeLong's Introduction to the Study of Insects,* 7th ed. Thomson Brooks/Cole.

Weygoldt, P. 1969. *The Biology of Pseudoscorpions.* Cambridge, Mass.: Harvard Univ. Press.

Wilson, E. O. 1971. *The Insect Societies.* Cambridge, Mass.: Harvard Univ. Press.

Winston, M. L. 1987. *The Biology of the Honeybee.* Cambridge, Mass.: Harvard Univ. Press.

Witt, P. N., and J. S. Rovner, eds. 1982. *Spider Communication: Mechanisms and Ecological Significance.* Princeton, N.J.: Princeton Univ. Press.

Search the Web

1. www.ucmp.berkeley.edu/arthropoda/arthropoda.html

   This site, maintained by the University of California at Berkeley, includes separate sections on arthropod systematics, ecology, and fossils. Good color images of representatives in each major arthropod group.

2. http://tolweb.org/tree/phylogeny.html

   This brings you to the Tree of Life site. Search on "Arthropoda."

3. www.mbl.edu

   This site contains photographs and an identification key to arthropods found in the region of Woods Hole, MA. To locate the taxonomic key, choose "Biological Bulletin" under Quick Links and then "Other Biological Bulletin Publications." Next select "Keys to

Marine Invertebrates of the Woods Hole Region," and then choose the arthropod group of interest. To locate the color photographs from the home page, choose "Publications" under Quick Links, and then under "Databases" choose "Marine Organisms Database." Then click on Arthropoda. Choose "Expand" to get a listing of all major marine arthropod groups and then select the group of interest to see a color image.

4. http://www.cals.ncsu.edu/course/ent425/

   This is the website from the General Entomology course at NC State.

5. www.who.int/ctd/

   This site is maintained by the World Health Organization. Search on "Malaria" to learn more about the role of mosquitoes in spreading this disease.

6. www.ent.iastate.edu/imagegallery/default.html

   This site provides an extensive gallery of insect images, provided by the Entomology Department of the Iowa State University.

7. www.ipmcenters.org

   This site, operated by the U.S.D.A., is a clearing house for information about the biological control of insect pests.

8. www.museum.vic.gov.au/crust

   This site provides an extensive collection of photographs and information about the marine crustaceans of southern Australia, offered through the Museum of Victoria in Melbourne, Australia.

9. www.slagoon.com

   Click on "Toons" to follow the daily adventures of Hawthorne, the hermit crab who lives in an empty beer can, and his friends. Drawn by Jim Toomey.

10. www.denniskunkel.com

    Superb, colorized scanning electron micrographs of some insect and arachnid species, provided by Dennis Kunkel.

11. www.crustacea.net

    A repository for information about all aspects of crustacean systematics, including taxonomic keys to crustaceans of the world.

12. http://www.cals.ncsu.edu/course/ent425/

    This informative website accompanies the General Entomology course at North Carolina State University.

13. http://www.myriapoda.org/

    Provided by the Biology Department, East Carolina University.

14. http://www.americanarachnology.org/AAS_information.html

    Information about arachnids, with excellent photographs, from the American Arachnological Society.

15

Two Phyla of Likely Arthropod Relatives: Tardigrades and Onychophorans

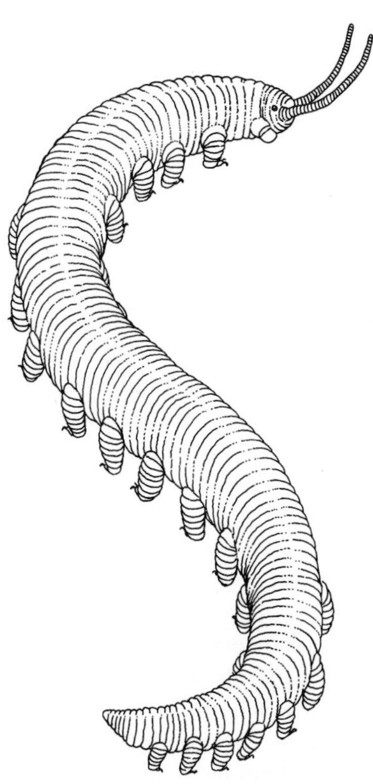

Introduction and General Characteristics

One would think that a group as stereotyped, and with as long a fossil record, as the Arthropoda would have a relatively uncontroversial phylogeny. The following features are considered arthropod characteristics: external, jointed exoskeleton; tracheae; compound eyes; Malpighian tubules; mandibles; heart with ostia. Nevertheless, there is considerable controversy regarding these animals' evolutionary history. In large part, the controversy centers on how many times these arthropod characteristics have independently evolved.

All of the arthropod subphyla listed in Chapter 14 are already represented among the earliest fossils of the Burgess Shale, formed early in the Cambrian period, about 530 million years ago. The group's ancestry is difficult to trace because arthropods most certainly are derived from soft-bodied ancestral forms poorly represented as fossils. Many biologists have long believed that arthropods are derived from annelids, or at least from annelid-like ancestors; indeed, arthropods and annelids are sometimes referred to collectively as the Articulata. In particular, both arthropods and annelids demonstrate clearly segmented coelomic spaces during embryogenesis; absence of this characteristic in the arthropod adult could easily be a modification of the ancestral condition. And as in annelids, arthropod segments arise from a specific zone at the rear of the animal. Although most molecular data do not support the Articulata hypothesis, the existence of animals possessing some characteristics of arthropods and some characteristics typical of annelids or other groups, or at least atypical for arthropods, presents tantalizing phylogenetic implications. The animals in the 2 phyla discussed in this chapter—Tardigrada and Onychophora—show this intriguing combination of arthropod and nonarthropod characteristics.

The phylogenetic position of both groups has been uncertain for some time. In particular, the tardigrades have often been associated with rotifers, nematodes, or other aschelminth groups (Chapters 10, 16–17), and the onychophorans have often been associated closely with annelids. There is growing recognition, however, that tardigrades and onychophorans are closely related to each other and to the arthropods. One characteristic shared between tardigrades and onychophorans is the possession of unjointed walking legs, called **lobopods,** that terminate in a number of sharp claws. These legs are also encountered among some fossils dating from the mid-Cambrian period, about 525 million years ago (Fig. 15.1). Some workers suggest that lobopods may have given rise to both arthropod and annelid appendages. A number of zoologists and paleontologists have suggested grouping the tardigrades, onychophorans, and their fossilized relatives in a new phylum, the Lobopoda . . . and that's where they may end up by the next edition of this book.

A close association between the lobopods and arthropods is supported by both anatomical and molecular studies, and the 3 groups are often referred to collectively as the Panarthropoda. Like arthropods, both tardigrades and onychophorans secrete chitinous cuticles that they molt periodically; whether the molting process has a common biochemical basis in the 3 panarthropod groups has yet to be determined.

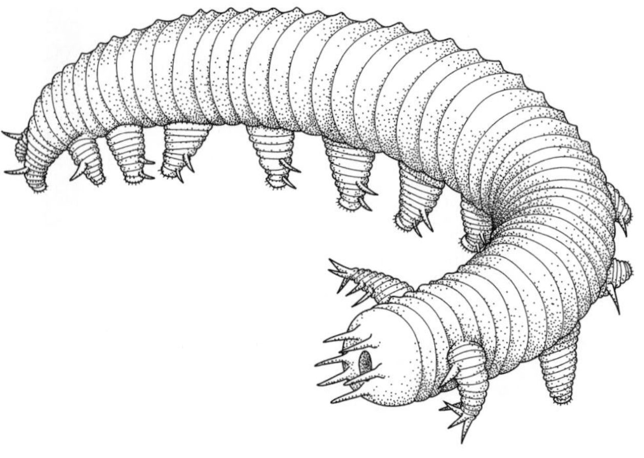

Figure 15.1
Reconstruction of the fossilized lobopod *Aysheaia,* from material preserved in the Burgess Shale of British Columbia, formed about 525 million years ago in the mid-Cambrian period. Note that "lobopod" can refer to the animal as well as the limb.
From S. Conway Morris. 1998. *The Crucible of Creation.* By permission of Oxford Univ. Press, NY.

Phylum Tardigrada

Phylum Tardi • grada
(Latin: slow walker)
tar´-di-grād-ah

Defining Characteristic:[1] Mouthparts include protrusible, oral stylets for piercing plant and, to a lesser extent, animal tissues

Members of the phylum Tardigrada (the water bears) have clear arthropod affinities, but don't quite seem to fit in as *bona fide* members of that phylum. About 900 tardigrade species have been described. All tardigrades are quite small, ranging between about 50 µm (micrometers) and 1200 µm in length. The typical tardigrade is only about 0.5 mm (millimeters) long. Most species live in surface films of freshwater on terrestrial plants, especially mosses and lichens. Some marine species have been described, many of them living in the spaces between sand grains (i.e., **interstitially**), as deep as 5000 m (meters). Although small, water bears may occur in impressively dense aggregations, up to several million per square meter of substrate surface. A few species are commensal or parasitic with other invertebrates.

The first fossilized tardigrades were found about 40 years ago, in Cretaceous amber (~100 million years old). But fossil tardigrades, or their early relatives, have now been reported from mid-Cambrian rocks in Siberia, formed about 520 million years ago. Thus, tardigrades, like the arthropods and onychophorans, were apparently present at or near the start of the Cambrian explosion.

Like the arthropods, tardigrades possess a complex, chitinous cuticle that is periodically molted and that not only covers the outside of the body, but also lines the foregut and hindgut. Moreover, tardigrades possess arthropod-like striated muscles, and their capacious body cavity appears to be a hemocoel, formed from spaces in the connective tissue. Because the tardigrades are small and their cuticle is highly permeable to water and gases, they are essentially restricted to moist habitats. Gas exchange occurs across the general body surface; tardigrades have no specialized respiratory structures. Like the arthropods, tardigrades lack motile cilia. All tardigrades possess 4 pairs of clawed appendages, with which the animals lumber over the substratum in bearish fashion (Fig. 15.2), but the appendages are never jointed.

The tardigrade nervous system is organized in arthropod manner, with a paired ventral nerve cord (Fig. 15.3). Also, several tardigrade glands look suspiciously like Malpighian tubules; their function in osmoregulation has recently been demonstrated. The mouthparts are a pair of stylets, which are mostly used to pierce plant cells; a few species are carnivorous. The mouth leads into a muscular sucking pharynx. Tardigrades have no specialized larval stages; offspring develop as miniature adults.

Several tardigrade characteristics are decidedly nonarthropod. As already mentioned, the appendages are segmented but not jointed. Moreover, the cuticle is never calcified. Tardigrade embryological development may

1. Characteristics distinguishing the members of this phylum from members of other phyla.

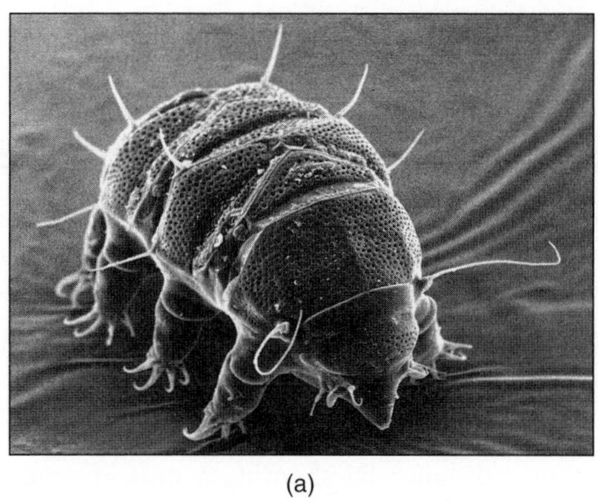
(a)

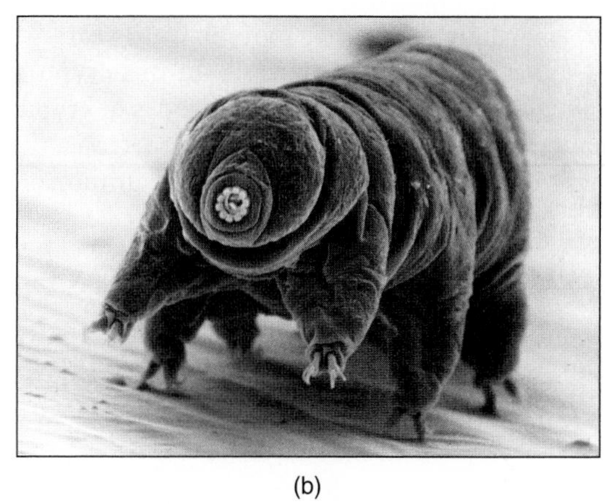
(b)

Figure 15.2
Scanning electron micrographs of (a) *Echiniscus spiniger* and (b) *Macrobiotus hufelandi,* two tardigrades.

(a,b) Courtesy of D. R. Nelson.

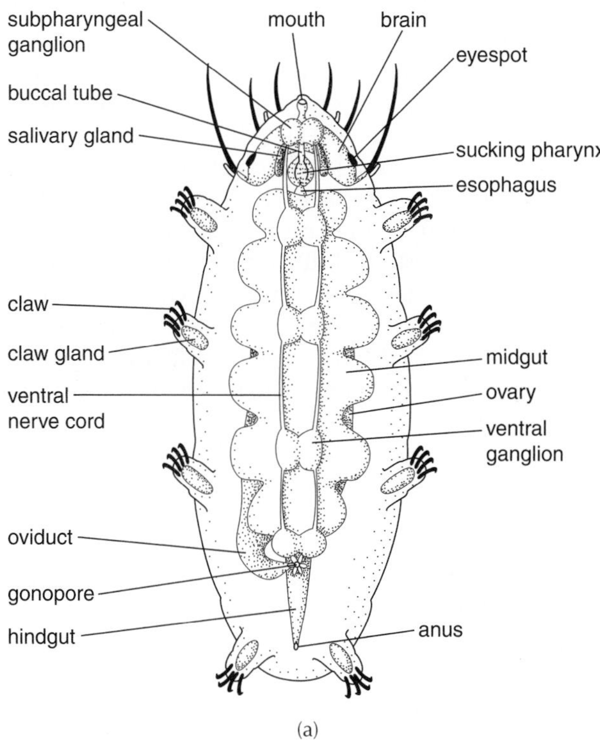

(a)

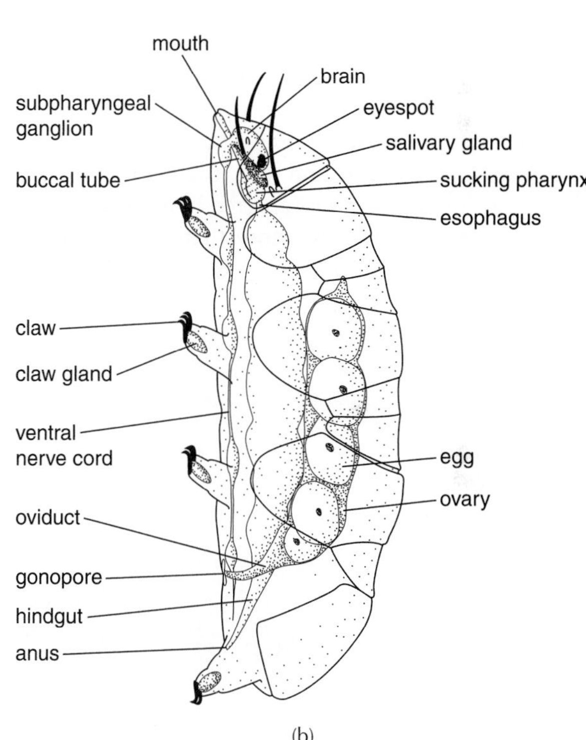

(b)

Figure 15.3
(a) Internal anatomy of a typical tardigrade (*Bryodelphax parvulus,* female). The claw glands secrete the claws. (b) Detail of the tardigrade nervous system. Note the large ganglia associated with each pair of appendages. (Note also that the nerve cord is ventrally located, as in annelids and arthropods.)

(a,b) From D. Nelson, 1982. In F. Harrison and R. Cowden, eds., *Developmental Biology of Freshwater Invertebrates;* Alan R. Liss, NY. Drawing by R. P. Higgins.

display at least one typically deuterostome characteristic (enterocoely—something never encountered among arthropods), although recent embryological studies contradict this conclusion. Similarly, indeterminate cleavage, typical of deuterostomes, has recently been reported for one tardigrade species[2]; more embryological work seems warranted. If tardigrades are indeed coelomates, that coelom—whatever its embryological origin—is reduced to a small pouch surrounding the gonads. But some workers have linked tardigrades to the nematodes, rotifers, and other pseudocoelomate aschelminths, based upon constancy of cell numbers in the tardigrade cuticle and the ultrastructural organization of the pharynx. Moreover, like nematodes and rotifers, tardigrades exhibit **cryptobiosis,** a bizarre ability to dehydrate and reduce metabolic rate to withstand extreme environmental conditions of low-temperature and desiccation

2. Hejnol, A., and R. Schnabel. 2005. *Development* 132:1349–61.

stress.[3] A tardigrade may live more than 10 years—more than 100 years in some instances!—in this cryptobiotic state; the total life span, including episodes of cryptobiosis, may thus be many decades in some species, although life spans of less than one year are more common.

Phylum Onychophora

Phylum Onycho • phora
(Greek: claw bearer)
on-ē-koh´-fer-ah

Defining Characteristics: 1) Second pair of appendages highly modified to form jaws surrounding the mouth; 2) third pair of appendages form stubby projections (oral papillae); 3) specialized glands (slime glands) discharge adhesive material through openings on the oral papillae; 4) subcutaneous hemal channels beneath the cuticle, forming part of the animal's hydrostatic skeleton

Onychophorans possess some characteristics that are clearly annelid in nature and some that are clearly arthropod in nature. All members are free-living and are clearly protostomous coelomates. More than 100 species have been described, with *Peripatus* being the best-known genus (Fig. 15.4).

3. See *Topic for Further Discussion and Investigation,* at the end of the chapter.

All modern onychophorans are terrestrial—although fossil onychophorans are primarily from marine sediments—and most are found in moist habitats in tropical environments and in southern temperate regions (e.g., New Zealand). Indeed, they appear to be restricted to such environments largely because they possess a thin, nonwaxy cuticle that does not deter evaporative loss of body water. Perhaps this danger of dehydration is also a factor in making onychophorans exclusively nocturnal (active only at night). Some species are carnivores (feeding particularly on various smaller arthropods), some are herbivores, and some are omnivores. Onychophoran predators attack their prey from a distance by shooting a proteinaceous glue from large "slime glands" opening at the tips of 2 specialized oral protuberances, the oral papillae. Once the victim is sufficiently entangled in the glue, the onychophoran bites through any protective coverings, secreting into the tissue substances that kill the prey and partially liquify the tissues (Research Focus Box 15.1). Onychophorans also eject glue as a defense against predators.

Individual onychophorans may grow to 15 cm (centimeters) in length. Onychophorans possess the following annelid-like characteristics: (1) body wall musculature is smooth and composed of longitudinal, circular, and diagonal elements (Fig. 15.4b); (2) a single pair of feeding appendages (jaws) is present; (3) no appendages are jointed; (4) a hydrostatic skeleton plays a role in locomotion (as described later); (5) one pair of nephridia is found in most segments; (6) light receptors are ocelli rather than compound eyes; (7) the outer body wall is deformable; and (8) spermatozoan morphology resembles that of

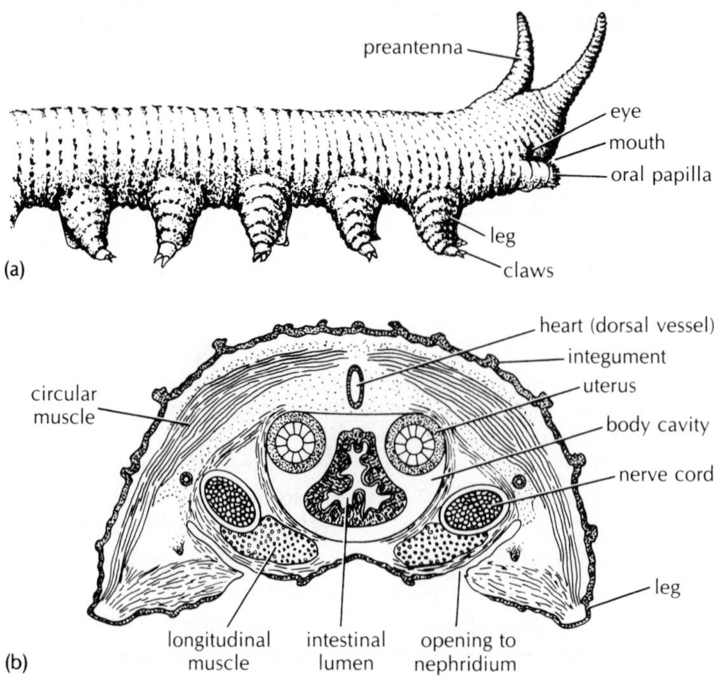

Figure 15.4
(a) The onychophoran *Peripatus,* in lateral view. (b) Same animal, seen in diagrammatic cross section.

(a,b) Sherman/Sherman, *The Invertebrates: Function and Form,* 2/e, © 1976, p. 169. Reprinted by permission of Prentice Hall, Upper Saddle River, New Jersey.

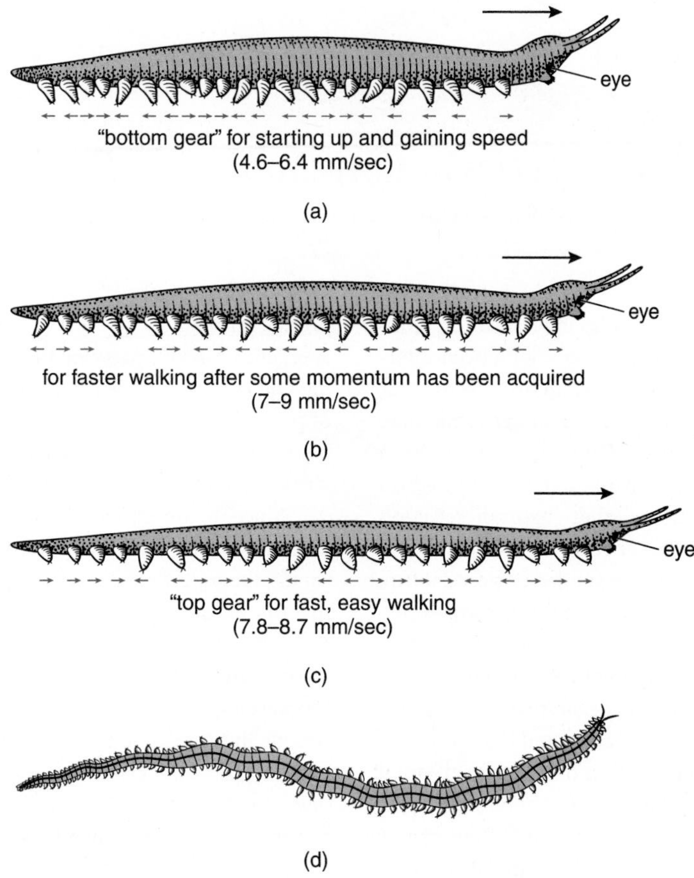

Figure 15.5

(a–c) The 3 most common gaits observed during locomotion of the onychophoran *Peripatopsis sedgwicki.* Shifts in walking speeds are associated with changes in body length and width. Sufficient speed must be obtained using gaits (a) and (b) before the animal can switch to gait (c) for prolonged, rapid walking. (d) Locomotion of the polychaete *Nereis diversicolor* (Annelida: Polychaeta), shown for comparison. The locomotory appendages (parapodia) project laterally, and waves of limb activity pass down the length of the body. Note that, for each segment, the left parapodium and right parapodium are not in the same phase of the activity cycle. Among polychaetes, waves of longitudinal muscle contraction aid in generating forward thrust.

(a,c) From Manton, in *Journal of the Linnean Society (Zoology)*, 41:529, 1950. Copyright © Academic Press, Ltd., London. Reprinted by permission. (d) Based on Gray, in *Journal of Experimental Biology*, 16:9, 1939.

oligochaetes and leeches. The following characteristics are arthropod-like: (1) the jaw musculature is striated; (2) the cuticle contains chitin; (3) the main body cavity is a hemocoel, not a true coelom; (4) gas exchange is achieved by spiracles opening into a tracheal system (the spiracles cannot be closed, however—another factor preventing the onychophorans from invading drier habitats); (5) mouth appendages are mandible-like; (6) the heart bears ostia; (7) legs are extended by hemocoelic pressure rather than by direct muscular contraction (as also found for the legs of arachnids and merostomes and the maxillae of butterflies); (8) excretory organs closely resemble the green glands (= antennal glands) of crustaceans; and (9) adhesive defense secretions are reminiscent of those produced by repugnatorial glands of the centipedes and millipedes (myriapods). The onychophoran nervous system is of the annelid/arthropod type: segmented, with a pair of ventral nerve cords.

A few characteristics separate the onychophorans from both the arthropods and the annelids. The head appendages consist of a pair of antennae, a pair of jaws, and a pair of oral papillae. These are followed by a series of unjointed walking legs that are quite unlike the parapodia of polychaete annelids, both structurally and functionally. Blood pigment is lacking.

Studies of several onychophoran species have shown onychophoran locomotion to be unique. Propulsion is generated directly by the musculature of the limbs themselves, with the body remaining rigid, serving as an anchor point against which the limbs can operate. Of course, in the absence of a solid skeleton, body rigidity is a function of the body wall musculature. The limbs (about 20 pairs are typically found) project ventrolaterally, elevating the body above the ground. This is quite different from polychaete parapodia, which project laterally, leaving the body surface in direct contact with the substrate.

As in polychaetes, waves of limb activity pass down the length of the onychophoran body, and several waves are generally progressing concurrently (Fig. 15.5). In the preparatory stroke for any given onychophoran limb, the tip of the leg is raised above the substrate and extended forward. On the backstroke (power stroke), the tip of the leg is

RESEARCH FOCUS BOX 15.1

The Energetics of Onychophoran Feeding

Read, V. M. St. J., and R. N. Hughes. 1987. Feeding behaviour and prey choice in *Macroperipatus torquatus* (Onychophora). *Proc. Royal Soc. London B* 230:483–506.

Most animals feed only on certain types of foods or on foods within a certain size range. Ecologists have long been interested in documenting the relative costs and benefits of feeding on particular foods in an effort to understand these species-specific food preferences. One explanation for a predator specializing on prey of particular size is based on the feeding-efficiency argument that natural selection should favor feeding on prey that provide the greatest caloric content for the amount of energy expended in prey capture and ingestion. Carnivorous onychophorans seem well suited to feeding-efficiency studies: They are slow-moving for their size (about 4 cm per minute) and rarely travel more than a meter from their burrows, which means that experiments can easily be performed in the laboratory and that prey capture must account for a good fraction of the animal's total daily energy expenditure. In addition, onychophorans typically eat only one victim per night and ingest the soft tissues completely, facilitating measurements of energy intake. Also, the primary cost of food collection is the proteinaceous glue needed to entangle the prey—a single, measurable entity.

Read and Hughes (1987) examined the energetic efficiency of food capture for a single species of onychophoran (*Macroperipatus torquatus*) found in the rain forests of Trinidad. To estimate the net amount of energy gained from feeding, the biologists first needed to determine the cost of prey capture, as estimated from the amount of glue expelled in capturing prey of different sizes. To do this, they weighed individual prey and put each into a small box with onychophorans. Once an onychophoran attacked and completely subdued its intended victim, the researchers scraped any excess glue from the bottom of the box and weighed it along with the glue-entangled prey to assess the amount of glue used in the attack. They also weighed each predator to determine how the use of glue varied with predator size. A single attack typically used up 40–50%, and as much as about 80%, of an individual's available glue supply.

Equating the amount of glue released with the cost of feeding may overestimate the cost of prey capture: In eating the prey, the predator also consumes some of the glue, so this protein is reclaimed, not lost. On the other hand, ignoring the energetic costs of *producing* the glue and of physically locating and consuming the predator may underestimate the cost of prey capture. On balance, the procedure may well produce a very reasonable estimate of capture costs, assuming that the 2 inaccuracies are approximately equal in magnitude.

Since the onychophoran must produce glue in order to eat, the amount of time required for an individual to replenish its glue supply after feeding determines how often the animal can indulge. To estimate the rate of glue replenishment, the biologists first had to determine how much glue a fully loaded onychophoran possesses and then how long it takes the onychophoran to produce that much glue. How would you go about making these determinations? The researchers chose a simple but clever approach. They induced individual onychophorans to squirt all of their glue at preweighed pieces of aluminum foil, and then they reweighed the foil. The onychophorans were then allowed to restore their glue supplies for different amounts of time, and the extent of restocking was again assessed by force-firing the animals at preweighed aluminum foil. The researchers found that these onychophorans needed more than 5 weeks to fully reload (Focus Fig. 15.1).

To measure the energetic benefits of feeding on arthropod prey of different sizes, Read and Hughes weighed prey before they were offered to an onychophoran of known weight, and then reweighed what remained of the prey after the onychophoran had concluded its meal; weight loss reflects food consumption. Typically, the remains consisted only of the prey's external cuticle.

A priori, it might seem advantageous for an onychophoran to attack any animal it can capture, but such is not the case. For one thing, more glue is needed to subdue a larger animal, so the cost of predation rises for larger prey. The larger animal also has a greater chance of escaping with some of the predator's precious glue and providing nothing in return. In addition, the predator has a gut of finite capacity; it does not make energetic sense to expel extra glue to capture something too big to fully ingest. Finally, the onychophoran that exhausts

applied to the substrate, and the leg is held straight and stiff. The tip of the limb remains stationary while the limb musculature contracts, sweeping the body past the point of contact between the limb and the substrate. As the body progresses past the stationary end of the leg, the leg shortens until it is essentially perpendicular to the substrate and then elongates as the portion of the body wall above the limb continues to move forward. Because of these well-timed changes in the limb length, the body itself shows little undulation, changing position neither laterally nor vertically with respect to the ground. If the legs could not be seen, the body would appear to be gliding. The jointed

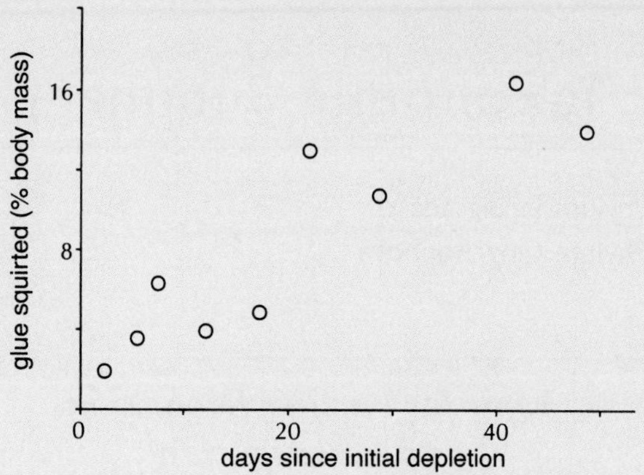

Focus Figure 15.1

Rate at which the onychophoran *Macroperipatus torquatus* replenishes its supply of glue. The animals were forced to expel all of their glue on day 0. The amount of glue held on subsequent days is shown here as a percentage of the animal's body mass.

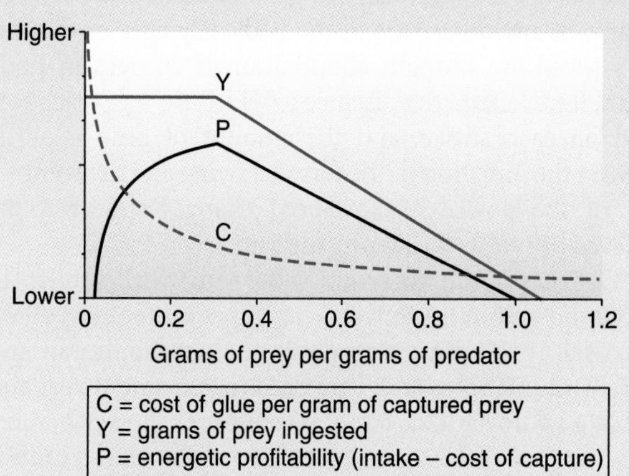

Focus Figure 15.2

The relationship between yield of prey and cost of capture for prey and predators of different sizes. Individual data points have been omitted for clarity.

Based on V. M. St. J. Read and R. N. Hughes, "Feeding Behaviour and Prey Choice in *Macroperipatus torquatus* (Onychophora)" in *Proceedings of the Royal Society of London B* 230:483–506, 1987.

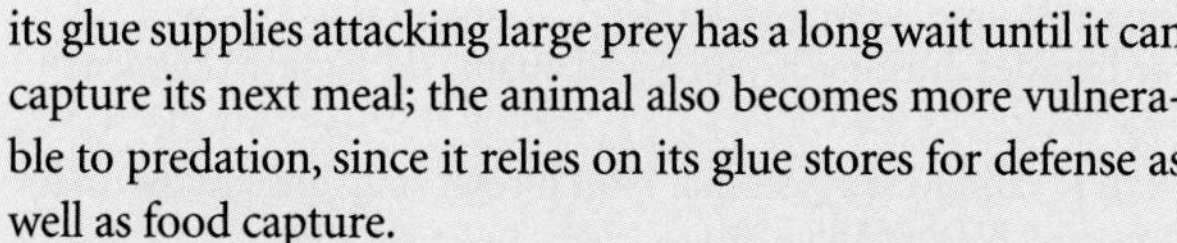

its glue supplies attacking large prey has a long wait until it can capture its next meal; the animal also becomes more vulnerable to predation, since it relies on its glue stores for defense as well as food capture.

Focus Figure 15.2 summarizes some of the key relationships concerning the profitability of prey capture. The *X* axis shows the weight of the prey relative to the weight of the predator. A value of 1.0 means that the predator and prey are equal in weight. A value of 0.5 indicates that the prey weighs half as much as the predator.

As shown in Focus Figure 15.2 (curve *C*), the amount of glue that must be expended to capture each gram of prey declines exponentially for larger and larger prey. Although it takes more glue to capture a larger animal, the amount of glue needed does not increase directly with the size of the prey; it increases more slowly. For example, if an onychophoran weighing 3 g (grams) requires 0.25 g of glue to subdue an animal half its size and 0.30 g of glue to subdue an animal of its own size, then the relative amount of glue required in the first case (with the smaller prey) is 0.25 g glue/1.5 g prey = 0.17 g glue used per gram of prey, but only 0.3 g glue/3.0 g prey = 0.1 g glue used per gram of prey in the second case, with the larger prey.

Curve *P* (Focus Fig. 15.2) shows the difference between the amount of energy extracted by a predator feeding on prey of different sizes and the approximate energy content of the glue expended in prey capture. The net energy obtained declines dramatically for heavier prey because of the predator's limited gut capacity; more glue is needed to capture larger prey, and excess prey tissue (curve *Y*) simply cannot be ingested. Curve *P* shows that it is energetically most advantageous for an onychophoran to feed on prey about 0.2 to 0.6 times its own weight. Attacking much larger or much smaller prey is not nearly as efficient.

Additional laboratory studies showed that these onychophorans did indeed grow more slowly when reared exclusively on very small or very large prey and that, when given a choice, they generally did not capture the smallest prey. It seems reasonable to assume that natural selection has encouraged a predilection for feeding on prey that return the greatest, or at least a substantial, amount of the energy invested in prey capture.

legs of arthropods also show such a shortening during movement, but the mechanism by which this is accomplished is necessarily quite different.

As so far described, the hydrostatic skeleton appears to have a rather passive role in onychophoran locomotion. However, a more active involvement has been demonstrated through analysis of film footage. As the animal moves, the body's length and width are continually changing; indeed, body dimensions rarely remain constant for more than a few seconds. Changes in body length are correlated with changes in speed and gait (i.e., the manner of walking) (Fig. 15.5). These alterations of body dimensions are brought about through the actions of the circular and longitudinal musculature, interacting through a constant-volume, internal hydrostatic skeleton (the hemocoel). Because onychophorans have no internal septa, contractions in one part of the body can bring about rapid distension, or rigidity, in any other part of the body. Changes in the animal's speed are brought about through changes in body length, limb length, the distance to which the legs reach forward on each stroke, and the amount of time required between the initiation of the forward swing and the completion of the power stroke. Speed changes appear to be achieved primarily by altering the gait.

Onychophoran locomotion is somewhat similar to that found within the Polychaeta, in that outfoldings of the body wall (projecting ventrally in the onychophoran and laterally in the polychaete) are used to generate thrust and in that a hydrostatic skeleton is implicated. However, there are important differences. First, the onychophoran limb is stiffened not by internal, rigid acicula, but rather by the intrinsic musculature of the limbs themselves and by hydrostatic forces generated by contraction of the body or limb musculature against the fluid-filled hemocoel. Moreover, the onychophoran body does not undulate; all thrust is generated by the limbs directly. In contrast, among polychaetes, contractions of the longitudinal body wall musculature may be used to generate waves of body undulation (Fig. 15.5d). These undulations transmit additional thrust against the substrate through the stationary parapodia. Indeed, when polychaetes are really making headway, most of the progress is attributable to contractions of the body wall musculature rather than to the parapodial elements directly. An additional difference between the locomotory mechanics of the 2 groups is that the polychaete limb does not change length during the propulsive cycle as does the onychophoran limb. These locomotory differences support the contention that if onychophorans arose from annelids or annelid-like ancestors, they probably did not arise from a polychaete ancestor.

The evolutionary history of the Onychophora is unknown. Fossils of marine, onychophora-like animals such as *Aysheaia* (Fig. 15.1) from the mid-Cambrian period (approximately 525 million years ago) have been described, but it is impossible to tell whether these fossilized animals were more like arthropods or more like annelids. Existence of these fossils suggests, at least, that the evolution of onychophoran-like limbs predated movement of the group to land; limb development can thus be viewed as a preadaptation for a terrestrial existence in this group. But little else about the evolutionary history of onychophorans is even hinted at in examination of other animals, either living or extinct. Certainly, there has been no consensus about the degree of relationship between the Onychophora and the Arthropoda; both morphological and molecular evidence argue increasingly that onychophorans are an early offshoot from an ancestor that eventually gave rise to both tardigrades and arthropods. Whether this ancestor, or a different lobopod ancestor, also gave rise to the annelids is a very much unsettled issue.

Taxonomic Summary

Phylum Tardigrada
Phylum Onychophora

Topic for Further Discussion and Investigation

Investigate the tardigrades' tolerance of low temperature, high pressure, and desiccation.

Crowe, J. H. 1972. Evaporative water loss by the tardigrades under controlled relative humidities. *Biol. Bull.* 142:407.

Crowe, J. H., and A. F. Cooper. 1971. Cryptobiosis. *Sci. Amer.* 225:30.

Jönsson, K. I., and L. Rebecchi. 2002. Experimentally induced anhydrobiosis in the tardigrade *Richtersius coronifer:* Phenotypic factors affecting survival. *J. Exp. Zool.* 293: 578.

Pigón, A., and B. Weglarska. 1955. Rate of metabolism in tardigrades during active life and anabiosis. *Nature (London)* 176:121.

Seki, K., and M. Toyoshima. 1998. Preserving tardigrades under pressure. *Nature* 395:853.

Westh, P., and H. Ramløv. 1991. Trehalose accumulation in the tardigrade *Adorybiotus coronifer* during anhydrobiosis. *J. Exp. Zool.* 258:303.

Taxonomic Detail

Phylum Tardigrada

The approximately 900 species are distributed among 3 classes.

Class Heterotardigrada

Although most species live in terrestrial or freshwater habitats, many are marine, often living interstitially in the spaces between sand grains. Nine families. About 345 species.

Family Echiniscidae. *Echiniscus* (a genus containing nearly 25% of all tardigrade species). All members of the family are terrestrial or freshwater inhabitants, and a few species occur only in South America at altitudes exceeding 1000 m. Many species are brightly colored: red, yellow, or orange.

Class Mesotardigrada

Thermozodium esakii. The single species in this class is known only from hot springs (65°C) in Japan. The hot springs were destroyed in an earthquake and the tardigrades have not been seen since.

Class Eutardigrada

This class includes about 435 species from marine, freshwater, and terrestrial habitats. Many of the marine species have been described over the past several years. Eutardigrades are divided between 2 orders based on differences in claw morphology.

Order Parachela

Macrobiotus, Minibiotus, Hypsibius. At least 6 families, containing about 430 species.

Order Apochela

Milnesium, Limmenius. These tardigrades are exclusively terrestrial. Some species are carnivorous, eating rotifers, nematodes, and other tardigrades inhabiting the mosses and lichens on which they live.

Phylum Onychophora

The 70 species are distributed among 2 families.

Family Peritopsidae. *Peripatoides, Peripatopsis. Peripatoides* occurs only in Australia and New Zealand, while other family members live in parts of Africa, South America, and New Guinea.

Family Peripatidae. *Peripatus.* These best-known onychophorans are widespread in subtropical lands.

General References about the Tardigrades and Onychophorans

Harrison, F. W., and M. E. Rice, eds. 1993. *Microscopic Anatomy of the Invertebrates, Vol. 12: Onychophora, Chilopoda, and Lesser Protostomata.* New York: Wiley-Liss.

Kinchin, I. M. 1994. *The Biology of Tardigrades.* London: Portland Press.

Nelson, D. R. 2002. Current status of the Tardigrada: Evolution and ecology. *Integr. Comp. Biol.* 42: 652–59.

Nichols, P. B., D. R. Nelson, and J. R. Garey. 2006. A family level analysis of tardigrade phylogeny. *Hydrobiol.* 558: 53–60.

Parker, S. P., ed. 1982. *Classification and Synopsis of Living Organisms, Vol. 2* (Tardigrada and Onychophora). New York: McGraw-Hill, 729–30, 731–39.

Pollock, L. W. 1975. Tardigrada. *Reproduction of Marine Invertebrates,* Vol. II. Giese, A. C., and J. S. Pearse, eds. New York: Academic Press, 43–54.

Thorpe, J. H., and A. P. Covich, eds. 2001. *Ecology and Classification of North American Freshwater Invertebrates,* 2nd ed. New York: Academic Press.

Wright, J. C., P. Westh, and H. Ramløv. 1992. Cryptobiosis in Tardigrada. *Biol. Rev.* 67:1–29.

Search the Web

1. www.tardigrades.org

   This is the website for the Edinburgh Tardigrade Project, established in December 2003.

2. http://animaldiversity.ummz.umich.edu/site/index.html

   Search using the terms "tardigrada" or "onychophora" to see drawings of particular species.

16

The Nematodes

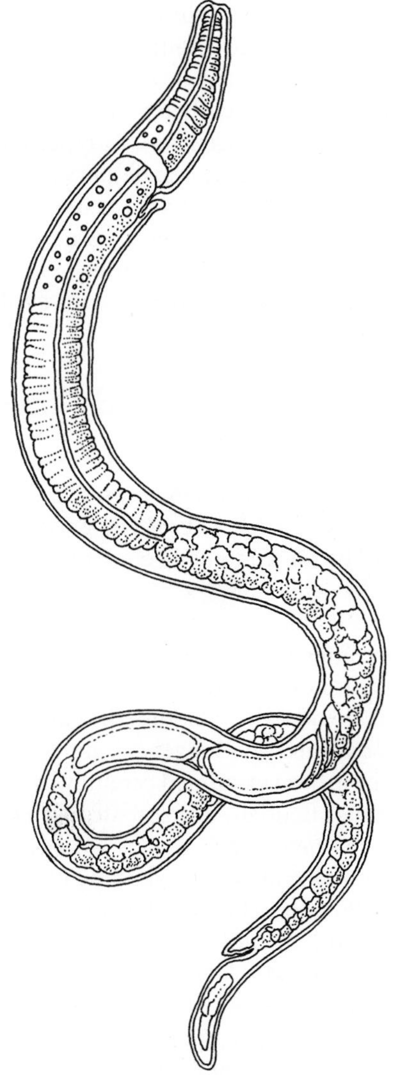

Introduction and General Characteristics

Phylum Nematoda
(Greek: thread)
nem-ah-tō´-dah

> "Nematodes are the haiku among multicellular animals, combining endless variation with a deceptively simple underlying anatomical pattern."
> Paul De Ley

Defining Characteristic:[1] Paired lateral sensory organs (amphids) on the head, derived from cilia and opening to the outside through a small pore

The nematodes are ubiquitous, unsegmented, acoelomate and pseudocoelomate worms. They are probably the most abundant multicellular animals alive today; nematode concentrations reach 1 million individuals per square meter in shallow-water sediments in both fresh- and salt water, and concentrations exceeding 4 million per square meter have been reported from some marine habitats. Someone once estimated about 90,000 nematodes in a single rotten apple. Other free-living nematode species are found on land, and still other nematode species are major parasites of vertebrates, invertebrates, and plants. About 16,000 species have so far been described, but authorities estimate that between about 100,000 and many millions of nematode species (mostly free-living species) may share our planet. The difficulty in describing and recognizing nematode species lies in the generally small size of these animals and in the considerable uniformity of both internal anatomy and external morphology. Often, species determinations must be based upon

1. Characteristics distinguishing the members of this phylum from members of other phyla.

Table 16.1 Proposed Relationships between Nematodes, Other Aschelminth Phyla, and the Arthropods

Phylum	"Aschelminths"	Cycloneuralia	Ecdysozoa
Rotifera	✓		
Acanthocephala	✓		
Gnathostomulida	✓		
Gastrotricha	✓	✓	
Nematoda	✓	✓	✓
Nematomorpha	✓	✓	✓
Kinorhyncha	✓	✓	✓
Priapulida	✓	✓	✓
Loriciferans ?	✓	✓	✓
Onychophora			✓
Tardigrada			✓
Arthropoda			✓

biochemical attributes or morphological details not readily visible, such as the size and placement of microscopic sensory structures or the morphological details of the male reproductive system.

In the literature, the Nematoda has sometimes been listed as a class within another phylum, the Aschelminthes ("cavity worms," in reference to their supposed possession of a pseudocoel). As noted earlier (Chapter 10), the major aschelminth groups (Nematoda, Rotifera, Gastrotricha, Kinorhyncha, Nematomorpha, Acanthocephala, Priapulida, and Gnathostomulida) are no longer believed to have a single ancestor in common, so each group has been elevated to phylum status.

The relationships among the various aschelminth phyla are still far from clear. Recent cladistic analyses of morphological data along with data from several gene sequences suggest that all of the aschelminths that molt their cuticles, including the nematodes, are more closely related to arthropods than to other aschelminths. It has been proposed, therefore, that Arthropoda, Nematoda, and 6 other phyla (Table 16.1) be considered together as the Ecdysozoa ("molting animals"). The molting aschelminth phyla had been grouped together previously, as the Cycloneuralia, but affiliating them with arthropods is a recent idea. A key implication of this arrangement is that the molting of a cuticle evolved only once in animal evolution. In support of this inference, ecdysone-like steroid hormones (which trigger molting among arthropods) have been reported from several nematode species, and the external application of the insect molting hormone, 20-hydroxyecdysone, is reported to stimulate molting in at least one nematode species. Additional studies along these lines should be forthcoming. Obviously it will be important to determine whether molting has essentially the same biochemical basis in all ecdysozoans. It will be equally important to examine the distribution of ecdysone-like steroid hormones outside of the ecdysozoan phyla, where they may play role(s) in physiology, development, or behavior unrelated to molting. For example, ecdysteroids have been detected in some parasitic flatworm species, where they apparently play a role in egg development and maturation (oogenesis). Ecdysteroids have also been found in some terrestrial snail species, but they seem to come from the diet rather than being synthesized by the snail. For the present, a close association between nematodes and arthropods seems to be generally accepted. We'll have to wait to see how well the association holds up under closer scrutiny.

Body Coverings and Body Cavities

The typical nematode is 1–2 mm long; shows no external segmentation; is tapered at both ends; and is covered by a thick, multilayered **cuticle** (a noncellular covering) of collagen secreted by the underlying epidermis (Fig. 16.1).

As pseudocoelomates, the members of at least some nematode species have a small internal body cavity lying between the outer body wall musculature and the gut, and the cavity is not lined with mesodermally derived tissue; the cavity may be derived from the embryonic blastocoel (p. 10), but this is no longer certain. Most nematode species that have been examined carefully lack a body cavity, and are thus essentially acoelomate. With or without a body cavity, the nematode's organs are never enveloped by **peritoneum** (the mesodermal lining of the coelomate body cavity). Pseudocoel fluid, when present, serves as the circulatory medium, and in some species it contains hemoglobin; nematodes lack a closed circulatory system of discrete blood vessels.

The epidermis of nematodes is often **syncytial;** that is, nuclei are not separated from each other by complete cell membranes. The cuticle of some nematode species is composed of a highly complex network of fibers that are virtually inelastic (Fig. 16.2a). The trellis-like arrangement of

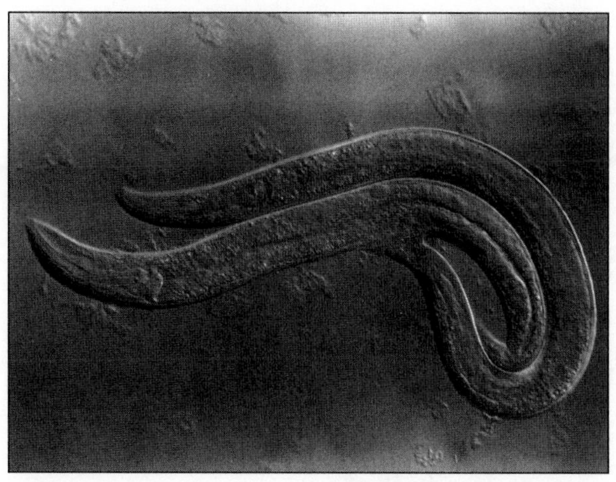

Figure 16.1
The free-living nematode *Caenorhabditis elegans*. Two individuals are shown mating.
Courtesy of Anne M. Villeneuve.

these fibers permits bending, stretching, and shortening of the cuticle (Fig. 16.2b). The cuticle is permeable to water and to gases, so gas exchange can take place across the entire body surface. On the other hand, since the cuticle offers little protection against dehydration, all free-living nematodes must live in water or at least in a film of water. The cuticle is also selectively permeable to certain ions and organic compounds, regulating the movement of these substances between the internal and external environment.

The cuticle is shed (molted) and resecreted 4 times during development from the juvenile to the reproductively mature adult. Within a species, the cuticle of each developmental stage may be structurally and biochemically distinct, providing one of several nematode models for studies of gene expression and its control during development. Unlike the situation in most arthropods (Chapter 14), nematodes continue to increase in size between molts and even after the final molt. But unlike

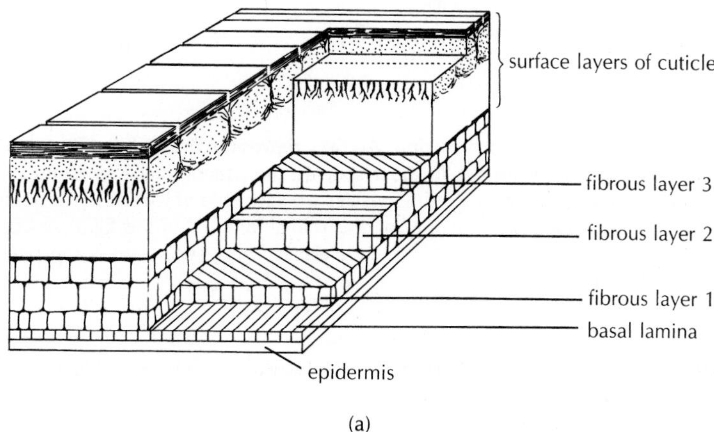

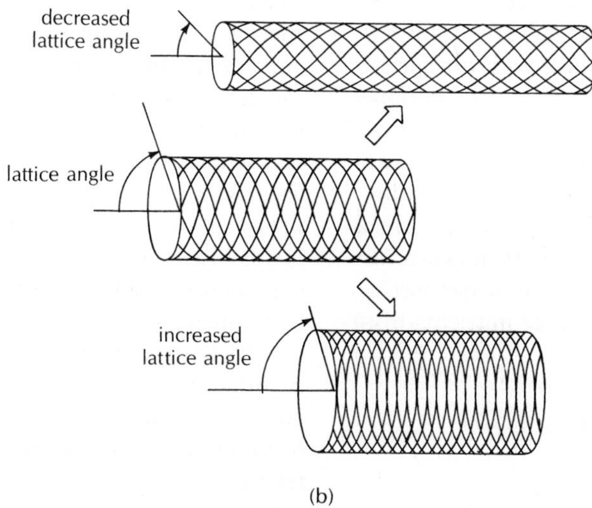

Figure 16.2
(a) Diagrammatic illustration of the multilayered nematode cuticle. Note the 3 layers of crossed fibers. (b) Although the collagenous fibers of the cuticle are individually inelastic, changes in the angle at which the fibers cross each other permit the animals to change shape.

(a) From A. F. Bird and K. Deutsch, "The structure of the cuticle of *Ascaris lumbricoides var suum*" in *Parasitology*, 47:319–28. Copyright © Cambridge University Press, New York. Reprinted by permission. 1957.

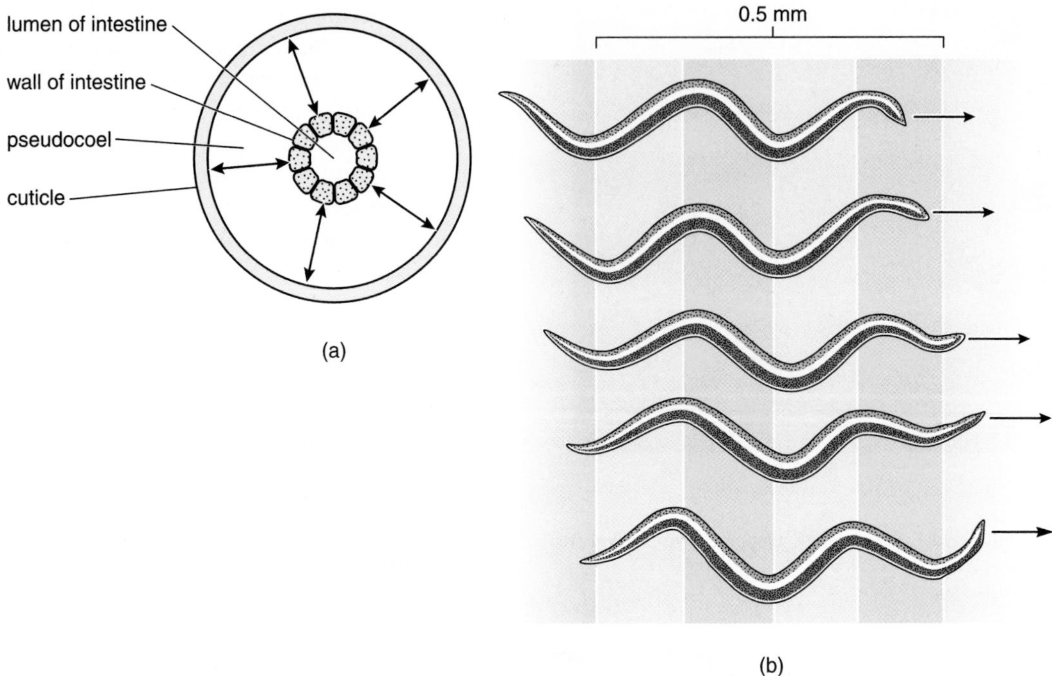

Figure 16.3

(a) Diagrammatic representation of a nematode in cross section. Arrows represent the high pressure within the pseudocoel, acting to maintain the rounded body shape and to collapse the gut. In most species, the pseudocoel, if present, is much smaller than illustrated. (b) Nematode locomotion on a solid surface. By contracting the muscles on each side of its body alternately, the animal forms a series of sinusoidal waves that thrust against the substrate surface, propelling the animal forward. In this figure, successive silhouettes of the nematode have been displaced below each other for greater clarity. The time elapsed between stages is approximately 0.33 seconds.

(b) Based on Gray and Lissman, in *Journal of Experimental Biology,* 41:35, 1964.

most other animals, nematodes grow mainly by increasing the size of individual cells rather than by increasing the number of cells; cell number in most tissues of adults is constant, a phenomenon referred to as **eutely.** Eutely is a classic "aschelminth" character, found also, for example, among rotifers, tardigrades, and gastrotrichs.

In some parasitic nematode species, the first 2 juvenile stages are free-living. In many of these parasitic species, the animal may become enclosed within 2 envelopes before completing the second molt. The outer envelope is termed the **sheath.** Escape from the sheath **(exsheathment),** the subsequent 2 molts, and development to adulthood often do not occur until the encapsulated form is eaten by a suitable host.[2] Once in the host, much of the cuticle may be lost and replaced by a microvillar surface, presumably facilitating uptake of soluble nutrients from the host.

Musculature, Internal Pressure, and Locomotion

The nematode body wall contains no circular muscles. This is most unusual for **vermiform** (worm-shaped) animals, and it places great limitations on their locomotory potential in that, for example, they cannot generate peristaltic waves of contraction.

Another obstacle to graceful locomotion is that the body is quite turgid in many species, due to substantial hydrostatic pressure within the pseudocoel (Fig. 16.3a). The internal pressure within nematodes of at least some species averages 70–100 mm Hg (millimeters of mercury), perhaps 10 times higher than pressures reported for most other invertebrates; recorded pressures within the common earthworm range from about 5–10 mm Hg, while those within a common polychaete annelid (*Nereis* sp.) are less than 1 mm Hg. Internal hydrostatic pressures as high as 225 mm Hg have been measured inside members of some nematode species. (For comparison, atmospheric pressure is approximately 760 mm Hg.)

At least 2 factors contribute to generating and maintaining high pressures within nematodes. First, the cuticle cannot expand to relieve pressure. Second, the musculature is always in a partially contracted state, trying to compress an incompressible fluid. The high internal pressure gives the nematode a very circular cross section. Thus, nematodes are commonly called "roundworms."

The rigid cuticle, high internal pressure, and lack of circular muscles preclude generation of pedal waves or peristaltic waves for locomotion. Nematodes, like

2. See *Topics for Further Discussion and Investigation,* no. 2, at the end of the chapter.

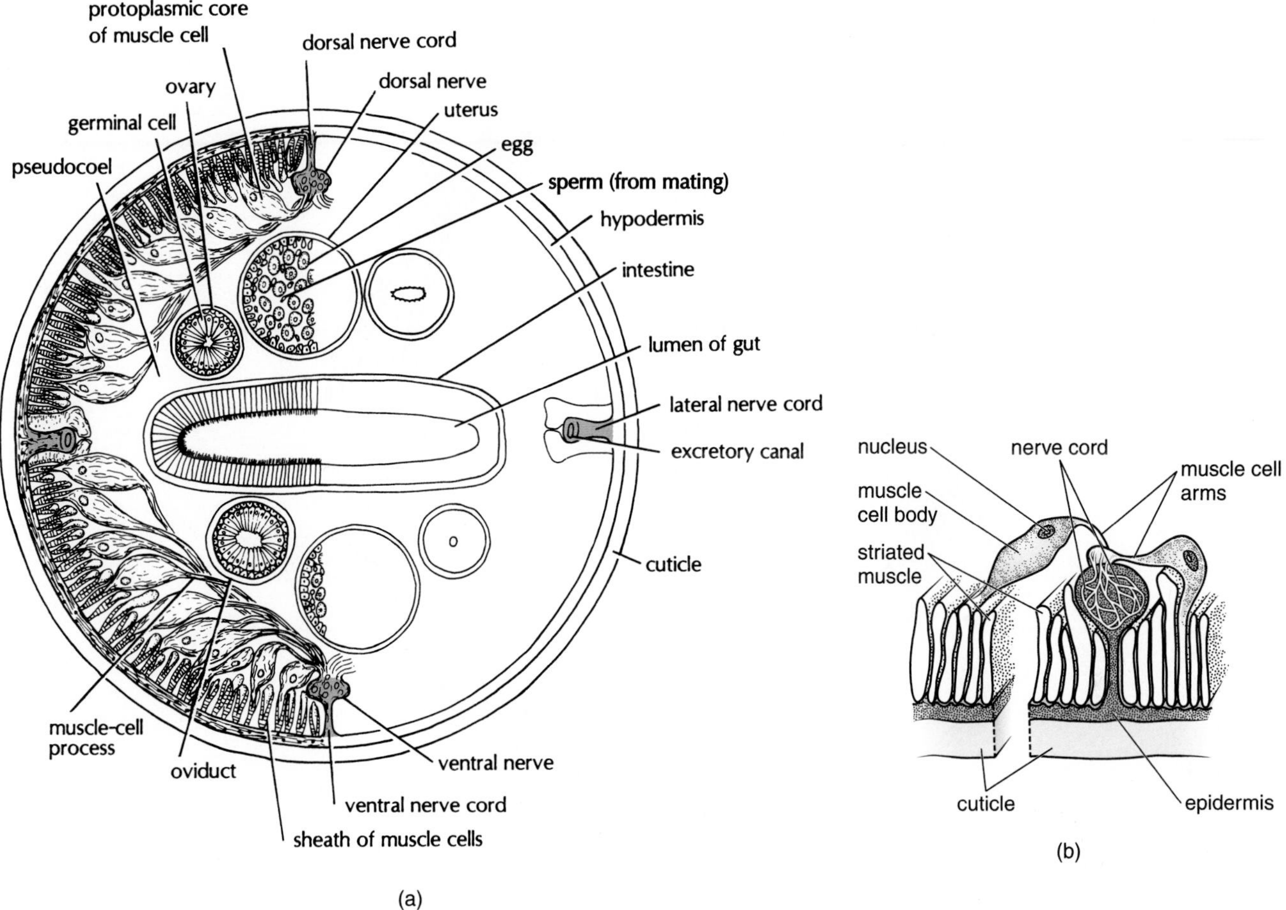

Figure 16.4

(a) *Ascaris lumbricoides* in cross section. The processes sent out from the muscle fibers to the nerve cords are clearly illustrated. For clarity, muscle cells are shown only in half of the figure. (b) Detail of nematode muscle cell from *Ascaris suum*. Note the long, noncontractile extension of the muscle cell leading from its position in the body wall to one of the major nerve cords.

(a) From Frank Brown, *Selected Invertebrate Types*. Copyright © John Wiley & Sons, Inc., New York. Reprinted by permission of Mrs. Frank Brown. (b) Based on J. T. Debell, in *Quarterly Review of Biology*, 1965, 40:233–51. 1965.

arthropods, completely lack locomotory cilia, so ciliary movement is also an impossibility. Instead, nematodes generally must move by undulating the body into sinusoidal waves, through alternating contractions of the longitudinal muscles on the dorsal and ventral surfaces of the body[3] (Fig. 16.3b). Contracting one set of muscles causes bending and stretches out muscles elsewhere in the body. Thus, muscles antagonize each other by means of pressure changes transmitted through the fluid skeleton of the pseudocoel (or of the internal cells themselves, for species lacking a pseudocoel), according to the basic principles of the hydrostatic skeleton (Chapter 5). Reextension of contracted muscles may be aided by the stiff cuticle surrounding the animal and the high pressure within the animal, both acting to spring the body back to a linear configuration when the muscles relax. Clearly, nematode design is not well suited to a free-swimming existence. Instead, most free-living nematodes live in soil, in aquatic sediments, in fruits, on surface films, and in other similar situations where either the substrate or the surface tension of a fluid at the air-water interface can provide resistance against which the animals can generate thrust.

Muscle contraction is controlled by a simple nervous system consisting of an anterior brain (nerve ring plus associated ganglia) and 4 or more major longitudinal nerve cords (ventral, dorsal, and at least one pair of lateral nerve cords) (Fig. 16.4a). Strangely, the nerve cords seem not to send out processes innervating the muscles. Rather, noncontractile extensions of the muscle fibers hook up to the nerve cords (Fig. 16.4b). While not unique to nematodes—a similar arrangement has been found in some flatworm and echinoderm species, for example—this pattern of innervation is certainly unusual, and it characterizes all nematode species examined to date.

3. See *Topics for Further Discussion and Investigation*, no. 1.

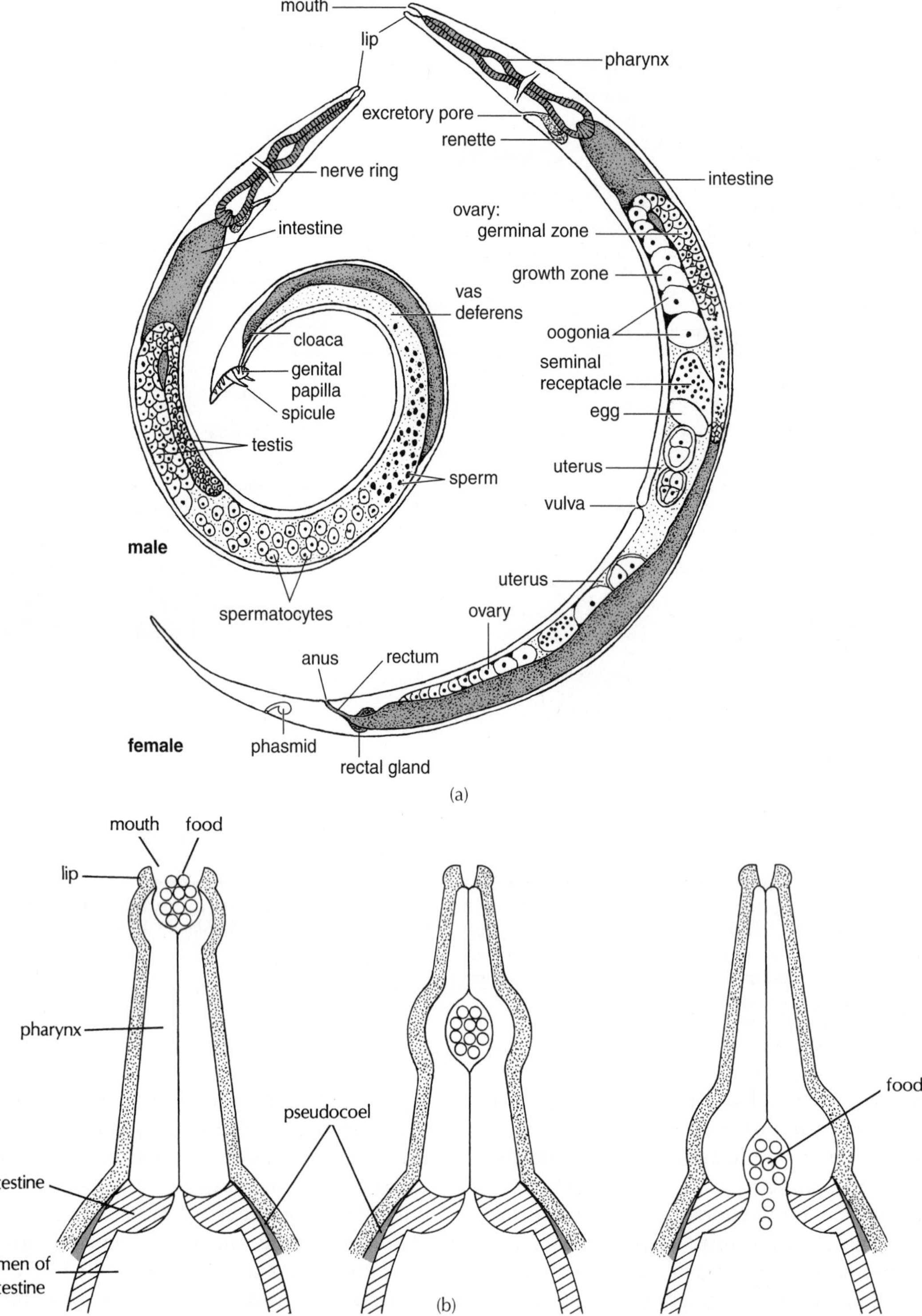

Figure 16.5

(a) Internal anatomy of a free-living nematode, *Rhabditis* sp. (b) Sequence of movements of the pharynx musculature during swallowing.

(a) From Cleveland P. Hickman, *Biology of the Invertebrates,* 2d ed. Copyright © 1973. The C. V. Mosby Company, St. Louis, Missouri. Reprinted by permission. (b) After Sherman and Sherman; after Clark.

Organ Systems and Behavior

High internal pressures pose potential digestive as well as locomotory difficulties. Nematodes have a linear digestive system, with a mouth **(stoma)** at the anterior end leading, in sequence, through a muscular pharynx, intestine, and rectum, and then an anus located near the body's posterior end (Fig. 16.5a). The difficulty in processing food lies in preventing the high pressure of the surrounding pseudocoel from collapsing the tubular, nonreinforced

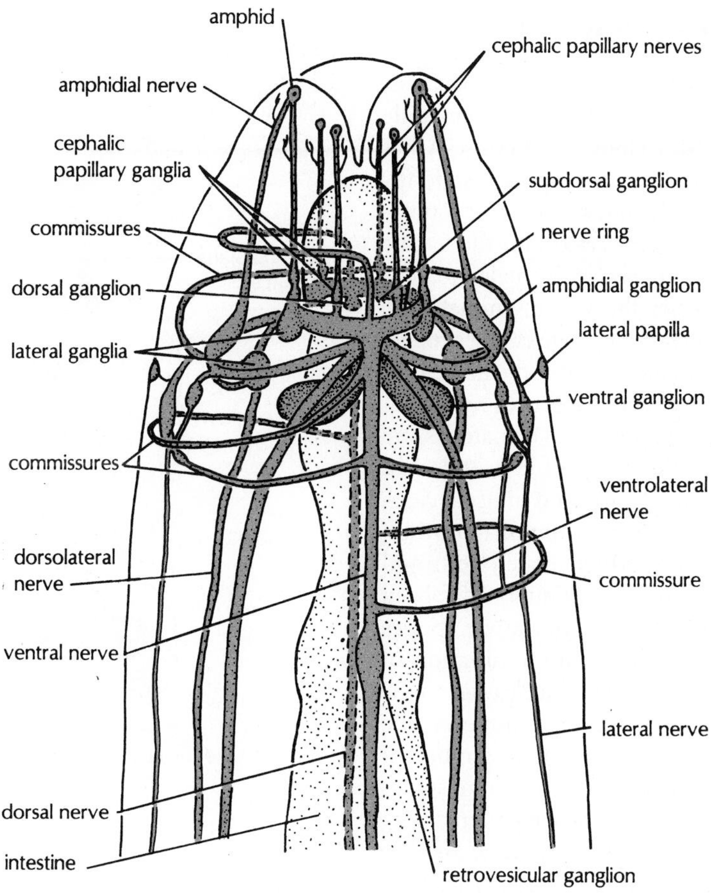

Figure 16.6

Nervous system of a nematode. Note in particular the brain ganglia; lateral, dorsal, and ventral nerve cords; and innervation of the amphids.

Modified from Brown, 1950. *Selected Invertebrate Types*. New York: John Wiley & Sons. After Hyman; after Chitwood and Chitwood.

digestive tract (Fig. 16.3a). The inner surface of the nematode gut is usually not lined by cilia. Indeed, ciliary beating probably would not be very effective in countering the high positive pressure exerted on the gut by the fluid in the pseudocoel. Instead, the gut is kept open by the activity of a highly muscular pharynx, which pumps in fluid at rates of up to 4 pulses per second (Fig. 16.5b). At the animal's posterior end, the high pressure in the pseudocoel keeps the terminal end of the digestive tract tightly closed. A dilator muscle at the anus opens to discharge wastes.

In addition to its role in keeping the gut lumen open against the high pressure of the pseudocoel, the pharynx (and associated glands) also adds lubricants and digestive enzymes to the food. Digestion is mainly extracellular, and nutrients are absorbed by the very thin wall (only one cell thick) of the intestine. Wastes are voided from the anus at intervals, about once every 1 or 2 minutes.

The respiratory and excretory systems are easily described: There are no specialized organs for gas exchange, no specialized circulatory system, and no nephridia. Metabolic wastes are apparently discharged along with other materials leaving the gut, or they diffuse across the body wall. A glandular system, the **renette,** or a modification of this system is present in most nematodes. The renette system varies considerably in complexity among species. It is often referred to as an excretory system, but its actual function has never been convincingly documented.

Despite their small size and limited locomotory abilities, nematodes are capable of fairly sophisticated behaviors. Various species are known to respond to temperature, light, mechanical stimulation, and a variety of chemical cues, including those produced by other individuals of the same species **(pheromones).** The general body surface seems to be light-sensitive in some species, possibly reflecting direct sensitivity of underlying nerve fibers. Many species have simple, pigmented light receptors **(ocelli)** as well. The major chemosensory organs, called **amphids** (Fig. 16.6), are anteriorly located pits lined with highly modified, nonmotile cilia **(sensillae).** Similar structures, called **phasmids,** are located at the posterior ends of some nematodes, and these are also thought to be chemosensory. The anterior and posterior ends of the body often have **cephalic** and **caudal papillae,** respectively, which also contain modified cilia. These structures are arranged around the mouth or anus, and are believed to be sensitive to mechanical stimulation. Many species possess external setae at various locations on the body; these are also thought to be **mechanoreceptors.**

Reproduction and Development

Although many species can reproduce by parthenogenesis, without the need for a mating partner, most nematode species are **gonochoristic** (= **dioecious**—G: 2 houses); that is, they have separate sexes. Individuals copulate, so fertilization is internal. Typically, males possess one to 2 copulatory spicules posteriorly that are inserted into the female gonopore during sperm transfer (Fig. 16.5a, male), to hold it open, but in at least one species, the male inseminates the female by piercing her body wall. Cleavage is determinate (mosaic), so cell fates are permanently set at the first cleavage, as in protostomes. Certain other aspects of reproduction and development, while not unique to nematodes, are highly unusual. For example, the sperm of nematodes are amoeboid, rather than flagellated. A second unusual characteristic is the phenomenon of **chromosome diminution,** in which specific regions of DNA are deliberately destroyed during development. Chromosome diminution has been observed in at least a dozen parasitic nematode species (members of the Ascarididae). Following fertilization of the egg, the zygote undergoes a normal first cleavage to the 2-cell stage. Prior to the second cleavage, however, the chromosomes of one of the cells fragment, and much of the chromatin from the ends of the original chromosomes is destroyed. The pieces of the chromosomes remaining after fragmentation and disintegration replicate, and are distributed in normal fashion to the 2 daughter cells of the subsequent division (Fig. 16.7). The other 2 daughter cells **(germ cells,** or **stem cells)** of the 4-cell embryo retain the full chromosome complement. Chromosome diminution occurs several more times during the next few cleavages. By the 64-cell stage, only 2 cells retain the complete genetic information present at fertilization. These 2 cells give rise to the gonad and produce the gametes for the next generation. The remaining cells, which retain as little as about 20% of the original genome, produce all of the somatic (i.e., nongamete producing) tissues (Research Focus Box 16.1). Loss of genetic information by chromosome diminution also occurs in some insects, crustaceans, ciliated protozoans, and plants.

Yet a third peculiar feature of nematode development is the constancy of cell numbers **(eutely)** mentioned earlier. Once organogenesis is completed, mitosis ceases in all of the somatic cells of most tissues. Thus, further growth of these tissues is due not to increases in cell numbers, but to increases in cell size.

Finally, nematodes are among the few groups of invertebrates that lack a morphologically distinctive free-swimming larval stage. The young animals emerge from their sturdy egg coverings as miniature adults. However, many free-living species can enter a state of developmental arrest, called the **dauer larva,** at the second molt. The dauer "larva" (a juvenile stage, really) is essentially inactive, has a very low metabolic rate, and cannot feed. It may, nevertheless, live for many months until environmental conditions improve, at which time normal development resumes and the animal completes the final 2 molts to adulthood. The genetic basis for entry into and departure from the dauer state has been much studied over the past 15 years.

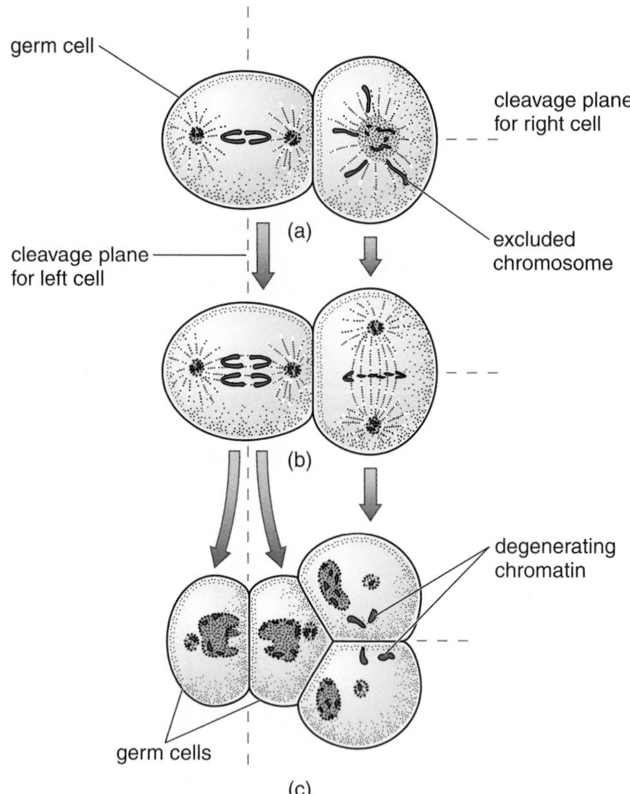

Figure 16.7

(a–c) Illustration of chromosome diminution in the parasitic nematode *Ascaris megalocephala.* In the second cleavage, only the chromosomes of the germ cell (= stem cell) undergo normal mitosis. Some of the chromosomal material in the other cell is excluded from participation in mitosis (a), and at the 4-cell stage (c) is seen to lie outside the nucleus. This material will later degenerate.

Based on *Biology of Developing Systems* by Philip Grant, 1978.

Parasitic Nematodes

Many nematodes are parasitic, and are modified accordingly. Nearly every major animal group, from sponges to mammals, plays host to some parasitic nematode species; nematodes of some species even parasitize members of other nematode species. According to a recent cladistic interpretation of gene sequence data,[4] parasitism arose independently among nematodes at least 7 times. Happily for biologists, much research on nematode biology has been driven by the need to control the potentially devastating impact of these parasitic species (Fig. 16.8). Nematodes parasitize humans, cats, dogs, and many domestic animals of economic importance, such as cows and sheep. Considerable

4. Blaxter et *al.* 1998. A molecular evolutionary framework for the phylum Nematoda. *Nature* 392:71–75.

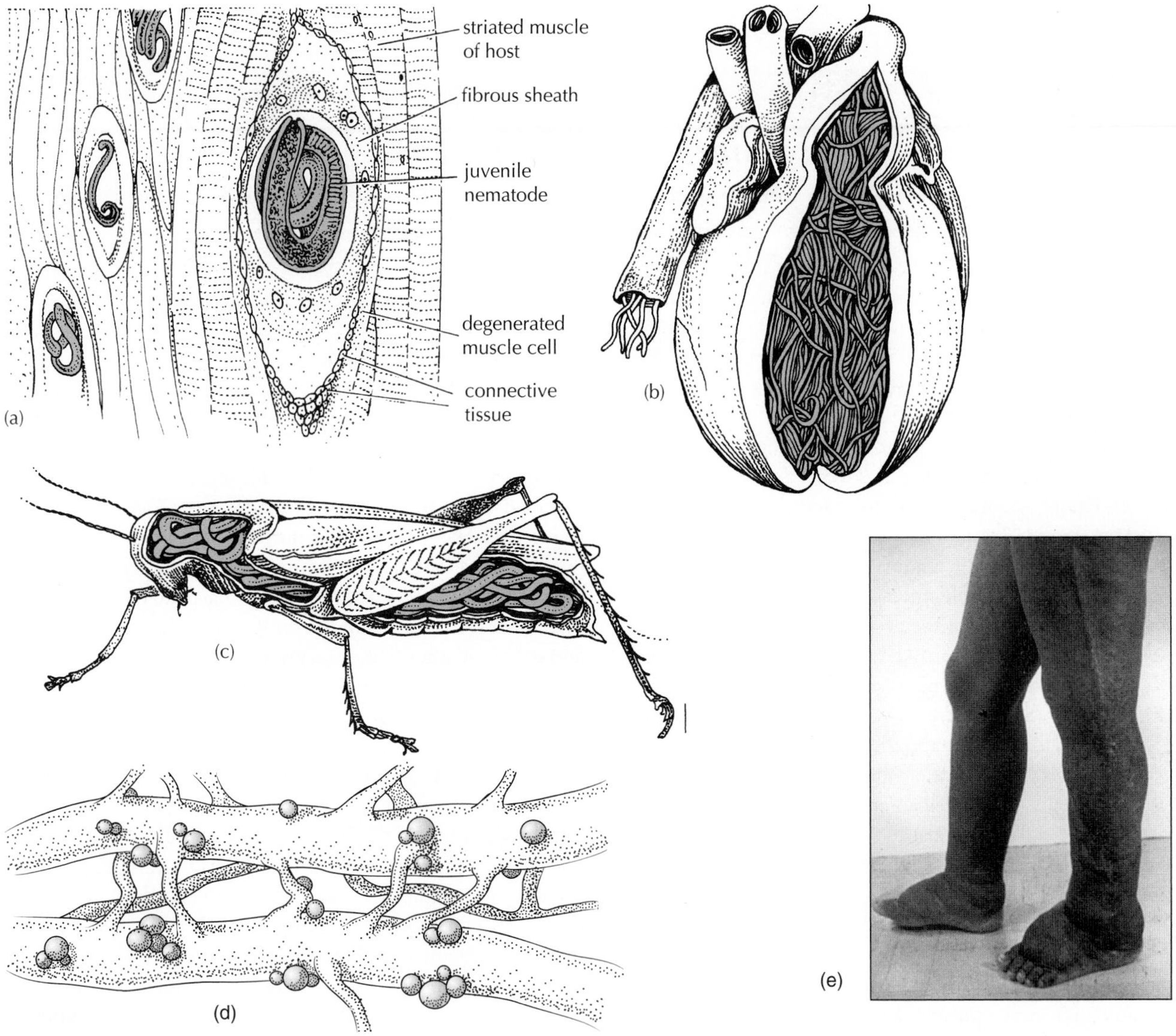

Figure 16.8
Examples of parasitism by nematodes and its effects on hosts. (a) *Trichinella spiralis* (responsible for trichinosis), shown encysted in vertebrate striated muscle. (b) Heart of a dog infested with *Dirofilaria immitis* (heartworm). (c) Cutaway view of a grasshopper infected with *Agamermis decaudata.* (d) Gall formation by nematodes (*Heterodera rostochiensis*) parasitic in plant roots. (e) Human male with elephantiasis of the legs.

(a) Modified from Brown, 1950. *Selected Invertebrate Types.* New York: John Wiley & Sons, and from Harmer and Shipley, *The Cambridge Natural History,* Macmillan. (b) From Harmer and Shipley. (c) From Hyman; after Christie. (d) Based on T. Cheng, *General Parasitology,* 2d ed. 1986. (e) Courtesy of the Centers for Disease Control, Department of Health and Human Services, Atlanta, GA.

veterinary and clinical research activity is now focused on how the various parasitic species suppress or otherwise outfox the host's immune response, and on why some individuals appear less susceptible to infection than others. Nematodes also parasitize the roots, stems, leaves, and flowers of plants, including species of great economic importance, such as soybeans, potatoes, oats, tobacco, onions, and sugar beets. Some of these parasitic nematodes attain great length, although they may be extremely thin. The largest nematode so far described is 9 m long and resides in the placenta of female sperm whales.

Hookworms and pinworms are 2 groups of nematodes well known to many humans (Fig. 16.9). The damage done

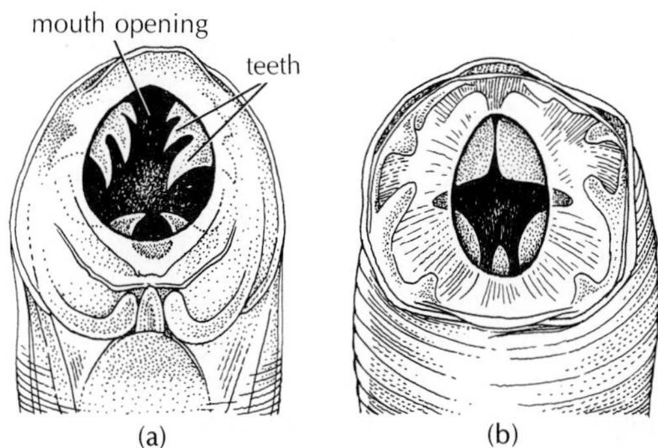

Figure 16.9
Hookworm mouth regions. (a) *Ancylostoma duodenale.* (b) *Necator americanus.*
From Chandler, *Introduction to Parasitology,* 8th ed. Copyright © 1949 John Wiley & Sons, New York. After Looss. Reprinted by permission.

RESEARCH FOCUS BOX 16.1

Nematode Development

Bennett K. L., and S. Ward. 1986. Neither a germ line-specific nor several somatically expressed genes are lost or rearranged during embryonic chromatin diminution in the nematode *Ascaris lumbricoides* var. *suum. Devel. Biol.* 118:141–47.

Perhaps a dozen parasitic nematode species are reported to undergo chromosome diminution during their embryonic development. In *Ascaris lumbricoides,* the somatic cells lose about 30% of their DNA by weight (dropping from 0.63 pg [picograms, 10^{-12} grams] of DNA per cell to 0.46 pg of DNA per cell), whereas the germ-line cells retain the full 0.63 pg of DNA per cell throughout their lives. What is the nature of the lost chromatin, and why is it lost from some cells but not others? More than 100 years ago, Theodor Boveri suggested that the somatic cells might eliminate genetic material needed only for germ-cell development—a very reasonable case of the somatic cells ridding themselves of unnecessary genetic baggage. Bennett and Ward (1986) attempted to determine whether DNA unique to the germ line is, in fact, completely eliminated from somatic cells as predicted. To do this, the researchers had to find a specific protein produced only by germ-line cells and then determine whether the genes coding for that protein were indeed absent from somatic cells.

Bennett and Ward chose to work with the Major Sperm Protein, a protein detected only in the sperm of both parasitic and free-living nematodes. The Major Sperm Proteins of different nematode species must be very similar, since antibodies prepared against the Major Sperm Protein from the free-living species *Caenorhabditis elegans* also react with the protein from *A. lumbricoides.*

The researchers obtained adult worms from a pig-slaughtering company and used standard procedures to isolate DNA from the sperm. This DNA is known to contain the gene coding for the Major Sperm Protein since the protein is synthesized by developing sperm. Bennett and Ward also isolated DNA from the oocytes; the first 2 larval stages, which lack significant gonadal tissue; and adult intestines, obtained by careful dissection. The trick then was to determine how germ-line DNA differed from that contained in the cells of the somatic tissues.

The DNA isolated from the various tissues was first chemically fragmented. The fragments from each tissue were then separated by size, using gel electrophoresis, and next denatured with a strong base, to separate the double-stranded DNA into separate strands. The DNA was then transferred from the gel to a nitrocellulose filter, by laying the filter atop the gel and allowing the DNA to blot onto the filter. The task then was to locate the particular DNA fragment coding for the Major Sperm Protein in testicular tissue and to see whether that same gene was present in the other tissues; if Boveri's hypothesis about gene elimination is correct, that gene would be absent from all samples except those prepared from sperm.

To locate the gene coding for the Major Sperm Protein, Bennett and Ward used purified messenger RNA that, when translated, produces that protein; they used this messenger RNA to synthesize a complementary DNA strand that was labeled with a radioactive tag. This synthesized DNA fragment will code for the Major Sperm Protein; making it radioactive enables researchers to recognize it again later. The radioactive DNA probe was then spread over the nitrocellulose filter, allowed to sit for a time, and then rinsed off; tissue DNA complementary to that radioactive DNA probe binds to it and will show up as a distinct band on X-ray film. If the X-ray shows such a band, the tissue DNA must contain the gene coding for the Major Sperm Protein; if no such band appears on the X-ray, the gene for the Major Sperm Protein must be absent from the sampled tissue. Focus Figure 16.1 clearly indicates that the Major Sperm Protein

by parasitic nematodes is generally indirect, resulting from competition with the host for nutrients. One hookworm, for example, may imbibe more than 0.6 ml (milliliters) of blood per day. Someone with a respectable infection of 100 hookworms would then be losing perhaps 60 ml of blood daily. Infections of 1000 hookworms per host are not uncommon. Other nematode species do most of their damage by becoming so densely packed in their preferred tissues that they block the flow of nutrients or fluids. Some ascarids (members of the genus *Ascaris*), for example, may block the host's intestine or bile duct. Other nematode species (e.g., *Wuchereria bancrofti* and *Onchocerca volvulus*) plug the lymphatic system, which sometimes results in a substantial buildup of fluid and subsequent dense growth of connective tissue in various regions of the body (Fig. 16.8e). The scrotum of one afflicted human male reportedly grew to a weight of about 18 kg (kilograms) before its removal by surgery. Less extreme forms of the disease (elephantiasis) presently afflict nearly 120 million people; elephantiasis is said to be one of the world's fastest-spreading diseases. In 1998, the World Health Organization and the pharmaceutical company Smithkline Beecham announced a collaborative attempt to eliminate elephantiasis by the year 2020.

Many nematodes that parasitize animals make extensive migrations during their development within the

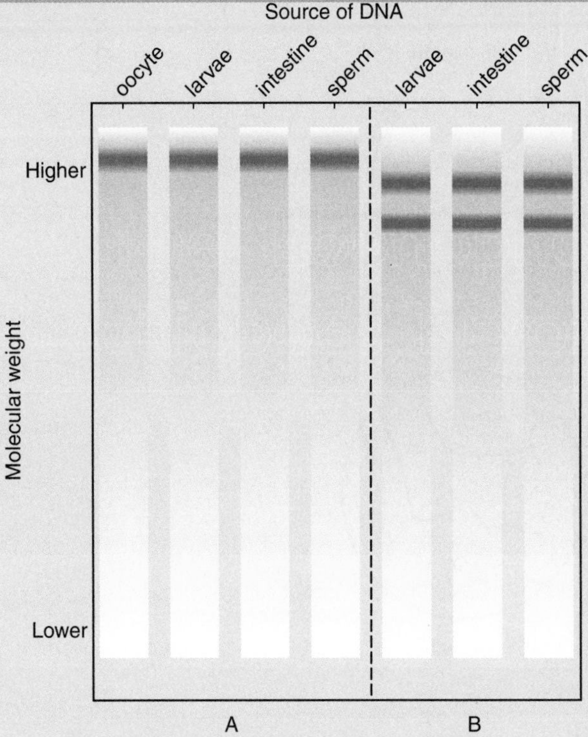

Focus Figure 16.1

Drawing made from X-ray film showing that the gene for nematode Major Sperm Protein is found in all nematode tissues examined. After the DNA fragments were transferred from an electrophoretic gel to a nitrocellulose filter, the filter was coated with radiolabeled DNA fragments coding specifically for the Major Sperm Protein. This radioactive DNA binds only to its complement on the nitrocellulose filter, and the unbound fragments are then washed off. If there are no complementary sequences on the filter, all the radioactive fragments wash off. Although the filter contains fragments of many different genes, only those coding for the Major Sperm Protein can become radioactive. The radioactive bands are located by placing X-ray film over the filter for about one week and then developing the film. The left and right sides (A and B) of the X-ray show results when the DNA from each tissue is fragmented using 2 different enzymes.

gene was present in all samples tested: oocyte, sperm, larvae, and adult intestine.

This work clearly demonstrates that the DNA coding for the Major Sperm Protein is present in all cells of the nematode, even after chromosome diminution. Moreover, the Major Sperm Protein genes in the different tissues fragmented into units virtually identical in molecular weight (Focus Fig. 16.1), suggesting that these genes are apparently not altered in any substantial way from one tissue to the next; DNA molecules differing markedly in sequence would have been cut by enzymes into fragments of different sizes. Thus, whatever function is served by chromosome diminution, the process does *not* seem to rid somatic cells of all DNA that functions only in gonadal tissues.

However, as Bennett and Ward note, perhaps *other* germ-cell specific genes *are* eliminated by chromosome diminution; the gene for the Major Sperm Protein may have been an unrepresentative choice. Bennett and Ward also acknowledge that the Major Sperm Protein might have some other, unknown function in somatic tissues; they have been unable to detect the Major Sperm Protein, or the messenger RNA coding for that protein, in somatic tissues, but it may nevertheless be present, in concentrations too low to measure. Obviously if the gene functions in somatic cells, it is not surprising to find it there. Until the sensitivity of available techniques improves, the functional significance of chromosome diminution must remain uncertain.

Based on Karen L. Bennett and Samuel Ward, "Neither a Germ Line-specific Nor Several Somatically Expressed Genes Are Lost or Rearranged during Embryonic Chromatin Diminution in the Nematode *Ascaris lumbricoides var. suum*" in *Developmental Biology*, 118:141–47, 1986.

host—from intestine, to liver, to heart, to lungs, to esophagus, and back to intestine, for instance. Various organs, including the intestinal wall, lungs, and eyes, can be damaged from these migrations. The larvae of *Onchocerca* spp. often migrate to the victims' eyes, causing blindness (so-called river blindness, or onchocerciasis); indeed, *Onchocerca* infection is one of the major causes of blindness in the world. About 18 million people are presently affected, mostly in western Africa.

Plant parasites also tend to do their damage indirectly, by (1) causing wounds, which are then susceptible to bacterial or fungal infection; (2) injecting plant viruses; or (3) damaging root, leaf, or stem transport systems.

The life cycles of parasitic animals are generally more complex than those of their free-living relatives, and the nematodes are no exception. Fecundities are enormous (a single female *Ascaris* releases some 200,000 fertilized eggs per day—that's 73 million per year!), and one or more intermediate hosts are often obligate in the life cycle; the life histories of some important nematode parasites are outlined in the *Taxonomic Detail* section at the end of this chapter. Here I will describe just a few examples.

The so-called American hookworm (*Necator americanus*), infecting some 1.3 billion people worldwide, has a rather simple life cycle with only a single host: humans. Once in the human host, the parasite undergoes an extensive

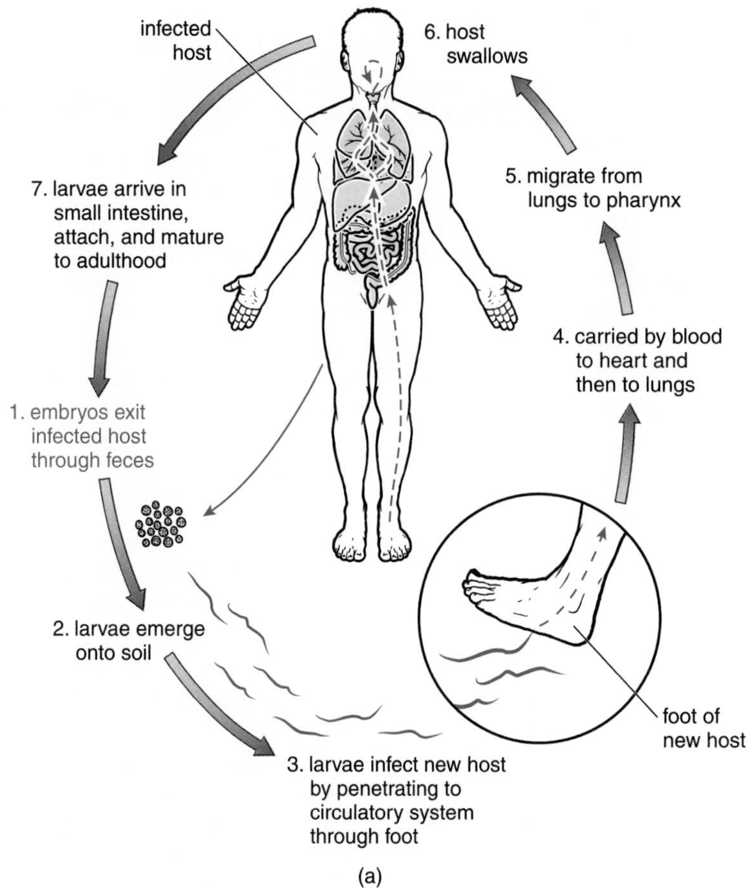
infected host
6. host swallows
5. migrate from lungs to pharynx
7. larvae arrive in small intestine, attach, and mature to adulthood
4. carried by blood to heart and then to lungs
1. embryos exit infected host through feces
2. larvae emerge onto soil
foot of new host
3. larvae infect new host by penetrating to circulatory system through foot
(a)

Figure 16.10

(a) Life cycle of the hookworm *Necator americanus*. The parasite matures in the host's small intestine and begins releasing fertilized eggs, which pass out with the host's feces. After hatching out of the egg case and molting twice, the parasite becomes infective to humans, typically lysing its way in through the foot. Adulthood is reached only after undergoing an extensive migration within the host's body.

internal migration before ending up in the host's intestine, where it matures. Fertilized eggs pass out in the host's feces and soon become infective to another human host upon reaching the third juvenile stage (Fig. 16.10a). Another globally distributed nematode parasite of humans is pinworm, infecting in temperate regions alone some 500 million people, particularly children; about 20% of U.S. children are infected. Again, the life cycle involves no intermediate host, and the juveniles often exhibit an impressive migration within the host before reaching their final location, this time in the host's large intestine and colon (Fig. 16.10b). Improving human sanitary practices is an essential step in limiting infection by both parasites.

The guinea worm, *Dracunculus medinensis,* presents us with a more complex and far more colorful example of a parasitic nematode life cycle; the generic name suggests that this nematode is up to no good. The female, only about 1 mm wide but generally 1 m or more long, lives just beneath the skin in humans, releasing an ulcer-producing secretion. When the skin comes into contact with water (as when the host bathes), the nematode protrudes its posterior end through the sore on the host's skin and ejects a considerable number of young from its uterus—up to about 1.5×10^7 offspring per day! These juveniles cannot reinfect humans directly but must first be ingested by a species of microscopic aquatic crustacean, a copepod (phylum Arthropoda). Humans become infected by drinking water containing these crustaceans. The parasite, liberated from the intermediate host during digestion, migrates through the intestinal wall of the primary host and into the host connective tissue. Male parasites, which are relatively small, die soon after inseminating the females.

Once females reach adulthood beneath the skin, the cure is charmingly simple. Commonly, an incision is made and the worm is rolled out on a stick, very slowly (only a few centimeters daily), to prevent breakage (Fig. 16.11).

Dracunculus medinensis is found only in Africa, South America, and Western Asia. The World Health Organization hopes soon to eliminate the problem worldwide, something it has previously accomplished for only one other disease, smallpox. The goal is far more feasible for the guinea worm than for most other parasites, since the guinea worm appears to have no alternate hosts. Thus, its elimination is mainly a question of keeping infected

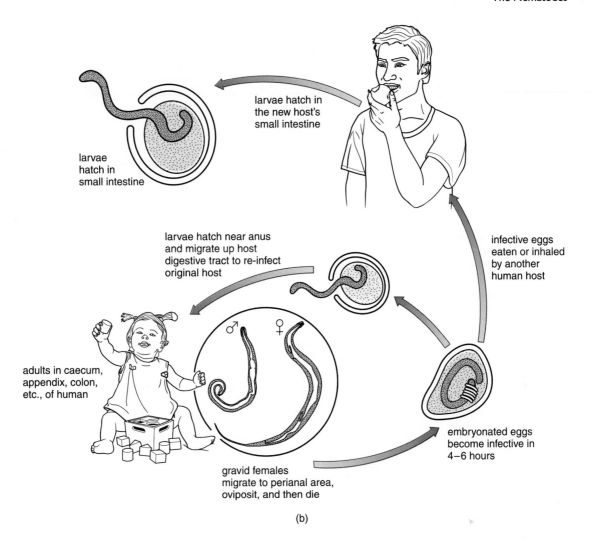

larvae hatch in the new host's small intestine

larvae hatch in small intestine

larvae hatch near anus and migrate up host digestive tract to re-infect original host

infective eggs eaten or inhaled by another human host

adults in caecum, appendix, colon, etc., of human

♂ ♀

embryonated eggs become infective in 4–6 hours

gravid females migrate to perianal area, oviposit, and then die

(b)

Figure 16.10 *Continued*

(b) Life cycle of the pinworm, *Enterobius vermicularis*. Mature females deposit their fertilized eggs at the host's anus and then die, releasing even more eggs. The embryos are transmitted to new hosts on the child's hands or through the air. Embryos remaining in the host's anal region emerge from their capsules after the second molt and migrate up the child's intestinal tract to reinfect the same host.

(b) Based on *Human Parasitology* by Burton J. Bogitsch and Thomas C. Cheng.

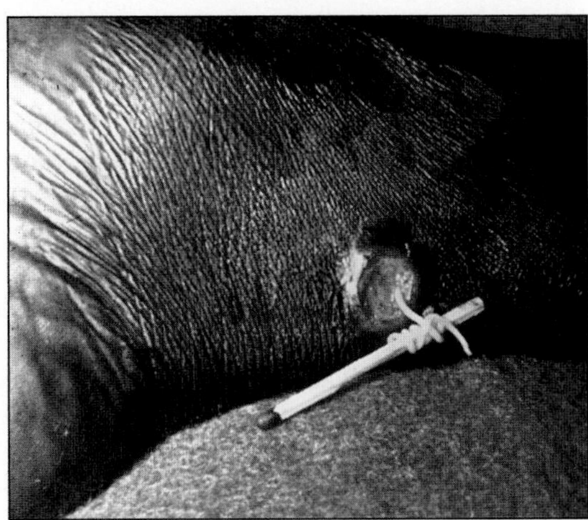

Figure 16.11

Using a matchstick to wind *Dracunculus medinensis* out of an infected human leg.

Courtesy of the Centers for Disease Control, Department of Health and Human Services, Atlanta, GA.

persons away from public water supplies and teaching people in target areas to boil water before drinking it. The number of infected people has been reduced by at least 95% since 1986.

Wuchereria, Loa loa, Brugia, Onchocerca, and other **filarial nematodes** (so named because they produce a characteristic infective stage called a **microfilaria**) also exhibit complex life cycles involving an arthropod intermediate host and extensive migration within the definitive host (Fig. 16.12).

In the past several years, it has become clear that most filarial nematodes form, in both adult and larval stages, symbiotic relationships with certain bacteria *(Wolbachia)*. The association is obligate—without the bacteria the nematode cannot develop, mate, or leave offspring—so that at least some filarial infections can now be successfully treated with tetracycline and some other antibiotics.[5]

5. See *Topics for Further Discussion and Investigation,* no. 4.

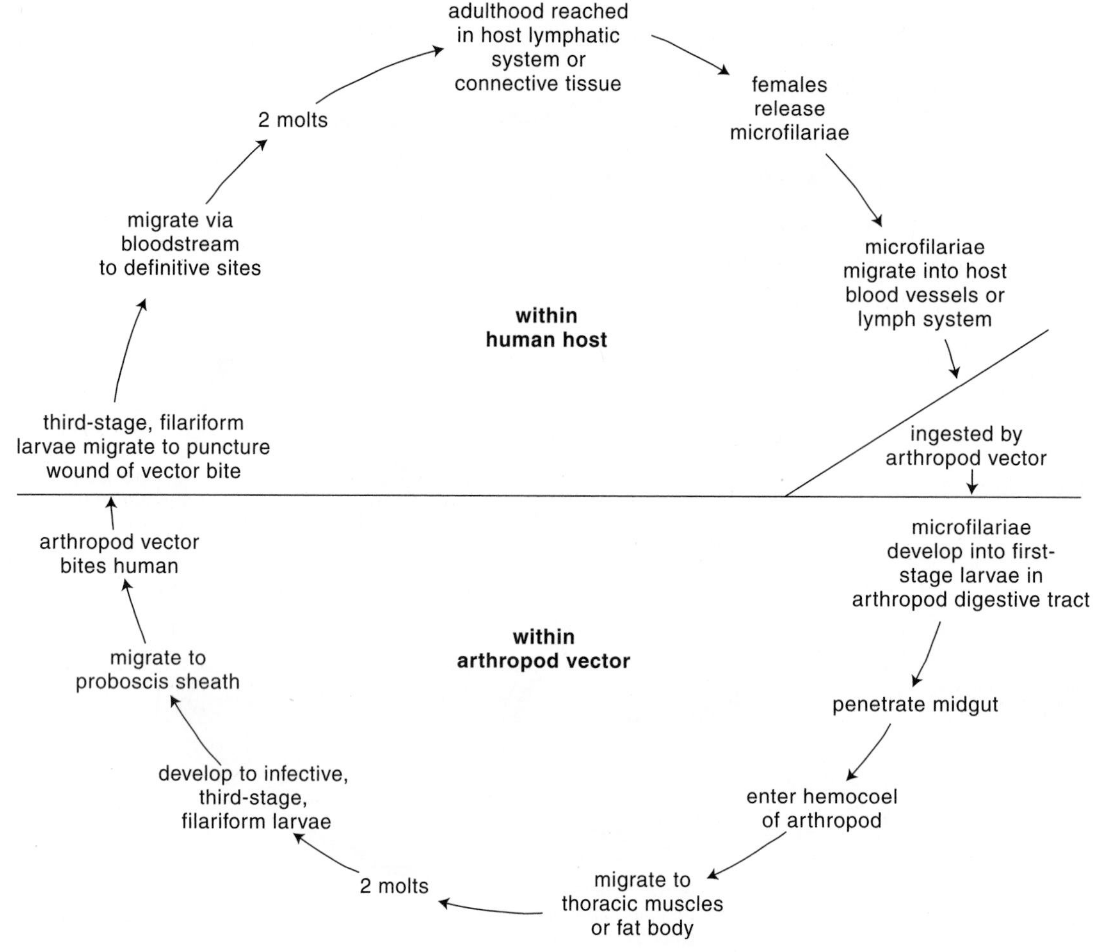

Figure 16.12

Generalized life cycle of *Wuchereria, Loa loa, Onchocerca,* and other filarial nematodes.

Figure from *Human Parasitology* by Burton J. Bogitsch and Thomas C. Cheng, copyright © 1990 by Saunders College Publishing, reproduced by permission of the publisher.

Beneficial Nematodes

It is unfortunate that parasitic species have given nematodes such a bad name. Species that require insects as intermediate hosts are increasingly being used to decrease population sizes of various insects, including mosquitoes and several major agricultural pests; the parasite often does mortal harm to its insect vector, particularly if infection densities are high. Besides, most nematode species are free-living, innocuous detritivores, not parasites. These species likely play valuable ecological roles in the cycling of nutrients and energy in a variety of environments.[6] Free-living nematodes may also have nascent commercial value as potential food sources in the aquaculture of some edible animals, such as penaeid shrimp and certain fish, and may soon be exploited as sensitive monitors of environmental contamination. In addition, much of what is currently known about aging, inheritance, the factors controlling gene expression and programmed cell death during development, and how nerve cells find their targets during development comes from the study of nematodes, particularly the free-living species *Caenorhabditis elegans* (Fig. 16.1). This animal has a small genome with only 6 chromosomes and some 100 million base pairs comprising nearly 20,000 genes. The entire genome was sequenced in 1998, making *C. elegans* the first animal to have its full set of genes completely described. The eventual role of every cell in development is now known, and the interconnections of all 302 neurons have been completely described, making these nematodes especially suitable for studying the molecular basis of differentiation, behavior, and learning. Genetic and developmental studies are further facilitated by the extremely short generation time in the laboratory—only about 3 days at room temperature—and by the ease with which the worms can be reared at high densities, up to about 10,000 individuals on a single petri dish. Molecular biologists can now implant foreign genes into the genome of *C. elegans;* these genes are not only expressed in their foreign environment, but are transmitted to succeeding

6. See *Topics for Further Discussion and Investigation,* no. 3.

generations, opening the door to important studies in genetic manipulation and gene therapy. At least 60 research laboratories worldwide are currently studying the development of *C. elegans*.

Topics for Further Discussion and Investigation

1. Discuss nematode locomotion.

Clark, R. B. 1964. Nematode locomotion. *Dynamics in Metazoan Evolution.* New York: Oxford University Press, 78–83.

Gray, J., and H. W. Lissman. 1964. The locomotion of nematodes. *J. Exp. Biol.* 41:135.

Harris, J. E., and H. D. Crofton. 1957. Structure and function in the nematodes: Internal pressure and cuticular structure in *Ascaris. J. Exp. Biol.* 34:116.

Wallace, H. R. 1959. The movement of eelworms in water films. *Ann. Appl. Biol.* 47:350.

Wallace, H. R., and C. C. Doncaster. 1964. A comparative study of the movement of some microphagous, plant-parasitic and animal-parasitic nematodes. *Parasitology* 54:313.

2. Investigate the factors that induce the hatching or exsheathment of parasitic nematodes.

Barrett, J. 1982. Metabolic responses to anabiosis in the fourth stage juveniles of *Ditylenchus dipsaci* (Nematoda). *Proc. Royal Soc. London Ser. B* 216:159.

Clarke, A. J., and A. H. Sheperd. 1966. Picrolonic acid as a hatching agent for the potato cyst nematode *Heterodera rostochiensis* Woll. *Nature (London)* 211:546.

Lackie, A. M. 1975. The activation of infective stages of endoparasites of vertebrates. *Biol. Rev.* 50:285.

Ozerol, N. H., and P. H. Silverman. 1972. Enzymatic studies on the exsheathment of *Haemonclius contortus* infective larvae: The role of leucine aminopeptidase. *Comp. Biochem. Physiol.* 42B:109.

Rogers, W. P., and R. I. Sommerville. 1957. Physiology of exsheathment in nematodes and its relation to parasitism. *Nature (London)* 179:619.

Wilson, P. A. G. 1958. The effect of weak electrolyte solutions on hatching rate of *Trichostrongylus retortaeformis* (Zeder) and its interpretation in terms of a proposed hatching mechanism of Strongylid eggs. *J. Exp. Biol.* 35:584.

3. Investigate the distribution, abundance, and ecological roles of the free-living nematodes found in shallow-water marine environments.

Bell, S. S., M. C. Watzin, and B. C. Coull. 1978. Biogenic structure and its effect on the spatial heterogeneity of the meiofauna in a salt marsh. *J. Exp. Marine Biol. Ecol.* 35:99.

Hopper, B. E., J. W. Fell, and R. C. Cefalu. 1973. Effect of temperature on life cycles of nematodes associated with the mangrove (*Rhizophora mangle*) detrital system. *Marine Biol.* 23:293.

Moens, T., and M. Vincx. 1997. Observations on the feeding ecology of estuarine nematodes. *J. Marine Biol. Ass. U.K.* 77:211.

Tietjen, J. H. 1977. Population distribution and structure of the free-living nematodes of Long Island Sound. *Marine Biol.* 43:123.

Warwick, R. M., and R. Price. 1979. Ecological and metabolic studies on free-living nematodes from an estuarine mud flat. *Est. Coast. Marine Sci.* 9:257.

4. Discuss the methods being investigated for control of parasitic nematodes and their biological basis.

Fanelli, E., M. Di Vito, J.T. Jones, and C. De Giorgi. 2005. Analysis of chitin synthase function in a plant parasitic nematode, *Meloidogyne artiellia*, using RNAi. *Gene* 349:87–95.

Hallem, E. A., M. Rengarajan, T. A. Ciche, and P. W. Sternberg. 2007. Nematodes, bacteria, and flies: A tripartite model for nematode parasitism. *Current Biol.* 17:898–904.

Higazi, T. B., A. Fillano, C. R. Katholi, Y. Dadzie, J. H. Remme, and T. R. Unnasch. 2005. *Wolbachia* endosymbiont levels in severe and mild strains of *Onchocerca volvulus. Molec. Biochem. Parasitol.* 141:109–12.

Taylor, M. J., H. F. Cross, and K. Bilo. 2000. Inflammatory responses induced by the filarial nematode *Brugia malayi* are mediated by lipopolysaccharide-like activity from endosymbiotic *Wolbachia* bacteria. *J. Exp. Med.* 191:1429–36.

Zhang, Y. H., J. M. Foster, L. S. Nelson, D. Ma, and C. K. S. Carlow. 2005. The chitin synthase genes chs-1 and chs-2 are essential for *C. elegans* development and responsible for chitin deposition in the eggshell and pharynx, respectively. *Devel. Biol.* 85:330–39.

Taxonomic Detail

Phylum Nematoda (= Nemata)

This phylum contains about 200 families of pseudocoelomate and acoelomate worms, with about 16,000 named species and perhaps many millions more waiting to be discovered and named. A number of cladistic analyses over past 10 years indicate that the previous division of nematodes into the 2 families Adenophorea and Secernentea is invalid: Molecular data suggest that members of the Secernentea evolved from within the Adenophorea, and that many of the morphological similarities among different nematode groups evolved through convergence. Nematodes are now divided into 3 major classes, the Enoplia, the Dorylaimia, and the Chromadoria, although many important relationships remain unresolved within those groups.[7]

7. Taxonomic hierarchy based largely on Meldal, B. H. M., *et al.* 2007. An improved molecular phylogeny of the Nematoda with special emphasis on marine taxa. *Molec. Phylog. Evol.* 42: 622–36. See also www.wormbook.org.

Class Enoplia
Free-living and predatory nematodes, including terrestrial, freshwater, and especially many marine species. None are parasitic. The largest of free-living nematodes belong to this group. At least 7 orders, and 25 families.

Family Tricodoridae. *Trichodoris.* An important parasite of plant roots, and one of the few known nematode vectors for plant viruses.

Class Dorylaimia
This large group includes many free-living species, a number of commercially important plant and animal parasites, and no marine species.

Order Dorylaimida
Longidoris. The group includes species (e.g., *Longidoris*) that are ectoparasitic on plants, and some large predatory species. Sixteen families.

Order Mermethida
Three families.

Family Mermethidae. *Hexamermis, Mermis, Romanomermis.* The adult stage in this order is typically free-living. However, the larval stages all parasitize invertebrates, especially insects, which can be killed by heavy parasite infestations. For this reason, considerable interest has been shown in using members of this order as biological control agents for target insects, including blackflies and mosquitoes, in which the larvae develop.

Order Trichinellida
Six families.

Family Trichuridae. *Trichuris*—whipworm. This family includes many parasites of mammals. *Trichuris trichuris* is one of the most common gastrointestinal worm parasites of humans, infecting the colons of perhaps 500 million people worldwide, and killing about 100,000 people yearly. One female parasite can produce over 45,000 eggs per day. The mammalian host becomes infected by ingesting embryos in contaminated water or food.

Family Trichinellidae. *Trichinella spiralis.* A parasite widespread in mammals and responsible for trichinosis in humans eating undercooked pork or some other meats. Approximately 40 million people are presently infected around the world. The male nematodes are only about 1.5 mm long, making them the smallest nematodes parasitizing humans. Oddly, there is no required intermediate host in the life cycle: Rather, juveniles and adults develop in different organs of a single host. Juveniles develop to adulthood intracellularly in muscle tissue, within special "nurse" cells that they themselves induce the host to make.

Class Chromadoria
This is a huge group of great diversity, containing at least 7 orders. It includes members of the former class Secernentea (mostly free-living terrestrial species), along with many marine and freshwater species, including parasites of both plants and animals. Members of some species live symbiotically on fish gills.

Order Rhabditida
This order contains all members of the former nematode class Secernentea. Most rhabditids ingest bacteria; the group includes *Caenorhabditis elegans,* the first animal to have its complete genome sequenced. Four suborders with at least 115 families.

Suborder Rhabditina
Members of this group possess a true dauer stage larva, something rarely seen in most other groups.

Infraorder Diplogasteromorpha
Ditylenchus, Heterodera, Pristionchus. In addition to free-living species, this group contains a variety of insect and plant parasites and is thus of considerable economic importance in agriculture and pest management. *Pristionchus pacificus,* a free-living species, is the first non-*Caenorhabditis* nematode to have its complete genome sequenced, and is the subject of growing interest for behavioral, ecological, and developmental studies. Six families.

Family Heteroderidae. *Heterodera.* This family contains some of the most important agricultural parasites in the world. The larvae invade underground roots, and the parasites spend their lives there, destroying root tissue.

Infraorder Rhabditomorpha
Family Rhabditidae. *Caenorhabditis elegans.* These small, free-living nematodes, less than 1 mm long when full-grown, have become very important animal models for studying the molecular basis of aging and development. Unlike most nematode species, members of *C. elegans* are simultaneous hermaphrodites, rather than gonochoristic. Individuals live about 3 weeks, and they can reproduce when only 3 to 4 days old.

Superfamily Strongyloidea
Superfamily Ancylostomatidae—hookworms. *Ancylostoma, Necator.*

Family Ancylostomatidae. As adults, all members parasitize the mammalian intestinal tract, feasting on host blood; the first 2 larval stages are generally free-living on feces. More than 20% of the world's human population presently may be infected with various hookworm species.

Necator americanus—the North American hookworm. Although the species occurs commonly in Asia, Africa, Central and South America, and the Caribbean, there are presently about 1 million human infestations in North America, primarily in southern states. Worldwide, the species infects some 1.3 billion people. Infestations with fewer than about 100 worms cause little trouble and may escape attention altogether. Infestations of more than 500 worms are highly pathological, due to extreme blood loss: Hosts are anemic and physically weak; periodically suffer severe abdominal pain, fever, and dizziness; and often experience a strong craving to eat dirt or wood. Fertilized eggs pass out with the host feces and hatch into the soil as long-lived, infective larvae. These larvae infect the next host (or reinfect a host) by penetrating the skin, often between the toes of the foot. The developing worm then undergoes an extensive, damaging journey through the host as described earlier for *Strongyloides* spp. and *Ascaris lumbricoides*. Once the host coughs and swallows the infective agent, the worm chomps down on the mucosa of the small intestine and begins its blood-sucking business. *Ancylostoma duodenale*—the Asian hookworm, the foreign equivalent of *Necator americanus*. Species are common parasites of dogs, cats, and humans in Europe, China, Africa, and India. Some 1 billion people probably harbor this parasite. Each worm typically ingests blood at twice the rate of its North American counterpart, so equivalent levels of pathology occur with smaller infestations.

Suborder Tylenchina

Twenty-eight families.

Infraorder Panagrolaimomorpha

Family Panagrolaimidae. *Turbatrix aceti*—vinegar eels. These free-living nematodes thrive in high-acidity liquid environments, such as vinegar, and feed on bacteria.

Family Strongyloididae. *Strongyloides stercoralis.* The complete life cycle is typically nonparasitic in soil or dung. Under unfavorable environmental conditions, however, the nematode larvae metamorphose into nonfeeding, infective filariform larvae. These may then penetrate the skin of a suitable mammalian host, traveling in the bloodstream to the heart and thence to the lungs. Juveniles then burst out of the lung capillaries, causing considerable damage in the process, and migrate up the host's throat. A cough and a swallow sends the parasite to its final resting place in the small intestine, where it dedicates its life to producing eggs by parthenogenesis. The embryos and larvae leave with the host feces and may then develop into free-living adults or infective filariform larvae, depending on environmental conditions. Humans get infected by contacting contaminated soil or feces or through contact with parasitized cats and dogs. Some 50 to 60 million people are currently infected, mostly in tropical and subtropical countries.

Suborder Spirurina

All members are parasitic. Fifty-eight families.

Family Ascarididae. *Ascaris, Parascaris.* Chromosome diminution is known only from members of this group, occurring in at least 12 species. All species are intestinal parasites of mammals. Ascarids are the most common human parasites. About 25% of the entire human population is infected with *A. lumbricoides* worldwide, with over 75% of the cases occurring in Asia. About 20,000 people die of the infection each year. The incidence of human infestation is perhaps 10 times higher in Europe than in the United States, but about 3 million cases are presently known in North America. The worms reside in the host's small intestine, feeding mostly on chyme and less frequently on blood sucked from the mucosal lining of the digestive tract. Their reproductive capacity is impressive: One female can produce over 25 million offspring in her short lifetime, at rates of up to several hundred thousand per day. Adult ascarids are among the largest of parasitic nematodes, reaching lengths of 30–40 cm. Fertilized eggs leave the host in feces and become infective after a time in soil. If the infective larval stage is swallowed, it hatches in the host's intestine and begins a long journey through the host, only to return finally to the intestine, where it reaches maturity. The migration through the host is similar to that previously described for *Strongyloides* spp., including passage through the heart and lungs, rupture from the alveolar capillaries, and an upward crawl to the throat, followed by the inevitable cough and swallow. Larvae may stray from this route and cause serious damage in hosts of species that are not normally infected.

Family Anisakidae. *Anisakis, Pseudoterranova.* All species are obligate parasites of aquatic vertebrates in both marine and freshwater habitats. Fish, including such popular types as sardines, salmon, flounder, and mackerel, typically act as intermediate hosts. The final host, usually a mammal and sometimes a human, contracts anisakiasis by eating raw or undercooked fish or squid. The parasite causes acute abdominal pain in the host. Infection can be avoided by cooking, salting, or freezing contaminated fish.

Family Oxyuridae. This family contains parasites of vertebrates and invertebrates. *Enterobius vermicularis*—pinworm. Gravid females, residing in the host's large intestine, migrate to the anus nightly and there deposit their fertilized eggs, creating a very itchy situation. When the sleeper scratches, the embryos collect on the fingers and under the fingernails, or become airborne. Other humans can then become infected (or reinfected) by breathing in or ingesting the fertilized eggs. Larvae hatch in the digestive tract and migrate directly to the large intestine, where they become adults. At least 400 million people are currently infected with *E. vermicularis* worldwide.

Family Filariidae. *Wuchereria, Brugia, Onchocerca, Loa.* Members of this family—the filarial parasites—parasitize all vertebrates but fish, and they cause a number of infamous human disorders, including elephantiasis (a consequence of infection by *Wuchereria bancrofti* or *Onchocerca volvulus*), loa loa (a consequence of infection by a nematode of the same name, *Loa loa*), and river blindness (a common consequence of infection by *Onchocerca volvulus*). More than 200 million people (and equal or even greater numbers of domesticated and wild animals) presently may be infected with filarial nematodes, particularly in India, the Philippines, South America, and the Caribbean. The family also includes *Dirofilaria immitis,* the agent of canine (and feline) heartworm; infection can be transmitted from pet to human through physical contact. All filarial species are parasitic throughout their lives. The first juvenile stage, the colorless and transparent **microfilaria** (which gives the group its common name), circulates in the host's blood, while the adults live a more sedentary existence in host tissues and lymphatic glands.

Wuchereria bancrofti requires a mosquito as an intermediate host, as do many other members of this family (including dog heartworm). In at least some strains of this parasite, infective microfilariae circulate in the host bloodstream mainly at night; mosquitoes feeding while the host sleeps ingest the microfilariae with their blood meal. The parasite then develops to the third juvenile stage within the mosquito's thoracic muscles and soon migrates to the mouth, where it accesses the definitive host the next time the mosquito feeds. If the human host houses more than about 5000 microfilariae per milliliter of blood, the indulging mosquitoes will likely die.

Loa loa, and other members of the genus, are restricted to African primates, including people. Over 13 million people currently may be infected. The worms are small and slender, perhaps 40–50 mm long and less than 0.5 mm wide. Rather than residing in any permanent location, adults usually migrate about the host's body subcutaneously; they are sometimes observed migrating under the conjunctiva of the eye. The adults secrete toxins that may cause 4–6 cm swellings on the host's body. The densities of microfilariae, living as always in blood, are highest in the daytime; transfer to a new host is therefore mediated by a daytime blood-sucking insect, the biting tabanid flies.

Onchocerca volvulus, the agent of river blindness and one cause of elephantiasis, requires blackflies (*Simulium* spp.) as intermediate hosts. Most of the serious damage to humans results from activities of the small microfilariae, rather than from the 50-cm-long, but merely annoying, adults. Lesions in the eye, resulting from the buildup of dead juveniles, eventually cause blindness. Nearly 40 million people currently may be blind or going blind from onchoceriasis and 20,000 to 50,000 die of the infection each year. As with *Wuchereria bancrofti,* heavy infestations may kill the insect vectors, primarily by damaging the midgut epithelium and Malpighian tubules. Some workers suggest placing these parasites in a separate family, the Onchocercidae.

Family Dracunculidae. *Dracunculus medinensis*—the guinea worm. This large, debilitating parasite probably infects about 1 million people worldwide, particularly in Africa and India. Adult females reach lengths of some 100 cm (about 3 ft) in their hosts, some 250 times larger than the largest males. Adults secrete substances that cause considerable itching and burning in the host. (See p. 442 for a complete life-cycle description.)

General References about the Nematodes

Aguinaldo, A. M. A., J. M. Turbeville, L. S. Linford, M. C. Rivera, J. R. Garey, R. A. Raff, and J. A. Lake. 1997. Evidence for a clade of nematodes, arthropods and other moulting animals. *Nature* 387:489.

Bird, A. F., and J. Bird. 1991. *The Structure of Nematodes,* 2d ed. New York: Academic Press.

Chen, Z. X., S. Y. Chen, and D. W. Dickson, eds. 2004. *Nematology: Advances and Perspectives. Vol 1: Nematode Morphology, Physiology and Ecology. Vol. 2: Nematode Management and Utilization.* CABI Publishing & Tsinghua University Press, China.

Cunha, A., R. B. R. Azevedo, S. W. Emmons, and A. M. Leroi. 1999. Developmental biology: Variable cell number in nematodes. *Nature* 402: 253.

De Ley, P., and M. Blaxter. 2004. A new system for Nematoda: Combining morphological characters with molecular trees, and translating clades into ranks and taxa. *Nematology Monographs and Perspectives,* 2: 633–53.

Desowitz, R. S. 1981. *New Guinea Tapeworms and Jewish Grandmothers: Tales of Parasites and People.* New York: W. W. Norton.

Harrison, F. W., and E. E. Ruppert. 1991. *Microscopic Anatomy of Invertebrates, Vol. 4. Aschelminthes.* New York: Wiley-Liss.

Heip, C., M. Vincx, and G. Vranken. 1985. The ecology of marine nematodes. *Oceanogr. Marine Biol. Ann. Rev.* 23:399–489.

Malakhov, V. V. (translated by W. D. Hope). 1994. *Nematodes: Structure, Development, Classification, and Phylogeny.* Washington, D.C.: Smithsonian Institution Press.

Meldal, B. H. M., et al. 2007. An improved molecular phylogeny of the Nematoda with special emphasis on marine taxa. *Molec. Phylog. Evol.* 42: 622–36.

Roberts, L. S., J. Janovy, Jr., and P. Schmidt. 2004. *Foundations of Parasitology*, 7th ed. Dubuque, Iowa: McGraw-Hill Publishers.

Schierenberg, R. 1997. Nematodes, the roundworms. In: S. F. Gilbert, and A. M. Raunio. *Embryology: Constructing the Organism.* Sunderland, MA: Sinauer Associates, Inc. Publishers, pp. 131–48.

Smythe, A. B., M. J. Sanderson, and S. A. Nadler. 2006. Nematode small subunit phylogeny correlates with alignment parameters. *Syst. Biol.* 55: 972–92.

Stevens, L., R. Giordana, and R. F. Fialho. 2001. Male-killing, nematode infections, bacteriophage infection, and virulence of cytoplasmic bacteria in the genus *Wolbachia. Ann. Rev. Ecol. Syst.* 32: 519–45.

Thorpe, J. H., and A. P. Covich. 2001. *Ecology and Classification of North American Freshwater Invertebrates,* 2nd ed. New York: Academic Press.

Search the Web

1. www.ucmp.berkeley.edu/aschelminthes/aschelminthes.html

   This address includes a brief discussion of all "aschelminth" groups. This site is offered by the University of California at Berkeley.

2. http://www.tolweb.org/tree/phylogeny.html

   This brings you to the Tree of Life site. Search on "Nematoda."

3. http://nematode.unl.edu/wormepns.htm

   This site, offered through the University of Nebraska at Lincoln, describes the uses of certain nematode species as agents for the control of agricultural insect pests, and includes links to related websites.

4. http://vm.cfsan.fda.gov

   This site is maintained by the U.S. Food and Drug Administration. Under the heading "Program Areas," select "Bad Bug Book" and click on entries for (a) *Ascaris lumbricoides,* (b) *Anisakis simplex* and (c) *Eustrongylides* for information about the diseases these nematodes cause among humans.

5. www.who.int/ctd/

   This site is maintained by the World Health Organization. Click on the entry for "Intestinal nematodes" to learn more about the effects of parasitic nematodes on human health. From the main menu, you can also access information on dracunculiasis, lymphatic filariasis, onchocerciasis, and schistosomiasis. Also, click on "International Travel and Health" to see the geographical distribution of diseases caused by filarial nematodes. This is a good map to view before making plans for vacation travel.

6. http://biodidac.bio.uottawa.ca

   Choose "Organismal Biology," "Animalia," and then "Nematoda" for photographs and drawings, including illustrations of sectioned material.

7. http://www.dpd.cdc.gov/DPDx/

   This site is maintained by the U.S. Government Centers for Disease Control and Prevention. It contains a section on "Parasites and Parasitic Diseases" along with a separate image library of causative agents.

8. www.nematodes.org

   This website contains detailed information about current research on nematode genetics and gene sequences, provided by the lab of Mark Blaxter.

9. www.wormbook.org

   A large collection of material on all aspects of research on the nematode *Caenorhabditis elegans.*

10. http://8e.devbio.com/chapter.php?ch=19

    Click on Topic number 1 to read about the mechanisms of chromosome diminution. This material accompanies Scott Gilbert's developmental biology textbook.

11. http://plpnemweb.ucdavis.edu/Nemaplex/index.htm

    This site provides detailed information concerning plant and soil nematodes, from the University of California at Davis.

17

Four Phyla of Likely Nematode Relatives: Nematomorpha, Priapulida, Kinorhyncha, and Loricifera

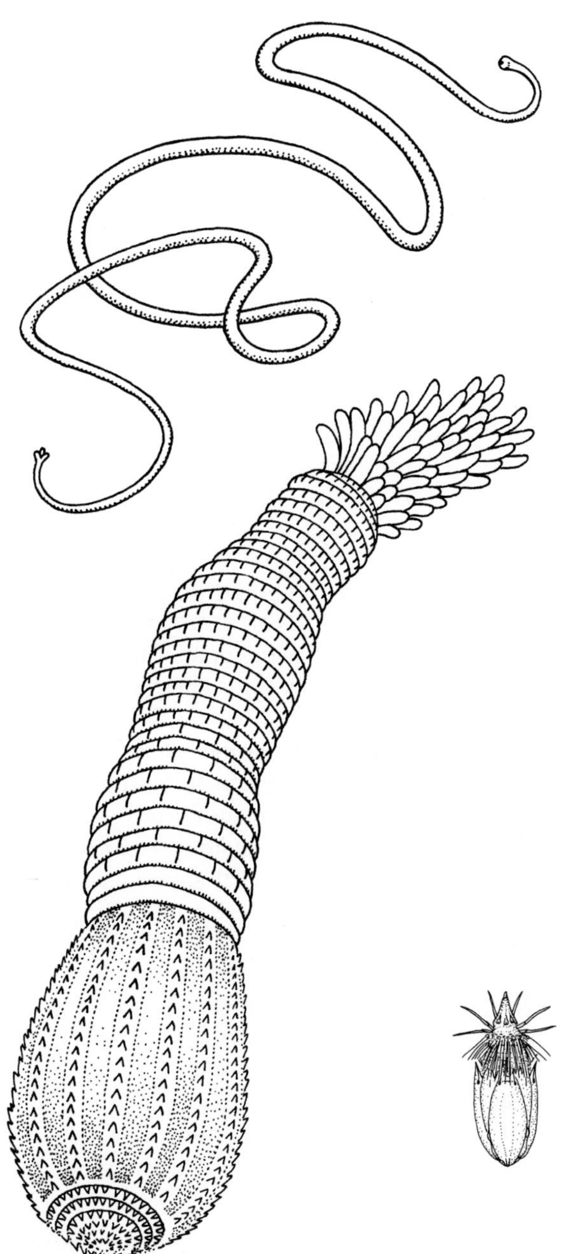

Introduction

Rotifers, nematodes, and other aschelminths have long been assigned to separate phyla, largely because their phylogenetic relationships have been uncertain. There is growing recognition, however, based on cladistic analyses of both morphological and molecular characters, that the aschelminths form 2 natural groupings: the Rotifera, Acanthocephala, and two smaller groups on the one hand (Chapter 10), and the molting animals on the other. Increasingly, the molting aschelminths are being placed together to form a larger grouping, the Cycloneuralia (*cyclo* = Greek: circular; *neura* = G: a nerve) (Table 16.1). The name Cycloneuralia refers to the fact that in all of these animals the brain encircles the pharynx like a collar. There is good reason to think that this will eventually be formalized as a phylum, with the nematodes and the 4 groups discussed in this chapter incorporated as separate classes.

Cycloneuralians have a number of similarities in addition to their circular brain shape. All members possess a cuticle of at least 3 layers, and that cuticle is molted at least once; the cuticle is collagenous in 2 groups, chitinous in the others (Table 17.1). Recently, cycloneuralians have been tentatively grouped with the arthropods, onychophorans, and tardigrades to form the Ecdysozoa, a group that includes all animals that periodically molt their cuticles (Table 16.1). Although many cycloneuralians are now believed to be acoelomate, some are either pseudocoelomate, or, in the case of nematodes, at least to include some pseudocoelomate species, with the fluid-filled body cavity formed from a persistent blastocoel. As with at least one nematode species, all of the animals discussed in this chapter exhibit, either as larvae or adults, an eversible anterior portion of the body with the mouth opening at the end—an **introvert;** thus, one can safely say that the animals discussed in this chapter are introverted.

Three of the groups discussed in this chapter (priapulids, kinorhynchs, and loriciferans) are linked by both

Table 17.1 Some Comparative Features of the Cycloneuralia*

Phylum	Cuticle	Body Cavity	Parasitism?	Species Distribution
Nematoda	Collagenous	Acoelomate, pseudocoelomate	Yes (adults)	All habitats
Nematomorpha	Collagenous	Pseudocoelomate	Yes (larvae)	Mostly freshwater, some marine
Priapulida	Chitinous	Pseudocoelomate? coelomate?	No	All marine
Kinorhyncha	Chitinous	Pseudocoelomate	No	All marine
Loricifera	Chitinous	Pseudocoelomate	No	All marine

*The name Nemathelminthes (G: "thread worms") has also been proposed, but that name has previously been used to encompass other phyla, including the Rotifera and Acanthocephala.

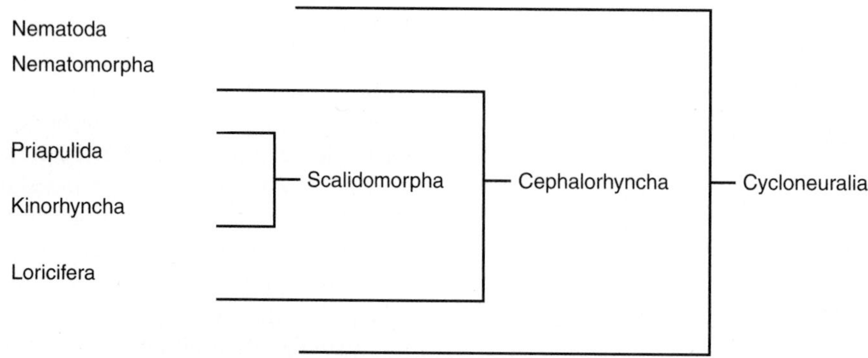

Figure 17.1
Distribution of animals among the Cycloneuralia and Cephalorhyncha.

morphological and molecular evidence, forming a larger taxonomic grouping within the Cycloneuralia called the Cephalorhyncha (Fig. 17.1); in all cephalorhynchans, the cuticle is chitinous, as are the introvert's scalids. Some workers have suggested making the Cephalorhyncha a phylum, in which case its 3 constituent members would become separate classes.

None of these phyla of likely nematode relatives includes more than about 325 species. Moreover, none of the 4 phyla is of substantial medical, veterinary, or agricultural importance, and none of their members have been exploited as model systems for the study of any basic biological phenomena; their biology is therefore not nearly as well studied as that of many nematode species. Still, the animals live fascinating lives and present interesting evolutionary puzzles.

Phylum Nematomorpha

Phylum Nemato • morpha
(G: thread body)
nem-at-ō-mor´-fah

Defining Characteristic:[1] Pseudocoelomates lacking a functional digestive tract in the adult stage

The least phylogenetically troublesome of the 3 phyla are the nematomorphs, also called "horsehair worms." About 320 species have been described from freshwater habitats and another 5 species from marine habitats. Although some molecular analyses affiliate nematomorphs with flatworms, most analyses align nematomorphs with nematodes. The adults are free-living, aquatic pseudocoelomates with a decidedly nematode external appearance: circular in cross section, long (commonly 0.5–1.0 m in length), thin (rarely more than 1 mm wide), lacking body segmentation, and enclosed in an external collagenous cuticle (Fig. 17.2a). Like most nematodes, nematomorphs are gonochoristic, and the eggs are fertilized internally. Also like nematodes, nematomorphs molt their collagenous external cuticle as they grow, possess only longitudinal body wall muscles, and lack locomotory cilia, as well as specialized circulatory and respiratory systems. They differ

1. Characteristics distinguishing the members of this phylum from members of other phyla.

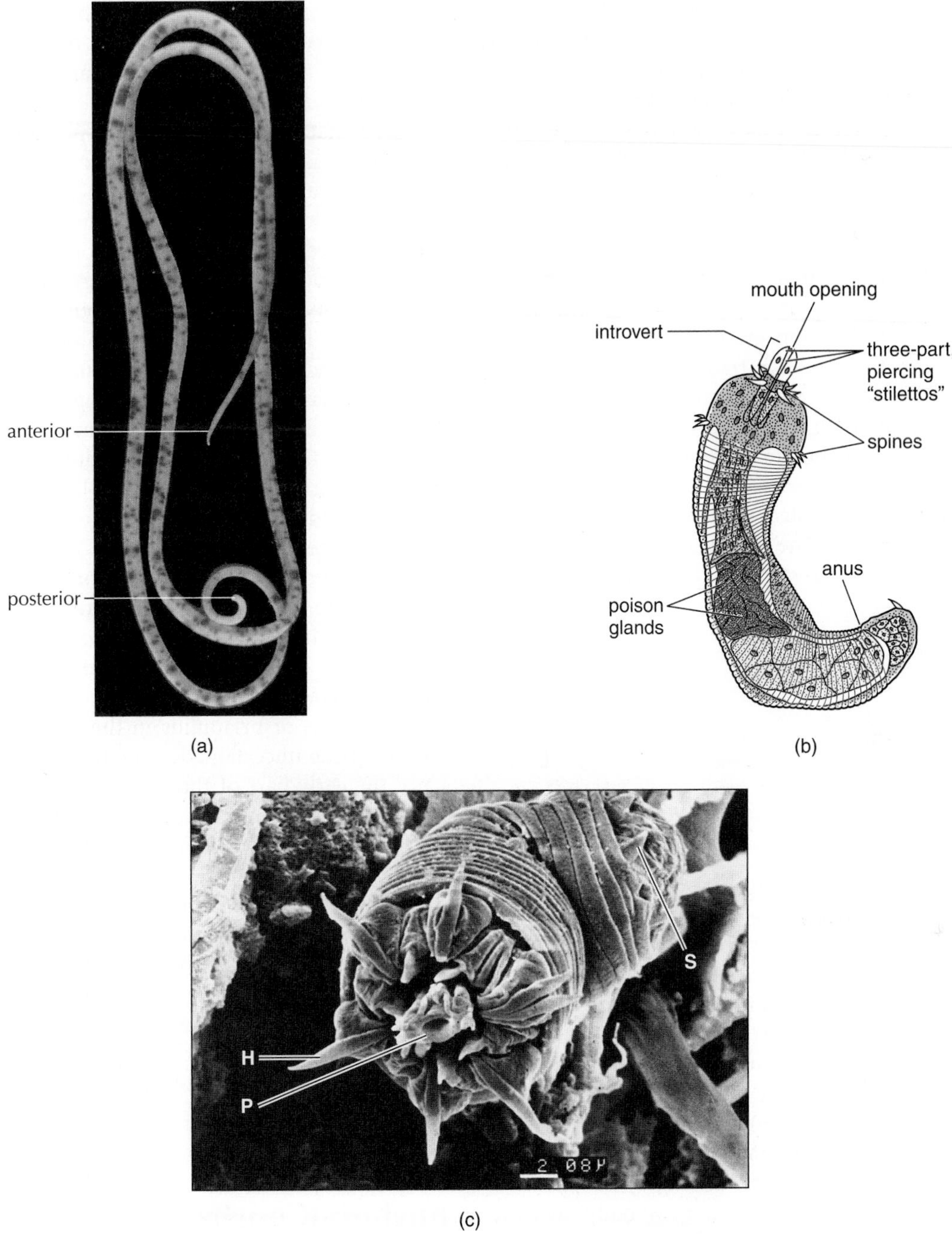

Figure 17.2
The males of this species are typically 10–30 cm long but may grow to more than 2 m. (a) Male horsehair worm (phylum Nematomorpha) from Tanzania, *Chorades ferox.* (b) The infective larva of the nematomorph *Gordius aquaticus*. (c) Larval stage of the nematomorph *Chordodes morgani*, seen in anterior and ventral view. Scale bar is 2.08 µm. P = proboscis (introvert); H = outer spine: S = tail spine. The mouth is visible at the center of the introvert, while the anus is just visible behind the labeled tail spine.
(a) Courtesy of A. Schmidt-Rhaesa. From A. Schmidt-Rhaesa. 2002. *Integr. Comp. Biol.* 42: 633–40. (b) Based on Gosta Jagersten, *Evolution of the Metazoan Life*. 1972. (c) Courtesy of Marion R. Wells. From Bohall, P. J., M. R. Wells, and C. M. Chandler. 1997. *Invert. Biol.* 116:26–29.

from nematodes primarily in lacking any excretory system and in the morphological details of their nervous and reproductive systems. Also unlike nematodes, they do not show constancy of cell numbers (eutely) in any of their tissues. Moreover, horsehair worms have nonfunctional digestive tracts that are degenerate at both the anterior and posterior ends. Indeed, the adults do not feed at all, but typically live rather inactive lives, metabolizing nutrients acquired as juveniles. Females are especially inactive, devoting most of their energy to egg production: A female nematomorph commonly releases more than 1 million fertilized eggs in her lifetime.

Juvenile nematomorphs also lack a functional digestive system; they live as internal parasites, mostly in insect hosts, exploiting dissolved nutrients in the host tissues and fluids. Life cycle details are still unknown for most species. In at least some species, the parasite enters the arthropod host as a small larva (about 100 μm), either through drinking contaminated water or eating infected prey, or by directly penetrating the intended host. The larvae possess a spined, eversible proboscis (introvert) (Fig. 17.2b, c), similar to that found in the 3 other phyla to be discussed in this chapter. Once in their unlucky host, the nematomorphs develop gradually to full adult size and shape, with the digestive tract degenerating in the process; they ultimately emerge into a suitable aquatic environment by bursting out, killing the host miserably in the process. In the days before horses gave way to cars, the sudden appearance of these large but slender worms into the external world suggested that they were hairs from horses' tails somehow come to life. Their actual life history is, perhaps, almost as startling.

Phylum Priapulida

Phylum Priapulida
(G: the penis)
prē-ap-´ū-lē´-dah

Defining Characteristic: Large body cavity contains amoebocytes and blood cells (erythrocytes)

The phylum Priapulida contains only about 18 described species, most of which live in muddy sediments and are rarely encountered. In contrast, what appear to be priapulids form a far more conspicuous part of the fauna fossilized some 525–540 million years ago, in the early to mid-Cambrian period. Like most nematodes, most priapulids are only a few millimeters long (although individuals in some species reach 20 cm in length) and, like the nematomorphs, they possess an extensive body cavity. It is not yet clear whether the body cavity is a pseudocoel or a true coelom; the necessary, careful embryological studies have never been done.

Like nematodes and nematomorphs, priapulids secrete an external cuticle and molt it periodically; however, this cuticle is chitinous, as in arthropods. Like that of both nematodes and nematomorphs, the priapulid body lacks internal septa and is circular in cross section, cylindrical, and not annulated. Like nematomorphs and most nematode species, priapulids are gonochoristic.

Like the 2 groups remaining to be discussed in this chapter, priapulids are all marine and free-living throughout their lives. Priapulids typically show external fertilization rather than copulation and internal fertilization. The body cavity fluid serves as a hydrostatic skeleton for locomotion within sediment and also acts as the circulatory medium. It contains cells bearing the unusual blood pigment hemerythrin, an iron-based oxygen-binding protein also found in such unrelated animals as polychaete worms and brachiopods.

Like that of larval nematomorphs and of the other 2 phyla discussed in this chapter, the priapulid body bears an eversible anterior introvert, with the mouth at its end (Fig. 17.3). Most of the larger priapulids and at least some of the others seem to be active predators, while many of the smaller priapulids apparently feast exclusively on detritus. The excretory system is conspicuous and includes protonephridia. There are no free-swimming larvae in the life history; developmental stages resemble adult priapulids (except for having a cuticular covering about the abdomen) and live in sediment. Individuals of many species develop a pronounced tail as they approach adulthood; the adult tail may be single or multiple (Fig. 17.3b).

Priapulids occur in both warm- and cold-water habitats from the intertidal to abyssal depths, but they are seldom found in high concentrations. The larger priapulids seem to do best in habitats intolerable to most other animals, such as anoxic muds and high-salinity pools. The smaller species tend to be found within thriving interstitial communities.

The place of priapulids in the evolutionary scheme has long been uncertain, despite their free-living lifestyle. Until the early part of the 20th century, priapulids were grouped with 2 other types of unsegmented worms, the Echiura and the Sipuncula (Chapter 13), to form a single phylum, the Gephyrea. However, unlike the members of those other 2 groups, and indeed unlike protostomes in general, priapulids show a radial cleavage pattern during early development. Their present placement among protostomes by most recent authors is a testament to the diminishing importance of cleavage patterns (and types of body cavities) in deducing phylogenetic relationships. Increasing evidence suggests a close relationship between priapulids and the next animals to be discussed, the kinorhynchs.

Phylum Kinorhyncha (= Echinoderida)

Phylum Kino • rhyncha (= Echino • derida)
(G: movable snout; G: hedgehog hide)
kī-nō-rink´-ah ē-kī-nō-deer´-id-ah

Defining Characteristic: Body consists of 13 segments

Kinorhynchs are *bona fide* pseudocoelomates, and the approximately 180 described species are exclusively marine. Like the other cycloneuralians, and like the arthropods, they lack external, motile cilia. Instead of gliding or swimming, they crawl through the mud in which they live, using the active, eversible head to thrust forward;

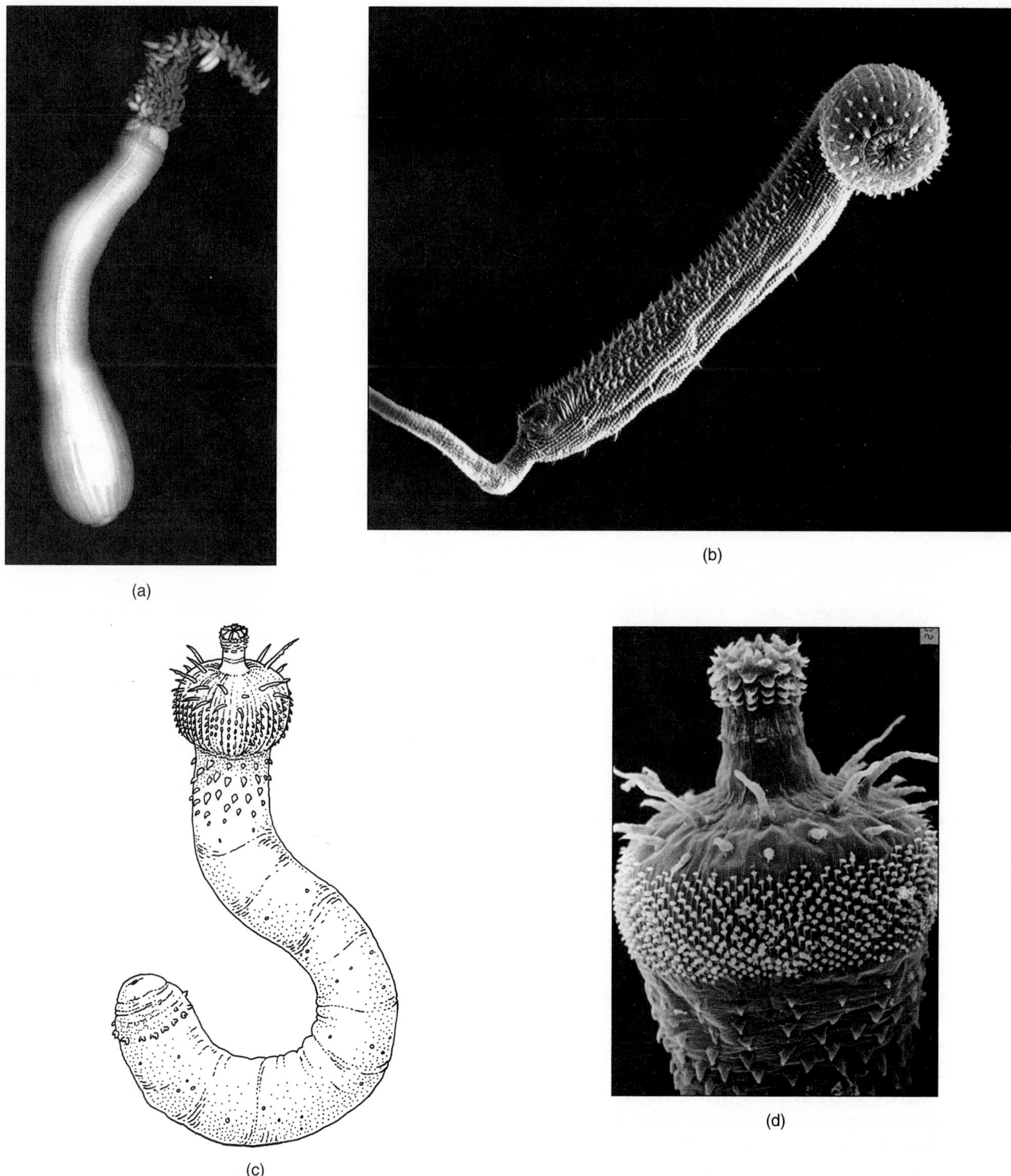

Figure 17.3

(a) *Priapulus caudatus.* The animal is 4 cm long, excluding the tail. The longitudinal nerve cord is visible on the ventral surface of the body. (b) Scanning electron micrograph of the interstitial priapulid *Tubiluchus corallicola* with introvert everted. The body is about 2 mm long, excluding the tail. (c) Diagram of the external morphology of *Meiopriapulus fijiensis,* an interstitial priapulid species. (d) Scanning electron micrograph of the anterior end of *M. fijiensis.*

(a,b) Courtesy of C. Bradford Calloway. From Calloway, 1975. (Springer-Verlag) *Marine Biology,* Vol. 31, pp. 161–74. "Morphology of the introvert and associated structures of the priapulid *Tubiluchus corallicola* from Bermuda." (Fig. 1, p. 163, in *Marine Biology.*) (c) From M. P. Morse, in *Transactions of the American Microscopic Society* 100: 239, 1981. Courtesy of Dr. M. P. Morse. (d) Courtesy of Dr. M. P. Morse.

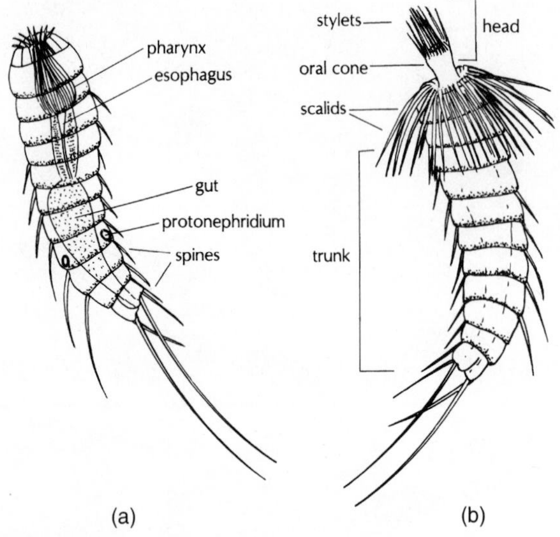

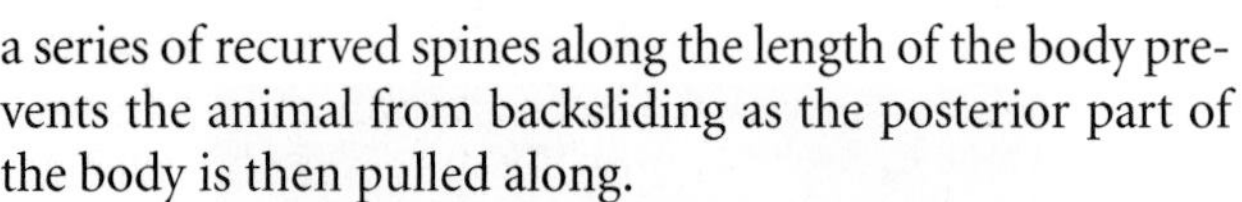

Figure 17.4
Typical kinorhynch (*Echinoderella* sp.), with head withdrawn (a) and protruded (b).
(a,b) After Hyman.

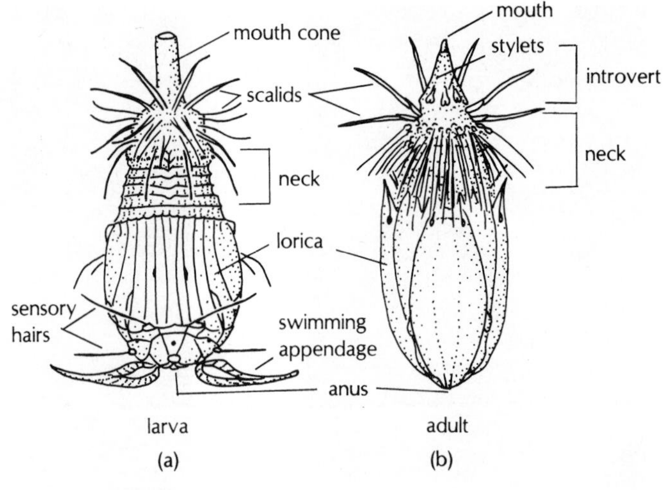

Figure 17.5
Nanaloricus mysticus, the first known member of the recently erected phylum Loricifera. (a) Higgins larva. (b) Adult.
From Kristensen, 1983. *Zeit. Zool. Syst. Evol.-forsch.* 21:163.

a series of recurved spines along the length of the body prevents the animal from backsliding as the posterior part of the body is then pulled along.

The tiny body (usually less than 1 mm long) is covered by an external, chitinous cuticle, as in priapulids. Kinorhynchs and priapulids may have an immediate ancestor in common, and are sometimes grouped together as scalidophorans (Fig. 17.1). However, the kinorhynch cuticle, musculature, and nervous elements are distinctly segmented. Such segmentation is unique among aschelminths. The first segment is the head, which displays several rings of curved spines called **scalids** and an additional, inner ring of piercing stylets surrounding the mouth opening (Fig. 17.4b). Since the anterior end bears the mouth and is eversible—giving rise to the most widely used phylum name (Kinorhyncha, meaning "movable snout")—it is properly termed an **introvert,** as discussed earlier. The second segment, called the "neck," typically bears numerous cuticular plates that seal the opening once the head has been retracted. However, in some kinorhynch species, both the head and neck are retracted, and the opening is sealed with plates on the third segment. The trunk, always consisting of 11 segments, bears numerous spines and adhesive tubes; these are important to the animal in locomotion, and to the systematist in species identification. As is so common for such small organisms, kinorhynchs lack specialized respiratory and circulatory systems. The excretory system consists of one pair of solenocytic protonephridia.

Kinorhynchs can occur at impressively high concentrations: Aggregations of about 2 million individuals per square meter of sediment have been reported. Kinorhynchs mostly eat bacteria and all have linear digestive tracts. All species are gonochoristic, with separate males and females, but developmental details are poorly known. Juveniles molt the cuticle periodically as they grow, passing through 6 stages before reaching adulthood. There are no free-living larvae in the life cycle.

Phylum Loricifera

Phylum Lorici • fera
(Latin: armor bearing)
lor-ih-sif′-er-ah

Defining Characteristic: Spinelike, chitinous scales (scalids) on the introvert are operated by individual muscles

One would think that all the animal phyla would have been discovered long ago. Even the kinorhynchs, for example, despite their small size and specialized habitat, have been well known for about 150 years. Yet, in 1983, a new phylum, the Loricifera, was erected to hold a number of newly discovered little animals, *Nanaloricus mysticus* (Fig. 17.5). More than 100 loriciferan species are now known, although most have not yet been formally described. All known loriciferans are marine. They are only about 50–500 µm long, and all live interstitially in subtidal sediments, clinging tightly to the surrounding sand grains. Despite their small size, each loriciferan is composed of thousands of cells and exhibits remarkable structural complexity. Loriciferans were discovered by accident, when newly collected sediments were rinsed with tap water instead of being relaxed in the normal fashion, with $MgCl_2$ in seawater; the osmotic shock apparently caused the animals to release their grips on the sand grains, bringing them to zoologists' attention for the first time.

Like the kinorhynchs just discussed, loriciferans begin with an anterior introvert surrounded by recurved spines, called **scalids;** like that of the kinorhynchs, the introvert is retractable and bears the mouth at its end (by definition). The mouth is encircled by piercing stylets. The neck is made of several segments, rather than the single segment of kinorhynchs, and bears numerous plates that likely protect the anterior end when it is withdrawn, as with kinorhynchs.

The posterior half of the body is covered by 6 overlapping plates comprising an external cuticle, or **lorica,** which is molted as the juveniles grow. Many rotifer species also exhibit a distinct lorica, making them superficially resemble loriciferans. However, like kinorhynchs (and unlike gastrotrichs and rotifers), loriciferans lack external cilia. The animals are gonochoristic and fertilization is probably internal, but few details of their sex lives or embryological development are yet known. Some species seem to be acoelomate throughout their lives, while others seem to have acoelomate larvae but pseudocoelomate adults. The preadult stage, called a **Higgins larva,** differs morphologically from the adult mainly in having a pair of posterior, unjointed swimming appendages and several pairs of locomotory spines on the ventral surface of the lorica (Fig. 17.5a). The Higgins larva shares certain morphological characteristics with the juveniles of priapulids, nematomorphs, and kinorhynchs, but also with rotifers. For the time being, at least, most systematists seem comfortable with a scenario in which the loriciferans, priapulids, kinorhynchs, and nematomorphs all derive from a common ancestor. The discovery of so many shared characteristics, including gene sequence comparisons, seems to be binding a good number of the various aschelminth groups more closely together again, possibly into a single phylum. Whether they retain their presumed close affiliation with arthropods remains to be seen.

Taxonomic Summary

Phylum Nematomorpha—horsehair worms
Phylum Priapulida
Phylum Kinorhyncha (= Echinoderida)
Phylum Loricifera

Topics for Further Discussion and Investigation

1. Until the early part of the 20th century, priapulids were grouped with 2 other types of marine worms, the Sipuncula and the Echiura (Chapter 13), in a single phylum, the Gephyrea. In what respects do the priapulids resemble sipunculans and echiurans?

2. Nematomorphs are unusual parasites in that the adults are nonfeeding and free-living, while the larvae parasitize a variety of arthropod hosts. What characteristics do nematomorphs share with the parasitic nematodes? What characteristics set the Nematomorpha apart from the Nematoda?

Taxonomic Detail

Phylum Nematomorpha

The approximately 325 species in this phylum are divided among 2 classes.

Class Nectonematoida

Nectonema. This small class of only 5 species contains the marine horsehair worms, which parasitize crabs and other decapod crustaceans.

Class Gordioida

Chordodes, Gordius. This class contains most nematomorph species. All live in freshwater or semiterrestrial habitats, and all are endoparasites of various insects, including beetles, grasshoppers, and cockroaches. Four families.

Phylum Priapulida

Maccabeus, Meiopriapulis, Priapulus, Tubiluchus. The entire phylum contains only 18 described species, divided among 3 families (the Priapulidae, Tubiluchidae, and Chaetostephanidae); a few more species are awaiting formal description, and another 11 species are known only as ancient fossils. All species are marine. Many priapulids are only a few millimeters long when fully grown, but the largest individuals, found in the genus *Priapulus,* reach lengths of up to 20 cm. Priapulids are found at all depths, from intertidal to abyssal, and in both warm and cold waters.

Phylum Kinorhyncha (= Echinoderida)

All approximately 180 species are marine. They are distributed among only 2 orders.

Order Cyclorhagida

Echinoderes. This order contains a highly diverse group of kinorhynchs. Some species are intertidal, while others are found only at depths of several thousand meters. Many species are free-living, while some appear to be commensal with such other invertebrates as sponges, bryozoans, or hydrozoans. The order contains slightly more than 60% of all kinorhynch species, and all are small, certainly less than about 500 μm long. Four families.

Order Homalorhagida

Pycnophyes, Kinorhynchus. This group contains the largest of the kinorhynchs, with some individuals growing to nearly 1 mm in length. Individuals commonly live at depths of several thousand meters. Two families.

Phylum Loricifera

Nanaloricus. All known loriciferans live interstitially in marine sediments. Twenty-two species have been formally described, and about 100 more await description.

General References about the Nematomorphs, Priapulids, Kinorhynchs, and Loriciferans

Bleidorn, C., A. Schmidt-Rhaesa, and J. R. Garey. 2002. Systematic relationships of Nematomorpha based on molecular and morphological data. *Invert. Biol.* 121: 357–64.

Garey, J. R., and A. Schmidt-Rhaesa. 1998. The essential role of "minor" phyla in molecular studies of animal evolution. *Amer. Zool.* 38:907–17.

Harrison, F. W., and E. E. Ruppert, eds. 1991. *Microscopic Anatomy of Invertebrates, Vol. 4: Aschelminthes.* New York: Wiley-Liss.

Higgins, R. P., and H. Thiel, eds. 1988. *Introduction to the Study of Meiofauna.* Washington, D.C.: Smithsonian Institution Press.

Hyman, L. H. 1951. *The Invertebrates,* Volume 3. *Acanthocephala, Aschelminthes, and Entoprocta.* New York: McGraw-Hill.

Kristensen, R. M. 2002. An introduction to Loricifera, Cycliophora, and Micrognathozoa. *Integr. Comp. Biol.* 42: 641–51.

Morris, S. C., and D. W. T. Crompton. 1982. The origins and evolution of the Acanthocephala. *Biol. Rev.* 57:85–115.

Nehaus, B., and R. P. Higgins. 2002. Ultrastructure, biology, and phylogenetic relationships of Kinorhyncha. *Integr. Comp. Biol.* 42: 619–32.

Nielsen, C. 2001. *Animal Evolution: Interrelationships of the Living Phyla,* 2nd ed. New York: Oxford University Press.

Parker, S. P., ed. 1982. *Classification and Synopsis of Living Organisms,* vol. 1. New York: McGraw-Hill, 857–77, and 931–44.

Schmidt-Rhaesa, A. 2002. Two dimensions of biodiversity research exemplified by Nematomorpha and Gastrotricha. *Integr. Comp. Biol.* 42: 633–40.

Thorpe, J. H., and A. P. Covich, eds. 2001. *Ecology and Classification of North American Freshwater Invertebrates. Nematomorpha,* 2nd ed. New York: Academic Press.

Willmer, P. 1990. *Invertebrate Relationships.* New York: Cambridge University Press.

18

Three Phyla of Uncertain Affiliation:
Gastrotricha, Chaetognatha, and Cycliophora

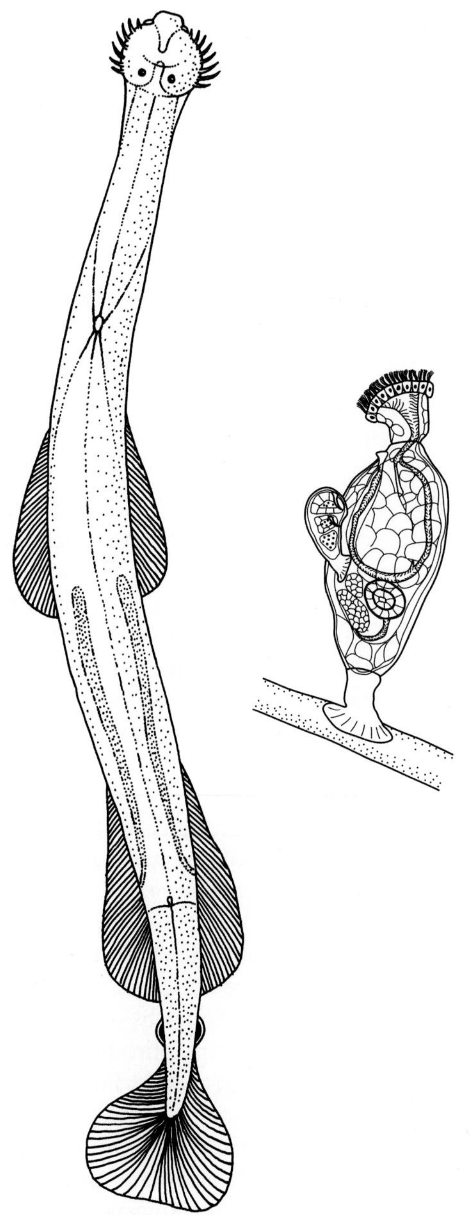

Introduction

The phylogenetic positions of the animals discussed in this chapter are uncertain at best, either because useful diagnostic characters have been lost through adaptation to small body size or because the animals show a baffling mixture of morphological and developmental characters. The animals discussed in this chapter are related only by the uncertainty of their relationships with other animal groups.

Phylum Gastrotricha

Phylum Gastro • tricha
(G: belly hair)
gas-tro-trik´-ah

Gastrotrichs (Fig. 18.1) are small—as tiny as 80 μm and rarely as large as 1 mm—acoelomate members of benthic (bottom-dwelling) fresh- and saltwater communities, occurring in concentrations of up to 100,000 individuals per m^2. A blastocoel forms during development but does not persist into adulthood. About 300 gastrotrich species live in freshwater, and nearly another 400 species are marine.

Gastrotrichs have long had an uncertain phylogenetic position. They have been allied in various studies with flatworms (Chapter 8), gnathostomulids and rotifers (Chapter 10), and nematodes and other ecdysozoans (Chapter 16, p. 432). Recent studies suggest that gastrotrichs should be included with nematodes and other animals discussed in the previous 2 chapters in the larger clade Cycloneuralia (p. 451). Like cycloneuralians, gastrotrichs have a secreted outer cuticle. However, unlike the cycloneuralians listed in Chapter 17, gastrotrichs do not molt. Neither do they have an introvert; moreover, unlike the animals discussed in the previous chapter, gastrotrichs have locomotory cilia.

Most gastrotrichs live interstitially in the spaces between sediment particles. The head bears a number of sensory bristles, and the body has numerous spines

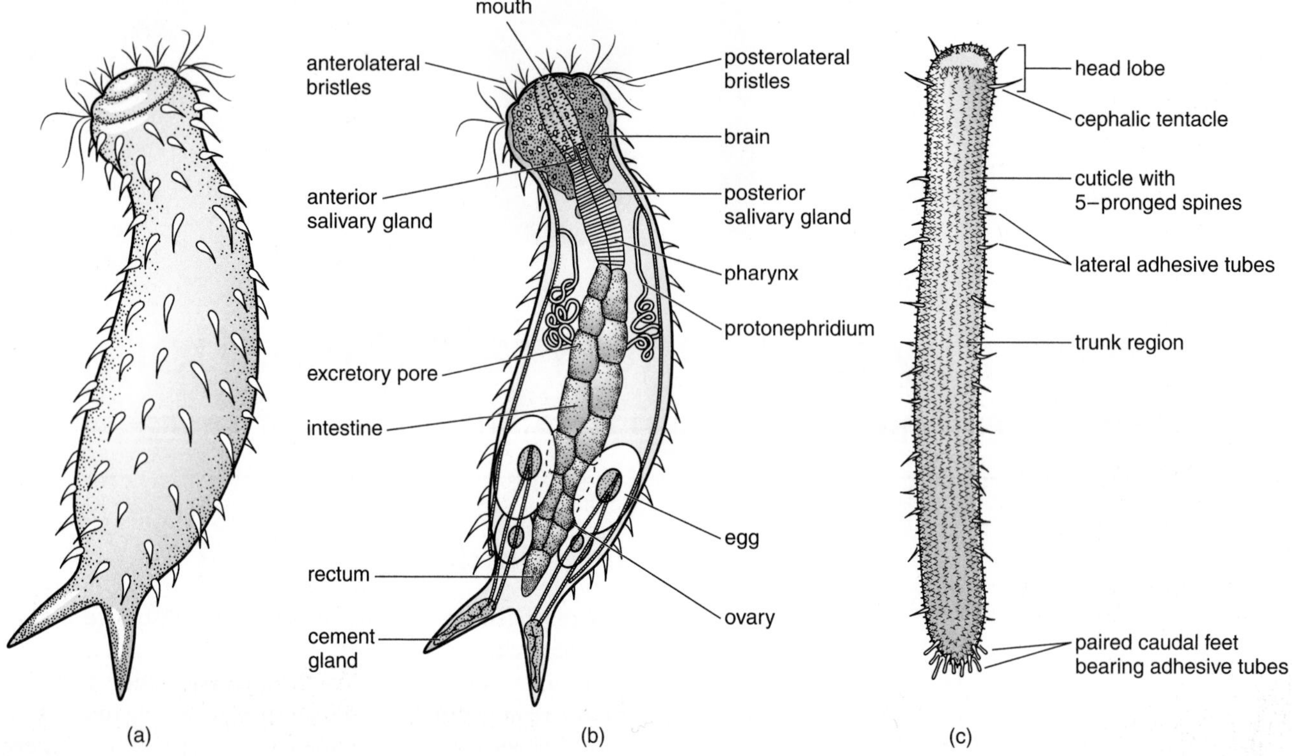

Figure 18.1

Typical gastrotrich (*Chaetonotus* sp.). (a) External appearance. (b) Internal anatomy. (c) *Tetranchyroderma* sp., a marine gastrotrich. The animal forms temporary attachments to sand grains using its adhesive tubes. The specimen illustrated was about 500 μm long.

(a) After Brunson; after Zelinka. (b) From Brown; after Remane. (c) Based on L. Margulis and K. Schwartz, *Five Kingdoms*, 2d ed. 1988.

(Fig. 18.1). Despite their superficial resemblance to rotifers (Chapter 10), gastrotrichs are clearly beasts of a different phylum: In particular, they lack the characteristic rotifer corona and mastax. In any event, all gastrotrichs possess both circular and longitudinal body wall muscles. The ventral surface is amply ciliated (as the name "gastrotrich" implies), so gastrotrichs can glide over a substrate and even swim short distances. Oddly, the locomotory cilia are covered by cuticle, unlike any other metazoans. Epidermal cells are monociliated in some species, a trait they share only with gnathostomulids, with a single cilium per cell, but multiciliated in others. The pattern of ciliation is an important taxonomic characteristic. Gastrotrichs also can form temporary attachments to solid surfaces, as can rotifers. More like the free-living flatworms, however, they possess a double-gland system in which one gland secretes the glue and the other secretes a de-adhesive to release the attachment. They cling so tightly to particles that zoologists must first anesthetize them with magnesium chloride ($MgCl_2$) to dislodge them for enumeration and further study.

There are no known parasitic species or carnivorous species; all gastrotrichs eat detritus, bacteria, diatoms, or protists. All have a linear digestive system, with an anterior mouth and a posterior anus.

Like other aschelminths, gastrotrichs lack any specialized respiratory or circulatory systems. Despite their small size, however, they do possess a discrete, protonephridial excretory system, which differs in morphological detail from those of both flatworms and rotifers; protonephridia are especially common in the freshwater species. In freshwater species, the protonephridia most likely function in maintaining osmotic concentration and body volume, in addition to their presumed function in removing soluble wastes.

Like nematodes and rotifers, gastrotrichs demonstrate **eutely,** with all adults of any given species having the same number of cells; cell numbers increase only in early development. Similar to that of nematodes but unlike that of rotifers, the gastrotrich body is covered with an external cuticle; the rotifer cuticle is typically intracellular. Also unlike rotifers, gastrotrichs—particularly the marine species—are mostly hermaphroditic (Fig. 18.2); no other aschelminth can so boast. Freshwater gastrotrichs commonly reproduce by parthenogenesis. In sexual reproduction, fertilization is always internal, and the fertilized eggs display determinate cleavage, as in rotifers and other aschelminths. However, the cleavage pattern is bilateral and radial rather than spiral. There are no free-living larvae in the life cycle.

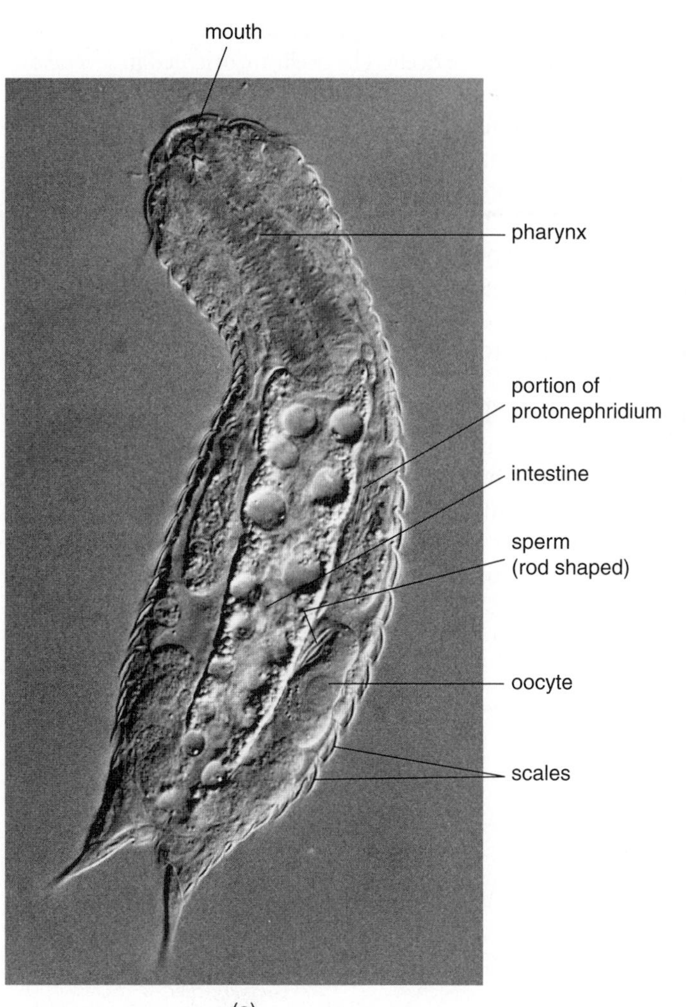

(a)

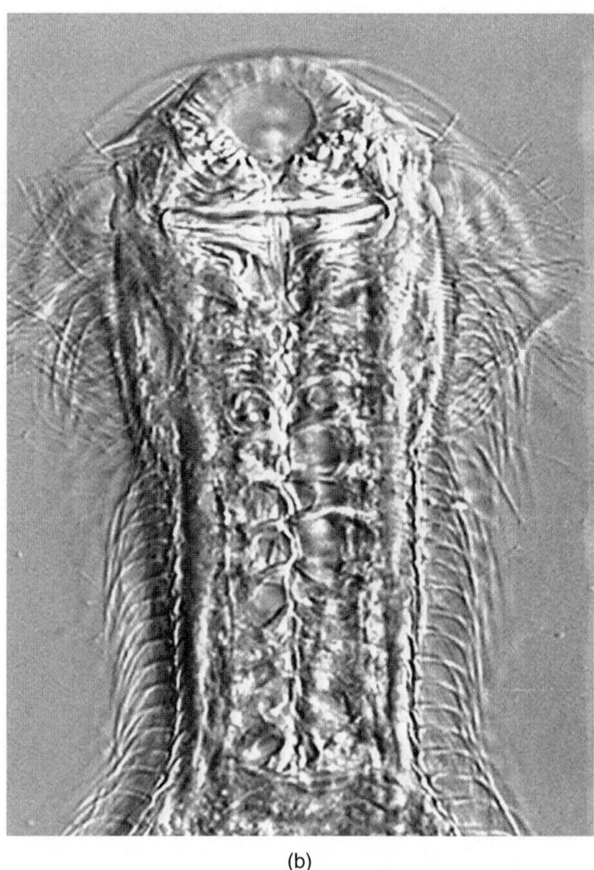

(b)

Figure 18.2
(a) Photograph of *Lepidodermella squamata,* a freshwater gastrotrich. (b) Anterior end of the Australian gastrotrich *Polymerurus nodicaudus.*

(a) Courtesy of Mitchell J. Weiss. (b) Courtesy of Rick Hochberg. From Hochberg, R. 2005. *Invert. Biol.* 124: 119–30.

Phylum Chaetognatha

Phylum Chaeto • gnatha
(G: bristle jaw)
kēt´-og-nath´-ah

Defining Characteristics: 1) A series of curved, chitinous, grasping spines on both sides of the head, for seizing prey; 2) lateral stabilizing fins, composed of ectodermal derivatives

General Characteristics and Feeding

Chaetognaths are all free-living, marine carnivores. The average chaetognath is only a few centimeters long, and even the largest individuals are no more than 15 cm long. Apparently, the ratio between surface area and volume is sufficiently large that gas exchange and excretion requirements can be met by diffusion across the general body surface; chaetognaths bear no specialized respiratory or excretory organs, and they lack a blood circulatory system. The inner lining of the adult body cavity is ciliated, however, so that gas exchange and nutrient transport are achieved by circulation of the coelomic fluid.

As befits an active carnivore, the chaetognath nervous system is fairly complex (Fig. 18.3). The anterior part of the intestine is encircled by a ring of nerve tissue bearing several ganglia. Posteriorly, there is a conspicuous ventral ganglion. Sensory and motor nerves extend from this and associated ganglia to the various light, tactile, and chemosensory systems; to the musculature of the trunk, tail, spines, and digestive tract; and to the cerebral ganglion located in the head. Scattered fan-like clusters of external cilia are sensitive to vibration, enabling chaetognaths to detect the presence and location of the copepods and fish larvae that constitute their principal foods.

Feeding behavior is impressive, both for the human observer and for the prey. Typically, the nearly transparent chaetognath[1] lies motionless in the water, sinking slowly until something edible comes its way, either in front of it or alongside it. The chaetognath then darts forward and grasps its prey with 2 rows of long, curved, stiff spines adjacent to the mouth (Fig. 18.4), or it flexes its body and captures its prey to the side. The entire prey-capturing maneuver takes only about 1/15 of a second in the laboratory. The large, grasping spines also may serve as mechanoreceptors, supplementing information provided by the external ciliary sensors.

Two rows of short teeth on either side of the mouth aid in holding prey during ingestion. These teeth also can puncture the prey exoskeleton and body wall; recent studies indicate that toxic secretions may exit from the vestibular pits, or from pores along the vestibular ridge, and enter prey through these punctures (Research Focus Box 18.1, p. 464).

Chaetognaths bear a pair of small eyes, each composed of 5 pigment-cup ocelli. The ocelli are oriented in several different directions within a single eye, so the chaetognath has a wide visual field; in fact, several ocelli point downward, so the chaetognath actually sees through its own, transparent body. The ocelli are probably not image forming, but they may enable the animal to detect motion and changes in light intensity.[2]

Chaetognaths are commonly called "arrowworms," since the body is substantially longer than it is wide and lacks appendages, although it does bear one or 2 pairs of delicate, **lateral fins** and a terminal **caudal fin** (Fig. 18.5). The lateral fins probably serve to reduce the rate at which the animal sinks while motionless, by increasing body surface area, and to stabilize the body when the animal swims. The caudal fin generates forward thrust for swimming and also must aid in flotation. The fins also seem to stabilize the body—acting as "sea anchors"—when the anterior end flexes to the side

1. See *Topics for Further Discussion and Investigation*, no. 3.

2. See *Topics for Further Discussion and Investigation*, no. 1.

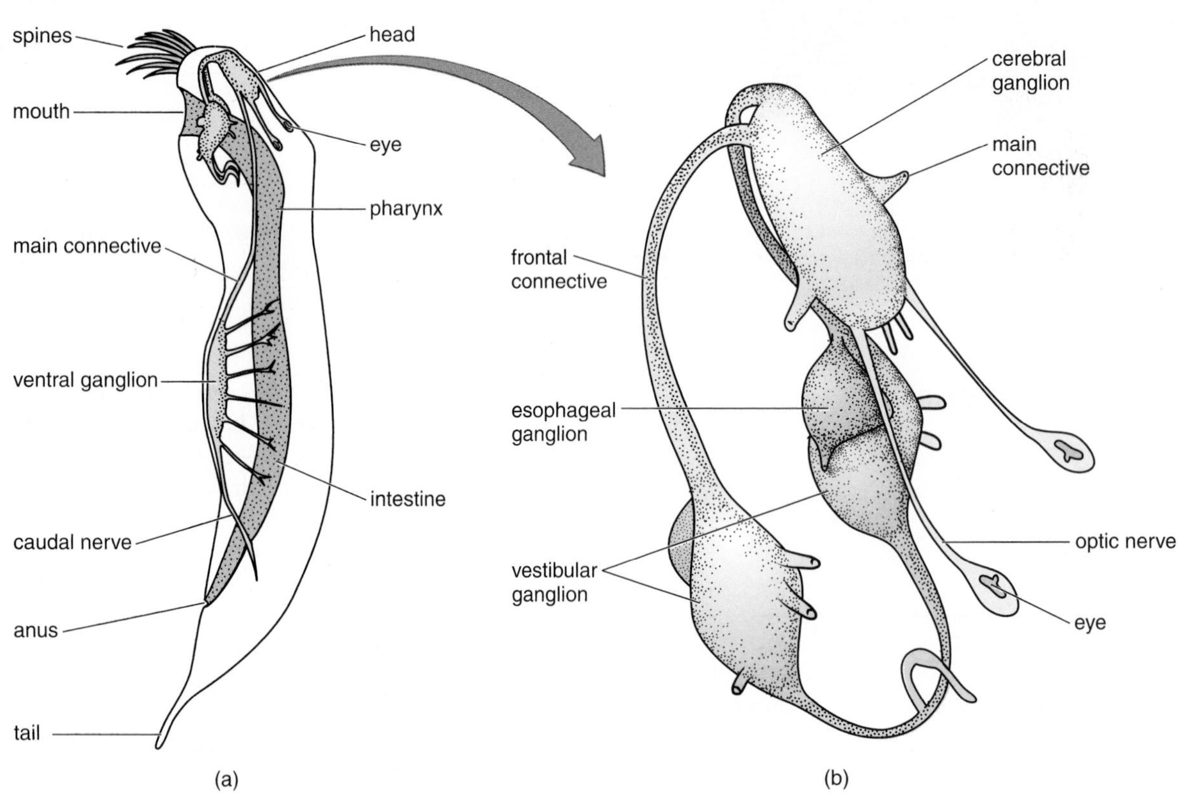

Figure 18.3
(a) Nervous system of a typical chaetognath. (b) Detail of nervous system in the head of a typical chaetognath.

(a,b) Based on T. Goto and M. Yoshida, in *Nervous Systems in Invertebrates*, M. A. Ali, 1987.

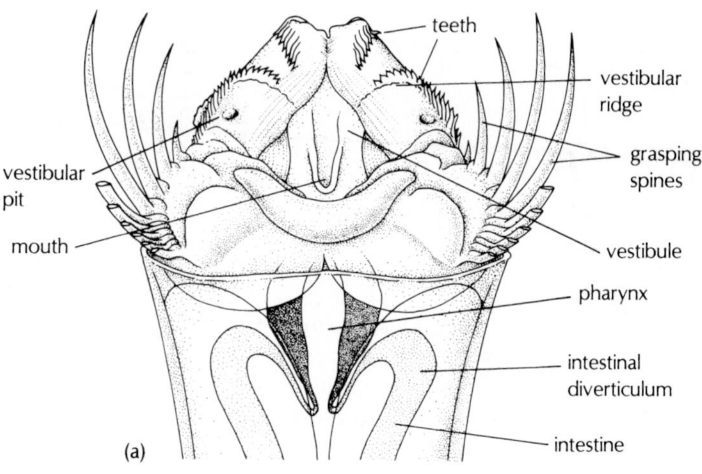

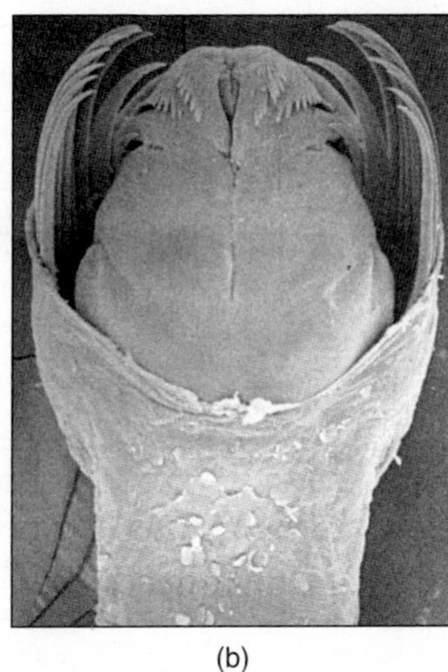

Figure 18.4

(a) The anterior end of *Sagitta elegans,* in ventral view. The eyes are located dorsally and are not visible in this orientation. The vestibular pits and the papillae of the vestibular ridges are probably secretory. The spines and teeth of the head are normally covered by a hood, which is retracted only for prey capture. (b) Scanning electron micrograph of *Sagitta setosa,* anterior end.

(a) From Hyman; after Ritter-Zahony. (b) Courtesy of Q. Bone et al., 1983. *Journal of the Marine Biological Association of the United Kingdom,* Vol. 63:929. Reprinted with the permission of Cambridge University Press.

during prey capture. The entire body is covered by a thin cuticle secreted by the underlying epidermis. The cuticle is not molted.

The body cavity of adult chaetognaths is compartmentalized. One septum isolates the head compartment from that of the trunk. A second septum divides the trunk compartment from that of the tail (Fig. 18.5). The chaetognath gut is linear and open at each end (Fig. 18.6). Digestion is extracellular and occurs within the intestine. There is no distinct stomach or digestive gland. Chaetognaths can repair wounds but cannot regenerate lost body parts.

Chaetognath Reproduction

All chaetognaths are simultaneous hermaphrodites (Fig. 18.7), with the male gonads generally maturing earlier than the female gonads. The ovaries are found in the coelomic compartment of the trunk, while the testes occupy the coelomic compartment of the tail. Self-fertilization occurs at least occasionally in some species. Chaetognaths do not copulate during cross-fertilization; instead, sperm are transferred indirectly, through an often mutual exchange of sperm-filled containers called **spermatophores.** The spermatophores are manufactured within a conspicuous pair of seminal vesicles and are eventually attached to the outside of the recipient's body. Once sperm escape from the spermatophores, they migrate along the body to the opening of the female reproductive tract. Fertilization is internal, after which the fertilized eggs are released and develop into miniatures of the adult. Chaetognaths do not have a morphologically distinct larval stage in their life history.

Chaetognath Lifestyles and Behavior

Most chaetognath species are planktonic, spending their entire lives being passively transported by ocean currents. Although, as planktonic animals, arrowworms cannot swim against any substantial water current, many species can and do undertake extensive vertical migrations (Fig. 18.8), sometimes traveling hundreds of meters daily. Many other planktonic animals, including pteropods (Gastropoda); copepods and cladocerans (Crustacea); and a variety of invertebrate larval stages display similar patterns of behavior. Generally, the animals migrate downward during the day and swim upward at night, although the timing and extent of the migrations often differ significantly among species and among different developmental stages of a given species. No single rationale for such **diurnal vertical migrations** has been demonstrated conclusively in laboratory or field experiments, although several reasonable adaptive benefits have been proposed: (1) avoiding visual predators, (2) increasing the energetic efficiency of feeding and digestion, (3) reducing competitive interactions both within and among species, (4) improving physiological well-being by experiencing varied temperatures and salinities, and (5) achieving dispersal by taking advantage of vertical differences

RESEARCH FOCUS BOX 18.1

Chaetognath Prey Capture

Thuesen, E. V., K. Kogure, K. Hashimoto, and T. Nemoto. 1988. Poison arrowworms: A tetrodotoxin venom in the marine phylum Chaetognatha. *J. Exp. Marine Biol. Ecol.* 116:249–56.

For many years, biologists have suspected that prey capture among chaetognaths is facilitated by paralytic toxins, but the animals' small size made the hypothesis impossible to demonstrate: If the entire animal is only perhaps a few centimeters long, how does one isolate, and then hope to characterize, what are likely to be minute amounts of toxic secretions? Thuesen et al. (1988) developed techniques sufficiently sensitive to conclusively demonstrate that at least 6 chaetognath species do indeed produce a potent neurotoxin, one that paralyzes victims by blocking sodium channels in cell membranes.

To isolate any toxins, the researchers collected adult chaetognaths, decapitated them, and then extracted potential toxins by bathing the heads in a solution of 0.1% acetic acid. Remaining tissues were discarded, and the extract was concentrated by evaporation. To bioassay for the presence of neurotoxins, the researchers used tissue cultures of mouse nerve cells (neuroblastomas). The approach was both clever and simple, and it made use of the known physiological effects of 2 other chemicals: ouabain and veratridine.

Ouabain disengages the sodium-potassium pump that maintains the normal resting potential of all eukaryotic cells. With the pump operating normally, sodium ions are actively expelled while potassium ions are simultaneously carried in. In the presence of ouabain, the pump is deactivated, and sodium ions flood into cells along their concentration gradient, a response that is further amplified by the second chemical, veratridine. Thus, when the 2 chemicals are added to the nerve cells in tissue culture, sodium ions flood into the cells, and the cells quickly swell and die. In the presence of a neurotoxin that blocks sodium channels, however, the cells should be spared since sodium ions will not be able to move across the cell membrane. Thuesen et al. could therefore show that the chaetognath head contains neurotoxins by demonstrating that the extract blocks the effects of ouabain and veratridine.

This was precisely the effect shown by head extracts of the 6 chaetognath species tested; in contrast, extracts of headless chaetognaths did not save the cells from the effects of ouabain and veratridine.

Do all chaetognaths produce neurotoxins? The chaetognaths tested in the Thuesen et al. study included both benthic and pelagic species from 3 different families and a wide range of geographical distributions (Focus Table 18.1), suggesting that the phenomenon is widespread. Secretion of neurotoxins certainly would explain the ability of many chaetognaths to capture larval fish as large as themselves.

Where is the toxin produced and secreted? The researchers suggest that the most likely sites are the papillae of the vestibular ridges adjacent to the mouth (Fig. 18.5), because venom secreted here would have ready access to piercing wounds made by the teeth. How could you demonstrate whether this is indeed the secretion site?

Focus Table 18.1 Habitat and Geographical Distribution of Chaetognaths Shown to Use Neurotoxins in Prey Capture

Family and Species	Habitat	Geographical Distribution
Eukrohniidae		
Eukrohnia hamata	Open ocean	Polar waters, deep tropical waters
Sagittidae		
Sagitta elegans	Coastal	Arctic waters
Flaccisagitta scrippsae	Open ocean	North Pacific
F. enflata	Open ocean	Tropical waters
Aidanosagitta crassa	Coastal	Japan
Spadellidae		
Spadella angulata	Benthic	Japan, Southeast Asia

From E. V. Thuesen, K. Kogure, K. Hashimoto, and T. Nemoto. 1988. Poison arrowworms: A tetrodotoxin venom in the marine phylum Chaetognatha. *J. Exp. Marine Biol. Ecol.* 116:249–56.

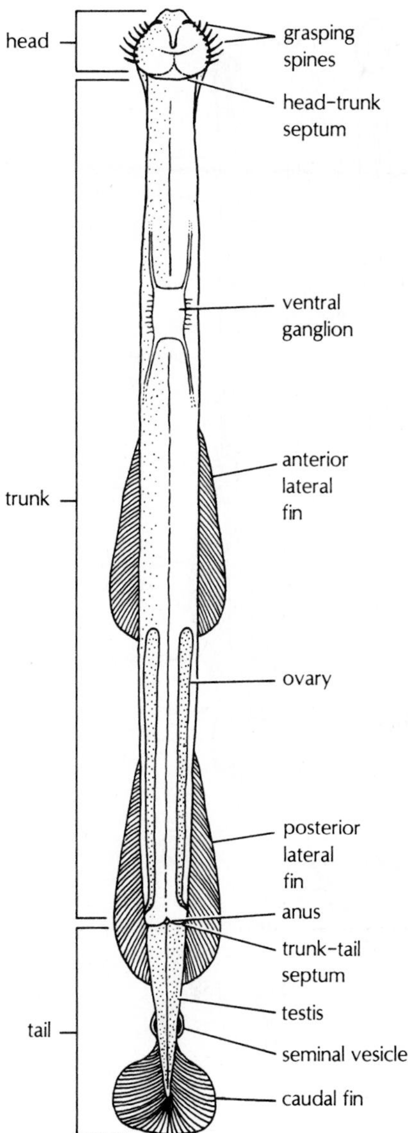

Figure 18.5

Sagitta elegans, seen in ventral view.
From Hyman; after Ritter-Zahony.

in speeds and directions of water currents.[3] The most convincing evidence presented to date (field studies on certain planktonic crustaceans) best supports the anti-predation hypothesis.

One family of chaetognaths (the Spadellidae) is entirely benthic. These arrowworms use specialized **adhesive papillae** to form temporary attachments to solid substrates, usually rocks and/or macroalgae. In the typical feeding posture the attachment is made posteriorly, with the rest of the body held elevated above the substrate. When the need arises, the attachment can be broken and the animal can dart off to a new location.

Only about 150 chaetognath species are known. This phylum is nevertheless an important one, ecologically and economically. Local concentrations as high as several hundred chaetognaths per cubic meter of seawater have been reported. As significant predators of fish embryos, fish larvae, and copepods (Fig. 18.9), chaetognaths must be important components of oceanic food chains and, in particular, must play a major role in determining the size of herring and other commercially exploitable fish populations.

Chaetognath bioluminescence was described for the first time in 1994, for one deep water species rarely found at depths shallower than 700 m. This species luminesces using the same chemical substrate found in other marine animals as disparate as cnidarians, squid, and crustaceans. Bioluminescence is known from no other chaetognath species.

Chaetognath Relationships

Chaetognaths are also intriguing for their enigmatic place in the phylogenetic scheme. A putative fossilized chaetognath, about 520 million years old, was recently described from Southern China,[4] making chaetognaths an ancient group. Yet, their relationship to other animal groups has long been uncertain (Fig. 2.12). The adult body cavity is lined with mesodermally derived peritoneum and is thus a true coelom, by definition. Studies of chaetognath embryology establish the arrowworms as deuterostomous coelomates in that cleavage is basically radial and indeterminant (i.e., cell fates are not irrevocably fixed following the first cell division) and the site of the blastopore gives rise to the anus; as in echinoderms and other deuterostomes, the mouth arises elsewhere (recall that *deuterostome* = G: second mouth). Moreover, the embryonic coelom arises from an archenteron, although, in detail, the method of coelom formation by chaetognaths differs significantly from the basic deuterostome plan. As discussed in Chapter 2, the standard deuterostome coelom arises from a symmetrical outpouching of the archenteron. In chaetognaths, on the other hand, the coelom is formed by invagination of the archenteron. Nevertheless, coelom formation is clearly enterocoelous, rather than schizocoelous, in nature.

In many other respects, however, chaetognaths are not convincing deuterostomes. Most marine deuterostomes have a distinctive ciliated larval stage in the life history; in contrast the morphology of the young chaetognath closely resembles that of the adult. The body cavity and musculature of adult chaetognaths are also unlike those of other deuterostomes. In fact, the only major, conspicuous morphological similarity with other deuterostomes is the division of the adult body cavity into 3 distinct compartments. The trunk and tail compartments are clearly illustrated in Figure 18.7; the head contains a small, additional coelomic space. For many years, it was believed that the embryonic

3. See *Topics for Further Discussion and Investigation*, no. 2.

4. Chen, J.-Y., and D.-Y. Huang. 2002. *Science* 298: 187.

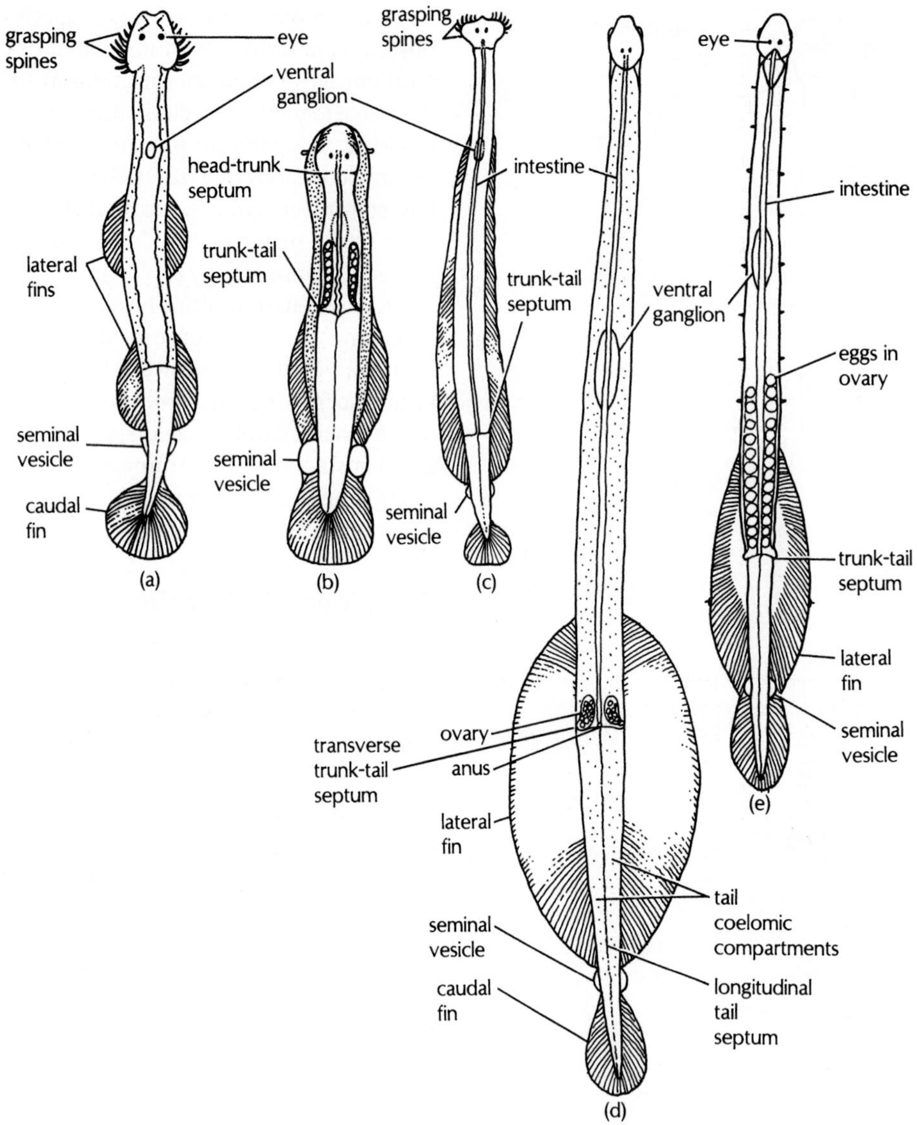

Figure 18.6

Chaetognath diversity. Note the differences in the number and extent of the lateral fins and in the external appearance of the seminal vesicles. (a) *Sagitta macrocephala.* (b) *Sagitta cephaloptera.* (c) *Eukrohnia fowleri.* (d) *Krohnitta subtilis.* (e) *Krohnitta pacifica.*
(d,e) From Hyman; after Tokioka.

coelom was obliterated during subsequent development, with the adult body cavity forming later and by an entirely different mechanism. It turns out, however, that the embryonic coelomic cavities are simply compressed for a time and later reopen in the juvenile; in this respect, at least, chaetognath development is not as strange as had been thought. Still, many have long doubted their deuterostome affinities. Indeed, the Chaetognatha has long been among the most phylogenetically isolated of animal phyla.

Some workers have suggested affliliating chaetognaths with aschelminths. In particular, chaetognath musculature bears some resemblance to that encountered within the phylum Nematoda. No circular muscles are found among the members of either phylum: The musculature of the chaetognath body wall is exclusively longitudinal and, like that of the nematodes, is arranged in discrete bundles. Alternating contraction of the ventral and dorsal musculature of the chaetognath trunk and tail provides the thrust for locomotion, the 2 sets of muscles antagonizing each other through the hydrostatic skeleton of the fluid-filled, compartmentalized body cavity. Unlike nematodes, however, chaetognaths do not generate sinusoidal waves of activity. Instead, muscle contractions are sporadic, so the animal darts forward intermittently. And, unlike many aschelminths, chaetognaths are not eutelic (i.e., chaetognaths grow primarily through increases in cell numbers, not increases in individual cell size).

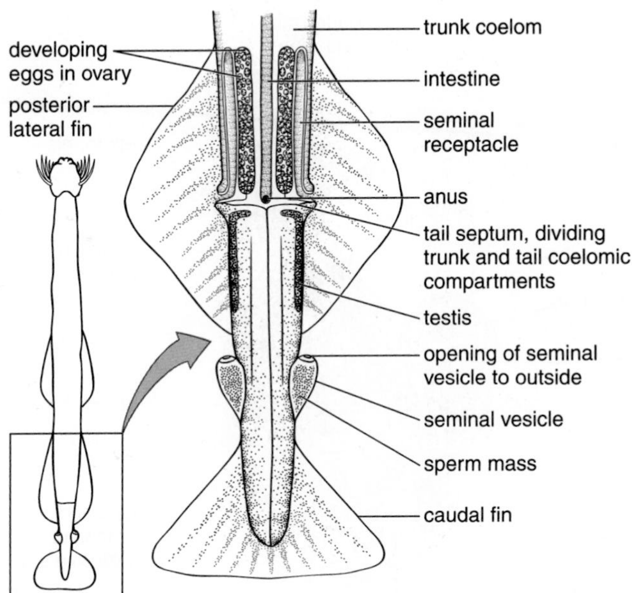

Figure 18.7

Ventral view showing reproductive system of *Sagitta elegans*. Chaetognaths are simultaneous hermaphrodites, possessing both male and female gonads. The sperm are packed into spermatophores within the seminal vesicles. Sperm received from another individual are stored in the seminal receptacles.
Based on Frank Brown, *Selected Invertebrate Types*.

Whether or not they are closely related to nematodes and other ecdysozoans, two independent molecular studies now place the chaetognaths either squarely among protostomes, or as the sister group to protostomes.[5] If chaetognaths are indeed protostomes, then protostomes most likely evolved from ancestors with deuterostome-like developmental characteristics.

Phylum Cycliophora

Phylum Cyclio • phora
(*cyclo* = G: a small wheel; *phora* = G: to carry)
sī´-klē-ō-for´-ah

Defining Characteristic: Ciliated larval stage (chordoid larva) with a mesodermal rod of muscle cells

"I didn't even know that lobsters had lips, but it turns out that they do, and these lips are the stomping ground of a tiny creature called *Symbion pandora* (literally, a 'couple of Greek words')."
Dave Barry

5. Summarized by Telford, M. J. 2004. *Nature* 431: 254–55.

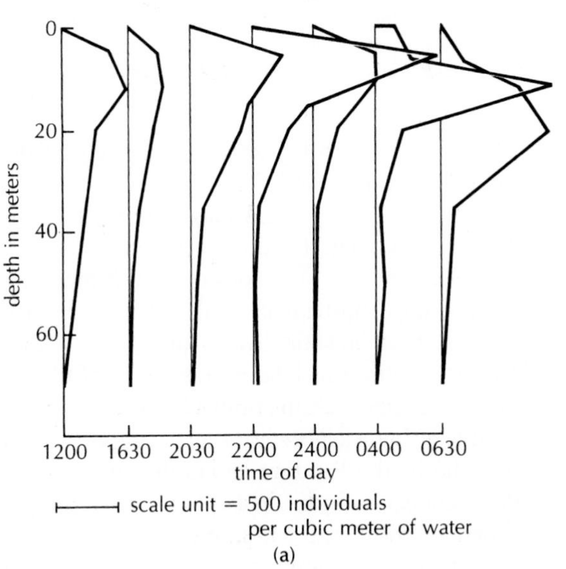

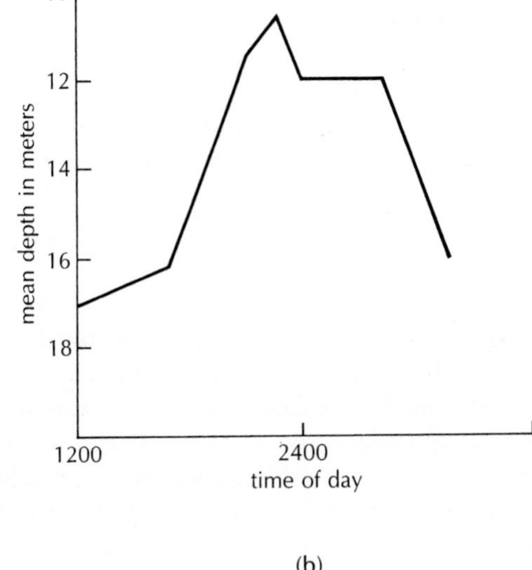

Figure 18.8

Diurnal vertical migration by the chaetognath *Sagitta elegans*. (a) Changes in numbers of individuals per unit volume of seawater at different depths over time. Note that a pronounced population peak begins to form toward the surface in the early evening and that the average depth of this peak shifts downward toward early morning. (b) The average depth for the population over time, from the data in (a). The population clearly migrates upward toward evening and downward in the early morning.
(a,b) From Froneman, P. W., E. A. Pakhomov, R. Perissinotto, and V. Meaton, 1998. *Marine Biol.* 131:95–101.

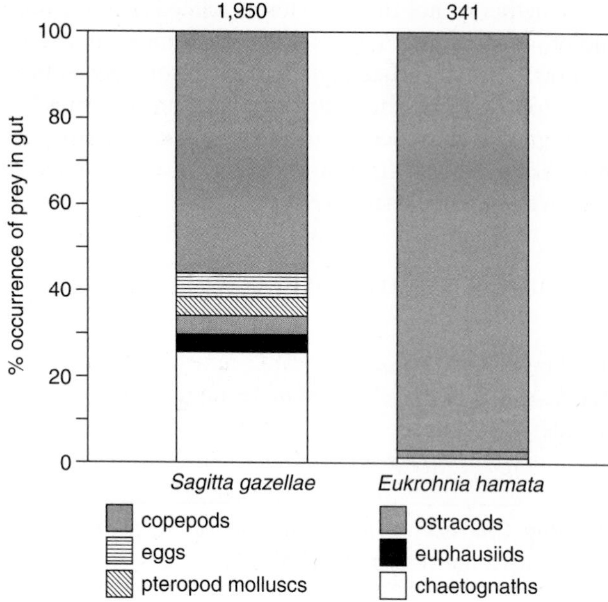

Figure 18.9

Gut content analysis of 2 chaetognath species collected at 4-hour intervals over 24 hours in the Indian Ocean southeast of Africa during April/May 1986. The number of digestive tracts opened by microdissection for each species is indicated above each bar. Only 9–12% of the animals had prey in their guts when examined. Of those individuals, copepods (blue shading) were clearly the major prey item.

From Froneman, P.W., E.A. Pakhomov, R. Perissinotto, and V. Meaton, 1998. *Marine Biol.* 131:95–101. (Springer-Verlag.)

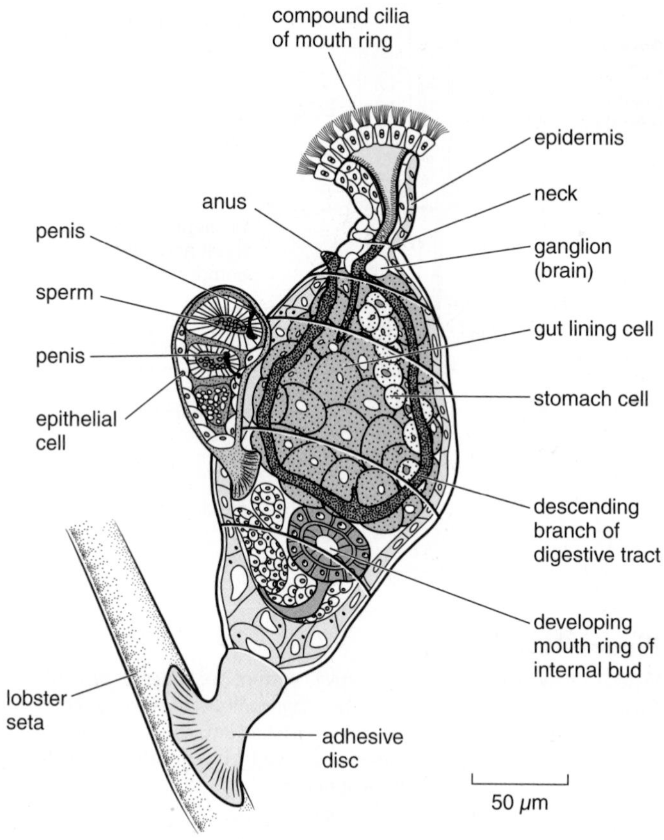

Figure 18.10

Feeding stage of the cycliophoran *Symbion pandora*, shown attached to one seta of a mouthpart appendage of the Norwegian lobster *Nephrops norvegicus*. The U-shaped digestive tract is shaded dark blue. Note the nonfeeding dwarf male attached to the outside of the feeding stage, which is probably developing a female stage internally. Also note the mouth ring and associated gut (shaded blue) developing internally near the base of the feeding individual; this will eventually replace the old feeding apparatus of this individual when it degenerates.

Based on Funch, P., and R. M. Kristensen, 1995. *Nature*, 378:711–14.

The microscopic, acoelomate animal known as *Symbion pandora* was first described in 1995 from the mouth appendages of the Norwegian lobster and placed in a newly erected phylum: the Cycliophora. Several more species are now known. The animals, which are less than 0.5 mm tall, seem to live exclusively on lobster mouth appendages, to which they attach by means of an **adhesive disc** at the base of a short stalk (Fig. 18.10). They are characterized by a ring of compound cilia (to which the phylum name refers) surrounding the mouth and accumulating food particles in the fashion of protostomes, by downstream collection (Chapter 2, p. 14). The digestive tract is lined with multiciliated cells and is U-shaped, with the anus located on the "neck," outside of the feeding funnel (Fig. 18.10). The body is covered by a thin cuticle. Periodically, the feeding apparatus and brain degenerate and are replaced by a new feeding apparatus and brain produced by internal budding (Fig. 18.10). But the most interesting part of the biology is yet to come.

The life cycle is bizarre, somewhat resembling that of colonial hydrozoans (Chapter 6) but perhaps most reminiscent of that found among some endoparasites, particularly the trematode flatworms (Chapter 8) and the causative agent of malaria (Chapter 3).[6] The animal illustrated in Figure 18.10 is an asexually reproducing feeding stage. At intervals it apparently releases an asexually produced swimming larva, called the Pandora larva, which contains within itself a developing feeding-stage animal (note the Pandora's Box analogy). The Pandora larva probably attaches to an appendage on the same lobster host; it then gives rise to a feeding-stage individual, which ultimately produces and releases another short-lived Pandora larva that probably attaches to the same lobster host. Many of the "individuals" on one lobster are thus probably modules **(ramets)** representing a single genotype, or **genet.** In the process of metamorphosis, the Pandora larva's nervous system and body degenerate; new individuals, with new nervous systems, are generated from buds within the larva.

But this can only go on for so long, because eventually the lobster will molt and the feeding ramets will be jettisoned with the lobster's cast off cuticle. As the lobster is preparing to molt, the feeding symbionts become sexually mature and produce, through internal differentiation, either a sexually mature swimming male or a sexually mature female, which contains a single egg (Fig. 18.11).

6. See *Topics for Further Discussion and Investigation*, no. 4.

The female is held within the feeding stage until her egg becomes fertilized. To bring this about, we presume that the male emerges from its feeding-stage progenitor, swims for a brief time, and attaches to a feeding stage that harbors a female. The small, nonfeeding, "dwarf" male either injects its sperm at this time (probably by hyperdermic impregnation) through what seem to be 2 penises (Fig. 18.10), or waits until the female emerges.

The female emerges and mates, or takes her already fertilized egg along with her. She then presumably attaches to the same lobster, which is still preparing to molt, and soon degenerates. The fertilized egg within the female, meanwhile, differentiates to form a ciliated **chordoid larva,** the name reflecting a conspicuous rod of muscle cells located ventrally in the larva (*chorda* = G: a string). The larva also bears a pair of protonephridia and may be a modified trochophore. The chordoid larva, for the first time in this story representing a new genet, presumably emerges before the lobster sheds its exoskeleton, and quickly disperses to another lobster host, where it produces a new feeding individual and proliferates the new genet asexually on that lobster, as described earlier.

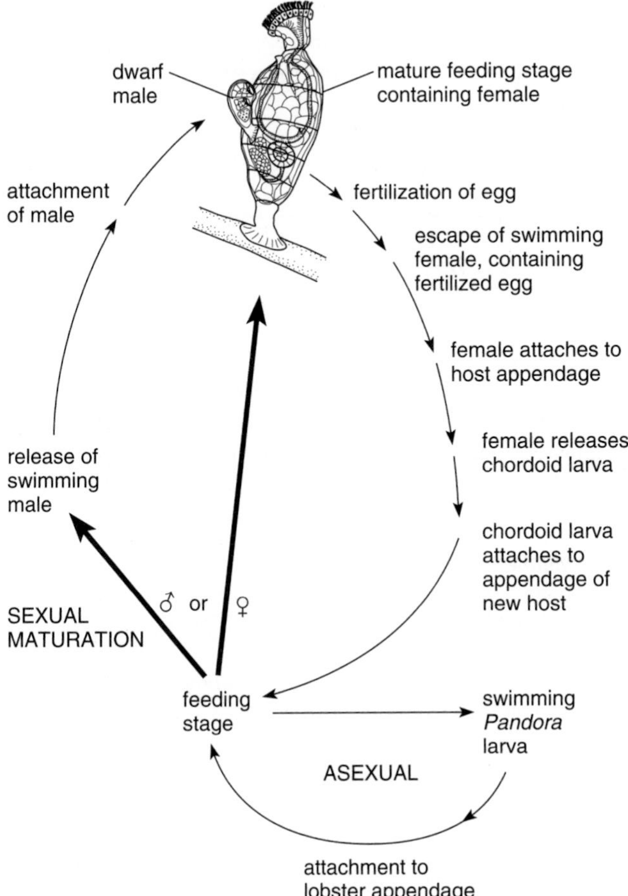

Figure 18.11
Life cycle of the cycliophoran *Symbion pandora,* hypothesized from laboratory observations of live animals and microscopic examination of what appear to be sequences of stages in preserved material. Based on Funch, P., and R. M. Kristensen. 1995. *Nature* 378:711–14.

Recent molecular analyses of 18S ribosomal RNA genes affiliate the cycliophorans with rotifers and acanthocephalans. In support of such an affiliation, both rotifers and cycliophorans exhibit dwarf males, hypodermic injection of sperm into females, and multiciliated epithelial cells. Also, the cuticle ultrastructure of *S. pandora* resembles that of some other aschelminths. But when originally described, cycliophorans were thought to be most closely related to certain other suspension-feeding animals to be discussed in the next chapter: the bryozoans and the entoprocts. In particular, both entoprocts and bryozoans exhibit modular growth by budding, and some aspects of cycliophoran brooding and their periodic regeneration of the feeding apparatus resemble phenomena exhibited among the Bryozoa; in marked contrast, budding and regeneration are unknown among rotifers, and extremely rare among other aschelminths. Moreover, the larval brain and nervous system degenerate at metamorphosis in all 3 groups, but never among the Rotifera.

A number of morphological characters are ambiguous. Protonephridia are encountered among both rotifers and entoprocts, and the cycliophoran U-shaped gut also characterizes rotifers and all of the phyla discussed in the next chapter, including the bryozoans and entoprocts. The absence of a mastax would seem to argue against a close relationship with rotifers, but the mastax could have been lost during cycliophoran evolution, as has apparently occurred among the acanthocephalans (Chapter 10). The proper placement of this enigmatic group of symbiotic animals is still uncertain.

Taxonomic Summary

Phylum Gastrotricha
Phylum Chaetognatha
Phylum Cycliophora

Topics for Further Discussion and Investigation

1. Discuss the evidence suggesting that chaetognaths can orient to light and to vibrations.

Feigenbaum, D., and M. R. Reeve. 1977. Prey detection in the Chaetognatha: Response to a vibrating probe and experimental determination of attack distance in large aquaria. *Limnol. Oceanogr.* 22:1052.

Goto, T., and M. Yoshida. 1983. The role of the eye and CNS components in phototaxis of the arrowworm, *Sagitta crassa* Tokioka. *Biol. Bull.* 164:82.

Newbury, T. K. 1972. Vibration perception by chaetognaths. *Nature* (London) 236:459.

2. What environmental cues appear to regulate the cycle of vertical migration of chaetognaths? Discuss the likely adaptive benefits and costs associated with this migratory behavior.

Pearre, S., Jr. 1973. Vertical migration and feeding in *Sagitta elegans* Verrill. *Ecology* 54:300.

3. Transparency must be greatly beneficial to planktonic animals, making it harder for them to be seen by potential prey and hiding them from potential predators. How can animals become transparent?

Johnsen, S. 2001. Hidden in plain sight: The ecology and physiology of organismal transparency. *Biol. Bull.* 201: 301–18.

4. Based upon lectures and your readings in this book, compare and contrast the life cycle of *Symbion pandora* with that of (a) *Plasmodium*, (b) colonial hydrozoans, and (c) trematode flatworms.

Taxonomic Detail

Phylum Gastrotricha

The approximately 700 species are divided among 2 orders.

Order Chaetonotida

Chaetonotus, Lepidodermella. Although most species live in freshwater (including freshwater bogs), representatives occur in all aquatic habitats. Individuals are always small, with members of most species less than 300 μm long. The order includes nearly 65% of all gastrotrich species, many of which reproduce mostly by parthenogenesis. Seven families.

Order Macrodasyida

Turbanella, Tetranchyroderma, Macrodasys, Dactylopodola. Members of this order occur in marine, brackish, and estuarine habitats; there are no freshwater species. This order contains the largest gastrotrichs, some of which reach lengths of about 3.5 mm. Most species are either simultaneous hermaphrodites or sequential hermaphrodites. Six families.

Phylum Chaetognatha

The approximately 150 chaetognath species are distributed among 2 orders within a single class. The division into orders is based on the presence (order Phragmophora) or absence (order Aphragmophora) of ventral, transverse muscles.

Class Sagittoidea

Two orders.

Order Phragmophora

Paraspadella, Spadella—all members of these genera are bottom-dwellers, mostly in shallow waters; *Eukrohnia*—species in this genus are planktonic.

Order Aphragmophora

Parasagitta, Sagitta. The group contains 50% or more of all chaetognath species. Its members are all planktonic and compose a major fraction of the animal biomass in the world's oceans.

Phylum Cycliophora

Symbion. Species have so far been described from the Norway lobster, *Nephrops norvegicus*, the American lobster, *Homarus americanus*, and the European lobster, *H. gammarus.* Recent molecular data suggest that at least 3 cryptic species, previously assumed to be a single species, live on the mouthparts of American lobsters. Previous reports of cycliophorans associated with non-lobster hosts are apparently erroneous.

General References about:

Gastrotrichs:

Harrison, F. W., and E. E. Ruppert, eds. 1991. *Microscopic Anatomy of Invertebrates, Volume 4: Aschelminthes.* New York: Wiley-Liss.

Higgins, R. P., and H. Thiel, eds. 1988. *Introduction to the Study of Meiofauna.* Washington, D.C.: Smithsonian Institution Press.

Hyman, L. H. 1951. *The Invertebrates, Volume 3. Acanthocephala, Aschelminthes, and Entoprocta.* New York: McGraw-Hill.

Morris, S. C., et al., eds. 1985. *The Origins and Relationships of Lower Invertebrates. Systematics Association, Special Volume 28.* Oxford: Clarendon Press, pp. 248–60.

Nielsen, C. 1996. *Animal Evolution: Interrelationships of the Living Phyla.* New York: Oxford University Press.

Parker, S. P., ed. 1982. *Classification and Synopsis of Living Organisms,* vol. 1. New York: McGraw-Hill, 857–77.

Schmidt-Rhaesa, A. 2002. Two dimensions of biodiversity research exemplified by Nematomorpha and Gastrotricha. *Integr. Comp. Biol.* 42: 633–40.

Thorpe, J. H., and A. P. Covich, eds. 2001. *Ecology and Classification of North American Freshwater Invertebrates,* 2nd ed. New York: Academic Press, (gastrotrichs).

Todaro, M. A., M. J. Telford, A. E. Lockyer, and D. T. J. Littlewood. 2006. Interrelationships of the Gastrotricha and their place among the Metazoa inferred from 18S rRNA genes. *Zool. Scripta* 35: 251–59.

Chaetognaths:

Alvarino, A. 1965. Chaetognaths. *Oceanogr. Marine Biol. Ann. Rev.* 3:115–94.

Ghirardelli, E. 1968. Some aspects of the biology of the chaetognaths. *Adv. Marine Biol.* 6:271–375.

Halanych, K. M. 1996. Testing hypotheses of chaetognath origins: Long branches revealed by 18S ribosomal DNA. *Syst. Biol.* 45:223–46.

Hyman, L. H. 1959. *The Invertebrates,* Vol. 5. *Smaller Coelomate Groups.* New York: McGraw-Hill.

Parker, S. P., ed. 1982. *Classification and Synopsis of Living Organisms,* vol. 2. New York: McGraw-Hill, 781–83.

Shinn, G. L. 1994. Epithelial origin of mesodermal structures in arrow worms (Phylum Chaetognatha). *Amer. Zool.* 34:523–32.

Cycliophorans:

Funch, P., and R. M. Kristensen. 1995. Cycliophora is a new phylum with affinities to Entoprocta and Ectoprocta. *Nature* 378:711–14.

Funch, P., and R. M. Kristensen. 1997. Cycliophora. In F. W. Harrison and R. M. Woollacott, eds., *Microscopic Anatomy of the Invertebrates,* vol. 13, pp. 409–74.

Search the Web

1. www.ucmp.berkeley.edu/aschelminthes/aschelminthes.html

   This address, operated through the University of California at Berkeley, includes a brief discussion of all "aschelminth" groups, which here includes the Chaetognatha and Gastrotricha.

2. http://www.microscopy-uk.org.uk/mag/indexmag.html

   Click on "Micscape Library" and then search on "Symbion" to see a popular article ("New Life Form") on the discovery and biology of *Symbion pandora,* published in the British online journal *Micscape.*

3. www.meiofauna.org

   This site is maintained by the International Association of Meiobenthologists.

4. http://www.microscopyu.com/moviegallery/pondscum/gastrotrich/chaetonotus/

   This website, offered by Nikon, includes videos of gastrotrich locomotion.

19

The "Lophophorates" (Phoronids, Brachiopods, Bryozoans) and Entoprocts

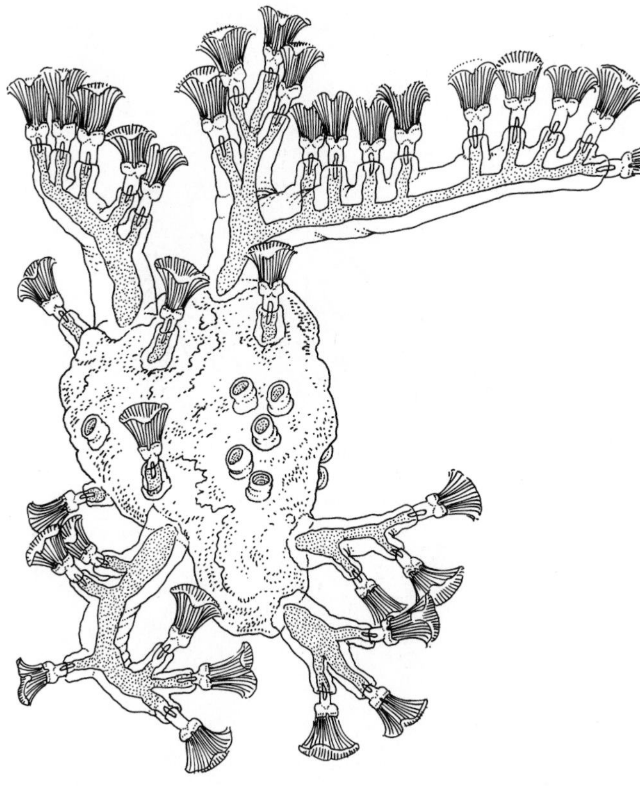

Introduction and General Characteristics

> "'Lophophorata' is thus a polyphyletic assemblage and the word should disappear from the zoological vocabulary, just as 'Vermes' disappeared many years ago."
> Claus Nielsen.

The 4 phyla discussed in this chapter—Phoronida, Brachiopoda, Bryozoa, and Entoprocta—have long had uncertain phylogenetic relationships to other animal phyla and to each other. We begin with the first 3 phyla, as they have in common one major anatomical feature that had long been thought to be homologous in the 3 groups: the **lophophore,** a ciliated organ used for both food collection and gas exchange. The lophophore is a circumoral (i.e., around the mouth) body region characterized by a circular or U-shaped ridge around the mouth. This ridge bears either one or 2 rows of ciliated, hollow tentacles. The internal space within the lophophore and its tentacles is always a coelomic cavity, and the anus always lies outside the circle of tentacles; both characteristics have come to be important parts of the definition of a lophophore. However, the bryozoan lophophore differs structurally from that of phoronids and brachiopods—tentacles on the bryozoan lophophore, for example, are multiciliated while those of phoronids and brachiopods are monociliated—and molecular data increasingly suggest that the bryozoan "lophophore" is indeed not homologous with that of phoronids and brachiopods.[1] I will refer here to the food-collecting organ in members of all 3 groups as a lophophore, but it is only for convenience.

1. Passamaneck, Y., and K. M. Halanych. 2006. *Molec. Phylog. Evol.* 40: 20–28.

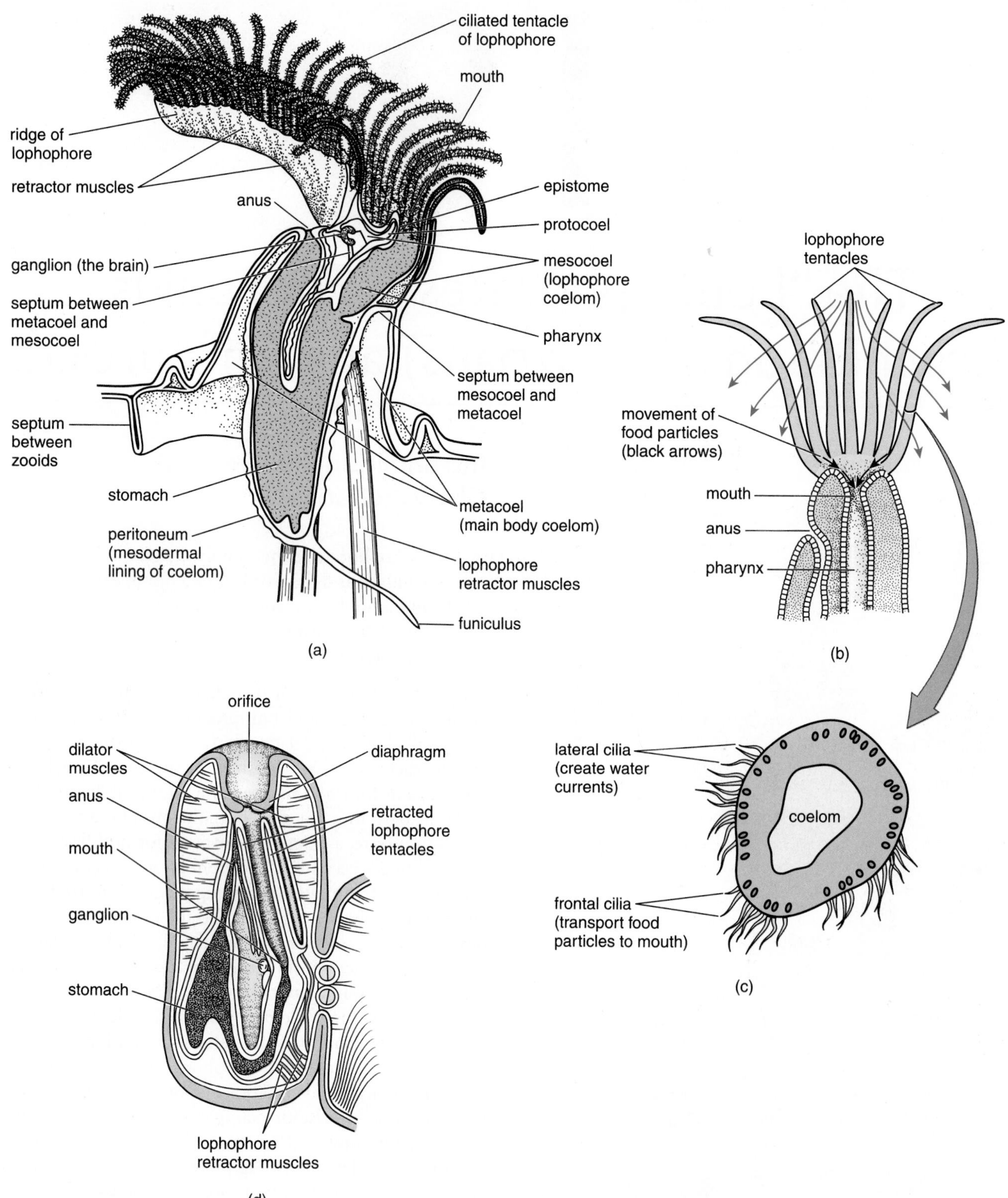

Figure 19.1

(a) Longitudinal section through the body of a lophophorate (the bryozoan *Cristatella* sp.). Note that the mouth opening lies within the ring of lophophore tentacles, but that the anus does not. (b) Movement of water through a lophophore. (c) Cross section through one lophophore tentacle, showing location and function of ciliary traits. (d) Illustration of a polypide retracted into a cystid. Anatomical details are discussed further in the section on phylum Bryozoa.

(a) From Hyman; after Cori. (b,c) Based on P. G. Willmer, *Invertebrate Relationships*. 1990. (d) From Hyman, after Marcus.

The 3 groups also differ with respect to their circulatory and excretory systems, the nature of the protective coverings they secrete, and important details of development. Indeed, their developmental patterns are unlike those reported for any other coelomate groups: They exhibit features characteristic of both protostome and deuterostome development, along with features that characterize development in neither group. Cleavage is basically radial, as in deuterostomes, and cell fates are not fixed—at least in the few species so far studied—until after the first several cleavages. Such indeterminate cleavage is a typical deuterostome characteristic. However, coelom formation is not distinctly enterocoelous (as it would be in a typical deuterostome) or schizocoelous (as it would be in a typical protostome). Both modes of coelom formation have been observed among "lophophorates," and in at least some phoronid species the coelom forms as a split (schizocoely) in endodermal tissue (enterocoely!) rather than mesodermal tissue. To further add to the intrigue, the bryozoan "coelom" forms by neither of these mechanisms! It arises not during embryogenesis, but only following the completion of larval life.

Among phoronids, the mouth forms from the blastopore, a distinctly protostome characteristic, and in 2 of the groups (the Phoronida and the Brachiopoda), typical protostome nephridia are found in larvae and adults. On the other hand, the tripartite coelom typically encountered among "lophophorates" has long been considered a uniquely deuterostome feature. On balance, morphological and embryological criteria link the lophophorates with deuterostomes.

On the other hand, a growing amount of molecular data, based on 18S rRNA, mitochondrial DNA, and *hox* gene sequences, consistently place them with annelids, molluscs, sipunculans, and the entoprocts to be discussed at the end of this chapter, forming a new clade of protostomes called the Lophotrochozoa.

What do the 3 "lophophorate" groups have in common? All members are sessile or sedentary suspension feeders, employing the lophophore cilia to capture phytoplankton and small planktonic animals, and none have a distinct head. Moreover, the pattern of water flow created by the tentacular cilia is the same in all species: Water is pulled down into the center of the lophophore by the action of lateral cilia and expelled between adjacent tentacles after frontal cilia have removed food particles (Fig 19.1b).

The lophophore's coelomic cavity, called the **mesocoel,** is physically separated by a septum from the larger, primary coelomic cavity, the **metacoel** (Fig. 19.1a). In some groups, this septum is perforated or incomplete, so the 2 coelomic cavities are interconnected. A third coelomic cavity, the **protocoel,** may be present anterior to the mesocoel. The protocoel is distinct but quite small in the adults of some lophophorate groups, but it is particularly conspicuous during the larval stage of phoronids. The absence of a conspicuous protocoel among brachiopods and bryozoans appears related to the evolutionary reduction of the head.

One feature common to the members of all 3 groups is the possession of very simple gonads, arising from a portion of the mesodermal lining of the main (trunk) coelomic compartment—that is, from the metacoel. Additionally, all species possess U-shaped digestive tracts, in which the anus (when present) terminates near the mouth (Fig 19.1), and all species secrete some form of protective covering about the body.

Finally, nearly all of the species discussed in this chapter are marine, except for a few bryozoan species that are commonly encountered in freshwater. None of the species is terrestrial, in any sense of the word; the sedentary, suspension-feeding lifestyle is not a likely preadaptation for life on land.

Phylum Phoronida

Members of the phylum Phoronida (for-ŏ′-nih-dah) conform closely to the same basic body plan. Only 14 phoronid species are known, and all of these are marine.

Most phoronids live in permanent, chitinous tubes implanted in muddy or sandy sediments or attached to solid surfaces. A few species burrow into hard, calcareous substrates, but nevertheless secrete a chitinous tube within the burrow. Adults do not move from place to place, although they can move within their tubes and, if artificially removed from these tubes, can burrow back into the sediment. A giant nerve fiber permits the animal to withdraw rapidly into its tube or burrow upon provocation.

Phoronids typically are about 12 cm long, with the form of an elongated, cylindrical sac (Fig. 19.2). They have no appendages, except for the anterior lophophore. A flap of tissue called the **epistome** covers the mouth (*epi* = Greek: around; *stoma* = G: mouth). The epistome is hollow, as it contains a remnant of the embryonic protocoel.

The lophophore is the phoronids' only prominent external structure. It consists of a conspicuous ring of tentacles, usually deeply indented to form a U-shape, and a less conspicuous, ciliated food groove (Fig. 19.3). Ciliary activity drives water into the ring of tentacles from the top of the lophophore and outward through the narrow spaces between the tentacles, as with brachiopods and bryozoans. In this manner, suspended food particles can be captured by the tentacular cilia and mucus, transferred to the cilia of the food groove, and conducted to the mouth for ingestion.

Internally, phoronids possess a pair of metanephridia, with the ciliated nephrostomes collecting coelomic fluid (and mature sperm or fertilized eggs) and the nephridiopores discharging urine near the anus (Fig. 19.3b). A blood circulatory system with hemoglobin (contained within blood corpuscles) is present in all species. There is no distinct heart, but the major trunk blood vessel is contractile. Blood circulates largely through a series of interconnected, discrete vessels. Each tentacle of the lophophore is serviced by a single, small vessel, through which the blood both ebbs and flows.

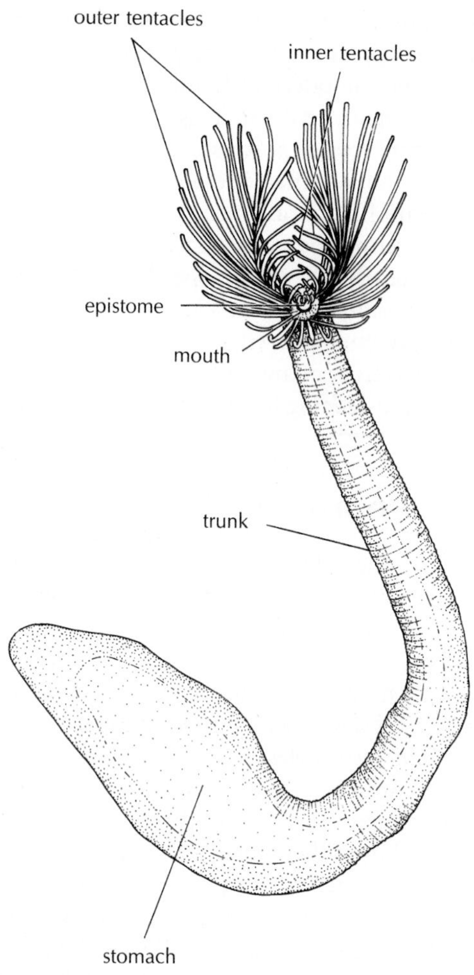

Figure 19.2
The phoronid *Phoronis architecta* removed from its tube.
From Hyman; after Wilson.

Recent analyses of molecular data suggest that phoronids have a recent ancestor in common with the group of animals discussed next, the brachiopods, or that they are themselves modified brachiopods.

Phylum Brachiopoda

Phylum Brachio • poda
(G: arm foot)
brak-ē-ō-pō´-dah

Defining Characteristic: Body enclosed within a 2-valved shell, with the shells oriented dorsally and ventrally

The external appearance of brachiopods, including the so-called "lampshells," is quite unlike that of the phoronids. Brachiopods superficially resemble the bivalved molluscs, with the body protected externally by a pair of convex, calcified shells (Fig. 19.4) that are coated with a thin layer of organic periostracum. Indeed, up until about 100 years ago, brachiopods were considered *bona fide* members of the phylum Mollusca. Only about 350 brachiopod species currently exist, but over 30,000 species, extending back nearly 550 million years, appear in the fossil record. This phylum has clearly seen better times (Fig. 19.5).[2]

As noted earlier, some molecular data[3] indicate that phoronids are modified brachiopods. If this proposal were to be generally accepted, the diagnosis of what it

2. See *Topics for Further Discussion and Investigation,* no. 2, at the end of the chapter.

3. Cohen, B. L. 2000. *Proc. R. Soc.* London B 267: 225–31; and A. Weydmann. 2005. *Org. Divers. Evol.* 5: 253–73.

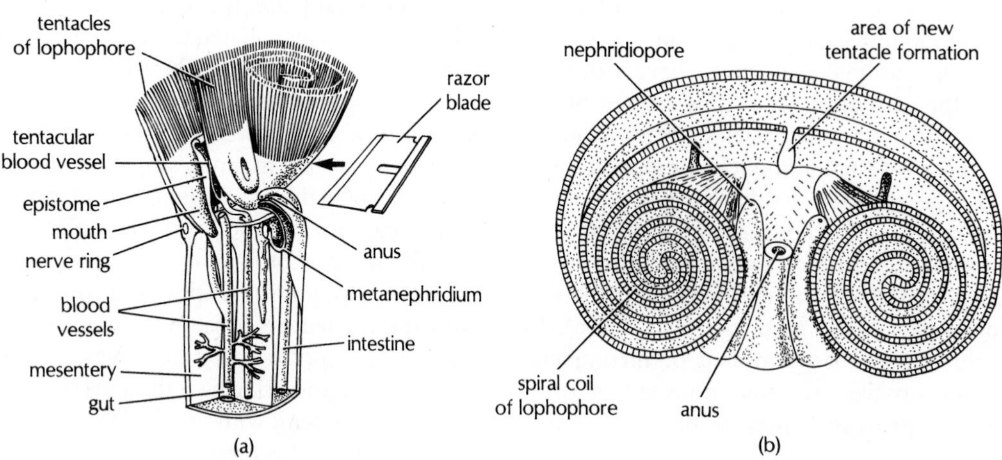

Figure 19.3
Phoronid anatomy. (a) Lateral view of *Phoronis australis,* cut vertically to reveal internal anatomy. (b) Looking down on the lophophore, after the tentacles have been chopped off in the plane indicated by the razor blade in (a). The tentacles would be coming out of the page if they had not been chopped off. The lophophore is basically U-shaped, although each end of the U is curled to form a spiral in this species.
From Hyman; after Shipley and Benham.

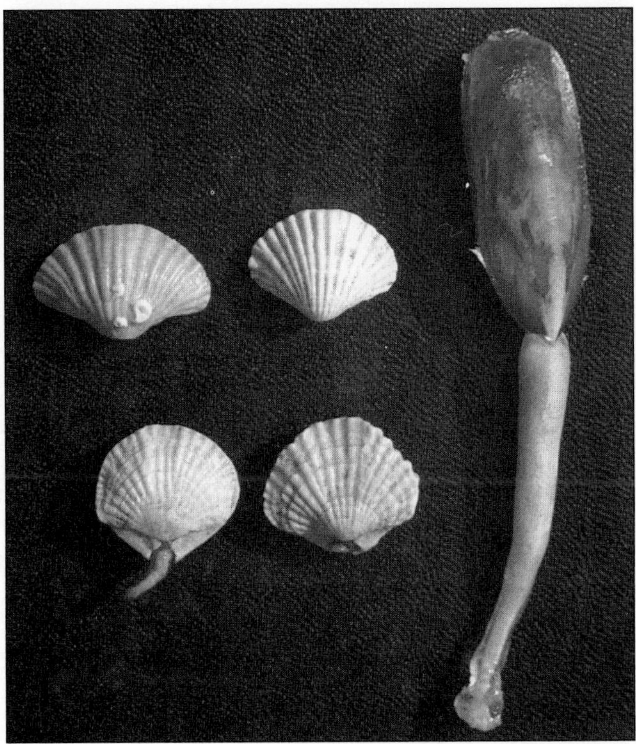

Figure 19.4

Photograph of an inarticulate brachiopod, *Lingula* sp. (right), and 4 specimens of an articulate brachiopod, the lampshell *Terebratella* sp. (left). *Lingula* sp. has a long muscular pedicle, which anchors the animal in a burrow in the sediment. In contrast, lampshells generally have a short, nonmuscular pedicle that attaches to rocks. The pedicle of *Terebratella* sp. is seen protruding from an opening in the ventral shell valve in one of the specimens. In a few articulate and inarticulate brachiopod species, the pedicle is completely lost; the valves cement directly to solid substrate.

Photograph by J. Pechenik and L. Eyster.

means to be a brachiopod would obviously have to be altered, since phoronids secrete multilayered tubes, not bivalved shells.

The brachiopod lophophore resembles that of the phoronids except that in most species it is drawn out into 2 arms (Fig. 19.6), increasing the effective surface area for food collection and gas exchange. Also, as in the Phoronida, metanephridia (1 or 2 pairs in brachiopods) serve as excretory organs. A circulatory system is also present, with blood being circulated by the action of a well-developed heart and from one to several contractile vessels associated with the main dorsal vessel. Subsequent to leaving this main vessel, the blood moves somewhat haphazardly through a system of interconnected blood sinuses. The blood contains no oxygen-binding pigments; a blood pigment is found in the coelomic fluid, but it is hemerythrin rather than hemoglobin. Hemerythrin, although it does contain iron, is structurally and functionally quite different from hemoglobin. Hemerythrin distribution among animals has long puzzled zoologists; it is found only in brachiopods, sipunculans, priapulids, and a few species of polychaetes, groups with no close phylogenetic connection. Hemerythrin is not found among any other lophophore-bearing animals.

Most brachiopods live permanently attached to a solid substrate or firmly implanted in sediment. Attachment is generally achieved by means of a stalk, called the **pedicle** (Latin: little foot), which protrudes posteriorly through a notch or hole in the ventral shell valve (Figs. 19.4 and 19.6b, c). The stalk is quite long and flexible in some species, perhaps serving to keep the body up above the substrate and in a zone of greater water flow.[4] This could benefit the brachiopod both in terms of increased gas exchange and increased rate of food capture. The pedicle is often muscular and hollow, housing an extension

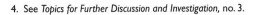

4. See *Topics for Further Discussion and Investigation,* no. 3.

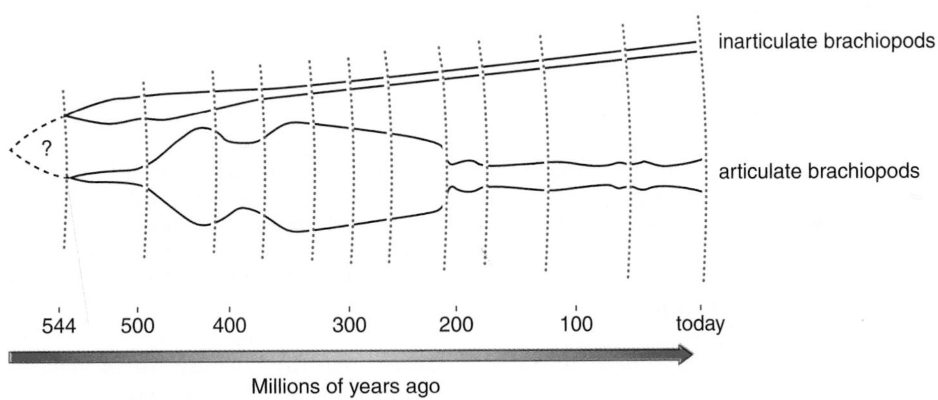

Figure 19.5

The occurrence of brachiopods in the fossil record. Decreases in width correspond to comparable decreases in the number of species. The distinction between articulate and inarticulate species is given in Table 19.1.

Based on Boardman et al., *Fossil Invertebrates.* Palo Alto: Blackwell Scientific Publications, Inc., 1987.

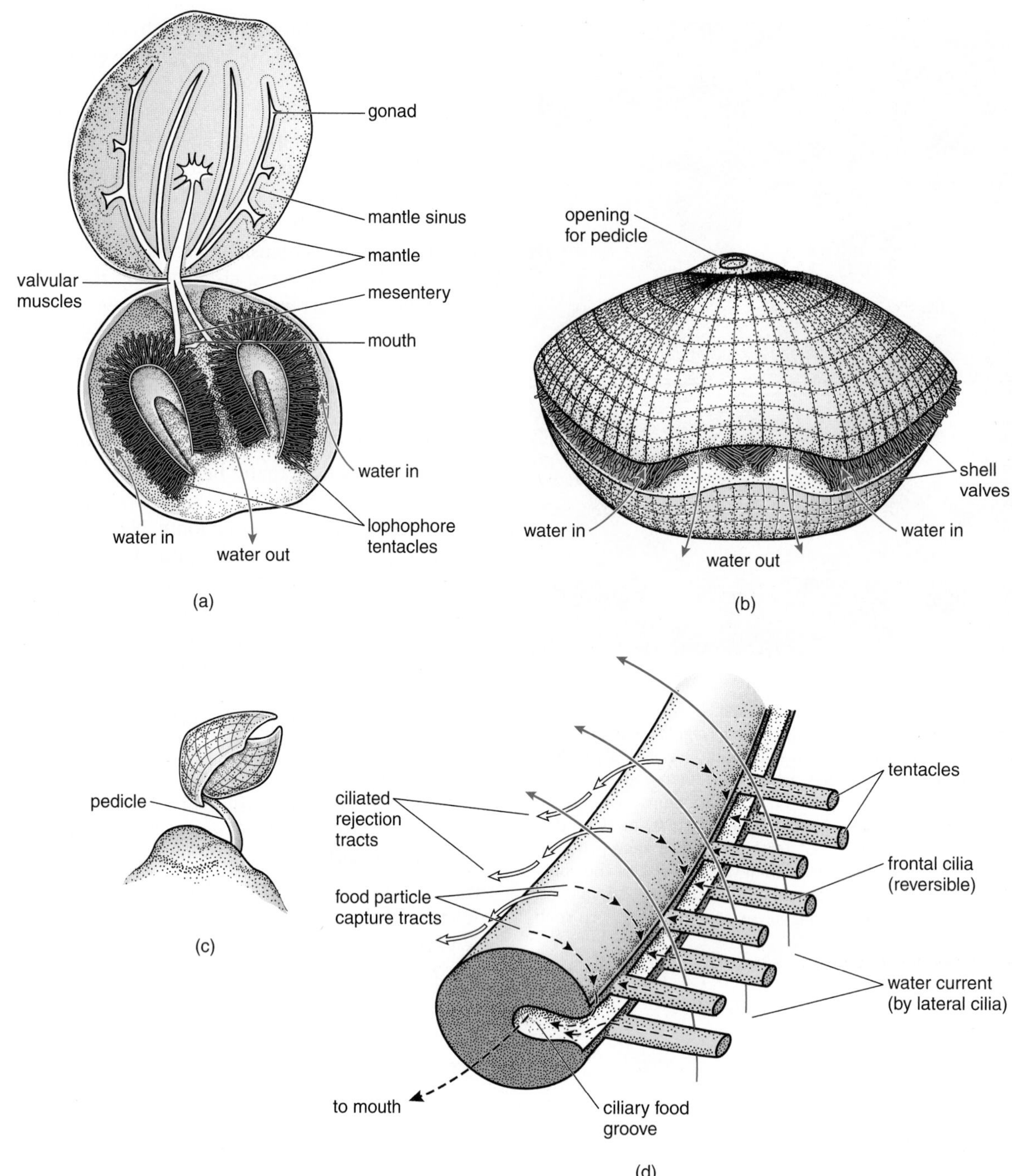

Figure 19.6

(a) Brachiopod with valves opened to show orientation of lophophore. (b) Path of water flow through the lophophore of a brachiopod. (*Terebratella* sp.) in the process of feeding. Note the opening through which the pedicle protrudes in life. (c) The brachiopod *Hemithyris psittacea* in its normal feeding posture. (d) Detail of one part of a lophophore arm. Dashed arrows show paths of captured food particles in transit to the mouth; open arrows show particles being rejected. Blue arrows indicate water flow through adjacent tentacles.

(a) Based on Beck/Braithwaite, *Invertebrate Zoology, Laboratory Workbook,* 3/e, 1968. (b) After Hyman; after Blochman. (c) After Hyman. (d) Modified after Russell-Hunter.

of the main coelomic cavity (the metacoel), through which coelomic fluid can circulate.

Brachiopod shells are composed of a protein matrix plus either calcium carbonate or calcium phosphate, and are secreted by 2 lobes of tissue referred to as the **mantle tissue.** This mantle tissue is by no means homologous with the shell-secreting molluscan tissue of the same name. Like the "mantle" of barnacles, the terminology simply reflects past ideas about the relationships between animal groups, ideas that have changed substantially over the past hundred years or so; both barnacles and brachiopods were once classified as molluscs.

The members of many brachiopod species secrete chitinous setae along the mantle edges, and tufts of

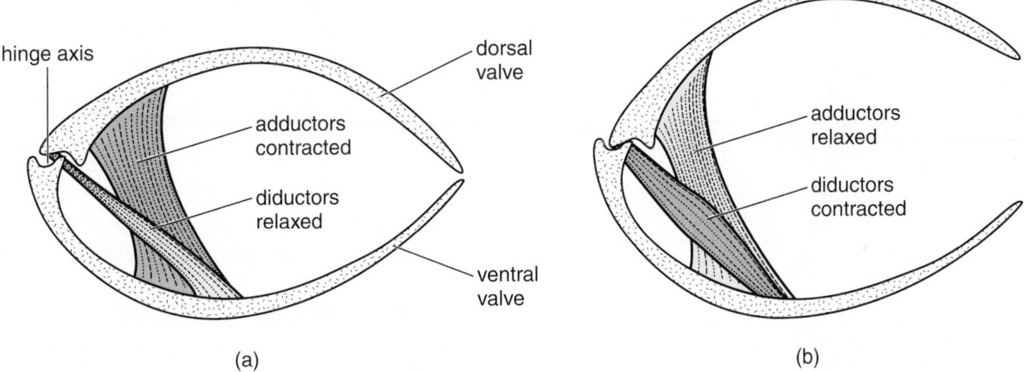

Figure 19.7
The use of adductor and diductor muscles in opening and closing the shell valves of a typical hinged brachiopod. (a) Contracting the adductor muscles while relaxing the diductor muscles closes the valves and stretches the diductor muscles. (b) Contracting the diductor muscles while relaxing the adductor muscles levers the valves apart and stretches the adductor muscles, preparing them for another contraction.
Based on Boardman et al., *Fossil Invertebrates*. 1987.

chitinous setae are also prominent in the larvae (Fig. 19.16b). Careful ultrastructural studies show that these setae are formed—in both adults and larvae—in precisely the same way as they are formed among polychaete annelids. Is this an example of convergent evolution? Or does it support the molecular data indicating that brachiopods are in fact protostomes, closely related to annelids?

Although brachiopods may be protostomes, there is nothing to indicate that shell formation in brachiopods and bivalves has a common evolutionary origin (i.e., that it is homologous).

The brachiopod body is oriented so that the shell valves are ventral and dorsal. In marked contrast, the shell valves of bivalves are to the left and right sides of the body. Brachiopod shells are typically less than 10 cm in any dimension.

In some brachiopod species, members of the class Inarticulata, the shell valves are held together entirely by adductor muscles. In members of the only other brachiopod class, the Articulata, the shell valves are hinged and articulate, as in bivalved molluscs; that is, the shell margins possess a series of interlocking teeth and sockets that prevent substantial sliding of one valve relative to the other.

In both brachiopod classes, the shell valves are brought together through the contraction of adductor muscles (Fig. 19.7), as in bivalved molluscs. However, there is no springy bivalve-style hinge ligament to force the valves open when the adductors relax. Instead, separating the valves from each other is an active process among brachiopods, dependent upon the contraction of an opposing set of muscles, the **diductor muscles.** The shell thus acts as a complete skeletal system; not only do the shell valves protect the soft body parts, but they also serve as the vehicles through which the 2 muscle groups, the adductors and the diductors, antagonize each other (Fig. 19.7).

Table 19.1 Comparison of the 2 Major Brachiopod Classes

Characteristic	Articulata	Inarticulata
Shell	Always calcium carbonate Tooth and socket hinge	Usually calcium phosphate No articulating hinges
Digestive tract	Ends blindly	Open at mouth and anus
Rigid internal support	Present in some species	Never present
Larval morphology	Lophophore develops at metamorphosis	Lophophore already present in larva

Members of the 2 brachiopod classes differ with respect to their shells' chemical composition and the morphology of their digestive tracts (Table 19.1). The shells of articulate brachiopods are all strengthened with calcium carbonate. In contrast, those of inarticulate species usually contain calcium phosphate. The digestive tract of the inarticulates is always U-shaped, with a mouth and separate terminal anus. The digestive tract of articulate brachiopods, on the other hand, terminates blindly (Fig. 19.8). Thus, the articulate brachiopods, sophisticated in so many other respects, make do without an anus. In addition, the lophophores of some articulate brachiopod species contain a rigid, calcified, internal support never encountered within the Inarticulata. Lastly, larval morphology also differs markedly between the 2 groups (see p. 489).

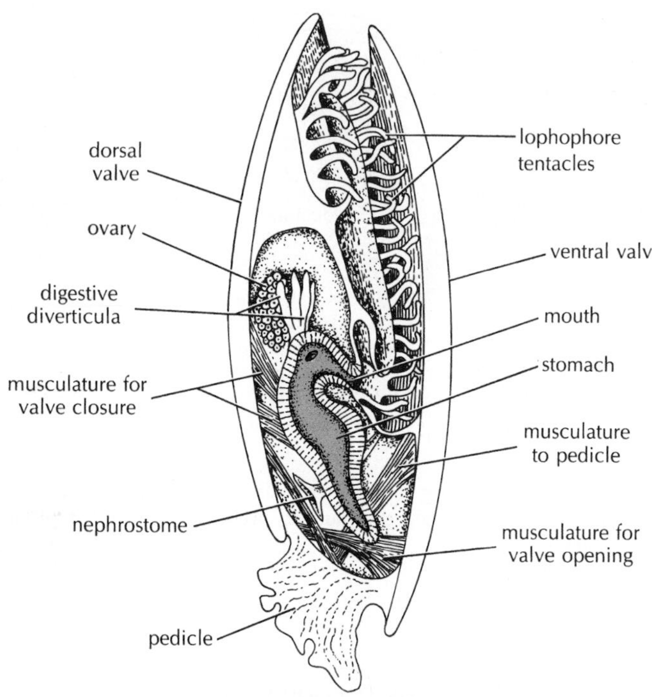

Figure 19.8
Diagrammatic illustration of the internal anatomy of an articulate brachiopod. Note the blind-ending intestine and the complex musculature operating the shell valves. Also note the conspicuous nephrostome.
After Harmer and Shipley.

Phylum Bryozoa (= Ectoprocta; = Polyzoa)

Phylum Bryo • zoa
(G: moss animals)
brī-ō-zō´-ah

All 3 names for the phylum Bryozoa are widely used in the literature. The name "Bryozoa" (moss animal) refers to the finely branched appearance of many common species. "Ectoprocta" emphasizes that the anus lies outside the ring of tentacles, as in other lophophorates (*ecto* = G: outside; *proct* = G: anus). Finally, "Polyzoa" (i.e., multiple animals) refers to the colonial nature of all bryozoan species: They always form clusters of asexually produced, physically interconnected units, or **modules.** The name Bryozoa is used in this text, as it seems to be the term of choice in recent publications. With the exception of one Antarctic species whose members apparently live a pelagic life in open water, bryozoans live attached to macroalgae, rocks, wood, sea turtle shells, and a variety of other hard and soft substrata.

As mentioned earlier, the relationship of bryozoans to phoronids and brachiopods is probably not a close one and the Lophophorata is probably doomed as a valid clade. I include the bryozoans in this chapter nevertheless, because it is not yet clear where else to place them.

Some marine bryozoans have been under intense study recently for their production of unique metabolites. Approximately 200 compounds have so far been isolated from a variety of species; in particular, the "bryostatins" show an exciting range of potential biomedical applications. Remarkably, the bryostatins are produced by specific bacterial symbionts, rather than by the bryozoans themselves.[5]

All bryozoans secrete a house around the body. The contents of the house—that is, the lophophore, gut, nerve ganglia, and most of the musculature—are referred to as the **polypide** (Fig. 19.9a). The house itself, plus the body wall that secretes it, constitutes the **cystid,** while the secreted, nonliving part of the house is termed the **zooecium** (*zoo* = G: animal; *oecus* = G: house). This rather confusing terminology evolved from the mistaken idea that the body wall of the house and the contents of the house were 2 separate individuals. That misconception was corrected quite some time ago, but the terminology had been used so widely that it has persisted. The body wall is actually attached to the zooecium, so the bryozoan is essentially glued to its house. One entire individual (cystid and polypide) is termed a **zooid.**

The bryozoan body wall has a unique developmental potential: The entire zooid can be generated, or regenerated, from the body wall of the cystid. Such generation and regeneration is, in fact, an integral part of the bryozoan life cycle.

During the life of an individual zooid, the entire polypide periodically degenerates into a dark-pigmented, spherical mass known as a **brown body.** In most species, a new polypide is subsequently produced from the cystid. In some species, the brown body remains conspicuously present in the coelomic space of the new polypide. In other species, the brown body is engulfed by the digestive system of the regenerating polypide and is then discharged unceremoniously through the anus. Bryozoan zooids may go through 4 or more such cycles of degeneration and rebirth during their lives. In some instances, brown-body formation also may provide a mechanism for bypassing unfavorable environmental conditions. Brown-body formation may also be a mechanism for eliminating insoluble waste products, by disposing of the entire polypide—a rather remarkable method of garbage disposal, to say the least. Bryozoans lack nephridia.

The septum dividing the metacoel from the mesocoel is very incomplete in bryozoans, so the coelomic fluid of the body cavity is continuous with that of the lophophore and tentacle cavities. The bryozoan lophophore is strikingly similar to that of phoronids, except that it can be retracted within the zooecium for protection and pro-

5. See *Topics for Further Discussion and Investigation,* no. 7.

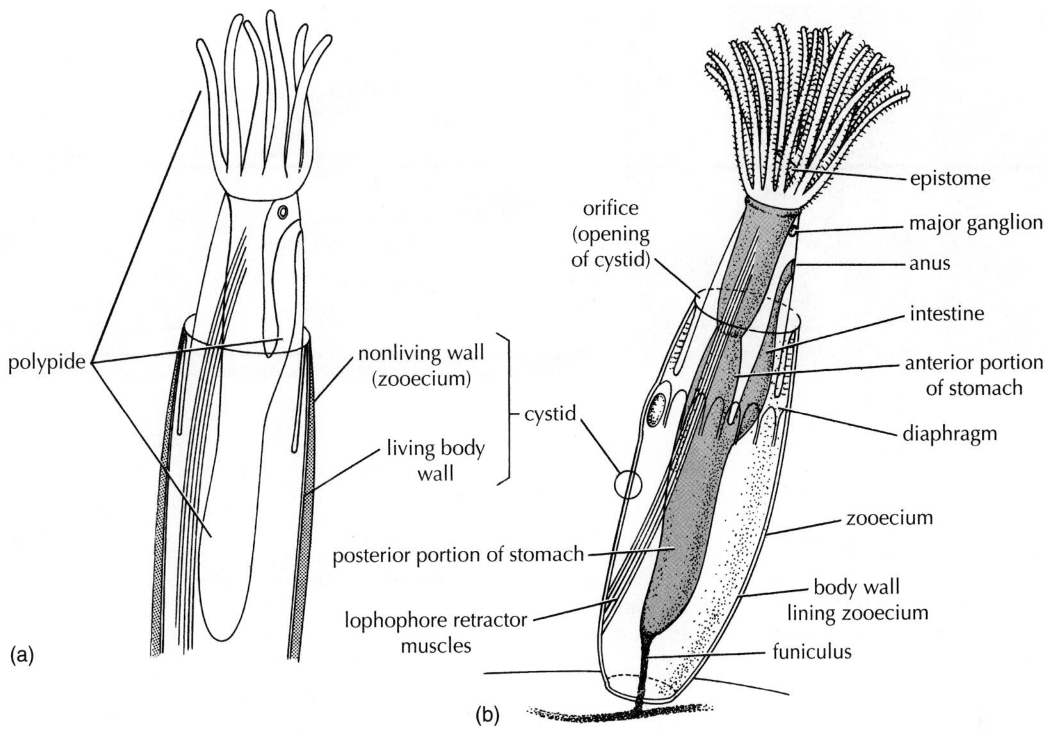

Figure 19.9
(a) Diagrammatic illustration demonstrating terminology of the polypide, cystid, and zooecium. (b) The freshwater bryozoan (class Phylactolaemata) *Fredericella,* showing details of the internal anatomy. From Hyman; after Allman.

truded for feeding and gas exchange. It protrudes through an **orifice** in the zooecium (Fig. 19.9). In many species, a muscular **diaphragm** is positioned just beneath this orifice. Protrusion of the lophophore is accomplished indirectly, by increasing the hydrostatic pressure within the main body cavity and dilating the diaphragm. The means of achieving the temporary increase in pressure within the body cavity differ among bryozoan species, as discussed later in this chapter.

Bryozoans show a great variety of external morphology. Moreover, all bryozoans are colonial; that is, as with corals, a single individual reproduces itself asexually to form a contiguous grouping of genetically identical individuals, with up to about 2 million zooids in a single colony (Fig. 19.10a). Each of these individual zooids is extremely small, usually less than 1 mm each, although an entire colony can exceed 0.5 m in length or circumference. Since the colonies grow by adding new modules, rather than by increasing the size of each unit, the aging and growth of bryozoans seem not to be subject to the sorts of physiological rules and limitations imposed on most other animals (Research Focus Box 19.1, p. 484).

Bryozoan colonies show a considerable variety of species-specific geometrical patterns. Colonies may be erect and branching or flat and encrusting. Great diversity of form is encountered within each of these 2 basic patterns.

In many bryozoans, thick mesenchymal cables—called **funicular cords**—form tissue connections between the individual colony members. A single cable is termed a **funiculus** (Figs. 19.1a and 19.9b). The funicular system of a single zooid may extend across the coelomic cavity from the stomach to a pore in the body wall, providing a mechanism for possible direct transfer of nutrients among adjacent zooids.[6] The funicular system may be homologous with the circulatory systems of other metazoans, including other lophophorates.

Most bryozoans (about 5000 species) are marine, but about 50 species are restricted to freshwater. These 50 bryozoan species constitute the only nonmarine lophophorates. Approximately 15,000 additional marine bryozoan species are known only from the fossil record. Like that of brachiopods, the bryozoan fossil record is long and interesting (Fig. 19.11), although it does not begin until about 70 million years after the Cambrian "explosion."

Bryozoans are divided among 2 major classes and one minor class, based largely on differences in the morphology of the lophophore, mechanism of lophophore protrusion, chemical composition of the external body covering, and the presence or absence of an epistome and of body wall musculature.

6. See *Topics for Further Discussion and Investigation,* no. 6.

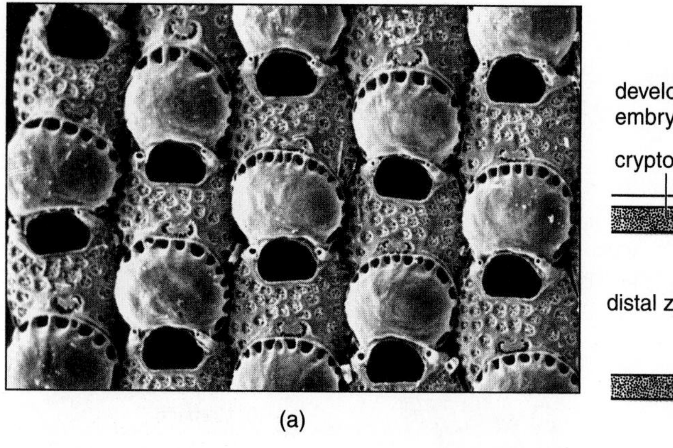

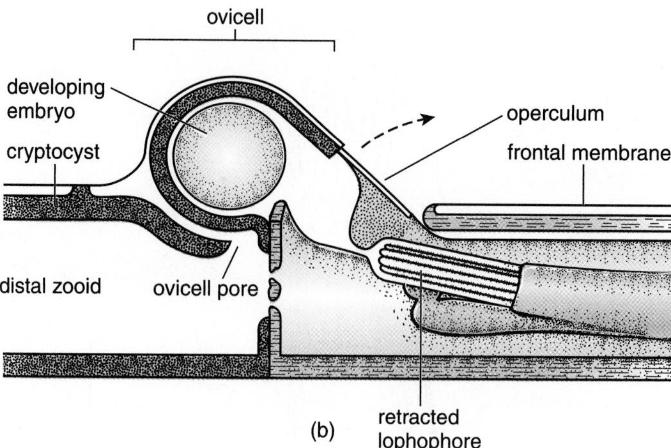

Figure 19.10
(a) Scanning electron micrograph of the marine bryozoan *Fenestrulina malusii.* Each zooid is associated with an ovicell, in which a single embryo is brooded after fertilization, as indicated in (b). The arrows adjacent to the operculum indicate the direction of opercular movement.

(a) From Claus Nielsen, in *Ophelia,* 9:209–341, 1971. Copyright © 1971 Ophelia Publications, Helsingør, Denmark. (b) Based on Claus Nielsen, in *Ophelia,* 9: 209–341, 1971.

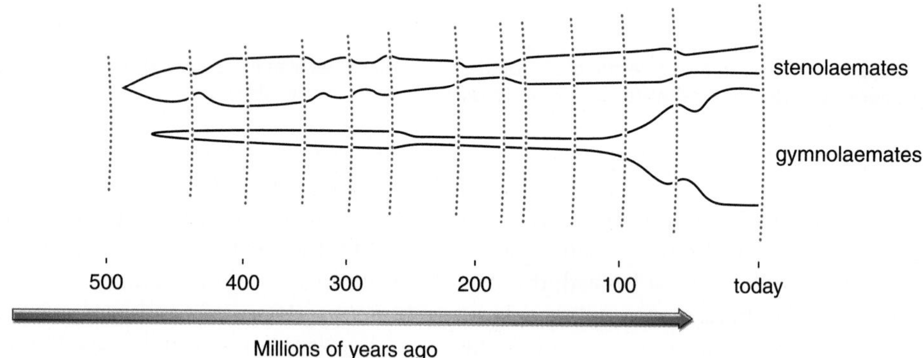

Figure 19.11
The occurrence of calcified bryozoans in the fossil record. Increases in width correspond to proportional increases in the number of species. The distinction between gymnolaemates and stenolaemates is given on pp. 484–488.
Based on Boardman et al., *Fossil Invertebrates.* Palo Alto: 1987.

Class Phylactolaemata

All members (about 50 species) of the class Phylactolaemata (fī-lak′-tō-lē-mah′-tah) are found in freshwater, and most (but not all) freshwater bryozoans are members of this class. The zooids of a given colony are morphologically identical; that is, the colonies are **monomorphic.** Some species secrete a chitinous outer covering, while others produce thick, gelatinous surroundings (Fig. 19.12a). In some species, the diameter of a single colony can exceed 50 cm. Colonies develop on a variety of submerged, solid surfaces, including shells, rocks, and the leaves and branches of freshwater vegetation. Most species are permanently affixed to these substrates and are incapable of locomotion. However, the colonies of a few species can move slowly from place to place, using the single muscular "foot" shared by all of the zooids in the colony.

The lophophore is U-shaped in all phylactolaemate species, as in phoronids. A flap of tissue, termed the **epistome** as in phoronids, hangs over the mouth. A protocoel is evident within the epistome of some species. As in the phoronids, the body wall contains both circular and longitudinal muscles. A pronounced, hollow funiculus extends from the stomach into the coelomic cavity shared by all of the zooids (Fig. 19.9).

Lophophore protraction is brought about by the contraction of muscles in the deformable body wall.[7] Because the coelomic fluid is essentially incompressible, this contraction increases the pressure within the metacoel and, since the septum between the 2 main body

7. See *Topics for Further Discussion and Investigation,* no. 4.

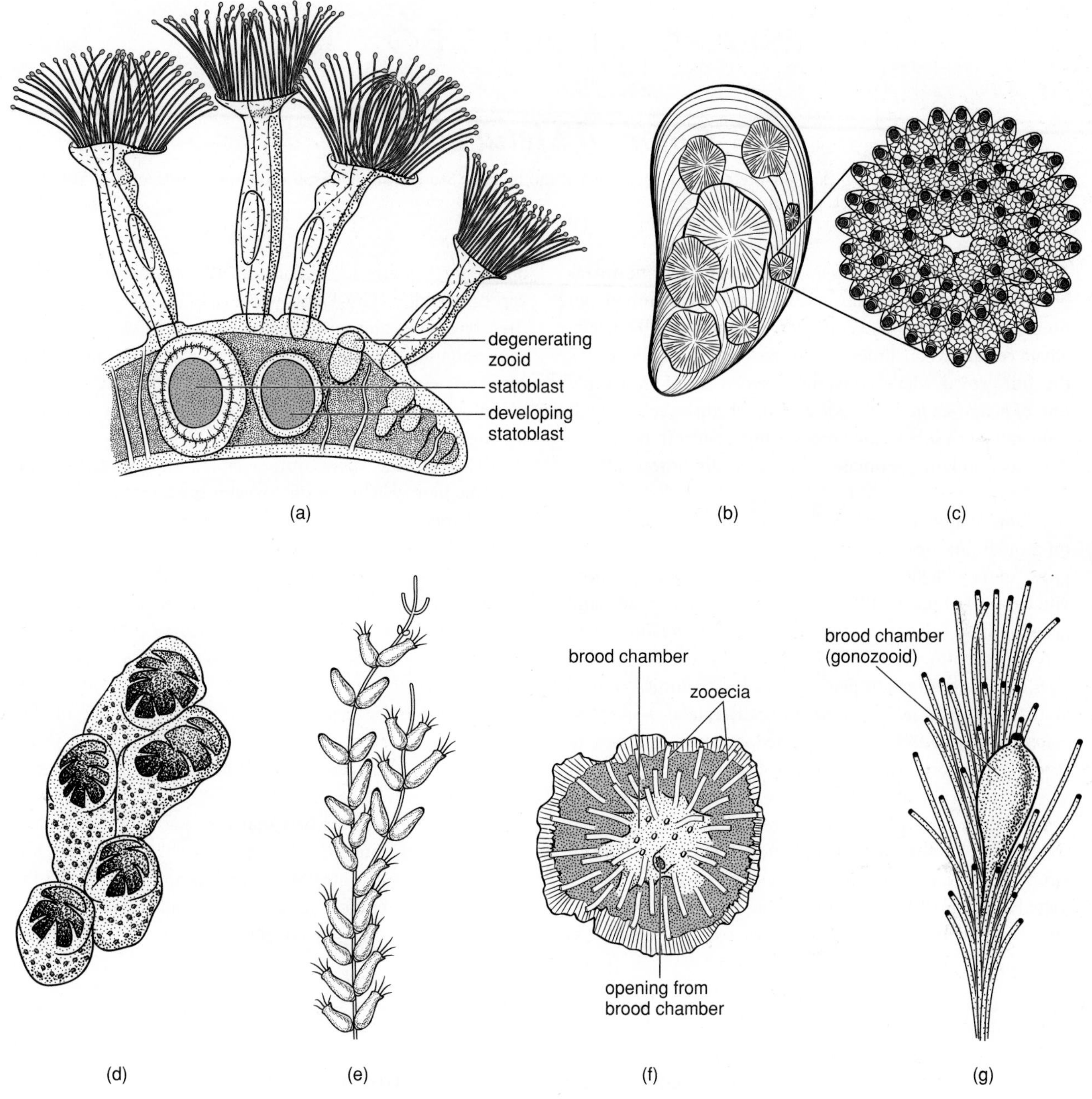

Figure 19.12

Bryozoan diversity. Phylactolaemata: (a) *Cristatella* colony. This species is found commonly encrusting lily pads in freshwater ponds. Note the gelatinous, creeping sole. The degenerating zooid is forming a brown body. Gymnolaemata: (b) Mussel shell valve encrusted with colonies of *Cryptosula,* shown in detail in frame (c). (d) *Electra pilosa.* Note the conspicuous spines guarding the frontal opening. (e) *Bowerbankia gracilis,* with tentacles withdrawn in several zooids. Stenolaemata: (f) *Lichenopora* colony. Note the communal brood chamber. (g) *Crisia ramosa,* showing tubular zooecium.

(a) After Hyman. (b,c) From Hyman; after Rogick. (d) Based on R. I. Smith, *Key to Marine Invertebrates of the Woods Hole Region.* 1964. (e) From Smith; after Rogick and Croasdale. (f) From Hyman; after Hincks. (g) After Harmer and Shipley.

cavities is incomplete, squeezes coelomic fluid from the metacoel into the mesocoel of the lophophore. Contraction of the body wall musculature thus inflates the lophophore tentacles and increases the hydrostatic pressure within the metacoel as well. Once the orifice diaphragm is opened by the contraction of specialized dilator muscles, the lophophore is forced out from the cystid by the elevated hydrostatic pressure within the metacoel. Retractor muscles extending from the body wall to the lophophore (Figs. 19.9 and 19.13) bring the lophophore back within the zooecium.

Perhaps the most intriguing feature unique to the phylactolaemates is the formation of **statoblasts.** These structures, produced seasonally, can withstand considerable desiccation and thermal stress. The statoblasts consist of a cell mass enclosed within a bivalved protective capsule of

RESEARCH FOCUS BOX 19.1

Colonial Metabolism

Hughes, D. J., and R. N. Hughes. 1986. Metabolic implications of modularity: Studies on the respiration and growth of *Electra pilosa*. *Phil. Trans. Royal Soc. London B* 313:23–29.

As most animals grow larger, their weight-specific metabolic rate (typically measured as oxygen consumed per hour per gram of tissue) declines; although a fully grown whale certainly consumes oxygen faster than a baby whale, the full-grown whale consumes oxygen more slowly *per unit of body weight.* Such a decline in weight-specific metabolic rate with increasing body volume results in most animals accumulating biomass at progressively slower rates as they grow.

Does coloniality offer any escape from this limitation on growth rate, or must colonial animals like bryozoans also grow more slowly with time? To answer this question, Hughes and Hughes (1986) measured metabolic rates and growth rates of the encrusting bryozoan *Electra pilosa.* Oxygen consumption rates of 37 colonies of widely different sizes were determined by placing individual colonies in small volumes of seawater and measuring the rate at which seawater oxygen concentration declined, at constant temperature. Oxygen concentrations were measured electronically.

At the end of each measurement, the researchers scraped each colony from its substrate, dried it at 60°C, and weighed it to the nearest milligram. This total weight would include inert calcium carbonate. To determine the actual amount of metabolizing tissue (i.e., **biomass**) present, Hughes and Hughes then cremated the samples at 500°C for 6 hours to combust all the organic material to carbon dioxide (CO_2) and water (H_2O). The researchers then reweighed the samples; the amount of weight lost in the cremation process reflects tissue weight, since only the inorganic "ash" is left behind. Tissue weights of the colonies tested in this experiment ranged from less than 0.1 mg to over 10 mg.

This "ashing" procedure enabled the researchers to examine how weight-specific oxygen consumption changed as the *E. pilosa* colonies grew. In contrast to the clear inverse relationship seen with noncolonial animals, weight-specific oxygen consumption did not decline with growth of the bryozoan colony (Focus Fig. 19.1). Moreover, monitoring the growth rates of 76 individual bryozoan colonies over a period of 2 weeks showed that, as colony size increased, so did the rate at which zooids on the growing edge (the peripheral zooids, located along the circumference of the colony) reproduced asexually (Focus Fig. 19.2). Thus, colonies of *E. pilosa* actually grew more rapidly as they aged, not more slowly.

In short, encrusting bryozoans appear to have a decided advantage over noncolonial organisms when competing for space on solid substrates, such as rocks and macroalgae: By proliferating an endless series of genetically identical modules, *E. pilosa* avoids the slowdown in growth inevitable in

species-specific morphology. They are produced along the funiculus of each zooid (Fig. 19.13), often in great numbers, and are released from the zooecium through a pore or, more typically, upon degeneration of the polyp in the late fall. When environmental conditions improve the following spring, the 2 valves of the statoblast separate along a preformed suture line, and a polypide soon emerges. Colony formation ensues by modular proliferation, through asexual budding of additional zooids. Phylactolaemates produce 2 types of statoblasts: **floatoblasts,** which are buoyant by means of gas-filled cells, and **sessoblasts,** which are firmly cemented to a solid substrate. Floatoblasts may be dispersed considerable distances by water, wind, or animals.

Statoblast formation by phylactolaemate bryozoans corresponds to the formation of gemmules and resting eggs by freshwater sponges and tardigrades, respectively. Clearly, formation of resting (i.e., **diapause**) stages is a common adaptation to the vagaries of freshwater existence.

Class Gymnolaemata

Most extant bryozoan species are assigned to the class Gymnolaemata (jim´-nō-lē-mah´-tah) (Fig. 19.12b–e), although they have dominated the phylum for only the past 70 million years or so (Fig. 19.11). The gymnolaemates are primarily marine. They are especially fascinating animals, even by bryozoan standards, displaying a broad range of morphological and functional diversity. Bryozoans are heavily preyed upon by turbellarians, polychaetes, insect larvae, crustaceans, arachnids (mites), gastropods, asteroids, and fish. To a great extent, the story of gymnolaemate evolution is one of increasing the degree to which the polypide is protected from these predators. In general, this is achieved by strengthening the zooecia, which is not so simply accomplished among bryozoans. Recall that lophophore protrusion depends upon the animal changing its shape to cause an elevation in coelomic pressure, and that the body wall is joined to the zooecium. The

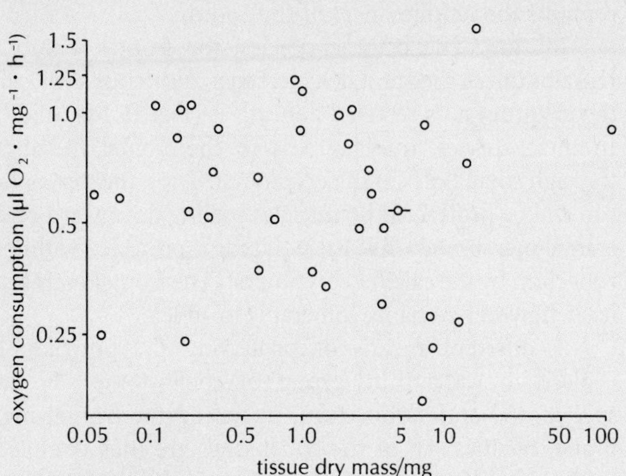

Focus Figure 19.1

The effect of colony size (measured as dry tissue weight) on weight-specific oxygen consumption (measured as microliters of oxygen consumed per milligram dry tissue weight per hour) in the encrusting bryozoan *Electra pilosa.* No obvious relationship is apparent; certainly there is no indication that weight-specific oxygen consumption declines with increasing colony size.

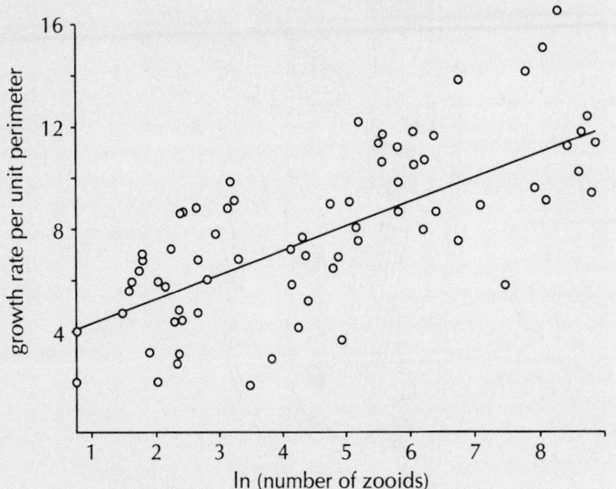

Focus Figure 19.2

The effect of colony size on the division rate of zooids along the edge of a colony in the encrusting bryozoan *Electra pilosa.* Increasing colony size is represented by the increasing number of zooids along the X-axis (plotted as natural log). In general, zooids along the edge of the colony divided faster in larger colonies.

organisms that grow by increasing their body volume. Indeed, growth actually accelerates with increased colony size, possibly because the feeding, nondividing zooids nearer the colony's center contribute nutrients to those actively dividing zooids along the edge. Can you design an experiment to determine whether the increased division rates of these edge zooids are, in fact, brought about by increased movement of nutrients from more central zooids?

trick, then, as described further on, is to strengthen the zooecium without losing the ability to generate elevated pressures within the metacoel.

Gymnolaemates differ from the phylactolaemates in a number of major respects. In phylactolaemates, the walls of the cystid are very incomplete, so adjacent zooids lack morphological boundaries (Fig. 19.13). Although each zooid protracts its lophophore through its own orifice, the polypides of a phylactolaemate colony essentially share a common metacoel. In contrast, the zooids of gymnolaemates are morphologically distinct individuals, although small **pore plates** allow the exchange of coelomic fluid between neighboring zooids (Fig. 19.14b). The 2 classes also differ with respect to lophophore morphology. Whereas the lophophore of phylactolaemate bryozoans has the deeply invaginated U-shape seen in phoronids, that of the gymnolaemates has a circular appearance.

In a few gymnolaemate species, the feeding zooids are borne on stolons—that is, on tubular extensions of the body wall. The stolons may be upright **(erect)** or flat against the substrate. In most gymnolaemate species, however, stolons are absent. Instead, zooids are contiguous, with the zooecium of one zooid supported by the zooecium of adjacent zooids (Figs. 19.10a and 19.12c), thereby strengthening the entire colony.

The colonies of most marine gymnolaemate species are composed of a large array of small, box-like or elliptical houses, typically encrusting on any submerged, solid surface. Zooids are arranged in species-specific patterns, determined by the pattern of asexual budding. The zooecia themselves are characterized by species-specific morphology and are therefore of great taxonomic value.

The Gymnolaemata contains 2 orders: the Ctenostomata (tēn´-ō-stō-mah´-tah) and the Cheilostomata (kēl´-ō-stō-mah´-tah). The ctenostomes are characterized by a flexible, chitinous zooecium, and lophophore protraction is achieved in much the same way as in the phylactolaemates. Muscle contractions draw the zooecium inward, and the resulting increase in hydrostatic pressure of the coelomic fluid forces the lophophore out through the orifice.

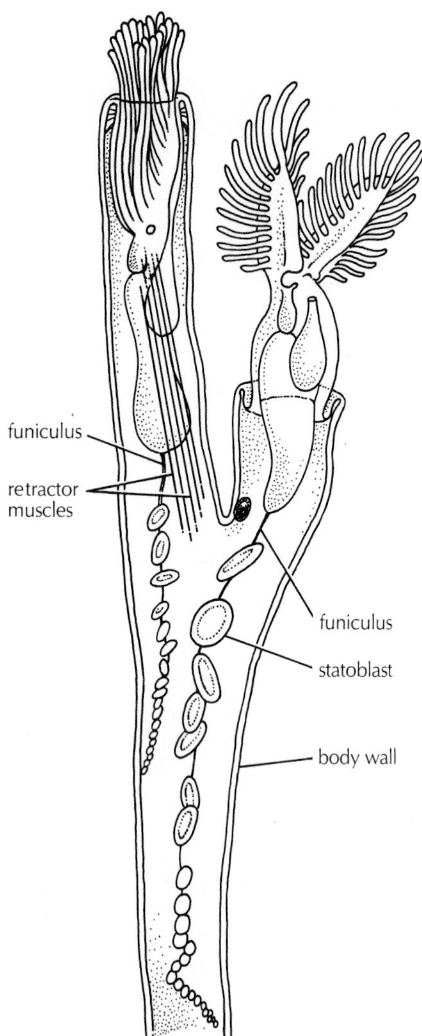

Figure 19.13
Statoblast formation in the freshwater bryozoan *Plumatella repens*. Note the association of the statoblasts with the funiculus and the lack of boundaries between adjacent zooids.
After Harmer and Shipley.

But most gymnolaemate bryozoans are cheilostomes, all of which calcify their zooecia to varying degrees. In many species, the orifice is sealed by a hinged, calcareous **operculum** when the polypide is fully withdrawn within the zooecium (Figs. 19.10b and 19.14). Calcium carbonate is also liberally deposited between the chitinous cuticle of the zooecium and the epidermis of the cystid of most species. The **frontal membrane,** however, often remains uncalcified, and muscles attach from the calcified body wall to this flexible surface (Fig. 19.14a, top); lophophore protraction is achieved by contracting these muscles, which increases the coelomic pressure. The body wall musculature is dramatically reduced or, more frequently, completely absent. Although calcification in these species certainly strengthens the colony structurally and probably provides some protection from predators, the zooid remains vulnerable through its uncalcified frontal membrane.

Often, the frontal membrane is partially shielded by spines projecting from the frontal margins of the zooecium (Fig. 19.12d) or, more rarely, from the frontal membrane itself. Nevertheless, the frontal membrane remains the Achilles' heel of the zooid.

In some cheilostome species, the frontal membrane remains uncalcified and flexible, but a calcareous shelf, called the **cryptocyst,** is secreted beneath it (Fig. 19.14a, middle). In most species, muscles pass to the frontal membrane through small holes in the cryptocyst. Thus, the lophophore can still be protracted by muscles pulling downward on the frontal membrane while the polypide is protected within the zooecium by the calcified cryptocyst. The frontal membrane itself, however, remains vulnerable to attack.

A different means of protecting the polypide has evolved in yet a third group of cheilostomes. In these species, the entire zooecium, including the frontal membrane, calcifies. All of the soft tissues are thus completely protected within the zooecium, but the frontal membrane can no longer serve as the vehicle through which internal hydrostatic pressure can be elevated. Instead, these species possess a new, uncalcified membrane that separates the metacoel from the calcified frontal surface. A space exists between the calcified frontal wall and this new membrane, forming a sac—specifically, a **compensation sac** or **ascus** (Fig. 19.14a, bottom). The compensation sac opens to the outside through a single pore, the **ascopore,** in the frontal surface (*ascus* = G: sac). Muscles extend from the calcified sides of the zooecium to the undersurface of the compensation sac. Contraction of these muscles pulls the compensation sac tissue inward, elevating the pressure within the coelomic cavity and forcing the lophophore out through the orifice of the zooecium. A vacuum does not form within the compensation sac during this process because seawater is drawn into the compensation sac through the ascopore.

The zooids described so far are the feeding and reproductive members of the colony, the **autozooids.** All gymnolaemate colonies contain some nonfeeding members as well. Thus, all gymnolaemate colonies are **polymorphic,** with individuals being morphologically or physiologically specialized for different functions, even though all individuals in the colony have a single genotype. The nonfeeding colony members, called **heterozooids,** take a variety of forms. The stolons and holdfasts of stoloniferous species are composed of a series of short heterozooids. Strangely enough, the pore plates permitting the exchange of coelomic fluid between adjacent zooids are also highly specialized heterozooids in some species. Other heterozooids, found only among the cheilostomes, are specialized for protecting and cleaning the colony. One such heterozooid is little more than a highly modified operculum, drawn out into a long, movable bristle. The continual sweeping movements made by these **vibracula** (Fig. 19.15d) presumably discourage invertebrate larvae from attaching to the colony and keep the colony surface free of debris. Other heterozooids form immobile, protective spines.

Probably the most intriguing heterozooids are the **avicularia.** Avicularia, in their least modified form, have the

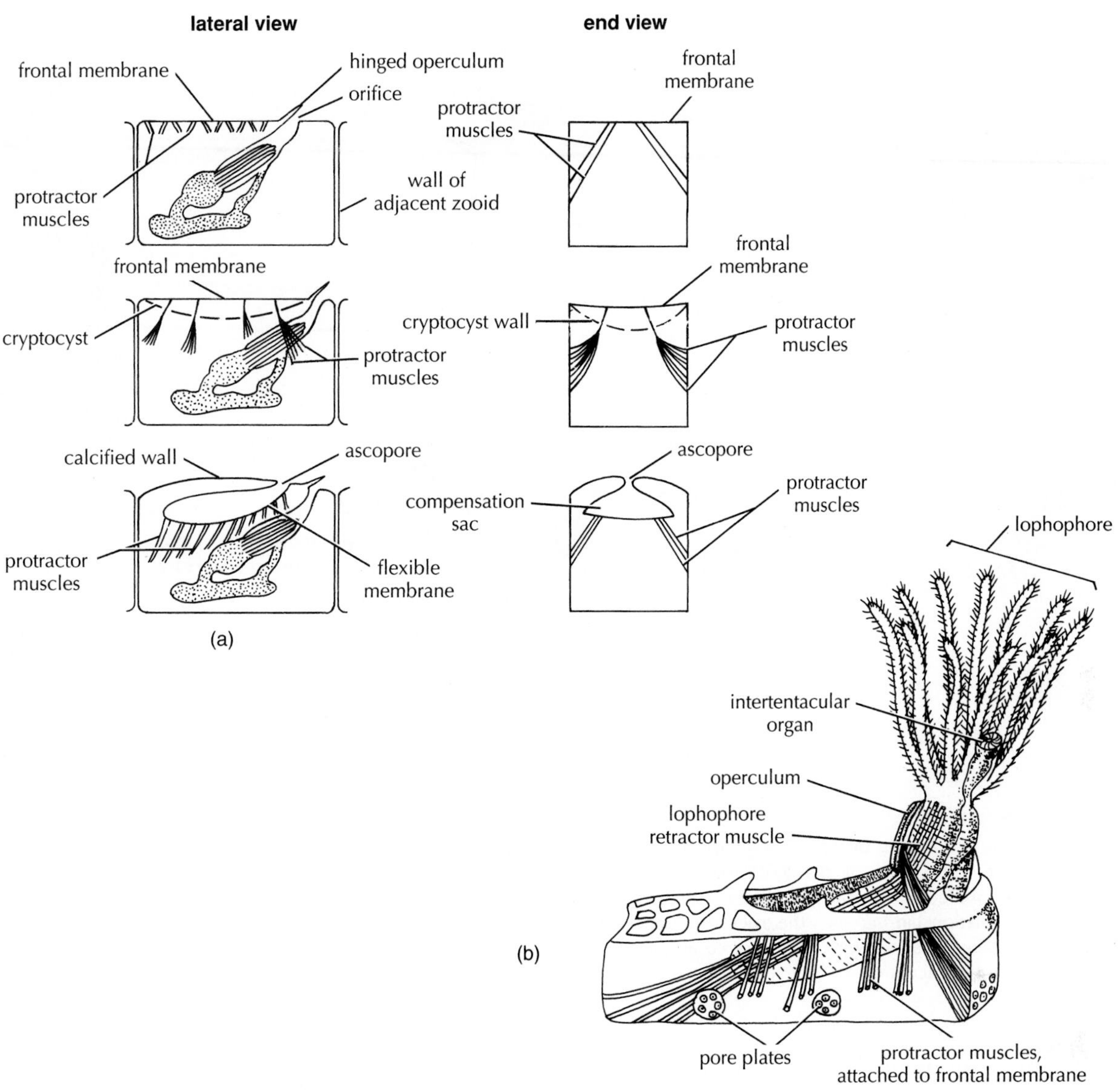

Figure 19.14

(a) Functional diversity among the gymnolaemates, showing different degrees of protection of the frontal membrane. (b) *Electra pilosa,* a species with a well-exposed frontal membrane, shown with lophophore protruded from the zooecium. The intertentacular organ serves as an exit for oocytes and as an entry point for sperm in some species.
(a) After Clark; after Harmer. (b) After Clark; after Marcus.

external appearance of a normal autozooid (Fig. 19.15c). Internally, the avicularia consist of little more than well-developed muscle fibers extending across a capacious coelom to the vicinity of the ventral, hinged operculum. The polypide is much reduced or completely absent. The operculum can be closed suddenly and with considerable force, discouraging, mutilating, or even killing potential predators. The avicularia of some species are highly modified for their tasks. In the most extreme form, the avicularia consist primarily of a modified opercular system, now referred to as a **mandible** (Fig. 19.15a–c). These avicularia may be mounted on a stalk, permitting some rotation. Some of these very specialized avicularia closely resemble a bird's head, a resemblance that gave rise to their name (*aves* = L: bird; thus, *avicularia* = little bird) (Fig. 19.15a, b).

In at least some species, adaptive shifts in zooid morphological development are induced by environmental factors, such as contact with potential predators and competitors.[8]

Class Stenolaemata

A small number of bryozoans are sufficiently dissimilar from the phylactolaemates and gymnolaemates to warrant placement into a third class, the Stenolaemata (stē´-nō-lē-mah´-tah). The members of this class are all marine, and

8. See *Topics for Further Discussion and Investigation,* no. 7.

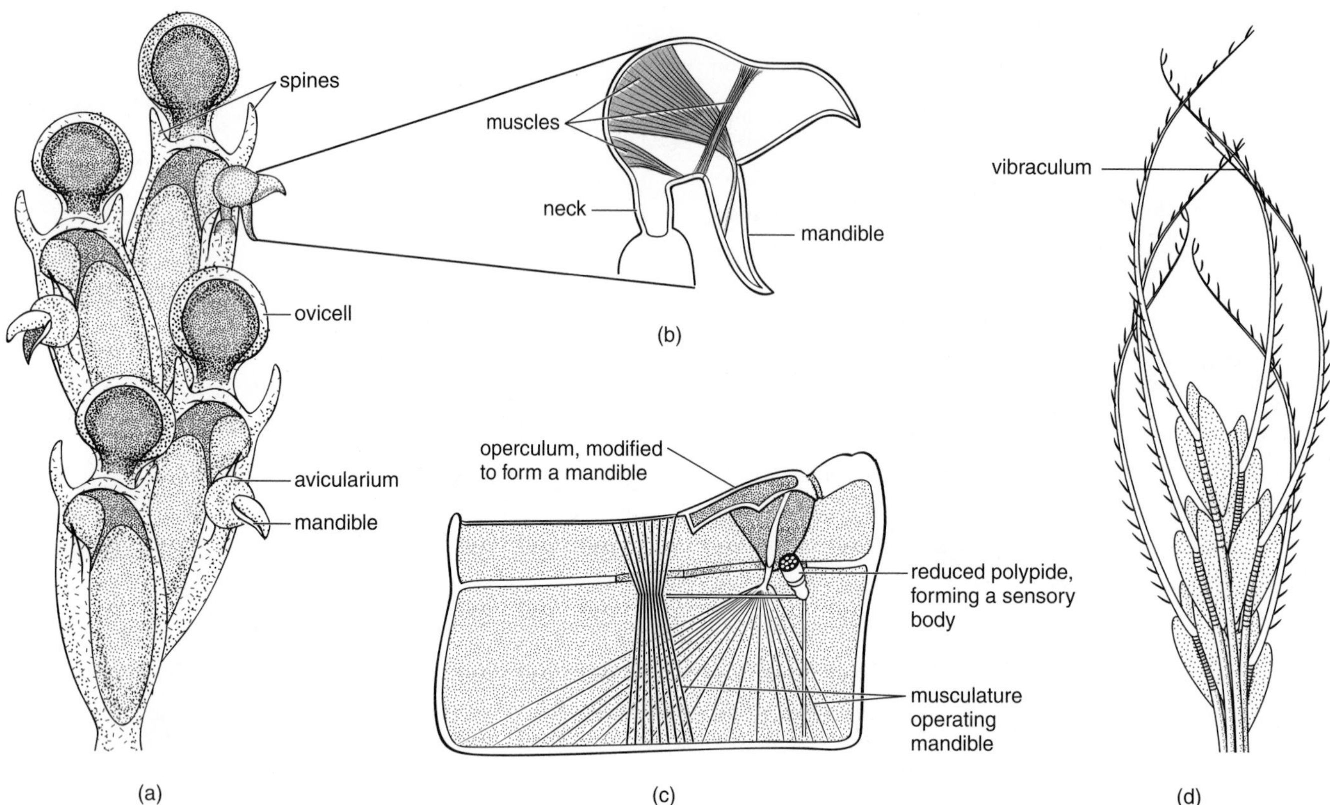

Figure 19.15

(a) *Bugula* sp., an erect, marine bryozoan with ovicells and highly modified avicularia. (b) Detail of the avicularia shown in (a). (c) Relatively unmodified avicularium of *Flustra foliacea,* showing more clearly the relationship between the avicularium and a normal zooid. (d) *Caberea ellisi,* showing part of a colony, each possessing a single, long vibraculum.

(a) From Hyman; after Rogick and Croasdale. (c) Based on Lars Silen, in Woollacott and Zimmer, eds., *The Biology of Bryozoa.* 1977. (d) From Smith.

the living species belong to a single order, the Cyclostomata (sī´-clō-stō-mah´-tah). Cyclostome zooids are always tubular and erect, and the zooecia are completely calcified (Fig. 19.12f, g). Not surprisingly, the cystid is nonmuscular, as in most cheilostomes. A thin, cylindrical membrane divides the main coelomic space into 2 compartments, one of which contains the polypide. Lophophore protraction is accomplished by the sideways, muscular displacement of this membrane. As in all bryozoans, the lophophore is drawn back into the zooecium by powerful lophophore retractor muscles.

Other Features of "Lophophorate" Biology

Reproduction

Asexual reproduction is most characteristic of the Bryozoa, in keeping with their habit of colony formation. Asexual reproduction by budding or fission is encountered in only one or 2 phoronid species, and reproduction is exclusively sexual in brachiopods. Brachiopods are also the exception when it comes to sexuality, with most species being gonochoristic. In contrast, most phoronids and bryozoan colonies are hermaphroditic. Gametes are formed by simple gonads—really just a cluster of germ cells in the mesodermal lining—within the metacoel. Phoronids and brachiopods usually discharge their gametes through the nephridia. Bryozoans, however, lack nephridia; instead, their sperm are released through the lophophore tentacles, and eggs are released through a special pore located between 2 of the tentacles. None of the animals copulate.

Phylum Phoronida

In phoronid reproduction, sperm are collected in the nephridia, packaged into discrete masses (**spermatophores**), and released through the nephridiopores of one individual to be captured from the seawater by another, neighboring individual. The embryos of all but one phoronid species develop into a characteristic, ciliated, feeding larval stage called an **actinotroch** (Figs. 19.16a and 24.15a). Metamorphosis to adult form involves a rapid and dramatic "turning inside out" of the larval body (Fig. 24.15a). Members of the one other species release crawling juveniles rather than free-swimming larvae. If

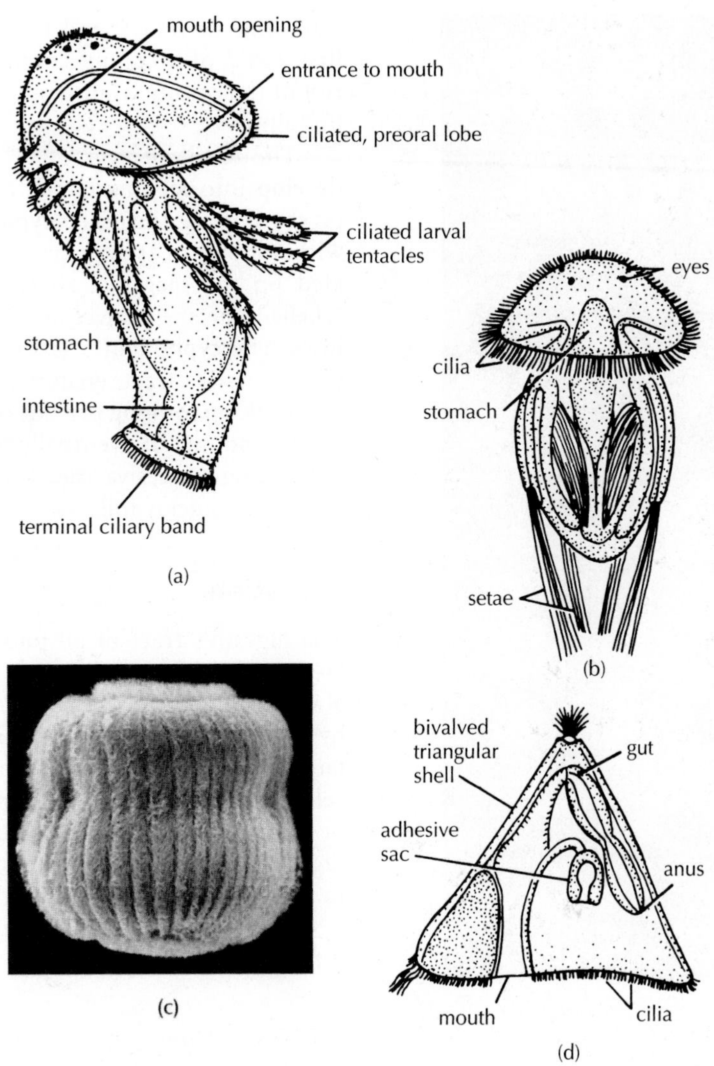

Figure 19.16

(a) Actinotroch larva of a phoronid. (b) Larva of the articulate brachiopod *Argyrotheca* sp. (c) Scanning electron micrograph of the coronate larva of the bryozoan *Bugula neritina.* (d) Cyphonautes larva of a bryozoan.

(a) From Hyman; after Wilson. (b) From Hyman; after Kowalevsky. (c) Courtesy of R. M. Woollacott and C. G. Reed. (d) After Hardy.

this reflects a secondary loss of larvae, then "development of a characteristic actinotroch larva" would be a valid Defining Characteristic of the phylum.

Phylum Brachiopoda

Among brachiopods, sperm and eggs are typically discharged through the nephridiopores with the eggs being fertilized in the surrounding seawater, although a few species brood their embryos. Articulate brachiopods have a unique larval form that does not resemble the adult (Fig. 19.16b). The larval stage of inarticulate brachiopods, in contrast, resembles a miniature adult, complete with shell valves and ciliated lophophore.

Phylum Bryozoa

Although all bryozoan colonies are hermaphroditic, an individual zooid may be of single sex. Sperm are released into the zooid's coelomic cavity and exit from openings in the lophophore tentacles. Neighboring individuals collect the sperm from the surrounding seawater. Once the eggs have been fertilized, a period of brood protection generally follows. Some species brood their embryos within the metacoel, but many other brooding sites are found within the phylum. Most cheilostomes possess specialized brooding chambers, called **ovicells,** at one end of the zooid (Fig. 19.10). As the embryos develop within the various brood chambers, the polypide of the parent usually degenerates.

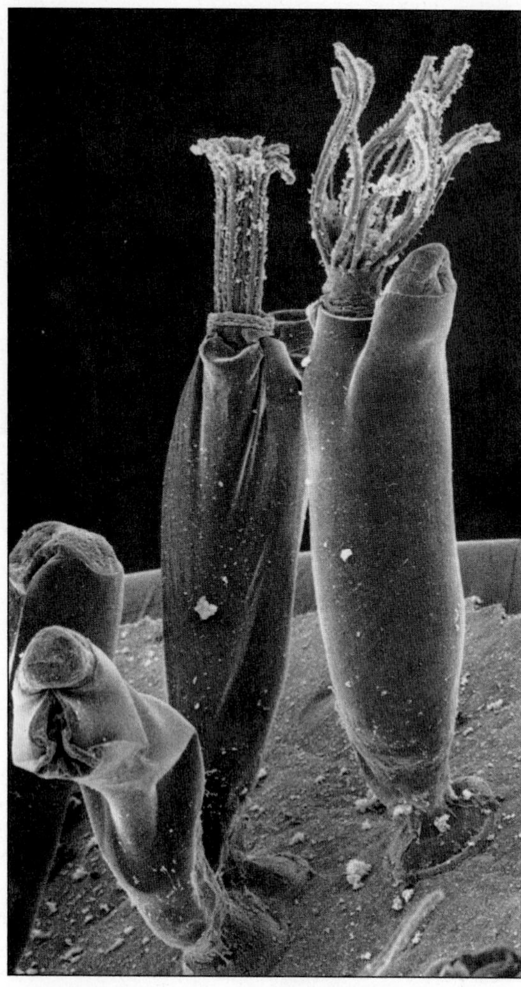

Figure 19.17
Four ancestrulae of *Bugula neritina,* newly metamorphosed from 4 coronate larvae. The ancestrulae are less than 1 mm tall, including the tentacles.
Courtesy of R. M. Woollacott and C. G. Reed.

This presumably reflects a transformation of parental tissues into nutrition for the offspring. The polypide of the parent may later be regenerated from the cystid.

In some brooding cheilostome species, the maternal zooid contributes substantial nutrients to the developing embryo in the ovicell. Because of this "placental brooding" system, the volume of the final larva as it escapes from the ovicell may be 15 to 500 times greater than the volume of the newly fertilized egg.

One or, more rarely, several nonfeeding, ciliated **coronate larvae** eventually emerge from each ovicell (Fig. 19.16c). After a limited period of dispersal away from the parent colony, the larvae attach to a substrate through the eversion of a sticky **adhesive sac,** and then they metamorphose to adult form (Fig. 19.17). During metamorphosis, all larval tissues move to the interior of the animal and are destroyed through a combination of phagocytosis and autolysis. Only the outer body wall remains intact, and this becomes the cystid. The first zooid of a colony is termed the **ancestrula.** The rest of the colony is subsequently generated via asexual budding.

The fertilized eggs of a few, nonbrooding species develop into feeding larval stages called **cyphonautes larvae** (Fig. 19.16d). The cyphonautes larva is enclosed within a pair of triangular, chitinous valves and is ciliated on the marginal surfaces not covered by these "shells." The shell valves are lateral, as in bivalved molluscs. The cyphonautes larva is equipped with a complete digestive tract, enabling it to feed and remain in the plankton for long periods of time, perhaps as long as several months. It eventually attaches to a substrate by everting an adhesive sac, and metamorphosis to the ancestrula quickly follows.

Digestion

The digestive tract of all phoronids and bryozoans is U-shaped, with a mouth and separate anus. The anus always lies outside the circle of lophophore tentacles, by definition. Digestion occurs within the relatively large stomach, and it has both extracellular and intracellular components. Brachiopods may (Inarticulata) or may not (Articulata) possess a U-shaped, one-way digestive system. In either case, the stomach connects to a large **digestive gland,** within which food material is broken down. Digestion is believed to be primarily intracellular.

Nervous System

The nervous system takes the form of a ring in all lophophorate species. This ring is found at the base of the lophophore in phoronids, encircling the esophagus in brachiopods, and adjacent to the pharynx in bryozoans. The ring may or may not be distinctively ganglionated. From the nerve ring, nerve fibers innervate the tentacles and musculature. An epidermal nerve network is commonly found in the body wall. Discrete sense organs are rare among lophophorates. Although a balance organ (**statocyst**) has been described in one species of inarticulate brachiopod, the sensory apparatus of other lophophorate species consists of scattered mechanoreceptor and chemoreceptor cells and/or the mantle setae.

Phylum Entoprocta (= Kamptozoa)

Phylum Ento • procta (= Kampto • zoa)
(G: within anus [G: flexible animal])

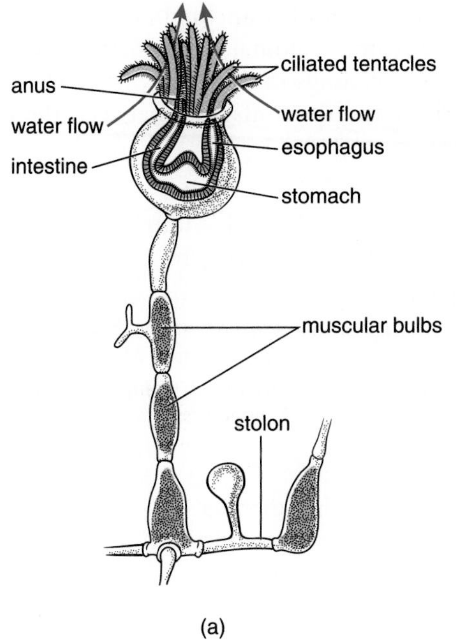

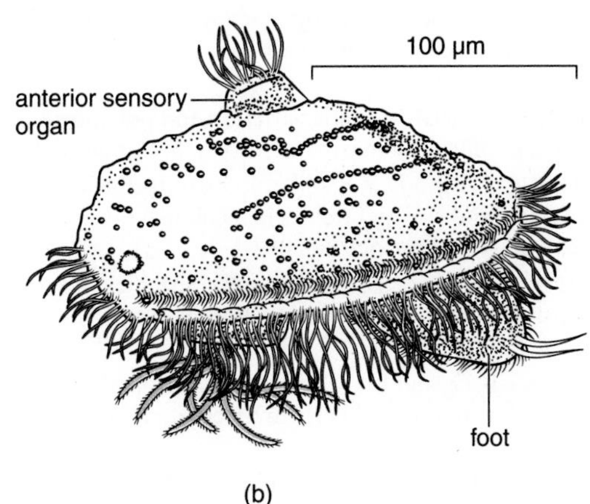

Figure 19.18

(a) An entoproct, *Anthropodaria* sp. Flow of water past the feeding tentacles of an entoproct. (b) Larval stage of the entoproct *Loxosomella harmeri,* viewed from the left side.

(a) From Hyman; after Nasonov. (b) Based on Claus Nielsen, in *Ophelia,* 9:209–341, 1971.

The Entoprocta is a phylum of benthic, solitary or colonial animals restricted to marine environments. About 150 species have been described, and all are suspension feeders. Like the other animals discussed in this chapter, entoprocts collect food particles using an anterior organ bearing numerous ciliated tentacles (Fig. 19.18). However, since the anus lies within this ring of tentacles (*ento* = G: within; *procta* = G: anus), the organ is not, by definition, a lophophore. Moreover, the pattern of water flow is opposite that exhibited among phoronids, brachiopods, and bryozoans: Among entoprocts, water flows into the center of the tentacular circle laterally, between the tentacles, and then passes upwards and outwards through the center (Fig. 19.18a; compare with Fig. 19.1b, p. 474). In several other respects, however, colonial entoprocts resemble bryozoans. Like bryozoan zooids, entoprocts are small, typically less than a few millimeters. Like bryozoans, entoprocts lack blood vessels and, as with phoronids, bryozoans, and many brachiopods, the entoproct digestive tract is U-shaped (Fig. 19.18).

However, unlike that of bryozoans, entoproct cleavage is determinate and spiral, and the embryos of some entoproct species develop into perfectly formed protostome-like trochophore larvae. This is in keeping with recent molecular studies of 18S rRNA genes, which suggest that entoprocts are protostomes.

Entoprocts have a small body cavity of sorts (filled with gelatinous material), but how it forms is still unclear. Consequently, some workers consider the entoprocts to be acoelomate, while others believe them to be pseudocoelomate. They are indeed a troublesome group at present. Some molecular data indicate no close relationship between entoprocts and bryozoans, while other molecular data suggest that those 2 groups should be placed in a single phylum.[9] On the other hand, some have argued that entoprocts are most closely related to the cycliophorans discussed at the end of the previous chapter, and have even suggested grouping entoprocts and cycliophorans in a single phylum.

Many entoproct species form extensive colonies through asexual replication, as do bryozoans, but entoproct colonies are never polymorphic; all individuals in an entoproct colony look the same and are capable of feeding and sexual reproduction. All entoprocts are apparently hermaphroditic.

Taxonomic Summary

Phylum Phoronida
Phylum Brachiopoda—the lampshells
  Class Inarticulata—the inarticulate brachiopods
  Class Articulata—the articulate brachiopods
Phylum Bryozoa (= Ectoprocta; = Polyzoa)—the moss animals
  Class Phylactolaemata
  Class Gymnolaemata
    Order Ctenostomata
    Order Cheilostomata
  Class Stenolaemata
    Order Cyclostomata
Phylum Entoprocta (= Kamptozoa)

9. See *Topics for Further Discussion and Investigation,* no. 1.

Topics for Further Discussion and Investigation

1. For a long time, entoprocts and ectoprocts (bryozoans) were grouped in a single phylum. Discuss the anatomical and functional similarities and differences between bryozoans and entoprocts, and the evidence for and against a close evolutionary relationship between these 2 groups.

Hausdorf, B., M. Helmkampf, A. Meyer, A. Witek, H. Herlyn, I. Bruchhaus, T. Hankeln, T. H. Struck, and B. Lieb. 2007. Spiralian phylogenomics supports the resurrection of Bryozoa comprising Ectoprocta and Entoprocta. *Molec. Biol. Evol.* 24: 2723–29.

Hyman, L. H. 1951. *The Invertebrates. Vol. III, Acanthocephala, Aschelminthes and Entoprocta.* New York: McGraw-Hill.

Mackey, L. Y., B. Winnepenninckx, R. De Wachter, T. Backeljau, P. Emschermann, and J. R. Garey. 1996. 18S rRNA suggests that Entoprocta are protostomes, unrelated to Ectoprocta. *J. Molec. Evol.* 42:552.

Mariscal, R. N. 1965. The adult and larval morphology and life history of the entoproct *Barentsia gracilis* (M. Sars, 1835). *J. Morphol.* 116:311.

2. Discuss the potential role of bivalved molluscs in bringing about the dramatic evolutionary decline of brachiopods.

Gould, S. J., and C. B. Calloway. 1980. Clams and brachiopods—ships that pass in the night. *Paleobiology* 6:383.

Sepkoski, J. J., Jr. 1996. Competition in macroevolution: The double-wedge revisited. In D. Jablonski, D. H. Erwin, J. H. Lipps, eds. *Evolutionary Paleobiology.* Chicago Univ. Press, Chicago, 211–55.

Stanley, S. M. 1968. Post-Paleozoic adaptive radiation of infaunal bivalve molluscs; a consequence of mantle fusion and siphon formation. *J. Paleontol.* 42:214.

3. Investigate the role of water currents in the feeding biology of phoronids, brachiopods, and bryozoans.

Gilmour, T. 1979. Ciliation and function of food-collecting and waste-rejecting organs of lophophorates. *Canadian J. Zool.* 56:2142.

LaBarbera, M. 1977. Brachiopod orientation to water movement. I. Theory, laboratory behavior, and field orientations. *Paleobiology* 3:270.

LaBarbera, M. 1984. Feeding currents and particle capture mechanisms in suspension feeding animals. *Amer. Zool.* 24:71.

Nielsen, C., and H. U. Riisgård. 1998. Tentacle structure and filter-feeding in *Crisia eburnea* and other cyclostomatous bryozoans, with a review of upstream-collecting mechanisms. *Marine Ecol. Progr. Ser.* 168:163.

Strathmann, R. R. 1973. Function of lateral cilia in suspension-feeding of lophophorates (Brachiopoda, Phoronida, Ectoprocta). *Marine Biol.* 23:129.

Strathmann, R. R., and Q. Bone. 1997. Ciliary feeding assisted by suction from the muscular oral hood of phoronid larvae. *Biol. Bull.* 193: 153–62.

4. Based upon your readings in this book, compare and contrast the operation of the bryozoan lophophore with the operation of the nemertine proboscis.

5. Some species of branching bryozoans superficially resemble some colonial hydrozoan colonies. Based upon your readings in this book, what are some of the similarities and dissimilarities between these 2 animal groups?

6. Discuss the evidence that nutrients are shared among the zooids of a bryozoan colony.

Best, M. A., and J. P. Thorpe. 1985. Autoradiographic study of feeding and the colonial transport of metabolites in the marine bryozoan *Membranipora membranacea. Marine Biol.* 84:295.

Carle, K. J., and E. E. Ruppert. 1983. Comparative ultrastructure of the bryozoan funiculus: A blood vessel homologue. *Z. Zool. Syst. Evol.-forsch.* 21:181.

7. Investigate the manner in which gymnolaemate bryozoans cope with competition, predation, and other environmental stresses.

Bayer, M. M., C. D. Todd, J. E. Hoyle, and J. F. B. Wilson. 1997. Wave-related abrasion induces formation of extended spines in a marine bryozoan. *Proc. R. Soc. London* 264:1605.

Harvell, C. D. 1986. The ecology and evolution of inducible defenses in a marine bryozoan: Cues, costs, and consequences. *Amer. Nat.* 128:810.

Harvell, C. D. 1998. Genetic variation and polymorphism in the inducible spines of a marine bryozoan. *Evolution* 52:80.

Harvell, C. D., and D. K. Padilla. 1990. Inducible morphology, heterochrony, and size hierarchies in a colonial invertebrate monoculture. *Proc. Nat. Acad. Sci.* 87:508.

Iyengar, E. V., and C. D. Harvell. 2002. Specificity of cues inducing defensive spines in the bryozoan *Membranipora membranacea. Marine Ecol. Progr. Ser.* 225: 205–18.

Jackson, J. B. C., and L. W. Buss. 1975. Allelopathy and spatial competition among coral reef invertebrates. *Proc. Nat. Acad. Sci.* 72:5160.

Lopanik, N. B., N. M. Target, and N. Lindquist. 2006. Ontogeny of a symbiont-produced chemical defense in *Bugula neritina* (Bryozoa). *Marine Ecol. Progr. Ser.* 327: 183–91.

Padilla, D. K., C. D. Harvell, J. Marks, and B. Helmuth. 1996. Inducible aggression and intraspecific competition for space in a marine bryozoan, *Membranipora membranacea. Limnol. Oceanogr.* 41:505.

Shapiro, D. F. 1992. Intercolony coordination of zooid behavior and a new class of pore plates in a marine bryozoan. *Biol. Bull.* 182:221.

Tzioumis, V. 1994. Bryozoan stolonal outgrowths: A role in competitive interactions? *J. Marine Biol. Assoc. U.K.* 74:203.

Taxonomic Detail

Phylum Phoronida

Phoronis. The 14 species in this phylum are found intertidally to depths of about 400 m. The smallest species are less than about 0.5 cm long, and the largest are about 50 cm long.

Phylum Brachiopoda

This phylum contains approximately 350 species distributed between 2 classes.

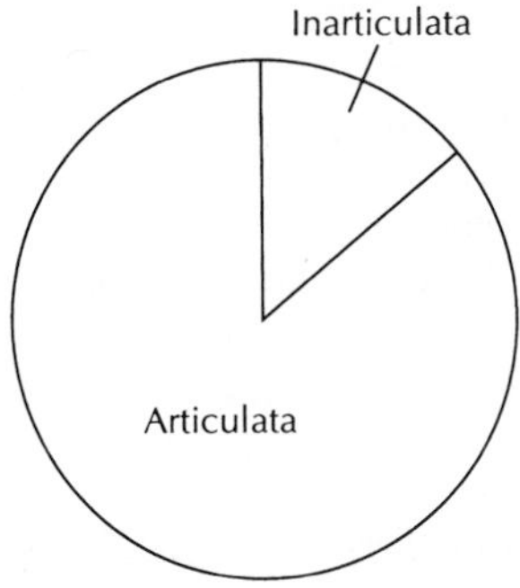

Class Inarticulata

The approximately 50 species are contained within 2 orders.

Order Lingulida

Lingula. Individuals are up to 44 mm in shell length. All species use their shells to form burrows in sediment, intertidally to depths of about 125 m. Some species are eaten by humans (in Australia and Japan). Fossils over 400 million years old closely resemble species of present-day *Lingula.* One family.

Order Acrotretida

Crania. Some of the species in this group range from the intertidal zone to depths exceeding 7600 m. Two families.

Class Articulata

The approximately 300 species are contained within 2 orders.

Order Rhynchonellida

These brachiopods live permanently attached to solid substrates from about 6 m deep to over 3000 m, in both tropical and cold waters. Approximately 30 species, distributed among 4 families.

Order Terebratulida

This group contains most living brachiopod species. The approximately 250 species are distributed among 12 families.

Family Terebratellidae. *Terebratella.* Shells are up to 91 mm long. One species, *Magadina cumingi,* uses its pedicle to move up and down in the substrate, unlike any other species of articulate brachiopod. Species occur from the intertidal zone to depths exceeding 4000 m.

Phylum Bryozoa

This phylum contains approximately 5000 species distributed among 3 classes.

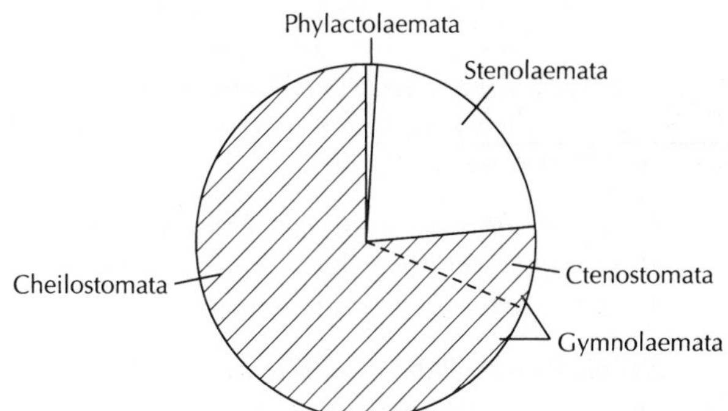

Class Phylactolaemata

These bryozoans are found only in freshwater. The approximately 50 species are distributed among 4 families.

Family Plumatellidae. *Plumatella.* Members of this family produce both sessile and free-floating, dispersive statoblasts.

Family Lophopodidae. *Pectinatella.* Colonies form large, gelatinous masses up to 0.5 m across, and the lophophores of some species bear up to 120 tentacles. All species produce floating statoblasts.

Family Cristatellidae. *Cristatella.* The polypides are confined to the upper surface of the rounded, unbranched colony. The colony's muscular lower surface enables the entire colony to move slowly, perhaps 10 cm per day, over a solid substrate. All species produce only floating statoblasts.

Class Stenolaemata

Crisia, Tubulipora. All species are marine and exhibit tubular, calcified zooids. Reproduction is unique among bryozoans in that each fertilized embryo may divide asexually, numerous times, giving rise to 100 or more genetically identical embryos within each reproductive individual (gonozooid). The approximately 900 species are distributed among about 20 families.

Class Gymnolaemata

This is the most diverse group of bryozoans and contains the greatest number of species (approximately 3000), most of which are marine. Two orders.

Order Ctenostomata

Zooid walls are not calcified but instead are membranous or gelatinous. Avicularia, opercula, and ovicells are lacking. The approximately 250 species are distributed among about 20 families.

Family Alcyonidiidae. *Alcyonidium.* This genus has a global distribution, being especially common in shallow coastal waters. All species are marine. Colonies are fleshy, rubbery, or gelatinous, and are erect in some species, flat and encrusting in others. Often members of this genus live symbiotically with shelled molluscs, crustaceans, hydrozoans, or other bryozoans. Remarkably, one northern circumpolar species lives unattached on soft substratum, while an Antarctic species apparently has a uniquely pelagic existence, living unattached in the open water.

Family Nolellidae. *Nolella, Aethozoon.* Members of the genus *Aethozoon* have zooids up to 8 mm long, the largest of any bryozoan species.

Family Paludicellidae. *Paludicella.* All species are restricted to freshwater.

Family Monobryozoontidae. *Monobryozoon.* The smallest bryozoan colonies are found in this group: The entire colony consists of a single, feeding autozooid. As new zooids are produced, they break free of the parent, which soon degenerates. All 3 species are marine and live in sandy sediments.

Family Hypophorellidae. *Hypophorella.* The branched stolons bore holes into sedentary polychaete tubes or into the calcareous walls of other bryozoans. All species are marine.

Family Penetrantiidae. These species secrete phosphoric acid, thereby boring into the calcareous shells of molluscs and barnacles. All species are marine.

Family Vesiculariidae. *Bowerbankia, Zoobotryon.* All species are marine, forming branching, typically erect colonies. Lophophores bear only 8 to 10 tentacles each. Members of one genus, *Zoobotryon,* attain colony lengths of up to 0.5 m.

Order Cheilostomata

Most species are marine and form highly polymorphic colonies with specialized reproductive zooids, attachment zooids, avicularia, vibracula, and ovicells. The zooids are often rectangular, and the walls are always at least partly calcified. This group contains the greatest diversity of bryozoan morphology. The approximately 2750 species are distributed among about 70 families.

Suborder Anasca

Approximately 30 families. Most or all of each zooid's frontal surface is membranous and virtually transparent.

Family Membraniporidae. *Membranipora.* These species lack avicularia and ovicells. Species may be encrusting, often forming almost circular colonies, or erect. All species produce feeding cyphonautes larvae with bivalved shells.

Family Electridae. *Electra.* Colonies lack avicularia or ovicells, and most species are encrusting. All species produce cyphonautes larvae.

Family Microporidae. *Micropora.* A number of the species in this group are free-living. One, *Selenaria maculata,* walks on the long setae of its avicularia at speeds of over 1.5 cm per minute.

Family Bugulidae. *Bugula, Dendrobeania.* These species form bushy, erect colonies. The zooids are only lightly calcified, and the entire frontal surface is membranous, so the internal organs are easily visible through a dissecting microscope. Colonies usually possess well-formed "bird's head" avicularia on stalks. *Bugula* is one of the largest and most widespread of all bryozoan genera. One member of the genus, *B. neritina,* is the source of bryostatin, a promising anticancer compound produced by a bacterial symbiont, particularly of the larval stage.

Family Cellariidae. *Cellaria.* These species are unusual in that the structure of the operculum and cryptocyst makes it impossible for parietal muscles to reach the flexible frontal membrane; the lophophore is apparently protruded by a different, and more complex mechanism (Perez and Banta, 1996. *Invert. Biol.* 115:162).

Suborder Ascophora

This is the largest and most diverse group of all living bryozoans. The frontal wall demonstrates a great variety of form and function. The lophophore is always operated through a compensation sac (= ascus). About 50 families.

Family Watersiporidae. *Watersipora.* The members of this group are all encrusting and lack avicularia and ovicells; zooids brood embryos internally.

Family Hippoporinidae. *Pentapora.* All species have ovicells, and many have avicularia. Colonies of *Pentapora* may become larger than 1 m in circumference, making them the largest of all bryozoan colonies.

Family Schizoporellidae. *Schizoporella.*

Phylum Entoprocta

Most of the approximately 150 species are marine, although a few live in freshwater. Four families.

Family Loxosomatidae. *Loxosoma, Loxosomella.* This family contains all of the solitary species; all other entoprocts are colonial. Some species are mobile, using a muscular, distal expansion of the stalk. All of the at least 110 species are marine.

Family Barentsiidae. *Barentsia, Urnatella.* All species are colonial. The few freshwater entoprocts are all placed in the genus *Urnatella.* Some species produce cysts, which tolerate adverse environmental conditions. Approximately 30 species.

General References about the "Lophophorates"

Boardman, R. S., A. H. Cheetham, and A. J. Rowell, eds. 1987. *Fossil Invertebrates.* Palo Alto, Calif.: Blackwell Scientific, 445–549.

Cohen, B. 2000. Monophyly of brachiopods and phoronids: Reconciliation of molecular evidence with Linnaean classification (the subphylum Phoroniformea nov.). *Proc. Royal Soc. London* B 267: 225–31.

Harrison, F. W., and R. W. Woollacott. 1997. *Microscopic Anatomy of Invertebrates, Vol. 13. Lophophorates and Entoprocta.* NY: Wiley-Liss.

Hyman, L. H. 1959. *The Invertebrates, Vol. 5. Smaller Coelomate Groups.* New York: McGraw-Hill.

James, M. A., A. D. Ansell, M. J. Collins, G. B. Currey, L. S. Peck, and M. C. Rhodes. 1992. Biology of Living Brachiopods. *Adv. Marine Biol.* 28:176–387.

Levin, H. L. 1999. *Ancient Invertebrates and Their Living Relatives.* New Jersey: Prentice Hall, pp. 157–75 (Bryozoa) and 176–202 (Brachiopoda).

Nielsen, C. 2002. The phylogenetic position of Entoprocta, Ectoprocta, Phoronida, and Brachiopoda. *Integr. Comp. Biol.* 42: 685–91.

Parker, S. P., ed. 1982. *Classification and Synopsis of Living Organisms, Vol. 2, Phoronids, Bryozoans, and Brachiopods.* New York: McGraw-Hill, 741, 743–69, 773–80.

Passamaneck, Y., and K. M. Halanych. 2006. Lophotrochozoan phylogeny assessed with LSU and SSU data: Evidence of lophophorate polyphyly. *Molec. Phylog. Evol.* 40: 20–28.

Ryland, J. S. 1970. *Bryozoans.* London: Hutchinson Univ. Library.

Sharp, J. H., M. K. Winson, and J. S. Porter. 2007. Bryozoan metabolites: an ecological perspective. *Nat. Prod. Rep.* 24: 659–73.

Thorpe, J. H., and A. P. Covich, eds. 2001. *Ecology and Classification of North American Freshwater Invertebrates,* 2nd ed. New York: Academic Press (bryozoans).

Woollacott, R. M., and R. Zimmer, eds. 1977. *Biology of Bryozoans.* New York: Academic Press.

Zimmer, R. L. 1996. Phoronids, Brachiopods, and Bryozoans, the Lophophorates. In: Gilbert, S. F., and A. M. Raunio. 1997. *Embryology: Constructing the Organism.* Sunderland, MA: Sinauer Associates, Inc. Publishers, 279–305.

General References about the Entoprocts

Harrison, F. W., and R. W. Woollacott. 1997. *Microscopic Anatomy of Invertebrates, Vol. 13. Lophophorates and Entoprocta.* New York: Wiley-Liss.

Hyman, L. H. 1951. *The Invertebrates, Vol. 3. Acanthocephala, Aschelminthes and Entoprocta.* New York: McGraw-Hill.

Nielsen, C. 1964. Studies on Danish Entoprocta. *Ophelia* 1:1–76.

Parker, S. P., ed. 1982. *Classification and Synopsis of Living Organisms,* vol. 2. New York: McGraw-Hill, 771–72.

Search the Web

1. www.ucmp.berkeley.edu/brachiopoda/brachiopoda.html

   www.ucmp.berkeley.edu/bryozoa/bryozoa.html

   These websites include information on the fossil records of bryozoans and brachiopods. The sites are maintained by the University of California at Berkeley.

2. www.mbl.edu

   Choose under Quick Links, "Biological Bulletin" and then "Other Biological Bulletin Publications." Next select "Keys to Marine Invertebrates of the Woods Hole Region," and then choose "Ectoprocta" or "Entoprocta." This brings up a taxonomic key for animals commonly found near Woods Hole, MA.

3. www.microscopy-uk.org.uk/mag/artmay01/bryozoan.html

   This brings you to "Flowers of the sea: Bryozoans and Cnidarians." This article includes wonderful photographs and an excellent video of bryozoan feeding and avicularial activity. The site is maintained by *Micscape,* an online magazine produced in the U.K.

4. http://paleopolis.rediris.es/Phoronida/

   World database on phoronids and phoronid research, provided by Christian C. Emig and Christian de Mittelwihr. Includes links to other websites concerning "lophophorates."

5. http://www.marinespecies.org/brachiopoda/

   World database on brachiopods and brachiopod research, provided by Christian C. Emig.

20

The Echinoderms

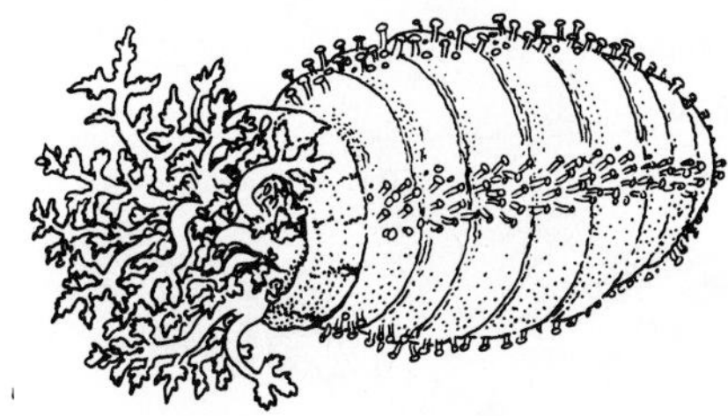

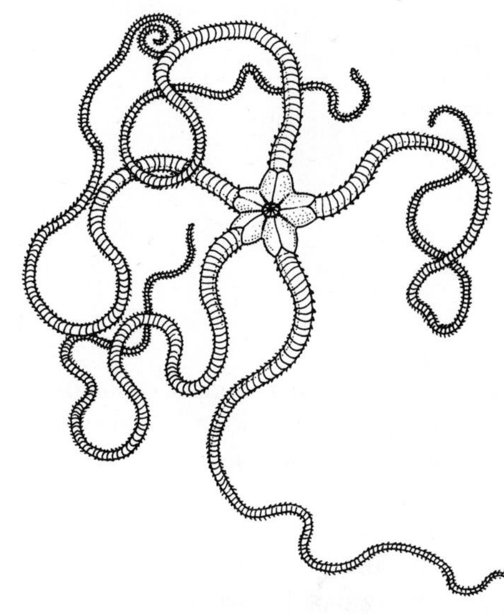

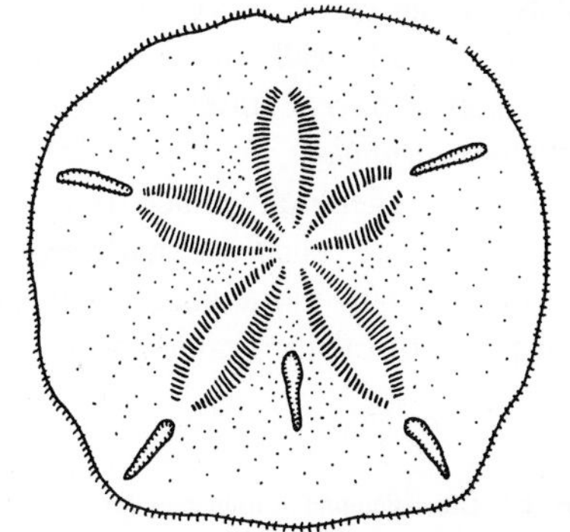

Introduction and General Characteristics

Phylum Echino • dermata
(Greek: spine skin)
ē-kīn´-ō-der-mah´-tah

Defining Characteristics:[1] 1) A complex series of fluid-filled canals (the water vascular system) derived from a pair of coelomic compartments and which service numerous flexible feeding and locomotory appendages (tube feet); 2) 5-pointed (pentamerous) radial symmetry in adults; 3) calcareous ossicles derived from mesodermal tissue forming an endoskeleton; 4) connective tissue is mutable: Its stiffness and fluidity can be rapidly and dramatically altered by the nervous system

Echinoderms are remarkable animals that include the sea lilies, feather stars, brittle stars, sea stars, sand dollars, sea urchins, sea biscuits, and sea cucumbers. Like us, echinoderms are deuterostomes (see Chapter 2). In late 2006, biologists finished sequencing the genome of the purple sea urchin (*Strongylocentrotus purpuratus*), the first such sequence for a nonchordate deuterostome; as it turns out, some of its approximately 23,300 genes are identical with those previously considered unique to vertebrates. Recent analyses indicate that echinoderms, together with hemichordates (Chapter 21), form the sister group (the Ambulacraria) to chordates (see Chapter 23, Fig. 23.10).

Nearly all of the approximately 6500 echinoderm species living today are marine; a few species are estuarine, but none live in freshwater. An additional 13,000 or so species, distributed among approximately 16 classes, are

1. Characteristics distinguishing the members of this phylum from members of other phyla.

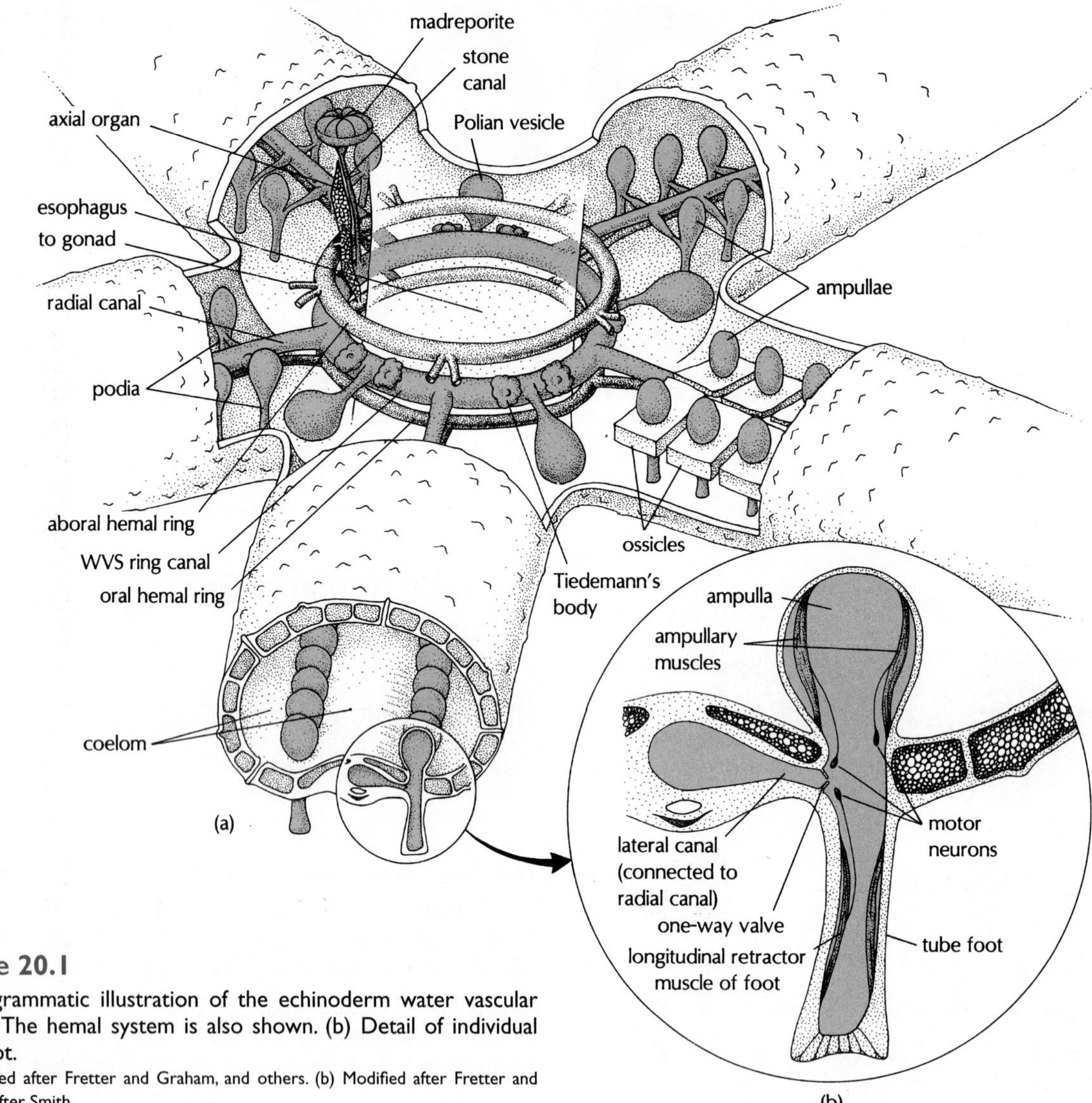

Figure 20.1

(a) Diagrammatic illustration of the echinoderm water vascular system. The hemal system is also shown. (b) Detail of individual tube foot.

(a) Modified after Fretter and Graham, and others. (b) Modified after Fretter and Graham; after Smith.

known from the fossil record; most of those classes have no living representatives. Remarkably, although their free-living larval stages are bilaterally symmetrical, most adult echinoderms show a basic 5-point (pentamerous) radial symmetry. As radially symmetrical animals, they lack cephalization. Thus, adult echinoderms generally do not have anterior and posterior ends. Instead, body surfaces are designated as being either **oral** (bearing the mouth) or **aboral** (not bearing the mouth).

Most echinoderms possess a well-developed internal skeleton composed largely (up to 95%) of calcium carbonate, with smaller amounts of magnesium carbonate (up to 15%), even lesser amounts of other salts and trace metals, and a small amount of organic material. The components of the echinoderm skeleton are individually manufactured within specialized cells originating from embryonic mesoderm. This is in sharp contrast to the method of shell production in molluscs and other invertebrate groups, in which minerals are deposited into an extracellular protein matrix.

The major unifying characteristic of the phylum Echinodermata is the presence of what is known as the **water vascular system** (often abbreviated as the **WVS**). The WVS consists of a series of fluid-filled canals derived primarily from one of 3 pairs of coelomic compartments (the hydrocoel) that form during embryonic development. These canals lead to thin-walled tubular structures called **podia,** or **tube feet** (*podium* = G: a foot) (Fig. 20.1a.) The podia are best visualized as tubular extensions of the WVS that penetrate the echinoderm body wall and skeleton in particular regions, known as **ambulacral zones,** or, in some groups, **ambulacral grooves** (*ambulacr* = Latin: walk). The system of internal WVS canals is generally linked to the outside seawater through a sieve plate called the **madreporite,**[2] which leads down a **stone canal** (so named because it is reinforced with spicules or plates of calcium carbonate) and then to a

2. See *Topics for Further Discussion and Investigation,* no. 9, at the end of the chapter.

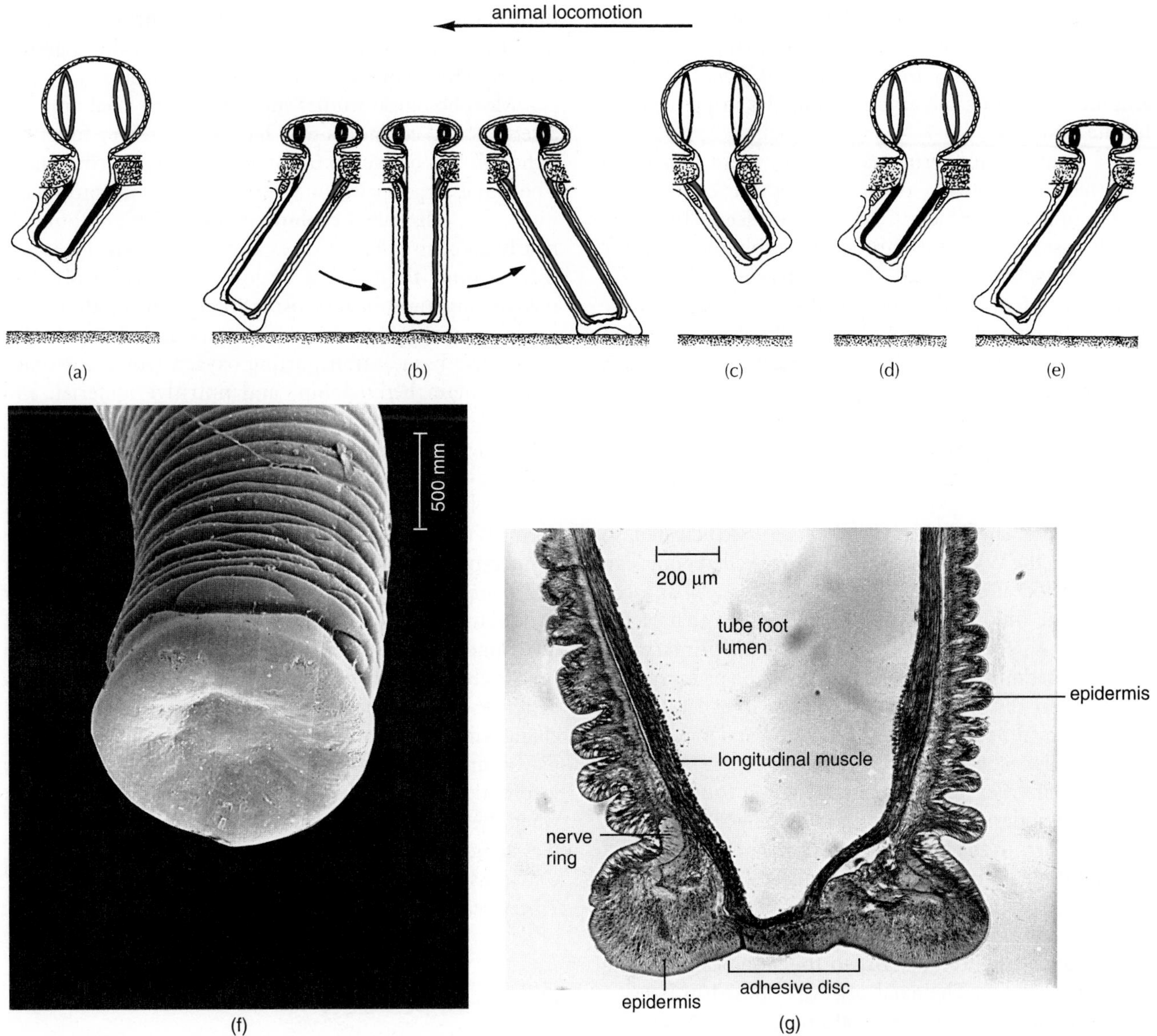

Figure 20.2

Tube foot locomotion over a solid substrate. Blue shading indicates muscles about to contract to produce the next step in the sequence. (a) Ampullar muscles about to contract, elongating the tube foot by hydraulic action. (b) Base of tube foot applied to substrate, movement achieved by contracting longitudinal muscles on one side of tube foot. (c–e) Tube foot being prepared for the next step; note that fluid is pumped back into the ampulla as longitudinal muscles are contracted. (f) Distal portion of tube foot from the sea star *Marthasterias glacialis,* showing terminal adhesive disc. (g) Longitudinal section through distal end of a tube foot from the sea star *M. glacialis.*

(f,g) Courtesy of P. Flammang. From P. Flammang et al., 1994. *Biol. Bull.* 187:35–47.

ring canal, which forms a ring around the esophagus in all but a few echinoderm species (Fig. 20.1a). Accessory fluid-storage structures called **Polian vesicles** and **Tiedemann's bodies** are often associated with the ring canal. In addition to storing fluid, the Tiedemann's bodies also serve to filter fluid from the water vascular system into the main body cavity (the **perivisceral coelom**), helping to maintain body turgor. Five (or some multiple thereof) **radial canals** radiate symmetrically from the ring canal. Pairs of bulb-shaped **ampullae** usually connect to these radial canals, with each ampulla servicing a single tube foot. Both ampullae and tube feet are supported by a system of calcareous ossicles, the **ambulacral ossicles.**

Tube feet often lack circular muscles (Fig. 20.1b) and so cannot extend themselves. In most species, fluid is pumped into the tube foot by contraction of the ampulla, extending the foot hydraulically (Fig. 20.2). A one-way valve at the juncture of the ampulla and radial canal ensures that fluid flows from ampulla to tube foot when the ampulla contracts, rather than to the radial canal.

The tube foot retracts when its longitudinal muscles contract. Again, the one-way valve ensures that fluid leaving the contracting tube foot is directed back into the ampulla, stretching the ampullar muscles in preparation for the next cycle.

A single echinoderm may possess more than 2000 tube feet. Locomotion often requires the coordinated protraction and retraction of all those feet.[3] Tube feet attach to solid substrates through a combination of ionic interactions, suction, and the activity of a duo-gland adhesion system. In the duo-gland system, one or more gland cells on each tube foot apparently secrete an adhesive that binds that tube foot temporarily to a substrate; adjacent gland cells then apparently release another chemical that somehow breaks those bonds. Similar duo-gland adhesive systems have been described from other animals as disparate as flatworms (p. 153) and scaphopod molluscs.

The inner surface of each tube foot is well ciliated, so the fluid in the WVS circulates. The thin-walled podia can thus function effectively in gas exchange as well as locomotion; the fluid of the WVS, together with that of the other coelomic compartments, serves as the primary circulatory medium. Tube feet may also be the primary sites of excretion (by simple diffusion), and in at least some groups, may function in chemoreception and food collection as well.

Specialized excretory organs are never found among adult echinoderms, although a cilia-driven nephridial system occurs in the larvae. A true heart is also absent.

Associated with the echinoderm WVS is a peculiar system of tissues and organs known as the **hemal system.** A major component of this hemal system is a spongy **axial organ** that lies adjacent to the stone canal of the WVS (Fig. 20.1a). The axial organ is housed in its own coelomic compartment, the **axial sinus,** and connects to 2 **hemal rings,** one oral and the other aboral. From the aboral hemal ring, strands of tissue, each contained within a coelomic, **perihemal canal,** extend outward to the gonads. Another series of strands radiates from the oral hemal ring to the tube feet.

The hemal system's functional significance has long been uncertain. There seem to be no direct connections between the hemal system and the WVS, or between the hemal and digestive systems. Recent studies with asteroids (sea stars) and holothurians (sea cucumbers) suggest that the hemal system transports nutrients from the coelomic fluid to the gonads. This conclusion is based in part on determining the sites of uptake of ^{14}C-labeled food material; radioactivity appeared sequentially in the digestive system, hemal system, and finally, in the gonads. In addition, concentrations of carbohydrates, lipids, proteins, and amino acids in hemal fluid may be 10 times greater than in other echinoderm fluids, indicating a high nutritional content. How nutrients move from the digestive system to the hemal system remains uncertain.

Morphological studies suggest that the axial organ of asteroids and echinoids may have an excretory function, although this has yet to be demonstrated experimentally. The axial organ also may be involved in producing intriguing cells called **coelomocytes,** which are found in nearly all echinoderm tissues and body fluids, including the coelomic fluid. These coelomocytes are involved in recognizing and phagocytosing foreign material, including bacteria;[4] synthesizing pigments and collagen (for connective tissue); transporting oxygen (some coelomocytes contain hemoglobin) and nutritive material; and digesting food particles. They also play a role in wound repair. Most echinoderms have great, and sometimes extraordinary, regenerative capabilities.[5]

Another feature that is apparently unique to echinoderms is their possession of what has been termed **mutable connective tissue,** or **catch tissue.** In some way, the details of which are still being worked out, nerve impulses can rapidly and dramatically alter the degree of stiffness and fluidity of the connective tissue—in some species, rock hard tissue can practically liquify in a fraction of a second, and revert back to extreme stiffness equally quickly. Such remarkable and rapid changes in the mechanical properties of the connective tissue play many roles in echinoderm biology, including feeding, locomotion, and the deliberate shedding of arms or viscera when the animals are attacked by predators. Specific examples are given as we discuss each class in the following text. Catch tissue occurs in all classes within the phylum.

At least 85 echinoderm species are known to be toxic or venomous, although few are deadly to humans.

Class Crinoidea

Class Crin • oidea
(G: lily-like)
krī-noy′-dē-ah

Defining Characteristic:[6] The main part of the body is supported above the substrate either by a long stalk or by a series of grasping claws (cirri)

The class Crinoidea is the oldest of the extant echinoderm classes, with a fossil record extending back nearly 600 million years; its members show many characteristics that, based upon studies of fossils, appear to be primitive. ("Primitive" characteristics are those that appear to show the least change

3. See *Topics for Further Discussion and Investigation,* nos. 5 and 8.

4. See *Topics for Further Discussion and Investigation,* no. 2, and Research Focus Box 20.1, p. 517.

5. See *Topics for Further Discussion and Investigation,* no. 1.

6. Characteristics distinguishing the members of this class from members of other classes within the phylum.

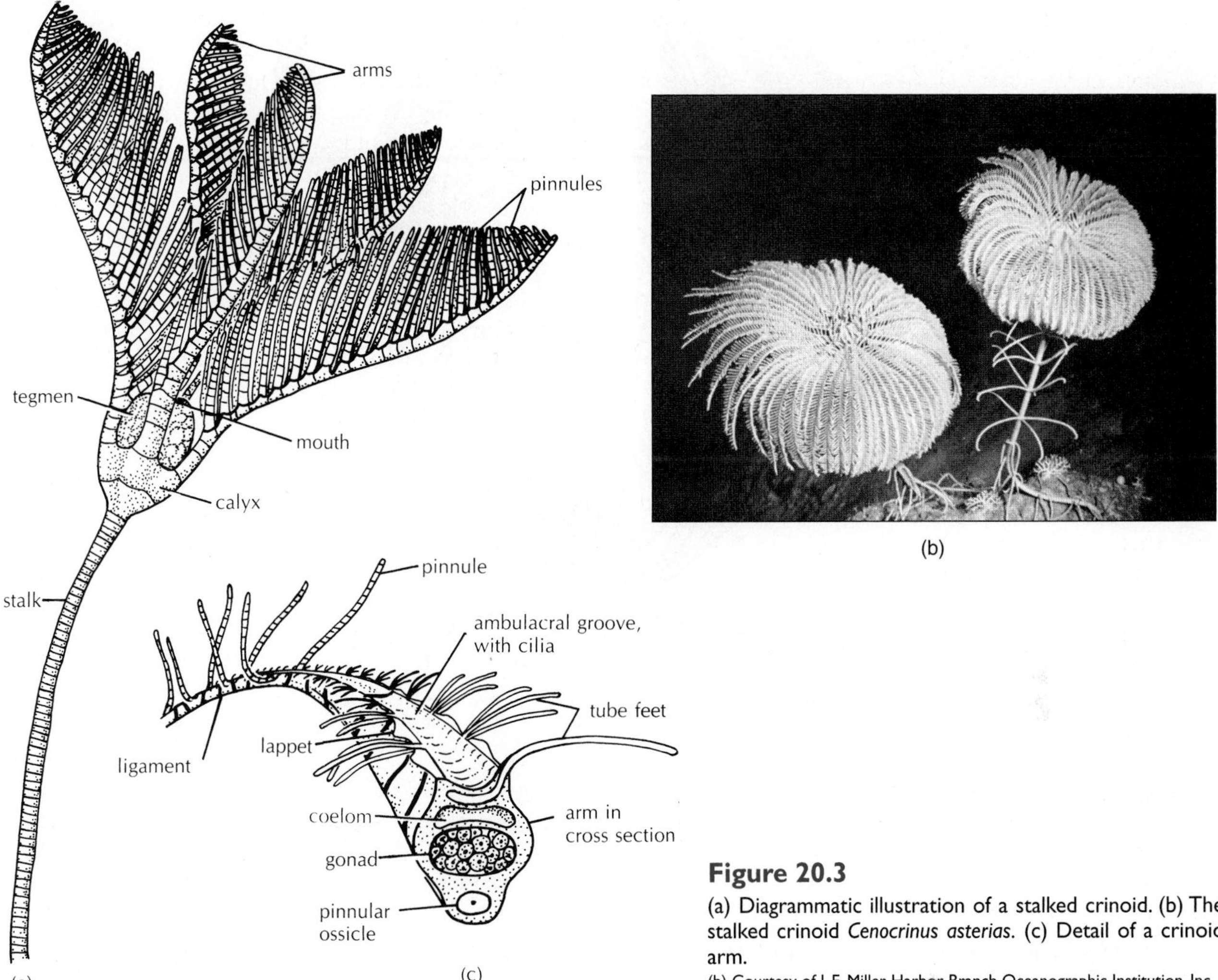

Figure 20.3
(a) Diagrammatic illustration of a stalked crinoid. (b) The stalked crinoid *Cenocrinus asterias*. (c) Detail of a crinoid arm.
(b) Courtesy of J. E. Miller, Harbor Branch Oceanographic Institution, Inc.

from the presumed ancestral condition; the word does not imply lack of complexity.) The Crinoidea is comprised of the stalked crinoids (the sea lilies, of which about 100 species now exist, all in deep water) and the nonstalked, motile comatulid crinoids (the feather stars), of which about 600 living species are known. Crinoids were far more successful long ago, and indeed they make up the majority of fossilized echinoderms. Today, the most diverse assemblage of crinoid species is found on the Great Barrier Reef surrounding Australia, where more than 50 different species co-occur in some areas. All crinoids are suspension-feeders.

The sea lilies, although a small group at present, were quite numerous 300 to 500 million years ago.[7] Most fossil crinoids were, as the sea lilies remain today, permanently attached to the substrate by a **stalk** (Fig. 20.3). The stalk is flexible, being composed of a series of calcareous discs (**columnals**) stacked one on top of the other and held together by connective tissue.

The feeding and reproducing part of the animal is situated at the top of the stalk. The digestive system is tubular, consisting of a mouth, intestine, and terminal anus, and is confined entirely to a calyx/tegmen complex; the **calyx,** a cup-shaped structure containing the complete digestive system (Fig. 20.3a), is covered by a lid-forming membrane (the **tegmen**) that bears the mouth. The mouth, and therefore the oral surface in general, is always directed away from the base of the calyx. Crinoids are unique among echinoderms in having the oral surface on the upper half of the body, a clear adaptation for suspension feeding. Both the calyx and tegmen generally bear numerous protective calcareous plates on the outer surface.

From 5 to more than 200 arms (usually in some multiple of 5) extend outward from the sea lily calyx, and all bear tube feet (podia). These arms, like the stalk, consist of a series of jointed calcareous ossicles, so that the arms can bend.

Two rows of tubular **pinnules** extend outward from each arm, one row on each side of the ambulacral groove, which runs the length of the arm (Fig. 20.3c). Thin, elongated podia, grouped in triplets, also flank the ambulacral groove of each arm. These podia are studded with mucus-secreting glands. Crinoids collect food by extending the arms, pinnules, and tube feet into the surrounding

7. See *Topics for Further Discussion and Investigation,* no. 7.

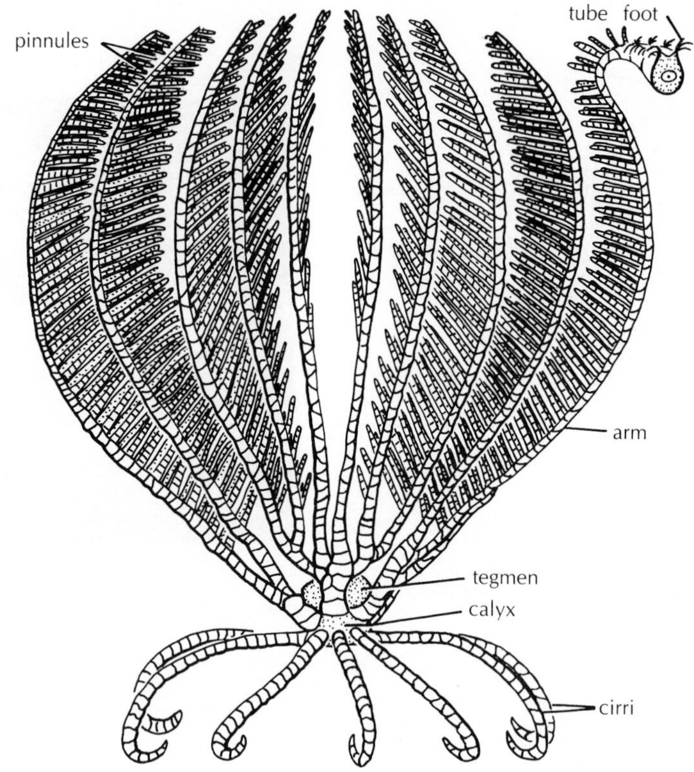

Figure 20.4
Diagrammatic illustration of a comatulid crinoid.
Modified from Hyman; after Clark.

water current,[8] the arms and pinnules being moved by contractions of extensor muscles and ligaments. As food particles in the water contact the podia, the particles are entangled in mucus and then flicked into the ambulacral grooves. The food particles are then transported to the mouth by ambulacral cilia. Note that crinoid tube feet have no locomotory function. Their use is limited to food collection, gas exchange, and, probably, the elimination of nitrogenous wastes by diffusion.

The sea lily stalk contains no musculature. Yet the animal can reorient its body quickly and repeatedly for more efficient food capture in response to temporary changes in current speeds and directions. Sea lilies accomplish this reorientation by rapidly altering the stiffness of the connective tissues holding the columnal discs of the stalk together. Conversely, crinoids are able to quickly discard (**autotomize**) one or more arms when attacked by predators, by liquifying appropriate connective tissue. The transition between solid and fluid states in the connective tissue is under direct nervous control, and it takes less than one second to accomplish.

Crinoid tube feet are never associated with ampullae, distinguishing the crinoids from most other echinoderms; crinoids protract their podia by contracting muscles in the radial canals. Crinoids also lack a madreporite, although stone canals opening into the coelom are numerous. The WVS opens to the outside through a large number of ciliated tubes penetrating the tegmen.

Feather stars, also referred to as **comatulids** after the name of the single order in which all are placed (order Comatulida), resemble the sea lilies from the calyx upward. However, in place of a long stalk, a series of jointed, flexible appendages called **cirri** occur near the base of the body (Fig. 20.4). The cirri are used to grasp solid substrates during periods of resting and feeding, which, for a comatulid, is most of the time. Comatulids are often observed perching atop sponges, corals, and other structures, living and nonliving; in this way, the feather stars extend their feeding appendages into the faster-moving water above the local substrate, increasing the frequency of food capture. Movement of the cirri, and to some extent of the arms themselves, is apparently mediated by contractile connective tissue rather than muscles in at least some species. Food collection is as described for stalked crinoids.

Feather stars locomote either by "snowshoeing" atop soft sediments using their cirri, or by swimming short distances above the substrate using forceful downward movement of the arms. A number of arms distributed uniformly around the calyx beat downward in unison; another group of arms then beats downward while the first group of arms makes its recovery stroke. Swimming

8. See *Topics for Further Discussion and Investigation,* no. 3.

thus involves a highly coordinated series of arm movements. The ability to move gives the comatulids a means of escaping or avoiding predators that is not available to the stalked crinoids.

Class Stelleroidea

Class Steller • oidea
(L: a star)
stel-er-oy´-dē-ah

Defining Characteristic: Arms (generally 5, or a multiple of 5) extend from a central disc

The class Stelleroidea contains all other armed echinoderms—namely, the brittle stars (ophiuroids) and sea stars (asteroids). As the class and common names imply, these animals differ from the crinoids in lacking stalks and in having the arms arrayed star-like around a flattened body. Despite the pronounced morphological differences between brittle stars and sea stars that will be discussed shortly, early fossil remains indicate a sufficiently close evolutionary relationship to warrant grouping these animals in this one class. Asteroids and ophiuroids also share a peculiar arrangement of mitochondrial DNA (a conspicuous multigene inversion), supporting other evidence that members of the 2 classes are closely related. In this taxonomic arrangement, brittle stars and sea stars are placed in separate subclasses within a single class.

Subclass Ophiuroidea

Subclass Ophiur • oidea
(G: snake-like)
ōf´-ē-yor-oy´-dē-ah

Defining Characteristics: 1) Well-developed ossicles in the arms form a linear series of articulating "vertebrae," joined together by connective tissue and muscles; 2) the oral surface bears 5 pair of invaginations (bursal slits), which may serve for gas exchange and as brood chambers for developing embryos

Most echinoderms are ophiuroids. The approximately 2100 species in this subclass are motile, as are all living echinoderms other than the stalked crinoids (sea lilies). As with the crinoids, arms extend from a central body, are built of jointed, vertebral ossicles **(vertebrae),** and are quite flexible (Fig. 20.5). The subclass is named in recognition of the snake-like movements made by these arms during locomotion. Even so, tube feet often play a role in ophiuroid movement, particularly in very young individuals, in species that remain small as adults, and in burrowing species.

Ophiuroids generally possess 5 long arms, radiating symmetrically from a small central disc (generally only a few centimeters in diameter). In some species (the basket stars), each arm branches several to many times (Fig. 20.5f). The discs of basket stars may reach 10 cm in diameter, and the outstretched arms may be about 1 m across! The common name for a typical ophiuroid is "brittle star," reflecting the tendency of the arms to detach from the central disc when provoked;[9] again, this autotomy is mediated through the mutable connective tissue. It is not uncommon to find 50% or more of the individuals in an ophiuroid population regenerating at least one arm, a process that requires several months to complete.

The oral surface of the ophiuroid body disc is often covered with a thin layer of minute, calcareous scales, while the body's aboral surface usually bears a number of protective calcareous plates, or **shields** (Fig. 20.5a, b), each composed of a single calcite crystal. Recent studies show that in at least some particularly light-sensitive species, the surface of each shield forms a very regular array of microscopic (40–50 μm) hills (Fig. 20.5b, c), which turn out to be lenses that focus light very precisely on underlying nerve fibers.

The arms are similarly encased in a series of endoskeletal ossicles. The bulk of each arm, however, consists of a series of thick, articulating, calcareous discs (the vertebral ossicles), with the tube feet (Fig. 20.5e) penetrating these "vertebrae" to the outside through a series of minute holes.

The ophiuroid digestive system, like that of the crinoids, is generally confined to the central disc. Unlike most other members of the Echinodermata, however, the ophiuroids possess only a single opening to the digestive system: A mouth is present, but an anus is lacking.

One final similarity between the crinoids and ophiuroids is that the podia are generally not operated by ampullae. In most other respects, the ophiuroid WVS follows the typical echinoderm pattern described earlier, except that a single ophiuroid may possess numerous madreporites, all of which open on the oral surface.

One feature found among most ophiuroids that distinguishes them from all other echinoderms is the occurrence of slit-shaped infoldings on the oral surface along the arm margins, adjacent to the arm shields. These 10 invaginations are known as **bursae** (Fig. 20.5d), and they project well into the coelomic space in the central disc. Seawater is constantly circulated through the bursae, presumably for gas exchange and, perhaps, for waste elimination. This circulation of external fluid is accomplished by cilia and, in some species, by muscular contractions. The bursae can also function in reproduction: In many species, the bursae serve as brood chambers within which the embryos develop.

9. See *Topics for Further Discussion and Investigation,* no. 1.

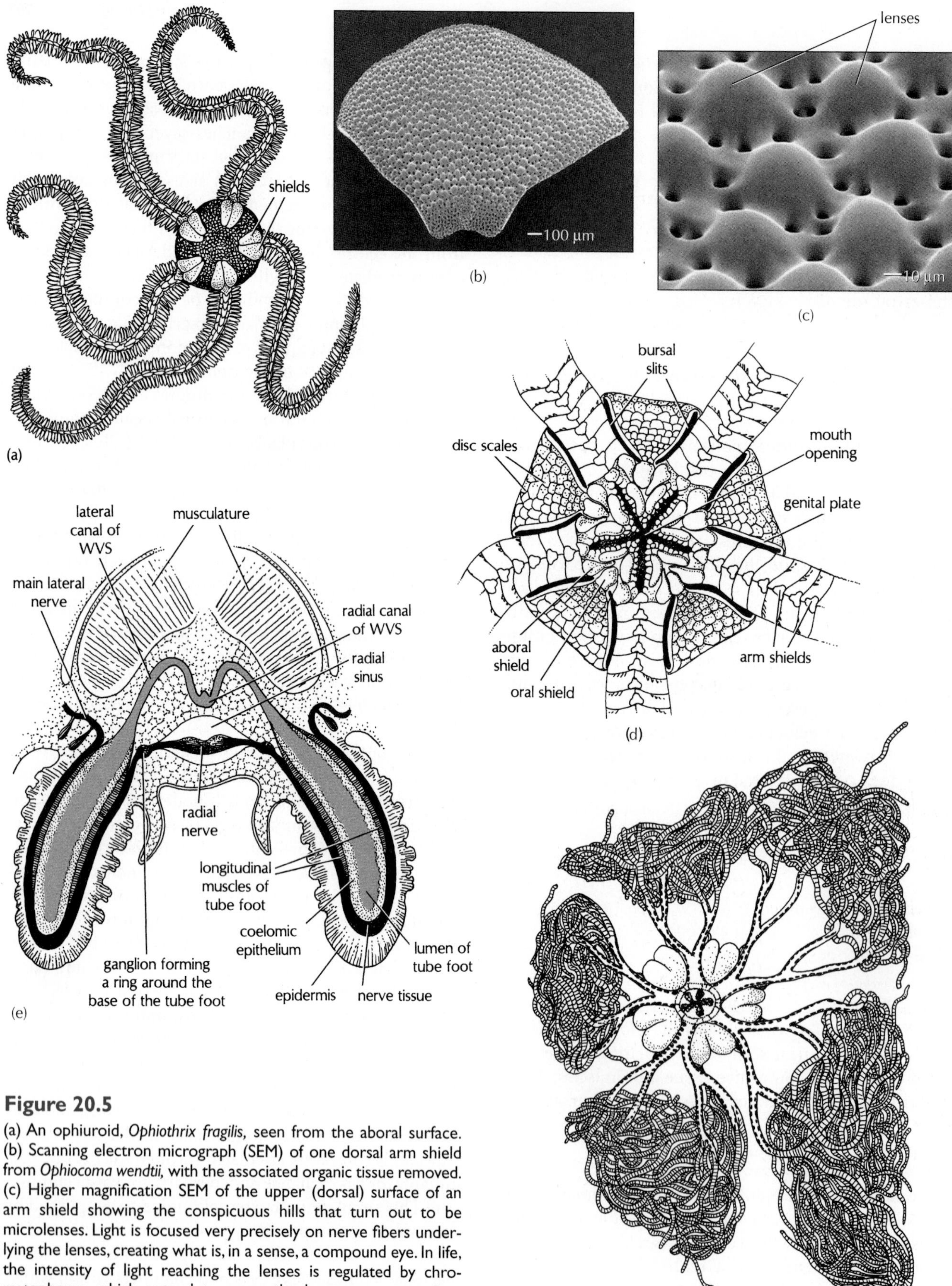

Figure 20.5

(a) An ophiuroid, *Ophiothrix fragilis,* seen from the aboral surface. (b) Scanning electron micrograph (SEM) of one dorsal arm shield from *Ophiocoma wendtii,* with the associated organic tissue removed. (c) Higher magnification SEM of the upper (dorsal) surface of an arm shield showing the conspicuous hills that turn out to be microlenses. Light is focused very precisely on nerve fibers underlying the lenses, creating what is, in a sense, a compound eye. In life, the intensity of light reaching the lenses is regulated by chromatophores, which extend to cover the lenses or retract into channels surrounding them. Chromatophore movements also alter the brittle star's surface color. (d) Oral view of the disc of *Ophiomusium* sp., showing the location of the mouth opening and bursal slits. (e) Cross section through an arm of *Ophiothrix* sp. Note the absence of ampullae. (f) The basket star, *Gorgonocephalus* sp.

(a) After Kingsley. (b,c) Courtesy of J. Aizenberg. From Aizenberg, J., A. Tkachenko, S. Weiner, L. Addadi, and G. Hendler. 2001. *Nature* 412:819–22. (e) From Hyman; after Cuenot. (f) After Pimentel.

Many ophiuroids are **deposit feeders,** ingesting sediment and assimilating the organic fraction. They also capture small animals in the sediment and ingest them individually. Some other brittle star species are **suspension-feeders,** filtering food particles from the water, while still other species function as carnivores or scavengers. Mutable connective tissue allows them to stiffen the arms and maintain them in filtering posture for long periods of time. Ophiuroids typically hide under rocks or in crevices during the day, emerging to feed only at night; feeding aggregations of several thousand individuals per square meter have been reported from some shallow-water habitats. Many species live in association with other invertebrates, especially sponges and sessile cnidarians. A few tropical ophiuroid species recently have been shown to exhibit diurnal alterations in body coloration that are mediated by light-sensitive chromatophores.

Subclass Asteroidea

Subclass Aster • oidea
(G: star-like)
as-ter-oy′-dē-ah

Defining Characteristic: The gonads and portions of the digestive tract extend into each arm

Approximately 1600 species of sea stars (often misleadingly termed "starfish") have been described from the living fauna, making the Asteroidea the second largest group (behind the Ophiuroidea) in the Echinodermata. This subclass now contains the aberrant concentricycloids, which I discuss separately later in this chapter (pp. 508–509). The discussion that follows pertains only to the "normal" sea stars.

The asteroids and the ophiuroids are superficially similar in that nearly all members of both groups possess arms and a basically star-shaped body (Fig. 20.6). However, the arms of the sea stars are not distinct from the central body disc, and they do not generally play an active, direct role in locomotion. Moreover, there are important morphological and functional differences in the digestive system and WVS. Finally, adult asteroids tend to be larger than adult ophiuroids. Few sea stars are smaller than several centimeters in diameter; individuals of most species are about 15–25 cm in diameter, and some are considerably larger.

Sea stars move slowly, through the highly coordinated activities of the tube feet that radiate out along the oral surface of each arm. The WVS is essentially as described at the beginning of this chapter, with the madreporite opening on the aboral surface. The tube feet lie in distinct **ambulacral grooves** on the oral surface (Fig. 20.6b). Such grooves are not encountered among ophiuroids (or in any other group of echinoderms, other than the crinoids). Each tube foot is individually operated by an ampulla and generally terminates in a small suction cup at the distal end (Figs. 20.2g and 20.7). A given tube foot is extended through contraction of the associated ampulla, and it is swung forward or backward by contraction of the longitudinal musculature on one side or the other of the podium (see Fig. 20.2a–e). The **ambulacral ossicles** apparently support the podia during these movements. Once the terminal portion of the podium contacts the substratum, slight contractions of the longitudinal muscles pull the central portion of the distal end of the podium upward, creating suction. Suction-cup locomotion seems best adapted for movement over firm substrates; the tube feet of asteroid species that move over or burrow into soft substrates do not terminate in suction cups.

The sea star mouth is directed downward, opens into a very short esophagus, and passes upward into a lower stomach (the **cardiac stomach**) (Fig. 20.8). This "lower" stomach is confined to the central body disc and is chiefly responsible for the digestion of food. Above the lower stomach is an upper or **pyloric stomach,** branches of which radiate out into each arm as **pyloric caeca.** The great surface area of the pyloric caeca, achieved through outfolding of the tissue, is in keeping with the primary functions of these organs: Secretion of digestive enzymes and absorption of digested nutrients. The pyloric caeca are also primary storage sites for assimilated food. The anus lies on the aboral surface, nearly in line with the mouth.

Asteroids typically prey on large invertebrates, including sponges, gastropods, polychaetes, bivalves, and other echinoderms.[10] A few species consume small fish. During feeding on large prey, the cardiac stomach of some species is actually protruded out of the body disc through the mouth and placed in contact with the prey's soft tissues. The cardiac stomach can be protruded through spaces as narrow as 1–2 mm (e.g., the gap between the shell valves of a nearly closed bivalve). Digestion in such species is frequently external, the resulting nutrient broth being transferred to the pyloric stomach by means of the ciliated channels of the cardiac stomach. No other echinoderms feed in this manner. On the other hand, if the prey is sufficiently small, the stomach will be retracted while it holds the victim, and digestion will proceed internally. Alternatively, the small prey may be ingested directly at the mouth, without the stomach having to be protruded at all. Some suspension-feeding asteroid species have also been described.

Many sea stars will sever, or **autotomize,** some of their arms if subjected to physical disturbance. As with crinoids and ophiuroids, this appears to be a form of escape, leaving potential predators with only a nutritious souvenir of the encounter. The response seems to be chemically mediated, since coelomic fluid from an

10. See *Topics for Further Discussion and Investigation,* no. 3.

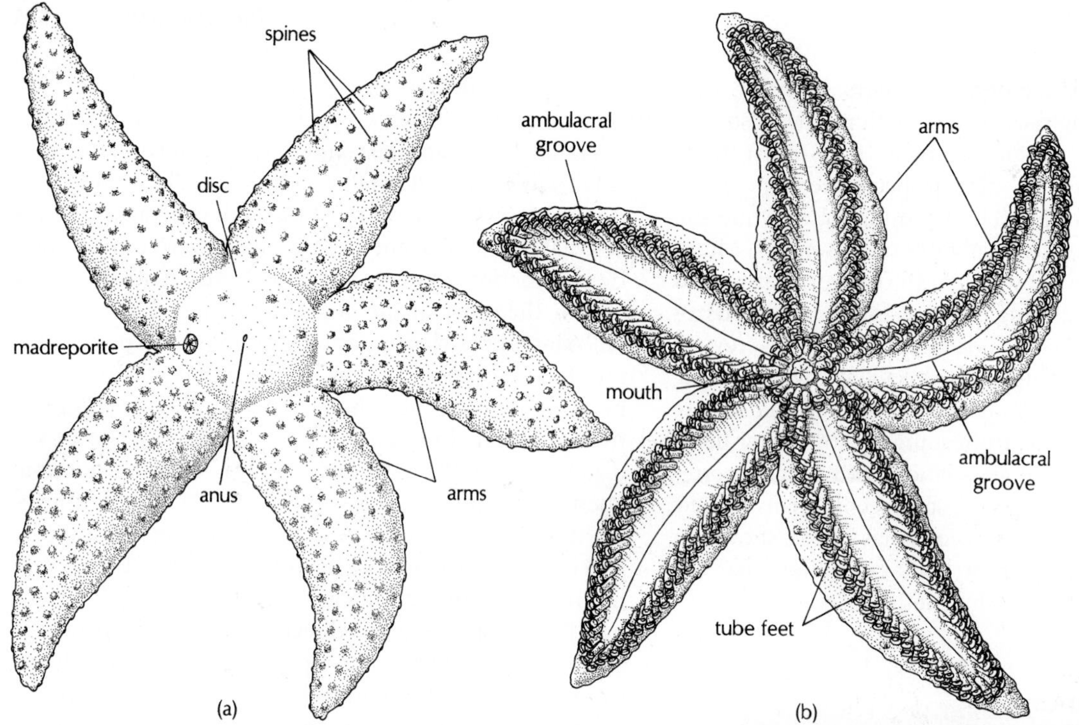

Figure 20.6

(a) A sea star, *Asterias vulgaris,* in aboral view. (b) Oral view of *Asterias vulgaris,* showing the tube feet in ambulacral grooves. (c) A deep-water sea star, *Rosaster alexandri.* Note the conspicuous tube feet extending from the 5 slender arms.

(a) After Hyman. (b) After Sherman and Sherman. (c) Courtesy of J. E. Miller. Harbor Branch Oceanographic Institution, Inc.

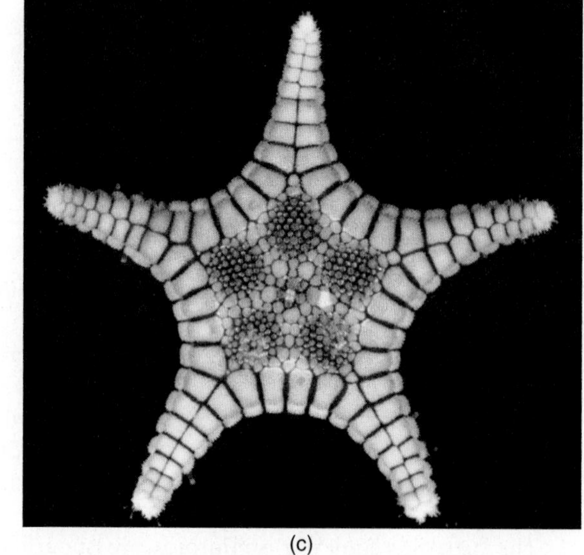

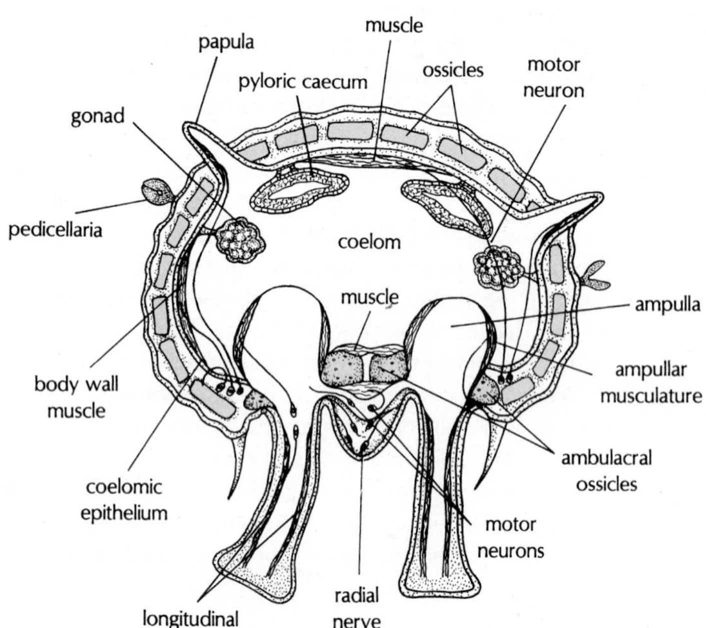

Figure 20.7

Cross section through the arm of an asteroid. Note that the podia protrude through minute pores in the ambulacral ossicles. Other anatomical features, including the papulae and pedicellariae, are discussed later in the chapter.

From Hyman; after J. E. Smith.

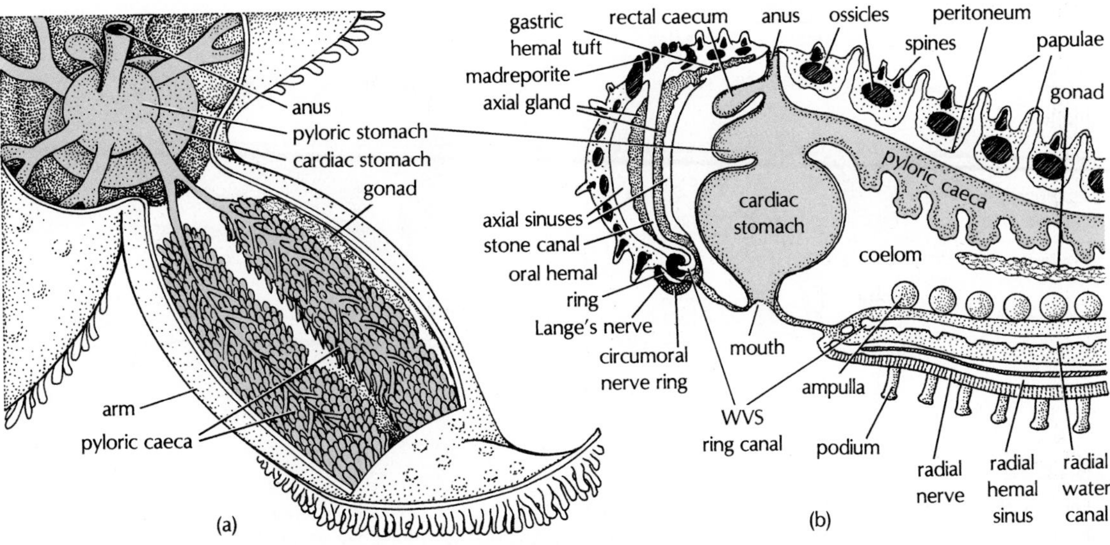

Figure 20.8

(a) Diagrammatic illustration of the digestive system of a sea star. (b) Diagrammatic illustration of a sea star digestive system, viewed laterally.

(b) After Chadwick.

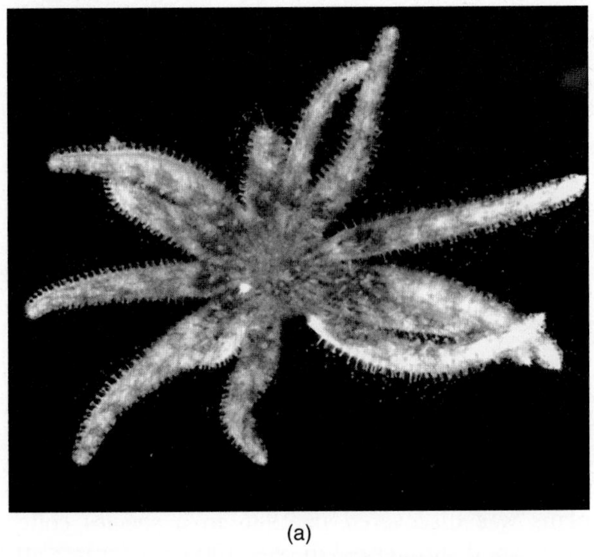

(a) (b)

Figure 20.9

Autotomy in the asteroid, *Pycnopodia helianthoides*. Coelomic fluid from an autotomizing individual was injected into the intact specimen shown. The animal was photographed (a) 25 seconds and (b) 60 seconds later.

(a,b) Courtesy of V. Mladenov, from Mladenov et al., 1989. *Biol. Bull.* 176:169–75. Permission of *Biological Bulletin*.

autotomizing sea star induces autotomy if injected into another sea star, as shown in Figure 20.9; again, the response involves the rapid liquifying of mutable connective tissue. The lost arms are eventually regenerated.

The sea star's calcareous skeleton takes the form of thousands of discrete rods, crosses, and plates embedded in connective tissue (Fig. 20.10). Ossicles other than those supporting the ambulacral grooves often bear outwardly directed calcareous spines that can be moved from side to side by muscles connecting to the underlying ossicles. The spines and outer surface of the skeletal ossicles are covered by a fairly thick cuticle that is secreted by the underlying, ciliated epidermis.

Asteroids possess 2 other types of appendage in addition to spines and podia. Thin, noncalcified outfoldings of the outer body wall serve a respiratory function.

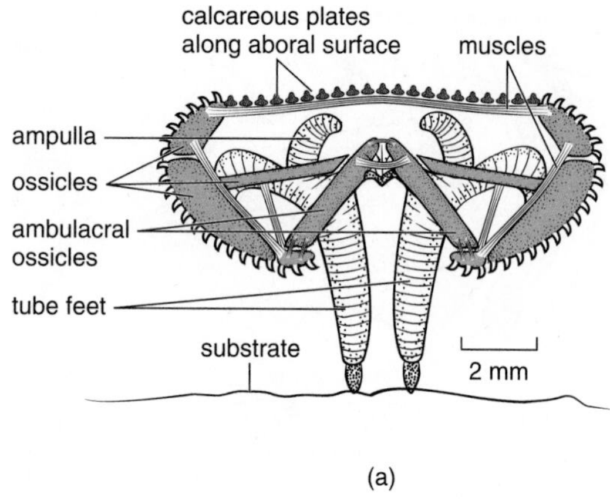

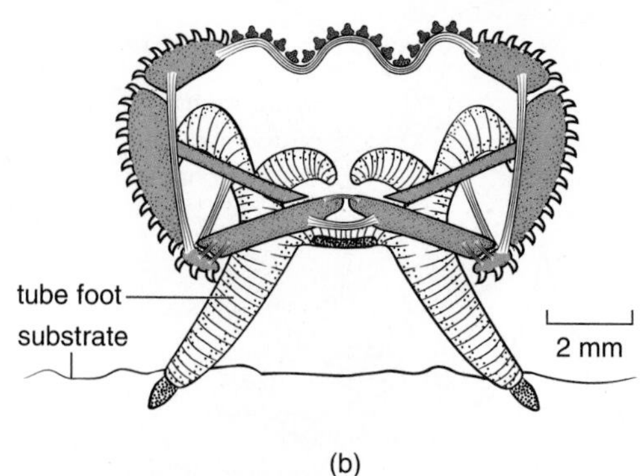

Figure 20.10

Transverse section through the arm of the asteroid *Astropecten irregularis,* showing the arrangement of calcareous ossicles and plates. The overlying cuticle has been omitted for clarity. In (a), the animal is walking over a substrate. In (b), the animal is burrowing. Note the changes in orientation of the ossicles, made possible by the elastic connective tissue that joins adjacent plates.

Based on D. Heddle, in *Symposium of the Zoological Society of London,* 20:125. 1967.

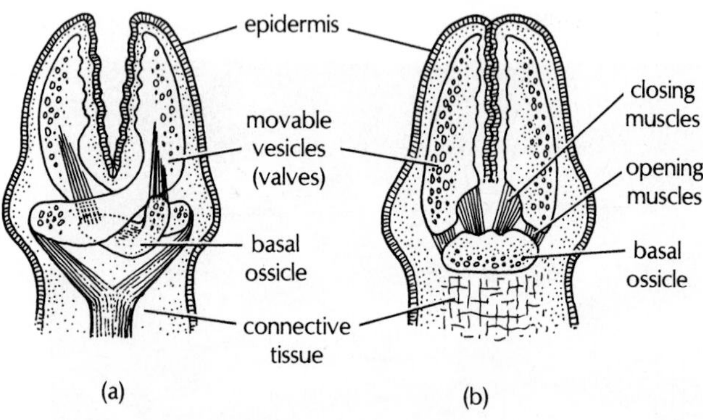

Figure 20.11

Two types of pedicellariae found on the body of *Asterias* sp. In (a), the movable ossicles (valves) cross, as in the blades of a pair of scissors. In (b), the valves are straight and nearly parallel, and operate in the manner of forceps. In both types of pedicellariae, the basal ossicle supports the valves but does not itself move.

(a,b) After Hyman.

These structures, called **papulae,** are found protruding between ossicles and are connected directly with the main coelomic cavity. The second type of appendage is much more dynamic. These appendages, called **pedicellariae,** consist of 2 (sometimes 3) calcium carbonate ossicles **(valves),** whose ends can be moved together or apart by muscles (Fig. 20.11). The 2 jaws are supported by a nonmovable, basal ossicle. The pedicellariae generally function in the removal of unwanted organisms and debris that contact the surface of the animal, and they also have been shown to capture living prey (including small fish!) in several asteroid species. Pedicellariae are found in only one other echinoderm class, the Echinoidea, which includes the sea urchins and sand dollars.

The Concentricycloids

The concentri • cycloids
(L: concentric ring)
kon-sen´-trē-si´-kloyds

Defining Characteristics: 1) The water vascular system includes what appear to be 2 concentric water vascular rings; 2) the tube feet are arranged in a circular pattern along the animal's periphery

"Many a flower was born to blush unseen. . . ."
(From Sir Thomas Gray, "*Elegy Written in a Country Churchyard*")

A marked departure from other echinoderm body plans was discovered in 1986 in a species collected from wood submerged in over 1000 m of water off the coast of New Zealand. A second species was later collected from wooden panels deliberately planted in the Bahamas, at a depth of about 2000 m, and a third species was described in 2006. To date, these small animals—less than 1 cm in diameter—are known exclusively from the crevices of submerged wood recovered from the deep sea. Each animal is circular, flat, and without arms, and more resembles a jellyfish (medusa) or a flower than an echinoderm (Fig. 20.12). The body exhibits no evidence of pentamerous radial symmetry, and recent studies show that their sperm—with the nucleus and mitochondrion drawn out into long, thin threads and the elongated acrosome divided into numerous distinct segments—are unlike those known from any other echinoderms.

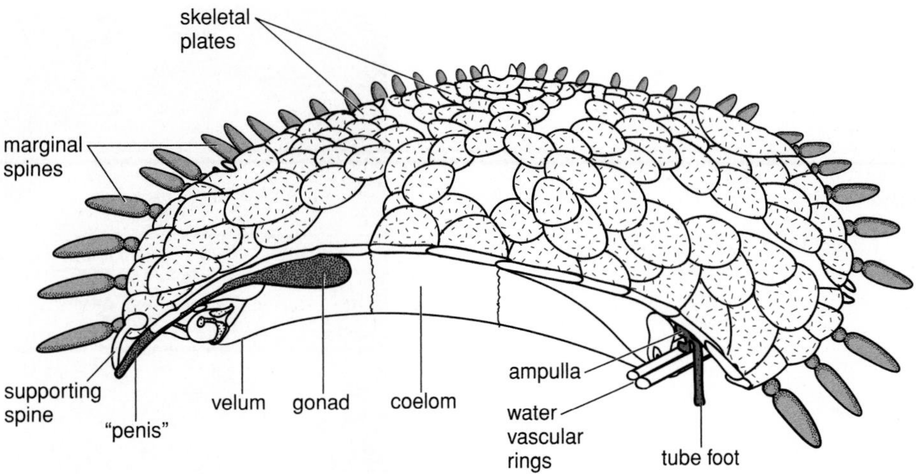

Figure 20.12

A male concentricycloid, *Xyloplax medusiformis,* with a portion cut away to show internal features.

Based on F. W. E. Rowe, in *Proc. Royal Soc., London* B. 233:431–59. © 1988.

The animals are called sea daisies, or concentricycloids. Originally placed in a separate class (the Concentricycloidea), most recent analyses argue for their inclusion within the Asteroidea, as highly modified sea stars. Their relationships to other sea stars, however, remain unclear. Molecular and developmental studies should help to resolve their phylogenetic position over the next several years.

Whatever their evolutionary history, these small but intriguing animals are clearly echinoderms. Like most other echinoderms, sea daisies have a spacious coelom, a water vascular system complete with unmistakable tube feet serviced by individual ampullae (Fig. 20.12), and a calcareous endoskeleton of distinct ossicles. Unlike the tube feet of any other known echinoderm, however, those of the sea daisies are arranged in a single circle along the periphery of the animal and are connected to a double-ringed water vascular system (Fig. 20.12), for which the class is named. The 2 concentric water vascular rings are connected at intervals by short radial canals. In all other echinoderms, radial canals bearing the tube feet radiate from a single ring canal encircling the esophagus, as discussed earlier.

The sea daisy body is supported by a series of overlapping skeletal plates (ossicles) arranged in concentric rings, and the daisy's "petals" (Fig. 20.12) are actually calcareous spines. The oral surface of one species is covered by a thin sheet of tissue, the **velum,** further increasing the animal's superficial resemblance to a hydrozoan medusa. In that species, and in one other species, the adults have no gut; they are presumed to subsist on dissolved organic matter. Members of the third species have a large mouth and a well-defined stomach, which is probably eversible and is presumed to function as in asteroids. The stomach ends blindly, as in ophiuroids; there is no anus. Neither species has any specialized gas exchange organs. Most details of the biology of these species have yet to be described.

Class Echinoidea

Class Echin • oidea
(G: spine-like)
ek-in-oy´-dē-ah

Defining Characteristics: 1) Ossicles are joined to form a rigid test; 2) podia pores pass through the ambulacral plates; 3) adults generally possess a complex system of ossicles and muscles (Aristotle's lantern) that can be partially protruded from the mouth for grazing and chewing

The last 2 classes of the Echinodermata remaining to be discussed consist of species that lack arms. The Echinoidea include the sea urchins, heart urchins, and sand dollars, somewhat less than 1000 species in total. The class is perhaps best represented by the sea urchins, which possess large numbers of long, rigid, calcium carbonate spines. The Greek word "*echinus*" means, literally, "a hedgehog." The spines serve for protection and, in some species, are actively involved in locomotion. Most sea urchins are free-living, roaming individuals, but a number of species bore into rock.

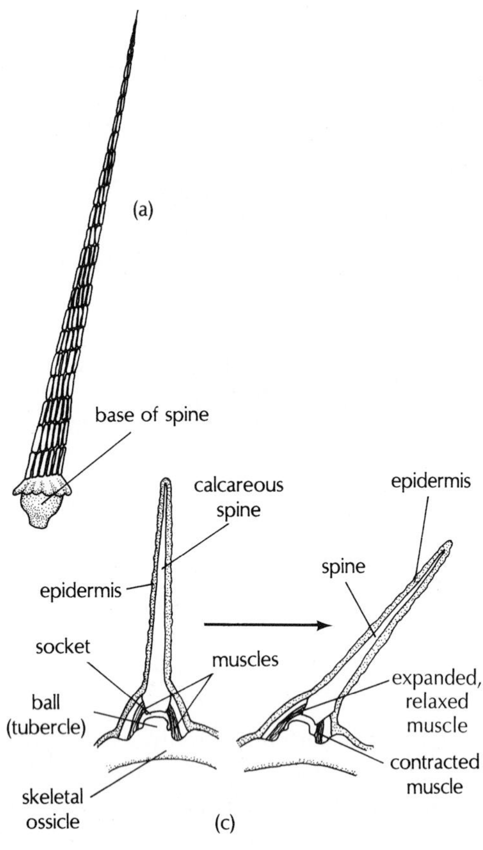

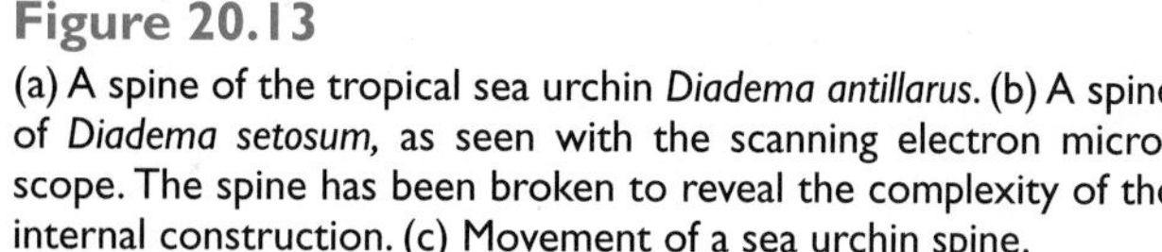

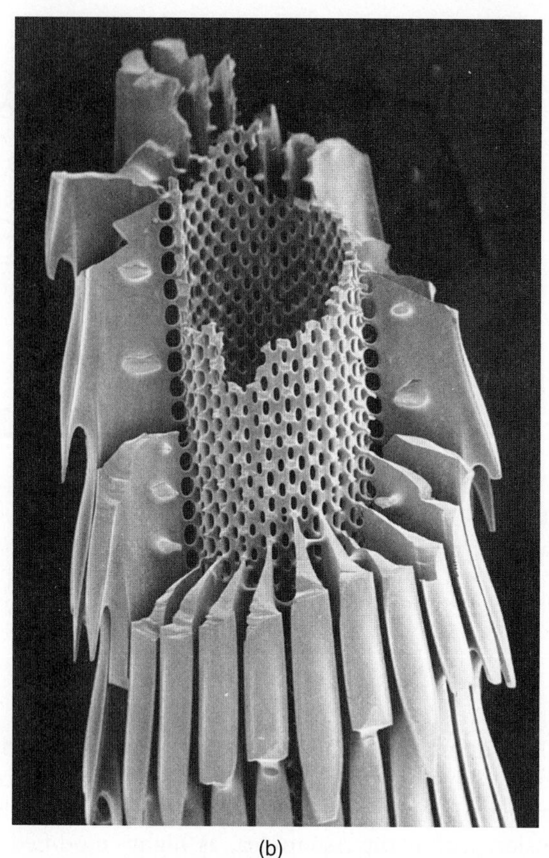

Figure 20.13

(a) A spine of the tropical sea urchin *Diadema antillarus*. (b) A spine of *Diadema setosum,* as seen with the scanning electron microscope. The spine has been broken to reveal the complexity of the internal construction. (c) Movement of a sea urchin spine.

(a,c) After Hyman. (b) Courtesy of K. Märkel from Burkhardt et al., 1983. *Zoomorphol.* 102:189–203. Courtesy © Springer-Verlag Co.

Echinoid spines attach to the underlying skeleton via ball-and-socket joints and can be declined rapidly in various directions by contracting specialized muscle fibers that connect between the ball **(tubercle)** and the spine (Fig. 20.13). The spines are often thin and sharp, but they are thick and blunt in a few species (Fig. 20.15a, b). They may function in bracing the animal when it wedges into crevices (stiffened by mutable connective tissue at the base of the spines), in gathering and manipulating food, and in defense. Toxins may be extruded through the spines or from glands associated with the spines in some species, as many an unwary, warm-water vacationer has discovered. If broken or damaged, the spines are replaced or repaired within a month or 2.

The ossicles comprising most echinoid skeletons are flat and joined together by collagenous ligaments, so ossicles cannot move relative to one another. In most species, the skeleton thus forms a solid, inflexible **test,** a feature that sets the typical echinoid apart from most other echinoderms. As the individual echinoid grows, the test is enlarged simultaneously in all directions. This size increase is accomplished through the addition of calcareous material at the margins of existing ossicles and by the secretion of new ossicles at the edges of the 5 "ocular plates" near the anus on the aboral surface (Fig. 20.14c).

Tube feet are widely distributed on the body, protruding through 5 double rows of pores in the **ambulacral plates** of the test (Fig. 20.14a, b). These pores are most easily seen if a denuded, empty, sea urchin test is held before a bright light. The body areas containing tube feet (i.e., the ambulacral plates) are distributed symmetrically about the body in strips extending orally/aborally. These regions are separated from each other by distinct **interambulacral** areas that are devoid of tube feet. The tube feet of echinoids are especially well developed and generally bear suction-cup ends; the podia commonly function in locomotion, as in asteroids.

Pedicellariae are also prominent echinoid appendages. However, they are generally borne on stalks and, unlike asteroid pedicellariae, are often equipped with calcareous support rods and bear 3 opposing jaws rather than 2 (Fig. 20.14e). Globular forms of pedicellariae, found in most

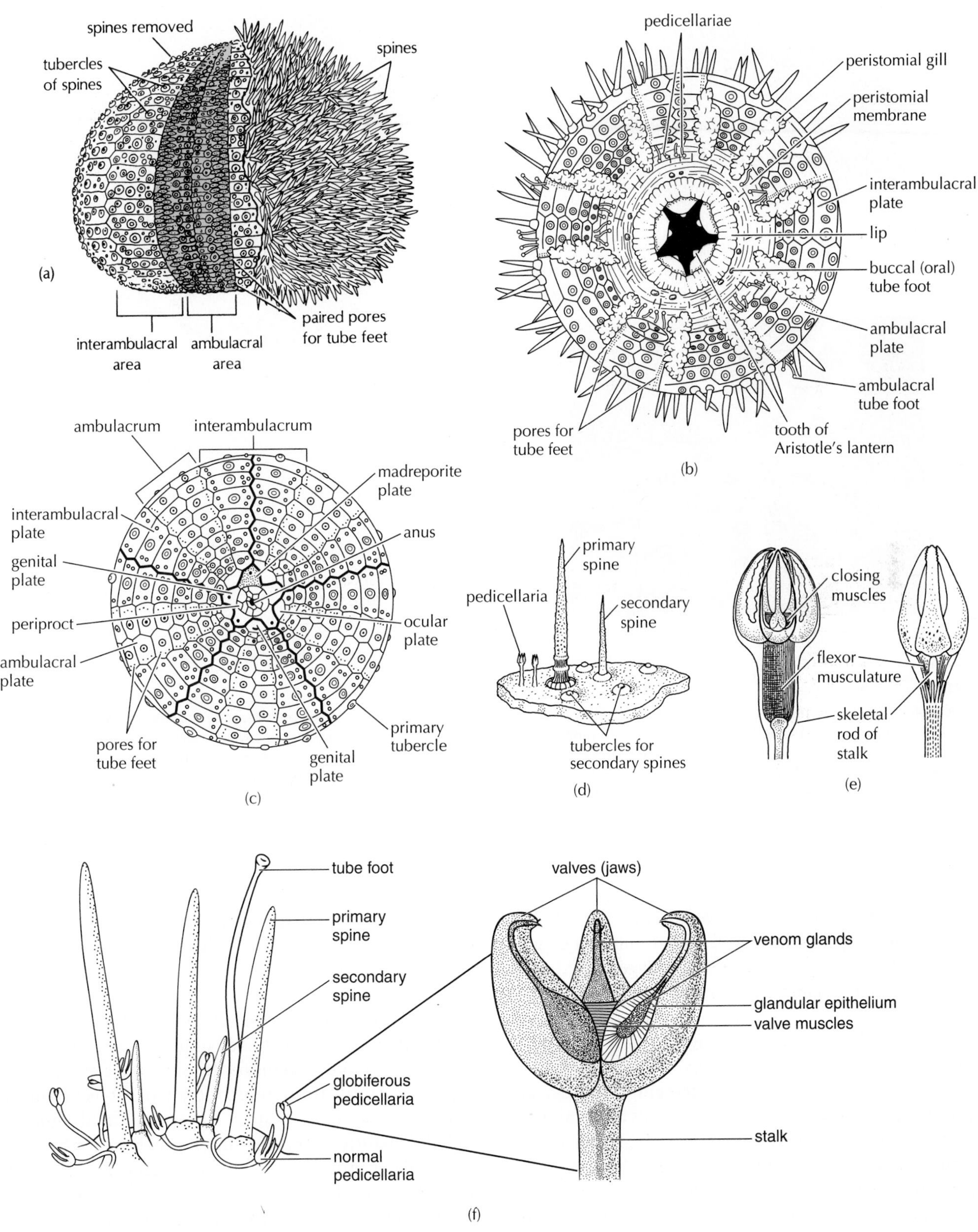

Figure 20.14

(a) A sea urchin, *Echinus esculentus,* with the spines removed from a portion of the test to reveal the ambulacral and interambulacral zones. Small pores, through which tube feet protrude in the living urchin, are seen in the ambulacral plates. Tubercles, upon which the spines pivot, are conspicuous on the interambulacral plates. The sea urchin test is comprised of 5 double rows of ambulacral plates and 5 double rows of interambulacral plates, for a total of 20 rows of plates. (d) Schematic illustration of an urchin test with appendages. (e) Pedicellariae from the echinoids *Strongylocentrotus droebachienis* (left) and *Eucidaris* sp. (right), showing the stalks and 3-part jaws. (f) Globiferous pedicellaria, showing poison glands.

(d) After Jackson. (e) From Hyman; after Mortensen. (f) Based on F. E. Russell, in J. H. S. Blaxter, F. S. Russell and M. Yonge, Eds., *Advances in Marine Biology,* 21:60–217, 1984.

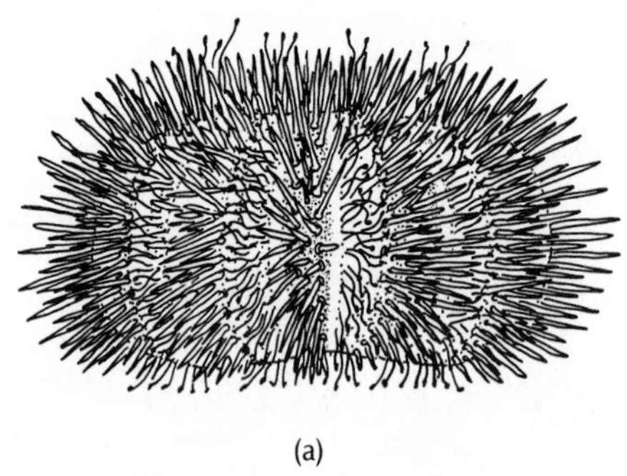

(a)

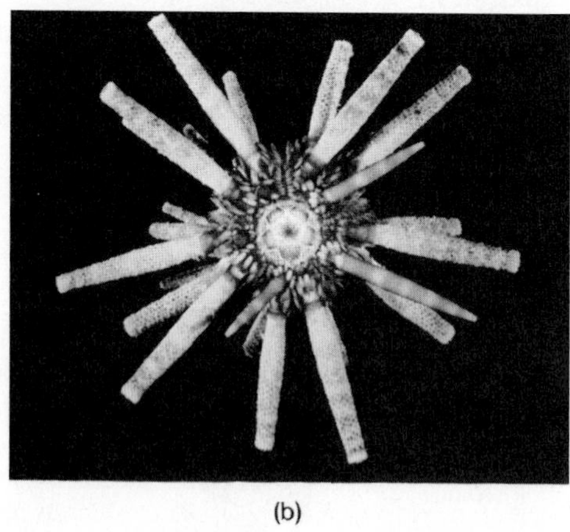

(b)

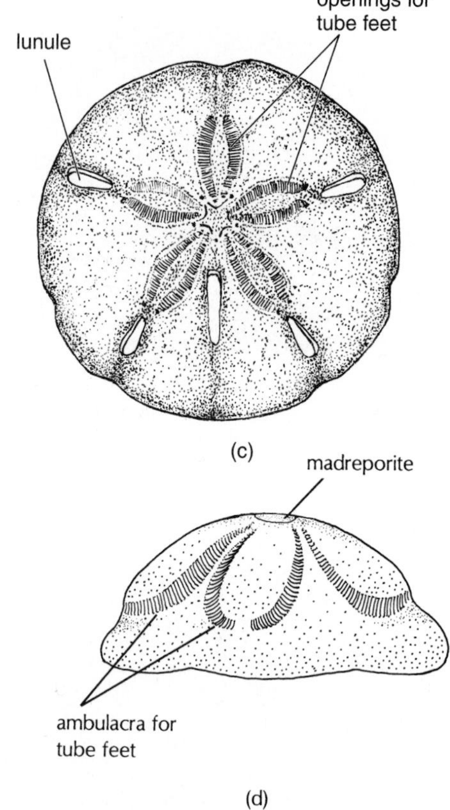

(c)

(d)

Figure 20.15

(a) A spiny urchin, a regular echinoid; see also Figure 20.17. (b) *Eucidaris tribuloides,* a tropical pencil urchin; the common name refers to the animal's unusually thick, cylindrical spines. (c) A sand dollar, an irregular echinoid. (d) A sea biscuit, another irregular echinoid. Regular echinoids (the sea urchins) are characterized by pentamerous symmetry, a globular test, and long spines. Irregular echinoids (including the heart urchins, sea biscuits, and sand dollars) have a somewhat-to-very flattened test and relatively short spines, and tend toward bilateral symmetry; the periproct with anus is displaced posteriorly, and the mouth may be displaced anteriorly. The ambulacra of irregular species often resemble flower petals in outline as in (c) and (d).
(a,c,d) After Brown. (b) Courtesy of J. E. Miller, Harbor Branch Oceanographic Institution, Inc.

urchin species, discharge defensive poisons (Fig. 20.14f). Some echinoid species have thin-walled outfoldings surrounding the mouth. These small appendages, called **gills,** presumably function in gas exchange.

Echinoids may be either regular or irregular (Fig. 20.15). **Regular urchins** have an almost perfect, spherical symmetry. All "sea urchins" fall into this group. Other echinoids are classified as **irregular** and display various degrees of bilateral symmetry. This bilateral symmetry may be associated with a lifestyle of burrowing through sand, mud, or gravel. In association with the burrowing habit, the tube feet of irregular urchins tend to lack terminal suckers. In all irregular echinoids, ambulacral areas (and thus the tube feet) are restricted to the oral and aboral surfaces, rather than extending in an unbroken line from the oral to the aboral surfaces. The ambulacral areas on the aboral surface form a conspicuous 5-pointed pattern, resembling the petals of a flower. The heart urchins have distinct anterior and posterior ends, with the mouth located anteriorly and the anus posteriorly. The spines of heart urchins are much more numerous than those of regular urchins, but they are also much shorter. Sand dollars (Fig. 20.15c) also bear very short spines, most likely an adaptation for burrowing. Unlike the heart urchins, sea biscuits, and regular urchins, in which the aboral surface is convex, the test of most sand dollars is greatly flattened to form a very thin disc.

The echinoid systems for feeding and digestion differ significantly from those of all other echinoderm species. A complex system of ossicles and muscles, called **Aristotle's lantern,** surrounds the esophagus in all regular echinoids and in some irregular species as well (Figs. 20.14b and 20.16); its absence in some echinoid species is believed to reflect a secondary loss. The teeth of Aristotle's lantern can be protruded from the mouth and moved in various directions to eat seaweed or to scrape food, especially algae, from solid substrates. Detritus-feeding deep-sea urchins use the lantern to scoop mud. Some urchin species also are known to routinely consume small bivalved molluscs and other invertebrates. Echinoids lacking Aristotle's lantern generally feed on small organic debris, which is

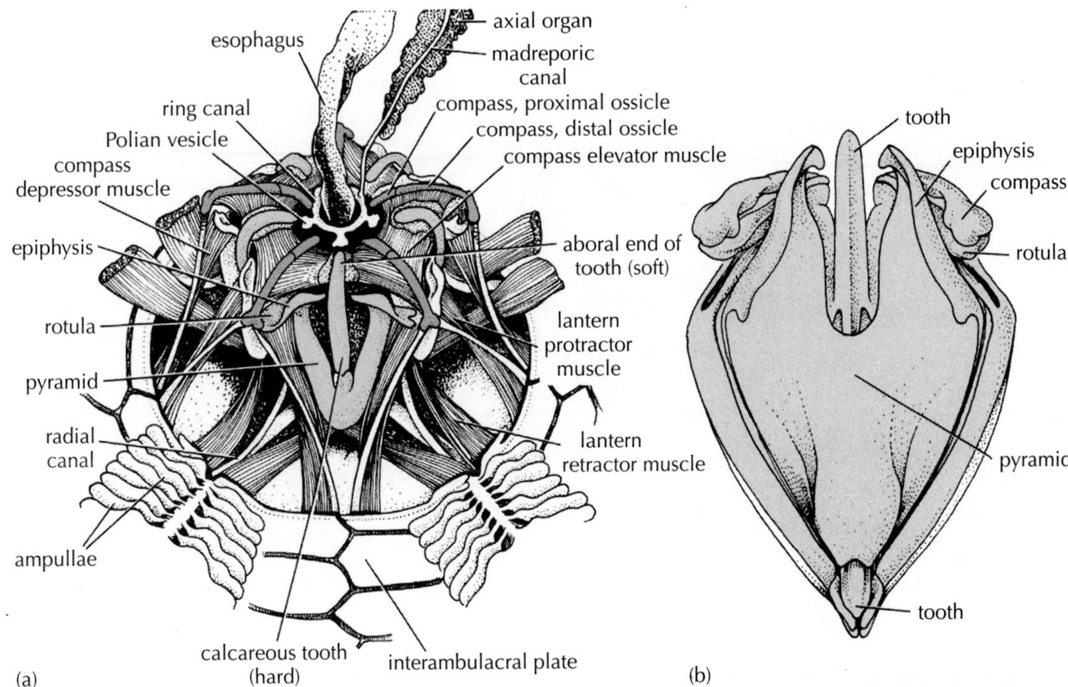

Figure 20.16

(a) Aristotle's lantern and its associated complex musculature, from the sea urchin *Arbacia punctulata.* (b) Detail of the lantern ossicles, with the musculature omitted for clarity. Lantern morphology differs considerably among echinoid species and is an important tool for species identification. The lantern consists of 5 bulky pyramids that support the 5 teeth; a series of bars (epiphyses) running along the aboral ends of the pyramids; a series of 5, thin compass ossicles aborally; and a series of 5 similar pieces, the rotulas, lying below the compass. Protractor muscles push the teeth outward; retractor muscles move the teeth apart and draw them back into the test. The teeth are very hard at the tips but soft at the aboral end; new tooth material is formed continually at the aboral end, compensating for tooth wear distally.
(a,b) After Brown.

often collected by means of modified tube feet, spines, and/or external ciliary tracts.[11]

The echinoid stomach is not protrusible. Indeed, no echinoid has a true stomach, the esophagus leading instead into a very long, convoluted intestine (Fig. 20.17), where the food is both digested and absorbed **(assimilated).** The anus is located aborally and is surrounded by a series of plates comprising the **periproct** (G: around the anus).

Assimilated food passes into the echinoid coelomic fluid. The echinoid coelomic space is immense, particularly in the regular urchins (Fig. 20.17). Coelomic fluid is the principal transporter of both food and wastes; that is, the coelomic fluid is the primary circulatory fluid. The inner surface of the mesodermal lining of the coelomic cavity is ciliated, maintaining a constant movement of the circulatory medium.

The echinoid WVS follows the archetypical pattern closely, with a single madreporite opening aborally, as in the asteroids (Fig. 20.17). The Japanese (and people living near the Mediterranean Sea) prize echinoid gonads as an edible delicacy, for which they are willing to pay more than $100 per pound. As a result of overfishing and pollution, local Japanese urchin populations no longer meet the culinary demand; canned echinoid gonads are increasingly imported from the United States.

Class Holothuroidea

Class Holothur • oidea
hōlo´-thur-oid´-ē-ah

Defining Characteristics: 1) The body is worm-shaped, being greatly elongated along the oral/aboral axis; 2) the calcareous ossicles are reduced in size and embedded individually in the body wall; 3) highly branched, muscular respiratory structures (the respiratory trees)—generally one pair—extend from the cloaca into the coelomic cavity

The class Holothuroidea contains about 1200 species, making it only slightly larger than the Echinoidea. Holothurians and echinoids resemble each other in lacking arms. In certain respects, the typical holothurian may be regarded as a flexible echinoid, whose morphological modifications reflect adaptation for a different lifestyle. Transforming an echinoid into a typical holothurian

11. See *Topics for Further Discussion and Investigation,* no. 3.

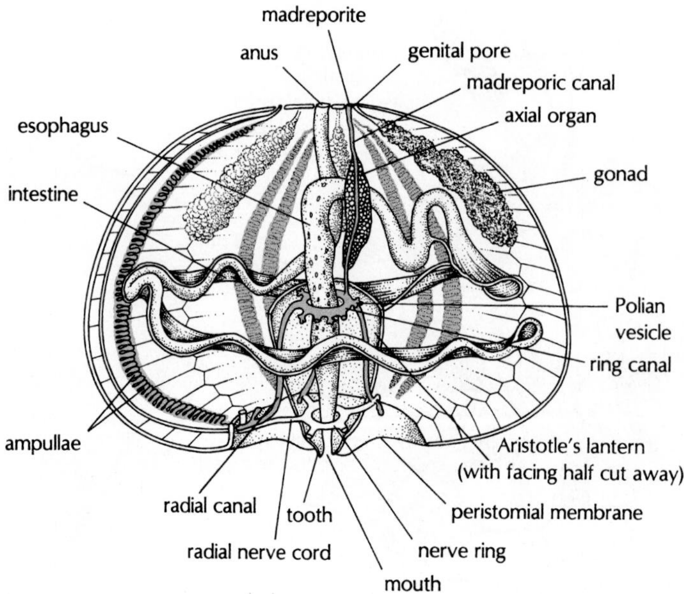

Figure 20.17
A sea urchin, *Arbacia punctulata,* seen laterally in diagrammatic section. Note the large coelomic space contained within the test.
After Brown; after Petrunkevitch.

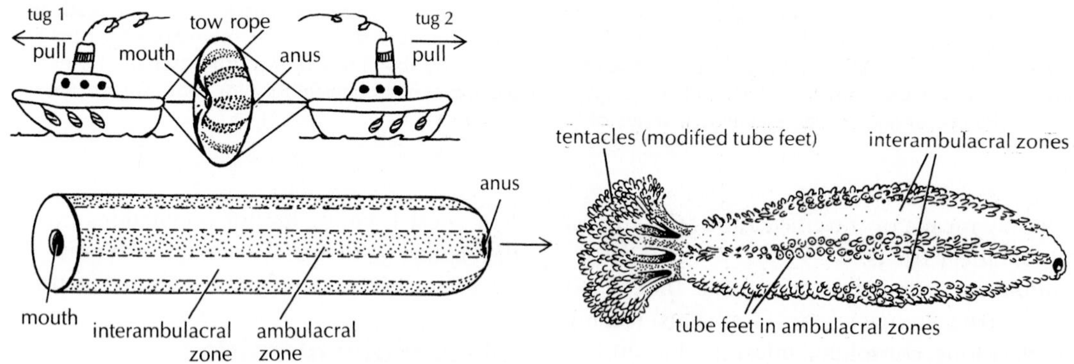

Figure 20.18
Relationship between the basic body plans of the echinoids and the holothurians. The holothurian shown at the end of the sequence is *Cucumaria frondosa.*
After Hyman.

would first require removing all spines and pedicellariae and then discarding Aristotle's lantern. (This is a thought experiment, not something that echinoderms actually do!) Next, the ossicles of the test would have to be separated from each other and greatly reduced in size. This mutual detachment and reduction in size of the ossicles would render the body wall stretchable, since it would now be composed largely of connective tissue. From this point to a finished holothurian would be largely a matter of stretching the imaginary animal, increasing the distance between the oral and aboral surfaces (Fig. 20.18).

Thus, holothurians are typically soft-bodied, bilaterally symmetric, vermiform creatures with distinct anterior and posterior ends and with podia generally confined to distinct ambulacral strips (as in the Echinoidea) (Fig. 20.18). The calcareous ossicles, so conspicuous in other echinoderm classes, are microscopic in holothurians and embedded in the body wall; these ossicles are often exquisitely shaped (Fig. 20.19). Calcareous ossicles compose up to 80% of the total dry weight of the body wall in some species, while a few species lack ossicles entirely. The outer body wall is often warty and dark colored. Some species actually resemble very closely the fruit from which they derive their common name, "sea cucumber." Adults range from several centimeters to over 1 m in length. Strangely enough, cephalization is not pronounced among holothurians, despite the presence of a distinct anterior end.

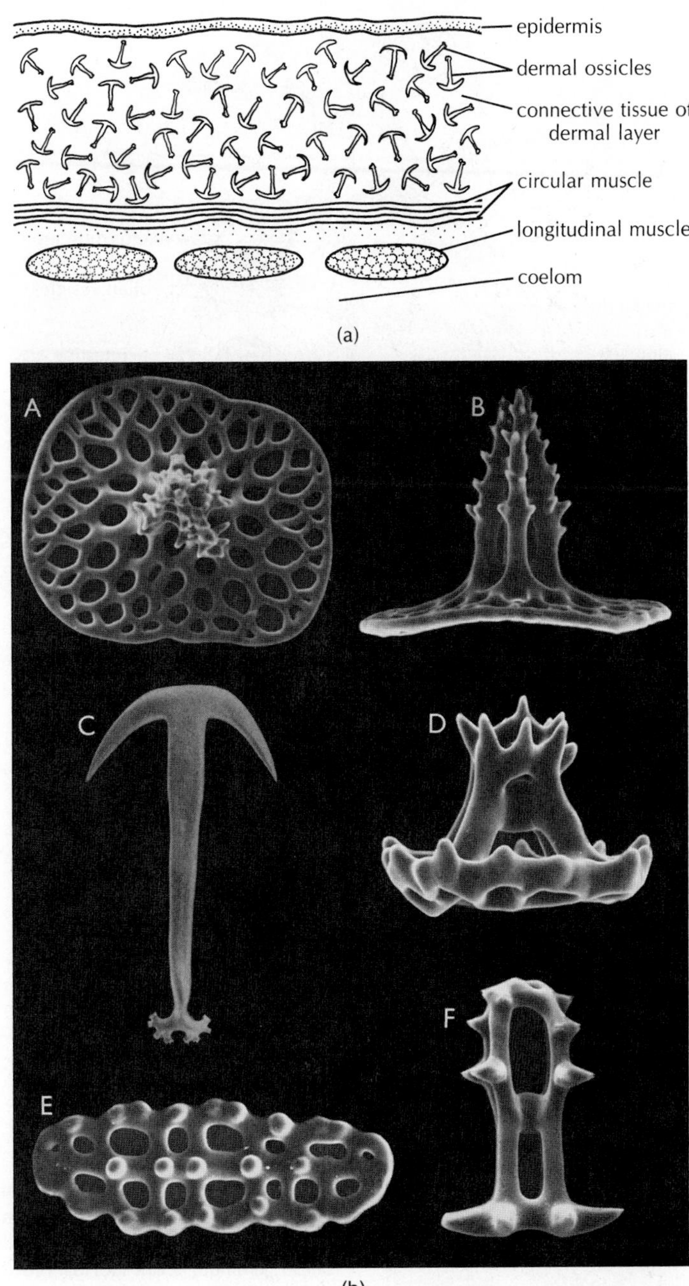

Figure 20.19

(a) Schematic cross section of the body wall of a typical holothurian, showing the arrangement of the musculature and location of the ossicles. Separation and size reduction of the ossicles permits holothurians to undergo major shape changes; the body wall, musculature, and coelom form a functional hydrostatic skeleton. (b) Scanning electron micrograph of the microscopic ossicles removed from the body walls of several holothurian species. Ossicle morphology plays a major role in species identification. (A,B) *Eostichopus regalis* (Cuvier), dorsal and lateral views; (C) *Euapta lappa* (Müller); (D) *Holothuria (Cystipus) occidentalis* Ludwig; (E) *Holothuria (Cystipus) pseudofossor* Deichmann; (F) *Holothuria (Semperothuria) surinamensis* Ludwig. These ossicles range from 60–400 μm (micrometers) in longest dimension.

(b) Courtesy of Harbor Branch Oceanographic Institution, Inc. Fort Pierce, Florida.

A holothurian's oral tube feet are modified as large, often feathery tentacles at the anterior end; these tentacles can be protracted from the mouth and used to capture food (Figs. 20.18 and 20.20). Each tentacle may be operated by a single, large ampulla. In some species, the tentacles are coated with sticky mucus to trap food particles from suspension and one species attaches to fish as an ectocommensal, but most sea cucumber species are deposit feeders, ingesting sediment and extracting the organic component. Some holothurians have been estimated to pass over 130 kg of substrate through their digestive systems per year. As deposit feeders, sea cucumbers thrive on the fine mud ooze that characterizes the deep-sea floor; they make up more than 90% of

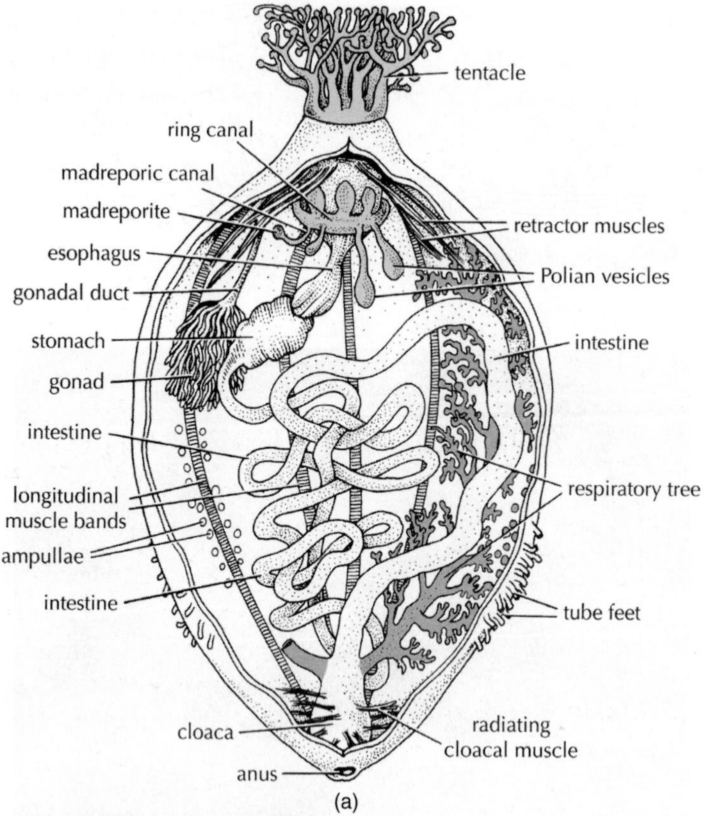

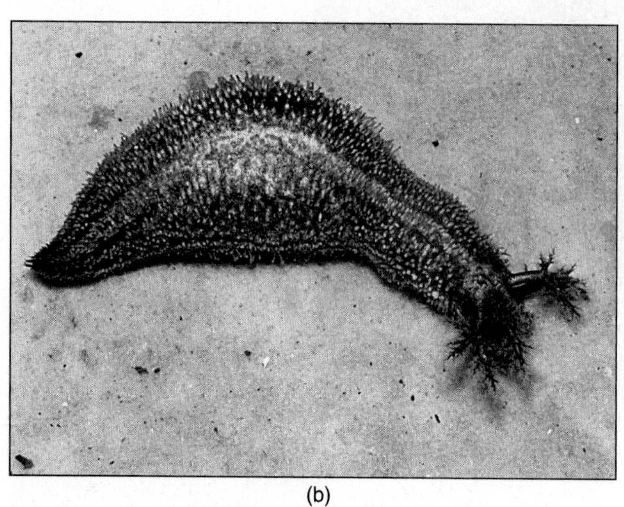

Figure 20.20

(a) Internal anatomy of the holothurian *Thyone briareus.* Note the extensive respiratory trees (only one shown), large coelomic space, and long intestine. (b) A mud-dwelling, burrowing holothurian, *Thyonella gemmata,* from Florida. Note the expanded tentacles, which are modified tube feet.

(a) After Hyman; After Coe. (b) Courtesy of Ralph Buschbaum.

the biomass in some abyssal habitats. Another group of deposit-feeders, the brittle stars (class Ophiuroidea), are also plentiful on the surface of such soft, deep-sea sediments.

The ambulacral tube feet of surface-living sea cucumber species generally bear suckers and are used in locomotion and attachment, as in the echinoids and asteroids.

The holothurian digestive system resembles that of the echinoids, except that it is greatly elongated (Fig. 20.20a). The WVS follows the typical echinoderm pattern, with the ring canal forming a ring about the esophagus, as usual. The ring canal is supported, however, by a calcareous ring, which may have an evolutionary origin in common with the Aristotle's lantern of echinoids. The madreporite generally lies free in the coelomic cavity (Fig. 20.20), so the WVS does not appear to be directly connected to the outside. The holothurian coelomic space is very large, as in the echinoids and asteroids, and the coelomic fluid is the primary circulatory medium. A few holothuroids also possess an extensive hemal system, with pulsatile hearts.

In contrast to other echinoderms, most holothurians have a body wall that contains well-developed layers of both circular and longitudinal musculature (Fig. 20.19a) and is considered by some to be an edible delicacy. As with other echinoderms, holothurians show a sophisticated immune response (see Research Focus Box 20.1).

RESEARCH FOCUS BOX 20.1

Influence of Pollutants

Canicatti, C., and M. Grasso. 1988. Biodepressive effect of zinc on humoral effector of the *Holothuria polii* immune response. *Marine Biol.* 99:393.

All animals, including the invertebrates, can distinguish between self and nonself. Among echinoderms, the immune response is mediated both by (1) humoral factors, which agglutinize (clump) or lyse (break apart) materials recognized as nonself, and (2) coelomocytes, which produce the lysins and also directly phagocytize or encapsulate foreign materials. Canicatti and Grasso (1988) set out to assess the effects of heavy metal pollution on the functioning of the self/nonself surveillance system. Heavy metals are major constituents of industrial effluents. Although numerous studies have considered the effects of various organic and inorganic pollutants on echinoderm survival, feeding, growth, reproduction, and development, the study by Canicatti and Grasso was one of the first to examine effects on the echinoderm immune response.

To conduct their study, the investigators drained the coelomic fluid from adult sea cucumbers (*Holothuria polii*) that were about 33 cm long. They then centrifuged the fluid to separate the liquid portion from the cellular constituents, including the coelomocytes. The pelleted cells were then opened by sonication, producing a "coelomocyte lysate." The ability of the centrifuged supernatant and coelomocyte lysate to agglutinate and lyse foreign cells was then assessed with and without the addition of zinc, cadmium, or mercury. Rabbit red blood cells (erythrocytes) served as foreign agents, to test the immune response.

Of the various heavy metals tested, only zinc affected the immune response, and the effect varied strikingly with concentration. Zinc concentrations of 1 mM (millimolar) or higher significantly depressed the lytic abilities of both the liquid and cellular fractions of the coelomic fluid (Focus Fig. 20.1). In contrast, lower zinc concentrations actually increased the lytic activity of the fluid fraction above control levels (Focus Fig. 20.1). Zinc concentrations up to 4 mM, the highest level tested, had no effect on the ability of coelomic fluid to agglutinate rabbit red blood cells; effects were limited to lytic activity.

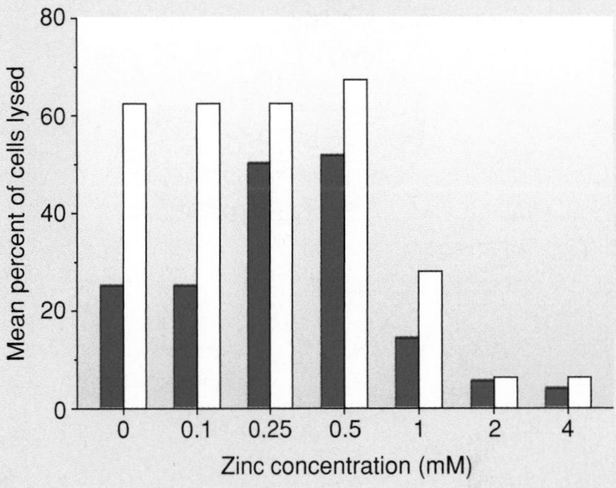

Focus Figure 20.1

Effect of dissolved zinc on the lytic activity of the liquid (shaded bars) and cellular (open bars) components of sea cucumber (*Holothuria polii*) coelomic fluid. Activity was tested against rabbit red blood cells, and each bar represents the mean of 5 replicates.
Based on C. Canicatti and M. Grasso, "Biodepressive effect of zinc on humoral effector of the *Holothuria polii* immune response," in *Marine Biology*, 99:393–96, 1988.

The data clearly indicate the ability of zinc pollution to alter a holothurian's immune recognition system. This raises a number of compelling questions: Would the researchers have seen the same depressive effect on the immune response if intact animals, rather than isolated cells and cellular products, were exposed to the same zinc levels? Do *organic* pollutants, such as fuel oils and pesticides, have comparable effects? Of the metals tested in this study, why did only zinc have an effect, and why did low concentrations of zinc *improve* the ability to recognize or attack foreign cells? How might suppression of the immune response affect the holothurian's survival in the field? Finally, does zinc have similar effects on the immune systems of other animals?

Clearly, holothurians have the characteristics necessary for operating a hydrostatic skeleton: a large, fluid-filled, constant-volume body cavity; a deformable body wall; and an appropriate musculature. It should not be surprising to learn, then, that members of many holothurian species burrow in sand and mud (Fig. 20.21a). The tube feet of such species are often much reduced, and some of the more specialized burrowing species lack tube feet entirely. Locomotion in these burrowing species is accomplished in part by using the tentacles to push substrate away, but it primarily results from waves of contraction of the circular and longitudinal muscles, earthworm style.

Holothurians are the only echinoderms to possess truly specialized, internal respiratory structures, called **respiratory trees.** Most holothurian coeloms hold a pair of these highly branched, muscular structures. The respiratory trees connect to the cloaca, which pumps water into the trees. Water is expelled through the cloaca by contraction of the respiratory tree tubules themselves.

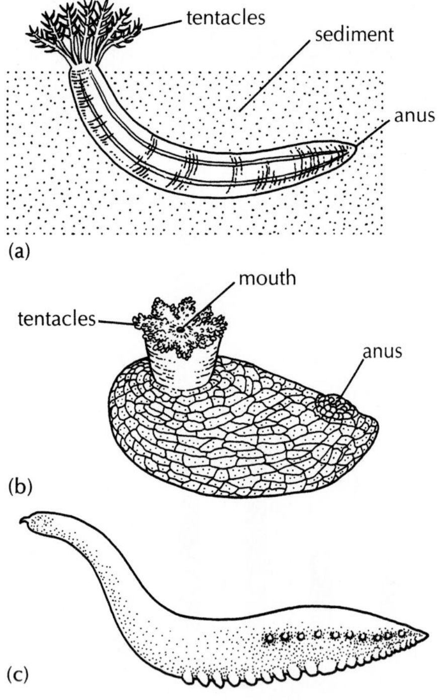

Figure 20.21

Holothurian diversity. (a) A burrowing holothurian, *Leptosynapta inhaerens,* in sand. This species is only a few centimeters long; has a very thin, flexible body wall; lacks respiratory trees; and except for the tentacles, lacks podia. (b) A highly modified holothurian, *Psolus fabricii.* The ventral surface is flattened, forming a "creeping sole." Podia are restricted to the flattened surface; the dorsal surface is covered with protective, calcareous scales. (c) A deep-sea holothurian, *Psychropotes longicauda,* with the tube feet modified as legs. Holothurians comprise one of the dominant elements of the deep-sea macrofauna. The specimen illustrated was obtained from a depth of perhaps 800 m; the body may reach nearly 0.5 m in length. Other deep-sea species, such as those in the genus *Scotoanassa,* are gelatinous and extremely fragile, and live several meters above the bottom.

(a,b) Modified from Fretter and Graham, 1976. *A Functional Anatomy of Invertebrates.* Orlando: Academic Press, Inc. (c) From Marshall, *Deep-Sea Biology.* Copyright © 1979. Garland STPM Press, New York. Reprinted by permission.

A lovely young starfish quite famous
Had a crush on a sea cuke named Amos.
She thought he was hot
'til she uttered in shock,
"My God! This guy breathes through his anus!"
(Courtesy of Chip Biernbaum, College of Charleston)

I have saved the best for last. Many holothurian species respond to a variety of physical and environmental factors by deliberately expelling internal organs. In some species, this is limited to the expulsion of incredibly sticky and/or toxic structures called **Cuvierian tubules,** which are attached to the left respiratory tree and apparently used only for discouraging and entangling potential predators. They are entirely separate from the animal's digestive system and are regenerated by the sea cucumber within a matter of weeks. In many other species, true **evisceration** occurs, in which the entire digestive system (the "viscera") may be expelled, along with the respiratory trees and gonads. Evisceration involves the liquifying and subsequent rupture of the connective tissue attaching the viscera to the inner body wall. All lost body parts are eventually re-formed, reflecting the substantial regenerative capabilities possessed by most echinoderms.[12]

Other Features of Echinoderm Biology

Reproduction and Development

Asexual reproduction is often encountered among echinoderms. Among adult asteroids and ophiuroids, the central disc separates into 2 pieces, and each piece proceeds to re-form the missing arms and organs (see Fig. 24.2). At least one asteroid species reproduces asexually from pieces of the arms alone and asexual replication has been reported for the larvae of some asteroid, echinoid, and ophiuroid species. Holothurians also exhibit asexual replication; in a few species, the adult body routinely breaks in half, transversely, with each half-cucumber regenerating its missing parts. Asexual reproduction is unknown among crinoids and echinoids.

Most echinoderm species reproduce only sexually, and the sexes are usually separate. With few exceptions (some ophiuroid species, for example), males and females are externally indistinguishable. All echinoderms except holothurians and crinoids bear multiple gonads. In the asteroids, at least one pair of gonads extends into each arm. Concentricycloids possess 10 gonads (5 pairs), which lie in the capacious coelomic cavity. In ophiuroids, from one to many gonads empty into each bursa. Echinoids usually have 5 gonads. The gametes of crinoids develop in tissues of the arms or pinnules, not in true gonads. Holothurians are unique among the Echinodermata in commonly possessing a single gonad. Gonadal development **(gametogenesis)** is controlled by steroid hormones in all echinoderms, as it is in vertebrates.

In most echinoderm species, gametes are liberated into the surrounding seawater, so fertilization is typically external. Concentricycloids seem to be a particularly notable exception to this general rule of external fertilization. The ducts leading out from the male testes continue beyond the margins of the circular body and are stiffened by peripheral, supportive spines (Fig. 20.12). These ducts may thus serve as copulatory organs, perhaps permitting internal fertilization.

With the exception of the concentricycloids, distinctive ciliated larval stages characterize each class, as illustrated in Figures 20.22, 20.23, and 24.14c, d, h. A delicate,

12. See *Topics for Further Discussion and Investigation,* no. 1.

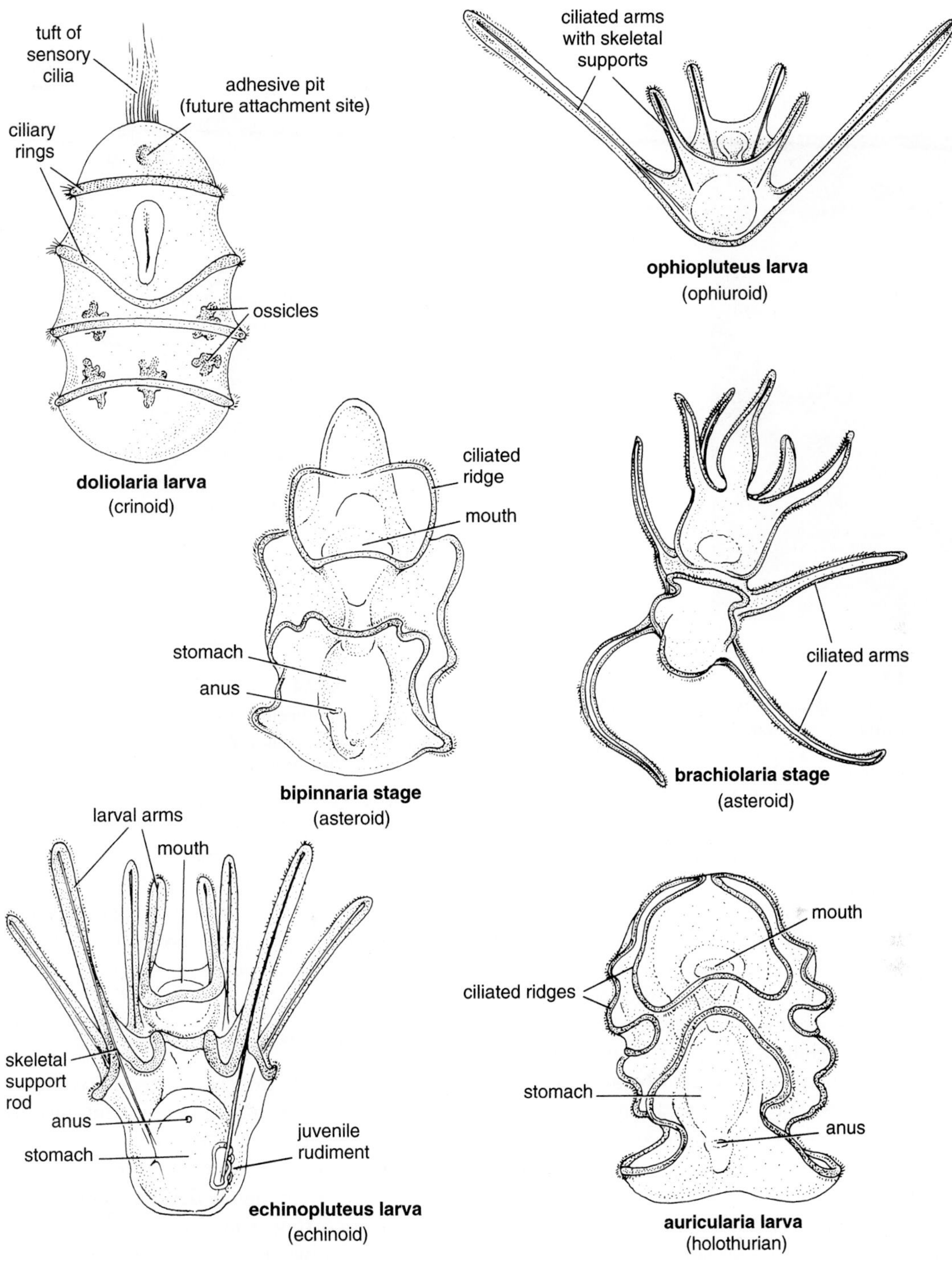

Figure 20.22

Larval forms typical of the various extant echinoderm classes. It has recently been found (Nakano et al., 2003. *Nature* 421:158–60) that in some crinoid species the doliolaria larva is preceded by a non-feeding auricularia stage.

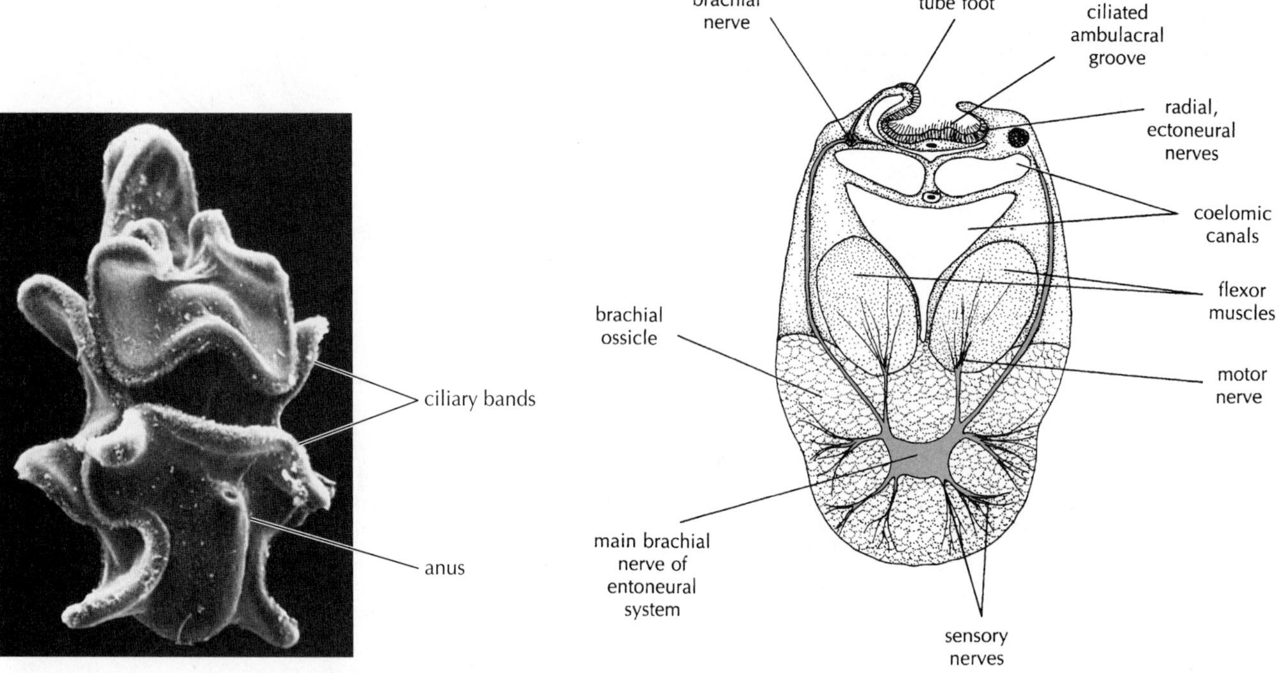

Figure 20.23
Scanning electron micrograph showing 10-day-old bipinnaria larva of the crown-of-thorns sea star, *Acanthaster planci.* This larva is about 750 μm long.
Courtesy of R. R. Olson.

Figure 20.24
Cross section through the arm of a crinoid, showing the arrangement of the nervous system.
From Hyman; after Hamann.

internal, calcareous skeleton supports the larval arms in echinoids and ophiuroids. Echinoderm metamorphosis to the adult body plan is often dramatic, in terms of speed and the complexity and magnitude of the morphological reorganization.[13] However, a number of species in most echinoderm classes exhibit severe reductions in the development of larval structures—particularly those associated with food collection—and an acceleration in the rate at which adult structures form, so embryos become adults more directly and often more quickly. Such modifications are often associated with some form of parental brood care. In some asteroid species, for example, embryos develop within the parental gonad and emerge as fully formed, tiny sea stars. These evolutionary departures from the standard larval morphology seem to have evolved independently a number of times even within the various echinoderm classes. Echinoderms are thus providing particularly good experimental material for studying the underlying molecular mechanisms through which developmental patterns have become modified by selective forces.[14] The development of prosobranch gastropods would also be well worth studying in this regard, but so far most of the action has been focused on the echinoderms and ascidians (Chapter 23).

Nervous System

Echinoderms do not have a centralized brain; nor are distinct ganglia generally found in them. Instead, the nervous system is composed of 3 diffuse nerve networks. An **ectoneural** system receives sensory input from the epidermis. This system, highly developed in all but the crinoids, consists of a ring around the esophagus with 5 associated nerves radiating outward. In species with arms, these radial nerve cords extend down each arm to the tube feet, ampullae (where present), and pedicellariae (where present). A second, **hyponeural,** system is exclusively concerned with motor function. It, too, consists of a circumoral nerve ring with 5 associated radial nerves, but it lies deeper within the animal's tissues. The hyponeural system is well developed only in ophiuroids and, to a lesser extent, in asteroids. In crinoids, the major nerve network is an **entoneural** system (Fig. 20.24), associated with the animal's aboral end. From a central mass in the calyx/tegmen complex, nerves radiate down the stalk of the crinoid to the cirri and up into each arm. The entoneural system is inconspicuous or entirely absent in the other echinoderm classes.

13. See *Topics for Further Discussion and Investigation,* no. 6.

14. See *Topics for Further Discussion and Investigation,* no. 10.

Taxonomic Summary

Phylum Echinodermata
- Subphylum Crinozoa
  - Class Crinoidea—the sea lilies and feather stars
- Subphylum Asterozoa
  - Class Stelleroidea
    - Subclass Ophiuroidea—the brittle stars
    - Subclass Asteroidea—the sea stars
      - The concentricycloids—the sea daisies
- Subphylum Echinozoa
  - Class Echinoidea—the sea urchins, heart urchins, and sand dollars
  - Class Holothuroidea—the sea cucumbers

Topics for Further Discussion and Investigation

1. Investigate the factors controlling the loss and regeneration of body parts among echinoderms.

Anderson, J. M. 1965. Studies on visceral regeneration in sea stars. III. Regeneration of the cardiac stomach in *Asterias forbesi* (Desor). *Biol. Bull.* 129:454.

Carnevali, M. D. C., E. Lucca, and F. Bonasoro. 1993. Mechanisms of arm regeneration in the feather star *Antedon mediterranea:* Healing of wound and early stages of development. *J. Exp. Zool.* 267:299.

Dobson, W. E. 1985. A pharmacological study of neural mediation of disc autotomy in *Ophiophragmus filograneus* (Lyman) (Echinodermata: Ophiuroidea). *J. Exp. Marine Biol. Ecol.* 94:223.

Lawrence, J. M., T. S. Klinger, J. B. McClintock, S. A. Watts, C.-P. Chen, A. Marsh, and L. Smith. 1986. Allocation of nutrient resources to body components by regenerating *Luidia clathrata* (Say) (Echinodermata: Asteroidea). *J. Exp. Marine Biol. Ecol.* 102:47.

Mladenov, P. V., S. Igdoura, S. Asotra, and R. D. Burke. 1989. Purification and partial characterization of an autotomy-promoting factor from the sea star *Pycnopodia helianthoides. Biol. Bull.* 176:169.

Pataruno, M., A. Smertenko, M. D. C. Carnevali, F. Bonasoro, P. W. Beesley, and M. C. Thorndyke. 2002. Expression of transforming growth factor β-like molecules in normal and regenerating arms of the crinoid *Antedon mediterranea:* Immunocytochemical and biochemical evidence. *Proc. Royal Soc. London* B 269:1741.

Pomory, C., and J. M. Lawrence. 1999. Energy content of *Ophiocoma echinata* (Echinodermata: Ophiuroidea) maintained at different feeding levels during arm regeneration. *J. Exp. Marine Biol. Ecol.* 238:139.

Pomory, C. M., and J. M. Lawrence. 1999. Effect of arm regeneration on energy storage and gonad production in *Ophiocoma echinata* (Echinodermata: Ophiuroidea). *Marine Biol.* 135: 57.

Sides, E. M. 1987. An experimental study of the use of arm regeneration in estimating rates of sublethal injury on brittle-stars. *J. Exp. Marine Biol. Ecol.* 106:1.

Smith, G. N., Jr. 1971. Regeneration in the sea cucumber *Leptosynapta.* I. The process of regeneration. *J. Exp. Zool.* 177:319.

Vandenspiegel, D., M. Jangoux, and P. Flammang. 2000. Maintaining the line of defense: Regeneration of Cuvierian tubules in the sea cucumber *Holothuria forskali* (Echinodermata, Holothuroidea). *Biol. Bull.* 198: 34.

Wilkie, I. C. 1978. Arm autotomy in brittle stars (Echinodermata: Ophiuroidea). *J. Zool., London* 186:311.

Zeleny, C. 1903. A study of the rate of regeneration of the arms in the brittle star *Ophioglypha lacertosa. Biol. Bull.* 6:12.

2. What is the role of coelomocytes in mediating the echinoderm immune response?

Bang, F. B. 1982. Disease processes in seastars: A Metchnikovian challenge. *Biol. Bull.* 162:135.

Beck, G., and G. S. Habicht. 1996. Immunity and the invertebrates. *Sci. Amer.* November 1996:60–71.

Canicatti, C., and A. Quaglia. 1991. Ultrastructure of *Holothuria polii* encapsulating body. *J. Zool., London* 224:419.

Dan-Sohkawa, M., J. Suzuki, S. Towa, and H. Kaneko. 1993. A comparative study on the fusogenic nature of echinoderm and nonechinoderm phagocytes *in vitro. J. Exp. Zool.* 267:67.

Hilgard, H. R., and J. H. Phillips. 1968. Sea urchin response to foreign substances. *Science* 161:1243.

Kaneshiro, E. S., and R. D. Karp. 1980. The ultrastructure of coelomocytes of the sea star *Dermasterias imbricata. Biol. Bull.* 159:295.

Service, M., and A. C. Wardlaw. 1984. Echinochrome-A as a bactericidal substance in the coelomic fluid of *Echinus esculentus* (L.). *Comp. Biochem. Physiol.* 79B:161.

Smith, L. C., C-S. Shih, and S. G. Dachenhausen. 1998. Coelomocytes express SpBf, a homologue of factor B, the second component in the sea urchin complement system. *J. Immunol.* 161: 6784.

Yui, M. A., and C. J. Bayne. 1983. Echinoderm immunology: Bacterial clearance by the sea urchin *Strongylocentrotus purpuratus. Biol. Bull.* 165:473.

3. Discuss the role of tube feet, spines, pedicellariae, cilia, and water flow in the collection of food by echinoderms.

Allen, J. R. 1998. Suspension feeding in the brittle-star *Ophiothrix fragilis:* Efficiency of particle retention and implications for the use of encounter-rate models. *Marine Biol.* 132:383.

Burnett, A. L. 1960. The mechanism employed by the starfish *Asterias forbesi* to gain access to the interior of the bivalve *Venus mercenaria. Ecology* 41:583.

Byrne, M., and A. R. Fontaine. 1981. The feeding behaviour of *Florometra serratissima* (Echinodermata: Crinoidea). *Can. J. Zool.* 59:11.

Chia, F. S. 1969. Some observations on the locomotion and feeding of the sand dollar, *Dendraster excentricus* (Eschscholtz). *J. Exp. Marine Biol. Ecol.* 3:162.

De Ridder, C., M. Jangoux, and L. De Vos. 1987. Frontal ambulacral and peribuccal areas of the spatangoid echinoid *Echinocardium cordatum* (Echinodermata): A functional entity in feeding mechanism. *Marine Biol.* 94:613.

Ellers, O., and M. Telford. 1984. Collection of food by oral surface podia in the sand dollar, *Echinarachnius parma* (Lamarck). *Biol. Bull.* 166:574.

Emson, R. H., and J. D. Woodley. 1987. Submersible and laboratory observations on *Asteroschema tenue,* a long-armed euryaline [*sic*] brittle star epizoic on gorgonians. *Marine Biol.* 96:31.

Emson, R. H., and C. M. Young. 1994. Feeding mechanism of the brisingid starfish *Novodinia antillensis. Marine Biol.* 118:433.

Fankboner, P. V. 1978. Suspension-feeding mechanisms of the armoured sea cucumber *Psolus chitinoides* Clark. *J. Exp. Marine Biol. Ecol.* 31:11.

Ghiold, J. 1983. The role of external appendages in the distribution and life habits of the sand dollar *Echinarachnius parma* (Echinodermata: Echinoidea). *J. Zool., London* 200:405.

Hendler, G. 1982. Slow flicks show star tricks: Elapsed-time analysis of basketstar (*Astrophyton muricatum*) feeding behavior. *Bull. Marine Sci.* 32:909.

Hendler, G., and J. Miller. 1984. Feeding behavior of *Asteroporpa annulata,* a gorgonocephalid brittle star with unbranched arms. *Bull. Marine Sci.* 34:449.

Holland, N. D., A. B. Leonard, and J. R. Strickler. 1987. Upstream and downstream capture during suspension feeding by *Oligometra serripinna* (Echinodermata: Crinoidea) under surge conditions. *Biol. Bull.* 173:552.

Lasker, R., and A. C. Giese. 1954. Nutrition of the sea urchin, *Strongylocentrotus purpuratus. Biol. Bull.* 106:328.

LaTouche, R. W. 1978. The feeding behavior of the feather star *Antedon bifida* (Echinodermata: Crinoidea). *J. Marine Biol. Assoc. U.K.* 58:877.

Leonard, A. B. 1989. Functional response in *Antedon mediterranea* (Lamarck) (Echinodermata: Crinoidea): The interaction of prey concentration and current velocity on a passive suspension-feeder. *J. Exp. Marine Biol. Ecol.* 127:81.

Macurda, D. B., and D. L. Meyer. 1974. Feeding posture of modern stalked crinoids. *Nature (London)* 247:394.

Mauzey, K. P., C. Birkeland, and P. K. Dayton. 1968. Feeding behavior of asteroids and escape responses of their prey in the Puget Sound region. *Ecology* 49:603.

Meyer, D. L. 1979. Length and spacing of the tube feet in crinoids (Echinodermata) and their role in suspension-feeding. *Marine Biol.* 51:361.

O'Neill, P. L. 1978. Hydrodynamic analysis of feeding in sand dollars. *Oecologia* 34:157.

Telford, M., and R. Mooi. 1996. Podial particle picking in *Cassidulus caribaearum* (Echinodermata: Echinoidea) and the phylogeny of sea urchin feeding mechanisms. *Biol. Bull.* 191:209.

4. Seawater contains fairly high concentrations (up to 3×10^3 g carbon/liter) of dissolved organic matter (DOM), especially in shallow coastal waters. To what extent are echinoderms capable of meeting their nutritional needs through the uptake of DOM directly from seawater?

Ferguson, J. C. 1980. The non-dependency of a starfish on epidermal uptake of dissolved organic matter. *Comp. Biochem. Physiol.* 66A:461.

Fontaine, A. R., and F. S. Chia. 1968. Echinoderms: An autoradiographic study of assimilation of dissolved organic molecules. *Science* 161:1153.

Hammond, L. S., and C. R. Wilkinson. 1985. Exploitation of sponge exudates by coral reef holothuroids. *J. Exp. Marine Biol. Ecol.* 94:1.

Manahan, O. T., J. P. Davis, and G. C. Stephens. 1983. Bacteria-free sea urchin larvae: Selective uptake of neutral amino acids from seawater. *Science* 220:204.

Shilling, F. M., and D. T. Manahan. 1990. Energetics of early development for the sea urchins *Strongylocentrotus purpuratus* and *Lytechinus pictus* and the crustacean *Artemia* sp. *Marine Biol.* 106:119–27.

Stephens, G. C., M. J. Volk, S. H. Wright, and P. S. Backlund. 1978. Transepidermal accumulation of naturally occurring amino acids in the sand dollar, *Dendraster excentricus. Biol. Bull.* 154:335.

5. How are individual echinoderm tube feet operated and integrated into the total behavior of the animal?

Binyon, J. 1964. On the mode of functioning of the water vascular system of *Asterias rubens* L. *J. Marine Biol. Assoc. U.K.* 44:577.

Flammang, P., S. DeMeulenaere, and M. Jangoux. 1994. The role of podial secretions in adhesion in two species of sea stars (Echinodermata). *Biol. Bull.* 187:35.

Kerkut, G. A. 1953. The forces exerted by the tube feet of the starfish during locomotion. *J. Exp. Biol.* 30:575.

Lavoie, M. E. 1956. How sea stars open bivalves. *Biol. Bull.* 111:114.

McCurley, R. S., and W. M. Kier. 1995. The functional morphology of starfish tube feet: The role of a crossed-fiber helical array in movement. *Biol. Bull.* 188:197.

Polls, I., and J. Gonor. 1975. Behavioral aspects of righting in two asteroids from the Pacific coast of North America. *Biol. Bull.* 148:68.

Prusch, R. D., and F. Whoriskey. 1976. Maintenance of fluid volume in the starfish water vascular system. *Nature (London)* 262:577.

Santos, R., S. Gorb, V. Jamar, and P. Flammang. 2005. Adhesion of echinoderm tube feet to rough surfaces. *J. Exp. Biol.* 208: 2555.

Smith, J. E. 1947. The mechanics and innervation of the starfish tube foot-ampulla system. *Phil. Trans. Royal Soc. B* 232:279.

Thomas, L. A., and C. O. Hermans. 1985. Adhesive interactions between the tube feet of a star fish, *Leptasterias hexactis,* and substrata. *Biol. Bull.* 169:675.

6. What major morphological changes take place as echinoderms metamorphose?

Cameron, R. A., and R. T. Hinegardner. 1978. Early events in sea urchin metamorphosis, description and analysis. *J. Morphol.* 157:21.

Emlet, R. B. 1988. Larval form and metamorphosis of a "primitive" sea urchin, *Eucidaris thouarsi* (Echinodermata: Echinoidea: Cidaroida), with implications for developmental and phylogenetic studies. *Biol. Bull.* 174:4.

Hardy, A. 1965. Pelagic larval forms. *The Open Sea: Its Natural History.* Boston: Houghton Mifflin, 178–98.

Hendler, G. 1978. Development of *Amphioplus abditus* (Verrill) (Echinodermata: Ophiuroidea). II. Description and discussion of ophiuroid skeletal ontogeny and homologies. *Biol. Bull.* 154:79.

Mladenov, P. V. M., and F. S. Chia. 1983. Development, settling behaviour, metamorphosis and pentacrinoid feeding and growth of the feather star *Florometra serratissima. Marine Biol.* 73:309.

Nakano, H. T., T. Hibino, Y. H. Oji, and S. Amemiya. 2003. Larval stages of a living sea lily (stalked crinoid echinoderm). *Nature* 421: 158.

Smiley, S. 1986. Metamorphosis of *Stichopus californicus* (Echinodermata: Holothuroidea) and its phylogenetic implications. *Biol. Bull.* 171:611.

7. The echinoderms have an extensive fossil record of 10,000 to 13,000 species, in keeping with their ancient origin, marine habitat, and solid skeleton. In fact, about 16 classes are known only from the fossil record, there having been no living representatives of these classes for tens of thousands of years. Most of these extinct echinoderms were sessile animals, permanently attached to a substrate. What are the morphological similarities and differences between these extinct species and the present-day sea lilies (class Crinoidea)?

Boardman, R. S., A. H. Cheetham, and A. J. Rowell, eds. 1987. *Fossil Invertebrates.* Palo Alto, Calif.: Blackwell Scientific Publications.

Clarkson, E. N. K. 1986. *Invertebrate Paleontology and Evolution*, 2d ed. Boston: Allen and Unwin.

Clausen, S., and A. B. Smith. 2005. Palaeoanatomy and biological affinities of a Cambrian deuterostome (Stylophora). *Nature* 438: 351.

David, B., A. Guille, J.-P. Féral, and M. Roux eds. 1994. *Echinoderms Through Time*. Rotterdam: A. A. Balkema.

Hyman, L. H. 1955. *The Invertebrates*, vol. IV. New York: McGraw-Hill.

Levin, H. L. 1999. *Ancient Invertebrates and Their Living Relatives*. New Jersey: Prentice Hall, 294–328.

Meyer, D. L., and D. B. Macurda, Jr. 1977. Adaptive radiation of the comatulid crinoids. *Paleobiology* 3:74.

Shu, D.-G., S. Conway Morris, J. Han, Z.-F. Zhang, and J.-N. Liu. 2004. Ancestral echinoderms from the Chengjiang deposits of China. *Nature* 430: 422.

8. Asteroids and echinoids resemble polychaete annelids in their highly coordinated use of numerous appendages in locomotion. Based upon lectures and your readings in this book, compare and contrast the structure and operation of polychaete parapodia with that of asteroid and echinoid tube feet.

9. The function of the echinoderm madreporite has been surprisingly difficult to document. What evidence indicates that it serves an important role in conducting seawater to the tube feet? What evidence suggests that it does not?

Binyon, J. 1984. A re-appraisal of the fluid loss resulting from the operation of the water vascular system of the starfish, *Asterias rubens*. *J. Marine Biol. Assoc. U.K.* 64:726.

Ellers, O., and M. Telford. 1992. Causes and consequences of fluctuating coelomic pressure in sea urchins. *Biol. Bull.* 182:424.

Ferguson, J. C. 1990. Seawater inflow through the madreporite and internal body regions of a starfish (*Leptasterias hexactis*) as demonstrated with fluorescent microbeads. *J. Exp. Zool.* 255:262.

Ferguson, J. C. 1996. Madreporite function and fluid volume relationships in sea urchins. *Biol. Bull.* 191:431.

Tamori, M., A. Matsuno, and K. Takahashi. 1996. Structure and function of the pore canals of the sea urchin madreporite. *Phil. Trans. Royal Soc. London B* 351:659.

10. How extensive are the changes in gene expression that seem to underlie shifts in developmental patterns among asteroids and echinoids?

Ferkowicz, M. J., and R. A. Raff. 2001. Wnt gene expression in sea urchin development: Heterochronies associated with the evolution of developmental mode. *Evol. Devel.* 3: 24.

Lowe, C. J., L. Issel-Tarver, and G. A. Wray. 2002. Gene expression and larval evolution: Changing roles of distal-less and orthodenticle in echinoderm larvae. *Evol. Dev.* 4: 111.

Wray, G. A., and R. A. Raff. 1991. The evolution of developmental strategy in marine invertebrates. *Trends Ecol. Evol.* 6:45.

11. What do you call a sea cucumber just after it has eviscerated?

12. How do echinoderms protect themselves from predators?

Bakus, G. J. 1968. Defensive mechanisms and ecology of some tropical holothurians. *Marine Biol.* 2: 23.

Bryan, P. J., J. B. McClintock, and T. S. Hopkins. 1997. Structural and chemical defenses of echinoderms from the northern Gulf of Mexico. *J. Exp. Marine Biol. Ecol.* 210: 173.

Iyengar, E.V., and C. D. Harvell. 2001. Predator deterrence of early developmental stages of temperate lecithotrophic asteroids and holothuroids. *J. Exp. Marine Biol. Ecol.* 264: 171.

Rosenberg, R., and E. Selander. 2000. Alarm signal response in the brittle star *Amphiura filiformis*. *Marine Biol.* 136: 43.

Taxonomic Detail

Phylum Echinodermata

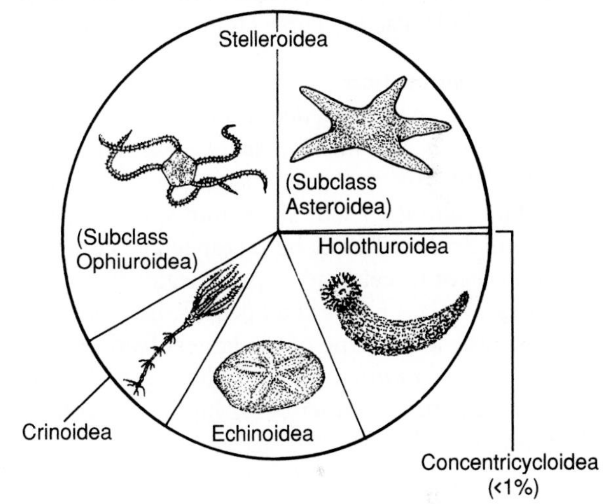

Subphylum Crinozoa

Representatives of most classes are extinct and are known only as fossils; existing species are contained within a single class. All are marine.

Class Crinoidea

The sea lilies and feather stars. The approximately 700 crinoid species are distributed among 5 orders, most of which contain only 1 to 3 families and 25 or fewer species.

Order Millericrinida

Ptilocrinus. These stalked sea lilies lack cirri and are mostly restricted to depths of 2000 m or more. One family.

Order Cyrtocrinida

Holopus. These sea lilies live attached to the substrate with a very short stalk, or none at all, and lack cirri. They are restricted to intermediate depths (several hundred meters) in the Caribbean. One family.

Order Bourgueticrinida

Bathycrinus, Rhizocrinus. These stalked sea lilies lack cirri and have 5 or 10 unusually short arms. The group includes the deepest known crinoids (members of the genus *Bathycrinus*), which extend to depths of nearly 10,000 m. Three families.

Order Isocrinida

Cenocrinus, Neocrinus, Metacrinus. These sea lilies have long stalks (up to 1 m). Many species live permanently attached to hard substrates, anchored at the base of their stalks. The other species attach to firm substrates using prehensile cirri found at intervals along the stalk; these cirri can release their grip on the substrate, allowing the animals to then walk or, in some species, swim to other sites. Most species are confined to depths of several hundred meters or more; all are located in the Caribbean and tropical Indo-Pacific. Three families.

Order Comatulida

The feather stars. *Antedon, Comantheria, Comanthina, Florometra, Heliometra, Nemaster.* This highly diverse and colorful group of crinoids contains about 88% of all crinoid species. They are widely distributed, being especially common both on coral reefs and in polar waters; representatives occur at all depths, from less than 1 m to the abyss. One of the largest feather star species, *Comanthina schlegelii*, has up to 200 arms, with the actual number being highly variable among individuals.

Subphylum Asterozoa

The approximately 3700 species are all marine and contained in a single class.

Class Stelleroidea

All but a few asterozoan species are included in this class. Three subclasses.

Subclass Somasteroidea

The members of this subclass share characteristics with both ophiuroids and asteroids, indicating a close relationship between these 2 major groups. Most representatives are known only as fossils.

Subclass Ophiuroidea

The brittle stars and basket stars. The approximately 2100 species are distributed among 3 orders.

Order Phrynophiurida

Five families.

Family Gorgonocephalidae. *Gorgonocephalus.* This group, which includes the basket stars (species with highly branched arms), is distributed from the tropics to the poles and from shallow water to the abyss. Individuals often occur in dense aggregations, perched on solid substrates with the oral surface facing up and the arms fully extended into the surrounding water; these animals are suspension feeders. Including the arms, some individuals grow to diameters exceeding 70 cm. About 100 living species are known.

Order Ophiurida

Amphiura, Amphipholis, Ophiura, Ophiomusium, Ophiocoma, Ophiomastix, Ophioderma, Ophiacantha, Ophiactis, Ophiopholis, Ophiostigma, Ophiothrix. This is the largest order of brittle stars, including over 2000 described species distributed among 11 families. Representatives are found in all oceans, at all depths. The embryos of some *Ophiomastix* species are brooded by adults of *Ophiocoma* spp., the only known example of what appears to be "brood parasitism" in ophiuroids (Hendler et al., 1999. *Invert. Biol.* 118:190–201), and probably the only known example among echinoderms.

Subclass Asteroidea

The sea stars. Nearly 1600 species have been described, distributed among 5 orders.

Order Platyasterida

Luidia. This is the most primitive extant group of asteroids, with a fossil record extending back nearly 500 million years. The stars in this order were formerly placed in the now abandoned order Phanerozonida. Most species are found on sandy substrates in shallow, tropical waters. Their tube feet are pointed and lack suckers. At least some species are euryhaline, capable of living in water of substantially reduced osmotic concentration. The free-swimming larvae produced by some species of *Luidia* are among the largest of all known invertebrate larvae, attaining lengths of 2 cm or more, and seem capable of remaining in the water for more than a year before metamorphosing; the larvae are unusual in reproducing more larvae by asexual fission while still planktonic. Some zoologists consider *Luidia* an aberrant member of another order, the Paxillosida. One family.

Order Paxillosida

Astropecten; Ctenodiscus—mud stars (they ingest sediment). This large order of nearly 400 described species is composed of primitive stars living in or on sediment. Not surprisingly for animals that do not walk on solid substrates, the tube feet lack suckers. Many species are exclusively abyssal, to depths of nearly 8000 m. *Luidia* may belong to this order (see Order Platyasterida). Five families.

Order Valvatida

Asterina, Goniaster, Linckia, Odontaster, Oreaster, Patiriella. Some species grow to diameters of only about 15 mm, while others reach diameters of nearly 50 cm. In some species, the tube feet are flat distally and lack suckers. The order contains one of the largest asteroid families, the Goniasteridae, which contains about 300 described species. Species in the genus *Linckia* show remarkable regenerative powers, even in comparison with most other asteroids: An entire animal can develop from a single arm. Nine families.

Order Spinulosida

Acanthaster—the crown-of-thorns sea star, a voracious predator on coral, with toxic spines and 9 to 23 arms; *Echinaster; Henricia*—blood stars; *Pteraster; Solaster*—sun stars, with as many as 15 arms. Twelve families.

Order Forcipulata

Heliaster—another group of sun stars, with up to 50 arms per animal; *Zoroaster*—rat tail stars. Four families.

Family Asteriidae. *Asterias, Leptasterias, Pisaster, Pycnopodia.* This is the largest of all asteroid families, containing more than 300 species, most of which live on hard substrates. Members of the genus *Pycnopodia* possess as many as 40,000 tube feet distributed among as many as 25 arms; these unusually large stars can attain diameters exceeding 1 m.

Order Brisingida

Brisinga, Novodinia. Sea stars in this order are common in the deep sea. Members of the genus *Brisinga* may exceed 1 m in diameter, including the arms; these stars possess unusually long tube feet, which they use for suspension feeding. Members of the genus *Novodinia* have been observed capturing prey using pedicellariae.

The concentricycloids

Xyloplax—the sea daisies. These deep-water echinoderms seem most closely related to members of the asteroid orders Valvatida or Spinulosida. Only a few species have been described to date, all associated with submerged wood. One family.

Subphylum Echinozoa

Two classes.

Class Echinoidea

The urchins, sea biscuits, and sand dollars—about 900 species. Thirteen orders, many with few surviving species.

Order Cidaroida

Eucidaris—pencil urchins. These widespread urchins are an ancient group, with fossilized representatives extending back over 300 million years. Each ambulacral plate is penetrated by a single tube foot, which sets these urchins apart from all others. All are nonselective scavengers. The nearly 140 species that have been described are distributed among 2 families.

Order Echinothuroida

These are all deep-water species (1000–4000 m depth), with a delicate, highly flexible test. They may reach 70 cm in diameter. All live on soft sediments, often in large aggregations. All species studied to date develop as highly modified, nonfeeding larvae. One family.

Order Diadematoida

Diadema, Echinothrix. These are common in warm, shallow waters, particularly on coral reefs. All representatives have extremely long, sharply pointed spines. The spines are not venomous, but you will sincerely regret stepping on them. Four families.

Order Arbacioida

Arbacia. One of the several dozen species in this group (*A. punctulata*) has been used widely in research by developmental biologists. One family (Arbaciidae).

Order Temnopleuroida

Lytechinus, Tripneustes. This group contains over 100 species, mostly in shallow tropical waters. One Pacific species (in the genus *Toxopneustes*) is venomous, although not deadly to humans; the pedicellariae are unusually large and capable of piercing the skin. Two families.

Order Echinoida

Echinus, Paracentrotus, Echinometra, Strongylocentrotus, Heterocentrotus. Several species of *Strongylocentrotus* have long been used in studies of developmental processes. This order includes the only known suspension-feeding urchin, *Derechinus horridus;* the animal remains stationary on rocks in the subantarctic, and grows only in height, projecting a large surface area into the water for particle capture. Four families, based largely on morphological differences in the pedicellariae.

Order Holectypoida

Echinoneus. This order contains some of the most widely distributed shallow-water tropical echinoid species. One family.

Order Clypeasteroida

Clypeaster—sea biscuits; *Dendraster, Echinarachnius, Encope, Mellita*—sand dollars. These echinoids possess numerous but very short spines with which they burrow into sand and mud substrates. All species deposit-feed and display varying degrees of bilateral symmetry, as in heart urchins. Because of this bilateral rather than radial symmetry, the members of this order, together with those in the order Spatangoida, are often referred to as **irregular urchins.** Nine families.

Order Spatangoida

Brissopsis, Echinocardium, Meoma, Moira, Paraster, Spatangus—heart urchins; *Pourtalesia*—bottle urchins. These deposit-feeding echinoids show a distinct bilateral symmetry, as in sand dollars, and spines and tube feet on different parts of the test are specialized for either burrowing or food collection. As with the other irregular urchins (contained in the order Clypeasteroida), the spines of heart and bottle urchins are very short, with more the appearance of fur than of spines. Fourteen families, including a number of poorly known deep-water groups.

Class Holothuroidea

The sea cucumbers. About 1200 species have been described, distributed among 6 orders.

Order Dendrochirotida

Most of the 400 or so species in this order are restricted to shallow waters, and all possess oral tentacles that are elaborately branched. The tentacles trap food particles from suspension in the surrounding water and push them into the mouth. Seven families.

Family Placothuriidae. *Placothuria.* These are strange cucumbers with many primitive features. The body is U-shaped and enclosed completely in a test of large, overlapping plates. Living representatives are known only off the coast of New Zealand.

Family Psolidae. *Psolus.* This is a widespread group, with representatives extending from the intertidal zone to depths of about 2800 m. As an adaptation for life on hard substrates, they have a distinct, flattened, ventral surface that lacks ossicles. The rest of the body is enveloped by a test of overlapping plates. About 80 species are known.

Family Cucumariidae. *Cucumaria.* The more than 150 described species in this group have thick-walled but flexible bodies that contain numerous small and separate calcareous ossicles. The group is especially well represented in temperate and cold waters, but it is distributed worldwide and at all depths.

Order Aspidochirotida

Holothuria, Stichopus, Parastichopus, Thelenota. Cuvierian tubules are found only among members of this order (some members of the family Holothuriidae). Most species are nonselective deposit feeders. Some are commercially important as a food source in the Far East, being fished and sold as "beche-de-mer" or "trepang"; only the body wall is eaten. Some individuals reach lengths of 1–2 m, with correspondingly impressive diameters, maintaining a cucumber physique. More than a third of the species in this order are restricted to deep waters, and at least a few (such as *Bathyplotes natans*) are good swimmers. Three families.

Order Elasipodida

Psychropotes, Pelagothuria. The hundred or so deposit-feeding species in this group are widespread in deep waters and, in fact, may constitute about 95% of the total biomass in some abyssal habitats. They are remarkable in a number of respects: None have respiratory trees; the bodies are typically gelatinous and extremely fragile; all show pronounced bilateral symmetry; and many species swim, either some or all of the time. Five families.

Order Apodida

Chiridota, Euapta, Leptosynapta. The several hundred species in this order are characterized by the complete absence of tube feet (*a* = G: without; *poda* = G: feet) and respiratory trees. Some representatives also lack body wall ossicles. In all, the body wall is very thin and worm-like. Most species live in shallow water, burrowed in sand or sequestered under rocks. The sticky tentacles, which are not serviced by ampullae, exhibit characteristic feeding movements in which they take turns carefully and deliberately wiping food particles into the mouth. Many species possess minute, attached, internal sacs (vibratile urnae) within the coelom that remove foreign particles from coelomic fluid. Three families.

Order Molpadiida

Molpadia, Caudina. These deposit-feeding cucumbers are mostly confined to shallow waters, although some species live at depths to about 8000 m. All have respiratory trees. Four families.

General References about the Echinoderms

Binyon, G. 1972. *Physiology of Echinoderms.* New York: Pergamon Press.

Blake, D. B., D. A. Janies, and R. Mooi (eds). 2000. Evolution of starfishes: Morphology, molecules, development, and paleobiology. *Amer. Zool.* 40:311–92.

Boardman, R. S., A. H. Cheetham, and A. J. Rowell, eds. 1987. *Fossil Invertebrates.* Palo Alto, Calif.: Blackwell Scientific, 550–611.

Gage, J. D., and P. A. Tyler. 1991. *Deep-Sea Biology: A Natural History of Organisms at the Deep-Sea Floor.* New York: Cambridge University Press.

Harrison, F. W., and F. S. Chia, eds. 1994. *Microscopic Anatomy of Invertebrates, Vol. 14. Echinoderms.* New York: Wiley-Liss.

Hendler, G., J. E. Miller, D. L. Pawson, and P. M. Kier. 1995. *Sea Stars, Sea Urchins, and Allies: Echinoderms of Florida and the Caribbean.* Washington, D.C.: Smithsonian Institution Press.

Hyman, L. H. 1955. *The Invertebrates, Vol. IV: Echinoderms.* New York: McGraw-Hill.

Janies, D. 2001. Phylogenetic relationships of extant echinoderm classes. *Can. J. Zool.* 79:1232–50.

Lawrence, J. M. 1987. *A Functional Biology of Echinoderms.* London: Croom Helm.

Levin, H. L. 1999. *Ancient Invertebrates and Their Living Relatives.* New Jersey: Prentice Hall, 294–328.

Motokawa, T. 1984. Connective tissue catch in echinoderms. *Biol. Rev.* 59:255–70.

Nichols, D. 1962. *The Echinoderms.* London: Hutchinson University Library.

Wray, G. A. 1997. Echinoderms. In: Gilbert, S. F., and A. M. Raunio. *Embryology: Constructing the Organism.* Sunderland, MA: Sinauer Associates, Inc. Publishers, 309–29. Chapter 20.

Search the Web

1. www.ucmp.berkeley.edu/echinodermata/echinodermata.html

   Click on "Systematics" for access to information about individual echinoderm classes, with accompanying images. This site is maintained through the University of California at Berkeley.

2. http://www.tolweb.org/Echinodermata

   This is the Tree of Life site. Note the separate, very detailed subsections concerning the crinoids and holothurians. Written by Gregory Wray.

3. http://www.calacademy.org/research/izg/echinoderm/classify.htm

   This site presents a complete classification of the extant Echinodermata, and includes links to images for representatives of some families.

4. www.mbl.edu

   This site is operated by the Marine Biological Laboratory at Woods Hole, MA. Under "Publications, Databases" select "Biological Bulletin" and then "Other Biological Bulletin Publications." Then choose "Keys to Marine Invertebrates of the Woods Hole Region" and select "Echinodermata." In addition to providing an identification guide to the echinoderms encountered near Woods Hole, MA, the site also includes an annotated listing of the echinoderm species found in the area.

   Websites concerning fossilized echinoderms:

5. http://www.museum.vic.gov.au/Infosheets/10140.pdf

   This information sheet, with photos, is provided by the Museum of Victoria in Australia.

6. http://www.ucmp.berkeley.edu/echinodermata/blastoidea.html

   This site introduces an extinct group of fossilized echinoderms, the Blastoidea, courtesy of University of California at Berkeley.

7. http://www.nearctica.com/paleo/inverts/echino.htm

8. http://www.fossilcrinoid.com/

9. http://www.uky.edu/KGS/fossils/echinos.htm

   This site is provided by the Kentucky Geological Survey at the University of Kentucky.

21

The Hemichordates

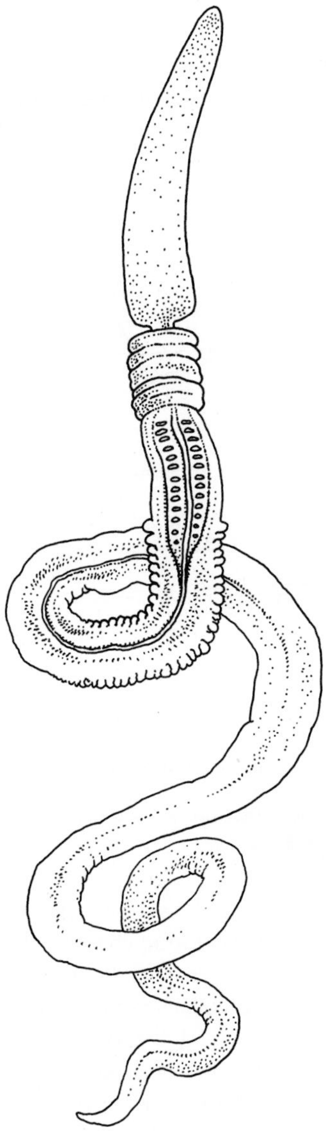

Introduction and General Characteristics

Phylum Hemi • chordata
(Greek: one-half string)
hem-ē-kor-dah´-tah

Defining Characteristic:[1] A conspicuous dorsal extension of the pharynx forms an anterior buccal tube, or stomochord

The hemichordates are a small group of marine worms with an intriguing relationship with both the Echinodermata and our own phylum, the Chordata (see Fig. 23.10). As typical deuterostomes, cleavage is radial and the coelom forms through enterocoely, as 3 distinct compartments (protocoel, mesocoel, and metacoel). Although hemichordates lack a notochord, excluding them from membership in the phylum Chordata, they do exhibit 2 other chordate characteristics: pharyngeal gill slits and, in some species, a dorsal, hollow nerve cord. Although some analyses have linked hemichordates to vertebrate chordates or to brachiopods and other "lophophorates," most of the recent evidence from 18S and 28S rDNA (ribosomal DNA) sequence analyses, certain patterns of tissue specific gene expression, and mitochondrial codon characteristics all suggest a closer link between hemichordates and echinoderms than between hemichordates and chordates. Indeed, hemichordates and echinoderms are now believed to be sister groups, forming the new taxon Ambulacraria within the Deuterostomia. Further supporting this proposition, the nerve cells of at least one hemichordate species react positively to antibodies raised against certain echinoderm neuropeptides;[2] such

1. Characteristics distinguishing the members of this phylum from members of other phyla.

2. Stach, T., S. Dupont, O. Israelson, G. Fauville, J. Nakano, T. Kanneby, and M. Thorndyke. 2005. *J. Mar. Biol. Ass. U.K.* 85: 1519–24.

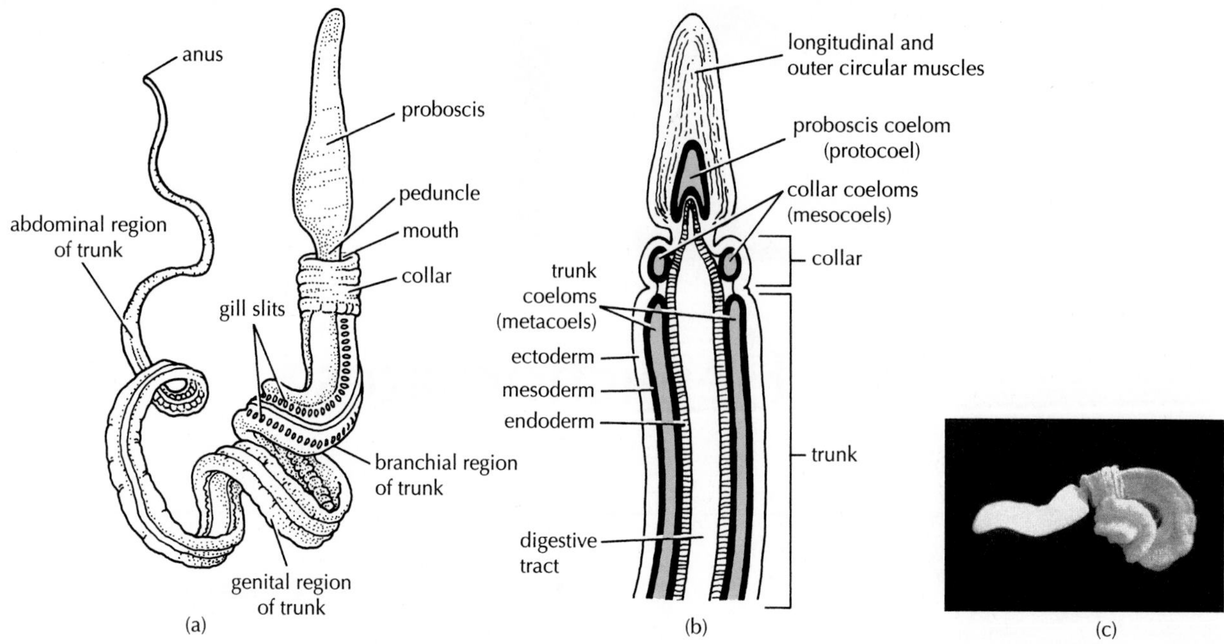

Figure 21.1

(a) The acorn worm, *Saccoglossus kowalevskii*, removed from its burrow to show the basic body structure. (b) Schematic illustration of the coelomic compartments in an enteropneust. (c) Individual of *S. kowalevskii* collected intertidally from Nahant, MA.
Photo by J. Pechenik.

positive immunoreactivity with these peptides is otherwise known only from the ciliated worm *Xenoturbella bocki* (see Chapter 22). Most of the approximately 100 hemichordate species are contained within a single class, the Enteropneusta.

Class Enteropneusta

Class Entero • pneusta
(G: gut breathing)
en-ter-op-nū´-stah

The enteropneusts are common inhabitants of shallow water, with most species forming mucus-lined burrows in sandy or muddy sediment. The body is long and narrow, and it is divided into 3 distinct regions corresponding to the tripartite compartmentation of the coelom (Fig. 21.1). The anteriormost section, called the **proboscis,** houses a single coelomic chamber, the **protocoel** (*proto* = G: first; *coel* = G: cavity). This anterior section of the body is generally conical in shape and has given rise to the common name for enteropneusts: "acorn worms." The proboscis is highly muscular and has major responsibility for burrowing and food collection. Indeed, the proboscis is the only truly active body part; enteropneusts are sedentary animals, rarely moving from place to place as adults. Locomotion is largely restricted to movements within burrows (Fig. 21.2).

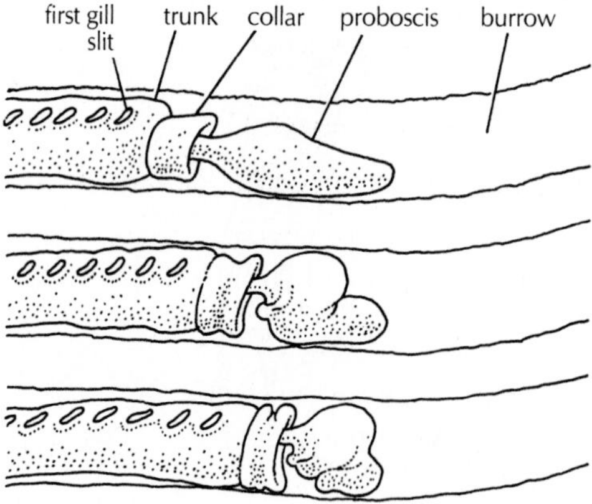

Figure 21.2

Use of the proboscis in locomotion within the burrow by an acorn worm, *Saccoglossus horstii.*
Adapted from R. B. Clark, 1964. *Dynamics in Metazoan Evolution.* New York: Oxford University Press. After Burdon-Jones.

The proboscis is followed by a narrow **collar** region, containing a pair of coelomic chambers derived from the embryonic mesocoel (*meso* = G: middle). The enteropneust mouth opens on the collar's ventral anterior surface. The bulk of the body, termed the **trunk,** contains a pair of coelomic compartments derived from the embryonic metacoel. External ciliation of the trunk may aid locomotion within burrows. The total body length is typically about 8–45 cm, although the members of one South American enteropneust species attain lengths of up to 2.5 m!

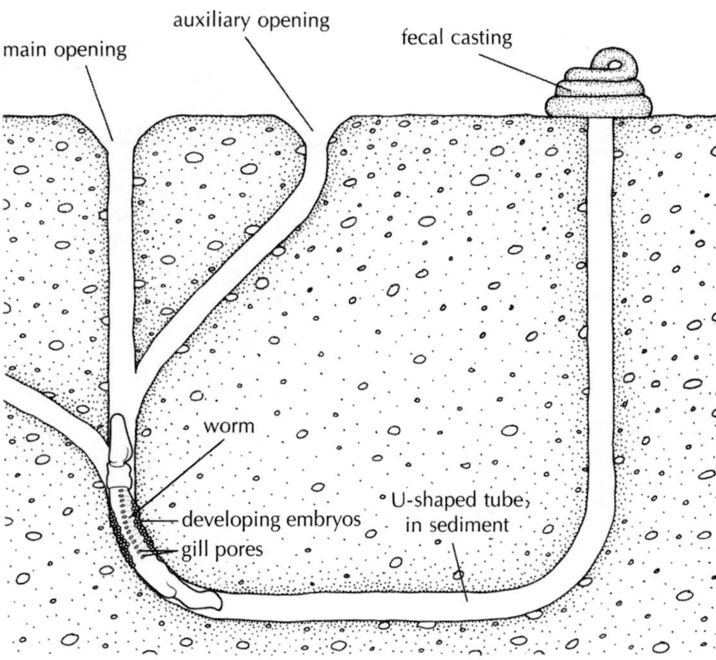

Figure 21.3
The U-shaped burrow of a typical enteropneust. Note the developing embryos in a cylinder around the animal's body, and the coil of fecal material that has been deposited at the burrow's posterior end.

Modified from Hyman; after Stiasny and after Burdon-Jones.

Many enteropneust species are **deposit feeders,** ingesting sediment, extracting the organic constituents, and extruding a coil (**casting**) of mucus-bound, organically deprived sediment from the anus (Fig. 21.3). Other species are **mucociliary feeders:** planktonic organisms and detritus adhere to mucus on the proboscis and are then conveyed along ciliated tracts to the mouth. At least one species is a true **suspension-feeder,** filtering and manipulating food particles from the incoming water using pharyngeal cilia (see below).

The enteropneust digestive tract is tubular, consisting of a mouth (located behind the proboscis, on the collar—see Figs. 21.1 and 21.4) leading into an esophagus (which compacts ingested particles into a mucus-bound rope), a pharynx, an intestine (the main site of digestion and absorption), and a terminal anus. Food is moved through the gut primarily by the action of ciliated cells lining the inner wall of the digestive tract.

An anterior extension of the pharynx forms a **buccal tube,** or **stomochord** (G: gut chord), within the collar of the animal (Figs. 21.1b and 21.4a), possibly serving to support the proboscis. For some time, this tube was thought to be a notochord, and the hemichordates were thus included within our own phylum, the Chordata. Although the enteropneust buccal tube and the chordate notochord are now generally believed to have had independent origins, current information does not rule out the possibility that the 2 structures are indeed homologous.

The pharynx opens to the outside through a series of lateral, paired gill slits (more than a hundred in some species). Cilia lining these gill slits beat in coordinated fashion so that water is drawn in at the mouth and discharged through the slits (Fig. 21.4). This flow of water is believed to serve for gas exchange, and the meaning of the name "enteropneust" then becomes apparent: The animals essentially breathe through a portion of their gut.

Acorn worms possess a true blood circulatory system. The blood, which lacks pigment, is circulated through dorsal and ventral blood vessels and associated blood sinuses by pulsations of the muscular blood vessels themselves; a distinct heart is never found. Evaginations of the blood sinuses are pronounced in the pharynx so that the gill slits are highly vascularized, in keeping with their likely function in respiration.

The enteropneust nervous system is echinoderm-like in that much of it is in the form of an epidermal nerve network. However, although there is clearly no brain, the nerve network is consolidated in some body regions to form longitudinal nerve cords; a midventral and a middorsal nerve cord are particularly well developed. Only the dorsal nerve cord extends into the collar region; moreover, this dorsal, collar nerve cord lies well below the epidermis and, in some species, is hollow (Fig. 21.4a). It might be homologous with the dorsal, hollow nerve cord of vertebrates and other chordates.

Acorn worms are **gonochoristic** (sexes are separate), with the gonads housed in the trunk, and they fertilize their eggs externally in the seawater. A free-living larval form is found among a number of enteropneust species (Fig. 21.5). This planktonic, **tornaria** larva is equipped with a series of sinuous, ciliated bands reminiscent of those encountered among the echinoderms.

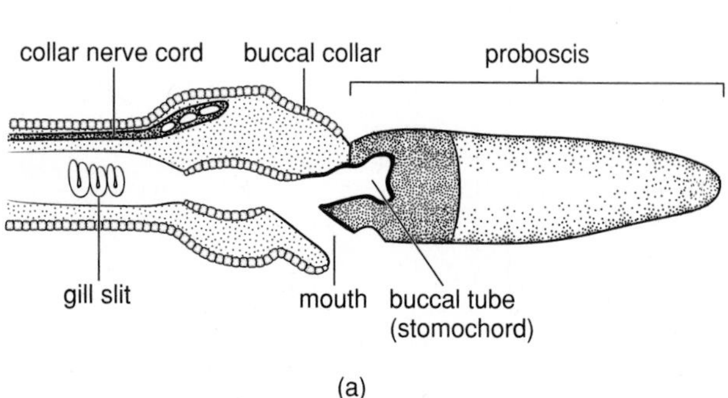

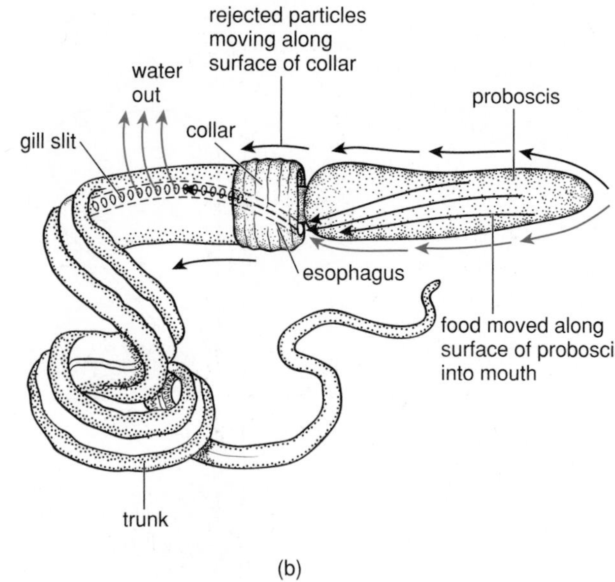

Figure 21.4

(a) Longitudinal section through anterior portion of an enteropneust, showing internal features. The proboscis is drawn intact (not sectioned). (b) Diagrammatic illustration of the respiratory and feeding pathways of a typical enteropneust. Fine food particles are trapped in mucus and transferred by cilia along the proboscis to the mouth, which is located in the collar; rejected particles are moved posteriorly over the collar surface. Ciliary activity draws water in at the mouth, through the pharynx, and out at the gill slits. The general body surface also plays a major role in gas exchange.

Based on George C. Kent, *Comparative Anatomy of the Vertebrates,* 6th ed. 1987. (b) After Russell-Hunter.

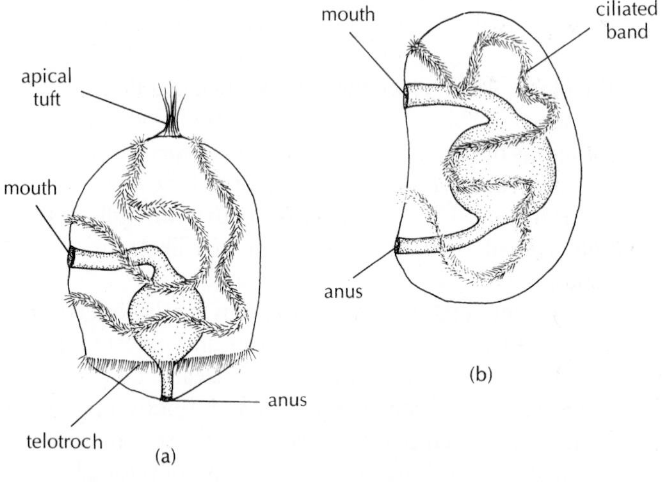

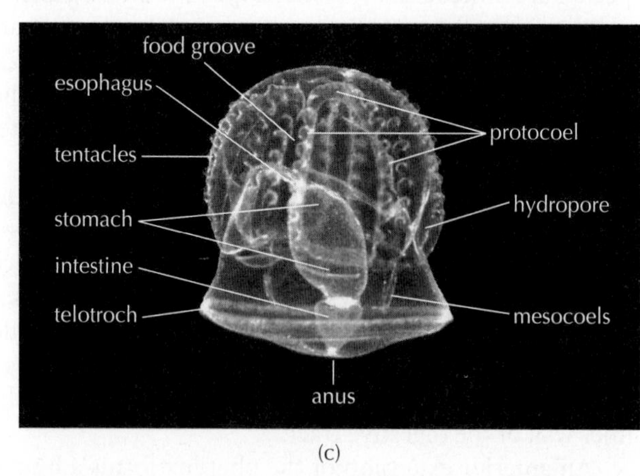

Figure 21.5

(a) The tornaria larva of *Balanoglossus* sp. The tornaria bears a conspicuous terminal ring of cilia (the telotroch), which is lacking in the auricularia larva of holothurian echinoderms (b). In other respects, ciliation patterns for larvae of the two groups are similar. (c) Tornaria larva of the hemichordate *Ptychodera flava,* in side view. The larva is approximately 4 months old and 3 mm long. Note that the ciliary bands (other than the telotroch) are outfolded to form complex "tentacles," a feature apparently unique to larvae in this genus.

(a,b) Adapted from Hardy, 1965. *The Open Sea: Its Natural History.* Boston: Houghton Mifflin Company. (c) Courtesy of Kevin Peterson. From Peterson, K. J., R. A. Cameron, J. Tagawa, N. Satoh, and E. H. Davidson. 1999. *Development* 126:85–95.

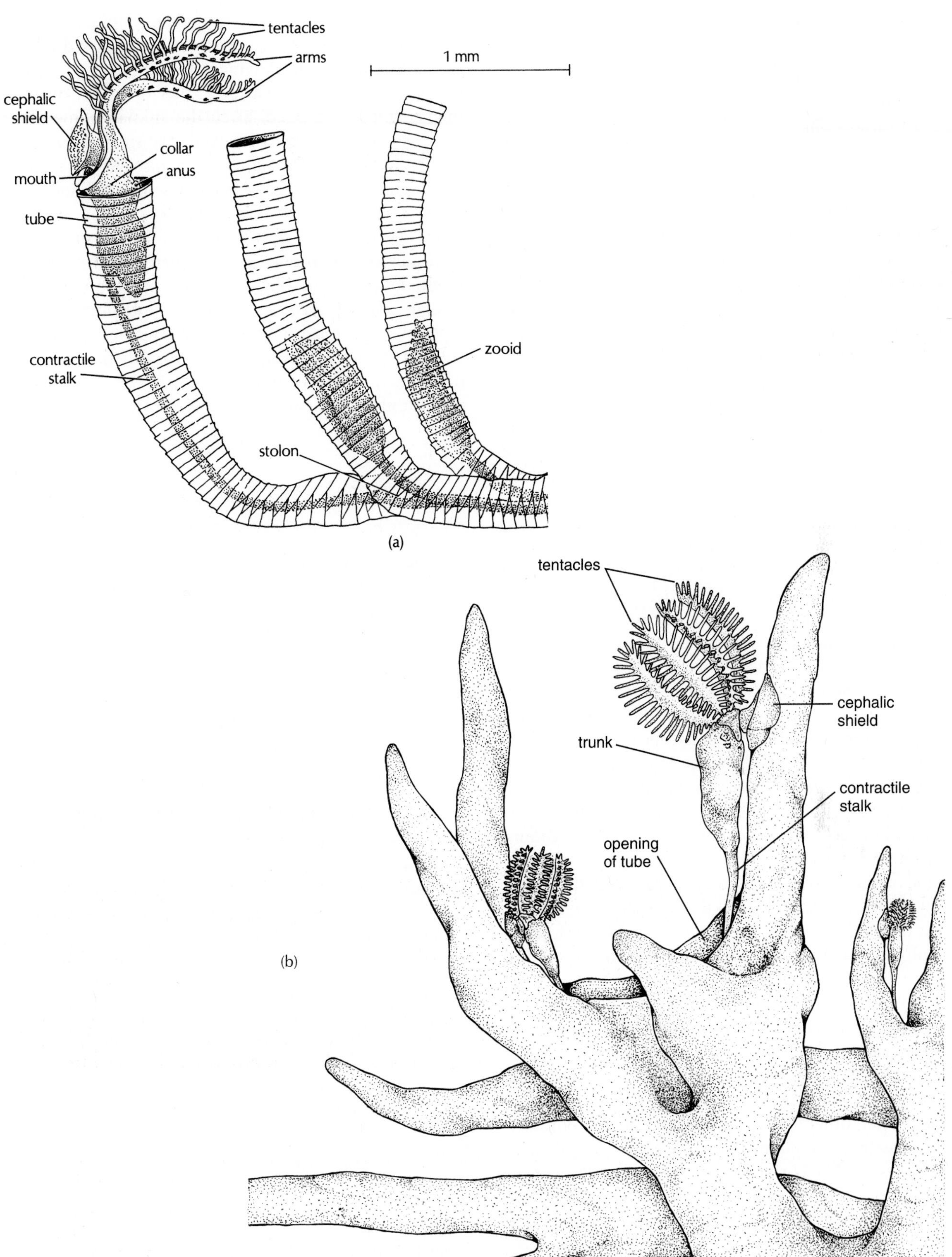

Figure 21.6

(a) A colony of the pterobranch *Rhabdopleura* sp. (b) Colony of *Cephalodiscus* sp., showing several individuals climbing about on their elaborate tubes. Each zooid (excluding the stalk) is only about 2 mm long.

(b) Based on S. M. Lester in *Marine Biology,* 85:263–68, 1985.

Class Pterobranchia

Class Ptero • branchia
[G: feather (or wing) gill]
ter-ō-brank´-ē-ah

A small number of hemichordates (about 25 species) are markedly dissimilar from the enteropneusts and are placed in a separate class, the Pterobranchia. Pterobranchs have been collected primarily by dredging in fairly deep water, especially in the Antarctic; until recently, they were considered inaccessible to the average invertebrate zoologist. However, over the past 25 years or so, several shallow-water populations have been found, particularly in the clear and inviting waters around Bermuda.

As with enteropneusts, pterobranchs exhibit a dorsal extension of the pharynx, forming an anterior stomochord. And as in enteropneusts, the pterobranch stomochord probably serves to support the surrounding tissue. However, unlike the enteropneusts, pterobranchs possess ciliated, anterior tentacles and a U-shaped gut. No pterobranch has more than one pair of pharyngeal gill slits, and some species have none. Also unlike enteropneusts, members of most pterobranch species occupy rigid tubes that they secrete (Fig. 21.6). Tube secretions are produced by glands on an anterior **cephalic shield** (Fig. 21.6). The cephalic shield also serves as an attachment organ and can be used as a muscular proboscis to crawl within the tube or even on solid substrates adjacent to the tube; at the slightest provocation, the animal withdraws into its tube with a rapid contraction of its muscular, highly extensible stalk (Fig. 21.6).

As illustrated in Figure 21.6, nearly all pterobranch species are colonial. Each individual, called a **zooid,** is only a few millimeters long, and colonies rarely exceed 20 mm in diameter. The zooids within one colony are produced by asexual budding, so all are descended from a single larva.

Current thinking suggests that pterobranchs may be direct descendants of an ancient, but now extinct group of colonial, often planktonic, marine animals called "graptolites."[3] Graptolites are very abundant as fossils; they flourished in the oceans about 300 to 500 million years ago. Some zoologists view pterobranchs, in turn, as likely ancestors of the enteropneusts just discussed. However, some recent 18S rDNA data imply the reverse: that pterobranchs have evolved from an enteropneust-like ancestor. Yet other biologists would link pterobranchs more closely to "lophophorates" (Chapter 19) than to enteropneusts. The U-shaped gut and the structure of the anterior region of the pterobranch body—complete with ciliated tentacles containing extensions of the mesocoel and surrounding the mouth but not the anus—does seem remarkably phoronid-like (p. 475). In addition, the pattern of ciliation, path of water flow, mechanism of particle capture, and mechanism of particle rejection are decidedly lophophore-like.[4] The enigmatic evolutionary position of pterobranchs has focused far greater attention on them than might be expected, given the animals' dimensions, their general inaccessibility, and the small number of species in the class. The rather recent discovery of shallow-water pterobranch species increases the likelihood that their evolutionary position will be clarified in time. Molecular data (18S rDNA analyses) to date support the inclusion of enteropneusts and pterobranchs within a single deuterostome phylum. If so, then the remarkably similar ciliated feeding organs of pterobranchs and lophophorates must have arisen independently in different ancestors, by convergence, an hypothesis made even more likely by recent indications that lophophorates are protostomes, not deuterostomes (Chapter 19).

3. See *Topics for Further Discussion and Investigation,* no. 2, at the end of the chapter.

4. See *Topics for Further Discussion and Investigation,* no. 4.

Taxonomic Summary

Phylum Hemichordata
 Class Enteropneusta—the acorn worms
 Class Pterobranchia

Topics for Further Discussion and Investigation

1. What morphological features of pterobranchs ally these animals with the enteropneusts?

Barrington, E. J. W. 1965. *The Biology of the Hemichordata and Protochordata.* Edinburgh, Scotland: Oliver & Boyd.

Hyman, L. H. 1959. *The Invertebrates. Vol. V, Smaller Coelomate Groups.* New York: McGraw-Hill.

Nielsen, C. 1995. *Animal Evolution: Interrelationships of the Living Phyla.* New York: Oxford University Press, 359–65, 387–95.

2. Discuss the evidence for and against a likely relationship between living pterobranchs and the extinct graptolites. What additional sorts of information might convince you one way or the other?

Bates, D. E. B., and N. H. Kirk. 1985. Graptolites, a fossil case-history of evolution from sessile, colonial animals to automobile superindividuals. *Proc. Royal Soc. London B* 228:207.

Dilly, P. N. 1988. Tube building by *Cephalodiscus gracilis. J. Zool., London* 216:465.

Levin, H. L. 1999. *Ancient Invertebrates and Their Living Relatives.* New Jersey: Prentice Hall, 329–43.

3. Some time ago, a separate class (the Planctosphaeroidea) was established to hold a single species of what was thought to be a planktonic hemichordate. It was subsequently determined that these organisms were unusual larval stages of a benthic adult. How would you determine whether a newly discovered animal was an adult or a larva? If the animal was a larval form, how could you determine the species to which it belonged?

4. Compare and contrast the feeding biology of pterobranchs and "lophophorates."

Halanych, K. M. 1993. Suspension feeding by the lophophore-like apparatus of the pterobranch hemichordate *Rhabdopleura normani*. *Biol. Bull.* 185:417.

Strathmann, R. 1973. Function of lateral cilia in suspension feeding of lophophorates (Brachiopoda, Phoronida, Ectoprocta). *Marine Biol.* 23:129.

Taxonomic Detail

Phylum Hemichordata

The approximately 100 species are distributed among 3 classes.

Class Enteropneusta

Balanoglossus, Ptychodera, Saccoglossus—acorn worms. *Balanoglossus gigas* can grow to a length of about 1.5 m. The approximately 70 species are distributed among 4 families and 15 genera.

Class Pterobranchia

Cephalodiscus, Rhabdopleura. The 25 species are distributed among 3 families.

Class Planctosphaeroidea

Planctosphaera pelagica. This animal—the only member of its class—is so far known only from its larval stage, which is large, gelatinous, and well-ciliated. If its adult identity is ever discovered, the class might be dissolved.

General References about the Hemichordates

Barrington, E. J. W. 1965. *The Biology of Hemichordata and Protochordata.* San Francisco: W. H. Freeman.

Bullock, T. H. 1946. The anatomical organization of the nervous system of the Enteropneusta. *Q. J. Microsc. Sci.* 86:55–111.

Cameron, C. B. 2005. A phylogeny of the hemichordates based on morphological characters. *Canadian J. Zool.* 83:196–215.

Cameron, C. B., J. R. Garey, and B. J. Swalla. 2000. Evolution of the chordate body plan: New insights from phylogenetic analyses of deuterostome phyla. *Proc. Natl. Acad. Sci.*

Halanych, K. M. 1996. Convergence in the feeding apparatuses of lophophorates and pterobranch hemichordates revealed by 18S rDNA: An interpretation. *Biol. Bull.* 190:1–5.

Harrison, F. W., and E. E. Ruppert, eds. 1996. *Microscopic Anatomy of Invertebrates, Vol. 15. Hemichordata, Chaetognatha, and the Invertebrate Chordates.* New York: Wiley-Liss.

Hyman, L. H. 1959. *The Invertebrates, Vol. 5: Smaller Coelomate Groups.* New York: McGraw-Hill.

Lambert, G. 2005. Ecology and natural history of the protochordates. *Canadian J. Zool.* 83:34–50.

Ruppert, E. E. 2005. Key characters uniting hemichordates and chordates: Homologies or homoplasies? *Canadian J. Zool.* 83:8–23.

Search the Web

1. www.mbl.edu

   Choose "Biological Bulletin" under "Quick Links," and then "Other Biological Bulletin Publications." Next select "Keys to Marine Invertebrates of the Woods Hole Region," and then choose "Protochordates." This article includes a description of the one hemichordate species (*Saccoglossus kowalevskii*) found near Woods Hole, MA.

2. http://www.ucmp.berkeley.edu/chordata/hemichordata.html

   This site provides an excellent introduction to all aspects of hemichordate biology and systematics.

3. http://tolweb.org/tree/phylogeny.html

   This brings you to the Tree of Life site. Search on "Hemichordata."

4. http://biodidac.bio.uottawa.ca

   Choose "Organismal Biology," "Animalia," and then "Hemichordata" for photographs and drawings.

5. https://www.webdepot.umontreal.ca/Usagers/cameroc/MonDepotPublic/Cameron/

   Chris Cameron's webpage (University of Montreal) contains useful information and color images of hemichordates.

6. http://faculty.washington.edu/bjswalla/Hemichordata/hemichordata.html

   More excellent images of hemichordates, from Billie J. Swalla (University of Washington).

22

The Xenoturbellids: Deuterostomes at Last?

Phylum Xenoturbellida

zē´-nō-ter-bell´-ih-dah
(*xeno* = G: strange)

> "*Xenoturbella* is merely a ciliated bag with epithelial epidermis and gastrodermis, a basiepidermal nerve plexus and a ventral mouth, but without an anus or any distinct organs."
> O. Israelsson and G. E. Budd (2005)

The idea that a small ciliated worm with few distinctive characteristics might have a phylum unto itself is a strange one, but it seems to be the case. A single species, *Xenoturbella bocki*, was first described in 1949 as an acoel flatworm (Chapter 8) and later claimed as either an early metazoan offshoot or a primitive deuterostome—possibly a neotenous hemichordate (Chapter 21).

It was recently affiliated with primitive bivalved molluscs, based upon a study of gamete development (oogenesis) and an analysis of sequence data from both 18S rRNA and mitochondrial genes. It turns out, however, that the bivalve gene sequences found were contaminants—*Xenoturbella* eats bivalves and bivalve embryos![1] New gene sequence data and new analyses, with the bivalve sequences ignored, mark the animals as deuterostomes and place them in a separate phylum as close relatives of echinoderms and hemichordates (Fig. 22.1). Supporting this proposition, an antibody raised against a particular neuropeptide (SALMFamide-2), known previously only from echinoderms, also reacts with tissues of both *Xenoturbella* and hemichordates and, apparently, with nothing else. A second species, *X. westbladi*, was described in 1999. Thus, we now have a new phylum of

1. Bourlat, S. J., H. Nakano, M. Åkerman, M. J. Telford, M. C. Thorndyke, and M. Obst. 2007. Feeding ecology of *Xenoturbella bocki* (phylum Xenoturbellida) revealed by genetic barcoding. *Molecular Ecology Notes* (Published online: 17-Aug-2007).

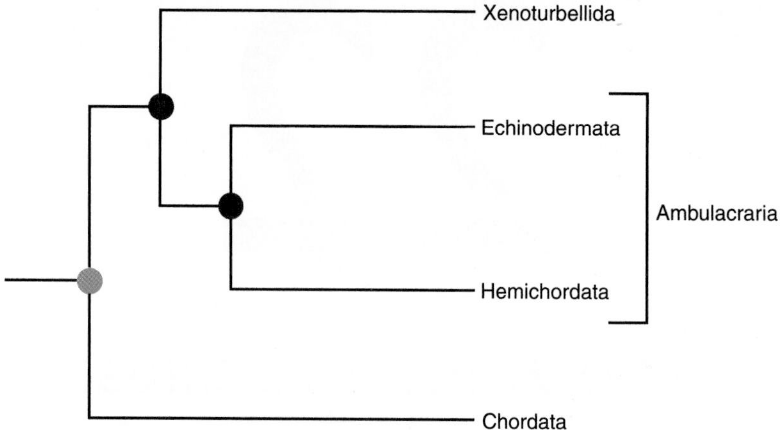

Figure 22.1
Apparent relationships among the 4 deuterostome phyla. The phylum Xenoturbellida was established in 2006. The node shaded blue represents the ancestral deuterostome. Modifed from several sources.

unlikely deuterostomes—the Xenoturbellida—containing only 2 species in a single genus. Oddly (for deuterostomes), individuals of *X. bocki* share with acoels and certain other flatworms (members of the Nemertodermatida, p. 173) some very distinctive features of their ciliary apparatus, and also a very peculiar tendency to withdraw and resorb worn epidermal cells, so that some workers still suspect a close connection to flatworms.

The animals are up to 4 cm long, **vermiform** (worm-shaped), and covered by locomotory cilia. They have no digestive tract, and indeed no organs at all. Their only conspicuous morphological feature, other than their cilia, is a statocyst for determining orientation. To date, they have been collected only off the coasts of Sweden and Scotland, in sediments at depths of 20–100 m.

Little is known about the reproductive biology of these animals, other than that the eggs and embryos are individually enclosed in double-layered follicles that float free in the space between the musculature and gastrodermis; there is no distinct ovary.[2]

Placement of *Xenoturbella* as a phylum in its own right and as sister group to echinoderms and hemichordates is likely to stimulate more research into the biology and development of this enigmatic and little-studied animal. Such research has the potential to tell us interesting things about deuterostome ancestry and evolution. Still, this promises to remain the shortest chapter in the book for a long time to come, should it in fact remain a separate chapter at all in future editions.

2. Israelsson, O., and G. E. Budd. 2005. Eggs and embryos in *Xenoturbella* (phylum uncertain) are not ingested prey. *Devel. Genes Evol.* 215: 358–63.

Topic for Further Discussion and Investigation

Members of the genus *Xenoturbella* are ciliated worms with a number of protostome characteristics. Discuss the evidence that they are in fact deuterostomes closely allied with echinoderms and hemichordates rather than to flatworms or other protostomes.

Bourlat, S. J., T. Juliusdottir, C. J. Lowe, R. Freeman, J. Aronowicz, M. Kirschner, E. S. Lander, M. Thorndyke, H. Nakano, A. B. Kohn, A. Heyland, L. L. Moroz, R. R. Copley, and M. J. Telford. 2006. Deuterostome phylogeny reveals monophyletic chordates and the new phylum Xenoturbellida. *Nature* 444: 85–87.

Bourlat, S. J., C. Nielsen, A. E. Lockyer, D.T. J. Littlewood, and M. J. Telford. 2003. *Xenoturbella* is a deuterostome that eats molluscs. *Nature* 424: 925–28.

Israelsson, O. 2006. Observations on some unusual cell types in the enigmatic worm *Xenoturbella* (phylum uncertain). *Tissue Cell* 38: 233–42.

Lundin, K. 2001. Degenerating epidermal cells in *Xenoturbella bocki* (phylum uncertain), Nemertodermatida and Acoela (Platyhelminthes). *Belg. J. Zool.* 131 (supplement 1): 153–57.

Stach, T., S. Dupont, O. Israelsson, G. Fauville, J. Nakano, T. Kanneby, and M. Thorndyke. 2005. Nerve cells of *Xenoturbella bocki* (phylum uncertain) and *Harrimania kupfferi* (Enteropneusta) are positively immunoreactive to antibodies raised against echinoderm neuropeptides. *J. Mar. Biol. Ass. U.K.* 85: 1519–24.

23

The Nonvertebrate Chordates

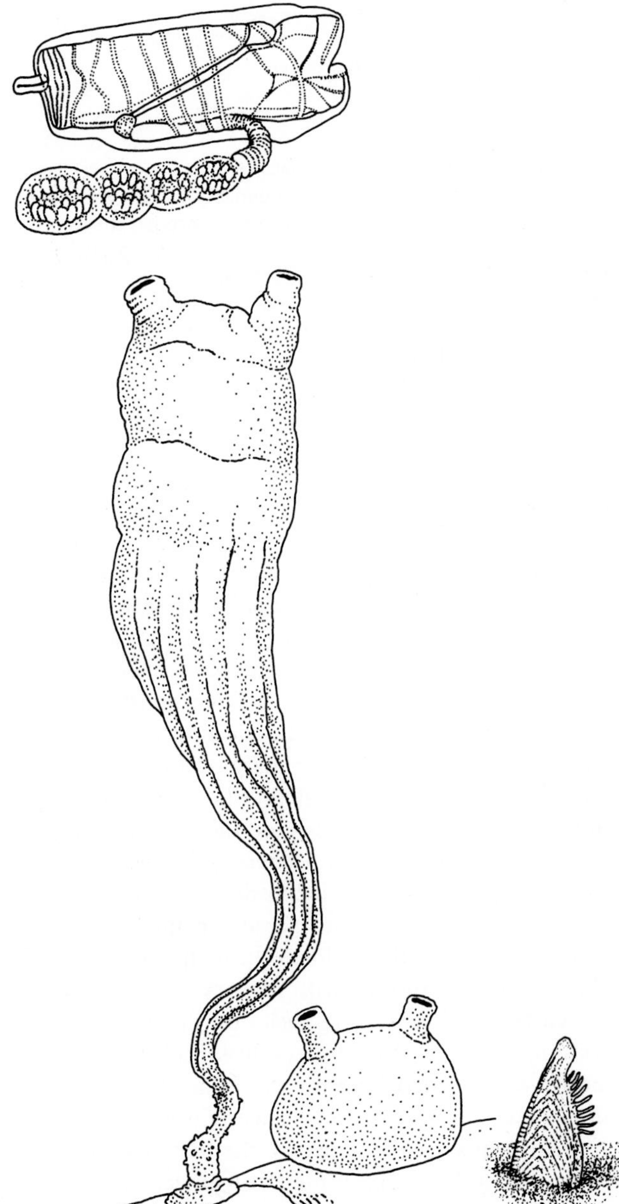

Introduction and General Characteristics

Phylum Chordata
(Greek: string)
kor-dah´-tah

Defining Characteristics:[1] 1) Nerve cord is dorsal and hollow; 2) the body is supported in at least some stage of development by a stiff rod (the notochord) formed from the roof of the archenteron and running the length of the animal ventral to the nerve cord, from anterior to posterior; 3) the pharynx is perforated with numerous ciliated slits (stigmata); 4) the body ends with a tail extending beyond the anus (a "post-anal" tail)

In addition to the 48,000 primates, cats, dogs, birds, and other familiar species contained in the subphylum Vertebrata, the phylum Chordata includes some 2200 invertebrate species in the subphyla Tunicata (=Urochordata) and Cephalochordata. The nonvertebrate chordates are all marine. In common with their vertebrate relatives, the invertebrate chordates generally show, at some point in their life histories, the following characteristics: pharyngeal gill slits; a hollow, dorsal nerve cord; a notochord; and a post-anal tail. The **notochord** (G: back string) consists of a linear series of cells, each of which contains a large, fluid-filled vacuole. In the vertebrates, the notochord is eventually surrounded by cartilaginous tissue or by calcareous vertebrae during **ontogeny** (i.e., development). About 90% of the invertebrate chordate species are contained within the subphylum Tunicata (=Urochordata), while the much smaller subphylum, the Cephalochordata, probably contains our nearest invertebrate relatives.

1. Characteristics distinguishing the members of this phylum from members of other phyla.

Subphylum Tunicata (= Urochordata)

Subphylum Tunic • ata
(L: tunic-bearing)
toon´-ih-cah´-tah

Subphylum Uro • chordata
(G: tail string)
ūr´-ō-kor-dah´-tah

Defining Characteristic: Notochord and nerve cord are found only in the larval stage, being reabsorbed at metamorphosis

The Tunicata is one of the few major taxonomic groups to contain no parasitic species. Members of this subphylum are commonly referred to as "tunicates," for reasons soon to become apparent. Most tunicates (also called urochordates) feed by straining small particles (especially phytoplankton) from the surrounding seawater, although some deep-sea ascidian species (see following paragraphs) are carnivorous. The method of generating the water current from which these food particles are obtained differs dramatically among the different urochordate classes. Tunicates are distributed among 3 major classes (Ascidiacea, Larvacea, and Thaliacea) and one small class of deep-water carnivores, the Sorberacea (see *Taxonomic Detail,* pp. 552–553). None of these animals have left a fossil record.

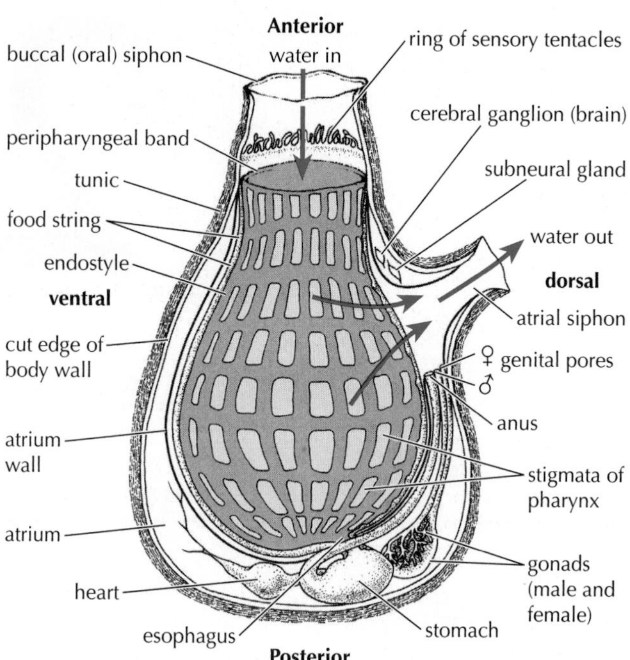

Figure 23.1
Diagrammatic illustration of the anatomy of a typical ascidian. A thin sheet of mucus, produced by the endostyle, coats the basket-like pharynx. All incoming water must pass through this mucous coating before reaching the atrium, so that food particles are strained out. Cilia move the sheet dorsally, where it compacts to form a discrete mucus-food string, ready for ingestion and digestion. The water current is generated by the action of ciliated cells along the margins of the stigmata.
Modified after Bullough and other sources.

Class Ascidiacea

Class Ascidia • cea
G: a little bag
a-sid-ē-ā´-sē-ah

The class Ascidiacea contains over 90% of all described tunicate species; its members are found throughout the world's oceans, in both shallow and deep water. Most adult ascidians, commonly known as "sea squirts," live attached to solid substrates (including boat hulls—greatly reducing fuel efficiency—and aquaculture cages, greatly reducing water flow for the animals being cultivated). Some species live anchored in soft sediments. With few exceptions, ascidian adults are **sessile** (i.e., incapable of locomotion). The body is bag-like and covered by a secretion of the epidermal cells (Fig. 23.1). This secreted, protective **test,** also called the **tunic,** is composed largely of protein and a polysaccharide (tunicin) that closely resembles cellulose. Amoeboid cells, blood cells, and in some species, blood vessels are found within the tunic. In some species, the tunic also bears numerous calcareous spines, hairs, or spicules (Fig. 23.2). Although lacking true nerves and muscles, the tunic of a few species contains elongated, multipolar cells capable of both nervous conduction and contraction. Tunic coloration varies from nearly transparent in some species to dramatically pigmented in others; reds, browns, greens, and yellows are common.

The ascidian digestive tract is especially noteworthy in that the pharynx is unusually wide and is perforated with numerous ciliated slits called **stigmata;** stigmata morphology differs substantially among species and is therefore an important taxonomic characteristic. The perforated pharynx thus forms a large **pharyngeal** (or **branchial**) **basket** (Fig. 23.1). In addition to the ciliated pharyngeal stigmata, the outer margin of the pharynx is lined by a double ring of ciliated cells—the **peripharyngeal band** (*peri* = G: around)—and the general surface of the pharynx is also ciliated; the lateral cilia create a flow of water through the ascidian for feeding and gas exchange and, in many species, for the release of gametes or larvae. Water is drawn into the ascidian through a **buccal (oral) siphon,** which is studded with sensory receptors on the outer surface and sensory tentacles on the inner surface. Incoming water is also apparently sampled by a **subneural gland** located beneath the brain; the subneural gland's precise function is still uncertain. In some deep-sea, carnivorous ascidian species, the buccal siphon can be expanded to form lips, for ingesting macroscopic prey.

Extending along the ventral surface of the pharynx is an elongated gland called the **endostyle** (Fig. 23.1). This

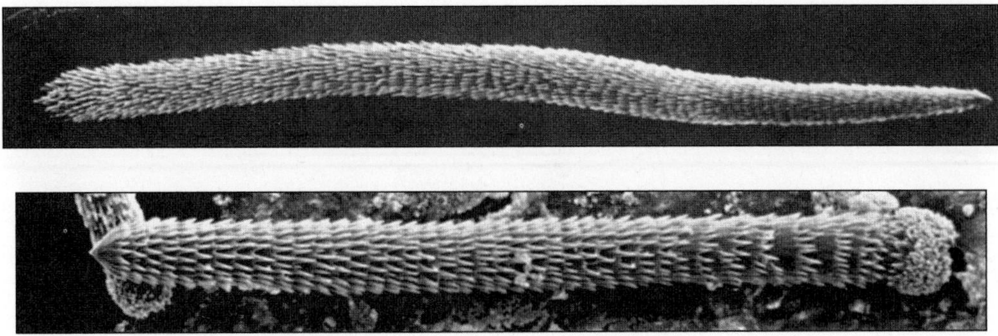

Figure 23.2
Calcareous spicules isolated from the test of the solitary, warm-water ascidian *Herdmania momus*. The spicules were soaked in a mild base to remove adhering tissue. Each spicule bears at least 100 rows of overlapping spines, as illustrated.

From G. Lambert and C. C. Lambert, *Journal of Morphology* 192:145–59, figs. 9a & b.

endostyle is generally thought to be homologous with the vertebrate thyroid gland, whose iodine-rich secretions regulate important aspects of development and metabolism. In all tunicates, the endostyle secretes a thin net of iodine-containing mucus that is then drawn across the pharynx by cilia; water passing through the stigmata must pass also through this mucous net. Food particles as small as approximately 1 μm (micrometer) thus can be filtered from the inflowing water. Frontal cilia on the pharyngeal basket continually move the particle-laden net of mucus dorsally (away from the endostyle) to form a mucus-food string. The string is then moved posteriorly (away from the siphons) and enters the esophagus and then the stomach for digestion. Digestion is extracellular, and nutrients are absorbed in the intestine. Food is moved through the gut by cilia lining the digestive organs, and solid wastes are discharged through the anus.

The water that has been relieved of its particles (and oxygen) during passage through the stigmata enters a saclike chamber, termed the **atrium,** which is enclosed by the tunic (Fig. 23.1). Water leaves the atrium through an **atrial siphon,** taking with it feces, excretory products, respired carbon dioxide, and any gametes that may have been discharged from the gonad. The buccal and atrial siphons can be closed off by contracting sphincter muscles ringing the incurrent and excurrent openings.

The ascidian body wall (not the tunic) contains both circular and longitudinal muscles. The animal can thus make major shape changes, although it cannot, with few exceptions, move from place to place. Rapid muscle contractions cause a characteristic squirting behavior, important in rejecting particles and dislodging anything stuck in the esophagus. A reminder: As with bivalved molluscs, water flow in ascidians is otherwise driven by ciliary activity, not by muscular action of the siphons.

Although many ascidian species are **solitary** or **unitary** (i.e., individuals are physically separated from each other), other species are **social** or **colonial.** As with colonial species in other phyla (e.g., hydrozoans and bryozoans), colonial and social ascidians are produced by asexual budding from a sexually produced founder, with the resulting modules remaining physically and functionally connected. Each colony thus represents a single genotype (a single **genet**). In social species, each module has its own incurrent and excurrent siphon. For true colonial species, however, the modules in the colony have separate buccal siphons and separate mouths, but share a single atrial siphon—the epitome of communal living (Fig. 23.3). Coloniality seems to have evolved many times within the subphylum. Chemicals extracted from some colonial species are showing considerable promise as anti-tumor agents for humans.

A small, tubular heart is located adjacent to the stomach. Remarkably, the heart reverses the direction of its pumping many times each hour; the opening to the heart through which circulatory fluid enters will be the opening through which fluid exits several minutes later, and what was a vein becomes an artery and vice versa.[2] The circulatory fluid itself is peculiar in that it contains **amoebocyte cells,** which accumulate and store excretory wastes, and **morula cells,** which accumulate vanadium (element no. 23 in the periodic table), sulfuric acid, or both; in some species, the concentration of vanadium in morula cells is more than 10 million times higher than its concentration in seawater! In many species, the morula cells deposit these substances in the tunic.[3] The ascidian circulatory fluid contains no oxygen-carrying respiratory pigments, so the blood's oxygen-carrying capacity is determined entirely by the solubility of oxygen in the fluid.

Although ascidians mostly release metabolic waste products (e.g., ammonia) in soluble form, some species (members of the genus *Molgula*) store uric acid and calcium oxalate internally in **renal sacs,** making them potential models for the study of human kidney stone formation.

The only diagnostic chordate characteristic encountered among adult ascidians is the presence of **pharyngeal gill slits (stigmata).** To see your chordate reflection in the ascidians, you must look to the larval stage, which is known

2. See *Topics for Further Discussion and Investigation*, no. 1, at the end of the chapter.

3. See *Topics for Further Discussion and Investigation*, no. 2.

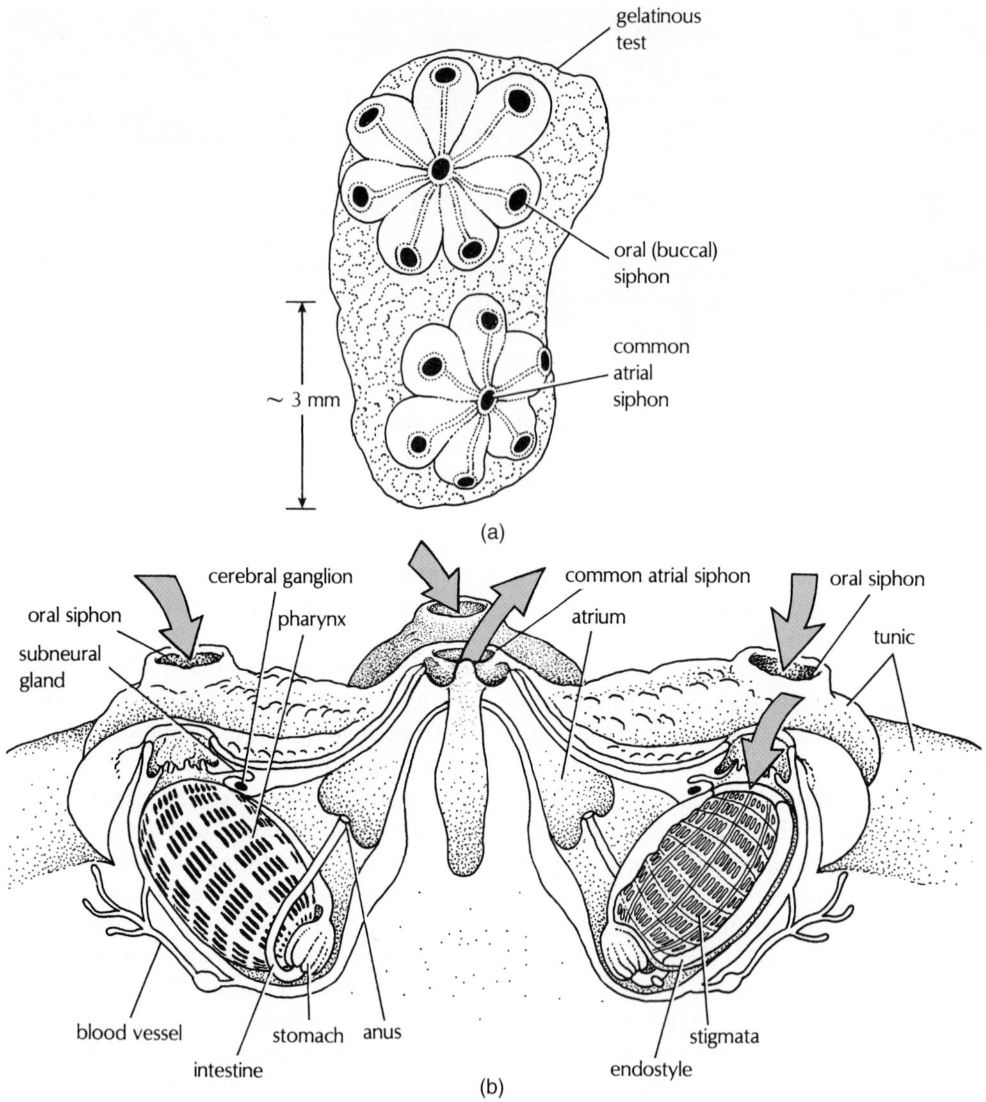

Figure 23.3

(a) A colonial ascidian, *Botryllus violaceus*. As many as a dozen individuals may share a single excurrent opening. (b) Diagrammatic section through a colony, showing the anatomy of the individual ascidians. Note that the circulatory system extends into the tunic; this is also true in solitary species.

(a) After Milne-Edwards. (b) After Delage and Herouard.

as a **tadpole** because of its superficial resemblance to the developmental stage of frogs (Fig. 23.4). The tadpole heart and digestive system are restricted to the "head" region. Tadpole larvae cannot feed, and the digestive tract becomes functional only following metamorphosis to the adult form. The tadpole's tail contains a conspicuous, dorsal, hollow nerve cord, a stiff notochord, and longitudinal muscles extending the length of the tail. The notochord can bend but can be neither elongated nor shortened. The longitudinal muscles on one side of the tail can thus antagonize the longitudinal muscles on the other side of the tail, enabling the tadpole to swim by flexing the tail from side to side. In the absence of the notochord, contracting the longitudinal musculature in the tail would merely cause the tail to buckle or shorten. Swimming of tadpole larvae absolutely depends upon the skeletal function subserved by the notochord.[4]

Attachment to a substrate by means of anterior suckers and a subsequent dramatic metamorphosis to adulthood generally follow a free-swimming larval life of less than one day. In particular, metamorphosis includes the resorption of the larval notochord and tail. Muscular activity thus plays a significant, but very short-lived, role in ascidian locomotion. The larval nervous system is also destroyed at metamorphosis, but is replaced with a new nervous system in the juvenile. Many ascidian species—both solitary and colonial—have now successfully invaded new geographical areas, where they are affecting community structure and displacing native species.[5]

4. See *Topics for Further Discussion and Investigation*, no. 8.

5. See *Topics for Further Discussion and Investigation*, no. 11.

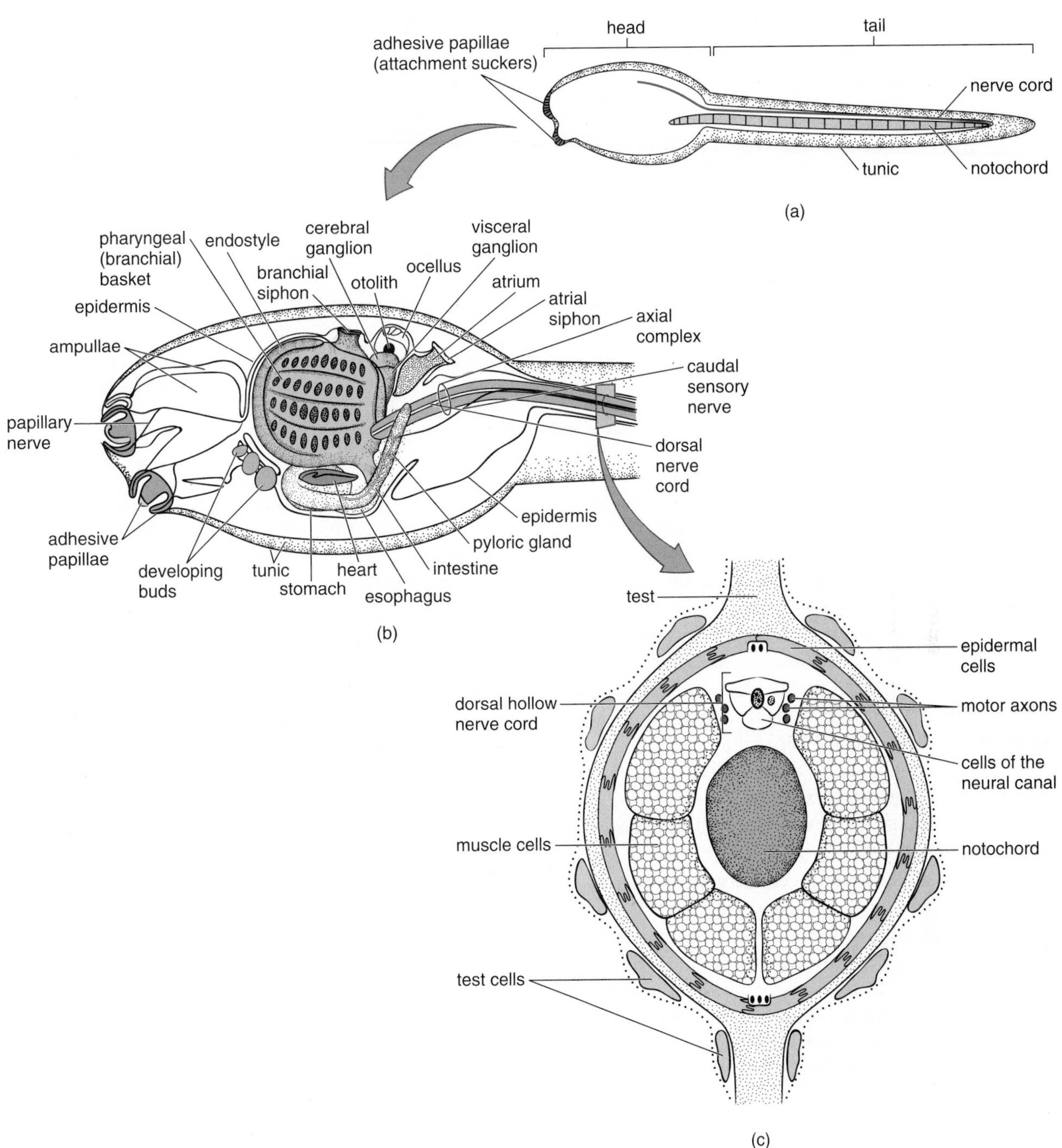

Figure 23.4

(a) Diagrammatic illustration of an ascidian tadpole larva, with the digestive system omitted for clarity. (b) Internal anatomy of an ascidian tadpole. Buds that will become additional zooids are already developing, indicating that this is a colonial ascidian species. (c) Diagrammatic cross section through the tail of an ascidian tadpole larva (*Ciona intestinalis*). Note the dorsal nerve cord above the conspicuous notochord, which is surrounded by a substantial musculature. Compare with the Cephalochordata cross section in Figure 23.12.

(b) Based on Cloney, in *American Zoologist,* 22:817, 1982. (c) From Q. Bone, et al., "Diagrammatic cross-section of mid-region of tail . . . in Ciona" in *Journal of the Marine Biological Association of the United Kingdom,* Vol. 72. Copyright © 1992 Cambridge University Press, New York. Reprinted by permission.

Class Larvacea (= Appendicularia)

The members of the class Larvacea (lar-vā´-sē-ah) may have evolved from ascidian ancestors through the process of **neoteny** (Fig. 23.5a), in which the rates at which **somatic** (i.e., nongonadal) body structures differentiate are slowed down relative to the rate of differentiation of reproductive structures.[6] Thus, the animal becomes sexually mature while retaining the larval morphology. The

6. See *Topics for Further Discussion and Investigation,* no. 10.

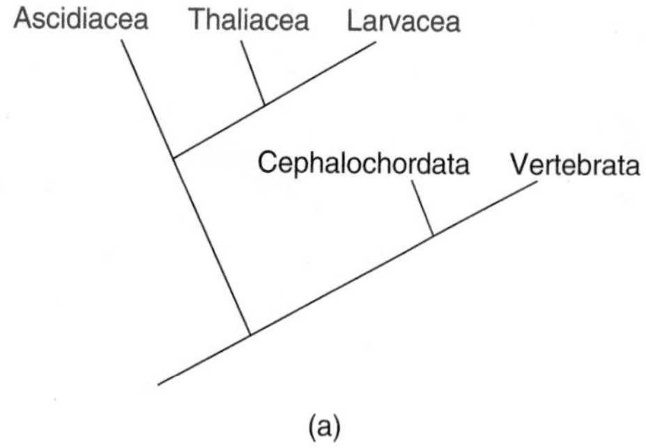

Larvacea
Thaliacea
Ascidiacea
Cephalochordata
Vertebrata
(b)

Figure 23.5
Two competing hypotheses regarding the phylogenetic relationships among urochordates (shaded blue). In hypothesis (a), thaliaceans and larvaceans evolved from ascidian-like ancestors. In hypothesis (b), based on 18S rRNA gene sequence analyses, the ancestral urochordate would have resembled a larvacean or an ascidian tadpole, and the benthic ascidian lifestyle would have evolved secondarily.
Modified from various sources.

former adult morphology is thereby deleted from the life cycle, and the new adult morph gradually becomes increasingly better adapted to its new lifestyle through natural selection. Another view, based on 18S rRNA gene sequence comparisons, suggests that larvaceans most closely represent the ancestral condition, and that current members of the other 2 urochordate classes—along with the cephalochordates to be discussed later in this chapter—evolved from something resembling a larvacean or an ascidian tadpole (Fig. 23.5b).

Adult larvaceans rarely exceed 5–6 mm in body length, although a few species are said to reach lengths of 100 cm. Larvaceans are encountered in most areas of the ocean, at all latitudes.

The larvacean heart and respiratory, digestive, and reproductive systems are confined to the head (Fig. 23.6a), as in the ascidian tadpole larva. Larvaceans, however, are feeding individuals. The mechanism of food collection differs considerably from that encountered among the Ascidiacea. Unlike the ascidian tadpole, the larvacean secretes a gelatinous house around itself (Fig. 23.6b). This house plays a role in both locomotion and food collection; the undulation of the larvacean tail—which, like that of the ascidian tadpole, contains a notochord—drives water through the house for locomotion and feeding. Water exits through a narrow-diameter opening that is partially occluded by a fine-mesh filter. The orientation of this opening determines the direction of movement of the house and animal resulting from the expulsion of fluid, as summarized poetically for members of the larvacean genus *Oikopleura* by Sir Walter Garstang:

> *Oikopleura,* masquerading as a larval Ascidian,
> Spins a jelly-bubble-house about his meridian:
> His tail, doubled under, creates a good draught,
> That drives water forward and sucks it in aft.

A typical individual with a trunk length of about 1.2 mm can move about 35 ml of water through its house each hour (about 0.8 liter per day), and it can jet along at about 1 cm • sec^{1} (centimeter per second).

Particles as small as 0.1 μm are filtered out as water passes through a mucous food-concentrating filter before leaving the house. The concentrated food suspension is then swept through a second, pharyngeal mucous filter and ingested. Very coarse particles, which might damage the delicate house, are screened out from the incoming water current by a coarser mesh positioned across the incurrent opening. The larvacean abandons its house periodically as the meshes become clogged and as the house becomes littered with feces. Before abandoning the old house, the larvacean secretes the material for a new one; it then inflates the new house within seconds of leaving the old one. The food-laden filters of the abandoned house may be important food sources for open-ocean fishes: A single larvacean may form and abandon 15 houses daily.

Class Thaliacea

The members of the class Thaliacea (thal-ē-ā´-sē-ah) are mostly free-living, planktonic individuals, but they achieve their mobility through modification of the adult ascidian body plan rather than through exploitation of the tadpole morph. They have two separate body forms, as described later (p. 547): a sexual form (the blastozooid) and an independent asexual form (the oozooid).

Thaliaceans, including the 3 orders of animals known commonly as pyrosomes, salps, and doliolids, are planktonic and nearly transparent. Transparency probably helps them avoid detection both by predators and by potential prey. Thaliaceans commonly attain concentrations of hundreds to thousands of individuals per m^3 (cubic meter) in subtropical continental-shelf waters. The buccal and atrial siphons are at opposite ends of the body, and bands of circular muscle are generally highly developed. These 2 features account for most thaliaceans' substantial locomotory capabilities.

The most primitive (least modified) of the thaliaceans, members of the genus *Pyrosoma,* are much like colonial ascidians except that the buccal and atrial siphons are at opposite ends of the body. Like colonial

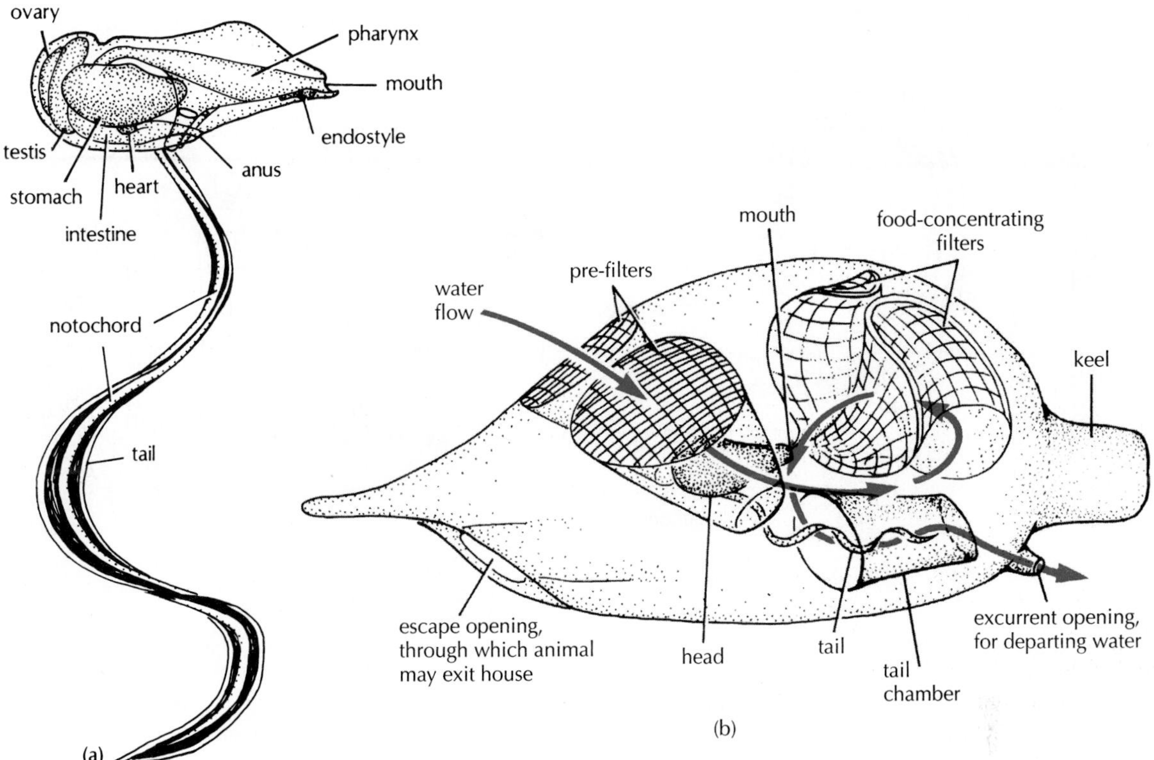

Figure 23.6

(a) Schematic illustration of a larvacean, removed from its gelatinous house. (b) The larvacean *Oikopleura* sp., generating water currents within its complex, disposable house. The house is about 5 cm long.

(a) After Pimentel. (b) After Hardy: after Lohmann.

ascidians, feeding is accomplished by means of cilia, the pharyngeal basket is well developed, and water is discharged through a common exit (Fig. 23.7a). Locomotion of *Pyrosoma* spp. is also achieved through ciliary activity.

Other thaliaceans move through the water by closing off the buccal aperture and contracting the thick circumferential bands of muscle underlying the test. This contraction deforms the test. The resulting decrease in internal volume forces water out of the atrial siphon (since water is incompressible), and the animal is jet-propelled forward. With a single such contraction, one doliolid only about 5 mm long can achieve instantaneous velocities of about 25 cm (50 body lengths) per second, albeit for only a fraction of a second. Following the contraction, the aperture of the atrial siphon is closed while the buccal aperture is opened, and the circular muscles relax. The volume of the animal increases as the elastic test quickly regains its resting shape. Water is thus drawn into the buccal siphon, and the animal is pulled forward as it prepares for the next power stroke. Note that the bands of circular muscle are antagonized by the test itself, rather than by longitudinal musculature.

There is presently some controversy about whether thaliaceans form a monophyletic group; both embryological and morphological evidence suggest that salps and doliolids may have evolved independently from different ascidian ancestors. The probable evolution of thaliaceans from ascidian-like ancestors appears to be a story of decreasing reliance on cilia and increasing development of musculature. In doliolids, the pharyngeal basket is reduced to a flattened plate, although cilia and stigmata are still conspicuous (Fig. 23.7b). The details of feeding have only recently been described for these animals.[7] Pharyngeal cilia seem to play a major role in food collection in both groups. Among the salps, however, the pharyngeal basket is reduced to a slender **branchial bar.** Water flows through a mucous bag suspended between the endostyle, branchial bar, and peripharyngeal bands; stigmata are absent (Fig. 23.7c). In salps (Fig. 23.8), muscular contractions thus play the dominant role in both feeding and locomotion.

Among doliolids (Fig. 23.7b, d), in contrast, feeding currents are apparently produced entirely by cilia lining the pharyngeal stigmata, and the animals can feed without moving appreciably.

Although most thaliacean species occur primarily as individuals that rarely exceed 5 cm in length, species in the

7. See *Topics for Further Discussion and Investigation*, no. 4.

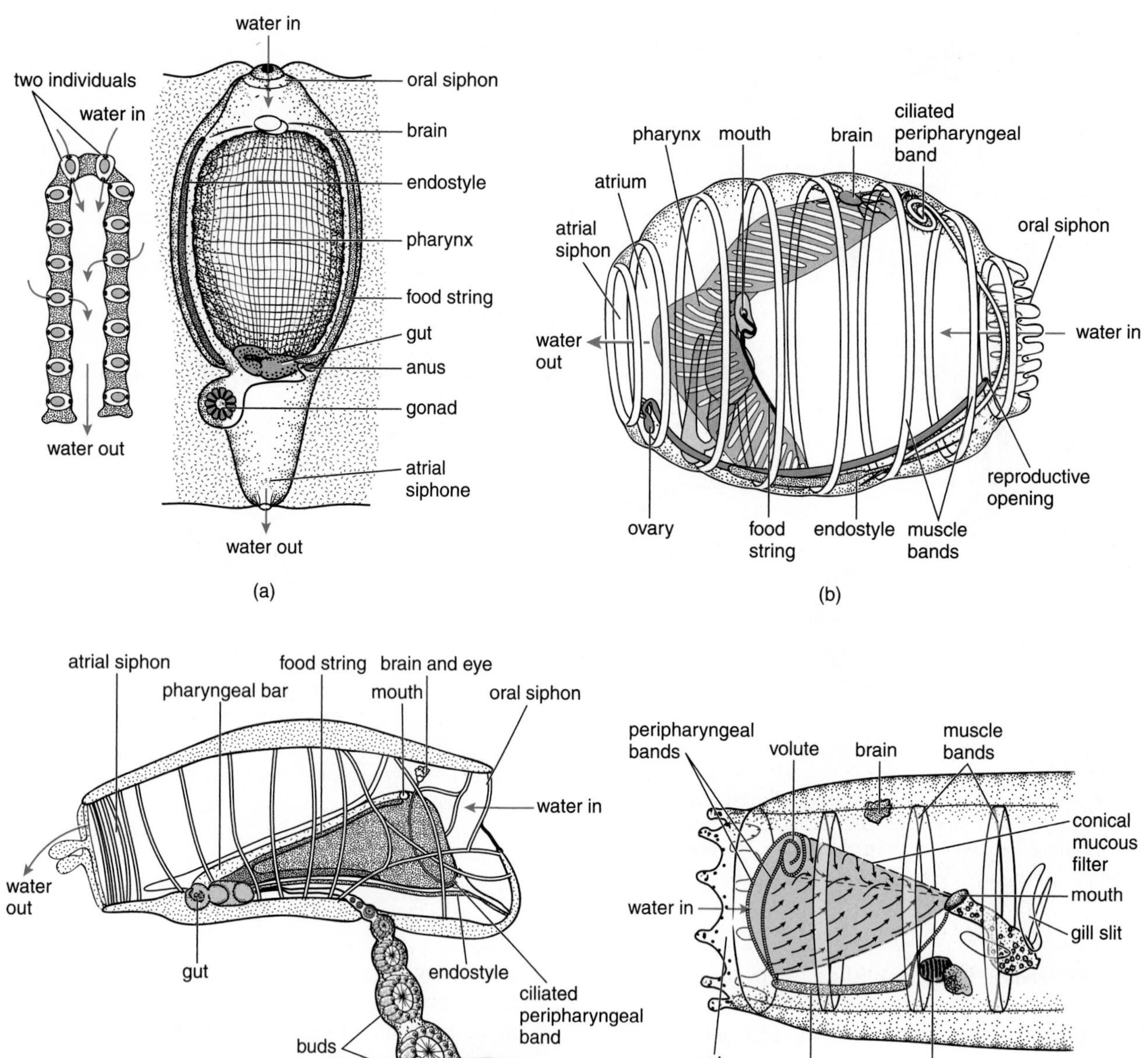

Figure 23.7

(a) *Pyrosoma atlanticum*. The general shape of a typical colony is shown on the left; the general morphology of an individual colony member is shown on the right. Although individuals are only a few millimeters long, a single colony may exceed 3 m in length. Pyrosome anatomy is much like that of ascidians, except that the atrial and oral siphons are directly opposite each other. Tinted areas indicate sheets of mucus. (b) Diagrammatic illustration of a doliolid, *Doliolina intermedia*. Body lengths rarely exceed 4–5 mm. Note the conspicuous rings of musculature that generate locomotory currents. Feeding currents are produced by pharyngeal cilia. (c) Diagrammatic illustration of a salp, *Cyclosalpa affinis*. The pharyngeal basket is reduced to a thin bar. The cilia on the pharyngeal bar function to compact the mucous net into a food string and in moving the food string toward the mouth; both locomotion and feeding are accomplished through contraction of the well-developed muscular bands. (d) Anterior end of a feeding doliolid, *Doliolum nationalis* (oozooid stage). Arrows on the mucous filter indicate the path taken by trapped food particles. Particle-laden mucus is wound into a string where the peripharyngeal bands meet (at the volute), which is then pulled downward toward the mouth. Gill slits have been largely omitted for improved clarity.

(a,c) Based on N. J. Berrill, *The Tunicata*. (b) Reprinted from Tokioka and Berner, in *Pacific Science*, 12:317–26, 1958, by permission of University of Hawaii Press, Honolulu, Hawaii. (d) Based on Q. Bone, J. -C. Branconnet, C. Carré, K. P. Ryan, 1997. *J. Exp. Marine Biol. Ecol.* 179:179–93.

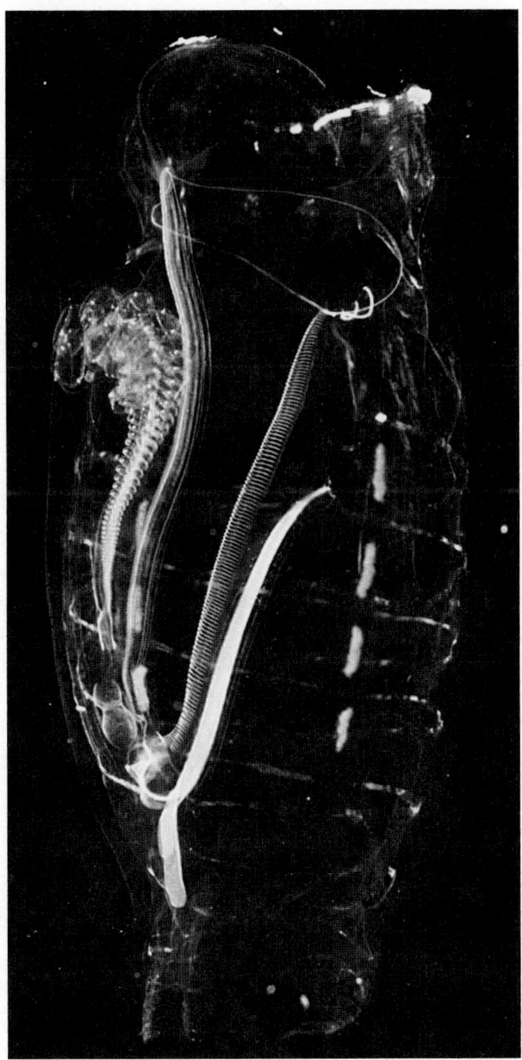

Figure 23.8
Salp photographed in the open ocean.
Courtesy of G. R. Harbison and J. Carleton.

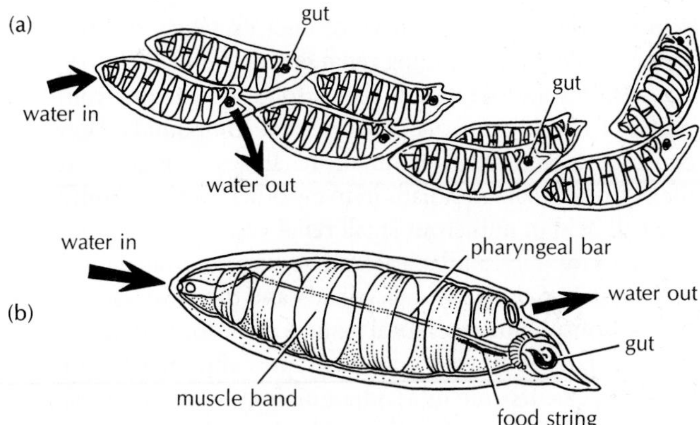

Figure 23.9
In the life cycle of many salps, chains of blastozooids (a) are budded off from a solitary oozooid (b). The blastozooids are released from the parent while still aggregated. After swimming for a time as a group, the individual blastozooids separate and reach sexual maturity. Note that the blastozooids are produced asexually from the oozooid, so that all of the individuals in an aggregate are genetically identical to each other and to the parental oozooid.
(a,b) After Hardy.

genus *Pyrosoma* are entirely colonial and may attain lengths of 8–10m. Most thaliaceans occur only in warm waters.[8]

Other Features of Urochordate Biology

Reproduction

Reproductive biology differs considerably among the various tunicates. Fertilization is almost always external in solitary ascidians and, probably, in all larvaceans, but it is almost always internal in colonial ascidians and thaliaceans. Gametes or larvae are typically discharged through the atrial siphon or, primarily in larvaceans and some colonial ascidians, through rupture of the body wall. Ascidians commonly exhibit asexual reproduction by budding, interspersed with bouts of sexual reproduction; most ascidians are simultaneous hermaphrodites. Most larvaceans are also hermaphroditic, but reproduction is exclusively sexual. Since asexual reproduction is not found within the class Larvacea, larvaceans are always solitary (i.e., there are no colonial species).

Thaliacean reproduction is both sexual and asexual, as with ascidians, but among the Thaliacea, the 2 forms of reproduction are allocated between 2 types of individuals. The union of sperm and egg gives rise to an asexually reproducing morph called the **oozooid.** The oozooid produces no gametes. Instead, individuals called **blastozooids** are budded off from the oozooid (Fig. 23.9).[9] These blastozooids develop hermaphroditic gonads, and their gametes ultimately give rise to the next generation of oozooids. Asexual replication combined with generation times as short as 1 or 2 days permit rapid buildups of thaliacean populations. Thaliaceans have been found in groups of more than 1000 individuals per cubic meter of seawater, over perhaps a hundred square kilometers of ocean. Larvaceans have been reported at even higher concentrations, sometimes exceeding 20,000 individuals per cubic meter of seawater despite their exclusively sexual reproduction.

Excretory and Nervous Systems

Specialized excretory organs are rare among tunicates. In most species, wastes are apparently removed by diffusion across general body surfaces and through the activities of scavenging amoebocytes. Several species within the ascidian

8. See *Topics for Further Discussion and Investigation,* no. 5.

9. See *Topics for Further Discussion and Investigation,* no. 9.

genus *Molgula,* however, have distinct excretory organs.[10] These ductless, blind-ending **renal sacs** accumulate nitrogenous wastes and deposit them as solid concretions composed primarily of uric acid and related compounds. These concretions accumulate within the renal sacs throughout the life of the ascidian. Ascidians in many other species accumulate uric acid in numerous small **renal vesicles** located near the digestive tract, and at least one species accumulates uric acid crystals in the body wall, gonads, and digestive tract.

A conspicuous cerebral ganglion is located in the urochordate body wall and innervates the muscles and sensory cells. Its role in ascidian biology seems minimal, since surgical removal of the cerebral ganglion causes little apparent change in activity. The brain's involvement in coordinating the activities of larvaceans and thaliaceans is likely to be much greater, since muscular activity plays a greater role in these animals' lives. In addition, salps have a prominent, horseshoe-shaped photoreceptor associated with the cerebral ganglion.

Subphylum Cephalochordata (= Acrania)

Subphylum Cephalo • chordata
(G: head string)
sef´-al-ō-cor-dah´-tah

A • crania
(G: without a cranium)
ā-krān´-ē-ah

Defining Characteristics: 1) The notochord extends beyond the nerve cord to the anterior end of the animal; 2) the notochord is contractile, formed as a longitudinal series of flattened discs containing thick myosin filaments

It's a long way from amphioxus,[11]
 it's a long way to us.
It's a long way from amphioxus
 to the meanest human cuss.
Goodbye to tails and gill slits,
 Hello: nails and hair!
It's a long, long way from amphioxus,
 but we came from there.

Cephalochordates are small (less than 10 cm long), laterally flattened, marine invertebrates that have much in common both with tunicates and with members of our own subphylum, the Vertebrata. Comparisons of gene sequences coding for 18S rRNA indicate that cephalochordates are indeed our closest nonvertebrate relatives (Fig. 23.10). Their fossil record extends back to the early Cambrian; what appear to be cephalochordate fossils have been found in the Burgess Shale deposits of western Canada and in even earlier deposits (about 525 million years old) discovered more recently in China. Like vertebrates, cephalochordates have a distinct notochord and a dorsal, hollow nerve cord (Figs. 23.11 and 23.12), and coelomic body cavities that form by enterocoely. There is, however, no sign of vertebrae in the cephalochordate body plan, and correspondingly the "brain" is not encased in a cranium (skull). Specific genes (*Hox* genes, a subset of homeobox genes) that apparently control differentiation along the anteroposterior axis during cephalochordate development appear to be homologous with gene sequences expressed during development of the vertebrate hindbrain. The temporal and spatial pattern of expression for these regulatory *Hox* genes during the development of vertebrate brain and cephalochordate nerve cord tissue suggests that the vertebrate brain has indeed evolved from an extensive portion of the cephalochordate nerve cord.[12]

But cephalochordate feeding biology most closely resembles that of tunicates: Microscopic food particles are captured on mucous sheets as water flows through gill slits in the ciliated pharynx. The captured food then enters the gut as a mucous string, to be digested enzymatically and by cellular phagocytosis. As in tunicates, the mucus is secreted by an endostyle and the pharynx functions in gas exchange as well as in food collection, with the pharyngeal gill bars being serviced by a *bona fide* circulatory system with well-developed veins, arteries, and capillary beds; colorless blood is circulated through the gill and to the rest of the body tissues by muscular contractions of the blood vessel walls. Wastes are eliminated through a well-formed nephridial system located primarily in the pharyngeal region adjacent to the gill slits. The paired excretory organs have a combination of protonephridial and metanephridial characteristics. One additional nephridium (**Hatschek's nephridium,** which is unpaired) is located in the head.

As with doliolids and ascidians, water flow is created by cilia on the pharynx, not by muscular activity. Water flows into the cephalochordate through a buccal opening, which is fringed with sensory tentacles, and exits through an atrial opening located anterior to the anus (Fig. 23.11). The buccal tentacles bear chemosensory and mechanoreceptor cells, and they also serve to prevent very large particles from entering the pharynx.

Unlike that of most urochordates, the cephalochordate notochord persists throughout life. But cephalochordates are clearly not vertebrates: The nerve cord does not elaborate anteriorly to form a brain; there is no protective cranium about the anterior part of the nervous system (not

10. See *Topics for Further Discussion and Investigation*, no. 3.

11. Pope, Phillip H. Sing to the tune of "It's a Long Way to Tipperary."

12. Holland, P. W. H., and J. Garcia-Fernàndez. 1996. *Devel. Biol.* 173:382–95.

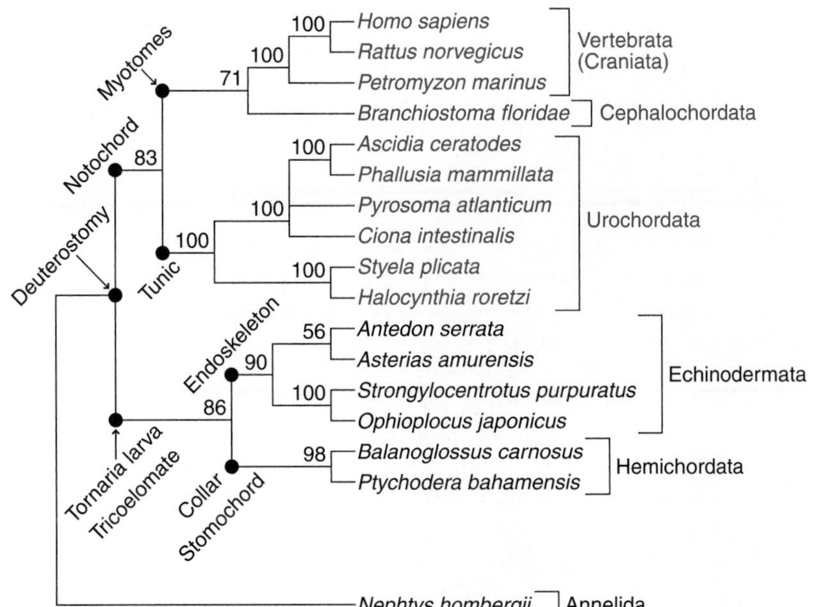

Figure 23.10

A leading hypothesis of the relationships among the different deuterostome groups, also showing the major differences in adult body plans. Members of the Chordata (the phylum containing humans) are shaded blue. In this analysis, an annelid species served as the outgroup (see Chapter 2).

Modified from Cameron, C. B., J. R. Garey, and B. J. Swalla. 2000. Evolution of the Chordata body plan: New insights from phylogenetic analyses of deuterostome phyla. *Proc. Nat. Acad. Sci.* 97:4469–74.

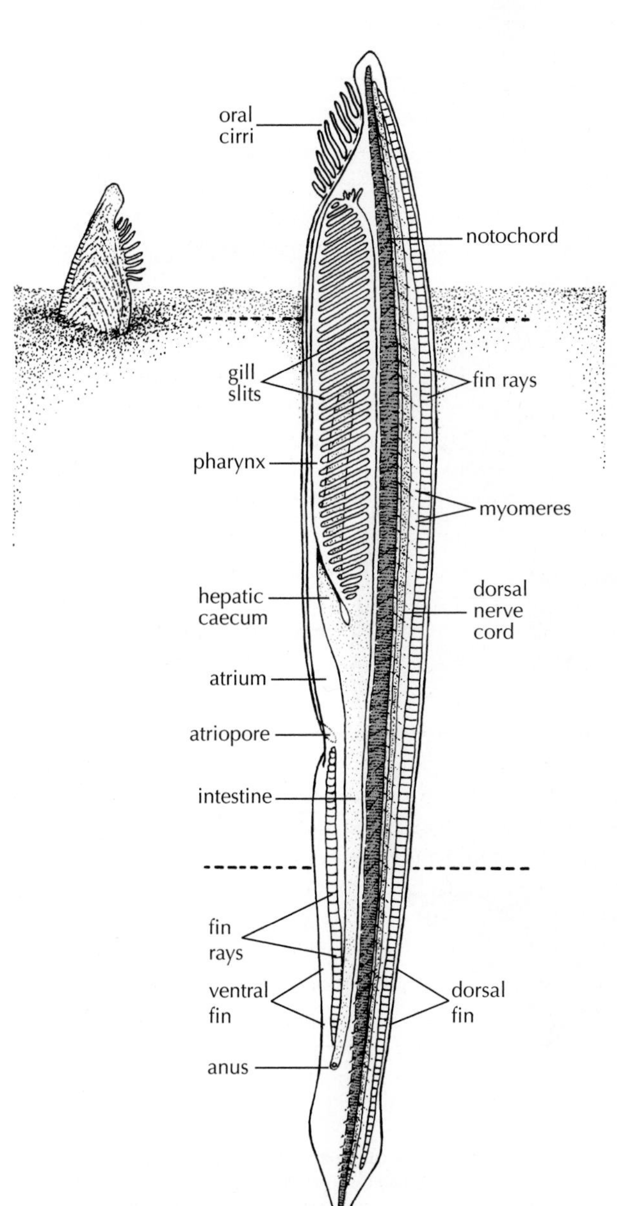

Figure 23.11

The lancelet *Branchiostoma* sp. burrowed into substrate, with its internal anatomy showing through the nearly transparent body. Dotted lines indicate regions from which the cross sections of Figure 23.12 were taken.

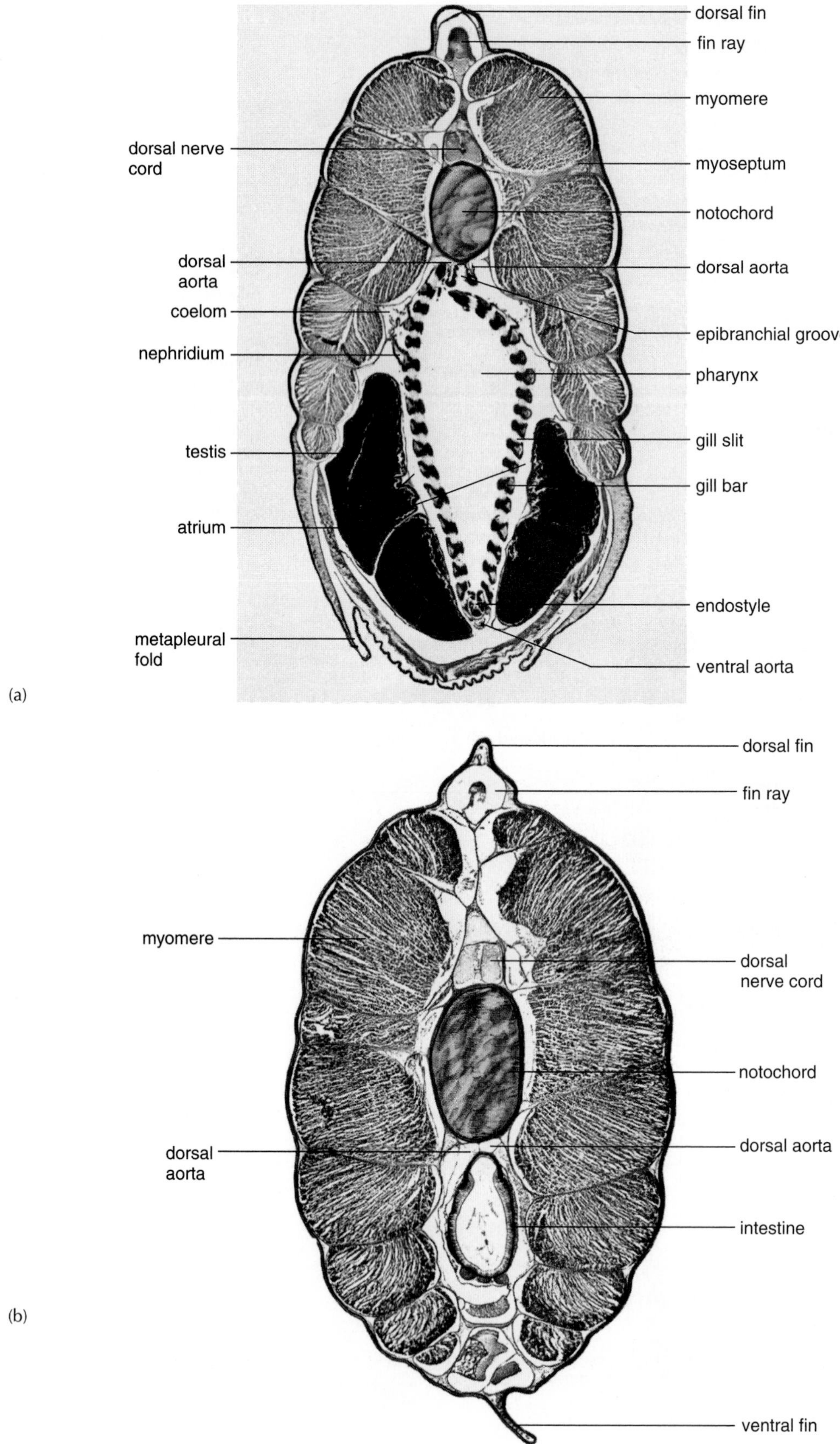

Figure 23.12

Cross sections taken through (a) the anterior region and (b) the posterior region of a male lancelet. Approximate areas from which the sections were taken are shown in Figure 23.11.

surprisingly, since there is nothing much to protect); the notochord extends the entire length of the animal, well into the head (hence, the name "cephalochordate;" *cephalo* = G: head); and the notochord is never replaced by a vertebral column. Specialized sensory organs are limited to the buccal tentacles and to a series of simple, pigmented light receptors (*ocelli*) distributed along the nerve cord.

Adult cephalochordates are nearly transparent and have fishlike bodies, earning them the common name of "lancelets" (little spears). The body being somewhat pointed at both ends, the members of this subphylum are also called by the general term "amphioxus," meaning "sharp at both ends" (*amphi* = G: on both sides; *oxys* = G: sharp). The body is covered neither by a test nor by scales, and adults spend most of their time feeding in sand, with the head protruding above the substrate (Fig. 23.11). When they need to find another feeding area or to escape predators, they move by using strong contractions of paired, lateral muscles. The muscles are arranged in a longitudinal row of about 60 distinct, V-shaped segments (**myomeres**) that are readily visible through the thin, nearly transparent epidermis. Muscle groups on opposite sides of the body (Fig. 23.12) antagonize each other through the flexible but incompressible notochord, as in larvaceans and ascidian tadpole larvae; the backbone serves this same function in fish. Unlike most tunicates, lancelets are almost always gonochoristic rather than hermaphroditic. Eggs are fertilized externally, and the embryos, after passing through a vertebrate-like "neurula" stage, develop into free-swimming larvae that metamorphose after spending several more weeks in the plankton.

Thirty lancelet species have been described. Lancelets are eaten by people in some parts of China, but they are perhaps of greater importance for what they may tell us about vertebrate origins, and about the underlying genetic changes that drove vertebrate evolution.

Taxonomic Summary

Phylum Chordata
- Subphylum Tunicata (=Urochordata)
  - Class Ascidiacea
  - Class Larvacea (= Appendicularia)
  - Class Thaliacea
    - Order Pyrosomida
    - Order Doliolida
    - Order Salpida
- Subphylum Cephalochordata (= Acrania)
- Subphylum Vertebrata

Topics for Further Discussion and Investigation

1. Investigate the significance of heartbeat reversal in tunicates and the mechanism by which it is accomplished.

Herron, A. C. 1975. Advantages of heart reversal in pelagic tunicates. *J. Marine Biol. Assoc. U.K.* 55:959.

Jones, J. C. 1971. On the heart of the orange tunicate, *Ecteinascidia turbinata* Herdman. *Biol. Bull.* 141:130.

Kriebel, M. E. 1968. Studies on cardiovascular physiology of tunicates. *Biol. Bull.* 134:434.

Ponec, R. J. 1982. Natural heartbeat patterns of six ascidians and environmental effects on cardiac function in *Clavelina huntsmani. Comp. Biochem. Physiol.* 72A:455.

2. Investigate the adaptive significance of vanadium and acid accumulation in the tests of ascidians.

Davis, A. R., and A. E. Wright. 1989. Interspecific differences in fouling of two congeneric ascidians (*Eudistoma olivaceum* and *E. capsulatum*): Is surface acidity an effective defense? *Marine Biol.* 102:491.

Pisut, D. P., and J. R. Pawlik. 2002. Anti-predatory chemical defenses of ascidians: Secondary metabolites or inorganic acids? *J. Exp. Marine Biol. Ecol.* 270:203.

Stoecker, D. 1978. Resistance of a tunicate to fouling. *Biol. Bull.* 155:615.

Young, C. M. 1986. Defenses and refuges: Alternative mechanisms of coexistence between a predatory gastropod and its ascidian prey. *Marine Biol.* 91:513.

3. Discuss the evidence for functional excretory systems among ascidians.

Das, S. M. 1948. The physiology of excretion in *Molgula* (Tunicata; Ascidiacea). *Biol. Bull.* 95:307.

Heron, A. 1976. A new type of excretory mechanism in tunicates. *Marine Biol.* 36:191.

Lambert, C. C., G. Lambert, G. Crundwell, and K. Kantardjieff. 1998. Uric acid accumulation in the solitary ascidian *Corella inflata. J. Exp. Zool.* 282:323.

Saffo, M. B. 1988. Nitrogen waste or nitrogen source? Urate degradation in the renal sac of molgulid tunicates. *Biol. Bull.* 175:403.

4. Discuss the similarities and differences in the feeding mechanisms of tunicates and bivalves.

Bone, Q., J.-C. Braconnot, C. Carré, and K. P. Ryan. 1997. On the filter-feeding of *Doliolum* (Tunicata: Thaliacea). *J. Exp. Marine Biol. Ecol.* 179:179.

Deibel, D. 1986. Feeding mechanism and house of the appendicularian *Oikopleura vanhoeffeni. Marine Biol.* 93:429.

MacGinitie, G. E. 1939. The method of feeding of tunicates. *Biol. Bull.* 77:443.

Young, C. M., and L. F. Braithwaite. 1980. Orientation and current-induced flow in the stalked ascidian *Styela montereyensis. Biol. Bull.* 159:428.

5. Discuss the potential ecological significance of planktonic urochordates in the open ocean.

Bochdansky, A. B., and D. Deibel. 1999. Measurement of in situ clearance rates of *Oikopleura vanhoeffeni* (Appendicularia: Tunicata) from tail beat frequency, time spent feeding and individual body size. *Marine Biol.* 133:37.

Bruland, K. W., and M. W. Silver. 1981. Sinking rates of fecal pellets from gelatinous zooplankton (salps, pteropods, doliolids). *Marine Biol.* 63:295.

Deibel, D. 1988. Filter feeding by *Oikopleura vanhoeffeni:* Grazing impact on suspended particles in cold ocean waters. *Marine Biol.* 99:177.

Flood, P. R., D. Deibel, and C. C. Morris. 1992. Filtration of colloidal melanin from sea water by planktonic tunicates. *Nature* 355:630.

Harbison, G. R., and R. W. Gilmer. 1976. The feeding rates of the pelagic tunicate *Pegea confederata* and two other salps. *Limnol. Oceanogr.* 21:517.

Madin, L. P. 1974. Field observations on the feeding behavior of salps (Tunicata: Thaliacea). *Marine Biol.* 25:143.

Morris, C. C., and D. Deibel. 1993. Flow rate and particle concentration within the house of the pelagic tunicate *Oikopleura vanhoeffeni. Marine Biol.* 115:445.

Wiebe, P. H., L. P. Madin, L. R. Haury, G. R. Harbison, and L. M. Philbin. 1979. Diel vertical migration by *Salpa aspera* and its potential for large-scale particulate organic matter transport to the deep-sea. *Marine Biol.* 53:249.

6. About 20 ascidian species invariably house algal symbionts. Evaluate the contributions made by these algae to their ascidian hosts.

Olson, R. R. 1986. Photoadaptations of the Caribbean colonial ascidian-cyanophyte symbiosis *Trididemnum solidum. Biol. Bull.* 170:62.

7. Investigate self-/non-self recognition among ascidians.

Findlay, C., and V. J. Smith. 1995. Antibacterial activity in the blood cells of the solitary ascidian, *Ciona intestinalis,* in vitro. *J. Exp. Zool.* 273:434.

Hirose, E., Y. Saito, and H. Watanabe. 1997. Subcuticular rejection: An advanced mode of the allogeneic rejection in the compound ascidians *Botrylloides simodensis* and *B. fuscus. Biol. Bull.* 192:53.

Kingsley, E., D. A. Briscoe, and D. A. Raftos. 1989. Correlation of histocompatibility reactions with fusion between conspecifics in the solitary urochordate *Styela plicata. Biol. Bull.* 176:282.

Raftos, D. 1996. Adaptive transfer of alloimmune memory in the solitary tunicate, *Styela plicata. J. Exp. Zool.* 274:310.

Raftos, D. A., and E. L. Cooper. 1991. Proliferation of lymphocyte-like cells from the solitary tunicate, *Styela clava,* in response to allogeneic stimuli. *J. Exp. Zool.* 260:391.

Raftos, D., and A. Hutchinson. 1997. Effects of common estuarine pollutants on the immune reactions of tunicates. *Biol. Bull.* 192:62.

Rinkevich, B. 2005. Rejection patterns in botryllid ascidian immunity: The first tier of allorecognition. *Canadian J. Zool.* 83:101–21.

Rinkevich, B., and I. L. Weissman. 1987. A long-term study on fused subclones in the ascidian *Botryllus schlosseri:* The resorption phenomenon (Protochordata: Tunicata). *J. Zool., London* 213:717.

Rinkevich, B., Y. Saito, and I. L. Weissman. 1993. A colonial invertebrate species that displays a hierarchy of allorecognition responses. *Biol. Bull.* 184:79.

8. Compare the locomotory mechanism of ascidian tadpoles with that of nematodes.

9. Compare the reproductive biology of thaliaceans with that of the scyphozoans.

10. Argue either for or against the proposition that larvaceans evolved from ascidian tadpole larvae by neoteny.

Berrill, N. J. 1955. *The Origin of Vertebrates.* Oxford: Oxford Univ. Press.

Bone, Q. 1992. On the locomotion of ascidian tadpole larvae. *J. Marine Biol. Assoc. U.K.* 72:161.

Garstang, W. 1928. The morphology of the Tunicata, and its bearing on the phylogeny of the Chordata. *Q. J. Micros. Sci.* 72:51.

Holland, L. Z., G. Gorsky, and R. Fenaux. 1988. Fertilization in *Oikopleura dioica* (Tunicata, Appendicularia): Acrosome reaction, cortical reaction and sperm-egg fusion. *Zoomorphology* 108:229.

Stach, T., J. Winter, J.-M. Bouquet, D. Chourrout, and R. Schnabel. 2008. Embryology of a planktonic tunicate reveals traces of sessility. *Proc. Nat. Sci. USA* 105:7229–34.

Wada, H., and N. Satoh. 1994. Details of the evolutionary history from invertebrates to vertebrates, as deduced from the sequences of 18S rDNA. *Proc. Nat. Acad. Sci. USA* 91:1801.

11. Investigate the impact of invasive ascidian species on native animals.

Whitlatch, R. B., and Stephan G. Bullard (eds.). 2007. Introduction to the Proceedings of the 1st International Invasive Sea Squirt Conference. *J. Exp. Marine Biol. Ecol.* 342: 1–2. (More than 20 research papers in this issue deal with the topic of invasive ascidian species.)

Taxonomic Detail

Phylum Chordata
Subphylum Tunicata (= Urochordata)

The approximately 2200 species are distributed among 3 classes.

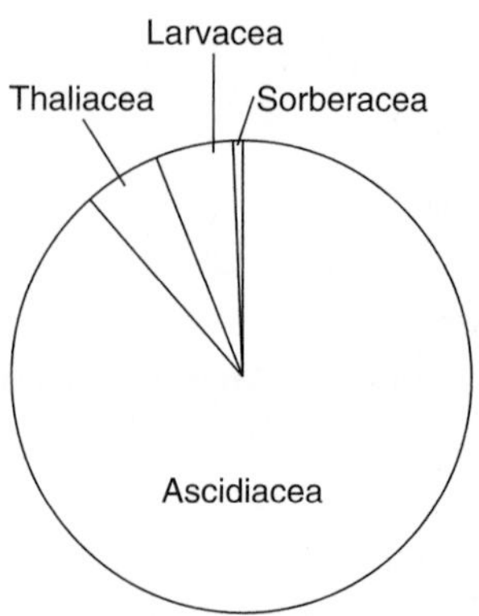

Class Ascidiacea

The approximately 2000 species are distributed among 3 orders, each of which has worldwide representation.

Order Asplousobranchia

Didemnum, Diplosoma, Clavelina, Distaplia, Trididemnum. All species form colonies by asexual budding; in some species, colony members are encased in a common gelatinous tunic. Didemnid colonies often contain symbiotic unicellular algae; they are also capable of locomotion, through gradual extension of the tunic in finger-like projections. Some species contain chemicals of possible value in cancer treatment, and at least one species in this order (*Clavelina minata,* from Japan) exhibits intracellular bioluminescence. Three families.

Order Phlebobranchia

Ciona, Corella, Ascidia, Phallusia, Ecteinascidia (colonial). Most species are solitary, but some are social or colonial; especially common in shallow water. The complete genome (150 million DNA base pairs) of *Ciona intestinalis* was sequenced in 2002. An alkaloid isolated from the tunicate *Ecteinascidia turbinata* is currently in phase II clinical trials in the United States and Europe for its potential as an anticancer drug.

Order Stolidobranchia

Boltenia, Halocynthia, Molgula, Styela, Polyandrocarpa (colonial), *Botryllus* (colonial). Representatives of this order occur worldwide, often in shallow water. Most species are solitary, with some solitary individuals reaching lengths of about 15 cm. Some members live interstitially in sand, and a few species are restricted to abyssal depths. Some species develop without a free-living larval stage. *Styela clava* is raised for food in Korea. Three families.

Class Sorberacea

This is a small class (about 12 species) of deep-water animals that resemble ascidians but have a huge "mouth" opening through which they ingest live prey (*Sorb* = L: to suck in). Also, adults have a dorsal nerve cord. The biology is not yet well known.

Class Larvacea (= Appendicularia)

Oikopleura, Fritillaria. Larvaceans occur in all oceans and are particularly common near the water's surface. Individuals can reach about 9 cm in length, although most are smaller than 1 cm. Some species of *Oikopleura* bioluminesce when poked or shaken. The class contains about 70 species.

Class Thaliacea

All species are colonial in at least part of the life cycle. About 75 species, distributed among 3 orders.

Order Pyrosomida

Pyrosoma. This group of colonial tunicates is best represented in warm water. All individuals exhibit striking bioluminescence, probably through the chemical activities of bacterial symbionts. Colonies commonly range from several centimeters to several meters in length.

Order Doliolida

Doliolum. Representatives live in all oceans, although they are most common in warmer waters. Doliolids are the fastest moving of all thaliaceans, despite their small size (usually only a few millimeters long).

Order Salpida

Salpa. Representatives live in all the world's oceans, although they are especially abundant in warm waters. Some species are found at depths exceeding 1500 m, although most salps live much closer to the surface. Salps reproduce rapidly; some species have generation times shorter than one day. Individuals may grow to lengths exceeding 24 cm. About 30 species have been described.

Subphylum Cephalochordata (= Acrania)

Branchiostoma—amphioxus, or lancelets. All 30 described species occur in shallow water and live partially buried in sandy substrates. Some species tolerate estuarine conditions, although most are fully marine. Two families, 2 genera.

Subphylum Vertebrata

Homo sapiens and other vertebrates. Of the approximately 48,000 vertebrate species with us today, nearly half are fishes and nearly 20% are birds. Only about 4000 species are mammals, and half of those are bats and rodents.

General References about the Nonvertebrate Chordates

Alexander, R. M. 1981. *The Chordates,* 2nd ed. New York: Cambridge Univ. Press.

Alldredge, A. 1976. Appendicularians. *Sci. Amer.* 235:94–102.

Barrington, E. J. W. 1965. *The Biology of Hemichordata and Protochordata.* San Francisco: W. H. Freeman.

Berrill, N. J. 1955. *The Origin of Vertebrates.* London: Oxford University Press.

Berrill, N. J. 1961. Salpa. *Sci. Amer.* 204:150–60.

Bone, Q., ed. 1998. *The Biology of Pelagic Tunicates.* Oxford University Press, USA.

Bone, Q., C. Carré, and P. Chang. 2003. Tunicate feeding filters. *J. Marine Biol. Ass. U.K.* 83:907–19.

Cameron, C. B., J. R. Garey, and B. J. Swalla. 2000. Evolution of the Chordata body plan: New insights from phylogenetic analyses of deuterostome phyla. *Proc. Nat. Acad. Sci.* 97:4469–74.

Conklin, E. G. 1932. The embryology of amphioxus. *J. Morphol.* 54:69–151.

Harrison, F. W., and E. E. Ruppert, eds. 1997. *Microscopic Anatomy of Invertebrates,* Vol. 15. Hemichordata, Chaetognatha, and the Invertebrate Chordates. New York: Wiley-Liss.

Jeffery, W. R., and B. J. Swalla. 1997. Tunicates. In: S. F. Gilbert and A. M. Raunio. *Embryology: Constructing the Organism.* Sunderland, MA: Sinauer Associates, Inc. Publishers, 331–64.

Johnsen, S. 2001. Hidden in plain sight: the ecology and physiology of organismal transparency. *Biol. Bull.* 201:301–18.

Millar, R. H. 1971. The biology of ascidians. *Adv. Marine Biol.* 9:1–100.

Monniot, C., F. Monniot, and P. I. Laboute. 1991. *Coral Reef Ascidians of New Caledonia.* Paris: *ORSTOM.*

Parker, S. P., ed. 1982. *Classification and Synopsis of Living Organisms,* vol. 2. New York: McGraw-Hill, 823–30.

Petersen, J. K. 2007. Ascidian suspension feeding. *J. Exp. Marine Biol. Ecol.* 342:127–37.

Sawada, H., H. Yokosawa, and C. C. Lambert, eds. 2001. *The Biology of Ascidians.* Tokyo: Springer-Verlag.

Stokes, M. D., and N. D. Holland. 1998. The Lancelet. *Amer. Scient.* 86:552–60.

Whittaker, J. R. 1997. Cephalochordates, the lancelets. In: S. F. Gilbert and A. M. Raunio. *Embryology: Constructing the Organism.* Sunderland, MA: Sinauer Associates, Inc. Publishers, 365–81.

Search the Web

1. www.ucmp.berkeley.edu/chordata/chordata.html

   Includes definitions of key terminology, and separate sections on life history and ecology, fossil record, morphology, and systematics. Enter the "Systematics" section to view color images of representative tunicates and cephalochordates along with information about each group. This site is maintained by the University of California at Berkeley.

2. http://tolweb.org/tree/phylogeny.html

   This brings you to the Tree of Life site. Search on "Chordata."

3. http://biodidac.bio.uottawa.ca

   Choose "Organismal Biology," "Animalia," and then "Urochordata" for photographs and drawings of adults and larval stages, including illustrations of sectioned material.

4. www.ascidians.com

   This site, maintained by Arjan Gittenberger, offers many color photographs of ascidians from around the world, and provides links to many other relevant websites.

5. http://depts.washington.edu/ascidian/

   The "Ascidian News." This is the place to go for the latest informal information about ascidians, provided by Gretchen and Charles Lambert. Includes abstracts from recent meetings and theses, and listings of recent publications.

6. www.mbl.edu

   Choose "Inside the MBL," then "Biological Bulletin," and then "Other Biological Bulletin Publications." Next select "Keys to Marine Invertebrates of the Woods Hole Region," and then choose "Protochordates." This brings up an identification key to the tunicate (urochordate) and cephalochordate species found near Woods Hole, MA.

7. http://workshop.molecularevolution.org/resources/amphioxus

   Here you will find the complete history of the Amphioxus song, with all the lyrics.

8. http://home.uchicago.edu/~egrey/ascidian_biology.htm

   Provided by Erin Grey. Includes a section on invasive species.

24

Invertebrate Reproduction and Development—An Overview

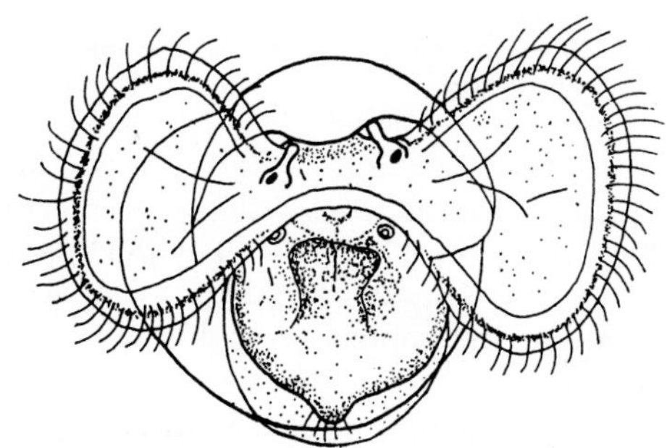

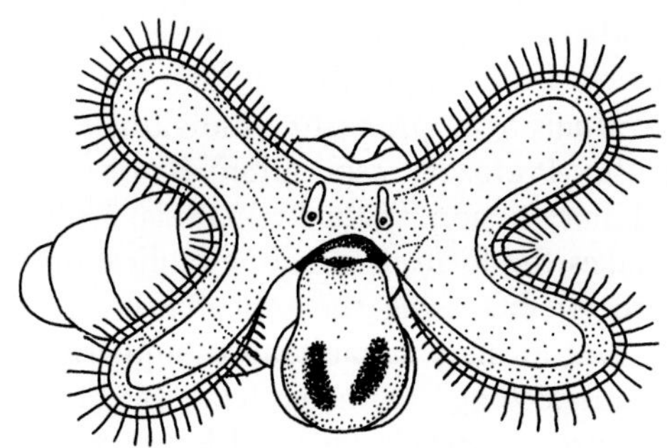

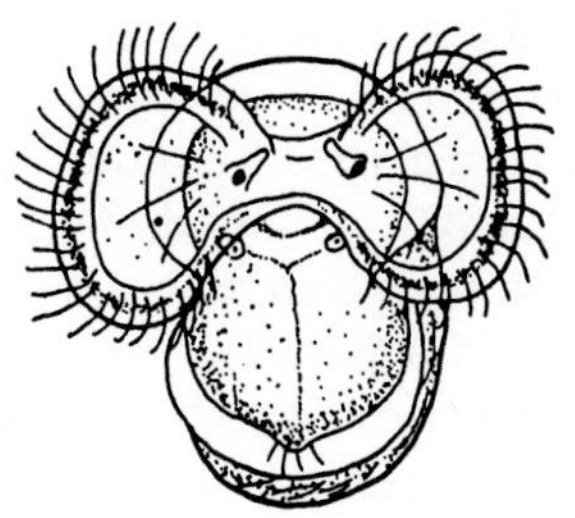

Introduction

The continued existence of a species depends upon the ability of individuals of that species to reproduce. Nearly every behavioral, morphological, or physiological adaptation characterizing a species may be presumed to contribute to its reproductive success, either directly or indirectly. In a sense, then, all organisms live to reproduce. Moreover, no matter how the process of reproduction begins or proceeds, differentiation (genetically controlled specialization of cells) is always involved; indeed, much of what is presently understood about the control of gene expression comes from studies of invertebrate development. The topic of reproduction and development thus seems ideal for uniting all of the phyla in the consideration of a single aspect of invertebrate biology. As you read through this chapter, I hope you will be able to take pride in recognizing many terms and the names of many organisms that were new to you when you began this book.

Invertebrates show a great diversity of reproductive and developmental patterns, a diversity far surpassing that encountered among the vertebrates. Most vertebrates fertilize internally and show some degree of parental care for the developing young. All vertebrates are deuterostomes, and thus cleavage is basically radial and indeterminate, and the mouth does not form from the blastopore. Variations on the basic deuterostome theme occur, of course, largely because of differing amounts of yolk in the eggs. The diversity of developmental patterns observed among invertebrates, however, is much more than variations on a theme. Among invertebrates, there are radical differences in

1. expression of sexuality;
2. site of fertilization (if present);
3. pattern of cell division;
4. stage at which cell fates become determined;
5. number of distinct tissue layers formed;
6. mechanism through which mesoderm (if any) is formed;
7. extent to which a body cavity develops;

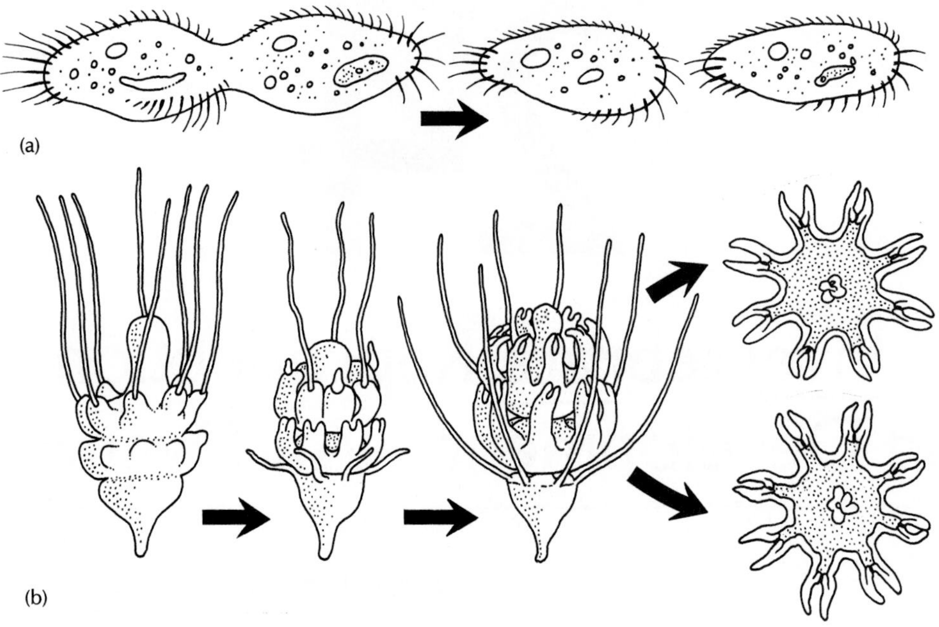

Figure 24.1
(a) Binary fission in a ciliated protozoan. (b) Strobilation in the scyphozoan *Stomolophus meleagris,* culminating in the release of numerous ephyrae.

(b) From D. R. Calder in *Biological Bulletin,* 162:149, 1982. Copyright © 1982 *Biological Bulletin.* Reprinted by permission.

8. mechanism through which a body cavity develops; and
9. origin of the mouth and anus (when present).

In this chapter, we survey invertebrate patterns of reproduction and development, identify some of the groups most closely associated with these different patterns, and consider the ecological significance—that is, the likely adaptive benefits—of the various patterns discussed. Cleavage patterns, modes of coelom formation, and other features of early metazoan embryology have already been summarized in Chapter 2. The goal of this chapter is to put reproduction and development into ecological and evolutionary contexts.

Asexual Reproduction

Reproduction among invertebrates may be either sexual or asexual. Sexual reproduction always involves the union of genetic material contributed by 2 genomes. Asexual reproduction, on the other hand, is reproduction in the absence of fertilization (i.e., without a union of gametes). The timing of both sexual and asexual reproductive events is controlled by both environmental and internal factors.[1]

Asexual reproduction is often a process of exact replication; in such instances, barring mutation, asexually produced offspring are genetically identical to the progenitor. This form of asexual reproduction (**ameiotic;** i.e., without meiosis) can add no genetic diversity to a population. On the other hand, through asexual reproduction a single individual can contribute to a potentially rapid increase in population size, excluding potential competitors and flooding the population with a particularly successful genotype.

Asexual reproduction need not involve egg production by a female. In sponges (poriferans), anthozoans, hydrozoans, scyphozoans, bryozoans, thaliaceans, and some ascidians and protozoans, for example, asexual reproduction is accomplished through the budding of new individuals from preexisting individuals (Fig. 24.1). Among the Protozoa, replication is often achieved through binary fission. In the Trematoda, asexual reproduction takes the form of ameiotic replication of larval stages, greatly increasing the probability that any one genotype will locate a suitable host. Similarly, larvae of many fungus-feeding gall midges (class Insecta) asexually produce more genetically identical larvae within themselves, increasing the likelihood that a given genotype will eventually locate a suitable fungus. In some other groups, such as the Anthozoa, Ctenophora, Turbellaria, Rhynchocoela, Polychaeta, Asteroidea, and Ophiuroidea, body parts may be detached from the adult (or even the larvae, in some asteroid, ophiuroid, and echinoid species)[2] and left behind to regenerate into new, morphologically complete individuals (Fig. 24.2).

Egg production is intimately involved in the ameiotic, asexual reproduction of many other invertebrate species. Among selected arthropods and rotifers, asexual

1. See *Topics for Further Discussion and Investigation,* nos. 1 and 2, at the end of the chapter.

2. See *Topics for Further Discussion and Investigation,* no. 14.

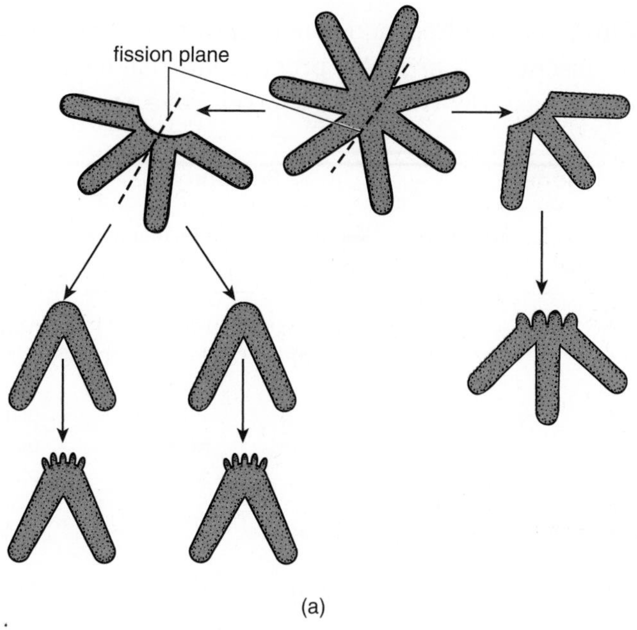

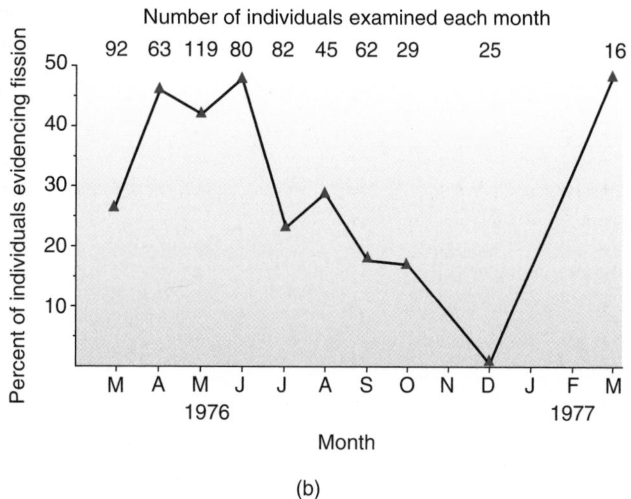

Figure 24.2

(a) Asexual reproduction in the asteroid *Nepanthia belcheri.* This species routinely undergoes fission, in which 6- or 7-armed adults divide to form primarily 2- or 3-armed individuals, which then proceed to regenerate the full complement of arms. (b) Percent frequency of fission varied between 0% and nearly 50% of the population over the course of one year. The number of individuals examined each month is shown at the top of the graph.

Based on Ottesen and Lucas in *Marine Biology,* 69:223, 1982.

reproduction takes the form of **parthenogenesis,** in which eggs develop to adulthood in the absence of fertilization. As simple as this sounds, there can be unusual complications. In some mites and ticks (class Arachnida), for example, the females cannot oviposit unless they first mate with a male, even though the eggs never become fertilized and the male makes no genetic contribution to the offspring. This is termed **pseudogamy** (false marriage). Something similar occurs in many beetles, except that males do not exist in some of these species. In such cases, the females mate with males of an allied species; although no union of gametes occurs, the eggs will not develop in the absence of contact with sperm.

In other invertebrate groups, asexual reproduction may involve meiosis, so that pairing and segregation of chromosomes occurs, and new genetic combinations can be generated despite the lack of genetic input by a second individual. This occurs in some protozoans and nematodes, both parasitic and free-living, but it is most commonly encountered among arthropods, especially insects and arachnids. As seen in Table 24.1, asexual reproduction is quite common among invertebrates. In fact, reproduction without fertilization is the primary reproductive mode in many species. Note that, with few exceptions, asexual reproduction requires the presence of only a single individual.

Table 24.1 Summary of Invertebrate Reproductive Patterns (+ = present; – = absent; ? = unknown; –? = probably absent; +? = probably present; internal fertilization = within the body of the animal, but not necessarily within the reproductive tract)

	Reproductive Mode			**Sexuality**			**Fertilization**	
Taxonomic Group	Asexual	Sexual	Spermatophore	Gonochoristic	Hermaphroditic	Larval Stage	Internal	External
Protozoa	+	+	–	Have mating strains		none	+	–
Porifera	+	+	–	+(rare)	+	parenchymula	+	–
Cnidaria								
Scyphozoa	+	+	–	+	+(rare)	planula	+(rare)	+
Hydrozoa	+	+	–	+	+(rare)	planula, actinula	+	+
Anthozoa	+	+	+(rare)	+	+(rare)	planula	+	+
Cubozoa	+	+	+(rare)	+	–	planula	+(rare)	+
Ctenophora	+(rare)	+	–	+(rare)	+	cydippid	+(rare)	+
Platyhelminthes								
Turbellaria	+	+	–	+(rare)	+	Müller's	+	–
Trematoda	+	+	–	–	+	miracidium, cercaria	+	–
Cestoda	+	+	–	+(rare)	+	none	+	–
Rhynchocoela	+	+	–	+	+(rare)	pilidium	+(rare)	+
Rotifera								
Seisonidea	–	+	–	+	–	none	+	–
Bdelloidea	+	–	–	+	–	none	–	–
Monogononta	+	+	–	+	–	none	+	–
Acanthocephala	–	+	–	+	–	acanthor	+	–
Loricifera	?	+	?	+	–	Higgins	?	?
Kinorhyncha	–	+	+	+	–	none	+	+
Gastrotricha	+	+	+	–	+	none	+	–
Priapulida	–	+	–	+	–	nameless (benthic)	+	+
Nematomorpha	–	+	–	+	–	nameless	+	–
Pentastomida	–	+	–	+	–	nameless (parasitic)	+	–
Gnathostomulida	–	+	–	–	+	none	+	– ?
Annelida								
Polychaeta	+	+	+	+	+(rare)	trochophore	–	+
Oligochaeta	+	+	–	–	+	none	–	+
Hirudinea	–	+	+	–	+	none	+	–
Echiura	–	+	–	+	–	trochophore	+(rare)	+
Sipuncula	+(rare)	+	–	+	+(rare)	trochophore, pelagosphera	–	+
Mollusca								
Gastropoda	+(rare)	+	+	+	+	veliger	+	+(rare)

Sexual Reproduction

Patterns of Sexuality

Although many invertebrates reproduce asexually, sexual reproduction, which requires the fusion of haploid gametes, is quite common. Two individuals are usually involved in bringing this about. Moreover, the genetic composition of the offspring is always dissimilar to that of either parent. Indeed, this seems to be a major selective advantage of reproducing sexually: by increasing genetic diversity within populations, sexual reproduction facilitates rapid adaptation to deteriorating environmental

Table 24.1 *Continued*

	Reproductive Mode			Sexuality			Fertilization	
Taxonomic Group	Asexual	Sexual	Spermatophore	Gonochoristic	Hermaphroditic	Larval Stage	Internal	External
Bivalvia	–	+	–	+	+	trochophore, veliger	+(rare)	+
Polyplacophora	–	+	–	+	+(rare)	trochophore	+(rare)	+
Cephalopoda	–	+	+	+	–	none	+	–
Scaphopoda	–	+	–	+	–	trochophore	–	+
Aplacophora	–	+	–?	+	+	trochophore	+?	+?
Monoplacophora	–	+	–	+	–	veliger?	–?	+?
Nematoda	+	+	–	+	+(rare)	none	+	–
Arthropoda								
Merostomata	–	+	–	+	–	trilobite	–	+
Arachnida	–	+	+	+	–	none	+	–
Chilopoda	–	+	+	+	–	none	+	–
Diplopoda	–	+	+	+	–	none	+	–
Insecta	+	+	+	+	+(rare)	larva, pupa	+	–
Crustacea	+	+	+	+	+	nauplius, cyprid, zoea, megalopa	+	+(rare)
Tardigrada	+	+	–	+	+	none	+	–
Onychophora	–	+	+	+	–	none	+	–
Bryozoa	+	+	–	+(rare)	+	coronate, cyphonautes	+	+(rare)
Phoronida	+	+	+	+(rare)	+	actinotroch	+	+(rare)
Brachiopoda	–	+	–	+	+(rare)	nameless	+	+(rare)
Chaetognatha	–	+	+	–	+	none	+	–
Echinodermata								
Asteroidea	+	+	–	+	+(rare)	bipinnaria, brachiolaria	+(rare)	+
Ophiuroidea	+	+	–	+	+	ophiopluteus	+	+
Echinoidea	–	+	–	+	+(rare)	echinopluteus	+(rare)	+
Holothuroidea	–	+	–	+	+(rare)	auricularia, doliolaria	–	+
Crinoidea	–	+	–	+	–	doliolaria	+	+
Hemichordata								
Enteropneusta	+(rare)	+	–	+	–	tornaria	–	+
Urochordata								
Ascidiacea	+	+	–	+(rare)	+	tadpole	+	+
Larvacea	–	+	–	+(rare)	+	none	–?	+?
Thaliacea	+	+	–	–	+	none	+	–
Cephalochordata	–	+	–	+	+(rare)	nameless	–	+

conditions.[3] The 2 parents are usually of different sexes, in which case the species is said to be **gonochoristic** or **dioecious.** Alternatively, a single individual may be both male and female, either simultaneously (**simultaneous hermaphroditism**) or in sequence (**sequential hermaphroditism**) (Table 24.2).

Hermaphroditism is common among invertebrates, as shown in Table 24.1. The East Coast oyster, *Crassostrea virginica,* is a good example of a species exhibiting sequential

Table 24.2 Forms of Sexuality Encountered among Invertebrates

Term	Form
Gonochoristic (dioecious)	♂ or ♀
Simultaneous hermaphrodite	♂ + ♀
Sequential hermaphrodite	
Protandric	♂ → ♀
Protogynous (relatively rare)	♀ → ♂

3. See *Topics for Further Discussion and Investigation,* no. 11.

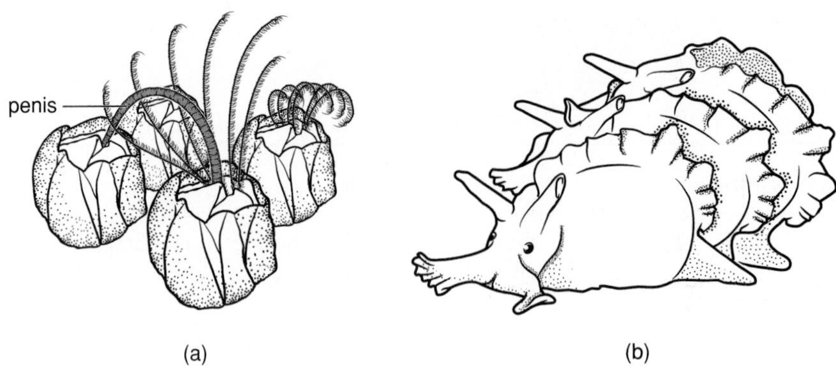

Figure 24.3

(a) Copulation among the Cirripedia. Barnacles are simultaneous hermaphrodites, a decided advantage for individuals that are incapable of locomotion as adults; any 2 adjacent individuals are potential mates. (b) Each sea hare, *Aplysia brasiliana* (Gastropoda: Opisthobranchia), bears a penis at the right side of the head and a vaginal opening posteriorly. Mutual insemination is common. The individuals in the middle of the chain illustrated are functioning simultaneously as males and females.

(a) After Barnes and Hughes. (b) Based on Purves and Orians, *Life: The Science of Biology,* 2d ed. 1983.

hermaphroditism. The young oyster matures as a male, later becomes a female, and may change sex every few years thereafter. Most sequential hermaphrodites change sex only once and usually change from male to female. This is known as **protandric hermaphroditism,** or **protandry** (*prot* = Greek: first; *andros* = G: male).

In contrast to species that change sex as they age, many invertebrates, including most ctenophores and cestodes, are simultaneous hermaphrodites (Fig. 24.3). Self-fertilization is rare among simultaneous hermaphrodites, although it can occur, as in the Cestoda and within some coral, polychaete, bryozoan, gastrotrich, cnidarian, and barnacle species. An advantage of simultaneous hermaphroditism is that a meeting between any 2 mature individuals can result in a successful mating. This is especially advantageous in sessile animals, such as barnacles (Crustacea: Cirripedia). Indeed, the benefits of simultaneous hermaphroditism are so conspicuous that one wonders why such reproductive patterns are rare among vertebrates. Possibly the reproductive systems and behaviors of most advanced vertebrates have become so complex and specialized that 2-sexed individuals are simply not feasible. This is perhaps unfortunate; many of the inequalities commonly encountered among human societies would be unlikely in a society of simultaneous hermaphrodites.

The advantages of sequential hermaphroditism are less evident. Age-dependent sex changes can deter self-fertilization (a rather extreme form of inbreeding) in hermaphroditic species. In addition, it may be energetically more efficient to be one sex when small and the other sex when larger. Certainly, a single spermatozoon requires less structural material and nutrients than does a single ovum and is therefore less costly to produce. The total cost of reproduction, however, will depend upon the total number of gametes produced, the cost of each individual gamete, and the amount of energy expended in obtaining a mate and protecting the young; it may or may not be cheaper to be a male, depending upon the species.

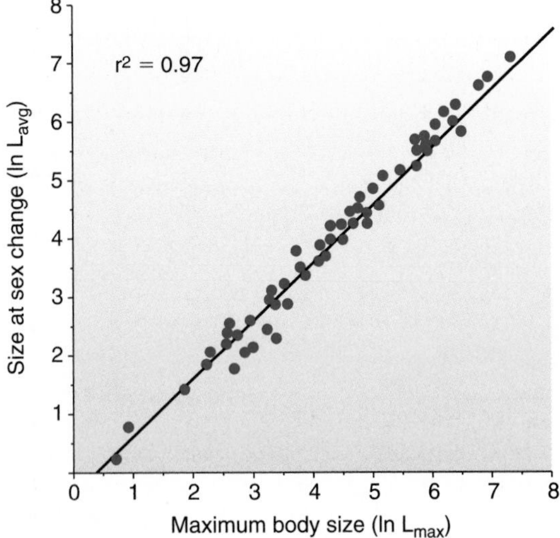

Figure 24.4

Sequential hermaphrodites generally change sex when they reach 72% of their maximum size. The figure shows data for 77 animal species representing molluscs, crustaceans, echinoderms, and chordates (fish). The r^2 value means that 97% of the variation in average size at sex change is explained by maximum body size. Based on Allsop, D. J., and S. A. West. 2003. Changing sex at the same relative body size. *Nature* 425: 783–84.

Remarkably, most individuals that switch sex—regardless of taxonomic group—do so when they reach about 72% of their maximum body size (Fig. 24.4). It is not yet clear why this characteristic is so widespread.

In some invertebrate species, sex and sex ratios are controlled by microscopic symbionts, such as microsporidians (Chapter 3) or bacteria in the genus *Wolbachia. Wolbachia* bacteria, which occur in many insect, arachnid, crustacean, and parasitic nematode species (see p. 443), are transmitted to the next generation of hosts exclusively through egg cytoplasm; sperm lack sufficient cytoplasm to accommodate the symbiont. Remarkably, the symbionts

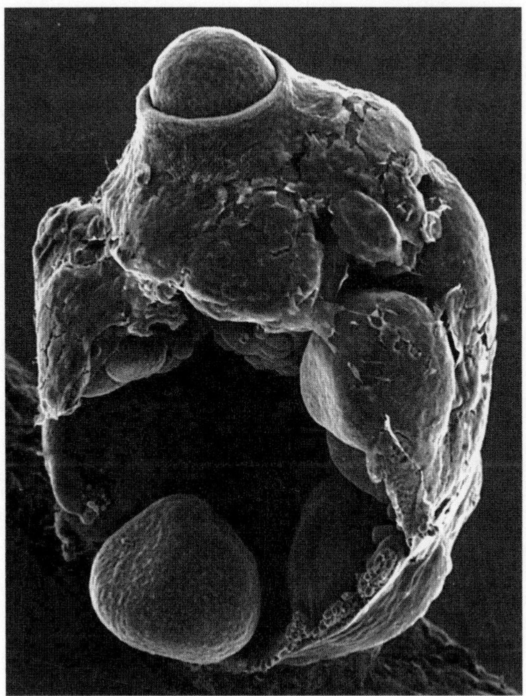

Figure 24.5

Ingestion of nurse eggs during prosobranch gastropod development. An embryo of the marine species *Searlesia dira* is shown in the process of ingesting a single nurse egg. A portion of the body wall has been torn away, revealing a number of previously ingested nurse eggs. Each nurse egg is approximately 230 µm (micrometers) in diameter.

Courtesy of Brian Rivest, from Rivest, 1983. *J. Exp. Marine Biol. Ecol.* 69:217. Elsevier Science Publishers, Physical Sciences and Engineering Division.

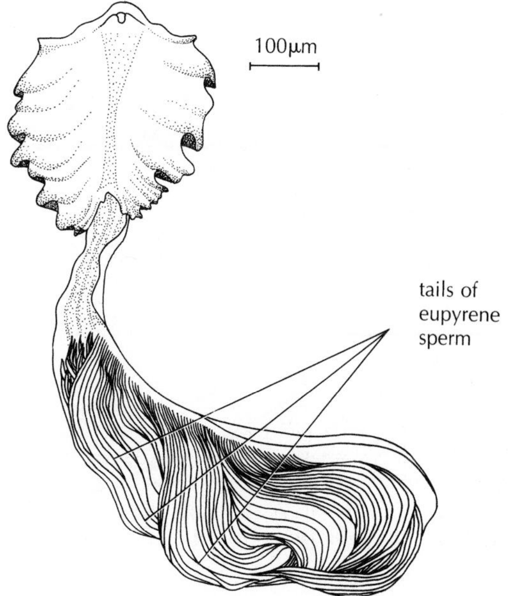

Figure 24.6

A giant, apyrene sperm of the marine gastropod *Cerithiopsis tubercularis,* with thousands of eupyrene sperm embedded in its tail. This form of apyrene sperm, called spermatozeugmata, may serve to transport eupyrene sperm to the egg.

From V. Fretter and A. Graham, *A Functional Anatomy of Invertebrates*. Copyright © 1976 Academic Press. Reprinted by permission of Harcourt Brace & Company Limited, London, England.

either kill male offspring, increasing the proportion of female hosts in the population, or somehow alter the hormonal machinery of host offspring so that males develop as functional females.[4]

Gamete Diversity

Invertebrate gametes show considerable diversity of structure and, often, of function. Some invertebrates produce a percentage of eggs that are incapable of being fertilized and/or of sustaining development following fertilization. These **nurse eggs** are ultimately consumed by neighboring embryos (Fig. 24.5). Nurse eggs are especially common among the Gastropoda. Sperm also show a high degree of functional diversity. Many invertebrate species produce only a percentage of normal sperm; that is, sperm that have a haploid DNA content and are capable of fertilizing an egg and promoting subsequent development of the embryo. These are **eupyrene** sperm. The other sperm can play no direct role in development because they have either an excess number of chromosomes or too few. In the extreme case, the abnormal sperm are without chromosomes entirely; that is, they are **apyrene.** Apyrene sperm production is especially common among gastropods and insects (Fig. 24.6). The functional significance of apyrene and other atypical sperm remains speculative at the present time.[5] Possibly they provide transportation or nutrition to eupyrene sperm that are associated with them.

Considerable morphological diversity exists among even normal invertebrate sperm (Fig. 24.7), and the arrangement of microtubules within the axoneme (see Chapter 3, pp. 42–43) often departs radically from the usual 9 + 2 arrangement (Fig. 24.8). Indeed, the sperm of some species lack flagella entirely, in which case the sperm are either incapable of any movement or move in amoeboid fashion. Aberrant sperm are encountered especially often within the Arthropoda. The non-flagellated sperm of one particular fruit fly species are nearly 6 cm long when uncoiled, more than 20 times the length of the fly itself!

Getting the Gametes Together

All sexual development begins with the fertilization of a haploid egg; the trick, then, is to get the eggs and sperm together. Invertebrates demonstrate a variety of ways of bringing this union about. On land and in freshwater, fertilization of the egg is, with few exceptions, internal, for reasons presented in Chapter 1 (these environments are generally too dry or osmotically stressful for survival of exposed gametes). Internal fertilization may be accomplished in several ways. Males of many invertebrate species

4. See *Topics for Further Discussion and Investigation,* no. 15.

5. See *Topics for Further Discussion and Investigation,* no. 6.

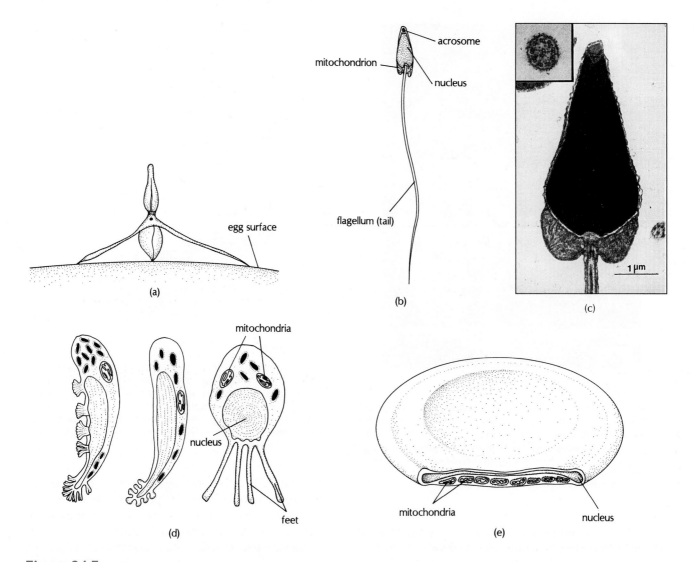

Figure 24.7

Gamete diversity. (a) The tripodlike sperm of the crustacean *Galathea* sp. The sperm is about to penetrate the egg. (b) Sea urchin sperm. (c) Transmission electron micrograph of a sea urchin sperm (head end only). (d) The aflagellate sperm of gnathostomulid worms (interstitial acoelomates). These sperm use the small feetlike processes to move. (e) The disc-shaped sperm of the insect *Eosentomon transitorium.*

(a) From Kume and Dan, 1968. *Invertebrate Embryology.* Washington, D.C.: National Science Foundation. After Kortzoff. (c) Courtesy of K. J. Eckelbarger. From Eckelbarger et al., 1989. *Biological Bulletin* 176:257–71. Permission of *Biological Bulletin.* (d) From Bacetti & Afzelius, "Biology of the Sperm Cell" in *Monographs in Developmental Biology,* No. 10. Copyright © 1976 S. Karger, AG, Basel, Switzerland. (e) From Bacetti and Afzelius; after Bacetti et al.

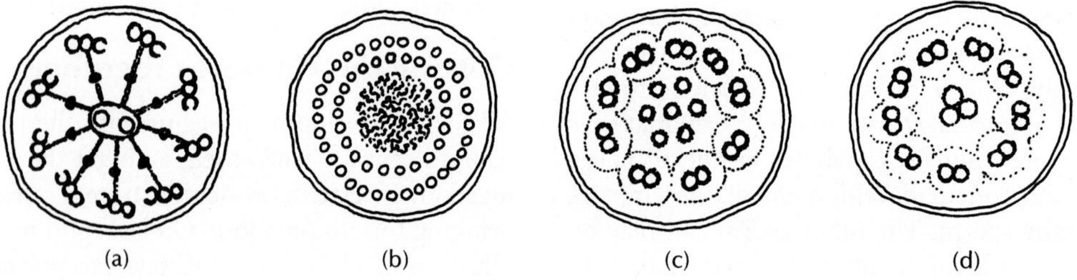

Figure 24.8

(a) Cross section through the sperm tail of a sea urchin. Note the 9 + 2 arrangement of microtubules in the axoneme. (b) Cross section through the axoneme of an insect sperm, *Parlatoria oleae.* (c) Axoneme of a caddisfly sperm, *Polycentropus* sp. The microtubules have a 9 + 7 arrangement. (d) 9 + 3 arrangement of microtubules in the axoneme of sperm from the spider *Pholeus phalangioides.*

From Bacetti and Afzelius, "Biology of the Sperm Cell" in *Monographs in Developmental Biology,* No. 10. Copyright © 1976 S. Karger, AG, Basel, Switzerland.

are equipped with a penis, through which sperm are transferred directly into the female's genital opening. In the case of hypodermic impregnation, as encountered among some turbellarians, leeches, gastropods, and rotifers, sperm are forcefully injected through the female's body wall. In both cases, sperm transfer is said to be **direct.**

Males of other species lack any such copulatory organs, and yet internal fertilization may still occur. Several means of achieving such **indirect sperm transfer** are encountered among invertebrates.[6] Commonly, the sperm are packaged in containers of varied complexity. These sperm-filled containers, called **spermatophores** (literally, "sperm carriers"), are secreted by specialized glands found only in the male. Among terrestrial invertebrates, spermatophores are typically employed by pulmonate gastropods, onychophorans, and terrestrial arthropods, including the insects, arachnids, centipedes, and millipedes (Fig. 24.9a, b). The spermatophore is transferred from the male to the female in diverse ways. In all collembellans (e.g., springtails) and some pseudoscorpions (Arthropoda: Arachnida), for example, no actual mating occurs. Males may deposit spermatophores onto suitable substrates without a female being present; the sperm capsules are located by the female either chemotactically or, in a few pseudoscorpion species, by following silk threads secreted by the male. Once the female has located a spermatophore, she inserts it into her genital opening, discharging the sperm from the container. This mechanism seems to be admirably suited for achieving internal fertilization in a species in which the proximity of one individual to another frequently prompts physical attack and even cannibalism. Some insects implant a spermatophore into the female through a copulatory organ, so this would be an example of direct sperm transfer even though it involves a spermatophore.

The pseudoscorpion example to the contrary, internal fertilization generally requires a high degree of cooperation between pairs of individuals, and it is often preceded by elaborate courtship displays. This is typically true when fertilization is accomplished by copulation, but it is also common when fertilization is achieved through use of spermatophores. In what appear to be the more highly evolved pseudoscorpions, for example, males again deposit spermatophores on the ground but only in the presence of females and only after a complex mating dance. The male then physically guides the female and positions her over the stalk of the attached spermatophore, whereupon she takes up the sperm packet through her genital opening. The 2 partners then quickly separate from each other. The female removes the capsule a short time later, after it has been emptied by osmotically generated pressure and the sperm have been safely stored within her.

This mode of sperm transfer closely resembles that observed in true scorpions. Here, too, the stalked spermatophore is deposited only in the presence of the female and its deposition is preceded by an intricate dance, during which time the male searches for a suitable substrate on which to cement the sperm-filled container. The scorpion spermatophore is quite complex, possessing a mechanically operated sperm-ejection lever (Fig. 24.9a). Again, the male positions the female so that her genital opening is over the tip of the capsule (Fig. 24.10). The capsule is then inserted into the genital opening just far enough to operate the lever of the spermatophore; the ejected sperm are then taken up and stored for later use.

In some other arachnids, the spermatophore's exact functional significance is dramatically apparent. The males of some species forcefully subdue a female, open her genital pore, deposit a spermatophore on the ground, pick it up with the chelicerae, insert the spermatophore into the female's genital opening, close the opening, and leave. Spermatophore transfer in centipedes and millipedes follows a somewhat similar script, except that the female voluntarily picks up and inserts the sperm packet following an often elaborate mating dance. Clearly, the spermatophore is a functional substitute for a copulatory organ in transporting spermatozoa to the female.

In bushcrickets and many other terrestrial arthropod species, the female feeds on a specialized part of the spermatophore, both during and after sperm transfer. In such cases, the "nuptial gift" occupies the female's attention for some time, increasing the likelihood that sperm transfer will be completed before the female removes the spermatophore, and also contributes nutrients that can increase female fecundity and offspring quality; either outcome will clearly increase the fitness of the donor male.

Spermatophores also are used by many marine and some freshwater invertebrates. Spermatophores are known to occur in monogonont rotifers, polychaetes, oligochaetes, leeches, gastropods (Fig. 24.9c), cephalopods (Fig. 24.9d), crustaceans (Fig. 24.11), phoronids, pogonophorans, and chaetognaths. In some species, the spermatophores are simply discharged into the sea and reach a female by chance. Species employing this mode of sperm transfer always live in close association (i.e., communally), so the loss of sperm is not so great as might be supposed. Floating spermatophores have been reported among the Gastropoda, Polychaeta, Phoronida, and Pogonophora. More commonly, males deliver the spermatophores directly.

Among the cephalopods, the complexity of the spermatophore, and of its mode of transfer to the female, rivals that encountered among terrestrial invertebrates. Males often use their chromatophores to perform species-specific color displays for the female as a prelude to sperm transfer. The spermatophore is a large (up to 1 m long!), cylindrical mass of sperm, incorporating a complex osmotically or mechanically activated sperm discharge mechanism (Fig. 24.9d). The spermatophores of some species contain up to approximately 10^{10} sperm! Large numbers of spermatophores are stored in a pouch, called **Needham's sac,** opening into the mantle cavity. Typically, at the appropriate moment, the male grabs one or more

6. See *Topics for Further Discussion and Investigation,* no. 3.

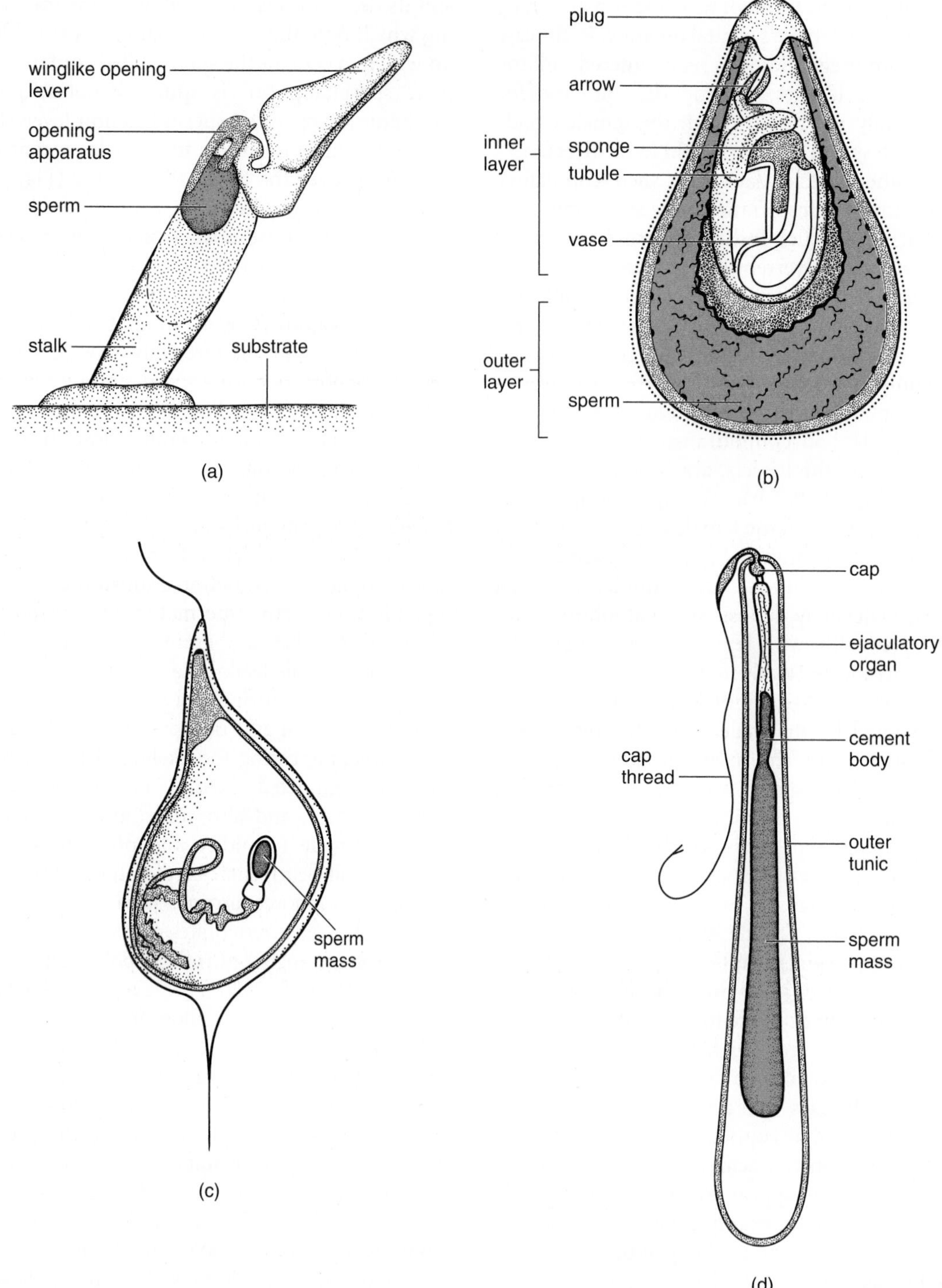

Figure 24.9

(a) The spermatophore of a scorpion. (b) Longitudinal section through the spermatophore of a tick (Arachnida). After transfer to the female, the outer layers of the spermatophore elongate over a one-minute period. The tip of the inner section of the spermatophore then opens, and within one additional second, the arrow, tubule, sponge, and vase are shot out from the spermatophore, discharging sperm. (c) Spermatophore of a vermetid gastropod, tentatively identified as *Dendropoma platypus.* (d) A cephalopod spermatophore.

(a) From Barnes, 1980. *Invertebrate Zoology,* 4th ed. Orlando, Florida: W. B. Saunders College Publishing. After Angermann. (b) Based on Feldman-Muhsam, in *Journal of Insect Physiology,* 29:449, 1983. (c) From Hadfield and Hopper in *Marine Biology,* 57:315, 1980. (d) from Brown, *Selected Invertebrate Types.* 1950. John Wiley & Sons. NY.

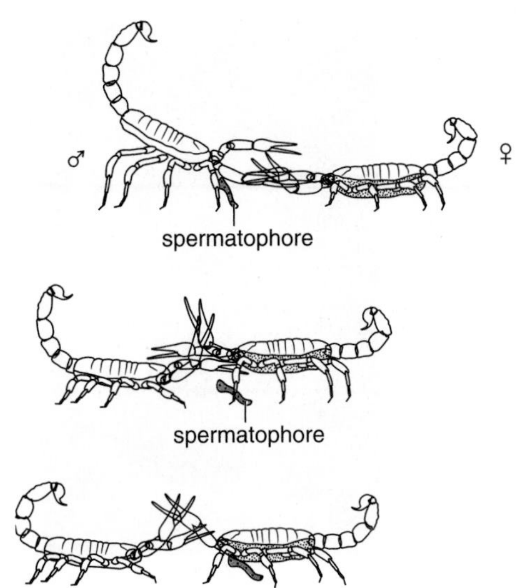

Figure 24.10

Exchange of spermatophores in scorpions. The male is shown on the left. After the male deposits the spermatophore, he guides the female over it.

Based on Angerman, in *Ziteschrift fur Tierpsychologie,* 14:276, 1957.

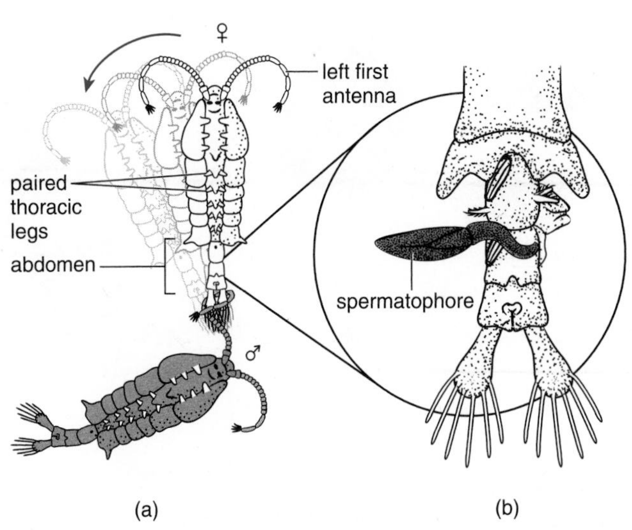

Figure 24.11

Spermatophore exchange in a marine copepod, *Centropages typicus* (as seen in ventral view). (a) The male grabs the female, using the hinged right first antenna. The male then swings the female around and grasps her around the abdomen, using his clawed right fifth thoracic appendage (Fig. 14.29a2). (b) The spermatophore is shown in position after its transfer to the female.

Based on Blades in *Marine Biology,* 40:57, 1977.

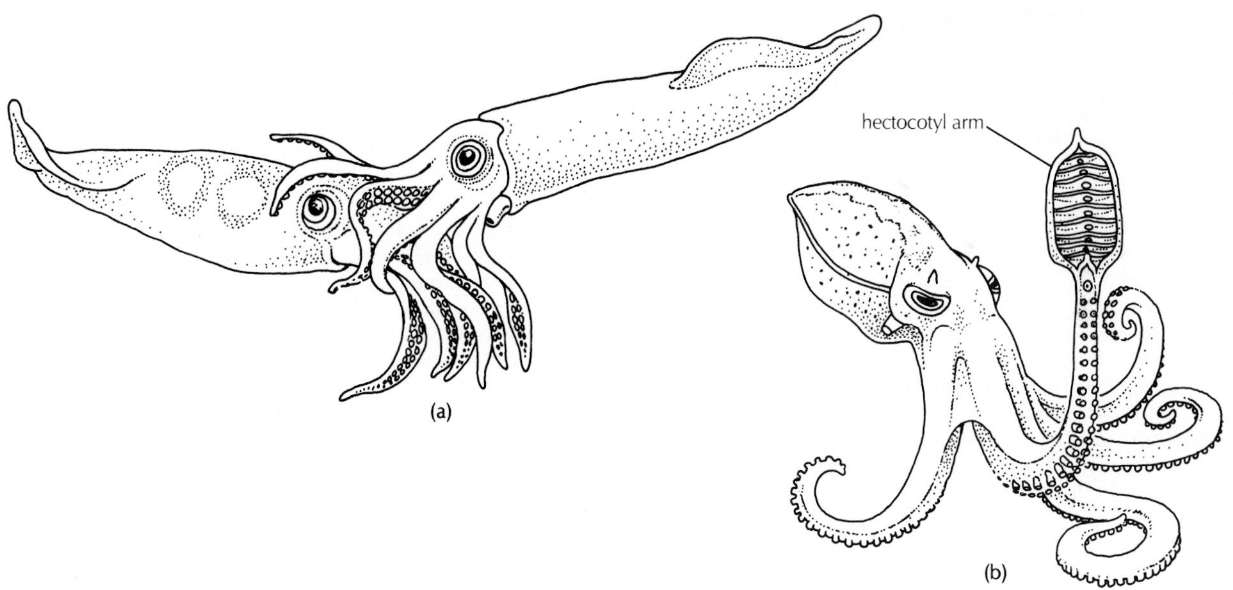

Figure 24.12

(a) Mating in the squid *Loligo.* (b) *Octopus lentus,* showing hectocotyl arm. The end of the arm forms a broad, cuplike depression that holds the spermatophores after the male has removed them from his mantle cavity.

(a) After Barnes; after R. F. Sisson. (b) After Huxley; after Verrill.

spermatophores from Needham's sac and inserts them into the female's mantle cavity, adjacent to the genital opening. The organ used to insert the spermatophores is an often highly modified arm called the **hectocotylus** (Fig. 24.12). In some cephalopod species, the hectocotylus breaks off from the male following its insertion into the female's mantle cavity. Before the reproductive biology of these species was understood, the disconnected hectocotylus was thought to be a parasitic worm, a not unreasonable assumption. In some cuttlefish species, males actively flush out the female's mantle cavity to eliminate spermatophores from previous matings before depositing their own.

The evolutionary movement of invertebrates to land and freshwater from the sea may well have been accompanied not by the development of entirely new systems of fertilization, but rather by the modification of already existing ones. Indeed, reproduction through the use of

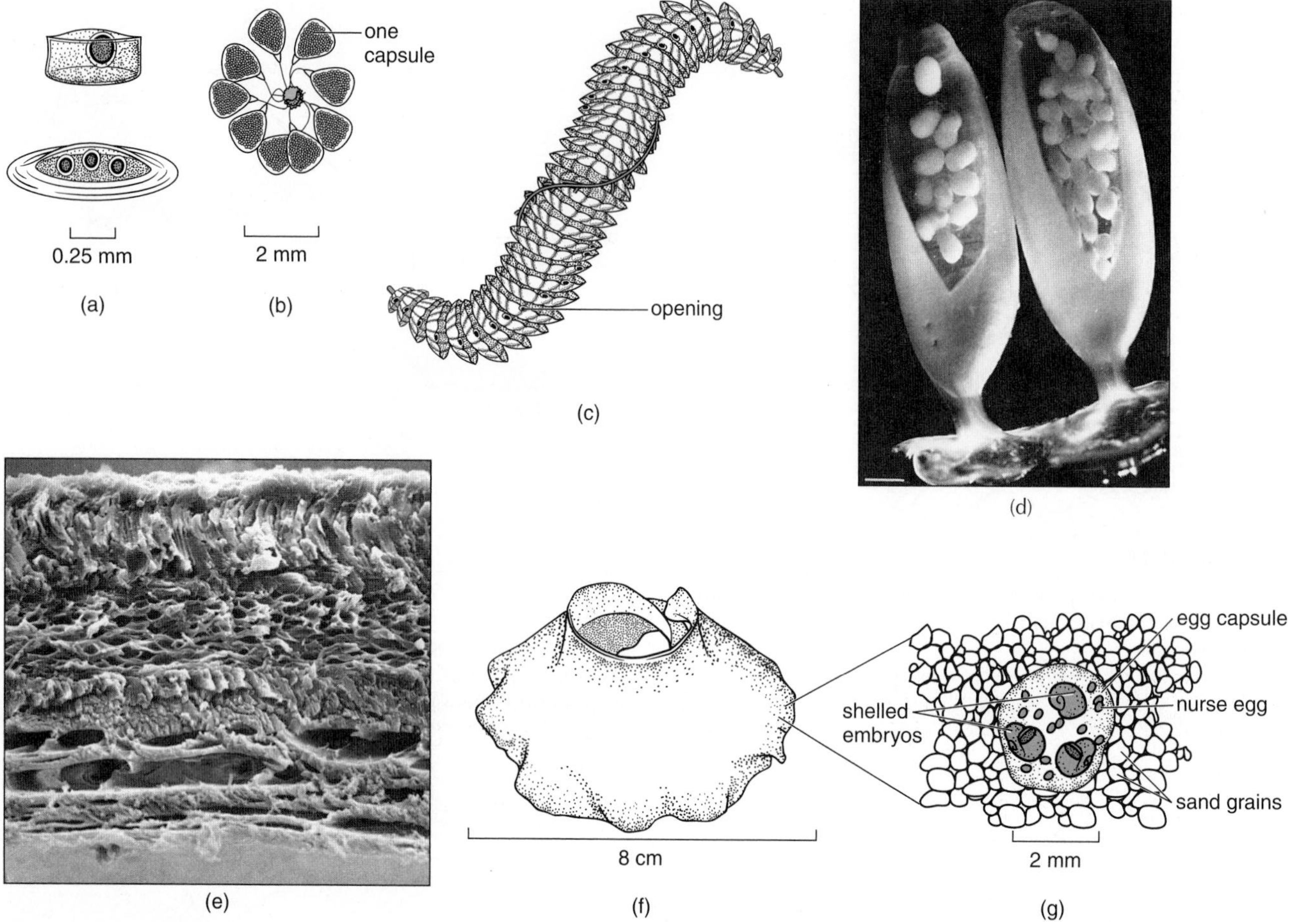

Figure 24.13

Representative egg capsules of marine prosobranch gastropods. (a) Planktonic egg capsules of 2 periwinkles, *Littorina littorea* and *Tectarius muricatus.* (b) Cluster of egg capsules of *Turritella.* (c) Necklace of egg capsules deposited by the whelk *Busycon carica.* Each capsule is about the size of a quarter. (d) Egg capsules of the marine gastropod *Nucella lapillus.* Part of the capsule wall has been cut away, revealing numerous yolky embryos within each capsule. The capsules are about 1 cm (centimeter) tall. Offspring emerge from the top of each capsule as miniature snails, having passed through the veliger stage within the capsules. (e) Cross section (scanning electron micrograph) of the egg capsule wall of *Nucella* (= *Thais*) *lima,* showing the complex, multilayered construction. The wall is approximately 55 μm thick. (f) Sand collar made by a female naticid gastropod. (g) Thousands of tiny egg capsules are embedded within the collar; many of the eggs (nurse eggs) are incapable of developing and serve as a food source for those that can.

(a,b) Based on "British Prosobranch Molluscs, Their Functional Anatomy and Ecology," in *British Prosobranch Molluscs* Vol. 161, The Ray Society. (d) Courtesy of R. Stöckman-Bosbach. From Stöckman-Bosbach et al., 1989. *Marine Biology* 102:283–89. Permission of Springer-Verlag. (e) From J. Pechenik, unpublished micrograph. (f) Modified from several sources. (g) From Fretter and Graham. 1994. *British Prosobranch Molluscs,* 2nd ed. Intercept Scientific, Medical and Technical Publications, U.K.

spermatophores or through copulation may be considered preadaptations for life on land; development of spermatophores may well have been one of the preadaptations that made the evolutionary transition from salt water to land or freshwater possible.

In marine animals, internal fertilization also may be accomplished in the absence of any form of physical copulation or sophisticated sperm packaging. Because the concentration of dissolved salts in seawater closely approximates that of most cells and tissues, sperm can be discharged freely into the seawater and transported to the female by water currents. Those species of bryozoans, echinoderms, bivalves, sponges, and cnidarians that have internal fertilization commonly employ this mechanism.

Once an egg has been fertilized internally, the embryos may develop within the female's body until released as miniatures of the adult—as, for example, in some brooding gastropods, bivalves, and ophiuroids.[7] Alternatively, the fertilized eggs may be packaged in groups within egg capsules or egg masses, which are then either protected by the female or affixed to, or buried within, a substrate and abandoned. These encapsulating structures are especially complex among members of the Gastropoda (Fig. 24.13).

Among marine invertebrates, union of gametes also may be accomplished in the absence of the structurally and behaviorally complex mechanisms generally associated with internal fertilization. In the ocean, fertilization may in fact be achieved by the coordinated release of eggs and sperm into the surrounding seawater. Such **external**

7. See *Topics for Further Discussion and Investigation,* no. 8.

fertilization is common in the marine environment, as seen in Table 24.1. On the Great Barrier Reef of Australia, for example, hundreds of coral species release gametes together on just several nights after a full moon in the spring. Remarkably, this highly synchronized spawning event seems to be coordinated through the same blue-light-sensitive photoreceptor proteins (cryptochromes) known to entrain the circadian rhythms of insects and mammals.[8] In any external fertilization event, the percentage of eggs that become fertilized varies with water current velocity and direction, position of individuals within a spawning aggregation, the size of the aggregation, egg diameter, the degree to which eggs chemically attract sperm, and the extent to which spawning is coordinated within any given population.[9]

Larval Forms

The product of an external fertilization generally develops into a free-living, swimming larva, an individual that grows and differentiates entirely in the water as a member of the plankton. Larval forms are also frequently produced by species that have internal fertilization, the larvae either emerging from the females after a period of brooding, or from egg capsules or egg masses. The larvae of most invertebrate species are ciliated, the cilia serving for locomotion and, in species with feeding larvae, for food collection as well. External ciliation is incompatible with the chitinous exoskeleton of larval arthropods; among such larvae, locomotion and food collection must be achieved using specialized appendages (Table 24.3). External ciliation is also lacking during the development of nematodes, which also are enclosed within a complex cuticle. Urochordates and chaetognaths constitute the final exceptions to the general rule of marine invertebrate larvae being ciliated.

Although free-living larvae—other than insect larvae—are relatively rare in freshwater habitats, they are produced by some sponge, rotifer, copepod, and even bivalve species (e.g., the troublesome zebra mussel, *Dreissena polymorpha*—see p. 265); the physiological mechanisms through which early developmental stages cope with the intense osmotic stress imposed by their freshwater surroundings would be well worth exploring.

The likely adaptive benefits of free-living larval forms for aquatic species that are slow moving or sessile as adults are (1) dispersal and genetic exchange between geographically separated populations of the same species, (2) rapid recolonization of areas following local extinctions, (3) minimal likelihood of inbreeding (mating with close relatives) in the next generation, and (4) lack of direct competition with adults for food or space during development. The last benefit is also of particular significance for terrestrial insects.

Through the course of evolution by natural selection, larval forms have become increasingly well adapted to their own niches and may little resemble the adults of their own species, either morphologically or physiologically (Fig. 24.14). Larvae and adults may be considered ecologically distinct organisms—often exploiting entirely different habitats, lifestyles, and food sources—that just happen to have a genome in common. This latter point is crucial. Despite their ecological dissimilarity, the success of the one form in its stage of the life cycle determines the very existence of the other. Reproductive patterns comprising 2 or more ecologically distinct phases are termed **complex life cycles.**

The transition between phases of a complex life cycle often takes the form of an abrupt morphological, physiological, and ecological revolution termed a **metamorphosis** (Fig. 24.15). The greater the degree of difference between the adult and larval lifestyles, and the greater the degree of adaptation to those different lifestyles, the more dramatic the metamorphosis.[10] Complex life cycles are generally believed to be the original condition for marine invertebrates. Such life cycles seem generally to have been lost in association with the invasion of land and freshwater, but they have re-evolved in at least one group, the Insecta. The percentage of insect species exhibiting **holometabolous development** (i.e., development involving a conspicuous metamorphosis) has increased from about 10% (325 million years ago) to about 63% (200 million years ago) to about 90% (presently). The adaptive benefits of complex life histories must indeed be considerable, making it an intriguing challenge to explain why larvae have apparently been lost within a number of marine invertebrate groups.[11]

The larvae of some invertebrates are **lecithotrophic** (*lecitho* = G: yolk; *trophy* = G: feeding); that is, they subsist on nutrient reserves supplied to the egg by the parent and are independent of the outside world for food. Lecithotrophic development is common among most groups of invertebrates, particularly among marine invertebrates living at high latitudes or in very deep water. It is much less common in shallow-water environments of the tropics and subtropics. In such shallow-water habitats, the larval stages typically develop functional guts and feed upon other members of the plankton, both plant (phytoplankton) and animal (zooplankton). Such larvae are said to be **planktotrophic.** Although there would seem to be many advantages to producing such planktotrophic larvae—for example, parents need to supply only minimal nutrients per egg, so a greater number of offspring can be produced from any given amount of stored yolk—many species have apparently switched from planktotrophic to lecithotrophic development in the course of their evolution. For example, Figure 24.16 shows one recent hypothesis regarding the evolutionary relationships among 12 sea star species in two

8. Levy, O., L. Appelbaum, W. Leggat, Y. Gothlif, D.C. Hayward, D.J. Miller, and O. Hoeugh-Guldberg. 2007. *Science* 318: 467–70.

9. See *Topics for Further Discussion and Investigation*, no. 9.

10. See *Topics for Further Discussion and Investigation*, no. 5.

11. See *Topics for Further Discussion and Investigation*, no. 8.

Table 24.3 Representative Larval Invertebrates

Phylum	Characteristic Larva	Adult	Page Reference
Porifera	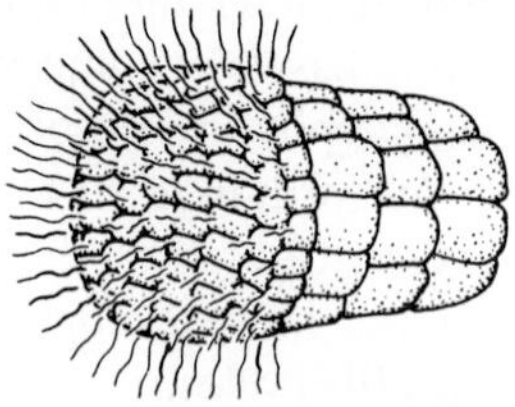 amphiblastula	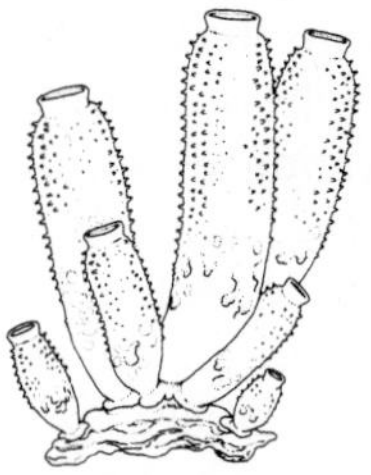 sponge	89
Cnidaria (= Coelenterata)	strobilating scyphistoma → ephyra	jellyfish (Scyphozoa)	108
	planula	hydroid (Hydrozoa)	112
	planula	anemone (Anthozoa)	118
Platyhelminthes	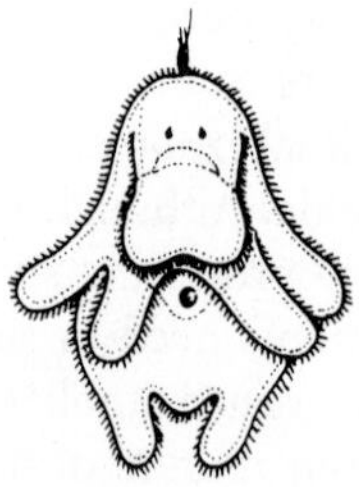 Müller's larva	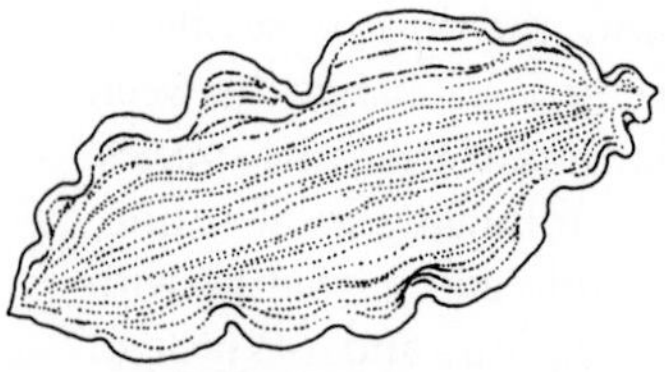 flatworm (Turbellaria)	154

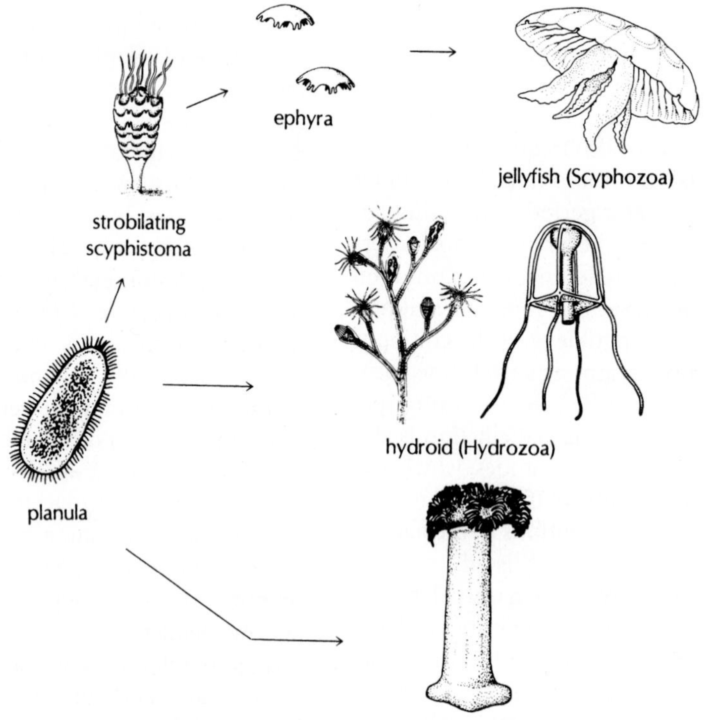

(*Continued on following page.*)

Table 24.3 *Continued*

Phylum	Characteristic Larva	Adult	Page Reference
Platyhelminthes (continued)	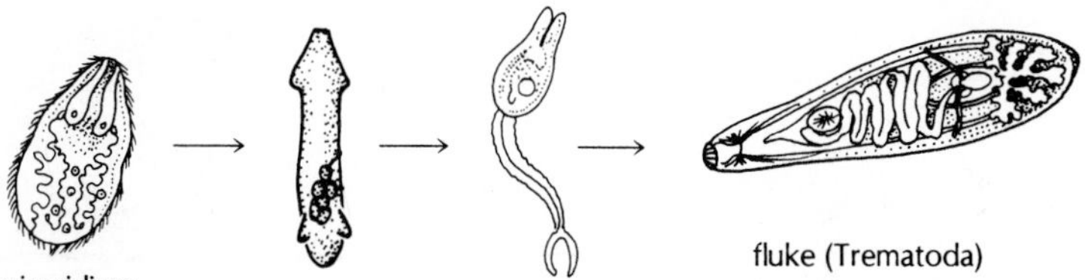 miracidium → redia → cercaria →	fluke (Trematoda)	161
Nemertea (= Rhynchocoela)	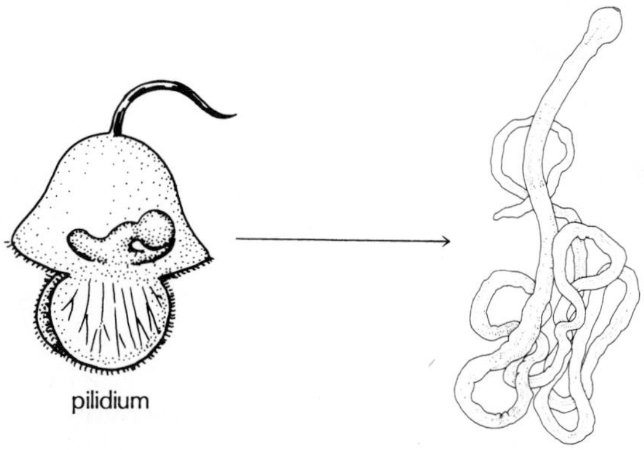 pilidium →	ribbon worm	210
Annelida	trochophore → setigerous larva →	polychaete worm (Polychaeta)	300, 310
	trochophore → pelagosphaera →	sipunculan	316

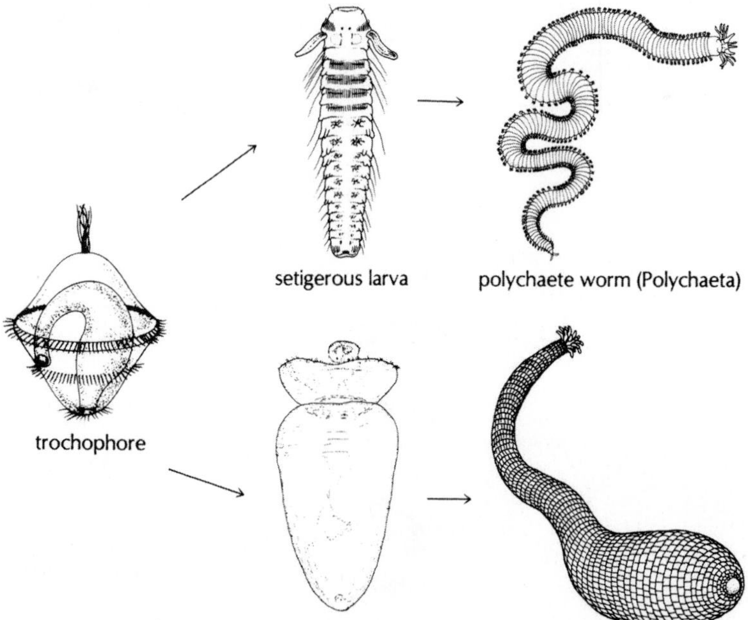

(Continued on following page.)

Table 24.3 *Continued*

Phylum	Characteristic Larva	Adult	Page Reference
Mollusca	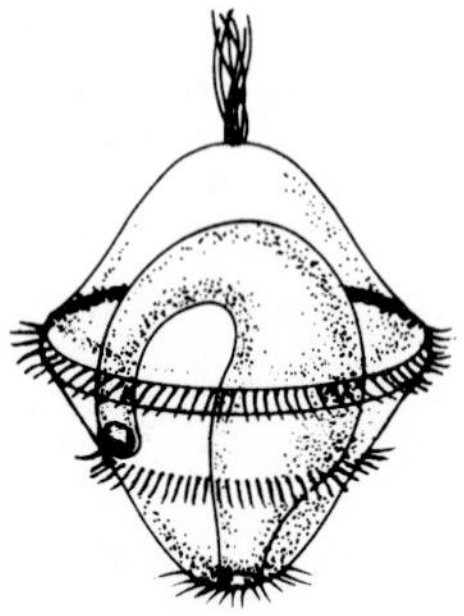 trochophore	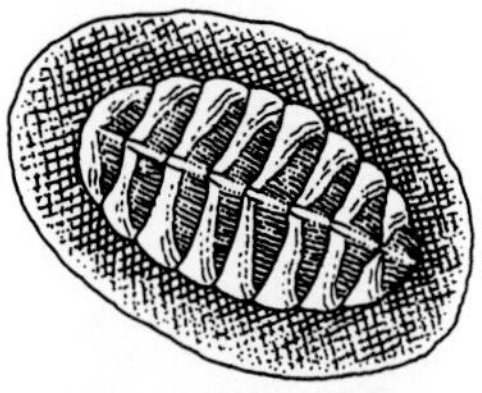 chiton (Polyplacophora)	265
	trochophore	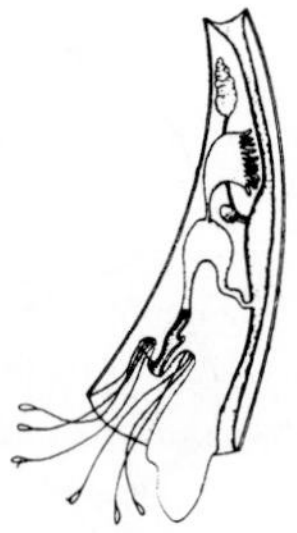 scaphopod (Scaphopoda)	265
	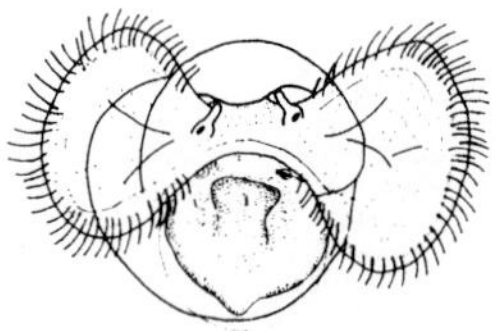 trochophore → veliger	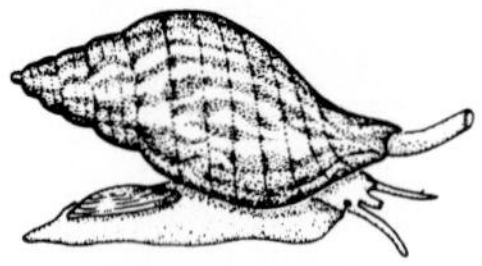 snail (Gastropoda)	265
	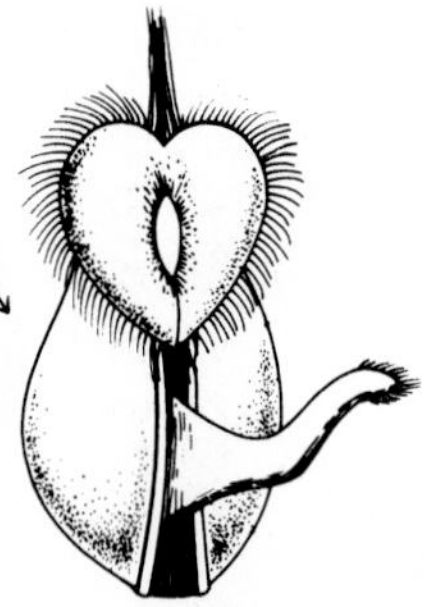 trochophore → veliger	 clam (Bivalvia)	265

(Continued on following page.)

Table 24.3 *Continued*

Phylum	Characteristic Larva	Adult	Page Reference
Arthropoda	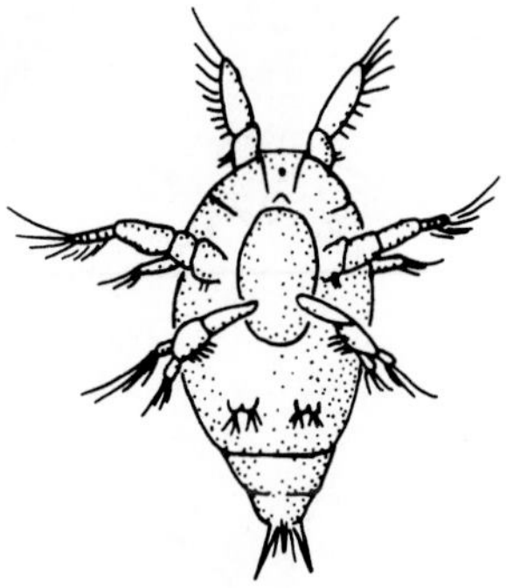 nauplius →	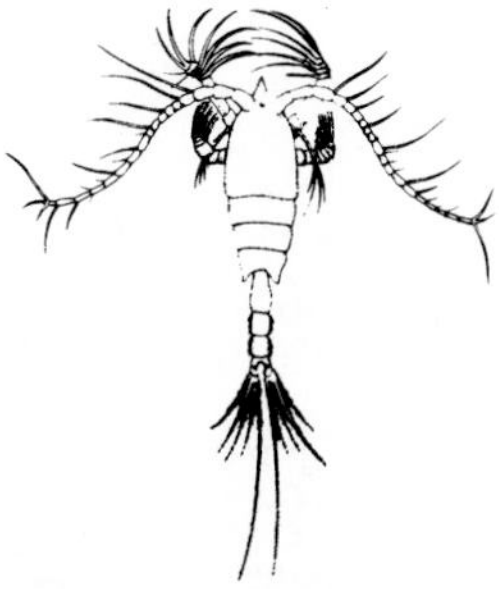 copepod (Crustacea, Copepoda)	389
	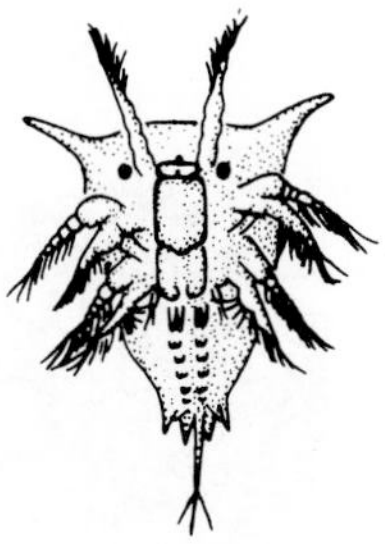 nauplius → 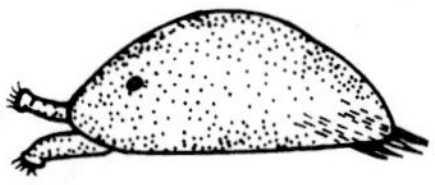 cyprid →	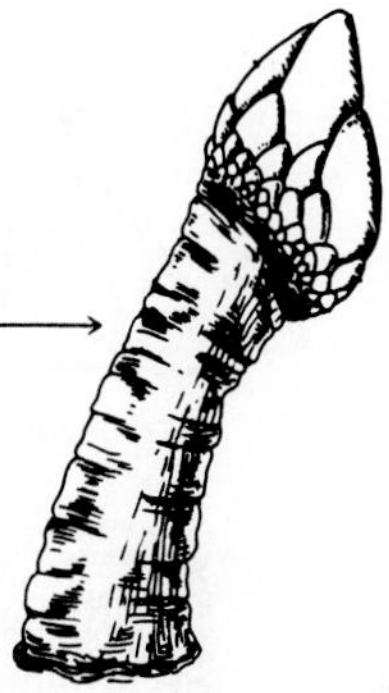 barnacle (Crustacea, Cirripedia)	389
	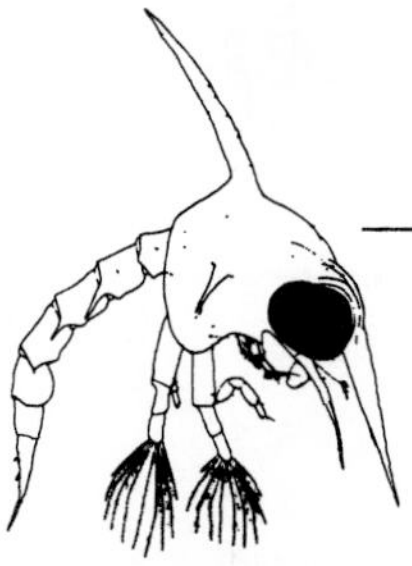 zoea → 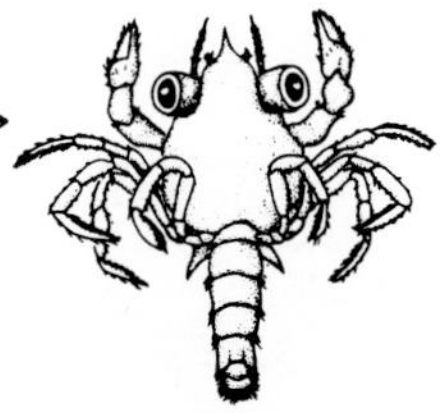 megalopa →	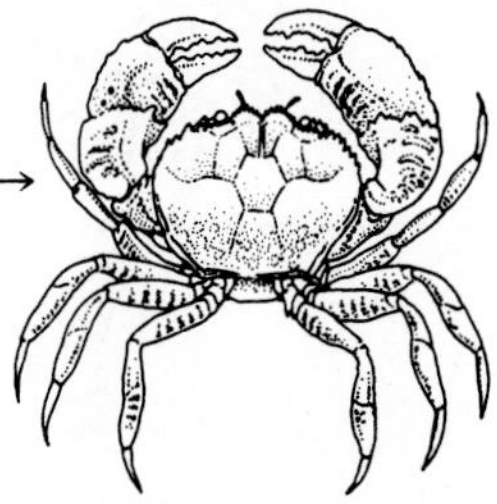 crab (Crustacea, Malacostraca, Decapoda)	391

(Continued on following page.)

Table 24.3 *Continued*

Phylum	Characteristic Larva	Adult	Page Reference
Arthropoda (continued)	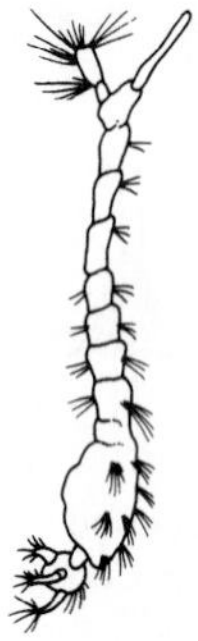 nymph, 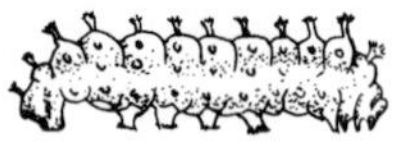 caterpillar,  pupa →	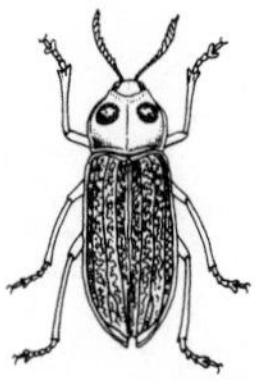 beetle,  butterfly, 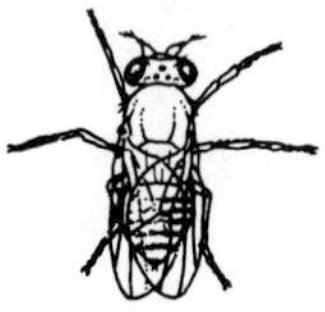 fly, 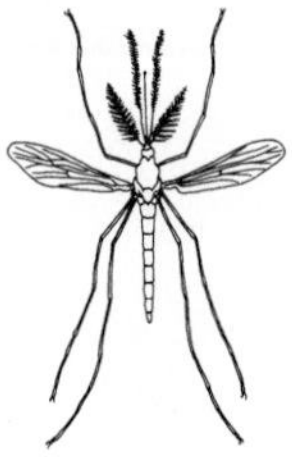 midge (Insecta)	371
Bryozoa	 cyphonautes 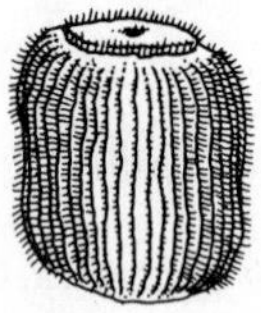 →, coronate larva →	  bryozoans	488
Phoronida	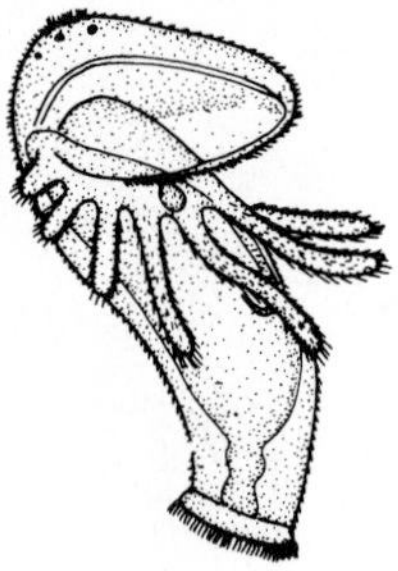 actinotroch →	 phoronid	488

(Continued on following page.)

Table 24.3 *Continued*

Phylum	Characteristic Larva	Adult	Page Reference
Brachiopoda	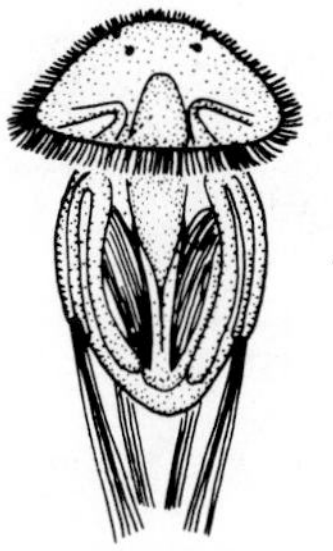 brachiopod larva	 lampshell	489
Echinodermata	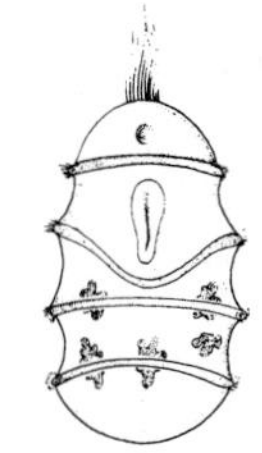 doliolaria	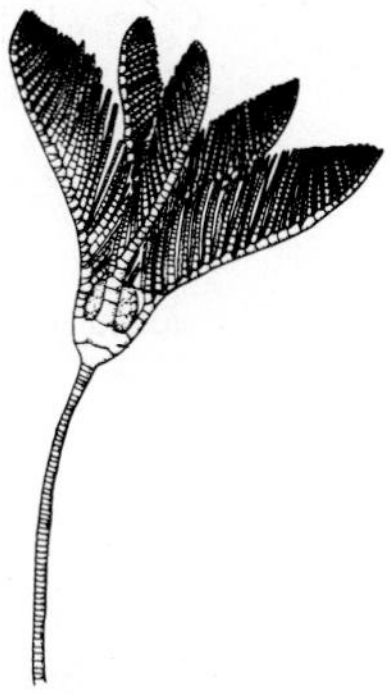 sea lily (Crinoidea)	518
	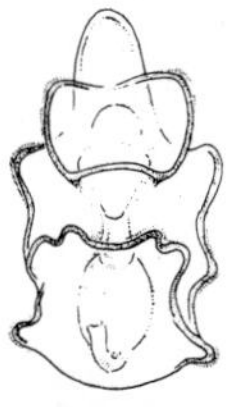 bipinnaria 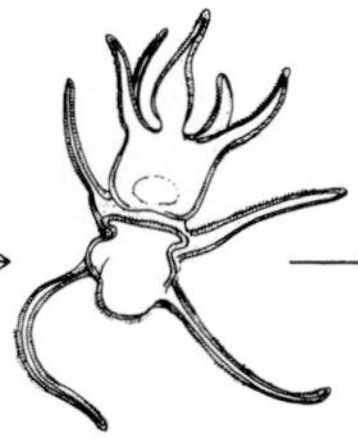 → brachiolaria	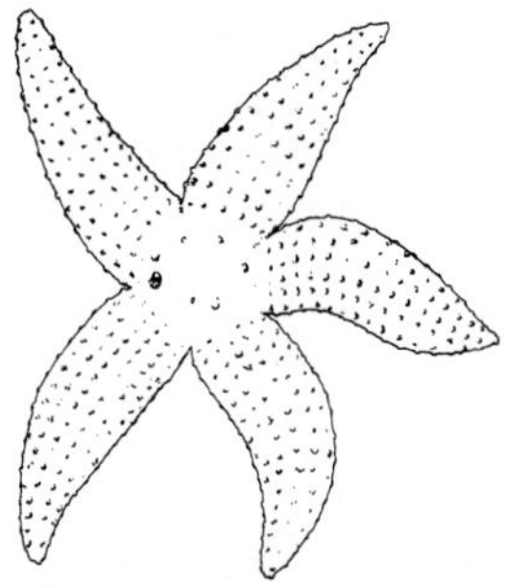 sea star (Asteroidea)	518
	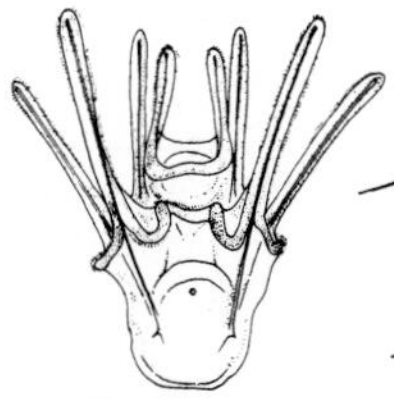 echinopluteus	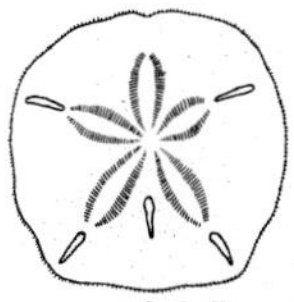 sand dollar,  sea urchin (Echinoidea)	518

(Continued on following page.)

Table 24.3 *Continued*

Phylum	Characteristic Larva	Adult	Page Reference
Echinodermata (continued)	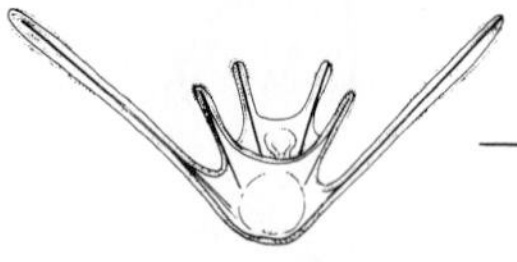 ophiopluteus	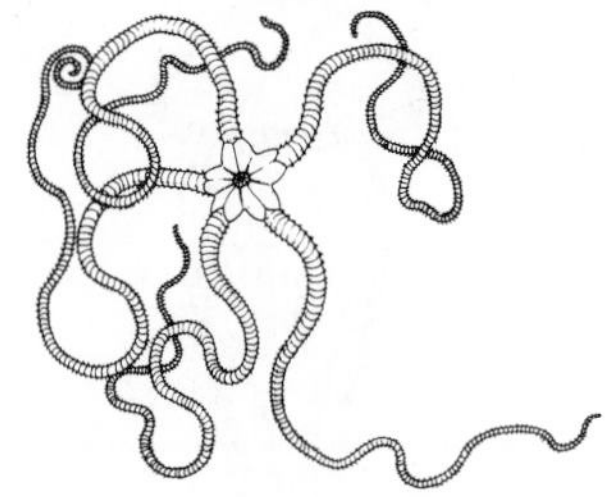 brittle star (Ophiuroidea)	518
	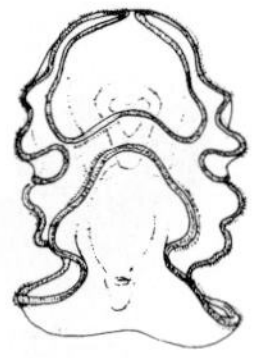 auricularia	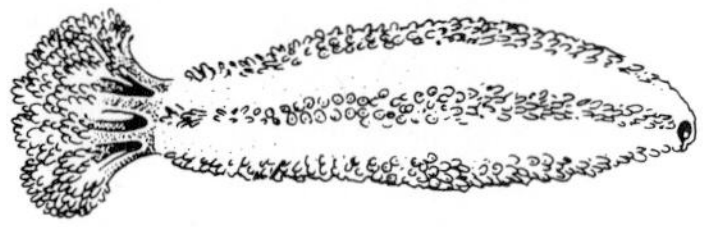 sea cucumber (Holothuroidea)	518
Hemichordata	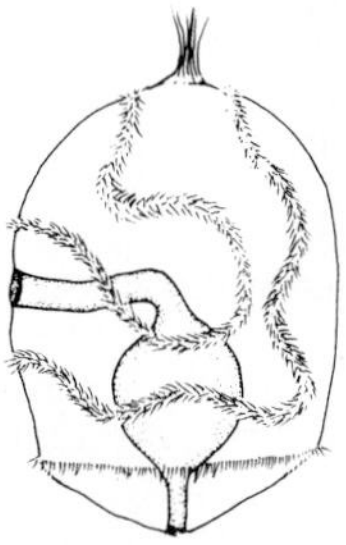 tornaria	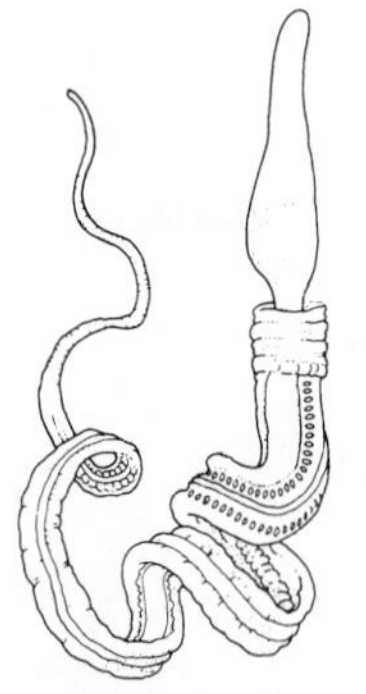 acorn worm	531
Chordata, Urochordata	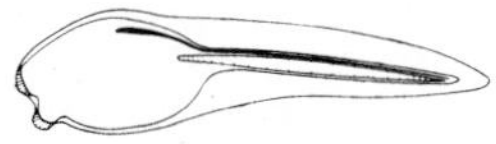 tadpole	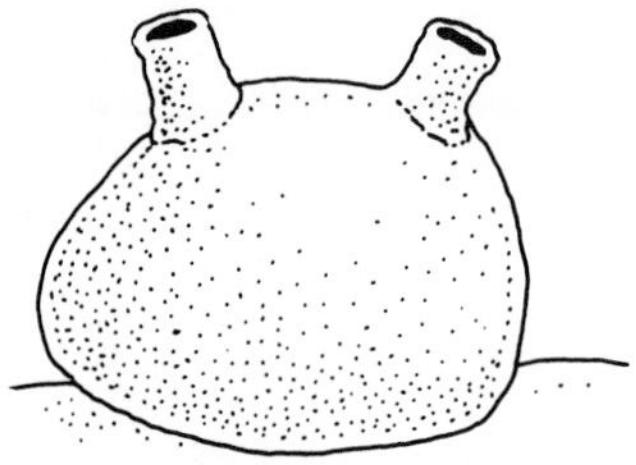 tunicate, sea squirt (Ascidiacea)	541–543

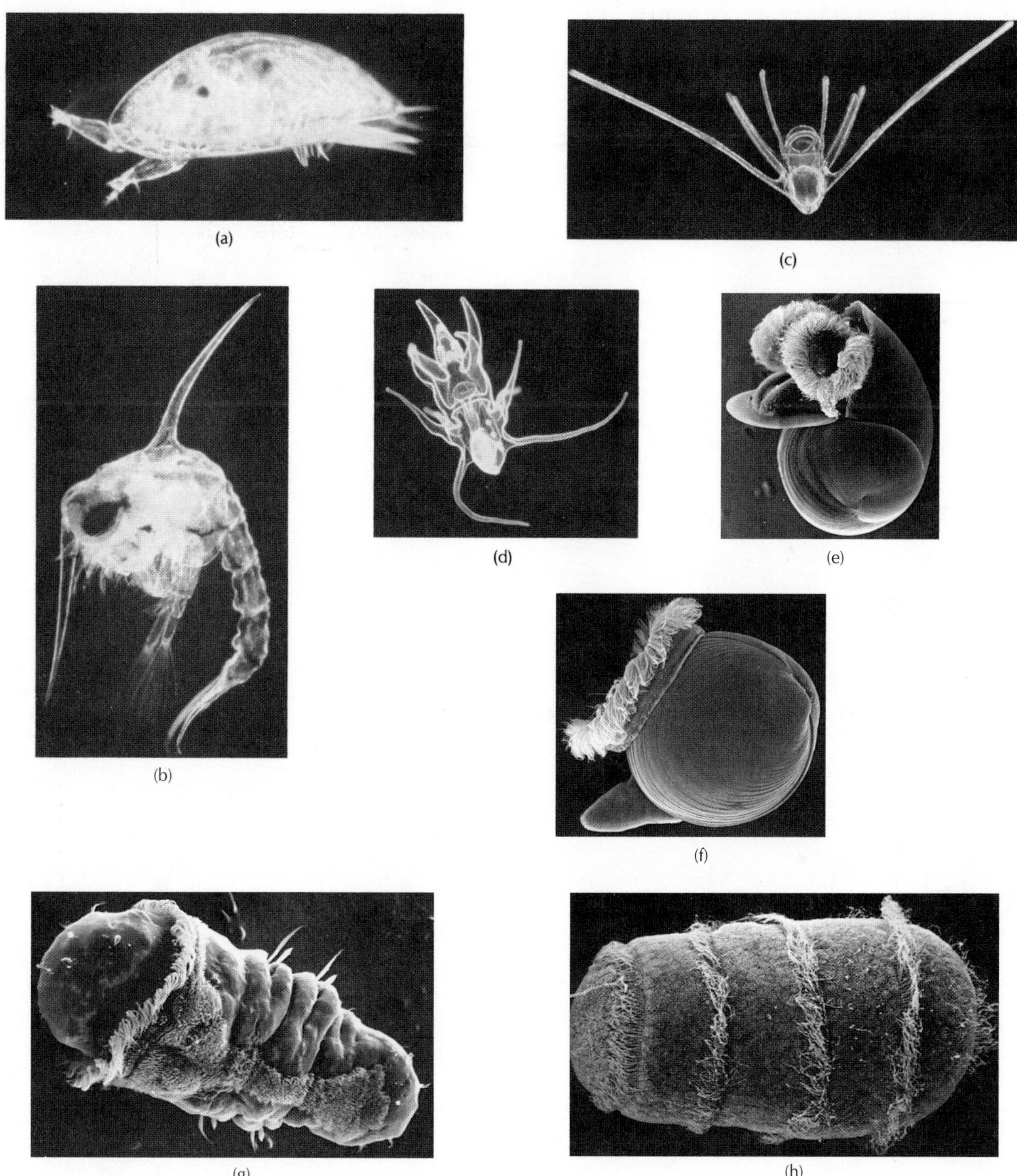

Figure 24.14

Invertebrate larvae. (a) Barnacle cyprid. (b) Zoea of a decapod crustacean. (c) Ophiopluteus of *Ophiothrix fragilis,* an ophiuroid. (d) Brachiolaria of the asteroid *Asterias vulgaris.* (e) Veliger of the opisthobranch *Rostanga pulchra.* (f) Veliger of the bivalve *Lyrodus pedicellatus,* showing shell, foot, and ciliated velum. Shell length is about 330 µm. (g) Advanced polychaete trochophore larva. Note that setae have already developed on 2 of the segments. (h) Doliolaria of the crinoid *Florometra serratissima.*

(a–d) Courtesy of D. P. Wilson/Eric and David Hosking. (e) Courtesy of F. S. Chia, from Chia, 1978. *Marine Biology* 46:109. Permission of Springer-Verlag. (f) Courtesy of C. Bradford Calloway. (g) Courtesy of F. S. McEuen, from McEuen, 1983. *Marine Biology* 76:301. Permission of Springer-Verlag. (h) Courtesy of Philip V. Mladenov, from Mladenov, P. V., and F. S. Chia, 1983. *Marine Biology* 73:309. Permission of Springer-Verlag.

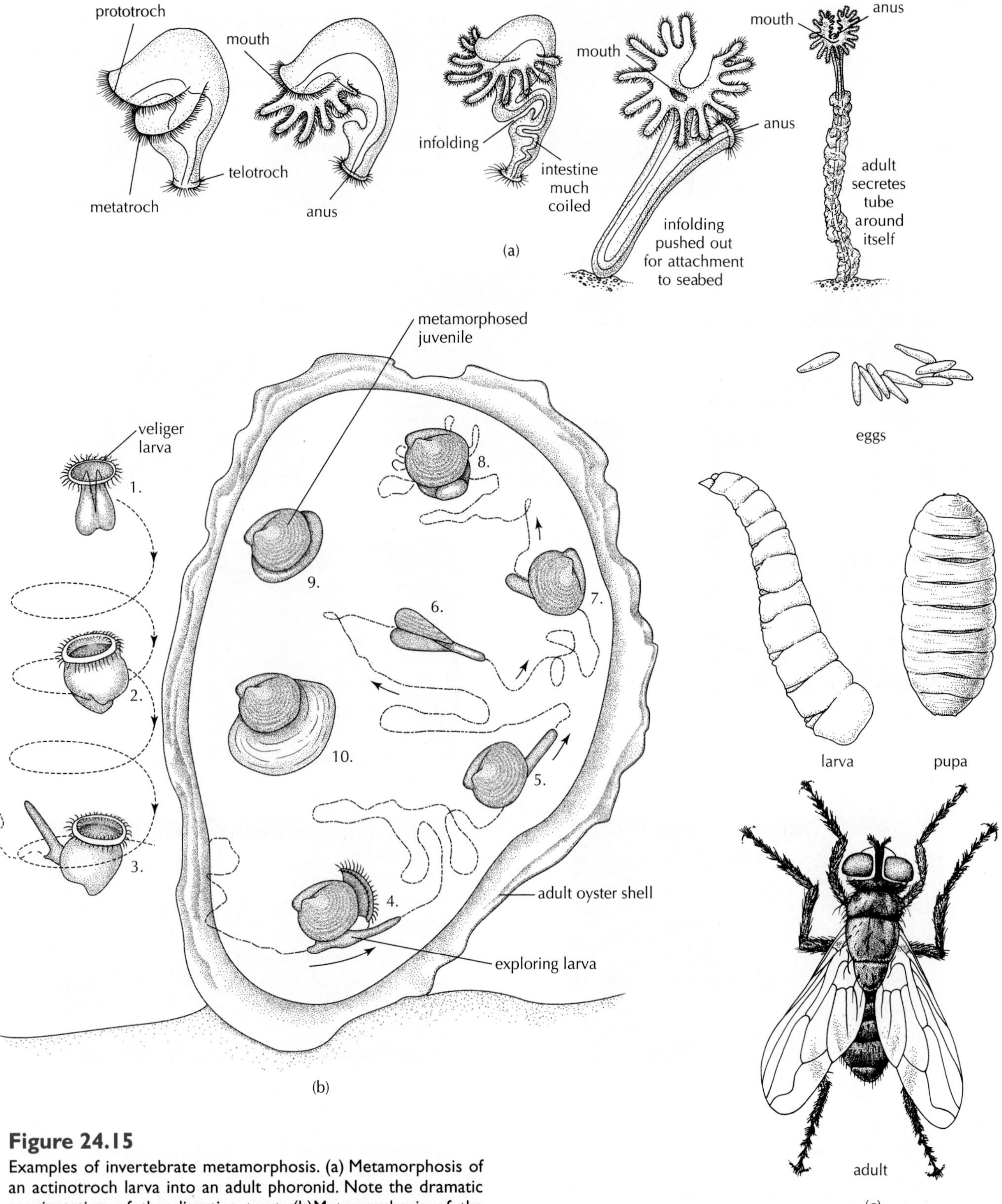

Figure 24.15

Examples of invertebrate metamorphosis. (a) Metamorphosis of an actinotroch larva into an adult phoronid. Note the dramatic reorientation of the digestive tract. (b)Metamorphosis of the oyster *Crassostrea virginica.* (c) Stages in the development of the common housefly. This animal undergoes 2 distinct metamorphoses: one from the larval to the pupal stage, and another from the pupa to the adult.

(a) Adapted from Hardy, 1965. *The Open Sea: Its Natural History.* Boston: Houghton Mifflin Company. (b) From H. F. Prytherch. 1934. Role of copper in the American oyster. *Ecolog. Monog.* 4:49–107. (c) After Engelmann and Hegner; after Packard.

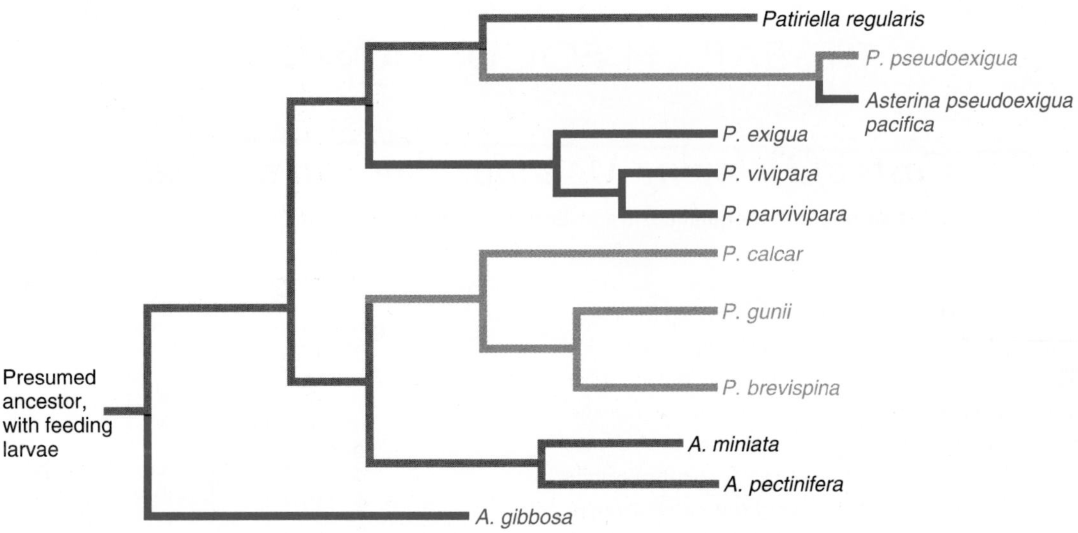

Figure 24.16

One proposed phylogenetic hypothesis for 12 species in the sea star genera *Patiriella* and *Asterina*. Species producing microscopic larvae that feed in the plankton are shown in gray. Species producing microscopic planktonic larvae that do not feed are shown in light blue. Species producing embryos that neither feed in the plankton nor disperse during development are shown in dark blue.
Based on Hart, M. W., M. Byrne, and M. J. Smith, 1997. *Evolution* 51:1848–61.

asteroid genera (Echinodermata). Some of the sea star species (shown in gray) produce microscopic, swimming larval forms that spend many days or weeks dispersing in the sea while feeding on unicellular protists before metamorphosing to adult form and habitat. Other species either release larvae that do not feed while dispersing (shown in light blue)—the larvae are endowed by the parent with enough yolk to fuel their entire planktonic development and the process of metamorphosis—or brood their offspring within or underneath the body (shown in dark blue), so that the developing young neither feed nor disperse; instead, they simply crawl out from the parent as miniature sea stars.

If the phylogeny in Figure 24.16 (one of several presented in the original paper) is correct, the ancestor to all of these 12 species had free-living, feeding larvae, and feeding larvae have been lost independently at least 4 times, once on the way to *Patieriella pseudoexigua* and *Asterina pseudoexigua pacifica*, and once on the way to *A. gibbosa*, for example. Similarly, dispersive larvae, whether they are feeding or not, have been lost during the evolution of these sea stars at least 3 times, once on the way to *A. pseudoexigua pacifica* and once on the way to *P. exigua*, *P. vivipara*, and *P. parvivipara*, for example.

The selective pressures responsible for such shifts in developmental mode are difficult to document convincingly, but certainly lecithotrophic larvae are independent of seasonal and spatial variations in phytoplankton concentration and also may benefit from more rapid development to the metamorphic stage.[12] In any event, once feeding larvae are lost from the life history of any particular species, they seem to be only rarely regained, particularly when that loss has been accompanied by extensive morphological simplification.

Dispersal as a Component of the Life-History Pattern

Most freshwater, marine, or terrestrial animals have a dispersal stage at some point in their life histories. For those marine invertebrates that exploit the properties of seawater by living a sedentary or even a sessile adult existence, dispersal is typically accomplished by a planktonic larval stage, as previously discussed. How much dispersal occurs depends on how long the larval form can be maintained and on the speed and direction of water currents in which the larvae live. In general, greater dispersal is associated with greater genetic homogeneity between geographically separated adult populations, greater ability to recolonize areas after localized extinction events, reduced rates of speciation, and greater species longevity (Fig. 24.17).[13] In recent times, larvae of many aquatic invertebrate groups have been dispersed enormous distances artificially by ocean-going ships, which take on ballast water in shallow coastal areas and release it when docking thousands of miles away. The long-term ecological consequences of such ship-mediated dispersal are likely to be substantial, as in the case of the freshwater zebra mussel, which has already spread throughout much of North America's lakes and rivers after its probable introduction from Russia in 1986.[14]

12. See *Topics for Further Discussion and Investigation*, no. 10.

13. See *Topics for Further Discussion and Investigation*, no. 12.

14. See *Topics for Further Discussion and Investigation*, no. 13.

RESEARCH FOCUS BOX 24.1

Costs of Delaying Metamorphosis in the Field

Wendt, D. E. 1998. Effect of larval swimming duration on growth and reproduction of *Bugula neritina* (Bryozoa) under field conditions. *Biol. Bull.* 195:126–35.

Aquatic invertebrate larvae must typically develop for a time in the plankton before they become competent to metamorphose. This is true for species with either planktotrophic or lecithotrophic development. Competent larvae often metamorphose in response to specific chemical or physical cues typically associated with conditions appropriate for juvenile and adult development. In the absence of those cues, competent larvae of many species delay their metamorphosis, thereby increasing the likelihood of eventually metamorphosing into a suitable adult habitat. However, recent laboratory studies have shown that delaying metamorphosis can impose costs, such as reduced juvenile survival, reduced juvenile growth rates, or reduced fecundity, so that the benefits of delaying metamorphosis may not be fully realized. Wendt (1998) sought to determine whether postponing metamorphosis of the bryozoan *Bugula neritina* would influence colony fitness in the field.

Conveniently, colonies of *B. neritina* and related species are readily stimulated to release larvae by keeping them in the dark for a time and then exposing them to bright light, and the larvae are brooded to an advanced stage of development before release: the larvae are competent to metamorphose within an hour after their release into the plankton. After obtaining hundreds of larvae from several colonies collected near Ft. Pierce, Florida, Wendt prevented some larvae from metamorphosing in the laboratory for 24 h, by keeping the seawater brightly lit and well stirred (at about 20°C). The following day, he stimulated the larvae to metamorphose onto plastic substrata, by increasing the potassium ion (K^+) concentration of the seawater they were swimming in by 10 mM (millimolar). At about the same time, he obtained an additional release of larvae, and triggered their metamorphosis onto plastic substrata immediately, again by increasing the K^+ concentration of seawater. Thus, juvenile bryozoans were about the same age after metamorphosis, whether metamorphosis had been postponed for 24 h or not.

Wendt then cut the plastic substrata into small strips, each holding one recently metamorphosed individual. The strips were then transplanted into the field and left for 14 to 17 days (water temperatures ~20°C). Wendt's experiments were designed to examine the influence of delayed metamorphosis, animal orientation, and competitive interactions on rates of colony growth and development, but here I will consider only the influence of delayed metamorphosis, for about 200 colonies.

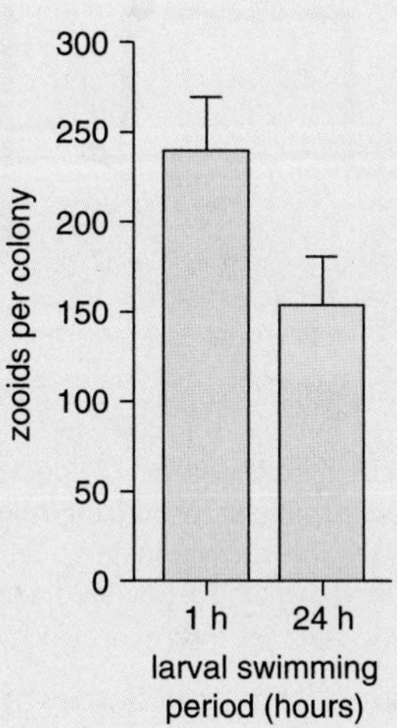

Focus Figure 24.1

Number of zooids in colonies of the bryozoan *Bugula neritina* after 14 days in the field, depending on whether or not metamorphosis was delayed. Each bar represents the average number of zooids in about 50 colonies; error bars represent the 95% confidence interval above the mean.

By the end of 2 weeks in the field, colonies were significantly smaller if they were initiated by individuals whose metamorphosis was delayed by 24 h; on average, such colonies contained nearly 40% fewer zooids (Focus Fig. 24.1). Just as impressive was the impact of the additional larval swimming period on the colonies' reproductive output.

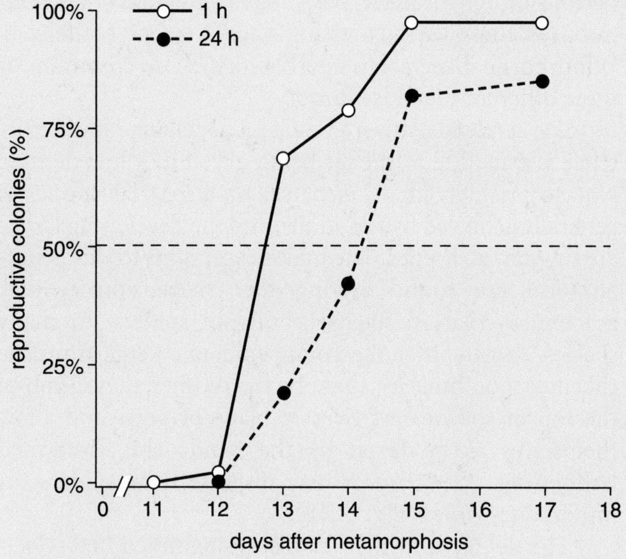

Focus Figure 24.2

The influence of delayed metamorphosis on time to first reproductive activity for bryozoan colonies in the field. Open circles represent data obtained when larvae were kept swimming for less than 1 h (at 20°C) before adding plastic substratum; solid circles represent data obtained when larvae were kept swimming for 24 h. Each curve is based on the maturation rates of 40 to 50 colonies.

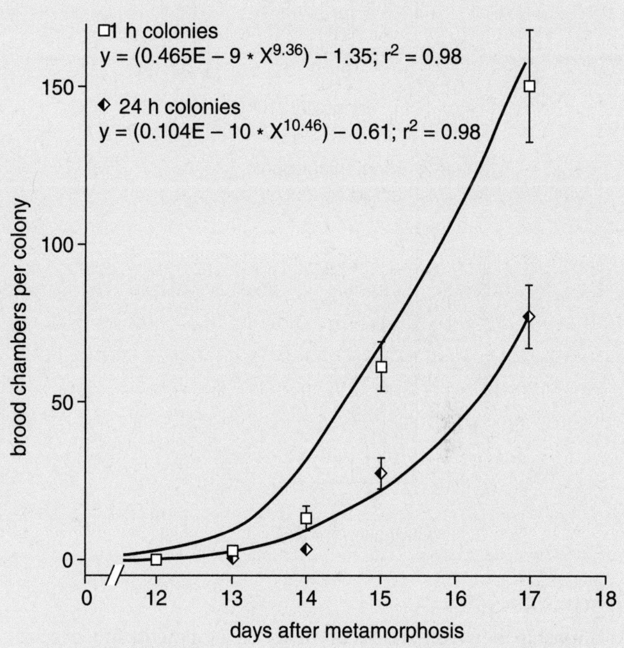

Focus Figure 24.3

The influence of delayed metamorphosis on rate of brood chamber production for bryozoan colonies in the field. Each curve is based on the development of about 90 colonies. Error bars represent 95% confidence intervals about the means.

All figures from D. E. Wendt, 1998. *Biol. Bull.* 195:126–35.

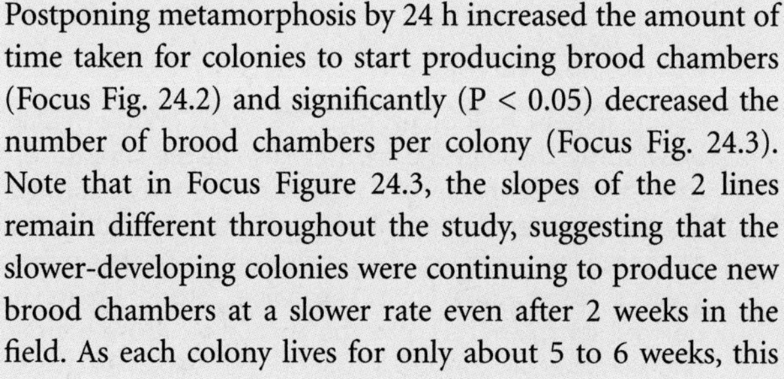

Postponing metamorphosis by 24 h increased the amount of time taken for colonies to start producing brood chambers (Focus Fig. 24.2) and significantly ($P < 0.05$) decreased the number of brood chambers per colony (Focus Fig. 24.3). Note that in Focus Figure 24.3, the slopes of the 2 lines remain different throughout the study, suggesting that the slower-developing colonies were continuing to produce new brood chambers at a slower rate even after 2 weeks in the field. As each colony lives for only about 5 to 6 weeks, this slower rate of development should translate into dramatically reduced total reproductive output.

One next question to ask might be, Do larvae of this species ever delay their metamorphosis in the field sufficiently long to bring about such dramatic reductions in growth and reproductive fitness? Can you think of a way to conduct such an investigation? What other animals might lend themselves to these sorts of field studies? What characteristics would they need to possess to make such studies feasible?

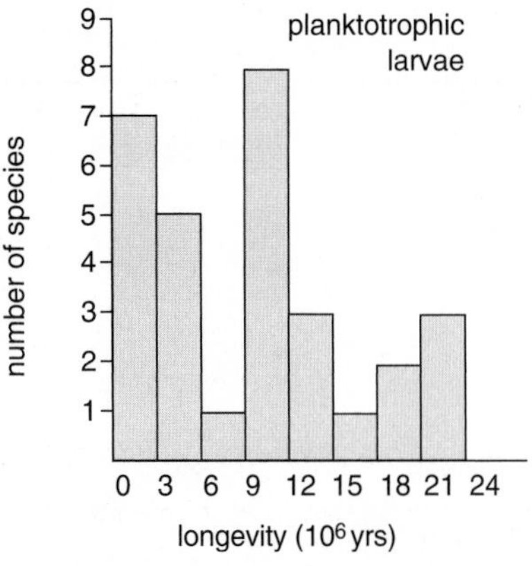

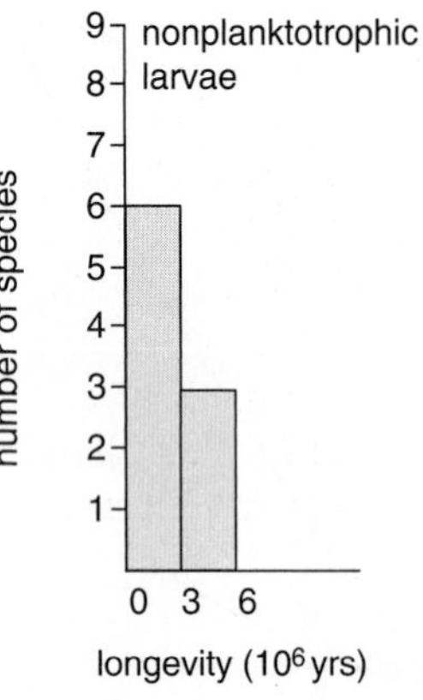

Figure 24.17

Relationship between mode of larval development and species duration in the fossil record, for 40 species in the gastropod genus *Nassarius*. Developmental mode is deduced from morphological characteristics of the premetamorphic shell, which is retained at the apex of juvenile and adult shells as the individual grows. No species with nonplanktotrophic larvae (limited dispersal potential) persisted beyond 4.5 million years, whereas many species with planktotrophic larvae (extensive dispersal potential) have fossil records extending from 9–24 million years.

Freshwater invertebrates must often deal with an ephemeral habitat. Some groups, such as the sponges, tardigrades, bryozoans, and a number of crustaceans and rotifers, often avoid the need to disperse away from unfavorable conditions by forming resistant stages of arrested development. Such stages take the form of gemmules in sponges, resting eggs in crustaceans and rotifers, and statoblasts in bryozoans. Rotifers, tardigrades, and many protozoans also may enter into a cryptobiotic state during periods of dehydration. When environmental conditions improve, these various resting stages become revitalized. Wind-borne dispersal to new habitats is also common for these different diapause forms.

For terrestrial invertebrates, a sedentary adult existence is rare, due to the air being dry and of low density. Not surprisingly then, dispersal of terrestrial species is generally achieved by the adults, and the developing larvae are the stay-at-homes. The major exceptions to this generalization are found among the "suspension-feeding" arachnids—that is, the web-building spiders. In many species, shortly after the young spiders emerge from the silken cocoon built for them by the mother, they climb to the top of the nearest twig or blade of grass and allow themselves to be taken by the wind. This is termed **ballooning.** Air currents may disperse baby spiders for hundreds of miles (Fig. 24.18).

The degree to which the dispersing individual selects the appropriate habitat for the sedentary phase of the life cycle obviously determines the probability that a given individual will indeed survive to reproductive maturity. Among marine invertebrates with free-living larvae, selective pressures acting against random metamorphosis to adult form and habitat must have been substantial. The dispersing larvae of many species are highly selective about where they will metamorphose and can delay their metamorphosis (up to many months for some species of molluscs, echinoderms, polychaetes and arthropods) if they fail to encounter the appropriate cues. The cues that

Figure 24.18

Young spiders ballooning. . . . "A warm draft of rising air blew softly through the barn cellar. The air smelled of the damp earth, of the spruce woods, of the sweet springtime. The baby spiders felt the warm updraft. One spider climbed to the top of the fence. Then it did something that came as a great surprise to Wilbur. The spider stood on its head, pointed its spinnerets into the air, and let loose a cloud of fine silk. The silk formed a balloon. As Wilbur watched, the spider let go of the fence and rose into the air.

"Good-bye!" it said, as it sailed through the doorway.

"Wait a minute!" screamed Wilbur. "Where do you think you're going?"

But the spider was already out of sight. . . . The air was soon filled with tiny balloons, each balloon carrying a spider.

Wilbur was frantic. Charlotte's babies were disappearing at a great rate."

Table 24.4 Gregarious Metamorphosis by Barnacle Larvae

	Control	*B. balanoides*	*B. crenatus*	*E. modestus*
Total number of larvae in experiment	120	120	120	120
Average number metamorphosing per dish (± 1 standard error about the mean)	0.4 ± 0.31	9.5 ± 0.81	3.7 ± 0.93	1.6 ± 0.85
Average percent metamorphosed (± 1 standard error about the mean)	3.3 ± 2.5%	79.2 ± 6.7%	30.8 ± 7.7%	13.3 ± 7.0%

Source: Data from Knight-Jones. 1953. *J. Exp. Biol.* 30:584.
Note: Groups of 12 cyprid larvae of *Balanus balanoides* were placed in each of 10 dishes of seawater (controls) or in dishes containing seawater plus the adults and shells of *B. balanoides* or of 2 other barnacle species, *B. crenatus* or *Elminius modestus*. The number of metamorphosed individuals was assessed after 24 hours.

trigger metamorphosis are often associated with some component of the adult environment: The adult food source or prey species, for example, or quite commonly, adults of the same species (Table 24.4).[15]

The extent to which larvae actually delay their metamorphosis in the field and the conditions under which they delay their metamorphosis have been difficult to explore. It is becoming clear, however, that there may be unanticipated costs associated with delaying metamorphosis in terms of postmetamorphic survival or growth rate (Research Focus Box 24.1); in such cases, the adaptive benefits of prolonging the competent period of larval development, so obvious and convincing in theory, may not be realized in practice.

Among the terrestrial insects, the larval stages are the sedentary eating machines, and the adults are the dispersal stages. Not surprisingly, the larvae of many species have become highly adapted for life on specific hosts, and the adults, in turn, often have evolved adaptations promoting the placement of eggs in the habitats supporting the best larval growth and survival.

Even though much of the existing diversity of reproductive pattern has not found its way into the preceding discussion, invertebrates clearly show a dazzling variety of ways to reproduce. Many of the patterns that seem strange to us—external fertilization, simultaneous and sequential hermaphroditism, and parthenogenesis, for example—are actually very common reproductive modes. Why are there so many different ways to reproduce?[16] The question is much discussed in the literature of evolutionary ecology but challenging to answer definitively. For one thing, the adaptive significance of any particular pattern is often difficult to demonstrate. Moreover, the selective pressures responsible for presently observed reproductive patterns may be quite different today from what they were hundreds of thousands of years ago. Lastly, the role of historical accident in shaping the reproductive patterns of various invertebrate groups is hidden among evolutionary interrelationships that are often unclear and that may never be satisfactorily unraveled.

15. See *Topics for Further Discussion and Investigation,* no. 4.

16. See *Topics for Further Discussion and Investigation,* nos. 8 and 11.

Nevertheless, discussions regarding the remarkable diversity of reproductive patterns encountered among invertebrates surely will continue long into the future, and our understanding surely will improve as new data on the costs and benefits associated with different patterns of reproduction and development are accumulated for more and more species, as more convincing phylogenies are constructed and tested, and as the mechanisms through which reproductive patterns evolve become clearer.

Topics for Further Discussion and Investigation

1. What are the roles of temperature and light in regulating the cycles of invertebrate reproductive activity?

Davison, J. 1976. *Hydra hymanae:* Regulation of the life cycle by time and temperature. *Science* 194:618.

De March, B. G. 1977. The effects of photoperiod and temperature on the induction and termination of reproductive resting stage in the freshwater amphipod *Hyallela azteca* (Sanssure). *Canadian J. Zool.* 55:1595.

Fell, P. E. 1974. Diapause in the gemmules of the marine sponge, *Haliclona loosanoffi,* with a note on the gemmules of *Haliclona oculata. Biol. Bull.* 147:333.

Hardege, J. D., H. D. Bartels-Hardege, E. Zeeck, and F. T. Grimm. 1990. Induction of swarming of *Nereis succinea. Marine Biol.* 104:291.

Hayes, J. L. 1982. Diapause and diapause dynamics of *Colias alexandra* (Lepidoptera: Pieridae). *Oecologia (Berlin)* 53:317.

Jokiel, P. L., R. Y. Ito, and P. M. Liu. 1985. Night irradiance and synchronization of lunar release of planula larvae in the reef coral *Pocillopora damicornis. Marine Biol.* 88:167.

Pearse, J. S., and D. J. Eernisse. 1982. Photoperiodic regulation of gametogenesis and gonadal growth in the sea star *Pisaster ochraceus. Marine Biol.* 67:121.

Rose, S. M. 1939. Embryonic induction in *Ascidia. Biol. Bull.* 77:216.

Schierwater, B., and C. Hauenschild. 1990. A photoperiod determined life-cycle in an oligochaete worm. *Biol. Bull.* 178:111.

Stanwell-Smith, D., and L. S. Peck. 1998. Temperature and embryonic development in relation to spawning and field occurrence of larvae of three Antarctic echinoderms. *Biol. Bull.* 194:44.

Stross, R. G., and J. C. Hill. 1965. Diapause induction in *Daphnia* requires two stimuli. *Science* 150:1462.

Vowinckel, C. 1970. The role of illumination and temperature in the control of sexual reproduction in the planarian *Dugesia tigrina* (Girarad). *Biol. Bull.* 138:77.

Walker, C. W., and M. P. Lesser. 1998. Manipulation of food and photoperiod promotes out-of-season gametogenesis in the green sea urchin, *Strongylocentrotus droebachiensis:* Implications for aquaculture. *Marine Biol.* 132:663.

Wayne, N. L., and G. D. Block. 1992. Effects of photoperiod and temperature on egg-laying behavior in a marine mollusk, *Aplysia californica. Biol. Bull.* 182:8.

West, A. B., and C. C. Lambert. 1976. Control of spawning in the tunicate *Styela plicata* by variations in a natural light regime. *J. Exp. Zool.* 195:263.

Williams-Howze, J., and B. C. Coull. 1992. Are temperature and photoperiod necessary cues for encystment in the marine benthic harpacticoid copepod *Heteropsyllus nunni* Coull? *Biol. Bull.* 182:109.

2. What aspects of reproductive activity appear to be under chemical control among invertebrates?

Abdu, U., P. Takac, H. Laufer, and A. Sagi. 1998. Effect of methyl farnesoate on late larval development and metamorphosis in the prawn *Macrobrachium rosenbergii* (Decapoda, Palaemonidae): A juvenoid-like effect? *Biol. Bull.* 195:112.

Adamo, S. A., and R. Chase. 1990. The "love dart" of the snail *Helix aspersa* injects a pheromone that decreases courtship duration. *J. Exp. Zool.* 255:80.

Bishop, J. D. D., P. H. Manriquez, and R. N. Hughes. 2000. Water-borne sperm trigger vitellogenic egg growth in two sessile marine invertebrates. *Proc. Royal Soc. London* B 267:1165–69.

Boettcher, A. A., and N. M. Targett. 1998. Role of chemical inducers in larval metamorphosis of queen conch, *Strombus gigas* Linnaeus: Relationship to other marine invertebrate systems. *Biol. Bull.* 194:132.

Coll, J. C., B. F. Bowden, G. V. Meehan, G. M. Konig, A. R. Carroll, D. M. Tapiolas, P. M. Aliño, et al. 1994. Chemical aspects of mass spawning in corals. Sperm-attractant molecules in the eggs of the scleractinian coral *Montipora digitata. Marine Biol.* 118:177–82.

Crisp, D. J. 1956. A substance promoting hatching and liberation of young in cirripedes. *Nature* 178:263.

Engelmann, F. 1959. The control of reproduction in *Diploptera punctata* (Blattaria). *Biol. Bull.* 116:406.

Forward, R. B., Jr., and K. J. Lohmann. 1983. Control of egg hatching in the crab *Rhithropanopeus harrisii* (Gould). *Biol. Bull.* 165:154.

Golden, J. W., and D. L. Riddle. 1982. A pheromone influences larval development in the nematode *Caenorhabditis elegans. Science* 218:578.

Golding, D. W. 1967. Endocrinology, regeneration and maturation in *Nereis. Biol. Bull.* 133:567.

Kanatani, H., and M. Ohguri. 1966. Mechanism of starfish spawning. I. Distribution of active substance responsible for maturation of oocytes and shedding of gametes. *Biol. Bull.* 131:104.

Kopin, C. Y., C. E. Epifanio, S. Nelson, and M. Stratton. 2001. Effects of chemical cues on metamorphosis of the Asian shore crab *Hemigrapsus sanguineus,* an invasive species on the Atlantic Coast of North America. *J. Exp. Marine Biol. Ecol.* 265:141–51.

Kuhlmann, H.-W., C. Brünen-Nieweler, and K. Heckmann. 1997. Pheromones of the ciliate *Euplotes octocarinatus* not only induce conjugation but also function as chemoattractants. *J. Exp. Zool.* 277:38.

Lindquist, N., and M. E. Hay. 1996. Palatability and chemical defense of marine invertebrate larvae. *Ecolog. Monogr.* 66:431.

Marthy, H., J. R. Hauser, and A. Scholl. 1976. Natural tranquilizer in cephalopod eggs. *Nature* 261:496.

Morse, D. E., N. Hooker, H. Duncan, and L. Jensen. 1979. γ-Aminobutyric acid, a neurotransmitter, induces planktonic abalone larvae to settle and begin metamorphosis. *Science* 204:407.

Painter, S. D., B. Clough, S. Black, and G. T. Nagle. 2003. Behavioral characterization of attractin, a water-borne peptide in the genus *Aplysia. Biol. Bull.* 205:16.

Reynolds, S. E., P. H. Taghert, and J. W. Truman. 1979. Eclosion hormone and bursicon titres and the onset of hormonal responsiveness during the last day of adult development in *Manduca sexta* (L.). *J. Exp. Biol.* 78:77.

Shorey, H. H., and R. J. Bartel. 1970. Role of a volatile sex pheromone in stimulating male courtship behaviour in *Drosophila melanogaster. Anim. Behav.* 18:159.

Snell, T. W., R. Rico-Martinez, L. N. Kelly, and T. E. Battle. 1995. Identification of a sex pheromone from a rotifer. *Marine Biol.* 123:347.

Takeda, N. 1979. Induction of egg-laying by steroid hormones in slugs. *Comp. Biochem. Physiol.* 62A:273.

Ting, J. H., and T. W. Snell. 2003. Purification and sequencing of a mate-recognition protein from the copepod *Tigriopus japonicus. Marine Biol.* 143:1.

Truman, J. W., and P. G. Sokolove. 1972. Silk moth eclosion: Hormonal triggering of a centrally programmed pattern of behavior. *Science* 175:1491.

Vaughn, D., and R. R. Strathmann. 2008. Predators induce cloning in echinoderm larvae. *Science* 319:1503.

Watson, G. J., F. M. Langford, S. M. Gaudron, and M. G. Bentley. 2000. Factors influencing spawning and pairing in the scale worm *Harmothoe imbricata* (Annelida: Polychaeta). *Biol. Bull.* 199:50.

Wheeler, D. E., and H. F. Nijhout. 1983. Soldier determination in *Pheidole bicarinata:* Effect of methoprene on caste and size within castes. *J. Insect Physiol.* 29:847.

Wigglesworth, V. B. 1934. The physiology of ecdysis in *Rhodnius prolixus* (Hemiptera). II. Factors controlling moulting and "metamorphosis." *Quart. J. Microsc. Sci.* 77:191.

Zimmer-Faust, R. K., and M. N. Tamburri. 1994. Chemical identity and ecological implications of a waterborne, larval settlement cue. *Limnol. Oceanogr.* 39:1075.

3. Discuss the morphological and behavioral adaptations for indirect sperm transfer.

Blades, P. I. 1977. Mating behavior of *Centropages typicus* (Copepoda: Calanoida). *Marine Biol.* 40:57.

Feldman-Muhsam, B. 1967. Spermatophore formation and sperm transfer in ornithodoros ticks. *Science* 156:1252.

Hsieh, H.-L., and J. L. Simon. 1990. The sperm transfer system in *Kinbergonuphis simoni* (Polychaeta: Onuphidae). *Biol. Bull.* 178:85.

Legg, G. 1977. Sperm transfer and mating in *Ricinoides hanseni* (Ricinulei: Arachnida). *J. Zool., London* 182:51.

Reeve, M. R., and M. A. Walter. 1972. Observations and experiments on methods of fertilization in the chaetognath *Sagitta hispida. Biol. Bull.* 143:207.

Vreys, C., E. R. Schockaert, and N. K. Michiels. 1997. Formation, transfer, and assimilation of the spermatophore of the hermaphroditic flatworm *Dugesia gonocephala* (Triclada, Paludicola). *Canadian J. Zool.* 75:1479.

Wedell, N. 1994. Dual function of the bushcricket spermatophore. *Proc. Royal Soc. London* B 258:181.

Weygoldt, P. 1966. Mating behavior and spermatophore morphology in the pseudoscorpion *Dinocheirus tumidus* Banks (Cheliferinea: Chernetidae). *Biol. Bull.* 130:462.

Yund, P. O. 1990. An in situ measurement of sperm dispersal in a colonial marine hydroid. *J. Exp. Zool.* 253:102.

4. How are the cues used by dispersal stages related to the habitat requirements of the sedentary or sessile stage in a complex life cycle?

Brewer, R. H. 1976. Larval settling behavior in *Cyanea capillata* (Cnidaria: Scyphozoa). *Biol. Bull.* 150:183.

Chew, F. S. 1977. Coevolution of pierid butterflies and their cruciferous food plants. II. The distribution of eggs on potential food plants. *Evolution* 31:568.

Cohen, L. M., H. Neimark, and L. K. Eveland. 1980. *Schistosoma mansoni:* Response of cercariae to a thermal gradient. *J. Parasitol.* 66:362.

Grosberg, R. K. 1981. Competitive ability influences habitat choice in marine invertebrates. *Nature* 290:700.

Highsmith, R. C. 1982. Induced settlement and metamorphosis of sand dollar (*Dendraster excentricus*) larvae in predator-free sites: Adult sand dollar beds. *Ecology* 63:329.

Hunte, W., J. R. Marsden, and B. E. Conlin. 1990. Habitat selection in the tropical polychaete *Spirobranchus giganteus.* III. Effects of coral species on body size and body proportions. *Marine Biol.* 104:101.

Hurlbut, C. J. 1993. The adaptive value of larval behavior of a colonial ascidian. *Marine Biol.* 115:253.

Jensen, R. A., and D. E. Morse. 1984. Intraspecific facilitation of larval recruitment: Gregarious settlement of the polychaete *Phragmatopoma californica* (Fewkes). *J. Exp. Marine Biol. Ecol.* 83:107.

Knight-Jones, E. W. 1953. Laboratory experiments on gregariousness during settling in *Balanus balanus* and other barnacles. *J. Exp. Biol.* 30:584.

MacInnes, A. J., W. M. Bethel, and E. M. Cornfield. 1974. Identification of chemicals of snail origin that attract *Schistosoma mansoni* miracidia. *Nature* (*London*) 248:361.

Olson, R. 1983. Ascidian-*Prochloron* symbiosis: The role of larval photoadaptations in midday larval release and settlement. *Biol. Bull.* 165:221.

Pawlik, J. R., C. A. Butman, and V. R. Starczak. 1991. Hydrodynamic facilitation of gregarious settlement of a reef-building tube worm. *Science* 251:421.

Raimondi, P. T. 1988. Settlement cues and determination of the vertical limit of an intertidal barnacle. *Ecology* 69:400.

Scheltema, R. S. 1961. Metamorphosis of the veliger larvae of *Nassarius obsoletus* (Gastropoda) in response to bottom sediment. *Biol. Bull.* 120:92.

Turner, E. J., R. K. Zimmer-Faust, M. A. Palmer, M. Luckenbach, and N. D. Pentcheff. 1995. Settlement of oyster (*Crassostrea virginica*) larvae: Effects of water flow and a water-soluble chemical cue. *Limnol. Oceanogr.* 1579.

Wallace, R. L. 1978. Substrate selection by larvae of the sessile rotifer *Ptygura beauchampi. Ecology* 59:221.

Williams, K. S. 1983. The coevolution of *Euphydryas chalcedona* butterflies and their larval host plants. III. Oviposition behavior and host plant quality. *Oecologia (Berlin)* 56:336.

Wood, E. M. 1974. Some mechanisms involved in host recognition and attachment of the glochidium larva of *Anodonta cygnea* (Mollusca: Bivalvia). *J. Zool.* 173:15.

5. Describe the morphological changes that take place during metamorphosis in one group of marine invertebrates.

Amano, S., and I. Hori. 2001. Metamorphosis of coeloblastula performed by multipotential larval flagellated cells in the calcareous sponge *Leucosolenia laxa. Biol. Bull.* 200:20–32.

Atkins, D. 1955. The cyphonautes larvae of the Plymouth area and the metamorphosis of *Membranipora membranacea* (L.). *J. Marine Biol. Assoc. U.K.* 34:441.

Berrill, N. J. 1947. Metamorphosis in ascidians. *J. Morphol.* 81:249.

Bickell, L. R., and S. C. Kempf. 1983. Larval and metamorphic morphogenesis in the nudibranch *Melibe leonina* (Mollusca: Opisthobranchia). *Biol. Bull.* 165:119.

Bonar, D. B., and M. G. Hadfield. 1974. Metamorphosis of the marine gastropod *Phestilla sibogae* Bergh (Nudibranchia: Aeolidacea). I. Light and electron microscopic analysis of larval and metamorphic stages. *J. Exp. Marine Biol. Ecol.* 16:227.

Cameron, R. A., and R. T. Hinegardner. 1978. Early events in sea urchin metamorphosis, description and analysis. *J. Morphol.* 157:21.

Cloney, R. A. 1977. Larval adhesive organs and metamorphosis in ascidians. *Cell Tissue Res.* 183:423.

Cole, H. A. 1938. The fate of the larval organs in the metamorphosis of *Ostrea edulis. J. Marine Biol. Assoc. U.K.* 22:469.

Emlet, R. B. 1988. Larval form and metamorphosis of a "primitive" sea urchin, *Eucidaris thouarsi* (Echinodermata: Echinoidea: Cidaroida), with implications for developmental and phylogenetic studies. *Biol. Bull.* 174:4.

Factor, J. R. 1981. Development and metamorphosis of the digestive system of larval lobsters, *Homarus americanus* (Decapoda: Nephropidae). *J. Morphol.* 169:225.

Lang, W. H. 1976. The larval development and metamorphosis of the pedunculate barnacle *Octolasmis mülleri* (Coker, 1902) reared in the laboratory. *Biol. Bull.* 150:255.

Reed, C. G., and R. M. Woollacott. 1982. Mechanisms of rapid morphogenetic movements in the metamorphosis of the bryozoan *Bugula neritina* (Cheilostomata, Cellularioidea). I. Attachment to the substratum. *J. Morphol.* 172:335.

Stricker, S. A. 1985. An ultrastructural study of larval settlement in the sea anemone *Urticina crassicornis* (Cnidaria, Actiniaria). *J. Morphol.* 186:237.

6. Discuss the possible adaptive significance of the apyrene sperm of insects and suggest ways to test the various hypotheses.

Silberglied, R. E., J. G. Shepherd, and J. L. Dickinson. 1984. Eunuchs: The role of apyrene sperm in Lepidoptera? *Amer. Nat.* 123:255.

7. What impact do UV irradiation and environmental pollutants have on the reproduction and development of marine invertebrates?

Bellam, G., D. J. Reish, and J. P. Foret. 1972. The sublethal effects of a detergent on the reproduction, development, and settlement in the polychaetous annelid *Capitella capitata. Marine Biol.* 14:183.

Bigford, T. E. 1977. Effects of oil on behavioral responses to light, pressure and gravity in larvae of the rock crab *Cancer irroratus. Marine Biol.* 43:137.

Bingham, B. L., and N. B. Reyns. 1999. Ultraviolet radiation and distribution of the solitary ascidian *Corella inflata* (Huntsman). *Biol. Bull.* 196:94.

Curtis, L. A., and J. L. Kinley. 1998. Imposex in *Ilyanassa obsoleta* still common in a Delaware estuary. *Marine Poll. Bull.* 36:97.

Heslinga, G. A. 1976. Effects of copper on the coral-reef echinoid *Echinometra mathaei. Marine Biol.* 35:155.

Holland, D. L., D. J. Crisp, R. Huxley, and J. Sisson. 1984. Influence of oil shale on intertidal organisms: Effect of oil shale extract on settlement of the barnacle *Balanus balanoides* (L.). *J. Exp. Marine Biol. Ecol.* 75:245.

Hovel, K. A., and S. G. Morgan. 1999. Susceptibility of estuarine crab larvae to ultraviolet radiation. *J. Exp. Marine Biol. Ecol.* 237:107.

Hunt, J. W., and B. S. Anderson. 1989. Sublethal effects of zinc and municipal effluents on larvae of the red abalone *Haliotis rufescens. Marine Biol.* 101:545.

Kobayashi, N. 1980. Comparative sensitivity of various developmental stages of sea urchins to some chemicals. *Marine Biol.* 58:163.

Macdonald, J. M., J. D. Shields, and R. K. Zimmer-Faust. 1988. Acute toxicities of eleven metals to early life-history stages of the yellow crab *Cancer anthonyi. Marine Biol.* 98:201.

Muchmore, D., and D. Epel. 1977. The effects of chlorination of wastewater on fertilization in some marine invertebrates. *Marine Biol.* 19:93.

Pechenik, J. A. 2006. Larval experience and latent effects—metamorphosis is not a new beginning. *Integr. Comp. Biol.* 46: 323–33.

Grosberg, R. K. 1991. Sperm-mediated gene flow and the genetic structure of a population of the colonial ascidian *Botryllus schlosseri. Evolution* 45:130.

LaMunyon, C. W., and S. Ward. 1998. Larger sperm outcompete small sperm in the nematode *Caenorhabditis elegans. Proc. Royal Soc. London* B 265:1997.

Levitan, D. R., M. A. Sewell, and F-S. Chia. 1992. How distribution and abundance influence fertilization success in the sea urchin *Strongylocentrotus franciscanus. Ecology* 73:248.

Meidel, S. K., and P. O. Yund. 2001. Egg longevity and time-integrated fertilization in a temperate sea urchin (*Strongylocentrotus droebachiensis*). *Biol. Bull.* 201:84–94.

Styan, C. A. 1998. Polyspermy, egg size, and the fertilization kinetics of free-spawning marine invertebrates. *Amer. Nat.* 152:290.

Yund, P. O. 1990. An *in situ* measurement of sperm dispersal in a colonial marine hydroid. *J. Exp. Zool.* 253:102.

Ziegler, K. S., and H. A. Lessios. 2003. 250 million years of bindin evolution. *Biol. Bull.* 205:8.

8. Many aquatic invertebrates release gametes or larvae into the environment, whereas many others show various degrees of protection and care for their developing young. What are the costs and benefits that might explain either (a) why so many species have evolved some form of direct parental care or (b) why so many species have not?

Gillespie, R. G. 1990. Costs and benefits of brood care in the Hawaiian happy face spider *Theridion grallator* (Araneae, Theridiidae). *Amer. Midl. Nat.* 123:236.

Menge, B. A. 1975. Brood or broadcast? The adaptive significance of different reproductive strategies in two intertidal seastars, *Leptasterias hexactis* and *Pisaster ochraceus. Marine Biol.* 31:87.

Milne, I. S., and P. Calow. 1990. Costs and benefits of brooding in glossiphoniid leeches with special reference to hypoxia as a selection pressure. *J. Anim. Ecol.* 59:41.

Pechenik, J. A. 1979. Role of encapsulation in invertebrate life histories. *Amer. Nat.* 114:859.

Pechenik, J. A. 1999. On the advantages and disadvantages of larval stages in benthic marine invertebrate life cycles. *Marine Ecol. Progr. Ser.* 177:269.

Swalla, B. J., and W. R. Jeffery. 1996. Requirement of the *Manx* gene for expression of chordate features in a tailless ascidian larva. *Science* 274:1205.

Thorson, G. 1950. Reproductive and larval ecology of marine bottom invertebrates. *Biol. Rev.* 25:1.

9. External fertilization is common among marine invertebrates. How can one assess fertilization success in the field and the biological and physical factors that contribute to such success?

Babcock, R. C., C. N. Mundy, and D. Whitehead. 1994. Sperm diffusion models and *in situ* confirmation of long-distance fertilization in the free-spawning asteroid *Acanthaster planci. Biol. Bull.* 186:17.

Denny, M. W., and M. F. Shibata. 1989. Consequences of surf-zone turbulence for settlement and external fertilization. *Amer. Nat.* 134:859.

Farley, G. S., and D. R. Levitan. 2001. The role of jelly coats in sperm-egg encounters, fertilization success, and selection on egg size in broadcast spawners. *Amer. Nat.* 157:626–36.

10. In many marine invertebrate groups, some species apparently have switched from having a feeding (planktotrophic) larva to a nonfeeding (lecithotrophic) larva. What selective pressures likely account for shifts in larval feeding type, what morphological changes accompany such shifts, what is the evidence that such shifts have occurred, and through what developmental mechanisms have such shifts been accomplished?

Amemiya, S., and R. B. Emlet. 1992. The development and larval form of an echinothurioid echinoid, *Asthenosoma ijimai,* revisited. *Biol. Bull.* 182:15.

Collin, R., O. R. Chaparro, F. Winkler, and D. Véliz. 2007. Molecular phylogenetic and embryological evidence that feeding larvae have been reacquired in a marine gastropod. *Biol. Bull.* 212: 83–92.

Hart, M. W., M. Byrne, and M. J. Smith. 1997. Molecular phylogenetic analysis of life-history evolution in asterinid starfish. *Evolution* 51:1848.

Jeffrey, W. R. 1994. A model for ascidian development and developmental modification during evolution. *J. Marine Biol. Assoc. U.K.* 74:35.

Krug, P. J. 1998. Poecilogony in an estuarine opisthobranch: Planktotrophy, lecithotrophy, and mixed clutches in a population of the ascoglossan *Alderia modesta. Marine Biol.* 132:483.

McEdward, L. R. 1992. Morphology and development of a unique type of pelagic larva in the starfish *Pteraster tesselatus* (Echinodermata: Asteroidea). *Biol. Bull.* 182:177.

McHugh, D., and G. W. Rouse. 1998. Life history evolution of marine invertebrates: New views from phylogenetic systematics. *Trends Ecol. Evol.* 13:182.

Raff, R. A. 1987. Constraint, flexibility, and phylogenetic history in the evolution of direct development in sea urchins. *Devel. Biol.* 119:6.

Sinervo, B., and L. R. McEdward. 1988. Developmental consequences of an evolutionary change in egg size: An experimental test. *Evolution* 42:885.

Strathmann, R. R., and D. J. Eernisse. 1994. What molecular phylogenies tell us about the evolution of larval forms. *Amer. Zool.* 34:502.

Strathmann, R. R., L. Fenaux, and M. F. Strathmann. 1992. Heterochronic developmental plasticity in larval sea urchins and its implications for evolution of nonfeeding larvae. *Evolution* 46:972.

Wray, G. A., and R. A. Raff. 1991. The evolution of developmental strategy in marine invertebrates. *Trends Ecol. Evol.* 6:45.

11. The widespread occurrence of sexual reproduction has long fascinated evolutionary biologists, partly because, as a result of meiosis, each adult gets to contribute only half of its genes to the next generation and partly because male offspring make no direct contribution to future reproduction; parthenogenesis would seem to be the more efficient reproductive pattern. What factors might select for sexual reproduction over parthenogenetic reproduction, and how can the various hypotheses be tested?

Barton, N. H., and B. Charlesworth. 1998. Why sex and recombination? *Science* 281:1986.

Bell, G. 1982. *The Masterpiece of Nature: The Evolution and Genetics of Sexuality.* Berkeley, Calif.: Univ. California Press.

Case, T. J., and M. L. Taper. 1986. On the coexistence and coevolution of asexual and sexual competitors. *Evolution* 40:366.

Ghiselin, M. T. 1974. *The Economy of Nature and the Evolution of Sex.* Berkeley, Calif.: Univ. California Press.

Goddard, M. R., H. C. J. Godfray, and A. Burt. 2005. Sex increases the efficacy of natural selection in experimental yeast populations. *Nature* 434:636–40.

Jaenike, J. 1978. An hypothesis to account for the maintenance of sex within populations. *Evol. Theory* 3:191.

Klatz, O., and G. Bell. 2002. The ecology and genetics of fitness in *Chlamydomonas.* XII. Repeated sexual episodes increase rates of adaptation to novel environments. *Evolution* 56:1743–53.

Lively, C. M., and S. G. Johnson. 1994. Brooding and the evolution of parthenogenesis: Strategy models and evidence from aquatic invertebrates. *Proc. Royal Soc. London* B 256:89.

Lloyd, D. 1980. Benefits and handicaps of sexual reproduction. *Evol. Biol.* 13:69.

Michod, R. E., and T. W. Gayley. 1992. Masking of mutations and the evolution of sex. *Amer. Nat.* 139:706.

Wuethrich, B. 1998. Why sex? Putting theory to the test. *Science* 281:1980.

12. To what extent does dispersal discourage or prevent genetic differentiation among adult populations?

Allcock, A. L., A. S. Brierley, J. P. Thorpe, and P. G. Rodhouse. 1997. Restricted gene flow and evolutionary divergence between geographically separated populations of the Antarctic octopus *Pareledone turqueti. Marine Biol.* 129:97.

Bossart, J. L., and J. M. Scriber. 1995. Maintenance of ecologically significant genetic variation in the tiger swallowtail butterfly through differential selection and gene flow. *Evolution* 49:1163.

Breton, S., F. Dufresne, G. Desrosiers, and P. U. Blier. 2003. Population structure of two northern hemisphere polychaetes, *Neanthes virens* and *Hediste diversicolor* (Nereididae), with different life-history traits. *Marine Biol.* 142:707.

Hamm, D. E., and R. S. Burton. 2000. Population genetics of black abalone, *Haliotis cracherodii,* along the central California coast. *J. Exp. Marine Biol. Ecol.* 254:235.

Hansen, T. A. 1983. Modes of larval development and rates of speciation in early Tertiary neogastropods. *Science* 220:501.

Hoskin, M. G. 1997. Effects of contrasting modes of larval development on the genetic structures of populations of 3 species of prosobranch gastropods. *Marine Biol.* 127:647.

Johannesson, K., and A. Tatarenkov. 1997. Allozyme variation in a snail (*Littorina saxatilus*)—Deconfounding the effects of microhabitat and gene flow. *Evolution* 51:402.

Lessios, H. A., B. D. Kessing, and D. R. Robertson. 1998. Massive gene flow across the world's most potent marine biogeographic barrier. *Proc. Royal Soc. London* B 265:583.

Palumbi, S. R., G. Grabowsky, T. Duda, L. Geyer, and N. Tachino. 1997. Speciation and population genetic structure in tropical Pacific sea urchins. *Evolution* 51:1506.

Untersee, S., and J. A. Pechenik. 2007. Local adaptation and maternal effects in two species of marine gastropod (genus *Crepidula*) that differ in dispersal potential. *Marine Ecol. Progr. Ser.* 347:79–85.

Williams, S. T., and J. A. H. Benzie. 1998. Evidence of a biogeographic break between populations of a high dispersal starfish: Congruent regions within the Indo-West Pacific defined by color morphs, mtDNA, and allozyme data. *Evolution* 52:87.

13. What factors are likely to determine whether an invasion will be successful; i.e., that invading animals will establish a reproductively successful population in the new habitat?

Carlton, J. T. 1996. Pattern, process, and prediction in marine invasion ecology. *Biol. Conservation* 78:97.

Cohen, A. N., and J. T. Carlton. 1998. Accelerating invasion rate in a highly invaded estuary. *Science* 279:555.

Lawton, J. H., and K. C. Brown. 1986. The population and community ecology of invading insects. *Phil. Trans. Royal Soc. London* B 314:607.

Pechenik, J. A., D. E. Wendt, and J. N. Jarrett. 1998. Metamorphosis is not a new beginning. *BioScience* 48:901.

Ruiz, G. M., J. T. Carlton, E. D. Grosholz, and A. H. Hines. 1997. Global invasions of marine and estuarine habitats by nonindigenous species: Mechanisms, extent, and consequences. *Amer. Zool.* 37:621.

Simberloff, D. 1989. Which insect introductions succeed and which fail? In: J. A. Drake, ed. *Biological Invasions: A Global Perspective.* New York: John Wiley & Sons Ltd.

14. Only about 15 years ago it was first reported that at least some planktotrophic echinoderm larvae replicate themselves naturally before metamorphosis. How does this phenomenon relate to the traditional view that cell fates are eventually fixed during echinoderm development?

Balser, E. J. 1998. Cloning by ophiuroid echinoderm larvae. *Biol. Bull.* 194:187.

Knott, K. E., E. J. Balser, W. B. Jaeckle, and G. A. Wray. 2003. Identification of asteroid genera with species capable of larval cloning. *Biol. Bull.* 204:246.

Vickery, M. S., M. C. L. Vickery, and J. B. McClintock. 2000. Effects of food concentration and availability on the incidence of cloning in planktotrophic larvae of the sea star *Pisaster ochraceus. Biol. Bull.* 199:298.

Vickery, M. S., M. C. L. Vickery, and J. B. McClintock. 2002. Morphogenesis and organogenesis in the regenerating planktotrophic larvae of asteroids and echinoids. *Biol. Bull.* 203:121.

15. How convincing is the evidence that bacterial and microsporidian symbionts can alter the sex ratios of host arthropod populations?

Dyson, E. A., M. K. Kamath, and G. D. D. Hurst. 2002. *Wolbachia* infection associated with all-female broods in *Hypolimnas bolina* (Lepidoptera: Nymphalidae): Evidence for horizontal transmission of a butterfly male killer. *Heredity* 88:166.

Kageyama, D., G. Nishimura, S. Hoshizaki, and Y. Ishikawa. 2002. Feminizing *Wolbachia* in an insect, *Ostrinia furnacalis* (Lepidoptera: Crambidae). *Heredity* 88:444.

Mautner, S. I., K. A. Cook, M. R. Forbes, D. G. McCurdy, and A. M. Dunn. 2007. Evidence for sex ratio distortion by a new microsporidian parasite of a Corophiid amphipod. *Parasitol.* 134:1567–73.

Terry, R. S., J. E. Smith, R. G. Sharpe, T. Rigaud, D. T. J. Littlewood, J. E. Ironside, D. Rollinson, D. Bouchon, C. MacNeil, J. T. A. Dick, and A. M. Dunn. 2004. Widespread vertical transmission and associated host sex-ratio distortion within the eukaryotic phylum Microspora. *Proc. Royal. Soc.* London B 271:1783–89.

Tinsley, M. C., and M. E. N. Majerus. 2006. A new male-killing parasitism: Spiroplasma bacteria infect the ladybird beetle *Anisosticta novemdecimpunctata* (Coleoptera: Coccinellidae). *Parasitol.* 132:757–65.

Weeks, A. R., K. T. Reynolds, A. A. Hoffmann, and H. Mann. 2002. *Wolbachia* dynamics and host effects: What has (and has not) been demonstrated? *Trends Ecol. Evol.* 17:257–62.

Weeks, A. R., R. Velten, and R. Stouthamer. 2003. Incidence of a new sex-ratio-distorting endosymbiotic bacterium among arthropods. *Proc. Royal Soc. London* B 270:1857.

General References about Invertebrate Reproduction and Development

Adiyodi, K. G., and R. G. Adiyodi, eds. 1983–1993. *Reproductive Biology of Invertebrates. Vol. I: Oogenesis, Oviposition, and Oosorption; Vol. II: Spermatogenesis and Sperm Function; Vol. III: Accessory Sex Glands; Vol. IV A and B: Fertilization, Development, and Parental Care; Vol. V: Sexual Differentiation and Behaviour; Vol. VI A and B: Asexual Propagation and Reproductive Strategies.* New York: Wiley Interscience.

Giese, A. C., and J. S. Pearse, eds. 1974–1979. *Reproduction of Marine Invertebrates. Vol. I: Acoelomate and Pseudocoelomate Metazoans; Vol. II: Entoprocts and Lesser Coelomates; Vol. III: Annelids and Echiurans; Vol. IV: Molluscs: Gastropods and Cephalopods; Vol. V: Molluscs: Pelecypods and Lesser Classes.* New York: Academic Press.

Giese, A. C., J. S. Pearse, and V. B. Pearse. 1987. *Reproduction of Marine Invertebrates. Vol. IX: General Aspects: Seeking Unity in Diversity.* Palo Alto, Calif.: Blackwell Scientific Publications.

Gilbert, L. I., J. R. Tata, and B. G. Atkinson. 1996. *Metamorphosis: Postembryonic reprogramming of gene expression in amphibian and insect cells.* New York: Academic Press.

Hosken, D. J., and P. Stockley. 2004. Sexual selection and genital evolution. *Trends Ecol. Evol.* 19:87–93.

Jägersten, G. 1972. *Evolution of the Metazoan Life Cycle.* New York: Academic Press.

McEdward, L. R., ed. 1995. *Ecology of Marine Invertebrate Larvae.* New York: CRC Press.

Nielsen, C. 1998. Origin and evolution of animal life cycles. *Biol. Rev.* 73:125–55.

Reverberi, G. 1971. *Experimental Embryology of Marine and Fresh-Water Invertebrates.* Amsterdam: North-Holland Publ. Co.

Wilbur, K. M., ed. 1983. *The Mollusca, Vol. 3: Development.* New York: Academic Press.

Williamson, D. I., and S. E. Vickers. 2007. The origins of larvae. *Amer. Scient.* 95:509–17. (This is a controversial but fascinating consideration of the topic. See also *Amer. Sci.* 96:91–92 for responses.)

Young, C. M., ed. 2002. *Atlas of Marine Invertebrate Larvae.* New York: Academic Press.

Glossary of Frequently Used Terms

Definitions of more specialized terms can be found by using the Index.

A

aboral The part of the body farthest from the mouth.

acoelomate Lacking a body cavity between the gut and the outer body wall musculature.

advanced characters Characters that have become modified from the ancestral condition (see "apomorphic" characters).

annulation External division of a worm-shaped body into a series of conspicuous rings.

apomorphic A modified ("derived") character state.

archenteron A cavity that eventually becomes the digestive tract of the adult or larva; formed during the development of a deuterostome embryo.

asexual reproduction Reproduction that does not involve the fusion of gametes; reproduction without fertilization.

B

basal lamina A thin collagenous sheet secreted by epithelial cells and on which they rest.

benthic Living on or within a substrate.

benthos The aquatic animals, plants, and algae living on or within a substrate.

binary fission Asexual division of one organism into two nearly identical organisms.

bioluminescence Biochemical production of light by living organisms.

biramous Two-branched.

blastocoel The internal cavity commonly formed by cell division early in embryonic development, prior to gastrulation.

brooding Parental care of developing young.

budding A form of asexual reproduction in which new individuals develop from a portion of the parent, as in all bryozoans and in many protozoans, cnidarians, and polychaetes.

C

cephalization Concentration of nervous and sensory systems in one part of the body, which becomes known as the "head."

chromatophore A pigment-containing cell that can be used by an animal to vary its external coloration.

cilium A thread-like locomotory organelle containing a highly organized array of microtubules; shorter than a flagellum.

cirri Among ciliated protozoans, a group of cilia that function as a single unit; among barnacles, the thoracic appendages, which are modified for food collection; among crinoids, the prehensile appendages located aborally that are used for walking and for clinging to solid substrates.

coelom An internal body cavity lying between the gut and the outer body wall musculature that is lined with derivatives of the embryonic mesoderm.

colony A cluster of genetically identical individuals formed asexually from a single colonizing individual.

conjugation A temporary physical association in which genetic material is exchanged between two ciliate protozoans.

convergent evolution The process whereby similar characteristics are independently evolved by different groups of organisms in response to similar selective pressures.

cuticle A noncellular, secreted body covering.

cyst A secreted covering that protects many small invertebrates, including some protozoans, rotifers, and nematodes, from environmental stresses, such as desiccation and overcrowding.

D

deposit feeding Ingesting substrate (sand, soil, mud) and assimilating the organic fraction.

desiccation Dehydration.

dioecious Characterized by having separate sexes; that is, an individual is either male or female, but never both. "Gonochoristic" means the same thing.

diploblastic Possessing only two distinct tissue layers during embryonic development.

E

ectoderm An embryonic tissue layer; forms epidermal, nervous, and sensory organs and tissues.

encystment The secretion of a protective outer covering that permits some small invertebrates to withstand exposure to extreme environmental stresses, such as desiccation and overcrowding, as in many protozoan, rotifer, nematode, sponge, and tardigrade species.

endoderm An embryonic tissue layer; forms wall of digestive system.

enterocoely Formation of a coelom through outpocketing of the inner portion of the archenteron in some animals (deuterostomes).

estuary A partially enclosed body of water influenced by both tidal forces and by freshwater input from the land.

eutely Species-specific constancy of cell numbers or nuclei; growth occurs through increased cell size rather than increased cell number.

exoskeleton A system of external levers and joints that permits pairs of muscles to act against, or antagonize, each other; the exoskeleton is also protective.

F

filiform Thread-like.

filter feeder An organism that filters food particles from the surrounding fluid.

flagellum A thread-like locomotory organelle containing a highly organized array of microtubules; longer than a cilium and often bearing numerous lateral projections.

flame cell Flagellated cell associated with protonephridia, as in flatworms, rotifers, and some polychaetes.

G

gametes The sex cells involved in fertilization.
gametogenesis The process by which gametes (sperm or eggs) are produced.
gastrovascular canals Fluid-filled canals opening at the mouth of cnidarians and ctenophores that function in gas exchange and in the distribution of nutrients.
gastrulation Creation of a new tissue layer by the movement of cells in the early embryo (blastula).
genet A unique genetic entity, generally formed through the fusion of egg and sperm.
gill A structure specialized for gas exchange in aquatic animals.
gonochoristic Having separate sexes; that is, an individual is either male or female, but never both. "Dioecious" means the same thing.

H

hermaphrodite A single individual that functions as both male and female, either simultaneously or in sequence.
homology Having identical evolutionary origins and developing through identical developmental pathways.
homoplasy Independent evolution of similar or identical character states through convergence or parallel evolution.
hydrostatic skeleton A constant-volume, fluid-filled cavity that permits muscles to be restretched after contraction, often through the mutual antagonism of muscle pairs.

I

imaginal discs Discrete masses of undifferentiated embryonic cells that become preprogrammed to form specific adult tissues and organ systems at metamorphosis, as in insect and nemertean development.
interstitial Living in the spaces between sand grains.
intertidal Living in the area between high and low tides and, thus, alternately exposed to the air and to the sea.
introvert A tubular, eversible extension of the head, bearing the mouth at its tip.

L

larva A free-living developmental stage in the life history of many invertebrate species.

M

meiofauna Small, interstitial animals living between sand grains.
mesenteries Infoldings of gastroderm and mesoglea extending into the gastrovascular cavity of cnidarians; sheets of peritoneum from which the digestive tract is suspended in coelomates.
mesoderm An embryonic tissue layer that gives rise to certain tissues and organs of the adult, including the muscles and gonads.
mesoglea Gelatinous layer found between the epidermis and the gastrodermis of cnidarians.
mesohyl Nonliving, middle gelatinous layer of sponges; living cells are often found within it.
metamerism Serial repetition of organs and tissues, including the body wall, nervous and sensory systems, and musculature.
metamorphosis A dramatic transformation of morphology and function occurring over a short period of time during development.
metanephridia Excretory organs collecting coelomic fluid through a ciliated open funnel and transporting modified fluid to the outside through a nephridiopore.
metanephridium An organ open to the body cavity through a ciliated funnel (nephrostome) and involved in excretion or in the regulation of water balance or salt content.
metazoan A multicellular animal.
microtubules Tubulin-containing cylinders characteristic of cilia and flagella.
module The functional unit of a colonial animal.
monoecious Characterized by the presence of both sexes in a single individual, either simultaneously or sequentially; hermaphroditic.
monophyletic groups Groups derived from a single, common ancestor and including all the descendants of that ancestor.

N

nematocyst Cnidarian cell type that explosively emits long threads specialized for defense and food capture.

O

ocellus A simple, pigment-containing photoreceptor found in a variety of unrelated invertebrates.
osmosis Diffusion of water across a semipermeable (permeable to water but not to solute) membrane along a concentration gradient.

P

paraphyletic groups Groups derived from a single, common ancestor that do not include all of the descendants of that ancestor.
parasitism A close association between species in which one member benefits at the expense of the partner.
parthenogenesis Development of an unfertilized egg into a functional adult.
pelagic organism An organism that lives above the bottom in the open sea.
peristalsis Progressive waves of muscular contraction passing down the length of an organism or organ system.
peritoneum The mesodermal lining of the body cavity of coelomates.
phagocytosis The process through which food particles are surrounded by cell membrane and incorporated into the cell cytoplasm, forming a food vacuole.
plankton Animals (zooplankton) and unicellular algae (phytoplankton) that have only limited locomotory capabilities and are therefore distributed by water movements.
plesiomorphic An ancestral ("primitive") character state.
polyphyletic groups Groups containing species that have evolved from two or more different ancestors rather than from a single, common ancestor.
preadaptation A trait that is adaptive only in a new set of physical or biological circumstances.
primitive characters Characters that are ancestral ("plesiomorphic"), reflecting the ancestral condition.
proboscis A tubular extension at the anterior of an animal, generally used for locomotion or food collection; may or may not be directly connected to the gut.
protandric hermaphroditism A pattern of sexuality in which a single individual functions as male and then female in sequence.
protonephridia Excretory organs that open to the outside through a nephridiopore, as in metanephridia, but which are capped with a fine mesh through which body fluids are ultrafiltered.
pseudocoel An internal body cavity lying between the outer body wall musculature and the gut; not lined with mesoderm and generally formed by persistence of the embryonic blastocoel.
pseudopodia Amorphous protrusions of cytoplasm involved in the locomotion and feeding of amoebae and related protozoans.

R

radial cleavage A form of early cell division in which all cleavage planes are perpendicular, so daughter cells come to lie directly in line with each other.
ramet A single functionally independent module produced by budding, fission, or another asexual process.

S

schizocoely Coelom formation accomplished by a split in the mesoderm during embryonic development of some animals (protostomes).

sedentary Bottom-dwelling and capable of only limited locomotion.

septa Peritoneal (mesodermal) sheets separating adjacent segments, as in annelids, or body divisions, as in chaetognaths.

sessile Bottom-dwelling and generally incapable of locomotion.

sexual reproduction Reproduction involving fusion of gametes.

spermatophore A container of sperm transferred from one individual to another during mating.

spicules Calcareous or silicious formations present in the tissues of some organisms and generally serving protective or supportive functions.

spiral cleavage Pattern of cell division in which cleavage planes are at 45° to the animal-vegetal axis of the egg.

statocyst A sense organ that informs the bearer of orientation of the body to gravity.

suspension-feeder An animal that feeds on particulates suspended in the surrounding medium; this may be accomplished by filtering or by other means.

synapomorphic Uniquely shared, derived (modified) character states; describes a group of species that can be defined as different from all others because they share some unique homologous character.

T

test Any hard external covering; may be secreted by the animal or constructed from surrounding materials.

triploblastic Exhibiting three distinct tissue layers during embryonic development.

V

vector Any organism that transmits parasites from one host species to another.

vermiform Worm shaped—that is, soft-bodied and substantially longer than wide.

Z

zooid A single member of a colony.

zooplankton The animal component of the plankton, having only limited locomotory powers.

Index

Note: Page numbers followed by *f* and *t* indicate figures and tables, respectively.

A

Abalone, 230, 234*f*, 278
Abarenicola, 334
Abdomen, 353, 374
ABO. *See* Accessory boring organ
Aboral surface, 498
Abralia veranyi, 261*f*
Abraliopsis, 261*f*
Acanthamoeba, 264
Acantharia (phylum), 63, 76
Acanthaster, 525
Acanthaster planci, 520*f*
Acanthobdella, 337
Acanthocephala (phylum), 196–198
  body characteristics, 196
  defining characteristics, 196
  evolution, 198
  life cycle, 196, 197*f*
  reproduction and development, 196–198, 558*t*
  taxonomic detail, 201–202
  taxonomic summary, 199
Acanthocephalus, 197*f*
Acanthochaetetes, 94
Acanthochitonidae (family), 277
Acanthocolla, 76
Acanthodbellida (order), 337
Acanthodesmia, 76
Acanthoeca, 77
Acanthometra, 76
Acanthometra elasticum, 67*f*
Acanthopleura, 277
Acanthopleura echinata, 221*f*
Acanthor, 196
Acanthoscurria, 404
Acanthospira, 76
Acanthostaurus, 76
Acari (order), 357, 404
Acartia, 416
Acartia clausi, 142
Acartia tonsa, 386–387, 387*f*, 388*f*
Accessory boring organ (ABO), 219*f*, 281
Acellular, 9
Achatina, 285
Achatina fulica, 285
Achatinella, 285
Achatinellidae (family), 285
Achatinidae (family), 285
Acholades, 174–175
Acholadidae (family), 174
Acicula, 297, 300
Acidity, seawater, 5, 44, 65, 122
Acmaea, 278
Acmaediae (family), 278
Acochlidiidae (family), 282
Acochlidioidea (order), 282
Acochlidium, 282
Acoela (order), 154, 172–173
Acoelomate, 10, 10*f*
Acontia, 119, 120*f*
Acorn barnacle, 390*f*
Acorn worms. *See* Enteropneusta (class)
Acotylea (suborder), 173
Acrania (subphylum). *See* Cephalochordata (=Acrania) (subphylum)
Acropora, 134
Acrorhagi, 119, 121*f*
Acrothoracica (order), 417
Acrotretida (order), 493
Acteon, 270*f*
Acteon tornatilus, 237*f*
Actin, 44, 45*f*
Actinia, 133
Actiniaria (order), 133
Actinophyrs, 77
Actinopus, 404
Actinosphaerium, 67*f*, 77
Actinophyridae, 77
Actinotroch, 488, 572*t*, 576*f*
Actinula larvae, 132
Actinulida (order), 133
Actophila (suborder), 284
Aculifera (subphylum), 276–277
Adductor muscles, 238, 251, 479, 479*f*
Adenophorea (class), 445
Adenoplana, 173
Adhesive disc, 468
Adhesive papillae, 465
Adhesive sac, 490
Advanced condition, 19
Aedes, 409
Aedes aegypti, 168*f*
Aeolidia, 284
Aeolidiidae (family), 284
Aeolidina (suborder), 284
Aequipecten, 287
Aequorea, 132
Aequorea victoria, 230, 230*f*
Aesthetes, 222
Aethozoon, 494
African sleeping sickness, 69
Agamermis decaudata, 439*f*
Agamete. *See* Axoblast cell
Agaricia, 134
Agaricia tenuifolia, 123*f*
Agnathiella, 202
Ahermatypic coral, 122, 124
Aidanosagitta crassa, 464*t*
Ailoscolecidae (family), 336
Aiptasia, 133
Air, *vs.* water, as environment for invertebrates, 1–4, 3*t*
Airfoil, 367, 367*f*
Alaria, 176
Alcadia, 278
Alcyonacea (order), 133
Alcyonaria. *See* Octocorallia (=Alcyonaria) (subclass)
Alcyonidiidae (family), 494
Alcyonidium, 494
Alcyonium, 133
Alectona wallichii, 83*f*
Algae, symbiotic, 112–113, 113*t*
Allogramia, 76
Allogromia laticollaris, 45*f*
Alloincompatibility, 80
Allopora, 132
Alpheidae (family), 413
Alpheus, 413
Alveolar sacs, 55
Alveolates, 46–61, 73, 75
Alveoli, 46, 48, 50*f*
Alvin (submersible), 308, 309*f*
Alvinellidae (family), 335
Amazonian leech, 325
Amber snails, 285
Amblypgi (order), 403
Ambulacral grooves, 498, 505
Ambulacral ossicles, 499, 505
Ambulacral plates, 510, 511*f*
Ambulacral zones, 498
Ambulacraria, 17*ff*, 497
Ameiotic replication, 556
Ameiotic reproduction, 556
American hookworm, 441–442, 442*f*, 447
American lobster, 413, 470
American oyster, 245, 249*f*, 267*f*
Ameritermes hastatus, 375*f*
Ametabolous species, 370
Amictic females, 190
Ammonia, 76
Ammonia, in aquatic *vs.* terrestrial environments, 3
Amoeba, 76
Amoeba proteus, 404*f*
Amoebic dysentery, 44, 62
Amoebocyte cells, 541
Amoeboflagellates, 72, 72*f*, 75
Amoeboid protozoans, 43–44, 62–68
  evolutionary relationship, 62
  locomotion, 42–44, 42*f*, 43*f*
  reproduction, 62
  taxonomic summary, 73
Amoebomastigotes, 72, 72*f*
Amoebozoa
  Arcellanida, 63, 63*f*, 73
  Gymnamoebae, 62–63, 63*f*
  Mycetozoa, 63–64, 76
  taxonomic detail, 76
  taxonomic summary, 73
Ampelisca, 412
Ampharetidae (family), 335
Amphiblastula, 90, 90*f*
Amphibola, 285
Amphibola crenata, 286
Amphibolidae (family), 285–286
Amphictenidae (=Pectinariidae) (family), 335
Amphids, 431, 437, 437*f*
Amphilina, 175
Amphineura, 222
Amphioxus. *See* Lancelet
Amphipholis, 524
Amphipoda (order), 374, 378*f*, 379, 391–392, 396, 412–413
Amphitrite, 335
Amphitrite ornata, 302*f*
Amphiura, 524
Amphoteric females, 194
Ampullae, 498*f*, 499
Ampullariidae (=Pilidae) (family), 279
Anachis, 281
Anagenesis, 27*t*
Anal pores, 140, 140*f*
Anal sacs, 313
Analogous characters, 19
Anasca (suborder), 494
Anaspidea (=Aplysiacea) (order), 237, 283
Anaspides, 411
Ancestral (derived) state, 26, 27*t*
Ancestral (primitive) state, 27*t*
Ancestrula, 490
Ancylostoma duodenale, 439*f*, 447
Ancylostomatidae (family), 447–448
Ancylostomatoidea (superfamily), 447–448
Animal pole, 12
Animal rubber, 342
Anisakidae (family), 447
Anisakis, 447
Anisoptera (suborder), 406
Annelida (phylum), 295–339
  circulatory system, 328
  Clitellata (class), 296, 318–322, 336–337
  coelomic cavities, 12
  comparison with Arthropoda, 341, 350, 395
  comparison with Sipuncula, 315, 317*t*
  defining characteristics, 295
  digestive system, 325
  Echiura (class), 297, 312–314, 315*f*, 337–338, 454, 558*t*
  evolution, 296
  excretory system, 296, 297*f*
  gas exchange, 286
  general characteristics, 285–286
  locomotion, 296
  musculature, 345*t*
  nervous system, 327–328
  Polychaeta (class). *See* Polychaeta (class)
  reproduction and development, 558*t*, 569*t*
  sense organs, 327–328
  taxonomic detail, 332–338
  taxonomic summary, 328
Anodonta, 247*f*, 269*f*, 287
Anodonta cyngea, 267*f*
Anomalodesmata (subclass), 239, 254
  defining characteristics, 254
  feeding mechanisms, 254
  taxonomic detail, 290
  taxonomic summary, 271
Anomia, 287
Anomiidae (family), 287
Anomopoda (suborder), 415
Anomura (infraorder), 414
Anopheles, 362, 409
Anopheles gambiae, 60*f*, 361*f*
Anopheles stephensi, 362, 363*f*
Anopla (class), 212
Anoplodium, 174
Anoplura (order), 407
Anostraca (order), 415
Ant lions, 408
Antagonize, and musculature, 97, 98*f*
Antedon, 524
Antennae, 374
Antennal glands, 396
Antennules, 374
Anthomedusae (suborder), 132
Anthopleura, 133
Anthopleura krebsi, 121*f*
Anthozoa (class), 118–125
  defining characteristics, 118
  digestion, 119
  Hexacorallia (=Zoantharia) (subclass), 120–124, 125*f*, 128, 133–134
  locomotion, 119–120
  musculature, 119, 122*f*, 345*t*
  Octocorallia (=Alcyonaria) (subclass), 124–125, 125*f*, 128, 133
  reproduction and development, 119, 556, 558*t*, 568*t*
  taxonomic detail, 133–134
  taxonomic summary, 128
Anthrax, 409
Anthropodaria, 491*f*
Antipatharia (order), 134
Antipathes, 134
Antonbrunnia viridis, 332
Antonbruuniidae (family), 332
Ants, 360, 370, 373, 374*f*, 410
Aphididae (family), 407
Aphids, 407
Aphonopelma, 404
Aphragmophora (order), 470
Aphrodita, 333
Aphroditacea (superfamily), 333
Aphroditidae (family), 333
Apical sense organ, 138–139, 139*f*
Apicomplexa (=Sporozoa) (phylum), 58–61
  control, 61
  defining characteristics, 58
  life cycle, 59, 60*f*, 61
  taxonomic detail, 75
  taxonomic summary, 73

Apidae (family), 411
Apis, 411
Aplacophora (class), 222, 223*f*
body characteristics, 222
defining characteristics, 222
evolution, 222
nervous system, 222
reproduction, 265, 559*t*
taxonomic detail, 278
taxonomic summary, 271
Aplysia, 236*f*, 270*f*, 283
Aplysia brasiliana, 560*f*
Aplysiacea (order). *See* Anaspidea (=Aplysiacea) (order)
Aplysiidae (family), 283
Apochela (order), 429
Apocrita (suborder), 410
Apodida (order), 526
Apomorphic condition, 19, 27*t*
Aporchis, 177
Appendicularia (class). *See* Larvacea (=Appendicularia) (class)
Apple snails, 279
Apposition eye, 347
Apterygota (subclass), 360, 361*f*
Apyrene sperm, 561, 561*f*
Aquatic environments
benefits of, 1–4, 3*t*
problems with, 3*t*, 4–5
Arabella, 333
Arabella iricolor, 299*f*
Arabellidae (family), 333
Arachnid silk, 357
Arachnida (class), 343, 352–357, 354*f*–356*f*
body characteristics, 353
digestive system, 393
dispersal, 580
excretory system, 395
reproduction, 557, 559*t*, 560, 563, 564*f*, 580
respiratory system, 353
taxonomic detail, 403–405
taxonomic summary, 396
web, 356*f*, 357
Araneae (order), 353, 403
Araneidae (family), 354*f*, 404
Araneus, 404
Arbacia, 525
Arbacia punctulata, 513*f*, 514*f*, 525
Arbaciidae (family), 525
Arbacioida (order), 525
Arcella, 76
Arcellanida, 63, 63*f*, 73, 76
Archaeocytes, 81–82
Archaeogastropods, 228*f*, 277–278
Archaeognatha (order), 405–406
Archegetes, 175
Archenteron, 12*f*
Archiacanthocephala (class), 201
Archiannelids, 336
Archidorididae (family), 284
Archidoris, 284
Architaenioglossa (order), 279
Architectonica, 282
Architectonicidae (family), 282
Architectonicoidea (superfamily), 282
Architeuthidae (family), 291
Architeuthis, 256, 291
Archivortex, 174
Arctica islandica, 241*f*, 288
Arcticidae (family), 288–289
Arenicola, 334
Arenicola marina, 299*f*, 301*f*
Arenicolidae (family), 334
Argiope, 354*f*, 404
Argonauta, 257, 291
Argonautidae (family), 291–292
Argonemertes, 213
Argopecten, 287
Argulus, 416
Argyronecta, 404
Argyronetidae (family), 404
Argyrotheca, 489*f*
Arhynchobdellae (order), 337
Arion fuscus, 240*f*
Aristotle's lantern, 511*f*, 512, 513*f*
Armandia, 334
Armillifer annulatus, 385*f*
Armina, 284
Arminidae (family), 284
Arminina (suborder), 284
Arms, 259
Armyworms, 410
Arrow squids, 291
Arrow worms. *See* Chaetognatha (phylum)
Artemia, 415
Artemia salina, 379, 380*f*
Arthropoda (phylum), 341–419
Arachnida (class). *See* Arachnida (class)
Chelicerata (subphylum), 352–357, 396, 403–405
circulatory system, 344–345, 346*f*
classification, 349–350
comparison with Annelida, 341, 345, 350, 395
comparison with Mollusca, 342, 345
Crustacea (class). *See* Crustacea (class)
defining characteristics, 341
digestive system, 392–394
evolutionary relationships, 349–350
excretory system, 395–396
exoskeleton, 342, 342*f*
eyes, 350–349
fossil record, 350, 357
general characteristics, 341–350
Insecta (class). *See* Insecta (class)
Mandibulata (subphylum), 358–392, 396, 405
Merostomata (class), 352, 392–393, 395, 396, 403, 559*t*
molting, 343, 344*f*
musculature, 344, 345*f*, 345*t*
Myriapoda (class), 358–360, 396, 405
nervous system, 344, 345*f*
Pycnogonida (=Pantopoda) (class), 357, 358*f*, 396, 405
reproduction and development, 349, 556, 557, 561, 563, 567, 569*t*, 571*t*–572*t*, 580
sense organs, 345–349
taxonomic detail, 402–418
taxonomic summary, 396
Trilobita (class), 351–352, 396
Trilobitomorpha (subphylum), 350–352, 396
Articulata (class), 477f, 479, 479*t*, 480*f*, 489, 489*f*, 490, 491, 493
Articulata hypothesis, 421
Asbestopluma, 94
Ascarididae (family), 447
Ascaris, 440, 441, 447
Ascaris lumbricoides, 434*f*, 440, 447
Ascaris megalocephala, 438*f*
Ascaris suum, 435*f*
Ascending limb, 245, 246*f*
Aschelminthes (phylum), 183
comparison with Chaetognatha, 466
comparison with Nematoda, 432, 432*f*
Aschemonella, 76
Ascidia, 553
Ascidiacea (class), 540–542, 540*f*, 541*f*, 542*f*
body characteristics, 540
circulatory system, 541
colonial, 541, 542*f*
development, 542
digestive system, 524–525
excretory system, 541–542, 547–548
musculature, 541
nervous system, 547–548
reproduction and development, 547, 556, 559*t*, 574*t*
taxonomic detail, 552–553
taxonomic summary, 551
Ascoglossa (order). *See* Sacoglossa (=Ascoglossa) (order)
Asconoid sponge, 85*f*, 86, 87, 87*f*, 94
Ascophora (suborder), 494
Ascopore, 486
Ascothoracica (order), 417
Ascus, 486
Asellus. See Caecidotea (=Asellus)
Asexual reproduction, 556–557
Anthozoa, 119, 556
Apicomplexa, 59, 60*f*, 61
Arachnida, 557
Arthropoda, 556–557
Ascidiacea, 556
Asteroidea, 556, 557*f*
Bdelloidea, 188
Bryozoa, 489–490, 556
Ciliophora, 52
Ctenophora, 556
Cubozoa, 109–110
Cycliophora, 469
Digenea, 169–70
Echinodermata, 518
Hydrozoa, 110, 5456
Insecta, 556
lophophorates, 488
Monogononta, 190
Nemertea, 210
Oligochaeta, 322
Ophiuroidea, 556
Orthonectida, 180
Placozoa, 91
Polychaeta, 300–301, 301*f*, 556
Porifera, 89, 556
Protozoa, 44, 46*f*, 556
Rhombozoa, 181–182
Rhynchocoela, 556
Rotifera, 188, 556–557
sarcodinids, 62
Scyphozoa, 108, 556, 556*f*
vs. sexual reproduction, 556
Thaliacea, 556
Trematoda, 556
Turbellaria, 158, 556
Urochordata, 547
Asian hookworm, 447
Asolene, 279
Aspidobothrea. *See* Aspidogastrea (=Aspidobothrea) (subclass)
Aspidochirotida (order), 526
Aspidogaster, 177
Aspidogastrea (=Aspidobothrea) (subclass), 151, 161, 170, 170*f*, 177
Aspidosiphonidae (family), 338
Asplanchna, 192–193, 193*f*, 193*t*, 201
Asplanchna girodi, 192, 193*f*, 193*t*
Asplanchna intermedia, 194*f*
Asplanchna priodonta, 187*f*
Asplanchna sieboldi, 187*f*
Asplanchnidae (family), 201
Asplousobranchia (order), 553
Assimilated food, 513
Astacidae (family), 413
Astacidea (infraorder), 413
Astacus, 413
Asterias, 508*f*, 525
Asterias vulgaris, 506*f*, 575*f*
Asteriidae (family), 525
Asterina, 524, 577*f*
Asteroidea (subclass), 505–508, 506*f*, 507*f*, 508*f*
body characteristics, 505, 507–508
coelom formation, 12*f*
defining characteristics, 505
digestive system, 505, 507*f*
feeding mechanisms, 505
locomotion, 505
reproduction and development, 518, 519*f*, 520*f*, 556, 557*f*, 559*t*, 573*t*, 575*f*, 577
taxonomic detail, 524–525
taxonomic summary, 521
Asterozoa (subphylum), 521, 524–525
Astraea, 278
Astraea pallida, 125*f*
Astrammina rara, 45*f*
Astrangia, 134
Astropecten, 524
Astropecten irregularis, 508*f*
Astrosclera, 94
Astrosphaera, 76
Asymmetrical body plan, 9, 9*f*
Asynchronous flight, 367
Atelocerata, 350
Athecate, 112, 115*f*
Athoracophoridae (family), 285
Atlanta, 280
Atlantidae (family), 280
Atoke, 301, 304*f*
Atrial siphon, 541
Atrium, 541
Aulacantha, 77
Aulacopleura konincki, 351*f*
Aulosphaera, 77
Aurelia, 107*f*, 131
Aurelia aurita, 108*f*, 109*f*, 131
Auricles, 146
Auricularia larva, 519*f*, 574*t*
Austrobilharzia, 177
Austrocochlea, 278
Austrodoris, 284
Autapomorphy, 27*t*
Autogamy, 52, 53*f*
Autolytus, 304*f*, 332
Autotomize, 502, 505, 507*f*
Autotrophs, 68
Autozooids, 486
Avagina, 172
Avicularia, 486–487
Axial cells, 181
Axial organ, 500
Axial sinus, 500
Axoblast cell, 181–182
Axoneme, 43, 42*f*
Axopodia, 44, 65, 76–77
Aysheaia, 422*f*, 428
Azygia, 177
Azygiida (order), 177

B

Babesia, 75
Bacterial symbiosis, 252–253, 253*f*, 274, 309, 310*f*, 311, 393, 442
Bacteriocytes, 252
Balancers, 138, 139*f*
Balanoglossus, 532*f*, 535
Balanus, 390*f*, 417
Balanus amphitrite, 1
Balanus balanoides, 581*t*
Balanus crenatus, 581*t*
Balanus improvisus, 390*f*
Ballooning, 580, 580*f*
Bamboo worms, 334
Bankia, 289
Bankivia, 278
Barentsia, 494
Barentsiidae (family), 494
Bark lice, 407
Barnacles. *See* Cirripedia (subclass)
Bartolius, 177
Basal animals, 149
Basal body, 42, 43, 47, 50*f*
Basal lamina, 83
Basement membrane, 84
Baseodiscidae (family), 212
Baseodiscus, 212
Basket shells, 281
Basket stars, 503, 524
Basommatophora (order), 285–286
Bathycalanus, 416
Bathychitonidae (family). *See* Ischnochitonidae (=Bathychitonidae) (family)
Bathycrinus, 524
Bathynella, 411
Bathynomus giganteus, 412
Bathyplotes natans, 526
Batillaria, 279
Bayesion inference, 27*t*
Bdelloidea (class), 188–190, 191*f*, 194*f*
reproduction, 558*t*
taxonomic detail, 201
taxonomic summary, 199
Bdellonemertea (order), 213
Bdelloura, 174
Bdellouridae (family), 174
Beachhoppers, 412
Bee flies, 410
Beef tapeworm, 176
Bees, 373, 410
Beetles, 361*f*
development, 371*f*, 572*t*
taxonomic detail, 408
Bell, 105, 106*f*, 107
Bellura, 371*f*
Bembicium, 279
Benthic
Chaetognatha, 465
Dinoflagellata, 55
Turbellaria, 151
Bernoulli, Daniel, 366
Bernoulli's principle, 366–367, 367*f*
Beroë, 138, 145f, 148
Beroida (order), 146, 148
Berthelinia, 282
Berthella, 283
Biceps, 97, 98*f*
Bilateral symmetry, 9, 9*f*
Binary fission, 556, 556*f*
Ciliophora, 52, 52*f*
Placozoa, 91
Protozoa, 44, 46*f*
sarcodinids, 62
Bioluminescence, 55, 141, 143, 258, 261*f*, 378, 465
Biomass, 343, 484
Biomphalaria, 166–167, 286
Biomphalaria glabrata, 165
Bipaliidae (family), 174
Bipalium, 174
Bipectinate, 218, 235, 241
Bipinnaria stage, 520*f*, 573*t*
Biradial symmetry, 138
Biramous appendages, 351, 351*f*, 358, 375*f*
Birgus, 414
Bithynia, 279
Bithyniidae (family), 279
Biting lice, 407
Bittium, 279
Bivalvia (=Pelecypoda) (class), 238–254
Anomalodesmata (subclass), 239, 254, 271, 290
burrowing, 251, 251*f*
defining characteristics, 238
lamellibranchs, 239, 241–254, 265, 271
musculature, 345*t*

Bivalvia (=Pelecypoda) (class)—*Cont.*
Protobranchia (subclass), 239, 241, 271, 286–290
reproduction and development, 558t, 265, 566, 567, 570t, 575f
shell, 238, 239, 240*f*
symbiotic relationships, 251, 252–253, 253*f*
taxonomic detail, 286–290
taxonomic summary, 271
Blaberus, 406
Blaberus giganteus, 406
Black corals, 134
Black Sea, 137, 147
Black widow spider, 404
Blackflies, 409, 448
Black-legged tick, 354*f*
Bladder, 196
Bladder worm, 158
Blastocoel, 10
Blastozooid, 547, 547*f*
Blatella, 406
Blattaria (order), 406
Blatteria, 406
Blepharisma, 75
Blood flukes, 161, 162*f*, 177
Blood pigments, 266, 328, 396
Blood stars, 525
Bloodletting, 324
Bloodworms, 332
Blue button, 133
Blue crab, 377*f*, 414
Blue mussel, 15f, 244f, 246*f*
Bodo, 73, 75
Body symmetry
classification by, 9
types of, 9, 9*f*
Bolinopsidae (family), 147
Bolinopsis infundibulum, 230*f*
Boltenia, 553
Bombardier beetles, 408
Bombus, 411
Bombycidae (family), 410
Bombyliidae (family), 410
Bombyx mori, 357, 410
Bomolochus, 416
Bonellia, 337
Bonellia viridis, 313, 313*f*, 315*f*
Bonellidae (family), 337
Book gills, 352
Book lice, 407
Book lungs, 353
Boonea, 282
Bootstrapping, 27*t*
Boring sponges, 93
Bosmina, 415
Botflies, 409, 410
Bothriocyrtum, 404
Bothrioplana semperi, 174
Bothrioplanidae (family), 174
Botryllus, 553
Botryllus violaceus, 542*f*
Bottle urchins, 526
Bougainvillia, 132
Bourgueticrinida (order), 524
Boveri, Theodor, 440
Bowerbankia, 494
Bowerbankia gracilis, 483*f*
Brachiolaria stage, 520*f*, 573*t*, 575*f*
Brachionidae (family), 201
Brachionus, 201
Brachionus calyciflorus, 192, 194*f*
Brachionus rubens, 191*f*
Brachiopoda (phylum), 475, 476–479, 477*f*, 478*f*, 479*f*
circulatory system, 477
defining characteristics, 476
digestive system, 490
fossil record, 476, 477*f*
musculature, 479, 479*f*
nervous system, 490, 491
pedicle, 477–478
reproduction and development, 489, 489*f*, 559*t*, 573*t*
shells, 478, 479
taxonomic detail, 493
taxonomic summary, 491
Brachyura (infraorder), 346, 414–415
Bracts, 113
Brain. *See also* Nervous system
Annelida, 327
Cephalopoda, 259, 262*f*
Nematoda, 435
Pogonophora, 327
Brain coral, 125*f*, 134
Branchial bar, 545
Branchial basket, 540
Branchial heart, 259
Branchial plume, 308, 308*f*
Branchinecta, 415
Branchiobdellida (order), 337
Branchiodrilus, 336
Branchiopoda (subclass), 379–381
development, 389
taxonomic detail, 415–416
taxonomic summary, 396
Branchiostoma, 549*f*, 553
Branchiostoma belcheri, 14*f*
Branchiura (subclass), 382, 416
Briareum, 133
Brine shrimp, 379, 380*f*, 396, 415
Brisinga, 525
Brisingida (order), 525
Brissopsis, 526
Bristletails, 370, 405
Brittle stars, 497, 503, 524, 574*t*
Brizalina spathulata, 66*f*
Brood parasitism, 524
Brown body, 480
Brugia, 443, 448
Bryodelphax parvulus, 423*f*
Broystatins, 480
Bryozoa (=Ectoprocta; = Polyzoa) (phylum), 473, 474*f*, 475, 480–488, 481*f*, 482*f*, 483*f*, 486f
body characteristics, 480–481
colonial, 482, 484–485, 485*f*, 490
digestive system, 490
dispersal, 580
fossil record, 481, 482*f*
Gymnolaemata (class), 484–487, 491, 493–494
Phylactolaemata (class), 482–484, 491, 493–494
reproduction and development, 489–490, 489*f*, 490*f*, 556, 559*t*, 560, 566, 572*t*, 578–579
Stenolaemata (class), 487–488, 491, 493
taxonomic detail, 493–494
taxonomic summary, 491
Bubble-shell snails, 282
Bubonic plague, 409
Buccal field, 190
Buccal mass, 218, 219*f*
Buccal siphon, 540, 540*f*
Buccal tube, 531
Buccinidae (family), 281
Buccinum, 281
Budding, 44, 91, 108, 112, 556
Buffalo gnats, 409
Bugula, 488*f*, 494
Bugula neritina, 489*f*, 490*f*, 494, 578, 578*f*
Bugulidae (family), 494
Bulinus, 286
Bulla, 237*f*, 282
Bumblebees, 411
Bunodactis, 133
Bunodosoma, 133
Buprestidae (family), 408
Burgess Shale, 7, 198, 421, 422*f*, 548
Bursa, 280
Bursae, 503
Bursidae (family), 280
Bursovaginoidea (order), 202
Bushcrickets, 563
Busycon, 226*f*, 233*f*, 281
Busycon carica, 566*f*
Butterflies, 361*f*, 410
development, 370, 572*t*
eye, 347, 348*f*
Buxtehudea, 77
Byssal threads, 248

C

Caberea ellisi, 488*f*
Caddisflies, 361*f*, 410
Cadmium, 320–321, 321*f*
Caecidae (family), 278
Caecidotea (=Asellus), 412
Caecum, 278
Caenogastropoda (superorder), 232, 271, 279–281
Caenorhabditis elegans, 433*f*, 440, 444, 446
Calamyzas amphictenicola, 332
Calanoida (order), 416
Calanus, 416
Calanus pacificus, 142, 143
Calcarea (class), 86, 87, 92, 94
Calciferous glands, 325
Calcium carbonate, 121–122
Caligus, 416
Caligus curtus, 385*f*
Callanira, 144*f*
Calliactis, 133
Calliactis parasitica, 32*f*
Callianassa, 414
Callianassidae (family), 414
Callianira, 147
Callinectes, 414
Callinectes sapidus, 377*f*, 414
Calliobdella, 337
Calliostoma, 278
Calliphoridae (family), 409
Callochiton, 276
Callyspongia, 94
Callyspongia diffusa, 80–81, 81*t*
Callyspongiidae (family), 94
Calotte, 182
Calyptogena, 289
Calyptraea, 279
Calyptraeacea (superfamily), 279–280
Calyptraeoidea (family), 279–280
Calyx, 501
Cambaridae (family), 413
Cambarincola, 337
Cambarus, 413
Cambrian explosion, 8, 31
Cameral fluid, 256
Campanularia, 114*f*, 132
Campodea staphylinus, 361*f*
Cancellus, 414
Cancer, 414
Cancer crabs, 414
Cancridae (family), 414
Candacia, 416
Caobangia, 335
Caobangidae (family), 335
Capitella, 334
Capitellida (order), 334
Capitellidae (family), 334
Caprella, 413
Caprella equilibra, 378*f*
Caprellidea (suborder), 413
Capsala, 176
Captacula, 255, 255*f*
Carabidae (family), 408
Carapace, 351, 352, 374, 378*f*, 379, 380*f*
Carausius, 406
Carcinonemertes, 213
Carcinonemertidae (family), 213
Carcinus, 414
Carcinus maenas, 393*f*
Cardiac stomach, 394, 505
Cardiidae (family), 288
Cardisoma, 414
Cardisoma guanhummi, 414
Cardita, 288
Carditidae (family), 288
Cardium, 244*f*, 288
Caridea (infraorder), 413
Carina, 389
Carinaria, 280
Carinaria lamarcki, 234*f*
Carinariidae (family), 280
Carinariodae (superfamily), 280
Carnivores, 32
Carybdea, 110*f*, 132
Caryophyllaeides, 175
Caryophyllidea (order), 175
Caspian Sea, 137
Cassiopea, 131
Castenella, 77
Casting, 531
Castrella, 174
Catch tentacles, 119
Catch tissue, 500
Catenula, 173
Catenulida (order), 173
Caterpillar, 572*t*
Cats, parasites in, 61
Cattle face fly, 409
Caudal fin, 462, 465*f*
Caudal papillae, 437
Caudina, 526
Caudofoveata (subclass). *See* Chaetodermomorpha (=Caudofoveata) (subclass)
Caulerpa, 283
Cavolina, 238*f*, 283
Cavoliniidae (=Cuvieriidae) (family), 283
cDNA. *See* Complementary DNA fragments
Cell number, classification by, 9
Cellana, 234*f*, 278
Cellaria, 494
Cellariidae (family), 494
Cellularia (subphylum), 93–94
Cellulose, 71
Cement glands, 196
Cenocrinus, 524
Cenocrinus asterias, 501*f*
Centipedes. *See* Chilopoda (order)
Central sheath, 42f, 43
Centrohelids, 67
Centropages, 416
Centropages typicus, 383*f*, 565*f*
Centruroides, 403
Cepaea, 285
Cephalaspidea (order), 282
Cephalic lobe, 304, 306, 306*f*
Cephalic papillae, 437
Cephalic shield, 534
Cephalization, 9, 259
Cephalobaena, 417
Cephalobaenida (order), 417
Cephalocarida (subclass), 411
Cephalochordata (=Acrania) (subphylum), 548–551, 549*f*, 550*f*
cleavage, 14*f*
defining characteristics, 548
feeding mechanisms, 548
fossil record, 548
musculature, 551
nervous system, 548, 550
reproduction, 559*t*
taxonomic detail, 553
taxonomic summary, 551
Cephalodiscus, 533, 535
Cephalopoda (class), 256–265
behavior, 259, 264, 264*f*
body characteristics, 256, 259
circulatory system, 259, 266
defining characteristics, 256
digestive system, 258*f*–259*f*, 259
diversity, 258–259
evolution, 257, 261, 263
fossil record, 257, 260*f*
gas exchange, 256
ink sac, 259
locomotion, 256–257
musculature, 256
nervous system, 259, 262*f*, 268–269, 271*f*
photophores, 258–259, 261*f*
reproduction, 265, 563, 559*t*, 564*f*, 565
sense organs, 259–261, 262*f*
shell, 256–257, 257*f*
taxonomic detail, 290–292
taxonomic summary, 271
Cephalothorax, 352, 374
Cephalothrix bioculata, 208*f*
Cerambycidae (family), 408
Cerata, 236
Ceratium hirundinella, 58*f*
Cercaria, 161, 163*f*, 164, 558*t*
Cerci, 363
Cercopidae (family), 407
Cercozoans, 68, 77
Cerebratulus, 205*f*, 212
Cerebratulus lacteus, 206
Ceriantharia (order), 134
Cerianthus, 125*f*, 134
Cerithidea, 279
Cerithiidae (family), 279
Cerithioidea (superfamily), 279
Cerithiopsis tubercularis, 561*f*
Cerithium, 279
Cestida (order), 145f, 146, 148
Cestoda (class), 32*f*, 157–159
defining characteristics, 157
feeding mechanisms, 157
life cycle, 158–159, 159*f*
reproduction and development, 158–159, 558*t*, 560
taxonomic detail, 175–176
taxonomic summary, 170
Cestodaria (subclass), 158, 158*f*, 170, 175
Cestum, 148
Cestum veneris, 145*f*
Chaetae, 295
Chaetoderma, 277
Chaetodermatidae (family), 277
Chaetodermomorpha (=Caudofoveata) (subclass), 277
Chaetogaster, 336
Chaetognatha (phylum), 461–467, 462*f*
classification, 465–467
defining characteristics, 461
digestive system, 463
feeding mechanisms, 462, 464, 464*t*
general characteristics, 461–463
lifestyles and behavior, 463, 465, 467*f*
nervous system, 461, 462*f*
reproduction and development, 463, 467*f*, 559*t*, 563, 567
sense organs, 461
taxonomic detail, 470
taxonomic summary, 469
Chaetonotida (order), 470
Chaetonotus, 460*f*, 470
Chaetopleura, 276
Chaetopterida (order), 334

Chaetopteridae (family), 334
Chaetopterus, 334
Chaetopterus variopedatus, 301*f*, 334
Chaetostephanidae (family), 457
Chambered nautilus. *See Nautilus*
Chaoboridae (family), 409
Chaoborus, 409
Chaos, 63*f*, 76
Charonia, 280
Cheilostomata (order), 485, 486, 489, 490, 491, 494
Chelicerae, 352, 353
Chelicerata (subphylum), 352
  Arachnida (class). *See* Arachnida (class)
  defining characteristics, 352
  Merostomata (class), 352, 392–393, 395, 396, 403, 559*t*
  Pycnogonida (=Pantopoda) (class), 357, 358*f*, 396, 405
  taxonomic detail, 403–405
  taxonomic summary, 396
Chelifer, 404
Chemoautotrophs, 247
Chemosynthesis, 311
Cherax, 413
"Cherry-stone" clams, 289
Chewing lice, 407
Chilaria, 352
Chilopoda (order), 350, 358–360, 359*f*
  cuticle, 359
  digestive system, 393
  excretory system, 395
  reproduction, 559*t*
  respiration, 359
  sense organs, 359
  taxonomic detail, 405
  taxonomic summary, 396
Chinese liver fluke, 162*f*, 166*f*, 177
Chiridota, 526
Chironex, 132
Chironex fleckeri, 132
Chironomidae (family), 409
Chitin, 342
Chiton, 277
Chitonidae (family), 277
Chitons. *See* Polyplacophora (class)
Chlamydomonas, 69*f*
Chlamys, 287
Chloragogen, 325
Chlorella, 112
Chlorocruorin, 328
Chloromyxum, 77
Chlorophyll, 58, 68
Chloroquine, 61
Choanocytes, 81, 82*f*, 94
Choanoflagellates, 68, 69*f*, 77
Chondracanthus nodosus, 385*f*
Chondrophora (order), 117*f*, 133
Chorades ferox, 453*f*
Chordata (phylum), 539–554
  Ascidiacea (class), 540–542, 547–548, 551, 552–553, 556, 559*t*, 574*t*
  Cephalochordata, 548–551, 553, 559*t*
  comparison with Hemichordata, 529
  defining characteristics, 539
  general characteristics, 539
  Larvacea (=Appendicularia) (class), 543–544, 547, 548, 551, 553, 559*t*
  reproduction and development, 574*t*
  taxonomic detail, 552–553
  taxonomic summary, 551
  Thaliacea (class), 544–547, 548, 551, 553, 559*t*
  Urochordata (=Tunicata) (subphylum), 539, 540–547, 551, 552–553, 559*t*, 574*t*
Chordodes, 457
Chordodes morgani, 453*f*
Chordoid larva, 469
Choricotyle louisianensis, 160*f*
Choristella, 278
Choristellidae (family), 278
Christmas-tree worm, 335
Chromadoria (subclass), 446
Chromatophores, 258, 260*f*, 343, 376, 375*f*
Chromista (kingdom), 38*f*, 72, 73, 77
Chromosome diminution, 438, 438*f*
Chrysaora, 131
Chrysomelidae (family), 408
Chthalamus, 417
Cicadas, 407
Cicadidae (family), 407
Cidaroida (order), 525
"Ciguatera poisoning," 55
Cilia
  Ctenophora, 138–139
  Echiura, 315*f*
  Enteropneusta, 532*f*
  Gastropoda, 229
  gill, 239, 242*f*, 245
  lamellibranchs, 241, 242, 244*f*, 246*f*, 247*f*
  metachronal beating, 47, 47*f*, 184
  Nemertea, 203
  Polychaeta, 301
  Protobranchia, 241, 242*f*
  Rotifera, 184, 184*f*, 186, 188*f*
  statocysts, 107
  structure and function, 42–43, 42*f*, 43*f*
Ciliary bands, 14, 15*t*, 16*f*
Ciliated protozoans, 39, 42–43
Ciliation, patterns of, 47, 48*f*–50*f*
Ciliophora (phylum), 46–54
  behavioral complexity, 54–55
  defining characteristics, 46–47
  excretory system, 48–49
  feeding, 53–54
  lifestyles, 52–54, 56*f*, 57*f*
  morphological features, 47–49, 50*f*
  patterns of ciliation, 47, 48*f*–50*f*
  phenotype plasticity, 54–55
  reproduction, 49–52, 53
  symbiotic, 53
  taxonomic detail, 75
  taxonomic summary, 73
Ciona, 553
Ciona intestinalis, 543*f*, 553
Circular muscles, 98–99, 98*f*, 119
Circulatory system
  Annelida, 328
  Arthropoda, 344–345, 346*f*
  Ascidiacea, 541
  Brachiopoda, 477
  Cephalopoda, 259, 266
  Cirripedia, 389
  Enteropneusta, 531
  lamellibranchs, 244*f*, 269*f*
  Mollusca, 216, 218, 217*f*, 266
  Nemertea, 204–205, 205*f*
  Prosobranch, 232, 233*f*
  Sipuncula, 318
Cirratulida (order), 334
Cirratulidae (family), 334
Cirratulus, 334
Cirratulus cirratus, 302*f*
Cirri, 389, 502, 502*f*
Cirripedia (subclass), 389, 390*f*, 391*f*
  circulatory system, 389
  defining characteristics, 389
  development, 389, 391
  feeding mechanisms, 389, 391*f*
  reproduction and development, 560, 560*f*, 571*t*
  shell, 389, 390*f*
  taxonomic detail, 417–418
  taxonomic summary, 396
Cirroteuthidae (family), 291
Cirrothauma, 291
Cirrus, 47, 49*f*, 50*f*
Cistena, 335
Cistenides. See Pectinaria (=Cistenides)
Cittarium, 278
Clade, 27*t*
Cladistics, 27–30, 31
  in action, 27–28, 28*f*
  appeal of, 31
  controversy, 31
  and molecular data, 29–30, 30*f*
  vocabulary associated with, 27*t*
Cladocera (order), 379, 415
Cladogenesis, 27*t*
Cladogram, 27*t*, 28
Cladorhiza, 94
"Cladorhizid" sponges, 88
Cladorhizidae (family), 94
Clam shrimp, 396, 415
Classical taxonomy, 26
Classification, 7–36
  by body symmetry, 9
  by cell number, 9
  by developmental pattern, 9–16, 17*f*
  by embryology, 9
  by evolutionary relationship, 16–31
  by habitat, 31–32
  by lifestyle, 31–32
Clathriidae (family), 94
Clausila, 285
Clausiliidae (family), 285
Clausilium, 285
Clavagella, 290
Clavagellidae (family), 290
Clavelina, 553
Clavelina minata, 553
Clavularia, 125*f*
Cleavage, 12–14, 26
Clevelandia ios, 314*f*
Clibanarius, 414
Click beetles, 408
Climacostomum, 49*f*
Clio, 283
Cliona, 93
Clione, 283
Clione limacina, 238*f*, 283
Clionidae (family), 93, 283
Clitellata (class), 296, 318–325
  comparison with Sipuncula, 317*t*
  defining characteristics, 318
  Hirudinea (subclass), 322–325, 328, 337, 558*t*
  nervous system, 327–328
  Oligochaeta (subclass), 318–322, 325, 328, 336–337, 558*t*
  sense organs, 327–328
  taxonomic detail, 336–337
  taxonomic summary, 328
Clitellio, 336
Clitellum, 318, 319*f*, 322, 323*f*, 325
Cloaca, 195, 196
Clonorchis sinensis, 177
Clymenella, 334
Clypeaster, 525
Clypeasteroida (order), 525
Clytia, 132
Clytia gracilis, 132
Cnidae, 102
Cnidaria (=Coelenterata) (phylum), 101–135
  Anthozoa (class), 118–125, 128, 133–134, 345*t*, 556, 558*t*, 568*t*
  vs. Ctenophora, 138, 140, 141, 142*t*
  Cubozoa (class), 108–110, 110*f*, 128, 132
  defining characteristics, 101
  evolution, 118, 119*f*
  feeding mechanisms, 102
  general characteristics, 102–104
  Hydrozoa (class), 110–115, 116*f*, 128, 132–133, 556, 558*t*, 568*t*
  musculature, 104, 105*f*
  nervous system, 102, 104*f*
  reproduction and development, 558*t*, 560, 566, 568*t*
  respiration, 104
  rotifer parasites of, 187*f*
  Scyphozoa (class), 104–108, 109*f*, 128, 131–132, 345*t*, 556, 558*t*, 568*t*
  symbiotic relationships, 112–113, 113*t*
  taxonomic detail, 131–134
  taxonomic summary, 128
Cnidoblasts, 102
Cnidocil, 102
Cnidocytes, 105*f*
Cocculina, 278
Cocculinella minutissima, 270*f*
Cocculinidae (family), 278
Cocculiniformia (superorder), 278
Cockles, 244*f*, 288
Cockroaches, 71, 75, 370, 406
Coconut crabs, 414
Codosiga, 77
Codosiga botrytis, 69*f*
Coelenterata. *See* Cnidaria (=Coelenterata) (phylum)
Coeloblastula, 90
Coelogynoporidae (family), 174
Coelom, 10
  advantage of, 10
  formation of, 10, 10*f*, 11*f*
Coelomates, 10–14
  advantage, 10
  ciliary bands, 14, 16*f*
  cleavage, 12–14
  coelom formation, 10, 11*f*
  coelomic cavities, 12
  cross section, 10*f*, 11*f*
  ideal characteristics, 15*t*
  mesoderm origin, 14
  mouth origin, 11
  polar lobe formation, 14, 15*f*
Coelomic cavities, 12, 212
Coelomic pouch, 11*f*
Coelomocyte lysate, 517
Coelomocytes, 500
Coeloplana, 147
Coeloplana mesnili, 145*f*
Coeloplanidae (family), 147
Coenobita, 414
Coenobitidae (family), 414
Coiled shell, 224, 225, 234*f*, 260*f*
Coleoidea (=Dibranchiata) (subclass), 290–292
Coleoptera (order), 408
Collar cells, 81
Collar region, 530
Collars, 347
Collembola (order), 405
Collisella. See Lottia (=Collisella)
Collisella scabra, 234*f*
Colloblasts, 139–140, 139*f*
Collotheca, 191*f*, 201
Collothecaceae (order), 201
Collothecidae (family), 201
Colonies
  Ascidiacea, 541, 542*f*
  Bryozoa, 482, 484–485, 485*f*, 490
  Entoprocta, 491
  Gymnolaemata, 486
  Hexacorallia, 121, 123*f*
  Hydrozoa, 112–113, 114*f*, 132
  Insecta, 372, 373
  Octocorallia, 124
  Phylactolaemata, 482–484
  Protozoa, 46
  sponges, 80
  zooflagellates, 68, 69*f*
Color changes, 343
Columbella, 281
Columbellidae (family), 281
Columella, 224
Columellar muscle, 224
Columnals, 501
Colus, 281
Comantheria, 524
Comanthina, 524
Comanthina schlegelii, 524
Comatulida (order), 502, 524
Comatulids, 502, 502*f*
Comb rows, 138, 139*f*
Comb-footed spiders, 404
Commensal, 32
Commensalism, 32
Common earthworm. *See Lumbricus terrestris*
Common garden spider, 354*f*
Common octopus, 259
Common wrasse, 126–127, 127*f*
Compensation sac, 486
Compensatory sacs, 316
Complementary DNA fragments (cDNA), 206
Complete mesenteries, 119
Complex life cycles, 567
Compound eye, 345–349, 346*f*, 351, 351*f*, 363, 374, 378*f*, 567
Concentricycloidea (class), 509
Concentricycloids, 508–509, 509*f*
  body characteristics, 508
  defining characteristics, 508
  evolution, 509
  reproduction and development, 518
  taxonomic detail, 525
  taxonomic summary, 521
Conchifera (subphylum), 277–292
Concholepas, 281
Conchophthirus, 48*f*
Conchostraca (order), 415
Conchs, 279
Cone shells, 281
Cone snails, 232
Conidae (family), 281–282
Conjugation, 50, 51*f*
Conjunctivitis, 409
Conocyema, 182
Conoidea (superfamily), 281–282
Contractile vacuoles, 40–41, 40*f*, 41*f*
Contractile vessels, 316
Conus, 232, 232*f*, 281
Conus abbreviatus, 267*f*
Convergence, 19
Convergent evolution, 39
Convoluta, 172
Copepoda (subclass), 381–382, 383*f*
  defining characteristics, 381
  feeding mechanisms, 382, 385*f*
  locomotion, 382
  reproduction and development, 389, 392*f*, 565*f*, 567, 571*t*
  taxonomic detail, 416
  taxonomic summary, 396
Copepodite, 389
Copepods, 142–143, 143*f*, 396
Coral gall crabs, 415
Coral reefs, 122, 126–127, 127*f*, 134
Corallimorpharia (order), 133
"Coralline sponge," 88*f*, 94
Corals
  ahermatypic, 122, 124
  hermatypic, 102, 122, 134
  horny, 124, 134, 125*f*
  true (stony), 123, 124*f*, 134
Corbicula, 289
Corbicula fluminea, 289
Corbiculidae (family), 289
Cordylophora, 132
Corella, 553

Cormidia, 114, 116*f*
Cornea, 347, 348*f*
Corolla, 283
Corona, 184, 184*f*, 185*f*, 186, 190
Coronadena, 173
Coronatae (order), 131
Coronate larvae, 490, 572*t*
Corophium, 412
Corpora allata, 370–371, 372*f*
Corpora cardiaca, 372*f*
Corticium, 94
Corynactis, 133
Coryphella, 284
Cotylea (suborder), 173
Cotylogaster occidentalis, 170*f*
Cotylurus, 175
Countercurrent exchange, 216, 217*f*, 218
Cowries, 280
Coxa, 379
Crab lice, 407
Crab spiders, 404
Crabs, 353*f*, 396
Cranchiidae (family), 291
Crane flies, 409
Crangon, 413
Crangonidae (family), 413
Crania, 493
Craspedacusta, 132
Crassostrea, 287
Crassostrea virginica, 216*f*, 249*f*, 267*f*, 559–560, 576*f*
Cratena, 284
Cratenemertidae (family), 213
Crayfish (crawfish), 376, 394*f*, 395*f*, 396, 411, 413
Crepidula, 279
Crepidula convexa, 17
Crepidula fornicata, 13*f*, 17, 234*f*
Crepidula plana, 17
Crickets, 406
Crinoidea (class), 500–503, 501*f*, 502*f*
  defining characteristics, 500
  digestive system, 501
  feeding mechanisms, 501–502
  fossil record, 500, 501
  locomotion, 502–503
  nervous system, 520, 520*f*
  reproduction and development, 518, 519*f*, 559*t*, 573*t*, 575*f*
  taxonomic detail, 523–524
  taxonomic summary, 521
Crinozoa (subphylum), 521, 523–524
Crisia, 493
Crisia ramosa, 483*f*
Cristatella, 474*f*, 483*f*, 493
Cristatellidae (family), 493
Crop, 325
Crown conchs, 281
Crown-of-thorns, 525
Crustacea (class), 350, 373–392
  Branchiopoda (subclass), 349, 379–381, 391, 396, 415
  Cirripedia (subclass), 389, 393, 411, 417–418, 560, 560*f*, 571*t*
  comparison with Insecta, 31
  Copepoda (subclass), 381–382, 386–387, 389, 391, 396, 416, 565*f*, 567, 571*t*
  cuticle, 342*f*
  defining characteristics, 373
  digestive system, 394, 394*f*
  dispersal, 580
  ecdysis, 343, 344*f*
  excretory system, 395–396
  eyes, 347
  Malacostraca (subclass), 374–379, 391, 394*f*, 396, 411–415, 571*t*
  Ostracoda (subclass), 381, 389, 396, 416
  Pentastomida (subclass), 382–385, 396, 417, 558*t*
  reproduction and development, 389–392, 392*f*, 559*t*, 560, 562*f*, 571*t*
  taxonomic detail, 411–418
  taxonomic summary, 396
Cryptobiosis, 189, 194, 423–424
Cryptobiotic state, 580
Cryptocercus, 406
Cryptochiton, 277
Cryptocotyle, 177
Cryptocyst, 486
Cryptodonta (subclass). *See* Protobranchia (=Paleotaxodonta =Cryptodonta) (subclass)
Cryptomonas, 192
Cryptomya californica, 314*f*
Cryptoplacidae (family), 277
Cryptoplax, 277
Cryptosula, 483*f*
Crystalline cone, 347, 348*f*
Crystalline style, 247, 249*f*
Ctene, 138
Ctenidia, 216, 232, 233*f*
Ctenodiscus, 524
Ctenophora (phylum), 137–148
  bioluminescent, 141, 143
  body characteristics, 138, 139*f*
  vs. Cnidaria, 138, 140, 141, 142*t*
  digestive system, 138
  feeding mechanisms, 139–140
  fossil records, 138
  general characteristics, 137–143
  impact of, 137
  locomotion, 138
  musculature, 138
  nervous system, 138
  Nuda (class), 145*f*, 146, 148
  reproduction and development, 140–141, 556, 558*t*, 560
  sense organs, 138–139, 139*f*
  taxonomic detail, 147–148
  taxonomic summary, 146
  Tentaculata (class), 144–146, 144*f*, 145*f*, 147–148
Ctenoplana, 147
Ctenoplanidae (family), 147
Ctenopoda (suborder), 415
Ctenostomata (order), 485, 491, 493–494
Cubomedusae, 108–109, 110*f*, 132
Cubozoa (class), 108–110, 110*f*, 128, 132, 558*t*
Cucumaria, 526
Cucumaria frondosa, 514*f*
Cucumariidae (family), 526
Culex, 409
Culicidae (family), 409
Cultellidae (family). *See* Pharidae (=Cultellidae) (family)
Cultellus, 288
Cultured pearls, 216
Cumacea (order), 412
Cup-and-saucer shells, 279
Curculionidae (family), 408
Cuspidaria, 290
Cuspidariidae (family), 290
Cuthona, 284
Cuthonidae (family). *See* Tergipedidae (=Cuthonidae) (family)
Cuticle, 342
  Chilopoda, 359
  Crustacea, 342
  Diplopoda, 360
  Gastrotricha, 459
  Insecta, 343, 363
  Kinorhyncha, 456
  Merostomata, 393
  molting, 343, 344*f*, 432
  Nematoda, 432–434, 433*f*
  Nematomorpha, 452
  Onychophora, 424
  Pentastomida, 382
  Priapulida, 454
  Tardigrada, 422
Cuttlebone, 260*f*, 291
Cuttlefish, 256, 260*f*, 271, 291, 565
Cutworms, 410
Cuvierian tubules, 518
Cuvieriidae (family). *See* Cavoliniidae (=Cuvieriidae) (family)
Cyamidae (family), 413
Cyanea, 131
Cyanea capillata, 103*f*
Cyclestherida (suborder), 415
Cycliophora (phylum), 20, 467–469, 468*f*, 469*f*
  defining characteristics, 467
  life cycle, 468–469, 469*f*
  reproduction and development, 469
  taxonomic detail, 470
  taxonomic summary, 469
Cyclomorphosis, 192–193, 193*f*, 193*t*, 380
Cycloneuralia, 17*t*, 21*f*, 183, 432, 432*t*, 452*t*
Cyclophoridae (family), 279
Cyclophyllidea (order), 176
Cyclopoida (order), 416
Cyclorhagida (order), 457
Cyclosalpa affinis, 546*f*
Cyclostomata (order), 488, 491
Cyclostremiscus beauii, 269*f*
Cyclotornidae (family), 410
Cydippid, 141
Cydippida (order), 144, 144*f*, 145*f*, 146, 147
Cymatiidae (family). *See* Ranellidae (=Cymatiidae) (family)
Cymatium, 280
Cymbulia, 283
Cymbulidae (family), 283
Cyphoma, 280
Cyphonautes larvae, 490, 572*t*
Cypraea, 280
Cypraeidae (family), 280
Cypraeoidea (superfamily), 280
Cyprid, 389, 571*t*, 575*f*
Cypridina, 416
Cypris, 389
Cypris, 416
Cyrtocrinida (order), 523
Cyst, 44, 190, 194*f*
Cysticercus, 158
Cystid, 480
Cystipus occidentalis. *See Holothuria (Cystipus) occidentalis*
Cystipus pseudofossor. *See Holothuria (Cystipus) pseudofossor*
Cytochrome-c oxidase, 310–311, 311*f*
Cytoplasm, 39
Cytoplasmic streaming, 44
Cytoproct, 49
Cytostome, 47

D

Dactylogyrus, 176
Dactylopodola, 470
Dactylozooids, 113, 114, 115*f*, 116*f*, 117
Daddy longlegs, 354*f*, 404
Dalyellia, 174
Dalyelliidae (family), 174
Dalyellioida (suborder), 174
Damselflies, 371*f*, 406
Danaidae (family), 410
Danaus plexippus, 410
Daphnia, 379, 415
Daphnia pulex, 380*f*
Darwin, Charles, 19, 25, 318, 336
Dauer larva, 438
Daughter sporocyst, 161, 167*f*
Daughters, 52
DDT, 61
Deadman's fingers, 133
Deafness, in cephalopods, 261
Decapoda (order), 375*f*, 376. *See also* Teuthoidea (=Decapoda) (order)
  development, 391, 393*f*, 571*t*
  taxonomic detail, 413–415
  taxonomic summary, 396
Deep-sea limpets, 270*f*, 278
Deerflies, 409
Definitive host, 58
  of Cestoda, 158, 159, 159*f*
  of Pentastomida, 385
  of Trematoda, 164, 166*f*–167*f*, 169
Delamination, 141, 141*f*
Delayed metamorphosis, 578–579, 579*f*, 580–581, 581*t*
Demibranchs, 241, 242*f*, 245
Demospongiae (class), 80, 86, 87–88, 90, 91, 93
Dendraster, 525
Dendrobeania, 494
Dendrochirotida (order), 526
Dendrocoelidae (family), 174
Dendrograms, 20, 21*f*–23*f*
Dendronephthya hemprichi, 111*f*
Dendronotidae (family), 284
Dendronotina (suborder), 236*f*, 284
Dendronotus, 284
Dendronotus arborescens, 236*f*
Dendropoma, 280
Dendropoma platypus, 564*f*
Density, of water, 3–4
Dentaliidae (family), 290
Dentalium, 255*f*, 290
Deposit feeders, 32, 241, 505, 531
Derechinus horridus, 525
Derived (ancestral) state, 27*t*
Derived condition, 1, 27*t*
Dermacentor, 404
Dermacentor andersoni, 354*f*
Dermaptera (order), 407
Dermatophagoides, 408
Dermestes, 408
Dermestidae (family), 408
Dero, 336
Deroceras, 285
Descending limb, 245, 246*f*
Determinate (mosaic) cleavage, 14
Detorsion, 229
Deuterostomes, 12–14, 17*f*, 21*f*
  vs. chaetognaths, 467
  ciliary bands, 14, 16*f*
  cleavage, 12
  coelom formation, 10, 11*f*
  coelomic cavities, 12
  ideal characteristics, 15*t*
  mesoderm origin, 14
  mouth origin, 11
Development. *See* Reproduction and development
Developmental pattern, classification by, 9–16, 17*f*
Dextral shell, 225, 227*f*
Diadema, 525
Diadema antillarus, 510*f*
Diadema setosum, 510*f*
Diadematoida (order), 525
Diadumene, 133
Dialula sandiegensis, 236*f*
Diapause, 194, 371, 484
Diaphanoeca, 77
Diapheromena, 406
Diaphragm, 481
Diaptomidae (family), 416
Diaptomus, 416
Diarrhetic shellfish poisoning, 55
Diaulula, 284
Diaululidae (family). *See* Discodorididae (=Diaululidae) (family)
Dibranchiata (subclass). *See* Coleoidea (=Dibranchiata) (subclass)
Dicrocoelium, 177
Dicrocoelium dendriticum, 168, 177
Dictyostelium, 68*f*, 76
Dicyema, 182
Dicyemida (order), 182
Didemnum, 553
Didinium, 75
Didinium nasutum, 53, 56*f*, 57*f*
Diductor muscles, 479, 479*f*
Dientamoeba, 75
Difflugia, 76
Difflugia gassowskii, 63*f*
Digenea (subclass), 161–170
  evolution of life cycles, 169–170
  life cycles, 161–170, 166*f*–167*f*
  taxonomic detail, 176–177
Digestive caeca, 269, 325
Digestive canals, 140
Digestive diverticula, 247
Digestive glands, 247, 269, 325, 490
Digestive system. *See also* Feeding mechanisms
  Annelida, 325
  Anthozoa, 119
  Arachnida, 393
  Arthropoda, 392–394
  Ascidiacea, 540
  Asteroidea, 505, 507*f*
  Cephalopoda, 258*f*–259*f*, 259
  Chaetognatha, 463
  Chilopoda, 393
  Cnidaria, 102
  Crinoidea, 501
  Crustacea, 394, 394*f*
  Ctenophora, 138
  Diplopoda, 393
  Echinoidea, 512–513, 513*f*
  Enteropneusta, 531, 532*f*
  Hirudinea, 325
  Holothuroidea, 516
  Hydrozoa, 112
  Insecta, 393–394
  lamellibranchs, 248, 249*f*–250*f*
  lophophorates, 490
  Merostomata, 392–393
  Mollusca, 269, 271
  Myriapoda, 393
  Nematoda, 436–437
  Nematomorpha, 454
  Nemertea, 204, 205, 208–209
  Oligochaeta, 325, 326*f*
  Ophiuroidea, 503
  Placozoa, 91
  Platyhelminthes, 150
  Polychaeta, 325, 326*f*
  Polyplacophora, 222
  Porifera, 81
  Protozoa, 46
  Pycnogonida, 357
  Rotifera, 195, 186
  Scyphozoa, 106, 107*f*
  Siboglinidae, 306
  Sipuncula, 316
  Turbellaria, 153, 154, 155*f*
Dilepididae (family), 176
Dimorphic colonies, 112
Dimorphic nuclei, 49
Dimorphic species, 313
Dinoflagellata. *See* Dinozoa (=Dinoflagellata) (phylum)
Dinophilus, 336
Dinozoa (=Dinoflagellata) (phylum), 54–58, 58*f*, 73, 75
Diodora, 278
Dioecious species. *See* Gonochoristic species
Dioecocestidae (family), 176

Dioecotaenia, 176
Diogenidae (family), 414
Diopatra, 333
Diophrys scutum, 57*f*
Diphyllobothriidae (family), 175
Diphyllobothrium, 175
Diphyllobothrium latum, 175
Diploblastic animals
  classification of, 9
  Cnidaria, 104
Diplocentrus, 403
Diplocotyle, 175
Diplogasteromorpha (infraorder), 446
Diploid eggs, 190
Diplomonads, 76
Diplopoda (order), 350, 358–360, 359*f*
  cuticle, 360
  digestive system, 393
  excretory system, 395
  reproduction, 559*t*
  taxonomic detail, 405
  taxonomic summary, 396
Diploria, 134
Diplosoma, 553
Diplospora, 75
Diplostraca (superorder), 415
Diplozoon, 176
Diplura (order), 405
Diptera (order), 361, 409–410
Dipylidium caninum, 176
Direct sperm transfer, 563
Direct waves, 229, 231*f*, 321
Dirofilaria immitis, 439*f*, 448
Discocelidae (family), 173
Discocotyle, 176
Discodorididae (=Diaululidae) (family), 284
Discodoris, 284
Dispersal, 577, 580–581, 580*f*
Dissolved organic matter (DOM), 306, 307
Distaplia, 553
Ditylenchus, 446
Diurnal vertical migrations, 463, 467*f*
DNA
  and evolutionary relationships, 20
  in macronucleus, 49
Dobson flies, 408
Dodecaceria, 334
Dogwinkles, 281
Doliolaria larva, 519*f*, 573*t*, 575*f*
Doliolida (order), 551, 553
Doliolids, 544, 545, 546*f*
Doliolina intermedia, 546*f*
Doliolum, 553
Doliolum nationalis, 546*f*
DOM. *See* Dissolved organic matter
Donacia, 371*f*
Donacidae (family), 288
Donax, 288
Door snails, 285
Dorididae (family), 284
Doridina (suborder), 283–284
Doris, 284
Dorylaimida (order), 446
Doto, 284
Dotoidae (family), 284
Douglas, Angela, 112–113
Dove shells, 281
Dracunculidae (family), 448
Dracunculus medinensis, 416, 442, 443*f*, 448
Drag, 4–5
Dragonflies, 406, 370
Dreissena, 289
Dreissena polymorpha, 242, 267, 567
Dreissenidae (family), 289
Drosophila, 361*f*, 409
Drosophila melanogaster, 348*f*, 367*f*
Drosophilidae (family), 409
Drupa, 281
Dufour's gland, 374*f*
Dugesia, 155*f*, 174
Dugesiidae (family), 174
Dung beetles, 408
Dung flies, 409
Duo-glands, 153, 153*f*
Duvaucellidae (family). *See* Tritoniidae (=Duvaucellidae) (family)
Dynein, 43
Dynein arms, 43
Dysentery, 409

E

Earthworms, 318, 322
Earwigs, 407
East Coast oyster, 559
Ecdysis, 343, 344*f*
Ecdysone, 370
Ecdysozoa (phylum), 17*f*, 20, 24*f*, 350, 432, 432*f*, 451
Ecdysteroid hormones, 343, 344*f*, 432
Echinarachnius, 525
Echinaster, 525
Echiniscidae (family), 428–429
Echiniscus, 428
Echiniscus spiniger, 423*f*
Echinocardium, 526
Echinococcus, 176
Echinoderella, 456*f*
Echinoderes, 457
Echinoderida (phylum). *See* Kinorhyncha (=Echinoderida) (phylum)
Echinodermata (phylum), 497–527
  comparison with Hemichordata, 529
  Crinoidea (class), 500–503, 518, 520, 521, 523–524, 559*t*, 574*t*, 575*f*
  defining characteristics, 497
  Echinoidea (class), 509–513, 518, 520, 521, 525–526, 559*t*, 573*t*
  general characteristics, 497–500
  hemal system, 500
  Holothuroidea (class), 13*f*, 513–518, 521, 526, 559*t*, 574*t*
  locomotion, 499–500
  mutable connective tissue, 500
  nervous system, 520, 520*f*
  reproduction and development, 518, 519f, 520, 559*t*, 566, 573*t*–574*t*
  Stelleroidea (class), 503–509, 521, 524–525
  taxonomic detail, 523–526
  taxonomic summary, 521
  tube feet, 498–500, 498*f*, 499*f*
  water vascular system, 498–499, 498*f*
Echinoida (order), 525
Echinoidea (class), 509–513
  body characteristics, 510, 511*f*, 514*f*
  defining characteristics, 509
  digestive system, 512–513, 513*f*
  feeding mechanisms, 512–513
  irregular, 512, 512*f*, 526
  regular, 512, 512*f*
  reproduction and development, 518, 519*f*, 559*t*, 573*t*
  taxonomic detail, 525–526
  taxonomic summary, 521
Echinometra, 525
Echinoneus, 525
Echinopluteus larva, 519*f*, 573*t*
Echinostelium, 76
Echinostoma, 164, 167*f*, 177
Echinostoma caproni, 72*f*
Echinostoma trivolvis, 163*f*
Echinostomida (order), 177
Echinothrix, 525
Echinothuroida (order), 525
Echinozoa (subphylum), 521, 525–526
Echinus, 525
Echinus esculentus, 511*f*
Echiura (class), 295, 312–314, 454
  body characteristics, 312
  comparison with Sipuncula, 317*t*
  defining characteristics, 312
  excretory system, 312, 313
  feeding mechanisms, 312
  gas exchange, 313
  nervous system, 315
  reproduction and development, 313–314, 315*f*, 558*t*
  taxonomic detail, 337–338
  taxonomic summary, 328
Echiura (order), 337
Echiuridae (family), 337
Echiuroidea (order), 337
Echiuris, 337
Echiurus echiurus, 313*f*
Ecteinascidia, 553
Ecteinascidia turbinata, 553
Ectocotyla paguris, 172
Ectoderm, 9
Ectognatha, 396, 405–411
Ectoneural system, 520
Ectoparasites, 160, 188, 333, 382
Ectoplasm, 40
Ectoprocta (phylum). *See* Bryozoa (=Ectoprocta; =Polyzoa) (phylum)
Ectosymbionts, 32
Ediacara, 7
Edwardsia, 133
Eimeria, 75
Elasipodida (order), 526
Electra, 494
Electra pilosa, 483*f*, 484–485, 485*f*, 487*f*
Electridae (family), 494
Elephantiasis, 439*f*, 440, 448
Elkhorn corals, 134
Ellobiidae (family), 284
Ellobium, 284
Elminius, 417
Elminius modestus, 581*t*
Elphidium, 76
Elphidium crispum, 66*f*
Elysia, 282
Elysiidae (family), 282
Elytra, 297
Emarginella, 278
Emarginula, 278
Embata, 201
Embiidina (order), 407
Embiids, 407
Embioptera (order), 28
Embryo fossil, 8, 8*f*
Embryogenesis, 383
Embryology, classification by, 9–10
Emerita, 414
Encephalitis, 357
Encope, 525
Encystment, 44, 62
Endoderm, 9
Endoparasites, 157, 164, 198, 382
Endoplasm, 40
Endopodite, 376, 375*f*
Endoskeleton, siliceous, 65
Endostyle, 540–541
Endosymbionts, 32
Enopla (class), 212, 213
Enoplia (class), 446
Ensis, 288
Entamoeba histolytica, 44
Enterobius vermicularis, 443*f*, 448
Enterocoely, 10, 11*f*, 423, 475
Enteromonas, 76
Enteropneusta (class), 530–531, 531*f*, 532*f*, 534, 535
  body characteristics, 530, 530*f*
  burrowing, 530, 530*f*, 531*f*
  circulatory system, 531
  digestive system, 531, 532*f*
  feeding mechanisms, 530–531, 531*f*
  locomotion, 530, 530*f*
  nervous system, 531
  reproduction and development, 531, 532*f*, 559*t*
  respiratory system, 531
  taxonomic detail, 535
  taxonomic summary, 534
Entobdella soleae, 160*f*
Entoconcha, 281
Entoconchidae (family), 281
Entognatha (class), 360, 396, 405
Entoneural system, 520
Entoprocta (=Kamptozoa) (phylum), 473, 490–491, 491*f*, 495
Entovalva, 288
Environments, air *vs.* water, 1–4, 3*t*
Eoacanthocephala (class), 202
Eogastropoda (subclass), 271, 277–278
Eostichopus regalis, 515*f*
Ephemera, 406
Ephemera varia, 371*f*
Ephemerella, 406
Ephemeroptera (order), 406
Ephydatia, 84*f*, 94
Ephydatia muelleri, 84*f*
Ephyra, 108, 109*f*, 568*t*
Epiboly, 141, 141*f*
Epicuticle, 342, 342*f*
Epidermis, syncytial, 157, 432
Epilabidocera longipedata, 230*f*
Epimenia babai, 223*f*
Epiphanes, 201
Epiphanes senta, 187*f*
Epistome, 475, 482
Epitheliomuscular cells, 104
Epitoke, 301, 304*f*
Epitoky, 301
Epitoniidae (family), 280
Epitonium, 280
Erect stolons, 485
Ergasilus, 416
Erpobdella, 337
Erpobdellidae (family), 337
Errant polychaetes, 299*f*, 300, 300*f*
Errantia (subclass), 300
Escargot, 238, 239*f*, 285
Eteone, 332
Euapta, 526
Euapta lappa, 515*f*
Eubranchipus, 415
Eucarida (superorder), 413–415
Eucestoda (subclass), 157, 170, 175–176
Euchaeta, 416
Euchaeta prestandreae, 388*f*
Eucidaris, 511*f*, 525
Eucidaris tribuloides, 512*f*
Eucladocera (suborder), 415
Eucoelomates, 10
Eudendrium, 132
Eudistylia, 335
Eudistylia vancouveri, 303*f*
Eudrilidae (family), 337
Eudrilus, 337
Euglena, 68, 69*f*, 75
Euglenida (order), 75
Euglenoidea (class), 75–76
Euglenozoa (phylum), 69, 75–76
Eukrohnia, 470
Eukrohnia fowleri, 466*f*
Eukrohnia hamata, 464*t*, 468*f*
Eukrohniidae (family), 464*t*
Eulalia, 332
Eulamellibranch gills, 245, 246*f*
Eulima, 281
Eulimidae (family), 281
Eulimoidea (superfamily), 281
Eumycetozoa, 63, 65, 73, 76
Eunapius, 94
Eunice, 333
Eunicida (order), 333
Eunicidae (family), 333
Eupagurus bernhardus, 32*f*
Euphausia, 413
Euphausia superba, 19
Euphausiacea (order), 376, 378*f*, 396, 413
Euplectella, 89*f*, 94
Euplectellidae (family), 94
Euplokamis, 147
Euplokamis dunlapae, 144*f*
Euplokidae (family), 147
Euplotes, 75
Eupolymnia, 335
Eupolymnia nebulosa, 305*f*
Eupulmonato (order), 284–285
Eupyrene sperm, 561, 561*f*
Eurhamphea vexilligera, 144*f*
European spiny lobster, 413
Eurypterida (order), 403
Eurytemora, 416
Eusocial species, 372
Euspira, 219*f*
Eutardigrada (class), 429
Eutely, 184, 185, 188, 196, 434, 438, 460
Euterpina, 416
Euthyneura (superorder). *See* Heterobranchia (=Euthyneura) (superorder)
Evadne, 415
Evisceration, 518
Evolution, 7–8
  convergent, 39
Evolutionary classification, 32
Evolutionary relationships
  classification by, 16–19, 18*f*
  deduction, 19–30
  determination, 25–30
  uncertainty about, 30–31
Evolutionary systematics, 26
Evolutionary trees, 22–25, 25*f*
Excavata, 38*f*, 69, 73, 75–76
Excretory system
  Annelida, 296, 297*f*
  in aquatic *vs.* terrestrial environments, 2
  Arachnida, 395
  Arthropoda, 395–396
  Ascidiacea, 541–542, 547–548
  Chilopoda, 395
  Ciliophora, 48–49
  Crustacea, 395–396
  Diplopoda, 395
  Echiura, 312, 313
  Insecta, 364, 366*f*, 395, 395*f*
  Merostomata, 395
  Mollusca, 271
  Nematoda, 437
  Nemertea, 204
  Phoronida, 490
  Platyhelminthes, 150–151, 150*f*
  Polychaeta, 300
  Porifera, 81
  Rotifera, 196
  Sipuncula, 318
  Urochordata, 547–548
Excurrent chamber, 241
Excurrent siphon, 242
Excystment, 44
Exogone, 332
Exopodite, 376, 375*f*
Exoskeleton, 342, 342*f*, 343
Exsheathment, 434
External fertilization, 566–567
Extinction, 7, 16
Extracapsular zone, 67
Extrusomes, 42, 42*f*

Eyes. *See also* Ocelli; Sense organs
  Annelida, 327
  Arthropoda, 345–349
  Cephalopoda, 261, 262f
  Chaetognatha, 461
  Copepoda, 382

F

Fabricia, 335
Fairy shrimp, 379, 396, 415
Falcidens, 223f, 277
Fanworms, 335
Fasciola, 177
Fasciola hepatica, 166f, 177
Fasciolaria, 281
Fasciolaria tulipa, 232f
Fasciolariidae (family), 281
Feather stars, 497, 501, 502, 502f, 521, 523, 524
Feather-duster worms, 335
Fecampia, 175
Fecampia erythrocephala, 175
Fecampiidae (family), 175
Feeding mechanisms. *See also* Digestive system
  Anomalodesmata, 254
  in aquatic environments, 2–3
  Arachnida, 393
  Asteroidea, 505
  Cephalochordata, 548
  Cestoda, 157
  Chaetognatha, 462, 464, 464t
  Chilopoda, 393
  Ciliophora, 53–54
  Cirripedia, 389, 391f
  Cnidaria, 102
  Copepoda, 382, 385f
  Crinoidea, 501–502
  Ctenophora, 139–140
  Diplopoda, 393
  Echinoidea, 512–513
  Echiura, 312
  Enteropneusta, 530–531, 531f
  Foraminifera, 65
  Holothuroidea, 515–516
  Insecta, 393
  lamellibranchs, 242, 244f, 245, 246f
  Larvacea, 544
  Merostomata, 392–393
  Mollusca, 218
  Myriapoda, 393
  Nemertea, 209
  Onychophora, 424, 426–427, 427f
  Ophiuroidea, 505
  Polyplacophora, 222
  Priapulida, 454
  Prosobranch, 232
  Protobranchia, 241
  Protozoa, 46
  Rotifera, 186, 187f
  Scaphopoda, 255, 255f
  Scyphozoa, 105–106
  Siboglinidae, 306
  Thaliacea, 545
Fenestrulina malusii, 482f
Fertilization
  in aquatic *vs.* terrestrial environments, 2–3
  external, 566–567
  internal, 566
Ficopomatus (=Mercierella) enigmaticus, 335
Fiddler crabs, 415
Filament, 241, 242
Filarial nematodes, 443, 448
Filariasis, 409
Filariidae (family), 448
Filibranch gills, 245, 246f
Filinia, 185f
Filinia longiseta, 188f
Filopodia, 44, 44f, 65
Filospermoidea (order), 202
Fingered sponge, 89f
Fingernail clams, 289
Fire coral, 114, 132
Fireflies, 408
Fission, 44, 119, 155
  binary. *See* Binary fission
  multiple, 44, 62
Fissurella, 278
Fissurellacea (superfamily), 278
Fissurellidae (family), 278
Flabelligera, 334
Flabelligera commensalis, 334
Flabelligerida (order), 334
Flabelligeridae (family), 334
Flabellina, 284
Flabellinidae (family), 284
Flaccisagitta enflata, 464t
Flaccisagitta scrippsae, 464t
Flagella, structure and function, 43
Flagellated protozoans, 39, 43, 68–72
  evolutionary relationship, 68
  phytoflagellated, 68, 69f
  taxonomic summary, 73
  zooflagellated, 68–72, 69f, 73
Flame cell, 150f, 151
Flamingo tongue snails, 280
Flatworms. *See* Platyhelminthes (phylum)
Fleas, 370, 409
Flies, 409
  compound eye, 346f
  development, 572t
  wings, 367f
Flight, Insecta, 365–367, 367f
Floatoblasts, 484
Florometra, 524
Florometra serratissima, 575f
Floscularia, 201
Floscularia ringens, 187f, 190f
Flosculariaceae (order), 201
Flosculariidae (family), 201
Flour beetle, 408
Flow cytometer, 387
Flukes. *See* Digenea (subclass)
Flustra foliacea, 488f
Folliculina, 75
Folliculinids, 52, 56f
Food cup, 62
Food vacuole, 46
Foot, 216, 224
Foraminifera (phylum), 65, 66f, 76
  defining characteristics, 65
  feeding, 65
  locomotion, 42–43, 42f, 43f
  symbiotic, 53, 76
  taxonomic summary, 73
Forcipulata (order), 525
Foregut, 392
Formica, 411, 374f
Formicidae (family), 411
Fossil record, 7–8, 8f, 19, 31
  Acanthocephala, 198
  ants, 360
  Arthropoda, 350, 357
  Brachiopoda, 476, 477f
  Bryozoa, 480f, 481
  Cephalochordata, 548
  Cephalopoda, 257, 260f
  Crinoidea, 500, 501
  Ctenophora, 138
  Foraminifera, 65, 66f, 76
  Insecta, 370, 373
  Lobopoda, 422, 422f
  Mollusca, 218, 220f
  Monoplacophora, 222, 225f
  Onychophora, 428
  Polyplacophora, 222
  radiolarians, 65
  sarcodinids, 62
  Scaphopoda, 255
  Tardigrada, 422
  Trilobita, 351, 351f
Foundry Cove, 320–321, 321f
Fragmentation, 91
Fredericella, 481f
Frenulata (subfamily). *See* Perviata (=Frenulata) (subfamily)
Freshwater environment, 5
Fritillaria, 553
Frog shells, 280
Froghoppers, 407
Frontal cilia, 241, 243f, 245, 246f, 252
Frontal membrane, 486
Fruit flies, 361f, 409
Fungia, 134
Funicular cords, 481
Funiculus, 481
Funnel, 256
Fusitriton, 280

G

Galathea, 414, 562f
Galatheidae (family), 414
Galeodes dastuguei, 354f
Galeommatidae (family), 288
Galiteuthis, 291
Gamete diversity, 561, 561f, 562f
Gametocytes, 59, 60f
Gametogenesis, 325, 518
Gammaridea (suborder), 412
Gammarus, 412
Ganesha, 148
Ganeshida (order), 148
Ganglia, 327
Garden spider, 354f
Garstang, Walter, 230, 544
Gas exchange. *See* Respiratory systems
Gastric caeca, 195
Gastric filaments, 106
Gastric glands, 195
Gastric mill, 394
Gastric pouches, 106
Gastric shield, 247
Gastropoda (class), 224–238
  circulation, 266
  defenses against predators, 225–226
  defining characteristics, 224
  evolution, 226
  locomotion, 229, 231f
  musculature, 224, 229, 231f, 345t
  nervous system, 270f
  Opisthobranchia, 235–237, 265, 271, 282–284, 560f, 575f
  Prosobranch, 229–235, 265, 271, 277–278, 561f, 566f, 580
  Pulmonata, 237–238, 265, 271, 284–286
  reproduction and development, 265, 267f–268f, 558t, 560f, 561, 561f, 563, 564f, 566, 566f, 570t, 580f
  shell, 224–225, 227f, 269f
  taxonomic detail, 277–286
  taxonomic summary, 271
  torsion, 226, 228f, 229, 230, 230f
Gastrotricha (phylum), 459–460, 460f
  reproduction, 460, 558t, 560
  taxonomic detail, 470
  taxonomic summary, 469
Gastrovascular canals, 105–106, 107f, 138, 140f
Gastrozooids, 112, 113, 114, 115f, 116f, 117f
Gastrulation, 141, 141f
Gates, 370
Gecarcinidae (family), 414
Gecarcinus, 414
Gel, 44
Gemma, 289
Gemmules, 82–84, 84f
Generic name, 17
Genet, 108, 109f, 468, 541
Geocentrophora, 173
Geoduck, 289
Geonemertes, 213
Geoplana, 174
Geoplanidae (family), 174
Georissa, 278
Germ balls, 161
Germ cells, 438, 438f
Germ layers, 9
Geukensia demissa, 216f
Ghost shrimp, 414
Giant clams, 288
Giant fibers, 268–269, 271f
Giant silk moths, 410
Giant squids, 256, 291
Giant tun, 234f
Giardia, 72, 72f, 76
Giardia lamblia, 72
Gibbulla, 278
Gigaductus, 75
Gigantocypris, 416
Gill cilia, 241, 243f, 247–248
Gill filament, 241, 243, 243f, 244f, 245
Gills, 1
  bacterial symbiosis, 252–253, 253f
  Echinoidea, 512
  Enteropneusta, 531
  lamellibranchs, 242, 244f, 245, 246f, 247
  Mollusca, 266
  Opisthobranchia, 236, 237f
  Prosobranch, 235
  Protobranchia, 241, 242f, 243f
  Pulmonata, 238
Girdle, 54, 218, 220f
Gizzard, 325, 392–393
Glass sponges, 80, 88, 92, 94
Glaucidae (family), 284
Glaucus, 284
Gleba, 283
Globigerina, 76
Globigerina bulloides, 66f
Globigerinoides ruber, 66f
Glochidium, 266, 267f
Glossina, 409
Glossiphonia, 323f, 337
Glossiphonia complanata, 323f
Glossiphoniidae (family), 337
Glossoscolecidae (family), 336
Glycera, 332
Glycera dibranchiata, 332
Glyceracea (superfamily), 332
Glyceridae (family), 332
Gnathifera, 17f
Gnathobase, 352
Gnathostomula, 202
Gnathostomulida jenneri, 198f
Gnathostomulida (phylum), 198–199, 198f, 202, 558t, 562f
Gnats, 409
Goby fish, 314f
Golfingia, 319f, 338
Golfingia minuta, 338
Golfingiidae (family), 338
Goniaster, 525
Gonionemus, 132
Gonochoristic species, 559, 559t
  Acanthocephala, 196
  Arthropoda, 349
  Digenea, 164
  Echiura, 313
  Enteropneusta, 531
  Kinorhyncha, 456
  Mollusca, 265
  Nematoda, 438
  Nematomorpha, 452
  Nemertea, 210
  Pentastomida, 382–383
  Polychaeta, 300
  Scyphozoa, 108
  Seisonidea, 188
  Siboglinidae, 311
Gonodactylus, 411
Gononemertes, 213
Gonozooids, 112, 113, 114, 115f, 116f, 117f
Gooseneck barnacles, 390f
Gordius, 457
Gordius aquaticus, 453f
Gorgonacea (order), 133
Gorgonia, 125f, 133
Gorgonians, 124, 125f, 126–127, 127f
Gorgonocephalidae (family), 524
Gorgonocephalus, 504f, 524
Graffilla, 174
Graffillidae (family), 174
Grantia (=Scypha), 94
Grantiidae (family), 94
Grapsidae (family), 414
Graptolites, 534
Grasshoppers, 361f, 370, 372f, 406
Great Barrier Reef, 122, 501
Green bottle fly, 409
Green crab, 393f, 414
Green glands, 396
Greenflies, 409
Gregarina, 75
Grillotia, 175
Growth lines, 239, 241f
Grylloblattaria (order), 406
Guinea worm, 442
Gymnamoebae, 62–63, 63f, 73
Gymnodimium, 75
Gymnolaemata (class), 484–487, 487f, 488f, 491, 493–494
Gymnophallidae, 177
Gymnosomata (order), 283
Gymnosphaera, 77
Gyrocotyle, 175
Gyrocotyle fimbriata, 158f
Gyrodactylus, 176

H

Habitat, classification by, 31–32
Haeckel, Ernst, 37
Haemadipsa, 337
Haemadipsidae (family), 337
Haemopis, 337
"Hairy girdle" syndrome, 276
Halammohydra, 133
Halichondria, 90f, 94
Halichondriidae (family), 94
Haliclona, 94
Haliclona oculata, 89f
Haliclona viridis, 86f
Haliclonidae (family), 94
Haliclystus, 131
Haliotidae (family), 278
Haliotis, 234f, 278
Haliotis kamtschatkana, 230, 230f
Haliplanella, 133
Halobates, 360, 407
Halocynthia, 553
Halteres, 362
Haminoea, 282
Hamster, parasites in, 72f
Hapalocarcinidae (family), 415
Haplognathia, 202
Haploid eggs, 190
Haplorchis, 177

Haplopoda (suborder), 415
Haplosporidium, 75
Haplotaxida (order), 336–337
Haptocysts, 54, 57*f*
Haptor, 160, 160*f*
Hard-shell clams, 246*f*, 249*f*–250*f*, 289
Harmothoe, 333
Harmothoe imbricata, 299*f*
Harpacticoida (order), 384*f*, 416
Harvestmen, 370, 404
Hatschek's nephridium, 548
Head, 374
Heart
branchial, 259
with ostia, 344, 346*f*
systemic, 259
Heart urchins, 509, 512, 512*f*, 521, 526
Heartworm, 439*f*, 448
Hectocotylus, 265, 265*f*, 565, 565*f*
Heliaster, 525
Heliastra heliopora, 125*f*
Helicidae (family), 285
Helicina, 278
Helicinidae (family), 278
Heliodiscus, 76
Heliolithium, 76
Heliometra, 524
Heliozoa (phylum), 67–68, 67*f*, 73, 77
Helisoma, 240*f*, 286
Helix, 239*f*, 285
Hemal rings, 500
Hemal system, 500
Hematodinium perezi, 58
Hemerythrin, 328, 334, 477
Hemichordata (phylum), 529–535
comparison with Chordata, 529
comparison with Echinodermata, 529
defining characteristics, 529
Enteropneusta (class), 530–531, 531*f*, 532*f*, 534, 535, 559*t*
general characteristics, 529–530
Pterobranchia, 533*f*, 534
reproduction and development, 559*t*, 574*t*
taxonomic detail, 534
taxonomic summary, 534
Hemimetabolous species, 370
Hemiptera (order), 407
Hemithyris psittacea, 478*f*
Hemiurus, 177
Hemocoel, 218, 232, 266, 269*f*, 342–343
Hemocyanin, 266, 343, 396
Hemoglobin, 39, 328, 396, 475
Henricia, 525
Hepatic ceca, 393
Hepatopancreas, 394
Herbivores, 32
Herdmania momus, 541*f*
Hermaphrodites, 559–560
Bryozoa, 489
Cestoda, 158
Chaetognatha, 463, 467*f*
Cirripedia, 560*f*
Clitellata, 318
Ctenophora, 140
Digenea, 168, 176
Gastrotricha, 460
Hirudinea, 323*f*, 325
Mollusca, 265
Nemertea, 210
Oligochaeta, 322, 323*f*
Platyhelminthes, 151
Porifera, 89
Rhombozoa, 182
Scyphozoa, 108
Urochordata, 547
Hermatypic coral, 102, 122, 134
Hermenia, 333
Hermissenda, 284
Hermit crabs, 32*f*, 115*f*, 116*f*, 132, 377*f*, 396, 414
Hesionidae (family), 332
Hesionides, 332
Hesperonoe adventor, 314*f*
Heterobranchia (=Euthyneura) (superorder), 271, 282
Heterocentrotus, 525
Heterocyemida (order), 182
Heterodera, 446
Heteroderidae (family), 446
Heterodonta (subclass), 287–290
Heteromeyenia, 94
Heteromyota (order), 337–338
Heteronemertea (order), 212
Heterophyes, 177
Heterophyidae (family), 177
Heteropods, 234*f*, 235, 280
Heterotardigrada (class), 428–429
Heterozooids, 486
Hexabranchidae (family), 284
Hexabranchus, 284
Hexacontium, 76
Hexacorallia (=Zoantharia) (subclass), 120–124, 125*f*
colonial, 121, 123*f*
solitary, 120
taxonomic detail, 133–134
taxonomic summary, 128
Hexactinellida (class), 86, 88, 89*f*, 92, 94
Hexapoda (class). *See* Insecta (class)
Hiatellidae (family), 289
Higgins larva, 456*f*, 457
Hindgut, 392
Hippa, 414
Hippidae (family), 414
Hippoporinidae (family), 494
Hippopus, 288
Hippospongia, 94
Hirudin, 325
Hirudinea (subclass), 322–325
blood-sucking, 324–325
body characteristics, 323–324, 323*f*, 324*f*
circulatory system, 323–324
defining characteristics, 322
digestive system, 325
locomotion, 324, 324*f*
musculature, 324, 324*f*
reproduction, 323*f*, 325, 558*t*
taxonomic detail, 337
taxonomic summary, 328
Hirudinidae (family), 337
Hirudo medicinalis, 324–325, 337
Histoincompatibility, 80–81, 81*t*
Histriobdella, 333
Histriobdellidae (family), 333
Holectypoida (order), 525
Holocoela. *See* Prolecithophora (=Holocoela) (order)
Holomastigotes, 75
Holomatabolous development, 567
Holometabola (superorder), 408
Holometabolous species, 370
Holopus, 523
Holothuria, 526
Holothuria (Cystipus) occidentalis, 515*f*
Holothuria (Cystipus) pseudofossor, 515*f*
Holothuria polii, 517, 517*f*
Holothuria (Semperothuria) surinamensis, 515*f*
Holothuriidae (family), 526
Holothuroidea (class), 513–518
body characteristics, 514, 514*f*
burrowing, 517, 518*f*
cleavage, 13*f*
defining characteristics, 513
digestive system, 516
evisceration, 518
feeding mechanisms, 515–516
internal anatomy, 514, 514*f*
pollution and, 517, 517*f*
reproduction and development, 518, 519*f*, 559*t*, 574*t*
taxonomic detail, 526
taxonomic summary, 521
Holozoic ciliates, 53
Holozoic flagellates, 68
Homalorhagida (order), 458
Homarus, 413
Homarus americanus, 377*f*, 470
Homarus gammarus, 470
Homo sapiens, 553
Homogammina, 76
Homogenizing, 112
Homokaryotic cells, 49
Homologous characters, 20, 26, 27*t*
Homology, 26, 27*t*
Homoplasy, 27*t*
Homoptera (order), 407
Honeybees, 367, 368*f*, 368–369, 369*f*, 411
Hookworm, 439–440, 439*f*, 441–442, 442*f*, 446–447
Hoplocarida (superorder), 411
Hoplonemertea (=Hoplonemertini) (order), 213
Hoplonemertini (order). *See* Hoplonemertea (=Hoplonemertini) (order)
Hoploplana, 173
Hoploplanidae (family), 173
Hornets, 373
Horny corals, 124, 125*f*, 134
Horse botflies, 410
Horseflies, 409
Horsehair worms. *See* Nematomorpha (phylum)
Horseshoe crabs, 346, 350, 352, 353*f*, 396, 403
Host, 32, 32*f*
definitive. *See* Definitive host
intermediate. *See* Intermediate host
transport, 196
Host penetration, 164, 164*f*
"Host release factor," 113
Houseflies, 406, 576
Hox genes, 20, 120, 475, 548
Human body lice, 407
Hutchinsoniella, 411
Hyaline cap, 44
Hyalophora cecropia, 410
Hydatid cyst, 158, 176
Hydatina, 282
Hydatina physis, 237*f*
Hydra, 101, 105*f*, 110, 110*f*, 111, 132
Hydractinia, 132
Hydractinia echinata, 116*f*
Hydranth, 111
Hydraulic action, 499, 499*f*
Hydrobia, 279
Hydrobia ulvae, 279
Hydrobiidae (family), 279
Hydrocena, 278
Hydrocenidae (family), 278
Hydrocorallina (order), 114–115, 128
Hydrogen sulfide, 308–309, 310–311, 311*f*
Hydrogenosomes, 75
Hydroida (order), 110–113, 114*f*, 115*f*, 128, 132–133
Hydroides, 335
Hydrostatic pressure, internal, 434
Hydrostatic skeleton, 97–99
Hydrotheca, 112, 114*f*
Hydrothermal vents, 308, 309*f*
20-hydroxyecdysone, 432
Hydrozoa (class), 110–115, 116*f*, 117*f*
colonial, 112–113, 114*f*, 132
digestion, 112
Hydrocorallina (order), 114–115, 128
Hydroida (order), 110–113, 114*f*, 115*f*, 128, 132–133
reproduction and development, 111, 111*f*, 112, 114*f*, 556, 558*t*, 568*t*
sense organs, 111
Siphonophora (order), 113–114, 117*f*, 128, 132
taxonomic detail, 132–133
taxonomic summary, 128
Hymenolepididae (family), 176
Hymenolepis diminuta, 176
Hymenoptera (order), 361, 410–411
Hyperia, 412
Hyperia gaudichaudii, 378*f*
Hyperiidea (suborder), 378*f*, 412
Hypermastigia (class), 71, 73, 75
Hyperosmosis, in freshwater, 4
Hyperparasites, 62
Hypoblepharina, 175
Hypoblepharinidae (family), 175
Hypobranchial gland, 233*f*
Hyponeural system, 520
Hypophorella, 494
Hypophorellidae (family), 494
Hypsibius, 429
Hyridella depressa, 267*f*

I

Ichneumonidae (family), 410–411
Iciligorgia schrammi, 126
"Ick," 53
Idiosepiidae (family), 291
Idiosepius, 291
Idotea, 412
Ikeda taenioides, 337
Illex, 291
Ilyanassa, 281
Ilyanassa obsoleta, 19
Imaginal discs, 211, 211*f*, 370
Immune response, 517, 517*f*
Inarticulata (class), 477*f*, 479, 479*t*, 489, 490, 491, 493
Incomplete mesenteries, 119, 120*f*
Incurrent chamber, 241
Incurrent siphon, 242
Indeterminate (regulative) cleavage, 14
Indirect sperm transfer, 563
Inella, 281
Infaunal siboglinids, 307
Infraciliature, 47, 48*t*, 49*f*
Infusoriform larvae, 182
Infusorigens, 182
Ingolfiellidea (suborder), 413
Ingression, 141
Ink sac, 259
Inner demibranch, 245
Innkeeper worm. *See Urechis caupo*
Insecta (class), 350, 360–373
body characteristics, 360–362
comparison with Crustacea, 31
cuticle, 343, 363
defining characteristics, 360
digestive system, 393–394
diversity, 360, 361*f*
evolution, 365, 366, 373
excretory systems, 364, 366*f*, 395, 395*f*
eyes, 347
flight, 365–367, 367*f*
fossil record, 370, 373
gas exchange, 363365, 365*f*
musculature, 367
reproduction and development, 370–371, 371*f*, 372*f*, 556, 559*t*, 560, 561, 562*f*, 563, 567, 572*t*, 581
sense organs, 362–363, 364*f*
social systems, 372–373, 374*f*
studies on, 360
taxonomic detail, 405–411
taxonomic summary, 396
water conservation, 363–365, 365*f*
Instars, 370, 406
Interambulacral areas, 510, 511*f*
Interfilamental ciliary junctions, 245, 246*f*
Interfilamental junctions, 241, 243*f*
Interfilamental tissue junctions, 245
Interlamellar junctions, 245, 246*f*
Intermediate host, 58
of Acanthocephala, 196
of Cestoda, 158, 159, 159*f*
of Pentastomida, 383
of Trematoda, 161, 162*f*–163*f*, 164, 166*f*–167*f*
Internal fertilization, 566
Internal pressure, Nematoda, 434, 434*f*
Interstitial animals, 184, 422
Interstitial cells, 105*f*
Intertidal animals, 31
Intertidal limpet, 234*f*
Intracapsular zone, 67
Introvert, 315–316, 317*f*, 451–452, 456
Invagination, 141, 141*f*
Inversion, 90
Iridocytes, 258
Irregular urchins, 512, 512*f*, *526*
Ischnochiton, 276
Ischnochitonidae (=Bathychitonidae) (family), 276
Isocrinida (order), 524
Isopoda (order), 374, 378*f*, 379, 396, 412
Isoptera (order), 372, 373, 406
Ixodes, 404
Ixodes dammini, 354*f*
Ixodes pacificus, 354*f*
Ixodes scapularis, 354*f*

J

Jackknifing, 27*t*
Janthina, 280
Janthina janthina, 133
Janthinidae (family), 280
Janthinoidea (superfamily), 280
Japanese beetles, 408
Jellyfish. *See* Scyphozoa (class)
Jet propulsion, 256, 269
Jewel beetles, 408
Jiggers, 409
Jingle shells, 287
Johanssonia arctica, 324*f*
Julia, 282
Juliidae (family), 282
Jumping plant lice, 407
Jumping spiders, 354*f*, 404
Juvenile hormone (JH), 370–371

K

Kalyptorhynchia (suborder), 175
Kamptozoa (phylum). *See* Entoprocta (=Kamptozoa) (phylum)
Katharina, 276
Katharina tunicata, 220*f*
Katydids, 406
Kentrogon, 389
Keratella, 201
Keratella slacki, 192–193, 193*f*, 193*t*
Keyhole limpets, 228*f*, 232, 278
Killer bees, 411
Kinbergonuphis, 333
Kinetic energy, 333
Kinetodesmata, 47, 49*f*
Kinetodesmos, 47

Kinetoplast, 69
Kinetoplastea (order), 73, 75–76
Kinetosome, 42, 42*f*, 47, 49*f*, 50*f*
Kinety, 49*f*
King crabs, 414
Kinorhyncha (=Echinoderida) (phylum), 452*t*, 454, 456, 456*f*
  body characteristics, 456
  defining characteristics, 454
  locomotion, 454, 456
  reproduction, 558*t*
  taxonomic detail, 457–458
  taxonomic summary, 457
Kinorhynchus, 458
Komarekiona, 336
Krill, 376–378, 377*f*, 396, 413
Krohnitta pacifica, 466*f*
Krohnitta subtilis, 466*f*
Kronborgia, 175

L

Labial palps, 2341
Labidognatha (suborder), 404
Labium, 360, 362, 363, 365*f*
Labrum, 351, 363, 365*f*
Lacewings, 408
Lacistorhyncus, 175
Lacuna, 279
Ladybugs, 408
Laevicardium, 288
Lamellae, 245
Lamellibrachia, 336
Lamellibranchs, 239, 241–254
  burrowing, 251, 251*f*
  circulation, 244*f*, 269*f*
  defining characteristics, 241
  digestive system, 248, 249*f*, 250*f*
  feeding mechanisms, 242, 244*f*, 245, 246*f*, 247
  gills, 241, 242, 244*f*, 246*f*, 247
  locomotion, 251
  musculature, 251
  reproduction, 265
  symbiotic relationships, 248, 252–253, 253*f*
  taxonomic summary, 271
Lampshells. *See* Brachiopoda (phylum)
Lampsilis, 287
Lampyridae (family), 408
Lancelet, 549*f*, 550*f*, 551, 553
Lancet liver fluke, 177
Land leeches, 337
Larvacea (=Appendicularia) (class), 543–544, 544*f*, 545*f*
  nervous system, 547–548
  reproduction and development, 547, 559*t*
  taxonomic detail, 553
  taxonomic summary, 551
Larvae, 370, 567, 568*t*–574*t*, 575*f*, 576*f*, 577
Lasaea, 288
Lateral cilia, 241, 243*f*, 245, 246*f*
Lateral fins, 462, 465*f*
Laterofrontal cilia, 245
Latrodectus, 404
Leaf beetles, 408
Leaf hoppers, 407
Leaf insects, 406
Leaf-cutter bee, 361
Leavicaudata (suborder), 415
Lecane, 201
Lecanidae (family), 201
Lecithoepitheliata (order), 173
Lecithotrophic larvae, 567, 577
Leeches. *See* Hirudinea (subclass)
Leiobunum, 404
Leishmaiasis, 76
Leishmania, 75
Leishmania donovani, 69–70
Leocrates, 332
Lepadella, 201
Lepas, 390*f*, 417
Lepidochitona, 276
Lepidochitona dentiens, 267
Lepidochitona hartweiqii, 267*f*
Lepidodermella, 470
Lepidodermella squamata, 461*f*
Lepidonotus, 333
Lepidopleuridae (family), 276
Lepidopleurus, 276
Lepidoptera (order), 410
Leptasterias, 525
Leptochiton, 276
Leptogorgia, 133
Leptomedusae (suborder), 132
Leptomonas, 75
Leptonemertes, 213
Leptopoma, 279
Leptosynapta, 526
Leptosynapta inhaerens, 518*f*
Leptoxis, 279
Lernaeocera branchialis, 385*f*
Lernaeodiscus, 417
Leuconoid sponge, 86, 86*t*, 87, 87*f*, 93, 94
Leucosolenia, 94
Leucosoleniidae (family), 94
Leucothea, 145, 147
Leucotheidae (family), 147
Libinia, 414
Lice, 407, 413
Lichenopora, 483*f*
Life cycles
  Acanthocephala, 196, 197*f*
  Apicomplexa, 59, 60*f*, 61
  Cestoda, 158–159, 159*f*
  complex, 567
  Cycliophora, 468–469, 469*f*
  Digenea, 161–170, 166*f*–167*f*
  Mesozoa, 179
  Monogenea, 161
  Monogononta, 195*f*
  Nematoda, 441–442, 442*f*, 443*f*
  Nematomorpha, 454
  Orthonectida, 180, 180*f*
  parasitic, 58, 59*f*
  Pentastomida, 384–385
  Rhombozoa, 181–182, 181*f*
  Scyphozoa, 108, 109*f*
  Trematoda, 163*f*
  zooflagellates, 68
Lifestyle, classification by, 31–32
Ligament, 238
Ligament sacs, 196
Light
  in aquatic *vs.* terrestrial environments, 3
  compound eyes, 345–346, 348*f*
Ligia, 412
Ligumia, 287
Limacidae (family), 285
Limacina, 283
Limacinidae (=Spiratellidae) (family), 283
Limax, 285
Limax flavus, 240*f*
Limifossor talpoideus, 223*f*
Limmenius, 429
Limnius, 191*f*
Limnocodium, 132
Limnodrilus, 336
Limnodrilus hoffmeisteri, 320–321, 321*f*
Limnognathia maerski, 199, 199*f*, 202
Limnomedusae (suborder), 132
Limnoria, 412
Limpets, 277
Limulus, 403
Limulus polyphemus, 352, 353*f*
Linckia, 525
Lineidae (family), 212
Lineus, 212
Lineus longissimus, 212
Linguatulidae (family), 417
Lingula, 477*f*, 493
Lingulata, 417
Lingulida (order), 493
Linnaeus, Carolus, 16
Lion's mane jellyfish, 131
Liriope, 132
Lithodes, 414
Lithodidae (family), 414
Lithophaga, 287
Litiopa, 279
Little heart shells, 288
Littorina, 279
Littorina littorea, 226*f*, 233*f*, 566*f*
Littorinidae (family), 279
Littorinimorpha (infraorder), 279–281
Littorinoidea (superfamily), 279
Loa, 448
Loa loa, 443, 444*f*, 448
Lobata (order), 144, 144*f*, 145*f*, 146, 147
Lobopoda (phylum), 422
Lobopodia, 43–44, 44*f*
Lobopods, 422, 422*f*
Lobsters, 377*f*, 396, 413, 469, 470
"Loco," 281
Locomotion
  Annelida, 296
  Anthozoa, 119–120
  in aquatic *vs.* terrestrial environments, 3
  Asteroidea, 505
  Cephalopoda, 256–257
  Copepoda, 382
  Crinoidea, 502–503
  Ctenophora, 138
  Cubozoa, 109
  Echinodermata, 499–500
  Enteropneusta, 530, 530*f*
  Gastropoda, 229, 231*f*
  Hirudinea, 324, 324*f*
  Kinorhyncha, 454, 456
  lamellibranchs, 251
  and musculature, 98–99, 99*f*
  Nematoda, 434–435, 434*f*
  Nemertea, 203–204
  Oligochaeta, 318, 320–321, 322*f*
  Onychophora, 425–428, 425*f*
  Opisthobranchia, 237
  Placozoa, 91
  Polychaeta, 297–300, 299*f*, 300*f*
  Polyplacophora, 218–219
  Protozoa, 42–44, 42*f*, 43*f*, 45*f*
  Rotifera, 188, 190*f*
  Scyphozoa, 105, 106*f*
  Sipuncula, 316
  Thaliacea, 545
  Turbellaria, 151–153, 152*f*, 153*f*
Locusts, 406
Loliginidae (family), 291
Loligo, 258*f*, 291, 565*f*
Loligo pealei, 262*f*
Lolliguncula, 291
Loma, 77
Longitudinal muscles, 98, 98*f*, 119
Looping, 152, 153*f*, 188, 189*f*
Lophomonas, 75
Lophophorates, 473–496
  digestive system, 490
  nervous system, 490
  reproduction and development, 488
Lophophore, 473
Lophopodidae (family), 493
Lophotrochozoa, 17*f*, 20, 21*f*, 150, 180, 350
Lophoura (=Rebelula) bouvieri, 385*f*
Lorica, 52, 190, 457
Loricifera (phylum), 20, 452*t*, 456–457, 456*f*
  defining characteristics, 456
  reproduction, 558*t*
  taxonomic detail, 458
  taxonomic summary, 457
Lottia (=Collisella), 378
Lottiidae (family), 378
Love darts, 265, 266*f*
Loxosoma, 493
Loxosomatidae (family), 495
Loxosomella, 495
Loxosomella harmeri, 491*f*
Loxothylacus, 417
Lubomirskiidae (family), 94
Lucifer, 413
Lucina, 287
Lucina floridana, 252
Lucina fosteri, 332
Lucinidae (family), 252, 287–288
Lucinoma, 287
Lucinoma aequizonata, 252, 253*f*
Lucinoma annulata, 252
Lugworms, 301*f*, 334
Luidia, 524
Lumbricidae (family), 336
Lumbricina (suborder), 336–337
Lumbriculida (order), 336
Lumbriculidae (family), 336
Lumbriculus, 336
Lumbricus, 336
Lumbricus terrestris, 318, 319*f*, 326*f*, 327*f*
Lumbrinereidae (family), 333
Lumbrinereis, 333
Lumbrinerides, 333
Lunatia, 280
Lycosa, 404
Lycosidae (family), 404
Lycoteuthidae (family), 291
Lycoteuthis, 291
Lyme disease, 254*f*, 357, 404
Lymnaea, 240*f*, 286
Lymnaeidae (family), 286
Lyratoherpia, 223*f*
Lyrodus pedicellatus, 570*f*
Lyses, 158
Lytechinus, 525

M

Maccabeus, 457
Macoma, 288
Macracanthorhynchus hirudinaceus, 201
Macrobdella, 337
Macrobiotus, 429
Macrobiotus hufelandi, 423*f*
Macrobrachium, 413
Macrocheira kaempteri, 414
Macrocilia, 146
Macrodasyida (order), 470
Macrodasys, 470
Macromeres, 13*f*, 14
Macronuclei, 49, 50, 52
Macroparesthia rhinoceros, 406
Macroperipatus torquatus, 426–427, 427*f*
Macrospironympha xylopletha, 71*f*
Macrostomes, 54–55, 54*f*
Macrostomida (order), 173
Macrostomum, 173
Macrothrix, 415
Macrotrachela multispinosus, 191*f*
Mactra, 288
Mactridae (family), 288
Madreporite, 498, 498*f*
Magadina cumingi, 493
Magelona, 334
Magelonida (order), 334
Magelonidae (family), 334
Maggots, 409
Maja, 414
Majidae (family), 414
Major Sperm Protein, 440–441, 441*f*
Malacobdella, 213
Malacostraca (subclass), 374–379, 375*f*, 377*f*
  body characteristics, 374, 375*f*, 376
  color changes, 376
  defining characteristics, 374
  development, 391, 571*t*
  digestive system, 394*f*
  taxonomic detail, 411–415
  taxonomic summary, 396
Malaria, 59, 60*f*, 61, 362–363, 409
Maldanidae (family), 334
Mallophaga (order), 407
Malpighian tubules, 364, 366*f*
Maltose release, and symbiotic relationships, 112–113, 113*t*
Manayunkia, 335
Mandibles, 358, 359, 360, 363, 364*f*, 373, 375*f*, 376, 487
Mandibulata (subphylum), 358–392
  Crustacea (class). *See* Crustacea (class)
  defining characteristics, 358
  Insecta (class). *See* Insecta (class)
  Myriapoda (class), 358–360, 393, 396, 405
  taxonomic detail, 405–418
  taxonomic summary, 396
Manduca sexta, 365*f*, 410
Mantids, 406
Mantis shrimp, 379, 411
Mantle, 216
Mantle cavity, 216
Mantle tissue, 478
Mantodea (order), 406
Mantophasmatodea (order), 406
Manubrium, 105–106, 112
Margarites, 278
Marine animals, 31
Marine environment, 4–5
Marine sandworm, 332–333
Marine snail
  cleavage, 13*f*
  polar lobe formation, 15*f*
  scientific names of, 17, 19
Marisa, 279
Marphysa, 333
Marsupium, 391
Martesia, 289
Marthasterias glacialis, 499*f*
Mastax, 186, 187*f*
Mastigamoeba, 64–65, 64*f*
Mastigamoeba, 64*f*, 76
Mastigamoebae, 64, 65, 76
Mastigamoebidae, 65, 73
Mastigella, 76
Mastigonemes, 43
Mastigophorans. *See* Flagellated protozoans
Mastigoproctus, 403
Mating types, 51–52
Maxilla, 353, 359, 363, 364*f*, 376
Maxillary glands, 396
Maxillipeds, 359, 376
Mayflies, 371*f*, 406
Mealy bugs, 407
Mechanoreceptors, 437
Mecopoda, 406
Mecoptera (order), 409
Medicinal leech, 324–325, 337
Mediomastus, 334
Medionidus, 287
Medusa, 101, 106*f*
  Cubozoa, 109
  Hydrozoa, 110–111, 111*f*, 113, 132
  Scyphozoa, 106–107, 106*f*, 107*f*, 131
Medusoid, 105
Megalopa, 391
Meganyctiphanes, 413

Meganyctiphanes norvegica, 379*f*
Megascolecidae (family), 336
Megascolides australis, 336
Meglitsch, Paul, 357
Mehlis's gland, 162*f*
Meiobenthos, 382
Meiopriapulus, 457
Meiopriapulus fijiensis, 455*f*
Meiosis, 52, 190, 194, 557
Melampus, 284
Melampus bidentatus, 240*f*
Melanella, 281
Melanoides, 279
Melarhaphe, 279
Melibe, 284
Melinna, 335
Mellita, 525
Melongena, 281
Melongenidae (family), 281
Membranelle, 47, 49*f*, 50*f*
Membranipora, 494
Membraniporidae (family), 494
Menippe, 414
Meoma, 526
Mercenaria mercenaria, 246*f*, 249*f*, 250*f*, 289
Mercenaria (=Venus), 289
Mercierella enigmaticus. See Ficopomatus (=Mercierella) enigmaticus
Mermis, 446
Mermithida (order), 446
Merostomata (class), 352, 353*f*
  body characteristics, 352
  digestive system, 392–393
  excretory system, 395
  reproduction, 559*t*
  taxonomic detail, 403
  taxonomic summary, 396
  use of, 352
Merozoites, 59
Mertensia, 147
Mertensidae (family), 147
Mesenchyme, 323
Mesenteries, 119, 120*f*, 121*f*
Mesocoel, 475
Mesoderm, 9, 14, 15*t*
Mesogastropods (=Taenioglossans), 278–281
Mesoglea, 101–102
Mesohyl, 81
Mesotardigrada (class), 429
Mesozoa (phylum), 179–182
  development, 179
  Orthonectida (class), 180
  Rhombozoa (class), 181–182
  taxonomic detail, 182
  taxonomic summary, 182
Metacercaria, 164
Metachronal beating, 47, 47*f*, 184
Metacoel, 175
Metacrinus, 524
Metallic beetles, 408
Metamerism, 295, 296*f*
Metamorphosis, 567, 567*f*, 580–581, 581*t*
  Arthropoda, 370–371
  delayed, 578–579, 579*f*, 580–581, 581*t*
  Nemertea, 211
Metanephridia, 271, 296, 297*f*, 300
Metazoans
  classification of, 9
  evolution of, 7–8
Metchnikovella, 77
Methane, 308–311
Metridium, 133
Metridium senile, 122*f*
Microciona, 94
Microciona prolifera, 94
Microconjugant, 53
Microcotyle, 176
Microcyema, 182
Microdalyellia, 174
Microfilaria, 443, 448
Microfilum, 77
Micrognathozoa, 20, 199, 202
Microhedyle, 282
Micromeres, 12, 13*f*
Micronuclei, 49, 50, 52
Micropilina, 277
Micropilina arntzi, 277
Micropora, 494
Microporidae (family), 494
Microsporidea (phylum), 59, 61, 73, 77
Microstomes, 54–55, 54*f*
Microstomum, 173
Microtubules, 42
Microvilli, 68
Micrura, 212
Mictic females, 190
Midges, 409, 572*t*
Midgut, 392
Millepora, 132
Milleporina (order), 132
Millericrinida (order), 523
Millipedes. *See* Diplopoda (order)
Milnesium, 429
Minibiotus, 429
Miracidium, 161, 163*f*, 164, 569*t*
Mites, 349, 353, 357, 362, 396, 404
Mitrocoma cellularia, 106*f*
Mitrocomella polydiademata, 111*f*
Mnemiopsis leidyi, 137, 138, 145*f*, 147, 148
Mnemiopsis macrydi, 144*f*
Mobile animals, 31
Modiolus, 287
Modules, 111, 480
Moina, 415
Moira, 526
Mole crabs, 414
Molecular relationships, and evolutionary relationships, 20
Molecular studies
  cladistics and, 29–30, 30*f*
  complications, 30–31
  controversy, 31
Molgula, 541, 548, 553
Mollusca (phylum), 215–293
  Aplacophora (class), 222, 265, 271, 277
  Bivalvia (class). *See* Bivalvia (=Pelecypoda) (class)
  Cephalopoda (class), 256–265, 269, 271, 290–292, 559*t*, 564*f*, 565
  circulation, 216, 217*f*, 218, 266
  comparison with Arthropoda, 342, 345
  defining characteristics, 215
  digestive system, 269, 271
  evolution, 218
  excretory system, 271
  feeding mechanism, 218
  fossil record, 218, 220*f*
  gas exchange, 216, 217*f*, 218, 266
  Gastropoda (class). *See* Gastropoda (class)
  general characteristics, 215–218
  Monoplacophora (class), 222–224, 271, 277, 559*t*
  nervous system, 266, 268–269, 270*f*, 271*f*
  Polyplacophora (class), 218–222, 271, 276–277, 559*t*, 570*t*
  reproduction and development, 265–266, 265*f*, 266*f*, 267*f*, 558*t*–559*t*, 570*t*
  Scaphopoda (class), 255, 265, 271, 290, 559*t*, 570*t*
  shell, 215–216, 216*f*
  taxonomic detail, 276–292
  taxonomic summary, 271
Molpadia, 526
Molpadiida (order), 526
Molting, 183, 343, 344*f*, 370, 432, 433
Monarch butterflies, 410
Monkey hoppers, 406
Monobryozoon, 494
Monobryozoontidae (family), 494
Monocelididae (family), 174
Monocelis, 174
Monocercomonoides, 76
Monocystis, 75
Monodonta, 278
Monoecious species, 108
Monogenea (class), 160–161, 170, 176
Monogononta (class), 190, 190*f*, 191*f*, 194
  life cycle, 195*f*
  reproduction and development, 190, 194, 558*t*
  taxonomic detail, 200–201
  taxonomic summary, 199
Monomorphic cells, 49
Monomorphic colonies, 482
Monopectinate, 235
Monophyletic groups, 16, 46, 62
Monophyletic taxon, 27*t*
Monopisthocotylea (suborder), 176
Monoplacophora (class), 222, 224
  defining characteristics, 222
  fossil record, 222, 225*f*
  nervous system, 225*f*
  reproduction, 559*t*
  shell, 224, 224*f*
  taxonomic detail, 277
  taxonomic summary, 271
Monosiga, 77
Monostilifera (suborder), 213
Monstrilla, 416
Monstrilloida (order), 416
Montacuta, 288
Montastraea, 134
Montfortula rugosa, 219*f*
Moon jelly, 131
Moon snails, 219*f*, 280
Mopalia, 276
Mopaliidae (family), 276
Morula cells, 541
Mosaic (determinate) cleavage, 14
Mosquitoes, 361*f*, 409
  and filarial parasites, 448
  and malaria, 59, 60*f*, 61, 362–363
Moss animals. *See* Bryozoa (=Ectoprocta; =Polyzoa) (phylum)
Mother sporocyst, 161
Moths, 371*f*, 410
Mouth, embryonic origin, 11, 15*t*
Mucociliary feeders, 531
Mud crabs, 414
Mud shrimp, 414
Mud stars, 524
Mudsnails, 281
Muggiaea, 116*f*
Mulinia, 288
Müller's larva, 154, 157*f*, 568*t*
Multicellular, 7, 9
Multiple fission, 44, 62
Murex, 281
Muricea pendula, 126
Muricidae (family), 281
Muricoidea (superfamily), 281
Musca automnalis, 409
Musca domestica, 409
Muscidae (family), 409
Musculature
  Annelida, 345*t*
  Anthozoa, 119, 122*f*, 345*t*
  Arthropoda, 344, 345*f*, 345*t*
  Ascidiacea, 541
  Bivalvia, 345*t*
  Brachiopoda, 479, 479*f*
  Cephalochordata, 551
  Cephalopoda, 256
  circular, 98–99, 98*f*, 119
  Cnidaria, 104, 105*f*
  Ctenophora, 138
  Gastropoda, 224, 229, 231*f*, 345*t*
  Gymnolaemata, 486
  Hirudinea, 324, 324*f*
  and hydrostatic skeleton, 97, 98*f*
  Insecta, 367
  lamellibranchs, 251
  longitudinal, 98, 98*f*, 119
  Nematoda, 434–435, 434*f*, 435*f*
  Nemertea, 203, 205, 208, 209
  Oligochaeta, 318
  Onychophora, 424
  Polychaeta, 297–300
  Rotifera, 185, 185*f*, 188
  Scyphozoa, 105, 345*t*
  Sipuncula, 316
  Tardigrada, 422
  Turbellaria, 152, 152*f*
Mushroom corals, 134
Mussels, 287
Mutable connective tissue, 500
Mutualism, 32
Mya, 289
Mya arenaria, 244*f*
Mycale, 94
Mycalidae (family), 94
Mycetozoa (phylum), 63–64, 73, 76
Myidae (family), 289
Myoida (order), 289–290
Myomeres, 551
Myosin, 43, 44
Myriapoda (class), 358–360
  Chilopoda (order), 350, 358–359, 393, 395, 396, 405, 559*t*
  digestive system, 393
  Diplopoda (order), 350, 359–360, 393, 395, 396, 405, 559*t*
  taxonomic detail, 405
  taxonomic summary, 396
Myrmica, 411
Mysella verrilli, 288
Mysia, 289
Mysid shrimp, 412
Mysidacea (order), 412
Mysis, 412
Mytilicola, 416
Mytilidae (family), 287
Mytilus, 287
Mytilus edulis, 15*f*, 244*f*, 246*f*, 249*f*, 287
Myxicola, 335
Myxidium, 77
Myxobolus, 118*f*, 133
Myxosomatidae (family), 133
Myxospora. *See* Myxozoa (=Myxospora) (phylum)
Myxozoa (=Myxospora) (phylum), 59, 61, 77, 115, 117, 118*f*, 133
Myzobdella, 337
Myzostomida (order), 335–336
Myzostomum, 335

N

Nacella, 278
Nacellidae (family), 278
Nacreous layer, 215, 216*f*
Naegleria gruberi, 72, 72*f*
Naiads, 370
Naididae (family), 336
Nais, 336
Naked ramicristate amoebae, 62
Nanaloricus, 458
Nanaloricus mysticus, 456, 456*f*
Nanophyetus, 177
Nanophyetus salmincola, 177
Narcomedusae (suborder), 132
Nasitrema, 177
Nasitrematidae (family), 177
Nassariidae (family), 281
Nassarius, 281, 580*f*
Nassarius reticulatus, 15*f*, 267*f*
Nassellaria, 76
Natica, 280
Naticidae (family), 280
Naticoidea (superfamily), 280
Natural pearls, 216
Nauplius larva, 389, 392*f*, 571*t*
Nautilidae (family), 290
Nautiloidea (subclass), 290
Nautilus, 256–257, 257*f*, 258–259, 262*f*, 291
Necator americanus, 439*f*, 441–442, 442*f*, 447
Neck, 158, 158*f*
Nectalia, 116*f*
Nectonema, 457
Nectonematoida (class), 457
Nectonemertes, 213
Nectophores, 113, 116*f*
Needham's sac, 563
Nemaster, 524
Nemata (phylum). *See* Nematoda (=Nemata) (phylum)
Nematoblasts, 102
Nematocysts, 102, 103*f*, 236
Nematoda (=Nemata) (phylum), 431–449, 452*t*
  behavior, 437
  beneficial, 444–445
  body coverings and body cavities, 432–434
  classification, 431–432, 432*t*
  comparison with Chaetognatha, 467
  defining characteristics, 431
  digestive system, 436–437
  excretory system, 437
  general characteristics, 431–432
  internal pressure, 434–435, 434*f*
  life cycle, 441–442, 442*f*, 443*f*
  locomotion, 434–435, 434*f*
  musculature, 434–435, 434*f*, 435*f*
  nervous system, 435, 435*f*
  parasitic, 432, 438–443, 439*f*, 442*f*
  reproduction and development, 438, 438*f*, 440–441, 557, 559*t*, 560, 567
  respiratory system, 437
  taxonomic detail, 445–448
Nematogen, 181–182
Nematomorpha (phylum), 452–454, 452*t*, 453*f*
  defining characteristics, 452
  digestive system, 454
  life cycle, 454
  reproduction, 558*t*
  taxonomic detail, 457
  taxonomic summary, 457
Nematostella, 133
Nematostella vectensis, 133
Nemertea (=Rhynchocoela) (phylum), 203–213
  cilia, 203
  circulation, 204–205, 205*f*
  classification, 210
  compared to flatworms, 203, 204–205
  cross section, 204*f*
  defining characteristics, 203
  digestion, 204, 205, 208–209
  evolution, 206–207
  excretion, 204
  feeding mechanisms, 209
  gas exchange, 204
  general characteristics, 203–210
  locomotion, 203–204

Nemertea (=Rhynchocoela) (phylum)—*Cont.*
  longitudinal section, 205*f*
  musculature, 23, 205, 208, 209
  nervous system, 204
  phylogenetic relationships, 206–207
  protection from predators, 210
  reproduction and development, 210–211, 556, 558*t*, 569*t*
  sense organs, 204
  taxonomic detail, 212–213
  taxonomic summary, 214
Nemertodermatida (order), 173
Nemertopsis, 208*f*
Neoblasts, 155
Neocrinus, 524
Neodermata, 151, 151*f*, 172
Neodiplostomum paraspathula, 162*f*
Neoechinorhynchus emydis, 197*f*
Neoechinorhynchus (=*Neorhynchus*), 193
Neogastropoda (infraorder), 281–282
Neomenia, 277
Neomenia carinata, 223*f*
Neomeniidae (family), 277
Neomeniomorpha (=Solenogastres) (subclass), 277
Neomphalidae (family), 278
Neomphalus, 278
Neopilina, 277
Neopilina galatheae, 225*f*, 277
Neopilinidae (family), 277
Neorhabdocoela (order), 173, 174–175
Neorhynchus. See Neoechinorhynchus
Neoteny, 543, 544*f*
Nepanthia belcheri, 557*f*
Nephridia, 296
Nephridiopore, 271
Nephrocytes, 395
Nephropidae (family), 413
Nephrops norvegicus, 468*f*, 470
Nephrostome, 296
Nephtyidacea (superfamily), 333
Nephtyidae (family), 333
Nephtys, 333
Nephtys scolopendroides, 175
Neptunea, 281
Neptune's goblet sponge, 89*f*
Nereididacea (superfamily), 332–333
Nereididae (family), 332–333
Nereis, 332, 434
Nereis diversicolor, 425*f*
Nereis virens, 299*f*, 300*f*, 326*f*, 332
Nerita, 278
Nerita undata, 219*f*
Neritidae (family), 278
Neritina, 278
Neritoidea (superfamily), 278
Neritopsina (superorder), 278
Nervous system
  Annelida, 327–328
  Aplacophora, 222
  Arthropoda, 344, 345*f*
  Ascidiacea, 548
  Cephalochordata, 548, 550
  Cephalopoda, 259, 262*f*, 268–269, 271*f*
  Chaetognatha, 461, 462*f*
  Cnidaria, 102, 104*f*
  Crinoidea, 520, 520*f*
  Ctenophora, 138
  Echinodermata, 520, 520*f*
  Echiura, 315
  Enteropneusta, 531
  Larvacea, 547–548
  lophophorates, 490
  Mollusca, 266, 268–269, 270*f*, 271*f*
  Monoplacophora, 225*f*
  Nematoda, 435, 435*f*
  Nemertea, 204
  Oligochaeta, 327*f*
  Onychophora, 425
  Pogonophora, 306
  Polyplacophora, 222
  Rotifera, 195–196, 195*f*
  Scyphozoa, 107, 107*f*
  Tardigrada, 422, 423*f*
  Thaliacea, 547–548
  Turbellaria, 151, 152*f*
  Urochordata, 547–548
Nests, 368
Neural cartridge, 347
Neuroptera (order), 408
Neurotoxins, 55, 210, 464, 464*t*
Nidamental glands, 258*f*
Nipponnemertes, 213
Noctiluca, 58, 75
Noctiluca scintillans, 58*f*
Noctuidae (family), 410
Node, 27*t*
Nolella, 494
Nolellidae (family), 494
North American hookworm, 447
Northern quahog, 289
Norwegian lobster, 468*f*, 470
Norwegian reindeer, 384
No-see-ums, 409
Nosema, 77
Notaspidea (order), 283
Notoacmea, 278
Notochord, 539, 540, 542, 543*f*, 534*f*, 544, 548, 549*f*, 551
Notomastus, 334
Notopala, 279
Notophyllum, 332
Notostraca (order), 415–416
Novodinia, 525
Nucella, 281
Nucella lapillus, 566*f*
Nucella (=*Thais*) *lima,* 566*f*
Nuchal organs, 315, 327
Nuclei, 49–50, 52
Nucula, 243*f*, 286
Nuculanidae, 286
Nuculidae (family), 286
Nuda (class), 145*f*, 146, 148
Nudibranchia (order), 236, 236*f*, 283–284
Numerical taxonomy, 26
Nurse eggs, 561, 561*f*
Nut shells, 286
Nutrient transfer, in symbiotic relationships, 112–113
Nutrition, in aquatic environments, 2–3
Nutritive-muscular cells, 104
Nymph, 572*t*
Nymphon, 405
Nymphs, 370

O

Obelia, 132
Obelia commissuralis, 115*f*
Obturaculum, 308
Obturata (=Vestimentifera) (subfamily), 328, 336
Ocelli
  Arthropoda, 345
  Chilopoda, 359
  Copepoda, 382
  Insecta, 363, 364*f*
  Nematoda, 437
  Scyphozoa, 107–108, 108*f*
Ocenebra, 281
Octocorallia (=Alcyonaria) (subclass), 124–125, 125*f*, 128, 133
Octopoda (order), 291–292
Octopodidae (family), 291
Octopus, 256 257, 262*f*, 291
Octopus lentus, 265*f*, 565*f*
Octopus vulgaris, 259, 264, 264*f*
Oculina, 134
Ocypodidae (family), 415
Ocyropsidae (family), 147
Ocyropsis, 147
Ocyropsis maculata, 144*f*
Ocythoe, 292
Ocythoe tuberculata, 292
Ocythoidae (family), 292
Odonata (order), 406
Odontaster, 525
Odontophore, 218, 219*f*
Odontophore complex, 218
Odontosyllis, 332
Odostomia, 282
Oekiocolax, 174
Oestridae (family), 410
Oikopleura, 544, 545*f*, 553
Oligobrachia, 336
Oligochaeta (subclass), 318–322
  circulatory system, 328
  digestive system, 325, 326*f*
  locomotion, 318, 320–321, 322*f*
  musculature, 318
  nervous system of, 327*f*
  pollution and, 318, 320–321, 321*f*
  reproduction and development, 322, 323*f*, 558*t*, 563
  taxonomic detail, 336
  taxonomic summary, 328
Olive shells, 281
Olivella, 281
Olivia, 281
Olividae (family), 281
Ommastrephidae (family), 291
Ommatidia, 346–347, 346*f*, 348*f*, 349*f*, 351
Onchidium, 286
Onchocerca, 441, 443, 444*f*, 448
Onchocerca volvulus, 440, 448
Onchocerciasis. *See* River blindness
Oncomiracidium, 160, 160*f*, 161
Oncorhynchus gorbuscha, 230*f*
Oncosphere, 158
O'Neal, W., 126–127
Ontogeny, 539
Onuphidae (family), 333
Onuphis, 333
Onychophora (phylum), 424–428
  body characteristics, 424–425
  cuticle, 424
  defining characteristics, 424
  evolution, 428
  feeding mechanisms, 424, 426–427, 427*f*
  locomotion, 425–428, 425*f*
  musculature, 424
  nervous system, 425
  reproduction, 559*t*, 563
  taxonomic detail, 429
  taxonomic summary, 428
Onychopoda (suborder), 415
Oocyst, 59, 60*f*
Oogenesis, 370
Ookinetes, 59, 60*f*
Oozooid, 547, 547*f*
Opalina, 76
Opalinids, 72, 73, 77
Open ocean animals, 31
Operculum, 227*f*, 232, 232*f*, 486
Ophelia, 334
Opheliida (order), 334
Opheliidae (family), 334
Ophiactis, 524
Ophiocantha, 524
Ophiocoma, 524
Ophiocoma wendtii, 504*f*
Ophioderma, 524
Ophiomastix, 524
Ophiomusium, 504*f*, 524
Ophiopholis, 524
Ophiopluteus larva, 519*f*, 574*t*, 575*f*
Ophiostigma, 524
Ophiothrix, 504*f*, 524
Ophiothrix fragilis, 504*f*, 575*f*
Ophiura, 524
Ophiurida (order), 524
Ophiuroidea (subclass), 503–505, 504*f*
  body characteristics, 503
  defining characteristics, 503
  digestive system, 503
  feeding mechanisms, 505
  reproduction and development, 518, 519*f*, 556, 559*t*, 566, 574*t*, 575*f*
  taxonomic detail, 524
  taxonomic summary, 521
Opiliones (order), 404
Opishaptor, 160
Opisthobranchia, 235–237
  compared to prosobranchs, 235–236
  defining characteristics, 235
  locomotion, 237
  nervous system, 270*f*
  reproduction and development, 265, 560*f*, 575*f*
  respiratory system, 236, 237*f*
  shell, 235–236, 237*f*
  taxonomic detail, 282–284
  taxonomic summary, 271
Opisthokonts, 73
Opisthorchiida (order), 177
Opisthorchiidae (family), 177
Opisthorchis, 177
Opisthorchis sinensis, 162*f*, 166*f*
Opisthorchis tenuicollis, 279
Opisthosoma, 306, 307*f*, 352, 353
Opisthoteuthidae (family), 292
Opisthoteuthis, 292
Opossum shrimp, 412
Oppia, 404
Oral lobes, 144
Oral siphon, 540, 540*f*
Oral surface, 498
Orb weavers, 404
Orchestia, 412
Orcula, 285
Oreaster, 525
Organelles, 40–42
Orifice, 481
Original condition, 19
Orthogastropoda (subclass), 271, 278
Orthognatha (suborder), 404
Orthonectida (class), 180, 182
Orthoptera (order), 406
Oscula, 84
Osmotic equilibrium, 5
Osmotic gradient, 41
Osmotic regulation, 41
Osphradium, 216, 217*f*, 232, 233*f*
Ostia, 84, 245, 344, 346*f*
Ostracoda (subclass), 381, 381*f*
  defining characteristics, 381
  development, 389
  taxonomic detail, 416
  taxonomic summary, 396
Ostrea, 287
Ostreidae (family), 287
Ototyphlonemertes, 213
Ototyphlonemertidae (family), 213
Outer demibranch, 245
Outgroup, 28
Ovalipes, 414
Ovarian balls, 196
Ovatella, 284
Ovicells, 489
Ovigers, 357
Oviposition, 370
Ovipositor, 370
Ovotestis, 265
Ovula, 280
Ovulidae (family), 280
Owenia, 335
Oweniida (order), 335
Oweniidae (family), 335
Oxymonads, 76
Oxymonas, 76
Oxyuridae (family), 448
Oyster drill, 281
Oyster leech, 173
Oysters, 271, 287
  development, 576*f*
  parasites in, 58
Ozobranchidae (family), 337
Ozobranchus, 337

P

Pachychilus, 279
Paguridae (family), 414
Pagurus, 414
Palaeacanthocephala (class), 202
Palaemonetes, 413
Palaemonidae (family), 413
Palaeonemertea (=Palaeonemertini) (order), 212
Palaeonemertini (order). *See* Palaeonemertea (=Palaeonemertini) (order)
Paleoheterodonta (subclass), 287
Paleotaxodonta (subclass). *See* Protobranchia (=Paleotaxodonta =Cryptodonta) (subclass)
Palinura (infraorder), 413
Palinuridae (family), 413
Palinurus, 413
Palola, 333
Palp proboscides, 241
Paludicella, 494
Paludicellidae (family), 494
Panagrolaimidae (family), 447
Panarthropoda, 17*f*
Pandalidae (family), 413
Pandalus, 413
Pandora, 290
Pandoridae (family), 290
Panopea, 289
Panopeus, 414
Pantopoda (class). *See* Pycnogonida (=Pantopoda) (class)
Panulirus, 413
Paper nautilus, 257, 265, 291
Paper wasps, 373
Papillae, 306, 313
Papillifera, 285
Papulae, 508
Parabasala (phylum), 70–71, 73, 75
Paracentrotus, 525
Paracerceis, 412
Parachela (order), 429
Paragonimus, 177
Paralithodes, 414
Paramecium, 42*f*, 53, 75
Paramecium aurelia, 51, 192
Paramecium caudatum, 51*f*, 57*f*
Paramecium multimicronucleatum, 56*f*
Paramecium sonneborni, 47*f*
Paramyxa, 75
*Paranemaa, 75
Paranemertes peregrina, 210*f*
Paraphyletic groups, 16, 27*t*
Parapodia, 237, 238*f*, 282, 283, 297, 298*f*, 299*f*, 300
Parasagitta, 470
Parascaris, 447

Parasites, 32, 32*f*
  Acanthocephala, 196, 198
  Acari, 357
  Apicomplexa, 58–61, 75
  Aspidogastrea, 151
  Bivalvia, 262
  Cestoda, 157–159
  Cirripedia, 389, 390*f*
  Copepoda, 382, 385*f*
  Digenea, 161–170
  Gastrotricha, 460
  Hirudinea, 324
  Kinetoplastea, 75
  Mesozoa, 179
  Microsporidea, 59
  Monogenea, 160
  Myxozoa (=Myxospora), 59, 77, 115
  Nematoda, 432, 438–443, 439*f*, 442*f*
  Nematomorpha, 454
  Orthonectida, 180
  Pentastomida, 382–383
  Pycnogonida, 357
  Rhombozoa, 181–182
  Rotifera, 187*f*, 188
  sarcodinids, 62
  Seisonidea, 188
  successful, 161
  Trematoda, 161, 164
  Turbellaria, 151
  zooflagellated protozoans, 68–72
Parasitic life cycle, 58, 59*f*
Parasitoids, 360
Paraspadella, 470
Parastacidae (family), 413
Paraster, 526
Parastichopus, 526
Paravortex, 174
Parazoa (subkingdom), 93–94
Parenchyma, 150, 204, 204*f*
Parenchymella, 90
Parenchymula, 90
Parlatoria oleae, 562*f*
Parsimony, 27*t*, 28
Parthenogenesis, 188, 190, 191*f*, 194, 322, 349, 557
Patella, 228*f*, 267*f*, 278
Patellidae (family), 278
Patellogastropoda (order), 234*f*, 271, 277–278
Patelloidea, 278
Patiriella, 525, 577*f*
Pauropoda (order), 350, 405
Pawlik, J. R., 126–127
Paxillosida (order), 524
PCR. *See* Polymerase chain reaction
Pea crab, 314*f*, 414
Peanut worms. *See* Sipuncula (phylum)
Pearl, 216
Pecten, 244*f*, 287
Pectinaria belgica, 302*f*
Pectinaria (=Cistenides), 302*f*, 335
Pectinariidae (family). *See* Amphictenidae (=Pectinariidae) (family)
Pectinatella, 493
Pectinidae (family), 287
Pedal laceration, 119
Pedal retractor muscles, 251
Pedal waves, 151–152, 152*f*, 218, 229, 231*f*
Pedalia mira, 191*f*
Pedicel, 353
Pedicellariae, 508, 508*f*
Pedicle, 477
Pediculus, 407
Pedipalps, 352, 353
Pelagica (tribe), 213
Pelagonemertes, 213
Pelagosphaera, 318
Pelagosphera, 318, 319*f*, 569*t*
Pelagothuria, 526
Pellicle, 47, 50*f*
Pen, 257
Pen shells, 287
Penaeidae (family), 413
Penaeidea (infraorder), 413
Penaeus, 413
Pencil urchins, 512*f*, 525
Penetrantiidae (family), 494
Pennaria, 132
Pennatula, 125*f*
Pennatulacea (order), 133
Pentagonia, 76
Pentapora, 494
Pentastomida (subclass), 382–385
  defining characteristics, 382
  life cycle, 384–385
  morphology, 382–383
  reproduction, 558*t*
  taxonomic detail, 417
  taxonomic summary, 396
Peracarida (superorder), 412
Pereopods, 376
Perforate, 298
Pericardium, 271
Perihemal canal, 500
Periostracum, 215, 216*f*
Peripatidae (family), 429
Peripatoides, 429
Peripatopsis, 429
Peripatopsis sedgwicki, 425*f*
Peripatus, 424, 424*f*, 429
Peripharyngeal band, 540, 540*f*
Periplaneta, 406
Periproct, 513
Perisarc, 111
Peristaltic waves, 318, 320–321, 322*f*
Peritoneum, 296, 432
Peritopsidae (family), 429
Peritrophic membrane, 393
Perivisceral coelom, 499
Periwinkles, 226*f*, 279
Perkinsus, 58, 75
Perviata (=Frenulata) (subfamily), 328, 336
Pesticides, 61
Petaloconchus, 280
Petricola, 289
Petricolidae (family), 289
Petrolisthes, 414
Pfiesteria piscicida, 55
PG. *See* Prothoracic glands
pH, of marine *vs.* freshwater environment, 5
Phaeodarea (class), 77
Phaeodina, 77
Phagocytosis, 58, 62, 72
Phagosome, 46
Phallusia, 553
Phantom midges, 409
Pharidae (=Cultellidae) (family), 288
Pharyngeal basket, 540
Pharyngeal gill slits, 541
Pharynx, protrusible, 154, 155*f*
Phascolion, 317*f*, 338
Phascolosoma, 338
Phascolosoma gouldi, 316*f*
Phascolosomatidae (family), 338
Phasmatoptera (order). *See* Phasmida (=Phasmatoptera) (order)
Phasmida (=Phasmatoptera) (order), 23, 25, 25*f*, 406
Phasmids, 437
Phenetics, 26, 32
Pheretima communissima, 323*f*
Pheromones, 437
Phestilla, 284
Phialidium gregarium, 230*f*
Phidiana, 284
Philippia, 282
Philodina, 201
Philodina roseola, 188*f*, 189*f*, 191*f*, 194*f*
Philodinidae (family), 201
Phlebobranchia (order), 553
Pholadidae (family), 289
Pholeus phalangioides, 562*f*
Pholoe, 333
Pholoidae (family), 333
Phoresy, 404
Phoronida (phylum), 475–476, 476*f*
  digestive system, 490
  reproduction and development, 488–489, 559*t*, 563, 572*t*, 576*f*
  taxonomic detail, 492
  taxonomic summary, 491
Phoronis, 492
Phoronis architects, 476*f*
Phoronis australis, 476*f*
Photinus, 408
Photoperiod, 371
Photophores, 258, 261*f*, 378, 379*f*
Photoreceptors. *See* Ocelli
Photosynthesis, 309, 311, 381
  dinoflagellates, 58
  Euglena, 68, 75
  and symbiotic relationships, 112–113
  zooxanthellae, 124
Photurus, 408
Phragmatopoma, 335
Phragmophora (order), 470
Phronima, 412
Phrynophiurida (order), 524
Phylactolaemata (class), 482–484, 483*f*, 486*f*, 491, 493
Phyllium, 406
Phyllochaetopterus, 334
Phyllodoce, 332
Phyllodocida (order), 332
Phyllodocidacea (superfamily), 332
Phyllodocidae (family), 332
Phyllopoda, 415
Phyllozooids, 113
PhyloCode, 19
Phylogenetic classification, 32
Phylogenetic systematics, 27–29
Phylogenetic trees, 20, 21–24, 25*f*, 207, 207*f*
Phylum, 16
Physa, 286
Physalia, 117*f*, 132
Physidae (family), 286
Phytoflagellated protozoans, 68–72
Phytoplankton, 4, 381, 567, 577
Piddocks, 289
Pieridae (family), 410
Pigment granules, 252
Pilidae (family). *See* Ampullariidae (=Pilidae) (family)
Pilidium larva, 210, 211*f*, 569*t*
Pillbugs, 378*f*, 379, 396, 412
Pinacocytes, 83, 85*f*
Pinacoderm, 83
Pinna, 287
Pinnate, 124
Pinnidae (family), 287
Pinnixa, 414
Pinnixia franciscana, 314*f*
Pinnixia schmitti, 314*f*
Pinnotheres, 414
Pinnotheridae (family), 414
Pinnules, 124, 303*f*, 501
Pinocytosis, 62
Pinworm, 439–440, 442, 443*f*, 448
Pisaster, 525
Piscicola, 337
Piscicolidae (family), 337
Pisidiidae (family). *See* Sphaeriidae (=Pisidiidae) (family)
Pisidium, 289
Pisionella, 333
Pisionidae (family), 333
Placiphorella, 276
Placobdella, 337
Placopecten, 287
Placothuria, 526
Placothuriidae (family), 526
Placozoa (phylum), 91, 91*f*
Plagiorchiida (order), 177
Plagiorchis, 177
Plagiorchis elegans, 168*f*
Plagiostomum, 173–174
Plakina, 94
Plakinidae (family), 94
Plakortis, 94
Planaria, 174
Planarians. *See* Tricladida (order)
Planariidae (family), 174
Planctosphaera pelagica, 535
Planctosphaeroidea (class), 535
Planktonemertes, 213
Planktonic animals, 31, 137
Planktotrophic larvae, 567
Planoceridae (family), 173
Planorbidae (family), 286
Planorbis, 286
Plant lice, 407
Planula, 108, 109*f*, 111, 111*f*, 568*t*
Plasmalemma, 39
Plasmodium, 63, 180
Plasmodium, 59, 60*f*, 61, 75, 362
Plasmodium berghei, 362, 363*f*
Plasmodium falciparum, 59
Plasmotomy, 44
Platyasterida (order), 524
Platyctenida (order), 145*f*, 146, 147
Platyhelminthes (phylum), 149–178
  Cestoda (class), 32*f*, 157–159, 170, 175–176, 558*t*, 560
  classification, 14
  compared to gnathostomulids, 198, 199
  compared to nemertines, 203, 204–205
  digestion, 150
  evolutionary relationships, 149–150, 151, 151*f*
  excretory systems, 150–151, 150*f*
  gas exchange, 150
  general characteristics, 149–151
  Monogenea (class), 70, 160–161, 160*f*, *176*
  reproduction and development, 150, 151, 558*t*, 568*t*–569*t*
  taxonomic detail, 172–177
  taxonomic summary, 170
  Trematoda (class), 161–170, 176–177, 558*t*, 569*t*
  Turbellaria (class), 151–155, 156*f*, 170, 172–175, 556, 568*t*
Platynereis, 332
Platynereis bicanaliculata, 210*f*
Platyzoa, 17*f*, 21*f*
Plecoptera (order), 407
Pleopods, 376
Plesiomorphic condition, 19, 27*t*
Pleurobrachia, 139*f*, 140*f*, 145*f*, 147
Pleurobrachia bachei, 142, 230*f*
Pleurobrachiidae (family), 147
Pleurobranchaea, 283
Pleurobranchidae (family), 283
Pleurobranchus, 283
Pleurocera, 279
Pleuroceridae (family), 279
Pleuroplaca, 281
Pleurotomarioidea (superfamily), 278
Plexaura, 133
Plexaurella, 133
Ploeotia, 75
Ploima (order), 201
Plumatella, 493
Plumatella repens, 486*f*
Plumatellidae (family), 493
Pneumatophores, 113, 116*f*, 117*f*
Pneumocystis carinii, 61
Pneumonia, 61
Pneumostome, 238, 239*f*
Pochella, 132
Pocillopora, 134
Pocillopora damicornis, 124*f*
Podarke, 332
Podia (tube feet), 498–500, 498*f*, 499*f*
Podocoryne, 132
Podocoryne carnea, 115*f*
Podocytes, 271
Podon, 415
Podon intermedius, 380*f*
Poecilostomatoida (order), 416
Pogonophora (family). *See* Siboglinidae (=Pogonophora) (family)
Polar capsules, 115, 118*f*
Polar filament, 115, 117
Polar lobes, 14, 15*f*
Polarity, 26, 27*t*, 28
Polarized light, 347
Polian vesicles, 498*f*, 499
Polinices, 280
Pollicipes, 390*f*, 417
Pollutants, 4
Pollution
  and Holothuroidea, 517, 517*f*
  and Oligochaeta, 318, 320–321, 321*f*
Polyandrocarpa, 553
Polyarthra, 201
Polybrachia, 306*f*, 336
Polycelis, 174
Polycentropus, 562*f*
Polyceratidae (family). *See* Polyceridae (=Polyceratidae) (family)
Polyceridae (=Polyceratidae) (family), 283
Polychaeta (class), 296, 297–304
  body characteristics, 297–298
  burrowing, 298, 299*f*, 300
  circulatory system, 328
  comparison with Sipuncula, 317*t*
  defining characteristics, 297
  digestive system, 325, 326*f*
  errant, 299*f*, 300, 300*f*
  excretory system, 300
  locomotion, 297–300, 299*f*, 300*f*
  morphology, 297
  musculature, 297–300
  nervous system, 327–328
  reproduction and development, 300–304, 304*f*, 305*f*, 556, 558*t*, 560, 563, 569*t*, 575*f*, 580
  sedentary, 300, 301*f*–302*f*
  sense organs, 327–328
  Siboglinidae (=Pogonophora) (family), 304–312, 328, 336, 543*t*, 548
  taxonomic detail, 332–336
  taxonomic summary, 328
Polycirrus, 335
Polycladida (order), 173
Polycycstinea (class), 76
Polydora, 334
Polydora ciliata, 305*f*
Polygonoporus, 175
Polygonoporus giganticus, 175
Polygordius, 336
Polymastimastix, 76
Polymerase chain reaction (PCR), and evolutionary relationships, 20
Polymorphic colonies, 112, 486
Polynoidae (family), 333
Polyodontes, 333
Polyodontidae (family), 333
Polyopisthocotylea (suborder), 176
Polyp, 101
  Anthozoa, 118
  Cubozoa, 109

Polyp—*Cont.*
Hydrozoa, 111, 115*f*, 112
Octocorallia, 124
Scyphozoa, 108, 109*f*
Polyphemus, 415
Polyphyletic grouping, 27*t*
Polypide, 480
Polyplacophora (class), 218–222
defining characteristics, 218
digestive system, 222
evolution, 222
feeding mechanism, 222
fossil record, 222
gas exchange, 218
locomotion, 218–219
nervous system, 222
reproduction and development, 265, 559*t*, 570*t*
sense organs, 222
shell, 218, 220*f*
taxonomic detail, 276–277
taxonomic summary, 271
Polystilifera (suborder), 213
Polystoma, 176
Polystomoidella oblongum, 160*f*
Polyzoa (phylum). *See* Bryozoa (=Ectoprocta; =Polyzoa) (phylum)
Pomacea, 279
Pomatias, 279
Pomatiasidae (family), 279
Pond snails, 286
Pontobdella, 337
Pontocypris, 416
Pontoscolex, 336
Porcelain crabs, 414
Porcellana, 414
Porcellanidae (family), 414
Pore plates, 485
Porifera (phylum), 20, 79–91
alloincompatibility, 80
body characteristics, 81, 82*f*, 86, 87*f*
Calcarea (class), 87, 92, 94
colonial, 80
defining characteristics, 79
Demospongiae (class), 87–88, 92, 93–94
digestion, 81
dispersal, 580
diversity, 86–88
evolutionary relationship, 80
excretion, 81
gemmule formation, 82–83, 84*f*
general characteristics, 81–86
Hexactinellida (class), 88, 89*f*, 92, 94
histoincompatibility, 80–81, 81*t*
reproduction and development, 89–91, 90*f*, 556, 558*t*, 568*t*
Sclerospongiae (class), 86, 88*f*, 94
taxonomic detail, 93–94
taxonomic summary, 92
tissue regression, 83
water flow through, 83, 84–85, 86*t*, 86*f*, 87*f*
Porites, 134
Pork tapeworm, 159*f*, 176
Porocephalida (order), 417
Porocephalus crotali, 385*f*
Porocyte, 88*f*
Poromya, 290
Poromyidae (family), 290
Porpita, 133
Portuguese man-of-war, 117*f*, 132
Portunidae (family), 414
Portunis sayi, 392*f*
Portunus, 414
Potamididae (family), 279
Poterion neptuni, 89*f*
Pourtalesia, 526
Power stroke, 43
Preadaptations, 6
Preying mantids, 406
Priapulida (phylum), 406, 452*t*, 454, 455*f*
body characteristics, 454
defining characteristics, 454
evolution, 454
feeding mechanisms, 454
reproduction, 558*t*
taxonomic detail, 457
taxonomic summary, 457
Priapulidae (family), 457
Priapulids, 198
Priapulus, 457
Priapulus caudatus, 455*f*
Primary mesenteries, 119, 120*f*
Primary nematogen, 181
Primary production, in water, 4
Primitive (ancestral) state, 27*t*
Primitive condition, 11, 19
Prismatic layer, 215, 216*f*
Pristina, 336
Pristionchus pacificus, 446
Proales, 201
Proales gonothyraea, 186*f*
Proalidae (family), 201
Proboscides, 41
Proboscis
Echiura, 312
Enteropneusta, 530, 530*f*
Nemertea, 205*f*, 208–209, 208*f*, 209*f*, 210*f*
Polychaeta, 298, 300
Rotifera, 188
Proboscis pore, 209
Proboscis receptacle, 196
Proboscis retractor muscle, 209
Procerodes, 174
Procerodidae (family), 174
Procuticle, 342, 342*f*, 343*f*
Proglottids, 158, 158*f*
Prohaptor, 160
Prolecithophora (=Holocoela) (order), 173–174
Promesostoma, 173
Pronuclei, 50, 53*f*
Prorhynchus, 173
Prorocentrum, 75
Proseriata (order), 174
Prosobranch, 229–235, 235*f*
circulatory system, 232, 233*f*
compared to opisthobranchs, 235–236
defining characteristics, 229
dispersal, 580
diversity, 235, 234*f*–235*f*
evolution, 232, 235
feeding mechanism, 232
gas exchange, 235
nervous system, 270*f*
reproduction, 265, 561*f*, 566*f*
shell, 232, 269*f*
taxonomic detail, 277–278
taxonomic summary, 271
Prosobranchia (subclass), 229
Prosoma, 352, 353
Prosorhochmidae (family), 213
Prosorhochmus, 213
Prostelium, 76
Prostoma, 213
Prostoma graecense, 208*f*
Prostomium, 297, 318
Protandric hermaphroditism, 265, 560, 559*t*
Protandric species, 210
Protandry, 560
Proteromonas, 77
Proterospongia, 68, 69*f*, 77
Prothoracic glands (PG), 343, 370, 372*f*
Prothoracicotropic hormone (PTTH), 370, 372*f*
Protista (kingdom), 37
Protobranchia (=Paleotaxodonta; =Cryptodonta) (subclass), 239, 241, 242*f*, 243*f*
defining characteristics, 239
feeding mechanisms, 241
gills, 241, 242*f*, 243*f*
taxonomic detail, 286–290
taxonomic summary, 271
Protocoel, 475, 530
Protodrilida (order), 335
Protodrillus, 336
Protodrilus, 335
Protogynous hermaphroditism, 560*t*
Protogyrodactylus, 176
Protonephridia, 150–151, 150*f*, 196, 198, 300
Protopodite, 376, 378*f*, 379
Protostomes, 12–14, 20, 21*f*
ciliary bands, 14, 16*f*
cleavage, 12–14
coelom formation, 10, 11*f*
coelomic cavities, 12
ideal characteristics, 15*t*
mesoderm origin, 14
mouth origin, 11
polar lobe formation, 14, 15*f*
Prototroch, 301
Protozoa (kingdom), 37–78
Acantharia (phylum), 65, 67, 76–77
amoeboid, 39, 43–44, 45*f*, 46*f*, 43–44
Apicomplexa (=Sporozoa) (phylum), 58–61, 73
body characteristics, 39
ciliated, 39, 42–43
Ciliophora (phylum), 46–54, 73
colonial, 46
defining characteristics, 38
Dinozoa (=Dinoflagellata) (phylum), 54–55, 58, 58*f*, 73
dispersal, 580
Euglenozoa (phylum), 75–76, 75*f*
evolutionary relationships, 38*f*, 39
feeding, 46
flagellated, 39, 43, 68–72, 73
general characteristics, 39–42
Heliozoa (phylum), 67–68, 67*f*, 73
locomotion, 42–44, 42*f*, 43*f*, 45*f*
Microsporidea (phylum), 61, 73
Mycetozoa (phylum), 63–64, 73, 76
Myxozoa (=Myxospora) (phylum), 59, 61, 77, 115, 117, 118*f*, 133
Parabasala (phylum), 70–71, 73, 75
Radiozoa (phylum), 65, 67, 67*f*, 73, 75
reproduction, 44, 46, 46*f*, 556, 558*t*
spore-forming, 39
taxonomic detail, 75–77
taxonomic summary, 73
transitional forms, 72
Protrusible pharynx, 154, 155*f*
Protura (order), 405
Proventriculus, 393
Provortex, 174
Provorticidae (family), 174
Psammetta, 76
Psammis, 416
Psammodrilida (order), 334
Psammodrilidae (family), 334
Psammodrilus, 334
Pseudergates, 373
Pseudicyema, 182
Pseudocalanus, 142, 143
Pseudoceridae (family), 173
Pseudoceros, 173
Pseudoceros crozieri, 155*f*
Pseudocoel, 10, 184
Pseudocoelomate, 10, 10*f*
Pseudodifflugia, 44*f*
Pseudofeces, 241, 245
Pseudogamy, 557
Pseudomicrothorax dubis, 42*f*
Pseudophyllidea (order), 175
Pseudoplasmodium, 63–64
Pseudopodia, 42, 44*f*, 62
structure and function, 43–44, 44*f*, 45*f*
Pseudopterogorgia, 133
Pseudoscorpion, 354*f*, 563
Pseudoscorpiones (order), 404
Pseudosquilla, 411
Pseudoterranova, 447
Pseudovermidae (family), 284
Pseudovermis, 284
Psocoptera (order), 407
Psolidae (family), 526
Psolus, 526
Psolus fabricii, 518*f*
Psychropotes, 526
Psychropotes longicauda, 518*f*
Pteraster, 525
Pteriomorphia (subclass), 287
Pterobranchia, 533*f*, 534, 535
Pteropods, 237, 238*f*, 280, 283
Pterotrachea, 280
Pterotrachea hippocampus, 234*f*
Pterotracheidae (family), 280
Pterygota (subclass), 406–411
Ptiliidae (family), 408
Ptilocrinus, 523
Ptilosarcus, 133
PTTH. *See* Prothoracicotropic hormone
Ptychocysts, 134
Ptychodactiaria (order), 134
Ptychodera, 535
Ptychodera flava, 532*f*
Pulmonata, 237–238, 239*f*, 240*f*
defining characteristics, 237
reproduction, 265
respiratory system, 238
shell, 238
taxonomic detail, 284–286
taxonomic summary, 271
Puncturella, 278
Pupa, 370, 572*t*
Pupilla, 285
Pupillidae (family), 285
Pupina, 279
Pupinidae (family), 279
Pycnogonida (=Pantopoda) (class), 357, 358*f*
body characteristics, 357
defining characteristics, 357
digestive system, 357
taxonomic detail, 405
taxonomic summary, 396
Pycnogonum litorale, 357
Pycnophyes, 458
Pycnopodia, 525
Pycnopodia helianthoides, 507*f*
Pyganodon, 287
Pygidium, 301, 325
Pyloric caeca, 505
Pyloric stomach, 394, 505
Pyralidae (family), 410
Pyramidella, 282
Pyramidellidae (family), 282
Pyramidelloidea (superfamily), 282
Pyrocystis, 75
Pyrosoma, 544–545, 547, 553
Pyrosoma atlanticum, 546*f*
Pyrosomes, 544, 546*f*
Pyrosomida (order), 553, 551
Pyrsonympha, 76

Q

Q fever, 357
Quadrigyrus nickolii, 197*f*
Quahogs, 289
Queen, 372, 373, 375*f*, 406, 411

R

Radial canal, 498*f*, 499
Radial cleavage, 12, 13*f*, 14*f*
Radial symmetry, 9, 9*f*, 138
Radiolarians, 65, 67–68, 67*f*, 73, 76–77
Radioles, 2303*f*
Radiozoa (phylum), 65, 67, 67*f*, 73, 75
Radula, 218
Radular sac, 218, 219*f*
Radular teeth, 218, 219*f*
Raillietiella mabuiae, 385*f*
Ramet, 108, 109*f*, 304, 468
Ram's horn, 290
Ram's horn snails, 286
Ranellidae (=Cymatiidae) (family), 280
Rangia, 288
Raptorial ciliates, 53
Rat tail stars, 525
Razor clams, 288
Rebelula bouvieri. *See Lophoura (=Rebelula) bouvieri*
Recluzia, 280
Recovery stroke, 43
Rectonectes, 173
Red Sea soft coral, 111*f*
Red spider mite, 354*f*
"Red tides," 55
Redia, 161, 163*f*, 569*t*
Regeneration, 155, 157*f*, 556
Regular urchins, 512, 512*f*
Regulative (indeterminate) cleavage, 14
Reighardia, 417
Remipedia (subclass), 20, 417
Renal pore, 271
Renal sacs, 541, 548
Renal vesicles, 548
Renette, 437
Renilla, 133
Replication, 556
Reproduction and development, 555–586
Acanthocephala, 196–198, 558*t*
Annelida, 558*t*, 569*t*
Anthozoa, 119, 556, 558*t*, 568*t*
Apicomplexa, 59, 60*f*, 61
Aplacophora, 265, 559*t*
in aquatic *vs.* terrestrial environments, 2, 3
Arachnida, 557, 559*t*, 560, 563, 564*f*, 580
Arthropoda, 349, 556, 557, 561, 563, 567, 569*t*, 571*t*–572*t*, 580
Ascidiacea, 547, 559, 559*t*, 574*t*
Asteroidea, 518, 519*f*, 520*f*, 556, 557, 559*t*, 573*t*, 575*f*, 577
Bdelloidea, 188–190, 558*t*
Bivalvia, 265, 558*t*, 566, 567, 572*t*, 575*f*
Brachiopoda, 489, 489*f*, 559*t*, 573*t*
Bryozoa, 489–490, 489*f*, 490*f*, 556, 559*t*, 560, 566, 578–579
Cephalochordata, 559*t*
Cephalopoda, 265, 559*t*, 563, 564*f*, 565
Cestoda, 158–159, 558*t*, 560
Chaetognatha, 463, 467*f*, 559*t*, 563, 567
Chilopoda, 559*t*
Chordata, 574*t*
Ciliophora, 49–52, 53
Cirripedia, 560, 560*f*, 571*t*
Cnidaria, 558*t*, 560, 566, 568*t*
Copepoda, 389, 392*f*, 565*f*, 567, 571*t*
Crinoidea, 518, 519*f*, 559*t*, 573*t*, 575*f*
Crustacea, 389–392, 392*f*, 559*t*, 560, 562*f*, 563, 571*t*
Ctenophora, 140–141, 556, 558*t*, 560
Cycliophora, 469
Decapoda, 391, 393*f*, 571*t*
Digenea, 161–170
Diplopoda, 559*t*
Echinodermata, 518, 519*f*, 520, 559*t*, 566, 573*t*–574*t*

Echinoidea, 518, 519f, 559t, 573t
Echiura, 313–314, 315f, 558t
Enteropneusta, 531, 532f, 559t
Gastropoda, 265, 267f–268f, 558t, 560f, 561, 561f, 563, 564f, 566, 566f, 572t, 580f
Gastrotricha, 459–460, 558t, 560
Gnathostomulida, 558t, 562f
Hemichordata, 559t, 574t
Hirudinea, 323f, 325, 558t
Holothuroidea, 518, 519f, 559t, 574t
Hydrozoa, 111, 111f, 112, 514f, 558t, 556, 568t
Insecta, 370–371, 371f, 372f, 556, 559t, 560, 561, 562f, 563, 567, 572t, 581
Kinorhyncha, 558t
Larvacea, 547
lophophorates, 488–490
Loricifera, 558t
Malacostraca, 391, 571t
Merostomata, 559t
Mesozoa, 179
Mollusca, 265–266, 265f, 266f, 267f, 558t–559t, 570t
Monogononta, 190, 194, 558t
Monoplacophora, 559t
Nematoda, 438, 438f, 440–441, 557, 559t, 560, 567
Nematomorpha, 558t
Nemertea, 210–211, 556, 558t, 569t
Oligochaeta, 322, 323f, 558t, 563
Onychophora, 559t, 563
Ophiuroidea, 518, 519f, 556, 559t, 566, 574t, 575f
Opisthobranchia, 265, 560f, 565f
Orthonectida, 180
Pentastomida, 558t
Phoronida, 488–489, 559t, 563, 572t, 576f
Placozoa, 91
Platyhelminthes, 150, 151, 558t, 568t–569t
Polychaeta, 300–304, 304f, 305f, 556, 558t, 560, 563, 569t, 575f, 580
Polyplacophora, 265, 559t, 570t
Porifera, 89–91, 90f, 556, 558t, 568t
Priapulida, 558t
Prosobranch, 265, 561f, 566f
Protozoa, 44, 46, 46f, 556, 558t
Rotifera, 185, 188, 556, 558t, 563, 567, 580
sarcodinids, 62
Scaphopoda, 265, 569t, 570t
Scyphozoa, 108, 109f, 556, 556f, 558t, 568t
Seisonidea, 188, 558t
Siboglinidae, 311–312
Sipuncula, 318, 319f, 558t, 569t
Tardigrada, 559t
Thaliacea, 547, 547f, 556, 559t
Trematoda, 556, 558t, 569t
Turbellaria, 154–155, 156f, 157f, 556, 558t, 568t
Urochordata, 547, 547f, 559t, 567, 574t
Reptentia (tribe), 213
Repugnatorial glands, 359
Resilin, 342
Respiratory plume, 308, 308f
Respiratory systems. *See also* Gills
Annelida, 296
in aquatic *vs.* terrestrial environments, 1–2
Arachnida, 353
Cephalopoda, 256
Chilopoda, 359
Cnidaria, 104
Echiura, 313
Enteropneusta, 531
Holothuroidea, 517
Insecta, 363–365, 365f
lamellibranchs, 242, 244f
Mollusca, 216, 217f, 218, 266
Nematoda, 437
Nemertea, 204
Opisthobranchia, 236, 237f
Platyhelminthes, 150
Polyplacophora, 218
Prosobranch, 235
Protobranchia, 241
Pulmonata, 238
Sipuncula, 318
Tardigrada, 422
Respiratory trees, 517
Resting eggs, 194, 194f, 195f
Reticulopodia, 44, 45f, 65, 66f
Retinular cells, 347
Retrograde waves, 229, 321
Retusa, 282
Reynolds numbers, 5
Rhabdites, 153, 153f
Rhabditida (order), 446
Rhabditidae (family), 446
Rhabditina (suborder), 446
Rhabditis, 436f
Rhabdocoela (order), 173
Rhabdoids, 153
Rhabdom, 347, 348f, 349f
Rhabdomeres, 347
Rhabdopleura, 533f, 535
Rhinophores, 236
Rhithropanopeus, 414
Rhizaria, 38f, 62, 65–68
Foraminifera (phylum), 65, 66f
Heliozoa (phylum), 67–68, 67f
Radiozoa (phylum), 65, 67, 67f
taxonomic detail, 76–77
taxonomic summary, 73
Rhizocephala (order), 393f, 417–418
Rhizocrinus, 524
Rhizostoma, 131
Rhizostomeae (order), 131–132
Rhodope, 282
Rhodopemorpha (order), 282
Rhodopidae (family), 282
Rhombogen, 181
Rhombozoa (class), 181–182, 181f
Rhopalia, 107, 107f, 108f
Rhopalura, 182
Rhynchobdellae (order), 337
Rhynchocoel, 205, 208, 208f, 209, 209f
Rhynchocoela (phylum). *See* Nemertea (=Rhynchocoela) (phylum)
Rhynchonellida (order), 493
Rhynchonympha tarda, 71f
Ribbon worms. *See* Nemertea (=Rhynchocoela) (phylum)
Ribosomal RNA (rRNA), 19–20, 206–207
Ricinulei (order), 404
Ridgeia, 308, 308f, 312, 336
Riftia, 336
Riftia pachyptila, 307f, 308, 308f, 309f, 310–311, 310f, 311f, 312
Ring canal, 499, 498f
Rissooidea (superfamily), 279
River blindness, 441, 448
RNA
in macronucleus, 49
ribosomal, 19–20, 206–207
Robber crabs, 414
Rocky Mountain spotted fever, 354f, 357
Rokopella, 277
Romanomermis, 446
Rosaster alexandri, 506f
Rostanga, 284
Rostanga pulchra, 575f
Rostroconchia (class), 218, 220f
Rostrum, 389, 374, 394f
Rotaria, 188f, 201
Rotatoria, 185f
Rotifera (phylum), 184–196
Bdelloidea (class), 188–190, 199, 201
cilia, 184, 184f, 186, 188f
comparison with Gastrotricha, 459
cyclomorphosis, 192–193, 193f, 193t
defining characteristics, 184
digestive system, 186, 195
dispersal, 580
excretory system, 196
feeding mechanisms, 186, 187f
general characteristics, 184–194
locomotion, 188, 190f
Monogononta (class), 190, 194, 195f, 199, 200–201
musculature, 185, 185f, 188
nervous system, 195–196, 195f
reproduction and development, 185, 188, 556, 558t, 563, 567, 580
Seisonidea (class), 188, 191f, 199, 201, 558t
sensory system, 195–196, 195f
taxonomic detail, 200–201
taxonomic summary, 199
water balance, 196
Roundworms, 434
Rove beetles, 408
Royal Society in London, 184
rRNA. *See* Ribosomal RNA
Runcina, 282

S

Sabella, 335
Sabella pavonia, 303f
Sabellaria, 335
Sabellaria alveolata, 302f
Sabellariidae (family), 335
Sabellida (order), 335
Sabellidae (family), 335
Saccocirrus, 335
Saccoglossus, 535
Saccoglossus kowalevskii, 530f
Sacculina, 389, 417
Sacculina carcini, 393f
Sacoglossa (=Ascoglossa) (order), 282–283
Sagartia, 133
Sagitta, 470
Sagitta cephaloptera, 466f
Sagitta elegans, 463f, 464t, 465f, 467f
Sagitta gazellae, 468f
Sagitta macrocephala, 466f
Sagittidae (family), 464t
Sagittoidea (class), 470
Salivary glands, 325
Salmincola, 416
Salmon gill maggot, 416
Salpa, 553
Salpida (order), 551, 553
Salps, 544, 545, 547f, 548, 553
Salt, in marine *vs.* freshwater environments, 4
Salticidae (family), 354f, 404
Sand crabs, 414
Sand dollars, 497, 508, 509, 512, 512f, 521, 525, 526, 573t
Sand fleas, 396, 412
Sand fly, 69
Sapphirina angusta, 384f
Sarcocystis, 75
Sarcodinids, 44f, 62, 63f
Sarcomastigophora (phylum), 62
Sarcostraca, 415
Sarsia, 132
Saturniidae (family), 410
Sawflies, 410
Saxitoxin, 55
Scalids, 456, 457
Scallops, 244f, 287
Scaphela, 234f
Scaphopoda (class), 255
defining characteristics, 255
feeding mechanisms, 255, 255f
fossil record, 255
reproduction and development, 265, 559t, 570t
shell, 255
taxonomic detail, 290
taxonomic summary, 263
Scarab beetles, 408
Scarabaediae (family), 408
Schistosoma, 177, 286
Schistosoma haematobium, 162f
Schistosoma japonicum, 167f
Schistosoma mansoni, 163f, 164f, 165, 165f, 166f
Schistosomatidae (family), 177
Schistosomiasis, 161, 162f, 165, 165f, 166f
Schizobranchia, 335
Schizocoely, 10, 11f, 475
Schizont, 59
Schizoporella, 494
Schizoporellidae (family), 494
Scientific name, 16
Scleractinia (order), 134
Scleractinian corals, 122, 123f
Sclerocytes, 82
Sclerospongiae (class), 86, 88f, 94
Sclerotization, 342
Scolelepis, 334
Scolex, 158, 158f
Scolopendra, 359f
Scolopendra gigantea, 405
Scorpion flies, 409
Scorpiones (order), 403
Scorpions, 354f, 396, 403, 563, 564f, 565f
Scotoanassa, 518f
Screwworm, 409
Scuta, 389, 390f
Scutellastra, 278
Scutigera, 405
Scutigera coleoptrata, 359f
Scutigerella, 405
Scypha. *See Grantia (=Scypha)*
Scyphistoma, 108
Scyphozoa (class), 103f, 104–108
defining characteristics, 104
digestion, 106, 107f
feeding mechanisms, 105–106
locomotion, 105, 106f
musculature, 105, 345t
nervous system, 107, 107f
reproduction and development, 108, 109f, 556, 556f, 558t, 568t
sense organs, 107, 107f
taxonomic detail, 131–132
taxonomic summary, 128
Sea anemones, 32f, 119, 121f, 122f, 128, 133
Sea biscuits, 512, 512f, 525
Sea butterflies. *See* Pteropods
Sea cucumbers. *See* Holothuroidea (class)
Sea daisies. *See* Concentricycloids
Sea fans, 125f, 128, 133
Sea feathers, 133
Sea hares. *See* Anaspidea (=Aplysiacea) (order)
Sea lilies, 498, 501–502, 501f, 503, 523, 573t
Sea monkeys, 379
Sea mouse, 333
Sea nettle, 131
Sea pansies, 128, 133
Sea pens, 128, 133
Sea slugs. *See* Nudibranchia (order)
Sea spiders. *See* Pycnogonida (=Pantopoda) (class)
Sea squirts. *See* Ascidiacea (class)
Sea stars. *See* Asteroidea (subclass)
Sea urchins, 497–500, 498f, 499f, 508, 509, 510, 510f, 511f, 512, 512f, 513f, 514f, 521, 525, 573t
Sea wasps, 109, 128, 132
Sea whips, 128, 133
Sea wolf, 333
Searlesia dira, 561f
Secernentea (class), 445
Second maxillae, 363
Secondary mesenteries, 120f
Secondary queen, 375f
Sedentaria (subclass), 300
Sedentary animals, 31
Sedentary polychaetes, 300, 301f–302f, 303f
Seed shrimp, 416
"Seep" communities, 309
Seison, 191f, 201
Seisonidae (family), 201
Seisonidea (class), 188, 191f
reproduction, 558t
taxonomic detail, 201
taxonomic summary, 199
Selenaria maculata, 494
Semaeostomeae (order), 131
Semibalanus balanoides, 391f
Semperothuria surinamensis. See Holothuria (Semperothuria) surinamensis
Senescence, 186
Sense organs. *See also* Eyes
Annelida, 327–328
Arthropoda, 345–349
Cephalopoda, 259–261, 262f
Chaetognatha, 461
Chilopoda, 359
Ctenophora, 138–139, 139f
Hydrozoa, 111
Insecta, 362–363, 364f
Nemertea, 204
Polyplacophora, 222
Rotifera, 195–196, 196f
Scyphozoa, 107, 107f
Sipuncula, 316
Turbellaria, 151
Sensillae, 437
Sensory lappets, 107
Sepia, 260f, 291
Sepiidae (family), 291
Sepioidea (order), 290–291
Sepioteuthis, 291
Septa, 119, 256, 256f, 296
Septibranch ctenidium, 254
Sequential hermaphroditism, 559–560, 559t
Sergestes, 413
Sergestidae (family), 413
Serpula, 335
Serpula vermicularis, 305f
Serpulidae (family), 335
Serpulorbis, 280
Sertularia, 132
Sesarma, 414
Sessile animals, 4, 2318
Ascidiacea, 540
Ciliophora, 52, 53
Rotifera, 188, 190f
Sessoblasts, 484
Setae, 295, 297
Setigerous larva, 569t
Sexual reproduction, 558–567, 577. *See also* Reproduction and development
vs. asexual reproduction, 556
Sexuality, patterns of, 558–561
Sheath, 434
Sheep liver fluke, 166f
Shell
Bivalvia, 238, 239, 240f
Brachiopoda, 478, 479
Cephalopoda, 256–257, 257f
Cirripedia, 389, 390f
coiled, 224, 225, 234f, 260f
dextral, 225, 227f
Gastropoda, 224–225, 227f, 269f

Shell—*Cont.*
- Mollusca, 215–216, 216*f*
- Monoplacophora (class), 224, 224*f*
- Opisthobranchia, 235–236, 237*f*
- Polyplacophora, 218, 220*f*
- Prosobranch, 232, 269*f*
- Pulmonata, 238
- Scaphopoda, 255
- sinistral, 225, 227*f*

Shell gland, 158*f*, 162*f*
Shell reduction or loss, 236, 257
Shellfish poisoning, 55
Shielding pigment, 347
Shields, 503
Shipworms, 251, 254*f*, 255*f*, 271, 289
Shrimp, 396, 411–416
Siboglinidae (=Pogonophora) (family), 304–312
- bacterial symbiosis, 309, 310*f*, 311
- body characteristics, 304, 306
- circulatory system, 328
- comparison with Sipuncula, 317*t*
- defining characteristics, 304
- digestive system, 306
- feeding mechanisms, 306
- general characteristics, 304–311
- reproduction and development, 311–312
- taxonomic detail, 336
- taxonomic summary, 328
- vestimentiferans, 307–309, 312

Siboglinum, 336
Sida, 415
Siderastraea, 134
Sigalion, 333
Sigalionidae (family), 333
Sige, 332
Sige fusigera, 299*f*
"Signature sequences," 20
Siliceous endoskeleton, 65
Silk, arachnid, 357
Silkworm, 357, 372*f*, 410
Silverfish, 360, 406
Simnia, 280
Simuliidae (family), 409
Simulium, 448
Simultaneous hermaphrodites, 151, 265, 559, 559*t*, 560*f*
Sinantherina, 201
Sinistral shell, 225, 227*f*
Sinularia, 133
Sinus gland, 376
Siphon, 227, 232, 232*f*, 233*f*, 256
Siphonaptera (order), 409
Siphonaria, 285
Siphonariidae (family), 285
Siphonoglyphs, 119, 120*f*
Siphonophora (order), 113–114, 117*f*, 128, 132
Siphonostomatoida (order), 416
Siphuncle, 256
Sipuncula (phylum), 314–318, 454
- body characteristics, 315–316
- burrowing, 315
- circulatory system, 318
- comparison with Annelida, 316, 317*t*
- defining characteristics, 314
- digestive system, 316
- excretory system, 318
- gas exchange, 318
- locomotion, 316
- musculature, 316
- reproduction and development, 318, 319*f*, 558*t*, 569*t*
- sense organs, 316
- taxonomic detail, 338
- taxonomic summary, 338

Sipunculidae (family), 338
Sipunculus, 327
Sipunculus nudus, 316*f*, 317*f*, 318*f*
Sipunculus polymyotus, 319*f*
Sister groups, 28
Skeleton, hydrostatic, 97–99
Skeleton shrimp, 413
Sleeping sickness, 69, 409
Slime glands, 424
Slime molds, 62*f*, 63–64, 64*f*, 76
Slipper shell snails, 234*f*, 279
Sludge-worms, 325
Smaragdia, 278
Smeagol, 286
Smithkline Beecham, 440
Snake flies, 408
Snapping shrimp, 413
Snow fleas, 409
Social species, 541
Social systems, Insecta, 372–373, 374*f*
Soft corals, 134
Soft-shell clams, 244*f*, 289
Sol, 44
Solaster, 525
Soldiers, 373, 375*f*, 406
Solemya, 286
Solemyidae (family), 286
Solenocytes, 151
Solenogastres (subclass). *See* Neomeniomorpha (=Solenogastres) (subclass)
Solenopsis, 411
Solifugae (=Solpugida) (order), 404
Solitary species, 116, 541
Solpugida (order). *See* Solifugae (=Solpugida) (order)
Somasteroidea (subclass), 524
Somatic structures, 543
Sorbeoconcha (order), 279–281
Sorberacea (class), 553
Sow bugs. *See* Pillbugs
Spadella, 470
Spadella angulata, 464*t*
Spadellidae (family), 464*t*, 465
Spasmoneme, 53
Spatangoida (order), 526
Spatangus, 526
Spathebothriidea (order), 175
Spathebothrium, 175
Species, 16
Species longevity, 577, 580*f*
Species name, 17
Specific name, 17
Sperm, 561, 561*f*
Spermathecae, 322
Spermatophores, 563, 564*f*, 565, 565*f*, 566
- Arthropoda, 349
- Chaetognatha, 463, 467*f*
- Hirudinea, 325
- Mollusca, 265, 265*f*
- Phoronida, 488

Spermatozeugmata, 561*f*
Sphaeriidae (=Pisidiidae) (family), 289
Sphaerium, 289
Sphaeroma, 412
Sphaeroma quadridentatum, 378*f*
Sphaerospora, 133
Sphingidae (family), 410
Sphinx moths, 410
Spicules, 81–82, 83*f*
Spider crabs, 414
Spider web, 356*f*, 357
Spiders. *See* Arachnida (class)
Spinnerets, 353
Spinicuadata (suborder), 415
Spinther, 333
Spintherida (order), 333
Spinulosida (order), 525
Spiny lobster, 413
Spiny urchin, 512*f*
Spio, 334
Spionida (order), 334
Spionidae (family), 334
Spiracles, 353, 363, 365*f*
Spiral cleavage, 12, 13*f*
Spiratella, 238*f*
Spiratellidae (family). *See* Limacinidae (=Spiratellidae) (family)
Spirillina, 76
Spirobranchus, 335
Spirometra, 175
Spironucleus, 76
Spironympha, 75
Spirorbidae (family), 335
Spirorbis, 335
Spirula, 290
Spirulidae (family), 290
Spirurina (suborder), 447–448
Spisula, 288
Spittlebugs, 407
Sponges. *See* Porifera (phylum)
Spongia, 94
Spongiidae (family), 94
Spongilla, 94
Spongillidae (family), 94
Spongin, 81
Spongiome, 40, 40*f*
Spongocoel, 81
Spongocytes, 82
Sporangia, 64
Spore-forming protozoans, 39
Sporoza. *See* Apicomplexa (=Sporozoa) (phylum)
Sporozoites, 59, 60*f*
Springs, 138
Springtails, 360
Spumella, 76
Spumellaria, 76
Spurilla, 284
Spurilla neapolitana, 236*f*
Squat lobsters, 414
Squids, 256–257, 258*f*, 259, 260*f*, 271, 291
Squilla, 411
Staghorn corals, 134
Stainforthia concava, 66*f*
Stalk, 501
Staphylinidae (family), 408
Star sand, 66*f*
Star shells, 278
"Starfish." *See* Asteroidea (subclass)
Statoblasts, 483
Statocysts, 107, 108*f*, 204, 490
Statolith, 107, 138, 139*f*
Stauracon, 76
Staurolonche, 76
Stauromedusae (order), 131
Stelleroidea (class), 503–509
- Asteroidea (subclass). *See* Asteroidea (subclass)
- concentricycloids, 508–509, 518, 521, 525
- defining characteristics, 503
- Ophiuroidea (subclass). *See* Ophiuroidea (subclass)
- taxonomic detail, 524–525
- taxonomic summary, 521

Stem cells, 438, 438*f*
Stem nematogen, 181
Stenolaemata (class), 487–488, 491, 493
Stenosemella, 56*f*
Stenostomum, 173
Stentor, 53, 57*f*, 75
Stentor coeruleus, 52*f*
Stephanoceros, 201
Stephanoceros fimbriatus, 193*f*
Stephanodrilus, 337
Stephanoeca campanula, 69*f*
Stephanoscyphus, 131
Stereoblastula, 90
Stichocotyle, 177
Stichopus, 526
Stick insects, 23, 25, 25*f*, 406
Stigma, 68
Stigmata, 540, 541, 542*f*
Stilifer, 281
Stoecharthrum, 182
Stolidobranchia (order), 553
Stolon, 111, 485
Stolonifera (order), 133
Stoma, 436
Stomatella, 278
Stomatopoda (order), 374, 378–379, 378*f*, 396, 411
Stomochord, 531
Stomodeum, 138
Stomolophus, 131
Stomolophus meleagris, 556*f*
Stomphia, 133
Stone canal, 498, 498*f*
Stoneflies, 407
Stony (true) corals, 123, 124*f*, 134
Streblospio, 334
Strepsiptera (order), 408–409
Streptaxidae (family), 285
Streptocephalus, 415
Stridulation, 406
Strigea, 176
Strigeidida (order), 176–177
Strobilation, 108, 109*f*, 556*f*
Stromatospongia, 94
Strombidae (family), 279
Stromboidea (superfamily), 279
Strombus, 279
Strongylocentrotus, 525
Strongylocentrotus droebochienis, 511*f*
Strongyloides, 447
Strongyloides stercoralis, 447
Strongyloididae (family), 447
Strontium sulfate, 65, 76
Styela, 553
Styela clava, 553
Stylactis, 132
Stylaster, 132
Stylasterina (order), 132
Style sac, 247, 249*f*
Stylet, 209
Stylochidea (family), 173
Stylochus, 173
Stylommatophora (suborder), 285
Stylonychia, 49*f*, 75
Stylonychia lemnae, 50*f*
Stylophora, 134
Subimago, 406
Subitaneous eggs, 190
Subneural gland, 540
Subtidal animals, 31
Succineidae (family), 285
Sucking lice, 407
Suctorians, 53–54, 57*f*
Sugar glands, 222
Sulcus, 54
Sulfides, 310–311, 311*f*
Sun stars, 525
Sundial shells, 282
Superposition eye, 347, 349*f*
Support system, in aquatic *vs.* terrestrial environments, 3
Surf clams, 288
Suspension feeders, 4, 32
- Ciliophora, 53–54
- Enteropneusta, 531
- Ophiuroidea, 505
- Rotifera, 187*f*

"Swimmer's itch," 162*f*, 177
Swimming crabs, 414
Sycon, 83*f*, 89*f*
Syconoid sponge, 86, 87*f*, 88, 89*f*, 94
Syllidae (family), 332
Syllis, 332
Symbiodinium, 123
Symbion, 470
Symbion pandora, 467, 468, 468*f*, 469, 469*f*
Symbionts, 32
- Anthozoa, 123
- Ciliophora, 53
- Dinozoa, 58
- diplomonads, 76
- flagellates, 75
- Foraminifera, 56, 73
- oxymonads, 76
- zoochlorellae, 113, 114*f*, 115*f*

Symbioses, 32
Symbiotic relationships, 32, 32*f*
- bacterial, 247, 252–253, 253*f*, 393, 442

Symnodinium, 58
Symphella, 405
Symphyla (order), 350, 362, 405
Symphyta (suborder), 410
Symphytognathidae (family), 404
Symplasma (subphylum), 94
Synalpheus, 413
Synalpheus regalis, 413
Synapomorphies, 27*t*, 28, 32
Synapta digitata, 13*f*
Syncarida (superorder), 411
Syncarids, 411
Synchaeta, 201
Synchaetidae (family), 201
Syncytial epidermis, 157, 432
Syndermata (phylum), 184, 198
Syngens, 51
Synkaryon, 50, 52
Systellommatophora (order), 286
Systemic heart, 259

T

Tabanidae (family), 409
Tadpole, 542, 543*f*, 574*t*
Tadpole shrimp, 381, 416
Taenia, 158, 158*f*, 176
Taenia solium, 32*f*, 159*f*, 176
Taeniarhynchus, 176
Taeniarhynchus saginata, 176
Taeniidae (family), 158, 176
Taenioglossans. *See* Mesogastropods (=Taenioglossans)
Tagmatization, 341
Tail-less whip scorpions, 403
Talorchestia, 412
Tanaidacea (order), 412
Tanning process, 342
Tantulocarida (subclass), 20, 417
Tapes, 289
Tapeworms. *See* Cestoda (class)
Tarantulas, 404
Tardigrada (phylum), 422–424, 422*f*, 423*f*
- cryptobiosis, 423–424
- cuticle, 422
- defining characteristics, 422
- development, 422–423
- dispersal, 580
- fossil record, 422
- gas exchange, 422
- musculature, 422
- nervous system, 422, 423*f*
- reproduction, 559*t*
- taxonomic detail, 428–429
- taxonomic summary, 428

Taxon, 16, 27*t*
Taxonomy
- classical, 26
- numerical, 26

Tectarius, 279
Tectarius muricatus, 566*f*
Tectura, 278
Tegmen, 501
Tegula, 278
Tegument, 151, 157

Tellina, 288
Tellinidae (family), 288
Telotroch, 301
Temnocephala, 175
Temnocephalida (suborder), 175
Temnopleuroida (order), 525
Temperature stability, of water, 4
Tenebrio, 408
Tenebrionidae (family), 408
Tenellia, 284
Tentacles
 catch, 119
 Cephalopoda, 259, 262*f*
 Cnidaria, 102
 Cubozoa, 109
 Polychaeta, 297
Tentaculata (class), 144–146, 144*f*, 145*f*
Tentillae, 114
Tephritidae (family), 409
Terebellida (order), 335
Terebellidae (family), 335
Terebralia, 279
Terebratella, 477*f*, 478*f*, 493
Terebratellidae (family), 493
Terebratulida (order), 492
Teredinidae (family), 289
Teredo, 289
Teredora malleolus, 254*f*
Terga, 389, 390*f*
Tergipedidae (=Cuthonidae) (family), 284
Termites, 372, 373, 406
 queen, 375*f*
 soldier, 361*f*, 375*f*
 worker, 375*f*
Terrestrial animals, 31
Terrestrial environments
 benefits of, 3*t*, 4–5
 problems with, 1–4, 3*t*
Tertiary mesenteries, 120*f*
Tertiary queen, 375*f*
Test, 52, 63, 63*f*, 65, 65*f*, 510, 511*f*, 540
Test-bearing amoebae, 63, 63*f*, 76
Tethydidae (family), 284
Tethys, 284
Tetrabothriidae (family), 176
Tetrahymena paravorax, 54*f*
Tetrahymena pyriformis, 48*f*
Tetrahymena thermophila, 54–55, 54*t*
Tetrahymena vorax, 54–55, 54*t*
Tetranchyroderma, 460*f*, 470
Tetraphyllidea (order), 176
Tetrastemma, 213
Tetrastemmatidae (family), 213
Tetrodotoxin (TTX), 210
Teuthoidea (=Decapoda) (order), 291
Tevnia, 336
Thais, 281
Thais haemastoma canaliculata, 217*f*, 219*f*
Thais lima, 566*f*
Thalassinidea (infraorder), 414
Thalassiosira weissflogii, 386–387, 386*t*, 387*f*, 388*f*
Thalassocalyce, 148
Thalassocalycida (order), 148
Thalassoma bifasciatum, 126–127, 127*f*
Thaliacea (class), 544–545, 546*f*, 547
 evolution, 544–545
 feeding mechanisms, 545
 locomotion, 545
 nervous system, 547–548
 reproduction and development, 547, 547*f*, 556, 559*t*
 taxonomic detail, 553
 taxonomic summary, 551
Thecate, 112, 115*f*
Thecosomata (order), 283
Thelenota, 526
Themiste, 338
Theridiidae (family), 404
Thermosbaena, 412
Thermosbaenacea (order), 412
Thermozodium esakii, 429
Thiara, 279
Thiaridae (family), 279
Thomisidae (family), 404
Thoracica (order), 417
Thorax, 374
Thorny corals, 134
Thrips, 407
Thyasira, 288
Thyasiridae (family), 288
Thyone briareus, 516*f*
Thyonella gemmata, 516*f*
Thyonicola, 281
Thysanoptera (order), 407
Thysanura (order), 406, 408
Ticks, 349, 353, 357, 396, 564*f*
Tiedemann's bodies, 498*f*, 499
Tiger beetles, 408
Timber beetles, 408
Tintinopsis parva, 56*f*
Tintinopsis platensis, 56*f*
Tipulidae (family), 409
Tisbe, 416
Tisbe furcata, 384*f*
Tissue regression, 83
Tobacco hornworm, 365*f*, 410
Todarodes, 291
Toes, 186, 188*f*
Tokophyra quadripartita, 57*f*
Tomopteridae (family), 332
Tomopteris, 332
Tongue worms. *See* Pentastomida (subclass)
Tonicella, 276
Tonicella lineata, 221*f*
Tonna, 280
Tonna galea, 234*f*
Tonnidae (family), 280
Tonnoidea (superfamily), 280
"Tooth shell," 256, 271
Tornaria, 531, 532*f*, 574*t*
Torsion, 226, 228*f*, 229, 230, 230*f*
Toxicysts, 41–42
Toxoplasma, 75
Toxoplasma gondii, 61
Toxopneustes, 525
Tracheae, 353
Tracheal system, 363, 365*f*
Trachelophis kahli, 57*f*
Tracheoles, 364, 365*f*
Trachylina (order), 132
Transport hosts, 196
Trapdoor spiders, 404
Trematoda (class), 161–170
 Aspidogastrea (subclass), 170, 177
 Digenea (subclass), 161–170
 life cycle, 163*f*
 reproduction and development, 556, 558*t*, 569*t*
 taxonomic detail, 176–177
 taxonomic summary, 170
Triadinium, 58*f*
Tribolium, 408
Triboniophorus, 285
Triceps, 97, 98*f*
Trichinella spiralis, 439*f*, 446
Trichinellidae (family), 446
Trichinosis, 439*f*, 446
Trichobilharzia, 177
Trichocysts, 40, 41, 42*f*, 48, 56*f*
Trichodoris, 447
Trichomonada (class), 73, 75
Trichomonas, 71, 75
Trichomonas vaginalis, 71
Trichonympha, 75
Trichonympha campanula, 71
Trichonympha collaris, 71*f*
Trichoplax adhaerens, 91, 91*f*
Trichoptera (order), 410
Trichotropis cancellata, 216*f*
Trichuridae (family), 446
Trichuris, 446
Trichuris trichuris, 446
Tricladida (order), 174
Tricodoridae (family), 446
Tridacna, 288
Tridacnidae (family), 288
Trididemnum, 553
Trigomonas, 76
Trilobita (class), 351–352
 body characteristics, 351–352
 fossil record, 351, 351*f*
 taxonomic summary, 396
Trilobitomorpha (subphylum), 350–352
 defining characteristics, 350
 taxonomic summary, 396
Triops, 416
Tripartite coelom, 12
Tripedalia, 132
Tripedalia cystophora, 110*f*
Triphora, 281
Triphoridae (family), 281
Triphoroidea (superfamily), 281
Triplecta, 76
Triploblastic animals, classification of, 9–10, 17*f*
Triplotaeniidae (family), 176
Tripneustes, 525
Tritonia, 284
Tritoniidae (=Duvaucellidae) (family), 284
Tritons, 280
Trivitellina, 176
Trochidae (family), 278
Trochoidea (superfamily), 278
Trochophore, 569*t*, 570*t*
 Echiura, 313, 315*f*
 Mollusca, 265, 267*f*
 Polychaeta, 301, 305*f*
Troglochaetus, 336
Trophi, 186, 187*f*
Trophosome, 306
Tropical (spiny) lobster, 413
Tropical sponge, 80–81, 81*t*
Tropocyclops prasinus, 192
True barnacles, 417
True bugs, 407
True (stony) corals, 123, 124*f*, 134
Trunk, 306, 307*f*, 312, 530
Trypanorhyncha (order), 175
Trypanosoma, 69–70, 73, 75
Trypanosoma brucei brucei, 70*f*
Trypanosoma lewisi, 70*f*
Trypanosomes, 69–70, 70*f*
Trypetesa, 417
Tsetse fly, 69, 409
TTX. *See* Tetrodotoxin
Tube feet, 498–500, 498*f*, 499*f*
Tube-dwelling polychaetes, 300, 301*f*
Tubercle, 510
Tubifex, 336
Tubificidae (family), 336
Tubificina (suborder), 336
Tubiluchidae (family), 457
Tubiluchus, 457
Tubiluchus corallicola, 455*f*
Tubipora, 133
Tubulanidae (family), 212
Tubulanus, 212
Tubularia, 132
Tubulin, 42
Tubulipora, 493
Tucca, 416
Tulip shells, 281
Tun shells, 280
Tunic, 540, 540*f*
Tunicata (subphylum). *See* Urochordata (=Tunicata) (subphylum)
Tunicin, 540
Turban shells, 278
Turbanella, 470
Turbatrix aceti, 447
Turbellaria (class), 151–155, 156*f*
 benthic, 151
 digestive system, 153, 154, 155*f*
 evolution, 154
 locomotion, 151–153, 152*f*, 153*f*
 nervous system, 151, 152*f*
 regeneration, 154–155, 157*f*
 reproduction and development, 154–155, 156*f*, 157*f*, 556, 558*t*, 568*t*
 sense organs, 151
 taxonomic detail, 172–175
 taxonomic summary, 170
Turbeville, James, 206
Turbinidae (family), 278
Turbo, 278
Turritella, 279, 566*f*
Turritellidae (family), 279
"Tusk shell," 256, 271
Tympanal organs, of insects, 362
Tympanum, of insects, 362
Typhloplana, 175
Typhloplanoida (suborder), 175
Typhloscolecidae (family), 332
Typhloscolex, 332
Typhlosole, 325
Typhoid, 409

U

Uca, 415
Uchidana parasita, 212
Ulophysema, 417
Ultraviolet light, 347
Umagillidae (family), 174
Umbo, 239
Umbonium, 278
Uncini, 336
Undulating membrane, 47
Undulipodia, 43
Unicellular, 9
Unio, 287
Unionidae (family), 242, 287
Uniramia (phylum), 350
Uniramous appendages, 358, 359, 360, 362, 363, 376
Unitary species, 541
Upogebia, 414
Upogebiidae (family), 414
Upside-down jellyfish, 131–132
Urechis, 312, 337
Urechis caupo, 312, 314*f*
Uric acid, 364–365, 395
Urnatella, 495
Urns, 318, 318*f*
Urochordata (=Tunicata) (subphylum), 539, 540–547
 Ascidiacea (class), 540–542, 547, 548, 551, 552–553, 556, 559*t*, 574*t*
 defining characteristics, 540
 excretory system, 547–548
 Larvacea (=Appendicularia) (class), 543–544, 547, 548, 551, 553, 559*t*
 nervous system, 547–548
 phylogenetic relationships, 544*f*
 reproduction and development, 547, 547*f*, 559*t*, 567, 574*t*
 taxonomic detail, 552–553
 taxonomic summary, 551
 Thaliacea (class), 544–545, 547, 548, 551, 553, 559*t*
Uropods, 376
Uropygi (order), 403
Urosalpinx, 281
Uterine bell, 196

Vaginulus, 286
Vallicula, 147
Valvatida (order), 525
Valves, 508
Vampire squids, 291
Vampirolepis nana, 176
Vampyromorpha (order), 291
Vampyroteuthidae (family), 291
Vampyroteuthis, 291
Vargula hilgendorfi, 381*f*
Vectors, 58
Vegetal pole, 12
Velamen, 148
Velella, 104*f*, 117*f*, 133
Veleropilina reticulata, 224*f*
Veliger, 265, 267*f*, 268*f*, 570*t*, 575*f*
Velum, 111, 111*f*, 265, 509
Vema, 277
Vema levinae, 225*f*
Veneridae (family), 289
Veneroida (order), 287–289
Venus. See Mercenaria (=Venus)
Venus's flower basket, 89*f*, 94
Venus's girdle, 145*f*
Veretoidea (superfamily), 280
Vermetidae (family), 280
Vermetus, 280
Vermicularia, 234*f*, 279
Vermiform, 117, 209, 222, 295, 434, 538
Vernalization, 82
Verruca, 417
Vertebrae, 503
Vertebrata (subphylum), 551, 553
Vesicomya, 289
Vesicomyidae (family), 289
Vesiculariidae (family), 494
Vestia, 285
Vestigastropoda (superorder), 278
Vestimentifera (subfamily). *See* Obturata (=Vestimentifera) (subfamily)
Vestimentiferans, 307–309, 312, 328
Vestimentum, 308, 308*f*
Vibracula, 486
Vibrio alginolyticus, 210
Villosa, 287
Vinegar eels, 447
Vinegar flies, 409
Violet snails, 280
Visceral mass, 224, 226, 226*f*
Viscosity, of water, 4–5
Viviparidae (family), 279
Viviparus, 279
Volume regulation, 41
Volutes, 234*f*
von Frisch, Karl, 368
Vorticella, 53, 57*f*, 75

Waggle dance, 368–369, 368*f*, 369*f*
Walking leaves, 406
Walking legs, 352, 353, 376
Walking sticks, 406
Wasps, 361, 364*f*, 370, 410
Water, *vs.* air, as environment for invertebrates, 1–4, 3*t*
Water balance, Rotifera, 196
Water bears. *See* Tardigrada (phylum)
Water conservation, Insecta, 363–365, 365*f*
Water fleas, 379, 380*f*, 396, 415
Water scorpions, 403
Water spiders, 404

Water vascular system (WVS), 398–499, 498*f*
Watering pot shells, 290
Waterpenny beetles, 408
Watersipora, 494
Watersiporidae (family), 494
Weevils, 408
Wentletraps, 280
Whale lice, 413
Whelks, 233*f*, 281, 566*f*
Whip scorpions, 403
Whipworm, 446
Whirligig beetles, 408
"White ants." *See* Termites
Wilson, E. B., 13*f*
Wind scorpions, 404
Winged insects, 405–411
Wingless insects, 405
Wings, Insecta, 365–367, 367*f*
Winter eggs. *See* Resting eggs
Wolbachia, 363, 443, 560
Wolf spiders, 404
Wood, J. G., 325
Wood roach, symbiotic relationships, 71, 71*f*, 75
Wood-boring bivalves, 251, 255*f*
Woodlice, 396, 412
Wood-storing caecum, 251
Workers, 372, 373, 374*f*, 375*f*, 406, 411
World Health Organization, 442
Worm leeches, 337
Worm-shell snail, 234*f*
Wuchereria, 443, 444*f*, 448
Wuchereria bancrofti, 440, 448
WVS. *See* Water vascular system

Xanthidae (family), 414
Xenophyophoreans, 76
Xenopneusta (order), 337
Xenoturbella, 537
Xenoturbella bocki, 530, 537, 538
Xenoturbella westbladi, 537
Xiphosura (order), 403
Xironodrilus, 337
X-organ, 343, 376
Xylophaga, 289
Xyloplax, 525
Xyloplax medusiformis, 509*f*

Yellow fever, 409
Yellow jackets, 373
Yoldia, 286
Yoldia eightsi, 224*f*, 243*f*
Yoldia limatula, 243*f*
Y-organ, 343

Z

Zebra mussels, 242, 289, 567, 577
Zelinkiella, 201
Zinc, 517, 517*f*
Zirphaea, 289
Zoantharia. *See* Hexacorallia (=Zoantharia) (subclass)
Zoanthidea. *See* Zoanthinaria (=Zoanthidea) (order)
Zoanthinaria (=Zoanthidea) (order), 134
Zoanthus, 134
Zoea, 391, 571*t*, 575*f*
Zooanthella, 75
Zoobotryon, 494
Zoochlorellae, 111, 112–113, 113*t*
Zooecium, 480, 481*f*
Zooflagellated protozoans, 68–72, 69*f*, 73
Zooids, 111, 480, 482*f*, 534
Zooplankton, 4, 186, 379, 382, 567
Zooxanthellae, 58, 106, 123, 124, 124*f*
Zoroaster, 525
Zygentoma (order), 46, 406
Zygocotyle lunata, 162*f*
Zygonemertes virescens, 210*f*
Zygoptera (suborder), 406